2nd International Conference on Green Design and Manufacture 2016 (IConGDM 2016)

AA004204

MATEC Web of Conferences Volume 78 (2016)

Phuket, Thailand
1 - 2 May 2016

Editors:

S. Sharif
M.M.A.B. Abdullah
S.Z. Abd Rahim
M.F. Ghazali
N. Mat Saad
M.M. Ramli
S.A. Ramli
S.A. Zainol Murad
S.S. Mat Isa

ISBN: 978-1-5108-3111-7

Printed from e-media with permission by:

Curran Associates, Inc.
57 Morehouse Lane
Red Hook, NY 12571

Some format issues inherent in the e-media version may also appear in this print version.

Printed by Curran Associates, Inc. (2016)

For additional information, please contact EDP Sciences – Web of Conferences at the address below.

EDP Sciences – Web of Conferences
17, Avenue du Hoggar
Parc d'Activité de Courtabœuf
BP 112
F-91944 Les Ulis Cedex A
France

Phone: +33 (0) 1 69 18 75 75
Fax: +33 (0) 1 69 28 84 91

contact-edps@webofconferences.org

Additional copies of this publication are available from:

Curran Associates, Inc.
57 Morehouse Lane
Red Hook, NY 12571 USA
Phone: 845-758-0400
Fax: 845-758-2633
Email: curran@proceedings.com
Web: www.proceedings.com

TABLE OF CONTENTS

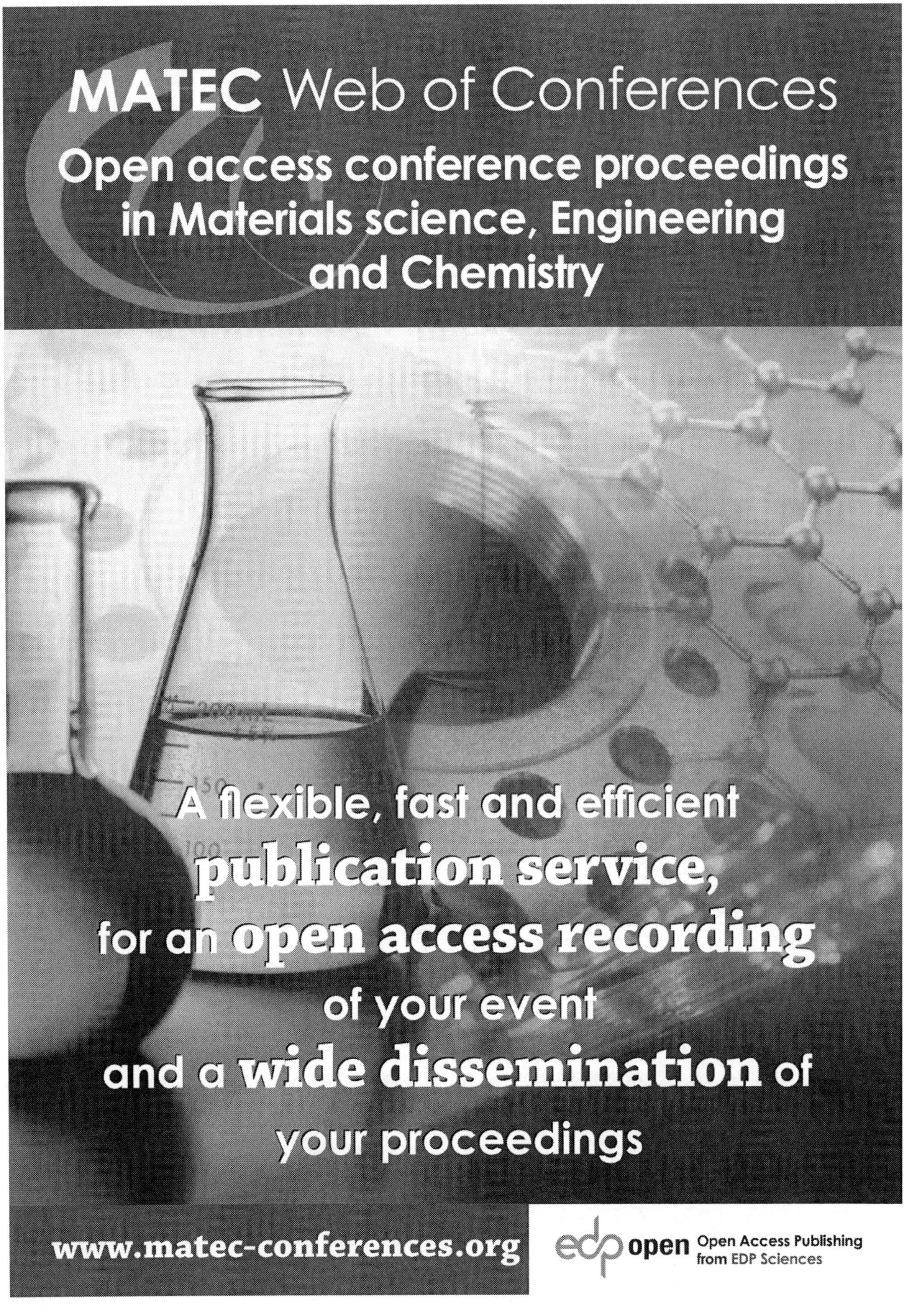
MATEC Web of Conferences
Open access conference proceedings
in Materials science, Engineering
and Chemistry
A flexible, fast and efficient
publication service,
for an open access recording
of your event
and a wide dissemination of
your proceedings
www.matec-conferences.org
edp open
Open Access Publishing
from EDP Sciences

ISSN (electronic edition): 2261-236X

Professor Dr. Safian Bin Sharif,
Universiti Teknologi Malaysia,
Johor, MALAYSIA

Publisher,
MATEC Web of Conferences,

July 5th, 2016

Dear Professor / Sir / Madam,

Statement of Peer Reviewing Process in IConGDM 2016

As the scientific editor appointed for the conference, I hereby witness that all papers accepted to the 2nd International Conference on Green Design and Manufacture 2016 (IConGDM 2016) are handled using a peer-reviewed system.

You may contact me or the secretariat committee should you need further information about the details of the peer-reviewed process handled in this conference.

Thank you.

Best Regards,

Prof. Dr. Safian Bin Sharif,
Department of Materials, Manufacturing & Industrial Engineering,
Universiti Teknologi Malaysia,
MALAYSIA
Email: safian@utm.my

Corrosion Characterisation of Al-Cu Reinforced In-Situ TiB_2

R. *Rosmamuhamadani*[1,*], S. *Sulaiman*[2], M.A. *Azmah Hanim*[3], M.I.S. *Ismail*[4], M.K. *Talari*[5], and *Sabrina* M. Yahaya[6]

[1,2,3,4]Mechanical and Manufacturing Department, Faculty of Engineering, Universiti Putra Malaysia, 43400 Serdang, Selangor, Malaysia.
[5]Materials Technology Programme, Faculty of Applied Sciences, Universiti Teknologi MARA, 40450, Shah Alam, Selangor, Malaysia.
[6]Applied Chemistry Programme, Faculty of Applied Sciences, Universiti Teknologi MARA, 40150, Shah Alam, Selangor, Malaysia

Abstract. Aluminium (Al) based in-situ metal matrix composites (MMCs) have better properties and performance compared to ex-situ MMCs. In this research, aluminium-copper (Al-Cu) alloy was reinforced with 3 and 6wt.% titanium diboride (TiB_2). Al-MMCs has been fabricated with salt route reaction process at 800 °C via potassium hexafluorotitanate (K_2TiF_6) and potassium tetrafluoroborate (KBF_4) salts. Hardness vickers tester and Gamry-Electrod Potentiometer were used to characterize the hardness properties and to determine the corrosion rate of Al-Cu alloys. From results obtained, increased TiB_2 contents will increase the hardness of Al-Cu alloys. Increased of TiB_2 contents also will increase the corrosion rate of Al-Cu alloys. Al-Cu with 3wt.%TiB_2 gave the good properties of corrosion when the wear rate recorded the lowest value compare to Al-Cu alloy itself and 6 wt.% TiB_2. The corrosion rate of Al-Cu with 3wt.TiB_2 was 16.15, while Al-Cu and Al-Cu-6wt%TiB_2 were 22.5 and 58.7 mm/y respectively.

1 Introduction

Al is the most popular matrix for MMCs. Al-MMCs has a wide application in the fields of aerospace, automobile and so on, because of their high specific strength, rigidity, wear resistance and good dimensional stability [1].

Recently, *in-situ* techniques have been developed to fabricate Al-based MMCs, which can lead to better adhesion at the interface and hence better mechanical properties. These *in-situ* routes provide many advantages such as the *in-situ* formed reinforcement phases are thermodynamically stable, disperse more uniformly in matrix, free of surface contamination and leading to stronger particle matrix bonding [2].

All metals and alloys undergo corrosion, which is defined as the destructive attack of a metal by the environment, by chemicals, or electrochemical processes [3]. The driving force is the free energy of reaction of the metal to form, generally, a metal oxide. Since

* Corresponding author: rosma614@salam.uitm.edu.my

corrosion reactions generally occur on the metal surface, they are called interfacial processes. The corrosion process takes place at the metal medium phase boundary and therefore is a heterogeneous reaction in which the structure and condition of the metal surface have a significant role. The corrosive medium must be transported to the surface and the corrosion products removed. Therefore, material transport phenomena, including free convection and diffusion into adjacent surface layers, must also be taken into account. Metallurgical factors that can affect corrosion in an alloy include: crystallography, grain size and shape, grain heterogeneity, impurity inclusions, and residual stress due to cold work.

The influence of nitric acid (HNO3-) on corrosion of the Al-Cu alloys, containing 1.3-30 at.% Cu have been studied [3]. They found the corrosion rates of the Al-1.3 at.% Cu and Al-2.7 at.% Cu alloys in HNO3- were be ~5 and 10 nm min−1 respectively. The corrosion of an Al-30 at.% Cu alloys was less uniform, with a local rate to ~13 nm s−1. Meanwhile [4] have studied the pitting corrosion of ClO4- on pure Al, Al-2.5 wt% Cu and Al-7 wt% Cu alloys in 1.0 M Na2SO4 solution at 25 °C. The susceptibility of the three Al samples towards pitting corrosion decreases in the order: Al>Al-2.5 wt% Cu>Al-7 wt% Cu.

Potentiostatic measurements showed that the rate of pitting initiation increases with increasing ClO4- ion concentration and applied step anodic potential, while it decreases with increasing Cu content. The corrosion behavior of pure Al, Al-6% Cu and Al-6% Si alloys in Na2SO4 solutions in the absence and presence of NaCl, NaBr and NaI were studied [5]. The corrosion resistance increases in the order Al<Al-6%Cu<Al-6%Si. The chloride ion showed the highest aggressiveness. The purpose of their work is to investigate the influence of Cu addition on the mechanical properties and corrosion resistance of commercially pure aluminum.

Cu addition on the mechanical properties and corrosion resistance of commercially pure Al have been investigated [6]. They studied the influence of Cu addition to commercially pure aluminum on microstructure, microhardness, grain size, impact energy, flow stress at 0.2 strain, mechanical behavior and corrosion resistance. Three different Al-Cu alloys of 3,6 and 9 wt.% Cu content were prepared and experimentally tested both mechanically and chemically. The results show the addition of Cu resulted in a linear increase of the hardness, and substantial reduction in the grain size, slight reduction the impact energy, substantial increase in the flow stress at 0.2 strains, and improve in the mechanical properties. The potentiostatic measurements showed that the susceptibility of the samples towards corrosion decreases in the order: Al>Al-3 wt% Cu>Al-9 wt% Cu>Al-6 wt% Cu. The corrosion rates of the 3, 6 and 9 wt% Cu alloys in HCl were found to be 0.29, 0.13 and 0.21 nm/s, respectively. The different properties, i.e. impact energy, flow stress at 0.2 strain, mechanical characteristics and corrosion resistance, showed that the 6 wt. % Cu is an optimal composition.

2 Methodology

2.1 Composite Fabrication

Al-6wt.% Cu was respectively melted at 720 °C in an induction furnace. The melts were homogenized for 15 minutes before added the salts which were K_2TiF_6 and KBF_4. These salts then were pre-heated to 250 °C for 1 hour. After melting of the base metals, the two salts were slowly added into the molten Al-6wt.% Cu in an atomic ratio in accordance with TiB_2 by using the stirring method.

2.2 Heat Treatment

The aging behaviour of the composites was studied by solutionising the samples at 540 °C for 2 hours followed by quenching in cold water and aging at 170 °C for different intervals. The composites subjected to age were tested their hardness properties within 48 hours regarding to [7] specifications.

2.3 Hardness Test

Vickers hardness is one of a method to measure the hardness of a material. Vickers test procedure as per [7] standard specifies making indentation with a range of loads using a diamond indenter which is then measured and converted to a hardness value.

2.4 Corrosion Characterization

Corrosion test was carried out on each alloy in the as-cast conditions. Potentiostatic polarization measurements were carried out using a Radiometer Analytical model PGZ 100 Potentiostat/Galvanostat with VoltaLab software. The working electrodes employed were the graphite and Al-Cu bars. Anodic and cathodic polarization curves were plotted. The investigated electrodes were cut as cylindrical rods, welded with Cu-wire for electrical connection to contact the test solution. The experiments were performed in a 250 ml volume Pyrex glass cell using Pt wire and a saturated calomel electrode as auxiliary and reference electrodes, respectively. All potentials given in this research are referred to this reference electrode. The experiments were carried out in 0.5 M NaCl solution. The NaCl solution was freshly prepared from analytical grade using doubly distilled water. For each run, a freshly prepared solution as well as a cleaned set of electrodes was used.

3 Results and Discussion

3.1 Hardness Properties

Hardness is one of the most important properties, which is commonly used to give a general indication of the strength and resistance to wear and scratching of a material. It can be defined as the ability of a material to resist permanent indentation or deformation when it is in contact with an indenter under load.

Table-1. Vickers hardness properties of un-aged and aged Al-Cu with different TiB_2 contents

Materials	Vickers Hardness Al-Cu alloy (Hv)		
	Un-aged	**Aged (540 °C, 2h)**	**Aged (150 °C, 48h)**
Al-Cu	125	136	174
Al-Cu-3wt.% TiB_2	134	142	175
Al-Cu -6wt.%TiB_2	141	152	150

For Vickers hardness, as references, the initial hardness of un-aged Al-Cu was 113 Hv for 5 N load. Similar trend can be shown in aged Al-Cu which was 136 Hv. Both of un-aged and aged-Al-Cu alloys, the value of hardness increased when TiB_2 contents increased.

 DOI: 10.1051/matecconf/20167801001

The hardness of un-aged Al-Cu-6wt.%TiB_2 were recorded 125 with 5 N load, while aged Al-Cu-6wt.%TiB_2 were 152 Hv. The addition of TiB_2 resulted in an increased of the hardness of Al-Cu alloy. The increase in hardness of the composites can be attributed to the decrease in the grain size to TiB_2 particles. It is evident that addition of TiB_2 leads to significant increment in the hardening by the precipitates, which can be attributed to additional strengthening due to coherency strains and misfit dislocations formed by the precipitation on TiB_2 particles. These results then are illustrated graphically in Table 1 and Figure 1 below.

The properties of a metal are strongly influenced by the grain size and the well-known Hall–Petch equation. It predicts an increase in yield strength with a decrease in grain size (d). This equation has been shown to be applicable to a wide variety of metals.

$$\sigma_y = \sigma_0 + kd^{-1/2} \tag{1}$$

Where σy, σo, and k is yield strength, friction stress and a constant respectively. In the as-cast condition, the hardening of the Al–Cu–TiB_2 composites can be attributed to the fine TiB_2 particles and reduction in grain size (due to grain refinement by TiB_2). In the case of solutionised composites, the hardness is governed by solid solution strengthening, in addition to Hall-Petch hardening and hardening from TiB_2 particles. The increase in the hardness of the solutionised composite with increase in fraction of TiB_2 particles clearly gives the contribution of TiB_2 particles to the hardness of the composites.

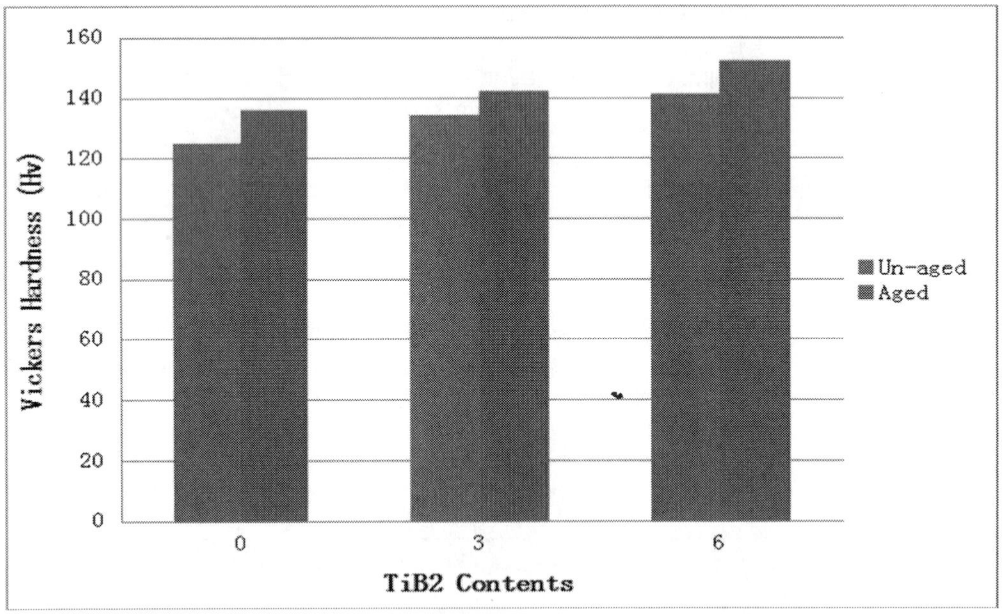

Fig. 1. Vickers hardness (Hv) of Al-Cu with different TiB_2 contents with 5N loads

3.2 Effect of ageing

Thus, the presence of higher amount of TiB_2 particles enhances the aging kinetics of Al–Cu alloy. The increase in aging kinetics in presence of TiB_2 particles could be attributed to the possible nucleation of the precipitates of Al2Cu on the basal planes of the TiB_2 particles. In addition, the difference in the coefficient of thermal expansion between Al (23.5×10^{-6} K−1) and TiB2 (7×10^{-6} K−1) causes the development of strain field at the interface between the reinforced particle and the matrix, which leads to an increase in the dislocation density. These dislocations can also act as nucleating sites during precipitation. From the research, tensile strength and Vickers hardness gave highest value compare alloy without aged.

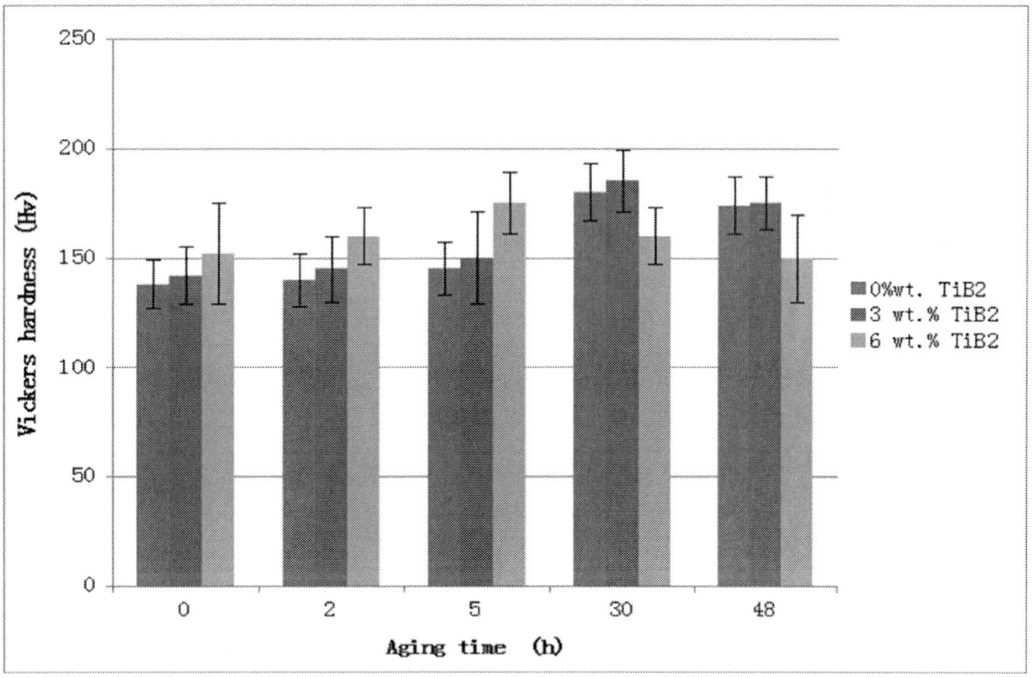

Fig. 2. Hardness of Al-Cu with different TiB_2 contents after aged for 48 hours

Fig. 2 above shows the age hardening response of the base Al-6Cu alloy and its composites. It is clearly evident from the figure that the peak hardness for Al-6Cu, Al-6Cu-3wt.%TiB2 and Al-6Cu-6wt.%TiB_2 were reached at 30, 30 and 5 hours, respectively. The figure also shows that the peak hardness increases with increasing reinforcement TiB_2 content (wt.%). The hardness of Al-6Cu alloy increases from 170 Hv in the base alloy to 188 Hv for the 6TiB_2 composite at optimum peak. Thus, the TiB_2 particles not only increase the hardness but also enhance the aging kinetics of Al-6Cu alloy.

3.3 Corrosion Characterization

Polarization is the displacement of an electrode potential from its equilibrium value and the magnitude of this displacement is the overvoltage which is expressed in terms of plus or minus volts (or mV) relative to the equilibrium potential. Potentiostatic polarization can be used to calculate the electrochemical parameters. Anodic and cathodic polarized potentials were measured to a reference sample and Al-Cu alloys in 0.1 M HCl solution in current density range of 2 to 6 mA/cm2 as shown in Figure 3. The electrochemical parameters;

corrosion potential (Ecorr), corrosion current (Icorr) and corrosion current density (in nm/s) were calculated from Figure 3 and summarized in Table 2. Corrosion potential decreases with increasing the TiB2 content in the alloys. The corrosion current decreases with increasing the TiB2 content reaching a minimum at 3 wt. %TiB_2 and then begin to increase with increasing the TiB2 content.

Table 2. Electrochemical parameters for Al-Cu with different TiB_2 alloys in 0.1 M HCl solution

TiB_2 [%]	E_{corr} [mV] (10^{-6})	I_{corr} [mA/cm²] (10^{-3})	CR [mm/y]
0	-568.0	1.942	22.50
3	-566.0	1.39	16.15
6	-574.9	5.06	58.7

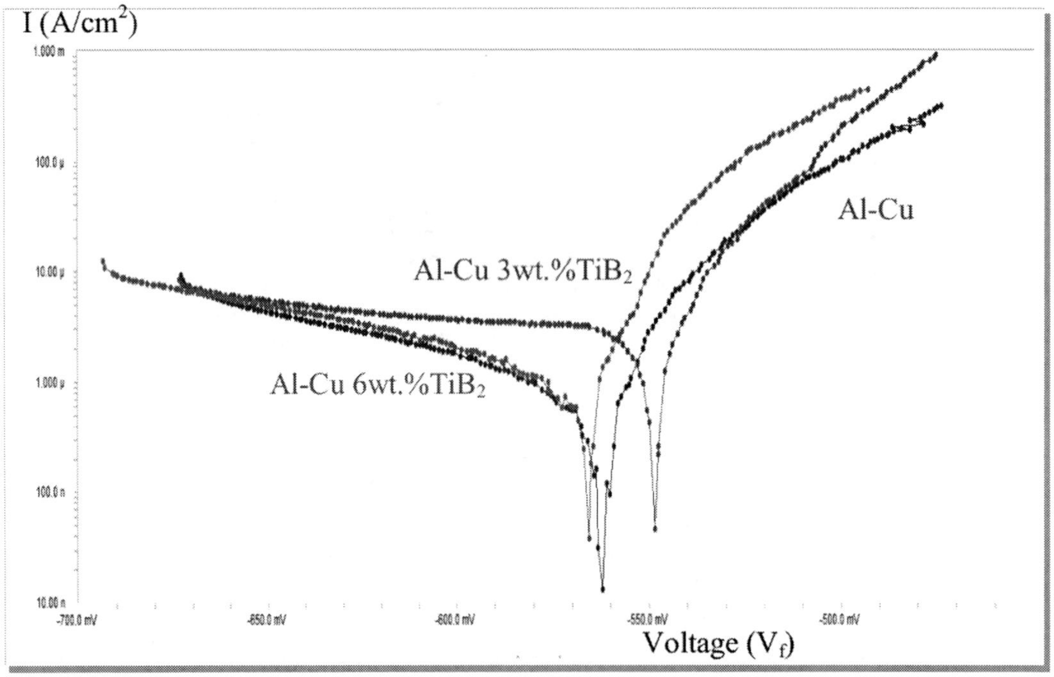

Fig. 3. Anodic and cathodic polarized curves of Al-6wt.%Cu alloys and Al-6wt.%Cu-TiB_2 in 0.5 M NaCl solution

The corrosion current and corrosion rate show an optimal value at 3 wt. %TiB_2. Addition of TiB_2 has a positive influence on the corrosion resistance in Al. This behavior can be attributed to the positive influence of TiB_2 addition on the grain size refinement. The fine grained materials have a more advantageous behavior with respect to corrosion and oxidation have been approved [8].

From Figure 3 above, Al-6wt.%Cu alloy with 3wt.%TiB_2 showed the best corrosion resistance compare to other composition of TiB2. The value of corrosion rate at 3 wt.% TiB2 was 16.15 x 10^{-3} mm/y, then decrease to 58.8 x 10^{-3} mm/y for Al-Cu with 6wt.%TiB2. The susceptibility of the Al-Cu alloys towards corrosion decreases in the order of:

$$\text{Al-Cu-3wt.\%TiB}_2 > \text{Al-Cu} > \text{Al-Cu-6 wt\%TiB}_2 \tag{2}$$

The lowest value of corrosion rate exhibit the alloy has a good property to withstand the corrosion.

4 Conclusions

From the research, we can conclude several conclusions which are;

i. *In-situ* Al-Cu alloy composites containing different weight fractions of particles of TiB_2 phase were synthesized successfully by the salt-metal reaction method and the particles were distributed evenly in the matrix of the composites.
ii. The susceptibility of the Al-Cu alloys towards corrosion decreases in the order of: Al-Cu-3wt.%TiB_2> Al-Cu> Al-Cu-6 wt%TiB_2.
iii. The composition of 3wt.%TiB_2 gave the best corrosion rate compare to cast Al-Cu alloy which were 16.15 and 22.50 x 10^{-3} mm/y.

This work was financially supported by FRGS from Ministry of Education (MOHE), Malaysia and Faculty of Applied Sciences UiTM, Shah Alam, Selangor.

References

1. S. Kumar, M. Chakraborty, V.S. Subramanya, B.S. Murty, Wear, **265**, 134 (2008)
2. H. Wang, L. Guirong, Z. Yutao, C. Gang, Mater. Sci. Eng., A **527,** 2881 (2010)
3. Y. Liu, M.A. Arenas, A. de Frutos, J. de Damborenea, A. Conde, P. Skeldon, G.E. Thompson, P. Bailey, T.C.Q. Noakes, Electrochim. Acta, **53**, 4454 (2008)
4. M.A. Amin, S.S. Abd. El-Rehim, S.O. Moussa, A.S. Ellithy, Electrochim. Acta, **53,** 5644 (2008)
5. S.S. Abdel Rehim, H.H. Hassan, M.A. Amin, Appl. Surf. Sci., **187**, 279 (2002)
6. A.E. Al-Rawajfeh, S.M.A. Al-Qawabah, Emirates Journal for Engineering Research, **14**, 47 (2009)
7. *Standard Test Method for Vickers Hardness of Metallic Materials*, **82** (ASTM E92, 2003)
8. K.T. Kashyap, T. Chandrashekar, Bull. Mater. Sci., **24**, 345 (2001)

Design and Development of a Solar Thermal Collector with Single Axis Solar Tracking Mechanism

Mogana *Theebhan*[1,*], Chua Yaw *Long*[2], *and* Tay Tee *Tiong*[3]
[1]Mechanical Engineering Department, College of Engineering, Universiti Tenaga Nasional, Malaysia
[2]Centre for Renewable Energy, Universiti Tenaga Nasional, Malaysia
[3]Faculty of Engineering and Built Environment, Tunku Abdul Rahman College University, Malaysia

Abstract. The solar energy is a source of energy that is abundant in Malaysia and can be easily harvested. However, because of the rotation of the Earth about its axis, it is impossible to harvest the solar energy to the maximum capacity if the solar thermal collector is placed fix to a certain angle. In this research, a solar thermal dish with single axis solar tracking mechanism that will rotate the dish according to the position of the sun in the sky is designed and developed, so that more solar rays can be reflected to a focal point and solar thermal energy can be harvested from the focal point. Data were collected for different weather conditions and performance of the solar thermal collector with a solar tracker were studied and compared with stationary solar thermal collector.

1 Introduction

Energy plays an important role in the development every country in the world. Though abundant energy sources are available, renewable energy is believed to be the key source of energy for the future. It is no doubt that renewable energy technologies also contribute to sustainable development.

Malaysia is a country which is located near the equator. This location provided Malaysia high solar energy potential. The daily solar energy potential was recorded to be around 4 kWh/m – 5 kWh/m. With an average sunshine duration of 4-8 h/day, Malaysia has a favorable climatic condition for the further development of solar energy. Solar thermal usage should be encouraged by the public.

Throughout the world, solar thermal collector had already been successfully used for various medium temperature applications. Solar thermal collectors were used to collect solar energy and convert it into another form of energy, namely thermal energy. The majority of these solar thermal collectors was installed fixed at a particular angle directed to the sun. As the sun moves from east to west from sunrise to sunset, and from north to south throughout the course of the year, this resulted in the solar thermal being unable to harvest the solar thermal energy to its maximum capacity. This drawback can be further overcome by applying a solar tracker system to the solar dish. The solar tracker system will change

[*] Corresponding author : chuayl@uniten.edu.my

the position of the collector dish automatically in accordance with the location of the sun in the sky.

2 Literature Review

Based on the literature review conducted, there are two main factors that affect the efficiency of the solar thermal collector. The first one is usage of solar tracking systems. The second factor would be the reflector used.

In 2012, Liene Kancevica, Henriks Putans, and Imants Ziemelis conducted a simple research and compared the efficiency of the solar collector with and without the presence of solar tracking system. According to their result, a solar collector dish equipped with a solar tracking system is proved to be able to produce 1.4 times more energy compared to the fixed angle solar collector. This experiment was carried out using a 1 m^2 collector dish. Besides, increasing the number of mirrors on the collector could also increase the overall efficiency [1].

In another research conducted by Fareed M. Mohamed et al, the research, talked about the operating process of solar thermal collector [3]. In this work, the researchers first designed and fabricated a solar dish concentrator according to corresponding theoretical data. The parabolic concentrator that the research team had come up with having high sunlight reflectivity (up to 76%). This high reflectivity potential had increased the outlet water temperature in receiver cavity. Consequently, it managed to increase the system operational efficiency. The researchers claimed that their product had increased the operational efficiency up to 30%.

In 2013, Deepthi et al. [6], presented a research paper that compared the performance of single axis solar tracking system and dual axis solar tracking system with the fixed mount solar system. Based on their research, it is known that the dual axis tracking system performed much better than the single axis tracker system. This better performance even included cloudy weather condition. This proved that the efficiency of dual-axis tracker system is higher. Even though the hardware complexity is higher in the dual axis tracker, it provided a higher efficiency and better performance when compared to a single axis tracker.

3 Methodology

The following flow chart in Fig. 1 illustrates the different phases in the research project.

The working principle for the solar dish to collect the thermal energy was identified in Phase 1 of the project. Two units of similar solar thermal collector dishes were designed using the parabolic dishes. One is meant for stationary collector, while another one will be equipped with the solar tracking mechanism. The dishes will be used to heat up a container that holds 1L of water at the focal point of the dish. The type of the solar tracker is later identified. Figure 2 illustrates the drawing of the designed solar thermal dish.

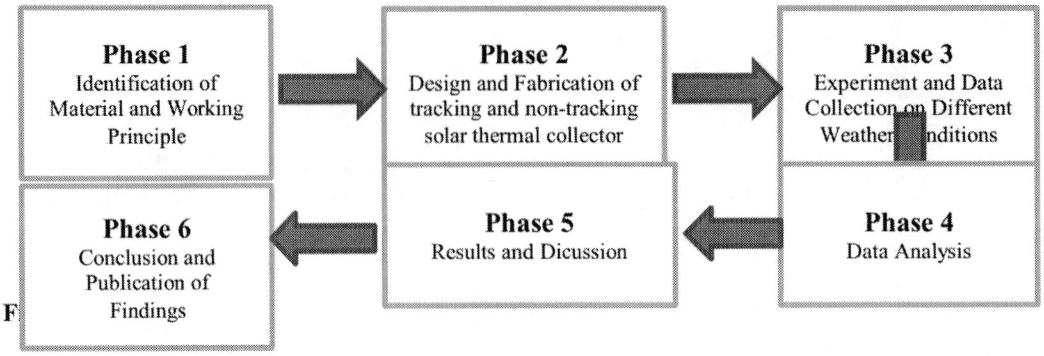

F

Fig. 2. Design of Solar Thermal Collector

In the second phase, the fabrication of the prototype was carried out. Appropriate solar tracker was acquired and installed in the second unit of the solar thermal dish.

An experiment to obtain the temperature difference between the solar thermal collectors equipped with solar tracker and non-solar tracker was carried out in the third phase. The data collection was carried out for 4 days in UNITEN Golf Range under the sunny conditions and also overcast conditions. The temperature of the water inside the container was measured using a digital thermometer placed in the water container. Reading was obtained with an interval of 60 minutes.

Fig.3. Stationary Solar Thermal Collector and Solar Tracking Solar Thermal Collector Collecting Data

Based on the results obtained from the experiment, the efficiency of solar concentrator is then calculated based on the solar tracking equipped device. First Law of Thermodynamics is used in calculating the efficiency of the overall system.

$$\text{Energy input} = \text{Energy loss} + \text{Energy Output} \tag{1}$$

The energy collected at the solar concentrator can be determined using the following equation:

$$E_t = I_t \times A_{SC} \tag{2}$$

Where:
E_i is the energy input in W
I_t is the total solar energy incident upon the plane of the solar air being heated in W/m^2
A_{sc} is the surface area of the solar cooker

The energy output from the solar concentrator can be found as shown below:

$$E_o = \frac{m_w c_{pw}(T_{wf} - T_{wi})}{t} \tag{3}$$

where:
E_o is the energy output in W
m_w is the mass of water in kg
c_{pw} is the specific heat of water in J/kgK
A_{sc} is the surface area of the solar concentrator in m^2

T_{wi} is the initial temperature of the water in K
T_{wf} is the final temperature of the water in K
t is the time in seconds

The efficiency of the solar concentrator can then be determined using the equation below:

$$\eta = \frac{Energy\,Output}{Energy\,Input} = \frac{E_o}{E_i} = \frac{[m_w C_{pw}(T_{wf} - T_{wi})]/t}{I_t \times A_{SC}} \quad (4)$$

4 Results

Figure 4 till Figure 8 illustrate the data obtained during the 4 days data collection using the solar dishes. The raw data shows that there were 2 days with sunny condition, 1 day with overcast weather condition while another day with sunny and overcast condition.

In the 4 days where data were collected, both tracking and non-tracking solar thermal collector were at work. The temperatures of the water at the focal point were collected, and increment in terms of efficiency is calculated and compared for all 4 days.

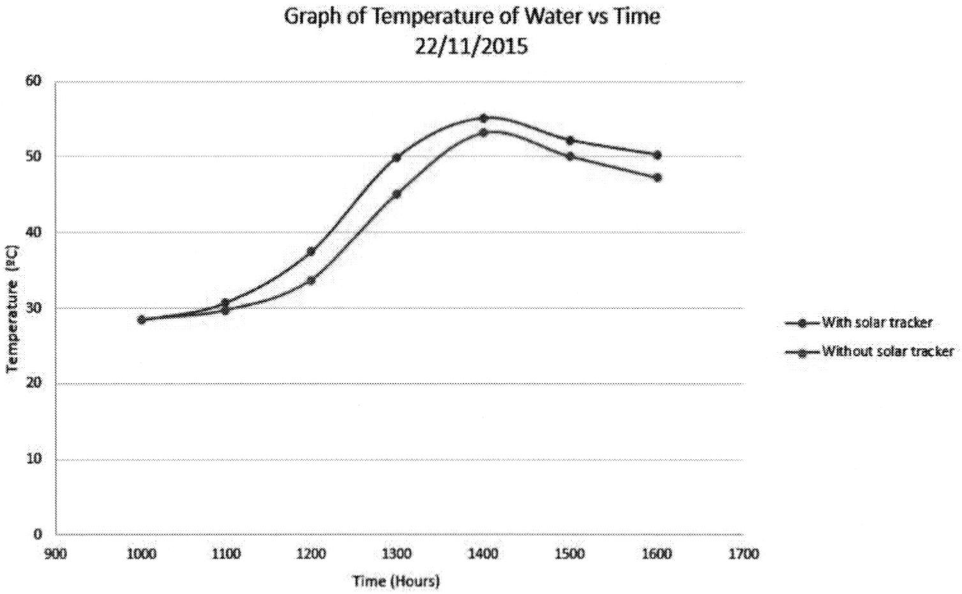

Fig. 4. Temperature Difference for Sunny and Overcast condition

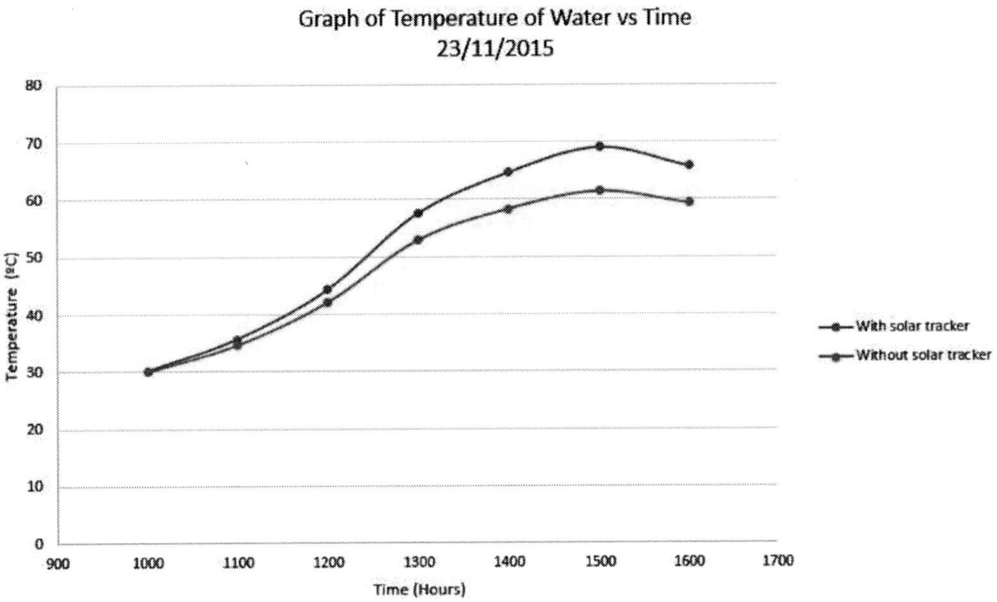

Fig. 5. Temperature Difference for Sunny Day Condition (Day 1)

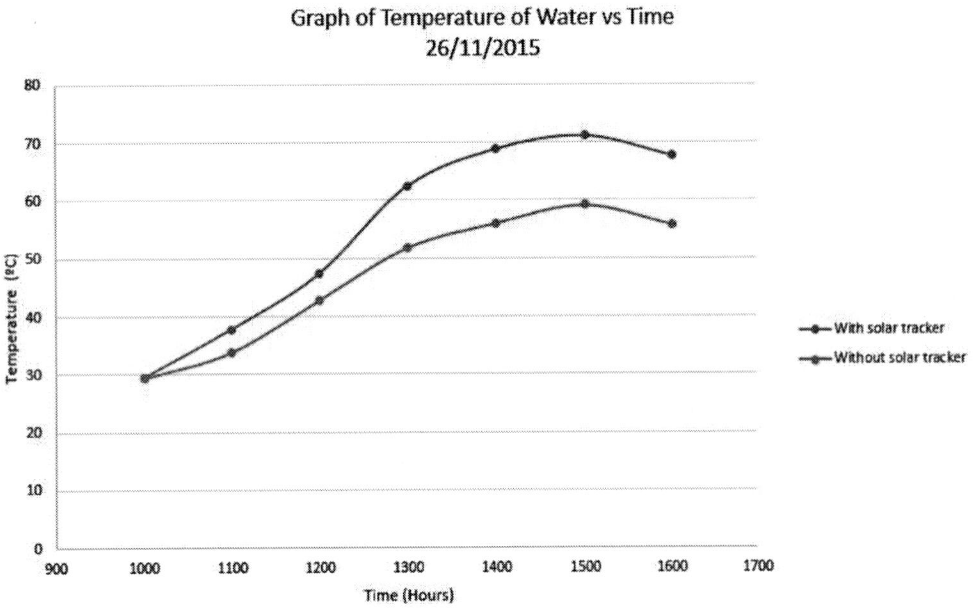

Fig. 6. Temperature Difference for Sunny Day Condition (Day 2)

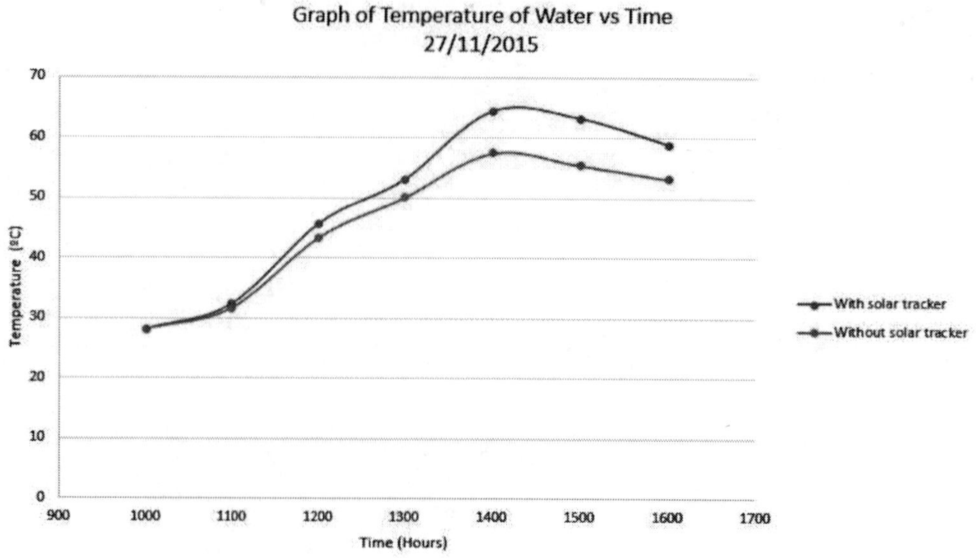

Fig. 7. Temperature Difference for Overcast Condition

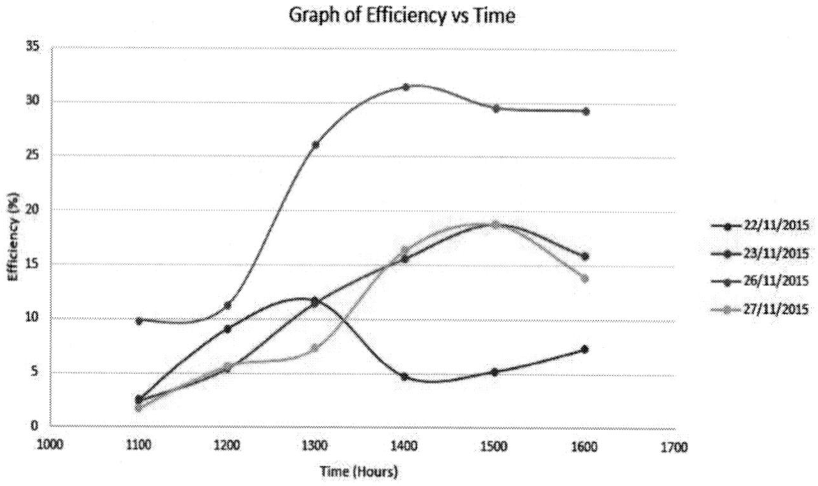

Fig. 8. Efficiency Increment

5 Discussion

From the results obtained, it is clear that the temperature of the water in the container is much higher for the solar thermal collector with tracking mechanism. Thus, the increment in efficiency is higher during sunny day's weather conditions where up to 31 % increment in efficiency.

The efficiency increment in both sunny and overcast, as well as overcast weather condition are lower compared to the sunny day condition. However the increment in efficiency can go up from 1.7 % to 18.76 %.

6 Conclusion

In this research project, a stationary solar dish concentrator as well as other solar dish concentrator equipped with had been designed and fabricated. The parabolic concentrator reflected the solar radiation at the focal point and increased water temperature in the container, and consequently arising system operational efficiency. Solar thermal collector, equipped with tracking system increased the operational efficiency to 31 %.

References

1. L. Kancevica, H. Putans, I. Ziemelis, *Proceeding of the International Scientific Conference "Renewable Energy and Energy Efficiency*, 190 (Jelgava, Latvia, 2012)
2. X. S. Jin, G.Q. Xu, R.J. Zhou, X. Luo, Y.K. Quan, Int. J. Clean. Coal. Energy., **2**, 16 (2013)
3. F.M. Mohamed, A.S. Jassim, Y.H. Mahmood, M.A.K. Ahmed, Int. J. Recent Res. Rev., **3**, 52 (2012)
4. M. Ghassoul, Int J of Emerging Tech. & Advanced Engineering, **3**, 453 (2013)
5. P.J. Hession, W.J. Bonwick, Solar Energy, **32**, 3 (1984)
6. C.S. Chin, A. Babu, W. McBride, Ren. Energy, **36**, 3075 (2011)
7. C. Philibert, *The present and future use of solar thermal energy as a primary source of energy,* (InterAcademy Council, 2005)
8. G.K.A. Sada, A.E.S. Aldeen, International Journal of Engineering and Innovative Technology (IJEIT), **4**, 147 (2015)

Innovation of Embedding Eggshell to Enhance Physical-Mechanical-Thermal Properties in Fired Clay Bricks via Extrusion Process

Nuchnapa Tangboriboon[1,*], *Sopita* Moonsri[1], *Atima* Netthip[1], and *Anuvat* Sirivat[2]

[1]Department of Materials Engineering, Faculty of Engineering, Kasetsart University ,Bangkok, 10900, Thailand.
[2]The Petroleum and Petrochemical College, Chulalongkorn University, Bangkok, 10330, Thailand.

Abstract. Eggshell functioned as a bio-filler and sintering aid added into earthenware clay compounds to make fire clay bricks via extrusion process. The main composition of eggshell is calcium carbonate ($CaCO_3$) while the main composition of earthenware is aluminosilcate ($Al_2O_3.2SiO_2.2H_2O$) and other oxide compounds. Calcium carbonate in eggshell can react to chemical substances in earthenware clay compound to form calcium alumnosilicate or anorthite and wollastonite or calcium silicate phase formation of the fired clay bricks. Adding 20%wt eggshell powder into clay bricks and fired at 1000°C for 5 hrs., affects to the physical-mechanical-thermal properties, i.e. high compressive strength and hardness, good thermal expansion coefficient, good true density, and low water absorption. In general, the theoretical compressive strength, hardness, and water absorption of desired refractory are more than 7.0 MPa, 6.0 HV, and less than 25%wt, respectively. Therefore, the obtained fired clay bricks added eggshell powder and fired at 1000°C for 5 hrs., are potential for use as bricks for construction and thermal insulation.

1 Introduction

Eggs have been used as foods, drugs, cosmetics, and foodstuffs. Unfortunately, eggshells are an egg product industry residue that contributes to environmental pollution as it favors microbial actions. The by-product eggshell represents approximately 11% of the total weight approximately 60 g of egg [1-2]. Eggshell is one kind of sources of calcium carbonate ($CaCO_3$) known as calcite which is the more stable polymorph at room temperature. An eggshell is a rich source of mineral salts, mainly calcium carbonate (94%), calcium phosphate (1%), magnesium carbonate (91%), and organic substances (4%) [2]. However, an important composition of eggshell is the calcium carbonate, which can be used in various material applications: dielectrics, dental and bone implantation, gypsum, catalysts, and also fillers used for in a variety industries such as rubber, ink, foodstuff, cosmetic, pharmaceutical, and fertilization.

Brick is a ceramic material mainly used in the building industry. There are divided into various groups on the basis of the mainly raw materials used, namely silica bricks, magnesia bricks, zirconia bricks, alumina bricks, mullite bricks, dolomite bricks, silicon carbide bricks, graphite bricks, and clay bricks [3-19]. Bricks are the major-construction and building material for a long time. The worldwide annual production of bricks is currently many thousand billion units and has trended the demand to increase every year. Conventional bricks are produced from clay and fired at high temperature. Earthenware is

* Corresponding author: fengnnpt@ku.ac.th

one kind of clay raw materials to produce clay brick because earthenware is the lowest price of raw material used, high percentage of iron oxide content, red or yellow color after firing, opaque, high plasticity, easily forming. Earthenware can be fired for biscuit at 1000°-1150°C. The most application of earthenware are the building bricks in block shape or rectangular building unit. There are many sources of earthenware in the provinces of Thailand such as Udon Thani (Ban Chiang), Sukhothai (Si satchanalai), Sawankhalok, Ayutthaya, Lopburi, Singburi, Kanchanburi, and Angthong which are high potential to make the good and strong traditional clay-bricks.

The annual growth rate of clay bricks using for construction, infrastructure, and industries, i.e. furnace for steel and jewelry making, materials, casting process, and others, increase continuously demand effect of the high rate of brick production for building enclosures. However, both economic and environmental problems during process need to concern on energy consuming, pollution, cost of production, and raw material sources. Therefore, one of the best ways of saving energy consumption and reducing eggshell waste is applied eggshell as a $CaCO_3$ filler and flux in clay bricks. In addition, recycled eggshell as a bio-filler can decrease the landfill pollution and contributes as an asset to the agriculture rather than being a waste material.

The objective of this work was to make fire clay bricks from earthenware clay mixed with eggshell ($CaCO_3$) powder. The obtained fired clay bricks which have good physical, mechanical, and thermal properties, lightweight, and low water absorption.

2 Experimentals

2.1 Materials and methods

Earthenware clay was obtained from Pa Mok district, Ban Norasing, Angthong province, Thailand. Hen eggshell was collected from the local cafeteria of the University. Eggshell was cleaned with tap water and let it dry in the air at room temperature for 1-2 days. The cleaned hen eggshell was ground with the rapidly porcelain ball mill to eggshell powder for 120 min.

2.2 Instruments

Extruder was used to prepare clay brick samples supplied by Petkasem Co.,Ltd., Thailand. Extruder composed of a screw to extrude the clay mixture from the hopper to the die. Barrel in the extruder can protect the mixture flow out the screw and get rid of the air containing in the mixture. Oven was used to dry the green products at 100°C for 24 hrs by using the WTB binder. Furnace was used to fire the clay bricks at 800°, 900°, 1000°, and 1100°C for 1, 3, and 5 hrs., at each firing temperature. The used furnace is Mudotemp, WW 51A model 1400°C.

2.3 Clay bricks preparation

The green products of clay brick samples were prepared according to the Table 1. The raw materials were mixed and kneaded to homogenous mixture and extrude pass through the die of extruder in a cylindrical shape having 5.00 cm diameter and 1.00 cm in thickness. The obtained clay samples or green products were dried at 100°C for 24 hrs., and then fired at 800°, 900°, 1000°, and 1100°C for 1, 3, and 5 hrs., at each firing temperature, respectively.

Table 1. Raw materials were mixed and needed to homogenous mixture and extrude pass

Samples	Earthenware clay (g)	Eggshell powder (%wt)	Water (%wt)
Formula 1	12,000	0	40
Formula 2	11,400	5	40
Formula 3	10,800	10	40
Formula 4	10,200	15	40
Formula 5	9,600	20	40

3 Results and discussion

3.1 Physical properties of clay bricks

The color of green products means the color of samples before firing as shown in Fig.1.

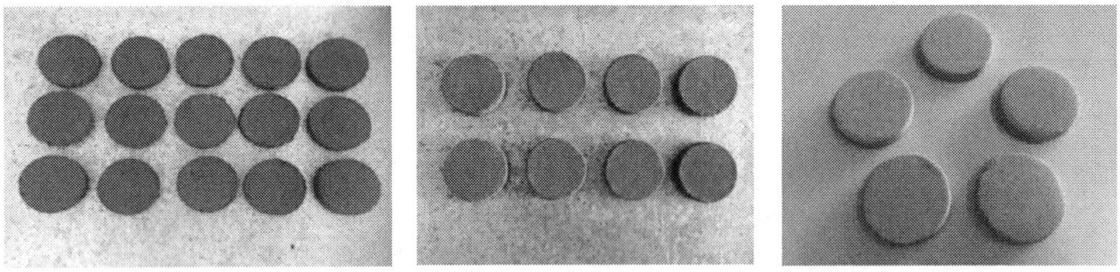

Fig. 1. Clay brick samples: a) the green products before drying; b) the green products after drying in an oven at 100°C for 24 hrs.; and c) the clay bricks after firing at 800°-1100°C.

Most color of clay bricks before firing or green products after kneading and extrusion are dark yellowish brown. After firing, the colors of clay bricks are reddish yellow to pale yellow.

The chemical composition of clay and chicken eggshell for clay refractory forming was measured by XRF as the data tabulated in Table2. The main composition of earthenware clay composed of 26.46%wt SiO_2, 22.32%wt Fe_2O_3, 20.66%wt Al_2O_3, and other oxide compounds. While the main composition of chicken eggshell composed of 98.101%wt $CaCO_3$ and the rest other oxide compounds.

Table 2. Chemical composition by XRF

Composition	Earthenware clay (%wt)	Hen eggshell (%wt)
SiO_2	26.46	0.073
Fe_2O_3	22.32	-
Al_2O_3	20.66	0.035
CaO	10.55	-
MgO	8.91	0.286
Na_2O	4.43	0.204
K_2O	4.39	0.038
TiO_2	2.28	-
$CaCO_3$	-	98.101
CuO	-	0.009
P_2O_5	-	0.443
SrO	-	0.013
SO_3	-	0.798
Total	100.00	100.00

"-" means not detected.

The physical properties (bulk density, true density, color, and water absorption) of clay brick samples added 0%, 5%, 10%, 15%, and 20% wt chicken eggshell powder fired at 800°C, 900°C, 1000°C, and 1100°C for 1, 3, and 5 hrs., are data tabulated in Table 3. Firing temperature and firing time are important factors to obtain the good clay bricks. The high firing temperature and firing time use, the better and higher strength of clay bricks obtained. The firing temperature is low less than 1000°C (800°C and 900°C), the samples are not good and low strength. However, if the firing temperature is too high more than 1000°C effect to crack on clay brick samples. Therefore, the suitable firing temperature of clay brick samples added eggshell powder is 1000°C for firing time 1, 3, and 5 hrs. Furthermore, if the percentage of chicken eggshell powder added in the clay bricks is high, the percentage value of water absorption is low and good appearance. The obtained clay brick samples added 20%wt eggshell powder are the optimum amount of hen eggshell adding. The average water absorption of adding 20%wt eggshell is equal to 14.36-14.88% consistent with the results obtained by Raut et al [9]. It means high densification effect to samples having high mechanical and good thermal properties. Furthermore, the obtained water absorption of clay brick samples has low values consistent with the theoretical values of the best refractory measured according to the Thai Industrial Standard Institute 243/2530 and the ASTM C 373-72 less than 25%wt water absorption.

Table 3. Physical properties of brick samples

Samples [a]	Bulk density [b] (g/cm^3)	True density [c] (g/cm^3)	Water absorption [d] (% wt)	Color of bricks
0-1000-1	1.80±0.04	2.05±0.01	16.69 ± 0.04	Red yellow
0-1000-3	1.83±0.01	2.07±0.02	15.79 ± 0.66	Red yellow
0-1000-5	1.83±0.03	2.12±0.07	15.37 ± 1.28	Red yellow
5-1000-1	1.71±0.03	1.97±0.05	N/A	Red yellow
5-1000-3	1.73±0.03	2.01±0.05	N/A	Red yellow
5-1000-5	1.75±0.01	2.05±0.02	N/A	Red yellow
10-1000-1	1.69±0.02	2.03±0.03	N/A	Red yellow
10-1000-3	1.67±0.01	2.02±0.03	N/A	Red yellow
10-1000-5	1.70±0.02	2.04±0.05	N/A	Red yellow
15-1000-1	1.63±0.02	1.97±0.01	N/A	Red yellow
15-1000-3	1.65±0.02	1.99±0.01	N/A	Red yellow
15-1000-5	1.64±0.03	2.00±0.02	N/A	Red yellow
20-1000-1	1.83±0.01	2.06±0.08	14.88 ± 0.89	Yellow
20-1000-3	1.70±0.02	1.98±0.02	14.69 ± 1.04	Yellow
20-1000-5	1.75±0.04	1.96±0.02	14.36 ± 0.80	Yellow

Remark: N/A means not measured.

[a] means 5 pieces of clay brick samples at each condition.

[b] means density including open and closed pores within the clay samples according to the ASTM B212-82.

[c] means density without open and closed pores within the clay samples due to boiled in water for 5 hrs., and let them in cool water for 24 hrs., before calculation according the ASTM B212-72 and ASTM C 373-72.

[d] means water absorption according to the ASTM C 373-72.

Encoded samples x-xxxx-x mean amount of eggshell powder adding-firing temperature-firing time i.e. 0-1000-3 (no adding (0%wt) eggshell powder -firing at 1000°C-firing time for 3 hrs).

When the clay bricks were mixed with the eggshell powder 0%, 5%, 10%, 15%, and 20%wt effect on the color of mixture changed according to the Munsell soil color chart from dark yellowish brown to very pale yellow as reported in Table 4. The clay brick samples were prepared by extrusion in a disk-like shape, having 5.00 cm diameter and 1.00 cm thickness. All clay brick samples were fired at 800°C, 900°C, 1000°C, and 1100°C for 1, 3, and 5 hrs., at each firing temperature. The colors of clay brick samples changed to reddish yellow or yellow as reported in Table 4 due to the amount of iron oxide (Fe_2O_3) having red color consistent with the results obtained by XRF and amount of hen eggshell powder adding.

The XRD peak patterns comparison of raw materials (earthenware clay and hen eggshell powder) and clay brick samples added 0% and 20%wt hen eggshell powder fired at 1000°C for 1, 3, and 5 hrs., are shown in Fig. 2. The XRD peak patterns of earthenware clay at 2Θ (26.832°, 50.079°, and 60.026°) consistent with the JCPDS nos. 01-089-8936 (quartz), 00-003-0447 (aluminum phosphate), and 00-019-0615 (iron oxide, Fe_2O_3) are hexagonal phase formation of SiO_2-Al_2O_3-Fe_2O_3. While the XRD peak patterns of chicken eggshell powder show rhombohedral of calcium carbonate ($CaCO_3$) or calcite consistent with the JCPDS nos. 01-072-1937 at 2Θ (29.369°, 48.452°, 47.450°, 39.370°, and 43.118°). The XRD peak patterns of fired clay brick without adding eggshell powder and fired at 1000°C for 1, 3, and 5 hrs., are shown only hexagonal crystalline phase formation of SiO_2-Al_2O_3-Fe_2O_3 consistent with the JCPDS nos. 01-089-8936 (quartz, SiO_2), 00-049-0134 (aluminum oxide, Al_2O_3), and 00-019-0615 (iron oxide, Fe_2O_3). The XRD peak patterns of fired clay brick samples added 20%wt eggshell powder and fired at 1000°C show sharp two peaks of anorthic phase formation of JCPDS nos. 00-003-0559 calcium aluminum silicate ($CaAl_2Si_2O_8$ or calcium feldspar) and calcium silicate ($CaSiO_3$ or wollastonite) at 2Θ

(26.881°-28.776°, 23.205°, 36.650°, and 43.473°) mixed with the cubic phase formation of calcium oxide (CaO) of JCPDS no. 00-048-1467 at 2Θ (37.360°, 53.860°, and 32.200°) and consistent with the results reported by Gökhan, G. et al. [13], Cultrone, G., et al. [15], Jordan, M.M et al.[20], Xia, W., et al. [21], and Wan, X. et al. [22]. Furthermore, the XRD peak patterns of fired clay bricks added 20%wt eggshell powder and fired at 1000°C show hexagonal phase formation of SiO_2-Al_2O_3-Fe_2O_3 consistent with JCPDS nos. 01-089-8936, 00-049-0134, and 00-019-0615. The sharp XRD peak at 2Θ 28.776° belongs to calcium aluminum silicate ($CaAl_2Si_2O_8$ or calcium feldspar) occurring obviously when the clay bricks were fired at 1000°C for 5 hrs., consistent with the results obtained by Jordan, M.M. et al. [20]. It means completely vitreous firing effect of high mechanical and good thermal properties of products.

Table 4. Color of clay brick before and after firing

Encoded samples (% wt eggshell-firing temperature)	Color of samples after firing (Munsell soil color chart)	Color of samples after firing
0 %wt-before firing	10YR5/6	dark yellowish brown
5 %wt-before firing	10YR5/6	yellowish brown
10 %wt-before firing	10YR7/4	very pale brown
15 %wt-before firing	10YR7/4	very pale brown
20 %wt-before firing	10YR7/4	very pale brown
0 %wt-800 °C	7.5YR7/8	reddish yellow
0 %wt-900 °C	7.5YR7/8	reddish yellow
0 %wt-1000 °C	7.5YR7/8	reddish yellow
0 %wt-1100 °C	7.5YR7/8	reddish yellow
5 %wt-800 °C	7.5YR7/6	reddish yellow
5 %wt-900 °C	7.5YR7/6	reddish yellow
5 %wt-1000 °C	7.5YR7/6	reddish yellow
5 %wt-1100 °C	7.5YR7/6	reddish yellow
10 %wt-800 °C	7.5YR7/6	reddish yellow
10 %wt-900 °C	7.5YR7/6	reddish yellow
10 %wt-1000 °C	7.5YR7/6	reddish yellow
10 %wt-1100 °C	7.5YR7/6	reddish yellow
15 %wt-800 °C	7.5YR7/6	reddish yellow
15 %wt-900 °C	7.5YR7/6	reddish yellow
15 %wt-1000 °C	7.5YR7/6	reddish yellow
15 %wt-1100 °C	7.5YR7/6	reddish yellow
20 %wt-800 °C	10YR8/6	Yellow
20 %wt-900 °C	10YR8/6	Yellow
20 %wt-1000 °C	10YR8/6	Yellow
20 %wt-1100 °C	10YR8/6	Yellow

The mechanical properties (compressive strength and hardness) of clay bricks are measured according to the Thai Industrial Standard Institute 243/2520. The data comparison of clay brick samples with and without 0%, 5%, 10%, 15%, and 20% wt eggshell powder fired at 1000°C for 1, 3, and 5 hrs., at each firing temperature are shown in Table 5. The high percentage of amount eggshell powder added in clay bricks, the better and higher compressive strength and hardness values of clay bricks are. The best condition of samples firing and eggshell adding to get the highest compressive strength and hardness clay brick is 20%wt eggshell powder adding and firing at 1000°C for 5 hrs., namely sample 20-1000-5. The compressive strength and hardness values of the best condition (20-1000-5) are 8.28 MPa and 8.79 ± 0.60 HV, respectively. While the clay brick sample without adding eggshell powder fired 1000°C for 5 hrs, has the compressive strength 7.42 MPa and hardness 7.63 ± 0.30 HV, respectively. The theoretical compressive strength and hardness values of the best fired clay brick according to the Thai Industrial Standard Institute 243/2530 and the ASTM C 373-72 should have the data more than 7.0 MPa and 6.0 HV, respectively.

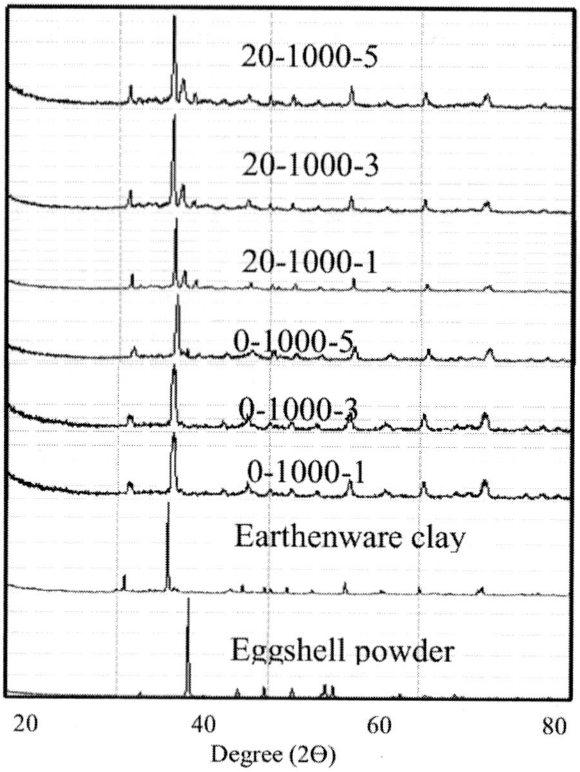

Fig. 2. The XRD peak patterns of samples: eggshell powder, earthenware clay, and fired clay bricks added 0 and 20%wt eggshell powder after firing at 1000°C for 1, 3, and 5 hrs., respectively.

The thermal expansion coefficient is one kind of the thermal properties of materials. The obtained fired clay bricks added eggshell powder and fired at 1000°C for 3 hrs., or 5 hrs., have low thermal expansion coefficient value less than those of fired clay bricks without adding eggshell powder at the same firing temperature and firing time. It means adding eggshell in fired clay bricks better than samples without eggshell adding. Thermal expansion coefficient values of clay bricks added eggshell powder and fired at 1000°C for 3 and 5 hrs., are 3.2605×10^{-6} and 2.3652×10^{-6} $(°C)^{-1}$. While thermal expansion coefficient values of fired clay bricks without adding eggshell powder and fired at 1000°C for 3 and 5 hrs., are 4.7846×10^{-6} and 3.6566×10^{-6} $(°C)^{-1}$. The low thermal expansion coefficient of materials has, the better thermal shock resistance is. Therefore, the fired clay brick sample added 20%wt eggshell powder and fired at 1000°C for 5 hrs., is the best condition to receive the good physical, mechanical and thermal properties.

Table 5. Mechanical properties of clay bricks

Samples*	Compressive strength (MPa)	Hardness (HV)
0-1000-1	7.05	7.27±0.20
20-1000-1	8.09	8.38±0.87
0-1000-3	7.37	7.47±0.60
20-1000-3	8.18	8.55±0.10
0-1000-5	7.42	7.63±0.30
20-1000-5	8.28	8.79±0.60

Table 6. Thermal expansion coefficient of clay bricks

Samples [a]	Thermal expansion coefficient [b] $(°C)^{-1}$
0-1000-3	4.7846×10^{-6}
0-1000-5	3.6566×10^{-6}
20-1000-3	3.2605×10^{-6}
20-1000-5**	2.3652×10^{-6}

Remark:
*Encoded of samples x-xxxx-x mean amount of eggshell powder adding-firing temperature-firing time i.e. 0-1000-3 (no adding (0%wt) eggshell powder -firing at 1000°C-firing time for 3 hrs.).
** means the best condition due to the low thermal expansion coefficient effect to good thermal shock resistance.

4 Conclusions

Adding chicken eggshell powder into clay bricks can increase physical-mechanical-thermal properties of products. The higher amount of adding eggshell powder into clay bricks is, the better physical-mechanical-thermal properties of clay bricks are due to calcium oxide content caused to calcium feldspar, calcium silicate, and wollastonite phase formation. The firing temperature and soaking time is an important factors affect to densification or vitrification due to low porosity and completion of the glassy phase formation. The best condition of fired clay bio-bricks are firing at 1000°C for 5 hrs. The fired clay bio-bricks added 20%wt eggshell powder and fired at 1000°C for 5 hrs., are high compressive strength and hardness (8.28 MPa and 8.79 ± 0.60 HV), good thermal expansion coefficient 2.3652×10^{-6} $(°C)^{-1}$, good true density 1.96 ± 0.02 g/cm^3, and low water absorption 14.36 ± 0.80%wt. The theoretical compressive strength, hardness, and water absorption values of fired refractory according to the Thai Industrial Standard Institute 243/2530 and the ASTM C 373-72 should have the values more than 7.0 MPa, 6.0 HV, and less than 25%wt, respectively. Therefore, the obtained fired clay bio-bricks added 20%wt eggshell powder and fired at 1000°C for 5 hrs., are potential to use as clay bricks for construction and thermal insulation.

The authors would like to thank the following: The Petroleum and Petrochemical College, and the Scientific and Technological Research Equipment Centre, at Chulalongkorn University, Thailand; the Department of Materials Engineering, at Kasetsart University for the use of their analytical equipment. We are grateful for the grant support from Agricultural Research Development Agency (Public Organization) or ARDA of Ministry of Agricultural and Cooperatives encoded PRP5605010080 of the fiscal year 2013-2014. We are also grateful for the grant support from Center for Advanced Studies in Industrial Technology and the Kasetsart University Research and Development of the fiscal year 2014. We also would like to acknowledge the financial support from Conductive and Electroactive Polymers Research Unit of Chulalongkorn University, the Thailand Research Fund (TRF-BRG), and the Royal Thai Government.

References

1. W.T. Tsai, J.M. Yang, C.W. Lai, Y.H. Cheng, C.C. Lin, C.W. Yeh, Bioresour. Technol., **97**, 488 (2006)
2. Y. Nys, J. Gautron, J.M. Garcia Ruiz, M.T. Hincke, C.R. Palevol., **3**, 549 (2004)
3. S. Ahmad, Y. Iqbal, F. Ghani, J. Pak. Mater. Soc., **1(2)**, 33 (2008)
4. L. Zhang. Constr. Build Mater., **47,** 633 (2013)
5. F.F. Singer, *Industrial ceramics* (Chapman Hall, London, 1979)
6. F. Karim, *Ceramics & Plastics: Testing of Engineering* (Ferozsons, Lahore, 1998)
7. J.B. Weems, Iowa Geological Survey Annual Report, **14**, 319 (1903)
8. *Technical Notes on Brick construction* (Brick Industry Association, 2006)
9. S.P. Raut, R.V. Ralegaonkar, S.A. Constr. Build Mater., **25(10)**, 4037 (2011)
10. I. Johari, S. Said, B. Hisham, A. Bakar, Z.A. Ahmad, Sci. Sinter., **42**, 245 (2010)
11. M. Sutcu, S. Akkurt, Ceram. Int., **35(7)**, 2065 (2009)
12. A. Ukwatta, A. Mohajerani, S. Setunge, N. Eshtiaghi, Constr. Build Mater., **91**, 86 (2015)
13. G. Gorhan, O. Şimşek, Constr. Build Mater., **40**, 390 (2013)
14. F. Kristaly, I. Kocserha, Mater. Sci. Forum, **659**, 43 (2010)
15. G. Cultrone, C. Rodriguez-Navarro, E. Sebastian, O. Cazalla, M.J. Delatorre, Eur. J. Mineral, **31,** 621 (2001)
16. *EN 832: Thermal Performance of Building Calculation of Energy Use for Heating Residential Buildings*, (CEN, Brussels, Belgium, 1998)
17. D.A. Oliveira, P. Benelli, E.R. Amante, J. Clean. Prod., **46**, 42 (2013)
18. F.S. Murakami, P.O. Rodrigues, C.M.T. Campos, M.A.S. Silva, Cienc. Tecnol. Aliment., **27(3)**, 658 (2007)
19. M.J. Trindade, M.I. Dias, F. Coroado-Rocha, Appl. Clay. Sci., **42(3)**, 345 (2009)
20. M.M. Jordan, T. Sanfeliu, C. de la Fuente, Appl. Clay. Sci., **20**, 87 (2001)
21. W. Xia, J. Chang, Micropor. Mesopor. Mat., **108(1-3)**, 345 (2008)
22. X. Wan, C. Chang, D. Mao, L. Jiang, M. Li, Mater. Sci. Eng. C, **25**, 455 (2005)

Effect of Differences Methanol Concentration and Extraction Time on the Antioxidant Capacity, Phenolics Content and Bioactive Constituents of Orthosiphon Stamineus Extracts

Noorhaslina Hashim[1,*], *Abdul Razak* Shaari[1] , *Awang* Soh Mamat[1] , and *Syarhabil* Ahmad[1]

[1] School of Bioprocess, Universiti Malaysia Perlis (UniMAP), Kompleks Pusat Pengajian Jejawi 3, 02600 Arau,Perlis, Malaysia.

Abstract: Bioactive compounds of *Orthosiphon stamineus* have been known to have the beneficial effects on health. Efficient extraction of these beneficial compounds is very important. The study was conducted to determine the efficiency of methanol as an extraction solvent for bioactive compound extraction of *O. stamineus* raw materials. Leaf sample was extracted in 25, 50 and 100% methanol and distilled water for 2, 4 and 8 hours at 40^{0}C. The free radical scavenging activity (FRSA) method was used to determine antioxidant capacity and Follin-Ciocalteu method was used to determine samples total phenolic compounds. The quantification of bioactive constituents by using the high performance liquid chromatography (HPLC) and the standard markers was used for this analysis such as rosmarinic acid (RA), sinensetin (SEN) and 3-hydroxy-5,6,7,4-tetramethoxyflavone (TMF). The current research shows that the yield of RA concentration was high in 50% methanol extracts at 2, 4 and 8 h of extraction. For the FRSA result showed that methanol at 50% and 100% was the most effective concentration levels inhibition between 45-55% at 8 hours' time extraction. For the total phenolic compound, the 50% methanol concentration in *O. stamineus* leaf had the highest value compared with 0%, 25% and 100% methanol concentration. But, it was observed that time extraction did not influence the TPC values for different concentrations of methanolic extracts. These indicated that time extraction had affected the total capacities of antioxidant and 50% methanolic extracts was the best concentration for indicated the TPC in the *O. Storminess* leaf.

1 Introduction

Orthosiphon stamineus (O.stamineus) contains 20 phenolic compounds, including two flavonol glycosides, nine lipophilic flavones and nine caffeic acid derivatives, such as

* Corresponding author: leein16@gmail.com

rosmarinic acid and 2,3-dicaffeoyltartaric acid. The core components of *O.stamineus* plant are the pharmacologically active polyphenols (the polymethoxylated flavonoids and caffeic acid derivatives) [1]. These elements have been reported to be actual in reducing oxidative stress by inhibiting the formation of lipid peroxidation products in biological systems [2].

Phenolics and polyphenolics compounds are considered by an aromatic or phenolics ring structure. These compounds are including the flavonoids, phenolics acids and lignans. Phenolics compounds are located in vacuole and have a tendency to be soluble in water or organic solvents [3]. In this study, the antioxidant activity of phenolic compounds was found to be mainly due to their scavenging and redox properties through counteracting and quenching of free radicals. Nevertheless, the extraction yield and antioxidant activity of the extracts highly depend on the solvent polarity, which determines both quantitatively and qualitatively the extracted antioxidant compounds. Usually, the constancy of different extracts from the same material depends on the extraction solvent used for removal of the polyphenol compounds, and it is apparent that extracts from the same plant material may vary widely with respect to their antioxidant concentrations and activities [4].

Hence, the objective of this study was to study the effects of different methanol concentration and different time extraction on the regaining of total phenolic content (TPC), antioxidant capacity by FRSA method of *O.stamineus* leaf extracts. Besides that, the effects of the extraction were also measured of bioactive compounds by HPLC instrument.

2 Materials and methods

2.1 Chemicals and reagents

Tetrahydrofuran (THF) [HPLC grade] and methanol (HPLC grade) were obtained from AR Alatan (K) Sdn Bhd. Sinensetin (SEN), eupatorin (EUP), Rosmarinic acid (RA) and 3'-hydroxy-5,6,7,4'-tetramethoxyflavone (TMF) were purchased from YKN Tech Ent. The other chemical such as Folin-ciocalteu reagent, caffeic acid, methanol (100% v/v), $NaNO_3$ and 2,2-diphenly-1- picrylhydrazyl (DPPH) were purchased from Sigma Aldrich.

2.2 Sample collection and preparation

In the early morning, the *O.stamineus* plant which grown at UniMAP Agrotechnology Research Station, Sg. Chucuh, Perlis was randomly harvested. Meanwhile, the parts of *O. stamineus* plant were dried at 40^0C for 5 days. The dried leaf of *O.stamineus* plant was balanced and stranded to powder form and used for the extraction process.

2.3 Extraction process

2.3.1 Differences of methanol concentration

Ten gram of *O.stamineus* dry powder was weighed and mixed with different methanol concentration. Actually, they are five methanol concentrations. For instance 0, 25, 50 and 100% methanol concentration. The methanol concentration was prepared by adjusting the composition of methanol and water. The extraction process was used in the conical flask, whereas the flask was closed with cotton wool and wrapped with aluminum foil to prevent spattering of mixture. Furthermore, the mixture was shaken at 150 rpm by incubator shaker at 40^0C.

2.3.2 Differences of time extraction

The samples with different methanol concentrations were extracted by using different time extraction for 2, 4 and 8 hours. Then, the temperature was fixing constantly at 40^{0}C. Furthermore, the extract was filtered into the storage bottle for further analysis. The extraction was conveyed out in replicates.

2.4 Free radical scavenging activity (FRSA)

The free radical scavenging activity (FRSA) method was used to determine the total antioxidant activity of O.stamineus leaf. The 1,1-diphenyl-2- picrylhydrazyl (DPPH) methanolic solution was prepared and two ml of the solution was mixed with 200 µl of sample. Meanwhile, the mixture was added with methanol solvent to make a final volume of 3 ml. Then, the fusions allowed to stance for 60 min, and the absorbance value was observed by using spectrophotometer at 517nm. The FRSA (%) of the samples were compared with a control. The formula for FRSA (%) as bellows:

$$FRSA = \left(\frac{(A_C - A_S)}{A_C} \text{ X } 100 \right)$$

Ac as an absorbance value for control; As as absorbance value for sample.

2.5 Total phenolic content (TPC) sssay

The total phenolic content (TPC) was determined by using Follin-Ciocalteu Reagent (FCR). The calibration curve was prepared with Caffeic Acid (CA) as a standard [5]. Nevertheless, the 0.2 ml of sample extract was mixed with 0.2 ml FCR. Then, the 1.58 ml water was added and mixed as a solution. Then, 1 ml of 20% $NaCo_3$ was added and allowed to stand at room temperature for two hours. Next, the mixture was restrained for the absorbance value at 760nm by spectrophotometer. The CA was used as a standard and the results were expressed as mg caffeic acid equivalent per 1 g of dry weight sample (CA mg/g DW).

2.6 Analysis of compounds by HPLC

The identification and quantification of bioactive compounds of *O.stamineus* leaf were determined by using HPLC instrument. Before that, the sample extract was filtered through a 0.45µm nylon membrane filter into a HPLC vial prior. The HPLC used was demoralized for analysis should equip with autosampler, column oven and UV/VIS detector. The HPLC column used was Merck Licrochart Purospher Start RP 18 column (250mm, 4.6 mm i.d, 5µm pore size). Then, the mobile phase used was prepared, whereas a mixture of water: methanol: tetrahydrofuran (50: 45: 5 v/v) (G. A. Akowuah et al.). The sample was evaluated at the mobile phase flow rate of 1 ml/min and detector wavelength of 340nm at 30°C for 40 min. However, the calibration curve was prepared by exploiting standard marker (Rosmarinic acid, TMF and Sinensetin) compounds purchased from Chromadex for SEN and Sigma Aldrich for RA.

 DOI: 10.1051/matecconf/20167801004

3 Results and discussions

3.1 Extraction by differences concentration solvent.

Extraction was done according to the method by [7]. Whereas, the combination of water and methanol were used for the extraction on this experiment. The characteristic of similar polarity solvents as the principle " like dissolve like" was suggested only extract those compounds which have same polarity [2]. We advocated that the methanol extract was found to be the good solvent for extraction of polyphenolic compounds from *O.stamineus* leaf and stem. Based on the situation of TPC and FRSA, methanolic extracts expressing good biological capacity. Besides that, in the extraction process, the polarity of solvents is very important to increase the solubility of antioxidant compounds [8]. Thus, it was significant to measure process to validate a good solvent in antioxidant compound extraction. This validation can be optimized to get the maximum antioxidant activity for a certain sample.

3.2 Free radical scavenging activity (FRSA)

Results of the FRSA of *O.stamineus* leaf extracts are presented in Figure 1. The results showed that 50 and 100% methanol concentration were the highest percentage of inhibition at different time extraction. However, the 0 and 25% methanol concentration were significantly declined the FRSA at 2 and 4 hours extraction.

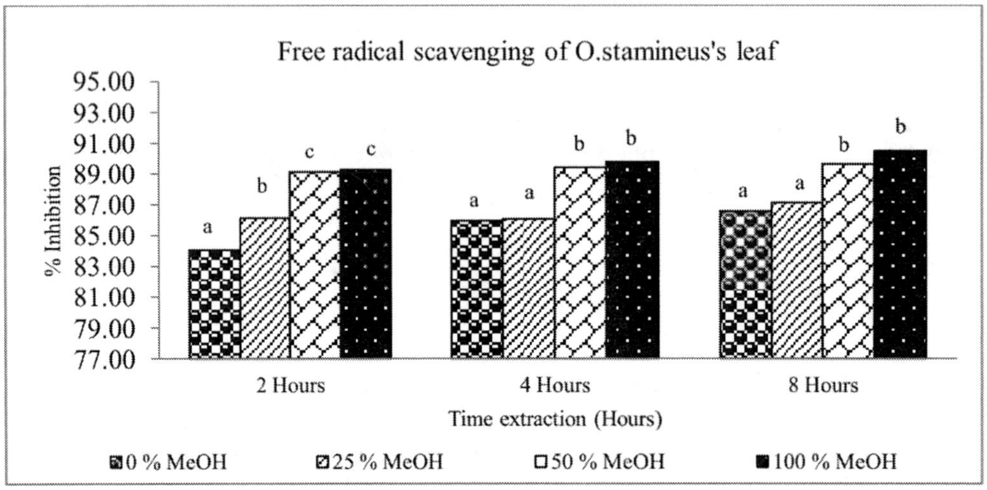

Fig. 1. Effect of different methanol extract at different time extraction on FRSA for *O.stamineus* leaf. Values with different letters are significantly different at $p < 0.05$.

Surprisingly, the FRSA for 100% methanol extract was slightly affected and not significantly different at 4 and 8 hours extraction. Based on previous studies, they were also indicated that sustained time extraction would enhance the amount of oxidation in extracting phenolic compound extract [6].

DOI: 10.1051/matecconf/20167801004

3.3 Total phenolic content (TPC)

The effects of different concentration of methanol extracts and different extraction time on TPC for *O.stamineus* leaf was presented in Figure 2. Results showed that the highest TPC was 50% methanol concentration extract and follows by 100% methanol concentration. However, the lowest value of TPC was 0% methanol concentration extract.

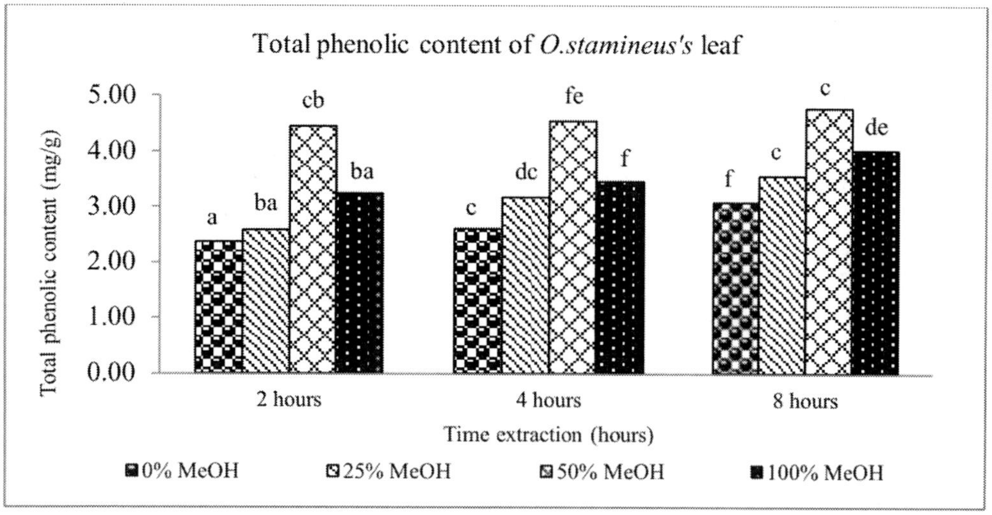

Fig. 2. Effects of different methanol extract at different time extraction on TPC for *O.stamineus* leaf. Values with different letters are significantly different at $p < 0.05$.

A similar situation was also observed in the previous study, it was reported that the long time extended for extraction would prime decreasing of phenolic compounds cause of the occurred extending and exposing to environment factors [9]. In our study, we suggested 50% methanol extract is the most suitable concentration extracts for the purpose of speed analysis.

3.4 HPLC Analysis

The results of the HPLC analysis are represented by its bioactive compound concentration (μg/g.dw). A result of the bioactive concentrations for O. stamineus leaf was presented in Table 1. The present study was indicated that the amount of RA, TMF and SEN concentration as the source of active constituents in *O.stamineus* leaf. Similarly to [2], they also indicated that the yield of RA was significantly affected in 50% methanol extract at 8 hours extraction. Despite the order of RA amounts were increased as 50% methanol extract> 100% methanol extract> 25% methanol extract> 0% methanol extract.

In contrast to TMF concentration, the TMF concentration was drastically increased after 8 hours extraction. It was observed that the 100% of methanol extract was significantly obtained high amount of TMF at 8 hours extraction. The sequence of TMF amounts were increased as 100% methanol extract>50% methanol extract>25% methanol extract>0% methanol extract. Similar situation was also observed for SEN concentration, whereas the amount of SEN compounds was significantly improved after 8 hours extraction in 100% methanol extract. Although, the SEN concentration increased in sequence as 100% methanol extract>50% methanol extract>25% methanol extract>0% methanol extract. Contrary to previous report, they were obtained 100% methanol extract did not influenced the TMF and SEN concentrations at different time extraction [7].

Table 1. The effects of different methanol extract at different time extraction on HPLC analysis of *O.stamineus* leaf.

Methanol	Time extraction	Concentration (μg/g.dw)		
Concentration (%)	(hrs)	RA	TMF	SEN
	2	4.799^{a}	0.302^{abc}	1.359^{a}
0	4	5.755^{a}	0.257^{bc}	9.848^{b}
	8	7.223^{a}	0.350^{abc}	1.681^{c}
	2	9.261^{a}	0.129^{c}	0.701^{d}
25	4	49.844^{b}	0.586^{a}	1.377^{a}
	8	26.500^{c}	0.451^{ab}	1.674^{c}
	2	208.764^{d}	0.415^{abc}	4.280^{e}
50	4	240.155^{e}	4.151^{d}	5.195^{f}
	8	235.153^{ef}	2.925^{e}	6.888^{g}
	2	197.205^{d}	4.653^{f}	7.135^{g}
100	4	225.379^{f}	4.891^{f}	7.946^{h}
	8	235.773^{ef}	5.307^{g}	8.272^{i}

Different letters for each treatment was stated as significant different values of means ($P < 0.05$).

4 Conclusions

As we can conclude that the *O. stamineus* leaf had the highest FRSA at 100% methanol extract and slightly changed at difference time extraction. Nevertheless, the 50% methanol extract was significantly influenced the TPC, but prolonged extraction did not significantly affected the TPC value. Therefore, the RA compounds was increased effectively in 50% methanol extract for four hours extraction. Differently with TMF and SEN compounds, it was significantly affected in 100% methanol extract at 8 hours time extraction.

This research was fully supported by Universiti Malaysia Perlis (Unimap), and also truly appreciated the UniMAP Agrotechnology Research Station, Sg. Chuchuh Perlis for their technical assistance.

References

1. G.A. Akowuah, I. Zhari, I. Norhayati, A. Sadikun, S.M. Khamsah, Food Chem., **87(4)**, 559 (2004)
2. G.A. Akowuah, Z. Ismail, I. Norhayati, A. Sadikun, Food Chem., **93(2)**, 311 (2005)
3. W.M. Maizura, M., Aminah, A. Wan Aida, Int. Food Res. J., **18**, 529 (2011)
4. K.K. Chew, M.Z. Khoo, S.Y. Ng, Y.Y. Thoo, W.M. Wan Aida, C.W. Ho, Int. Food Res. J., **18(4)**, 1427 (2011)
5. S. Kaur, P. Mondal, J. Microbiol. Exp., **1(1)**, 1 (2014)
6. G.A. Akowuah, I. Zhari, I. Norhayati, A. Sadikun, S.M. Khamsah, Food Chem., **87(4)**, 559 (2004)
7. K.K. Chew, M.Z. Khoo, S.Y. Ng, Y.Y. Thoo, W.M. Wan Aida, C.W. Ho, Int. Food

Res. J., **18(4)**, 1427 (2011)
8. S.I. Abdelwahab, S. Mohan, M.M. Elhassan, N. Al-Mekhlafi, A.A. Mariod, A.B. Abdul, M.A. Abdulla, K.M. Alkharfy, M. Mohamed Elhassan, Evid. Based. Complement. Alternat. Med., **2011**, 1 (2011)
9. N. Dewi, Y. Muhammad, **12(2)**, 293 (2013)

MATEC Web of Conferences **78**, 01005 (2016) DOI: 10.1051/matecconf/20167801005
IConGDM 2016

Optimum Design for Multi-angle Kitchen Grater Mechanism for Biodegrading Kitchen Waste

Zol Bahri Razali[1,*], *Abdul Rahim* Abdul Hasim[1] , and *Mohd Hisam* Daud[1]

[1]Robotics and Automation Technology, Faculty of Engineering Technology, Universiti Malaysia Perlis, 02100 Padang Besar, Perlis, Malaysia

Abstract. The study presented in this report is regarding on the conceptual design of the grater machine which meant to shred food wastes into small and fine size particle. This study will focus on the waste management factor of the food waste produced in the home kitchen. A study is conducted to compare the most suitable material and geometric shape of the grating blades by comparing with similar existing devices in the market. The chosen material and blade design are to be evaluated for its characteristics and performance by using the FEM method. The analysis is taking all the constraints regarding the design as well as its advantages are to be considered in designing a new grating blade for the optimum end product making. This design derived through in the consideration for the home application rather than domestic use. This brings the meaning that the compact and small size design. With the detailed evaluation of the stress reaction on the blades to provide the safety and sustainability factor of the device as well as the people that is using the device.

1 Introduction

The "Grater" machine (also known as a shredder) is a kitchen utensil used to grate foods into fine pieces [1]. This study will focus on the waste management factor of the food waste produced in the home kitchen. The project is to analyse the optimum design of Multi-angle Kitchen Waste Grater machine. For this intention, several types of existing grater blade design in the market are to be evaluated for its characteristics. All the constraints regarding the design as well as its advantages are to be considered in designing a new grating blade for the optimum fineness end product and the speed of grating capacity. This design should consider for the application in the house which mean it should be smaller size and the safety precaution factor of the operator and other peoples in surrounding area must be placed at the top priority.

The main objective is about the kitchen waste management where it is to be recycled and reuse for the purpose of agriculture. Today's growth of our country shows the rapid increment of waste producing amount. The waste producing rate is 19,000 tons of garbage a day and the number getting increase and estimated will approach 30,000 tons per day by year 2020 [2]. This scenario has put a major constraint in managing the resources for the landfill. The government has launched the Solid Waste Management Act and Public

* Corresponding author: zolbahri@unimap.edu.my

Cleansing 2007 (Act672) which had been implemented since 1st September seen tremendously a great effort to encourage the public in adopting recycling. Other developed countries like the United States, Japan and most of the European countries has long been practiced the recycling method to reduce the environmental loadings and natural resources consumption [3]. It is the time where recycling practices of waste produced at home become a new branch of an engineering equipment production to be explored in line with current needs [4].

The analysis consists of mincing mechanism design force analysis and the simulation of the grater. For this purpose, several samples of existing grater blades will be analysed. The main focus is where it should begin by mean of analyse the design of grate mechanism material and propose the new design which offer the improvement in the qualities and durability. The mechanical properties of the food variants will be taken in consider for determining the required strength of the greater material or the blades [5]. This includes every scope of human consumption varies from organic based product such as vegetables along with its wastes which have the lower hardness property for livestock material such as meat and bones which are the hardest materials in human food classification.

2 Literature review

Utilizing or recycling the food waste materials have inspired the invention of a grater machine to decomposition the kitchen waste to a suitable and fast dissolve to become agricultural fertilizer. A good blade design to shreds the organic materials is important to result a fine material for composting. It is also meant that the blade should also bear in working with the varied types of waste ranging from plants to livestock products. The market nowadays has many types of grater machine and this is just another innovation which may vary the choices available.

The design analysis will be based on the varied approaches that designer used, and using the thinking that have been arisen in design research. The logic will gives the lead in describing the reasoning in design. A 'sparse' description which comes from logic will opens the way to explore the variants of the design to be differ from other point of features and advantages [2]. The reasoning patterns that usually used in problem solving of the a set of unknowns and known are described as such equation:

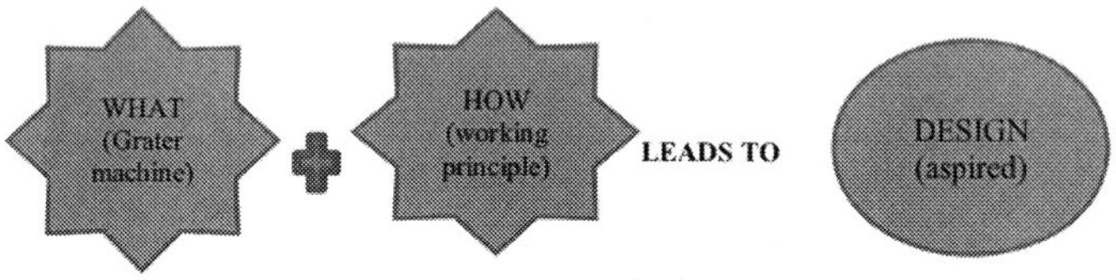

Fig. 1. The transformation of the problem solving method with regards to the current situation

2.1 Basic concept of food grater.

The existing products of grater only concerning to the small application in the kitchen, for cutting edges, used for grating cheese and other foods [5]. To expand the usage for home application of handling food waste, the similar equipment used in certain huge applications such as waste management center should be taken in consider. The grater machine should have the high capacity rate of processing and must be able to withstand the greatest hardness of the food material which are represented by the poultry products such as animal bones.

2.2 Grater blades materials.

When handling with food based material, the best and most common material used is the stainless steel. Referring to Table 1, stainless steel-430 type is commonly used for applications such as equipment enclosures and housings. For cutting mechanism applications, the hardened martensitic stainless steel-316 is likely chosen. The Stainless Steel-316 is the one classified as the "Food Grade" type. Where the applications involve with high temperatures, the non-hardened ferritic and austenitic types of stainless steel which having the higher duplex strength is the most suitable to be used [6].

The material selection is crucial to avoid weaknesses in the design. In this case it does majorly refer to the corrosion from being occur and the stand ability to what it purposed. Better surface finish will influence the functional performance. It also makes the device easy to clean, thus keeping away the risk of corrosion. The several types of stainless steel usually used in food processing equipment is as shown below together with its typical applications.

Table 1. Types of Stainless Steel and its typical applications

Type	Applications
420 (Martensitic)	Spatulas, Professional knives, Cooks, etc.
430 (Ferritic)	Moderately corrosive environment (e.g. dry foods, fruits, vegetables, drinks, etc.), panel (i.e., components that require weldability or little formability), table surfaces or cladding equipments.
304 (Austenitic)	Pipework, machinery parts (i.e. components requiring some formability or weldability), vats and bowls,. Greater corrosion resistance to 430.
316 (Austenitic)	More corrosive foods (e.g. meat/blood, foods with moderate salt contents) components, which requires frequently clean, not under excessive stress and without stationary solids.
14539 (Austenitic)	Corrosive foods (e.g. hot brine with solids that act as crevice forms, stagnant and slow moving salty foods).
14662 (Duplex)	Corrosive foods (e.g. hot brine in solids, stagnant and also slow carrying salty foods). Higher strength when compared with austenitics. High stress corrosion cracking resistance within salt merchandise at elevated temperatures.
6%Mo.Types (Austenitic)	Corrosive foods (e.g. hot brine in solids, that act equally crevice formers, stagnant in addition to slow taking salty foods). Good stress corrosion cracking resistance within salt merchandise at elevated temperatures. Used in steam heating and hot performs circuits, hot water boilers, etc.

2.3 Grater blades design.

For soft solid surfaces separating work, the local speed of the material passing through the cutting blades will be determined by the normal and tangential forces of an element of a blade material. The ratio of 'slice' and 'push' velocities, which is defined by ξ. Lower the ξ, the greater the force reacts at the blade tip and will increase the damage to the cut surfaces. The forces of reaction also vary depends to the blade curvature which will be discussed in more details later. The analysis of existing blade design will be concluded to establish a new design which improves the constraints found [7].

Wherever the intention is, the slicer is made from steel rather from any other materials and for this case, the circular plate with plural shredding blades formed by bending a small spaced section up through the plate by a punching form method which create a hole right behind each blade. The shredder will strips the food product into slices where then will pass through the hole to another side of the plate (to the discharge outlet).

This kind of device (Figure 2) seems to be effective to a soft solid shredding like vegetable. For a harder solids like meat and bones, it does not suitable and may cause damage to the machine during operation.

Fig. 2. Rotating Shredder/Slicer Plate [2]

The tougher and heavier duty food shredder blades come in a rotating knife blade form. This shredder assembly (Figure 3) includes the freeing jams-impact mechanism. This mechanism stores the energy in its rotating elements which sourced from the driven motor and being transmitted to the cutting blades which reacts with the materials that are going to be shredded. The cutting process will continue going to the food reached the desired size of particles which been determined through the hole size of the screen plates at the bottom of the chamber. The desired size particles will pass through the screen once meet the size to the outlet of the machine where the bucket is provided.

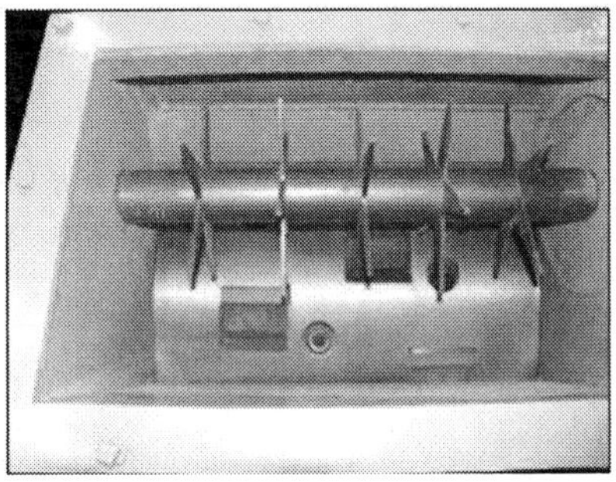

Fig. 3. Rotating Shredder/Grater Blade Types [2]

3 Methods and materials

3.1 Properties of food material.

The data of various foods modulus of elasticity, as shown in Table 2 has been collected to determine the ultimate forces that the grater blades will handle. The variants are from the soft solid like vegetables and to the hardest possible materials like animal bones. The table shown next give the values of the specific limits of forces required for varying types of foods.

Table 2. Variety of food material mechanical properties

Food Material	Modulus of Elasticity, E (MPa)
Onions	0.000284
Corn kernels	219
Apple	0.00014
Meat	3.15
Fish (frozen)	0.005
Potato	5.2
Wheat	239
Carrot	0.004
Bones	275.8

3.2 Residual stresses in grater blades.

The material structural components and engineering properties such as distortion, fatigue life, dimensional stability, the resistance to corrosion and brittle fracture are significantly influenced by residual stresses [7]. The design process indeed requires to accounting the residual stress analysis as compulsory to produce the final result of optimum reliability under actual working conditions. In cyclic loading fatigue (N > 106 cycles), the residual stresses effect is comparable to the stress concentration effect. Because the residual stresses do not cause by the loads, it has to be globally balanced and can be expressed in the equation as follow;

$$\int \sigma dA = 0 \tag{1}$$

3.3 Stress analysis and result.

The development and evaluation of finite element models of grating mechanism for the cutting and rotating effects will be considered in the thesis. To calculate the original stress, the finite element analysis and modelling where the displacement boundary conditions would be the smoothed data attained from the measurements. This data can be easily described as the input to the finite element modelling.

The theories relate upon the basic assumption that the energy required to produce a change in its length, dL for a particle of a specific size dimension of L, is given as a simple power function of L. With the consideration of Kick's and Rittinger's Law, leads to:

$$E = E\text{i}\,(5/L2)1/2[1 - (1/q^{1/2}) \tag{2}$$

3.4 Finite element analysis modeling.

To visualize the model for the finite element analysis, a simple model of a rotating solid disc and an annulus with the same dimensions as the grater blade to be analysed. Both the solid disc and an annulus make the model exactly represent the grater blade. It will also be modeled asymmetrically with 3-D eighth model. For radial stress calculation, the term can be defined by such expression;

$$\sigma rr = \frac{(3+\upsilon)}{8}(\rho b^2\omega^2)\left[1 - \frac{r^2}{b^2}\right] \tag{3}$$

where, υ = poisson ratio for the blade material.

ρ = the mass density of the blade material

b = disc outer radius
ω = angular velocity (rad/sec)
r = arbitrary radial position

3.5 Rotating solid disc analysis.

Here the illustration the will be applied by open up the hand calculations used for the finite element analysis results validation, then the parameters and the computer model are shown and discussed. Finally, the axisymmetric model results will be compared with the hand calculations. For solid rotating disc, the exact hand calculations are as cited in Boresi and Schmidt [8]. The hand calculations represent the displacements and stresses of a rotating solid disc at about its center. Then, the results obtained from these hand calculations are compared with the results of computer model as an axisymmetric model. The following Eq. 4 is used to define the radial stress, σ_{rr}, along any arbitrary radial position, r, in a solid disc rotating about its center.

$$\sigma_{rr} = \frac{(3+v)}{8}\left[1-\frac{r^2}{b^2}\right]\rho b^2 \omega 2 \quad (4)$$

and the arbitrary radial position of the radial displacement is found as the Eq. 5:

$$u = \frac{(1-v^2)}{8E}\rho r^3 \omega^2 + C_I r + \frac{1}{r} C_{II} \quad (5)$$

Where C_I and C_{II} area products of integration constants and this is the application for a solid disc only.

4 Results and discussions

4.1 Nominal dimensions and parameters of a typical cut-off grater/cutting blade.

There are many dimensions and parameters of a typical cut-off cutting blade which can be seen in Figure 3. Because the dimensions and parameters can be divided into two criteria, the parameters are given as:

i) Dimensions of the cutting blade as a whole.
ii) Dimensions of the cutting blade's rim or outer area.

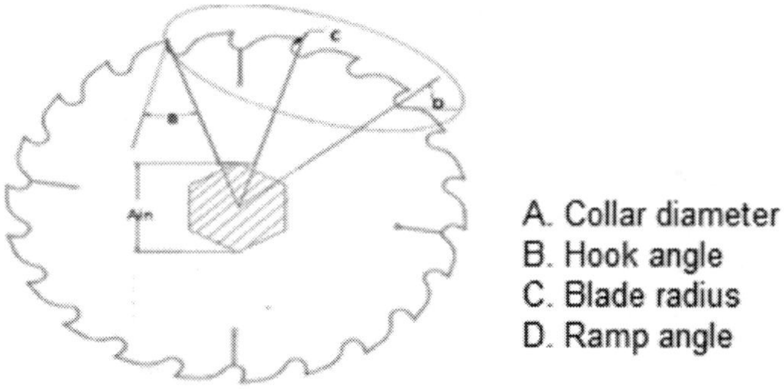

Fig. 4. Typical nominal parameters of a typical circular cut-off cutting blade

Dimensions of the cutting blade as a whole. Cutting Blade as a whole refers to the Figure 4 above where "A" is the collar diameter. This is where to fix the cutting blade into the shaft or the arbor of the grating machine. The arbor is a 12 inch length which can fix a number of 6 blades with a spacer in between of each blade, hexagonal shape as to improve the driven force to the blade. There are two rows of blades facing each other and the blade arrangement are alternately in between the opposite blades. Makes the whole area are included in the cutting mode during the operation. The end of both shafts will be extended out from the box. This section is to be attached to the gearing mechanism to enable the same angular velocity for both shafts. But the gearing mechanism and the driver will not be included in this design as the main objective only focus to the grater mechanism only. The arrangement of this grater blade only together with the shaft or arbor is as shown in Figure 5.

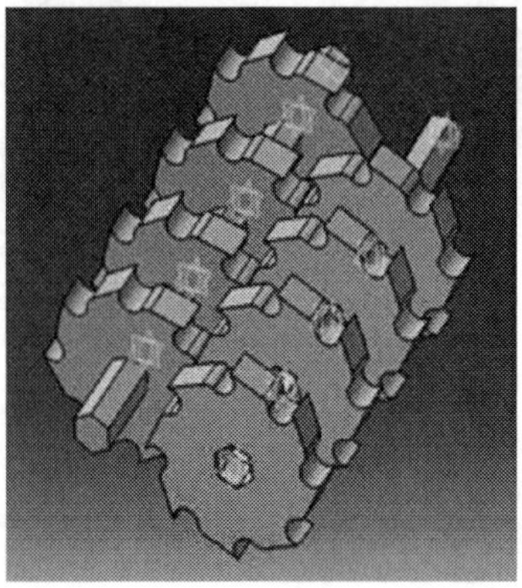

Fig. 5. The concept of grater blades arrangements

4.2 Grater blade outer area.

It is regarded the outer rim area and it represents the dimensions of the cutting teeth, gullets, shoulders, cutting teeth pockets, and stop hole radii. Dimension "A" in Figure 4 is the length of the cutting teeth, measured from a tooth's bottom side to the outermost point on the cutting blade. Dimension "B", is the gullet radius, which brings the meaning of the radius of the designed area to cope the cutting action dust during the cutting operation and releasing the dust after that when the cutting tooth passed through the material. Dimension "C" is the shoulder relief angle. Furthermore, the shoulder relief angle provides the rise to the dimension "D", which is the shoulder radius that define the fast dropping off-slope radius behind the shoulder that leads towards the preceding gullet.

4.3 Axisymmetric model of solid disc.

The results of the radial stress computer axisymmetric-model are shown in Figure 6. Here, the radial stress is displayed in colours which are defined by the key legend at the top left of the picture. This is the figure that represents an axisymmetric model of a solid rotating disc at about its center.

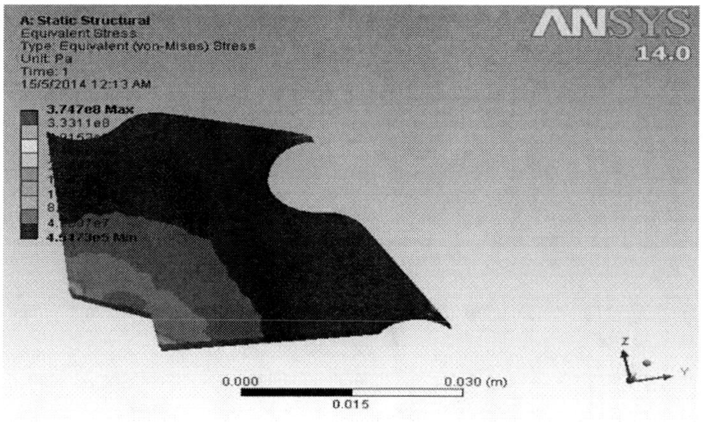

Fig. 6. Axisymmetric model showing radial stress (S11) in solid disc.

4.4 Comparing radial stress.

Here, the hand calculated results for the radial stress is to be comparable to the computer results, the two results did match quite well alike. For the signal-to-noise ratio between the 40 axisymmetric model results and the hand calculations of the radial stress is 3747000 Pa. Thus, the maximum error for the both results is very low at 9.18E-05 MPa and the error standard deviation is calculated at 0.0001585 for a maximum stress of 374.7 MPa. These statistical analysis values are obtained from the hand calculations and the computer results comparison proves that the computed results are very accurate and extremely well represent a rotating solid disc. The plots of the results of hand calculation results and the finite element analysis are as shown in Figure 7.

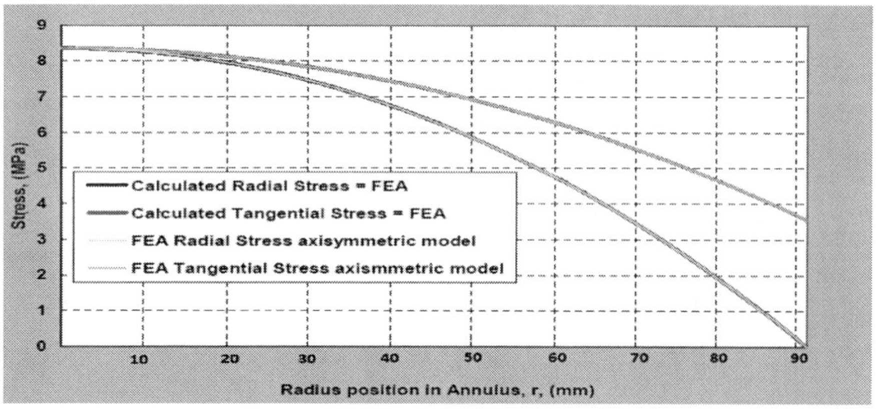

Fig. 7. The radial and tangential stress in a 4-in (100 mm) radius solid disc

4.5 Comparing radial displacement.

The radial displacement results between the hand calculations and the axisymmetric model results are compared. The radial displacement signal-to-noise ratio is 117,500 with a maximum error between the two cases of 3.30E-08 mm and a maximum displacement of 0.004 mm. The estimated error standard deviation is 3.40E-08. As seen in Figure 8, the solid disc radial displacement of a four inches diameter flats off at its outer edge. The difference between the axisymmetric results and the hand calculation for radial displacement on the plot is negligible.

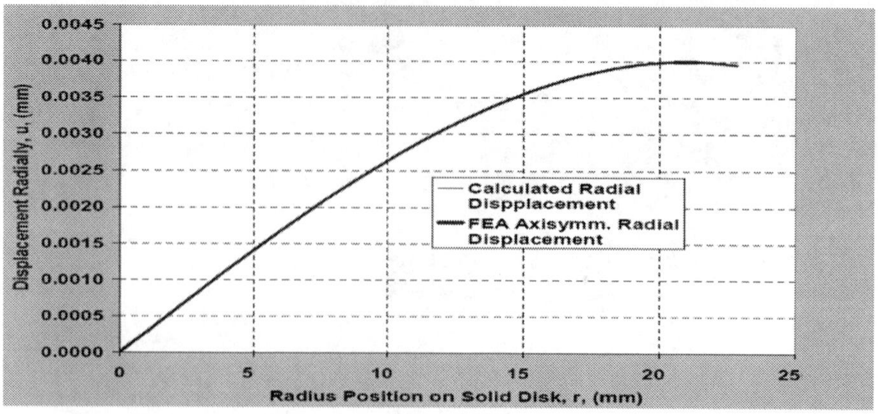

Fig. 8. The hand calculation of the radial displacement

4.6 Results of spinning grater blade.

The results obtained from the spinning blade body in Ansys shows the value of stresses occurred in the model of the solid grater blade disc.. The maximum and minimum stresses are as seen in the table gives the value of 374.7 MPa and 0.4547 MPa. By comparing these values to the grater blade material properties, which is given by the 250 MPa. It is assumed that this grating mechanism could reach the optimum working performance and will not get failed upon the most extreme food material in its operations. The contour of the equivalent stress occurs in the solid blade element as in Figure 8, only shows a small gradient of high stress locating at the center of the bore which is precisely designed in hexagonal shape to capture the driven inertia force transmitted from the driving device such as motor very effectively and with high durability.

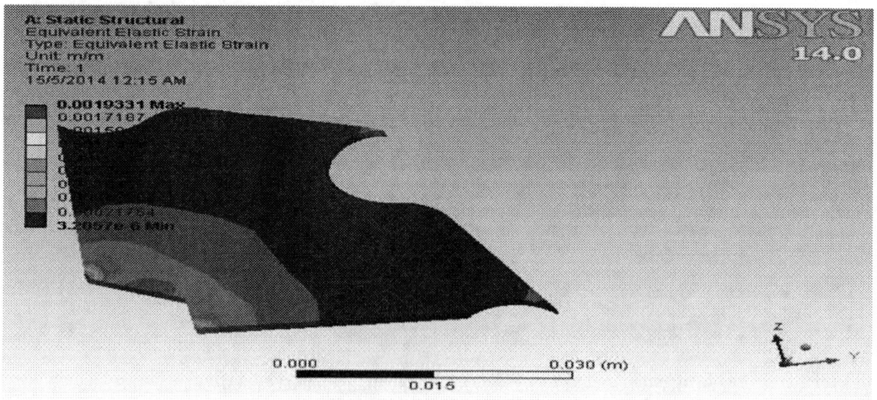

Fig. 9. The Ansys Elastic Strain Occurred Results of the Model during operations

5 Conclusion

The research overall has resulted in better understanding of a grating blade mechanism. This paper has given the detail views about the associated stresses that occur in the element. This will lead to the main objective which to create a new food waste grating device for home application. The radial stresses acting to the model is that obtained through computer analysis then compared to another method of hand calculation to see its discrepancies as well to validate the results.

The study conducted is able to compare the most suitable material and geometric shape of the grating blades by comparing with similar existing devices in the market. The proposed material and blade design have evaluated their characteristics and performance by using the FEM method. By taking all constraints regarding the design as well as its advantages are to be considered in designing, a new grating blade for the optimum multi-angle end product is able to be proved. The problem of solid disc that associates with the rotation and cutting loads, the length of the specimen to be cut and the residual stress has also been solved in this research.

This work was financially supported by the Universiti Malaysia Perlis Seed Money (9014-00034).

References

1. M. Fujii, Resour. Conserv. Recy., **49,** 564 (2012)
2. Z.B. Razali, N.A. Mat Nawi, *Automated Segregating Device for Reuse and Recycle Waste* (LAP, Germany, 2013)
3. http://en.wikipedia.org/wiki/Grater Retrieved on 24 October 2013.
4. Z.B. Razali, G.B. Lee, *High-End Technology for Effective Glass Manufacturing Industry* (LAP, Germany, 2013)
5. P.W. Jeffrey, *Designing an Appropriate Technology Shredder in a Developing Country,* Master thesis (University of St. Thomas, Minnesota, 2005)
6. D. Kees. Design Stud., **32**, 521 (2011)
7. N.S. Rossini, Mater. Design, **35**, 572 (2011).
8. B. Ponton, *Finite Element Analysis of Industrial Circular Sawblade with Respect to Tensioning, Rotating, Cutting, and Expansion Slots*, Master thesis (Vir. Poly. Institute and State U. 2007).
9. G.D. Saravacos, *Mechanical processing equipment* (Kluwer Acad. Pub, 2002)
10. A. Boresi, R. Schmidt, *Advanced Mechanics of Materials* (Wiley, New York, 2003)

MATEC Web of Conferences **78**, 01006 (2016) DOI: 10.1051/matecconf/20167801006
IConGDM 2016

Optimum Design of Multi-Function Robot Arm Gripper for Varying Shape Green Product

Zol Bahri Razali[1,*], *Mohammad Haniff* Othman[1] and *Mohd Hisam* Daud[1]

[1] Robotics and Automation Technology, Faculty of Engineering Technology, Universiti Malaysia Perlis, 02100 Padang Besar, Perlis, Malaysia

Abstract. The project focuses on thorough experimentally studies of the optimum design of Multi-function Robot Arm Gripper for varying shape green product. The purpose of this project is to design a few of robot arm gripper for multi-functionally grip a green product with varying shape. The main character of the gripper is that it can automated adjust its finger to suit with the shape of the product. An optimum design of multi-function robot arm gripper is verified through experimental study. The expected result is a series of analytical results on the proposal of gripper design and material that will be selected for the gripper. The analysis of the gripper design proposal by using ANSYS and CATIA software is described in detail in this paper.

1 Introduction

In a robotic system, the one of the most important part is the robotic gripper. The unit between the robot and the work piece is the gripper. That is the reason in an automated framework, the choice of the gripper is extremely imperative [1]. There are a lot variety kinds of grippers, the most common types of grippers are jaw type, vacuum and magnetic grippers.

In the industries, the type of mechanical gripper that most commonly used is jaw type, it have a fingers that move parallel. Depending on the jaw design, two-jaw gripper consists with two gripping fingers that apply pressure internally or externally on the object [2]. The gripper hold the object by uses two or more fingers that move toward each other.

In this era, a lot of the gripper has been desiged or created. The gripper is often designed according to the needs and desires of users. In the market, the users difficult to search the good material and design for the gripper robotic arm for varying shape product. A survey was conducted for studied about fundamental and experimentally studies of the behavior and mechanical properties for optimum design for multi-function robot arm gripper for varying shape product.

* Corresponding author : zolbahri@unimap.edu.my

2 Literature Review

Gripper. In a mechanical framework, the one amongst the most vital part is the automated gripper. The unit between the robot and the work piece is the gripper. That is why in a robotic system, the selection of the gripper is very important. There are a lot variety kinds of grippers, the most common types of grippers are jaw type, vacuum and magnetic grippers. It also can be categorized into three main groups, single surface gripper, clamping grippers and flexible grippers [1].

Jaw type gripper. Jaw type grippers are related to clamping gripers. The design is so simple, so it is not surprising the gripper clamp is one of the cheapest on the market. In the jaw type grippers, the simplest type is two-jaw gripper. Two jaw gripper types are most widely used in industry because it requires a lower cost to make it. This is because, the gripper is the easiest type to designed. Two jaw gripper includes with two gripping finger pressure imposed externally or internally on an object depends on the jaw design. It could be used for objects that are large and small. Mechanics for jaw-finger movements can include linkage, cam, pinion, actuators, pneumatic and hydraulic cylinder. When the object is too complex and difficult to grasping by two jaw gripper, three jaw gripper is the best choice for more complex objects. The three jaw grippers (Figure 1) using pressure as two jaw grippers but consist of three grasping fingers. The three jaw grippers are more expensive and therefore complicated than two-jaw grippers [3].

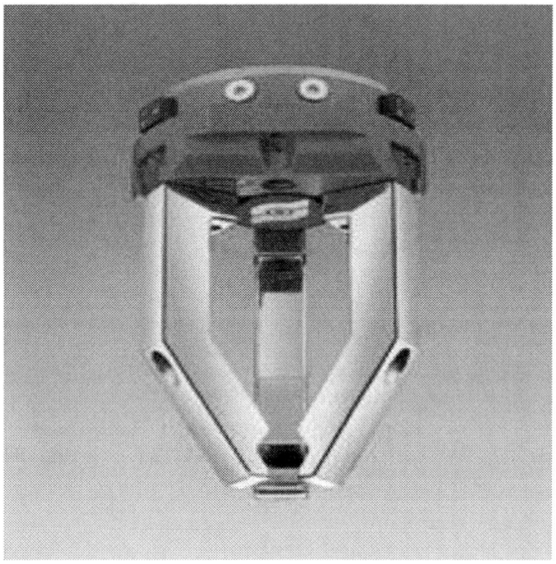

Fig. 1. Type of jaw gripper.

Magnetic gripper. Magnetic gripper is in the main group of single surface gripper, when the grippers have only one surface of component that available. These types are suitable for pick up the product that has a light and heavy weight and flat component when it difficult to controlled by the others main group of grippers. Magnetic grippers are divided into two types; electromagnet and permanent magnet. Magnetic gripper only suitable for grasping, and pick up types of ferrous materials. For permanent Magnets, the materials will be used that's from the nature of the magnetic. Permanent magnets require a mechanism to release the gripped object [4]. In addition to permanent magnet, the magnetic field cans also be generated from electricity. It will be created by a wire wounded into a coil. The point when power through the wires, an attractive field will be produced and will be lost when there is no electricity. Mostly electromagnetic are often used to take various types of metal and steel waste. The electromagnetic grippers can pick and release the ferrous object in a few second and easy to control [5]. Other benefits with electromagnetic grippers are that they can resist high forces [4].

Vacuum gripper. Vacuum grippers are also in the main group of single surface grippers. Vacuum gripper consists of a suction cup as pick up devices. The suction cup is created out of rubber. The tubes are associated through suction cups with under pressure devices for picking up item with air is sucked and releasing item with air is pumped out into the suction cups. The suction cups are mostly made by polyurethane or rubber that can stand by temperatures between -50 and 200°C. There are different types of suction cups, depth suction cup, universal suction cups, suction cups with bellow and flat suction cups with bars as shown in Figure 3 [6].

Universal suction cup is always applied to the surface is slightly curved and flat surface. In the market, the one of the cheapest suction cups is universal suction cups but at the type of suction cups, there have a several disadvantages. Under the pressure is too high, the vacuum gripper should use larger suction cup. For flat or flexible items that require assistance when lifted, flat suction cup with a bar is the best choice. These kinds of suction cup provides a small opening under load and keep the area under the pressure acting, this reduces wear on the flat with a suction cup bar, this makes the movement becomes faster and more secure. For suction cups with bellow, mostly it is used for curved surface. It is suitable for the separation of an object, grip smaller items and requires shorter movement. The depth suction cups are suitable for lifted over and edge, for curve and irregular surface. Vacuum gripper has a problem to a product that has a rough surface and a product that has holes in its surface. It will be difficult when the product is porous because the suction cups cannot sucked the surface [6].

Flexible grippers are new in the market. These main groups are difficult to handle and design. It consists with a few linkages on each finger and that have a two or more finger to pick up the product. These grippers are like a human hand. It is called multi finger gripper. This type of gripper can pick up products that have a complex shape because the gripper can control individually of the linkage in the fingers.

3 Methods and Materials

Propose of Design. In this project, some designs will be proposed for a different concept, developed and analysed by using ANSYS and CATIA software. As the proposal, three designs in different concept will be proposed, based on an observation and detail study about the application and characteristice of grippers. Figure 2, 3 and 4 show the proposed design with its green design concept.

Design 1: Human-hand concept. This is a robotic peripheral that is designed for industrial applications. Its design makes it a unique robotic end of arm tool to pick, place and handle a large range and volume of parts of varying sizes and shapes. This gripper constructed using stainless steel. In the grippers industries, stainless steel is suitable for lift the heavy weight product. The design is most complicated, but it suitable for pick a heavy weight product. The advantage for this design is it can pick any shape of product like a human hand. This design only has a three finger for pick and lifts the product. Figure 2 shows the reason why this design has only a three finger.

Fig. 2. Design 1- Human-hand concept.

Fig. 3. Design 2- Ball sucker concept.

Design 2: Ball-sucker concept. Gripping and holding of objects are key tasks for robotic manipulators. The development of universal grippers able to pick up unfamiliar objects of widely varying shape and surface properties remains, however, challenging. Most current designs are based on the multi-fingered hand, but this approach introduces hardware and software complexities. These include large numbers of controllable joints, the need for force sensing if objects are to be handled securely without crushing them, and the computational overhead to decide how much stress each finger should apply and where. Here we demonstrate a completely different approach to a universal gripper. Individual fingers are replaced by a single mass of granular material that, when pressed onto a target object, flows around it and conforms to its shape. Upon application of a vacuum the granular material contracts and hardens quickly to pinch and hold the object without requiring sensory feedback. This advance opens up new possibilities for the design of simple, yet highly adaptive systems that excel at fast gripping of complex objects. Tasks that appear simple to humans, such as picking up objects of varying shapes, can be vexingly complicated for robots. Secure gripping not only requires contacting an object, but also preventing potential slip while the object is moved. Slip can be prevented either by friction from contact pressure or by exploiting geometric constraints, for example by placing fingers around protrusions or into the opening provided by the handle of a cup. For reliable robotic gripping, the standard design approach is based on a hand with two or more fingers, and typically involves a combination of visual feedback and force sensing at the fingertips. The advantage of the design is can pick and place varying shape product through the object has a complex shape.

Design 3: Inflatable-pockets concept. In this design, following an extensive review and discussion on different gripper types and design issues in the first section, a flexible gripper design based on the use of compliant materials and internal pressure of inflatable rubber pockets. This type of grippers conforms to the shape of an object by means of elastic gripping elements and pressurization with active degrees of freedom. The advantage of this design is can lift any shape of product without damaged the surface of the product.

Fig. 4. Design 3 – Inflatable pockets concept.

4 Comparison of design

Analytic Hierarchy Process. Complete criteria comparison matrix [7] using 1–9 ratings described in Table 1. The table shows the comparison between operation, maintenance, durability and design that which most important for this design.

Table 1. The comparison matrix between characteristic (matric [C]).

Criteria	Operation	Maintenance	Durability	Design	Total
Operation	1.00	5.00	3.00	5.00	14.00
Maintenance	0.20	1.00	9.00	7.00	17.20
Durability	0.333333333	0.111111111	1.00	3.00	4.444444444
Design	0.20	0.142857143	0.333333333	1.00	1.676190476
TOTAL	1.733333333	6.253968254	13.33333333	16.00	

Table 2 shows the result of comparison matrix between characteristic. The rating at table 4.2 will be multiply with total of rating. The consistency of characteristic can be measure in this process.

Table 3 shows the comparison of characteristic of three design of gripper for varying shape product. This table helps to give a score for the comparative advantage of each design. All score will be referring in this table.

Table 2. The consistency measure of characteristic.

Criteria	Operation	Maintenance	Durability	Design	Total	Average	Consistency measure
Operation	0.58	0.80	0.23	0.31	1.91	0.48	5.96
Maintenance	0.12	0.16	0.68	0.44	1.39	0.35	5.48
Durability	0.19	0.02	0.08	0.19	0.47	0.12	4.11
Design	0.12	0.02	0.03	0.06	0.23	0.06	4.27
TOTAL	1.00	1.00	1.00	1.00		1.00	19.82

Table 3. The design characteristic of gripper for varying shape product.

Characteristic	Parameter	Design 1	Design 2	Design 3
Operation	rating	very good	very good	good
Maintenance	month	6	3	3
Durability	material	stainless steel	rubber	stainless steel and rubber
Design	shape	most complicated	simplest	not complicated

Table 4 shows the comparison of operation for each design. The result show design 1 and 2 get a same score. If it is able to get a little bit score, it shows that it is a good design.

Table 4. Comparison between operation of design 1, 2 and 3.

Operation	Rating	Design 1	Design 2	Design 3
Design 1	very good	1.00	1.00	3.00
Design 2	very good	1.00	1.00	3.00
Design 3	good	0.333333333	0.333333333	1.00
sum		2.333333333	2.333333333	7.00

Table 5 is used to get the average score for critical measure of rating for operation in each design after get the data from Table 4.

Table 6 shows the comparison of maintenance for each design. The result show design 2 and 3 get a same score. But this result show the design 1 is most good from another design. Design 1 gets a highest score because this design takes a long time to be maintenance. It shows that this design is not easily damaged.

Table 5. Critical measure of rating between design 1, 2 and 3.

Operation	Rating	Design 1	Design 2	Design 3	Total	Average	Consistency measure
Design 1	10	0.42857143	0.428571429	0.428571	1.285714	0.428571	1.00
Design 2	13	0.42857143	0.428571429	0.428571	1.285714	0.428571	1.00
Design 3	15	0.14285714	0.142857143	0.142857	0.428571	0.142857	1.00
sum		1.00	1.00	1.00		1.00	3.00
						CI	-1.33
						RI	0.52
						CR	-2.56

Table 6. Comparison between maintenance of design 1, 2 and 3.

Maintenance	Month	Design 1	Design 2	Design 3
Design 1	6	1.00	2.00	2.00
Design 2	3	0.50	1.00	1.00
Design 3	3	0.50	1.00	1.00
sum		2.00	4.00	4.00

Table 7 is used to get the average score for critical measure of rating for maintenance in each design after get the data from table 6. This table shows design 1 get the highest score from another design.

Table 8 shows the comparison of maintenance for each design. The result show the design 1 is most good from another design. Design 1 gets highest score because it only used a stainless steel. Stainless steel has a strong durability than a rubber.

Table 7. Critical measure of maintenance between design 1, 2 and 3.

maintenance	month	Design 1	Design 2	Design 3	Total	average	consistency measure
Design 1	4	0.50	0.50	0.50	1.50	0.50	1.00
Design 2	3	0.25	0.25	0.25	0.75	0.25	1.00
Design 3	3	0.25	0.25	0.25	0.75	0.25	1.00
sum		1.00	1.00	1.00		1	3.00
						CI	-1.33
						RI	0.52
						CR	-2.56

Table 8. Comparison between durability of design 1, 2 and 3.

durability	material	Design 1	Design 2	Design 3
Design 1	Stainless Steel	1.00	5.00	3.00
Design 2	Rubber	0.20	1.00	0.60
Design 3	Stainless Steel and Rubber	0.333333333	1.666666667	1.00
sum		1.533333333	7.666666667	4.60

Table 9 shows the comparison of design (shape) for each design. The result shows the design 2 is most good from another design. Design 2 gets highest score because this design is simplest to build compare with each design.

Table 9. Comparison between design and design 1, 2 and 3.

Design	Shape	Design 1	Design 2	Design 3
Design 1	most complicated	1.00	0.111111111	0.333333333
Design 2	simplest	9.000000009	1.00	1.00
Design 3	not complicated	3.000000003	1.00	1.00
sum		13.00000001	2.111111111	2.333333333

5 Finite Elemen Analysis.

It is seen from Figure 5, the moments about joints are;

$$M1 = F * (L1 + L2 + L3) \quad M1 = 70F \tag{1a}$$

$$M2 = F * (L2 + L3) \quad M2 = 41F \tag{1b}$$

$$M3 = F * (L3) \quad M3 = 20F \tag{1c}$$

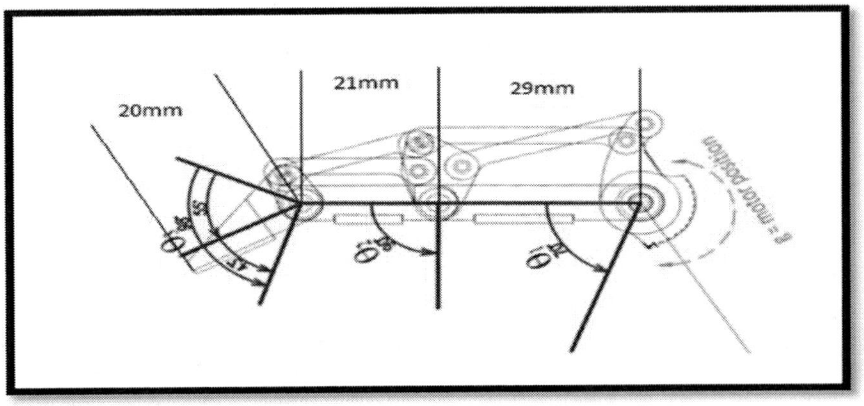

Fig. 5. Schematic Representation of Finger.

The critical moment that creates the tension is M1. After definition of M as reaction moment on joint 1 then it is written as;

$$M1 = M \tag{2}$$

Where

$$M = T * rp \tag{3}$$

Here, T is the tension force on rope and rp is the radius of pulley at joint 1. From Eq.2 and Eq.3,

$$70\,F = T * 6$$

$$T = 11.667\,F \tag{4}$$

After determination of tension force on rope the torque value of the actuator is calculated from;

$$Ta = T * ra \tag{5}$$

Where ra is radius of actuator gear and ra=10.5mm

$$Ta = 11.667\,F * 10.5 \quad Ta = 123\,F\ [\text{N.mm}] \tag{6a}$$

$$Ta = 0.123\,F\ [\text{N.m}] \tag{6b}$$

For 6 bar pressure, torque value for actuator is 2 [N.m]. Then; using this value and equation 4.6b, maximum gripping force for the tip point of the finger is calculated as;

$$\frac{1}{123F\ [\text{N.mm}]} \text{ x } 2 = 0.01626 \text{ kN}; \qquad F = 16.26\ [\text{N}] \tag{7}$$

By taking the coefficient of friction as 0.5 [8] (covering material), Maximum carrying capacity of one finger can be calculated as;

$$\text{m} = \text{F} * \mu / \text{g} \qquad \text{m} = 16.26 * 0.5 / 9.81 \qquad \text{m} = 0.8287\ [\text{kg}] \tag{8}$$

From the result calculated in Eq.8, the gripping capacity of the hand for three fingers and palm can be calculated as;

$$\text{mG} = 3 * \text{m} \quad \text{mG} = 2.4862\ [\text{kg}] \tag{9}$$

Finally, it is necessary to define an equation for torque-pressure relation for the actuator of the gripper in order to determine finger force for an open loop gripping force control [8]. The actuators used in the gripper are rotary vane type pneumatic actuators. So, torque value of the actuator is linearly proportional to applied pressure and defined as;

$$\text{Torque Factor x Pressure} = \text{Torque Output} \tag{10}$$

Using Eq.10, torque factor for actuator is found as;

$$\text{Torque Factor} = 2\ [\text{N.m}] / 600000\ [\text{N/m2}] = 0.334\text{x}10^{-5}\ [\text{m}] \tag{11}$$

Then, torque-pressure relation for the actuator is defined as;

$$Ta = 0.334\text{x}10^{-5}\ P \tag{12}$$

In this equation, unit of Ta is [N.m] and P is [N/m2]. Using Eq. 6b and Eq.12, the relationship between pressure and the gripping force is obtained as;

$$F = 1.662\text{x}10^{-5}P \tag{13}$$

Here, unit of F is [N] and P is [N/m2]. Eq.13 is valid for pressure ranges between 2 and 8 bars which is the operating range of the actuator.

6 Analysis of Design

From the force Analysis for fingers result calculated in Eq. 8, Maximum carrying capacity of one finger is 0.8287 kg. The gripping capacity of the hand for three fingers and palm is 2.4862 kg**.** In the theoretical, the maximum carrying capacity for this design is 2.4862 kg by length of the fingers. The result will be used to analysis the design. 2.4862 kg is equal to 24.4 N. For this analysis, 3 0N force will be applied to finger to get the data of total deformation, equivalent elastic strain and equivalent stress. In the creation of the gripper, the most commonly used materials are aluminum and stainless steel. In the industries of gripper, aluminum is mostly used for lift light objects and stainless steel is used to lift extreme weight objects. For this model, the parts of gripper will be constructed from stainless steel to make a gripper able to stand in the long time.

Deformation in continuum mechanics is the transformation of a body from a reference configuration to a current configuration. **Figure 6** shows the analysis of total deformation of finger. The maximum deformation is 3.3215 mm.

 DOI: 10.1051/matecconf/20167801006

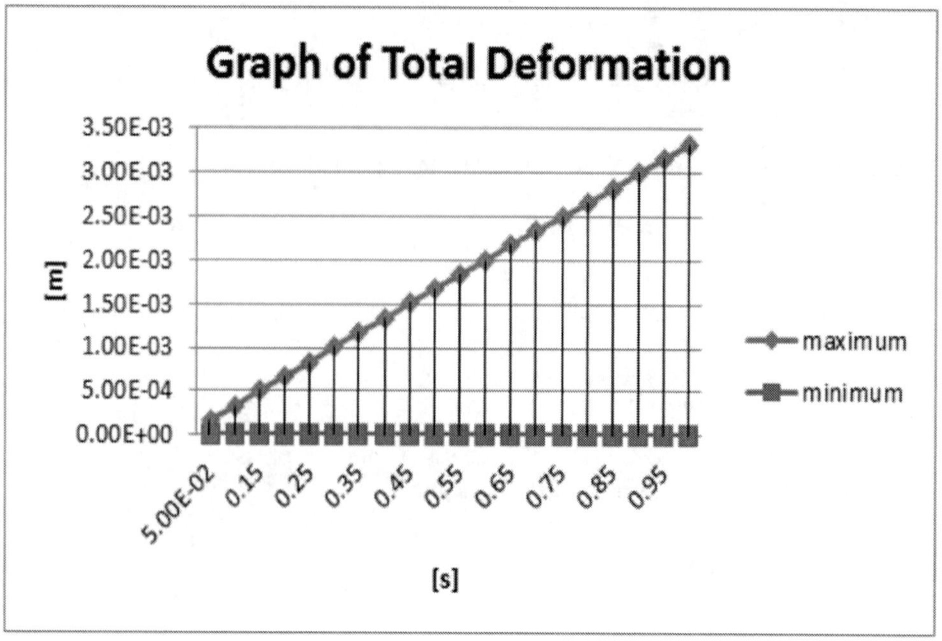

Fig.6. Graft of total deformation.

Figure 7 shows the result of equivalent elastic strain. Equivalent elastic strain is a form of strain in which the distorted body returns to its original shape and size when the deforming force is removed. The maximum strain is 0.044578 m/m.

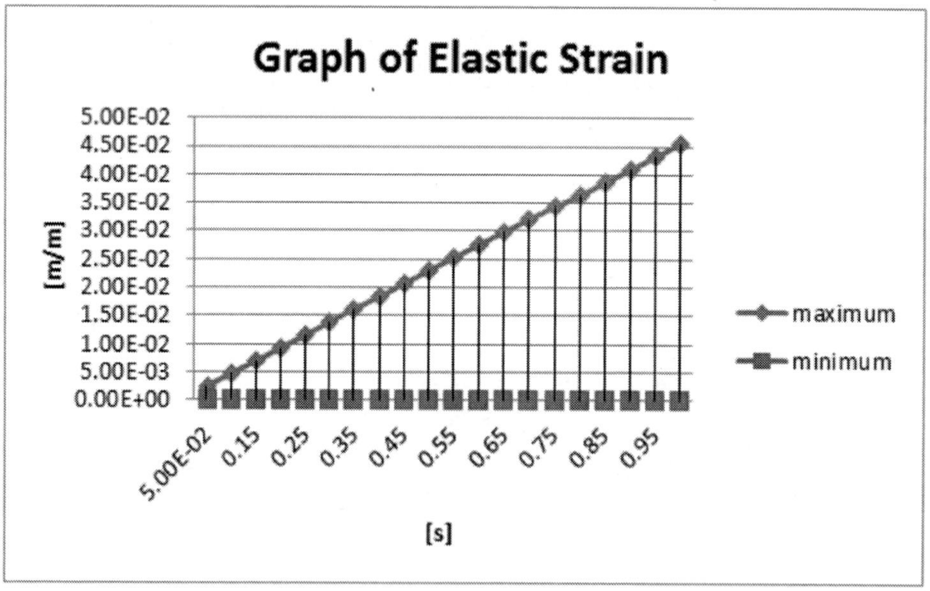

Fig. 7. Graft of elastic strain.

7 Conclusion

To design a gripper tool for an industrial robot has been very instructive. In this paper, detail method of designing an optimum robotic gripper is described and implemented. The gripper has been shown to be able to handle a variety of shapes and sizes. The objective for determining, selecting and designing the proposed green robot gripper is achieved. As the main objective is to optimally design multifunction robot arm gripper for varying shape of green product, is done by using CATIA software. Other than that, the comparison of the gripper design to choose the best design also completed by using the Analytic Hierarchy Process (AHP). For the analysis of FEA (finite element analysis), the result was get by using ANSYS software. The result shows that the selected design can stand by the pressure that was applied. For the conclusion, the optimum design of this gripper can be highlighted so that it can be marketed in the industries.

This work was financially supported by the Universiti Malaysia Perlis Seed Money (9014-00034).

References

1. Z.B. Razali, N.A. Mat Nawi, *Automated Segregating Device for Reuse and Recycle Waste* (Lambert, Chicago, 2013)
2. L.U. Odhner, L.P. Jentoft, M.R. Claffee, N. Corson, Y. Tenzer, R.R. Ma, M. Buehler, R. Kohout,R.D. Howe, A.M. Dollar, Int. J. Rob. Res., **33**, 736 (2014)
3. D.T. Pham, E. Tacgin, Int. J. Mach. Tool Manu., **32**, 349 (1992)
4. G.J. Monkman, S. Hesse, R. Steinmann, H. Schunk, *Robot Grippers,* (WILLEY-VCH, Germany, 2007)
5. P.V.P. Reddy, V.V.N. Satya Suresh, Int. J. Mech. Eng. Robot. Res., **2**, 255 (2013)
6. Z.B. Razali, G.B. Lee. *High-End Technology for Effective Glass Manufacturing Industry* (Lambert, Chicago, 2013)
7. T.L. Saaty, L.G. Vargas, *Models, Methods, Concepts & Applications of the Analytic Hierarchy Process* (Springer, New York, 2012)
8. S.H. Cho, S.D. Eppinger, *ASME 2001 Design Engineering Technical Conferences and Computers and Information in Engineering Conference* (2001)

MATEC Web of Conferences **78**, 01007 (2016) DOI: 10.1051/matecconf/20167801007
IConGDM 2016

Optimum Identification Method of Sorting Green Household Waste

Mohd Hisam Daud[1,*] and *Zol Bahri* Razali[1]
[1] Robotics and Automation Technology, Faculty of Engineering Technology, Universiti Malaysia Perlis, 02100 Padang Besar, Perlis, Malaysia

Abstract. This project is related to design of sorting facility for reducing, reusing, recycling green waste material, and in particular to invent an automatic system to distinguish household waste in order to separate them from the main waste stream. The project focuses on thorough analysis of the properties of green household waste. The method of identification is using capacitive sensor where the characteristic data taken on three different sensor drive frequency. Three types of material have been chosen as a medium of this research, to be separated using the selected method. Based on capacitance characteristics and its ability to penetrate green object, optimum identification method is expected to be recognized in this project. The output capacitance sensor is in analogue value. The results demonstrate that the information from the sensor is enough to recognize the materials that have been selected.

1 Introduction

The project aims to develop a material to be removed, in order to improve the concept od reduce, reuse and recycle. The aim is also to design a system to separate household waste objects to separate them from the main waste stream. In particular, it serves to distinguish some types of waste directly or automatically. A detector will be placed somewhere on the conveyor, the sensor is transmitted through the object to be distinguished, thresholding circuits or software to distinguish the objects located either above or below the level sensor which will decide what type of the detected object.

This research is to find the best identification method to sort co-mingled household waste material. Generally, the process of separating household waste of recyclable materials such as plastic, concrete, paper, forest green and others. However, there is some sort of technology is used, such as an infrared sensors and infrared laser to separate paper, concrete, plastic, and other. The techniques is used to analyze the waste materials based on size and shape of objects and in a recycling center facilities there is some of the techniques can be implemented. However, to meet current needs, vision systems or any other latest technique which can recognize different types of recycle material in household waste can be used as an alternative method. Optimum identification technique must be used to recognize the household waste material.

* Correcponding author : mhisam.daud@gmail.com

2 Literature Review

At present, Malaysia is running out of space to dispose of municipal solid waste generated each day from anthropogenic consumption. Based on a study by the Ministry of Housing and Local Government, the landfill in this country will be filled by 80% within a few years [1]. In fact, many serious enviromental problems that produce form landfills themselves that effect both for us and for future generations.

Malaysian daily average litter is 0.8kg per person per day. Waste generated by Malaysia also noted the high rate among the countries of 15000 tons of garbage every day. If we collect all the waste and put in one place, we can fill with garbage KL twin towers in just 9.5 days. Even expected amount of waste will increase by 2% per year, depending on the activity of the population, economic activities and waste disposal methods [1].

The definition of 'waste' is taken from the Concise Oxford Dictionary refers to "useless remains" or value or lack of use is the product of human activities. Physically, useful products we can find the same ingredients but different in terms of the lack of useful production.

Various schemes could be classified for the waste: the physical state (liquid, solid, gas) and then in solid waste by the original use (food waste, packaging waste, etc.), materials (paper, glass, etc.) [2], with the physical characteristics (burning, composting, recycling), the origin (commercial, domestic, industrial, agricultural, etc.) or with the level of security (dangerous, not dangerous). Household and commercial waste is often referred to as Municipal Solid Waste (MSW) [3]. Household waste is one of the most difficult to manage the waste effectively. It consists of various types of materials (metal, paper, plastic, glass, organic) at all mixed together. MSW composition is also variable, both geographically and seasonally from city to rural areas and from country to country[2].

Green waste is biodegradable waste materials which may consist of a park or garden waste, such as flower or grass cuttings and hedge trimmings, as well as commercial and domestic food waste. The differentiation green identifying it as high in nitrogen, is contrary to brown waste, which is mainly carbon. Green waste is often collected through private waste management contractors business or municipal curbside collection schemes and subject to an independent audit. Biofuel can be produced from biodegradable green waste that captured from Biogas. non-food crops also can be used to produce cellulosic ethanol that from green waste [4].

Currently, optical sorters is used by the facility to sort the waste form green household waste, have what it takes to pre-shredded green waste streams, automatically sort and clean contaminated. The combination of mechanical pre-processing systems (air classification, size, grinding, etc.) are use in the technology to remove prohibitive contaminants (metals, plastics) from the green waste automatically, and generates a contaminant-free green waste product of composting or anaerobic digestion [5]. There are two main areas in green waste processing facility where optical sorters are in a position to provide added value to customers. The automated removal of any metals and plastics are two of the most common preferred application option. If there are needs to remove plastic contaminants from the food, there is a sorting facility features with a full spectrum near infrared (NIR) sensor to classify polymers and analyze and other organic materials at lower volumes [6]. This latest generation separator, with a pixel resolution of up to 16x optical sorters else, really provide optical isolation in high definition.

3 Methods and Materials

3.1 Process Flow Charts.

Figure 1 shows a flow chart to describe the step to finish this project. The project starts with finding problems related to sorting household waste and what happens at present. Then study on the current project in sorting system, process, equipment such as sensors that. can detect material and any rules or principles involved in preparing the material that has been done. Review of the literature on the requirements and available methods or methods of concept will guide researchers to the idea of optimal design or the optimal method of preparing a variety of material systems. This process will continue by selecting materials from household waste as a medium in this study.

After selecting at least five types of waste as a medium of research, analysis, basic electrical properties of materials that have been done. Based on the results expected, an experimental study of the proposed methods will continue. After the results are compiled, comparing the identification method to choose the ingredients needed to make a difference in each method and then analyzing the optimum circuit identification method..

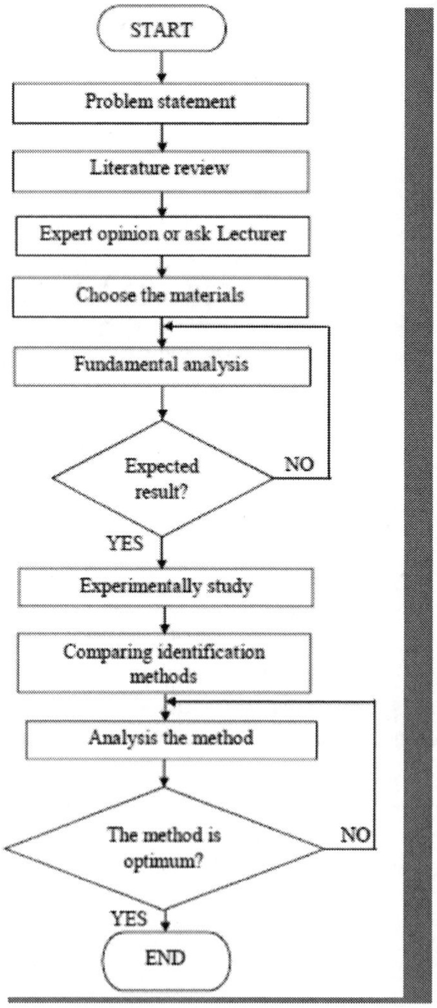

Fig.1. Flow Charts.

3.2 Analyze the Properties of Household Waste Materials.

Outside, there is a lot of household waste, therefore for the purposes of this study, the authors only focused on the type of waste that can be separated into three types such as: metal, concrete and green forests. type of material has the properties of mechanical, chemical or electrical different [7]. Therefore, to run some analysis on materials properties is important, as it will cause some impact on the sensor used in this project.

3.3 Selection of the Sorting Methods.

The main part of the project is to find the best methods for distinguishing sorting waste materials are different. It refers to the best method that can detect and differentiate forest green, plastic, metal, concrete and paper at once. It can also be said that this technology will lead to a good choice for the development of the economic sustainability of recycling sorting facility [1].

Five methods were shortlisted. Basic system studied methods to determine its relationship with material properties [8]. It is important to understand their relationship because of the properties of the sensor to be used in the sorting system has the effect of material properties.

Based on data regarding the correlation of these traits, one of the best methods will be selected for further study on methods like, how it will be implemented, what event, what is the advantage compared to others. When a complete system is known, we will propose a redesign sorting system, with the system to be as optimal isolation system for the project, such as a system that can be compiled for different materials.

3.4 Redesign the Sorting System.

For this case, the focus is on the circuit that follows the order of the sorting process. For example, when the sensor detects the green woods, it returns a signal to the rest of the sorting bins and the waste products should be sorted into the correct bins.

4 Results and Discussions

4.1 Analysis the Method of Sorting Using Capacitive Sensor.

The data can be collected by measure the analog signals that detect from the capacitive sensor. The output of capacitive sensor must be in analog. When it detect present objects or the materials, the voltage value (analog signal) will show the different value depent on the capacitive objects. Microcontroller only could read the digital signal, so that the analog signal data must be convert to digital signal first and process the data. This process also known as analog to digital converter (ADC). Three type of frequency will be used in this analysis are 104.7 khz, 96.6 khz and 89.7 khz to be tested within 20 cm of distance between the materials.

4.2 Results of Different Materials.

Table 1 and Figure 2 show the relationship between different materials at a frequency of 89.7 kHz with different distances. proximity sensor with surface materials up to 20 cm wide. The value that is different from the production of metal, wood, and concrete can be seen at a distance of 6 cm to 20 cm.

Table 1. Different Materials at 89.7 kHz.

	Type of material		
Distance (cm)	**Concrete**	**Metal**	**Wood**
1	0	0	0
2	0	0	0
3	0	0	0
4	2	155	77
5	231	272	238
6	352	332	310
7	398	363	338
8	416	380	336
9	429	388	366
10	434	400	351
11	443	406	361
12	445	408	370
13	447	407	357
14	448	411	360
15	446	410	368
16	445	409	371
17	446	408	364
18	445	414	379
19	444	405	372
20	444	401	369

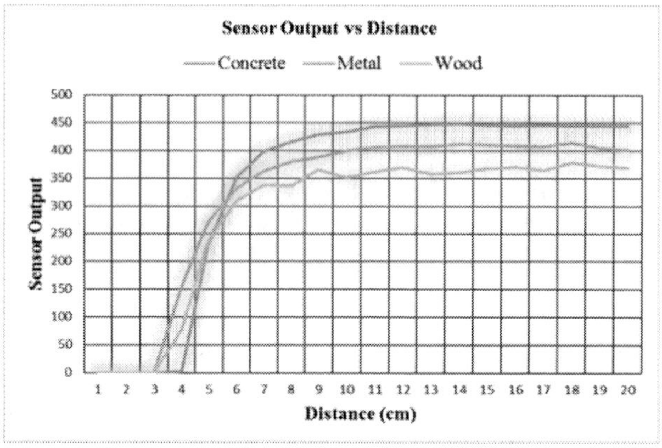

Fig. 2. Sensor output vs. Distance at 89.7 kHz.

Table 2 and Figure 3 shows the relationship between different materials at a frequency of 96.6 kHz with a difference. Using the driving frequency of the sensor, the sensor output readings are almost equal to each other starting from a distance of 5 cm with a sensor surface of the material.

Table 3 and Figure 4 show the relationship between different materials on the frequency of 104.7 kHz with a difference. Using the driving frequency of the sensor, the sensor output readings are almost equal to each other starting from 4 cm distance sensor with the material surface.

Table 2. Different Materials at 96.6 kHz.

	Type of material		
Distance (cm)	**Concrete**	**Metal**	**Wood**
1	0	0	0
2	0	2	0
3	216	648	49
4	671	722	746
5	738	746	745
6	760	742	745
7	764	749	749
8	760	753	745
9	758	748	746
10	762	749	743
11	760	750	747
12	759	749	744
13	756	750	743
14	764	748	747
15	762	754	744
16	757	755	743
17	758	753	740
18	755	754	750
19	762	753	747
20	760	757	742

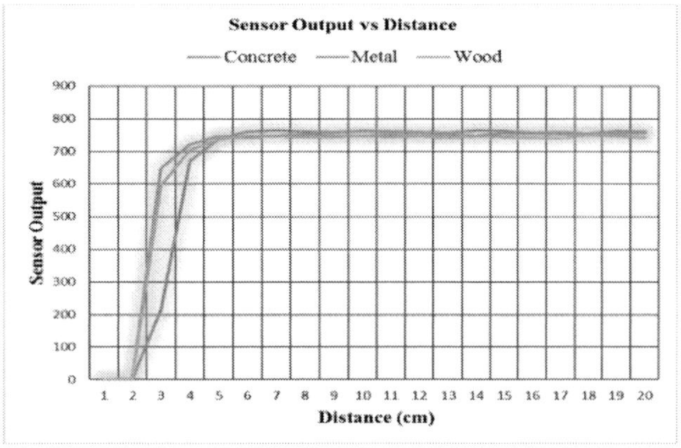

Fig. 3. Sensor output vs. Distance at 96.9 kHz.

4.3 Comparing the methods.

The comparison between all the methods is to determine which one is the optimum and the best method. Table 4 shows the different method that could recognize multi materials. This analysis come out with different result from different type method. Eddy current separator is only could sort the permanent magnet materials [9] and it can recognize ferrous and nonferrous materials (metallic material). Machine vision system can recognize the materials based on size, shape and color. The picture of materials will be taken to analyze it. Both capacitive proximity and capacitve sensor could detect all concrete, wood and metal but capacitive proximity sensor is not the best choice cause this sensor only detect the materials but could not recognize the materials because of the signal is digital value ('0' and '1'). The capacitve sensor is the best choice cause this sensor is the detect based on the analog value.

Means different type of materials will show the different value of data. Therefore, the best method to sort and recognize the materials is capacitive sensor and it have potential to be the optimal method.

Table 3. Different Materials at 104.7 kHz.

	Type of material		
Distance (cm)	**Concrete**	**Metal**	**Wood**
1	0	0	0
2	0	686	373
3	712	850	819
4	859	878	871
5	881	884	881
6	885	889	881
7	912	885	886
8	893	887	882
9	893	884	887
10	887	885	881
11	888	889	883
12	885	886	878
13	886	888	880
14	911	884	883
15	892	885	880
16	892	882	886
17	887	883	881
18	887	888	883
19	884	884	880
20	885	887	881

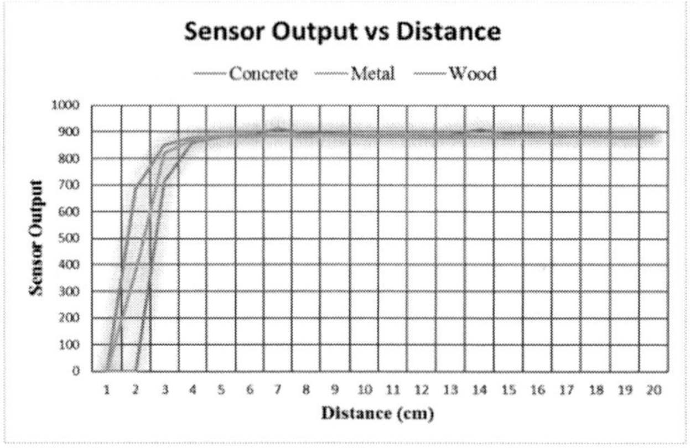

Fig. 4. Sensor output vs. Distance at 104.7 kHz.

Table 4. Recognizing various of materials.

	Materials		
Method	**Concrete**	**Metal**	**Green wood**
Eddy current separator	-	✓	-
Capacitive proximity sensor	✓	✓	✓
Machine vision	-	-	-
Capacitive sensor	✓	✓	✓

5 Discussions

In general, sorting waste systems will sort thr recyclable waste resulting from household and in the same time it could solve the problem regarding to landfills that require a big space to dispose the garbage. Especially in urban areas that produce more waste compare to the village area or rural area. The sorting system required suitable method to make it applicable to all types of household waste or various type of waste. From this study, there are several methods available and can recognize the type, size, shape and color. More importantly, you want to use the methods that should be able to recognize and classify the type of waste in certain materials.

Image processing in machine vision methods are not suitable for use in sorting this material to identify various types of waste as image processing can only recognize by shapes, sizes, and colors. This system requires an application that can recognize the various types of waste to classify material following classes of materials.

Capacitive sensors are more suitable than other methods because the sensor can recognize the various types of materials based on the electrical properties of the material. Each material will produce different electrical property values. Capacitive sensors using eddy current principle that makes it react just like the concept of eddy currents. Eddy current principle also react when there is any material having electrical characteristics close to the investigation that resulted in the electromagnetic field.

6 Conclusion

This paper presented the method that can recognize three materials of household waste (concrete, metal and green woods). The capacitive sensor is the best identification method based on the characteristic of the sensor to sorting multi-material of household waste. This sensor can recognize three of materials using various frequencies. Based on the results, the frequency of 89.7 kHz is the best to reach out to the capacitive sensor to analyze the types of waste materials to complete the rest of the household.

Data obtained using three household waste (concrete, metal and forest green). Preferred materials have different electrical properties in which the sensor can recognize and classify it. The large gap between the sensor output to concrete, metal and wood, green materials to make classification easier. If we use a capacitive proximity sensor, the sensor can detect each of the materials, however, because of different distances on the surface of the material; digital signal must give an inaccurate value.

Finally, the capacitive sensor is the most effective and efficient method to compare with others that can be used to separate the various materials as the method can recognize a variety of materials.

This work was financially supported by the Universiti Malaysia Perlis Seed Money (9014-00034).

References

1. Z.B. Razali, N.A. Mat Nawi, *Automated Segregating Device for Reuse and Recycle Waste* (Lambert, Chicago, 2013)
2. Z.B. Razali, G.B. Lee, *High-End Technology for Effective Glass Manufacturing Industry* (Lambert, Chicago, 2013)
3. P.C. Nicholas, *Handbook of Solid Waste Management and Waste Minimization Technologies* (Butterworth-Heinemann, Boston, 2003)
4. http://www.magsep.com/optical-sorting-applications/municipal-solid-waste-msw-sorting/green-waste-sorting/ Retrieved on 24 November 2013.

5. N. Kirchner, D. Hordern, D. Liu, G. Dissanayake, Sensors and Actuators A. Phys., **148**, 96 (2008)
6. M.O. Rahman, A. Hussain, E. Scavino, H. Basri, M.A. Hannan, Expert Syst. with Appl., **38**, 10398 (2011)
7. D. Fletcher, R. Gerber, IEEE Trans. Magn. **29**, 3255 (1993)
8. L. Zitouni, R. Beheim, Huez, F. Belloir, IEEE Sens. J., **6**, 1580 (2006)
9. M. Brojboiu, L. Mandache, V. Ivanov, *8th International Symposium on Advanced Topics in Electrical Engineering* (2013)

UniMAP e-Lab for Electrical Engineering Technology: Future Online Laboratory Classes

Mohd Hisam Daud[1,*], and *Zol Bahri* Razali[1]

[1]Robotics and Automation Technology, Faculty of Engineering Technology, Universiti Malaysia Perlis, 02100 Padang Besar, PERLIS, MALAYSIA

Abstract. This paper will describe a proposed design and approaches to the future provision of laboratory experience using communication and control technology. This approach provides broad access for students who can not attend conventional laboratory to laboratory work. The experimental system online (e-Lab) is under development for the students to carry out various experiments in engineering and technology education assessment system. In the field of engineering technology, one important element is laboratory work, although there are limitations in terms of space laboratories, distance learning provision. UniMAP e-Lab project aims to address many of the existing constraints. Beyond the educational goals, UniMAP e-Lab system enables experimental knowledge in a particular field of engineering technology and experimental results of the research are disseminated and exploited effectively. Solution design of hardware and software as well as the characteristics of education discussed.

1 Introduction

This paper discussed the development of UniMAP e-Lab system (or known as remote laboratory) to enable students to carry out experiments as an extension of the real world required a lot of distance learning systems and computer-based learning. The objective of this system is to provide a high quality learning experience in engineering technology education with a teaching laboratory for students [1], provides flexibility in terms of location, special needs and time. The project which is developing an online experimentation system, will allow students to conduct a range of experiments to demonstrate the utility of such an approach for engineering technology education [2].

For instance, currently the PEARL system will allow students to work together while at a distance from the laboratory site using a range of synchronous and asynchronous communications tools. Experiments developed include an implementation of a remote electron microscope, a spectrometer, visual inspection of printed circuit boards and a digital electronic bench. A key design feature of this web-based system is that the interface should be compatible with assistive technology [3].

Remote access laboratories have been widely reported by researchers and education innovators since the internet started to provide reliable data communications and easy access for large numbers of users. However, most reports in the contemporary literature

* Corresponding author: mhisam.daud@gmail.com

describe technical implementations without extended and detailed evaluation of student learning advantages. So far, it would appear, no detailed cost benefit studies have been reported. It would seem that the technology has stalled [4]. Early implementations, such as Telelabs at the University of Western Australia (UWA), are in regular operation for large classes of students [4]. Yet the technology does not seem to have been widely adopted, nor have there been reports of serious attempts to negotiate industry standards that would be a necessary precursor for large-scale adoption.

Why has such a promising technology not been adopted so readily? Are the benefits as great as proponents have claimed in the past?

2 Literature Review

2.1 Effectiveness of the e-Lab.

When evaluating the effectiveness of the learning process in engineering technology education, practical experience is one of the outcomes. Hands-on laboratories play an important role not only with respect to this practical aspect but also when talking about developing student's creativity and problem solving skills [5]. So that, in the past 10 years, many research involved in remote laboratory research to measure the experience of students.

A report by Trevelyan and his colleague [4] on a fundamental investigation of some learning phenomena in conventional hands-on laboratories that arose from the question *"What do students really learn in laboratory classes or remote lab experiences?"*. To explore this issue, the authors started a project in order to understand more about the results of practical learning in the traditional classroom lab. The notion of a group of researchers, to settle practical matters related to the specific domain it depends on the ability of someone in 'practical intelligence' (PI). Psychologists develop PI measurement instruments as part of the whole debate continued discipline predict performance on-the-job who use the results of psychometric tests. The study found that in laboratory-duty pliers, they can measure the significant gains in PI and these techniques can be used. Conventional evaluation (lab reports, tests, exams, training tutorials) obtained from the results of students not associated with PI. Even more interesting, perhaps through PI, fault diagnosis tasks to be perform can be predicted by the ability of students and their found this evidence.

For the first time, therefore, the researchers demonstrated that there are real advantages inherent in hands-on laboratory classes and we can measure this advantage [2]. What can be learned from this work is that students' learning in laboratory classes is not what the lecturer have come to expect. There is still much to discover and the report provided some research tools to enable others to follow similar investigations.

There are significant implications for developers of UniMAP e-Lab system and engineering technology education researchers from this work.

2.2 e-Lab: Evaluating Students Learning Experience.

In 2000, UWA decided to build on the Telelabs project to combine automated assessment and performance monitoring techniques with internet-connected remotely operable laboratory equipment to enhance student learning and provide a real return on investment through savings in laboratory demonstrator time [4].

This research shows how through the dial-up phone and internet connection, one or more users can use low-bandwidth connections simultaneously operated successfully by a complex machine. This research project and other projects that have stimulated using an internet connection for a whole generation of remote control equipment. The industrial automation and control equipment is now standard for most remote internet control for the hardware and software equipment.

Online access to lab equipment have given more time for students to interact with our equipment: conventional laboratory classes usually allows each student only a few minutes real time operating equipment [5]. By allowing students to operate the equipment for an hour or more at a time, and to collect data by means of semi-automatic, we have significantly improved opportunities for student learning.

At the same time, to improve the students learning technique, the researchers have also attempted an incentives to measure this through develop assessment techniques. Of course, real equipment should always be expose to students [5]. The question still needs to answer is "how much time needed for supervision in the real laboratory with real equipment can be replaced with remote access without supervised?"

In evaluating the engineering technology laboratory class effectiveness, student evaluations or opinions may also be included [6]. Explicit learning outcomes, usually written in a handout guide for a mechatronics laboratory class, create intentionally: students perform mechatronics laboratory exercises in order to learn these. Implicit learning, on the other hand, may not be intentional.

For example, in Robotics and Automation laboratory class setting, students will learn more than just the stated objectives of the class. Through their experience they will learn many other things that may not seem to be important for gaining marks, but nevertheless are helpful in achieving a commonly desired objective for students: completing the class as quickly as possible [2]. In an electronics laboratory, for example, this learning might include convenient ways to strip connecting wires, ways to make reliable electrical connections, etc. This is what we refer to as practical knowledge, and one way to measure this learning is through the notion of practical intelligence [4].

Measuring the hands-on components of laboratory classes might be valuable for engineering technology educators. Before moved from hands-on to e-laboratories, engineering technology educators need to justify the experiences of certain students when the measurement is validated.

Through a platform of assessment in the e-laboratory, learning component can be distinguished by measuring and assessing it. The educator could be more confident in predicting what students will gain in a given laboratory class [1]. These measuring instruments would then provide a powerful new means to separate between e-laboratories and hands-on and also to assess the effectiveness of engineering technology laboratory classes in UniMAP.

3 Methods and materials

3.1 Existing technology for e-Lab.

Nowadays the e-Laboratories technologies already been used. The e-Laboratory (or remote laboratory) is defined as a computer-controlled laboratory that can be accessed and controlled through some communication medium. Different architectures are used to support the e-learning environments. Some are based on software solutions such as LabVIEW, while others are supported by open-source software such as PHP, JavaScript, Java, Pearl, Python, AJAX.

e-lab allows the user to perform tasks and laboratory experiments without having to be or using real equipment via the Internet. In a traditional laboratory, consumers use equipment directly through physical action (eg manipulate by hand, switch button, pressing the button) and receive sensory feedback (visual, audio and tactile). In the e-lab, the same interactions occur just like being in the lab either in the wall or around the world, via the remote infrastructure. Remote infrastructure is a new layer of communication between the users of laboratory equipment. It is responsible for receiving sensory information from the equipment and convey the user's actions.

Therefore, students will have access to laboratory facilities without restrictions on time and location. Remote access is the application of laboratory in which students connect to a remote laboratory equipment, carry out experiments and get the results of that experiment as a response [7]. It introduces students to the physical remote laboratory provides physical and environmental practices that students can use the web interface from any computer [8].

So in this section is discussed a few types of e-laboratories with different architecture and tools used to develop them

3.2 LabVIEW web UI builder.

Unlike traditional web programming languages, LabVIEW Web UI Builder enables engineers and scientists to use their existing knowledge of LabVIEW graphical programming in order to develop web applications. National Instrument Web UI Builder is an online editor based on Microsoft's Silver Light Framework that provides the user a LabVIEW like experience for developing remote application based on the LabVIEW RESTful web services. Web services are available as a development option for LabVIEW application from version 8.6 and were introduced to address the growing need for an open standard way to interact with LabVIEW application over the Web. Thin client application created with LabVIEW Web UI Builder, in combination with LabVIEW web services, provide remote monitoring and control solutions that are based on modern, standard web technologies. These solutions do not depend on the LabVIEW Run-Time Engine, and all network communication performed through the standard Hypertext Transfer Protocol (HTTP) [8].

Furthermore, remote control applications can be created using LabVIEW Web UI Builder. Remote control easily implemented when only one end user via a thin client that will connect and use the system. during the development of thin clients, it is necessary to take into account if the end user is more than one. Solutions that LabVIEW Web UI Builder users can perform include requiring verification as part of the login or use a web service to track the number of users currently communicating with the system.Figure-1 shows an appearing monitor of LabVIEW Web UI Builder [8].

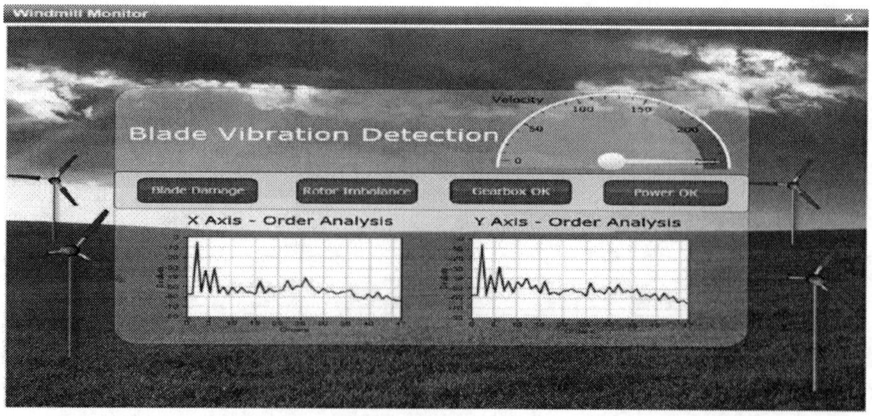

Fig. 1. LabVIEW Web UI Builder

3.3 Smartphone and cross-platform communication toolkit (SCCT).

Smartphone and Cross-Platform Communication Toolkit (SCCT) is a programming toolkit supported on a wide range of platforms and programming languages that manages communication among application transparently and speeds up your delivery time. SCCT create open solutions capable of exchanging data among a wide range of systems from smartphones to Java apps and UNIX workstations. The SCCT transparently manages communication between applications so you can create multiple simultaneous connections without changing a single line of code. Data is converted into a platform-independent format and optimizes data packing to get the best data transfer performance. It also supported on windows, android, Linux, iOS, and phone7 for different programming languages [9].

The publisher-subscriber architecture Remote Lab is realized using the library provided by SCCT Communication in LabVIEW. library publisher is to create a complete publisher, which confirms the customer in, check the status of a connection and send data to all active publishers and pass their requests to the application. Publishers maintain server side communication system. Client library is to create customers who handle all the details of communication with the publisher. It received the package data and present them in front end of the application according to their data type.

SCCT facilitate the exchange of data between applications and handle a lot of connections at a time and SCCT facilitate the exchange of data between applications and handle many connections at the time and publish data in different formats according to the client's ability. Furthermore, SCCT provides high-performance data communications on conventional platforms such as HMTL5, JavaScript, Android, LabVIEW, Java, [9], thus making it a multi-platform approach.

3.4 WebLab-deusto.

University of Deusto had developed the remote lab management system that call Weblab-Deusto. In both versions the beginning it was an ad hoc remote laboratories [10] which is far above the lab system that can be built different laboratories using different software technology and currently supports federal laboratories far. The WebLab-Deusto architecture (see Figure 2) is a client–server architecture. Thus, this the intensity of light is the factor of the other fireflies to move toward the other firefly.

The average student interacts with server login for confirmation and with server core to during operation. Server core will operate truth, tabulation and information storage sent by students, and submit a all requests to laboratory that is correct. Behind core server, WebLab-Deusto architecture defined two more waiter is lab server and experiment server.

The first is to be placed in a physical laboratory where various experiments Servers expected. The purpose is mainly to provide security Experiment Servers, and also periodically check if it still works, and acts as a wrapper. Experiments Server is a server that physically interact with the hardware. Library to build Experiment Server is available for different platforms (C ++, C, Java, Python, Labview). All interaction between the client and the server are sent via a web service. Either SOAP, JSON over HTTP and XML-RPC can be used. Two types of laboratories can be built using WebLab-Deusto laboratory managed and unmanaged

In managing the laboratory, all interaction between students and laboratory is managed through all the layers described. Therefore, the savings made on Server Core and Server Experiments must not be common at the network level. However, protocols based on directives must be implemented in each laboratory. In laboratory unmanaged, Experiment existing server to manage other platforms [10] as a Virtual Machine or LabVIEW Remote

Panel. Thus, the student is still interacting with all levels, but the main interaction is controlled by external platforms (eg Virtual Machines with VNC or Remote Desktop), and it is not stored or securized by WebLab-Deusto.

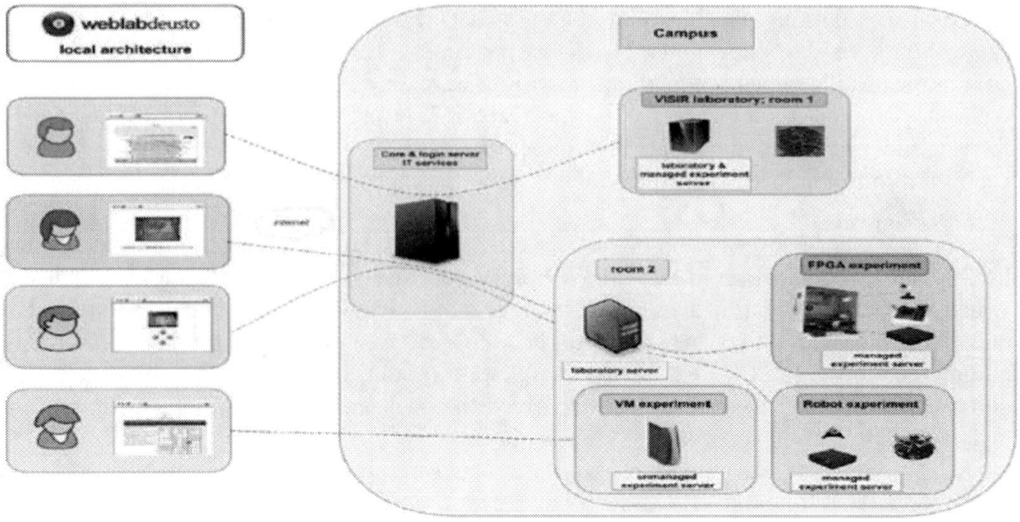

Fig. 2. The WebLab-Deusto Architecture

4 Results and discussions

4.1 Method of developing e-Lab.

This section has described the method of developing the e-laboratory system. The system divided into software and hardware. The most problems in developing the system are software developers (develop server: web server, database server and application server). It also needs a skill in programming such as C, C++, Java, NET, Python, etc.

Figure 3 shows that the general of e-laboratory architecture that divides into a few parts. In this architecture, clients connect to the core and login servers, using common HTTP. The users may use a personal computer (PC), laptop, hand phone (android, iOS, etc.), etc. to control the lab as long as there has internet connection. A web server is a program that uses client or server model and the World Wide Web HTTP serves files that make up a web page for web users (the computer containing the HTTP client that submitted their request).

Every computer needed a web server program to connected to the internet with a website. Web servers often come as part of a larger package of Internet and intranet-related programs to serve e-mail, download requests to the File Transfer Protocol (FTP) file, and building and publishing a website. Consideration in choosing a web server including how it works with the operating system and other server, the ability to control the program-server, security features, and publications, search engines, and tools of building a website that might come with it. This server manages authentication, authorization, user tracking, Federal (share) and scheduling. From there, the system forwards the request to the lab servers (application servers), which pointed to the end of the experiment.

For application waiter have been exposed business logic for customer application through various protocols, probably include HTTP. Although web server especially related to send HTML for display in web browser, application waiter providing access to business logic to be applied by client applications program. Application programme can use this logic just because it will call method in object (or function in procedure world). the application waiter customer will include GUI (graphical user interface) that walk in PC. Information that commute between application waiter and the client is unlimited to markup easy display. On the other hand, information that is logical programme.

Since the logic takes the form of data and method calls and not static HTML, the client can use the exposed business logic however it wants. In most cases, the server exposes this business logic through a component API, such as an EJB (Enterprise JavaBean) component model found on J2EE (Java 2 Platform, Enterprise Edition) application server. In addition, the application server manages its own resources. The task of keeping the door including security, transaction processing, resource gathering, and messages.

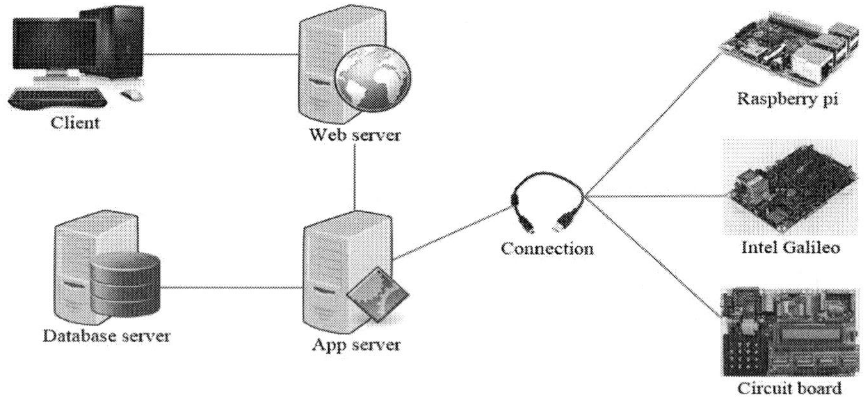

Fig. 3. The UniMAP e-Lab Architecture

The connection between the application server with instrument (electronic circuit, control system, etc.) is depending on the system. Some system uses serial communication cable (RS232), direct cable (such as Raspberry pie that use the USB cable and connect to the computer), etc. The instrument here is equipment that use in the laboratory and also the experiment that the student will be held. The signal that users send from the client will be sent to this equipment.

5 Discussions

Online access to lab equipment have given more time for students to interact with lab equipment: conventional laboratory classes usually allows each student only a few minutes away of equipment operating time. By allowing students to operate the equipment for an hour or more at a time, and to collect data by means of semi-automatic, has improved student learning opportunities significantly.

Apart from the advantage that it provides much more opportune time for students to operate equipment for themselves, the main strength of the UniMAP e-Lab project has been as a centrepiece for the Robotics and Automation Technology degree course. Students in the course have learned mechatronic systems integration and the development of techniques to develop and expand e-Lab system at all faculties of engineering technology. Students in the second and third using this system for practical work because it combines many different elements of robotics and automation.

Although the rapid risen of communication technology, the technology has not been widely adopted for the purpose of education, such as the development of e-Lab, either in our own faculty or elsewhere. Nearly all engineering laboratory classes still follow traditional patterns, as do lecture and tutorial classes. Therefore, it is worth asking why the adoption of such an attractive technology has been so much slower than expected. Many of the original researchers that pioneered remote laboratory implementations have moved on to other research or have returned their original research interests.

In our research, early in the project the authors were concerned that a complex LabVIEW-based software application for the server development might require extensive technical support as it migrated through the operating system and computer hardware upgrades. While significant effort has been needed on occasions. Gradually, however, some other significant limiting factors have become apparent.

6 Conclusion

Based on the concern about unadopted latest and current communication technologies and opportunities, the authors start to work on the UniMAP e-Lab (online Lab or remote Lab), with the hope to use and get benefit from the technologies. Through our hands-on working and observation, UniMAP e-Lab seem to provide more time for students or repeatly to interact with equipment: of actual time operating equipment. By allowing students to operate the equipment repeatedly, and collect data in the semi automated fashion, have significantly improved opportunities for student learning.

For students Robotics and Automation Technology degree course in UniMAP, apart from the advantage that it provides much more opportunity in laboratory sessions. By extending and develop the e-Lab system, users could learned electronic, electrical and mechanical integration and also system development techniques. Students in the second and third years using this system for practical work because it combines many different elements of robotics and automation.

The e-Lab system, when completed, will be embedded into UniMAP e-Learning LMS (Learning Management System) through e-Learrning platform. However to achieve the goal is still too far: we start from the scratch.

This work was financially supported by the Universiti Malaysia Perlis - Fundamental Research Grant Scheme (9003-00442).

References

1. Z.B. Razali, L.M. Kamarudin, *Blended Learning and Flipped Learning in Malaysia*, **9**, (UKM, 2014)
2. Z.B. Razali, *Advancements in Robotics and Mechatronics*, **34** (2015)
3. C. Colwell, E. Scanlon, M. Cooper, Comput. Educ., **38**, 65 (2002)
4. J.P. Trevelyan, Z.B. Razali, *Internet Accessible Remote Laboratories: Scalable E-Learning Tools for Engineering and Science Disciplines*, 416 (Engineering Science Reference, 2012)
5. Z.B. Razali, J.P. Trevelyan, Procedia Soc. Behav. Sci., **56**, 42 (2012)
6. L.D. Feisel, J.R. Albert, J. Eng. Educ., **94**, 121 (2005)
7. D. Cmug, T. Mutapcic, F. Zoino, *Remote Versus Classical Laboratory in Electronic Measurements Teaching - Effectiveness Testing*, XVIII Imeko World Congress Metrology for A Sustainable Development, Rio de Janerio, (2006)
8. P. Sladek, L. Pawera, J. Valek, Procedia Soc. Behav. Sci., **12**, 164 (2011)
9. http://www.toolsforsmartminds.com/products/SCCT.php rec: 21 April 2015
10. P. Orduna, L. Rodriguez Gil, D. Lopez de Ipina, J. Garcia Zubia, *Sharing the remote laboratories among different institutions: A practical case*, 9th International Conference on Remote Engineering and Virtual Instrumentation (2012)

Oxidation Resistance Analysis Of Metallic (FeCrAl Foil) Catalytic Converter Developed By Ultrasonic Approach

A.M Leman[1 *], *Dafit* Feriyanto[1], *N. Farhana*[1], *R.* Rahmat [1], *B.A.* Bakar[1], and *I.* Baba[1]

[1] Faculty of Engineering Technology, Universiti Tun Hussein Onn Malaysia (UTHM), Parit Raja, Batu Pahat, 86400 Johor, Malaysia.

Abstract. Mobile sources contribute about 44% of outdoor toxic emissions, approximately 50% of cancer risk and at around 74% of non-cancer risk health problems. Catalytic converter is quite needed in removing the pollutant and in preventing a health problem. The main problem in the catalytic converter is low oxidation resistance when operated at high temperature. Therefore, this paper aimed to develop catalytic converter material in high-temperature operation at around 1100 °C using FeCrAl foils as a metallic catalytic converter which coated by γ-Al2O3. This research is conducted using 3 various techniques such as ultrasonic bath for 3, 4, and 5 hours, Nickel (Ni) electroplating for 30, 45 and 60 minutes and the combination of ultrasonic bath and electroplating technique. Oxidation resistance analysis was conducted using tube furnace under argon gas for 60 hours in 3 cycles. Mass changes analysis of treated samples is showed by degradation mass. Lowest mass change of by ultrasonic bath samples is 0.3 wt%, for a combination of ultrasonic and electroplating samples is 0.3 wt% shown by UT 3 hours as well as 0.6 shown by EP 30 min. Parabolic rate constant is obtained by the time calculation based on the mass change of treated and untreated samples. It shown that UB 3 h is lowest parabolic rate constant of 2.258 x10-20 g2 cm-4s -1 and UB 5 h is 1.13 x10-20 g2 cm-4s -1. Lowest mass gain and lowest parabolic rate constant are become an indicator that the samples and that technique are recommended to fabricate the catalytic converter.

1 Introduction

Nowadays, there are many problems in our environment contributed by transportation sector especially from motor vehicle [1]. According to Malaysia Environmental Quality Report there are many types of pollutants which release by normal engine operation condition such as Carbon Monoxide (CO, 0.5 vol. %), Unburned Hydrocarbons (HC, 350 vppm), Nitrogen Oxides (NO_x, 900 vppm), Hydrogen (H_2, 0.17 vol. %), Water (H_2O, 10 vol. %), Carbon dioxide (CO_2, 10 vol. %), Oxygen (O_2, 0.5 vol. %). According to these data, HC, CO and NO_x are the major pollutants that influence to the outdoor toxic emission,

* Corresponding author : mutalib@uthm.edu.my

cancer risk and non-cancer risk health problem [2,3]. They reaction are mentioned in Eq 1, 2 and 3 [4].

$$2CO + O_2 \rightarrow 2CO_2 \quad (1)$$

$$C_xH_{2x}+2 + [(3x+1)/2]O_2 \rightarrow xCO_2 + (x+1)H_2O \quad (2)$$

$$2NO_x \rightarrow xO_2 + N_2 \quad (3)$$

These three reactions occur most efficiently when the catalytic converter receives the exhaust gas from an engine running slightly above the stoichiometric point. In below of stoichiometric point shows the necessary mixing rate of the most common fuels [4].

The technology of vehicle emission reduction can be categorized into two major parts, namely primary and secondary method. For the primary method, it depends on fuel used by vehicle based on air treatment and combustion process. In the meantime, the use of a catalytic converter in the exhaust system of vehicles is the secondary method [4]. These studies focus on the catalytic converter as vehicle emission reduction by new treated material. There are 2 major material types for catalytic converter which are ceramic and metallic material. Generally, physical parameter of the catalytic converter is listed in Table 1.

Table 1. Physical Parameters of the Converter [5].

Physical parameters of the converter	
Wall density	2500 kg/m
Wall thermal conductivity	2 W/mK
Wall heat capacity	1400 J/kgK
Elective wall thickness	0.25mm
Thickness of washcoat layer	0.05mm
Porosity of washcoat	0.41
Channel hydraulic diameter	1.4mm
Channel open frontal area	1.96mm
Porosity of monolith	72%

First catalytic converter is developed using ceramic honeycomb substrate which based on cordierite ($2MgO \cdot 2Al_2O_3 \cdot 5SiO_2$ - 14% MgO, 35% Al_2O_3 and 51% SiO_2). That material is selected because they it has many advantages in softening points with higher attrition resistance, high melting, and lower pressure losses, low thermal expansion, high shock resistance as compared to pallet converter [6]. start from the 1970s, honeycomb substrate with ultra thin foil, corrugated and laid up is developed by ferritic steel which called by FeCr alloy. It consists of Iron (Fe), Chromium (Cr), Aluminum (Al) and Yttrium (Y). It protected by Alumina (Al_2O_3) in order to achieve the withstand high temperature and corrosive condition in exhaust [7]. Metallic and ceramic catalytic converter is shown in Figure 1.

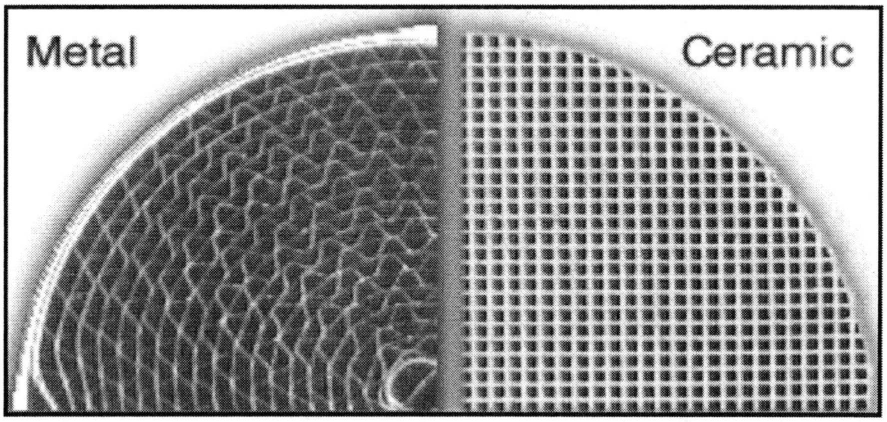

Fig. 1. Ceramic and metallic substrate converter [7].

The metallic catalytic converter has many promising properties such as high melting, high cell densities, high thermal shock resistance and low-pressure losses. But it has more advantages as compared to ceramic material in physical properties i.e. low specific heat and cheap (low-cost fabrication) [1]. Nowadays, FeCrAl is the promising metallic catalytic converter because it has excellent oxidation resistance at a high temperature of 1300 °C [8] and it commonly coated by γ-Al_2O_3 powders as corrosion and oxidation protective [9]. It used as coated material because the low price, good chemical stability, high micro-hardness and wear resistance at high temperature [10]. Other material such as Pt, Rh and Pd is not used in this research because it sensitive to thermal degradation at high temperature, CeO_2–Mn because it trapping of sulfur which led to corrosion [12].

Therefore, In this research is conducted using FeCrAL foil as substrate and γ-Al_2O_3 powders as wash coat material in order to achieve high oxidation resistance treated by ultrasonic approach. That treatment is chosen because during the ultrasonic process, high-speed bubbles producing a jet of liquid moving at around 100 m/sec. High velocity a jet potential to applied for cleaning, removing impurities from spent catalyst, electroplating or surface treatment process [13], and predicted to catalyst activation after and during coating process.

2 Methodology

Coating activity in this research is conducted using an ultrasonic approach which divided into three signed methods which are ultrasonic bath, nickel (Ni) electroplating and the combination of ultrasonic bath and electroplating technique. The material used in this research such as FeCrAl foil, γ-Al2O3, Ni plate, Nickel Sulphamate ($Ni(SO_3NH_2)2.4H_2O$), nickel chloride ($NiCl{\cdot}6H_2O$), boric acid (H_2BO_3), and sodium Lauryal Sulphate ($\underline{C_{12}H_{25}SO_4Na}$) Natrium Hydroxide (NaOH).

2.1 Ultrasonic Bath Process

Ultrasonic bath technique is purposed to the preliminary coating process of γ-Al_2O_3 to FeCrAl foil. FeCrAl foil is being cut in the size of 30 x 15 mm. Ethanol with a concentration of 20g/l is used as coating media in order to accelerate the γ-Al_2O_3 ebbed into the FeCrAl foil. Ultrasonic bath process is conducted using frequency of 35 kHz in various times of 3, 4, and 5 hours. The mechanism of the ultrasonic bath is shown in Figure 2.

 DOI: 10.1051/matecconf/20167801009

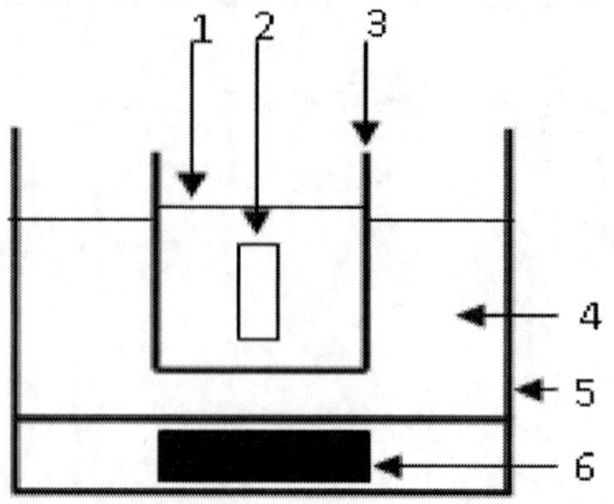

Fig. 2. Schematic diagram of ultrasonic cleaning bath (1) Ethanol; (2) FeCrAl; (3) Beaker; (4) Water; (5) Bath; (6) Ultrasonic source.

2.2 Nickel (Ni) Electroplating Process

In electroplating process, there are several components that are needed to be concern such as electrolyte (sulphamate type), anti-pitting agent, anode and cathode. The electrolyte is prepared by employing distilled water at a constant temperature of 60 °C and the adjustment of the pH value around 2.5-5.4 by manipulating the HCl and NaOH as the reagent. Ni plate is positioned as an anode and FeCrAl as the cathode with distance approximately of 25 mm with the current density of 2 A/dm^2 and total surface area of 450 mm^2. Illustration of the schematic diagram of the electroplating process is shown in Figure 3 and electroplating reaction is shown in Figure 4.

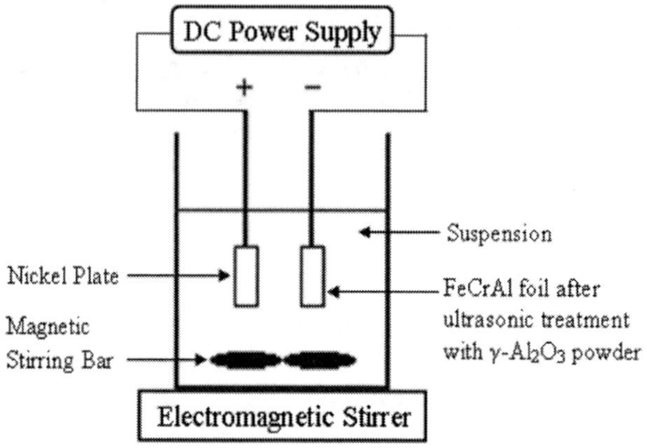

Fig. 3. Schematic diagram of electroplating process.

Fig. 4. The reaction occurs during electroplating.

2.3 Oxidation test

Tube furnace is held under argon gas environment at the temperature of 1100°C with heating and cooling rate of 5 °C/min, argon gas flow rate of 51 ml/min and flow density of 3.98 g/ml. The experiment was designed at 3 cycles and each cycle was carried out during 20 h of saturated period. Prior to, the material mass was measured by an electronic microbalance with the precision of 0.001g. After oxidation test, the data is measured by Eq. 3 and Eq. 4.

$$\Delta W = \frac{W_1 - W_1}{A_1} \tag{4}$$

Where;

Wt = Sample weight after each cycles of the oxidation process.
Wi = Initial weight sample
Ai = The surface area of the sample before oxidation process.

The calculation for the parabolic rate constant (kp) using parabolic rate law

$$\left(\frac{\Delta W}{A_1}\right)^2 = k_p t \tag{5}$$

Where;

t = The oxidation time (s)
k_p= The parabolic rate constant ($g^2\, cm^{-4} s^{-1}$)

3 Result and Discussion

3.1 Mass Changes

Mass change analysis is used to observe the oxidation resistance or thermal stability of the material. the effectiveness of the method to improve the oxidation resistance is signed by small mass change. Mass change of the ultrasonic bath and combination of ultrasonic and electroplating technique samples is listed in Table 2. It shows that the smallest mass change is signed by UB 5 hours in 3rd cycles with no mass change observed compared to UB 3 and 4 hours with 0.3 wt% mass change in every step. For UT samples show that different pattern with UB samples where the UT 5 hours shows the highest mass change of 2.167 wt% as compared to UT 3 hours of 0.3wt% and UT 4 hours of 0.9 wt%. Longer time of ultrasonic bath technique is led to the higher amount of γ-Al_2O_3 which embed in FeCrAl foil. γ-Al_2O_3 layer is proposed heat resistance before it spread into the FeCrAl foil. That phenomenon is not valid when the ultrasonic bath is combined with electroplating technique. It maybe caused by different liquid media where the ultrasonic bath use ethanol and electroplating use sulphamate type as solution. Different concentration is given high impact to the process of γ-Al_2O_3 to embed into FeCrAl foil.

Table 2. Mass changes of treated and untreated samples.

The mass of sample after oxidation test for ultrasonic bath sample (UB)			
	Mass change of UB for 3 Hours	Mass change of UB for 4 Hours	Mass change of UB for 5 Hours
Initial	0.319g	0.325g	0.32g
Cycle 1	0.319g	0.325g	0.32g
Cycle 2	0.318g	0.324g	0.319g
Cycle 3	0.317g	0.323g	0.319g
The mass of the sample after the oxidation test for ultrasonic bath followed by electroplating sample (UT)			
Initial	0.311g	0.327g	0.323g
Cycle 1	0.310g	0.327g	0.323g
Cycle 2	0.309g	0.324g	0.316g
Cycle 3	0.308g	0.321g	0.314g

The main coating process of this research is using Nickel electroplating technique. It aimed to improve the surface properties of FeCrAl by coating γ-Al2O3 such as to improve wear resistance, corrosion protection and oxidation resistance. This research is focused on developing high oxidation resistance of the treated FeCrAl sample which approved by small mass change after electroplating like shown in Figure 5.

Mass change analysis of treated samples shows the improvement properties high-temperature operation of 1100 °C. Electroplating (EP) for 30 minutes shown highest oxidation resistance by smallest mass changes of 0.6 wt% as compared to EP 45 min of 15.34 wt% and EP 60 h of 13.92 wt%. in these case, the mass changes increase linearly with the electroplating times. EP 30 min have higher deposition rate which provides uniformity of coating thickness and compact deposit are synthesized in template-based structure.

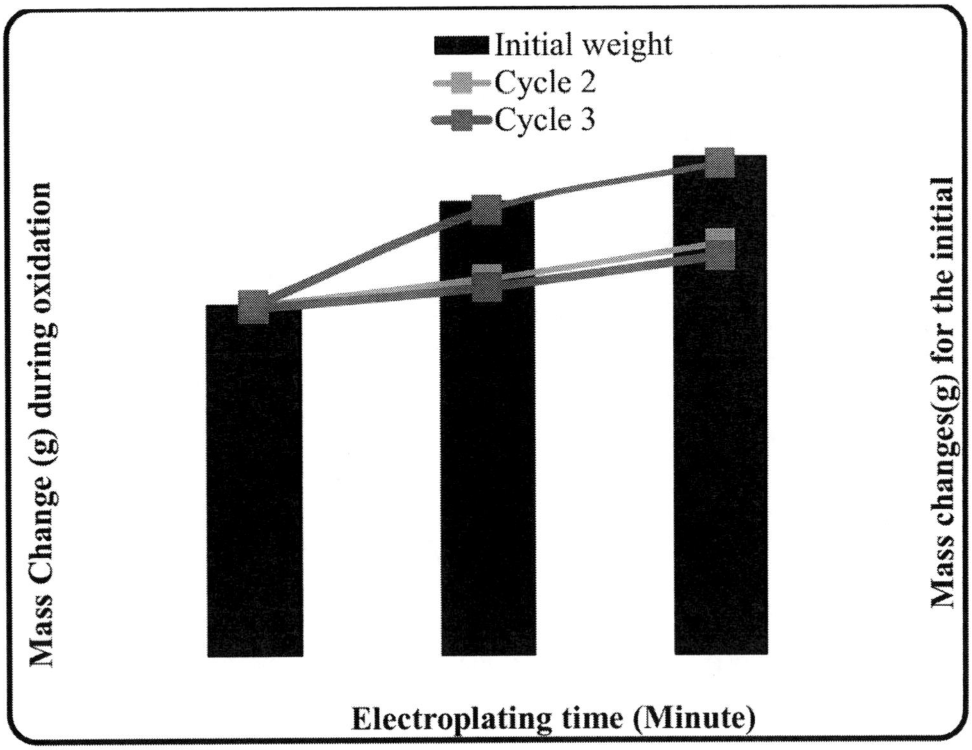

Fig. 5. Mass change of treated and untreated samples after electrodeposition process.

3.2 Parabolic Rate Constant

The parabolic rate constant is an intrinsic property of an oxidation resistant and normally used to measure its oxidation resistance. The parabolic rate constant of treated and untreated samples is listed in Table 3. Preliminary coating by UB 3 hours is obtain 2.258 x10-20 g^2 $cm^{-4}s^{-1}$ which compared to UB 4 (9.14 x10-09 g^2 $cm^{-4}s^{-1}$) and 5 hours (1.13 x10-20 g^2 $cm^{-4}s^{-1}$). The samples after a combination of ultrasonic bath and electroplating which signed by UT samples shown inline phenomena with UB samples. Where the UT 3 h shown lowest parabolic rate constant of 3.387 x10-20 g^2 $cm^{-4}s^{-1}$) as compared to UT 4 h (1.829 x10-18 g^2 $cm^{-4}s^{-1}$) and UT 5 h of (2.729 x10-17 g^2 $cm^{-4}s^{-1}$). Smallest mass change of EP 3 h effect to the lower parabolic rate constant for 7.968 x10-14 g^2 $cm^{-4}s^{-1}$) and higher kp shown by EP 4 h (4.27 x10-13 g^2 $cm^{-4}s^{-1}$) and EP 5 h (1.89 x10-12 g^2 $cm^{-4}s^{-1}$). All the results shows that 3 hours treatment times is most effective by ultrasonic bath, electroplating and combination of ultrasonic and electroplating technique.

Parabolic rate constant is remarkably reduced in 3 hours treatment times by spinel protection coating because it provided by γ-Al_2O_3 oxide scale. From the results of oxidation test that coated material shown an excellent resistance against the spallation and cracking which resulted by lower mass change and lower parabolic rate constant. Spalling and cracking phenomena commonly due to the oxide of coefficient of thermal expansion is mismatching [16].

DOI: 10.1051/matecconf/20167801009

Table 3. The parabolic rate constant for an ultrasonic bath.

Treatment	Average ($g^2\ cm^{-4}s^{-1}$)
UB 3Hours	2.258 $x10^{-20}$
UB 4Hours	9.14 $x10^{-09}$
UB 5Hours	2.258 $x10^{-20}$
UT 3Hours	3.387 $x10^{-20}$
UT 4Hours	1.829 $x10^{-18}$
UT 5Hours	2.729 $x10^{-17}$
EP 3Hours	7.968 $x10^{-14}$
EP 4Hours	4.27 $x10^{-13}$
EP 5Hours	1.89 $x10^{-12}$

3.3 Statistical analysis

Statistical analysis is performed for all method in order to investigate the significance of parameter which applied in this research. Analysis of Variance (ANOVA) single factor is used to analyze the data. For electroplating results (Table 4) shows that the Fcrit of 4.066181 is bigger than F of 1.041539 means that this method is significant in changing the properties. Current density, sulphamate type, pH, and electroplating times give high effect to the improving oxidation resistance in high temperature of 1100 °C.

Table 4. Statistical analysis of after electroplating process (EP).

Source of Variation	*df*	*MS*	*F*	*F crit*
Between Groups	3	0.002496	1.041539	4.066181
Within Groups	8	0.002397		
Total	11			

Ultrasonic bath (Table 5) and the combination of ultrasonic bath and electroplating (Table 6) shown that the parameter conducted in those process is a significant effect to the oxidation resistance properties. It approved by statistical analysis which mentions that value of Fcrit>F. Ultrasonic bath obtains Fcrit of 4.120312 and F of 0.130165 meanwhile for UT samples obtain Fcrit of 4.066181 and F of 0.411606. When the ultrasonic bath is conducted, shock out wave through the liquid cause high-speed collision which promote micro coating occur on the FeCrAl foil.

Table 5. Statistical analysis of after ultrasonic bath technique (UB).

Source of Variation	*df*	*MS*	*F*	*F crit*
Between Groups	3	1.5E-06	0.130165	4.120312
Within Groups	8	1.15E-05		
Total	11			

Various groups in this research is divided into 4 such as Initial, cycle 1, cycle 2 and cycle 3 groups which have the same value of the df of 3 and 8. The standard deviation of all data in every group and every method show the differences which indicated that the fluctuate condition when oxidizing in temperature of 1100 °C occurs. The SD of UB initial is 0.0019, cycle 1 is 0.0019, cycle 2 is 0.0019 and cycle 3 is 0.0018. The SD for UT samples initial is 0.0048, cycle 1 is 0.0051, cycle 2 is 0.0043 and cycle 3 is 0.0038. The SD for EP samples initial is 0.0382, cycle 1 is 0.0362, cycle 2 is 0.0159 and cycle 3 is 0.0131. High standard deviation means that the data that resulted by the parameter is more significant effect as compared to lower standard deviation. However, all methods is recommended to produce the catalytic converter material with high oxidation properties.

Table 6. Statistical analysis of after combination of ultrasonic bath and electroplating (UT).

Source of Variation	*df*	*MS*	*F*	*F crit*
Between Groups	3	2.54E-05	0.411606	4.066181
Within Groups	8	6.18E-05		
Total	11			

4. Conclusion

FeCrAl was coated by γ-Al_2O_3 using 3 various methods and it can be concluded that the coated material has higher oxidation resistance as compared to the uncoated material. The smallest mass change of coated material is 0.3 wt% for UB 3 h and UT 3 h and 0.6 wt% for EP 3 h. Parabolic rate constant is linear with mass change value. The significance data is approved by statistical analysis for all the data shown that Fcrit>F. Consistency data is established that all parameter and all procedure that conducted in this research is in-line with the fundamental methods.

The authors would like to thank the Ministry of Higher Education Malaysia and Universiti Tun Hussein Onn Malaysia (UTHM) through the funding supported FRGS grant under No Vot. 1216 and then Centre for Graduate Studies – UTHM.

References

1. H. Santos, M. Costa, Energ. Convers. Manage., **49**, 291 (2008)
2. H.S. Gandhi, G.W. Graham, R.W. McCabe, J. Catal., **216**, 433 (2003)
3. A. Firdianto. *Ultrasonic Treatment with Nickel Electroplating Combined with Oxidation for Developing Gamma-Alumina Washcoat on Fe-Cr-Al Substrate* (Diss. UTHM, Batu Pahat, 2012)
4. J. Kaspar, P. Fornasiero, N. Hickey, Catal. Today, **77**, 419 (2003)
5. K. Reszka, D. Murzynski, J. Rakoczy, Vacuum, **83**, S224 (2009)
6. Shi Jin Shuai, Jian Xin Wang, Chem. Eng. J., **100**, 95 (2004)
7. Bode Hans, *Material Aspect in Automotive Catalytic Converter.* (Wiley-VCH, Chicago, 2002)
8. J.P. Dong, H.G. Kim, J.Y. Park, Y. Jung, J.H. Park, Y.H. Koo, Corros. Sci., **94**, 459 (2015)
9. D.H. Kim, B.Y. Yu, P.R. Cha, W.Y. Yoon, J.Y. Byun, S.H. Kim, Surf. Coat. Technol., **209**, 169 (2012)
10. N.A. Badarulzaman, S. Purwadaria, A.A. Mohamad, Z.A. Ahmad, Ionics, **15**, 603 (2009)

11. J. Liwei, M. Shen, J. Wang, Surf. Coat. Technol., **201**, 7159 (2007)
12. Z. Bajram, M. Hodnett, A. Carroll, Ultrasonics, **44**, 73 (2006)
13. A. Ryosuke, M. Atobe, T. Fuchigami, Ultrason. Sonochem., **13**, 19 (2006)
14. E. Hadi, Z. Morteza, Indian J. Eng. Mater. S., **18**, 314 (2011)

MATEC Web of Conferences **78**, 01010 (2016) DOI: 10.1051/matecconf/20167801010
IConGDM 2016

Consumer Awareness in Energy Efficiency for Residential Houses in Peninsular Malaysia

K.A Rahman[1,*], *A.M* Leman[2], *M.Z.M* Yusof[1], *and M.M.N* Salleh[3]

[1]Faculty of Mechanical & Manufacturing Engineering, Universiti Tun Hussein Onn Malaysia, 86400 Batu Pahat, Johor, Malaysia
[2]Faculty of Engineering Technology, Universiti Tun Hussein Onn Malaysia, 86400 Batu Pahat, Johor, Malaysia
[3]Faculty of Computer Science & Information Technology, Universiti Tun Hussein Onn Malaysia,86400 Batu Pahat, Johor, Malaysia

Abstract. Most energy efficiency (EE) measures implemented (or yet to be implemented) in Malaysia involve technological interventions, but it will rely on consumer behaviour including their awareness in energy consumption. This paper highlight the energy-saving awareness among consumers for residential houses which involves two main components: i) knowledge and ii) practice. The method for this study is using the questionnaires of 408 respondents from four zones in Peninsular Malaysia. Pearson Correlation analysis technique was used to examine the relationship of several variables with electricity saving awareness. The results of the analysis found that the both components, knowledge and practice have a significant relationship with consumer awareness. It shows that both components are very important to ensure the awareness of EE will be implemented by the consumer.

1 Introduction

Global is energy consumption has risen 30% in the last 25 years. As a developing country, Malaysia faces the same situation regarding this issue. Residential sector in Malaysia has been recorded to emit 2,347,538 tonnes of CO_2 and will rise up to 11,689,308 by 2020 [1]. Based on the report by Energy Commission of Malaysia, 20.6 % of electricity were from domestic users [2]. There were about 7.8 million registered as electricity consumers with the largest electricity utility in Malaysia, Tenaga Nasional Berhad (TNB). Figure 1 shows the distribution of electricity users in Peninsular Malaysia and from total 7.8 million, 82 % is domestic users (6.4 million), 17 % commercial, 0.4 % industrial and 0.6% from other types [3].

* Corresponding author : khairunnisa7482@gmail.com

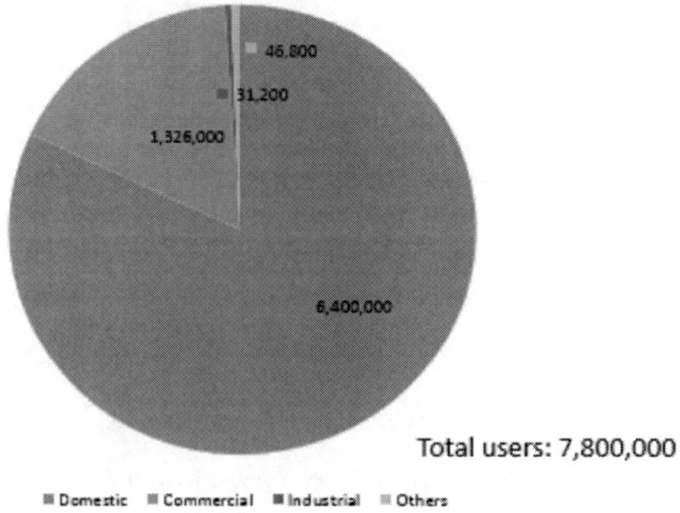

Fig .1 Distribution of electricity users in peninsular Malaysia for 2013.

According to AWER [4], in this scenario, improvements in energy efficiency are regarded as the solution to increase the economic growth, and at the same time decrease the energy use. According to Noor Hanita [5], three major issues that are associated with energy efficiency were architectural (design) issue, appliances/services (technology) efficiency issue, and the human (behavioural) issue. Households have a direct connection between their energy efficiency behavior and the monthly or quarterly cost of energy. Furthermore, the potential for savings through behaviour change is thought to be greater in the domestic environment because there is more direct control over energy consumption [6]. Thus, this study is directed at residential buildings because more than one-third of the world's energy is used in buildings especially in residential building [7].

2 Literature Review

According to Fragidis [8], energy awareness is refer to the knowledge about how much energy is consumed. Consumers need to reduce their energy consumption but they do not know how to do it without the knowledge. More importantly, consumers need to know details about their energy consumption, such as how much electricity they use, when they use it, and how much it costs in time to appropriately impact their monthly electricity bill. On the other hand, the more knowledgeable they are, the more interested they are adopting energy saving practices. In the beginning, energy consumers' awareness is a first step in adopting energy saving behaviours [8]. Yet even with adequate knowledge of how to save energy and a professed desire to do so, many consumers still fail to take noticeable steps towards EE and conservation. Energy awareness towards energy saving requires an appropriate attitude and human behaviour of the house inhabitants [1]. It seems rational to assume that increasing consumers' awareness about their EE may influence their consumption behavior.

Based on the study that have been done in Nigeria, it focused on the issues to determine the factors that affect energy efficiency on pre-determined housing samples. The results of the study indicated that there is a strong correlation between attitude and behaviour to EE in the residential housess. In addition, the same researchers has done a social survey with the same issues but focused on the implication of the energy efficiency delivery specific in Bouchi town, Nigeria [9]. The result reveals a much lower level of EE practice by the occupants due largely to lack of awareness and the requisite knowledge of EE. Moreover, a study has been conducted by Syed Hussain [10] to survey energy-saving awareness among students at Universiti Utara Malaysia (UUM), Malaysia. This research involves three main components which is knowledge, views and actions. The results found that the level of awareness about energy saving among UUM students is low and this has caused a great burden on the university management in terms to pay the cost of electricity bills. Therefore, they suggested the further concerted effort must be put in place towards improving the level electricity saving awareness among UUM students. In the same way, according to Noor Hanita [5] factors such as awareness and social status may have implication on EE awareness but not necessarily determine it. They also suggested that the needs of EE practice in the households to reduce the energy consumption. In line with that, the results from the study found that the levels of awareness and education of the individual influence the EE practice. As well as, awareness and practice also can be implemented by using the star rated electrical appliances [11]. In addition to consumer awareness, the EE awareness campaign were found to have a positive impact on decreasing levels of electricity consumption [12].

3 Approaches and Methods

This study focuses on the implications of human behaviour in household energy consumption. The objective is to determine the relationship of a combination of two behaviours: knowledge and practice of consumer awareness. According to Syed Hussain [10], knowledge is one of the main components that need to be considered when implementing energy saving awareness. Subsequently, Noor Hanita [5] highlight the needs of EE practice in the households to implement energy saving behaviour. Furthermore, the theoretical background has provided the dependent variable of awareness (level of efficiency practice by usage of energy star rating appliances) with the independent variables of knowledge and practice as the basic parameters of the study. A quantitative approach was adopted via questionnaire administration on the target respondent. According to Creswell [13], the appropriate sample size is determined by the level of confidence, level of precision and the type of test that will be used in data analysis. Hence, the formula by Fowler [14] for estimating the sample size and a table to determine the same were used. By using a table that specifies the required sample size that would ensure a good decision model (Table 1), the sample size was greatly reduced. This formula is based on three major scenarios: a) Percentage of sampling characteristic: The first row shows the percentage of the sample that will have the desired characteristics, ranging from 5/95 to 50/50 (an equal chance). As this research seeks to equally divide the sample as much as possible, 50/50 was selected. b) Sampling error: Sampling error is the amount of error that a study can tolerate. 4% to 6% is a common choice. In this research 6% of sampling error has been selected means the sample mean will differ from the true population mean only 6% of the time. c) Confidence level: The confidence level that has been choose is 95%.

Table 1. Sample size table (adapted from Fowler, [14]).

Sample size	Percentage of Sample with Characteristic				
	5/95	10/90	20/80	30/70	50/50
35	7	10	14	15	17
50	6	8	11	13	14
75	5	7	9	11	12
100	4	6	8	9	10
200	3	4	6	6	7
300	3	3	5	5	6
500	2	3	4	4	4
1000	1	2	3	3	3
1500	1	2	2	2	2

Based on the above characteristics, the recommended sample size for this research is 300 but for this study 500 questionnaires were distributed. This to prevent not significant data due to the reject questionnaire that not fulfil the study requirement. From 500 questionnaires, only 408 questionnaires were fully answered. The questionnaires were distributed to respondents in four main zones in West Malaysia including North Malaysia (Kedah, Perlis and Penang), South Malaysia (Negeri Sembilan, Johor and Malacca), East Malaysia (Terengganu, Pahang and Kelantan) and West Malaysia (Perak, Selangor and Wilayah Persekutuan). Figure 2 shows the fragments of the study area where the highest number of respondent comes from the South Zone (191) followed by West Zone (108), East Zone (55) and North Zone (54).

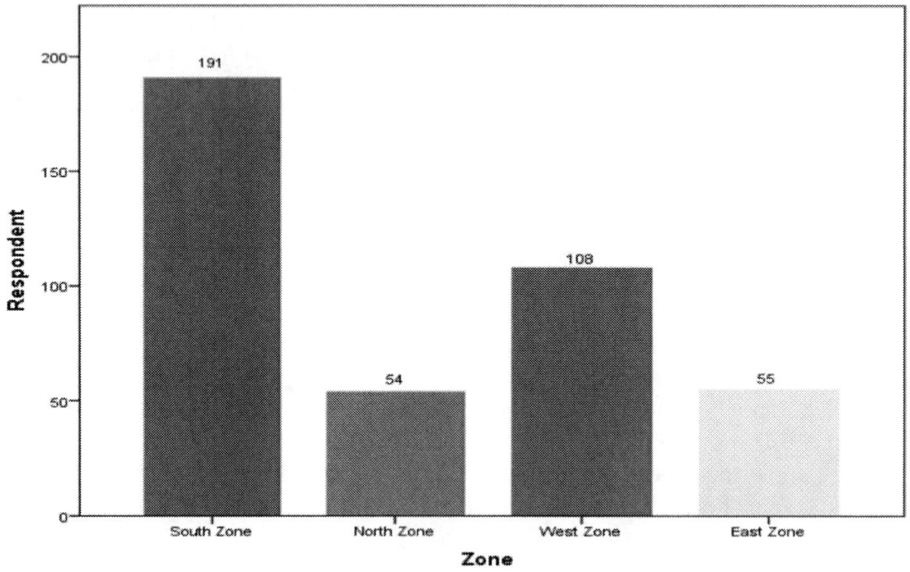

Fig. 2. Study area and the number of respondents for the energy survey.

The dependent variable of awareness was subjected to a five-point scale measure (one star - five star), while the independent variables of knowledge and practice were subjected to a four-point "Likert" scale (strongly disagree - disagree - agree - strongly agree). The reason to use only 4-point Likert Scale rather than 5-point Likert scale is to avoid the social desirability bias [15]. The questionnaire was developed based on two factors that influence the consumer awareness on energy efficiency in residential houses, which is knowledge (10 items) and practice (25 items).

Correlation analysis is used to describe the strength and direction of the linear relationship between two variables [16]. The correlation coefficient is a measure of linear association between two variables. The values of the correlation coefficient are always between -1 and +1 [17]. There have two types of correlation which is Pearson Correlation and Spearman Correlation. Both are used to explore the strength of the relationship between two continuous variables, but it depend to the normality of the data distribution. The normality test for this data have been done and it found that the data is normally distributed based on the inspection of the normal probability plots (labelled Normal Q-Q Plot). In this plot, the observed value for each score is plotted against the expected value of the normal distribution. Pallant [16] suggested that a reasonably straight line consider as a normal distribution.

4 Result and Discussion

4.1 Demographic Profile

The demographic questions contained herein was developed from various sources, including the International Energy Agency (IEA), TNB, Malaysia Energy Comission (ST) and Ministry of Energy, Green Technology and Water of Malaysia (KeTTHA). This demographic profile divided into two sections which is Section A for Respondents Information and Section B for Residence Information. Section A is to gather the information about the respondent, such as the number of residents, age, type of occupation of the head of the family, and monthly household income and for Section B is for estimated size of the house, number of rooms in the house and the average energy bill of the month which may then be giving an overview of respondents and their house. Table 2 shows the distribution of the respondent base on the demographic question.

This section gathered the information about the number of people living in the house as shown in the Table 2. Four people in a house were the popular number for 78 families for this item compared to two families which have more than 11 members living in the house. Total number of people living in the house for this study is 2024. Next, this item also use of determined distribution of people according to their group of age. The highest number of people is 816 that come from age 22 to 40 years old. However the lowest is from 61 years and above which is 61 people from total 2024 peoples which come from 408 families (N = 408). Other than that, the other information is about home leader occupation. It shows 38.97% of the respondent working in the government sector from various types of occupation related and the lowest come from respondents from industrial sector which stated about 4.41%. The other respondents come from other types of occupation, such as private sector (25.25%), self-employee (16.67%), and retired (5.39%). In addition, 9.3% respondents did not assign to one of the five main categories, but ticked the option 'others'. In this case, respondents come from other types of occupation or did not have any job for the time being. Next, respondents were asked about their monthly household income. The results show 136 respondents have income between RM4001 and above. This is because many of the respondents have more than one person that have income rather than only one from the family that have income.

Table 2. Distribution of the respondent base on the demographic questions.

	Demographic variables	Description	Freq	%
A	Respondents Information			
1	Number of People Living in the House	1 people	7	1.7
		2 peoples	47	11.5
		3 peoples	55	13.5
		4 peoples	78	19.1
		5 peoples	69	16.9
		6 peoples	63	15.4
		7 peoples	31	7.6
		8 peoples	29	7.1
		9 peoples	18	4.4
		10 peoples	9	2.2
		More than 10 peoples	2	.5
		Total	408	100
2	Age Group of the Family	Toddler under 3yrs	128	6.3
		4 to 12yrs	268	13.2
		13 to 21yrs	375	18.5
		22 to 40yrs	816	40.3
		41 to 60yrs	376	18.6
		61 yrs and above	61	3.0
		Total	2024	100
3	Home Leader's Occupation	Government Sector	159	39.0
		Private Sector	103	25.2
		Industry	18	4.4
		Self-Employee	68	16.7
		Retired	22	5.4
		Others	38	9.3
4	Monthly Household Income.	0 (no income)	1	0.2
		Below RM1000	46	11.3
		RM1001 - RM2000	61	15.0
		RM2001 - RM3000	76	18.6
		RM3001 - RM4000	88	21.6
		RM4001 and above	136	33.3
		Total	408	100

	Demographic variables	Description	Freq	%
B	Residence Information			
1	House Estimation Size	below 700sqf	12	2.9
		700sqf - 749sqf	52	12.7
		750sqf - 799sqf	110	27.0
		800sqf - 1000sqf	145	35.5
		1001sqf and above	89	21.8
		Total	408	100
2	Number of rooms	1 room	10	2.5
		2 rooms	34	8.3
		3 rooms	173	42.4
		4 rooms	139	34.1
		5 rooms	44	10.8
		Others	8	2.0
		Total	408	100.0
3	Average electricity bill	Below RM20	9	2.2
		RM20 - RM50	82	20.1
		RM51 - RM100	134	32.8
		RM101 - RM150	89	21.8
		RM151 - RM200	47	11.5
		RM201 and above	47	11.5
		Total	408	100.0
C	Level of awareness	1 star		
		2 star		
		3 star		
		4 star		
		5 star		
D	Energy efficiency knowledge and practice rating (Likert Scale)	Strongly disagree		
		Disagree		
		Agree		
		Strongly Disagree		

Moreover, for Residence Information section, there were 88 respondents have MYR3001- MYR4000 monthly household income and followed by MYR2001- MYR3000 (N=76), MYR1001- MYR2000 (N=61), and below MYR1000 (N=46). However, from all the respondents, there was one respondent that has no income at all. This has been identified that the respondents were very old and live lonely in the house and the daily expenses comes from his children every month. Meanwhile, the estimation size for housing for this item divided into five categories, which is 800sqf to 100sqf have a higher value (35.5%) followed by 750sqf to 799sqf (27.0%), 1001sqf and above (21.8%), 700sqf to 749sqf (12.7%) and below 700sqf is the lowest (12.7%).

Next category examined was the distribution of number of rooms in the house. The category has been broken down into six categories from one room to five rooms and followed by others (more than five rooms). The highest percentage was for respondents that have three rooms in their house (42.4%) and followed by four rooms (34.1%), five rooms (18.8%), two rooms (8.3%), one room (2.5%) and others (2.0%). Finally, the average electricity bill for respondents were examined. This category was divided into six categories which is below MYR20 (0-92kWh), MYR20 to MYR50 (93-220kWh), MYR51 to MYR100 (221-361kWh) MYR101 to MYR150 (362-462kWh), RM151 to MYR200 (463-563kWh) and the last categories is the respondent that has an average bill from RM201 and above (>563kWh). The highest average bill is comes from MYR51 to MYR100 category (N=134) and the lowest is for below RM20 category (N=9). It is more easier to collect the information about the electricity usage in Malaysian Ringgit (MYR) rather than usage by (kiloWatt/hour) kWh due to the background of the respondent that come from various background and some of them may not familiar with the technical term. The current electricity tariff in Malaysia as shown in Table 3 [18] and the price rate is not included of 6% of Goods and Service Tax (GST). GST only implement to those who have more than 300kWh of electricity usage.

Table 3. Electricity tariff in Malaysia.

Tariff Category	Unit	Current Rate (MYR) (effective 1st Jan 2014)
Tariff A – Domestic Tariff		
For the first 200 kWh (1 - 200 kWh) per month	cent/kWh	21.80
For the next 100 kWh (201 - 300 kWh) per month	cent/kWh	33.40
For the next 300 kWh (301 - 600 kWh) per month	cent/kWh	51.60
For the next 300 kWh (601 - 900 kWh) per month	cent/kWh	54.60
For the next kWh (901 kWh onwards) per month	cent/kWh	57.10

**The minimum monthly charge is MYR3.00*

5 Variables Measurement

The results of the measurements that have been made against both variables of awareness which is knowledge and practice of EE. Through this descriptive data, it found that the mean, standard deviation and variance of each factor affecting the consumer awareness on EE. Table 1 shows data mean, variance and standard deviation of each factors.

Table 4 shows the mean value of the knowledge is 3.41 and 2.75 for practice. Robinson [19] suggested for four point Likert scale that means score 1 for very low priority, 2 for low priority, 3 for high priority, and 4 for very high priority. This means that the majority of respondents said that the knowledge and practice can caused them to implement energy efficiency although for the factor practice have low priority. The mean score for knowledge is high priority (3.41) that indicates the knowledge on EE among consumer is high.

However, consumers' awareness towards practice the EE in residential houses is low based on the mean score (2.75). It means, they know and have the knowledge about EE but they do not or less practice in their daily life. Meanwhile, these factors has 0.39 standard deviation and 0.16 variance for knowledge and 0.42 standard deviation and 0.17 variance for practice.

Table 4. Mean, variance and standard deviation.

	N	Mean	Std. Deviation	Variance
Knowledge	408	3.4105	0.39332	0.155
Practice	408	2.7527	0.41693	0.174
Valid N	408			

5.1 Correlation between Knowledge and Practice

A Pearson product-moment correlation or Pearson's r was run to determine the relationship between knowledge and practice to implement energy efficiency approach in their daily life. The data showed no violation of normality and linearity. There was a positive correlation between knowledge and practice, which was statistically significant ($r = 0.348$, $n = 408$, $p < 0.05$) as shown in Table 5. This correlation means an increase in knowledge was increasing the practice & application and both variables were important for consumer to implement the energy efficiency of electrical appliances in residential houses.

Table 5. Correlation of knowledge and practice.

		Knowledge	Practice
Knowledge	Pearson Correlation	1	0.348**
	Sig. (2-tailed)		0.000
	N	408	408
Practice	Pearson Correlation	0.348**	1
	Sig. (2-tailed)	0.000	
	N	408	408

**. Correlation is significant at the 0.01 level (2-tailed).

5.2 Correlation of Awareness and Knowledge

The same analysis, Pearson's r was computed to assess the relationship between knowledge of energy efficiency and awareness that can determine from the usage of star rated appliances. From the analysis, it found that there has positive correlation between the two variables, *are* = 0.120, $n = 408$, $p = 0.015$ as shown in Table 6.

Table 6. Correlation of awareness and knowledge.

		Awareness	Knowledge
Awareness	Pearson Correlation	1	0.120*
	Sig. (2-tailed)		0.015
	N	408	408
Knowledge	Pearson Correlation	0.120*	1
	Sig. (2-tailed)	0.015	
	N	408	408

*. Correlation is significant at the 0.05 level (2-tailed).

5.3 Correlation of Awareness and Practice.

Another relationship that has been determined from the analysis is the relationship between awareness and consumers' practice taken from the usage of star rating appliances as shown in Table 7. The data show that there was a positive correlation between consumer's practice & application and the usage of star rating appliances, $r = 0.81$, $n = 408$, $p = 0.00$. Cohen [20] suggests that the correlation have a strong relationship if the value of Pearson's r is from 0.50 to 1.00. This result means increased in practice of energy efficiency can reduce the electricity bill for the month.

Table 7. Correlation of Star Rating appliances and practice.

		Awareness	Practice
Awareness	Pearson Correlation	1	0.814**
	Sig. (2-tailed)		0.000
	N	408	408
Practice	Pearson Correlation	0.814**	1
	Sig. (2-tailed)	0.000	
	N	408	408

**. Correlation is significant at the 0.01 level (2-tailed).

6 Conclusion

This study has revealed that knowledge and practice is a main component of awareness to implement the EE approach for residential houses. Both components must be implemented together to enhanced the awareness of energy saving among consumers. However, the mean score of the study found that the level of practice in implementing EE among consumers is low and this has caused the wastage of energy in residential. Meanwhile, the results revealed that the level of knowledge is good. It means consumers have good knowledge about EE and if they implemented both main components, it will enhance the awareness among consumers. Awareness is the challenge of implementing the EE which is majority consumers have the knowledge about EE but they do not practice the knowledge. Therefore, addressing the issue of energy efficiency practice could play a valuable role in guiding the consumer to reduce energy consumption. Other than that, the government should further initiate regular information and awareness programms for the general public on energy efficiency to boost the awareness level and subsequently energy efficiency practice.

The author would like to express her thanks and gratitude to Universiti Tun Hussein Onn Malaysia (UTHM) and the Ministry of Education Malaysia for HLP Scholarship for the study.

References

1. N.H.A. Majid, M.S. Salehudin, Z.A. Rahim, R. Othman, Procedia - Soc. Behav. Sci., **170**, 527 (2015)
2. Energy Commission, Peninsular Malaysia Electricity Supply Industry Outlook 2013 (Energy Commission, Putraya, 2013)
3. TNB, Tenaga Nasional Berhad (TNB, 2014)
4. S. Mekhilef, A. Safari, W.E.S. Mustaffa, R. Saidur, R. Omar, M.A Younis, Renewable and Sustainable Energy Reviews (2012)
5. N.H. Abdul Majid, H. Ibrahim Udale, Third International Conference on Applied Energy (2011)

6. A. Barbu, N. Griffiths, G. Morton, European Environment Agency-Copenhagen (2013)
7. D. Wulfinghoff, How to Build & Operate a Super-Efficient house (Wulfinghoff Energy Servics, Inc. , Maryland, 2003)
8. G. Fragidis, D. Olschewski, Consumer awareness and engagement for energy efficiency solutions (Cleopa GmbH, Berlin, 2015)
9. I.U. Hussaini, N.H. Abdul Majid, International Journal of Energy Sector Management **8**, 230 (2014)
10. T.P.R. Syed Hussain, H. Ismail, M.K. Md Noh, PERKEM Proceedings 2 (2013)
11. K.A Rahman, M. Zainal, M. Najib, A.M. Leman, Chem. Eng. Res. Des., **45**, 1663 (2015)
12. A. Kemp. Hesterman, S. Glick, J.E. Cross, J. Oncol. Manag., **12**, 4 (2014)
13. J.W. Creswell, New Jersey: Upper Saddle River (2012)
14. F.J. Fowler, Survey Research Methods (Sage Publication, Los Angeles, 2014)
15. M.S. Matell, J. Jacoby, Educ. Psychol. Meas., **31**, 657 (1971)
16. J. Pallant, SPSS SURVIVAL MANUAL : A step by step guide to data analysis using SPSS (McGraw Hill, New York, 2007)
17. A. Field, Discovering Statistics Using SPSS (SAGE Publications, Los Angeles, 2009)
18. Tenaga Nasional Berhad, Electricity Tariff (TNB, 2014)
19. P. Robinson, R. Shepard, J., Ext 49 (2011)
20. J. Cohen, Statistical power analysis for the behavioral sciences (Lawrence Erlbaum Associates, New Jersey, 1988)

MATEC Web of Conferences **78**, 01011 (2016) DOI: 10.1051/matecconf/20167801011
IConGDM 2016

Profiling of Indoor Plant to Deteriorate Carbon Dioxide Using Low Light Intensity

Mohd Mahathir Suhaimi Shamsuri[1,†], *A.M.* Leman[1], *Azian* Hariri[2] , *K.A* Rahman[2], *M.Z.M* Yusof[2], and *Azizi* Afandi[2]

[1] Faculty of Engineering Technology, Universiti Tun Hussein Onn Malaysia (UTHM), 86400, Parit Raja, Batu Pahat,Johor Malaysia.
[2] Faculty of Mechanical and Manufacturing Engineering, Universiti Tun Hussein Onn Malaysia (UTHM), 86400, Parit Raja, Batu Pahat,Johor Malaysia.

Abstract. Reasonable grounds that human needs the plants because their abilities reduce carbon dioxide (CO_2). However, it is not constantly human with the plants, especially in the building. This paper intends to study the abilities of seven plants (Anthurium, Dumb Cane, Golden Pothos, Prayer Plants, Spider Plant, and Syngonium) to absorb CO_2 gas. The research was conducted in chambers (one cubic meter) with temperature, lux intensity and CO_2 concentration at 25±10C, 300 lux, and 450±10 ppm. Before experimental were carried out, all plants selected should be assimilated with an indoor setting for performance purpose, and the experiment was conducted during daytime (9 am-5 pm). The experiments run in triplicate. Based on the results that are using extremely low light that ever conducted on plants, only Spider Plants are not capable to absorb CO_2, instead turn up the CO_2 rate during respiration. Meanwhile, Prayer Plant is the most plant performed with CO_2 reduction is 7.62%, and this plant also has equivalent results in triplicate study based on an ANOVA test with significant value at 0.072. The conclusions of this research, only Spider Plant cannot survive at indoor condition with extremely low light for plants live and reduce CO_2 concentration for indoor air quality (IAQ). The rate of 300 lux is a minimum light at indoor that are set by the Department of Occupational Safety and Health (DOSH, Malaysia).

1 Introduction

Carbon dioxide (CO_2) is one of a contaminant when elevated at higher concentration [1]. Based on the recommendations of ASHRAE (American Society of Heating, Refrigerating, and Air-Conditioning Engineers), limit CO_2 concentration in indoor is 1000 ppm [2]. Normally, CO_2 in the building due to human respiration, and this gas has a significant effect on human [3]. If the CO_2 concentration is not controlled, it will cause a headache, loss of focus, and unconsciousness on human [4]. T[‡]herefore, indoor air quality (IAQ) should be preserved wisely. Many efforts are made to achieve IAQ for both technical and

[†]Corresponding author: mohd_mahahir@rocketmail.com

management fields [5]. However, every study conducted to enhance air quality must be concerned about energy usage. Around 40% of the total global energy consumed by buildings, and it is expected to be increased by 50% in the 2030 [6]. Meanwhile, buildings in Malaysia (residential and commercial building) use about 48% of the total energy that produced by responsible bodies (Tenaga Nasional Berhad) [7].

Based on Malaysian Ministry of Energy, Green Technology and Water (KeTTHA) in 2015 strongly recommended to reduce the energy consumption for improving the quality of life and preserving natural resources and the environment. Indoor plants seem an alternative solution in response to the suggestion. Through to studies by former researchers claim that indoor plants are capable to serve as bio-filtration [8,9,10,11,12,13]. It is interesting to purify the indoor air using a minimum or without energy, and this would be very beneficial to humans.

However, indoor plants originally came from the forest, and the environments are different compared to the indoor. The significant difference of environment between indoor and outdoor, that effect on plant growth is lighting and temperature. Based on DOSH (Department of Safety and Health Malaysia), propose that lighting limit for indoor building is ranging from 300 lux to 700 lux [14], meanwhile, light intensity on daylight is more than 1000 lux. Besides that, the allowable temperature by ICOP (Industry Code of Practice) in an office building is 230°C to 260°C [15], while Malaysian temperature at noon exciding 300°C [16].

Light is the energy source for the plant to do photosynthesis process [17]. Devkota claim that certain plant needs a medium light (not too high or too low) to get optimum growing [18]. Moreover, the right concentration of light imposed on the plant will cause photosynthesis process perform preferable [19]. Other than that, blue and red light is the color that plants strongest response for photoperiod growth and sprout, yellow and green light will cause the color reflected back into space air [20].

In the meantime, the temperature is also one of factors that can impact on plant growth. High temperature is usually stress for plants, restricting growth, and will damage the photosynthetic system [21]. Whilst, low temperature also makes the plant in photoinhibition condition, limiting the productivity and geographical distribution of many species, including important agricultural crops [22,23]. Because of that, optimum temperature should be imposed on the plant environment to maximize the photosynthetic rate [24].

Nevertheless, to meet the desire to improve air quality using plants, aptitude tests in the building to plant life should be carried out. Based on former researchers such as Quero and Torpy, they had allowed the plants assimilated at indoor first (for certain periods) before the next test conducted [25,26]. Assimilation process will specify whether the plant will survive in the building or not, since rate of light and temperature at indoor very low compare to outdoor. Certain plants are able to down-regulate their photosynthetic apparatus when they are assimilated to yet lower light and temperature conditions [19]. Figure 1 show types of indoor plant in this study.

2 Method And Materials

2.1 Indoor Plant Material

Indoor ornamental plants were selected based on the existence in the local study area (Batu Pahat, Johor). The choice of plants is also through to the recommendation by former researchers such as Wolverton and Omar-Hor [27,28]. All plants grow in a pot-sized of 17

cm diameter and 25cm height, and pot mix ration is 2:2:1 (garden soil, compost, and perlite). Other than that, all plant ages that involved in this study is one year.
Before experimental were conducted, all plants involved must be assimilated with the indoor environment for at least two months, as recommended by Drozak and Torpy [29, 26]. In assimilation process, allowing plants to tolerate with indoor environment, where lighting and temperature is different compared to their original habitation. Light intensity during assimilated process is 300 lux, because this is minimum rates that are prescribed by the Department of Occupational Safety and Health [14]. Meanwhile, the temperature was fixed at 250C due to direction from the Prime Minister Office [30].

Fig. 1. Types of Indoor Plant [28, 29].

2.2 Method

Basically the method for this study based on report that is published by Australian Horticulture in the year 2011. In addition, this study also refers to other former researchers [10, 31, 32] as a method guideline, to ensure there is no doubt on the results. Every plant will be tested individually in one cubic meter chamber to analysis their performance to reduce CO_2 level. The reason why this control test is carried out because the plant also produces CO_2 [33], and before plants used in real situations, the researcher must determine the abilities of every single plant to reduce CO_2 by its own (without any other source of CO_2).

Besides that, adhesive foam-plastic insulation tape was used to provide airtight seal on the top of the chamber, and 12V DC fan inside the chamber to promote complete mixing. The temperature during the test was kept at 250C ± 10C to comply with Industrial Code Of Practice On Indoor Air Quality and Prime Minister Office, Malaysia [15, 30]. Meanwhile, the intensity of light that are provided along the tests is 300 lux, because researcher want to investigate the abilities of plants to reduce CO_2 gas (after plants was assimilated at indoor with the same rate of 300 lux) in the chamber.

Artificial lighting was provided by fluorescent bulbs that are placed at outside the chamber, about 30 cm from center of the plant. Apart from that, a portable IRGA TSI IAQ meter also used to monitor the CO_2 concentration in the chamber, and was set to record CO_2 reading at 5 minute intervals. All whole potted plant chamber trials were performed at ambient of CO_2 level at 450±10 ppm, where this being normal rate of indoor. Plants were tested individually during office hours from 9.00 am to 5.00 pm, and this study involving three plants for each species (triplicate study).

2.3 Data Analysis

Data were analyzed by One Way Analysis of Variance (ANOVA) using the Statistical Program for Social Sciences (SPSS) versions 20 for the significant analysis at α=0.05. Mean of data also important to analyze to support the analysis.

3 Results and Discussion

Figure 2 till Figure 8 shows all finding results based on graph illustration. Each graph is based on triplicate experiments. The findings also use the lowest light intensity that ever used on plants to reduce CO2 level. Generally, output graphs (CO2 concentration) for all types of plant are fluctuated, it is because plants also produce CO2 during respiration process [24]. Respirations on plants occur for both dark and light condition, and do not like photosynthesis, which only requires the presence of light [33]. Photosynthesis and respiration process can be attributed with light intensity. According to Irga, found that photosynthesis will more perform rather than respiration during high light intensity, that mean CO2 absorption will improve without fluctuate [26]. This research uses low light intensity (300 lux) because to comply the minimum of light intensity at indoor [14].

Table 1 shows details about CO2 reduction by each plant species. Prayer Plant is a plant that most perform to reduce CO2 concentration by 7.62%, followed by Dumb Cane (7.42%), Golden Pothos (6.5%), Kadaka Fern (4.65%), Syngonium (3.1%) and Anthurium (1.1%). In order to support percentage reading of CO2 reduction, average of CO2 reading also needed. This average is the result from 96 point reading of CO2 (5 minute interval from 8 hours test duration) which was taken in the chamber along with the plant.

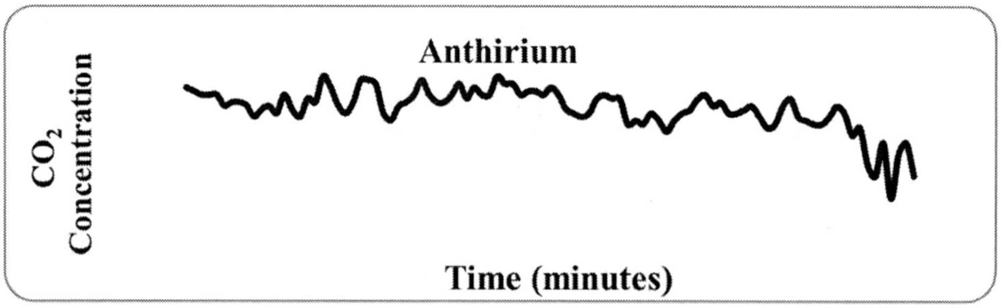

Fig. 2. Graph of CO_2 absorption by Anthurium.

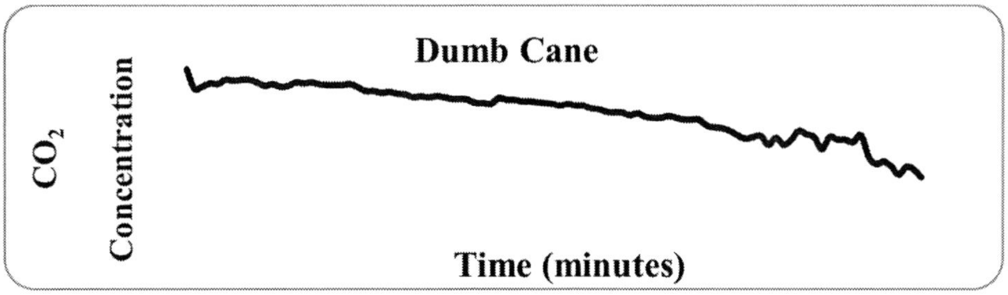

Fig. 3. Graph of CO_2 absorption by Dumb Cane.

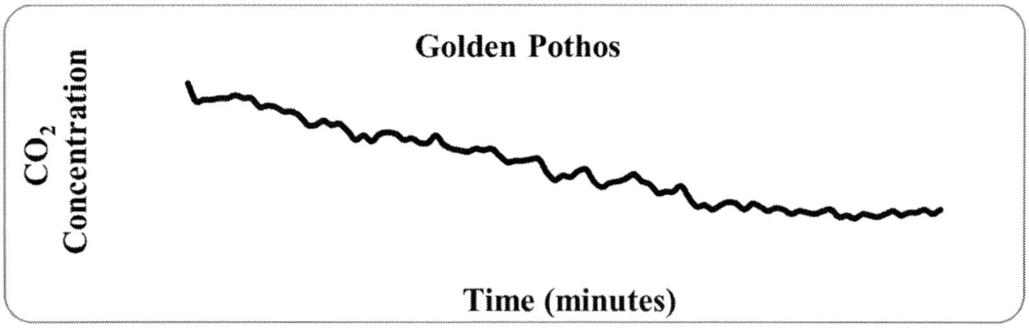

Fig. 4. Graph of CO_2 absorption by Golden Pothos.

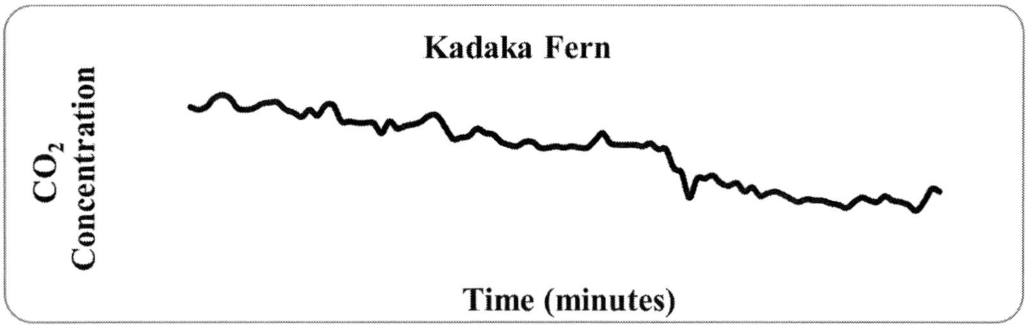

Fig. 5. Graph of CO_2 absorption by Kadaka Fern.

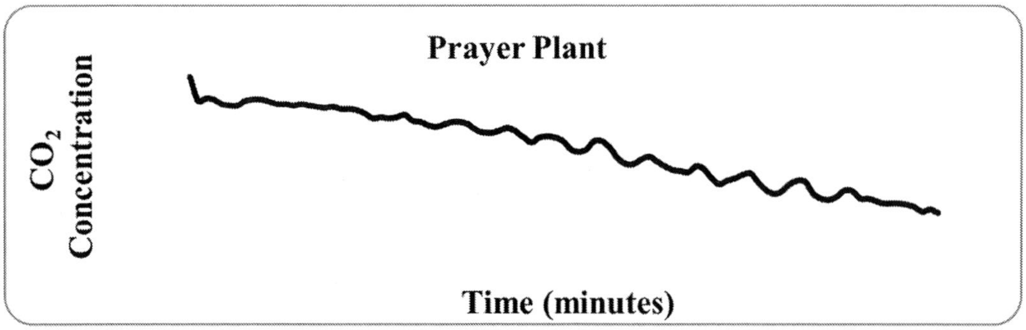

Fig. 6. Graph of CO_2 absorption by Prayer Plant.

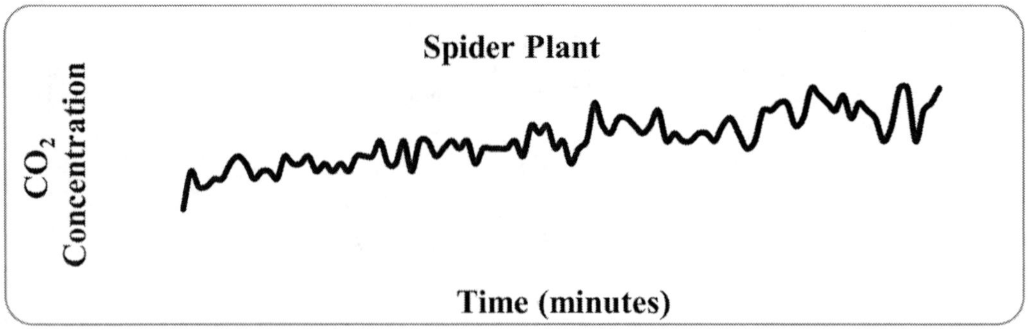

Fig. 7. Graph of CO_2 absorption by Spider Plant.

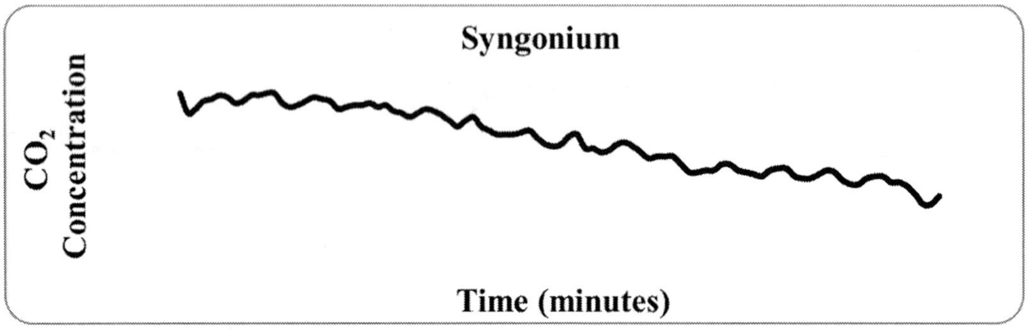

Fig. 8. Graph of CO_2 absorption by Syngonium.

All average value except the Spider Plant is lower than initial reading, and that means it supports the fact that the plants in this study were able to reduce CO2 levels even in low light conditions. CO_2 reduction by Anthurium is the fewest compare to others, and 300 lux light intensity seems the value of light compensation point (LCP) for the Anthurium whole potted plant. LCP is light level that photosynthesis and respiration occur at the same rate (have no CO2 reduction) [34]. Meanwhile, CO2 concentration in Spider Plant during this study is increasing around 1.1%. Failure Spider Plant to reduce levels of CO2 is due to lack of concentration levels of light that should have been imposed on the plant. If the intensity of light applied to a plant does not help in the process of photosynthesis, the CO2 concentration will increase [19].

Table 1. Details of reducing CO2 by each plant species.

Type of plant	Initial CO_2 reading (ppm)	Final CO_2 reading (ppm)	Average CO_2 reading for 96 point (ppm)	CO_2 reduction (%)	ANOVA sig. level at α =0.05
Anthurium	452	447	449.86	1.1	0.061
Dumb Cane	453.33	419.67	439.91	7.42	0.003
Golden Pothos	454	425.67	434.76	6.5	0.056
Kadaka Fern	458.67	437.33	447.2	4.65	0.064
Prayer Plant	454.67	420	434.05	7.62	0.072
Spider Plant	455.33	458.33	458.25	-1.1	0.000
Syngonium	454	435	444.1	3.1	0.045

In this study, the researcher uses three plants for every species to conduct the experiments. By doing so, researcher must know about the significance for every plant in the same species through to change rate of CO2 during this study. Because of that, researcher was used ANOVA technique in SPSS to find the significant for every plant in the same type. This solves problem recommended by Liu [32]. In ANOVA, to determine the significant between several groups, it depends on alpha (α) value. If alpha is greater than 0.05, it means that each group has a significant data; otherwise, if alpha is smaller than 0.05, it means that each group has no significant data [35]. Based on the Table 1, only Anthurium, Golden Pothos, Kadaka Fern, and Prayer Plant have significant value more than 0.05, where the value is 0.061, 0.056, 0.064, and 0.072. From this study that use low light level, can be seen through the data, Prayer Plant performs well, with the highest CO2 reduction, and have a higher level of significance. Unfortunately, Dumb Cane and Syngonium have significant value that lowers than 0.05, where the value is 0.003 and 0.045, and this is what is being said by Schornack and Wang that the testing of microbiology plant is intractable [36]. Even Dumb Cane and Syngonium have no significant data, but, each of plant from both species capable to reduce CO_2 (through to data finding). Meanwhile, significant value for Spider Plant is zero, and that means every plant for this species really different. From observations of researchers, found that Spider Plant has difficulty living in the building, where, in the low light, it makes the plant wither and fall leaves.

4 Conclusion

The focus of this study is to identify seven plant species that might function well to absorb CO_2 at indoor condition. Nevertheless, there may be constraints by plants to perform due to light factor, where indoor lighting is much lower compared to planting origin. This clearly when looking at Spider Plant that had no CO_2 reduction on it, instead increase the value. However, certain plants can tolerate with indoor condition after assimilation process is conducted on plants. As evidence, types of plants such as Anthurium, Dumb Cane, Golden Pothos, Kadaka Fern, Prayer Plant, and Syngonium capable to survive under indoor condition (with 300 lux light intensity), and even able to slightly decrease the concentration of CO_2 in the chamber. Overall in this experiment, it was found type of Prayer Pant is the most perform plant compare to others (based on the amount of CO_2 reduction, and the significant between plants).

The authors would like to thank the Ministry of Higher Education Malaysia through the funding supported (MyBrain15). Thank you to Universiti Tun Hussein Onn Malaysia (UTHM) and the Centre for Graduate Studies – UTHM.

References

1. M. Mahathir, J. Sch. Health, **12**, 2 (2015)
2. Standard, A. S. H. R. A. E., *55: Thermal Environmental Conditions for Human Occupancy American Society of Heating,* (Atlanta, USA, 1992)
3. P.N. Bierwirth, www.researchgate.net (2015).
4. Occupational Safety and Health Administration (OSHA), *Indoor Air Quality in Commercial and Institutional Buildings,* (Washington, 2011)
5. A.J. Respir, *Am. J. Respir. Crit. Care Med.*, **180**, 8 (1997)
6. R.M. Zin, J. Teknol., **70**, 7 (2014)
7. R. Saidur, H.H. Masjuki, Int. J. Mech. Mater. Eng., **3**, 1 (2008)
8. Z. Wang, Thesis, Syracuse University (2011)
9. Mahathir, A.M. Leman, H. Shafii, *The 3rd Scientific Conference on Occupational Safety and Health- Sci-Cosh* (2014)

10. U. Shome, Thesis, The University of Guelph (2004)
11. B.C. Wolverton, *Interior Landscape Plants For Indoor Air Pollution Abatement,* (Washington, 1989)
12. J. Zhou, F. Qin, J. Su, J. Liao, H. Xu, *J. Food, Agric. Environ.*, **9**, 1012 (2011)
13. H. Kim, J. Yang, J. Lee, J. Park, K. Kim, B. Lim, G. Lee, S. Lee, D. Shin, Y. Lim, *Environ. Health Toxicol.*, **29**, 1 (2014)
14. Department of Safety and Health, *Guidelines On Occupational Safety And Health For Working With Video Disply Units* (2003)
15. ICOP, *Industry Code Of Practice On Indoor Air Quality* (2010)
16. Jabatan Meteorologi Malaysia, *Batu Pahat Monthly Max TT Min TT Max RH Min RH 2011* (2013)
17. P.J. Aphalo, The Plant Photobiology Notes, **1**, 39 (2006)
18. A. Devkota, P.K. Jha, Middle-East J. Sci. Res., **5**, 230 (2010)
19. M.D. Burchett, F. Torpy, L. De Filippis, J. Brennan, P.J. Irga, University of Technology (2011)
20. Argus, Argus Control Systems Ltd. (2010)
21. D.R. Taub, J.R. Seemann, J.S. Coleman, Plant Cell Environ., **23**, 649 (2000)
22. G. Oquist, V.M. Hurry, N. Huner, Plant Physiol., **101**, 1 (1993)
23. D.J. Allen, D.R. Ort, Trends Plant Sci., **6**, 1 (2001)
24. W. Yamori, K. Hikosaka, D.A. Way, Photosynth. Res. **119**.1 (2014)
25. J.L. Quero, R. Villar, T. Marañón, R. Zamora, D. Vega, L. Sack, Funct. Plant Biol., **35**, 8 (2008)
26. F.R. Torpy, P.J. Irga, M.D. Burchett, Urban For. Urban Green., **13**, 2 (2014)
27. B.C. Wolverton, *How to Grow Fresh Air: 50 Houseplants that Purify Your Home Or Office* (Penguin Books, 1997)
28. K. Omar-Hor, *1001 Plants in Singapore*, (2006)
29. A. Drozak, E. Romanowska, Biochim. Biophys. Acta - Bioenerg., **1757**, 11 (2006)
30. Jabatan Perdana Menteri Malaysia, *Surat Pekeliling Am Bilangan 2 Tahun 2014* (2014)
31. M. Dela Cruz, J. H. Christensen, J.D. Thomsen, R. Muller, Env. Sci Pollut Res, **21**, 13909 (2014)
32. Y.J. Liu, Y.J. Mu, Y.G. Zhu, H. Ding, N. Crystal Arens, Atmos. Environ., **41**, 3 (2007)
33. D. Whiting, M. Roll, L Vickerman. Colorado State University (2014)
34. F.J. Sterck, R.A. Duursma, R.W. Pearcy, F.Valladares, M. Cieslak, M. Weemstra, , J. Ecol., **101**, 4 (2013)
35. A. Field, *Discovering Statistics Using IBM SPSS Statistics: And Sex and Drugs and Rock 'n' Roll* (2013)
36. S. Schornack, E. Huitema, L.M. Cano, T.O. Bozkurt, R. Oliva, M. Van Damme, S. Schwizer, S. Raffaele, A. Chaparro-Garcia, R. Farrer, M. E. Segretin, J. Bos, B. J. Haas, M.C. Zody, C. Nusbaum, J. Win, M. Thines, S. Kamoun, Mol. Plant Pathol.,**10**, 6 (2009)

A New Invention Method to Determine the Reduction Factor for Low Fabric Tension Properties for Head Garment Fabrication

A.F Aiman[1,*], *M.N* Salleh[2], *K.A* Ismail[1]

[1]School of Manufacturing Engineering, Universiti Malaysia Perlis, 02600 Arau, Perlis, Malaysia
[2]School of Technology Management and Logistic, UniversitiUtara Malaysia, 06010 Sintok, Kedah, Malaysia

Abstract. This paper proposes a new method to determine the reduction factor for producing a head garment with specified targeted pressure output. Pressure garment fabric mostly supplied to the local hospitals with no information of the material properties and the fabrication method generally used a single reduction factor at various body segments. Reduction factor defined as the percentage of reducing the garment size from the original circumference of the body part which contributes to the compression. The objective of this study is to compare the fabrication method of head garment using reduction factor equation from previous research with the new proposed method. The equation to predict the reduction factor requires the parameter of the fabric tension which is obtained from tensile test and radius of curvature of the human body parts. In the new proposed method, a 3D scanning was used for data acquisition to obtain the geometry of the head area. The pressure outputs are measured by a pressure measurement system developed from Flexiforce sensor and Arduino circuit board. By using the equation, the result shows the calculated reduction factor produced an extremely tight head garment compared to the conducted experiments which manage to produce an adequate reduction factor with a targeted pressure output of 20mmHg. The result of the experiment indicates that the reduction factor ranging from 17% to 23% compared to the equation which produces 20% to 47% of reduction factor. As an additional, the proposed experimental method can be used for different type of pressure garment fabrics in order to obtain the relationship between the reduction factor and the circumference of the body parts.

1 Introduction

Pressure therapy derives from an early surgical principle, in which the use of bandages and splints more than 40 years ago since 1970s [1, 2]. Methods of applying pressure have been documented and it included, among others, the method of splints, bandaging, and elastic garments. Pressure garment (PG) mostly used to treat hypertrophic scars because it is an easy device to apply, minimally disruptive to activity, and the availability in terms of

* Corresponding author : aiman@unimap.edu.my

commercially or through skilled technician. Regarding the skilled technician, mostly in medical or hospital departments they are known as an Occupational Therapist (OT). Besides that, PG manage to correct the deformity and eventually decrease the need for corrective surgery.

Even though there are still debates among researchers about the ideal pressure exerted by the pressure garment, most researchers agreed that the best pressure exerted from the compression of pressure garment is ideally in the range of 15mmHg to 24mmHg or 1999.85Pato 3199.74Pa [1, 3, 4]. It has been reported that pressure greater than 40mmHg may result other complications [5]. Nevertheless, most pressure garments fitting were assessed using ''sight, touch and experience"[2, 6]. In order to predict the pressure generated by pressure garment, Laplace`s Law has been widely used by other previous studies as given in Eq. 1 [7-12]. The parameters involved in the equation are the fabric tension and the radius of curvature. Both parameters can be obtained through tensile test and the dimension of the circumference of human body parts.

$$Pressure\ , P(Pa) = \frac{Tension\ in\ the\ cylinder\ wall\ ,T\ \ (Nm^{-1})}{Radius\ of\ the\ cylinder\ ,r(m)} \quad (1)$$

2 Problem statements

The Laplace`s Law was well known used in predicting the pressure output from the usage of the pressure garment. The important parameters known are the tension of the fabric and the circumference of the body parts. The tension of the fabric is determined from the standard tensile test and circumference is measured from the human body parts. Regarding the fabric used for this study, five specimens were tested by tensile test using MS ISO 13934-1:2003 standard. The modulus of elasticity calculated from the result of the tensile test was 246 N/m.

In order to obtain the circumference of head area, a 3D scanner was used to capture the geometrical of a mannequin head. The head segment divide to 20 layers of section-cut given by 10mm spacing in horizontal direction as shown in Fig. 1.

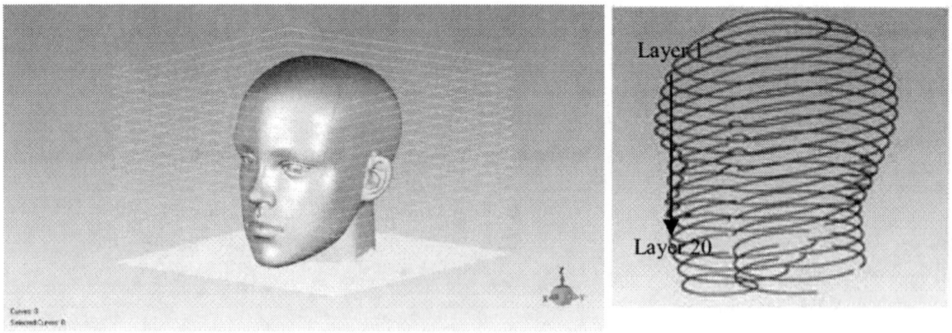

Fig. 1. 3D scanning for head segment using the mannequin`s head.

As for the reduction factor, which refers to the percentage of the reduced dimension from the original circumferences in order to deliver the exerted pressure, the prediction calculated by using Eq. 2 [13, 14].

$$Re = \frac{1}{1+\frac{E}{Pr}} \quad (2)$$

Based on the Eq. 2, there are three main parameters needed to calculate the reduction factor (Re), fabric tension in N/m (E), the pressure to be exerted (P) in Pascal and the radius of curvature (r) in meter. Using Equation (2), the reduction factor calculated value ranging from 20% to 47% of each section. The results indicate that the reduction factor produced was extremely big and resulted a tight head garment to be fitted to the mannequin's head. Reduction factor calculated using the fabric tension from the tensile test cannot be applied for the larger circumferences. According to previous studies, the reduction factor used for the most body part are in the range from 10% to 25% only[6, 15, 16]. Therefore, a new method is proposed to determine the relationship of reduction factor and circumference to achieve certain target pressure output.

3 Method

3.1 Experiment using a cylinder to determine the reduction factor for head segment

An experiment using several different sizes of round Polyvinyl Chloride (PVC) cylinder pipes with different circumference dimension range from 28cm to 64cm designed to investigate the reduction factor use to achieve a certain pressure output. The outer diameter of the cylinder was selected based on the range of the circumference size of a human head. In this study, the pressure output was set to 20mmHg or 2666.4Pa. A pressure sensor measurement system was developed from Flexiforce sensor and Arduino UNO board to measure the pressure output. The sensor is placed under the fabric and then monitored to achieve the targeted pressure output and the length which refer to the reduction factor is marked.

Three samples were used for each cylinder and the average of the reduced length were taken to make a validation sample. The validation sample is stitched and measured to validate either the pressure obtain as targeted. For each cylinder, fabric with 10mm in width used for the experiment. The length of each cylinder depends on the circumference size of each cylinder. Each sample is marked with the original circumference and labelled as original circumference. A scale with 5mm distance was drawn before the fabric being stretched as shown in Fig. 2 (a).

A jig was designed to replicate the stitch area located at the PVC area and the specimen will be inserted through the hole and wrap through the PVC and went through again in the same hole as the intake as shown in Fig. 2 (b). This procedure is done carefully to make sure the alignment according to the PVC circumference. A stopper was made at the intake of the fabric to ensure the fabric is locked in position when the fabric being pulled from the other direction. While the fabric being pulled and the Flexiforce sensor measure up to the targeted pressure, the measurement was being monitored continuously. Once the reading achieved the targeted pressure, the specimen locked using a clamp and the reading was monitored until the pressure targeted reading stabilized and marked the reduction length on the scale made previously. The length for the reduction factor recorded as shown in Fig. 2 (c). The procedures are repeated to all of the fabric specimens.

In order to validate the results, a validation sample using the mean value of the reduced length of the three samples. The mean value of the reduced length will be fabricated by stitching the fabric and measured to validate the pressure output achieved the specified targeted pressure output which is 20mmhg as shown in Fig. 2 (d).

 DOI: 10.1051/matecconf/20167801012

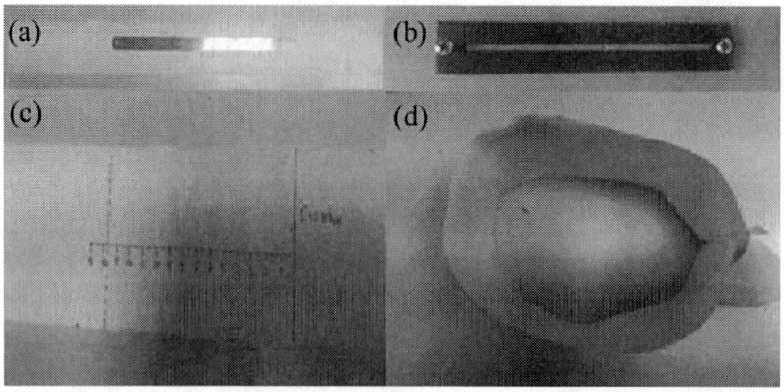

Fig. 2.(a) Sample of specimen (b) Jig for specimen (c) Length measured for the reduction factor (d) Sample from mean data stitched to verify with the pressure output.

3.2 Pressure measurement system (PMS)

Pressure measurement system was developed using Flexiforce sensor from Tekscan, USA which have been used in previous studies[17–19]. The pressure measurement used 1lb low force sensor. Arduino board was used as a combination to flexi force sensor for data acquisition. A 10g to 50g standard weight used to determine the linearity of the sensor output. The amount of the weight was selected due to the maximum pressure output calculated based on the formula of pressure equal to the force over area. The weight in Newton (0.49N) divides by the sensor active area, 0.000713 in square meter produced pressure of 6877.03Pa or 51.58mmHg. The output from the pressure measurement system converted to mmHg units because of the units are generally used in rehabilitation department. Fig. 3 (a), (b), and (c) show the diagram, sensor, and experiment setup.

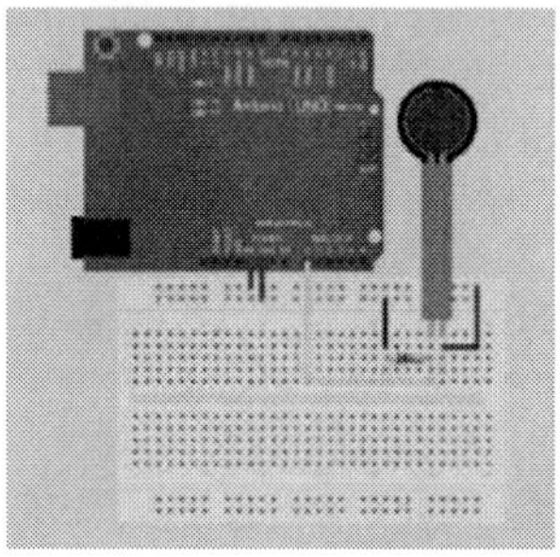

(a)

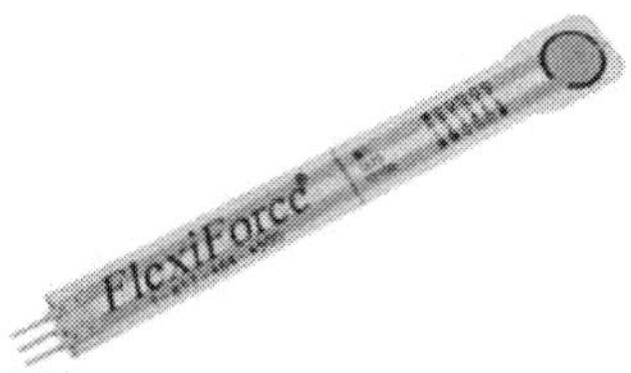

(b)

(c)

Fig. 3. (a) Fritzing wiring diagram (b) Flexiforce A201 1lb pressure sensor (c)Experiment setup.

4 Result and discussion

Table 1 shows the results of the length for the reduction gained by each specimen. A graph was plotted and polynomial curve fit was used to determine the equation that will be used to predict the reduction factor for certain circumferences. Fig. 4 shows the graph of the reduction factor that should be used to achieve the specified pressure output. As for the validation sample, the results of the pressure output show the reading of 20mmhg with tolerance of +/- 1mmhg.

Table 1. Reduction factor for achieved specified pressure output.

	Length reduction(cm) to achieve 20mmhg						
Cylinder Circumference (cm)	**Specimen 1**	**Specimen 2**	**Specimen 3**	**SD**	**Mean**	**cv**	**Reduction factor (%)**
27.91	3.50	3.60	3.20	0.21	3.43	6.10	12.30
35.89	7.40	7.20	7.30	0.10	7.30	1.40	20.34
43.87	10.60	10.70	10.10	0.32	10.5	3.10	23.86
52.84	13.70	13.50	13.60	0.10	13.6	0.70	25.74
63.80	17.10	16.80	17.20	0.21	17.0	1.20	26.70

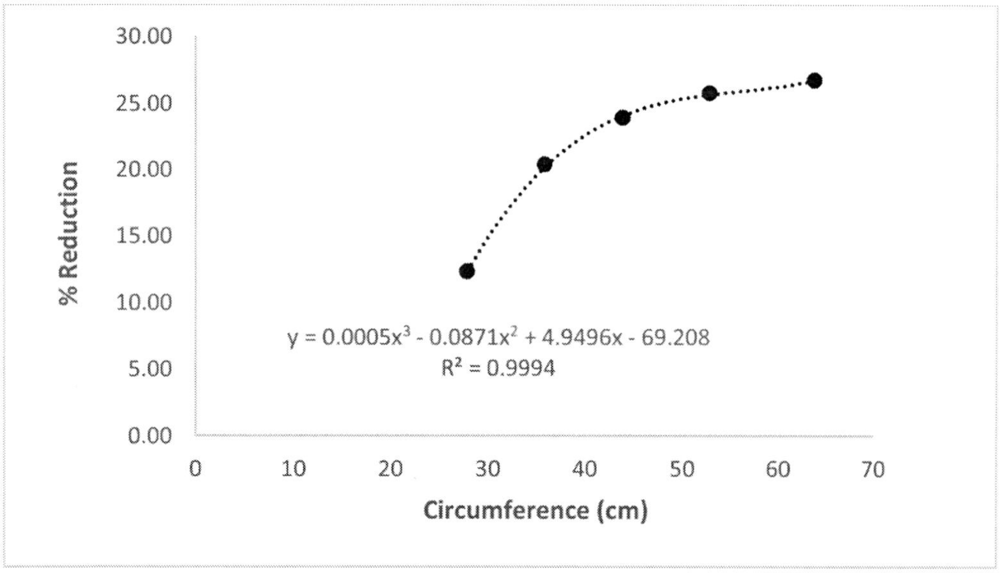

Fig. 4. A 3rd order Polynomial equation of the reduction factor for achieved 20mmHg pressure output.

Then, using the equation derived from the curve fitting using a 3rd order polynomial equation, the reduction factor is calculated for each section from the scanning data and compared with the prediction of the reduction factor calculated from the Eq. 2 as shown in Table 2. Comparison of the reduction factor is depicted in Fig. 5.

Table 2. Comparison of reduction from tensile method and variable cylinder circumference method

Section /layer	Original Circumference(cm)	Reduction factor calculated from Eq.(2)	Reduction factor from experiment
1	14.74	0.20	-0.14
2	33.23	0.36	0.17
3	41.61	0.42	0.22
4	46.39	0.44	0.23
5	49.50	0.46	0.23
6	50.91	0.47	0.23
7	51.81	0.47	0.23
8	52.15	0.47	0.23
9	51.11	0.47	0.23
10	48.32	0.45	0.23
11	47.27	0.45	0.23
12	46.52	0.45	0.23
13	45.29	0.44	0.23
14	44.09	0.43	0.23
15	42.26	0.42	0.22
16	42.38	0.42	0.22
17	41.88	0.42	0.22
18	40.92	0.41	0.22
19	40.24	0.41	0.22
20	39.65	0.41	0.21

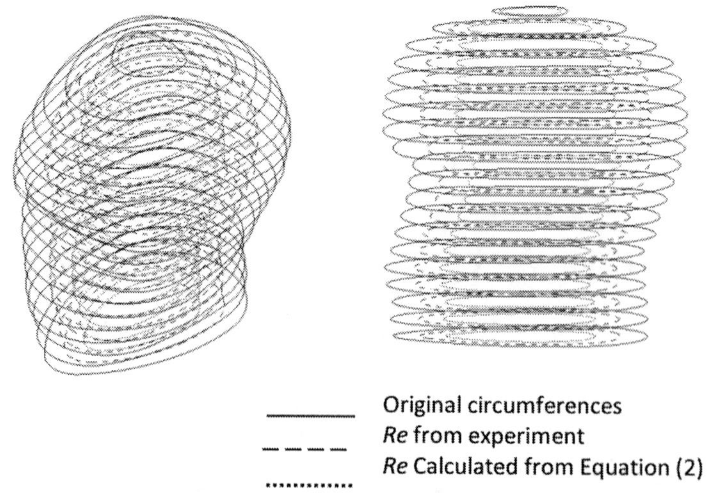

Fig. 5. Comparison of the Reduction Factor (Re) between Equation (2) and experiment.

In order to verify the polynomial equation, a different cylinder circumference from the experiment was selected. A cylinder with a 37.5 cm circumference was selected and the reduction factor calculated was 20.28%. A sample was fabricated and based on this reduction factor, the pressure generated was measured and the result showed the pressure output at 20mmHg with fluctuations of 1mmhg to 2 mmHg due to the drift of the sensor shown in Fig. 6.

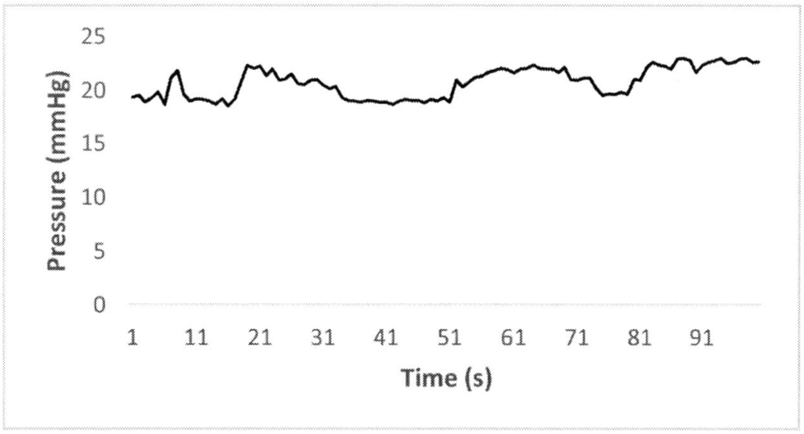

Fig. 6. Validating the polynomial curve equation using cylinder with a 37.5 cm circumference.

Based on Fig.5 and Table 2, the cylinder experiment indicates the reduction factor is in the range of 17-23%, while the range is 20-47% by using the Eq. 2. However, the cylinder experiment shows that at small circumference, at layer 1, indicating a negative value which refers that the reduction factor is too large to apply at the small circumference. Therefore, no reduction factor should be applied to the small circumference below 30cm.This result supported the finding from previous research that reported the circumference below 30cm does not provide an accurate results by using prediction calculated from the Laplace Law [4]. Even though the reduction factor indicates 20% of the reduction factor at layer 1, the result of pressure measurement using sensor indicates a high value exceeding 20mmHg. However, in this study, the tension of the fabric is slightly low, value of 246N/m which resulted a bigger reduction factor compare to the previous research that used a fabric with higher value of fabric tension which is more than 400N/m [7, 20]. Therefore, the experiment using cylinders can be used as an alternative method to calculate the reduction factor for a low fabric tension parameter. Based on the experiment results, the reduction factor from the cylinder experiment indicate an acceptable reduction factor to fabricate the head garment.

5 Conclusion

The experiments using cylinder provide a result that indicates the reduction factor should be used in producing head garment is about 17% to 23%. It is shown that the result can be used to produce a head garment that will fit in the mannequin`s head compared to the reduction factor calculated from the equation which is up to 47% of reduction factor that produce a very tight head garment which does not fit for the head. By using the cylinders, the equation from the polynomial graph can be used to predict the reduction factor for different circumferences. As an additional, any type of pressure garment`s fabric that's being received by local hospitals with no information about the material properties should be tested by this method in order to determine the reduction factor with a targeted pressure output.

We thank the Occupational Therapist staff from the rehabilitation unit of Hospital Tuanku Fauziah, Perlis for providing professional consultation and producing the head garments. Thanks to the Ministry of Higher Education Malaysia for supporting this research under Prototype Research Grant Scheme (PRGS).The authors would like to thank School of Manufacturing, Universiti Malaysia Perlis for providing laboratory facilities for 3D scanning.

References

1. L. Macintyre, M. Baird, Burns, **32**,10 (2006)
2. P. Joanne, W. Gill, *Pressure Garments: A manual on their design and fabrication*, (Butterworth-Heinemann, Oxord, 1995)
3. L.H.Y. Candy, L.T.W.P. Cecilia, Z.Y. Ping, Burns, **36**,1234(2010)
4. L. Macintyre, Burns, **33**, 579 (2007)
5. L.H. Engrav, D.M. Heimbach, F.P. Rivara, M.L. Moore, J. Wang, G.J. Carrougher, B. Costa, S. Numhom, J. Calderon, N.S. Gibran, Burns, **36**, 975 (2010)
6. L. Macintyre, M. Baird, Burns, **31**,11 (2005)
7. W.Y. Leung, D.W. Yuen, S.P. Ng, S.Q. Shi, J. Burn Care Res., **31**, 716 (2010)
8. T. O., E. Ashayeri, B. M., S. A., F. Alam, S. Marteau, Procedia Eng., **2**, 2823 (2010)
9. B.M.E. Valentinuzzi, A. J. Kohen, IEEE Pulse, **3**, 66 (2012)
10. K.D. Nguyen, A.A. Simard. Meilleur, C. Berthiaume, R. Godbout, L. Mottron, S.F.F. Ng, C.L.P. Hui, N. Yildiz, L. Wang, M. Felder, J.Y. Cai, Burns, **33**,59 (2007)
11. N. Yildiz, Burns, **33**,59 (2007)
12. L. Macintyre, R. Ferguson, Burns, **39**, 1073 (2013)
13. M.N. Salleh, M. Acar, N.D. Burns, *2011 4th International Conference on. IEEE* (2011)
14. M.N. Salleh, H.M. Lazim, S.N. Othman, A.F. Merican, *International Conference on Advacanced Mechantronic Systems* (2012)
15. S.F. Ng, C.L. Hui, Int J Eng Res Appl. **71**, 275 (2001)
16. L.T. Nguyen, N.J. Massman, B.J. Franzen, D.H. Ahrenholz, N.W. Sorensen, W.J. Mohr, L.D. Solem, J. Burn. Care. Res., **25**, 485 (2004)
17. M. Ferguson. Pell, S. Hagisawa, D. Bain, Med. Eng. Phys., **22**, 657 (2000)
18. M. Burke, B. Murphy, D. Geraghty, *IEEE SENSORS 2014 Proceedings* (2014)
19. J. Li, H. Liu, Y. Wang, L. Shi, F. He, Measurement, **45**, 2114 (2012)
20. S.F.F. Ng, C.L.P. Hui, Int. J. Cloth. Sci. Technol., **11**, 251 (1999)

MATEC Web of Conferences **78**, 01013 (2016) DOI: 10.1051/matecconf/20167801013
IConGDM 2016

A study of electrode wear ratio on EDM of Ti-6AL-4V with copper-tungsten electrode

Nurezayana Zainal[1,*] , *Azlan* Mohd Zain[1] and *Safian* Sharif[2]

[1]Soft Computing Research Group, Faculty of Computing, Universiti Teknologi Malaysia, 81310 Skudai, Johor, Malaysia.
[2]Faculty of Mechanical Engineering, UniversitiTeknologi Malaysia, 81310 Skudai, Johor, Malaysia.

Abstract. In this work, a study of electrode wear ratio (EWR) on the die-sinking electrical discharge machining (EDM) ofTi-6AL-4V titanium alloy with copper-tungsten (Cu-W) electrode has been carried out. Pulse on time (ON), pulse of time (OFF), peak current (V) and servo voltage (SV) were seen as the machining parameters. The experiments were run according to the design of experiments (DOE), which is two levels of full factorial with added centre points. The experimental results reveal that pulse on time and peak current are statistically significant parameters for affecting EWR with the p-value of 0.0013 and 0.0012 respectively. Moreover, based on ANOVA, we recognized peak current as the most significant parameters which contribute 31.75%, followed by pulse on time, servo voltage and pulse on time which contribute 30.99%, 8.68% and 0.72%, respectively.

1 Introduction

Machining process can be determined as the cognitive operation of shaping the workpiece as desired and it is commonly done using machine and cutting tools [1]. In the early 20th century, the machining process is recognized as non conventional machining process which includes milling, bending, grinding, drilling and drilling. Since the coming of novel technologies, there has been modern machining process such as electrochemical machining (ECM), electrical discharge machining (EDM), abrasive water jet (AWJ) and ultrasonic machining (USM) [2, 3, 4].

Electrical discharge machining (EDM) is unitary of the well known modern machining process used in machining materials that are mainly hard to be machined. EDM is widely being used in innovative manufacturing industry purposely used to transfer material from the work piece by controlling erosion through §a series of electric sparks occurring between the electrode and the work piece immersed in dielectric fluid [5, 6]. Generally, EDM produces complex shapes and permits high precision machining of any strong, difficult-to-machine, breakable or thin materials [7]. In addition, EDM is used for completing parts for aerospace and automotive industry and surgical components [8].

* Corresponding author: ezayana@gmail.com

In late years, titanium has become an important material which has widely applied in the aerospace industries. Basically, titanium has a high melting temperature and low thermal conductivity where it belongs, to the group of difficult-to-cut materials which is not suited for traditional machining [9]. Moreover, titanium has a low modulus of elasticity which causes significant spring back after deformation under the cutting load [10]. Hence, EDM process can effectively machine titanium alloy.

Copper-tungsten electrodes provide better surface finish and longer life than the graphite electrodes [7]. The thermal properties of copper–tungsten alloy render it suitable for utilization as an EDM electrode in this field. More often than not, the low melting point of Cu component reduces the resistance to electrode wear. Moreover, there is porosity in liquid phase sintered electrode due to the insolubility between Cu and W due to its high tungsten content [11]. In addition, Cu–W electrode has a good wear resistance properties and good thermal conductivity [12]. Therefore, it is desirable for the machining of small holes with flat beds, where high-precision machining is the primary consideration.

In the present study, the effect of different machining parameters (pulse of time, pulse of time, peak current and servo voltage) towards electrode wear ratio (EWR) of Ti-6AI-4V titanium alloy with copper-tungsten electrode is studied. Then, the result of these machining parameters on EWR is analyses using analysis of variation (ANOVA).

2 Experimental Procedure

The experimental study was conducted with an objective to identify the machining parameters that significantly affect the EWR. The experiments are carried out utilizing a numerical control programming, electrical discharge machine known as AG40L Sodick. The EDM has the provisions of movement in three axes which are longitudinal (X-axis), lateral (Y-axis) and vertical (Z0-axis). In this effort, titanium alloy Ti-6AL-4V was selected as the work piece and copper-tungsten (Cu-W) electrode was used for machining the work piece. The diemeter of the electrode is 8mm. Then, the dielectric fluid used is oil. The specimens were EDMed at 1.5mm depth of hole. Figure 1 illustrates the machining setup of EDM process. Table 1 presents the design factors and their levels.

Fig. 1. Experimental setup of electrical discharge machining.

Table 1. Factor and level for EDM parameters of titanium alloy.

Symbol	Name	Units	Level	
			Low (-1)	High (+1)
ON	Pulse on time	μs	150	230
OFF	Pulse off time	μs	60	90
IP	Peak current	Ampere	10	12
SV	Servo voltage	Volt	30	60

Two levels of full factorial design with added center points were used to carry out a total number of 20 experiments. In this study, EWR is defined as the ratio of the volume of material removed from the electrode to the volume of material removed from a workpiece. A digital balance (Ohaus - PA 214) was utilized to quantify the weight of the electrodes and work piece required. EWR was calculated by the formula as expressed in (1) and (2):

$$EWR = \frac{volume\ of\ material\ removed\ from\ electrode}{volume\ of\ material\ removed\ from\ work\ piece}\ x\ 100\% \qquad (1)$$

$$EWR = \frac{E_{WB} - E_{WA}}{WP_{WB} - WP_{WA}}\ x\ 100\% \qquad (2)$$

where, E_{WB} is the volume of the electrode before machining (g), E_{WA} is the volume of electrode after machining (g), WP_{WB} is the volume of electrode before machining (g) and WP_{WA} is the volume of electrode after machining (g).

3 Result and Discussion

The effect of each machining process; pulse on time, pulse off time, peak current and servo voltage on electrode wear ratio (EWR) in EDM of Ti-6AI-4L titanium alloy is discussed in this section. The results for the experiments are further analyzed using analysis of variance (ANOVA) for identifying the significant factors affecting EWR. Design Expert software was employed to carry out the ANOVA approach. Moreover, the electrodes after EDM process can be viewed in Figure 2.

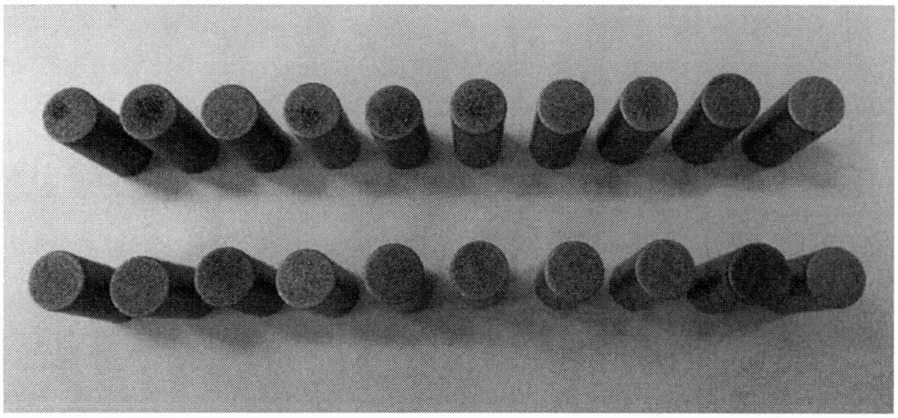

Fig. 2. The electrodes after EDM process.

From the experimental results plotted in Figure3, it can be seen that run No. 14 gave the highest reduction of EWR during the EDM process (50.04%). On the other hand, run No. 2 gave the lowest reduction of EWR, which is 16.19 %. Moreover, Figure 4 illustrates the effect of different machining parameters on EWR. Figure 4 (c) and Figure 4 (d) show that EWR increases as peak current and servo voltage increases. It can be concluded that high current and high servo voltage results in high EWR. This is imputable to the fact that the higher current will generate a stronger spark which producing more rate of heat energy. Furthermore, Figure 4 (b) shows that EWR increases sharply after the decrement of pulse off time. However, as can be seen from Figure 4 (a), EWR decreases as pulse on time increases from 150μs to 230μs.

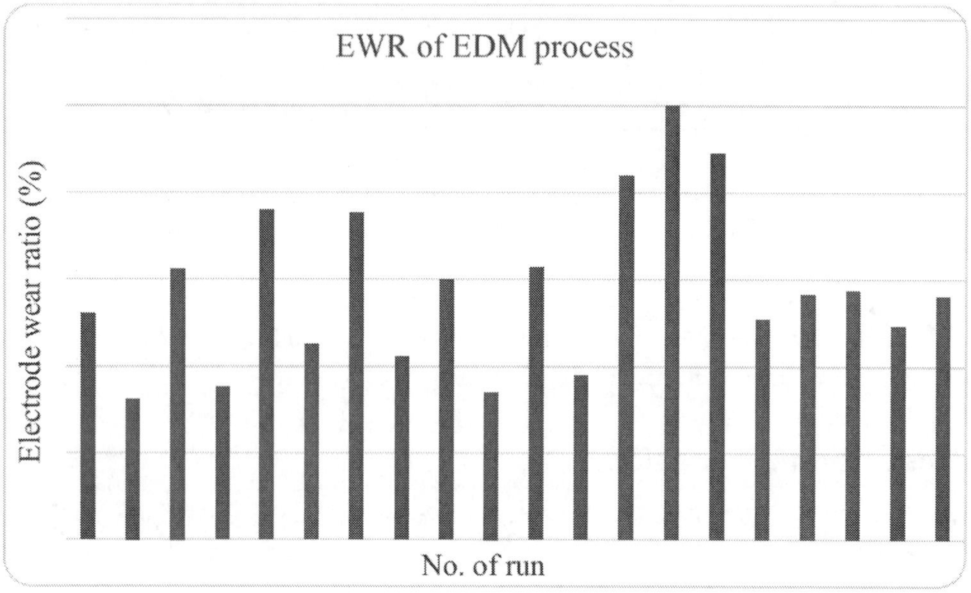

Fig. 3. Analysis of EWR.

Than, the experimental results obtained are evaluated by using analysis of variance (ANOVA) approach. The results of experimental data from ANOVA are given in Table3. This analysis is carried out for a significance level of α=0.05, for confidence level of 95%. Agreeing to the ANOVA technique, the values of "p-value" less than 0.05 will indicate the machining parameters are statistically significant. While the "p-value" greater than 0.05 will indicate the machining parameters are statistically insignificant. According to Table 3, we can observe that pulse on time and peak current are statistically significant parameters for affecting EWR with the p-value of 0.0013 and 0.0012 respectively. Besides, pulse off time and servo voltage appears as insignificant parameters for affecting EWR. Furthermore, it can be determined that the peak current is the most effective machining parameters with a 31.75 % contribution as presented in Table 3. It is followed by pulse on time with contributions 30.99%. The less affecting machining parameters are pulse off time with contributions 0.72% only.

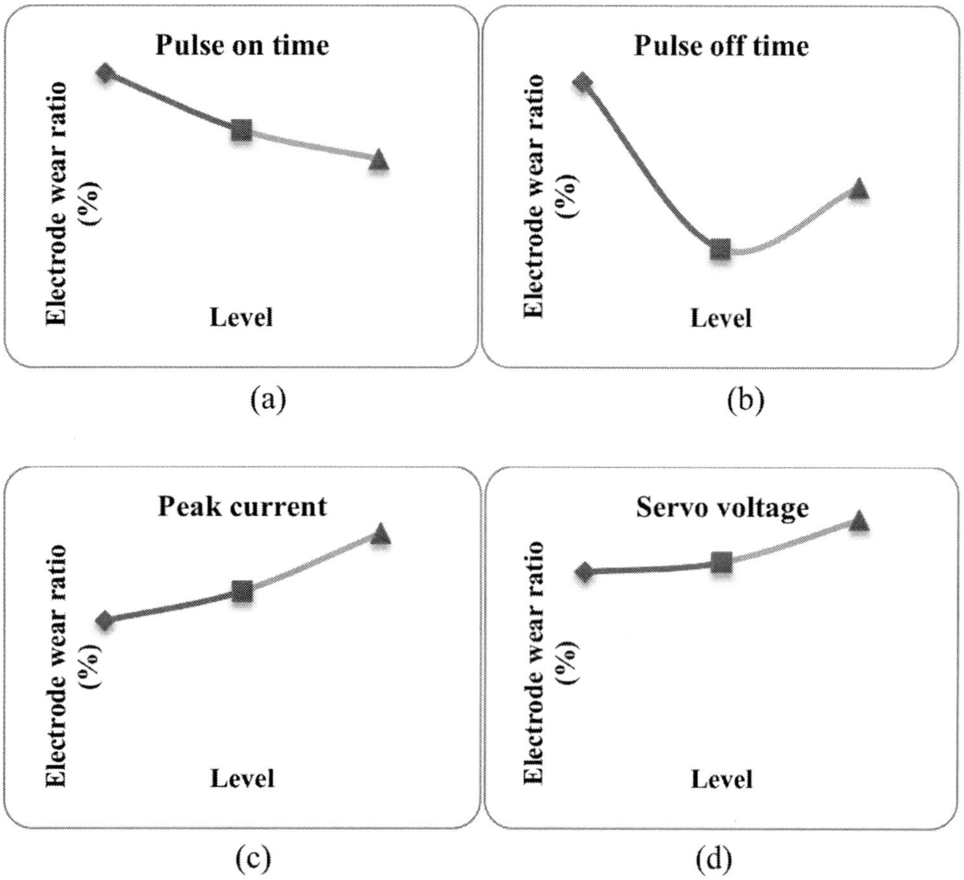

Fig. 4. Analysis of effect of machining parameters on EWR. (a) Effect of pulse on time on EWR, (b) Effect of pulse off time on EWR, (c) Effect of peak current on EWR and (d) Effect of servo voltage on EWR.

Table 3. ANOVA for EWR.

Source	Sum of square	df	Mean square	f-value	p-value		Contribution (%)
Model	0.1228	4	0.0307	9.2898	0.0007	Significant	
A-Pulse on time	0.0528	1	0.0528	15.9612	0.0013		30.99
B-Pulse off time	0.0012	1	0.0012	0.3709	0.5523		0.72
C-Peak current	0.054	1	0.054	16.3563	0.0012		31.75
D-Servo voltage	0.0148	1	0.0148	4.471	0.0529		8.68
Curvature	0.0012	1	0.0012	0.3494	0.5639	Not significant	
Residual	0.0463	14	0.0033				
Lack of Fit	0.0452	11	0.0041	11.5739	0.0340	Significant	
Pure Error	0.001	3	0.0004				27.86
Cor Total	0.1702	19					100

DOI: 10.1051/matecconf/20167801013

4 Conclusion

The results of EWR of Ti-6AL-4V titanium alloy with copper-tungsten electrode on the different machining parameters have been presented. A higher current accounts for more EWR followed by servo voltage. Initially, pulse off time, minimize EWR from 60μs to 75 μs, and then its increase the EWR sharply. However, pulse on time account for lower EWR from 150μs to 230μs. According to ANOVA technique, we can see that pulse on time and peak current are statistically significant parameters for affecting EWR with the p-value of 0.0013 and 0.0012 respectively. Then, we recognized peak current as the most significant parameters which contribute 31.75%, followed by pulse on time, servo voltage and pulse on time which contribute 30.99%, 8.68% and 0.72%, respectively.

Special appreciation to reviewer(s) for useful advices and comments. The authors greatly acknowledge the Research Management Centre, UTM and Ministry of Higher Education Malaysia (MOHE) for financial support through the Fundamental Research Grant Scheme (FRGS) vot. No. R.J130000.7828.4F721.

References

1. N. Yusup, A.M. Zain, S.Z.M. Hashim: Procedia Eng., (2012)
2. N. Zainal, A.M. Zain, N.H.M. Radzi, M.R. Othman: J. Intell. Manuf., (2014)
3. A.M. Zain, H. Haron, S. Sharif, Expert Syst. Appl., **37** (2010)
4. A.M. Zain, H. Haron, S. Sharif, Mach. Sci. Technol., **14** (2010)
5. M.B. Ndaliman, A.A. Khan, M.Y. Ali, Int. J. Mech. Mater. Eng.,**6**, 3 (2011)
6. S. Padhee, N. Nayak, S.K. Panda, P.R. Dhal, S.S. Mahapatra, Sadhana **37**, 2 (2012)
7. M.A.R. Khan, M.M. Rahman, K. Kadirgama, Int. J. Adv. Manuf. Tech., **77** (2014)
8. N.M. Abbas, D.G. Solomon, M.F. Bahari, Int. J. Mach. Tool Manu., **47**, 7 (2007)
9. S.L. Chen, B.H. Yan, F.Y. Huang: J. Mater. Process. Technol., **87** (1999)
10. B. Kuriachen, J. Mathew, Mater. Manuf. Processes, **30**, 8 (2015)
11. L. Li, Y.S. Wong, J.Y.H. Fuh, L. Lu, J. Mater. Process. Technol., **113** (2001)
12. H.T. Lee, F.C. Hsu, T.Y. Tai, Mater. Sci. Eng., **364** (2004)

MATEC Web of Conferences **78**, 01014 (2016) DOI: 10.1051/matecconf/20167801014
IConGDM 2016

Computational Approach for Multi Performances Optimization of EDM

Yusliza Yusoff[1,*], *Azlan* Mohd Zain[1], *Mohd Salihin* Ngadiman[1]

[1]Department of Computer Science, Faculty of Computing, 81310 UTM Skudai, Johor, Malaysia

Abstract. This paper proposes a new computational approach employed in obtaining optimal parameters of multi performances EDM. Regression and artificial neural network (ANN) are used as the modeling techniques meanwhile multi objective genetic algorithm (multiGA) is used as the optimization technique. Orthogonal array L256 is implemented in the procedure of network function and network architecture selection. Experimental studies are carried out to verify the machining performances suggested by this approach. The highest MRR value obtained from OrthoANN – MPR – MultiGA is 205.619 mg/min and the lowest Ra value is 0.0223μm.

1 Introduction

The high operational cost, long experimental period and requirement of an experienced machinist resulted to the aggressive research development in the area of machining optimization of EDM. Based on the literatures, there are four major issues in EDM which are monitored and control, machining, development, handling of parts and parameters optimization [1, 2].

In this paper, we are focusing on the computational approaches to optimize and improve the multiple machining performances of EDM. Traditionally, machining parameters are selected manually based on the engineer and operator experiences [3]. The selection of parameters in order to obtain the optimum machining performances is very complex due to the best parameter combination is unknown especially to the beginners and non machining expertise. Improper variables guesswork has contributed to a long development period, miss achieve of target date and loss of competitiveness. In general, during the early stage of development, the engineers and operators have a very limited processing knowledge to complete the machining experiments within the time given resulted in a very high development cost. To overcome difficulties in obtaining optimal solutions in a fast manner and low cost, new intelligent modeling and optimization techniques are proposed. There are many intelligent techniques proposed in optimizing machining operations as described by Rao and Kalyankar (2014) [4].

There are many modeling techniques proposed by previous researchers such as response surface methodology (RSM), regression, ANN, support vector machine (SVM) [5], fuzzy logic (FL) [6] etc. For example, Padhee et al. (2012) [7] adopted RSM to model, and NSGA

[*] Corresponding author: yusliza@utm.my

to optimize the machining parameters; dielectric fluid, pulse on time, duty cycle and peak current of powder mixed EDM. Regression is also considered as one of the most well known modeling technique employed by many researchers to overcome the machining optimization problems as mentioned by Zain et al. (2012) [8]. Zain et al. (2012) [8] in their study, applied regression and ANN model to optimize the surface roughness value of an end milling machining. In this paper, we are focusing on combining regression, ANN and multiGA to optimize multiple performances of EDM. Different from existing literatures which uses variable numerical experimentation [9] or random selection [10-12], we proposed orthogonal array as a technique to select the network parameters of ANN modeling in this study. Basically, the used of orthogonal array L256, is targeted to reduce the trial and error guesswork during the network parameters and network architecture selection. The regression and ANN model are then executed with multiGA in order to obtain the optimal machining solutions. Although there are many new optimization algorithms, such as cuckoo search algorithm [13], Levy flight algorithm [14], glowworm swarm algorithm [15], firefly algorithm [16], GA has taken a wide range of monopoly in solving the machining parameters optimization problems as surveyed by Yusup et al. (2012) [17] due to its advantages; mainly in reducing operational cost and production time besides maximizing the production rate and quality. In terms of multiple objectives, the journey of multiGA in machining optimization is still untapped and a continuous initiative on how to obtain optimal multiple machining performances need to be determined.

2 Experimental

The experimental data used in this study are obtained from experiments conducted by Kanagarajan et al. (2008) [18] in an Electronica die sinking EDM. The electrode material used by the authors is an electrolytic grade copper with the size of 12mm diameter. Kerosene is used as the dielectric fluid circulated by the jet flushing and the composites of the workpiece materials consist of 70% tungsten carbide and 30% cobalt. The machining performances considered are material removal rate (MRR) and surface roughness (Ra); meanwhile the parameters are rotation (S), current (T), pulse on time (U), flushing pressure (V).

ANN. The difficulty to predict expected output characteristic of EDM using mathematical equation resulted in the idea of using computational integration of ANN and MPR. The special logical characteristic of ANN which is easily constructed based on the input and output training data; the integration of these two models, ANN-MPR is expected to predict better optimal machining parameters. Besides that, the ability to completely detect complex nonlinear relationships between dependent and independent multiple variables is aimed to support MultiGA in generating considerable optimal solutions. The only problem heading with ANN is when there will be too much guesswork in choosing the best network functions. To avoid time consuming and trial error experimentation, we include orthogonal array L256 in the process of selection combination. The network functions are arranged based on the L256 combinatorial design. From the orthogonal array results, it is observed that 4-8-2 layer recurrent neural network (LRNN) architecture as depicted in Fig. 1 is found to be the most accurate and reliable with very good prediction accuracy. 4-8-2 LRNN, or generally prescribed as OrthoANN showed the lowest percent error; 1.74% for MRR and 1.80% for Ra with the configuration of Bayesian Regulation training function, hyperbolic tangent sigmoid transfer function, mean squared error performance function and gradient descent weight and bias learning function. The predicted results of OrthoANN and the percent error compared to real machining data are given in Table 2.

MPR. A statistical regression equation, known as multiple polynomial regression (MPR) is employed as part of the modeling technique of this study. The basic MPR equations of this study are given as Eq. 1(MRR) and Eq. 2 (Ra) below:

$$MRR = M1S + M2T + M3U + M4V + M5S2 + M6T2 + M7U2 + M8V2 + C \quad (1)$$

$$Ra = N1S + N2T + N3U + N4V + N5S2 + N6T2 + N7U2 + N8V2 + C \quad (2)$$

Where, M1 ~ M8 and N1 ~ N8 are the coefficients, S, T, U, V are the considered parameter variables and C is the constant value.

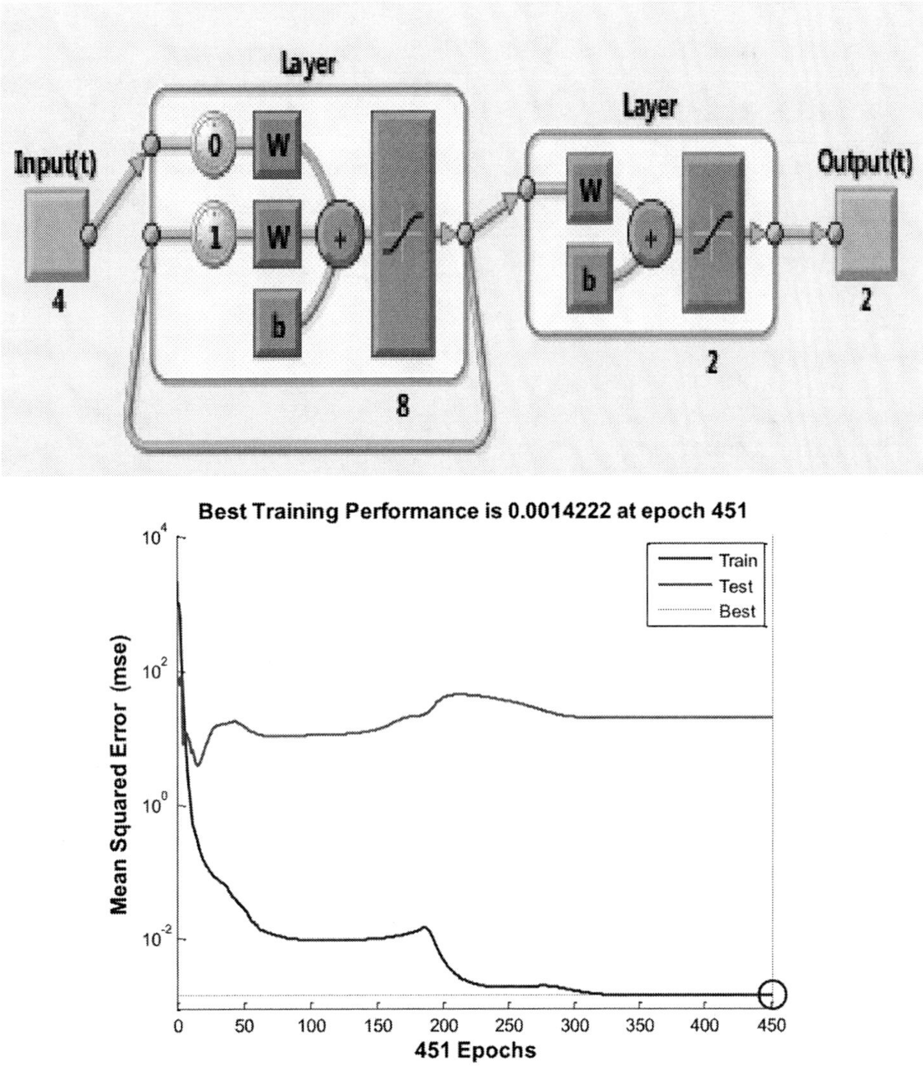

Fig. 1. 4-8-2 LRNN architecture and training performance.

Table 2. Real machining results and OrthoANN results.

Run	Real EDM Experiment (Kanagarajan et al, 2008)						OrthoANN Prediction Resutls			
	Rotational speed, rpm (S)	Current, A (T)	Pulse Time, µs (U)	Flushing pressure, kg/cm^2(V)	MRR, mg/min	Ra, µm	MRR, mg/min	Percent error, %	Ra, µm	Percent error, %
1	250	5	200	0.5	38.19	3.94	38.19838	0.00022	3.92347	0.004195
2	250	5	200	1	46.05	2.84	39.87993	0.133986	2.978744	0.048853
3	250	5	200	1.5	51.37	2.35	51.36999	2.78E-07	2.386609	0.015578
4	250	10	500	0.5	46.5	8.83	46.49825	3.77E-05	8.831211	0.000137
5	250	10	500	1	56.07	6.37	56.07072	1.28E-05	6.374118	0.000646
6	250	10	500	1.5	62.56	5.27	61.25816	0.020809	4.407064	0.163745
7	250	15	1000	0.5	49.31	14.7 4	49.31042	8.54E-06	14.56681	0.01175
8	250	15	1000	1	59.45	10.6 4	59.4498	3.44E-06	10.6538	0.001297
9	250	15	1000	1.5	66.33	8.8	66.33009	1.37E-06	8.793185	0.000774
10	500	5	500	0.5	37.77	3.89	42.02465	0.112646	3.561823	0.084364
11	500	5	500	1	45.54	2.81	45.53809	4.2E-05	2.82346	0.00479
12	500	5	500	1.5	50.81	2.39	50.81059	1.16E-05	2.377658	0.005164
13	500	10	1000	0.5	49.84	8.24	49.84137	2.75E-05	8.246744	0.000818
14	500	10	1000	1	60.09	5.94	60.08888	1.87E-05	5.927781	0.002057
15	500	10	1000	1.5	67.05	4.91	67.05043	6.42E-06	4.915196	0.001058
16	500	15	200	0.5	121.07	7.56	121.0698	1.38E-06	7.559584	5.5E-05
17	500	15	200	1	145.98	5.45	145.9803	2.22E-06	5.443669	0.001162
18	500	15	200	1.5	162.87	4.51	162.8678	1.33E-05	4.516236	0.001383
19	1000	5	1000	0.5	40.48	3.63	40.48258	6.38E-05	3.643697	0.003773
20	1000	5	1000	1	48.81	2.62	38.94707	0.202068	2.905432	0.108944
21	1000	5	1000	1.5	54.46	2.37	54.4608	1.47E-05	2.371298	0.000548
22	1000	10	1000	0.5	122.37	4.22	122.3699	1.22E-06	4.221237	0.000293
23	1000	10	1000	1	147.55	3.05	147.5507	4.45E-06	3.032144	0.005854
24	1000	10	1000	1.5	164.62	2.52	164.445	0.001063	2.479962	0.015888
25	1000	15	1000	0.5	119.74	7.47	119.7401	8.68E-07	7.472861	0.000383
26	1000	15	1000	1	144.38	5.39	144.38	2.48E-10	5.397557	0.001402
27	1000	15	1000	1.5	161.09	4.46	161.0933	2.05E-05	4.456306	0.000828
	Average percent error							1.74%		1.80%

MultiGA. GA is able to search for better solutions to linear and nonlinear problems through mutation, crossover and selection operations. The goals of multi objectives optimization are to find set of solutions near to the Pareto optimal front and should uniformly distributed. MultiGA which uses controlled elitist GA, a variation of NSGA-II [19] is used to generate the Pareto frontier of this study. MultiGA acts on individuals with better fitness value that can help to increase the diversity of the population even if they have a lower fitness value. It is very important to preserve the diversity of population for convergence to an optimal Pareto front by controlling the elite members of the algorithm. The optimization algorithm is run in Matlab 2012 software. The objective functions of this study are to maximize MRR and minimize Ra based on basic equation as shown by Eq. 1 and Eq. 2. The optimization trials have been conducted accordingly based on selected algorithm parameters and limitations. The algorithm parameters considered are: population size of 60, generation size of 800, tournament selection value of 2, intermediate crossover value of 1 and adaptive feasible mutation. 4-8-2 LRNN OrthoANN – MPR – MultiGA results are presented in Fig. 2. The results obtained at 542 iterations.

	S	T	U	V	MRR	Ra
1	963.6018	12.08074	284.3871	1.495271	200.9486	0.4469
2	963.5556	7.260906	284.3676	1.499418	159.2746	2.2040
3	963.5586	7.330306	284.3754	1.499594	160.1874	2.1901
4	963.5552	8.0441	284.3736	1.499758	169.0233	2.0270
5	963.7007	14.03597	284.3882	1.499999	205.6192	0.7086
6	963.6673	12.23692	284.376	1.499622	201.6796	0.3690
7	963.6251	10.32191	284.3798	1.499721	190.8441	1.2747
8	963.5767	11.24355	284.3815	1.498811	196.8875	0.8685
9	963.5697	6.85572	284.3693	1.499627	153.7904	2.2796
10	963.6431	12.6483	284.3824	1.497549	203.0627	0.1394
11	963.5703	10.08538	284.3764	1.499978	189.0299	1.3692
12	963.5481	7.678752	284.3741	1.498529	164.5937	2.1135
13	963.568	11.77626	284.379	1.498892	199.6722	0.6083
14	963.6917	11.51879	284.3901	1.498984	198.4084	0.7372
15	963.6251	13.9854	284.3799	1.499762	205.5783	0.6759
16	963.5802	9.357353	284.377	1.499701	182.7857	1.6362
17	963.6667	10.87795	284.3876	1.499459	194.6941	1.0372
18	963.6508	12.85612	284.3837	1.499299	203.7050	0.0223
19	963.5453	6.458783	284.3677	1.5	148.1197	2.3429
20	963.5712	8.461756	284.3775	1.499725	173.7502	1.9155
21	963.6909	13.71296	284.39	1.499849	205.3549	0.4991

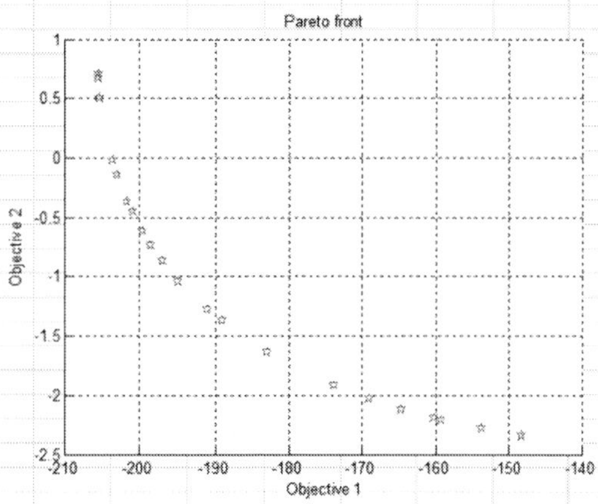

Fig. 2. OrthoANN – MPR – MultiGA optimal solutions.

3 Discussion and conclusion

From the experimental results of OrthoANN – MPR – MultiGA, presented in Fig. 2, none of the solutions is absolutely better than any other solution, the best solution can be selected based on the individual product requirements. The option of one solution over the other depends on the requirement. If the requirement is a better surface roughness or higher material removal rate, a suitable arrangement of variables can be decided. As for this study, the objectives are to maximize the MRR value, and minimize Ra value. From real EDM experimental results presented in Table 2, the highest MRR value obtained is 164.62 mg/min for solution no. 24, meanwhile the lowest Ra value is 2.35μm from solution no. 3. Contrast to OrthoANN – MPR – MultiGA results, most of the MRR solutions are higher than the value obtained by experiment. The highest MRR value is 205.6192mg/min from solution no. 5 and the lowest MRR value obtained is 148.1197mg/min from solution no. 19. As depicted in Fig. 2, only solutions 2, 3, 9, 12, 19 showed lower values than the real experimental MRR highest value. The lowest Ra solution from Fig. 2 is 0.0223μm which is much lower than the real experimental values.

Special appreciation to reviewer(s) for useful advices and comments. The authors greatly acknowledge the Research Management Centre, UTM and Ministry of Higher Education Malaysia (MOHE) for financial support through the Fundamental Research Grant Scheme (FRGS) vot. No. R.J130000.7828.4F721

References

1. K. Ho, S. Newman, Int. J. Mach. Tool. Manu., **43**, 1287 (2003)
2. K.H. Ho, S.T. Newman, S. Rahimifard, R.D. Allen, Int. J. Mach. Tool. Manu., **44**, 1247 (2004).
3. J.T. Huang, Y.S. Liao, Int. J. Prod. Res., **41**, 1707 (2003)
4. R.V. Rao, V. Kalyankar, Int. J. Adv. Manuf. Technol., **73**, 1159 (2014)
5. A.M. Deris, A.M. Zain, R. Sallehuddin, Procedia Eng., **24**, 308 (2011)
6. M.R.H. Mohd Adnan, A. Sarkheyli, A.M. Zain, H. Haron, Artif. Intell. Rev., **43**, 345, (2013)
7. S. Padhee, N. Nayak, S.K. Panda, P.R. Dhal, S.S. Mahapatra, Sadhana-Acad. P. Eng. S., **37**, 223 (2012)

8. A.M. Zain, H. Haron, S.N. Qasem, S. Sharif, Appl. Math. Model., **36**, 1477 (2012)
9. S.N. Joshi, S.S. Pande, Appl. Soft. Comput., **11**, 2743 (2011)
10. K.M. Tsai, P.J. Wang, J. Mater. Process. Tech., **117**, 111 (2001)
11. S. Assarzadeh, M. Ghoreishi, Int. J. Adv. Manuf. Tech., **39**, 488 (2008)
12. G. Krishna Mohana Rao, G. Rangajanardhaa, D. Hanumantha Rao, M. Sreenivasa Rao, J. Mater. Process. Tech., **209**, 1512 (2009)
13. A. Mohamad, A.M. Zain, N.E.N. Bazin, A. Udin, J. Intell. Manuf., **26**, 1247 (2013)
14. A.F. Kamaruzaman, A.M. Zain, S.M. Yusuf, A. Udin, Appl. Mech. Mater., **421**, 496 (2013)
15. N. Zainal, A.M. Zain, N.H.M. Radzi, M.R. Othman, J. Intell. Manuf., **27**, 797 (2016)
16. N.F. Johari, A.M. Zain, N.H. Mustaffa, A. Udin, Appl. Mech. Mater., **815**, 268 (2015)
17. N. Yusup, A.M. Zain, S.Z.M Hashim, Expert. Syst. Appl., **39**, 9909 (2012)
18. D. Kanagarajan, R. Karthikeyan, K. Palanikumar, J.P. Davim, Int. J. Adv. Manuf. Technol., **36**, 1124 (2008)
19. K. Deb, A. Pratap, S. Agarwal, T. Meyarivan, IEEE Trans. Evol. Comput., **6**, 182 (2002)

Experimental study of surface roughness in Electric Discharge Machining (EDM) based on Grey Relational Analysis

Ashanira Mat Deris[1,*], *Azlan* Mohd Zain[1] , *Roselina* Sallehuddin[1] , and *Safian* Sharif[2]

[1]Soft Computing Research Group, Faculty of Computing, UTM, Johor, Malaysia
[2]Faculty of Mechanical Engineering, UTM, Johor, Malaysia

Abstract. Electric Discharge Machining (EDM) is one of the modern machining which is capable in handling hard and difficult-to-machine material. The successful of EDM basically depends on its performances such as surface roughness (Ra), material removal rate (MRR), electrode wear rate (EWR) and dimensional accuracy (DA). Ra is considered as the most important performance due to it role as a technological quality measurement for a product and also a factor that significantly affects the manufacturing process. This paper presents the experimental study of surface roughness in die sinking EDM using stainless steel SS316L with copper impregnated graphite electrode. The machining experimental is conducted based on the two levels full factorial design of design of experiment (DOE) with five machining parameters which are peak current, servo voltage, servo speed, pulse on time and pulse off time. The results were analyzed using grey relational analysis (GRA) and it was found that pulse on time and servo voltage give the most influence to the Ra value.

1 Introduction

Electric disharge machining (EDM) which is firstly introduced in manufacturing industries in late 1940s, is one of the earlier non-conventional machining processes. EDM is well known as one of the most widely used and successfully applied machining processes for the geometrically complex parts, hard materials and difficult-to-machine materials [1]. The successful of EDM depends on its performances such as surface roughness (Ra). Surface roughness is considered as one of the most important performances in machining process that refers to the product quality [2,3]. According to Keskin et al. [4], surface roughness is the most measured performance in EDM compared with other performances such as tool wear and material removal rate [5]. In most cases, in the industries, surface roughness is a technical requirement needed for mechanical products. Practically, for the machine parts, surface roughness plays an important role in fatigue strength, wear resistance, tensile, and ductility [6]. For achieving optimal solution of surface roughness, a proper selection of machining parameters is essential. Surface roughness value generally influenced by many factors such as machining parameters, work piece properties, cutting phenomena and type

* Corresponding author : ashanira.md@gmail.com

of cutting tools [7]. Several machining parameters that have been widely studied by researchers are pulse on time, pulse off time, duty factor, servo voltage and peak current. The value of surface roughness is controlled by sparks frequency during the machining process, which is the number of discharges per second. The higher energy applied leads to the higher amount of material removed which can be also causing a rougher surface finish.

Ramirez and Franco [8] reports the experimental study of Ra in EDM process using AISI 316 stainless steel with copper electrodes. Two machining parameters were considered which are depth of penetration and current intensity. The experiment results proved that both considered parameters gave the high influence on surface roughness value. Bose and Mahapatra [9] investigated surface roughness in die sinking EDM process on AISI H31 tool steel. Discharge gap current, spark gap pulse on time and pulse off time were evaluated and analyzed in order to find which parameters give the most influence to the surface roughness. The experiment design based on Taguchi method L27 orthogonal array. The result showed that pulse on time gives the most influence to Ra value which contributes 47.24%. Surface roughness decreased when pulse on time value increased.

Mir et al. [10] applied Response Surface Methodology (RSM) to analyze machining parameters of surface roughness in powder mixed EDM. The experiment is conducted based on central composite design (CCM) while RSM is used to develop, improve and optimized the experiments. Discharge current, pulse on time and concentration of Aluminum powder added into dielectric fluid were chosen as machining parameters with H11 steel as a work piece. The results showed that powder concentration and peak current are the most significant parameters to the surface roughness. Tang and Du [11] investigated surface roughness value in green EDM process which tap water is used as dielectric fluid. The experiment is conducted using Taguchi L9 orthogonal array with four machining parameters which are pulse on time, discharge current, gap voltage and lifting height. The work piece used is Ti-6Al-4V alloy. The results found that the optimal values of machining parameters are 11A discharge current, 30 V gap voltage, 3mm lifting height, 30% duty factor and 30 μs pulse duration. Surface roughness decreased from 2.37μm to 1.93μm with the improvement of 18.9% when tap water is used as a dielectric fluid.

Wired EDM is also one of the EDM types. Ekici et al. [12] evaluates the surface roughness in wire electric discharge machining (WEDM) of high density Al/B_4C metal matrix composites using Taguchi method. Five machining parameters which are pulse on time, wire speed, pulse off time, reinforcement percentage and wire tension were considered as the control factors. The experiment design was conducted based on Taguchi L18 orthogonal array. The results showed that pulse on time is the most significant factors that affects the surface roughness value. Ikram et al. [13] investigated WEDM machining parameters which are wire feed velocity, wire tension, pulse on time, dielectric pressure, pulse off time, open voltage and servo voltage, to the surface roughness value using Taguchi approach. The experiment was conducted based on Taguchi L18 orthogonal array and the experiment results were analyzed based on analysis of variance (ANOVA) and signal-to-noise (S/N) ratio analysis. The results showed that pulse on time gives the most effects to the surface roughness value.

The main objective of this study is to perform the machining experiment of EDM and to investigate the machining parameters that give the most effect to the surface roughness. The experiment was conducted on stainless steel 316L work piece with copper impregnated graphite electrode in die sinking EDM. The experiment result has been collected and analyzed using GRA in order to investigate the significant parameters to the surface roughness value.

2 Design of Experiments and Experiment Procedure

The proper selection of EDM machining parameters and experimental design are important in order to get successful machining process with desired output performance. In this study, five parameters were considered. They are peak current, pulse on time, servo voltage, pulse off time and servo speed. The design setting of the machining parameters which acts as input parameters need to be determined prior to the machining process. In this study, the experiment is conducted based on two levels full factorial design. The selection of the range value for each input factors are based on manual EDM handbook and also based on the previous researchers. Table 1 shows the range value of EDM input parameters.

Table 1. The range value of EDM parameters

Variables	Unit	Levels	
		1	2
Peak Current (IP)	[A]	5.7	10.5
Pulse on time (ON)	[μs]	100	200
Pulse of time (OFF)	[μs]	50	70
Servo Voltage (SV)	[V]	30	90
Servo Speed (S)	-	74	92

The experiment design consists of two level full factorial design. The input parameters; peak current, pulse on time, pulse off time, servo voltage and servo speed are classified into two levels, high and low which labels as '1' and '2'. The design indicates 32 number of experiments. The machining experiment is conducted using AG40L die sinking EDM process as shown in Fig. 1. The material used in this experiment is stainless steel SS316L and the electrode is copper impregnated graphite. The general setting of EDM setup is shown in Table 2.

Fig. 1. AG40L model of die sinking EDM.

Table 2. General EDM setup.

Variables	Setup
Work piece	Stainless Steel 316L (100mm x 50mm x 10mm)
Electrode	Copper impregnated graphite (8mm)
Depth of hole	2mm
Polarity	Positive
Flushing	On
Dielectric Fluid	Oil

3 Results and Analysis

This section discusses the experimental result and analysis of surface roughness value using Grey Relational Grade (GRA). Surface roughness values were measured using surface roughness tester and the result is analyzed based on GRA method. GRA which is firstly introduced by Professor Deng Julong in Grey sytem is one of the analysis technique [14]. The first step in GRA is data preprocessing. Normalization of the data is calculated based on the characteristic of the input parameters which is also known as sequence data. This study applies "lower the better" which is used to compare levels in GRA, the formula is given in Eq.1:

$$x^*_i(k) = \frac{max x_i^o(k) - x^*_i(k)}{max x_i^o(k) - \min x_i^o(k)} \tag{1}$$

Relationship between the ideal and normalized experimental results is express using grey relational coefficient (GRC) which is calculated using Eq.2:

$$\xi_i(k) = \frac{\Delta \min + \xi \cdot \Delta max}{\Delta_{0i}(k) + \xi \cdot \Delta max} \tag{2}$$

Where ξ is distinguishing or identification coefficient, ξ = [0, 1] and $\Delta_{0i}(k)$ is the deviation sequence of the reference sequence and the comparability sequence, which can be expressed in Eq.3:

$$\Delta_{0i}(k) = ||x^*_0(k) - x^*_i(k)||$$

$$\Delta \min = \min_{\forall j \in i} \min_{\forall k} ||x^*_0(k) - x^*_j(k)||$$

$$\Delta \max = \max_{\forall j \in i} \max_{\forall k} ||x^*_0(k) - x^*_j(k)|| \tag{3}$$

where $x^*_0(k)$ is reference sequence and $x^*_j(k)$ is comparability sequence. For the identification coefficient, the value of ξ = 0.5 is generally used [15]. After the GRC value is calculated, the grey relational grade (GRG) is needed to derive. Normally, the average of GRC is taken to calculate the GRG which is defined in Eq.4:

$$\gamma_i = \frac{1}{n}\sum_{k=1}^{n} \xi_i\ (k) \qquad (4)$$

However, the Eq.4 is modified since in a real application, the effect of each factor on the system is varies. The modified grey relational grade is given in Eq.5:

$$\gamma_i = \frac{1}{n}\sum_{k=1}^{n} w_k \cdot \xi_i\ (k) \sum_{k=1}^{n} w_k = 1 \qquad (5)$$

where w_k refers to the normalized weights. GRG is calculated to show the relationships between the sequences. The value of GRG will be 1 if the two sequences are identity. Practically, GRG value with less than 0.6 will considered as least influential factors to the reference sequence. Table 3 shows the experimental result and also GRG value for each parameters.

Based on Table 3, it can be seen that the most minimum surface roughness is when pulse time is 100µs, pulse off time is 50µs, peak current is 5.7A, servo voltage 30A and servo voltage 74µs. GRA identifies the significant parameters for surface roughness in EDM process based on the GRG value. With the GRG values of 0.6502 and 0.6254, which are the first and second highest, it can be concluded that pulse on time and servo voltage are most significant parameters to the surface roughness value in EDM process.

Table 3. Surface Roughness value of EDM.

Run	ON	OFF	IP	SV	S	Ra
1	100	50	5.7	30	74	1.3974
2	100	50	10.5	30	74	4.501
3	100	50	5.7	90	74	4.0317
4	100	50	10.5	90	74	1.886
5	100	50	5.7	30	92	1.4589
6	100	50	10.5	30	92	4.1478
7	100	50	5.7	90	92	1.5578
8	100	50	10.5	90	92	3.8676
9	200	50	5.7	30	74	1.7793
10	200	50	10.5	30	74	5.4328
11	200	50	5.7	90	74	2.474
12	200	50	10.5	90	74	2.5733
13	200	50	5.7	30	92	1.9979
14	200	50	10.5	30	92	2.8592
15	200	50	5.7	90	92	15.1741
16	200	50	10.5	90	92	2.1229
17	100	70	5.7	30	74	1.8791
18	100	70	10.5	30	74	2.3766
19	100	70	5.7	90	74	1.5366
20	100	70	10.5	90	74	4.0896
21	100	70	5.7	30	92	1.6486
22	100	70	10.5	30	92	5.5439
23	100	70	5.7	90	92	1.7243
24	100	70	10.5	90	92	2.9212
25	200	70	5.7	30	74	1.6429
26	200	70	10.5	30	74	2.706
27	200	70	5.7	90	74	17.7071
28	200	70	10.5	90	74	2.5712
29	200	70	5.7	30	92	11.4499
30	200	70	10.5	30	92	4.1105
31	200	70	5.7	90	92	1.4481
32	200	70	10.5	90	92	2.4101
GRG	**0.650208**	0.619907	0.602967	**0.625414**	0.615313	-

4 Conclusion

The machining experimental of stainless steel 316L and copper impregnated graphite is conducted in die sinking EDM process in order to investigate the significant parameters of surface roughness value. Two levels full factorial experiment design is considered in this study. Based on the analysis of GRA, it was found that pulse on time and servo voltage are the most significance to the surface roughness value compared with pulse off time, peak current and servo speed.

Special appreciation to reviewer(s) for the useful advice and comments. The authors greatly acknowledge the Research Management Centre, UTM for financial support through Research University Grant (GUP) Vot. No. Q.J130000.2528.11H72

References

1 P.M. George, B.K. Raghunath, L.M. Manocha, A.M. Warrier, J. Mater. Process. Tech., **145,** 66 (2004)
2 A.M. Zain, H. Haron, S. Sharif, Expert Syst. Appl., **37,** 4650 (2010)
3 A.M. Zain, H. Haron, S. Sharif, Expert Syst. Appl., **37,** 1755 (2010)
4 Y. Keskin, H.S. Halkaci, M. Kizil, Int. J. Adv. Manuf. Tech., **28,** 1118 (2006)
5 N. Yusup, A.M. Zain, S. Z. M. Hashim, Expert Syst. Appl., **39,** 9909 (2012)
6 M.Y. Wang, H.Y. Chang, Int J. Mach. Tools. Manuf., **44,** 51 (2004)
7 A.M. Zain, H. Haron, S. Sharif, Machining Sci. Technol., **14,** 43 (2010)
8 C.M. Ramirez, P. Franco, Procedia Eng, **132,** 679 (2015)
9 G.K. Bose, K.K. Mahapatra, Adv. Prod. Eng. Management, **9,**168 (2014)
10 M.J. Mir, K. Sheikh, B. Singh, N. Malhotra, Int. J. Eng. Sci. Technol., **4**(3), 45 (2012)
11 L. Tang, Y.T. Du, Int. J. Adv. Manuf. Technol., **70,** 469 (2014).
12 E. Ekici, A.R. Motorcu, A. Kus, J. Comp. Mater., **1** (2015)
13 A. Ikram, N.A. Mufti, M.Q. Saleem, A.R. Khan, J. Mech. Sci. Technol., **27** (7), 2133 (2013)
14 A.M. Deris, A.M. Zain, R. Sallehuddin, Meccanica, **48,** 1937 (2013)
15 R. Sallehuddin, S.M. Shamsuddin, S.Z.M. Hashim, Int. Conf. Comp. Model. Simul., **80** (2010)

Process Characterization of 32nm Semi Analytical Bilayer Graphene-based MOSFET

Z.A. Noor Faizah[1,*], *Ibrahim* Ahmad[1], *P.J.* Ker[1], and *P.S.* Menon[2]

[1]Centre for Micro and Nano Engineering (CeMNE) College of Engineering, Universiti Tenaga Nasional (UNITEN) 43009 Kajang, Selangor, Malaysia
[2]Institute of Microengineering and Nanoelectronics (IMEN) Universiti Kebangsaan Malaysia (UKM) 43600 Bangi, Selangor, Malaysia

Abstract. This paper presents an inclusive study and analysis of graphene-based MOSFET device at 32nm gate length. The analysis was based on top-gated structure which utilized Hafnium Dioxide (HfO_2) dielectrics and metal gate. The same conventional process flows of a transistor were applied except the deposition of bilayer graphene as a channel. The analytical expression of the channel potential includes all relevant physics of bilayer graphene and by assuming that this device displays an ideal ohmic contact and functioned at a ballistic transport. Based on the designed transistor, the on-state current (I_{ON}) for both GNMOS and GPMOS shows a promising performance where the value is 982.857uA/um and 99.501uA/um respectively. The devices also possess a very small leakage current (I_{OFF}) of 0.289578nA/um for GNMOS and 0.130034nA/um for GPMOS as compared to the conventional SiO_2/Poly-Si and high-k metal gate transistors. However, the devices suffer an inappropriate subthreshold swing (SS) and high value of drain induced barrier lowering (DIBL).

1 Introduction

The never ending process challenge of scaling down to cramming more transistor in a single chip which was coined by Gordon E. Moore has led to an invention of new device design and proposed materials. Problems such as transport degradation, the increasing of parasitic effect and aggravated on-off transition attributed from the miniaturization of the channel length. From conventional SiO_2/Poly-Si to high-k metal gate, advanced silicide and multi-gate transistors; all has been adopted to overcome the problems. Further ahead, the carbon-based material is proposed as an alternative to SiO_2 and now is rising in nano-electronic design due to low cost of manufacturing and its outstanding properties such as extremely high mobility together with a promising ability to be scaled down to a smaller gate length. The development of Carbon Nanotubes (CNT) transistor is first introduced but due to significant gap in nanoribbons which can only be obtained at widths close to 1-2 nm [1] makes the CNT prohibited for fabrication on the scale of integrated circuit. Graphene was then introduced theoretically and experimentally [1-5]. The outcome shows that

* Corresponding author: faizahz@uniten.edu.my

graphene demonstrated to be thermodynamically stable and the 2-D nature of graphene has been confirmed by experimental observation of the quantum hall effect.

In addition, the fact that graphene is a planar form draws a major advantage over CNT, which mostly permits for highly developed top-down CMOS-compatible process flow [6]. In this research, a 32nm gate length of NMOS and PMOS bilayer graphene-based transistor were virtually design and tested for its functionality where the performance parameters were set to be in line with the International Technology Roadmap of Semiconductor, ITRS 2011 prediction for which the threshold voltage (V_{TH}) is 0.103 ± 12.7% V, on-state current (I_{ON}) greater than 100nA/um and leakage current (I_{OFF}) lower than 150nA/um. The simulation tools of ATHENA and ATLAS in Silvaco TCAD Tools was utilized during the simulation process.

2 Transistor virtual fabrication for simulations

In this experiment, both GNMOS and GPMOS transistor are virtually fabricated and analyzed separately. This is to simplify the simulation process and comparison of the device performance. The fabrication processes for both devices were similar except the types the dopant use; which is differs at its substrate, and its density of dopant. A single crystal of silicon wafer (100) with a heavily doping concentration was initially prepared. It was then thermally oxidized to a SiO_2 layer to create a p- and n-wells for NMOS and PMOS transistor. Graphene is then atomically deposited onto the SiO_2 layer where the thickness and the number of vertical grid spacing are practically specified according to the established simulation and fabrication results [2, 6-9].

Next, a thickness of 0.67 nm high-k material, Hafnium Dioxide (HfO_2) is then deposited onto graphene layers followed by 38nm thickness of different metal gate for GNMOS and GPMOS transistor. The metal gate that is used for GNMOS and GPMOS transistor is Tungsten Silicide (WSi_2) and Titanium Silicide ($TiSi_2$) respectively. The high-k and metal gate materials are etched precisely to produce a gate length of 32 nm (± 0.1nm) transistor. Then, Halo and Source-Drain (S/D) structures using heavily doped n-type and p-type material are implant to accumulate the Schottky tunneling source and drain. The function is to provide a low resistive path for only electrons or holes to be inoculated into the graphene channel and thus achieved a unipolar conduction. Compensation implantation of Phosphor dopant at 10^{12} cm^{-3} took place right after the growth of 0.015um Borophosphosilicate Glass (BPSG). Aluminum layer is then deposited to form the metal contact of the transistor which also dictates that the process is accomplished and ready for performance measurement through its electrical characterization. The complete design of 32 nm planar NMOS bilayer graphene transistor is as shown in Fig. 1.

 DOI: 10.1051/matecconf/20167801016

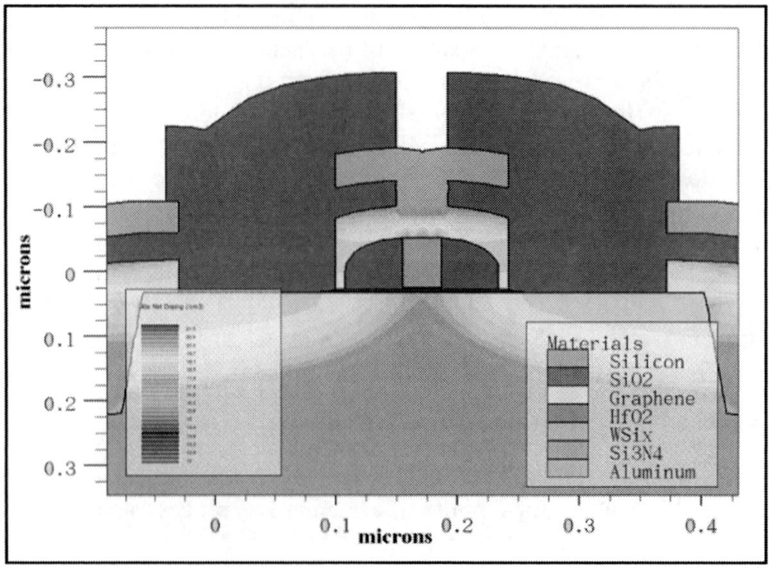

Fig. 1. Doping Profile of 32nm NMOS Bilayer Graphene Transistor

3 Physical characterization of graphene transistor

The complete design of virtual transistors was then ready for their electrical characterization using ATLAS simulator. At this stage, the performance of the transistors was measured in terms of threshold voltage (V_{TH}), leakage current (I_{OFF}), and on-state current (I_{ON}) which was specified in International Technology Roadmap of Semiconductor (ITRS) 2011. In this research, bilayer graphene was modeled as a semi-metal with a bandgap of approximately 0.55eV [3], permittivity of 2.4 [6], charge carrier mobility of top-gated graphene measured at room temperature and effective field of Eeff=0.4MV/cm [6]. The radiative recombination rate of electron and holes in graphene is set at 100ns [9]. The graphene's 2-Dimensional (2D) electron and hole densities of states are measured at 300K and obtained from equations below [9]:

$$Nc = \frac{8\pi m_e kT}{h^2} \ln\left(1 + e^{-(Ec-Ef)/kT}\right)$$

$$Nv = \frac{8\pi m_h kT}{h^2} \ln\left(1 + e^{-(Ef-Ev)/kT}\right) \quad (1)$$

where the effective masses of electrons and holes in graphene in this research was set at me $\approx$0.06 mo and mh$\approx$0.03mo while mo is the free electron mass according to paper in [8]. This is due to the transport properties of graphene attributed to a single spatially quantized sub band which is populated by electrons and holes at that value. The number of electrons and holes per unit area are experimentally measured to be ~10^{12}cm^{-2} at room temperature which also satisfies (1) [9].

4 Results and analysis

The electrical characteristic curves of a graphene transistor with a heavily doped n- and p-type Silicon S/D were shown in Fig. 2 and Fig. 3. Fig. 2 shows the I_{DS}-V_{GT} characteristics of GNMOS at different V_{DS} and Fig. 3 shows the I_{DS}-V_{GT} characteristics of GPMOS at different V_{DS}. The device's performance is measured at fixed threshold voltage, V_{TH} = 0.103 V ± 12.7% as guided by ITRS 2011 by variating the dopant density and annealing temperature. From, ITRS 2011 prediction, the on-state current (I_{ON}) is aimed to be greater than 100nA/um and leakage current (I_{OFF}) lower than 150nA/um.

In this experiment, the I_{ON} of graphene-based transistor was closer to the ITRS 2011 prediction and slightly higher than conventional SiO_2/Poly-Si [10] and high-k metal gate [11, 12] transistors due to low effective mass and less energy required to go through the outlawed gap between source and drain [3]. Leakage current (I_{OFF}) of graphene-based transistor possess a good value not only lower than the conventional SiO_2/Poly-Si and high-k metal gate transistor but so much lower than the ITRS 2011 prediction. The high value of I_{ON} and lower value of I_{OFF} will boost up the switching capability of a transistor which was proven in this simulation, $I_{ON}/I_{OFF} \approx 10^6$. However, due to the minority carrier (hole) concentration in graphene was very high, this device suffers a large subthreshold swing (SS) and drain-induced barrier lowering (DIBL) [9]. The performance results for both GNMOS and GPMOS were summarized in Table 1.

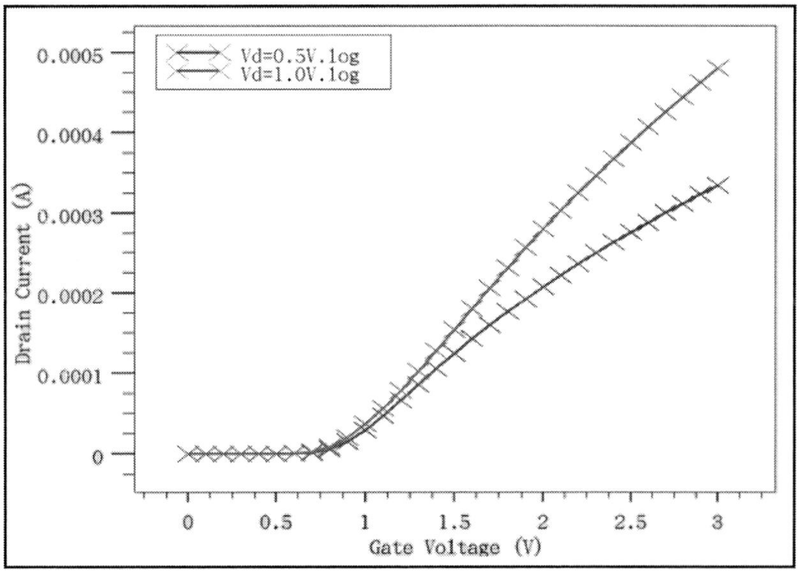

Fig. 2. I_{DS}-V_{GT} of GNMOS at different V_{DS}

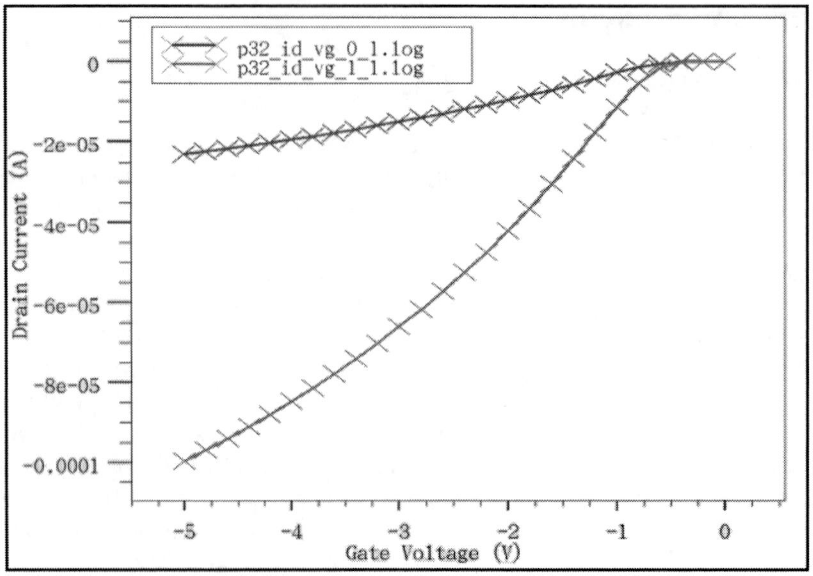

Fig. 3. I_{DS}-V_{GT} of GPMOS at different V_{DS}

Table 1. Performance Analysis of 32nm Graphene-based Transistor

Parameter	GNMOS	GPMOS
High-K/Metal Gate	HfO_2/WSi_2	$HfO_2/TiSi_2$
Threshold Voltage, V_{TH} (V)	0.103444	0.103184
On-state current, I_{ON} (uA/um)	982.857	99.501
Leakage current, I_{OFF} (nA/um)	0.289578	0.130034
Sub-threshold Swing, SS (mV/dec)	94.38	119.92
Drain-Induce Barrier Lowering, DIBL (mV/V)	134.6	30.112
On-Off ratio (I_{ON}/I_{OFF})	3.394×10^6	0.765×10^6

5 Conclusion

The performance of 32nm gate length of planar graphene-based transistor were analyzed in the paper. The utilization of highly-doped Silicon S/D which forms a Schottky tunneling junction resulted in high I_{ON} while the application of top-gated high-k metal gate material modulates the drain current and limits the carrier mobility. The results show a very high I_{ON} with such a low value of I_{OFF}. This confirms that the utilization of graphene layer is possible to be implemented with an obligatory to enhance the device performance by bandgap tuning method.

This project is sponsored by Ministry of Higher Education (MOHE) under grant project 20140123FRGS.

References

1. Martina Cheli, Gianluca Fiori, Giuseppe Iannaccone, IEEE T. Electron Dev., **56(12)**, 2979 (2009)
2. J. Clerk Maxwell, *A Treatise on Electricity and Magnetism. 3rd Edition* (Oxford, Clarendon, 1892)
3. P. Nemes Incze, Z. Osvath, K. Kamaras, L.P. Biro. Carbon **46(11)**, 1435 (2008)
4. A. Hamzah, A.S. Azman, R. Ismail, Z. Johari, *IEEE Regional Symposium*, 1 (2015)
5. K. Tamersit, F. Djeffal, D. Arar, M. Meguellati, *World Congress on Engineering*, **2**, (2013)
6. B. Guo, L. Fang, B. Zhang, J.R. Gong, Insciences J., **1(2)**, 80 (2011)
7. M.C. Lemme, T.J. Echtermeyer, M. Baus, H. Kurz, IEEE Electron Device Lett., **28(4)**, 282 (2007)
8. A.C. Ferrari, J.C. Meyer, V. Scardaci, C. Casiraghi, M. Lazzeri, F. Mauri, S. Piscanec, D. Jiang, K.S. Novoselov, S. Roth, A.K. Geim, Phys Rev Lett, **97(18)**, 187401-1 (2006)
9. K.S. Novoselov, A.K. Geim, S.V. Morozov, D. Jiang, Y. Zhang, S.V. Dubonos, I.V. Grigorieva, A.A. Firsov, Science, **306(5696)**, 666 (2004)
10. J. Zhu, J.C. Woo, *37th European Solid State Device Research Conference*, (IEEE, 2007)
11. Husam Ahmed Elgomati, *Optimization of Process Parameters for 32nm CMOS Device*, PhD Thesis, (Universiti Kebangsaan Malaysia 2012)
12. Noor Faizah Zainul Abidin, Ibrahim Ahmad, A.H. Afifah Maheran, *Proceeding of the 2nd National Graduate Conference, Universiti Tenaga Nasional, Malaysia*, 62 (2014)
13. A.A. Maheran, Z.N. Faizah, P.S. Menon, I. Ahmad, P.R. Apte, T Kalaivani, F. Salehuddin, *IEEE International Conference*, 232 (2014)

MATEC Web of Conferences 78, 01017 (2016) DOI: 10.1051/matecconf/20167801017
IConGDM 2016

Process Parameters Optimization of 14nm MOSFET Using 2-D Analytical Modelling

Z.A. Noor Faizah[1,*], *Ibrahim* Ahmad[1], *P.J.* Ker[1], *Y.* Siti Munirah[1], *R.* Mohd Firdaus [1], *E.* Md. Fazle[1], and *P. S.* Menon[2]

[1]Centre for Micro and Nano Engineering (CeMNE) College of Engineering, Universiti Tenaga Nasional (UNITEN) 43009 Kajang, Selangor, Malaysia.
[2]Institute of Microengineering and Nanoelectronics (IMEN) Universiti Kebangsaan Malaysia (UKM) 43600 Bangi, Selangor, Malaysia.

Abstract. This paper presents the modeling and optimization of 14nm gate length CMOS transistor which is down-scaled from previous 32nm gate length. High-k metal gate material was used in this research utilizing Hafnium Dioxide (HfO_2) as dielectric and Tungsten Silicide (WSi_2) and Titanium Silicide ($TiSi_2$) as a metal gate for NMOS and PMOS respectively. The devices are fabricated virtually using ATHENA module and characterized its performance evaluation via ATLAS module; both in Virtual Wafer Fabrication (VWF) of Silvaco TCAD Tools. The devices were then optimized through a process parameters variability using L9 Taguchi Method. There were four process parameter with two noise factor of different values were used to analyze the factor effect. The results show that the optimal value for both transistors are well within ITRS 2013 prediction where V_{TH} and I_{OFF} are 0.236737V and 6.995705nA/um for NMOS device and 0.248635 V and 5.26nA/um for PMOS device respectively.

1 Introduction

The gate length of MOSFET has continually been scaled down through last few decades to create smaller and smaller device in order to fabricate high density chips. Reduced power dissipation, faster switching and low cost were the main concern for this scaling. As MOSFET is scaled down it provides various kind of technical challenges and this become worsen when it comes to the nano-meter dimensions. This is because of the problems such as Short Channel Effects (SCE), increased leakage current, and lack of pinch off [1]. The utilization of SiO_2 to overcome the problems is no more effective today. Researchers has come up with a new solution to overcome the scaling impact by replacing the gate dielectric from SiO_2 to high permittivity materials and innovations to new device structure such as Finfet and double-gate. In this research, the virtual fabrication was simulated through ATHENA module and its electrical characteristic was simulated via ATLAS module where both modules can be found in Silvaco TCAD Tools. These modules are important in designing and optimizing the process parameter [2] of a semiconductor device. As part of

* Corresponding author: faizahz@uniten.edu.my

optimization process, Taguchi Method were used to observe the variance and mean effect of process parameter. With added noise factor called signal-to-noise ratio (SNR), this method becomes more reliable. The aim of this research is to virtually design and optimize a 14nm high-k metal gate CMOS transistor with threshold voltage, V_{TH} of 0.2301V and leakage current, I_{OFF} lower than 100nA/um. The values are set by ITRS 2013 specification.

2 Experiment description

2.1 Virtual transistor simulation recipes.

For simplicity, the transistors were designed separately in ATHENA module. The summarized simulation recipes were shown in Table 1. The complete virtual design of 14nm NMOS transistor is as shown in Fig. 1. The same structure was used to design the PMOS transistor except the utilization of metal gate which is Titanium Silicide ($TiSi_2$).

Table 1. Simulation Recipe

Process step	Parameters
Silicon substrate	• 7×10^{14} cm^{-3} Boron • <100> orientation
Retrograde well implantation	• 200Å oxide screen by 970°C, 20 min of dry oxygen • 4.55×10^{11} cm^{-3} Phospor • 30 min, 900°C diffused in nitrogen
STI isolation	• 130Å stress buffer by 900°C, 25min of dry oxygen • 1500Å Si_3N_4, applying LPCVD • 1.0 um photoresist deposition • 15 min annealing at,900 °C
Gate oxide	Diffused dry oxygen for 0.1 min, 815 °C
Vt adjust implant	• 1.8×10^{11} cm^{-3} Boron difluoride • 5KeV implant energy, 7° tilt • 20 min annealing at, 795 °C
High-k metal gate deposition	• 0.002 um HfO_2 • 0.1 um WSi_2 (NMOS) • 0.1 um $TiSi_2$ (PMOS) • 17 min, 900 °C annealing
LDD implantation	• 7.553×10^{13} cm^{-3} Phosphor (NMOS) • 5.05×10^{13} cm^{-3} Boron (PMOS) • 20° tilt
Sidewall spacer	• 0.047 um Si_3N_4
S/D implantation	• 0.99×10^{14} cm^{-3} Arsenic (NMOS) • 9.20×10^{12} cm^{-3} Boron (PMOS) • 10KeV implant energy, 7° tilt
PMD deposition	• 0.3 um BPSG • 25 min, 850 °C annealing
Metal 1	• 0.04 um Aluminium
IMD deposition	• 0.005 um BPSG • 15 min, 950 °C annealing
Metal 2	• 0.12 um Aluminium

DOI: 10.1051/matecconf/20167801017

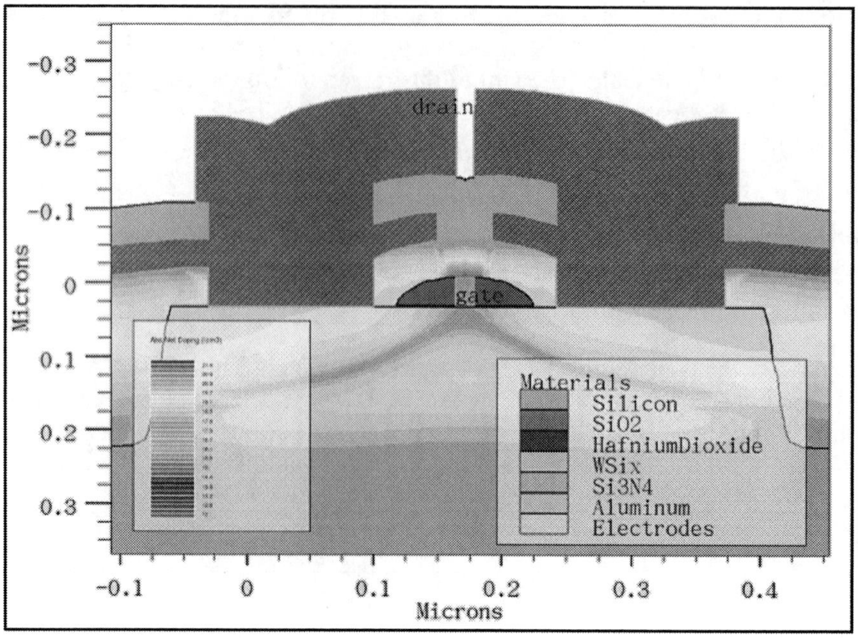

Fig. 1. Doping Profile of 14nm NMOS Transistor

2.2 Taguchi L9 orthogonal array method.

The optimization of the transistors was done using Taguchi Method by variation of the process parameters. The parameters that has been selected in this research includes Halo Implantation Dose (A), Halo Tilting Angle (B), Source Drain Implantation (C), and Compensation Implantation (D). Noise factors on the other hands includes Sacrificial Oxide Layer Temperature, BPSG Temperature and V_T Adjust Implant Temperature. Table 2 shows the value of each parameter at different level and Table 3 shows their noise factors for both type of device.

Table 2. Control Factors and Their Levels

Device	Symbol	Unit	Level 1	Level 2	Level 3
NMOS	A	atom/cm^3	7.5530×10^{13}	7.5535×10^{13}	7.5540×10^{13}
	B	°	31	32	33
	C	atom/cm^3	0.99×10^{14}	1×10^{14}	1.1×10^{14}
	D	atom/cm^3	$0.675 10^{14}$	0.68×10^{14}	0.685×10^{14}
PMOS	A	atom/cm^3	5.05×10^{12}	5.10×10^{12}	5.15×10^{12}
	B	°	19.7	19.8	19.9
	C	atom/cm^3	9.00×10^{12}	9.10×10^{12}	9.20×10^{12}
	D	atom/cm^3	0.4×10^{11}	1.1×10^{11}	1.8×10^{11}

Table 3. Noise Factors and Their Levels

Device	Symbol	Noise Factor	Unit	Level 1	Level 2
NMOS	X	Sacrificial Oxide Layer Temperature	°C	900	910
	Y	BPSG Temperature	°C	850	852
PMOS	X	Sacrificial Oxide Layer Temperature	°C	900	910
	Y	Vt Adjust Implant Temperature	°C	795	800

3 Results and analysis

Taguchi method was applied in this research to optimize the threshold voltage (V_{TH}) and leakage current (I_{OFF}). By following the L9 Orthogonal Array template, all four control factors are varied accordingly using the dopant values for which produced the output near to the ITRS 2013 prediction. In this experiment V_{TH} belongs to nominal-the-best quality characteristic while I_{OFF} belongs to lower-the-best quality characteristic. The initial results of V_{TH} and I_{OFF} after optimization are described in Table 4 and Table 5 for NMOS transistor and PMOS transistor respectively.

Table 4. V_{TH} and I_{OFF} Results for NMOS

Exp No.	Threshold Voltage, V_{TH} (V)				Leakage Current, I_{OFF} (nA/um)			
	X1, Y1	X1, Y2	X2, Y1	X2, Y2	X1, Y1	X1, Y2	X2, Y1	X2, Y2
1	0.493733	0.485195	0.493934	0.485400	5.104570	5.650100	5.091970	5.636180
2	0.242297	0.230865	0.242553	0.231122	70.55158	78.12330	70.35110	77.94000
3	0.430243	0.438392	0.430030	0.438201	1599.700	16709.90	1597.930	1669.170
4	0.248580	0.239892	0.248834	0.240148	66.82820	72.31010	66.67090	72.14010
5	-0.169930	-0.169712	-0.169712	-0.178187	2029.080	2181.740	2025.230	2177.640
6	0.165606	0.156053	0.165873	0.156322	138.1950	149.9200	137.8750	149.5730
7	0.163056	0.171313	0.162839	0.171095	1912.970	2054.570	1909.310	2050.680
8	0.181527	0.172296	0.181792	0.172562	120.8040	131.0810	120.5230	130.5060
9	0.228356	0.219083	0.228613	0.219342	79.66340	86.54280	79.47660	86.34030

Table 5. V_{TH} and I_{OFF} Results for PMOS

Exp No.	Threshold Voltage, V_{TH} (V)				Leakage Current, I_{OFF} (nA/um)			
	X1, Y1	X1, Y2	X2, Y1	X2, Y2	X1, Y1	X1, Y2	X2, Y1	X2, Y2
1	0.24358	0.243365	0.24315	0.242937	52.217	52.422	52.6363	52.8428
2	0.237851	0.237633	0.237386	0.237168	58.3183	58.5485	58.8322	59.0647
3	0.231531	0.231313	0.23103	0.230812	65.9534	66.2154	66.5826	66.8475
4	0.250575	0.250366	0.250189	0.249976	45.2084	45.3838	45.5318	45.7087
5	0.233934	0.233719	0.233523	0.23331	65.0706	65.324	65.5665	65.8223
6	0.262981	0.26277	0.262545	0.262332	35.6784	35.818	35.9749	36.1159
7	0.244912	0.2447	0.244515	0.244302	52.3622	52.5649	52.7469	52.9514
8	0.274721	0.274513	0.274348	0.274136	52.217	52.422	52.6363	52.8428
9	0.258453	0.258243	0.258061	0.257852	58.3183	58.5485	58.8322	59.0647

The next step is to determine the dominant and the adjustment factor for both transistors. The results were summarized in Table 6 for both V_{TH} and I_{OFF}. For NMOS device, it can be seen that Halo Implantation (**A**) shows the highest influence on V_{TH} (40%) and hence it was set to be the dominant factor. Halo tilting angle (**B**) on the other hand displays the lowest impact of 13% on variance but score the highest on Mean (29%). Thus, parameter **B** was set to be the adjustment factor. Parameter **B** was then varied between 31° to 33° to get the V_{TH} value near to ITRS 2013 prediction. For I_{OFF}, S/D Implantation (**C**) gives the highest impact (79%) to the device and thus it was set as a dominant factor.

The S/N response for PMOS device shows parameter **A** as the dominant factor because of the highest influence on V_{TH} (52%) and parameter **B** as the adjustment factor due to highest score on mean (56.23%) and lowest score on variance (32%). Parameter **C** and **D** was neglected and considered null as the percentage scores below 10%. It is because any changes made to parameter **C** and **D** will not influences the device performance. Parameter **B** was swept between 19.7° to 19.9° to get the V_{TH} within the ITRS 2013 prediction. The same response for I_{OFF} as parameter **A** possess the highest value on Mean (46%) and thus was set to be the dominant factor.

Table 6. S/N Response and Annova

Device	Process Parameter	V_{TH}					I_{OFF}			
		S/N Ratio (dB)			Factor Effect (%)		S/N Ratio (dB)			Factor Effect (%)
		L1	L2	L3	Variance	Mean	L1	L2	L3	Variance
NMOS	A	29.9	21.6	29.7	40	46.00	137.6	131.2	131.2	1
	B	24.6	30.0	26.6	13	29.00	140.9	131.4	124.7	11
	C	29.6	29.4	22.1	32	18.00	146.7	142.5	107.8	79
	D	30.4	25.0	25.7	15	7.00	140.2	131.2	125.5	9
PMOS	A	56.7	57.6	58.1	52	56.23	56.7	57.6	58.1	46
	B	58.0	57.5	57.0	32	40.56	58.0	57.5	57.0	14
	C	57.3	57.5	57.7	4	3.01	57.3	57.5	57.7	38
	D	57.1	57.5	57.8	12	0.19	57.1	57.5	57.8	2

The best setting parameters for both devices based on the Taguchi analysis are shown in Table 7. The best setting of parameter values was simulated again at different noise factor to verify the accuracy of Taguchi analysis. The result of the confirmation experiment based on different noise factors are shown in Table 8. Final results were then compared to ITRS 2013 prediction for conclusion of the optimization process experiment which is shown in Table 9.

Table 7. S/N Response and Annova

Device	Process Parameter	Unit	Best Value	
			V_{TH}	I_{OFF}
NMOS	A	atom/cm^3	7.5530x10^{13}	7.5530x10^{13}
	B	°	32	31
	C	atom/cm^3	0.99x10^{14}	0.99x10^{14}
	D	atom/cm^3	0.675x10^{14}	0.675x10^{14}
PMOS	A	atom/cm^3	5.15 x 10^{12}	5.15 x 10^{12}
	B	°	19.7	19.7
	C	atom/cm^3	9.20 x 10^{12}	9.20 x 10^{12}
	D	atom/cm^3	1.8 x 10^{11}	1.1 x 10^{11}

Table 8. Results of Confirmation Experiment

Device	Performance Parameter	X1, Y1	X1, Y1	X1, Y1	X1, Y1	Average Value
NMOS	V_{TH} (V)	0.242397	0.230766	0.242653	0.231132	0.236737
	I_{OFF} (nA/um)	6.704570	7.150100	6.891970	7.236180	6.995705
PMOS	V_{TH} (V)	0.24892	0.24873	0.24856	0.24833	0.24864
	I_{OFF} (nA/um)	5.23622	5.25649	5.27469	5.29514	5.26564

Table 9. ITRS Prediction vs. Optimization Results

Device	Performance Parameter	ITRS Prediction	Optimization Results
NMOS	V_{TH} (V)	0.23	0.236737
	I_{OFF} (nA/um)	<100	6.995705
PMOS	V_{TH} (V)	0.230	0.248635
	I_{OFF} (nA/um)	<100	5.258260

4 Conclusion

The modeling and optimization of 14nm gate length high-k metal gate CMOS transistor was successfully presented. The optimization of threshold voltage (V_{TH}) and leakage current (I_{OFF}) through Taguchi analysis was also fully utilized together with Silvaco TCAD Tools for virtual design simulations. The results were reported to be well within ITRS 2013 prediction.

We would like to thank Universiti Tenaga Nasional (UNITEN) and the Malaysian Ministry of Education for sponsoring the publication of this work using grant 20140123FRGS.

References

1. F. Matsuoka, H. Iwai, H. Hayashida, K. Hama, Y. Toyoshima, K. Maeguchi, *IEEE Trans. Electron Devices*, **37(6)**, 1487 (1990)
2. K. Goel Ashok, M. Merry, K. Arkenberg, E. Therkildsen, E. Chiaburu, W. Standfest, Silvaco International, (1995)
3. Z.N. Faizah, I. Ahmad, P.J. Ker, P.A. Roslan, A.A. Maheran. *IEEE Regional Symposium*, 1 (2015)
4. H.A. Elgomati, B.Y. Majlis, A.A. Hamid, P.M. Susthitha, I. Ahmad, *Sixth Asia Modelling Symposium*, 40 (2012)
5. A.A. Maheran, P.S. Menon, I. Ahmad, H.A. Elgomati, B.Y. Majlis, F. Salehuddin, 10th *IEEE International Conference*, 173 (2012)
6. I. Ahmad, Y.K. Ho, B.Y. Majlis, Quantum Electronics & Optoelectronics, **9(2)**, 40 (2006)

Integrated Index in Consideration of Appropriate Plastic Recycling System in Waste Bank Operation

Noorhan Firdaus Pambudi[1,2,*], *Kiyoshi* Dowaki[1], and *Akbar* Adhiutama[2]

[1]Tokyo University of Science, 2641 Yamazaki, Noda-shi, Chiba-ken 278-8510, Japan
[2]Institut Teknologi Bandung, Jalan Ganesha No. 10 Bandung 40132, Indonesia

Abstract. Several appropriate technology had been developed to maintain plastic waste in society according to minimize environmental impact. Landfill is no longer appropriate to maintain plastic waste based on the environmental impact that might be occurred for instance. However in developing countries such as Indonesia, although plastic recycling technology have been promoted by maintain waste bank policy for support community willingness to exchange their recyclable waste with certain monetary values, there is no guarantee that community will fully accept plastic recycling technology. This research aims to assess the performance of plastic recycling in environmental and social aspects as its integrated index. From that assessment, appropriate strategies in plastic recycling will be delivered in this research. Environmental aspects will be assessed by using life cycle assessment (LCA) through MiLCA software and selected by using data envelopment analysis (DEA). Social aspects will be analyzed by using qualitative and quantitative methodology such as observation, interview, secondary data, and questionnaires. Simulation and modelling will also developed by using agent-based modelling (ABM) to describe social dynamic of community in supporting waste bank policy. The appropriate system of plastic recycling will be promoted as expected results for this research.

1. Research Background

Increasing the number of population in Indonesia, the serious problem of amount of waste generated in the community has been caused. The government of Indonesia has proposed several policies to solve this problem. For instance, one strategy is the promotion of waste bank. In the waste bank, the waste transaction is executed then the provider can obtain the monetary value as mentioned in "Peraturan Menteri Lingkungan Hidup No. 13" [1]. This system needs participation from community to sustain in the household waste management system. Also, the success or failure of the waste bank operation would be affected by the different preferences of residential people.

Waste bank is one policy for maintaining appropriate technology which represented by plastic recycling technology and community empowerment which represented by

* Corresponding author: noorhan.firdaus@sbm-itb.ac.id

community willingness to exchange their waste to waste bank. There are two aspects in integrated index in this research, social aspects and environmental aspects. Social aspects lie on community acceptance and participation in supporting plastic recycling. Environmental aspects lie on technological appropriateness in perform better impact to environment for a plastic waste treatment. This study aims to develop and measure both of indexes in one integrated index to find appropriateness of the policy in waste bank in terms to develop plastic waste recycling technologies.

For plastic waste treatment, this research develops several scenarios or strategies to maintain plastic waste in Bandung including current treatment of plastic waste which uses landfill. Proposed scenario in this research is promoting plastic waste recycling in community with several modifications for each strategy. There are 5 scenarios proposed in this research,

- Scenario 1 : all plastic waste will be sent to landfill
- Scenario 2 : 50% of plastic waste will be sent to landfill, the rest will be incinerated with energy recovery
- Scenario 3 : all plastic waste will be incinerated with energy recovery
- Scenario 4 : all plastic waste will be recycled, scrap of remanufacturing process will be sent to landfill
- Scenario 5 : all plastic waste will be recycled, scrap of remanufacturing process will be incinerated with energy recovery

Therefore, this study will answer following research questions:

RQ1 How is currently plastic waste management performed in Bandung?
RQ2 What policy should be necessary taken by government to support plastic recycling in plastic waste management system?
RQ3 What kind of appropriate system or technology should be implemented to support plastic recycling system based on its opportunities and challenges?

2. Research Boundaries

The study will be focused on plastic waste management system in Bandung (Figure 1). The study also reach each value chain of plastic waste which is started from plastic waste generated by community in household, plastic waste collected by local government within or without waste bank, and up until plastic waste remanufacturing process. It also focused on household waste management system specifically in plastic waste. Since waste in Bandung categorized into organic and non-organic waste, so this study will covered non-organic waste and its stakeholder who connected with it. Stakeholders who will be considered lie from local government, community, and industry.

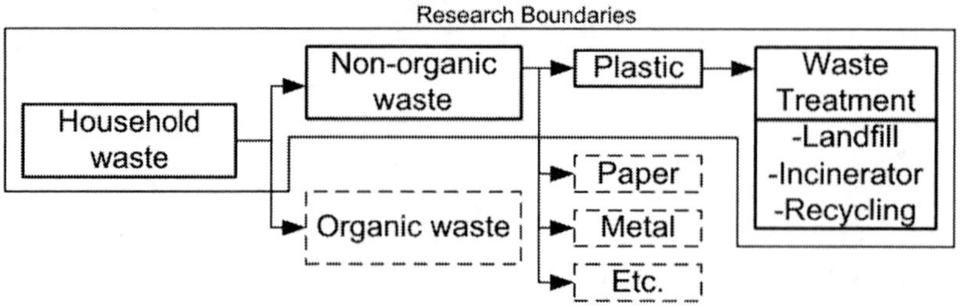

Fig. 1. Research Boundaries.

3. Research Methodology

All scenarios will be assessed by using LCA through MILCA software. The most effective scenario will be selected by using DEA. This measurement is applied for maintaining environmental performance of each proposed strategy. Decision making will also consider social performance of each strategy by analyzing community empowerment using qualitative methodology through observation or interview and quantitative methodology through questionnaire. Observation will be conducted in waste collection process from waste bank administrators to analyze community empowerment in supporting waste bank operation. Interview will be targeted to community and several cleaning service officer to develop needs in community to support plastic recycling in Bandung.

4. State of the Art and Research Gap

4.1 Life cycle assessment of plastic waste management system

By using LCA modelling, previous study conducted by Manfredi and Christensen [2] compared the environmental performance of six landfill technologies which conventional landfill had greater impact if there is no innovation to reduce environmental impact. The study from Yay [3] conclude that landfill and incineration have been confirmed as the worst waste final disposal alternatives, composting and material recovery showed better performance. Thus, landfill and incinerator without any combination of another waste treatment technology could give a bad impact to environment. Other previous study conducted by Fikri et al. [4] took an object in Indonesia especially in household hazardous waste management in Semarang had been conducted. The result showed that recycling HHW is the best scenario based on the LCA approach. Another study conducted by Arena et al. [5] defined through the analytical comparison between the six scenarios showed that the recycling option always environmentally preferable. Another study from Perugini et al. [6] confirmed that recycling scenarios are always preferable to those of non-recycling. Thus, recycling is one appropriate technology which could be considered to avoid greater impact to environment.

Previous study conducted by Matsuda et al. [7] was also focused on food waste which was resulted that prevention of food losses enhanced by separate collection lead to a significant reduction in GHG emissions. Another study conducted by Scortar and Suciu [8] covered that recycling rate is increasing in several countries especially in well-developed country such as OECD countries. Recycling idea is becoming one strategy that implemented in several countries to avoid greater impact in environment.

4.2 Social simulation in environmental aspect of community

Throughout the previous study conducted by Shi et al. [9] concluded that even when the participation rates increased, bottleneck facilities were found to be one of the primary obstacles when increasing recycling rates in several countries. Therefore, for those cases, strategic planning of capital projects, including both the bottleneck facilities' expansion and new facilities' construction is recommended in those studies. Without any improvement in infrastructure which was supported recycling system, the strategy would not sustain in the long terms. Another latest paper conducted by Suvageau and Frayret [10] proposed an agent-based simulation model to study and analyze the performance of various procurement and production policies in the recycled paper industry. Social simulation become one methodology to see the behaviour occurred in community related to support environmental friendly technologies or strategies. Previous study conducted by Liu and Ye [11] was also explored the evolvement of firm's environmental behaviour and influencing factors. There are several type of behaviour should be defined to conduct social simulation in related to environmental behaviour in our community.

4.3 Waste bank and plastic waste recycling system

Previous study conducted by Thanh et al. [12] about potential plastic recycling of household solid waste in Vietnam which household income and household size become existing correlation to plastic waste generation in detailed composition. Thus study concluded that household habits and behaviours of plastic waste discharge and the aspect of environmental impacts and resource consumption for plastic waste disposal alternatives were also evaluated. From other study conducted by Scortar and Suciu [8] said that the current waste management policies highlighted the importance of an integrated approach to waste management, which includes construction of facilities for waste disposal, together with measures to prevent waste generation and recovery measures, in accordance with the waste management hierarchy: waste prevention, quantity minimization, waste recovery and safe ultimate storage where there is no possibility of recovery. Another study conducted by Barterlings and Sterner [13] found that economic incentives, although important, are not the only driving force behind the observed reduction in municipal waste. Given the proper infrastructure that facilitates recycling, people are willing to invest more time than can be motivated purely by savings on their waste management bill. Another important result was that perceptions about the difficulties related to recycling and composting are important determinants of the effort people are prepared to make. Economic value of plastic waste and community occurred to evaluate plastic recycling technologies or systems. Waste bank strategies in this study conclude that aspect to evaluate the current system of plastic recycling.

5. Data Collection

5.1 Plastic waste volume

Secondary data which support to describe current situation of plastic waste recycling system will be gathered. This data will support the process of Life Cycle Assessment and Data Envelopment Analysis to give current performance in environmental aspect of plastic waste recycling system. The data also collected by using questionnaire to describe daily consumption of plastic waste in community. Table 1 show the specific percentage of waste type from 1981 until 2001 which have been collected by Deputi Bidang Analisa Sistem BPPT [14]. The data show that organic waste was dominated the number of waste generated by community and plastic waste percentage increasing year by year.

Table 1. Percentage of Waste 1981-2001.

Component	Unit	Year									
		1981	1983	1985	1989	1997	1998	1999	2000	2001	2002
Organic	%	79.49	77.90	73.97	79.37	74.60	75.38	75.18	74.99	74.60	74.22
Paper	%	7.97	6.70	8.28	8.57	10.18	10.50	10.71	10.93	11.15	11.37
Wood	%	3.65	2.97	3.94	0.75	0.98	0.39	0.20	0.02	0.02	0.02
Textile	%	2.40	1.98	3.05	0.79	1.57	1.20	1.13	1.06	1.00	0.93
Rubber	%	0.47	0.94	0.52	0.33	0.55	0.41	0.39	0.37	0.35	0.33
Plastics	%	3.67	5.13	5.64	6.51	7.88	8.11	8.30	8.50	8.69	8.88
Metal	%	1.37	1.93	2.04	1.45	2.04	1.89	1.89	1.90	1.90	1.90
Glass	%	0.50	0.65	1.55	1.57	1.75	1.93	1.99	2.05	2.10	2.16
Battery	%	0.48	1.80	0.97	0.48	0.29	0.01	0.01	0.01	0.01	0.01
Others	%				0.18	0.18	0.18	0.18	0.18	0.18	0.18
Total	%	100	100	100	100	100	100	100	100	100	100

Table 2 shows the specific proportion of plastic waste type in Surabaya and Jakarta from report conducted by Deputi Bidang Analisa Sistem BPPT [15]. HDPE, LDPE, and PP are the most contributed plastic waste which generated in both cities. Lesson learned from Surabaya showed that there is a decreasing number of waste tonnage sent to landfill according to an increasing number of waste bank developed. The data from previous research conducted by Wijayanti and Suryani [16] showed that the reduce of waste tonnage has been achieved, up to 7,14 tons per week. In 2005, the number of population was 2.740.490, and waste sent to landfill is 1.819 m^3. This amount had been decreased until 2011. In accordance, the number of waste bank in Surabaya increased from 2010 to 2012 by more than 50% and 30% from 2012 to 2013. All data were concluded that waste bank improvement will impact to decreasing number of waste which will be sent to landfill. Improvement of developing new waste bank and improvement of performance of waste bank is necessary to be conducted in terms to avoid bad impact of landfill as an option of waste management system.

Table 2. Proportion of Plastic Waste in Surabaya and Jakarta.

Component	Surabaya	Jakarta
LDPE	1.01	0.78
PP	2.64	2.03
HDPE	3.97	3.05
PVC Bottle	0	0
PVC Film	0.15	0.12
PET	0.09	0.07
Styrofoam	0.08	0.07
Others	0.54	0.41

5.2 Observation and interview

Observation conducted during waste bank operated in collecting plastic waste in community. This research followed and analyzed several phenomena occurred during waste collection phase of waste bank. This study defines several characteristic of community who is involved in collecting plastic waste and exchange plastic waste into certain monetary value. Interview conducted to several respondents who involved in waste bank operation, waste management system, and plastic recycling industries. This research aims to analyzed problems and limitations that occurred in promoting plastic waste recycling system. Thus, appropriate system will be promoted while several evaluation and improvement proposed into current situation of plastic waste recycling system.

6. Proposed Integrated Index in Plastic Waste Management

There two parts of indexes will be discussed in this research, environmental and societal aspects of plastic recycling in waste management system. Both of them will be integrated in terms to find the opportunities and challenges of plastic recycling implementation in current waste management system. Environmental aspects will be covered by using life cycle assessment which will be comparing five scenarios which already explained in chapter 1. Meanwhile, societal aspects will be observed by using simulation and modelling approach specifically using agent-based modelling since the context is lie on community involvement and acceptance of plastic recycling technology.

First scenario, plastic waste will be delivered to landfill as their final waste treatment which is occurred in business as usual. Second scenario, half of plastic waste will be sent to landfill and the rest will be incinerated with energy will be recovered. Third scenario, all plastic waste will be incinerated with energy will be recovered. Fourth scenario, all plastic will be recycled and production waste will be sent to landfill. Last scenario, all plastic will be recycled and production waste will be incinerated with energy will be recovered. The model of each scenario can be seen in Figure 2.

Societal aspects that will be included in this research is cost of maintaining plastic recycling in community and willingness of community to adopt the system which support plastic recycling. Cost of maintaining recycling in community contains of cost of waste collection, cost of waste transportation, and cost of waste storage. Cost of waste collection is occurred in collecting phase of waste started from community to temporary storage of waste or waste bank. The model of the simulation can be seen in Figure 3. Several data from qualitative research will support the result in environmental and societal aspect. The result will be considered to maintain more sustainable and appropriate system in community.

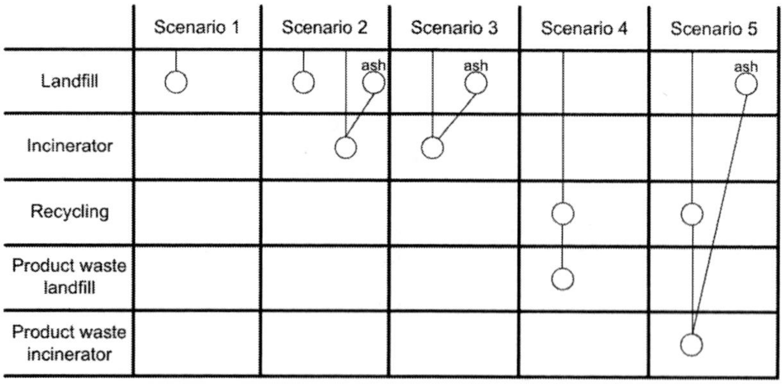

Fig. 2. Scenario of Life Cycle Assessment.

The model of simulation which showed in Figure 3 is explain the changing behaviour in community based on the changing of cost (c) according to monetary value of waste (p). Community will compare between this two variables to maintain their decision of willingness to adopt the system which support plastic recycling. In respond to that difference, community will be classified into three types which are adapting community, careless community and arguing community. Initial state of community will be called as common-community who not decides their willingness yet according to condition of differences between cost and monetary value still unrecognized. Adapting communities are peoples who willing to adopt the system in every single condition of differences between cost and monetary value. Arguing community are people who not willing to adopt the system according the monetary value is less than cost that must be spent by them, this kind of community can change their decision if the differences of monetary value and cost changing (monetary value higher than cost). Careless community are people who will not adopt the system even the monetary value is higher than cost that will be spent by them.

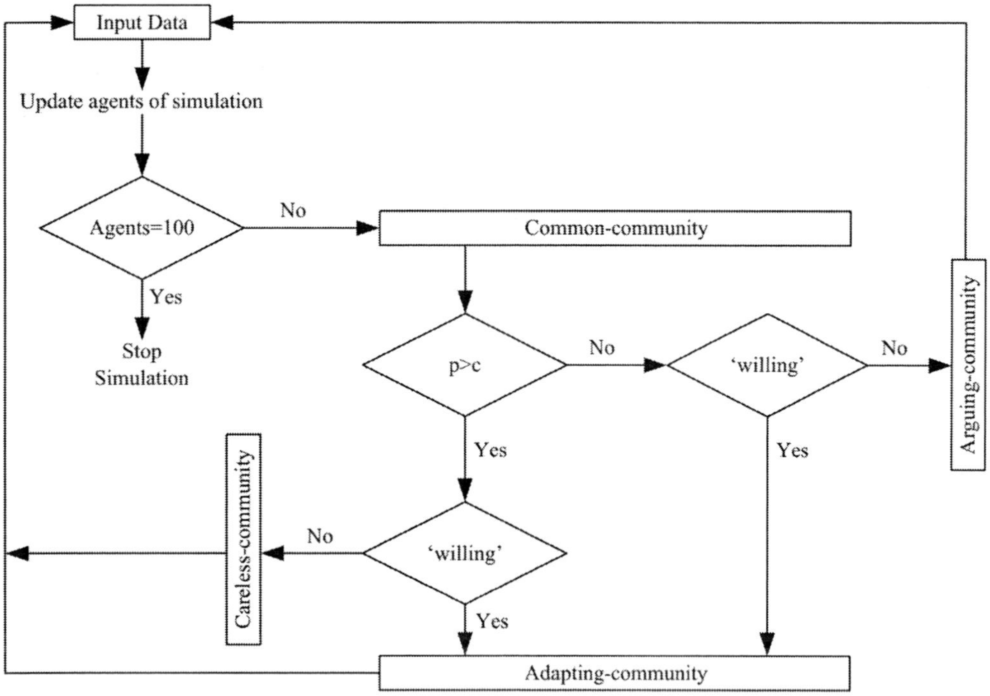

Fig. 3. Model of Societal Aspects using Agent-based Modelling.

7. Expected Result and Discussion

7.1 LCA and DEA Result

Through life cycle assessment, this research covered emission from each scenario as described in Table 3. From each emission which is occurred, scenario 1 and 2 has the worst impact in all emissions. This calculation assessed from green house gas emission in life cycle assessment using MiLCA. Scenario 3, 4, and 5 had has better result in green house gas emission. This study conclude that plastic waste can be treated using landfill which is still used in business as usual in Bandung and several city in Indonesia. Recycling and incinerator should be considered as alternative in treat plastic waste in Indonesia. Even scenario 4, which use landfill to treat production waste from recycling activity, still have a better performance in environmental impact especially in green house gas emission than landfill strategy (represented by scenario 1 and 2).

Table 3. Emission of each Scenario.

Elementary flow	Scenario 1	Scenario 2	Scenario 3	Scenario 4	Scenario 5	Unit
carbon dioxide (biogenic)	0.039103965	0.029360454	0.019616944	0.020835714	0.020839432	Kg
carbon dioxide (fossil)	0.038399083	0.000153408	0.000153187	0.000153867	0.000153922	Kg
methane	0.044838288	0.025911523	0.007030841	0.007431663	0.007413497	Kg
methane (fossil)	5.92E-07	1.90E-09	1.90E-09	1.91E-09	1.91E-09	Kg
nitrous oxide	3.12E-07	1.90E-09	1.90E-09	1.91E-09	1.91E-09	Kg
nitrous oxide	0.000491472	0.000486289	0.000481791	0.000535709	0.000535008	Kg
PFC-14	3.62E-09	3.44E-09	3.27E-09	4.20E-09	4.15E-09	Kg
sulfur hexafluoride	6.93E-12	6.82E-12	6.72E-12	7.99E-12	7.97E-12	Kg

Green house gas emission is considered as output in this research, this study also consider several resources which will be needed. From this input and output, DEA will assess which scenario is performs better than other scenarios. This study considers water resource that used from each scenario such as freshwater, ground water, and sea water. From DEA, scenario 3, 4, and 5 show the most effective and efficient of input and output. This result strengthens the result from green house gas result. The study pointed out that landfill is no longer appropriate to maintain plastic waste in the future. Recycling and incinerator system will be needed to avoid environmental impact in long terms.

7.2 Simulation Result

Simulation from the model which described in Figure 5 conclude that rate of adapting community increasing according to the decreasing cost which will be spent by community. Number and rate of arguing community decreasing following to the increasing number and rate of adapting community. There is changing behaviour from arguing community to adapting community. Number and rate of careless community increasing in first half of simulation which caused by number of arguing community are higher than adapting community.

However, rate of careless community decreasing in second half of simulation which caused by number of adapting community started to be higher than arguing community. This simulation concluded that changing cost is important key factor to maintain an increasing number of communities who will adopt the system which support plastic recycling. The change in cost can be maintained by analyzed the possible cost variables that influence the cost which will be spent by community. One key action that can be taken is government's subsidy in order to decrease the cost. This simulation using agent-based modelling and simulation by using SOARS software. The limitation of this simulation is there are cost variables that have not been added into the changing process of cost in the general term which totally spent by community.

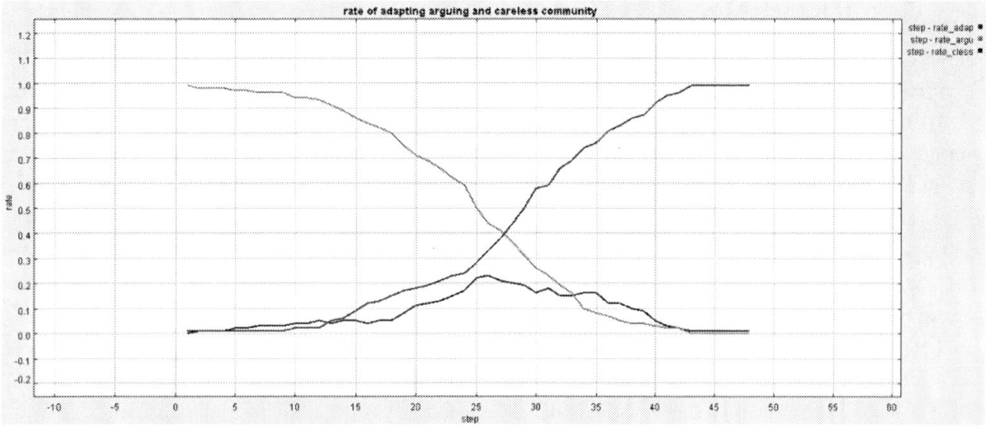

Fig. 5. Rate of Adapting, Arguing, and Adapting Community.

Table 4. Observation findings.

Findings	Observation	Interpretation
Economic class is not directly affected to community participation in separating their plastic waste	Several communities who collected their plastic waste located in area where only accessible by walking or motorcycle. Several communities who collected their plastic waste working as blue-collar labour or subcontract worker or small stall owner. There is a waste bank's customer live in a well developed house, however in her complex' area, she is the only one customer.	The area of household, type of community's occupancies is show that the community live in a low-middle economic class. There are customers live in a middle economic class but they are minority.
Educational background is not directly affected to community participation in separating their plastic waste.	Several of waste bank administrators especially in waste bank collector team and several waste banks' customer are senior high school educated. Community in hospital are controlled by one administrator who higher educated.	Lower or higher educational background can be generalized to be supporting factor of community participation in separating their plastic waste.
Knowledge is promising factor of community participation in separating their plastic waste.	Administrator in hospital, who control community there, is higher educated but also know well about waste management.	Knowledge can be influenced by educational background. Knowledge is more necessary to influencing community participation in separating their plastic waste than only based on educational background.

7.3 Qualitative Data Result

From observation which gathered in the waste bank, this study pointed out several results. First finding is economic class and educational background is not directly affected to community participation in separating their plastic waste. Other findings is educational background and knowledge in plastic waste management is not directly related, community with better knowledge in waste bank, plastic waste, or recyclable waste is actively participate in separating their plastic waste even their educational background are lower. Thus mean that promotional activities about knowledge in waste bank, plastic waste, or recyclable waste are necessary to be conducted. Several strategies could be conducted through social media, local newspaper, or other city information infrastructure. This observation result can be seen in Table 4.

From interview which targeted to several respondents who work as scavengers and cleaning services in local public places, this study pointed out several results and specific needs to be considered in the system. Sustainability and availability of waste separation infrastructure must be maintained in order to support plastic recycling system in community. One respondent pointed out the case of trash basket in Bandung area which are easy to be disappeared or destructed or stolen because built from metal which very valuable for some people (metal's monetary value is high in Bandung or other area in Indonesia). Ease of reassemble or collecting or separating plastic waste also considerable to support a better performance in plastic recycling system. Other respondent claimed that collecting or separating plastic waste from other kind of waste in Bandung is still difficult and take a long time. Cleaner and drier condition of waste storage also will support the quality of plastic waste which ready to be recycled. Respondent who work as cleaning service noted that because of plastic waste still mixed with other kind of waste, they often find that plastic waste is already contaminated by water or any liquid from other waste.

8. Conclusion

Plastic waste management in Bandung recently maintained through landfill. This study actually showed that landfill strategy is no longer appropriate to maintain plastic waste. Recycling and incinerator is preferable than landfill strategy. Recycling strategy with landfill as their production waste strategy is considerable than only rely on landfill. From LCA, ABM, and qualitative data collection there are several policies should be necessary to be taken by government. First, plastic recycling and incinerator policy should be considered as main strategy for plastic waste management. Second, subsidy to reduce cost in collecting plastic waste in community should be implemented by maintain sustainable and ease to use infrastructure from household level, waste transportation, and waste storage system. Third, promotional activities in waste bank and plastic recycling system are necessary to be conducted by using social media, local newspaper, and/or other information infrastructure in Bandung area. Plastic waste still increasing and it is promising to reduce waste which will be sent to landfill in order to life cycle of landfill infrastructure. Further study can talk in investment environment to support plastic recycling and incinerator in Indonesia or other city in Indonesia.

References

1. Hendri, Resyia, Harapan Tua, *Jurnal Online Mahasiswa Bidang Ilmu Sosial dan Ilmu Politik,* (2012)
2. S. Manfredi, T. H. Christensen, Waste Manag., **29(1)**, 32 (2009)
3. A.S.E. Yay, J. Clean. Prod., **94**, 284 (2015)
4. E. Fikri, P. Purwanto, H. R. Sunoko, Procedia Environ. Sci., **23**, 123 (2015)
5. U. Arena, M.L. Mastellone, F. Perugini, Int. J. Life Cycle Assess, **8(2)**, 92 (2003)
6. F. Perugini, M.L. Mastellone, U. Arena, Environ. Prog. Sustain. Energy, **24**(2), 137 (2005)
7. T. Matsuda, J. Yano, Y. Hirai, S.I.Sakai, Int. J. Life Cycle Assess., **17(6)**, 743 (2012)
8. L.M. Scortar, S. Lazar, L.E. Suciu, Managerial Challenges of the Contemporary Society. Proceedings, **249** (2009)
9. X. Shi, A.E. Thanos, N. Celik, Resources, Conservation and Recycling, **92**, 190 (2014)

10. G. Sauvageau, J.M. Frayret, Eur. J. Oper. Res. **242(3)**, 987(2015)
11. Y. Liu, H. Ye, J. Clean. Prod., **37**, 278 (2012)
12. N.P. Thanh, Y. Matsui, T. Fujiwara, Environ. Monit. Assess., **175(1-4)**, 23 (2011)
13. H. Bartelings, T. Sterner, Environ. Resour. Econ., **13(4)**, 473 (1999)
14. F.L. Sahwan, *Penelitian Komposisi Sampah* (Jurnal Teknologi Lingkungan, Jakarta, 1989)
15. F.L. Sahwan, *Penelusuran Pola Sistem Pengolahan Plastik Bekas di Jakarta dan Surabaya* (Jurnal Teknologi Lingkungan, Jakarta, 1989)
16. D.R. Wijayanti, S. Suryani, Procedia Soc. Behav. Sci., **184**, 171 (2015)

MATEC Web of Conferences **78**, 01019 (2016) DOI: 10.1051/matecconf/20167801019
IConGDM 2016

Influence of Optimization of Process Parameters on Threshold Voltage for Development of $HfO_2/TiSi_2$ 18 nm PMOS

N. Atan[1,*] , *Ibrahim* Ahmad[1] , *B.Y.* Majlis[2] , and *M.F* Azle[1]

[1]Centre of Micro and Nano Engineering (CeMNE) Universiti Tenaga Nasional (UNITEN), 43000 Kajang, Selangor, Malaysia
[2]Institute of Microengineering and Nanoelectronics (IMEN) Universiti Kebangsaan Malaysia (UKM) 43600 Bangi, Selangor, Malaysia

Abstract. Manufacturing a 18-nm transistor requires a variety of parameters, materials, temperatures, and methods. In this research, HfO_2 was used as the gate dielectric ad TiO_2 was used as the gate material. The transistor $HfO_2/TiSi_2$ 18-nm PMOS was invented using SILVACO TCAD. Ion implantation was adopted in the fabrication process for the method's practicality and ability to be used to suppress short channel effects. The study involved ion implantation methods: compensation implantation, halo implantation energy, halo tilt, and source–drain implantation. Taguchi method is the best optimization process for a threshold voltage of $HfO_2/TiSi_2$ 18-nm PMOS. In this case, the method adopted was Taguchi orthogonal array L9. The process parameters (ion implantations) and noise factors were evaluated by examining the Taguchi's signal-to-noise ratio (SNR) and nominal-the-best for the threshold voltage (VTH). After optimization, the result showed that the VTH value of the 18-nm PMOS device was -0.291339.

1 Introduction

"Metal–oxide–semiconductor field-effect transistor" or "MOSFET" is a very prevalent type of transistor. It is unique because it entails very little current to turn on (less than 1.0 mA) yet delivers a much higher current to a load (10.0mA to 50.0A or more). This transistor is cheap yet flexible and reliable; it can be considered the key active component in all modern electronics.

Over a billion of individually packaged MOSFETs (known as discrete) are manufactured every year in order to meet the increasing demands for ICs from electronic companies. This in turn reduces the price of every chip. Smaller-sized ICs allow more chip transistor per wafer to be produced. Scaling down the metal-oxide-semiconductor, according to Moore's law, have been improving at sensational level [1], leading to massive improvement in the operation and performance of MOSFET. The transistor's faster digital switching can save time, energy, and power, improving the operation of transistors. Additionally, the reduced size of MOSFET has not only proportionally reduced the

* Corresponding author: norani@uniten.edu.my

dimensions of all devices dimensions in terms of the channel length, channel width, and oxide thickness but has also affected the transistor's doping.

However, smaller-sized transistors are more difficult to develop. The process requires prolonged hours, concentration, and consistency to enable the transistor to function efficiently. To control the threshold voltage is the main task in produce a nanoscale transistor. If the optimum VTH value is not achieved, doping fluctuations will occur and the whole system of the device will be affected [2].

In this research, dopant ions such as phosphorus, boron and arsenic are used for ion implantation. Many process parameters can influence an 18nm-PMOS fabrication such as halo implantation, halo implantation energy, halo tilt, source/drain implantation, VTH adjustment implantation, compensation implantation, and VTH adjustment energy. Manipulating the dose, energy, and rotation of the implants will alter the profile and electrical characteristics of the devices [3]. The aim was to obtain one of the most optimum VTH value for MOSFET. In this research, HfO_2 was used as gate oxide. Combination HfO_2 and $TiSi_2$ is to obtain the highest value of ION/IOFF ratio and the lowest value of sub-threshold leakage current (IOFF). The purpose is to render this device suitable for low power application [4]. Nevertheless, we attempted to discover the best combination of four process parameters that can yield the optimal value of VTH.

This Taguchi's orthogonal array method evaluates the entire process parameters with only a few experiments and noise factor in order to get the optimal process parameters [5]. The aim of the current work is to meet the threshold voltage in the ITRS 2011 specification for a 18-nm gate length PMOS transistor with a VTH value of 0.302V [6]. The Taguchi's robust design is a powerful statistical tool often used in industrial process optimization and analysis. The method can be applied to the design of high quality systems without increasing costs, and it can also reduce the time for the development process [7].

2 Method and materials

2.1 Process steps and device

The development of HfO2/$TiSi_2$ 18-nm PMOS transistor was fabricated using Virtual Wafer Fabrication (VWF) SILVACO TCAD. A p-type (boron doped) silicon bulk sampled is used with <100> alignment and a doping concentration of 7 x 10^{14} atoms/cm^2. The next process was to develop retrograde N well. In this process a 200 Å oxide layer was growing on the top of the substrate at high temperature of 970°C for 20 minutes. The following step was to generate a shallow trench isolation (STI) of 130-Å thickness. This annealing process takes 25 minutes at temperature 900°C in dry oxygen. By applying a low pressure chemical vapour deposition process (LPCVD), a 1500-Å nitride layer was deposited on top sample of the oxide layer. Photoresist layer was then deposited on the wafer and through a reactive ion etching (RIE) process the unnecessary part at the top of the STI area was etched away. After the desired depth of STI was achieved, an oxide layer was grown on the trench sides to eliminate any impurities from entering the silicon substrate. To eliminate any extra oxides on the substrate surface, the Chemical mechanical polishing (CMP) was then applied. The STI was then annealed for 15 minutes at a temperature of 850°C to develop a phosphor silicate glass (PSG) at the top of the substrate followed by oxide and nitride pad. The sacrificial oxide layer was formed once the STI was completely annealed and etched accordingly to remove defects on the surface.

The next step was the crucial process of growing the gate oxide. The gate oxide layer was grown by exposing the silicon wafer to dry oxygen for a very short time of 10ms at 850°C. This condition served to ensure that a 1.1-nm thickness of gate oxide (TOX) was

DOI: 10.1051/matecconf/20167801019

grown. Once the substrate was annealed, boron difluoride (BF2) was implanted at the N-well active area to adjust the VTH of the device. The dosage of boron applied was 1.6757777x10^7 atoms/cm^2 whereas the energy used was 5KeV with a tilt angle of 7°. Among the channel regions, on the top of the silicon bulk was the deposition process of Hafnium Dioxide (dielectric permittivity HfO_2, εopt = 22) as dielectric material. The length of the high-K material was scaled and adjusted to get 18nm same value as the gate length of the transistor. Thus, the process done for Titanium Silicide ($TiSi_2$) deposited on the top of HfO_2, and followed by a Halo implantation process with implanted phosphor as a dopant at a dose of 5.581 x10^{13} atom/cm^2 with an angle tilted at 30° and halo energy of 290KeV.

The next step was to form the sidewall spacer using a chemical vapour deposition (CVD) process, a 0.047-µm silicon nitride layer was deposited and was etched for the same deposition thickness. Then, a source-drain implantation process was conducted at 7° with boron as a dopant at a density of 5.556666 x10^{13} atom/cm^2, with 11KeV implantation energy. The next step was to develop a 0.3 µm layer of boron phosphor silicate glass (BPSG). This layer acted as the pre metal dielectric (PMD). In the same way, the annealing process was performed at the wafer to strengthen the structure at 850°C. The final process of wafer was to implant the compensation a phosphor dose of 2.5 x 10^{13} atom/cm^2 at 60KeV implantation energy. The beam was tilted at 7°. Finally, the aluminum layer was etched accordingly after it was deposited on the top of structure to form metal contacts for the source and drain. This point completed the production of the 18-nm PMOS transistor (Figure 1 and Figure 2).

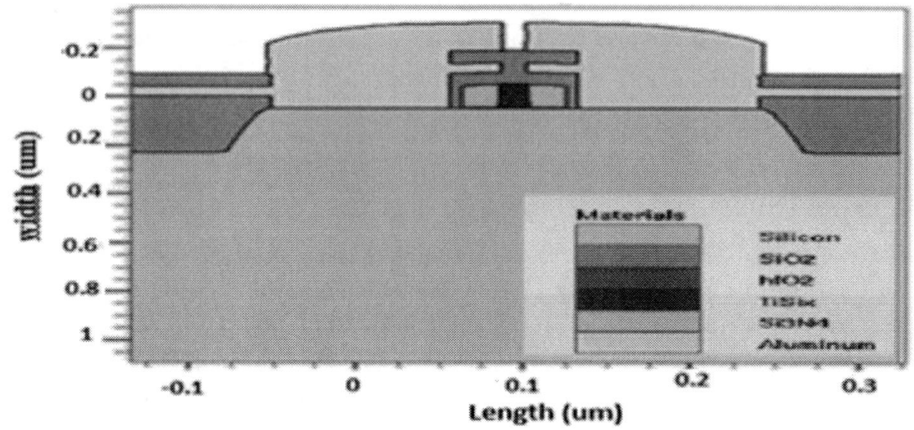

Fig. 1. A complete HfO_2/$TiSi_2$ 18nm PMOS.

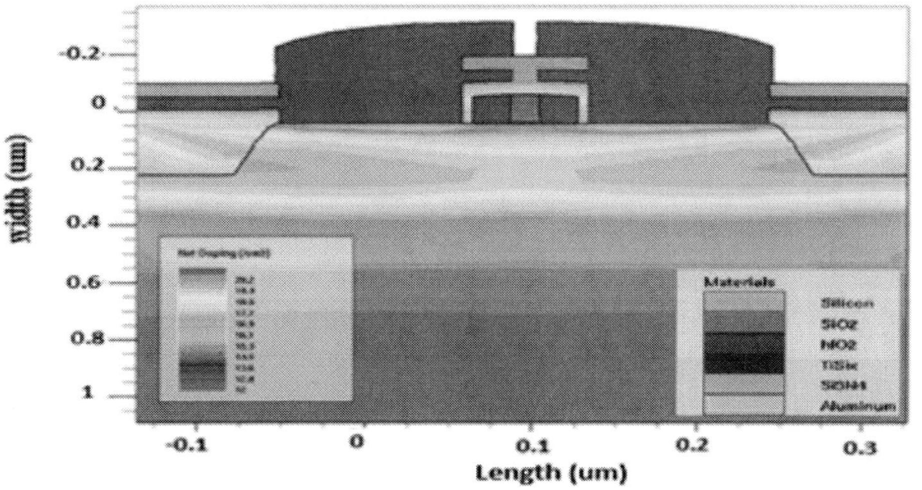

Fig. 2. A complete $HfO_2/TiSi_2$ 18nm PMOS doping profile.

2.2 Taguchi orthogonal L9 array method

In this L9 Orthogonal, the process parameters were Compensation Implantation, Halo Implantation Energy, Halo tilt and Source-Drain Implantation. Whereas, the temperatures of phosphorus silicate glass (PSG) and boron phosphorus silicate glass (BPSG) were the noise factors. In this study, we carried out three experiments of different combination of process parameters but with the same noise factors. The purpose was to obtain the best combination of process parameter with a good VTH value. The following section reports the best result of process parameters.

3 Result

From the result of the three experiments, this experiment was selected because it shows a good VTH value from the combination of different four process parameters and noise factors as tables below.

Table 1. Process Parameters and Levels.

Symbol	Process Parameter	Unit	Level 1 [L1]	Level 2 [L2]	Level 3 [L3]
A	Compensation Implantation	atom/ cm^3	14.871000e13	14.875000e13	14.879000e13
B	Halo Implantation Energy	KeV	294	295	296
C	Halo Tilt	Degree	30	32	34
D	Source-Drain Implantation	atom/ cm^3	5.394000 e13	5.400000 e13	5.406000 e13

Table 2. Noises Factor and levels.

Symbol	Noise Factor	Unit	Level 1	Level 2
X	PSG	°C	850	851
Y	BPSG	°C	850	851

3.1 Analysis of 18-nm PMOS device

The Taguchi Orthogonal L9 array projected thirty-six simulation runs from the nine experiments. The completed result for VTH data is shown in Table 3. Two control factors were considered: dominant factor and an adjustment factor. A signal-to-noise ratio (SNR) analysis was used to identify the optimal process parameters in this experiment. A larger S/N ratio corresponds to better performance [8].

Table 3. VTH Values for 18-nm PMOS Device.

EXPERIMENT NO	VTH1 X_1Y_1	VTH2 X_1Y_2	VTH3 X_2Y_1	VTH4 X_2Y_2
1	-2.51E-01	-2.39E-01	-2.49E-01	-2.37E-01
2	-3.16E-01	-3.04E-01	-3.14E-01	-3.02E-01
3	-5.85E-01	-5.42E-01	-5.76E-01	-5.84E-01
4	-2.77E-01	-2.66E-01	-2.75E-01	-2.64E-01
5	-4.12E-01	-3.98E-01	-4.10E-01	-3.95E-01
6	-3.56E-01	-3.43E-01	-3.54E-01	-3.41E-01
7	-1.27E-01	-1.31E-01	-1.28E-01	-1.33E-01
8	-2.85E-01	-2.74E-01	-2.83E-01	-2.72E-01
9	-3.62E-01	-3.54E-01	-3.61E-01	-3.52E-01

In this research, the VTH of the 18-nm device indicates the characteristics of nominal-the-best quality [8]. This characteristic was selected to obtain a VTH closer to -0.302V for 18-nm transistor which is the value predicted by International Roadmap of Semiconductor (ITRS) 2011. The formula of SNR (nominal-the-best), η can be expressed as [9]:

$$\eta = 10\mathrm{Log}_{10}\left[\frac{\mu^2}{\sigma^2}\right] \quad (1)$$

and

$$\sigma^2 = \frac{\sum_{i=1}^{n}(Y_i - \mu)^2}{n-1} \quad (2)$$

Where n is the number of tests, Y is the experimental value of the VTH, μ is mean, and σ is variance. The dominant factor and the adjustment factors need to be considered in SNR of category Nominal-the-Best. By applying the formulae (1) and (2), the SNR (η) for the PMOS device was calculated and the results are shown in Table 4.

Table 4. Mean, Variance and S/N Ratio for PMOS Device.

Experiment No.	Mean	Variance	SNR (Mean)	SNR (Nominal-the-best)
1	-2.44E-01	4.68E-05	-1.22E+01	3.11E+01
2	-3.09E-01	4.60E-05	-1.02E+01	3.32E+01
3	-5.72E-01	4.22E-04	-4.86E+00	2.89E+01
4	-2.70E-01	4.28E-05	-1.14E+01	3.23E+01
5	-4.04E-01	7.21E-05	-7.88E+00	3.35E+01
6	-3.49E-01	5.66E-05	-9.15E+00	3.33E+01
7	-1.30E-01	7.83E-06	-1.77E+01	3.33E+01
8	-2.79E-01	4.77E-05	-1.11E+01	3.21E+01
9	-3.57E-01	2.38E-05	-8.94E+00	3.73E+01

The mean SNR for the nine experiments were calculated and plotted in Figure 3. The larger the SNR, the better the quality characteristic for the VTH. From Figure 3, we can identify which level of the process parameters is the dominant factor and adjustment factor for this experiment. The dominant factor has the biggest effect on SNR and the adjustment factor poses the largest effect on mean and the smallest effect on SNR. From the graph, we can see that the dominant factor appears for compensation implantation (A3), followed by halo tilt (C2), source-drain implantation (D1), and halo implantation energy (B3), whereas halo implantation energy is the adjustment factor.

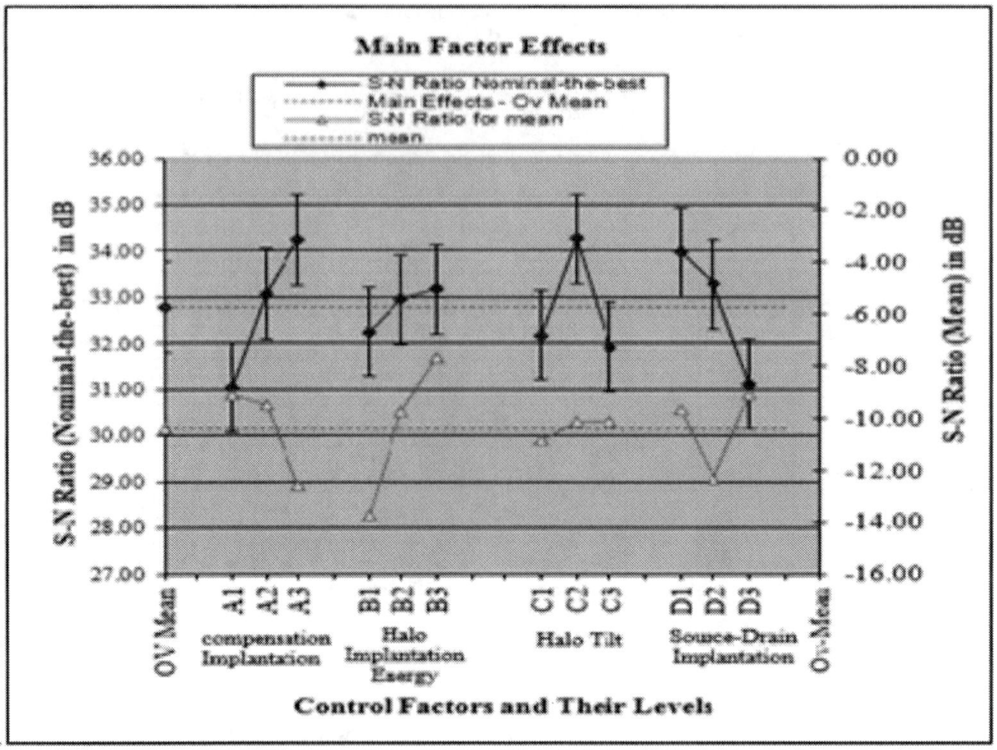

Fig. 3. Graph of S/N ratio (Nominal-the-Best).

3.2 Analysis of variance (ANOVA)

ANOVA is one of the most common statistical techniques to determine the percentage contribution of the optimum combination of the process parameters for the device. This analysis can also be used to identify which of the control factor has the most effective to the VTH. The highest percentage of SNR has the greatest effect on the stability of the VTH and hence the responsible parameter will be the dominant factor. From Table 5, compensation implantation has the most effect on the VTH of 18-nm PMOS (39% SNR). Hence, compensation implantation is the dominant factor. The source-drain implantation poses a factor effect on SNR of 33%, which is the second largest effect after compensation implantation. Halo tilt poses a percentage of SNR of 25% and 4% of SNR for halo implantation energy. We can also identify the adjustment factor from this table, which is the halo implantation energy. It has the largest effect on mean, (58.62%) and the smallest effect on SNR of (4%).

Table 5. Result of ANOVA for 18nm PMOS Device.

Symbol	Process Parameter	Factor Effect on SNR (%)	Factor Effect on Mean (%)
A	Compensation Implantation	39	22.29
B	Halo Implantation Energy	4	58.62
C	Halo Tilt	25	0.90
D	Source-Drain Implantation	33	18.20

3.3 Confirmation of optimum factor

The above results shows that halo implantation energy is suitable for the optimal design of the device because the parameter serves as the adjustment factor. The dose can be adjusted from 294KeV to 296KeV until we achieved the threshold voltage closest to the nominal or target value, which is -0.302 as required by ITRS 2011 for 18 PMOS device. Again, the final experiment must be conducted using the final value of process parameters as shown in Table 6. This time only one simulation was run with the same noise factors such as PSG temperature and BPSG temperature.

Table 6. Best Setting of Process Parameters.

Symbol	Process Parameter	Unit	Level	Best Value
A	Compensation Implantation	atom/cm^3	3	14.879000e13
B	Halo Implantation Energy	KeV	1	294
C	Halo Tilt	Degree	2	32
D	Source-Drain Implantation	atom/cm^3	1	5.394000e13

From the result of the best setting the process parameters, we obtained the values as shown in Table 7. Table 7 compares the VTH before optimization and after optimization, as well as the ITRS value. After optimization under X2Y2, the value of VTH was -0.291339. This value was selected as the final value of VTH after optimization because the value is within to ITRS 2011 requirements of -0.302 ± 12.7% [10, 11].

Table 7. Final Result of VTH after Confirmation Experiment.

Experiment No	Before Optimization (Volt)	After Optimization VTH 4 (Volt) X_2Y_2	ITRS 2011 VTH (Volt)
1	-0.32047	-0.291339	-0.302

4 Summary

The optimum solution in achieving the best level of process parameters in the development of VTH for $HfO_2/TiSi_2$ 18 nm PMOS transistor was successfully predicted by using Taguchi L9 Orthogonal array. The compensation implantation dose was identified to be the dominant factor, and halo implantation energy was identified as the adjustment factor in this device. Both process parameters are the most effective parameters with respect to a threshold voltage (VTH) of -0.291339V.

The authors would like to thank to Ministry of Higher Education (MOE), Institute of Micro engineering and Nano electronics (IMEN) Universiti Kebangsaan Malaysia (UKM), and Centre of Micro and Nano Engineering (CeMNE) Universiti Tenaga Nasional (UNITEN) for financial, facilities and moral support throughout the project.

References

1. G. E. Moore, *IEEE International Electron Devices Meeting*, **11**, (1975)
2. A. K. Goel, http://www.silvaco .com/products/descriptions, (1995)
3. S. Ninomiya, , Y. Kimura, , T. Kudo, , A. Ochi, , F. Sato, M.Tsukihara, ... & S.Shibata, *International Workshop on IEEE*, (2009)
4. N.Atan, B. Y. Majlis, I. Ahmad, *Semiconductor Electronics (ICSE), IEEE International Conference*, 56 (2014)
5. U. Esme, Arab J. Sci Eng., **34**, 2 (2009)
6. G.P. Syros, J. Mater. Process. Technol., 68 (2003)
7. N.V.R. Naidu, *Proc. International Conference on Flexible Automation and Intelligence Manufacturing*, 811 (2008)
8. A. Salehuddin, I. Ahmad, F.A. Hamid, A. Zaharim, H.A. Elgomati, B.Y. Majlis, P.R. Apte, World Academy of Science, Engineering and Technology, 1137 (2011)
9. Phadke, Madhav S, *Quality Engineering Using Robust Design, Pearson Education, Inc.* (1998)
10. Information on http: www.ITRS2011.net
11. I. Ahmad, Y.K. Ho, and B.Y. Majlis, *Semicond. Phys. Quantum Electron. Optoelectron*. **9**, 40, (2006)

MATEC Web of Conferences 78, 01020 (2016) DOI: 10.1051/matecconf/20167801020
IConGDM 2016

Drying Curves and Colour Changes of Cassia Alata Leaves at Different Temperatures

Norawanis Abdul Razak[1,*], *Abdul Razak* Shaari[1], *Maizura* Jolkili[1], and *Lee Yit* Leng[1]

[1]Faculty of Engineering Technology, Universiti Malaysia Perlis (UniMAP), Campus of UniCITI Alam Sungai Chuchuh, Padang Besar, 02100 Perlis, Malaysia.

Abstract. Cassia alata plant has been commonly used as medicinal herbs for treating fungal infection such as ringworm and eczema. Nevertheless, a few studies on drying characteristics have focused on C. alata. This work aims to investigate the drying characteristics of C. alata leaves at different temperatures. Moreover, the effects of different drying temperatures on colour qualities of C. alata leaves had been determined. The leaves were dried at different temperatures (40, 50, 60, 70, 80 and 90 °C) using a laboratory oven. Dried and fresh leaves were analyzed for the moisture content and colour analysis. The EMC of C. alata leaves decreased from 0.30 to 0.10 g water/dry matter as the temperature increased from 40 °C to 90 °C. The colour qualities of C. alata leaves were lower with increasing drying temperature.

1 Introduction

Drying process is very important to prolong the shelf life of the product and it is also important to preserve the chemical and physical properties of the products. Drying of herbs inhibits microbial growth and forestalls certain biochemical changes. Sun drying is commonly used by small-scaled entrepreneurs to dry the herbs [1]. In spite of the low cost in using this method, it has some great disadvantages to it such as contamination and also from the hygienic aspects. Furthermore, the final moisture content of the material is always relatively high after being sun dried. Alternative methods in drying process are oven drying, hot air drying and freeze drying. These methods are performed in a closed system which ensures better air circulation and controlled conditions while drying.

Nevertheless, drying variables, especially temperature and the drying method, can alter the appearance and chemical constituents of herb quality. The total polyphenols and antioxidants of Capsicum chinense Jacqui (Cumari chilli pepper) was reported decreased when dried at 65°C compared to 55 and 45°C in a fixed moisture [2]. Colour, is nevertheless an important characteristic to be sustained as it can provide the first impression of the dried product quality. Colour is related to consumers' appreciation of a product as they tend to associate product colour and other visual properties with its taste, hygienic security, shelf life, nutritional value and personal satisfaction [3]. Furthermore, colour is directly linked to aroma and taste of the product.

* Corresponding author: norawanis@unimap.edu.my

Oven drying is found out as a better way compared to sun drying in maintaining the colour of herbs [4]. However, there is lack of information on the effects of different oven drying temperatures on colour changes of C. alata leaves. Thus, the aims of this study were to determine the drying curve and colour changes of C. alata leaves at different temperatures of oven drying (40, 50, 60, 70, 80 and 90 °C). The hypothesis of this study is that the equilibrium moisture contents and colour qualities of C. alata leaves decrease with the increasing temperature. Therefore, we expected to determine which drying temperature is best for preserving the colour quality of C. alata during industrial scale production.

2 Materials and methods

2.1 Plant cultivation

C. alata plants were grown at the Institute of Sustainable Agrotechnology (INSAT), Sg. Chucuh, Perlis, Malaysia. The plants were grown from cuttings and maintained according to the standard practiced adopted by the majority growers in Malaysia.

2.2 Samples preparation

Fresh leaves were harvested, washed and rinsed. The leaves were then immediately removed from the twigs for subsequent drying treatments.

2.3 Drying process

About 10 g of C. alata leaves were oven dried at six different temperatures (40, 50, 60, 70, 80 and 90°C) by using laboratory oven until constant weight was obtained. The samples were replicated five times for each temperature and the data were recorded.

2.4 Colour properties determination

Colour properties (L*, a* and b*) of C. alata leaves were determined by using Minolta Chroma meter CR 400 (Minolta Co., Osaka, Japan).

2.5 Statistical analysis

All measurements were carried out in triplicate and the results were statistically analyzed using JMP pro 11 package to determine the average value and standard error.

3 Results and discussion

3.1 Drying curve of cassia alata leaves at different temperatures

Drying curves (Fig. 1) were experimentally obtained by plotting the moisture content versus drying time. A considerable time period is required to achieve complete drying and to reach equilibrium moisture content (EMC). At initial stage of the drying process, the moisture contents of C. alata leaves were high due to the availability of free moisture samples at the initial stage. The moisture contents of leaves were decreasing with increasing temperature. This finding agreed with work of Seremet et al. (2016), who also found that drying removes the water from the food and consequently inhibits the development of microorganisms, improves food preserving as well as proper storage [5]. Fig. 2 shows that C. alata took about longest duration (7 hours) to remove its free moisture at 40°C, while shortest duration (30 minutes) to remove the free moisture at 90°C. The moisture desorption rate increased with higher temperature of oven drying as demonstrated by a lower EMC and shorter time to reach the EMC was consistent with those reported in literature [6]. The EMC of C. alata decreased from 0.30 to 0.10 g water/dry matter as the temperature increased from 40°C to 90°C.

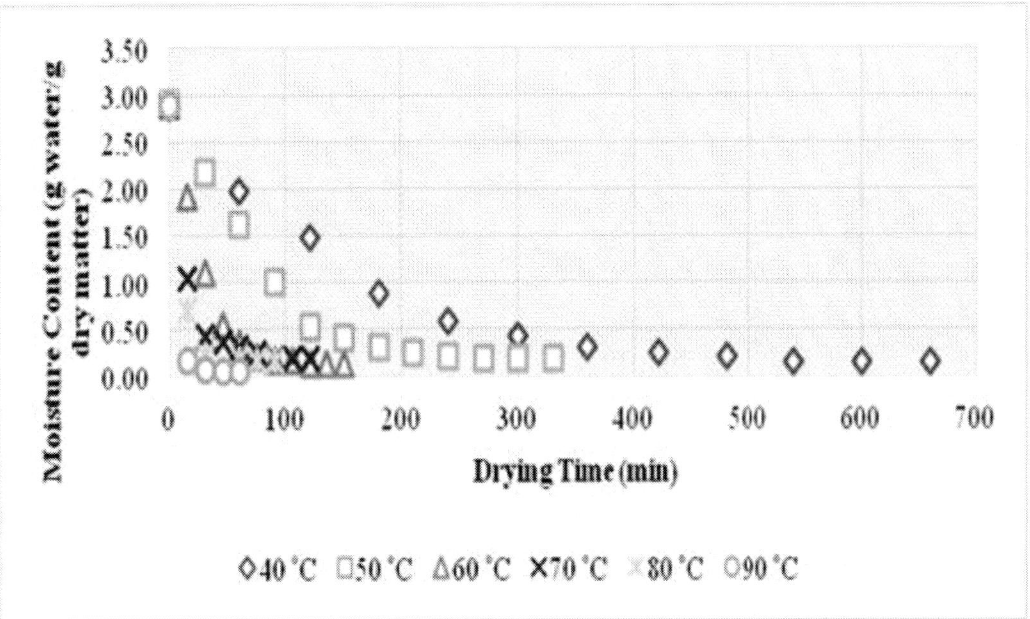

Fig. 1. Drying curves of C. alata leaves at different temperatures

The drying curves agreed with the theory of drying [7]. Initially, sensible heat was transferred from the oven drying to the fresh leave and the moisture was removed from the leave. The rate of evaporation increased during this period. Then, the moisture kept being removed from the saturated leaves surface. During this period, the rate of evaporation was the highest and it changed very little as the moisture content was reduced. As illustrated in Fig. 1, the actual drying rate was seen to fluctuate within this period; it could be attributed to the phase transition of water from the leave surface to the surrounding air. The falling-rate period coincided with the last stage of drying. The material surface in the falling rate period was no longer saturated with water and the drying rate was controlled by diffusion of moisture from the interior of solid to the surface [8].

3.2 Colour changes of C. alata leaves f at different temperatures

The determination of the colour quality is based on brightness (L*), greenish to reddish (a*), and blue to yellowish (b*) colour. For L*, the scale is given as 0 which is black and 100 as white. For a*, negative value indicates greenness and positive value indicates reddish. Meanwhile for b*, negative value means blue and positive indicates that the yellowish. The effect of drying temperatures to the colour quality of C. alata leaves is as shown in Table 1. The results showed that the L* values of the leaves decreased as the temperature increased. However, the L* values of C. alata were not significantly different when dried at 40 and 50°C compared to the fresh leaves. This demonstrated the colour of leaves became dark green as they were dried at higher temperature. The decreased in L* value when dried at all drying temperatures which was possibly due to drying time and drying temperature [9].

Table 1. Colour changes of C. alata leaves at different temperatures

Drying Temperature (°C)	Color Parameters		
	L*	a*	b*
Fresh	55.72 ± 0.59	-13.51 ± 1.01	28.02 ± 0.59
40	54.10 ± 0.75	-12.10 ± 1.51	27.88 ± 1.49
50	51.83 ± 2.15	-7.67 ± 1.17	23.84 ± 0.66
60	47.36 ± 0.93	-6.04 ± 0.90	23.08 ± 3.60
70	46.08 ± 3.01	-5.75 ± 1.12	22.80 ± 2.48
80	40.76 ± 5.29	-3.82 ± 0.94	20.46 ± 3.07
90	36.10 ± 3.00	-2.88 ± 1.71	20.44 ± 2.83

Mean values ± standard deviation

The value a* represents the greenness (negative value) to reddish (positive value) of the leaves. Table 1 shows an increase in a* values of the leaves with increasing the temperature (-13.51, -12.10, -7.67, -6.04, -5.75, -3.82 and -2.88 respectively). The green colours of leaves decreased as they were dried at higher temperature. The change of colour was probably due to heat degradation of chlorophyll. This finding was similar with work, which also found that the value of a* showed more positive as the drying temperature increased which indicated a change in colour from green to reddish tone [10]. The natural green colour of leaves is due to mixture of chlorophyll which is directly related to magnesium. During drying, the magnesium molecules were changed to pyropheophytin and pheophytin [11]. Therefore, greenness was reduced at higher temperatures. The degree of greenness is an important characteristic for customers to determine the final dried products [10]. Some literatures reported that the reduction of green shade in the leaves was also probably due to the undesirable enzymatic browning reaction during drying process [12]. This reaction was possibly due to the activity of polyphenol oxidase (PPO), which oxidizes polyphenols to produce quinines in the leaf tissues.

The b* values indicates the blue negative value) and yellow (positive value). The increase in temperature did not seem to affect much on the b* value of C. alata leaves (Table 1). The range of b* values was from 28.02 to 20.44 which indicated gradual colour changed from yellow towards blue. However, the change was not very noticeable through the difference of temperatures. A decrease in the yellowness (b* values) for C. alata leaves was probably due the decomposition of carotenoid compound [13].

4 Summary

Equilibrium moisture content (EMC) is a significant factor in biomass drying as it can provide a guideline for expediting drying and for terminating the drying process in a timely manner to save energy. With higher temperature of oven drying, the moisture desorption rate increased as demonstrated by a lower EMC and shorter time to reach the EMC. However, the colour qualities of C. alata leaves were lower with increasing drying temperature. This study suggests that the colour quality of C. alata was the best at the temperature of 40 °C. Low oven-drying temperature was better in preserve the colour quality of C. alata leaves.

This study was financially supported by the Ministry of Higher Education Malaysia (Fundamental Research Grant Scheme, FRGS); the research facilities of the Institute of Sustainable Agrotechnology (INSAT) and Faculty of Engineering Technology, UniMAP.

References

1. E.K. Akpinar, Energy Convers. Manage., **51(12)**, 2407 (2010)
2. R.C. Reis, V.C. Castro, I.A. Devilla, C.A. Oliveira, L.S. Barbosa, R. Rodovalho, Braz. J. Chem. Eng., **30(2)**, 337 (2013)
3. F. Pedreschi, J. Len, D. Mery, P. Moyano, Food. Res. Int., **39(10)**, 1092 (2006)
4. Y. Soysal, Biosyst. Eng., **89(2)**, 167 (2004)
5. L. Seremet, E. Botez, O.V. Nistor, D.G. Andronoiu, G.D. Mocanu, Food. Chem., **195**, 104 (2016)
6. X. He, A.K. Lau, S. Sokhansanj, C.J. Lim, X.T. Bi, S. Melin, T. Keddy, Biomass Bioenergy, **57**, 161 (2013)
7. I. Doymaz, J. Food Eng., **74(3)**, 370 (2006)
8. L.M. Diamante, P.A. Munro, Sol. Energy, **51(4)**, 271 (1993)
9. G. Rudra Shalini, U.S. Shivhare, S. Basu, J. Food Eng., **85(1)**, 147 (2008)
10. N. Therdthai, W. Zhou, J. Food Eng., **91(3)**, 482 (2009)
11. M.A. Ali, Y.A. Yusof, N.L. Chin, M.N. Ibrahim, S.M.A. Basra, Agric. Sci. Procedia, **2**, 394 (2014)
12. M.K. Krokida, V.Oreopoulou, Z.B. Maroulis, D. Marinos-Kouris, J. Food Eng., **49(4)**, 347 (2001)
13. D.S. Sogi, M. Siddiq, K.D. Dolan, Lwt-Food Sci. Technol., **62(1)**, 564 (2015)

Mechanical Properties of 3 Ply Plywood Made from *Acacia Mangium* Veneers and Green Starch-based Adhesives

Kang Chiang Liew[1,*], and *Genevieve* En Yee Chin[1]

[1]Faculty of Science and Natural Resources, Forestry Complex, Universiti Malaysia Sabah, Jalan UMS, 88400, Kota Kinabalu, Sabah, Malaysia.

Abstract. Recently, starch has attracted great consideration as a raw material on wood adhesives in the wood industry. Cassava and sago starch-based adhesive are renewable, biodegradable and environmental friendly product when compared with other petroleum-based adhesives. In this study, different starches-based adhesive has been produced. Mechanical properties of plywood made from *Acacia mangium* veneers with different starches-based adhesives (cassava and sago) as binder cured at different curing temperatures (100°C, 120°C and 140°C) has been determined. All materials (starch, vinegar, water and glycerol) were cooked and stirred until the mixture reached 70°C - 80°C which become sticky and whitish. After that, starch-based adhesives were applied on the veneers by using spreader, and the plywood were pre-pressed for 30 minutes with 20 kg load before hot-press. Cassava starch-based adhesive showed the highest Modulus of Elasticity which was 12410.56 N/mm² than sago starch-based adhesive, while Modulus of Rupture of the cassava starch-based adhesive at 100°C showed highest mean value at 74.19 N/mm². Sago-starch based adhesive at 140°C showed the highest shear strength with 1.11 N/mm². In short, cassava and sago starch-based adhesives gave good performance in mechanical properties such as bending for pressed temperature (100°C and 120°C), and shear at 140°C pressed temperature.

1 Introduction

Adhesive is used and applied on any surface of the material, glued together and resist separation. Most of the wood adhesives are derived from petroleum source, called as synthetic adhesive or petroleum-based synthetic adhesive. Petroleum-based adhesives are non-renewable resource and limited such as melamine formaldehyde, urea formaldehyde resin, phenol formaldehyde and ethylene vinyl acetate based adhesive. These types of adhesive are widely available in different type of material, good in performance, excellent in durability and resistance in moisture and heat. In manufacturing and end use of these synthetic adhesives, low molar mass of toxic compounds were emitted largely. Negative effects on the human health and environment, for example, end product made from

* Corresponding author: liewkc@ums.edu.my

plywood in the interior, formaldehyde will be emitted to the surrounding and effects on the human health [1].

There have been renewed interests in more environment-friendly adhesive such as bio-based adhesive or "green" adhesive from biological sources. There are some natural substances, which was extracted and derived from variety of organic or inorganic sources such as starch, casein, soy bean, protein, and blood [2]. In short, wood industries continuously looking for green product which renewable, biodegradable and non-toxic gas emitted into the environment and low price for adhesive. Therefore, the aim of this study was to evaluate the mechanical properties (bending test and shear test) of plywood made from *Acacia mangium* veneers with different starches-based adhesives (cassava and sago) as binders cured at different curing temperatures (100°C, 120°C and 140°C). Results expected that the starch-based adhesives would give comparable properties to the conventional adhesive of urea formaldehyde.

2 Materials and Methods

Acacia mangium was selected as the raw material where their logs were peeled and veneers cut into size at 19.5 cm in width and 20 cm in length. Starch-based adhesives from cassava flour and sago flour, vinegar, glycerol and water were cooked on hot plate. During the cooking process, the mixture was constantly stirred until it became sticky and whitish, which took around 30 minutes, at 70°C - 80°C. Urea formaldehyde adhesive was used as Control in this study.

Gluing process was performed after the completion of the process of preparation of veneers and adhesives. Starch-based adhesives and urea formaldehyde with 16.79±2 g of adhesives were applied onto each side of the veneers manually. Arrangement of the veneers was glued together at 90° on each layer prior to pre-pressed for 30 minutes with 20 kg of load before hot press. Hot press process was carried out by using temperature of 100°C, 120°C, and 140°C for a duration 30 minutes with pressure 0.5 MPa, for each 3-ply plywood. After hot press process, plywood was conditioned (relative humidity: 50±2 %, temperature: 23±2°C for a period of 7 days) to obtain constant moisture content before cutting into test pieces.

Bending test (JAS No. 233:2003) was done to determine Modulus of Elasticity and Modulus of Rupture between plywood and adhesive. The dimension of test pieces was 170 mm in length X 35 mm in width. Shear test (JAS No. 233:2003) with modification on the shear speed with 1 mm/s was done to determine the strength of adhesion between the layers of plywood. The test pieces were 79 mm in length X 25 mm in width. The groove were cut by using bandsaw at the top and back of the test piece with 2 mm in width. Test pieces were immersed into the water bath for 3 hours at 60°C before putting on towel to absorb excessive water. Then, Universal Testing Machine was used to test shear of test piece. Figure 1 showed the experimental design.

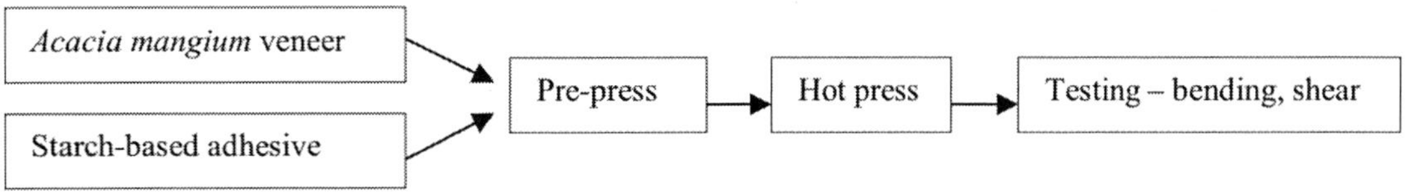

Fig. 1. Experimental design of study.

3 Results and Discussion

There were two mechanical test done throughout this study, which is bending test and shear test. The average of the mechanical test for Modulus of Rupture of plywood with different pressed temperatures recorded in the range 56.37 N/mm^2 to 78.95 N/mm^2, whilst the Modulus of Elasticity of in the range 7779.53 N/mm^2 to 12410.56 N/mm^2 and the shear test 0.28 N/mm^2 to 2.8 N/mm^2.

Figure 2 presented average of Modulus of Elasticity of the starch-based adhesives. Cassava starch-based adhesive with 100°C of hot press temperature in making plywood showed the highest Modulus of Elasticity among all starch-based adhesives and temperatures, noted as 12410.56 N/mm². This is due to structure of cassava at 100°C is good in performance as a binder and high elasticity compared with others starch-based adhesives in plywood. It also showed that structure and elasticity of the cassava starch-based is being destroy when temperature increasing to 140°C. Uthumporn *et al.* [3] stated that different starches have different gelatinization temperature and while the starch under alkali and acid treatment with increasing temperatures could influence gelatinization temperature of starch. According to Yu *et al.* [4], granules morphological changes is effected by heating temperature than compared with heating time, while increase in alkali concentration, gives a strong impression on the swollen granules than neutral and acid conditions.

At the same temperature, sago starch-based adhesive in plywood showed the lowest Modulus of Elasticity recorded as 7779.23 N/mm². On the other hand, at 120°C of hot press temperature showed the better results in starch-based adhesives in plywood when compared with urea formaldehyde. Sago starch-based adhesive showed the highest Modulus of Elasticity recorded as 12376.34 N/mm² which was higher than cassava starch-based adhesive at 57.78 N/mm² and urea formaldehyde about 2010.06 N/mm².

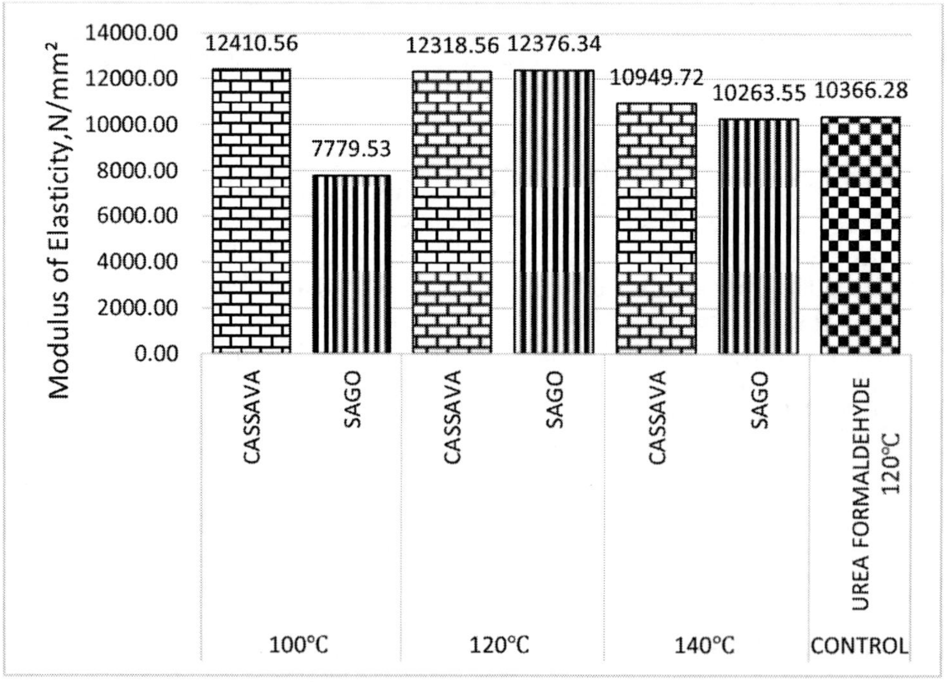

Fig. 2. Graph shows Modulus of Elasticity (N/mm²) for different starch-based adhesives and urea formaldehyde as binder in plywood pressed at different temperatures.

Figure 3 presented cassava starch-based adhesive in plywood with hot press temperature at 100°C recorded highest mean value of Modulus of Rupture which was 74.19 N/mm². At the same hot press temperature, sago starch-based adhesive in plywood showed the lowest mean value of Modulus of Rupture 56.37 N/mm². By comparison of mean value of Modulus of Rupture between the different adhesives with hot press temperature at 120°C, sago starch-based adhesive showed slightly lower than urea formaldehyde about 5.36 N/mm². In short, cassava and sago starch-based adhesives in plywood at 120°C of hot press temperature showed lower mean value of Modulus of Rupture when compared with urea formaldehyde. This was probably due to pure starch-based adhesives do not give good performance in strength and bonding between veneers and starch-based adhesive. According to Akbari *et al.* [5], the same testing was used for medium density fiberboard where there was no increment in loading at higher starch content due to lowest composition of natural rubber latex in the starch-based adhesive. Hence, pure starch-based adhesive has no ability to resist the loading strength except adding natural latex which tends to reduce brittle failure of the blend [6].

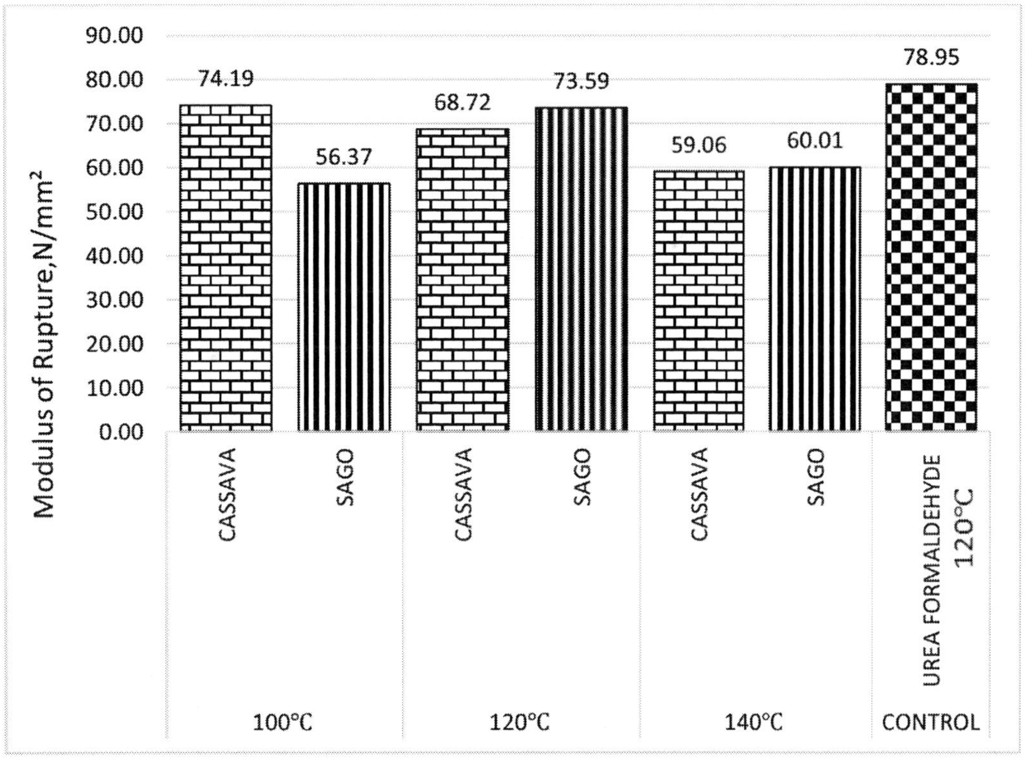

Fig. 3. Graph shows Modulus of Rupture (N/mm²) for different starch-based adhesives and urea formaldehyde as binder in plywood pressed at different temperatures.

Referring to Figure 4, sago starch-based adhesive in plywood with different hot press temperature at 140°C showed the highest mean value of shear strength which recorded 1.11 N/mm², while the lowest mean value of shear strength noted as 0.28 N/mm², which was starch-based adhesive at 120°C. As showed in bar graph, as the hot press temperature increased, the sago starch-based adhesive in plywood also increased in strength. With this situation, the structure of starch granules was broken and separated molecules of starch which is able to create an excellent adhesion to the veneer surface. According to Mohamed *et al.* [7], a high heating rate was achieved and the final temperatures was high enough to limit α-amylase activity. Hence, firmness of the starch gel was not reduced and a good cohesion was achieved. In contrast, bar graph showed fluctuation when cassava starch-based adhesive in plywood increased with hot press temperature, bond strength decreased first, then increased again. Yu *et al.*[4] mentioned that treatment temperature of starch-based adhesive was increased and bond (shear) strength of starch-based adhesive between veneers also increased first, and then decreased in bond strength. This is because of the amount of free hydroxyl groups was presented on starch molecular chain.

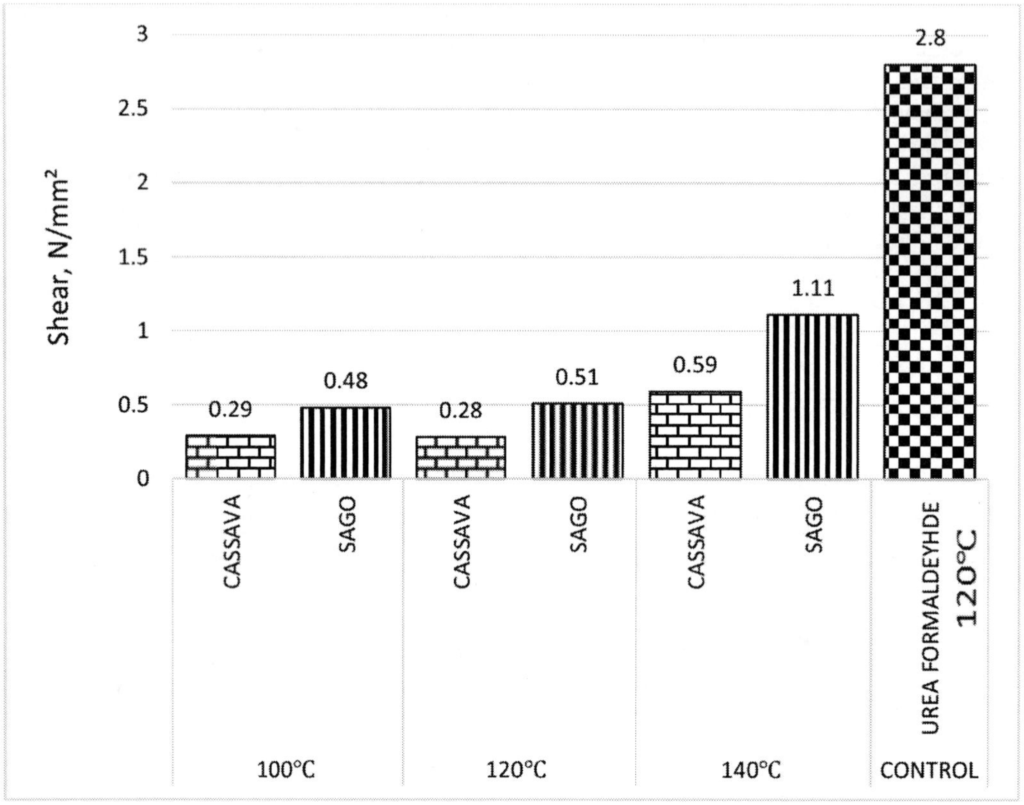

Fig. 4. Graph shows shear strength (N/mm²) for different starch-based adhesives and urea formaldehyde as binder in plywood pressed at different temperatures.

4 Conclusions

Mechanical testing were carried out with bending test and shear test to study bonding of starch-based adhesives with veneers. Cassava and sago starch-based adhesive in plywood with 120°C had higher elasticity and more elastic than urea formaldehyde as Control about 12318.56 N/mm² and 12376.34 N/mm². By comparison with starch-based adhesives and urea formaldehyde as Control in plywood at 120°C hot press temperature, sago starch-based adhesive showed slightly lower than urea formaldehyde was about 5.36 N/mm². At 120°C hot press temperature, cassava and sago starch-based adhesives showed mean value in shear strength as 0.28 N/mm² and 0.51 N/mm² when compared with urea formaldehyde recorded as 2.8 N/mm² due to pure starch-based adhesives were no good performance in shear test. In short, starch-based adhesive is an environmental friendly product, but in the process of making starch adhesive is expensive than urea formaldehyde. Hence, for the consumer who is consider environmental friendly can buy and use starch-based adhesive of plywood for making furniture or for floor and wall.

The authors sincerely thanked and acknowledged the Faculty of Science and Natural Resources, Forestry Complex, Universiti Malaysia Sabah for kindly supporting and providing the assistance and equipment throughout this research.

References

1. A.J. Kinloch, *Adhesion and Adhesives: Science and Technology* (Springer Science and Business Media, 1987)
2. L.F.M. da Silva, (ed.), *Handbook of Adhesion Technology: with 97 tables* (Springer Science & Business Media, 2011)
3. U. Uthumporn, Y.N. Shariffa, A. Fazilah, A.A. Karim, Colloid and Polym. Sci., **290**, 15 (2012)
4. H. Yu, Y. Cao, Q. Fang, Z. Liu, Bioresources, **10**, 2 (2015)
5. S. Akbari, A. Gupta, T.A. Khan, S.S. Jamari, N.B.C. Ani, P. Poddar, J. Ind. Acad. Wood Sci., **11**, 2 (2014)
6. J. Shey, S.H. Imam, G.M. Glenn, W.J. Orts, Ind. Crop. Prod., **24**, 1 (2006)
7. S.A. Mohamed, A.L. Al-Malki, T.A. Kumosani, Aust. J. Basic & Appl. Sci., **3**, 3 (2009)

Notched Strength of Woven Fabric Kenaf Composite Plates with Different Stacking Sequences and Hole Sizes

Anders Hans Romayne[1,*] and *Hilton* Ahmad[1]

[1]Department of Structures and Materials Engineering, Universiti Tun Hussein Onn Malaysia, 86400 Parit raja, Batu Pahat, Johor, Malaysia.

Abstract. Advantages of using kenaf fibres over synthetic fibres in composites manufacturing are relatively cheap, less abrasive and hazardous during handling, and renewable materials. Current work investigates parametric effects on notched strength of woven fabric kenaf polymer composites plates with variation of lay-up types, notch sizes and plate thickness. Testing coupons are prepared using hand lay-up technique and circular notch were drilled prior to mechanical testing. Stress concentration at the notch edge promotes micro-damage event as tensile loading was applied leading to crack initiation and propagations across the plate width. It is suggested that woven fabric kenaf polymer composites are potentially used in low and medium load bearing applications.

1 Introduction

Applicability of Carbon Fibre Reinforced Polymer (CFRP) and Glass Fibre Reinforced Polymer (GFRP) in Malaysia is relatively limited as compared to other engineering materials and current supplies are generally imported, production of composite materials made from local-produced fibres (e.g. kenaf fibres) may reduce CFRP and GFRP dependency and wider its applicability to a greater extent. Kenaf fibres are relatively cheaper, lighter, renewable, recyclable, biodegradable and less abrasive compared to synthetic fibres. Excellent tensile strength and modulus of kenaf fibres were reported by previous researchers [1]. Previous works [2] reported promising results of un-notched strength of woven fabric kenaf composite plates. Research works on other natural fibres are aggressively ongoing in searching for utilizing natural fibre in composite materials that are renewable and sustainable. Tenth Malaysian Plan (RMK10) has highlighted to activate agricultural sector with high driven value and biotechnology in agricultural research leading to larger production of agriculture in near future. Woven fabric fibre composites had attracted much attention from engineers and scientist in various sectors due to several advantages such as drape ability, a woven fabric layer is equivalent to two unidirectional layers and excellent resistance against impact and fatigue. However, compared with unidirectional composites counterpart, existing of crimping region in woven fabric composites leading to strength and stiffness degradation but offset by the ability of woven

* Corresponding author: hilton@uthm.edu.my

fabric to arrest crack growth and excellent fracture absorption at the tow crossover points. Combinations of weaving patterns, tow architectures and lay-up configurations lead to complexities of associated damage morphologies.

Comprehensive review of published literatures concerned with notched plates in composite materials and associated experimental observations has been presented [3-6]. Plates with discontinuities such as notch introducing stress concentrations at the vicinity of notch edge and different fracture and failure mechanisms are observed as compared with unidirectional (non-woven) lay-ups counterparts. Crack is initiated as the composite fracture toughness is exceeded and propagated along net-tension plane in self-similar manner until certain distance prior to catastrophic failure. The distance ahead the notch edge prior to catastrophic failure is known as process zone length where micro-damage event takes place. In a notched woven fabric GFRP composite subjected to remote tension, it has been shown experimentally using a model (transparent) GFRP system that prior to failure an intense zone of damage develops in the region of maximum tensile stress [4]. Damage within that zone comprises of matrix cracking, fibre-matrix splitting and fibre tow fracture. This approach is important to further the exploratory of plates associated with stress concentrations such as in bolted and welded joints [7-8].

Most of structures parts assembly requires discontinuity in the form of notches and cut-outs for number of reasons. The failure mechanisms in composite materials are rather complex, experimental work need to be carried out to study in-depth the behaviour of these material when subjected to loading. Notched plates response to remote tensile loading is somewhat complex (even more complex with woven fabric fibres due to existing of crimp region) as micro-failure involved ahead of the notch tip, and strength prediction on notched plate had shown quite limited success [5]. There is no reported literatures on notched strength of woven fabric kenaf composites, current work concentrates on parametric study of notched strength woven fabric kenaf composite plates with variation of lay-up types, hole size and plate thicknesses under quasi-static loading. The experimental methodology includes woven fabric kenaf coupon preparation as designated in testing series, drilling of circular hole and conducting mechanical testing.

2 Experimental works

Figure 1 showed the flow chart of current experimental framework.

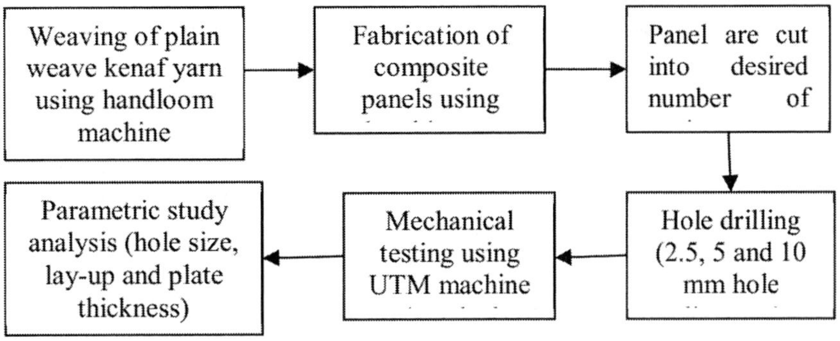

Fig. 1. Flow chart of experimental framework

Woven fabric layers were prepared using handloom weaving machine to required numbers as specified to testing series. Kenaf fibres are wrapped to avoid hairy fibre yarn as exposed to air and kept in UTHM Textile Laboratory. Current work use epoxy resin as matrix binder and fabricated using hand lay-up technique to produce composites lay-up panels. All fabricated panels were cut into testing coupon size using water-cooled saw and carefully cut to prevent any cracking failure along the coupon edge. Each testing coupons has 130 mm gauge length and plane width of 25 mm with varying notch sizes (i.e., open hole diameters), *d*, of 2.5, 5.0 and 10.0 mm giving *d/W* ratios of 0.1, 0.2 and 0.4 respectively. End tab is prepared to provide better gripping and to avoid coupon slipping during mechanical testing. Hole drilling were carefully conducted with low helix and high speed and all notch length was inspected visually to ensure there was no defects were seen.

3 Testing series

Testing series of current work were given in Table 1 which comprised of variation combinations of hole size, lay-up types and plate thicknesses. Three testing coupons were prepared for each testing configurations, notched strength are averaged and given as σ_N = peak load (P_{max})/[thickness, t x plate width, W]. Mechanical testing were carried out using Instron 1175 machine 3369 with a crosshead speed of 0.5 mm/min and 50 kN load cell.

Table 1. Testing series under investigation in current work

Weave type	Orientation	Lay-up	Designation
Plain Weave (PW)	Cross-ply	(0/90)	PX1
		(0/90 | 90/0)	PX2
		(0/90 | 90/0 1 0/90 / 90/0)	PX4
	Quasi-isotropic	(0/90 | +45/-45)	PQ2
		(0/90 | +45/-45 | -45/+45 | 90/0)	PQ4
		(60/-60 | -60/60)	PT2
		(+45/-45 | -45/-45)	PS2

4 Results and discussions

Figure 2(a) showed load-displacement profiles taken from data-logger that showed ultimate load of PX4 lay-up plates with different hole size. From the graph, the notched plate of woven fabric kenaf composite demonstrated a linear elastic behaviour until ultimate load and failure in a brittle manner. It is assumed that plastic deformation in tensile testing of notched plate is negligible, as found in other composite systems (Belmonte [4] and Manger [6]). Experimental observation found that a screeching sound were heard after few minutes testing, associated with matrix cracking and delamination occurred at its notch edge. Thereafter, fibre fracture occurred is an indicator of reaching peak load prior to catastrophic failure. Although this study did not measure the damage zone length (due to limited scope of current work), the existence of damage zone at vicinity of the notch has been reported by previous researchers [4-6].

Figure 2(b) showed an example of 5 mm hole size testing coupon at failure. As tensile force is applied, a crack has grown to approximately constant through the plate thickness across the net-tension plane (parallel to plate width). When the crack reached a critical length (associated with process zone length), catastrophic failure occurs ahead of the critical length. The size of damage zone length is approximately a size of radius hole. The notched strength decreased as hole size increased due larger remaining unloaded plate width, as expected. This also associated with smaller area of damage zone length developed at the hole vicinity prior to catastrophic failures as reported by Belmonte [4].

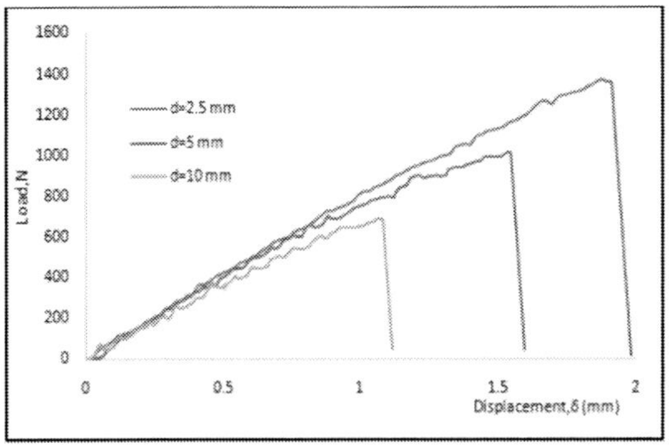

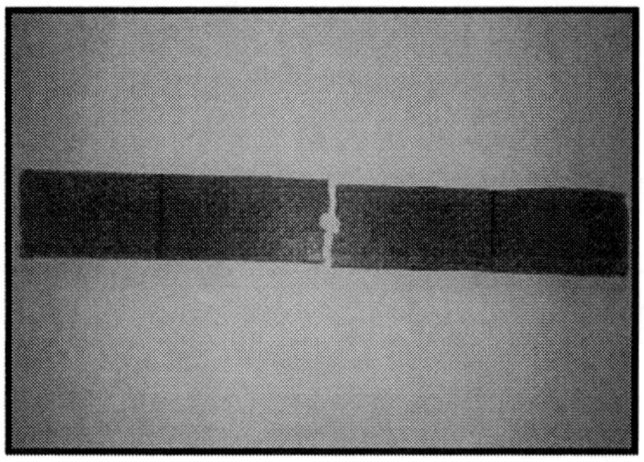

(a) (b)

Fig. 2. (a) Load-displacement profiles of cross ply lay-up for PX4 for 2.5, 5 and 10 mm hole size Fig. (b). Photograph showing a laminate coupon of cross-ply lay-up containing 5 mm circular hole at failure.

As mentioned previously, increased notched strength with smaller hole size is expected due to larger unloaded region from the vicinity of the notch. It is also associated with the formation of damage zone length which formed perpendicularly to the loading direction. As composite material displayed brittle failure types, occurrence of damage formation at the hole vicinity can help the explanation of low strength on larger hole size. Belmonte [4] concluded interrupted test to confirm the occurrence of matrix cracking, delamination and numbers of tow fracture under scanning electron microscope (SEM). However, interrupted test and determination of damage zone length is not covered in the current scope. Notice that the hole size increments were doubled from 2.5 mm to 5 mm and correspondingly 5 mm to 10 mm. About 10-20% strength reduction were observed with every increment of hole size. From PX series lay-up, it is also found that larger strength reductions were found in thicker plates. Similar trends were observed in all other lay-ups.

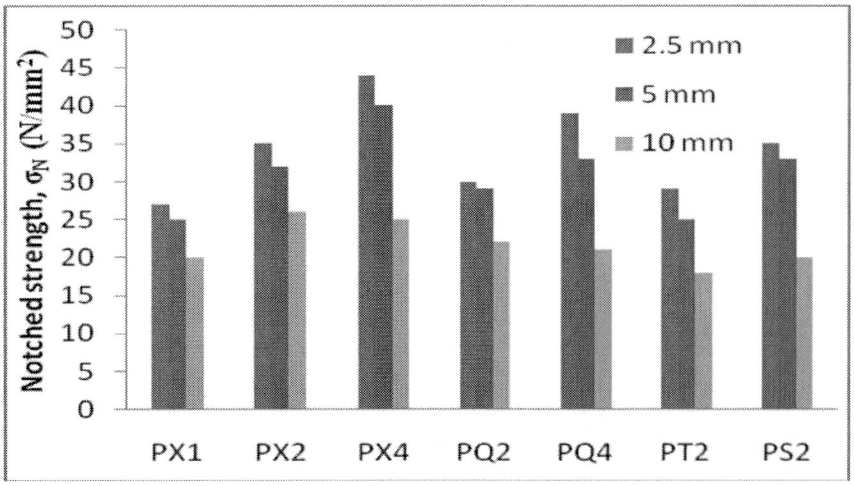

Fig. 3. Notched strength of all testing series conducted

It is obvious that cross-ply lay-up demonstrated higher notched strength as compared with equivalent quasi-isotropic lay-up. It is important to note that 0° is the optimum fibre direction to carry loading stress as the plate is subjected to remote tensile loading. Bear in mind that each woven fabric layer consists of two sets of orthogonal plies which equivalent to two unidirectional plies. Cross-ply lay-up consists of more 0° plies compared to equivalent quasi-isotropic lay-up of similar plate thickness. Therefore, higher notched strength in cross-ply lay-up is perhaps not surprising. The average measured plat thickness of PX1, PX2 and PX4 are 0.7 mm, 1.8 mm and 3 mm respectively. About 10% - 25% increment of notched strength were observed with increment of woven fabric layer.

Better notched strength with larger plate thickness is contributed by extra resistance from interfacial strength from neighbouring layers to reduce occurrence of damage event from delamination. Again, the study of delamination damage event is outside the scope of current work, but high resistance to delamination as plate thickness increases was reported by previous researchers (Belmonte, [4]). He also found that constant micro-damage event occurred in the form of matrix cracking, delamination and fibre fracture through the plate thickness, portrays that higher resistance to damage as plate thickness increases. In reality, most composite components in civil engineering works are relatively thicker than any other engineering sectors. Therefore it is beneficial to study wide range of notched plate thicknesses to explore their structures response upon loading and applicability of these materials in structures application. It may exhibit more complex failure mode as the plates are getting thicker.

It was found that woven fabric CFRP [4] and GFRP [6] systems exhibited larger notched strength as compared with KFRP system. Kenaf fibre exhibited much lower modulus and tensile strength as compared with synthetic fibre. On the other hand, the carbon fibres and to the less extent glass fibre has higher stiffness, high tensile strength, low weight and high chemical resistance. Therefore, lower notched strength of woven kenaf fabric composite system were seen as compared to woven fabric synthetic fibres systems is much expected. However, as notched strength gives the value in the range if 18 – 44 N/mm^2 is acceptable in low and medium load bearing application system. Although strength properties in carbon fibre system are much stronger but carbon fibres are much expensive in as it is considered as imported good. On the other hand, kenaf fibres are renewable, sustainable and much cheaper alternatives to synthetic fibres as it is grown locally. Bear in mind that the available notched strength datasets for synthetic fibres composites were conducted in large manufacturing factory scale, current study used in-house fabrication techniques with limited resources, therefore finishing plates may not as good as in factory-manufactured plates.

5 Conclusions

It was found that better notched strength was exhibited by cross-ply lay-ups due to larger volume fraction of 0° plies compared to equivalent quasi-isotropic lay-ups series. As expected, notched strength decreased with larger hole size due to larger unloaded region width prior to catastrophic failures. This is associated with process zone length where micro-damage event in the form of matrix cracking, delamination and fibre fracture occurred. Similar trends were found for plate thickness parametric study which notched strength were increased with thicker plates. Woven fabric kenaf fibres are renewable, sustainable and cheap materials gives alternatives to material designer to opt in low and medium load bearing applications.

The authors would like to be obliged to Universiti Tun Hussein Onn Malaysia for providing laboratory facilities and financial assistance under project Vot No. U245.

References

1. V. Nair, P. Khosia, M. Ramachandran, Res. J. Pharm., Biol. Chem. Sci., **7(1)**, 2001 (2016)
2. L.S. Yee, H. Ahmad, *MATEC web conf.*, **47**, 1 (2016)
3. H. Ahmad, ARPN Journal of Engineering and Applied Sciences, **11**(3), 1577 (2016)
4. H.M.S. Belmonte, C.I.C. Manger, S.L. Ogin, P.A. Smith, R. Lewin, Compos. Sci. Technol., **61**, 585 (2001)
5. P.P. Camanho, P. Miami, *16th International Conference on Composite Materials, Kyoto, Japan*, (2007)
6. C.I.C. Manger, *Failure of Notched Woven GFRP Composites: Damage Analysis and Strength Modelling*, PhD Thesis, (University of Surrey,Guildford 1999)
7. N.A. Rahman, W. Tizani, Adv. Mat. Res. **1025-1026**, 950 (2014)
8. N.A. Rahman, W. Tizani, T. Pitrakkos, J. Constr. Steel. Res., **93**, 1 (2014)

Manufacturing of Fire Resistance Geopolymer: A Review

Ikmal Hakem Aziz[1,*], *Mohd Mustafa Al Bakri* Abdullah[2] , *Heah* Cheng Yong[1] , *Liew* Yun Ming[1] , *Kamarudin* Hussin[1,2] , *Aeslina* Abdul Kadir[3] and *Emy Aizat* Azimi[1]

[1]Center of Excellence Geopolymer and Green Technology, School of Materials Engineering, Universiti Malaysia Perlis (UniMAP), P.O. Box 77, D/A Pejabat Pos Besar, 01000, Kangar, Perlis, Malaysia
[2]Faculty of Engineering Technology, Universiti Malaysia Perlis (UniMAP), P.O. Box 77, D/A Pejabat Pos Besar, Kangar, Perlis 01000, Malaysia
[3]Department of Water and Environmental Engineering, Faculty of Civil and Environmental Engineering, University Tun Hussein Onn Malaysia, Batu Pahat Johor, Malaysia

Abstract: Protection against fire using inorganic polymer is a new application of engineering technology. Even though, there are varieties of fire-protection materials, there is always a need for the development of new materials with improved thermophysical properties and low cost. Geopolymer composites materials are promising from this point of view. Granulated blast furnace slag, boiler ash and fly ash have been used as the prime materials for forming geopolymers composites. Geopolymers have been studied due to its unique properties such as a good fire resistance. Geopolymer offers an innovative for application associated with the high thermal application. This paper summarizes on the potential of alkali-activated materials over the past decades along with outlines of the manufacturing of geopolymer composites for fire resistance application.

1 Introduction

Ordinary Portland cement (OPC) is the topmost volume engineering materials in use today. It has been used as a binding agent for concrete due to its exceptional thermal performance, mechanical properties and durability [1]. However, there is an argument in environmental issues on the enormous energy consumption and emissions of carbon dioxide (CO2) during its manufacture [2].The manufacturing of concrete in industry is one of the major contributors of the global warming. It has been reported that 13,500 million tons of CO2 are produced from the process worldwide, accounting approximately 7% of the annually produced greenhouse gases [3]. Practically, geopolymers are increasing interest as ecologically friendly fireproof building materials, sound heat insulators, materials for encapsulating hazardous wastes and low energy consumption during production from raw materials [4, 5]. Geopolymer are well known as having great properties including high

[*] Corresponding author: ikmalhakem@gmail.com

compressive strength, high temperature stability, low thermal conductivity, and high thermal engineering applications [6-8].

Thermal conductivity, strength retention after heat treatment, non combustibility, and temperature capability are used as indicators for fire endurance of geopolymer materials. Geopolymer has been proposed for very good heat resistance having thermal stability up to 800 – 1200°C [9]. Nevertheless, it is reported that geopolymers panels made of granulated slag filler and metakaolin exhibited an improved fire resistance behavior. The temperature on the cold side of the panels was 350°C after exposure up to 1100°C for 35 min [10]. Others, Zhang et al. [9] found that geopolymer derived from alkali activation showed excellent strength retention after heating to temperature from 100 to 800°C, and the post-cooling compressive strength increased by as much as 100% after exposure 800°C due to densification and phase transformation as shown in Figure 1. This paper review demonstrate that geopolymer composites are promising fire resistance material for engineering applications.

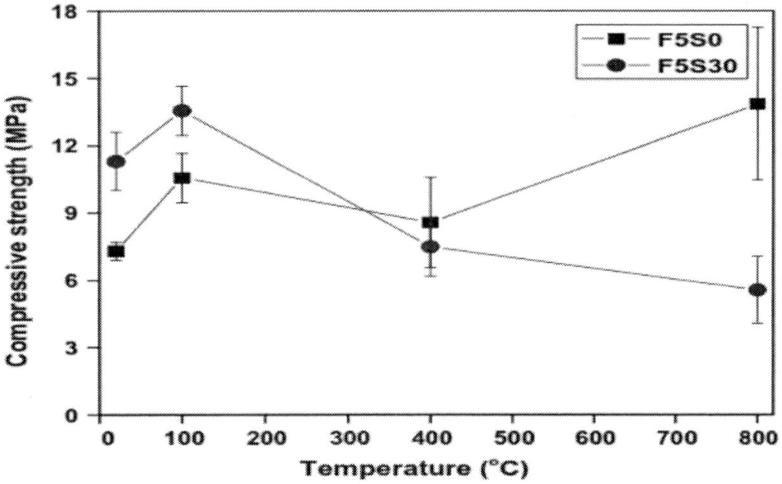

Fig. 1. Compressive strength of Geopolymer after exposure to different temperature [9].

2 Sources materials

Granulated blast furnace slag (GBFS) potential industrial waste used as raw material in geopolymerization process. Based on the study by Khale et al. [11], slag containing a wide range of natural Al-Si minerals, could be developed as potential sources materials for the synthesis of geopolymers. The chemical composition of the slag materials is shown in Table 1 which demonstrates major composition in the source material of geopolymer is SiO2.

Table 1. XRF analysis data of Blast furnace slag and boiler slag composition [12, 13].

Materials	SiO_2	Al_2O_3	Fe_2O_3	TiO_2	CaO	MgO	Na_2O
Blast Furnace Slag	34.39	14.47	0.63	0.53	41.6	6.49	0.22
Boiler Slag	49.29	25.39	7.65	N/A	0.63	0.53	N/A

Fly ash also becomes another potential waste material that contains a high percentage of amorphous silica and alumina precursors [14]. Geopolymerization process with fly ash happened via a mechanism involving the dissolution of aluminum and silicon species from the surface of fly ash as the surface hydration of undissolved waste particles, the polymerization of active group, gel formation and subsequently hardening of geopolymer structure [15]. Table 2 present one of the general chemical composition of fly ash obtained by XRF analysis.

Table 2. XRF analysis data of fly ash composition [16].

Chemical	Fly ash
SiO_2	52.11 %
Al_2O_3	23.59 %
Fe_2O_3	7.39 %
TiO_2	0.88 %
CaO	2.61 %
MgO	0.78 %
Na_2O	0.42 %

3 Alkaline solution

The fundamental of geopolymerization are the dissolution of solid aluminosilicate oxide M-OH solution where M is the class of alkali metal (mostly Na and K), dissolution of dissolved Al and Si complexes to an inter-particle space, then the formation of a gel phase by polymerization between silicate solution and Al and Si complexes, and the hardening formed in gel phase at the finishing steps [17]. The mixture of sodium hydroxide (NaOH) and sodium silicate (Na2SiO3) is the alkaline solution that often used [18]. Alkali hydroxide is needed for the dissolution of aluminosilicate sources while sodium silicate solution acts as a binder, plasticizer and alkali activator [19]. Alkaline solution also encourages a few amount of Si and Al atoms to dissolve the aluminosilicate sources from monomer in solutions, and then polycondense to form a rigid framework [20]. The rate of polymerization reaction increases as the sodium silicate is the preferred activating solution owing to its soluble silicate content. The specific surface composition of the initial raw materials, the composition, type, and quantity of alkali activator influencing the properties of geopolymers [21]. Besides, the various ratios of solid-to-liquid (S/L) ratio and Na2SiO3-to-NaOH ratio have significant effect on compressive strength [22].

3.1 Geopolymerization reaction and mechanism

Geopolymerization is a complex process and it is involving three steps as shown in Figure 2; (Eq.1) dissolution in alkaline solution; (Eq.2) reorganization and diffusion of dissolved ions with formation of small coagulated structures; (Eq.3) polycondensation of soluble species to form hydrated products [6-7,19]. A chemical reaction covering the polycondensation of orthosialiate (hypothetical monomer) that proposed by Davidovits [23].

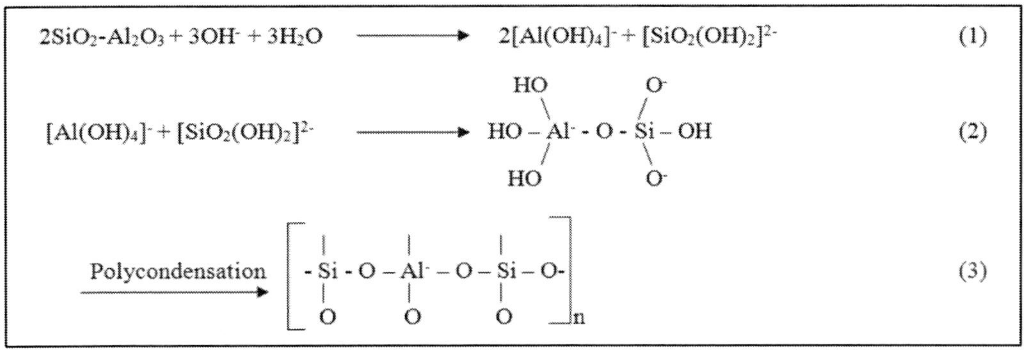

Fig. 2. Three steps of geopolymerization process [19].

3.2 Manufacturing geopolymer

Figure 3 presents the steps in manufacturing of geopolymer. There are involving a few technological parameters to achieve their certain properties of geopolymer such as fire resistance. Practically, the mixing of concentrated aqueous alkali hydroxide or silicate solution with solid aluminosilicates form a synthetic alkali aluminosilicates. Furthermore, filler may add in the preparing of geopolymer materials in order to achieve the required properties. Temuujin et al. [24] conducted research on the calcining of metakaolin at temperature 750°C for 24 hours. The calcined kaolin with the composition of Si:Al ratio 2.5 and Na:Al ratio of 1 was prepared as the optimum ratio to produce geopolymer. Water acts as an additional factor on the preparing of geopolymer material. The unexpected expansion with temperature of MK 2.5 provided a level of matching with the thermal expansion of the substrate resulting in the maintenance of the structural integrity of the resulted product. Calcining of bulk geopolymer with Si: Al = 2.5 to 1000°C resulted in formation of sodium aluminosilicate.

Mustafa et al. [25] study on fly ash based geopolymer for coating application at high temperature. The fly ash geopolymer coating was prepared by mixing alkaline activator solution 12M NaOH with the aluminosilicates. The ratio of fly ash/alkaline activator to Na2SiO3/ NaOH was standardized at 2.5. After compounding process, the slurry paste formed then applied to ceramic substrate by dipping method with the coating thickness ranging between 0.3-05mm. The ceramic plates were cured at 70°C for 24 hours. In order to improve thermal properties and strength of the obtained geopolymer, samples was sintered for 3 hours at temperature ranged between 600°C and 1500°C with heating and cooling rates 5°C/min. The result obtained that the highest compressive strength of 40 MPa improve proportional with the sintering temperature, 1500°C [25].

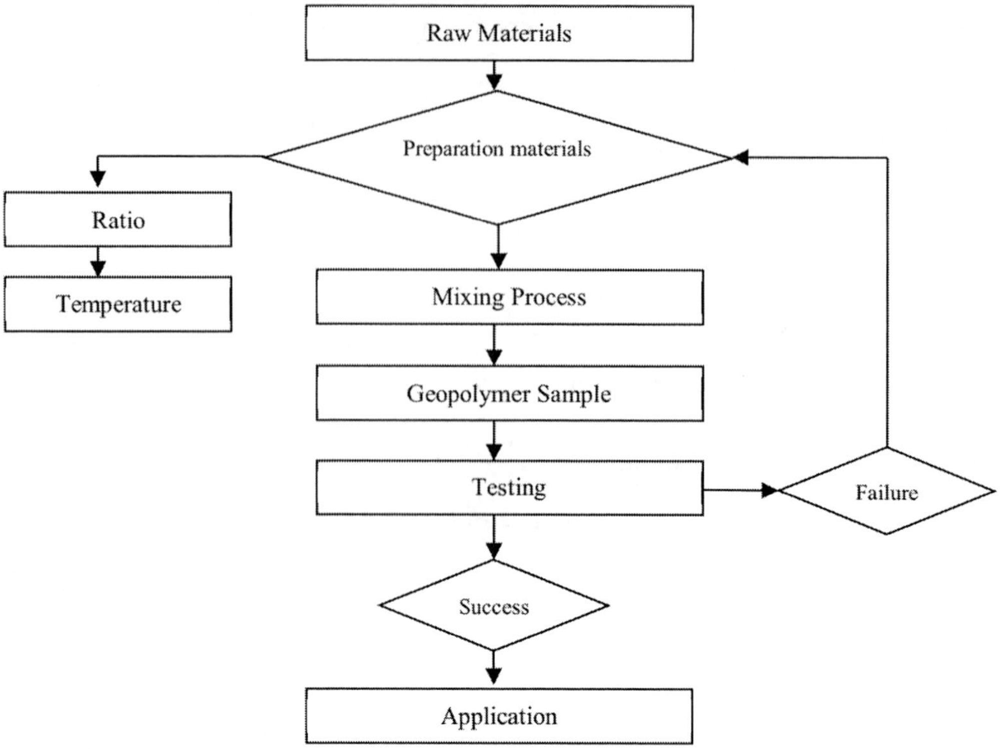

Fig. 3. General process in manufacturing of geopolymer

4 Conclusion

This paper summarized the manufacturing of geopolymer material for fire resistance applications. Furthermore, geopolymer consists great mechanical and physical properties which offer economic and environmental friendly as an alternative to Ordinary Portland cement (OPC). There is not much study being carried out on the application of geopolymer material at high temperature. Thus, the development of geopolymer for future research in this field will lead a new era of material engineering in fire fighting/ safety precautions area.

The authors gratefully acknowledge Center of Excellence Geopolymer and Green Technology (CEGeoGTech) and School of Materials Engineering , UniMAP for their expertise and support. The authors would also like to thank for the funding support from the Fundamental Research Grant Scheme (FRGS-9003-00540) under Ministry of Education Malaysia (MOE).

References

1. H. Djwantoro, E.W. Steenie, M.J.S. Dody, B.Vijaya Rangan, Aci. Mater. J., **101(6)**, 467 (2004)
2. V. Malhotra, Advances in Concrete Technologies in the Middle East, (2008)
3. A.M. Mustafa Al Bakria, H. Kamarudin, M. BinHussain, I. Khairul Nizar, Y. Zarina, A.R. Rafiza, Physics Procedia, **22**, 286 (2011)
4. R.E. Lyon, P.N. Balaguru, A. Foden, U. Sorathia, J. Davidovits, M. Davidovics, Fire Mater., **21(2)**, 67 (1997)

5. H. Khater, MATEC web conf., **11**, (2014)
6. D. Dimas, I. Giannopoulou, D. Panias, J. Mater. Sci., **44(14)**, 3719 (2009)
7. S. Alonso, A. Palomo, Cement Concrete Res., **31(1)**, 25 (2001)
8. Y.J. Zhang, Y.C. Wang, S. Li, Mater. Sci. Eng., **527(24)**, 6574 (2010)
9. Z. Zhang, J.L. Provis, A. Reid, H. Wang, Cement Concrete Comp., **62**, 97 (2015)
10. L. Vickers, A. Van Riessen, W.D.A Rickard, *Fire-Resistant Geopolymers: Role of Fibres and Fillers to Enhance Thermal Properties*, (Springer 2015)
11. D. Khale, R. Chaudhary, J. Mater. Sci., **42(3)**, 729 (2007)
12. T. Cheng, J. Chiu, Miner. Eng., **16(3)**, 205 (2003)
13. M.A. Villaquiran-Caicedo, R.M. de Gutierrez, S. Sulekar, C. Davis, J.C. Nino, Appl. Clay Sci., **118**, 276 (2015)
14. U. Rattanasak, P. Chindaprasirt, Miner. Eng., **22(12)**, 1073 (2009)
15. H. Xu, J. Van Deventer, G. Lukey, Ind. Eng. Chem. Res., **40(17)**, 3749 (2001)
16. M. Al Bakri, A. Mohd, H. Kamarudin, O.A. Abdulkareem, C.M.R. Ghazali, A.R. Rafiza, M.N. Norazian, Applied Mechanics and Materials, **110**, 734 (2012)
17. Y.M. Liew, H. Kamarudin, A.M. Mustafa Al Bakri, M. Binhussain, M. Luqman, I. Khairul Nizar, C.M. Ruzaidi, C.Y. Heah, Phys Procedia, **22**, 312 (2011)
18. N. Lloyd, B. Rangan, *2nd International Conference On Sustainable Construction Materials And Technologies, Ancona, Italy*, (2010)
19. K. Komnitsas, D. Zaharaki, Miner. Eng., **20(14)**, 1261 (2007)
20. H. Kamarudin, Y.M. Liew, A. Mohd Mustafa Al Bakri, M. Luqman, I. Khairul Nizar, C.Y. Heah, Aust. J. Basic & Appl. Sci., **5(9)**, 441 (2011)
21. P. Rovnanik, Constr. Build Mater., **24(7)**, 1176 (2010)
22. D.L. Kong, J.G. Sanjayan, K. Sagoe-Crentsil, Cement Concrete Res., **37(12)**, 1583 (2007)
23. J. Davidovits, J. Therm. Anal. Calorim., **37**, 1633 (1991)
24. J. Temuujin, W. Rickard, M. Lee, A. Van Riessen, J. Non-Cryst. Solids, **357**(5), 1399 (2011)
25. A.M.A.B. Mohd, L. Jamaludin, K. Hussin, M. Binhussain, C.M.R. Ghazali, A.M. Izzat, Adv. Mat. Res., **686**, 227 (2013)

Optimization of Machining Parameters for Minimization of Roundness Error in Deep Hole Drilling using Minimum Quantity Lubricant

Anis Farhan Kamaruzaman[1,*], *Azlan* Mohd Zain[1], *Noordin* Mohd. Yusof[2], and *Farhad* Nadjarian[2]

[1]Soft Computing Research Group, Faculty of Computing, Universiti Teknologi Malaysia, 81310 UTM, Skudai, Johor, Malaysia
[2]Faculty of Mechanical Engineering, Universiti Teknologi Malaysia, 81310 UTM, Skudai, Johor, Malaysia

Abstract. This paper presents an experimental investigation of deep hole drilling using CNC milling machine. This experiment investigates the effect of machining parameters which are spindle speed, feed rate and depth of hole using minimum quantity lubricant on the roundness error. The experiment was designed using two level full factorial with four center point. Finally, the machining parameters were optimized in obtaining the minimum value of roundness error. The minimum value of roundness error for deep hole drilling is 0.0266 at the spindle speed is 800 rpm, feed rate is 60 mm/min, depth of hole is 70 mm and minimum quantity lubricant is 30ml/hr.

1 Introduction

The demands in industrial nowadays need for the application of machining process to produce a particular product compared to the use of manpower in the past. Therefore, various types of machines have been introduced so far. Prior conventional machining was applied as turning [1, 2], milling [3, 4] drilling [5, 6], grinding [7] and so on. By its desire to meet the needs and demands of the present time, non-conventional machining or modern machining was introduced as abrasive water jet (AWJ) [8, 9], electrochemical machining (ECM) [10, 11], electric beam machining (EBM) [12, 13] and etc. Each machining process applied is based on the products to be produced. Thus, for the production of deep holes, a particular machining process is needed namely deep hole drilling.

The difference between drilling and deep hole drilling is based on the concept of deep hole drilling itself. Deep hole drilling is defined by its depth to diameter ratio (D: d) where the ratio is greater than 10: 1. In other definition, deep hole machining means machining of holes deeper than ten times the diameter of the holes. Deep hole drilling can be applied to a wide range of materials from aluminium to the super-alloy.

However, in order to produce a high-quality product range of issues that may affect the quality of the product is noteworthy. Furthermore, deep hole drilling process is a new

* Corresponding author: anisfarhankmi@yahoo.com

machining process was introduced and there is not much research being done to improve the quality of the resulting holes. Deep hole drilling process involves the production of a deep hole where any machining parameters and constraints during the machining process is carried out can affect the quality of the hole.

The importance of deep hole drilling can be seen in some areas such as in the area of aerospace [14], oil and gas [15], mold and die [16] and others where the quality of the hole resulting the effectiveness of the product. Therefore, it is necessary to study and run the experiments to identify the effect of machining parameters on hole quality.

For this purpose, the effect of different combinations of spindle speed, feed rate, depth of hole and MQL was examined. The optimum machining parameters were specified and the effect of cutting parameters on the roundness error was investigated.

2 Experimental Setup

The deep hole drilling experiments were carried out on a Maho Deckel Model MH500E CNC milling machine. The workpiece material used in the experiment was steel alloy (Fig. 1). The tools use was twisted drills (Fig. 2). There are four machining parameters involved in conducting this experiments which are spindle speed, feed rate, depth of holes and minimum quantity lubricant (MQL). The four machining parameters and constraints involved in this experiment are shown in Table 1 below.

Table 1. Machining parameters and constraints.

Machining parameters	**Constraints**
Spindle speed (*V*)	700-900 rpm
Feed rate (*f*)	50-70 mm/min
Depth of hole (*d*)	65-75 mm
Minimum quantity lubricant (*l*)	20-40 l/hr

Fig. 1. Workpiece.

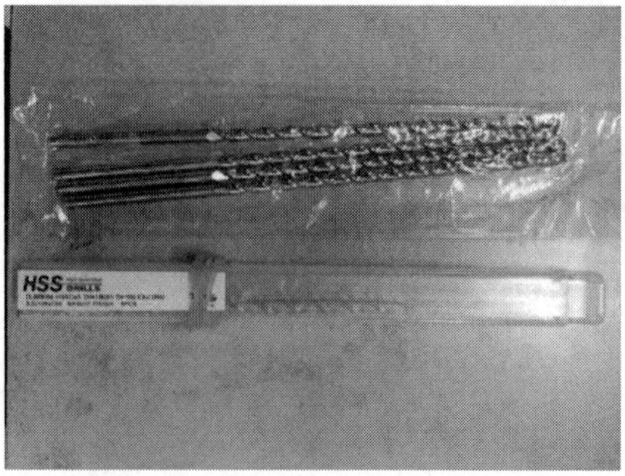

Fig. 2. Twist Drill.

The machine needs to be setup such as installing a workpiece, tool and MQL system. In the preliminary experiment, design of experiment (DOE) has been made using Minitab 17 software. The experiment was designed using two level full factorial with four center point. There are twenty tests conducted based on DOE. All experiments were performed based on the DOE. Therefore, this experiment involves twenty experiments to identify the effect of machining parameters on roundness error while identifying the optimum machining parameters to produce a minimum roundness error. The output which is roundness error was measured by a coordinate measuring machine (CMM).

3 Results and discussion

The variation of the roundness error of the drilled holes in relation to machining parameters involved was analyzed in this research. The result of experiments for roundness error is shown in Table 2 as shown below.

The result for optimization of machining parameters of deep hole drilling for roundness error is 0.0266 mm. The optimal machining parameters are 800 rpm for cutting speed (*V*), 60 mm/min for feed rate (*f*), 70 mm for deep of hole (*d*) and 30 l/hr for minimum quantity lubricant (*l*). It was found that minimum roundness error obtained at hole 20.

The ANOVA has been used to obtain the relative effect of the different factors by the decomposition of variance. By using ANOVA, the relative effect of the machining parameters with respect to diameter error was investigated to determine the optimal combination of the machining parameters. The results of ANOVA for this experiment was generated by Design Expert 7.

Table 2. The result of experiments for roundness error.

	Spindle speed (*V*)	**Feed rate (*f*)**	**Depth of hole (*d*)**	**Minimum quality lubricant (*l*)**	**Roundness error**
Hole 1	700	50	65	20	0.0327
Hole 2	700	70	65	20	0.0400
Hole 3	700	50	75	20	0.0550
Hole 4	700	70	75	20	0.0563
Hole 5	800	60	70	30	0.0527
Hole 6	700	50	65	40	0.0565
Hole 7	700	70	65	40	0.0639
Hole 8	700	50	75	40	0.0545
Hole 9	700	70	75	40	0.0605
Hole 10	800	60	70	30	0.0678
Hole 11	900	50	65	20	0.0683
Hole 12	900	70	65	20	0.0668
Hole 13	900	50	75	20	0.0467
Hole 14	900	70	75	20	0.0625
Hole 15	800	60	70	30	0.0457
Hole 16	900	50	65	40	0.0340
Hole 17	900	70	65	40	0.0366
Hole 18	900	50	75	40	0.0502
Hole 19	900	70	75	40	0.0524
Hole 20	800	60	70	30	0.0266

In Table 3, the results indicate that all the parameters are not significant for roundness error value as the α value used in ANOVA is 0.05. If $p < 0.05$, it shows that the observed different within four machining parameters are significant. However, all the machining parameters have $p>0.05$ which refer to the insignificant. This situation could indicate that this effect was certainly masked by the action of uncontrolled factors. These results would lead us to believe that some machining parameters need for improving this process.

Table 3. ANOVA for roundness error.

ANOVA for selected factorial model						
Analysis of variance table [Partial sum of squares - Type III]						
Source	**Sum of Squares**	**df**	**Mean Square**	**F Value**	**p-value Prob > F**	
Model	0.002047044	4	0.000511761	0.37576955	0.8221	not significant
A-Spindle Speed	0.000607178	1	0.000607178	0.445831272	0.5152	
B-Feed Rate	2.90946E-05	1	2.90946E-05	0.021363237	0.8859	
C-Depth of Hole	9.95027E-05	1	9.95027E-05	0.07306162	0.7909	
D-MQL	0.001311268	1	0.001311268	0.962822072	0.3431	
Curvature	3.39506E-05	1	3.39506E-05	0.024928854	0.8768	not significant
Residual	0.019066614	14	0.001361901			
Lack of Fit	0.015154451	11	0.001377677	1.056457169	0.5485	not significant
Pure Error	0.003912163	3	0.001304054			
Cor Total	0.021147608	19				

4 Conclusion

The experiment was conducted under three machining parameters and minimum quantity lubricant have managed to find a minimum value of roundness error. The minimum value of roundness error is 0.0266 at the spindle speed is 800 rpm, feed rate is 60 mm/min, depth of hole is 70 mm and MQL is 30ml/hr. The validation for the results of the experiment was performed using ANOVA. However the results were exhibited by ANOVA showed that the machining parameters and MQL is not significant to the roundness error. Therefore, the enhancements to the experiment need to be done to produce a better result.

Special appreciation to reviewer(s) for useful advice and comments. The authors greatly acknowledge the Research Management Centre, UTM and Ministry of Higher Education Malaysia(MOHE) for financial support through the Fundamental Research Grant Scheme (FRGS) No.R.J130000.7828.4F721.

References

1. T. Liu, C.J. Kim, Science, **346**, 6213 (2014)
2. A.R. Yildiz, Inform. Sciences, **220**, 399 (2013)
3. A.M. Zain, , H. Haron, S. Sharif, Expert Syst. Appl., **37(2)**, 1755 (2010)
4. A.M. Zain, H. Haron, S. Sharif, Expert Syst. Appl., **37(6)**, 4650 (2010)
5. J. Prasanna, L. Karunamoorthy, M.V. Raman, S. Prashanth, D.R. Chordia, Measurement, **48**, 346 (2014)
6. J.S. Nam, D.H. Kim, H. Chung, S.W. Lee, J. Clean. Prod., **102**, 428 (2015)
7. B. Ping, Y.C. Fu, , Z.W. Zhang, Z.C. Zhao, Mater. Sci. Forum, **770**, 263 (2014)

8. A.M. Zain, H. Haron, S. Sharif, , Expert Syst. Appl., **38(7)**, 8316 (2011)
9. A.M. Zain, H. Haron, S. Sharif, Appl. Soft. Comput., **11(8)**, 5350 (2011)
10. B. Ghoshal, B. Bhattacharyya, Int. J. Mach. Tool. Manu., **64**, 49 (2013)
11. F. Klocke, M. Zeis, S. Harst, A. Klink, D. Veselovac, M. Baumgartner, *Procedia CIRP*, **8**, 265 (2013)
12. M. Cronskar, M. Backstrom, L.E. Rannar, Rapid Prototyping J., **19(5)**, 365 (2013)
13. H.K. Rafi, N.V. Karthik, H. Gong, T.L. Starr, B.E. Stucker, J. Mater. Eng. Perform, **22(12)**, 3872 (2013)
14. O. Yilmaz, A.T. Bozdana, M.A. Okka, Int. J. Adv. Manuf. Tech., **74(9-12)**, 1323 (2014)
15. S. Agarwal, T.X. Phuoc, Y. Soong, D. Martello, R.K. Gupta, Can. J. Chem. Eng., **91(10)**, 1641 (2013)
16. M.C. Cakir, O. Irfan, K. Cavdar, Robot Cim-Int. Manuf., **21(2)**, 175 (2005)

Effect NaOH Concentration on Bagasse Ash Based Geopolymerization

Saloma[1,*], *Hanafiah*[1] , and *Kartika* Ilma Pratiwi[1]

[1]Sriwijaya University, Jl. Palembang-Prabumulih, KM.32, Inderalaya, Ogan Ilir, South Sumatra, 30662, Indonesia

Abstract. Geopolymer is a natural adhesive material which can be developed as a substitute for cement. The natural ingredients which want to use should contain silica and alumina. This paper uses bagasse ash as a basic material of mortar geopolymer. As an adhesive, the bagasse ash should be mixed with water and another activator alkali such as sodium hydroxide (NaOH) and sodium silicate (Na_2SiO_3). The NaOH's molarity variation are 8, 10, 12, 14 and 16 M with Na_2SiO_3/NaOH = 1,0 sand/bagasse ash = 2,75 and activator/bagasse ash = 0,42. This research use 50 x 50 x 50 mm cube sized specimen and conduct a compressive strength test with 3, 7, 14, 21 and 28 days. The fresh mortar test result showed that the use of NaOH's molarity variation influences the slump value and time setting. The bigger NaOH molarity variation that been used, the smaller slump value. But, the time setting is increased. While the result for density and compressive strength shown that the bigger NaOH molarity variation, the bigger density and the compressive strength. Maximum compressive strength resulted from the mixture of mortar geopolymer with 16 M concentration.

1 Introduction

Mortar geopolymer is an environmentally friendly material, which can be developed as a substitute for future construction material. While it can be used from industrial waste, the making process of geopolymer cement did not use as much energy as *Portland* cement. Portland cement needs 1.000°C temperature for clinker burning process. Those burning processes produce CO_2 which pollutes the air. The geopolymer making process could reduce the greenhouse gasses emission up to only 20% left.

The main ingredient for geopolymer is a natural ingredient which contains silica (Si) and Aluminum (Al). The silica and aluminium are important for polymerization bond. These basic ingredients then reacted with alkali activator solvents such as NaOH and Na_2SiO_3, with NaOH concentration about 8-16 M [1]. The rice husk ash has a huge potential to become the main ingredient of mortar geopolymer. The research has shown that after 28 days, the maximum compressive strength was 45 MPa with NaOH concentration for about 10 M [2].

* Corresponding author: saloma_571@yahoo.co.id

The basic ingredient used in this geopolymer research is bagasse ash waste, which is the fuel for sugar cane mill. Sugar cane is the raw material to produce sugar. This plant only grows in a tropical territory such as Indonesia. The age of this plant, from the plantation until the harvest, is around 1 year. In Indonesia, sugar cane usually grows in Java and Sumatera island. After harvest, the sugar cane processed into 5% sugar, 5% water and 90% the bagasse (sugar cane waste). The bagasse then used in a burning process within a steam boiler. The steam boiler is the main tools to stir the sugar cane mill.

Previous research on the use of bagasse ashes as a mixture of concrete material shown a satisfying result. [3] shown that the conventional concrete with a bagasse ash mixture could bear a 30-40 MPa compressive strength. [4-5] shown that the compressive, tensile and flexural strength a concrete with bagasse ash mixture, are 10% higher than an original concrete. Besides improving mechanical characteristic, bagasse ash also could improve workability. So, it does not need to use additional superplasticizer material.

2 Material

The basic material for mortar geopolymer making should contain high silica and alumina. In this research, the basic material is bagasse ash waste, which is the main fuel for sugar-cane mill. The bagasse ash comes from PT Perkebunan Nusantara X (Persero) Surabaya, East Java. Figure 1 shows the result of bagasse ash's scanning electron microscope (SEM) test. Based on this test, the bagasse ash shape has a spherical form with ± 20 µm maximum diameter. Figure 2 is the result of XRD analysis which shown that this material contains high silica and alumina.

Fig. 1. SEM bagasse ash.

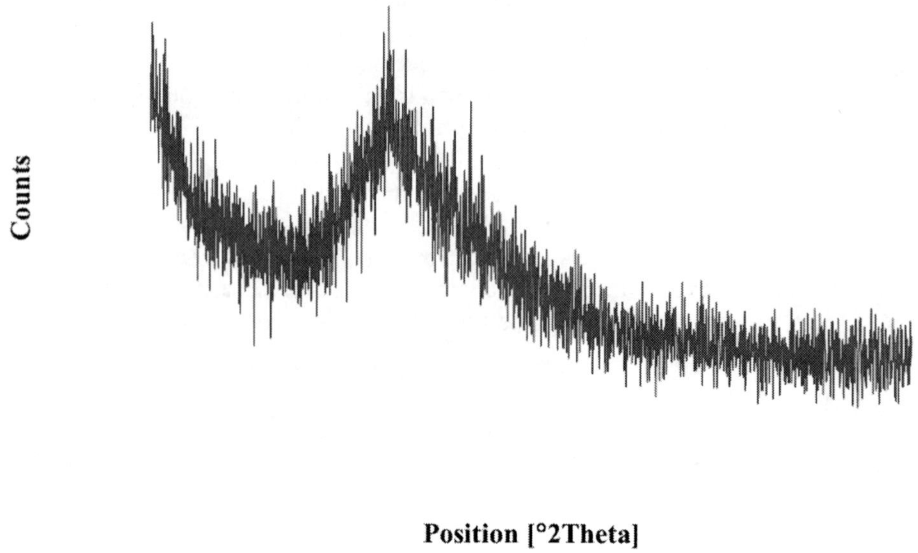

Fig. 2. XRD bagasse ash.

Activator that been used in this research is Na_2SiO_3 and NaOH. Na_2SiO_3 contain 58% gel-formed sodium silicate. Meanwhile, the powder-formed NaOH contain 98% purity level. NaOH mixed with water until it formed 8 M, 10 M, 12 M, 14 M, and 16 M concentration. This liquid then mixed with Na_2SiO_3. The comparison value between NaOH and Na_2SiO_3 is 1. Before it used to create mortar, this activator liquid bleached for about 1 day until it reaches a normal room temperature. The fine aggregate (sand) that used in this research comes from Tanjung Raja's rivers and OKI, South Sumatera. The ratio between sand/bagasse ash = 2,75 and the ratio between activator/bagasse ash = 0,42.

3 Laboratory test

This research conducted in Laboratory of Concrete and Material, Civil Engineering Faculty of Engineering, Universitas Sriwijaya. The standard for mortar mixed material refers to American Standard Testing and Material (ASTM C109). The experiment that has been conducted were workability, setting time, density and compressive strength test. The NaOH's molarity variation are 8, 10, 12, 14 and 16 M with Na_2SiO_3/NaOH = 1,0. sand/bagasse ash = 2,75 and activator/bagasse ash = 0,42. The workability test conducted by test the flow table to know the diameter distribution. While the compressive strength test conducted using a 50 x 50 x 50 mm cube-sized material for 3, 7, 14, 21 and 28 days period.

4 Result and discussion

4.1 Test result of slump flow

The test result for slump flow showed in Figure 3. Based on Figure 3, the higher NaOH concentration, the lower slump flow value. This happens due to the bigger NaOH concentration that been used, the bigger its density in a 1-liter liquid water. So, it makes the

NaOH liquid become thicker. Therefore, the use of small NaOH concentration makes the mortar mixed become flow and produced a broader diameter.

The decreasing percentage of slump flow mortar geopolymer value shown in Table 1 which shows that the slump flow percentage decreasing congruent with the increasing of NaOH concentration. The decreasing percentage are 10.93%, 33.72%, 40.82%, and 46.81%. This shown that the increasing concentration of NaOH can decrease mortar geopolymer workability.

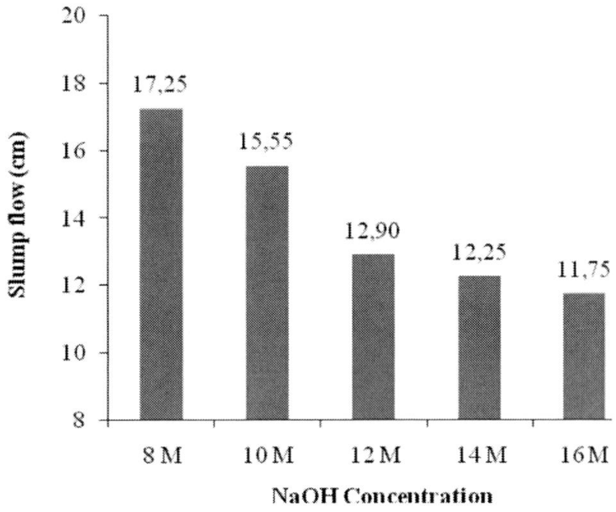

Fig. 3. Slump flow.

Table 1. Difference percentage of mortar geopolymer's slump flow.

NaOH concentration	Slump flow (cm)	Increasing percentage (%)
8 M	17.25	0
10 M	15.55	10.93
12 M	12.90	33.72
14 M	12.25	40.82
16 M	11.75	46.81

4.2 Setting time

The result of setting time test can be shown in Figure 4. The result showed that the fastest bonding-time happens in the mortar with 16 M NaOH concentration, with 105 minutes time and 165 minutes latest bonding-time. Meanwhile, the longest initial setting time happens in a mortar with 8 M NaOH concentration for 161,25 minutes and 225 minutes final time. Based on this research, it can be concluded that the bigger NaOH concentration, the faster the setting time.

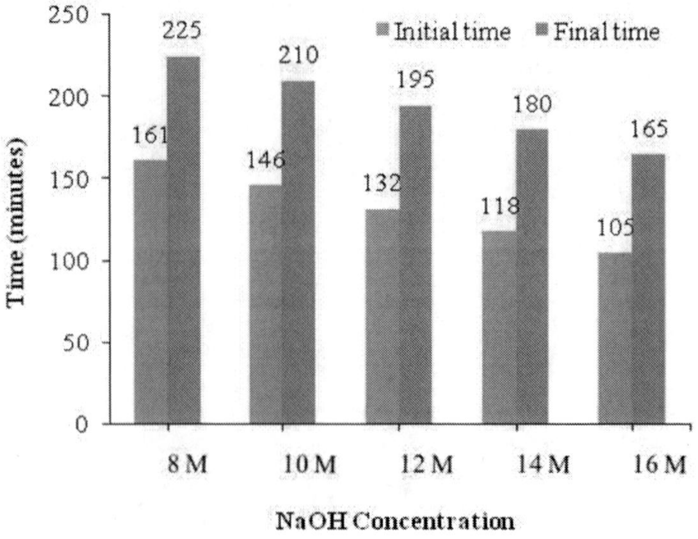

Fig. 4. Setting time.

4.3 Density

The result for density test can be shown in Figure 5. The maximum density which resulted from the mixture of 16 M NaOH is 1,79 gram/cm. The minimum density resulted from the mixture of 8M NaOH is 1,69 gram/cm. Based on this result, the more vary the NaOH, the higher its density. In a long period, the density will decrease until it reached the constant number. This happens because the water that contained in the mortar becomes evaporated. The increasing density percentage of mortar geopolymer can be shown in Table 2. Based on the data in Table 2, the density increaseds along with the increasing of NaOH concentration. The increasing density percentage on 10 M, 12 M, 14 M and 16 M in a row is 1.74%, 2.87%, 4.51% and 5.58%

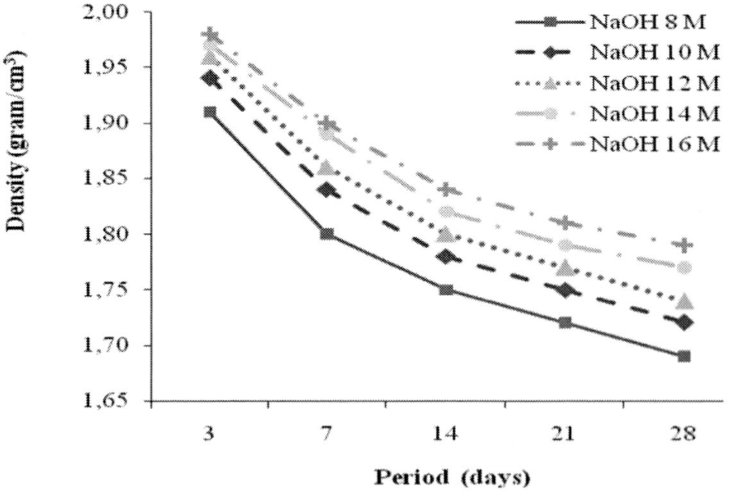

Fig. 5. Density.

Tabel 2. Difference percentage of mortar geopolymer's density.

NaOH concentration	Density (cm)	Increasing percentage (%)
8 M	1.69	0
10 M	1.72	1.74
12 M	1.74	2.87
14 M	1.77	4.51
16 M	1.79	5.58

4.4 Compressive strength

Figure 6 shown the development of compressive strength for 3–28 days period with NaOH concentration including 8 M, 10 M, 12 M, 14 M. In the age of 3 to 7 days, all mortar have the same Compressive strength development. Furthermore, from the age of 14 days, mortar with 14 M and 16 M NaOH concentration experienced a faster development of compressive strength compare to mortar with 8 M, 10 M and 12 M NaOH concentration. The fast improvement of compressive strength in 14 M and 16 M concentration NaOH happens due to the role of NaOH in the creation of zeolite formation. Na_2SiO_3 in an activator liquid also plays a role in the fast improvement of compressive strength. This is due to Na_2SiO_3 accelerating reaction in the polymerization process. Table 3 indicates the increasing number of Compressive strength, along with the improvement of NaOH concentration. The increasing Compressive strength percentage in a 10 M, 12 M, 14 M and 16 M mortar concentration in a row are 30.18%, 48.97%, 42.49% and 66.19%.

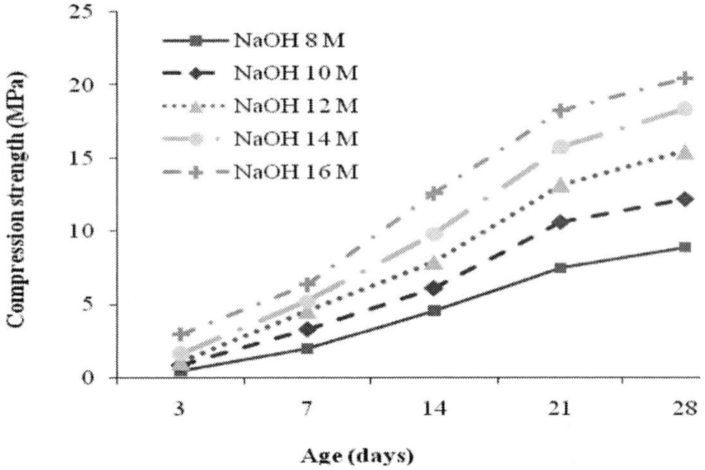

Fig. 6. Compressive strength.

Table 3. Percentage of mortar geopolymer's increasing compressive strength.

NaOH percentage	Compressive strength (MPa)	Increasing percentage (%)
Mortar type N	6,2	0
8 M	8.88	30.18
10 M	12.15	48.97
12 M	15.44	42.49
14 M	18.34	66.19
16 M	20.44	69.67

4.5 Compressive strength vs density

Graphic for analysis of mortar geopolymer's compressive strength and density can be shown in Figure 7. The result for analysis of mortar geopolymer's compressive strength and density with 8 M, 10 M, 12 M, 14 M and 16 M NaOH concentration shown that the coefficient determination value (R^2) is 0.9936 with Equation 1:

$$f_c' = 116.99\,\gamma - 188.75 \tag{1}$$

where:
f_c' = compressive strength (MPa)
γ = density (gram/cm^3)

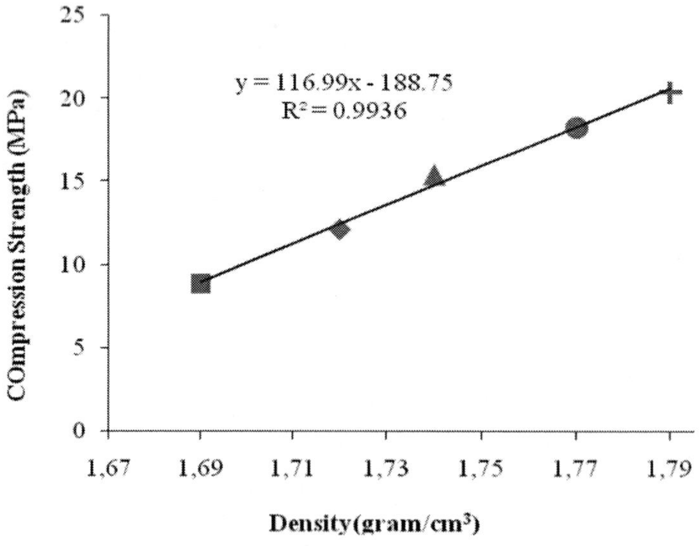

Fig. 7. Compressive strength vs density in 28th day.

4.6 Compressive strength vs NaOH concentration

Analysis of compressive strength regression and NaOH concentration can be shown in Figure 8. The analysis found that coefficient determination value (R^2) is 0.9962 with Equation 2:

$$f_c' = 8.7174\, M^{0.526} \tag{2}$$

where:
f_c' = compressive strength (MPa)
M = molarity (M)

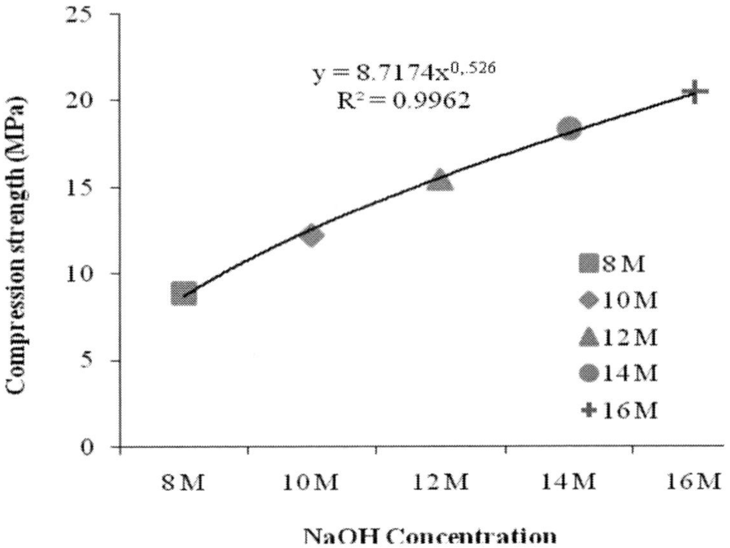

Fig. 8. Compressive strength vs NaOH concentration in 28th day.

4.7 NaOH concentration vs density

Analysis of NaOH concentration regression and density is shown in Figure 9. On the 28th day, the test found that the coefficient determination value (R^2) is 0.9962 with Equation 3:

$$\gamma = 0.025\, M + 1.667 \tag{3}$$

where the:
γ = density (gram/cm^3)
M = molarity (M)

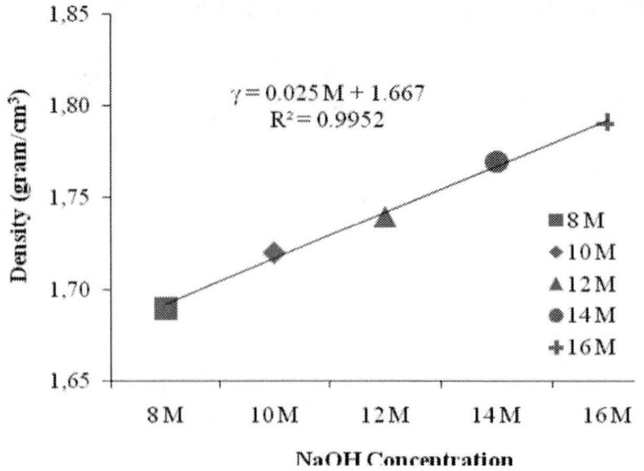

Fig. 9. Density vs NaOH concentration in 28th day.

4.8 Slump flow vs NaOH concentration

Analysis for slump flow regression and NaOH concentration is seen in Figure 10.

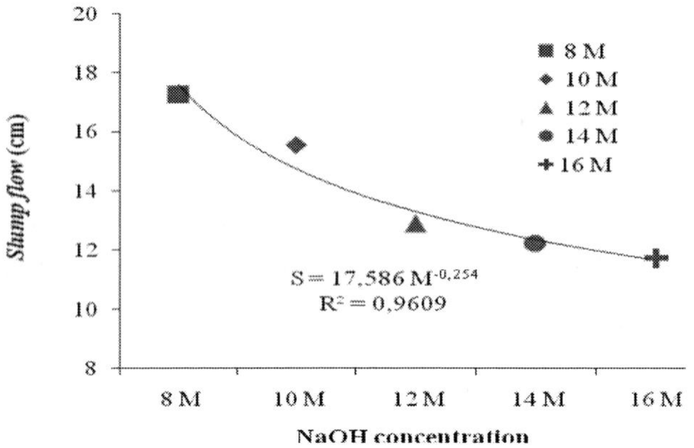

Fig. 10. Slump flow vs NaOH concentration in 28th day.

The analysis found that the coefficient determination value (R^2) is 0.9609, with Equation 4:

$$S = 17.586\, M^{-0.254} \tag{4}$$

where the:
S = *slump flow* (cm)
M = molarity (M)

5 Conclusion

Conclusions of this research are:

- The test result for slump flow mortar geopolymer shows that NaOH concentration influences the workability of the mixture. This happens due to the bigger NaOH concentration, the bigger density in a 1 Litre mortar geopolymer liquid. So, the more viscous the NaOH liquid.

- The test result for setting time mortar geopolymer show that the fastest bonding time resulted from 16 M NaOH concentration's which need 105 minutes. And for the last bonding, it need165 minutes. Based on this result, it can be concluded that the bigger NaOH concentration, the faster the bonding time.
- The result of density test shown that in each variant of NaOH concentration, the density for bagasse ash's mortar geopolymer is 1.69-1.79 gram/cm^3.
- The result of Compressive strength test showed that the maximum compressive strength for the 28th day mortar is 20.44MPa, with 16M NaOH concentration. The minimum compressive strength resulted from 8 M NaOH concentration, which is 8.88 MPa. It can be concluded that the bigger NaOH variant concentration, the bigger compressive strength that resulted.

This work was financially supported by the DIPA Sriwijaya University 2015.

References

1. A. Motorwala, V. Shah, R. Kammula, P. Nannapaneni, D. Raijiwala, International Journal of Emerging Technology and Advanced Engineering, **3**, 159 (2013)
2. Y, Kim, B.J. Lee, V. Saraswathy, S. Kwon, Scientific World Journal, **2014**, 1 (2014)
3. U.R. Kawade, V.R. Rathi, V.D. Girge, Int. J. Innov. Res. Sci. Eng. Technol., **2(7)**, 2997 (2013)
4. R. Srinivasan, K. Sathiya, International Journal for Service Learning in Engineering, **5(2)**, 60 (2010)
5. R. Srinivasan, and Sivakumar, J. Eng. Appl. Sci. (Asian Res. Publ. Netw.), **7(11)**, 1436 (2012)

Geopolymer Mortar with Fly Ash

Saloma[1,*], *Anis* Saggaff[1], *Hanafiah*[1], and *Annisa* Mawarni[1]

[1]Sriwijaya University, Jl. Palembang-Prabumulih, KM.32, Inderalaya, Ogan Ilir, South Sumatra, 30662, Indonesia.

Abstract. The cement industry accounts for about 7% of all CO_2 emissions caused by humans. Therefore, it is necessary to find another material in order to support sustainable material. An alternative way is replacing cement material with alternative material as fly ash. Fly ash as binder need to be added alkaline activator in the form of sodium silicate (Na_2SiO_3) or potassium silicate (K_2SiO_3) and sodium hydroxide (NaOH) or potassium hydroxide (KOH). The purpose of this research is to analyze the effect of activator liquid concentration on geopolymer mortar properties and to know the value of compressive strength. Molarity variation of NaOH are 8, 12, 14, and 16 M with ratio of Na_2SiO_3/NaOH = 1.0. Ratio of sand/fly ash = 2.75 and ratio of activator/fly ash = 0.8. The cube-shaped specimen 50 x 50 x 50 mm is cured by steam curing with a temperature of 60°C for 48 hours. The experimental result of fresh mortar reported that the molarity of NaOH affect the slump flow and setting time, higher of NaOH produces the smaller value of slump and the faster time of setting. The experimental of density results reported that the increase of specific gravity when the molarity of NaOH increased. The experimental results of the compressive strength are showed that the maximum compressive strength of geopolymer mortar 14 M is 10.06 MPa and the lowest compressive strength produced by geopolymer mortar 8 M is 3.95 MPa. Testing the compressive strength of geopolymer mortar 16 M produces compressive strength lower than 14 M geopolymer mortar is 9.16 MPa.

1 Introduction

Nowadays, construction world has been shown a fair progress with the development of materials technology. One of the construction progress that had been developed was mortar. Mortar is the material used for building construction that is a mixture of fine aggregate, a binder, and water. Generally, the binder uses in mortar is cement.

Fly ash is wasting material which is produced by the power plant and it can be used as an alternative replacement of cement material. To be a binder, fly ash should be added with an alkaline activator as a catalyst of fly ash. Alkaline activator that commonly used in geopolymerization is a combination of sodium silicate (Na_2SiO_3) or potassium silicate (K_2SiO_3) and sodium hydroxide (NaOH) or potassium hydroxide (KOH) [1]. The ratio of K_2SiO_3 to KOH in order to produce high strength of geopolymer mortar is in the range 0.8

* Corresponding author: saloma_571@yahoo.co.id

to 1.5 [2]. The molarity of NaOH or KOH that can be used as an activator is in the range 8 M to 16 M [3, 4].

Based on the previous research, the writer has conducted a manufacture of geopolymer mortar with alkaline activator sodium silicate (Na_2SiO_3) and sodium hydroxide (NaOH) as the catalyst of fly ash. This research is also conducted to determine the properties of geopolymer mortar.

2 Material

This research was conducted in the Laboratory of Materials and Concrete, Civil Engineering Department, Faculty of Engineering, University of Sriwijaya. The specimen of this research was a cube-shaped size 50 x 50 x 50 mm with steam curing at a temperature of 60°C for 48 hours. The variation of molarity, NaOH, which is used were 8, 12, 14, and 16 M with a ratio of Na_2SiO_3 / NaOH = 1.0. The ratio of sand / fly ash = 2.75 and the ratio of activator / fly ash = 0.8.

The material used in this research consists of fly ash where it came from power plant Tanjung Enim, South Sumatera, fine aggregate from Tanjung Raja, sodium hydroxide (NaOH) 98% purity flake, sodium silicate (Na_2SiO_3) with 58% of solid gel purity, and water. Material testing of this research are the physical properties of fine aggregate test and chemical analysis test of fly ash. The testing of chemical analysis was conducted in the Laboratory of PT.Semen Baturaja Palembang. The chemical composition of the chemical analysis of fly ash can be seen in Table 1. From the result, it showed that the compound of fly ash was classified into class F containing of $SiO_2 + Al_2O_3 + Fe_2O_3$ = 87.54% and the CaO content = 3.75%.

Table 1. The chemical composition of the fly ashes.

Chemical composition	SiO_2	Al_2O_3	Fe_2O_3	CaO	MgO	SO_3	LOI	IR
Percentage	82.04	3.70	1.80	3.75	0.01	0.08	3.50	78.97

Before producing mortar, first, it needed an activator solvent of NaOH and Na_2SiO_3 based on molarity needed. The solvent is made manually using a measuring cup and spatula. Then the solvent was allowed to stand for about 24 hours to achieve a homogenous solvent. The producing of fresh mortar is done by mixing fly ash and sand for 3 minutes. Then the activator solvent is added into the mixture of fly ash-sand and it is mixed for 10 minutes. Next, fresh mortar is filled into the mold in size 50 x 50 x 50 mm which has been smeared with lubricant oil. The filling process is done in two stages. Each stage of filling, 25 mm of concrete mix is filled then it's compacted. Then fresh mortar that has been moulded is cured with steam curing at temperature 60°C for 48 hours.

3 Laboratory Test

This research conducted in Laboratory of Concrete and Material, Civil Engineering Faculty of Engineering, Universitas Sriwijaya. The standard for mortar mixed material refers to American Standart Testing and Material (ASTM C109). The experiment that has been conducted were workability, setting time, density and compressive strength test. The NaOH's molarity variation are 8, 12, 14 and 16 M with a ratio of Na_2SiO_3/NaOH = 1.0. Ratio of sand/fly ash = 2.75 and ratio of activator/fly ash = 0.8. The cube-shaped specimen 50 x 50 x 50 mm is cured by steam curing with a temperature of 60°C for 48 hours. The

DOI: 10.1051/matecconf/20167801026

workability test conducted by test the flow table to know the diameter distribution. While the compressive strength test conducted using a 50 x 50 x 50 mm cube-sized material for 3, 7, 14, 21 and 28 days period.

4 Result and Analysis

4.1 Slump Flow

The result of slump flow test has a different value for each composition based on the molarity of NaOH. The result of slump flow can be seen in Figure 1. Based on the result of slump flow test it can be concluded that the greater molarity value of NaOH, the smaller slump value will be resulted. It causes due to the influence of molarity on the viscosity of activator solvent can influence the workability of the mixture.

4.2 Setting time

The result of setting with any variations of NaOH molarity can be seen in Figure 2. The result of setting time showed that the molarity of NaOH is greatly affected the setting time of geopolymer mortar. The greater value of NaOH molarity, the faster setting time is needed and vice versa the smaller value of NaOH molarity, the longer setting time is needed.

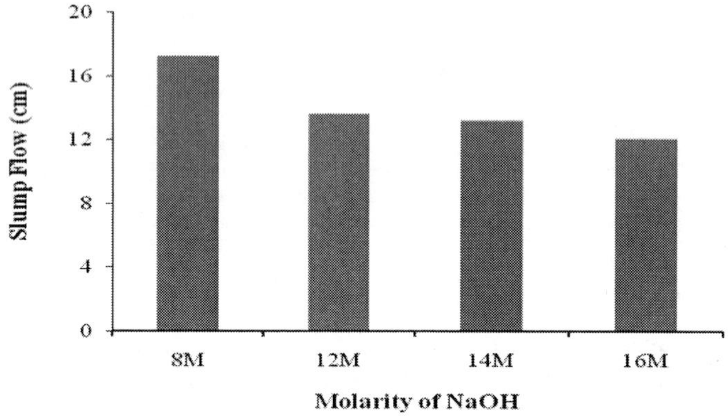

Fig. 1. The result of slump flow.

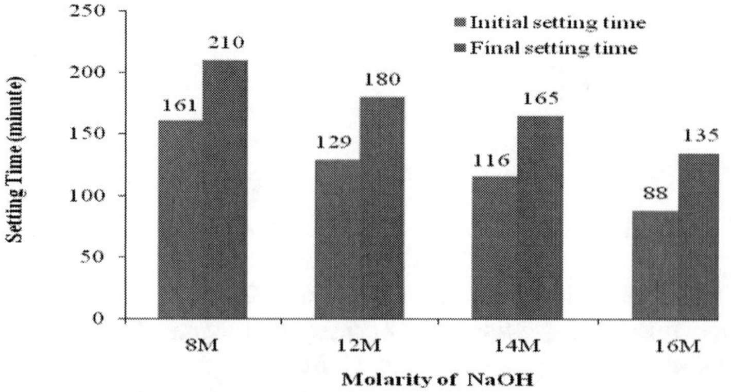

Fig. 2. The result of setting time.

DOI: 10.1051/matecconf/20167801026

4.3 Density

The test of density is tested before the test of compressive strength conducted. The data is obtained by weighing the weight of the specimen. The result of mortar geopolymer density can be seen in Figure 3. Based on the test, it can be explained that the greater value of NaOH molarity, the greater value of density is obtained. Mortar with the lowest molarity has more porous so it has a lower density than mortar with higher molarity. The result of density geopolymer mortar has a lower density than cement mortar is between range 2.17 to 2.19 gr/cm^3 [5, 6].

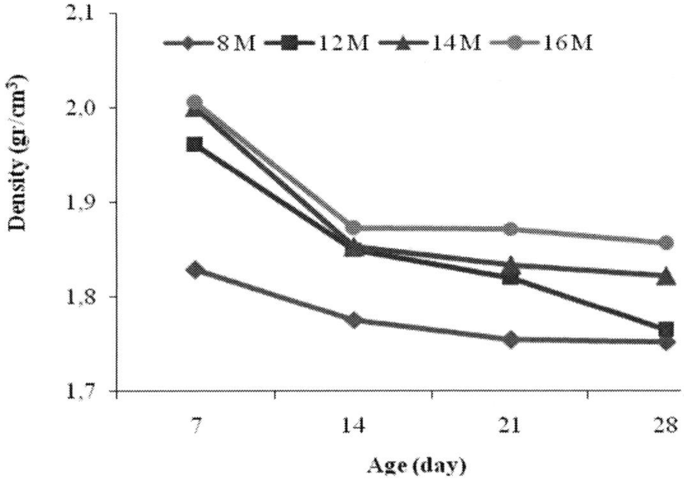

Fig. 3. The result tests of density.

4.4 Compressive Strength

The result of testing on each variation of NaOH molarity for aged 7, 14, 21, and 28 days can be seen in Figure 4. The minimum result of mortar compressive strength for aged 28 days is obtained by 8 M of mortar geopolymer with 3.95 MPa. The maximum compressive strength for 28 days is produced by mortar 14 M in the amount of 10.06 MPa. The compressive strength of geopolymer mortar 16 M resulting lower than geopolymer mortar NaOH 14 M.

The maximum compressive strength which is produced by geopolymer mortar 14 M is lower than the compressive strength of cement mortar type S according to ASTM C1329-03. The compressive strength of cement mortar type S which is required by ASTM C 1329-03 for 7 days and 28 days are 9.0 MPa and 14.5 MPa. The difference of compressive strength between geopolymer mortar and cement mortar is 30,62%.

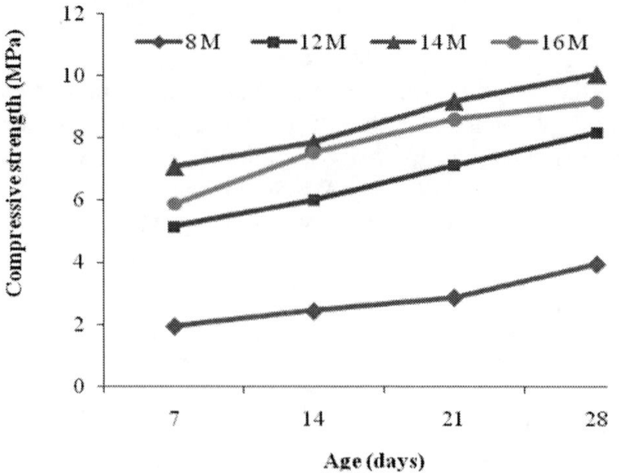

Fig. 4. The result of mortar compressive strength.

4.5 Regression Analysis of Mortar Compressive Strength

Regression analysis used the maximum result of compressive strength generated by geopolymer mortar 14 M. Regression analysis of the compressive strength of geopolymer mortar for various age testing refers to the equations recommended by ACI 209.2R-08. Regression analysis of the compressive strength of geopolymer mortar based on the compressive strength of mortar ACI 209.2R-08 can be seen in Figure 5.

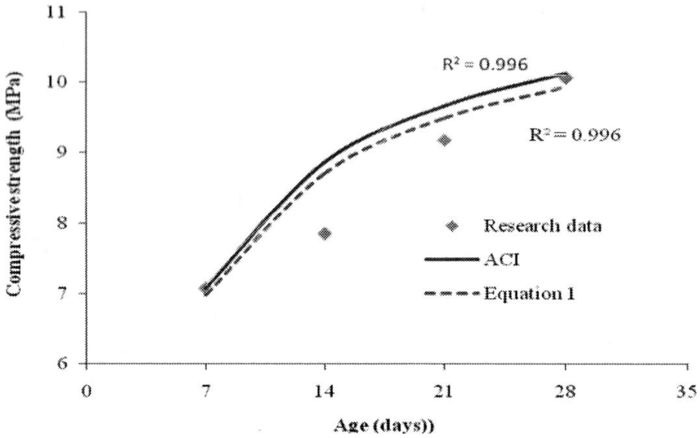

Fig. 5. Regression analysis of mortar compressive test.

The function used to determine the compressive strength for various age of this research can be seen in Equation 1.

$$f_c{'}_t = \left[\frac{t}{4{,}35 + 0{,}88.t}\right] f_c{'}_{28} \tag{1}$$

The coefficient of determination R^2 = 0.9966. The value of R^2 which is close to 1 indicates that the regression analysis approach is good at accuracy.

5 Conclusion

The conclusion that can be drawn based on the results of research on the effects Na_2SiO_3 and NaOH on the properties of geopolymer mortar are as follows

- The greater the molarity of NaOH, the smaller value of slump flow obtained, thereby it reduces the workability of the mixture. The smaller value of molarity NaOH, the greater value of slump flow obtained that improve the workability of the mixture.
- The greater the molarity of NaOH, the faster setting time needed.
- The greater the molarity of NaOH, the greater the density of geopolymer mortar obtained
- The result of compressive strength showed that the maximum value is obtained on geopolymer mortar 14 M in the amount of 10.06 MPa and the minimum value is obtained on geopolymer mortar 8 M which is equal to 3.95 MPa. In mortar mixture of 16 M, it decreases the compressive strength equal to 9.16 MPa. Based on these result, it can be concluded that the optimum mixture to get the maximum strength is used NaOH 14 M.
- The result of regression analysis is based on ACI 209.2R-08 can be drawn that the coefficient of determination R^2= 0.9966. Based on R^2 it can be concluded that the testing of geopolymer mortar is fairly reached good accuracy.
- The result of compressive strength of geopolymer mortar is lower than cement mortar type S according to ASTM C 1329-03.

This work was financially supported by the DIPA Sriwijaya University 2015.

References

1. D. Hardjito, Journal-American Concrete Institute, **101(6)**, 467 (2004)
2. D. Hardjito, S.S. Fung, Modern Applied Science, **4(1)**, 44 (2010)
3. D. Hardjito, S.S.E. Asia, Modern Applied Science, **2**, 3 (2008)
4. D. Hardjito, M. Tsen, *The 3rd International Conference ACF/VCA*, **2003**, 144 (2008)

Calcium Lactate addition in Bioconcrete: Effect on Compressive strength and Water penetration

J.M Irwan[1,*], *L.H* Anneza[1], *N.* Othman[2], *A. Faisal* Alshalif[1], *M.M.* Zamer[1], and *T.* Teddy[1]

[1] Jamilus Research Centre for Sustainable Construction (JRC), Faculty of Civil and Environmental Engineering,University Tun Hussein Onn, Malaysia.
[2]Micropollutant Research Centre (MPRC),Faculty of Civil and Environmental Engineering,University Tun Hussein Onn, Malaysia.

Abstract. This paper presents compressive strength and water penetration of bioconcrete with addition of calcium lactate. Bioconcrete has higher engineering concrete properties and durability compared to normal concrete but the natural production of calcium carbonate is limited to the calcium content in cement. Therefore, additional calcium is added as an additional calcium source to study the influence towards compressive strength and water penetration. The bacteria used in this research are *Enterococcus faecalis* and *Bacillus sp*. Calcium lactate was added into concrete mix in concentrations of 0.001mol/l, 0.005mol/l and 0.01mol/l of liquid used. The concentration of bacteria added into the mix is by partial replacement of water used in casting, which are 3% for *Enterococcus faecalis*and 5% for *Bacillus sp*. Both compressive strength and water penetration test used cubes of 150mm × 150mm × 150mm. The cubes were tested after 28 days. The result of compressive strength for control is 36 MPa while partial replacement of bacteria yields 38.2 MPa for 3% *Enterococcus faecalis*and 37.0 MPa for 5% *Bacillus sp*. Calcium lactate with 0.005 mol/L has the best performance with 42.8 MPa for *Enterococcus faecalis* and 39.6 MPa for *Bacillus sp*. Whereas for water penetration, the best concentration of calcium lactate which yielded the lowest water penetration is 0.01 mol/l for both *Enterococcus faecalis* and *Bacillus sp* which are 8.7 cm and 8 cm respectively. The addition of calcium lactate into bioconcrete is quite promising for improvement of concrete properties and durability.

1 Introduction

Over the years, many researchers had studied the addition of waste material into concrete [1-3]. Another green alternative for improving concrete that is currently explored by many researchers is bioconcrete. Bioconcrete is a combination of biological micro-organisms and concrete. The use of micro-organisms in concrete is due to the ability to produce calcium carbonate [4]. Different bacteria from various sources are used by researchers studying bioconcrete. Bacteria that was known as *Shewanella* was isolated from the mud of the hot springs in West Bengal, India. This bacteria is used to improve concrete properties [5]. The

[*] Corresponding author: irwan@uthm.edu.my

use of *Bacillus pasteurii* and *Pseudomonas aeruginosa* which are endospore-forming soil micro-organism are used in Ramachandran [6]. A significant decrease in water uptake resulted by the addition of bacteria in concrete. The surface deposition of calcium carbonate crystals decreased the water absorption by 85%, depending on the porosity of specimens [7]. The bacteria used in this study is *Enterococcus faecalis* and *Bacillus sp*, both of this bacteria are isolated and enriched to suit the concrete environment. The isolation, enrichment and the amount of bacteria added as a partial replacement is based on a previously conducted study by Irwan [8, 9]. Adding calcium source in the concrete helps in increasing the amount of precipitated calcium carbonate and therefore increases strength and improve durability. The amount of precipitated calcium carbonate depends on in the type of calcium source added [10].

2 Experimental Investigation

2.1 Material

The total amount of materials that was used in this research is shown in Table 1. Ordinary Portland cement, coarse and fine aggregates were used in the concrete mix. The aggregates were prepared according to BS 882.1992. Both *Enterococcus faecalis* (*Ent.f*) and *Bacillus sp* (*B.sp*) were added as partial replacement of water, 3% and 5% respectively. The concentration of calcium lactate added is based on 0.001 mol/l, 0.005 mol/l and 0.01 mol/l of the amount of water. The calculation for calcium lactate is based on the chemical formula ($C_6H_{10}CaO_6$) with molar mass (218g/mol) of calcium lactate. Therefore, concentration are used as 0.22g/L, 1.09 g/L and 2.18 g/L for 0.001mol/l, 0.005 mol/l and 0.01 mol/l respectively.

Table 1. Amount of materials used for fabrication.

Batch	Cement (kg)	Fine aggregate (kg)	Coarse aggregate (kg)	Water (kg)	Bacteria (kg)	Calcium lactate (g)
Control	7.9	13.9	22.6	4.3	-	-
3% *Enterococcus faecalis* (*Ent.f*)	7.9	13.9	22.6	4.2	0.13	-
3% *Ent.f*+ 0.001 mol/l	7.9	13.9	22.6	4.2	0.13	1
3% *Ent.f*+ 0.005 mol/l	7.9	13.9	22.6	4.2	0.13	4.7
3% *Ent.f*+ 0.01 mol/l	7.9	13.9	22.6	4.2	0.13	9.4
5% *Bacillus sp* (*B.sp*)	7.9	13.9	22.6	4.1	0.23	-
5% *B.sp* + 0.001 mol/l	7.9	13.9	22.6	4.1	0.23	1
5% *B.sp*+ 0.005 mol/l	7.9	13.9	22.6	4.1	0.23	4.7
5% *B.sp*+ 0.01 mol/l	7.9	13.9	22.6	4.1	0.23	9.4

2.2 Preparation of bacteria and calcium lactate

Both bacteria are prepared with the same method. Which is taking a colony of bacteria, adding it into sterilize nutrient broth and shaken for ten days. The number of days for optimum growth was determined by plotting bacteria growth curve using daily turbidity data. The bacteria are added into concrete mix as partial replacement of water.Calcium lactate is prepared by dissolving in water before adding into the mix.

2.3 Fabrication of concrete and testing

Fabrication of concrete is done according to BS 1881-125:2013. After the mix is complete, the bioconcrete is poured into plastic moulds of 150mm × 150mm× 150mm then left to hardened for 24 hours. After 24 hours, the concrete is removed and left for dry curing. Compressive strength test and water penetration test are conducted after 28 days of maturity. The testing is done using three samples to get the average results. The average results of the test are presented.

3 Results and discussion

3.1 Compressive strength

The compressive strength had considerable improvement with concrete specimens containing bacteria and calcium lactate shown at Figure 1. The highest increment of compressive strength for both bacteria is with calcium lactate of 0.005 mol/l concentration. The increment of concrete specimen strength with Ent.f and Ent.f with calcium lactate of 0.001 mol/l, 0.005 mol/l and 0.01 mol/l concentration compared to control are 6.11%, 6.9%, 18.9% and 8.3% respectively. Whereas, the increment of strength for concrete specimens containing B.sp and B.sp with calcium lactate of 0.001 mol/l, 0.005 mol/l and 0.01 mol/l concentrations compared to control are 2.7%, 6.9%, 10% and 5% respectively.

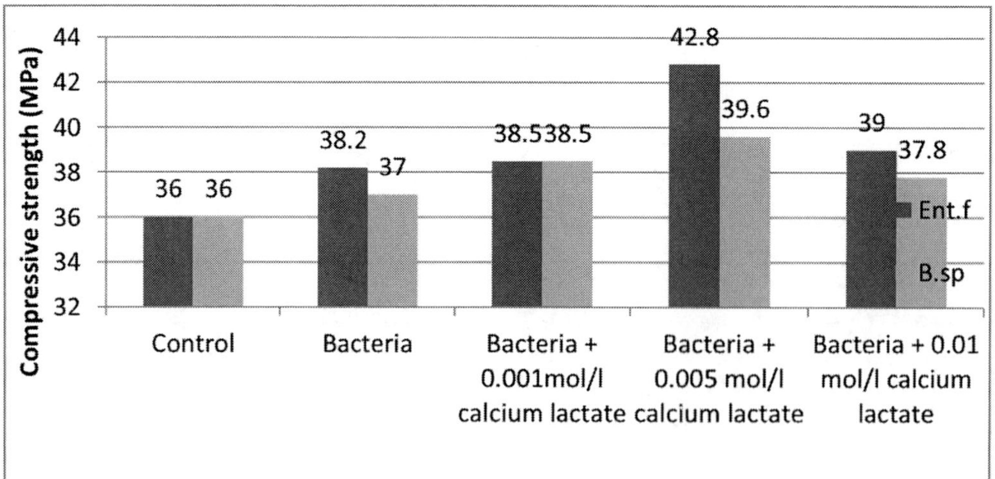

Fig. 1. Comparison *Enterococcus faecalis* and *Bacillus sp* with various concentration of calcium lactate.

Irwan et al., has stated that materials added into concrete which does not participate in hydration process with the binder could attribute to decrease in compressive strength. Bacteria added into concrete does not directly involved in the hydration process itself but used the by-product of the reaction to produce calcium carbonate. The slight reduction of compressive strength with higher concentration of calcium lactate shows that over production of calcium carbonate would affect the compressive strength. Similar result is reported by Faiz and Steve [11], which stated that minimal amount of calcium carbonate was found most effective in increment of strength. Researchers have studied that certain concentration of bacteria increases compressive strength compared to control [4-7]. However, the concentration of bacteria depends on the bacteria itself. As every bacteria used are of different origins. Thus, may have different reactions. In Xu *et al.* [10] and Abo-El-Enein [12] studies, calcium source were added into bioconcrete. After which compressive strength were tested and compared. It was found that different calcium source precipitates different amount of calcium carbonate. Different calcium source influence the rate of bacteria precipitation and amount.

The precipitation of calcium carbonate that fills the pore within the concrete matrix is the factor for increment of compressive strength. Addition of *Ent.f* in concrete has considerable impact on the compressive strength compared to *B.sp*. The obvious difference for both bacteria is with 0.005 mol/l concentration of calcium lactate. *Ent.f* has shown a tremendous increase of compressive strength compared to *B.sp*. This is due to *Ent.f* reaction to the addition of calcium source. The addition of calcium source allows the bacteria to increase precipitation which benefited the concrete. Calcium lactate with concentration of 0.01 mol/l has caused an over production of calcium carbonate which cause the strength to reduce slightly. This is similar for both bacteria. The increment of strength with addition of bacteria are due to deposition of calcite facilitated by calcium nutrient [10,14].

3.2 Water penetration

Bioconcrete specimens with different concentration of calcium lactate shows promising results, as seen in Figure 2. The lowest water penetration is with 0.01 mol/l of calcium lactate for both bacteria. Concrete specimens with *Ent.f* and *Ent.f* with 0.001 mol/l, 0.005 mol/l and 0.01 mol/l has decrease water penetration by 22%, 20%, 22.9% and 26.3% respectively compared to control. Whereas, concrete samples with *B.sp* and *B.sp* with 0.001 mol/l, 0.005 mol/l and 0.01 mol/l decreased water uptake by 20.3%, 25%, 26.3% and 32.2% respectively compared to control.

The presence of bacteria reduces the water penetration of concrete by filling the pores with deposited calcium carbonate [4,7,16]. The addition of calcium source acts as a catalyst to further deposit calcium carbonate.

The increment of precipitation of calcium carbonate by calcium source in concrete containing bacteria plugs pores within the concrete matrix. This blocks the pathway for water to enter concrete which reduces the water penetration of concrete. With the addition of calcium lactate, the decrement of water penetration for *B.sp* is better than *Ent.f*. This is due to the reaction between the bacteria and calcium lactate within the concrete pores. The calcium lactate crystallizes within the pores which caused the pores to block up thus reduces the water penetration. Previous study has reported that calcite deposition in pores caused reduction in water uptake [4,7]. Xu and Yao [10] had reported that additional calcium source in bioconcrete reduced the water penetration and increased the concrete durability.

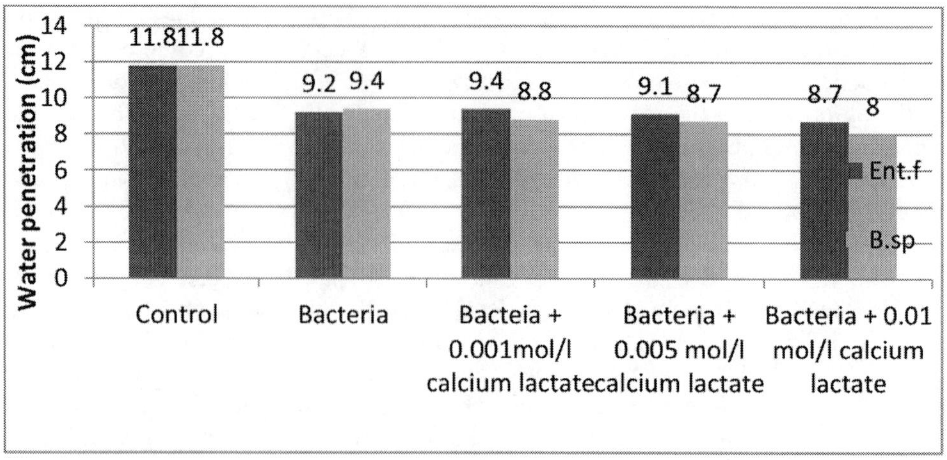

Fig. 2. Comparison *Enterococcus faecalis* and *Bacillus sp* with various concentration of calcium lactate.

4 Conclusion

The results obtained in this research showed that concrete added with calcium lactate and bacteria has its advantages in improving concrete strength and durability. In terms of compressive strength, the optimum percentage of calcium lactate that is added is 0.005mol/l for both bacteria. While, water penetration shows that 0.01 mol/l of calcium lactate is optimum for both bacteria.

This research was supported by UniversitiTun Hussein Onn Malaysia (UTHM) and The Ministry of Higher Education Malaysia through Fundamental Research Grant Scheme (FGRS), Vot 1211.

References

1. S.K. Faisal, J.M. Irwan, N. Othman, M.H. Wan Ibrahim, *Flexural Toughness of Ring-Shaped Waste Bottle Fiber Concrete* (EDP Sciences, 2016)
2. S.A. Radziah, W.W.Z.A. Raziman, S.Shahiron, *Physical and Mechanical Properties of Compressed Earth Brick (CEB) Containing Sugarcane Bagasse Ash* (EDP Sciences, 2016)
3. J. Norwati, A.H. Farhan, M.W.I. Haziman, P.J. Ramadhansyah, M.A. Fadzil, Z.A. Norul Ernida, D. Nurol Huda, *Fresh Properties and Flexural Strength of Self-Compacting Concrete Integrating Coal Bottom Ash (*MATEC Web of Conference, IConCEES, 01010, **47** 2016)
4. N. Chahal, R. Siddique, Constr. Build. Mater., **49,** 161 (2013)
5. P. Ghosh, S. Mandal, Indian Journal Of Experimental Biology, **44(4),** 336 (2006)
6. S.K. Ramachandran, V. Ramakrishnan, S.S. Bang, ACI Material Jurnal, **98(1)**, 3 (2001)
7. W.D. Muynck, K. Cox, N.D. Belie, W. Verstraete, Constr. Build. Mater., **22(5),** 875 (2008)
8. J.M. Irwan, A. Faisal Alshalif, N. Othman, R.M Asyraf, *Isolation of Ureolytic and Sulphate reduction bacteria: Acclimitize to concrete environment* (Interntional conference on civil, Biological and Environmental Engineering (CBEE), Istanbul, Turkey 2014)

9. J.M. Irwan, A. Faisal Alshalif, N. Othman, L.H. Anneza, *Effect of Ureolytic Bacteria on Compressive strength and Water Permeability on Bio-concrete* (International Conference on Civil, Architectural, Structural and Construction Engineering 2015).
10. J. Xu, W. Yao, Cement Concrete Res. **64**, 1 (2014)
11. U.A. Faiz, Shaikh, W.M.S. Steve, Constr. Build. Mater., **70,** 309 (2014)
12. S.A. Abo-El-Enein, A.H. Ali, Fatma, N. Talkhan, H.A. Abdel-Gawwad, Housing and building national research center journal, **9(1)**, 36 (2013)

Metal Removal and Antimicrobial Properties of Watermelon rind modified with clove

N. Othman.[1,*] , *N.* Azhar[1] , *Megat* Abdul Rani[1], P. S, and H. *Mohamed Zaini* [1]

[1]Micropollutant Research Centre, Faculty of Civil and Environmental Engineering, University Tun Hussein Onn Malaysia.

Abstract. The current rapid development of industrial activity indirectly discharged pollutant into the local water stream. One of the harmful industrial wastes that enter public drainage is heavy metal owing to its toxic, non-biodegradable and persistent in nature. Improper treatment of domestic waste also will discharged high amount of microbial. Various types of technology were developed for removing pollutant in wastewater, but most of the technology employed to address on removing organic content in wastewater. Other pollutant namely, heavy metals and microbial indirectly treated at low concentration of pollutant loading. Among various available technologies for water treatment adsorption process is considered a promising technology as compared to other methods because of convenience, easy operational and low cost of treatment. The aim of this study is to investigate potential used of local fruit rind modified with herb as adsorbent material. The rind present strong potential due to its high content of cellulose, pectin, hemicellulose and lignin as active binding sites for metal sorption. Modified rind improves absorption by having anti microbial properties to kill pathogenic organisms. The concentration of heavy metal ions was analysed using ICP-MS. *E. coli* and total colifrm were plated out using chromocult agar. The results showed significant reductions of heavy metals and microbial concentration after wastewater was treated with clove modified rind.

1 Introduction

Wastewater contains several types of contaminants that indicate its quality. The quality of wastewater depends on level of pH, biodegradable and non degradable organics, inorganic, pathogens and nutrients. Left untreated, these contaminants would enter our waterways excessively, affecting the source of water for consumption and fisheries and reducing the recreational value of the waterways as well.

Inorganic minerals, metals and compounds are common in wastewater from residential and non-residential sources. The inorganics such as sodium, potassium, calcium, magnesium, cadmium, copper, lead, nickel, zinc and mercury have originated from industrial or commercial sources and runoff water [1]. Inorganics are stable, persistent and

* Corresponding author: norzila@uthm.edu.my

non degradable. The inorganic produce effects to aquatic living things and the environment as well. Thus wastewater treatments to remove the inorganic are very important to reduce the risk of harmful caused by inorganic activities.

Greywater and blackwater from typical homes contain enough pathogens to pose a risk to public health. Other likely sources in communities include hospitals, schools, farms, and food processing plants. Some illnesses from wastewater-related sources are relatively common. Gastroentritis can result from varieties of phatogens and other important wastewater-related diseases include Hepatitis A, thypoid, polio, cholera and dysentery [2].

Currently, a number of methodologies with varying degrees of success have been developed to manage water pollution. Some of them involve chemical precipitation, chemical coagulation, ion-exchange, solvent extraction electrochemical, adsorption using activated carbon, biosorption and membrane filtration [3]. However, several of these methods have their own shortcomings and limitations. Among various available technologies for water pollution, adsorption process is considered a promising technology as compared to other methods because of convenience, easy operation and low cost of treatment [4].

Several suitable agricultural by-products which contain lignocellulosics have been investigated as adsorbent material for removal of heavy metals from contaminated aqueous systems [5]. Naturally, herbs contain active compound such as eugenol, carvacrol, linalol and carsonic acid that have been identified as antimicrobial activity by a degradation cell wall or disruption of cytoplasmic membrane [6]. The objective of this study is to investigate heavy metal removal and antibacterial properties from watermelon rind modified with herb.

2 Materials and method

2.1 Sampling of wastewater.

The wastewater was sampled using grab sampling. The samples were collected over dry weather and in the morning for ensuring peak flow. The sample was collected and stored in a clean bottle and kept in a box with ice. The sample was stored in a refrigerator at approximately 4°C.

2.2 Preparation of biosorbent.

The watermelon rind was collected from the local fruit stall and washed with tap water. The rind was dried in sunlight until dry. The biosorbent was then soaked in nitric acid (HNO_3) with concentration of 5% for 24 hours for removing metals and transferred to deionized water (H_2O) for 24 hours. The rind was then washed with 60^{0}C hot water several times to ensure pH in the range of 6.5-7.5. The rind was dried in an oven at 60°C until constant dry.

2.3 Preparation of herb.

The herb used in this study was clove. It was soaked in deionized water for 24 hours and air dry until reach constant weight. Dried herbs were grinded. Three different weights of herbs were placed in a round bottom flask. About 50ml of water was added for distillation purpose. After distillation the sample was concentrated using a rotary evaporator.

2.4 Preparation of rind with clove extract

The dried rind was boiled with clove extract. The prepared rind with the herb was air dried and was pulverized, using laboratory ball mill 150 μm particle sizes.

2.5 Design Experiment

All experiments were conducted using the batch method of varying mass of modified watermelon rind into three different masses namely, 3g, 5g and 7g. 100mL of wastewater was mixed with the modified biosorbent and the mixture was shaked at 125rpm at room temperature in aerobic condition for 24 hours. The mixture was then filtered. The biosorbent was analysed using XRF, SEM-EDX and FTIR. The wastewater was analysed using ICP-MS. *E. Cole* and total coliform were checked through out the study.

4 Results and Discussion

4.1 Characterization of metals in domestic wastewater

The results from ICP-MS show that zinc (Zn), ferum (Fe) and lead (Pb) were found in the domestic wastewater. Table 1 shows heavy metal concentration.

Table 1. Concentration range of heavy metals.

Heavy Metals	Concentration (mg/L)
Zinc (Zn)	1.50 – 2.50
Ferum (Fe)	0.90 – 0.85
Lead (Pb)	0.09 - 0.08

The results show that certain concentrations of heavy metals present in significant concentration. Even, in low concentration heavy metals can cause severe effects to human health and the environment due to non-degradable and persistent [5].

4.2 XRF Analysis.

The result from XRF before and after the biosorption is shown in Table 2. Concentration of silica oxide (SiO_2) was 78.80%. The highest value of silica compares to other chemical compound is triggering the capability of watermelon rind in metal adsorption. The results also show that loaded of Zn, Fe, Pb is 24.40%, 1.34% and 0.05% respectively after biosorption. These results indicate the capability of modified fruit rind to adsorb metals, namely Zn, Fe and Pb. The presents of a very high silica content in the watermelon rind as a precursor for metal removal [1]. This is due to the capability of silica in ion exchange process over metals removal.

Table 2. Result of XRF before and after biosorption.

Formula	Concentration before	Concentration After
ori-g	7	7
added-g	3	3
CO_2	0.10%	0.1
SiO_2	78.80%	1.18%
SO_3	8.32%	0.93%
CaO	2.18%	0.15%
Al_2O_3	1.61%	0.26%
Fe_2O_3	1.00%	1.34%
P_2O_5	0.99%	0 < LLD
K_2O	0.87%	ND
Cl	0.82%	ND
MoO_3	0.30%	ND
ZnO	ND	24.40%
PbO	ND	0.05%

4.3 SEM-EDX Analysis.

SEM-EDX analyses that are presented in this paper are based on the results using watermelon rind as biosorbent in removing specific target heavy metals. Figure 1 shows native biosorbent. The figure shows that watermelon rind gave a porous surface texture which enables the adsorption of metal ions onto the surface. Besides, it also shows that the surface morphology of watermelon rind consisted of two kinds of regions which were lighter and darker [7]. The lighter shades represented inorganic component, while the darker shades were organic components. Silica is present in the spectrum as shown in Figure.1 and 2. It is triggered the capability of watermelon rind in removing metals. This result agrees with XRF result in the appearance of silica oxide in certain concentrations. The morphology of watermelon rind underwent a physical change after biosorption metal ions. Lump-like deposits or shiny particle indicated the presence of metal ions as the light spotted in that area (Figure 2).

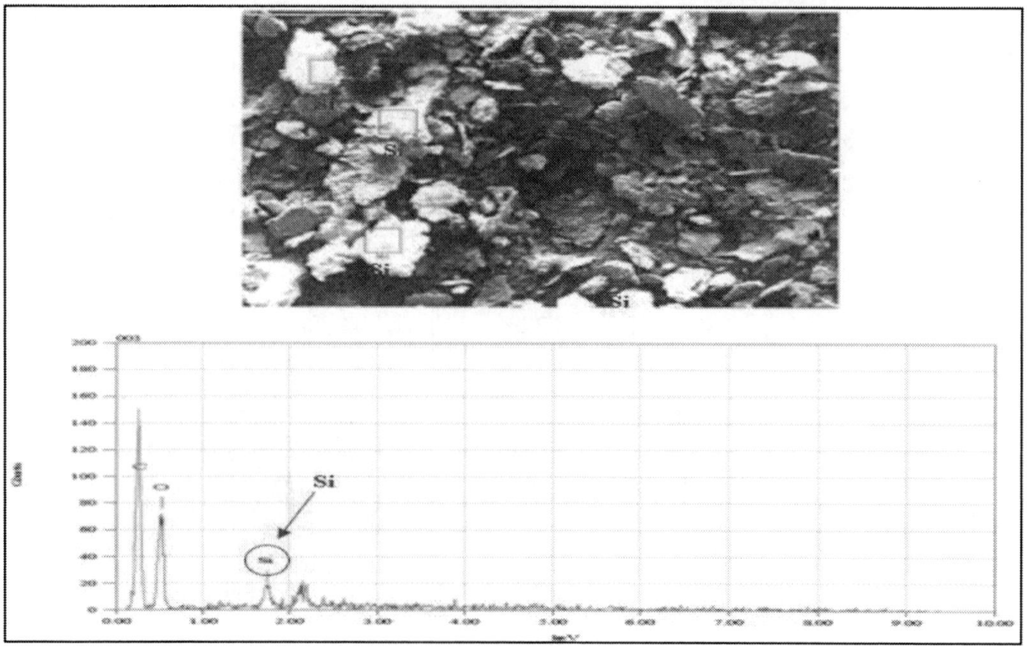

Fig. 1. SEM-EDX before biosorption.

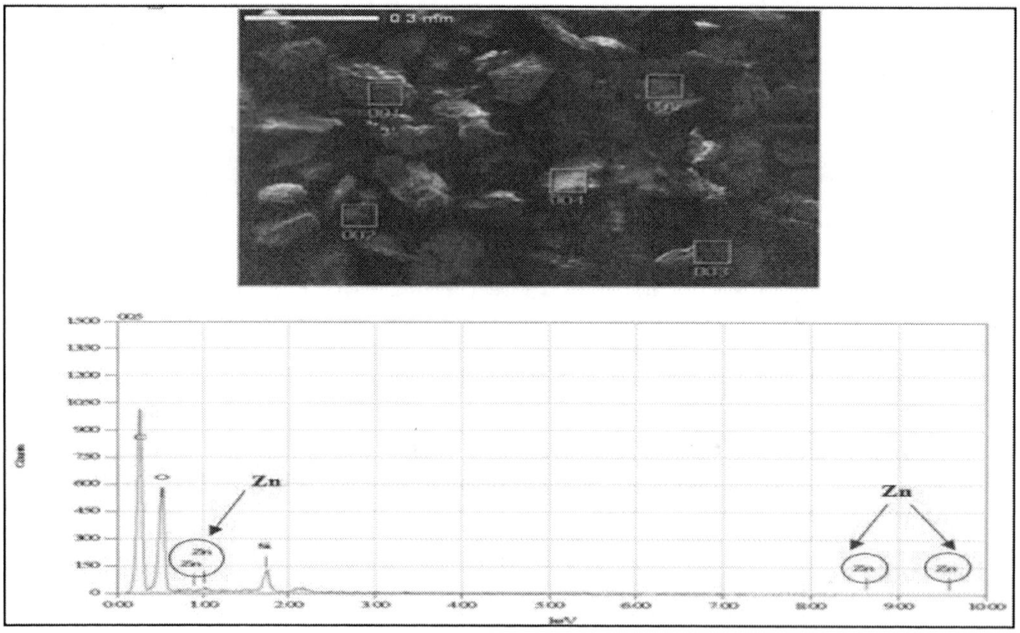

Fig. 2. SEM-EDX after biosorption.

4.4 FTIR Analysis

The characteristic band region at 1017 cm^{-1} to 1028 cm^{-1} suggests the presence of alcohol group; while alkene group present in band region at 1607 cm^{-1} to 1622 cm^{-1} as shown in Table 3. Both alcohol and alkene stretching show major shifting of the peak with more than 10 cm^{-1} that confirmed the function of both groups in helping heavy metal and microbial removal. The appearance of specific functional groups in this study is well agreed with previous studies by Ahmad et al [1] but the number of wavelengths shifted vary according to the type of adsorbent and media. The appearance of alcohol group shows potential of modified watermelon rind as antibacterial and high capability in removing *E. coli* and total coliform.

Table 3. FTIR Analysis of watermelon after modifying with herb.

Wavenumber (cm^{-1})		Identified groups	Remarks
Watermelon	Watermelon + clove		
1017.81	1028.42	C-O	Alcohol groups responsible for watermelon + clove– major shift of wave number ≥10 cm-1
1607.45	1622.84	C=C	Alkene groups responsible for watermelon + clove– major shift of wave number ≥10 cm-1

4.5 Biosorption

The data from Table 4 revealed that maximum removal of heavy metals can be achieved up to 80%, 77% and 67% for zinc, ferum and lead respectively in the actual domestic wastewater.

Table 4. Concentration of metals before and after treatment with modified watermelon rind.

Heavy Metals	Concentration before treatment (mg/L)	Concentration before treatment (mg/L)
Zinc (Zn)	1.50 – 2.50	0.30-0.36
Ferum (Fe)	0.90 – 0.85	0.20-0.35
Lead (Pb)	0.09 - 0.08	0.03-0.04

The maximum removal refers to the state where the metal bindings reach their equilibrium state. In this condition, the treatment became stable and further adsorption could not take place because the surface area has reached the saturation point [8-10]. Nevertheless, percentage removal also depends on concentration of heavy metal and other pollutant that present in the wastewater sample and also type of media being used in the study. It is proven by previous studies on using other biosorbent such as durian peel [11] and banana peel [12]

4.6 E. coli and Total coliform

The removal of both *E. coli* and total coliform increases as mass of modified watermelon rind increases. Othman [13] also revealed that mass of adsorbent among factors that affecting biosorbent. After treating with 7g modified watermelon rind both *E. coli* and total coliform are not detected. This is due to the presence of the active compound in alcohol group that have been identified as antimicrobial activity by a degradation cell wall or disruption of cytoplasmic membrane [6]. This result in line with the FTIR spectrum analysis that proved the appearance of alcoholic group in the clove extract that assisted as antimicrobial properties. Both results confirm the performance of clove modified watermelon rind in removing *E. coli* and total coliform from wastewater. Both *E.coli* and total microbial are important microbial need to be removed due to the detrimental effect towards health and normally ingested through foot exposed to polluted water. The results also show that both E.coli and total colifm treated with clove modified watermelon rind resulted to save level of water sample below than the standard, Interim National Water Quality Standard (INWQS).

Table 4. *E. coli* and Total coliform before and after treatment with modified watermelon rind

Mass herbs (g)	**Before treatment (x 10^6)**	**Total coliform after treatment**	***E.coli* after treatment**	**INWQS Total coliform (Standard)**	**INWQS *E.coli* (Standard)**
5	total coliform = 89 E.coli = 68	3gm rind = 10 x 10^2 5gm rind = 2 x 10^2 7gm rind = 3 x 10	3gm rind = 2 x 10^2 5gm rind = 4 x 10 7gm rind = ND	50 x 10^3	5 x 10^3
7	total coliform = 89 E.coli = 68	3gm rind = ND 5gm rind = ND 7gm rind = ND	3gm rind = ND 5gm rind = ND 7gm rind = ND	50x 10^3	5x 10^3
8		3gm rind = ND 5gm rind = ND 7gm rind = ND	3gm rind = ND 5gm rind = ND 7gm rind = ND	50x 10^3	5x 10^3

5 Conclusion

The watermelon rind modified with herb show high capability in removing metals such as zinc, ferum and lead in domestic wastewater. It is also shown antibacterial properties by reducing *E. coli* and total coliform.

This research was funded by UTHM under Centre of Research and Commercialization.

References

1. M. A. Ahmad, R. Alrozi, Chem. Eng. J., ***165***, 883 (2012)
2. M. Bilal, J.A. Shah, T. Ashfaq, S.M.H. Gardazi, A.A. Tahir, A. Pervez, Q. Mahmood, J. Hazard Mater., **263**, 322 (2013)

3. A.K. Minocha, Indian J. Chem. Techn., **13(3)**, 203 (2006)
4. M. Nikazar, N. Afshari, Afinidad, **62(518)**, 326 (2005)
5. N. Othman, S. Mohd Asharudin, Adv. Mat. Res., **93**, 266 (2013)
6. S. Sethi, S. Dutta, B. Gupta, B.S. Gupta, International Journal of Pharmacy and Pharmaceutical, **55**, 260 (2012)
7. N. Othman, A.S. Che Azhar, A. Suhaimi, Appl. Mech. Mater.,**680**, 150 (2014)
8. N. Othman, F.H. Azizul Rahman, Appl Mech Mater , **680**, 146(2014)
9. A. Abdul Kadir, N.M. Azmi, International journal of sustainable construction, **3(2)** 70 (2012)
10. R.M. Mohamed, A.H.Kassim, Journal of sustainable development, **9(1)**, 35 (2014)
11. W.W. Wong, F.M.A. Liong, M.E. Azhar, Food Res. Int., **15(3)**, 363 (2008)
12. P. Kaewsarn, W. Saikaew, S. Wongcharee, *Dried biosorbent derived from Banana peel : A potential biosorbent for removal of cadmium ions from aqueous solution* (The 18h Thailand Chemical Engineering and Applied Chemistry Conference October 20-2 2008, Pattaya, Thailand. 2008)
13. N. Othman, S. Mohd-Asharudin, Appl. Mech. Mater., **389**, 29 (2011)

Effect of Variant End of Injection Period on Combustion Process of Biodiesel Combustion

Amir Khalid[1,*], *Noorpuspadilla* Baharuddin[1] , *Azwan* Sapit[1] , *Azahari* Razali[1] , and *Norrizam* Jaat[2]

[1]Combustion Research Group (CRG), Centre for Energy and Industrial Environment Studies (CEIES), Faculty of Mechanical and Manufacturing Engineering, Universiti Tun Hussein Onn Malaysia, 86400 Batu Pahat, Johor, Malaysia
[2]Automotive Research Group (ARG), Centre for Energy and Industrial Environment Studies (CEIES), Faculty of Mechanical and Manufacturing Engineering, Universiti Tun Hussein Onn Malaysia, 86400 Batu Pahat, Johor, Malaysia

Abstract. Biodiesel is an alternative fuel as a replacement to the standard diesel fuel in combustion diesel engine. The biodiesel fuel has a significantly influences throughout the combustion process and exhaust emission. The purpose of this research is to investigate the combustion process behavior during the End of Injection (EOI) period and operates under variant conditions using Rapid Compression Machine (RCM). Experimental of RCM is used to simulate a combustion process and combustion characteristics of diesel engine combustion. Three types of biodiesel blend which are B5, B10 and B15 were tested at several injection pressures of 80 MPa, 90 MPa and 130 MPa under different ambient temperatures, 750 K to 1100 K. The results of this study showed that the ignition delay slightly reduced with increasing the content of biodiesel blends from B5, B10 and B15 and became more shorten as the injection pressure been enhanced. As the injection pressure increased, the behavior of combustion pressure at end of injection is reduced, radically increased the NOX emission. It is noted that the process of combustion at the end of injection increased as the ambient temperature is rising. In fact, higher initial ambient temperature improved the fuel atomization and mixing process. Under the biodiesel combustion with higher ambient temperature condition, the exhaust emission of CO, O2, and HC became less but increased in NOX emission. Besides, increased in blends of biodiesel ratio are found to enhance the combustion process, resulted a decreased in HC emissions.

1 Introduction

Nowadays, the major issues of automotive industry facing are to control fuel consumption, increasing the price of standard diesel fuel and pollutant emissions. Due to this major problem, biodiesel is one of the potential alternative fuels for diesel fuel, which could keep

* Corresponding author: amirk@uthm.edu.my

the high efficiency in diesel engines. Basically, biodiesel [1-3] is found as the alternative fuels to reduce emission of hydrocarbon (HC), carbon monoxide (CO), and particulate matter (PM) but it affects to the increasing of oxides of nitrogen (NOX) emission. The main usage of biodiesel is lead to the highest reduction in HC, CO and PM emissions without any power losses, but it also increased in fuel consumption and NOX emission on diesel engine with no modification or special fuelling equipment [4]. Although the regulation of NOX emission is very strict now, but biodiesel still can eligible to satisfy the stringent future emissions policies and the lack of fossil fuels [5-7]. In diesel engine, ignition delay is a major factor to determine the rapid pressure rise at the initial burning stage and subsequent combustion stage. The ignition delay, which is meant the duration or time period between the start of injection and ignition of the combustion. The process in ignition delay divided into physical and chemical processes. Physical delay processes correspond to the mixture formation that involving atomization, penetration, entrainment and vaporization. Meanwhile, the chemical delay to the time necessary to get an exponential increase in the chemical reaction rate which consist processes such as fuel decomposition and oxidation [8]. Increase in ignition delay has been often longer than the fuel injection duration, which causes ignition dwell, defined as the duration between the EOI and the start of combustion. With positive ignition dwell, mixing process occurring near or after the EOI dominate the following combustion and emission formation [9-10].

H. M. Pated [11] studied the fuel injection system in diesel engine and found that the fuel injection system can be improved the engine operation characteristics. The injection pressure increased influence the spray atomization and mixture preparation process of biodiesel fuel [12]. While the injection duration dependence on the fuel properties such as bulk modulus, viscosity and density [13]. The influence of high injection pressure can help attain better combustion of fuel which will result in low HC emission, but increased in the values of NOX emission [14]. The behavior of biodiesel fuel at the end of injection is poorly understood, yet it has important implications regarding to the diesel engine emissions. Recent researches have shown that at the end of injection, an entrainment wave is created which caused the fuel spray to entrain rapidly with the ambient gas. This rapid entrainment creates a dilute mixture of fuel that may be a source of unburned hydrocarbon emissions [15-16]. Unburned hydrocarbons (UHC) are fuel and lubrication particles that did not burn completely and smaller non-equilibrium particles of partially burned fuel. Besides, the emission also consists of carbon monoxide (CO) due to the incomplete air-fuel mixing which caused by the very short engine cycle time. In this research, the characteristics of the combustion process are investigated with focused on the behavior at the end of injection in terms of pressure with changing ambient condition and injection parameters. The experimental study was performed in a Rapid Compression Machine [16]. The detail behavior of combustion at the end of injection was collected by PICO Scope 3000 series.

2 Methodology

The biodiesel from Crude Palm Oil (CPO) is then blended with diesel in various concentrations for prepared the biodiesel blends of 5vol% (B5), 15vol% (B10) and 15vol% (B15). During blending process, the mixture was stirred at 70oC for 1 hour with the rotating blade speed maintained at 270 RPM. The properties of biodiesel are listed as illustrates in Table 1. The experimental apparatus can be divided into several systems such as rapid compression machine, single-shot common rail system, data acquisition system

DOI: 10.1051/matecconf/20167801029

and exhaust emission measurement system. Moreover, the exhaust emission was measured by gas analyzer.

Table 1. Properties of Biodiesel

Fuel Type	Properties		
	Density (g/m^3)	Kinematic Viscosity (Cp)	Water Content (ppm)
B5	0.837	3.0	120.1
B10	0.838	2.9	158.6
B15	0.840	3.0	219

A schematic diagram of Rapid Compression machine is shown in Figure 1, together with an outline of the fuel injection system. A free piston is placed at the left side of the cylinder paneled by an orifice plate and partitioned by a diaphragm as a start of the experiment. Then, the driver chamber and the pressure reservoir are filled with nitrogen gas to allow the diaphragm could withstand the pressures in order to avoid the damage of diaphragm. Once the valve is opened, the pressure inside the driver increased immediately and caused the diaphragm burst and the piston is travelled onto a tapered stop ring installed in front of the combustion chamber. Fuel was injected after a fixed interval from the time the piston stopped. The measured air temperature and pressure at the start of injection were used to identify the combustion behavior at EOI by the Pico Scope 3000 series. Table 2 summarizes the experimental conditions in this research.

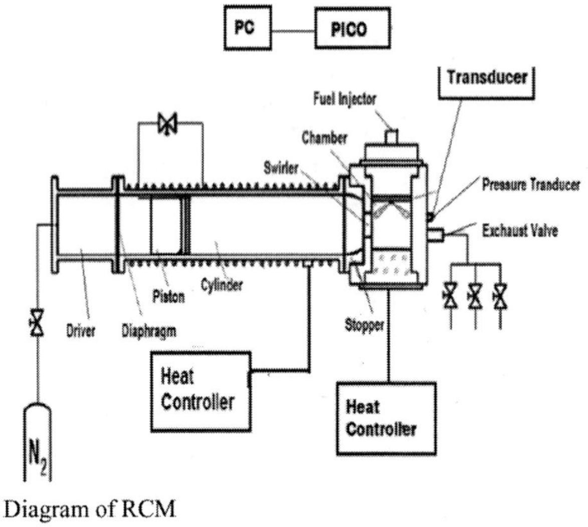

Fig. 1. Schematic Diagram of RCM

Table 2. Specifications of experimental

Fuel	Injector type	6 holes, Ø = 0.16 mm
	Fuel type	B5, B10, B15
	P_{inj} [MPa]	80, 90 and 130
	q_i [ml]	0.04
Ambient Condition (inside chamber)	T_i [°C]	70, 80
	r_s [m/s]	19
	ρ [kg/m^3]	16.6
	O_2 [vol%]	21

3 Result and discussion

The aim of this study was to investigate the influence of the combustion behavior during the End of Injection (EOI) under different parameters with biodiesel blends in the Rapid Compression Machine (RCM). Regarding with this purpose, a comparison between three types of blended biodiesel was investigated in this experiment. For the clarity purposed, the data experiments included emissions under variant conditions in EOI were discussed first. Effects of pressure at EOI on injection duration under variant injection pressure with biodiesel blended with Diesel used were investigated. The initial charging pressure investigated was Pc = 0.1 MPa, which corresponded to the ambient densities of ρ=16.6 kg/m3. The others parameter were kept constant with swirl velocity rs=19 m/s. Fuel was injected by a 6-hole injector with hole-diameter of 0.16 mm and variant injection pressures of Pinj=80 MPa, 90 MPa and 130 MPa.

Figure 2 to Figure 4 shows the characteristics of variant injection pressures at EOI with three different percentages of biodiesel blends. The measurement pressure at EOI of B5 against the duration injection, in which the time elapsed from begin of injection as shown in Figure 2. Fuel injection began at a predetermined time after the piston is stopped in front of the combustion chamber. A decreased in pressure can be identified due to the fuel vaporization, fuel vapor heated and heat transferred until combustion occurred, while the pressure reached the peak after the EOI.

It showed that the pressure inside the combustion chamber at EOI with the low injection pressure, 80 MPa and 90 MPa was higher compared to the high injection pressure at 130 MPa for B5 biodiesel blends.

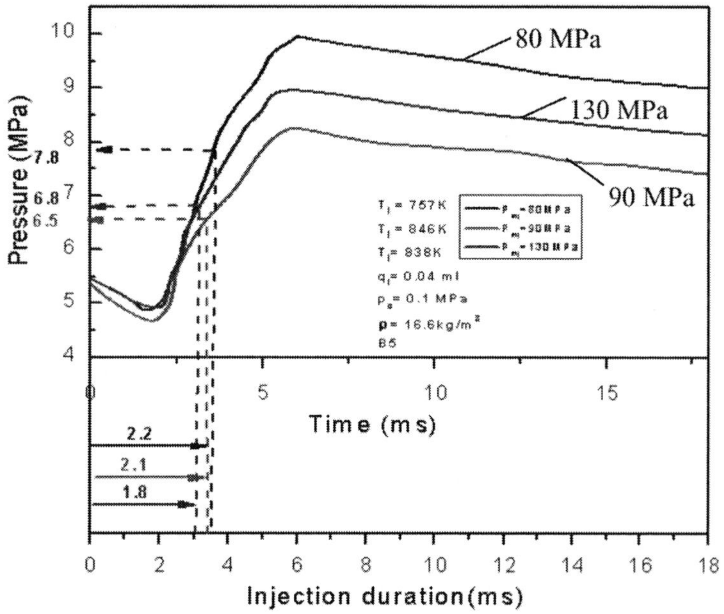

Fig. 2. Variation of Injection duration of B5 blend diesel at different Injection Pressure

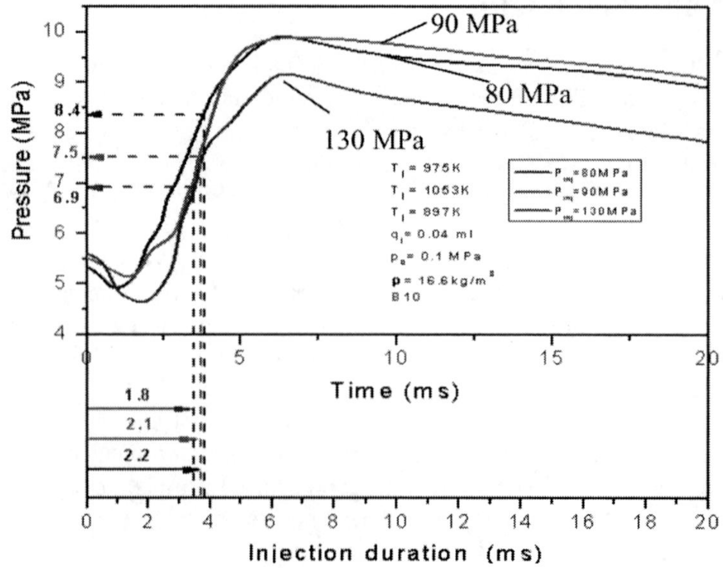

Fig. 3. Variation of Injection duration of B10 blend diesel at different Injection Pressure

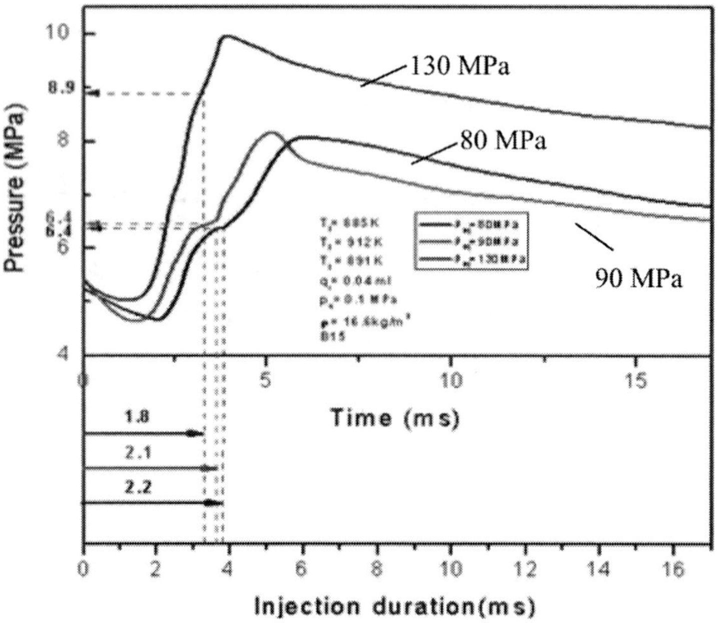

Fig. 4. Variation of Injection duration of B15 blend diesel at different Injection Pressure

B10 biodiesel did not show major changes along with the increased of injection pressure at EOI as illustrated in Figure 3. It can be seemed that, as the injection pressure decreased, the pressure at the EOI became promoted. On the contrary, Figure 4 clearly shows that higher injection pressure at EOI with 130 MPa under B5 of biodiesel blend, which is assumed as the reason of value of viscosity. On top of that, the result showed the injection duration for B5, B10 and B15 are slightly reduced with increased the fuel injection pressure. Possible reason predicted is the higher biodiesel blends (B15) has higher density due to the higher density of biodiesel. High injection pressure promoted mixture formation during the ignition delay period and large amount of combustible mixture is formed. Therefore, dark flame developed widely at spray tip region and between sprays before the appearance of bright flame after the ignition. Furthermore, bright flame has less intensity due to well mixing. Figure 5 illustrates the exhaust emissions of the experiment measured by an exhaust gas analyzer for different injection pressures and biodiesel blends. It can be seen that the higher the injection pressure, the more NOx is observed and less HC exhaust emission. Consistent with previous research, the NOx emission for biodiesel blends fuel is gradually increasing with the fuel injection pressure.

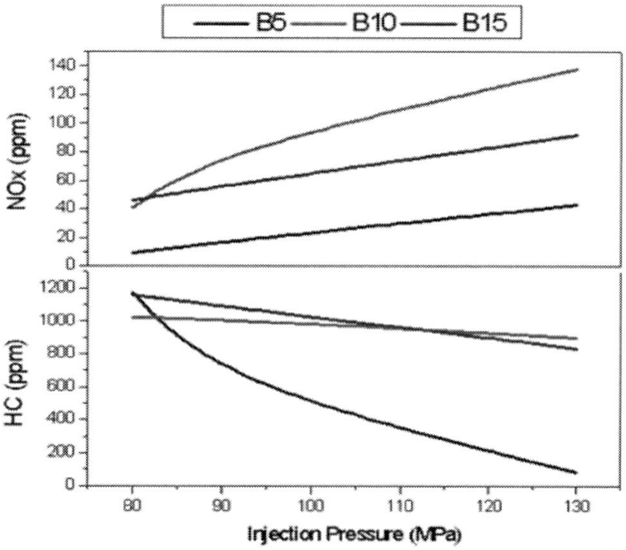

Fig. 5. Effect of pressure at EOI with variant injection temperature towards NOx and HC exhaust emission

4 Conclusion

In this paper, a rapid compression machine has been developed to study the basic aspects of the diesel combustion phenomenon. A satisfactory performance has been obtained, a study of combustion behavior at EOI toward the emission. From a series of tests carried out on this compression machine, the following conclusions are obtained:

- The injection duration for B5, B10 and B15 depends with the fuel injection pressure. Dependence of the injection duration, due to the relatively lower fuel properties. Increasing injection pressure and injection duration influences to the NOX emission. Nevertheless, a change of injection pressure is seemed to have higher impact on NOX emission than change in injection duration.
- From the visualization of combustion and emission by gas analyzer with the highest variant ambient temperature towards to the chamber pressure at EOI, CO, O2, and HC reduction are proven but the increase in NOX emission.

The authors also would like to thank the Ministry of Higher Education Malaysia for supporting this research under Exploratory Research Grant Scheme (ERGS) vot. E032.

References

1. A. Khalid, N. Jaat, A. Sapit, A. Razali, B. Manshoor, I. Zaman, A.A. Abdullah, International Journal of Automotive and Mechanical Engineering, **11**, 2447 (2015)
2. B. Tesfa, R. Mishra, F. Gu, A.D. Ball, *Combustion Characteristics of CI Engine Running with Biodiesel blends*, International Conference on Renewable Energies and Power Quality, Las Palmas de Gran Canaria (2011)
3. A. Khalid, N. Tamaldin, M. Jaat, M. F. M. Ali, B. Manshoor, I. Zaman, Procedia Eng., **68**, 225 (2013)
4. S. Chandrasekar, R. Niranchan, V. S., J.S. Leon, *Effect of Cooled EGR in Combustion Characteristics of a Direct Injection CI Engine Fuelled with Biodiesel Blend*, International Journal of Mechanical, Aerospace, Industrial, Mechatronic and Manufacturing Engineering, **73**, 109 (2013)
5. A. Demirbas, Energ. Convers. Manage., **50**, 2239 (2009).
6. M. Jaat, A. Khalid, B. Manshoor, S.M. Basharie, H. Ramsy, Appl. Mech. Mater., **465**, 265 (2014)
7. M.P.B. Musculus, T. Lachaux, L.M. Pickett, C. Idicheria, *Hydrocarbon Emissions in Low-Temperature- Combustion Diesel Engines*, **724**, 776 (SAE International, 2007)
8. S. Kook, L.M. Pickett, M.P.B. Musculus, SAE Int. J. Engines, **2**, 1194 (2009)
9. A. Khalid, M.D. Anuar, Y. Ishak, B. Manshoor, A. Sapit, M. Leman, I. Zaman, *Emissions characteristics of small diesel engine fuelled by waste cooking oil*, 4th International Conference on Production, Energy and Reliability, **13**, 6 (2014)
10. O.A. Kuti, J. Zhua, K. Nishidaa, X. Wang, Z. Huang, Fuel, **104**, 838 (2013)
11. H.M. Patel, *Design and Optimization of Fuel Injection System in Diesel Engine Using Biodiesel – A Review*, International Journal of Innovations in Engineering and Technology, **2**, 70 (2013)
12. S. Lee, S. Park, Y. Daisho, P. I. Mech. Eng. D-J. Aut., **218,** 1317 (2005)
13. A. Kastengren, Christopher F. Powell, Z. Liu, S. Moon, J. Gao, X. Zhang, J. Wang, *End-of-Injection Behavior of Diesel Sprays Measured With X-Ray Radiography.* (ASME, 2010)
14. K. Wislocki, I . Pielecha, D. Maslennikov, J. Czajka, Kones, **18,** 542 (2011)
15. A. Khalid, *Effect of Ambient Temperature and Oxygen Concentration on Ignition and Combustion Process of Diesel Spray*, Asian Journal of Scientific Research, **6**, 434 (2013)
16. A. Khalid, K. Hayashi, Y. Kidoguchi, T. Yatsufusa, *Effect of air entrainment and oxygen concentration on endothermic and heat recovery process of diesel ignition*, (SAE International, 2011)

Investigation of Flame Characteristics of Ethanol-Gasoline Blends Combustion Using Constant Volume Chamber

Nuriati Baddu[1,*], *Amir* Khalid[1] , *Dahrum* Samsudin[1] , *Izzuddin* Zaman[2] , and *Bukhari* Manshoor[2]

[1]Combustion Research Group (CRG), Centre for Energy and Industrial Environment Studies (CEIES), Faculty of Mechanical and Manufacturing Engineering, Universiti Tun Hussein Onn Malaysia, 86400 Batu Pahat, Johor, Malaysia

Abstract. The stricter on emission legislation and growing demands for lower fuel consumption require significant efforts to improve combustion efficiency while satisfying the demand of emission quality. Gasoline blend with ethanol provided advantages due to its self-sustaining concept and high octane number which can be influenced to the combustion process. The purpose of this study is to investigate the effect on the emission and flame propagation of gasoline under the different blending ratio of ethanol fuel in a constant volume chamber. In this paper, the flame propagation of ethanol blends with gasoline in a constant volume chamber has been carried out. The real image is observed by schlieren optical visualization technique and image processing technique. The parameters in this study also included different types of injection pressure, characteristics of flame propagation and different blending ratio of ethanol in gasoline fuels. As a result, the burn rate increases when the content of ethanol percentage in gasoline increased. Therefore, the higher flame propagation area for 0.2 MPa, 0.3 MPa, 0.4 MPa were headed by E15 followed by E10, E5 and gasoline fuels respectively. It is subjected that the increasing of ethanol percentage blends with gasoline could lead to the reduction of CO, HC and NOX emissions. However, the CO2 emission is higher due to the combustion process was enhanced.

1 Introduction

The significant growth of the population and changes in lifestyles had caused the consumption of energy increased. Major portion of energy supplies are presently based on fossil-fuels petroleum [1-3]. However, the energy sources are limited on the earth, thus the replacement of gasoline by liquid fuel produced from renewable sources is high priority in many countries at twentieth century [4-5]. Ethanol is widely promoted as a clean-burning, renewable fuel that could help wean the nation from oil, would likely cause more ozone-related health problems than gasoline [6]. Bioethanol has attracted attention worldwide due

* Corresponding author: yatiepytie@gmail.com

to its potential to use as an alternative automotive fuel [7]. Alcohol fuels have been used in internal combustion engines for many decades. In 1826, Samuel Morey developed an engine that can be used by ethanol fuels [8]. The use of ethanol blends with diesel was a subject of research in the 1980s. At this time, it showed that the ethanol blends were technically acceptable as a fuel for existing engines [9]. Indeed, ethanol showed powerful in some of the first internal combustion engines designed by Otto in the middle of the 19th century [10]. Besides, ethanol can be produced from ligno-cellulosic material like wood, agricultural, forest residues and municipal waste which has potential to be valuable substitute for gasoline fuel. Moreover, ethanol has positive influence on engine performance as well as exhaust emissions since its characteristic and properties are possessed easily [1]. Recently, ethanol has been used extensively as a fuel additive or an alternative fuel in spark ignition (SI) engines due to high octane number that can affect higher compression ratio or advance on sparks timing and clean-burning fuel [10]. The vapor pressure can be increased by using ethanol in gasoline which also can make the mixture up to the volatility standard. Combustion characteristics are playing a significant role to ensure the fuels will combust completely inside the combustion chamber. Meanwhile, it is also important to know the impact of exhaust emission to the air pollution such as carbon monoxide (CO), hydrocarbon (HC), carbon dioxide (CO2), oxides of nitrogen (NOx) by using ethanol blend with gasoline. In this research, the characteristics of combustion process are investigated which focusing on the effects of flame propagation areas and exhaust emission with various concentrations of ethanol content blends with the gasoline fuel (E0, E5, E10 and E15) and different injection pressures (0.2 MPa, 0.3 MPa, 0.4 MPa). This experiment was performed by schlieren optical visualization technique based on sampling time 7.6 ms, 15.2 ms, 32.8 ms, 40.4 ms, 48.0 ms and 55.6 ms.

2 Experimental setup

The effects of flame propagation and exhaust emission were investigated with different concentrations of ethanol 5vol% (E5), 10vol% (E10), 15vol% E(15) and pure gasoline at the injection pressure 0.2 MPa, 0.3 MPa, and 0.4 MPa. All parameters were experimenting with the same base conditions, ambient temperature 345K, fuel quantity 0.15 ml and equivalence ratio $\phi = 1.1$. However, flame propagation and combustion characteristics are essential investigation to the application of ethanol-gasoline blends in spark ignition (SI) engines. The physical and chemical properties indicate the quality of fuel to combust in an SI engine. The engine combustion quality, performance and emission characteristics are fully dependent on these properties. Nevertheless, the properties of ethanol blends with gasoline are shown in the Table 1.

Table 1. Properties of ethanol and petrol [10]

Property	E0	E5	E10	E15
Density (kg/m^3)	757.5	759.1	760.8	776
Reid Vapor Pressure	53.7	59.3	59.6	58.8
Research octane number	95.4	96.7	98.1	98.5
Sulfur (wt%)	0.0061	0.0059	0.0055	0.0063
Distillation temperature (°C)				
a) Initial boiling point	35.5-38.8	36.5	37.8	37.9
b) 10 vol %	54.5-56.1	49.7	52.9	51.7
c) 50 vol%	94.4-109.6	88	71.1-95.8	72.6
d) 90 vol%	167.3-206.3	167.7	157-166.4	165.3
e) End point	197.0	202.5	197.5-208.4	198.1
Heating Value (MJ/kg)	42.58-42.7	40.55-41.78	39.79-41	41.61

Table 2. Experimental specification

Experimental Specification	
Temperature	345 K
Type of fuel	Gasoline, E5, E10, E15
Injection Pressure	0.2 MPa, 0.3 MPa, 0.4 MPa
Injection system	Electronic Control

Table 3. Combustion Chamber specification

Combustion Chamber	
Material	Mild steel
Volume	400 cc
Dimension	160 mm × 160 mm × 160 mm
Bore	75 mm

Flame development and combustion characteristics are important to the application of ethanol-gasoline blends in DISI engines due to their unique physical properties. Several physical properties of ethanol-gasoline blends are enabled to have distinguishable combustion behavior due to their flame propagation characteristics. Therefore, the schlieren optical visualization technique is corresponded to make an analysis on flame propagation area and also act as an effective way to understand the combustion of ethanol-gasoline blends on direct injection spark ignition. Table 2 shows the experimental conditions throughout all the experiments.

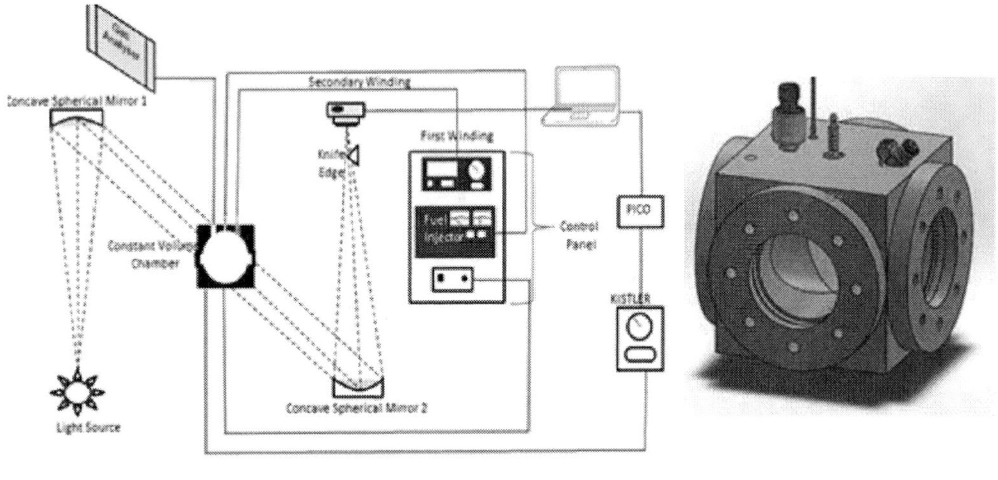

Fig. 1. Schematic Diagram of Schlieren optical visualization technique

Figure 1 shows the schematic of the experiment schlieren setup. The schlieren apparatus consists of two concave mirrors with a focal length 1.524 meter each. A light-emitting diode (LED) light source was projected to a concave mirror with a focal length of 1.524

meters. The light was deflected by the mirror to an acrylic placed in front of the chamber. Here, a straight and equal intensity of light is maintained by the light-emitting diode (LED) light source. The light was then passed through the test subject and reached the 2nd concave mirror. Then, the parallel light of (light-emitting diode) LED light will deflected again by 2nd concave mirror. A knife edge with a diameter of 2.5 mm was used as a schlieren stop which was set at the focal point of the light-emitting diode (LED) light deflected by 2nd concave mirror. After the knife edge cut the LED light, it was reached to camera lens and the image was taken. Figure 2 and Table 3 shows the schematic diagram of constant volume combustion chamber and its specification. In addition, exhaust emission was measured by a SPTC LTD, Automobile Exhaust Gas Analyzer.

3 Result and discussions

Experiments were performed with different blending ratio of ethanol-gasoline fuels and different injection pressures in a constant volume combustion chamber at 345 K temperature with 0.8 MPa initial filling chamber pressure. First of all, the effect of 0.4 MPa injection pressure on the flame propagation area was investigated. Table 4 shows the flame propagation of injection pressure 0.4 MPa, it is clearly seen that the fastest combustion rate are led by the highest additive ethanol in this experiment E15 follow by E10, E5 and gasoline fuels.

Table 4. Flame propagation for bioethanol E15

Injection Pressure (MPa)	Fuels	Time (ms)						
		0	7.6	15.2	32.8	40.4	48.0	55.6
0.4	E0			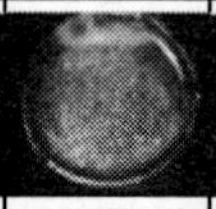				
	E5							
	E10						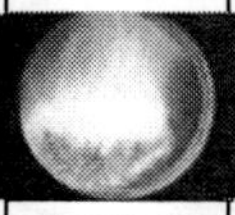	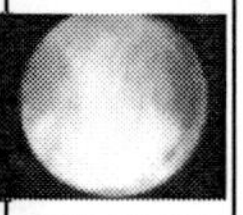
	E15							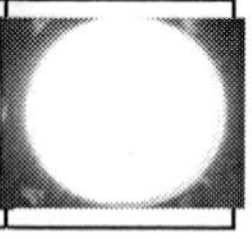

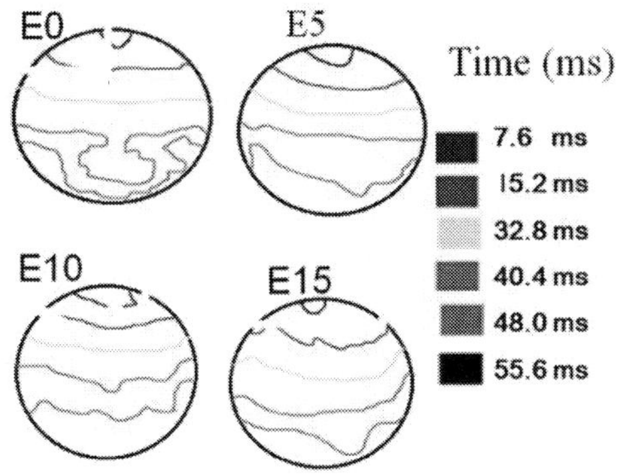

Fig. 3. Flame propagation pattern for injection pressure 0.4 MPa

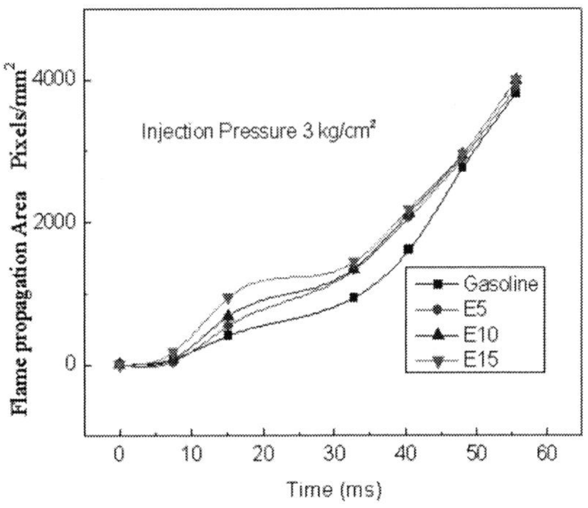

Fig. 4. Graph of flame propagation area against time for injection pressure 0.4 MPa

Moreover, Figure 3 and Figure 4 shows the analysis data referred from Table 4, it is clearly illustrated that the flame propagation area under E15 was higher compared to the pure gasoline, E5 and E10 achieved 173.037 pixels/mm2 and 941.455 pixels/mm2 respectively from time 7.6 ms to 15.2 ms. It was suggested that the spark discharge ionizes the surrounding ignitable mixture, thus an initial flame kernel form. The downstream evaporating fuel generated a favorable locally air-fuel mixture that sustained a subsequent flame propagation. With the fuel being ignited continuously, the flame quickly expanded to the mixture edges and a shock wave is produced like appearance in ambient, especially at the downstream flame tip.

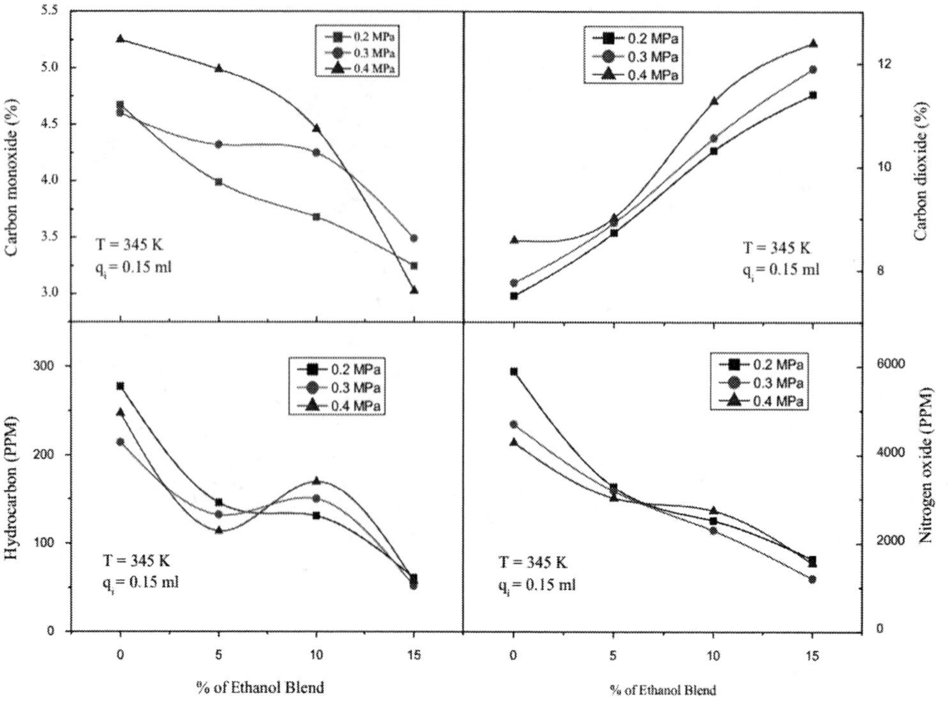

Fig. 5. Effects of ethanol blend on emissions

Figure 5 illustrates the pollutant emissions of combustion with ethanol-gasoline blends fuel. The experimental investigation showed that using ethanol-gasoline blends fuel could decrease NOx emission due to the higher heat vaporization of ethanol compared to gasoline which caused the combustion temperature decreased. Besides, the particle diameter of the accumulation mode became smaller when the concentration of ethanol content increased. Apart from that, the aerosol number concentration was found a decreased with increasing the ethanol concentration while it is predicted to lead to complete combustion. For this reason, CO and NOx emission decreased with increases in ethanol concentration. HC emission was reduced as the oxygen content of ethanol caused the reaction and move towards the complete combustion. Meanwhile, the CO2 emission increased because of the combustion process is improved.

4 Conclusion

This study was done by schlieren optical visualization technique to investigate the flame propagations and exhaust emissions for each types of the ethanol 5vol% (E5), 10vol% (E10), 15vol% E(15) and pure gasoline under variant fuel injection pressures of 0.2 MPa, 0.3 MPa, and 0.4 MPa. The following conclusions were obtained:

- Ethanol-gasoline blends influenced the combustion and flame propagation compared to the pure gasoline. The combustion or burn rate increased with increasing the content of ethanol percentage in gasoline. As a result, the highest flame propagation area for 0.4 MPa was led by E15 followed by E10, E5 and gasoline fuels respectively.
- Reduction of CO, HC and NOx emission were achieved, but CO2 emission increased when the ethanol percentage blends with gasoline increased.

The authors also would like to thank the Ministry of Higher Education Malaysia for supporting this research under Exploratory Research Grant Scheme (ERGS) vot. E032.

References

1. B.M. Masum, H.H. Masjuki, M.A. Kalam, I.M.R. Fattah, S.M. Palash, M.J. Abedin, Renew. Sust. Energ. Rev., **24**, 209 (2013)
2. A. Khalid, N. Jaat, A. Sapit, A. Razali, B. Manshoor, I. Zaman, A. A. Abdullah, International Journal of Automotive and Mechanical Engineering, **11**, 2447 (2015)
3. A. Khalid, K. Hayashi, Y. Kidoguchi, T. Yatsufusa, *Effect of air entrainment and oxygen concentration on endothermic and heat recovery process of diesel ignition*, (SAE International 2011)
4. F.T. Ansari, A.P. Verma, *Experimental determination of suitable ethanol-gasoline blend for spark ignition engine*, International Journal of Engineering Research & Technology, **1**, 1 (2012)
5. A. Khalid, N. Tamaldin, M. Jaat, M.F.M. Ali, B. Manshoor, I. Zaman, Procedia Eng., **68**, 225 (2013)
6. A. Khalid, *Effect of Ambient Temperature and Oxygen Concentration on Ignition and Combustion Process of Diesel Spray*, Asian Journal of Scientific Research, **6**, 434 (2013)
7. A. Ganguly, P.K. Chatterjee, A. Dey, Renew. Sustainable Energy Rev., **16**, 966 (2012)
8. H.O. Hardenberg, Samuel Morey, *Atmospheric Engine* (SAE International, Warrendale, 1992)
9. B. Kovarik, Automotive History Review, **32**, 7 (1988)
10. A. Keskin, Energy Sources. Part A: Recovery, Utilization and Environmental Effects, **32**, 1851 (2010)

Synthesis of Chitosan /Alginate/ Silver Nanoparticles Hydrogel Scaffold

Roslinda Hani Ramli[1,3,*], *Chin* Fhong Soon[1,2], and *Anika Zafiah* Mohd Rus[3]

[1]Biosensor and Bioengineering Laboratory, MiNT-SRC, Universiti Tun Hussein Onn Malaysia, 86400 Parit Raja, Johor, Malaysia.
[2]Faculty of Electric and Electronic Engineering, Universiti Tun Hussein Onn Malaysia, 86400 Parit Raja, Johor, Malaysia.
[3]Faculty of Mechanical and Manufacturing Engineering, Universiti Tun Hussein Onn Malaysia, 86400 Parit Raja, Johor, Malaysia.

Abstract. This work reports the preparation of silver nanoparticles (AgNPs) and synthesis of natural based hydrogel scaffold with an inclusion of AgNPs, chitosan/alginate/silver nanoparticles. The synthesised hydrogel scaffolds were characterised by using Fourier Transform Infrared Resonance Spectroscopy (FTIR). The FTIR result revealed that the shifting of the three peaks of 3252.95 cm^{-1} (–OH and $-NH_2$ stretching), 1591.33 cm^{-1} (C=O stretching) and 1411.88 cm^{-1} (N–H stretching) of chitosan/alginate/silver nanoparticles in compared to chitosan/alginate hydrogel indicating the presence of electrostatic interaction of $-NH^{3+}$ in chitosan reacted with the –COO– group of alginate and binding of the silver (Ag). These results indicated that chitosan/alginate/silver nanoparticles were consolidated in the composite system.

1 Introduction

The ill effects of tissue harms and organ disappointment as a consequence of mishaps or sickness was experienced by a large number of individuals worldwide [1, 2]. Be that as it may, in spite of numerous advances in this field, counterfeit embeds regularly cannot supplant the lost organ totally [3]. Synthetic biomaterials research by integrating the engineering field was conducted due to the necessity for supplanting the damaged or tissue failure either for short or long term [4]. The goal of tissue designing is to deal with the repair of tissue inside the body while coordinating science with building principle [5, 6].

Firstly, a little biopsy of tissue from the patient or a benefactor was included in tissue designing methodology. The transplanted cells then gathered from the tissue, and extended to a sought number in vitro (a manufactured situation outside a living being) [7]. The fuse of the cells include the three-dimensional polymer frameworks that carry on as analogs. In human tissue, there are natural extracellular frameworks, polymer platforms has distinctive structure to this natural extracellular networks yet comparable in capacity. The cells was

* Corresponding author: anika@uthm.edu.my

conveyed to the fancied site in the patient's body, space for regenerative tissue was given, and the structure control too built tissue capacity was controlled by the polymer platforms [2].

The restoration and reparation of a collection of tissues, for instance, ligament, bone and vasculature was repaired and aided by the polymer frameworks. The skeletal and practical similarities to the natural extracellular matrices (ECM) makes hydrogel an alluring tissue platforms [9]. For the most part, the hydrogels utilized for tissue designing ought to be (1) biocompatible and biodegradable; (2) take into account the free dissemination of supplements and metabolites; (3) furthermore have the capacity to give mechanical support over a drawn out timeframe [9].

Alginate is classified as a natural polymer and primarily found as a structural component of marine brown seaweed and furthermore as capsular polysaccharides in some dirt microbes [10]. Amazing biocompatibility and bioactivity properties required straightforward gelation with divalent cations, for example, Ca^{2+} was the advantages of alginate. Alginate can be effortlessly handled in hydrophilic solvents, for example, water and known as non-dangerous [11]. Competent connection will be formed by the negative charge of carboxyl groups of alginate with chitosan positive charge. However, due to the loss of divalent ions, alginate hydrogels typically have uncontrollable degradation kinetics [10].

Chitosan is a polycationic polymer, is acquired from scavanger shells by fractional and full antacid deacetylation. β (1-4) connected 2-acetamido-2-deoxy-D-glucose and 2-amino-2-deoxy-D-glucose units was construct chitosan structures [13]. The essential amine bunches existence make chitosan a cationic polyelectrolyte thus permit chitosan to bind to different biomolecules especially with functional groups of hydroxyl thus able to dissolve in acidic solvent [14]. The biodegradability and nontoxicity were the upsides of chitosan, however, chitosan alone was unstable and weak due to excessive swelling in solution.

Natural materials such as chitosan and alginate have been utilized to frame hydrogel frameworks [15]. Polysaccharides are ordinarily biocompatible and have alluring regular properties yet wild degradation [15, 16]. Chitosan originated from crustacean shells is a decent culture bed for cells however our research work experienced issue with microbes pollution. Thus, we proposed the consideration of silver nanoparticles inclusion in the composite hydrogel framework to solve the issue of microbial pollution. Silver nanoparticles have been comprehensively reported with high antimicrobial activity against broad assortment of microbes [17]. The temperature stability and low votality of silver will enhance the function of synthesised hydrogel [17]. The objectives of this work were to synthesis the natural based hydrogel scaffold with the inclusion of silver nanoparticles and to investigate the functional group present in the chitosan/alginate/silver nanoparticles.

2 Materials and methods

2.1 Preparation of silver nanoparticles

Firstly, 100 ml of polyvinyl pyrrolidone (PVP) (Qrec, Asia) solution at 0.3 % (v/v) was prepared by dissolving PVP in distilled water. This PVP solution was divided into two separate polystrene tubes, each with 95 ml PVP at 0.3 % (solution A) and 5 ml PVP at 0.3 % solution (solution B), respectively. Subsequently, 0.039g of silver nitrate ($AgNO_3$) (Qrec, Asia) was added to solution A and allowed to dissolve by stirring constantly on a magnetic stirrer plate (Thermo Scientific, US) for 10 minutes. For the preparation of solution B, borohydride solution was prepared by dissolving 0.17 mg of sodium borohydride ($NaBH_4$) (Qrec, Asia) in solution B. After the complete dissolution of AgNO3, solution B was added drop wise to solution A at a rate of 10 µl/sec. The reaction mixture was mixed vigorously at room temperature for 15 minutes. Then, silver nanoparticles were ready to be used for the subsequent experiment [18].

2.2 Synthesis of chitosan/alginate/silver nanoparticles

Experiment flowchart of chitosan/alginate/silver nanoparticles are as in Figure 1. Chitosan/alginate hydrogel was prepared firstly by dissolving 1g of chitosan in 25 ml of 1 % (v/v) acetic acid solution. A magnetic stirrer (Thermo Scientific, US) was used to stir and blend the solution until homogeneous (chitosan solution). 25 ml of distilled water (DI) then was heated up to 100 °C. 1 g of sodium alginate was then dissolved in the hot water prepared. The stirring was continued until the solution form homogeneous solution (alginate solution). All steps were repeated and sodium alginate solution finally was added to chitosan solution to form chitosan/alginate solution. The entire solution was stirred and blended for 20 minutes until homogeneous. All chitosan, alginate and chitosan/alginate solution was aged at room temperature for 2 hours and placed in a 15 ml polystyrene tube. Hydrogel then were freeze dried using a Labconco Freezone 18 (Labconco, US) freeze dryer. Then, 1% (v/v) calcium chloride ($CaCl_2$) solution was used to polymerised freeze dried alginate and chitosan/alginate for 10 minutes to form framework of hydrogel. Chitosan, alginate and chitosan alginate hydrogel scaffold were cut with thickness ~5 mm and 1 cm diameter. Next, chitosan/alginate hydrogel was soaked in yellow silver colloidal solution AgNPs for 24 hours. This composite of is denoted as Chi/Alg/AgNPs hydrogel. Similar procedures were repeated for the control without an addition of AgNPs and it was denoted as Chi/Alg hydrogel. All chitosan, alginate, chitosan/alginate and chitosan/alginate/silver nanoparticles were ready for subsequent experiment.

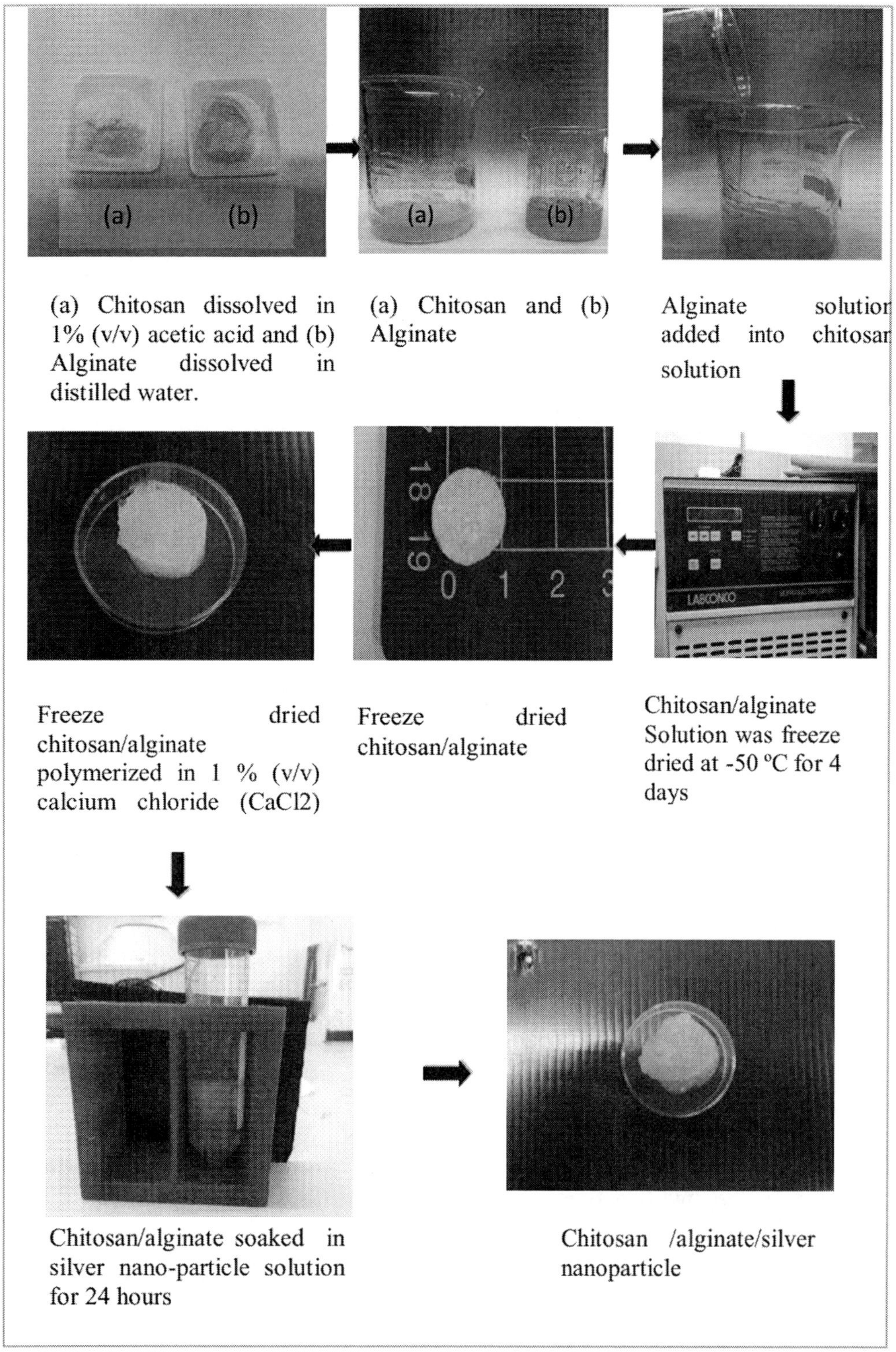

Fig. 1. Experiment Flowchart for the synthesis of chitosan/alginate/silver nanoparticles.

2.3 Characterising functional groups of the composite

Chitosan, alginate, chitosan/alginate and chitosan/alginate/silver nanoparticles scaffold were fabricated by freeze-drying method. Synthesised hydrogel scaffolds were observed with a pale brown color, spongy and inelastic. Chitosan, alginate, chitosan/alginate and chitosan/alginate/silver nanoparticles functional groups were studied by Fourier Transform Infrared Resonance Spectroscopy (FTIR). The Fourier Transform Infrared Resonance Spectroscopy (FTIR) spectra of the hydrogel scaffold was recorded using ATR mode in wavelength ranging from 4000 down to 600 cm^{-1} by using a FTIR spectrophotometer (Perkin Elmer Spectrum 1000, US). The FTIR spectrsocpy experiments were repeated three times.

3 Results and discussion

3.1 Fourier transform infrared resonance spectroscopy (FTIR)

Polymer classification and visualisation of groups-end as well branching of chain was mainly determined by Fourier Transform Infrared Resonance (FTIR) in a polymer study [19]. The connections of unit in a polymer and functional groups existence of chitosan, alginate, chitosan/alginate and chitosan/alginate/silver nanoparticles were studied and analysed by FTIR and are described as in Figure 2.

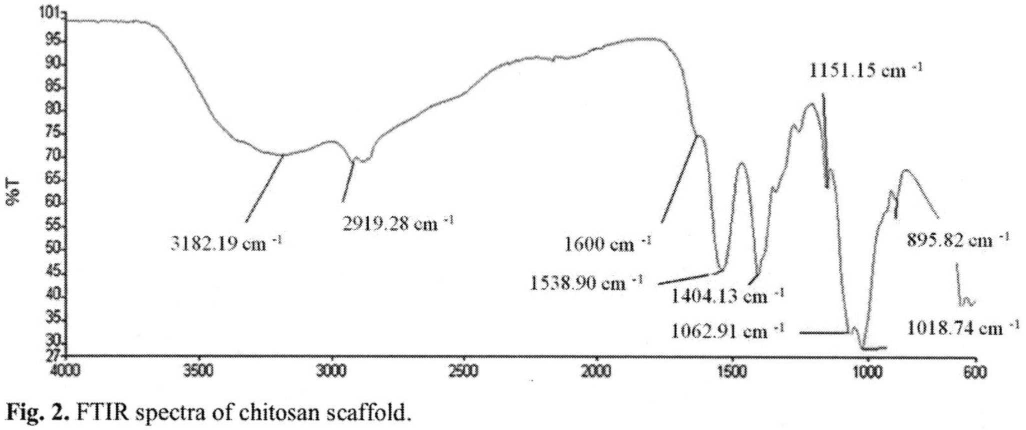

Fig. 2. FTIR spectra of chitosan scaffold.

In view of the IR spectra of chitosan platform in Figure 2, the expansive band at 3182.19 cm^{-1} conformed to the gathers of amine and hydroxyl; 2919.28 cm^{-1} point was prompted by –OH strain; while around 1600cm^{-1} and 1538.90 cm^{-1}, it involved the vibrations of carbonyl (C=O) amide I band (secondary amide), and N–H (amide II band) respectively. 1404.13 cm^{-1} point was ascribed to the stretching of N–H (ether and amide) and stretching of N-H (amide III band). This demonstrated the formation of chitosan/acetate salt in the system [20]. The structural tremor of C–O was characterized by 1062.91 cm^{-1} curve point [21].

Figure 3 illustrated the alginate frameworks demonstrates 1601.86 cm^{-1} curve that related to the (C=O) carboxylate bunch. Furthermore, 3350.99 cm^{-1} solid and strong peak was associated to the bunch of –OH. The extending recurrence at 1425.25 cm^{-1} and 1079-

1024 cm^{-1} was correlated by regularity carboxylate and irregular extending recurrence respectively [21].

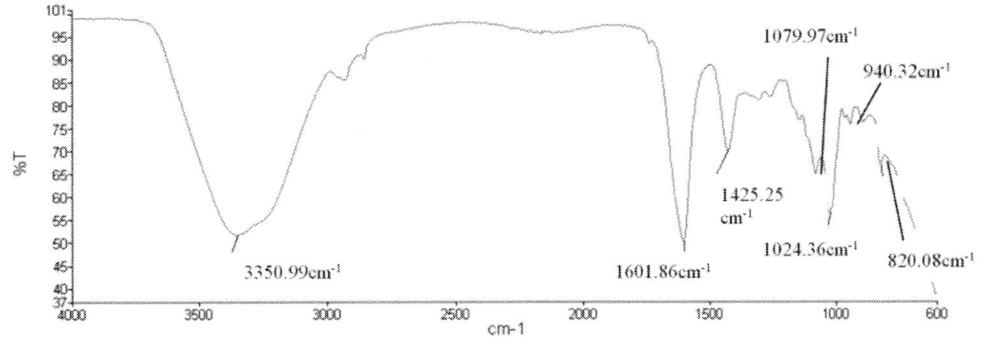

Fig. 3. FTIR spectra of alginate scaffold.

In correlation of the IR spectra of chitosan or alginate platforms to chitosan/alginate hydrogel framework, regular and irregular stretching of chitosan and alginate –COO– bunches has moved to 1591.33 cm^{-1} and 1411.88 cm^{-1}, separately on spectra of chitosan/alginate. Carbonyl (C=O) of secondary amide (amide I band) of chitosan at 1600cm^{-1} also observed with the reduction of retention band as appeared in Figure 4. Moreover, the extending vibrations of –OH and $-NH_2$ of chitosan (3182.19 cm^{-1} and 2919.28 cm^{-1}) and alginate (3350.99 cm^{-1}) both moved to 3252.95 cm^{-1} and get wider [22]. Polyelectrolyte complex was shaped by electrostatic connection of the carboxylic gatherings of alginate with ammonium gatherings of chitosan has caused the two polymers has framed a competent connections [23, 24]. The indication of –COO– of alginate was responded to the $-NH^{3+}$ of chitosan is likewise upheld with N–H twisting vibration of essential amines of chitosan (1538.90 cm^{-1}) and regular as well irregular stretching of –COO– at 1425.25 cm^{-1} of alginate was displaced [24].

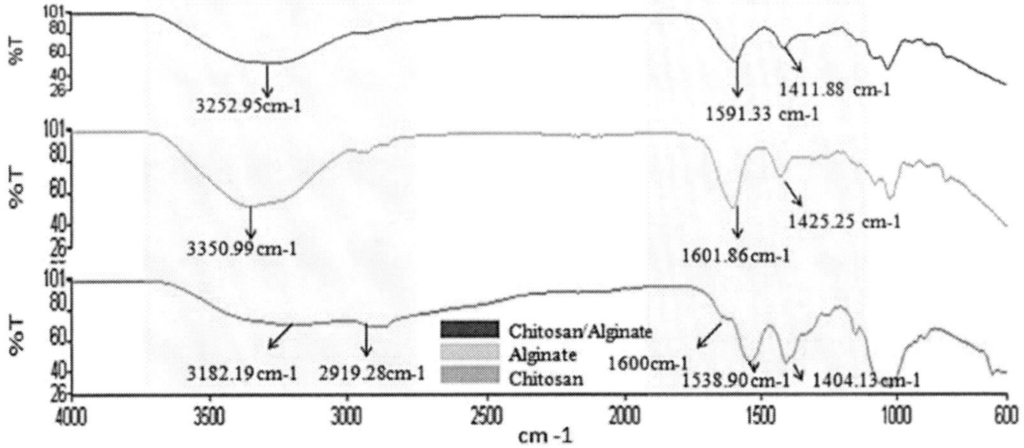

Fig. 4. FTIR spectra of chitosan/alginate, chitosan and alginate scaffold.

Solid retention band at 1635.89 cm^{-1}, that associated with amide carbonyl (C=O) of PVP was observed in IR spectrum of silver nanoparticles as in Figure 5 [18, 25]. The –OH gathering of H_2O and N-H extending of amine gathering were related to curve point at 3313.06 cm^{-1} og AgNPs [18].

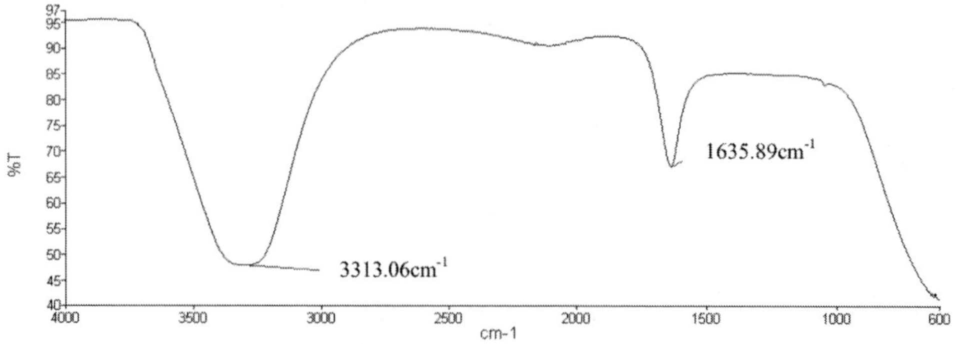

Fig. 5. FTIR spectra of silver nanoparticles.

The C=O motions in amino gatherings explained point of 1600 cm^{-1} of chitosan (Figure 2) thus give a confirmation of acetyl amino bunches existence. However, according to Figure 4, the absence of this point in chitosan/alginate IR spectra and the moving crest at 1591.33 cm^{-1} in chitosan/alginate (NH_2 amino bunches properties) in Figure 4 additionally demonstrated its inclusion in the development of polyelectrolyte complex. These progressions likewise happened in IR spectra of chitosan/alginate/silver nanoparticles (Figure 6), in which the silver (Ag) nanoparticles tie to the N-H obligation of chitosan [23, 26, 27]. Due to the reduction of ~1400 cm^{-1} absorbance in chitosan/alginate/silver nanoparticles, CH_2OH gatherings existence is relevant. The diminishing of this crest at chitosan/alginate/silver nanoparticles (Figure 6) was relative to Ag^+ particles expand adsorption [26].

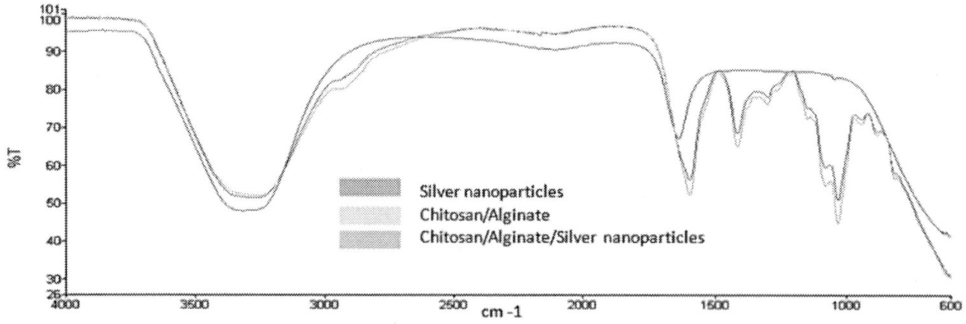

Fig. 6. FTIR spectra of silver nanoparticles, chitosan/alginate and chitosan/alginate/silver nanoparticles.

4 Summary

In this research work, a natural based hydrogel scaffold with an inclusion of silver nanoparticles, (chitosan/alginate/silver nanoparticles) was successfully synthesised and the functional group of the biomaterials composite has been studied. The competent connection which also been known as polyelectrolyte complex formation between two polymers, chitosan and alginate was due to the interaction of ammonium gatherings of chitosan with carboxylic bunches of alginate by electrostatic connection was confirmed by the displacement stretching of regular and irregular of –COO– groups of alginate, –OH– and –NH_2 of chitosan. The absence of C=O motions in amino group of chitosan and moving crest at ~1590 cm^{-1} (NH_2 amino bunches properties) in chitosan/alginate/silver nanoparticles also illustrates Ag nanoparticles has binded to N-H bond of chitosan. Thus, based on the preliminary result on IR spectra of chitosan/alginate/silver nanoparticles, Ag binding within the polyelectrolyte complex of alginate and chitosan were detected and synthesised successfully.

We would like to thank Universiti Tun Hussein Onn Malaysia (UTHM) and research financial support (ERGS E018) awarded by Universiti Tun Hussein Onn. The author also would like to thank Mr. Fazlannuddin Hanur Bin Harith, Mr. Sulaiman Bin Ngadiran and Mr. Mohd Faizal b. Mohamad @ Mohd Annuar for their technical supports.

References

1. H.C. Arca, S. Senel, J. Pharm. Sci., **33**, 211 (2008)
2. K.Y. Lee, D.J. Mooney, Chem. Rev., **101**, 1869 (2001)
3. F.P. Brandl, *PHD dissertation* (Universitat Regensburg, 2010)
4. J.B. Park, Med. Oral. Patol. Oral., **16(1)**, 115 (2011)
6. J.D. Enderle, J.D. Bronzino, *Introduction to biomedical engineering* (Elsevier Academic Press, Burlington, 2012)
7. S. Dumitriu, *Polysaccharides: structural diversity and functional versatility* (CRC Press, Boca Raton, 2004)
9. J.A. Burdick, R.L. Mauck, *Biomaterials for tissue engineering applications: a review of the past and future trends* (Springer New York, 2010)
10. N. Cao, *Master dissertation* (University of Saskatchewan, 2011)
11. K. Baysal, A.Z. Aroguz, Z.Adiguzel, B.M. Baysal, Int. J. Biol. Macromol., **59**, 342 (2013)
13. B.J. Lawrence, *PHD dissertation* (Oklahoma State University, 2008)
14. N.L. Francis, *PHD dissertation* (Drexel University, 2012)
15. L. Lum, J. Elisseeff, *Injectable hydrogels for cartilage tissue engineering* (Topic in tissue engineering, 2003)
16. J.P. Fisher, A.G. Mikos, J.D. Bronzino, *Tissue engineering* (CRC Press, Boca Raton, 2007)
17. G. Cirillo, U.G. Spizzirri, F. Iemma, *Functional Polymers in Food Science: From Technology to Biology, Volume 1: Food Packaging* (John Wiley & Sons, 2015)
18. D. Malina, A. Sobczak-Kupiec, Z. Wzorek, Z. Kowalski, Dig. J. Nanomater. Bios., **7**, 1527 (2012)
19. J. Scheirs, *Compositional and failure analysis of polymers: a practical approach* (John Wiley & Sons, 2000)
20. J. Tanigawa, N. Miyoshi, K. Sakurai, J. Appl. Polym. Sci., **110**, 608 (2008)
21. J. Venkatesan, I. Bhatnagar, S. K. Kim, Marine Drugs, **12**, 300 (2014)
22. V.K. Malesu, D. Sahoo, P. Nayak, Int. J. Appl. Biol. Pharm., **2(3)**, 402 (2011)
23. L.N. Thwala, *PHD dissertation* (University of Johannesburg, 2012)
24. S. Honary, M. Maleki, M. Karami, Trop. J. Pharm. Res., **8(1)**, 53 (2009)
26. P. Kaur, A. Choudhary, R. Thakur, Int. J. Sci. Eng. Res., **4**, 869 (2013)

Characterisation of silver nanoparticles produced by three different methods based on Borohydride reducing agent

Roslinda Hani Ramli[1,3] , *Chin Fhong* Soon[1,2, *], *Anika Zafiah* Mohd Rus[3]

[1]Biosensor and Bioengineering Laboratory, MiNT-SRC, Universiti Tun Hussein Onn Malaysia, 86400 Parit Raja, Johor, Malaysia
[2]Faculty of Electric and Electronic Engineering, Universiti Tun Hussein Onn Malaysia, 86400 Parit Raja, Johor, Malaysia
[3]Faculty of Mechanical and Manufacturing Engineering, Universiti Tun Hussein Onn Malaysia, 86400 Parit Raja, Johor, Malaysia

Abstract. This work reports the preparation and characterisation of silver nanoparticles (AgNPs) through chemical reduction method with three different procedures denote as modified Mulfinger's Method, Mulfinger's Method and Malina's Method. The yellow colloidal solution were characterised and analysed using Field Emission Scanning Electron Microscopy (FESEM), Energy Dispersive X-ray Spectroscopy (EDS), X-Ray Diffraction (XRD). FESEM and EDS analyses confirmed that the formation of AgNPs in Malina's Method with highest yield of silver (Ag), 67.39 % weight as compared to Mulfinger's and modified Mulfinger's Method. The peaks in XRD pattern for Malina's Method are in good agreement with face-centered-cubic form of metallic silver while modified Mulfinger's Method and Mulfinger's Method XRD pattern suggest the halite/sodium chlorate and silver chlorate peaks respectively. Malina's Method is found to be a suitable method to study AgNPs.

1 Introduction

As of late, nanotechnology has gotten to be a standout amongst the most essential and energizing cutting edge fields in physics, material science, engineering, biology and also chemistry [1]. Nanotechnology is connected with the field of manufacture, characterisation, manipulative and utilisation of structures on a nano scale. Because of the nano-scaled size (1-100 nm), nanoparticles have altogether enhanced biological, chemical, and physical properties [2]. Various changes in physical properties are because of the transformation from smaller scale particles (micro-sized) to the nanoparticles [1]. The expansion in the

[*] Corresponding author : soon@uthm.edu.my

surface-to-volume proportion is a steady movement as the molecule gets smaller, induce the expanding strength and response of atoms on the surface of the molecule [1].

Wet chemical synthesis is normally used to incorporate metallic nanoparticles in which the chemicals utilized are regularly toxic and combustible. Moreover, biosynthesis could likewise be connected to integrate metallic particles yet the utilizing of this strategy still faces the challenge to isolate the nanoparticles, decreased purity and inconsistent response or behaviour of particles [3]. Noble metal nanoparticles has pulled in much consideration, because of their prevalent chemical and physical characteristic [4]. Silver nanoparticles (AgNPs) have occupied the consideration of the specialists because of its novel substance, physical, and natural properties, particularly their adequacy against microorganisms, used to reduce inflammation and wound curing ability [5].

There are various utilisations of nanoparticles, for example, for biological, physical and pharmaceutical applications [6]. The imperative properties of the silver-based composite are non-toxic, antibacterial and inorganic characteristic [7]. Public places, for example, railway stations and lifts in China, silver nanoparticles has been utilised widely as antimicrobial agents [6]. It was accounted in 1889, M. C. Lea firstly synthesised the silver colloid-stabilized by citrate with a normal breadth of 7-9 nm [8].

The most frequent and most applied method to prepare the silver nanoparticles with stable properties is by the chemical reduction method. Borohydride ($NaBH_4$) and citrate are reducing agents that frequently used to reduce silver-containing salt in the midst of production of nanosilver particles [6]. Polyvinylpyrrolidone (PVP) was utilised as a stabiliser in the synthesis of the silver nanoparticles [9]. Biocompatibility, universal soluble both in water and polar solvents and has fantastic wetting properties are the unique properties of Polyvinylpyrrolidone (PVP) [9].

Colloidal silver with molecule breadths of a few nanometers were yielded by the reduction of silver ions (Ag^+) in aqueous solution [6]. The reduction of different edifices with Ag^+ particles leads to the formation of silver atoms and followed by agglomeration of silver nanoparticles into oligomeric bunches [6]. These bunches in the long run prompted the development of colloidal silver particles [6].

The objectives of this work were performed to study the properties and compare the yield of Silver (Ag) produced using different methods such as (a) modified Mulfinger's method, (b) Mulfinger's method and (c) Malina's method. The summary of the three methods is as listed in Table 1. Malina's method does not involve the usage of the ice bath during the experiment and polyvinyl pyrrolidone (PVP) was added as a stabiliser before the formation of yellow silver colloidal solution. While the difference between modified Mulfinger's method and Mulfinger's method is that Mulfinger's method originally does not involve the addition of sodium chloride (NaCl) to check the stability of nanoparticles and difference of volume $AgNO_3$ and PVP added. These differences may produce different yields and composition of silver. Thus, this research was conducted to investigate, compare the three methods and identifying which method produces the highest yield of Ag.

Table 1. Comparison of methods for producing silver nanoparticles.

Modified Mulfinger's method [10]	**Mulfinger's method [11]**	**Malina's method [12]**
Involve the usage of the ice bath to reduce the decomposition $NaBH_4$	Involve the usage of the ice bath to reduce the decomposition $NaBH_4$	Mixture prepared at room temperature, 25 °C.
PVP (stabiliser) was added after the formation of yellow silver colloidal solution	PVP (stabiliser) was added after the formation of yellow silver colloidal solution	PVP (stabiliser) was added before the formation of yellow silver colloidal solution
Involves an addition of sodium chloride (NaCl) to check the stability of AgNPs	Not involve the addition of sodium chloride (NaCl)	Not involve the addition of sodium chloride (NaCl)
Produced particles with range 1-10 nm	Produced particles with range 10-14 nm	Produced particles with non uniform size distribution
Energy Dispersive Xray Spectroscopy (EDS) confirms the presence of AgNPs	No elemental analysis has been conducted in paper	Energy Dispersive Xray Spectroscopy (EDS) confirms the presence of AgNPs

2 Materials and methods

2.1 Preparation of silver nanoparticles (modified Mulfinger's method)

Firstly, 30 ml of 0.002 M of sodium borohydride ($NaBH_4$) (QRec, Asia) was added to a conical flask. The flask was placed in an ice bath and a solution was stirred with magnetic stirrer (Thermo Scientific, US) for 1 minute. To reduce the decomposition during the experiment, an ice bath was used in the experiment. Then, 2 ml of 0.001 M silver nitrate ($AgNO_3$) (QRec, Asia) was added into the $NaBH_4$ solution at approximately 10 µl per second. The stirring was stopped as soon as all of $AgNO_3$ was added into $NaBH_4$ solution. Yellow colloidal suspension of silver nanoparticles should be formed at this stage. The reflection of a laser beam from the particles indicates the presence of a colloidal suspension. Next, 1 ml of the yellow colloidal solution of silver was transferred into a glass test tube. Approximately 100 µl of 1.5 M sodium chloride (NaCl) (Qrec, Asia) then was added to the test tube. NaCl solution was added to check the stability of silver nanoparticles formed. If the nanoparticles were unstable and has formed aggregates, further addition of NaCl will cause yellow colloidal solution to turn darker yellow, then gray. Subsequently, 200 µL of PVP (QRec, Asia) at 0.3 % (v/v) was added into the test tube to prevent aggregation. Silver nanoparticles are ready to be used for the next experiment [10].

2.2 Preparation of silver nanoparticles (Mulfinger's method)

For preparation of silver nanoparticles, 30 ml of 0.002 M of $NaBH_4$ was added to a conical flask. The solution was stirred with magnetic stirrer for 20 minutes flask and placed in an ice bath. 10 ml of 0.001 M $AgNO_3$ was added into stirring $NaBH_4$ solution at approximately 10 µl per second. The stirring was stopped as soon as all of $AgNO_3$ was added into $NaBH_4$ solution. At this stage, yellow colloidal suspension of silver nanoparticles should be formed. The reflection of a laser beam from the particles indicates the presence of a colloidal suspension. Next, 5 ml yellow colloidal solution were transferred into a test tube. Then 20 µL of PVP 0.3% solution was added into the test tube to prevent aggregation [11]. Silver nanoparticles are ready to be used for the next experiment. The difference between modified Mulfinger's method and Mulfinger's method is that Mulfinger's method originally does not involve NaCl to check the stability of nanoparticles and difference of volume $AgNO_3$ and PVP added. However, the concentration of the borohydride and silver nitrate solutions are similar.

2.3 Preparation of silver nanoparticles (Malina's method)

Firstly, 100 ml PVP 0.3 % (v/v) solution was prepared by dissolving PVP in distilled water. This PVP solution was divided by two in separated polystyrene tubes, each with 95 ml PVP 0.3 % (solution A) and 5 ml PVP 0.3 % solution (solution B) respectively. Then, solution A was added to 0.039 g of AgNO3 and allowed to dissolve by stirring constantly on a magnetic stirrer plate for 10 minutes. While in another polystyrene tube, 0.17 mg NaBH4 was dissolved in solution B to produce borohydride solution. Solution B was added drop wise to solution A at a rate of 10 µl/sec after the complete dissolution of silver nitrate. The reaction mixture was mixed vigorously at room temperature for 15 min [12]. Then, silver nanoparticles were ready to be used for the subsequent experiment.

2.4 Physical characterisation

Characterisation of the yellow colloidal solution was performed by Field Emission Scanning Electron Microscopy (FESEM), Energy Dispersive X-ray Spectroscopy (EDS), and X-ray Diffraction (XRD). To examine the physical structrues of the silver nanoparticles produced, the samples were first prepared by dispersing the synthesised yellow silver colloidal solution on a glass substrate and left air dried at room temperature, 25 °C. Images of the silver colloids were obtained using the Field Emission Scanning Electron Microscopy (FESEM, JEOL JSM-7600F) and elements of the silver colloids were investigated using EDS detector of the FE-SEM. The X-ray Diffraction (XRD) analysis of the silver colloids was studied using a X'pert PRO PANalytical diffractometer. The samples of silver colloids were investigated using C-Kα X-rays at a wavelength (λ) of 1.54060 Å and the detected data was taken in a 2Θ range of 10° to 120° with a step size of 0.02°. All the characterisation experiments were repeated three times.

3 Results and discussion

3.1 The physical structures and elements of the silver nanoparticles

The specimens arranged by modified Mulfinger's, Mulfinger's and Malina's method created comparative yellow, clear and transparent aqueous solution in the vicinity of PVP, in which PVP functions as a stabilising agent in preventing aggregation [11]. In any case, this work recognized that the yield, creations and physical properties of the silver nanoparticles delivered utilizing the three techniques are for sure distinctive. The micrographs of FESEM depicted as in Figures 1 and 2. As shown in Figure 1 (a, b), observed that there are large crystals on 800 × magnifications where the presence of the NaCl crystals is apparent and it is confirmed by EDS elemental analysis on modified Mulfinger's method. While as shown in Figures 1 (d) and 2 (b), silicon (Si) and oxygen (O) peaks are associated with the glass slide that was used as substrate to hold the sample during the microscopy. This also explains the presence of O in EDS result of modified Mulfinger's method. Besides, modified Mulfinger's method shows sodium (Na) and chlorine (Cl) peak is dominant due to the addition of NaCl in the system. In contrast, elements of Na, magnesium (Mg) and calcium (Ca) were identified in EDS data by Mulfinger's method, whereas Cl was found in EDS data by Malina's method. However, for Malina's method, based on FESEM and EDS analysis of samples produced, Malina's method indicated highest level of silver element at a weight 67.39%.

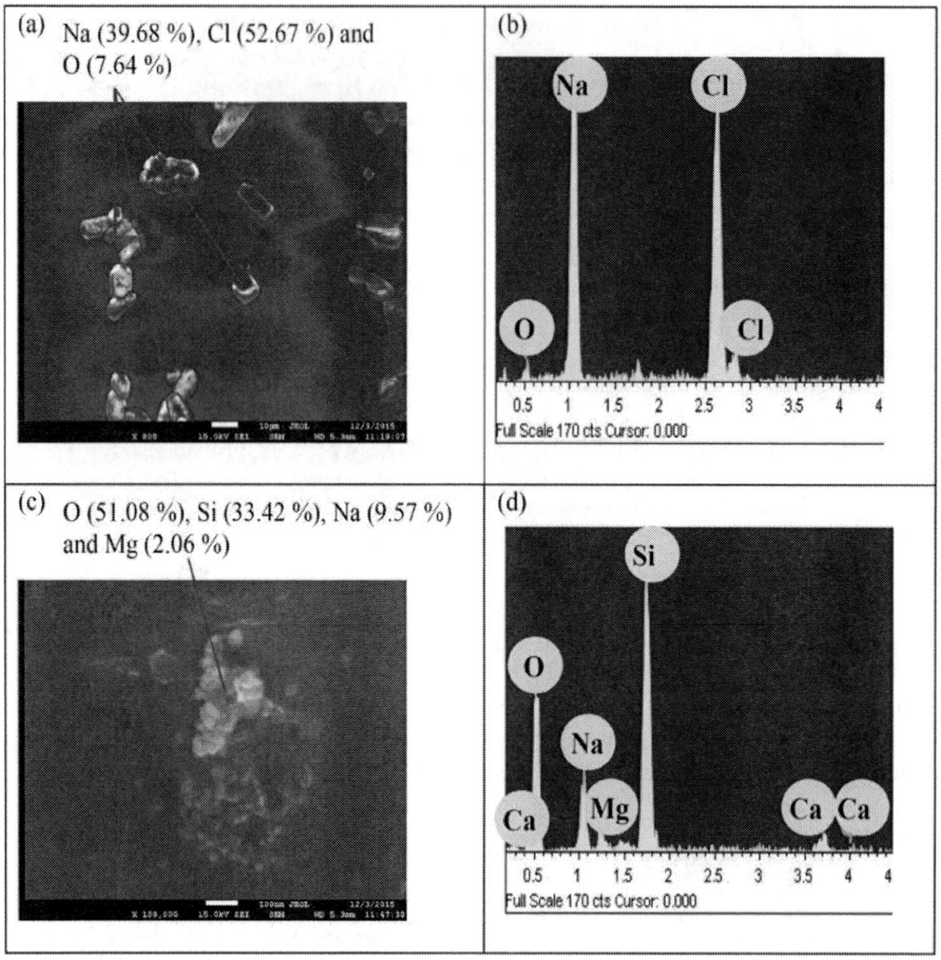

Fig. 1. FESEM image and EDS result of yellow colloidal solutions produced by modified Mulfinger's Method (a,b), Mulfinger's Method (c,d).

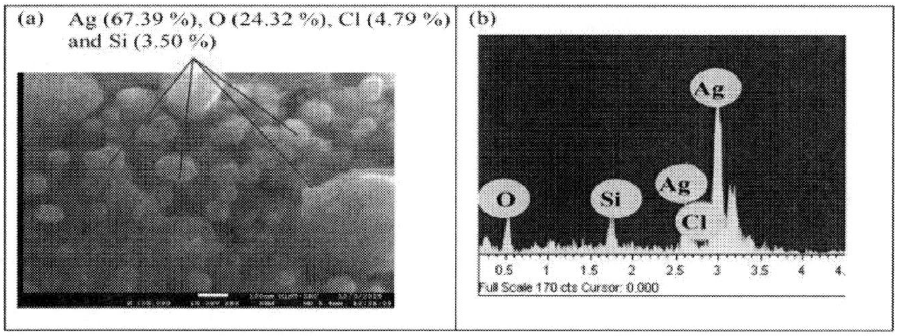

Fig. 2. FESEM image and EDS result of yellow colloidal solutions produced by Malina's Method (a,b).

FESEM images, elemental analysis with EDS confirmed the presence of silver nanoparticles in Malina's method. Modified Mulfinger's and Mulfinger's method indicated the absence of silver (Ag) peaks.

3.2 X-ray diffraction analysis (XRD)

According to Figure 3, the XRD result of modified Mulfinger's method, the peaks at 31.57° (200) and 54.05 ° (231) peaks are associated with sodium chlorate (ICSD no: 98-008-0341), while the peaks of 31.74 ° (200), 53.88 ° (311) and 75.34 ° (420) are associated with halite (ICSD no: 98-005-2232) . There were no pure silver element found for the modified Mulfinger's method. However, peak 38.05 ° (111) of silver chlorate and silver (ICSD no: 98-018-0878) were identified for Mulfinger's method. Nonetheless, more encouraging results were found for Malina's method in which all peaks at 38.1 ° (111), 44.3 ° (200), 64.55 ° (220) and 77.56 ° (311) are associated with the pure silver (ICSD no: 98-005-3761) [4]. Hence, the XRD peaks confirmed the presence of pure silver peaks in Malina's method as compared with modified Mulfinger's and Mulfinger's method.

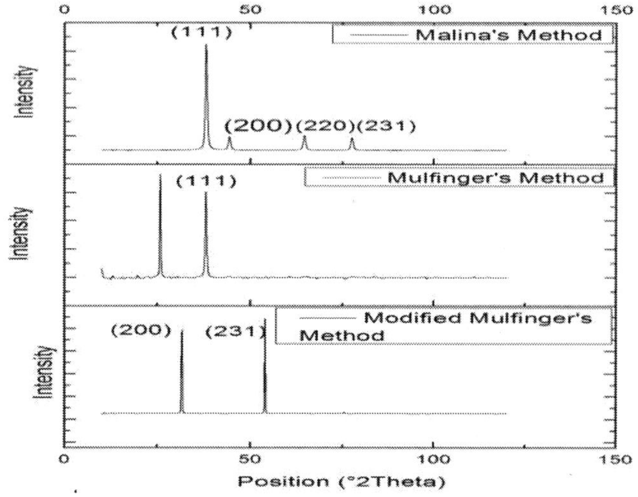

Fig. 3. XRD data of yellow colloidal solutions of modified Mulfinger's ,Mulfinger's and Malina's Methods.

4 Summary

In the experiments connected with the present work, colloidal silver nanoparticles stabilised with PVP was successfully prepared by oxidation-reduction reaction in sodium borohydride and silver nitrate in aqueous solution. Based on FESEM and EDS result, modified Mulfinger's method show Na and Cl peak is dominant due to the addition of NaCl in the system. In contrast, elements of Na, Mg and Ca were indentified in EDS data of Mulfinger's method. However, for Malina's method, EDS analysis of samples indicated highest level of silver element at a weight 67.39%. In addition, the peaks in XRD pattern for Malina's method are in good agreement with face-centered-cubic form of metallic silver while modified Mulfinger's method and Mulfinger's method XRD pattern suggest the halite/sodium chlorate and silver chlorate peaks respectively. Malina's method is found to be a suitable method to study AgNPs. In conclusion, the properties and Ag yield of yellow colloidal solution of modified Mulfinger's method, Mulfinger's method and Malina's method were succesfully investigated in this research paper. Malina's method proved to produce high yield of Ag content and found to be a suitable method to study AgNPs.

We would like to thank Universiti Tun Hussein Onn Malaysia (UTHM) and the authors are grateful to the research financial support (ERGS E018) awarded by Universiti Tun Hussein Onn Malaysia. We also would like to thank Miss Isrehetty and Madam Faezahana for their technical supports.

References

1. S.M. Landage, A.I. Wasif, P. Dhuppe, International Journal of Advanced Research in Engineering and Applied Sciences, **3**, 14(2014)
2. V. Sarsar, M.K. Selwal, K.K. Selwal, Journal of Saudi Chemical Society **19**, 682 (2015)
3. T. Theivasanthi, M. Alagar, Nano Biomed. Eng., **4** (2011)
4. M.A.M. Khan, S. Kumar, M. Ahamed, S.A. Alrokayan, M.S. Alsalhi, Nanoscale Res. Lett., **6**, 1 (2011)
5. A. Haider, I.K. Kang, Advances in Materials Science (2015)
6. S. Prabhu, E.K. Poulose, Int. Nano Lett., **2**, 1 (2012)
7. P. Rauwel, E. Rauwel, S. Ferdov, M.P. Singh, Advances in Materials Science (2015)
8. B. Nowack, H.F. Krug, M. Height, Environ. Sci. Technol. Lett., **45**, 1177 (2011)
9. N. Giri, R.K. Natarajan, S. Gunasekaran, S. Shreemathi, Arch. Appl. Sci. Res. **3**, 624 (2011)
10. M.F. Alsayed, A.A. Shoeib, A.A. Hindi, M.A. Awad, K.M.O. Ortashi, Dig. J. Nanomater. Biostruct., **10** (2015)
11. L. Mulfinger, S.D. Solomon, M. Bahadory, A.V. Jeyarajasingam, S.A. Rutkowsky, C. Boritz, J. Chem. Educ., **84**, 322 (2007)
12. D. Malina, A. Sobczak. Kupiec, Z. Wzorek, Z. Kowalski, Dig. J. Nanomater. Bios., **7**, 1527(2012)

Photostability Characterization of Wood Polymer Composites of Polyvinyl Chloride and Rice Husk to Ultra-Violet Irradiation Exposure

Farhana Hazwanee M.Jais[1,*], *Nurzhatul Aziemah* A. Omar[1], and *Anika Zafiah* M. Rus[1]

[1]Sustainable Polymer Engineering (E1), Advanced Manufacturing and Materials Center (AMMC), Faculty of Mechanical and Manucfacturing Engineering,University Tun Hussein Onn Malaysia, Malaysia

Abstract. In this study, the photostability of wood polymer composites (WPC) was assessed by using ultra-violet accelerated weathering test. Meanwhile, the mechanical properties of WPC show strength improvement at prolonged UV irradiation exposure with 43.21 MPa and maximum strain of 4.08 % at 5000 hours and 1000 hours UV irradiation exposure respectively. The colour stability of WPC was improved by addition of Ultraviolet (UV) stabilizer during pre-mixing process which shows positive effect on the colour stability and prevented chalking of the composites for external use. Generally, discolourations of WPC during test exposure were caused by degradation of both wood and plastic. Therefore, incorporation with pigments and other additives gives improvement to the photostability of WPC. This is based on the increment of WPCs mechanical property while the morphological fracture surface of dumbbell test specimens revealed the pull out rice husk fiber which contributed to the distribution of load in the WPC samples.

1 Introduction

Polymers are high molecular weight substances consisting of molecules that are approximately multiples of simple units [1]. The word polymer comes from the Greek *poli*, which means many, and *meros*, which means parts [1]. A polymer is called a plastic when it has other materials such as stabilizers, plasticizers, or other additives are in it [1].

WPCs are produced starting from rice husk (up to 50 %) in the form of fibres, mixed with a polymeric matrix polyvinyl chloride and additives such as coupling agents, plasticizer, UV stabilizer and lubricant to improve the physical and mechanical characteristic of polymer [2]. Interfacial adhesion between rice husk and thermoplastics is usually very weak due to the high interfacial tension which leads to undesirable properties of the composites. The process of modifying interfacial properties of an incompatible polymer blend leading to the production of polymer blend is called compatibilization [3].

* Corresponding author: farhana.irmb@gmail.com

In WPCs, the composites of wood and plastic basically contain several materials are added in small amounts to affect the performance and processing. Coupling agents, calcium zinc stabilizer, plasticizer, polyethylene wax and also titanium dioxide as an additives are all used to some extent in improving physical properties.

Coupling agents are used to improve the bonding between the hydrophilic wood fiber and hydrophobic matrix. This has been the most active area of research in employing an additive to provide moisture resistance to WPCs. Coupling agents can reduce the amount of moisture WPCs absorb by reducing gaps at the wood matrix interface and by reducing the number of hydroxyl groups available for hydrogen bonding with water. However, the decreases in water absorption can be minor. Steckel et al. [4] examined the moisture sorption properties of several WPCs using a full factorial design and determined that wood flour content had the largest influence on equilibrium moisture content, followed by removal of the original polymer rich composite surface.

The inclusion of a coupling agent and reducing particle size had a much smaller effect on equilibrium moisture content. Panthapulakkal et al. [5] examined water absorption of 65 % rice husk filled HDPE composites. After immersion in water for 1608 hours the control composites absorbed 14% water. Composites containing a coupling agent absorbed either 12 % or 9 % depending on the coupling agent used. However, the use of a lubricant as a processing aid negated any positive effect of the coupling agent. Lin et al. [6] reported an 18% decrease in moisture absorption when 2 % coupling agent was added to 15 % wood flour filled PP composite. Coupling agents may be more effective at mitigating changes in strength and modulus after exposure to cyclical freeze-thaw testing. The increase in weight change of rice husk filled HDPE composites was 9.6 % after 12 freeze-thaw cycles. Adding a coupling agent resulted in increases of 3.3-3.5 % of polymer strength [5].

UV irradiation exposure test can be used to study the polymer photostability, and can provide data regarding polymer life time. This test is important due to the photostability characteristic of polymer will influence the polymer commercial success. Furthermore, UV irradiation is one of the factors that can affect the polymer degradation process [7]. The two most popular UV sources for photo-degrading materials have been the xenon arc and fluorescent lamps. Both the xenon arc and fluorescent light are gas discharge tubes and are usually line source.

2 Methodology

The experiment of characterization and analysis were conducted on virgin polyvinyl chloride (VPVC), rice husk (RH), plasticizer, coupling agents, calcium zinc stabilizer, polyethylene wax, and titanium dioxide (TiO_2). Virgin PVC and RH were prepared based on addition of thermoplastic doped with Titanium Dioxide (TiO_2) and additives consist of coupling agents, plasticizer, lubricant and stabilizer known as dry blend. The dry blend was then mixed using high speed hot mixer at certain temperature and discharged to cooling mixer to bring down the dry blend's temperature. The dry blend was then fed into pelletizing feeder to produce WPCs pellet at certain parameter setting.

The WPCs pellets produced was then run on conical twin screw extruder to fabricate WPCs profile for further analysis. The WPCs sample fabricated was analyzed under scanning electron microscope and tensile strength samples preparation is using Injection moulding machine. The setting parameter of Injection moulding machine with nozzle temperature is 200°C, middle is 180°C, front is 195°C, rear 1 is 175°C, rear 2 is 165°C and feeds is 55°C as shown in Figure 1.

The WPCs samples were exposed under UV light in the UV Weatherometer to examine its tensile properties which examine the photostability of WPC. Higher improvement of WPC strength property is indicating higher stability of the samples under harsh environmental condition after UV weatherometer. Meanwhile the fracture morphological surface of tensile test sample of WPCs was analyzed using scanning electron microscope (SEM).

3 Tensile Test of Ultra Violet Irradiation of Wood Polymer Composites

The mechanical property of WPC is conducted such as to measure its strength of polymers and its filler. Tensile test is used to determine the tensile strength, modulus of elasticity and elongation at break. The tensile test samples are prepared according to ISO 527-2 (5A) as Figure 1. The test is conducted by using Universal Tensile Machine AG-I, Shimadzu, with 10 kN of force. The plotted results in Figure 2 revealed that the tensile strength were increased after uv irradiation exposure at 1000 hours but started to decrease at 5000 hours.

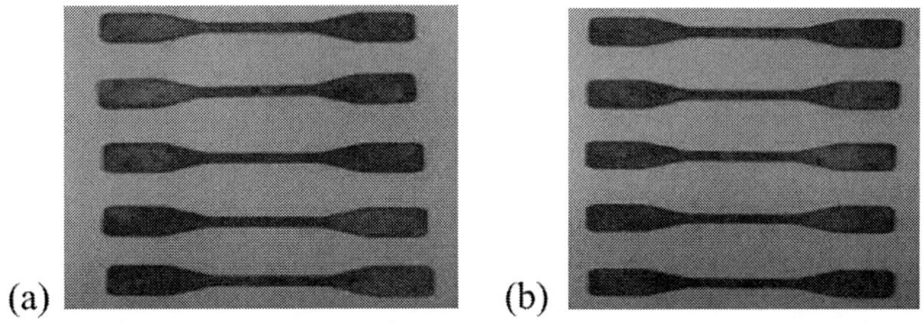

Fig. 1. Dumbell specimen of WPC (a) unirradiated WPC samples, (b) WPC of 5000 hours UV irradiated.

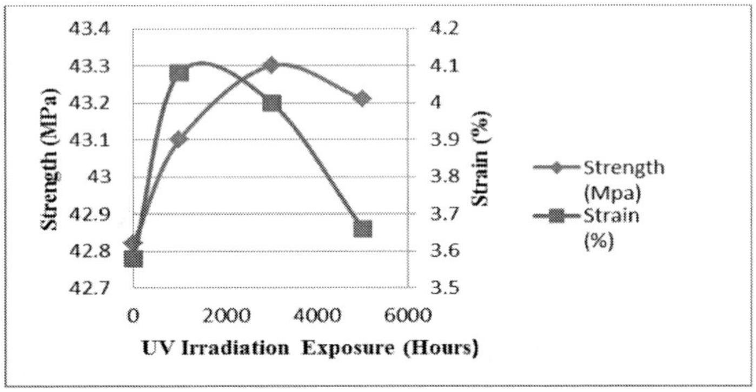

Fig. 2. Mechanical strength and strain of WPC samples of UV irradiation exposure (hours).

It can be seen that the nonlinear tensile behaviour to break is highly sensitive to structural changes in the polymer due to UV light exposure. The ultimate mechanical properties, and particularly strain at break of various polymer systems, react to the photo-oxidative degradation much earlier and in a more pronounced way than in the usual physicochemical analytic methods. Thus strain at break can be utilized in practice as a sensitive indicator of the degree of degradation of photo-oxidized polymers by chemi-crystallization.

The photo-oxidation causes molecular chain scissions in the amorphous regions and these parts of cut chains rearrange into a crystalline phase. Thus, the reduction of mechanical properties is related to the increase of the crystallinity fraction. Additionally, the strain value of polymer compound was increased with the increasing of UV irradiation exposure hours as shown in Figure 2. It is observed that the strain has increased with prolonged UV irradiation exposure with the maximum value of strain is 4.08 %. It is believed that the polymer content, cross-linker and UV irradiation have influenced the bond strength of the polymer. Thus, the variations in strength and strain may be due to the changes in chain-scission/cross-linking ratio that occurred during UV irradiation exposure of the polymer compounds.

4 Scanning Electron Microscopy (SEM) of Wood Polymer Composites of Ultra-Violet Irradiation Exposure

SEM is used to evaluate the changes in fracture surface morphology of polymer compound of WPC. The cross-section of fracture surfaces of samples are coated with platinum by using JFC-1600 auto fine coater and result as Table 1. The tensile specimens' fractures are observed by using scanning electron microscope of model Jeol JFM-6380 located at Material Characterization Laboratory, FKMP, UTHM. The SEM micrograph fracture surfaces of WPC shows the elongation break of filler appears after pulling out when stress is applied. The morphology of miscible polymer compound depends on the components, ratios and processing conditions. The PVC samples indicated that this polymer is contributed to high tensile strain.

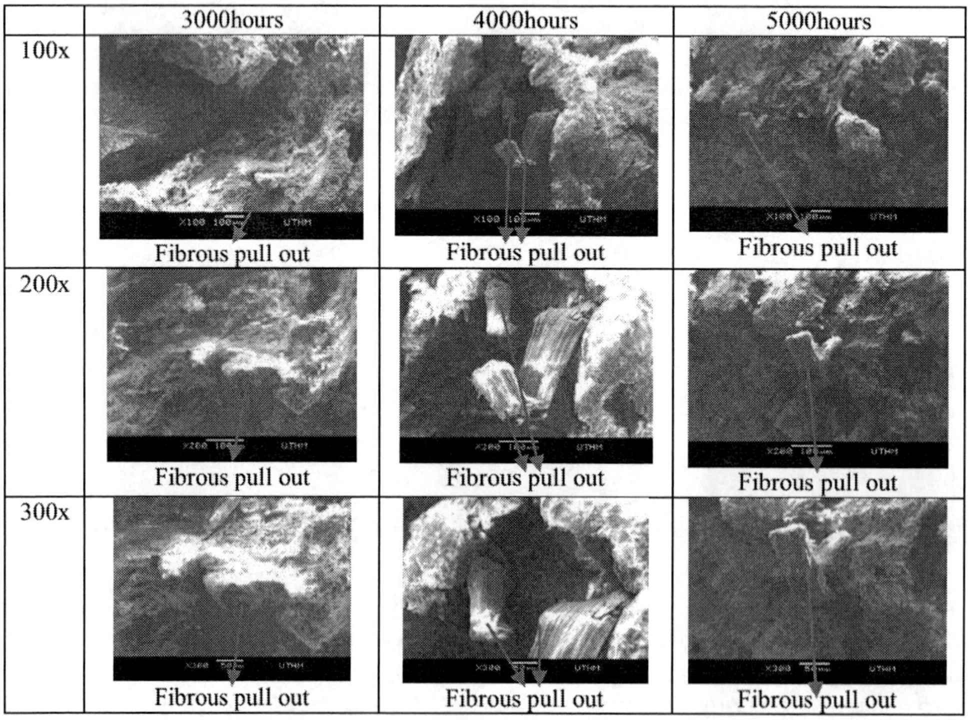

Fig. 3. Comparison between SEM cross section of fracture surface of WPC tensile specimens samples for 3000 hours, 4000 hours and 5000 hours.

Figure 3 shows fibrous pull out from fracture surface of WPC tensile specimen which contributed to stress distribution through the tensile specimen of the wood fiber and polymer matrix. These specimens experienced brittle interfaced failure. Weak interlayer bonding or some amount of interlayer porosity was evident in the failure of the specimens and appeared to be the cause of layer delamination along to the WPC during loading [8]. This is also contributed from the miscible polymer compound which depends on the materials composition ratios and processing condition during the fabrication process.

5 Conclusions

Wood polymer composites of polyvinyl chloride (PVC) with rice husk have good mechanical and physical properties. The addition of the rice husk in PVC was affecting the physical and mechanical properties of WPC during UV irradiation exposure. The WPC shows increment in strength gradually at 5000 hours of UV irradiation exposure. This is due to the crosslinking of polymer and increased interphases connection between polymer, rice husk fibres and other additive.

The authors would like to thank the Malaysian Government and Universiti Tun Hussein Onn Malaysia (UTHM), Johor, for supporting this research study under Exploratory Researched Grant Scheme, ERGS Vot E018. Not to forget, the authors express deep gratitude to ZFH Industries Sdn Bhd for supplying the pellet used in this research study.

References

1. C. Clemons, *Wood-Polymer Composites* (Woodhead Publishing Limited, 2008)
2. D. Ndiaye, E. Fanton, S.M. Therias, L. Vidal, A.Tidjani, J.L. Gardette, Compos. Sci. Technol., **68**, 2779 (2008)
3. H. Nasution, *Green Building Materials and Civil Engineering* (CRC Press, Taylor & Francis Group, London, 2015)
4. V. Steckel, C.M. Clemons, H. Thoemen, J. Appl. Polym. Sci., **103**, 752 (2007)
5. S. Panthapulakkal, S. Law, M. Sain, J. Thermoplast. Compos. Mater., **18**, 445 (2015)
6. Q. Lin, X. Zhou, G. Dai, J. Appl. Polym. Sci., **85**, 2824 (2002)
7. J. Tocháček, Z. Vrátníčková, Polym. Test., **36**, 82 (2014)
8. C. Ziemian, M. Sharma, S. Ziemi, *Mechanical Engineering* (Science, Technology and Medicine, 2012)

MATEC Web of Conferences **78**, 01034 (2016) DOI: 10.1051/matecconf/20167801034
IConGDM 2016

Electromagnetic Energy Absorption due to Wireless Energy Transfer: A Brief Review

A. Syafiq[1,*], *M. H.* Mat[2], and *H. A.* Rahim[3]

[1]School of Electrical System, UniMAP, Perlis, Malaysia
[2]Centre for Diploma Studies, UniMAP, Perlis, Malaysia
[3]School of Computer & Communication Engineering, UniMAP, Perlis, Malaysia

Abstract. This paper reviews an implementation of evaluating compliance of wireless power transfer systems with respect to human electromagnetic exposure limits. Methods for both numerical analysis and measurements are discussed. The objective is to evaluate the rate of which energy is absorbed by the human body when exposed to a wireless energy transfer, although it can be referred to the absorption of other forms of energy by tissue. An exposure assessment of a representative wireless power transfer system, under a limited set of operating conditions, is provided in order to estimate the maximum SAR levels. The aim of this review is to conclude the possible side effect to the human body when utilizing wireless charging in daily life so that an early severe action can be taken when using wireless transfer.

1 Introduction

Wireless power transfer (WPT) or wireless energy transmission is the transmission of electrical power from a power source to a consuming device without using any of discrete conductors. Nowadays, there is recent interest regarding battery charging wireless power transfer technology are now being seen by the ubiquity of cellular phone phones, automotive and the other type of mobile communication devices. Recently the concern about wireless electromagnetic waves are being concerned with communication devices. Wireless charging, as well known as inductive charging, which is suitable to shoot up your smartphone. Wireless chargers emit an alternating current by a transmitter coil, which induces a voltage in the receiver coil found in the device. Hence, the pace was quantified by the power absorbed with a limited arena of the body tissue in a standard measurement of watts per kilogram (W/kg). It can be an average measure over the intact body, or a small sample amount such at the hand by the Specific Absorption Rate (SAR). The Specific Absorption Rate (SAR) measures exposure within range of the frequency 100 kHz and 10 GHz, such as laptops, wireless charger, Wi-Fi routers and cell phones.

[*] Corresponding author: aliffsyafiqjamaluddin@yahoo.com

2 Specification absorption rate

Specific absorption rate (SAR) signifies a measurement for an electromagnetic energy that absorbed by the biological tissue in case by the different type of exposure such as wireless, transmitter, antennas, and phones [1-2]. The SAR limit were specified in the IEEE C95.1 : 1999 are 1.6W/Kg in a SAR1gm averaging mass while that specified in IEEE C95.1 : 2005 has been updated to 2W/Kg in a 10gm averaging mass [3]. This new SAR limit specified in IEEE C95.1 : 2005 is comparable to the limit specified in the International Commission on Non-Ionizing Radiation Protection (ICNIRP) guidelines. That's why the Specific Absorption Rate (SAR) will be calculated and be measured in this project to achieve the safety using wireless power transfer to human mankind. The basic property for assessing local RF exposure from wireless devices is the Specific Absorption Rate (SAR). So, the Specific Absorption Rate (SAR) testing is conducted by according to international standards regularly as part of the product type approval.

$$\mathrm{SAR} = \frac{\sigma}{\rho} E^2 \left(\frac{W}{kg}\right) \tag{1}$$

The Specific Absorption Rate (SAR), which can cause the heating of tissue were measured in Watts per kilogram and its measure of electromagnetic energy absorption by the body [5]. By σ is the electric conductivity of the tissue in S/m, ρ is the mass density in Kg/m^3, and E is the root-mean-square (rms) magnitude of the electric field strength in (V/m) inside the material [6]. Then, we can calculate the SAR by using this equation. The SAR of a human body exposed to RF-EMF which it is very difficult to measure directly because the SAR includes the internal electric field strength and conductivity of the body, so the equation above will be used for measure [7]. Therefore, Specific Absorption Rate (SAR) be situated estimated by numerical methods using computational models of human bodies.

3 Electromagnetic theory

The acronym EMF (Electromagnetic Fields) is commonly used to express the exposure of humans to electromagnetic fields. The exposure to EMF is covers a wide frequency range from (0 Hz - 300 GHz). Electromagnetic fields happens everywhere in our environment such as the earth, sun and ionosphere where this all natural sources of EMF.

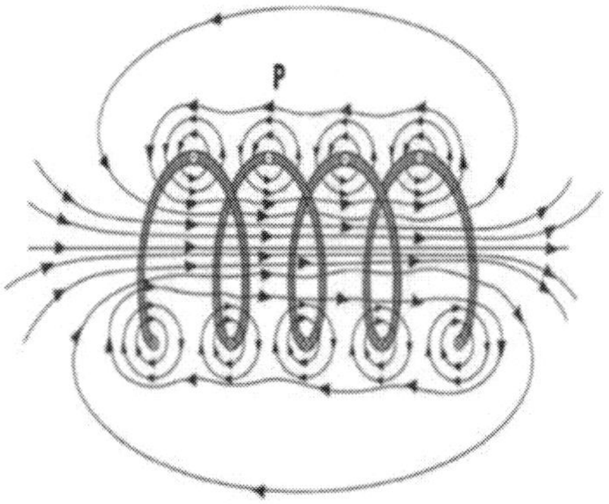

Fig. 1. Current through the coil (inductive charging) [9]

From the Figure 1, the implanted coil directly provides the charging current for the pacemaker battery through transcutaneous electromagnetic induction between the coils [10]. The idea breaks on inductive coupling that uses an electromagnetic field where its transfers energy from the transmitter to the receiver [10]. There's an alternating current that passes through the coil in the charger, which creates a fluctuating electromagnetic field. This technique can powerfully transfer power over a relatively long distance using the resonators by having the high quality factor [11]. So, the coil in the contrivance is placed in this electromagnetic field where the current is induced in the coil and next charges the contrivance's battery. But, for charging, phones are required to keep near the charging plate, so it will using inductively coupled power transfer [12]. This is reason why wireless charging is sometimes kenned as inductive charging.

4 Type and designation coil

There many type of coils with a winding material (copper and iron), different size, and conductor cross-sectional. Simply, the coils with the ferromagnetic wire differ substantially from the coils with the diamagnetic material which their inductance and resistance are significantly higher, and it's depend substantially on the signal frequency. In a coil testing, with the frequency increase from 20 kHz to 150 kHz inductance of the ferromagnetic coil decreased by 9.86 %, while the copper wire coil inductance doesn't change [13]. Hence, the type of copper coil is use in this project.

5 Wireless energy/power transfer and operation

The very essential of any magnetic Wireless Power Transfer system are the magnetically coupled transmitting and receiving coils [14]. The wireless power transfer (WPT) system using magnetically coupled coils has been discovered by Kurs in 2007 [15-16]. When the alternating current is passed through transmitter coil, it's generate the magnetic field.

The receive coil and loop (Rx) are identical to the transmit coil and loop [17]. It will receive the transmit power from the (Tx) coil. The strong reactive near-fields of wireless power systems may induce high fields in the body tissue of persons in their closest vicinity. Wireless charging can be as efficient as a wired charging when receiver implements direct charging [18].

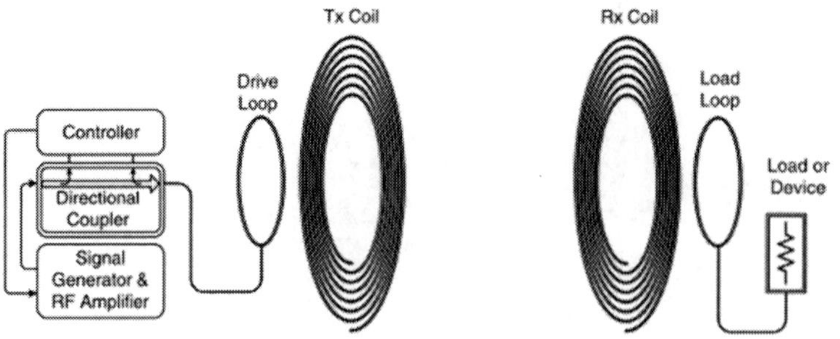

Fig. 2. Schematic block diagram for magnetically coupled resonant wireless power system [17]

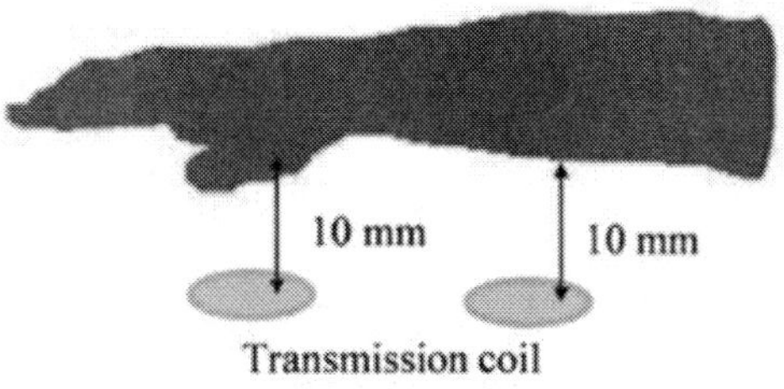

Fig. 3. Computational condition for the arm [19].

Figure 2 shows the exposure condition of the arm. The part of body human, arm was selected to perform in this simulation. Two positions of the transmitting coil in proximity to the hand and forearm are considered. This is because our hand was near when we charge our phone via wireless charging. The coil is located below the Palmar side of the hand or forearm at a distance of 10 mm.

6 Expected Results

For the best result, the wireless transfer and human hand has been consider to confirm the magnetic field distribution generated by the transfer system. For the effects of this results, the wireless power transfer as charging phone has been used to work out the SAR exposure on the human torso. Note that the magnetic field distribution without the magnetic sheet is symmetrical. So, the disturbance in the field is mainly attributed to the insertion of the magnetic sheet.

7 Summary

In a nutshell, overall studies have investigated the SAR and the electric field in anatomically based human-body models of the magnetic field emitted from the wireless power transfer system with induction coupling. Thus, it is agreed that probably that the wireless power transfer has an effect in the human tissues due to an electromagnetic absorption.

References

1. C. Lazarescu, V. Dafinescu, V. David, *International Conference and Exposition on Electrical and Power Engineering*, 25 (2012)
2. R. Butet, Y. Toutain, and S. Le Dall, *2013 IEEE Antennas and Propagation Society International Symposium (APSURSI)*, 2026 (2013)
3. I. International and E. Safety, *IEEE Standard for Safety Levels with Respect to Human Exposure to Radio Frequency Electromagnetic Fields, 3kHz to 300 GHz*, (IEEE Standards, 2006)
4. S. Akimoto, S. Kikuchi, T. Nagaoka, K. Saito, S. Watanabe, M. Takahashi, K. Ito, *IEEE transactions on microwave theory and techniques*, **58**, 3859 (2010)
5. B.M. Hochwald, D. J. Love, S. Yan, P. Fay, J. Jin, and B. M. Hochwald, *IEEE Communications Magazine*, **52**, 126 (2014)
6. B.A. Christ, M. Douglas, S.M. Ieee, J. Nadakuduti, M. Ieee, N. Kuster, F. Ieee, *Proceedings of the IEEE*, **101**, 1482 (2013)
7. T. Nagaoka and S. Watanabe, *Electromagnetic Compatibility (EMC EUROPE), 2013 International Symposium on. IEEE*, 891 (2013)

8. S. Nishizawa, S. Member, and O. Hashimoto, *IEEE transactions on microwave theory and techniques*, **47**, 277 (1999)
9. S. Garg and F. Khan, Int. J. Multidiscip. Res. Dev., **2**, 650 (2015)
10. J. C. Lin, *IEEE Antennas and Propagation Magazine*, **55**,251 (2013)
11. S. Park, H. Kim, J. Cho, and E. Kim, *2014 International Symposium on Electromagnetic Compatibilit*, 1190 (2014)
12. P. Kumar, International Journal of Engineering Science and Innovative Technology, **2**, 215 (2013)
13. A. Laizans, A. Galins, V. Osadcuks, A. Pecka, A. Rubenis, *Proceedings of the international scientific conference* (Latvia University of Agriculture, 2015)
14. R. Pinto, M. Bertoluzzo, V. Lopresto, S. Mancini, C. Merla, G. Pede, A. Genovese, and G. Buja, *2015 9th European Conference on Antennas and Propagation (EuCAP)*, 1 (IEEE, 2013)
15. A. Kurs, A. Karalis, R. Moffatt, J. D. Joannopoulos, P. Fisher, and M. Soljac, Sci., **317**, 83 (2007)
16. J. Chakarothai, K. Wake, and S. Watanabe, *General Assembly and Scientific Symposium (URSI GASS)*, (IEEE, 2014)
17. A. Christ, M.G. Douglas, J.M. Roman, E.B. Cooper, A.P. Sample, B.H. Waters, J.R. Smith, N. Kuster, *IEEE Trans. Electromagn. Compat.,* **55**, 265 (2012)
18. T. Bijalwan, R. Parveen, P. Uniyal, Kavita Panwar, A. Bagwari, ISTP Journal of Research in Electrical and Electronics Engineering, 212, 2014.
19. T. Sunohara, A. Hirata, I. Laakso, T. Onishi, Phys. Med. Biol., **59**, 3721 (2014)

Rutting Prediction in Asphalt Pavement Based on Viscoelastic Theory

Mohammed Hadi Nahi[1,*], *Ibrahim* Kamaruddin[1], and *Madzlan* Napiah[1]

[1]Civil and Environmental Engineering Department, Universiti Teknologi Petronas, Bandar Seri Iskandar 32610, Tronoh, Perak, Malaysia

Abstract. Rutting is one of the most disturbing failures on the asphalt roads due to the interrupting it is caused to the drivers. Predicting of asphalt pavement rutting is essential tool leads to better asphalt mixture design. This work describes a method of predicting the behaviour of various asphalt pavement mixes and linking these to an accelerated performance testing. The objective of this study is to develop a finite element model based on viscoplastic theory for simulating the laboratory testing of asphalt mixes in Hamburg Wheel Rut Tester (HWRT) for rutting. The creep parameters C1, C2 and C3 are developed from the triaxial repeated load creep test at 50°C and at a frequency of 1 Hz and the modulus of elasticity and Poisson's ratio determined at the same temperature. Viscoelastic model (creep model) is adopted using a FE simulator (ANSYS) in order to calculate the rutting for various mixes under a uniform loading pressure of 500 kPa. An eight-node with a three Degrees of Freedom (UX, UY, and UZ) Element is used for the simulation. The creep model developed for HWRT tester was verified by comparing the predicted rut depths with the measured one and by comparing the rut depth with ABAQUS result from literature. Reasonable agreement can be obtained between the predicted rut depths and the measured one. Moreover, it is found that creep model parameter C1 and C3 have a strong relationship with rutting. It was clear that the parameter C1 strongly influences rutting than the parameter C3. Finally, it can be concluded that creep model based on finite element method can be used as an effective tool to analyse rutting of asphalt pavements.

1 Introduction

Rutting is one of the wide common deformations for flexible pavement roads. For proper pavement design and better understanding of materials behaviour it is necessary to predict asphalt pavement rutting [1]. Rutting phenomenon is the longitudinal depressions along the wheel paths caused by the movement of asphalt pavement materials under high traffic loading based on consolidation or plastic flow [2,3]. Excessive amount of asphalt cement is one of reasons for plastic flow [4,5]. Rutting in asphalt pavement is mainly occurring due to its nonlinear viscoelastic nature [6,7]. The high wheel loads and high tire pressures are the main cause for Increasing of asphalt pavement rutting phenomenon [8,9]. Magdy, (1996) presents three rutting mechanisms wear rutting, structural rutting and instability rutting

* Corresponding author: mohammednahi2@gmail.com

[10]. This study deals with the last type, which is the most dangerous type of surface rutting. Sue and Wong (2009) has studied the air voids effects on the behaviour of rutting under wheel loadings. An eight-node, three-dimensional stress, displacement element was used in the finite element analysis. A study in (2003) developed a finite element model to simulate the laboratory testing of asphalt mixes in Asphalt Pavement Analyzer (APA) test for rutting and to relate the test results to basic material properties, a visco-elasto-plastic model (creep model) has been used to represent the time-dependent characteristics of the asphalt mixture [11, 12]. The objective of this study is to develop a finite element model to simulate the laboratory testing of asphalt mixes in (HWRT) for rutting. The nonlinear viscoelastic behaviour analysis of asphalt mixes was adopted in this study. It was found that creep model parameters C3 and C1 have a strong relationship with rutting. However, between these two the parameter C1 strongly influences rutting than the parameter C3 [8].

2 Asphalt viscoelastic behaviour

The hot-mix asphalt materials plays a major role in asphalt pavement rutting. Most paving materials are not elastic, but if the load is small compared to the strength of the material and is repeated for a large number of times, the deformation under each load repetition is nearly completely recoverable and can be considered elastic [13, 14]. The recoverable (elastic and viscoelastic) and nonrecoverable (plastic and viscoplastic) strains components are determined in the analyses. The viscoplastic strain component is the major contributor to the asphalt mixture rutting and only this component is used in the development of the material creep parameters [8]. If the creep model is used to describe the time dependant material behaviour, the repeated loading and continuous loading have the same effect on the predicted creep strain as long as the total loading times are the same.

3 Hamburg wheel rut tester

The Hamburg Wheel Tracking Device (HWTD) is well known in Canada as the Hamburg Wheel Rut Tester (HWRT). Ludomir Uzarowski 2006 employed The HWRT for asphalt mix accelerated performance testing in terms of its rutting resistance. The wheels can be either steel (47 mm wide) or solid rubber (50 mm wide), and the load applied to the wheels is (710 N). The customary temperature for the HWRT test in Canada and the United States is 50°*C*; this temperature is also typically used in Europe for a climate close to a Superpave high temperature PG of 58. The test path is (230 mm) long and the average speed of each wheel is approximately 1.1 km/h (53 ± 2 wheel passes per minute). All the HWRT testing was carried out at a temperature of 50°C. Fig. 1, shows the diagram of the Rutting Resistance Testing in the HWRT.

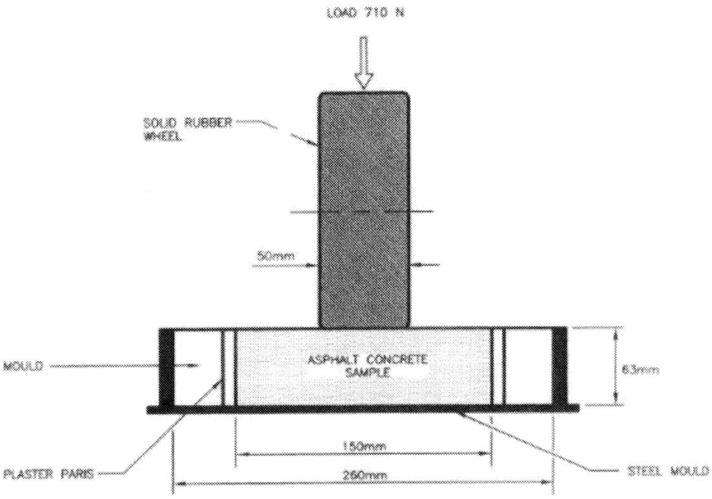

Fig. 1. Diagram of the Rutting Resistance Testing in the HWRT

Five asphalt mixes were used in this study [12]. These mixes present a wide range of applications, from high traffic to low volume roads. All five mixes were obtained from paving projects in Ontario [8]. In total, about one ton of asphalt mixes was used in the research and the designs for all five mixes were provided by the contractors who supplied the mixes. The mixes incorporated the aggregate types and asphalt cements listed in Table 1.

Table 1. Asphalt Mix Ingredients (12).

Mix	Aggregate		Additives	Asphalt Cement Grade
	Coarse	Fine		
HL 3	Crushed gravel (40%)	Asphalt sand (45%) and screening (15%)	-	PG 58-28
SMA L	Traprock (79%)	Traprock (13%) and mineral filler (8%)	Cellulose fibre (0.3%)	PG 70-28 PM*
SMAG	Traprock	Traprock and mineral filler	Cellulose fibre (0.3%)	PG 70-28 PM*
SP 19 D	Crushed rock (39%)	Screenings (10%) and manufactured sand (51%)	-	PG 64-28
SP 19 E	Crushed rock (63%)	High stability sand (37%)	-	PG 70-28

4 Creep model parameters

Five material parameters may affect the rutting predicted by the finite element method. These parameters include three creep model parameters (*C1*, *C2* and *C3)* and two elastic parameters (modulus of elasticity, and Poison's ratio). However, predicted rut depth is not sensitive to modulus of elasticity and Poison's ratio since these two factors only define the elastic behaviour, which is not directly related to the permanent deformation. Therefore, the sensitivity study was kept limited to the creep model parameters [9]. In this analysis, the modulus of elasticity obtained from the triaxial repeated load creep test based on Uzarowski, L. [11] work has been used. The triaxial repeated load creep test done at a temperature of 50°C and at a frequency of 1 Hz. The Poisson's ratio is calculated using the Eq. 1. given in MEPDG [8]:

$$\mu_{ac} = 0.15 + \frac{0.35}{1 + e^{-1.63+(3.84*10^{-6})E_{ac}}} \quad (1)$$

Where:
μ_{ac}= Poisson's ratio of asphalt mix at a specific temperature
E_{ac} = modulus of asphalt mix at a specific temperature (psi)
The creep parameters *C1, C2* and *C3* have been calibrated based on the rutting measured in the HWRT. The *C3* parameter is stress related. The rut resistance testing in the HWRT is conducted at a constant loading stress of 500 kPa, the *C3* parameter is fixed at the initial level shown in Table 2. Parameter *C1* is the value of the y-axis intercept while parameter *C2* is related to the slope of the strain-time relationship curve in a log-log scale [4]. The elastic and final creep parameters used in the simulation are shown in Table 2.

Table 2. Asphalt mix elastic and creep parameters.

	Material Parameter				
	Elastic		Creep		
Mix Type	**Modulus kPa**	**Poisson's Ratio**	***C1* (x10-8)**	***C2***	***C3***
HL3	950,000	0.41	41	1.48	-0.63
SMAL	800,000	0.42	509	1.04	-0.78
SMA G	800,000	0.42	138	1.31	-0.79
SP 19 D	1,600,000	0.39	66	1.20	-0.68
SP19 E	1,400,000	0.40	72	1.20	-0.64

5 ANSYS modelling

ANSYS is a suite of powerful engineering simulation based on the finite element method. It has an extensive library of finite elements and an extensive list of material models. Analysis in ANSYS includes two stages starting level and processing level, starting level include converting a mechanical problem for finite element model (FEM). Processing level includes three steps Pre-processing, Solve and Post-processing. The geometry of problem, material properties, element type and boundary conditions has been defined in Pre-processing step. While in the solve step, the type of analysis has been chosen. Finally the result of solving the problem has been observed in the Post-processing step (ANSYS user's manual-version 9.0 2004). There are lots of equations embedded in commercial finite element analysis software (ANSYS). In order to model primary and secondary stages of creep characteristics, the presented creep model used in this research can be written in the form of creep strain rate. The material is assumed to be isotropic and the basic solution technique used is the initial-stiffness Newton–Raphson method as in Eq. 2.

$$\dot{\varepsilon}_{cr} = C_1\, \sigma^{c_2}\, t^{c_3} \quad (2)$$

where $\dot{\varepsilon}_{cr}$ is the creep strain rate, *r* is the equivalent stress, *t* is the time at the end of sub-steps, *C1, C2* and *C3* are the parameters related to material properties. As the temperature in the analysis is fixed at 50°*C*, the parameters *C1, C2* and *C3* are developed from the triaxial repeated load creep test at 50°*C* and the modulus of elasticity and Poisson's ratio determined at the same temperature. The elastic parameters modulus of elasticity and Poisson's ratio and creep parameters *C1, C2* and *C3* are required for the finite element simulator (ANSYS) in order to calculate the rutting for various mixes under a uniform loading pressure of 500 kPa.

6 Finite element simulation

The structure and the auxiliary conditions in the finite element simulation have to be correctly modelled to obtain reasonable results. This includes the finfite element model's components such as include element type, boundary conditions, load conditions, material properties and geometry of the model As shown in Fig. 2, 150 mm x 300 mm (in width) and 63 mm (in height) rectangular bituminous mixture structure was constructed to simulate the Hamburg Wheel Rut Test.

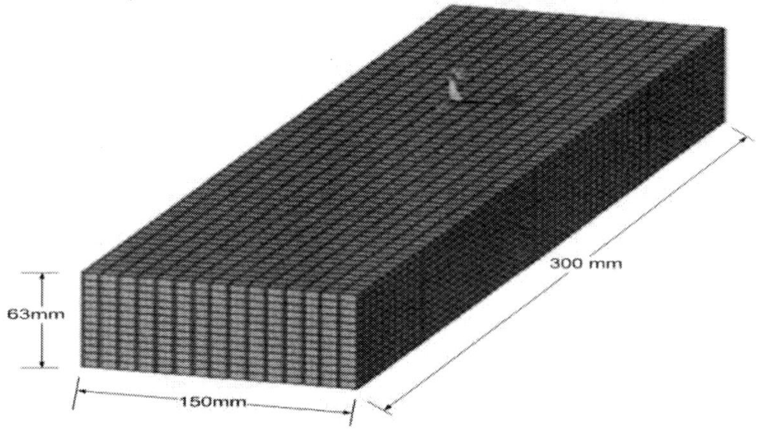

Fig. 2. Finite element mesh of wheel tracking test specimen.

The dual wheel loading configurations in the HWRT testing are symmetrical as shown in Figure 3, to reduce the element number and the required time for each analysis, only one wheel and half of the HWRT sample was included in the finite element modelling. The load magnitude was 710 N with 500 kpa tire pressure. The material creeps model discussed above was used to characterize the permanent deformation properties of the bituminous mixture. The parameters used in the creep model were derived from the triaxial repeated load creep test result. The bottom and the four surrounding vertical boundaries were set to be confined with restricting displacement in all directions except the axisymmetric side which is surrounding is an x direction only. SOLID45 is used for the 3-D modelling of solid structures and this element is defined by eight nodes having three degrees of freedom at each node translation in the nodal x, y, and z directions. The element has plasticity, creep, swelling, stress stiffening, large deflection, and large strain capabilities.

Fig. 3. General view of the Hamburg Wheel Rut Tester.

7 Result and discussion

The measured rut depth versus time relationship with rutting predicted in ANSYS and ABAQUS at the time of loading of about 1000, 1800, 2600, 3300 and 4200 sec has been shown in the Figures 4,5,6,7 and 8. The results show that there is a reasonable agreement on the difference between the predicted rutting values and the laboratory measured rutting values. The comparison between the obtained results from ABAQUS and the results obtained from ANSYS shows that rutting values from ANSYS are closer to the laboratory measured values.

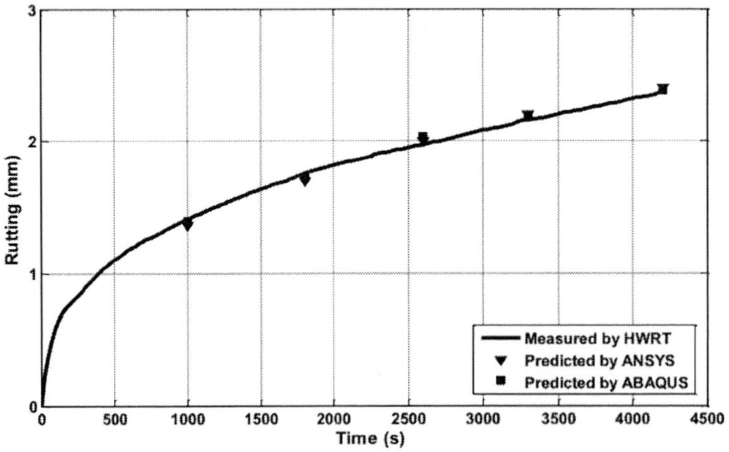

Fig. 4. Rutting measured in HWRT and predicted in ABAQUS and ANSYS for the HL 3 Mix.

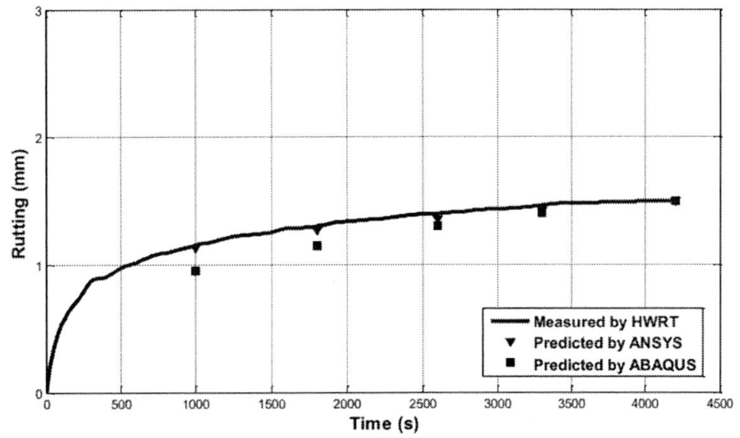

Fig. 5. Rutting measured in HWRT and predicted in ABAQUS and ANSYS for the SMA L Mix.

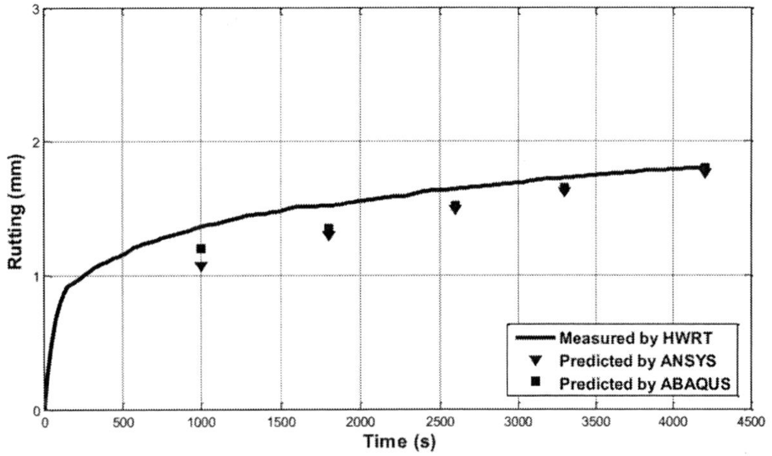

Fig. 6. Rutting measured in HWRT and predicted in ABAQUS and ANSYS for the SMA G Mix.

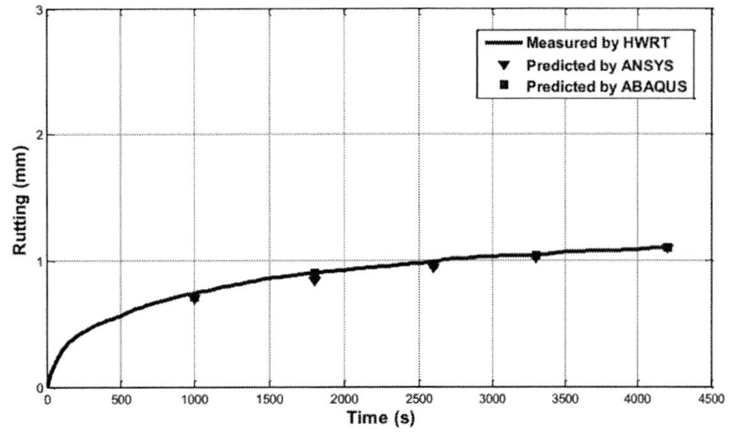

Fig. 7. Rutting measured in HWRT and predicted in ABAQUS and ANSYS for the SP 19 D Mix.

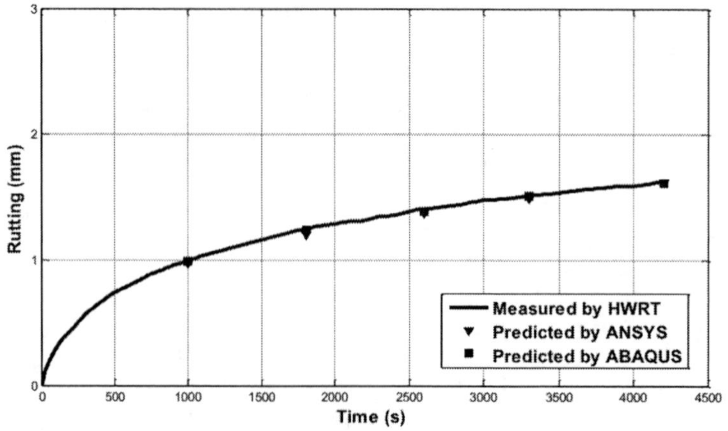

Fig. 8. Rutting measured in HWRT and predicted in ABAQUS and ANSYS for the SP 19 E Mix.

There are a total of five possible factors, which may affect the predicted model response. They are creeping model parameters *C1, C2* and *C3,* modulus of elasticity and Poisson's Ratio. The result of this work simulation shows that the predicted rut depth is not sensitive to modulus of elasticity and Poisson's Ratio. This expected since these two factors only define the elastic behaviour which is not directly related to permanent deformation. It was clear that the parameter *C1* strongly influences rutting than the parameter *C3* As can be seen from Fig. 9, predicted rut depth increases with increasing *C1* value. As can be seen from Figure 10, predicted rut depth increases with increasing *C3* value (noted that *C3* is in the range from -1 to 0) The relationship of *C3* and the predicted rut depth is non-linear as shown in fig. 10.

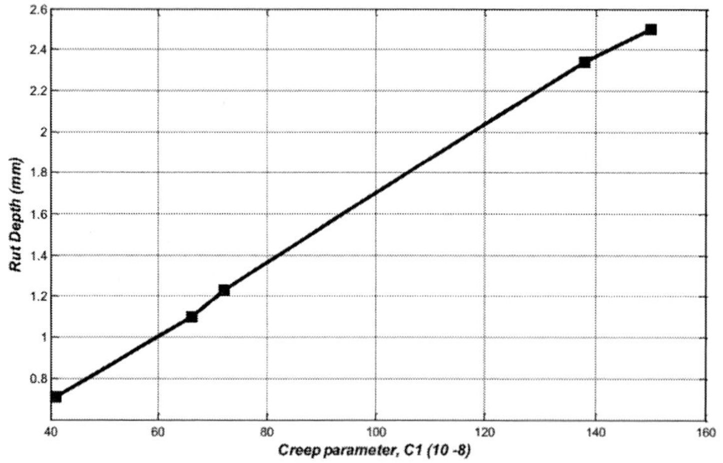

Fig. 9. The effect of varying Creep Model Parameter *C1* on Rut Depth.

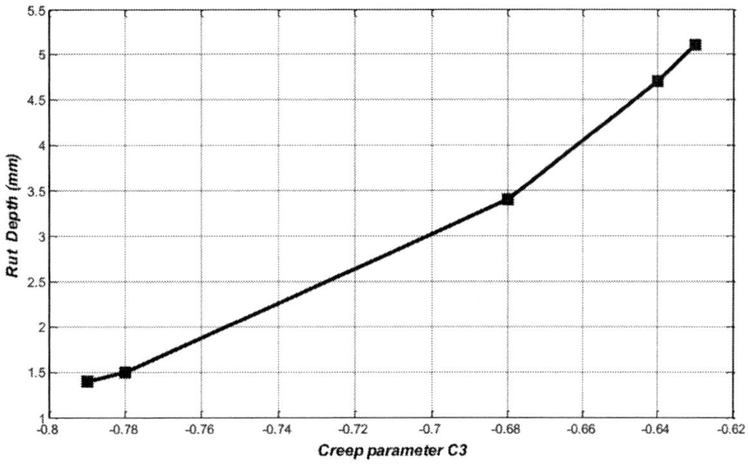

Fig. 10. The effect of varying Creep Model Parameter *C3* on Rut Depth.

8 Conclusion

The finite element model developed for HWRT tester was verified by comparing predicted and measured rut depths and also by comparing the rut depth with ABAQUS result from literature. Reasonable agreement was found between the predicted and measured rut depths furthermore, it was found that creep model parameters *C1* and *C3* have a strong relationship with rutting and that the parameter *C1* strongly influences rutting than the parameter *C3*. It was clear that the predicted rut depth increases with increasing *C1* value and with increasing *C3* value. The parameters *C1* and *C3* are closely related to the amount of asphalt content and the air void in hot mix asphalt. Finally, it can be concluded that creep model based on finite element method can be used as effective tools to analyse rutting of asphalt pavements.

References

1. M.E. Abukhettala, *The relationship between marshall stability, flow and rutting of the New Malaysian hot-mix asphalt mixtures*, (Universiti Teknologi Malaysia, Johor Bahru, 2006)
2. M.H. Nahi, A. Ismail, A.K. Anffm, J. Appl. Sci., **11**, 2562 (2011)
3. B.A. Feyissa, *Analysis and modeling of rutting for long life asphalt concrete pavement,* (TU, Darmstadt, 2009)
4. U.S. Department of Transportation, *Hamburg wheel tracking device*, (Federal Highway Administration, Turner-Fairbank Highway Research Center, McLean, VA, 2002)
5. M.A.A. Zahw, *Development of testing framework for evaluation of rutting resistance of asphalt mixes*, (Faculty of Engineering, Al-Azhar University, 1996)
6. M.H. Nahi, I. Kamaruddin, A. Ismail, R.A.E. Al Mansob, J. Appl. Sci., **14**, 2730 (2014)
7. NCHRP, *Guide for mechanistic-empirical design of new and rehabilitated pavement structures*. (National Cooperative Highway Research Program, Illinois, 2004)

8. S. Pirabarooban, M. Zaman, R.A. Tarefder, *Evaluation of Rutting Potential in Asphalt Mixes Using Finite Element Modeling*, Annual Conference and Exhibition of the Transportation Association of Canada (2003)
9. M.H. Nahi, I. Kamaruddin, M. Napiah, Appl. Mech. Mater., **567** (2014)
10. Texas DOT. *Pavement construction specifications.* Texas Department of Transportation, (2003)
11. L. Uzarowski, *The development of asphalt mix creep parameters and finite element modeling of asphalt rutting*, (University of Waterloo, Canada, 2007)
12. D. Walker, M. Buncher, *Intersection Strategy*, (Asphalt Institute, Kentucky, 1999)
13. D.Walker, M. Buncher, A. Institute, *High performance hot mix asphalt intersection strategy* (Asphalt Institute, Kentucky, 2004)

MATEC Web of Conferences **78**, 01036 (2016) DOI: 10.1051/matecconf/20167801036
IConGDM 2016

An experimental result of surface roughness machining performance in deep hole drilling

Azizah Mohamad[1,*], *Azlan* Mohd Zain[1], *Noordin* Mohd Yusof[2], and *Farhad* Najarian[2]

[1] Soft Computing Research Group, Faculty of Computing, Universiti Teknologi Malaysia, 81310 UTM Skudai, Johor, Malaysia
[2] Department of Materials, Manufacturing and Industrial Engineering, Faculty of Mechanical Engineering, Universiti Teknologi Malaysia, 81310 UTM Skudai, Johor, Malaysia

Abstract. This study presents an experimental result of a deep hole drilling process for Steel material at different machining parameters which are feed rate (*f*), spindle speed (*s*), the depth of the hole (*d*) and MQL, number of drops (*m*) on surface roughness, R_a. The experiment was designed using two level full factorial design of experiment (DoE) with centre points to collect surface roughness, R_a values. The signal to noise (S/N) ratio analysis was used to discover the optimum level for each machining parameters in the experiment.

1 Introduction

Deep hole drilling is a complex machining process in which the ratio of hole depth to hole diameter exceeds 10 [5]. Deep hole drilling is a relatively complex drilling production process due to the high hole diameter to length ratio which makes the tool shaft prone to vibration and results typically in low quality holes from geometry and surface roughness viewpoints [1]. Deep hole drilling processes are used to generate drill holes with a high length to diameter ratio. A good holes, which are holes with better straightness, better roundness and smoother surface roughness, there are many factors need to be considered such as material, spindle speed, feed rate and tool diameter [9]. According to Hayajneh (2001), the important machining parameters that effect hole quality is spindle speed and feed rate [6]. Surface roughness, R_a is an important aspect in evaluating the quality of products. R_a is widely used as an index of product quality and in most cases a technical requirement for mechanical products [3, 12]. Normally, R_a value is influenced by many factors such as machining parameters, cutting phenomena, workpiece properties and cutting tool properties [15, 16]. Based on previous literature, R_a is one of the machining performance measurements frequently considered by researchers [4, 11, 13, 14].

This study focuses on an experimental result of R_a in the deep hole drilling process. The flow of the experiment and analysis of result also discussed.

* Corresponding author: azizahbtmohamad@yahoo.com.my

DOI: 10.1051/matecconf/20167801036

2 Experimental procedure and equipment

The deep hole drilling experiment was carried out at Production Laboratory using CNC 3 Axis Milling Machine (MAHO MH500E2). Figure 1 shows the CNC 3 Axis Milling Machine. This machine was used to produce the holes on the steel for conducting the deep hole drilling process.

Fig. 1. CNC 3 Axis Milling Machine.

The tool used for this study is High Speed Steel, (HSS) with twist drill bits and workpiece material is Steel. The HSS diameter is 5.0 mm and the illustration of HSS is shown in Figure 2. Table 1 shows the mechanical properties of the HSS based on the book catalogue of Drilling Tools (YG-1 Drilling Tools Catalog 2009/2010).

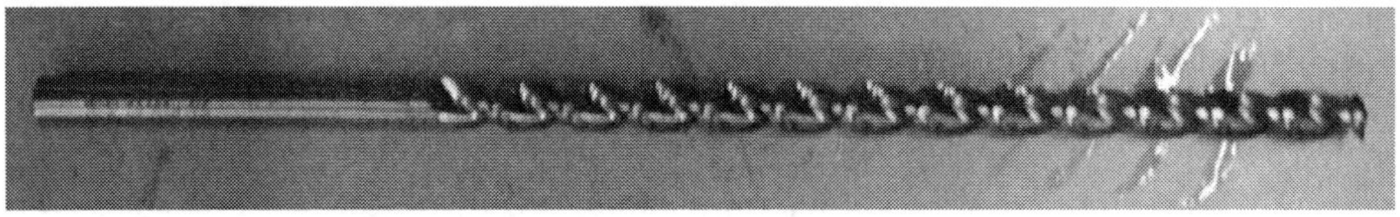

Fig. 2. HSS with twist drill bit.

Table 1. Mechanical properties of the HSS used in the experiment.

Characteristics	Properties
Standard	DIN 1869/1
Tool materials	HSS Co5 (Cobalt 5% is used in the tool material)
Helix angle	38°
Tolerance of toll diameter	H8
Point angle	130°
Drill diameter, d1	5.0 mm
Overall length,L1	195 mm
Flute length, L2	135 mm

3 Deep hole drilling parameters and experimental design

Design of experiment (DoE) is a collection of powerful statistical analysis techniques used for modelling, developing, improving, and optimizing various manufacturing and other useful processes [1]. It also for structuring and organizing method for determining the relationship between input parameters, which affect the process and the output of the process. DoE is an important step in the selection of machining parameters and their levels within an effective and suitable experimental run [4].

The Two Level Full Factorial Design with centre points is used as DoE in this study. Twenty experimental numbers were executed, which consists of sixteen data of two levels, which 2^k full factorial analysis where k is referred to number of machining parameters involved in these experiments with four centre points.

In this study, there are four machining parameters that had been put into consideration which is feed rate, f (mm/min), spindle speed, s (rpm), depth of hole, d (mm) and MQL, number of drops m (ml/hour). The machining parameters and their levels are shown in Table 2. The R_a value of the machined work piece was measured using the Handysurf Profile Meter.

Table 2. Levels of machining parameters for deep hole drilling.

Machining parameters	Units	Level		
		-1 (LOWER)	0 (CENTER POINT)	+1 (UPPER)
Feed rate, f	mm/min	65	75	85
Spindle speed, s	rpm	900	1000	1100
Depth of hole, d	mm	50	55	60
MQL, number of drops, m	ml/hour	20	30	40

Full Factorial Design is used widely because they are easy to design and analyse, efficient to run and full of information [7]. Then it is simplest and most common type of factorial design [2].

4 Results and discussion

The experimental results for R_a along with their computed S/N ratio values are shown in Table 3.

From the table, the minimum value obtained is 2.8 μm. The optimal values of machining parameters that contribute to minimum R_a are 85 mm/min for feed rate which in level +1, 1100 rpm for spindle speed which in level +1, 60 mm for depth of the hole which in level +1 and 40 ml/ hour for number of drops of MQL which in level +1. The lower the better (LB) category of S/N ratio has been selected for R_a and used to identify the optimal level of machining parameters [8]. The calculation of LB category can be referred in Eq.1.

$$S/N_{LB} = -10 \log \left(\frac{1}{n}\sum_{i}^{n} {y_i}^2\right) \quad (1)$$

where,
y_i= the observed data (R_a)
n= the number of observations

Table 3. Experimental results for R_a and S/N ratio.

Exp No.	Feed rate, *f* (mm/min)	Spindle speed, *s* (rpm)	Depth of hole, *d* (mm)	MQL, number of drops, *m* (ml/hour)	Surface roughness, (μm)	S/N ratio for R_a
1	-1	-1	-1	-1	4.4	-12.8691
2	-1	+1	-1	-1	3.8	-11.5957
3	+1	-1	-1	-1	4.0	-12.0412
4	+1	+1	-1	-1	5.0	-13.9794
5	0	0	0	0	4.5	-13.0643
6	-1	-1	+1	-1	4.0	-12.0412
7	-1	+1	+1	-1	4.1	-12.2557
8	+1	-1	+1	-1	3.3	-10.3703
9	+1	+1	+1	-1	4.1	-12.2557
10	0	0	0	0	5.3	-14.4855
11	-1	-1	-1	+1	4.4	-12.8691
12	-1	+1	-1	+1	3.1	-9.8272
13	+1	-1	-1	+1	4.5	-13.0643
14	+1	+1	-1	+1	3.7	-11.3640
15	0	0	0	0	4.1	-12.2557
16	-1	-1	+1	+1	3.9	-11.8213
17	-1	+1	+1	+1	4.2	-12.4650
18	+1	-1	+1	+1	4.4	-12.8691
19	+1	+1	+1	+1	2.8	-8.9432
20	0	0	0	0	4.6	-13.2552

The highest of S/N ratio value corresponds to the better machining parameters [10]. The mean response table of S/N ratio for each level of the machining parameters is summarized in Table 4.

Table 4. The mean S/N response table for R_a.

Level	Feed rate, *f* (mm/min)	Spindle speed, *s* (rpm)	Depth of hole, *d* (mm)	MQL, number of drops, *m* (ml/hour)
-1	-11.9680	-12.2432	-12.2013	-12.1760
0	-13.2652	-13.2652	-13.2652	-13.2652
+1	-11.8609*	-11.5857*	-11.6277*	-11.6529*

* Indicate the optimum level

Based on the Table 4, R_a value has been identified as minimum at the Level +1 of feed rate(*f*), Level +1 of spindle speed (*s*), Level +1 of depth of hole (*d*) and Level +1 MQL (*m*) respectively. The symbol of '*' corresponded to the highest value of S/N ratio among their levels and then it is identified as the optimum level. Consequently, the level that has a higher value of S/N ratio determines the optimum level of each machining parameters. In Table 4 for feed rate level +1 (f_{+1} = -11.8609), spindle speed level +1 (s_{+1}=-11.5857), depth of hole level +1 (d_{+1} = -11.6277) and MQL level +1 ($m+_1$ = -11.6529) has the highest S/N ratio value, which indicated that the machining parameters at that level produced minimum value R_a. As a result, the optimum combination to get the minimum R_a is $f_{+1}s_{+1}d_{+1}m_{+1}$as shown in Table 5.

Table 5. Optimum value of machining parameters for deep hole drilling.

Machining parameters	Level	Value
Feed rate, f (mm/min)	+1	85
Spindle speed, s (rpm)	+1	1100
Depth of hole, d (mm)	+1	60
MQL, m, number of drops (ml/hour)	+1	40

From the result, it was found that the combination value of R_a is acceptable which is the level obtained in real machining is the same level in the S/N ratio analysis which is $f_{+1}s_{+1}d_{+1}m_{+1}$.

5 Conclusion

This study presents an experimental result of R_a in deep hole drilling process. The results were analysed using S/N ratio. From the S/N ratio analysis, the optimal machining parameters for R_a is determined as $f_{+1}s_{+1}d_{+1}m_{+1}$, which are 85 mm/min for feed rate (level +1), 1100 rpm for spindle speed (level +1), 60 mm for depth of the hole (level +1) and 40 ml/ hour for number of drops of MQL (level +1). The results show that the combination level obtained from the S/N ratio analysis is the same level with the machining parameters that produced minimum value R_a. It was found to be statistically significant each other.

Special appreciation to reviewer(s) for useful advices and comments. The authors greatly acknowledge the Soft Computing Research Group (SCRG), Research Management Centre (RMC), UTM and Ministry of Higher Education Malaysia (MOHE) for financial support through the Fundamental Research Grant Scheme (FRGS) vot No R.J130000.7828.4F721.

References

1. T. Aized, M. Amjad, Int. J. Adv. Manuf. Technol., **69**, 9 (2013)
2. M. Buragohain, C. Mahanta, Applied Soft Computing, **8(1)**, (2008)
3. P.G Benardos, G.C Vosniakos, Int, J, Mach, Tool, Manu., **43(8)**, (2003)
4. U. Caydas, S. Ekici, J. Intell. Manuf., **23(3)**, 639 (2012)
5. C.S Deng, J.H. Chin, Int. J. Adv. Manuf. Technol., **25(5-6)**, 420 (2005)
6. M.T. Hayajneh, Mater. Manuf. Process., **16(2)**, 147 (2001)
7. R. Mee, *A comprehensive guide to factorial two-level experimentation* (Springer Science & Business Media, New York, 2009)
8. M. Nalbant, H. Gokkaya, G. Sur, Mater. Des., **28(4)**, 1379 (2007)
9. A.A. Rahman, A. Mamat, A. Wagiman, Modern Applied Science, **3(5)**, 221 (2009)
10. M. Sarıkaya, A. Gullu, J. Clean. Prod., **65**, 604 (2014)
11. D.R. Salgado, F.J. Alonso, I. Cambero, A. Marcelo, Int. J. Adv. Manuf. Technol., **43(1)**, 40 (2009)
12. N. Yusup, A.M Zain, S.Z.M Hashim, Expert Syst. Appl., **39**, 9909 (2012)
13. A.M. Zain, H. Haron, S. Sharif, Expert Syst. Appl., **37(2)**, 1755 (2010)
14. A.M. Zain, H. Haron, S. Sharif, Expert Syst. Appl., **37(6)**, 4650 (2010)
15. A.M. Zain, H. Haron, S. Sharif, Int. J. Prod. Res., **50(1)**, 191 (2012)
16. A.M. Zain., H. Haron, S.N. Qasem, S. Sharif, Appl. Math. Model., **36(4)**, 1477 (2012)

Renewable Polymer/ Thermoplastics Polyethylene Blended with Enhanced Mechanical and UV Stability Properties

Nurul Syamimi M. Salim [1,*], and M. Rus *Anika Zafiah* [1]

[1]Faculty of Mechanical and Manufacturing Engineering, University Tun Hussein Onn Malaysia (UTHM), Parit Raja, 86400, Batu Pahat, Malaysia.

Abstract. Blends of Renewable Polymer (RP) and thermoplastic polyethylene (LDPE and HDPE) may contribute to make recycling more economically attractive. In this study, the monomer is mixed with flexible isocynate as a crosslinker, these mixture is called Renewable Polymer. Renewable polymers are mixed in a Low-density polyethylene (LDPE) and High-density polyethylene (HDPE) with a ratio of 5%, 10%, 15%, 20%, 25% and 30%. The aim of this work to make LDPE/RP and HDPE/RP blends injected via injection molding and to evaluate their mechanical properties via tensile test. Accelerated weathering test up for 500 hours, 1000 hours, 1500 hours, 2000 hours, 2500 hours and 3000 hours. The blends yielded tensile strength and maximum elongation at break curves very dependent on their composition, especially regarding the presence of necking. The tensile strength increase at 500 hours, while maximum elongation at break were found to decreased with increase of UV irradiation hours. In conclusion, RP content and UV irradiation time play significant roles in controlling mechanical properties of the RP-blended with LDPE and HDPE synthetic polymer, thus providing the opportunity to modulate polymer properties.

1 Introduction

As the economy achieves global status, many factors regarding the competitiveness of a nation come under investigation. More recently, together with important areas such as technology advancement and technology transfer, issues related to sustainable development and environment preservation are receiving increasing attention from the world community.

Advantages of the mechanical recycling of polymers include reduction of oil and energy consumption compared with the synthesis of virgin polymers, reduced disposal of plastic waste in municipal garbage and generation of employment and income. However, recycling of municipal plastic waste is often an arduous task due to the fact that is material is usually a mixture of several polymer, which makes processing more difficult and also limits the number of potential application [1].

* Corresponding author : nurulsyamimisalim@gmail.com

The term of vegetable oils includes all vegetable oil such as olive oil, soybean oil, corn oil, coconut, palm oil and many others. The main constituents of vegetable oils are triacylglycerols, which are the product of esterification of glycerol with three fatty acids. Fatty acids account for 95% of the total weight of triglycerides and their content is characteristic for each vegetable oil [2]. In polymer industry, waste vegetable oils which represent a major potential source of chemicals have been utilized as an alternative feedstock for sustainable monomers.

Polyethylene (PE) is widely used material in outdoor applications, so its life time is determined by various environmental factors as solar radiation, temperature, thermal cycling, humidity, weather, pollutants but most importantly by ultra-violet (UV) radiation [3]. LDPE offer resistance to stress cracking and excellent impact properties, while HDPE provides stiffness, chemical resistance and barrier properties.

In this study, the evaluation of tensile properties of LDPE/RP and HDPE/RP blends was carried out to investigate the composition range for better mechanical performance and also to define the impact of UV irradiated exposure.

2 Experimental

Low density polyethylene (LDPE) density = 0.922 g/cm3, (melt index = 5 g/10 min, , vicat softening point = 93 °C), high density polyethylene (HDPE) density = 0.957 g/cm3, (melt index = 4 g/10 min, vicat softening point = 124 °C), flexible Polymethane Polyphenyl Isocyanate (Modified polymeric-MDI) (viscosity at 25 °C = 120-160 cps, specific gravity at 25 °C = 1.18-1.20 g/ml, NCO content, % wt = 26.3-27.3) and renewable monomers (renewable polyol). The monomer conversions from waste vegetable oils are started with their-house catalyst preparation to generate the epoxies from the unsaturated fatty compound. The acid-catalyst ring opening of the epoxides is to form renewable polyols.

Sample have been prepared in the form of dog-bone by injecting melted LDPE/RP and HDPE/RP in a mold as indicated in the ISO 527-2 (5A) standard. The injection machine is Nissei Horizontal Screw Type Injection Molding NP7 whose operating conditions were: temperature in nozzle = 185 °C (LDPE), 205 °C (HDPE), temperature of first cylinder = 175 °C (LDPE), 180 °C (HDPE), temperature of second cyclinder = 165 °C (LDPE and HDPE), injection pressure = 43.3 MPa (LDPE), 96.6 MPa (HDPE).

Artificial aging undergone by sample was carried out in an accelerated wheatherometer. This type of testing uses UV irradiation simulating the solar irradiation while the other environmental parameters, such as temperature and relative humidity remain under control (35-40 °C and 50% respectively).The UV machine used is Haida International Equipment Co., LTD and is equipped with a UV Lamp Tested Model HD-703. Samples were radiated over variable times of 500, 1000, 1500, 2000, 2500 and 3000 hours.

Tensile test have been carried out to determine the evolution of tensile strength and percentage of breaking deformation versus percentage of renewable polymer. To this purpose, the ISO 527-2 (5A) has been applied using a Universal Tensile Machine AG-I, Shimadzu. Testing speed was 5 mm/min and the load was 10 kN.

3 Result and Discussion

3.1 Mechanical Properties of LDPE/RP blends after UV irradiation exposure.

Figure 1 shows the trend of change for tensile strength of LDPE with blended of RP under UV irradiation exposure by UV wheatherometer. It showed in pure LDPE there's decreased in strength when under UV irradiation exposed but the decreasing is not lower than strength value of pure LDPE without UV irradiation exposed. With the addition of RP blends, it is observed that for LB5 and LB10, the initial 500 hours the tensile strength is increased from without UV irradiation exposed until 3000 hours. Meanwhile, for LB15 showed increasing tensile strength from without UV irradiation exposed to 500 hours until 1500 hours. At 2000 hours the strength decreased slightly until 3000 hours. It can be observed that LB20 and LB25 showed a similarity trends with increasing in tensile strength from without UV irradiation exposed to 500 hours and declined in tensile strength at 1000 hours to 3000 hours. For LB30, the tensile strength increased upon UV irradiation exposure from 500 hours until 1000 hours. When at 1500 hours it showed decreased strength until 3000 hours. On the one hand, upon UV irradiation exposed for each hour from 500 hours until 3000 hour the tensile strength decreased with addition of RP blends.

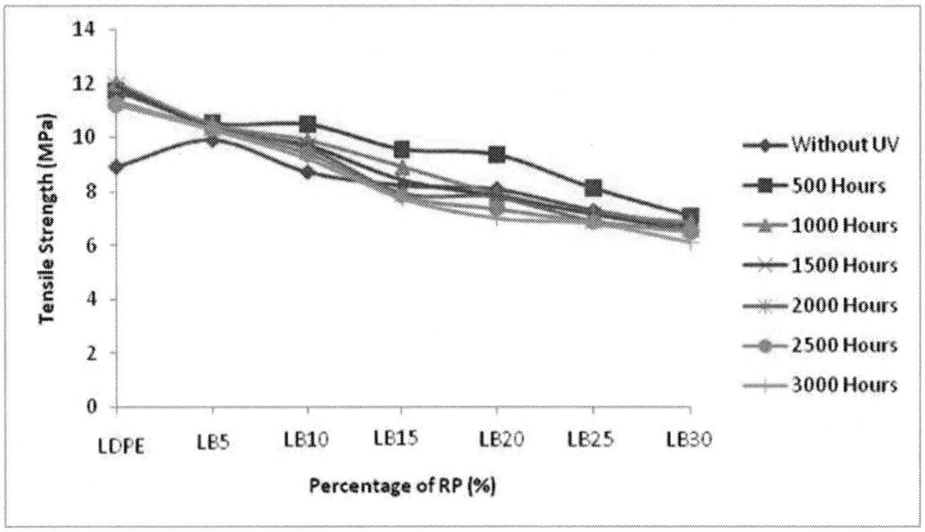

Fig. 1. Tensile Strength of LDPE/RP blends after UV irradiation Exposure.

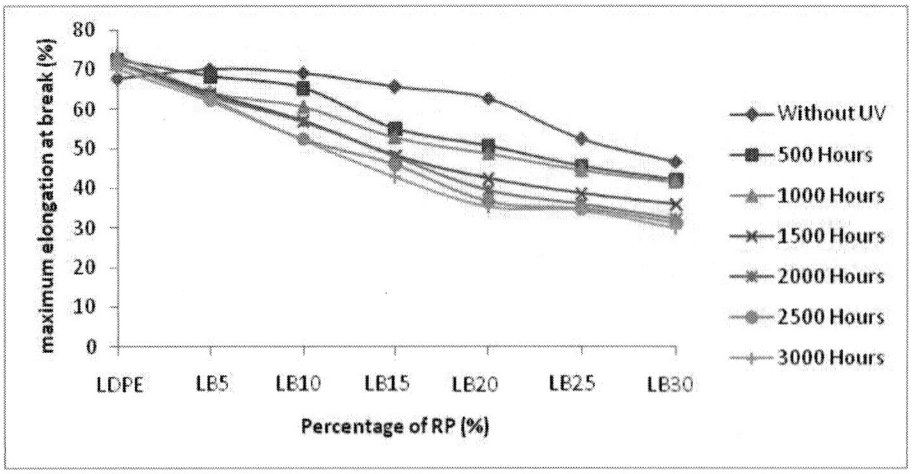

Fig. 2. Maximum elongation at break of LDPE/RP blends after UV irradiation Exposure.

Figure 2 shows a graph of the percentage of maximum elongation with ultraviolet exposure time for each percentage of LDPE/RP blends. The graph pattern changes also decreased and increased in the percentage of maximum elongation. It is observed that the percentage of maximum elongation at break decrease with addition of RP under UV irradiation exposed. The increased of percentage maximum elongation at break of pure LDPE when under UV irradiation from 500 hours. For LB5, the percentage of maximum elongation at break decreased when under UV irradiation. It is showed same trends in LB10 until LB30. When UV irradiation exposed it showed decreased in percentage of maximum elongation at break with addition of RP. It gave same trends from 500 hours until 3000 hours.

Increasing the strength is likely by the reaction cross connection (crosslink) in the polymer chain as a result of exposure to ultraviolet. Previous studies also noted that the disclosure of ultraviolet is one method of crosslinking the polymer polyethylene [4]. The decline in tensile strength at the time of exposure may be due to the occurrence of the chain breaking process (chain scission) as a result of photo-oxidation. These include energy intensity and ultraviolet, foreign objects, particle size and free radicals formed during the synthesis of specimen processing [5].

3.2 Mechanical Properties HDPE/RP blends after UV irradiation Exposure.

For the variation of the tensile strength upon UV irradiation exposure, we can see that the tensile strength increasing under UV irradiation for pure HDPE. It observed from Figure 3 that for HB5, the tensile strength increased at initial 500 hours until 1000 hours and decreased at 1500 hours until 3000 hours. These trends have similar to HB10 and HB25. However, for HB15 the tensile strength increased upon UV irradiation exposed at 500 hour until 2000 hours. There were decreases in the value of tensile strength after 2500 hours of UV irradiation till 3000 hours of irradiation. Other than that, for HB20 and HB30 had same trends of increased in tensile strength at 500 hours until 1500 hours and then declined at 2000 hours until 3000 hours. This result was similar as found in the previous study which stated that tensile strength at yield point is remaining stable or increased slightly till 100 hours of UV radiation and after that time a gradual reduction took place in all samples [5]. Meanwhile, for each UV irradiation hours from 500 hours until 3000 hours showed decreased with the addition of RP.

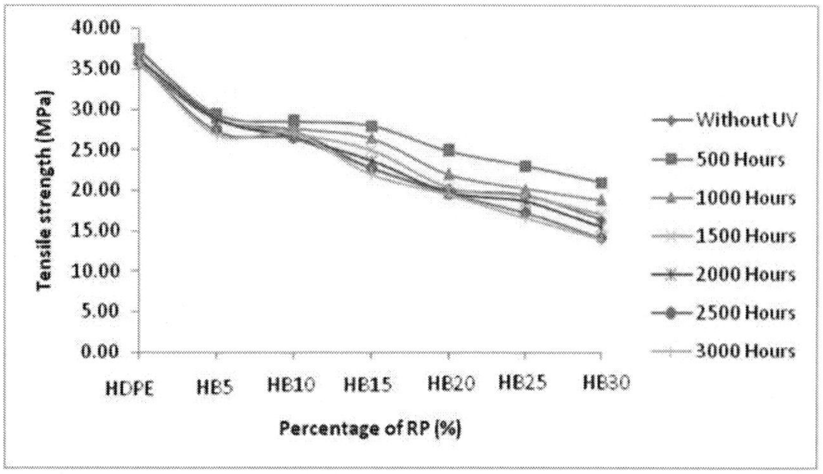

Fig. 3. Tensile Strength of HDPE/BP blends after UV irradiation Exposure.

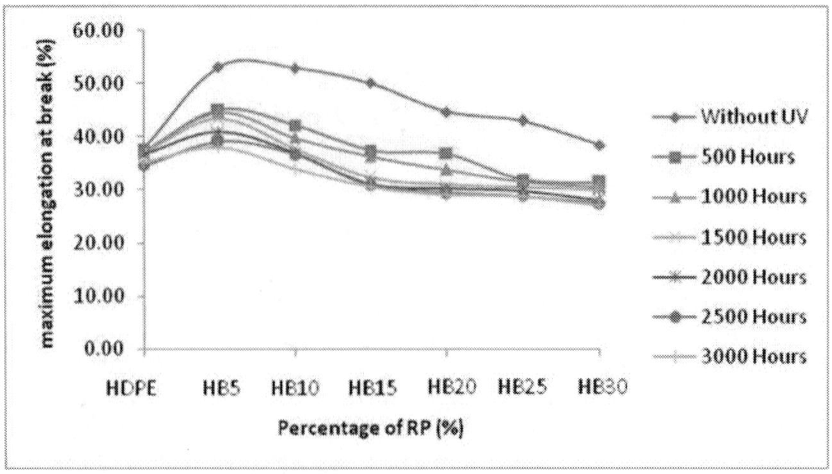

Fig. 4. Maximum elongation at break of HDPE/RP blends after UV irradiation Exposure.

With the UV irradiation there increased in percentage of maximum elongation at break with addition hours of UV irradiation exposed at pure HDPE as shown in Figure 4. For 500 hours, the percentage of maximum elongation at break increased at HB5, HB10 and HB15 and decreased at HB20, HB25 and HB30. At 1000 hours until 2000 hours, these trends similar that showed increased in percentage of maximum elongation at break at HB5 and HB10. However, it decreased in percentage of maximum elongation with addition of BP at HB15, HB20, HB25 and HB30. For 3000 hours, only HB5 increased in percentage of maximum elongation at break and then decreased at HB 10, HB15, HB20, HB25 and HB30. On the other hand, the percentage of maximum elongation at break decreased with addition of UV irradiation time for each addition of BP blends.

Based on previous study, during UV irradiation macromolecular chain scission is taking place that reduces the molecular weight of the polyethylene [7]. However, beyond chain scission, crosslinking reactions are, also, taking place simultaneously [6]. Crosslinking and accompanying stiffening may compete against chain scission and the resultant loss in molecular weight [8]. Thus, crosslinked macromolecules could increase initially the stiffness of HDPE while after a certain time a substantial reduction is caused to the HDPE molecular weight, due to chain scission reactions, reducing therefore its mechanical properties [6].

4 Conclusion

The renewable polymer (RP) was successfully blended with thermoplastic polyethylenes which are LDPE/RP and HDPE/RP and utilizing some modification processes like, UV irradiation. The overall process provides the freedom to develop a large variety of polymers with various mechanical properties. The characterization results indicate that the variations of renewable polymer (RP) content and UV irradiation are effective means to modulate the polymer's mechanical properties. The polymer strength decreased with the increase of renewable polymer (RP) content, while ductility of the polymer increased with the increase of the RP content. The UV irradiation of the polymer was enhanced with the increase of RP content. In conclusion, the renewable polymer/thermoplastic blended properties could be modulated by manipulating the RP content and UV irradiation exposure.

The author would like to thank the Malaysian Government and University Tun Hussein Onn Malaysia (UTHM), Johor and Postgraduate Incentive Research Grant (GIPS) Vot U222.

References

1. R. Strapasson, S.C. Amico, M.F.R. Pereire, T.H.D. Sydenstricker, Polym. Test, **24**, 468 (2005)
2. M.D.E. Lucas, M.A.R. Meier, Eur. Polym. J., **47(5)**, 837 (2010)
3. I.A. Husein, Polym. Degrad. Stabil, **92(11)**, 2026 (2007)
4. S.M. Tamboli, S.T. Mhaske, D.D. Kale, Indian Journal of Chemical Techonology, **11(6)**, 853 (2004)
5. M.A. Salem, Egypt J. Sol., **24(2)**, 141 (2001)
6. I. Grigoriadou, Compos. Part B-Eng., **55**, 407 (2013)
7. J.L.S. Angulo, H.O. Ortega, S.S. Valdes, J. Appl. Polym. Sci., **53(7)**, 847 (1994)
8. J. Khan, S. Hamid, J. Polym. Mater., **15(2)**, 177 (1998)

Mechanical Properties of Renewable Polymer with Thermoplastics Endurance to Ultraviolet irradiation Exposure

Nurul Syamimi M. Salim[1,*], *Anika Zafiah* M. Rus[1], and *Masnadia* Mashuri[1]

[1]Faculty of Mechanical and Manufacturing, University Tun Hussein Onn Malaysia (UTHM), Parit Raja, 86400, Batu Pahat, Malaysia

Abstract. At present the disposal of waste tyre rubber (WTR) has become a major waste management problem in the world. Therefore in this study, polymer blended based on Polyethylene which is Low Density Polyethylene (LDPE) or High Density Polyethylene (HDPE), with Renewable Polymer (RP) and waste tyre rubber (WTR) is prepared via injection molding. Blended polymer such as LDPE/RP/WTR and HDPE/RP/WTR is known as LRT and HRT respectively. The preparation of polymer blend steps start with the preparation of RP. The RP is prepared by crosslinking the renewable monomer with Polymethane Polyphenyl Isocyanate (MDI) at composition ratio of 1:0.5. The second steps involved by adding 10 gm of liquid RP prepared earlier on with fixed amount of LDPE and HDPE of 100 gm. Then the blended LDPE/RP or HDPE/RP namely as LR or HR respectively is further added with WTR with different percentages ratio of 5 %, 10 % and 15 %. The manually blended polymer mixture and filler is then melt mixing using injection moulding to fabricate the tensile specimen for mechanical tensile test and physical determination such as density, distribution of WTR in polymer blend and surface fracture morphology using scanning electron microscope. The samples were then exposed to UV irradiation exposure in UV Accelerated Weathering for 500, 1000, 1500, 2000, 2500 and 3000 hours to evaluate the photostability of the polymer blends. The optimum amount of WTR ratio composition is at 5 % for both LRT and HRT blends which indicate the stability of polymer blends towards UV irradiation exposure at 1000hours.

1 Introduction

At present most of industrial uses only synthetic polymer from petroleum and natural gas resources such as polyethylene, polypropylene and polyvinyl chloride. However, this type of polymer materials that is not renewable. In addition, the polymer of this type has the disadvantage because it can cause environmental pollution, congestion landfills and depend entirely on limited natural resources such as petroleum [1]. Renewable polymer is a renewable material because its ability to undergo degradation (decomposition) in the biosphere that is associated with micro-organisms, enzyme or under a state of

* Corresponding author : nurulsyamimisalim@gmail.com

nature[2].While, renewable resources for the conversion of renewable polymer is originated from vegetable oils. The main constituents of vegetable oils are triacylglycerols, which are the product of esterification of glycerol with three fatty acids. Fatty acids account for 95% of the total weight of triglycerides and their content is characteristic for each vegetable oil [3]. In polymer industry, waste vegetable oils which represent a major potential source of chemicals have been utilized as an alternative feedstock for sustainable monomers.

New technologies, processes and resources are currently under research due to the current worldwide energy panorama. Dramatic growth in the number of used tyres around the globe was recorded due to increasing number of vehicles on the road. Approximately 800 million tyres are discarded around the globe annually [4]. This figure is estimated to increase by 2% every year. The annual global production of tyres is about 1.4 billion unit, which corresponds to an estimated 17 million tonnes of used tyres each year [5].

Waste tyre particle (WTP) falls into this kind of solid residues and it is managed as waste even if it is going to be recycled or re-manufactured. This is a burden that adds significant cost over disposal and in many cases acts as a barrier to improve resource efficiency [6]. The scientific community's effort in finding ways to reduce tyre waste has leaded to intense research on rubber recycling. The technology of polymer blending has emerged as a useful tool in tailoring polymers to the needs of the end users. Accordingly, blending with waste rubber is important both from the point of view of disposal of waste and the reduction in the product cost [7].

The specimen of renewable polymer (RP) from monomer based on palm oil is reinforced with synthetic polymer, Low Density Polyethylene (LDPE) or High Density Polyethylene (HDPE) and mixed with waste tyre rubber (WTR) as filler. The specimen of HDPE/RP, LDPE/RP, WTP/HDPE and WTP/LDPE were irradiated under ultraviolet (UV) radiation exposure. The tensile test tested to study the effect of UV irradiation exposure on the mechanical properties of specimens the LDPE/RP, HDPE/RP, WTR/LDPE and WTR/HDPE.

2 Experimental

The raw materials involved for the experimental study were high density polyethylene (HDPE) density = 0.957 g/cm3, (melt index = 4 g/10 min, vicat softening point = 124 °C), low density polyethylene (LDPE) density = 0.922 g/cm3, (melt index = 5 g/10 min, , vicat softening point = 93 °C), flexible and rigid Polymethane Polyphenyl Isocyanate (Modified polymeric-MDI) (viscosity at 25 °C= 120-160 cps, specific gravity at 25 °C= 1.18-1.20 g/ml, NCO content, % wt = 26.3-27.3), renewable monomers and waste tyre rubber (WTR). The monomer conversions from waste vegetable oils are started with in-house catalyst preparation to generate the epoxies from the unsaturated fatty blends. The liquid monomer is mixed with crosslinker and with HDPE and LDPE until solidifies. The proportion ratio of crosslinker and monomer is 1:0.5. This process is repeated with different ratio of renewable-polymer (RP) and WTR of 5 %, 10 % and 15 %.

LDPE/RP, HDPE/RP, WTR/LDPE and WTR/HDPE were fed in the injection molding machine (Nissei Horizontal Screw Type Injection Molding NP7) equipped with mold of tensile specimens according to ISO 527 (5A). The processing temperature for nozzle, front, middle, rear 1 and rear 2 for LDPE were 185 °C, 180 °C, 175 °C, 170 °C and 165 °C while, for HDPE were 205 °C, 195 °C, 180 °C, 175 °C and 165 °C. The injection pressure for LDPE and HDPE were 43.3 and 96.6 MPa respectively. Dumbbell shape samples produced from injection moulding was exposed to the UV light in UV Lamp Tested Model HD-703 (Haida International Equipment Co., LTD) at different exposure time at 50 °C. Different UV exposure time was conducted for each group of 5 sample to calculate the average value of mechanical testing. The UV exposed time set 500, 1000, 1500, 2000, 2500 and 3000 hour.

3 Result and discussion

3.1 Mechanical Properties of LDPE/RP and HDPE/RP blends.

The mechanical properties of the blended were analyzed based on their tensile strength and strain or elongation at break as shown in Figure 1a and Figure 1b. The HDPE provides ductility that exhibit brittle behaviour with a subsequent loss of toughness to the blended materials. The addition of RP generally increases the toughness and decreases the strength of the blends. This could be due to the soft properties of the RP. Meanwhile, the elongation at break increased. The higher tensile strain of HDPE/RP blends compared to neat HDPE may also be due to plasticization effect. Based on these results, when the RP is added with crosslinker, it may act as filler as it improves the tensile strain values.

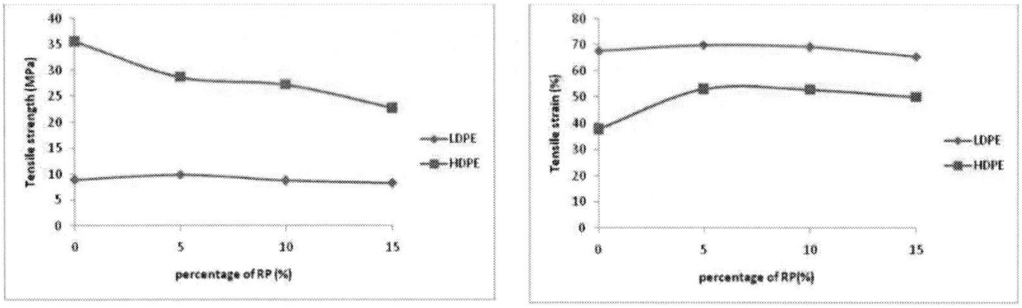

(a) (b)

Fig.1. LDPE/RP and HDPE/RP blends after UV Exposure with (a) Tensile Strength (b) Tensile Strain.

Besides that, the tensile strain of the LDPE/RP blends is observed to be higher than the HDPE/RP blends. The high value of elongation at break for the LDPE/RP blends is due to the more flexible properties. This allows it to significantly yield before breaking, leading to a high value of tensile strain. Therefore, LDPE/RP is comparatively more flexible than HDPE/RP.

3.2 Effects of UV irradiation exposure on tensile strength.

Figure 2a and Figure 2b shows the graph of tensile strength of LDPE/RP and HDPE/RP with consistent ratio of RP of 10% and waste tyre rubber (WTR) of 5%, 10% and 15%. The samples is namely as, L, LR, LRT5, LRT10, LRT15, H, HR, HRT5, HRT10 and HRT15 blends. Where the abbreviation of L is for LDPE, H for HDPE, R for renewable polymer, T is for waste tyre rubber with proportion ratio of 5, 10 and 15%. The presence of the WTR ratio did not improve the strength properties of the LDPE samples. Meanwhile WTR contributed better strength on the H and HR. It is clearly shows that the HRT5 strength values were approaching the neat HDPE from the Figure 2a. It is also noted that the UV irradiation exposure also give influence to the crosslink process between RP and the WTR.

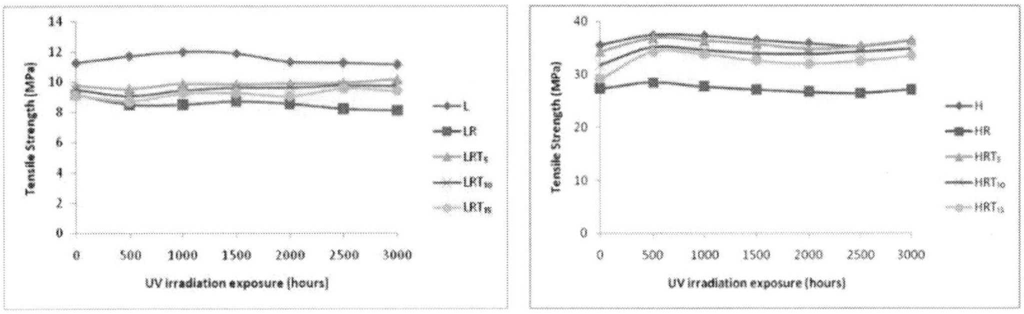

(a) (b)

Fig. 2. Effects of UV irradiation exposure on tensile strength for (a) LDPE and (b) HDPE.

H, HR, HRT5, HRT10 and HRT15 show the highest value of the tensile strength compare to the L, LR, LRT5, LRT10 and LRT15. However the trend starts to decreases after HRT5. The increase of the WTR percentages content will decrease the mechanical properties of HDPE and LDPE blends.

At the beginning the HR facing the decrement of 23.35 %, then the small increment of 3.35 % from HRT5 and then decrement start to begin from HRT10 and HRT15 with value of 10.80 % and 17.95 % respectively. After UV irradiation exposure the decrement trend are much slower than before the UV irradiation exposure of LDPE or HDPE blends.

It can be seen that by increasing the WTR content the tensile strength of LR, LRT5, LRT10, LRT15, HR, HRT5, HRT10 and HRT15 was decrease due to the attributed of WTR content. The WTR still remain as the dispersed in the blend while the LDPE or HDPE was bearing the tensile strength alone. Previous studies [8], stated several reasons for the decrease in tensile strength due to the decreasing of the blends rigidity.

It is well known that the WTR is quite hard to dispose and highly resistance to the degradation due to the various different chemical components on it. In conjunction with that, WTR particle induces to the photostability of the LRT5, LRT10, LRT15, HRT5, HRT10 and HRT15 blend because WTR particle did not give any reactions during the UV irradiation exposure process. UV irradiation exposures not lead the crosslinking between RP, WTR and the matrix. However, due to the lack of adhesion and low interfacial bonding, WTR particle lead to the decreasing on mechanical properties. The addition of WTR in the polymer blend is an advantages in order to recycle the scrap tyres. However, according to Pilar [9], WTR use as an additive should be at low amounts which are less than 10% wt.

3.3 Effects of UV irradiation exposure on tensile strain.

Figure 3a and 3b shows the graph of strain (%) for samples of polymer blends. Every group of polymer blends samples show fluctuate trend. It is noted that the presence of the WTR influence the strain properties. The fluctuate trend for all the samples of LRT5 show the highest improvement if compare to the sample of LR, LRT10 and LRT15. 1.54 % difference from comparison of unirradiated and 3000 hour on UV irradiation exposure. The LRT5 show the most significant improvement of strain (%). At 500 hour UV irradiation exposure for LRT10 show the biggest decrement with 13.17 %. However the tensile strength of HRT5, HRT10 and HRT15 slowly increased as shown in Figure 3b. The strain trend shows almost the same for all samples. However, lowest strain of H, HR, HRT5, HRT10 and HRT15 as compared to the L, LR, LRT5, LRT10 and LRT15.

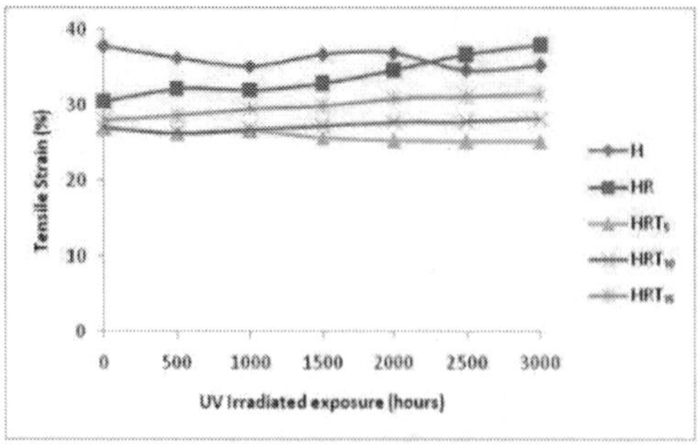

(a)

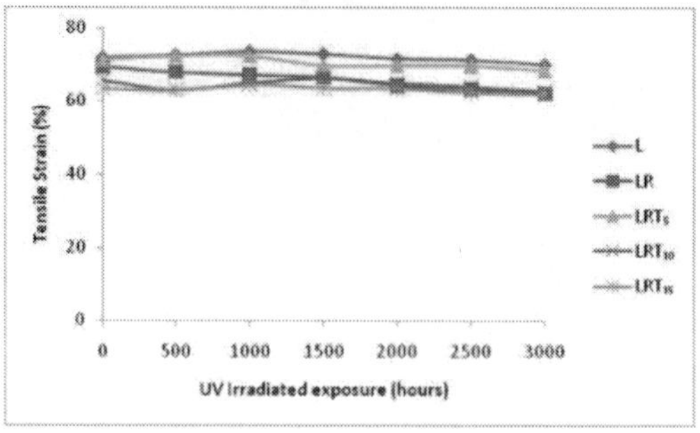

(b)

Fig. 3. Effects of UV irradiation exposure on tensile strain for (a) LDPE and (b) HDPE.

Neat LDPE still have highest strain compare to those samples added with WTR as shown in Figure 3a. LRT15 show the lowest value on strain. However, LRT5 shows the better influence of WTR. 5 % of WTR enhance the strain on LDPE approaching the neat LDPE value.

Figure 3b show the effects of UV irradiation exposure on strain for H, HR, HRT5, HRT10 and HRT15. The trend shows that there are slight increments of strain for HRT5, HRT10 and HRT15. Meanwhile, at 1000 hours until 3000 hours, the tensile strain of HR increases. It is well known that HDPE have high crystallinity that will make it hard to deform and cause to break in low strain.

From the previous study [10], it is also proven that by adding the WTR on HDPE blend will lead the strain to decrease. However, in this study show that there are some increments on the strain after UV irradiated exposure. It is believe, this situation happen due to the increase of adhesion between the interfacial of WTR and the matrix LDPE or HDPE. Thus this condition lead to WTR and HDPE blends become miscible blends, which reducing two separate phases on the blend that causing strain to increase slightly.

4 Conclusion

The processing conditions of polymer blends of LDPE and HDPE by injection molding were determined and analyzed. The initial processing conditions of the standard LDPE and HDPE thermoplastic were used as references. It was concluded that the processing conditions by injection molding were exactly the same to that of neat LDPE. For the HDPE/RP blends, increases in the parameters used in injection molding were required. Tensile strength and elongation at break increased with increasing RP content at low compositions and started to decrease at high RP content for the LDPE/RP blends. However, the tensile strength and strain of the LDPE/RP blends were generally better than the neat LDPE. Therefore, not only does the presence of RP in the blends provide sustainable characteristics, but it also improves the mechanical properties. These results comply with the properties of pure LDPE and HDPE. Apart from the changing ratios of RP to LDPE or HDPE, the processing temperature was also considered to have an influence on the mechanical properties of the blends. The effects of sustainable polymer to the LDPE or HDPE have become more significant as the polymer blends has been exposed to the UV irradiation.

The author would like to thank the Malaysian Government and University Tun Hussein Onn Malaysia (UTHM), Johor and Postgraduate Incentive Research Grant Vot U222 and Malaysian Technical University Centre of Excellence (MTUN CoE) for supporting this research study under research grant Vot 1481.

References

1. S. Wiwin, G.S. Monica, T. Wega, Triyono, A. Ria, I.F.F. Iip, Procedia Environmental Sciences , **20**, 215 (2014)
2. R. Chandra, R. Rustgi, Prog. Polym. Sci., **23(7)**, 1273 (2000)
3. M.D.E. Lucas, M.A.R. Meier, Eur. Polym. J., **47(5)**, 837 (2010)
4. P.J.H.V. Beukering, M.A. Janssen, Resour. Conserv. Recycl., **33(4)**, 235 (2001)
5. M. Sienkiewicz, J.K. Lipka, H. Janik, A. Balas, Waste Manage., **32(10),** 1742 (2012)
6. S. Ramarad, M. Khalid, C.T. Ratnam, A.L. Chuah, W. Rashmi, Prog. Mater. Sci., **72**, 100 (2015)
7. M.M. Hassan, N.A. Badway, A.M. Gamal, M.Y. Elnaggar, E.S.A. Hegazy, Nucl. Instrum. Meth. B., **268(9)**,1427 (2010)
8. H.H.R. Cui, M.R. Kessler, Polym. Degrad. Stabil, **98(11)**, 2357 (2013)
9. M.A. Salem, Egypt J. Sol., **24(2)**, 141 (2001)
10. I. Grigoriadou, Compos. Part B-Eng., **55**, 407 (2013)
11. S.M. Tamboli, S.T. Mhaske, D.D. Kale, Indian J. Chem. Techn., **11(6)**, 853 (2004)

The Economic Benefits of Generation Revenue Assessment in Pool-Based Market Model for Restructured Electricity Supply Industry

Zuraidah Ngadiron[1,*], *N.H* Radzi[1] and *M.Y.* Hassan[2]

[1]Faculty of Electrical and Electronic Engineering, Universiti Tun Hussein Onn Malaysia, Batu Pahat, Malaysia
[2]Centre of Electrical Energy Systems (CEES), Faculty of Electrical Engineering, Universiti Teknologi Malaysia, Johor Bahru, Malaysia

Abstract. The electricity supply industry had undergo deregulation and restructuring toward becoming a more transparent and competitive electricity market environment. The pool market model is amongst the most preferred electricity market model. Even though it is a safe option to be more competitive and transparent electricity supply industry, there are issues on the welfare of the generators involved. This paper addresses the pricing issue in the pool market by extending the capacity payment mechanism in the single auction power pool. In the proposed model, the approach of minimum capacity payment involving the efficiency of the generators is introduced. A case study is conducted to illustrate the proposed model. An economic analysis is performed to highlight the merits of the proposed model with the pure pool in term of generation revenue.

1 Introduction

In the competitive electricity market, participation of generation companies and large consumers in bidding methodologies is for their own advantages. By strategically bidding, the electricity generating companies expect to maximize their profits and control the market price. Even though, they will suffer in great losses of market efficiency because their offers will deviate from the true marginal costs. Furthermore, incur much more payments from the consumers than really required. In order to achieve a maximum return, it is very important for the generation companies to formulate optimal bidding strategies with risk terminology before step into the electricity market, as the market clearing price (MCP) in this landscape is flexible [1]. The pool market model is a standout amongst the most preferred power market model actualized in many developing countries. Regardless of being the sensible and safe choice for a more competitive and straightforward power supply industry, there are issues on the welfare of the generators included. As reported by Zou et al. [2], which utilized of the analytical paradigm of economic mechanism design theory is to deduce and design a customized pool-based market mechanism, in order to fulfill three major properties; incentive compatibility, individual rationality, and payment cost minimization.

* Corresponding author: zuraidahn@gmail.com

One of pool market model major benefits is that it provides a market for generators unable to sell their output via contract to a specific customer. This study is focusing on the economic aspect from the perspective of the generators. Therefore, the proposed model is being designed in order to overcome several disadvantages of the pure pool market such as price fluctuation and market power exercise.

2 Improvement of Pool Based Market Model for Malaysia Electricity Supply Industry (MESI)

The reformed of single buyer is reinforced by implementation of competitive bidding, while the Incentive Based Regulation (IBR) will ensure the efficient operation [3]. Yet still single buyer model lack of competition due to long term of Power Purchase Agreements (PPAs). Nevertheless, there are few issues of introducing the pure pool market model such as price fluctuation and market power exercises. Thus, it is important to modify the existing pure pool model, so it can provide a fair market to supplier and user. It is reported at low load condition, the generators would gain more revenue under single buyer model compared to pool market model, whereas at high load demand, the generators would gain more revenue under pool market model [4]. Aifa [5] proposed an alternative electricity market model, hybrid model for MESI, which combines the pure pool market and pro-rata base load profile. In the meantime, Nuryasmin [6] proposed a minimum generation model that guaranteed income for all IPPs, and also generates profit for Tenaga Nasional Berhad (TNB). An improvement of hybrid model by Nurehan et. al [7,8] successfully overcomes the shortcomings of the pool market model, in the context of guaranteed revenue remuneration for the generator with the introduction of based load demand sharing approach known as hybrid model 1 (HM 1), and hybrid model 2 (HM 2). The introduction of the base load sharing is to ensure the expensive generator will get the remuneration, in order to overcome the problem in the pool market model. This is because there is no revenue during low electricity demand for the expensive generator. However, all these models did not consider on the efficiency, and the electricity price offered by generators during base load sharing. Furthermore, the base load power plants are designated based on their low cost generation, efficiency, and safety at rated output power levels.

3 Methodology of Proposed Market Model

Malaysia load profile curve in Fig. 1(a) is used to present the pure pool, HM 1, HM 2, and the proposed market model. Fig. 1(b) shows the aggregated generation curve and the system marginal price (SMP) at 30%, 50%, 80%, and 100% demand, which proven the pool market price is extremely high due to high SMP, during high electricity demand. Two load profile curves for Sunday, and Monday are used in this research. Assuming the generators bid the same price for the 24 hours, Table 1 shows the details of the sixteen generators selected around Malaysia. Instead of open cycle gas turbine (OCGT), only combine cycle gas turbine (CCGT) and thermal plant type are chosen due to the efficiency and price offered by the generator. This is because generation costs of OCGT are much higher, and the low load-factor of the peak-load services [9]. As the monetary values involved in this paper are confidential, estimated values in used instead.

 DOI: 10.1051/matecconf/20167801039

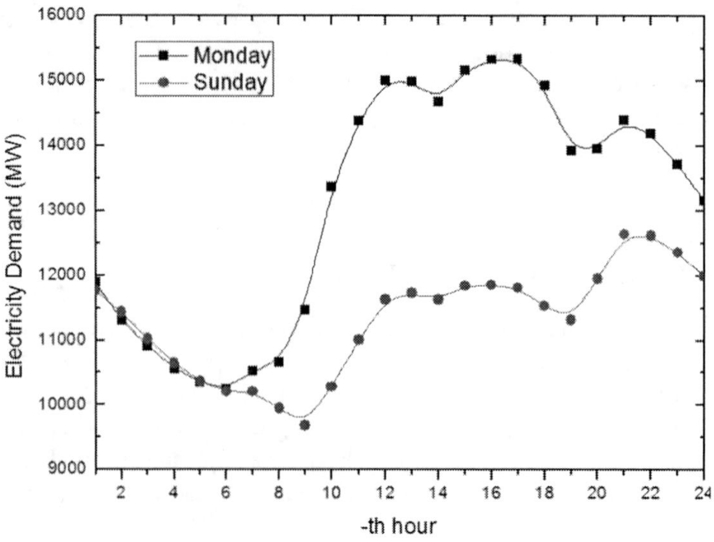

(a)

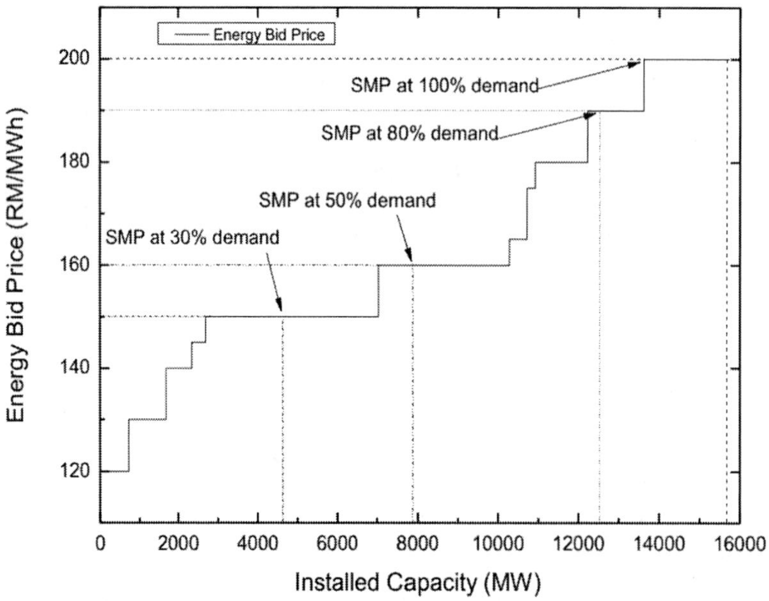

(b)

Fig. 1. (a): Diagram of Malaysia load profile curves; (b): The aggregated generation curve and the SMP at 30%, 50%, 80%, and 100% demand.

This proposed model is an improvement of the pool based marketing model. It comprises of electricity demand sharing and capacity payment approaches, in order to educate the generators to apply low prices for electricity generated. Thus, Independent Power Producers (IPPs) can contribute power to the power purchaser at all times. In addition, the efficiency of the generators is taken into account. This proposed market model has categorized the electricity demand into two areas, identified as a high demand area and low demand area. The low and high demand areas are determined from the daily electricity demand curve, constructed using the demand forecasting data. A line is drawn on the hourly electricity demand curve at 80% of the peak demand, which is used as the reference line to distinguish the low demand and high demand areas on the electricity demand curves. The electricity demand sharing is applied under high demand area. Meanwhile, the capacity payment is added for low demand area.

Table 1. Details on plant types, MW installed capacity, bid price, capacity price, and efficiency of 16 generators in Malaysia.

Gen	Plant Type	Installed Capacity (MW)	Energy/Bid Price (RM/MW/h)	Capacity Price (RM/MW/month)	Efficiency (%)
1	CCGT	720	120	30000	46.91
2	CCGT	640	130	30000	40.82
3	CCGT	334	130	35000	43.64
4	CCGT	650	140	35000	43.64
5	CCGT	350	145	35000	43.64
6	Thermal	2420	150	20000	39.03
7	CCGT	1170	150	30000	45.62
8	CCGT	762	150	45000	44.22
9	CCGT	1136	160	30000	40.74
10	Thermal	2100	160	30000	20.91
11	CCGT	440	165	35000	43.64
12	CCGT	100	175	35000	43.64
13	Thermal	100	175	30000	35.99
14	CCGT	1303	180	40000	42.09
15	Thermal	1400	190	25000	35.91
16	Thermal	2100	200	55000	25.82

3.1 The Low Demand Area.

The low demand area consists of the hourly electricity demand below the 80% of the peak demand value, and will be traded through bidding competition as in the pool market. Nevertheless, pool market cannot guarantee continuous remuneration for IPPs when the electricity demand is low. Consequently, numerous of high costs IPPs will experience low or even zero revenue during low demand period. Therefore, capacity payment with minimum generation is added during low electricity demand, in order to ensure continuous remuneration for all IPPs regardless their submitted energy bid prices and the fluctuating electricity demand. This minimum generation is based on the efficiency of the generator generates the electricity. Under the same bidding system as applied in the pool market model, the payment for power generated will depend on the SMP, which determined from the last IPPs being dispatched at that period of time. The purchasing price in low demand area, C_{PP} is paid to the IPPs for the power generated in an hour.

3.2 The High Demand Area.

The electricity demand under high demand area will be traded through demand sharing and bidding competition. The 80% reference line will separate electricity demand in the high area into two parts; i) the electricity demand below the 80% reference line will be traded equally among the IPPs through demand sharing approach, and ii) the IPPs will compete against each other to supply the remaining demand in the area above the 80% reference line. The distribution of the shared electricity demand depends on the IPPs available capacity. IPP with the highest capacity will have the biggest share of the sharing electricity demand. The payment during electricity demand sharing will be determined by two factors; the SMP of the sharing demand and the amount of electricity demand shares of each IPP. C_{PP80} is the sharing price with the SMP value at the total shared demand (80% of peak electricity demand). The upper part of the high demand area will be put under bidding competition using the energy bid prices submitted by the IPPs. During this time, all IPP are involved, from the cheapest to the expensive generators.

4 Results and Discussion

In order to be selected in the generation dispatch, IPPs must win in the energy bidding. The IPPs with the lower energy bid price compared to the hourly SMP will be selected to fulfil the electricity demand for that hour. The SMP at 100% demand was RM200 followed by RM190, RM160 and RM150 at 80%, 50%, and 30% demand respectively. All participated IPPs in the electricity dispatch are paid according to the pool price which is determined from hourly SMP. In this case study, the power contribution is chosen at 80% from the peak demand for the load demand sharing, reducing the SMP to RM190. Consequently, the production cost of electricity will be lower, in order to control the price fluctuation during high demand. Therefore, the non-participated IPPs are exposed to the risk of losing their revenue, because only at 100% demand, all IPPs received the revenue. It means that, more IPPs will lose their revenue as the electricity demand keeps reducing. However, the IPPs revenue in the pool market continues to strike as the demand keeps increasing due to most IPPs are paid at a higher price compared to their own bid price. For example, IPP 1 is paid with RM190 for every MW power produced at 80% demand, which is RM70 higher compared to its initial bid price. Fig. 2 (a) and (b) illustrate the comparison of generator's revenue on Sunday, and Monday respectively. The figures indicate that from low to middle price, the generators received the highest revenue under pure pool, but the revenue decreased for the expensive generator. This is because the generators were stacked from the lowest to the highest prices offered by generator. Furthermore, a cheaper generator will get the chance to win in the bidding competition. The generator's revenue of the proposed model is increasing compared to HM 1 and HM 2 during low to medium price generator.

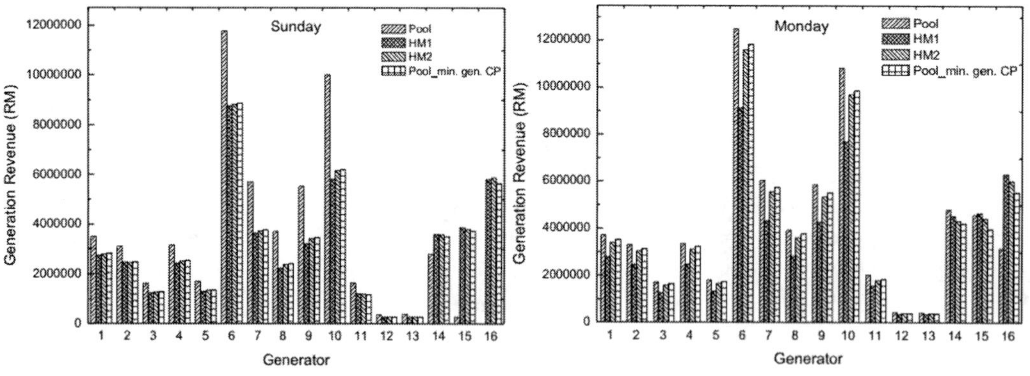

(a) (b)

Fig. 2. (a) and (b): Comparison of generator's revenue for Sunday and Monday.

It is reasonable to introduce minimum generation capacity payment based on the generator efficiency, instead of paying full capacity payment as compensation to the generators, because the possibility of expensive generators not selected in the bidding competition is high. Nonetheless, as a reward, this same type of capacity payment also will be given to the IPPs which have won in the bidding competition as an incentive, in order to educate the IPPs to bid and sell their electricity produced at a lower price. For proposed method, the reduction of revenue for expensive generators does happen compared to HM 2, due to this change.

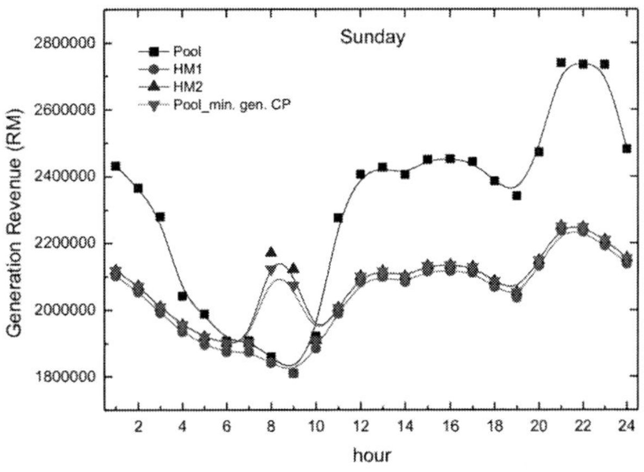

(a)

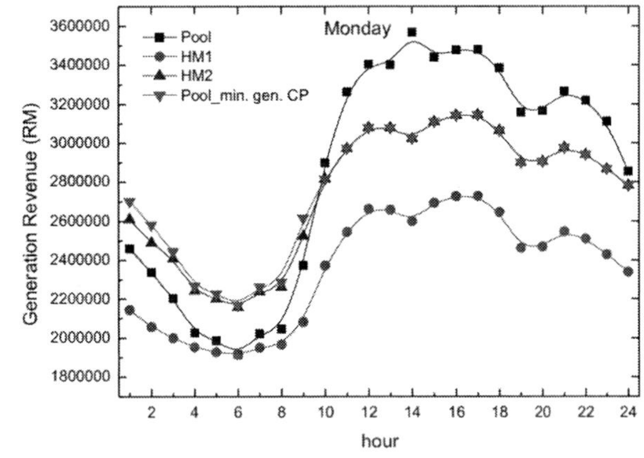

(b)

Fig. 3. (a) and (b): Comparison of hourly generation revenue for Sunday and Monday.

Fig. 3 (a) and (b) show the hourly generation revenue. HM 1 and pure pool model is the lowest and the highest revenue, respectively. For HM 2 and the proposed method, the revenue almost the same for the high demand. Meanwhile, revenue during low demand is slightly higher for the proposed method compared to HM 2, due to minimum generation capacity payment execution for participated and non-participated generator in the bidding competition.

5 Conclusion

Worldwide experience with the single buyer model has raised concern on how this model backfires due to poorly constructed such as lack of transparency and fairness, poor system planning and non-competitive. Several observers also voiced concerns over a profit-seeking single buyer, citing conflict of interest, duplication of cost and tariff hikes. Conversely, the pool market model offers full competitive model and based on uniform price scheme. The most expensive generators might not be able to get any revenue at all. Thus, will force each of them to bid for the cheapest energy price most of the time, and this will create competition. The proposed model seems to be very effective. As it gives the opportunity more to the least cost generators to participate not only in base load demand, but also in peak load demand. However, minimum generation capacity payment also given as an incentive, and compensation for the least cost and expensive generator, which participate and lost in bidding competition respectively. As a result, it creates win-win situations between the generators as the seller and distributor as the buyer.

References

1. K. Asokan, R. Ashok Kumar, *International Journal of Soft Computing and Engineering* (2013)
2. Peng Zou, Qixin Chen, Qing Xia, Chang He, Chongqing Kang, Appl. Energy, **158**, 508 (2015)
3. N.Z. Mohd Zamin, N.Z. Zainol Abidin, Joon B. Ibrahim, *In TENCON Spring Conference, 2013 IEEE* (2013)
4. M.Y. Hassan, M.P. Abdullah, A.S. Arifin, F. Hussin, M.S. Majid, *IEEE International Conference In Power and Energy* (2008)
5. A.S. Arifin, *IEEE International Conference In Power and Energy* (2008)
6. Nor Yasmin Binti Ismail, *Electricity Market Model For Deregulated Electricity Supply Industry* (Diss. UTM, Skudai, 2011)
7. Nurehan Othman, Mohammad Yusri Hassan, Faridah Hussin, Md Pauzi Abdullah, International Journal of Integrated Engineering, **5**, 26 (2013)
8. Nurehan Binti Othman, *Improving Electricity Market Model For Malaysia Electric Supply Industry* (Diss. UTM, Skudai, 2014)
9. J.D. Gifford, IEA, **02** (2011)
10. N. Othman, M.Y. Hassan, F. Hussin, *IEEE International Conference In Power and Energy* (2012)
11. A. Khalid, N. Jaat, A. Sapit, A. Razali, B. Manshoor, I. Zaman, International Journal of Automotive and Mechanical Engineering, **11**, 2447 (2015)
12. A. Khalid, N. Tamaldin, M. Jaat, M.F.M. Ali, B. Manshoor, I. Zaman, Procedia Engineering, **68**, 225 (2013)

MATEC Web of Conferences **78**, 01040 (2016) DOI: 10.1051/matecconf/20167801040
IConGDM 2016

The Influence of Moisture on the Performance of Polymer Fibre-Reinforced Asphalt Mixture

Ibrahim Kamaruddin[1,*], *Madzlan* Napiah[1], and *Mohammed Hadi* Nahi[1]

[1]Civil and Environmental Engineering Department, Universiti Teknologi Petronas, Bandar Seri Iskandar 32610, Tronoh, Perak, Malaysia

Abstract. A number of researches have been done worldwide to evaluate the damage caused by water in bituminous pavements. The use of the retained strength ratios obtained from laboratory moisture damage tests is a useful tool in making quantitative predictions of the related damage caused by water. This study involved laboratory work on the effect of water on the performance of bituminous mixtures. Comparisons are made between the performances of Hot-rolled Asphalt (HRA) bituminous mixtures containing base bitumen of 50pen grade to that of a polymer-fibre reinforced HRA mixture. Two types of polymer fibre were studied, namely polypropylene and polyester and these fibre were added in different concentrations in the bituminous mixtures. Changes in both the cohesive properties of the bitumen and the adhesion of the bitumen to the aggregate surface were observed as a result of exposing the bituminous mixtures to moisture. The effect of polymer fibre reinforcement in bituminous mixtures helps reduce the level of moisture damage. This was evident in the lower moisture susceptibility achieved in the polymer fibre reinforced bituminous mixtures as compared to the control mixture. The additional bitumen in the fibre reinforced mixtures also afforded an increased film thickness on the aggregate particles, thus affording additional protection of the mixtures from moisture. The reinforcement of polymer fibres in bituminous mixtures also acts to decrease the moisture sensitivity of the bitumen to aggregate bonding. This may be due to the strengthening of the wetted binder matrix that helps promote both adhesion and cohesion retention.

[*] Corresponding author: ibrakam@petronas.com.my

1 Introduction

The damaging effects of moisture on the physical properties and mechanical behaviour of bituminous mixtures have been the focus of study for many years. Many laboratory tests have been developed in order to evaluate and quantify the amount of damage that is caused by water on bituminous mixtures. The most widely used laboratory methods in conducting these tests appear to be the immersion-mechanical tests which measure the changes in mechanical properties of the bituminous specimens after exposure to moisture. Typically the results are reported in terms of percentage retained strength of the specimens. This paper is based on some laboratory work and addresses the damaging effect of moisture in polymer fibre-reinforced bituminous mixtures [1].

2 Stripping in bituminous mixtures

The unfilled void spaces in compacted bituminous mixtures are also able to hold sufficient quantities of water that can cause distress and damage and reduce mix performance. The volume of these voids is a variable dependent on the nature of the voids, characteristics of the mix and the degree to which they are compacted in the pavement [2]. The damaging effect of water will give rise to stripping or de-bonding of the bitumen from the aggregate surface. This is brought about by the loss of cohesion and softening of the binder due to the action of the moisture within the bitumen or bituminous mix. Both stripping and softening can occur in the same mix [3]. Stripping is the physical separation of the bitumen and aggregate produced by the loss of adhesion between the bitumen and the aggregate surface primarily due to the action of water or water vapour [4]. It is an aggregate interfacial phenomenon and induces the loss of stability in the bituminous mix that in turn promotes failure [1, 5]. The phenomenon can be further aggravated by the presence of aggregate surface coatings and by aggregates with a smooth texture surface. Stripping is primarily an aggregate problem, but that the type of bitumen used is also important [6]. Softening is the general loss of stability of a mixture that is brought about by a reduction in cohesion within the bituminous mix matrix due to the action of moisture. The resulting damage brought about by water can be assessed quantitatively by mechanical tests in which such properties as loss of tensile strength or decrease of resilient and stiffness moduli have been measured. Moisture induced damage is usually easy to identify when stripping is evident. Where a loss of pavement stiffness or moduli occurs without the visual evidence of stripping, the cause of the problem is less easily recognised [7].

There are a number of laboratory methods and tests that are cited in the literatures that enable the determination of the moisture susceptibility and thus the proneness of bituminous mixtures to stripping or de-bonding. Based on previous work [8], it appears that vacuum saturation of specimens followed by conditioning and testing using the Indirect Tensile test show promise for introducing moisture into the specimens and measuring their strength to predict the moisture susceptibility of the mixtures. Specimens were moisture conditioned before the determination of their structural strengths as those mixtures most susceptible to moisture damage would have the lowest structural strengths following conditions [9]. The retained strength is determined by comparing the dry tensile strength to the wet conditioned tensile strength of the bituminous mixtures.

2 Materials used in the investigation mineral aggregates, filler and bitumen

Limestone aggregates and Ordinary Portland cement (OPC) filler and a binder of nominal penetration Grade 50 were used in this study. Some relevant properties of the material used are shown in Table 1.

Table 1. Properties of the mineral aggregate, filler and bitumen used in the investigation.

Material	Percentage by Weight (%)	Relative Density
Coarse Aggregate	35	2.75
Sand	55	2.65
Filler (Ordinary Portland Cement)	10	3.15
	Penetration (0.1 mm)	Softening Point (°C)
Bitumen	52	48.5

3 Synthetic fibres

Two types of synthetic fibres polypropylene and polyester were used in this study. The fibres were used as a partial replacement of the filler; on an equal volume basis; at two concentrations of 0.5 % and 1 % by weight of the mix. The chopped fibres were the by-products of the textile industry and thus their potential use was desirable on environmental grounds [10].

Table 2. Characteristics of fibres used in the study.

	Specific Gravity	Denier	Length (mm)	Average Diameter (μm)	Degradation Temperature (°C)
Polypropylene	0.91	6	6	22*	160
Polyester (POL)	1.41	3	6	17*	250

*Values obtained from 20 readings using a light microscope at 400X magnifications.

Some characteristics of the fibres used are shown in Table 2. In order to maintain thermal stability when using the polypropylene fibres, it was decided that the mixing temperature when preparing the Hot Rolled Asphalt (HRA) mixture will not exceed 140°C and compaction be done at 130°C.

4 Degree of saturation

The degree of saturation gives a measure of the amount of water that is absorbed by the specimen into its permeable voids. The degree of saturation is thus defined as the ratio of the volume of water in the wet specimen to the total volume of voids in the specimen. The creation of a degree of saturation in the laboratory high enough without damaging the specimens and that the retained strength can be determined involves a moisture conditioning process. Static soaking seems to provide ideal conditions for stripping to occur while both pressure and vacuum saturation procedures may create damage to the specimens. For practical laboratory purposes, static soaking may require too much time. Saturation by partial vacuuming for a short period of time has therefore been used in moisture damage studies on bituminous mixes. If the volume of absorbed water exceeds the volume of voids, the specimen has been supersaturated and damage and should be discarded. A number of researchers have come up with various regimes for vacuuming and saturating the bituminous specimens. The specimens to hot water immersion (at 60°C) for up to 14 days and testing the specimens at different immersion period to determine the retained strength of the specimens.

Table 3. Concrete strain change with temperature.

Temperature (°C)	Concrete strain (m/m) x 10^{-6}
10	1.0
20	1.6
40	3.2

5 Void structure in bituminous mixtures

The moisture conditioning process attempts to allow water to penetrate and occupy the air voids in the specimen. An appreciation of the void structure in bituminous mixtures is thus very vital. A study to examine the influence of asphalt film thickness, voids and permeability on asphalt hardening in asphalt mixtures and come up with a hypothetical model of the air voids system in a compacted bituminous mixture [11, 12]. Different water saturation techniques were employed in their study that included a 24 hours soaking and vacuuming at different absolute pressures. The model divides the air voids system into three categories; through passage accessible air voids, dead end accessible air voids and non-accessible air voids. The 24 hours soaking allows water to only occupy the through passage accessible voids. The 24 hours soaking and hand pumping allowed the water to occupy the through passage accessible voids as well as a small portion of the dead end accessible voids. Vacuuming with an absolute pressure of 2.5 cm mercury allowed the water to occupy the through passage accessible voids and most of the dead end accessible voids [13].

6 Experimental procedures

As the water susceptibility of the mixes was to be determined, the specimens were tested in both the dry and wet condition. The dry conditioning involves curing the specimens at room temperature for two days prior to testing. The wet conditioning involved subjecting the bituminous samples to a combination of air vacuum, vacuum saturation and static soaking. Air vacuuming is to evacuate all the accessible pores from air and water. The objective of vacuum saturating the specimens was designed to accelerate the moisture damage process. It is extremely important that the process of vacuum saturating the specimens do not result in a degree of saturation greater than 100 %, which is indicative that more water was introduced into the voids than there are void spaces, making comparison between samples no longer valid. Fig. 2 is a schematic diagram of the vacuum saturation apparatus. In this study, the specimens were placed in a thick-walled desiccator jar. Valves (W) and (A) are closed and valve (V) which led to the vacuum pump is opened. The air vacuuming process takes about half an hour to drive out all the air trapped in the accessible voids. Distilled water was then used to fill the jar to about 2 cm above the specimens and about 3 cm below the top rim of the jar. This was followed by a 1-hour vacuum saturation period at 1 atmospheric pressure, which was considered a pre-treatment of moisture condition and a means of water-saturating the specimens, during which time the jar surfaces was gently agitated. After one hour of vacuuming, the vacuum was removed and the inside of the jar was allowed to reach ambient atmospheric pressure and the specimens undergoing static soaking for a period of 24 hours for the purpose of achieving a constant weight of the specimens (fully saturated condition).

Fig. 3 shows the relationship between the degree of saturation and bitumen content for the control Hot-Rolled Asphalt (HRA) mix in comparison with the fibre modified mixes. The general trend is in the degree of saturation to decrease with increasing bitumen content. The addition of fibres appears to bring about an increase in the degree of saturation, this increase being more pronounced at the higher fibre concentration. This may be the result of the higher porosity that is associated with the higher fibre content mixes. The polypropylene mixes appear to exhibit better result than the polyester fibre mixes. For all the mixes, the saturation lines obtained appear parallel to one another with a somewhat similar slope. This is indicative that the degree of saturation in all the mixes decreases consistently with increasing bitumen content.

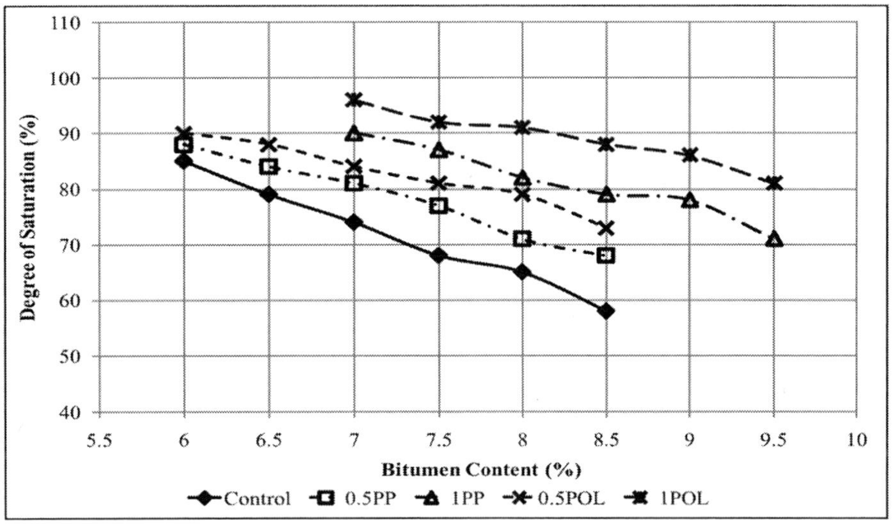

Fig. 3. Degree of Saturation vs. Bitumen Content.

Fig. 4 and Fig. 5 show the variation between the calculated and measured porosity between the control and the polypropylene fibre modified mixes and the polyester fibre modified mixes respectively. The calculated porosity gives a measure of all the voids in the specimens that include both the accessible and non-accessible voids while the measured porosity as was obtained from the moisture conditioning process determined only the accessible voids. The calculated porosity therefore is always greater than the measured porosity as shown in the figure. The general trend is the porosity decreases with increasing bitumen content. The result of adding fibres to the HRA mix resulted in higher porosity in the resulting mix. The relationship between the measured and calculated porosity is plotted graphically in Fig. 6 resulting in a linear relationship between the two. The point of intercept of the lines with the calculated porosity axis gives an indication of the percentage of non-accessible or unconnected pores in the respective mixes. The addition of fibres appears to reduce the percentage of non-accessible pores in the mix, this reduction was seen to be more pronounced in mixes with greater fibre content. A general trend shown from the figure suggests that increasing the bitumen content resulted in an increase in the unconnected voids as the bitumen fills up the void space or continuous channels in the specimen. Subsequently, this reduces the connectivity of the voids. The presence of the fibres also suggests an increase in the connectivity of the voids as the porous nature of the fibres gave a continuous channel (path) of void space. It must also remember that the control mix gave lower porosity than the fibre incorporated mixes, thus justifying its behaviour as in Fig. 6.

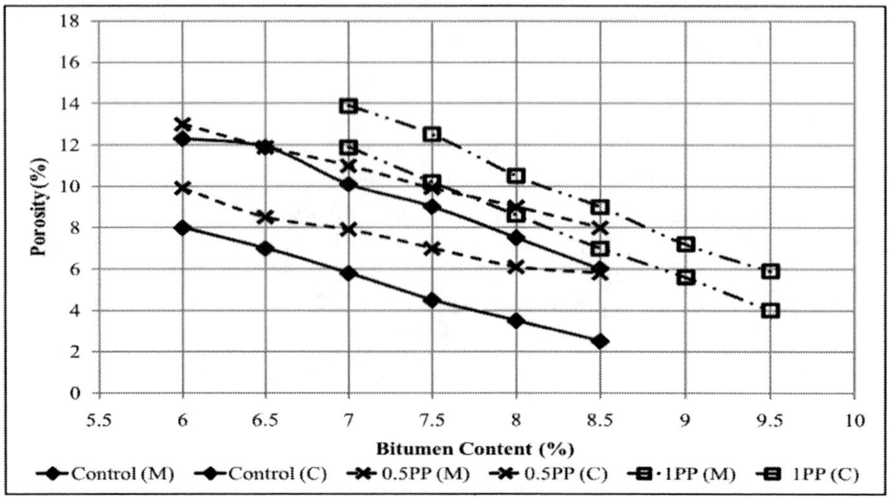

Fig. 4. Calculated and Measured Porosity vs. Bitumen Content for Control and PP Fibres.

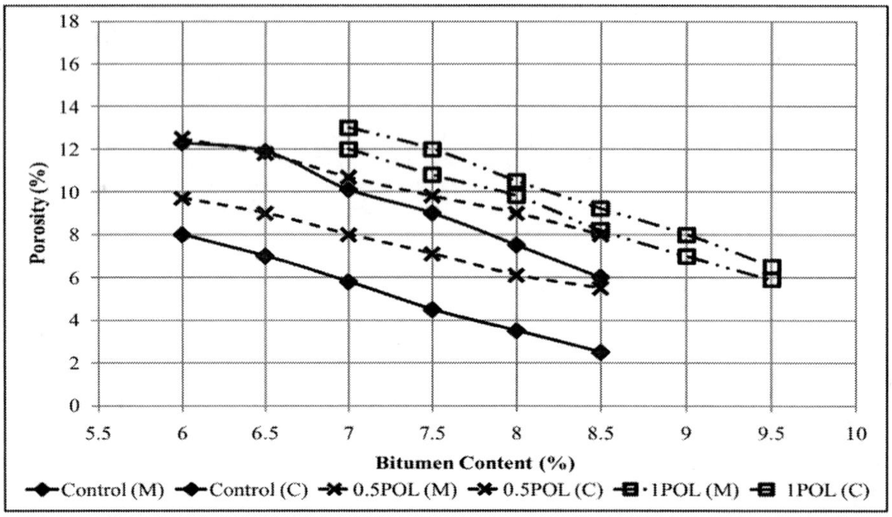

Fig. 5. Calculated and Measured Porosity vs. Bitumen Content for Control and POL Fibres.

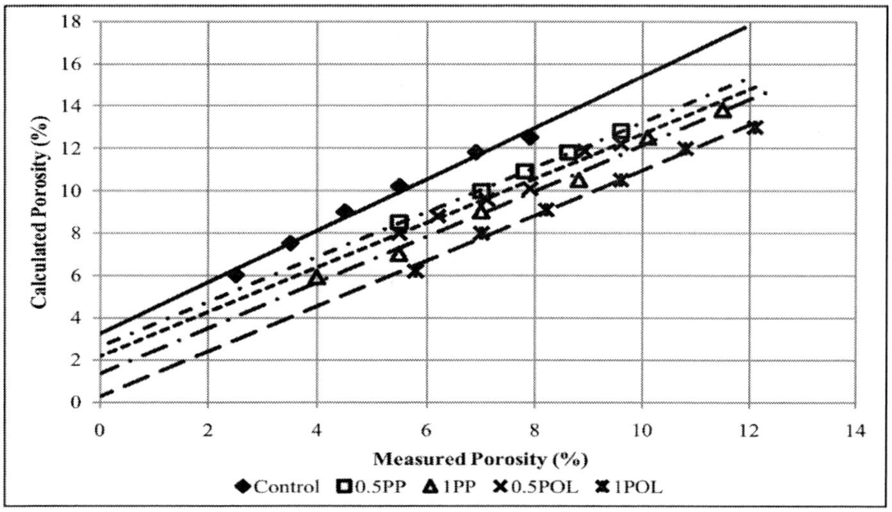

Fig. 6. Calculated vs. Measured Porosity for Different Mixes.

The indirect tensile strength (ITS) ratio is effectively an indication of the amount of strength loss due to the effect of water. The variation of the indirect tensile strength ratio and the degree of saturation is shown in Fig. 7. The ITS ratio shows it decreasing with increasing degree of saturation for all the mixes. The lines obtained are rather parallel to one another indicative that the decrease is somewhat similar. The fibre-modified mixes exhibited higher ITS ratios of around 11-25% over the control mix.

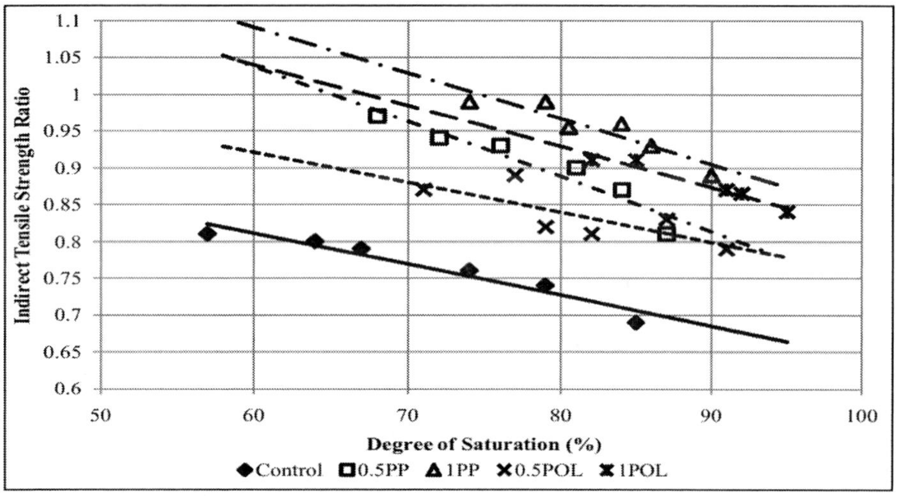

Fig. 7. Calculated vs. Measured Porosity for Different Mixes.

It is appropriate to be reminded that the fibre incorporated mixes had higher porosity and permeability than the control mixes that will permit easier access to water and increase the potential for stripping. It may thus appear that the more viscous binder of the fibre-incorporated mixes had a better cementing and adhesive property at the binder-aggregate interface that resulted in a reduction in stripping. It is believed that de-bonding may not have been solely responsible for the decrease in wet tensile strength values but other moisture damaging factors such as binder matrix softening may have been responsible as well.

7 Conclusion

Based on this study, the following conclusions can be drawn:

i. Changes in both the cohesive properties of the bitumen and the adhesion of the bitumen to the aggregate surfaces may occur as a result of exposing the bituminous mixtures to moisture. Polymer fibre incorporation into bituminous mixtures helps reduce the high level of moisture damage that was noted from the control mix. The polyester fibre modified mixes also showed lower moisture susceptibility than those of the polypropylene mixes at the same fibre concentration. However, the 0.5% fibre concentrated mixes showed better resistance to water damage than that at the 1% concentration.

ii. It is important to remember that mixes with polymer fibres had greater bitumen content and yet greater void contents than the control mix. Regarding resistance to moisture damage, these two parameters would be expected to oppose each other. Suffice to say that the additional bitumen in the fibre mixes increased the film thickness on the aggregate particles thus affording additional protection from moisture. The 0.5% fibre concentration may have provided enough reinforcement across the plane of failure in the mixtures while the 1% fibre concentration may have far too high void contents that allowed for more water penetration into the mixtures.

iii. The incorporation of polymer fibres in bituminous mixtures also acts to decrease the moisture sensitivity of the bitumen to aggregate bonding. This may be due to the strengthening of the wetted binder matrix which promote both adhesion and cohesion retention.

References

1. J. Ahmad, N.I.M. Yusoff, M.R. Hainin, M.Y.A. Rahman, M. Hossain, Constr. Build. Mater., **50**, 567 (2014)
2. M.Y. Aman, Z. Shahadan, M. Tamin, N. Raihan, Research Journal of Applied Sciences, Engineering and Technology, **9**, 797(2015)
3. A. Diab, Z. You, R. Ghabchi, M. Zaman, J. Mater. Civil. Eng., **27**, 04014254 (2014)
4. M.H. Nahi, I. Kamaruddin, M. Napiah, Appl. Mech. Mater., **567**, 539 (2014).
5. H.F. Haghshenas, A. Khodaii, M. Khedmati, S. Tapkin, Constr. Build. Mater., **75**, 488 (2015)
6. P. Jaskula, J. Judycki, Proceeding Advanced Characterization of Asphalt and Concrete Materials, 76 (ASCE, 2014)
7. Y.R. Kim, I. Pinto, S.W. Park, Constr. Build. Mater., **29**, 386 (2012)
8. J.L. Lambert, *Investigation into Asphalt Concrete Material and Volumetric Properties that Promote Moisture Damage* (University of Manitoba, Winnipeg, 2013)
9. M.H. Nahi, I. Kamaruddin, A. Ismail, R.A.E. Al Mansob, J. Appl. Sci., **14**, 2730 (2014)
10. J.F. Su, J. Qiu, E. Schlangen, Polym. Degrad. Stabil., **98**, 1205-1215 (2013)
11. R. Xiong, S. Chen, B. Guan, Journal of Highway and Transportation Research and Development, **30**, 28 (2013)
12. J. Yi, S. Shen, D. Wang, D. Feng, Y. Huang, J. Test. Eval., **44**, 856 (2016)
13. M.H. Nahi, A. Ismail, A.K. Anffm, J. Appl. Sci., **11**, 2562 (2011)

Fracture Behaviours in Compression-loaded Triangular Corrugated Core Sandwich Panels

N.Z.M. Zaid[1,*], *M.R.M.* Rejab[1], *A.F.* Jusoh[1], *D.* Bachtiar[1], *J.P.* Siregar[1], and *Zhang* Dian ping[2]

[1]Faculty of Mechanical Engineering, Universiti Malaysia Pahang, 26600 Pekan, Pahang, Malaysia
[2]School of Mechanical Engineering, Ningxia University, Yinchuan 750021, China

Abstract. The failure modes occurring in sandwich panels based on the corrugations of aluminium alloy, carbon fibre-reinforced plastic (CFRP) and glass fibre-reinforced plastic (GFRP) are analysed in this work. The fracture behaviour of these sandwich panels under compressive stresses is determined through a series of uniform lateral compression performed on samples with different cell wall thicknesses. Compression test on the corrugated-core sandwich panels were conducted using an Instron series 4505 testing machine. The post-failure examinations of the corrugated-core in different cell wall thickness were conducted using optical microscope. Load-displacement graphs of aluminium alloy, GFRP and CFRP specimens were plotted to show progressive damage development with five unit cells. Four modes of failure were described in the results: buckling, hinges, delamination and debonding. Each of these failure modes may dominate under different cell wall thickness or loading condition, and they may act in combination. The results indicate that thicker composites corrugated-core panels tend can recover more stress and retain more stiffness. This analysis provides a valuable insight into the mechanical behaviour of corrugated-core sandwich panels for use in lightweight engineering applications.

1 Introduction

A corrugated-core sandwich panel consist of a corrugated sheet positioned between two thin skins is its high strength-to-weight ratio [1]. The corrugated-core serves to keep the skins apart as well as stabilising the component by resisting vertical forces. This design also enables the complete structure to act as a single thick plate as a virtue of its high shear strength [2]. The second feature shows outstanding ventilation characteristics, avoiding humidity retention that is common in cellular core materials (e.g. polymeric foams, honeycombs) [3-5].

The failure mechanisms studied in composites can vary greatly from thin to thick ones. For composites, the notion of being thick should be described in a manner more than that of uniform materials. Even though the former governs the probability of global buckling at the structural level, the later represents the number of plies in the laminate and would effect its failure behaviour at a microscopic scale, which is more unpredictable and unavoidable. This can be known from the fact that thick composites frequently improve more complex fracture modes than those in thin composites. The thick composite is laminated by the relative number of plies. Both the number of ply interface and the whole interfacial area are

* Corresponding author: nzakiahmz@gmail.com

raised in proportion to the thickness, both increasing the chance of interfacial defects and delamination. From a statistical phenomenon point of view, the possibility of containing more and larger defects increases when a composite becomes thick. The compressive strength tends to decrease with an increased thickness, despite the fact that an accurate experimental validation is difficult due to the wide scatter in the results [6]. Compressive failure is a result of fibre instability sensitivity to fibre misalignment. Thicker composites contain more layers of fibre and a higher chance of greater misalignment [7].

In this paper, corrugated-core sandwich panels based on three varying materials are studied under the compression loading. Particular attention will focus on establishing the fracture modes during compression loading.

2 Corrugated-core sandwich panels

Three types of material have been investigated in the study, an aluminium alloy 2024-O (AL), a woven carbon fibre reinforced plastic (CFRP) and a fabric-type glass fibre reinforced plastic (GFRP). In order to get a repeatable and uniform corrugation, the sheets were prepared using a special 45° triangular profile with an effective surface of fabrication of 210 mm by 240 mm. Fig. 1 shows the mould, which was made from mild steel, with a nominal cell height of 10 mm height and cell length of 20 mm.

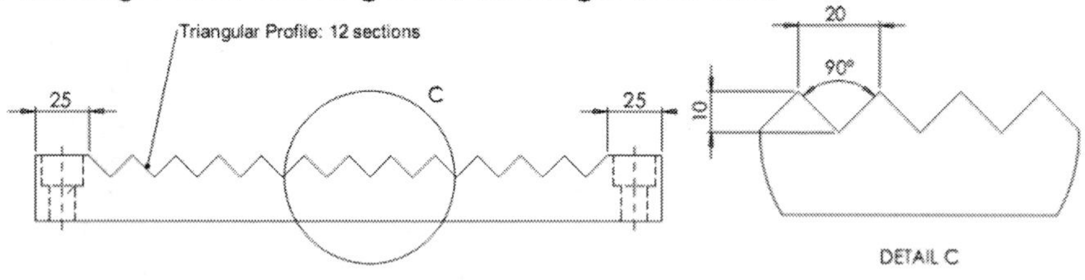

Fig. 1. Assembly drawing of the corrugated mould.

The composite corrugation was fabricated using a corrugated mould by curing it using compression mould machine. The sample was removed from the mould when the hot press cooled to a temperature below 60°C. The corrugated core was then bonded between the two skins using a strong epoxy adhesive (Araldite 420 A/B). The corrugated sandwich sheet was then heated in an oven at a temperature of 120°C for approximately 1 hour in order to cure the adhesive. The structure was cut into test samples with dimensions of 100mm (length) and 25mm (width).

Fig. 2 shows a typical corrugated-core sample consisting of five repetitions of a triangular unit cell. In this investigation, the unit cell was based on a triangular profile. The geometric parameters plotted in the Figure 2 are as follows: θ and β are the internal angle of the unit cell; T is the height of the core; H_U and H_L are the upper and lower thicknesses of the skins, correspondingly; H is the average thickness of the inclined core members, i.e. the wall thickness; x is the length of the unit cell; and w is the width of the sample. In this study, the value of x is 20 mm length while θ and β are set to 45° and 90°, respectively. The width of all of the specimens was 25 mm.

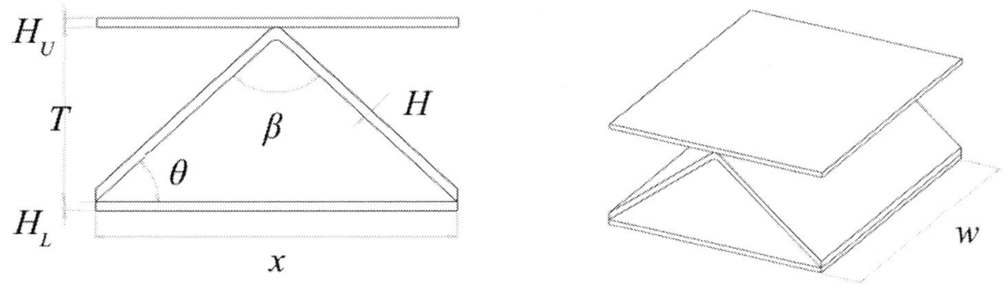

Fig. 2. Geometry of the corrugated-core sandwich panel.

3 Experimental work

Compression tests on the corrugated-core sandwich panels were operated using an Instron series 4505 testing machine. All the test specimens were make in a rectangular form (~ 25mm width x 100mm length), with different web thicknesses.

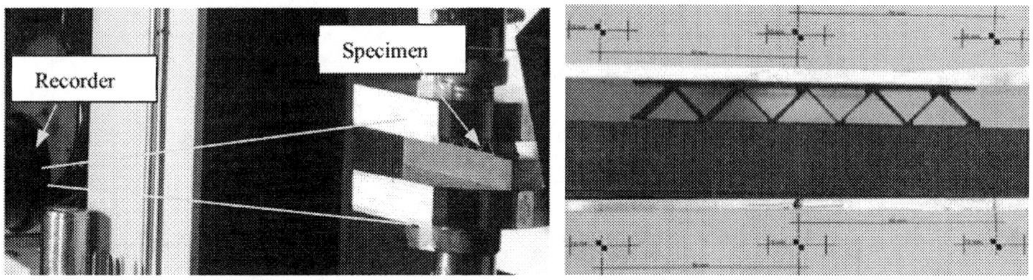

Fig. 3. Compression test set-up with the displacement measurement system.

Fig. 3 displays a specimen positioned between the platens of the test machine, with uniform lateral compression at a quasi-static loading rate of 1 mm/minute. A high-resolution displacement measurement system was used to gather displacement data as well as capturing the resulting images. The load-displacement trace was recorded until the specimen was entirely crushed.

4 Results and discussion

The experiments showed that there is a fundamental difference in the behaviour of the different panels and their respective responses are considered separately below.

4.1 Aluminium alloy corrugated-core.

A typical load-displacement trace for an AL corrugated-core sandwich panel is showed in Fig. 4. The deformed shapes were recorded by high speed video camera. Upon loading, the specimen shows a nonlinear response during the initial loading stage. This may partly be attributed to effects associated with the machine compliance and, perhaps more significantly, to the fact that closer inspection revealed that both skins were not fully parallel to each other. As a result, the initial deformation of the panel is influenced by the flattening of these skins. After this initial "stiffening", the specimen responds in a linear fashion (as in point A) to the first peak in the load, with deformation being symmetrical about the axis of loading. After reaching the maximum peak load, one of the struts in the corrugated-core was partially bent and as a result, the overall stiffness of the specimen decreased, the load required to further deform the panel steadily decreases due to the transmission of localised buckling in the corrugated-core. The response then becomes progressively nonlinear (point B). This is followed by a sudden drop in the applied load as the panel loses stability due to plastic buckling.

At point C, the corrugated-core takes on a trapezium shape, and the applied load increases, due to the interaction between the surfaces of the deformed struts and the flat skins. Finally, point D shows the corrugated-core has been completely densified, where both the skins and corrugated-core are flattened and in some cases the sides of edge are debonded.

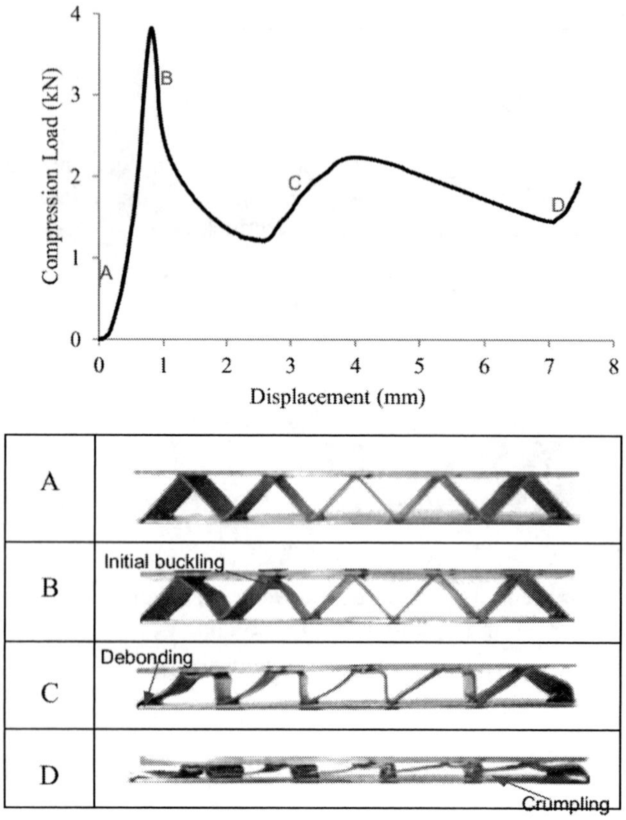

Fig. 4. Typical load-displacement curve showing progressive damage development in an AL corrugated-core sandwich panel with five unit cells.

4.2 Glass fibre-reinforced plastic

The GFRP corrugated-core sandwich panel exhibited a brittle behaviour, involving crushing under the compressive load as presented in Fig. 5.

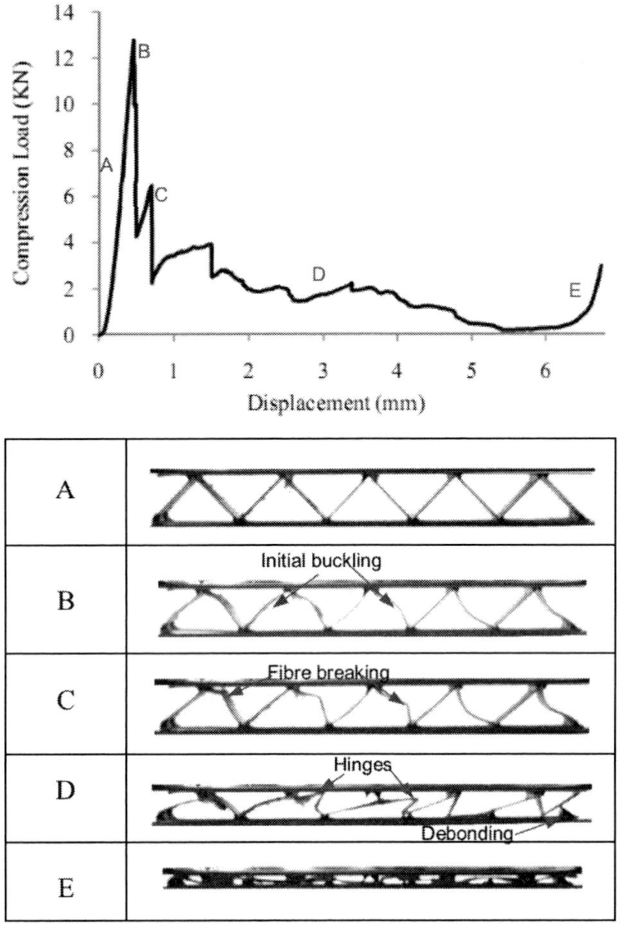

Fig. 5. Typical load-displacement curve with progressive damage development in a five unit cell GFRP corrugated-core sandwich panel.

At the point B, initial cell wall buckling was visible, followed by a compression fracture at the peak load. The drop in the load-displacement diagram is steeper than in the AL sample due to the different failure mechanisms involved (points C and D): here, failure involves the formation of hinges in the middle of the cell wall after fibre fracture, debonding of the adhesive between the surfaces and continuous crushing after failure at the cell wall from the upper to lower skins. Finally, at point E the corrugated-core is completely crushed, where both the skins and the core are flattened.

Fig. 6 shows the relationships between the compression strength and stiffness for differe nt GFRP cell wall thicknesses.

DOI: 10.1051/matecconf/20167801041

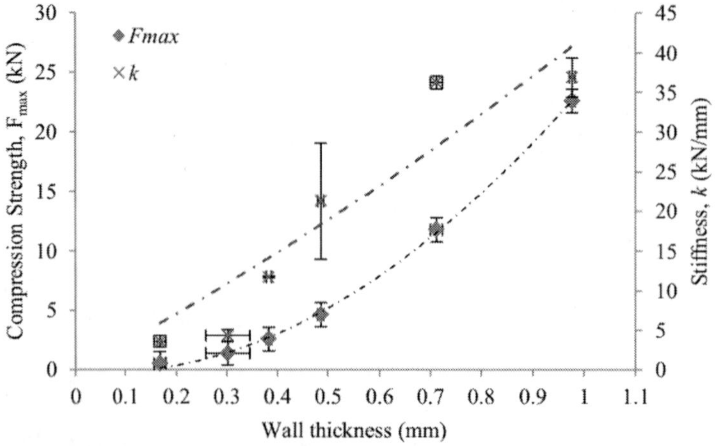

Fig. 6. Compression strength and stiffness responses to cell wall thickness of GFRP corrugated-core sandwich panels.

Table 1 displays that the failure modes and stiffness's of the 10-ply GFRP panel are quite different compared with those in the thinner specimens. The stiffness of the 10-ply sample is similar to 7-ply specimen, where the changes of the corrugation shape from sharp edges (triangular profile) to sinusoidal were observed following the fabrication process. Initial delamination damage on the wall edge associated with the demoulding process also may have contributed to the stiffness reduction.

Table 1. Post-failure examinations of GFRP corrugated-core in different cell wall thickness using optical microscope.

Observation	Magnification (20x)	Magnification (40x)
3-ply of GFRP with an average *H*=0.30mm, failure starts with buckling followed by fibre breakage	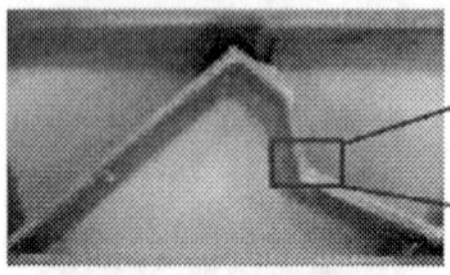	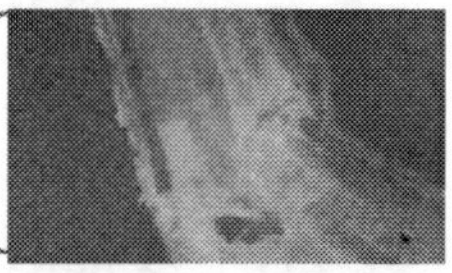
5-ply of GFRP with an average *H*=0.49mm, failure starts with buckling followed by hinges and fibre breakage	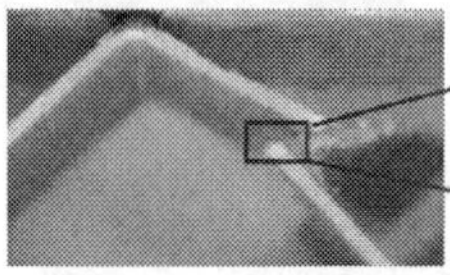	
7-ply of GFRP with an average *H*=0.71mm, failure starts with buckling followed by fibre breakage and delamination		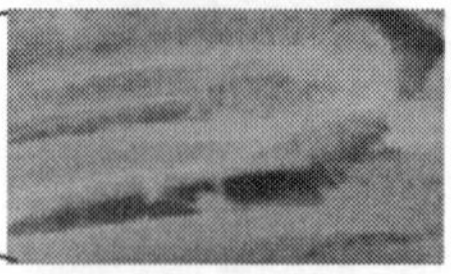
10-ply of GFRP with an average *H*=0.98mm, failure mode is dominated by delamination damage from the wall edges	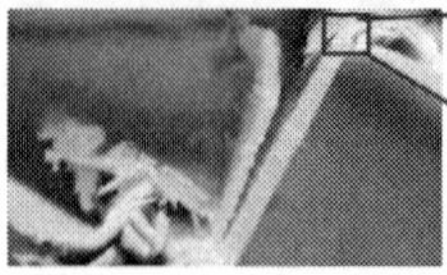	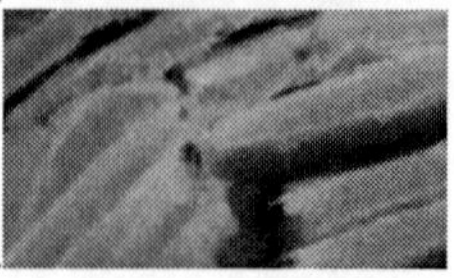

4.3 Carbon Fibre-Reinforced Plastic

The failure mechanisms in the CFRP corrugated-core under compression loading are shown in Fig. 7 where initial failure was dominated by cell wall buckling. The load-displacement plot is steeper than that for the AL corrugated-core and the post-damage failure modes are similar to the GFRP corrugated-core.

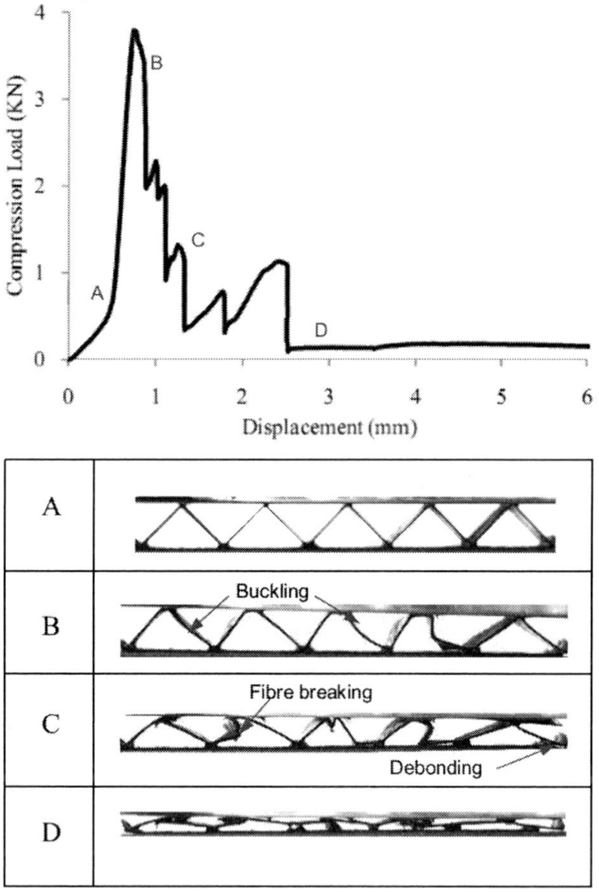

Fig. 7. Typical load-displacement curve with progressive damage development in five unit cells of CFRP corrugated-core sandwich panel.

Fig. 8 shows the relationship between the compression strength and stiffness for the different cell wall thicknesses. Regarding the stiffness response, there is considerable scatter in the data due to significant variations in the cell wall thickness. This reduced the rigidity of the strut members.

Table 2 shows the post-failure images of the fractured CFRP specimens where it can be seen that in addition to the fibre instability failure mode, other damage mechanisms such as matrix cracking and splitting are present. In most specimens, especially the thicker samples, final fracture (delamination) was located near the wall edge. In the 6-ply CFRP, initial debonding affected the global stiffness of the panels. It is believed that defects during the fabrication process, such as fibre waviness, variations in the fibre volume fraction may also have contributed in reducing the stiffness of the panels.

 DOI: 10.1051/matecconf/20167801041

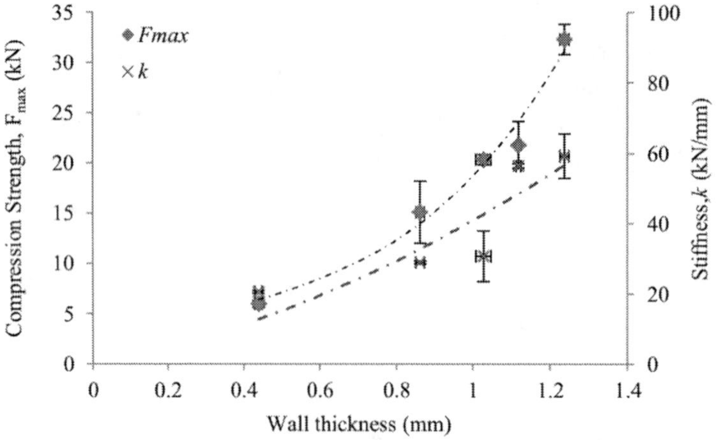

Fig. 8. Compression strength and stiffness responses of CFRP corrugated-core sandwich panels as a function of cell wall thickness.

Table 2. Post-failure examinations of CFRP corrugated-core in different cell wall thickness using optical microscope.

Observation	Magnification (20x)	Magnification (40x)
2-ply of CFRP with an average *H*=0.44mm, failure starts with buckling followed by hinges and fibre breakage		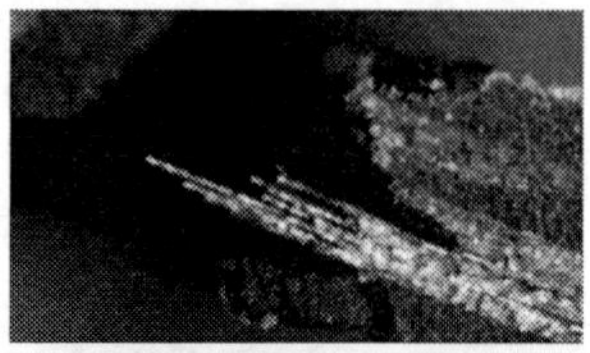
4-ply of CFRP with an average *H*=1.03mm, failure starts with buckling followed by delamination at the edges	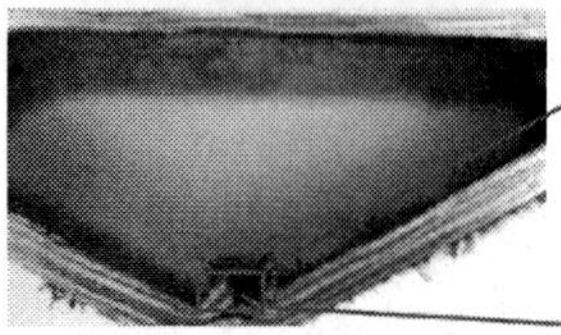	
6-ply of CFRP with an average *H*=1.24mm, fibre delamination occur in cell wall and debonding suddenly takes place.		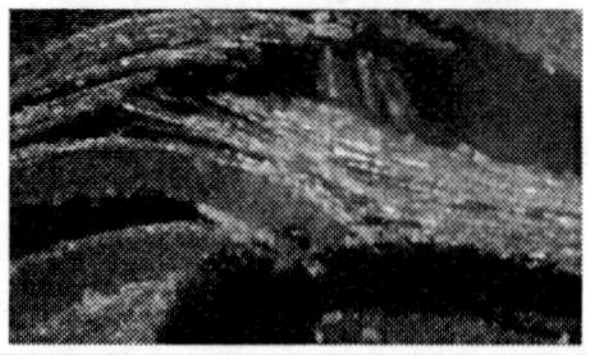

5 Conclusions

The compressive responses and failure mechanisms of corrugated-core sandwich panels with three different materials subjected to uniform lateral compression were investigated both experimentally and numerically. Failure modes for different cell wall thickness of GFRP and CFRP corrugated-core panels also have been examined in this study. From the observations, it is found that initial failure in the panel was dominated by the instability of the struts when the cell wall starts to buckle. Four modes of failure were described in the results: buckling, hinges, delamination and debonding. Each of these failure modes may dominate under different cell wall thickness or loading condition, and they may act in combination. The results indicate that thicker composites corrugated-core panels tend can recover more stress and retain more stiffness. However, results also showed that thicker panels (10-ply GFRP and 6-ply CFRP) are less stiff. This is not because of the thickness effect but mainly because of the geometry changes at the core edges from the triangular shape to sinusoidal during the lamination process.

The authors are grateful to the Ministry of Education Malaysia for funding this research (FRGS/1/2014/TK01/UMP/02/3).

References

1. H.G. Allen, *Analysis and design of structural sandwich panels* (Pergamon, Oxford, 1969).
2. F. Cote, V. S. Deshpande, N. A. Fleck, A. G. Evans, Int. J. Solids Struct., **43**, 6220 (2006).
3. M.R.M Rejab, W.J. Cantwell, Compos. Part B-Eng., **47**, 267 (2013)
4. Y.C. Zhang, S.L. Zhang, Z.L. Wang, Adv. Mat. Res., **217-218**,1584 (2011)
5. G.V. Movchan, Russian Aeronautics, Iz. VUZ, **50**, 439 (2007)
6. S. Fisher, K. Drechsler, S. Kilchert, A. Johnson, Compos. Part A-Appl. S., **40**, 1941 (2009)
7. J. Lee, C. Soutis, Compos. Part A-Appl. S., **36**, 213 (2005)

Investigation of Thermal Behavior for Natural Fibres Reinforced Epoxy using Thermogravimetric and Differential Scanning Calorimetric Analysis

F.A. Fauzi[1,*], *Z.* Ghazalli[1], *J.P.* Siregar[1], and *C.* Tezara[2]

[1]Faculty of Mechanical Engineering, Universiti Malaysia Pahang, 26600 Pekan, Pahang, Malaysia.
[2]Department of Mechanical Engineering, Faculty of Science, Technology, Engineering and Mathematics, INTI International University,71800 Nilai, Negeri Sembilan, Malaysia.

Abstract. This paper presented the research works on the investigation of the thermal behavior of the natural fibres; i.e. pineapple leaf fibre, kenaf fibre and mengkuang fibres reinforced epoxy. The thermogravimetric analysis and differential scanning calorimetric analysis were used to measure the thermal behavior of the treated and untreated pineapple, kenaf and mengkuang fibres reinforced epoxy. The samples for both analysis were subjected to maximum temperature 600°C at the heating rate of 10°C/min. The results showed that the treated fibres show higher maximum peak temperature as compared to the untreated fibres. Additionally, the glass transition temperature showed a lower value for all treated fibre. It can be concluded that investigation of thermal properties of these natural fibres could improve the utilization of natural fibre composites in various applications i.e. sports applications.

1 Introduction

Natural fibre reinforced epoxy is a relatively new material in the sports and recreation use as compared to the other synthetic materials. However, the usage of natural fibre composites are relatively high in automotive and construction applications [1]. Due to the increasing interest in green technology, the usage of natural fibres in polymer composites in the form of sports and recreation gaining its popularity [2, 3]. Researchers conducted various studies on many natural fibres such as hemp, jute, sisal, kenaf and pineapple leaf fibre (PALF) [4-6]. The differences in chemical composition make it essential to determine their thermal properties [7, 8]. One of the methods employed in studying the thermal properties of polymeric materials is thermogravimetric analysis (TGA). Additionally, differential scanning calorimetric (DSC) analysis is critical in determining the melting point and glass transition of a polymer composite.

Many researchers presented their works on the degradation of natural fibres with TGA. The thermal stability of various materials i.e. natural fibres has been identified by their degradation temperature [7, 9, 10]. Besides that, the DSC is also one of the necessary

* Corresponding author: farahayunifauzi@gmail.com

analysis to determine the chemical properties of the materials. According to Vilaplana et al. [11], glass transition somehow shows the transition between the glassy and the elastic state of a polymer. PALF and kenaf fibre have been widely discussed by many researchers [12-14]. However, least studies have been conducted on mengkuang fibre (*Pandanus Tectorius*) [15]. 12 different varieties of pineapple leaf fibre used in a study by Neto et al. [6] showed mechanical properties is inversely proportional to the diameter of the fibre bundle and lignin content. Yousif et al. [16] have studied the flexural properties of untreated and alkali treated kenaf fibre which results in improvement for the treated kenaf fibre.

In this work, the objective is to investigate the thermal properties of PALF, kenaf fibre, and mengkuang fibre reinforced epoxy under TGA and DSC analysis. A thorough study of the thermal properties of the untreated PALF, kenaf fibre, and mengkuang fibre reinforced epoxy are presented. Results of thermal properties of the PALF, kenaf and mengkuang composites were addressed. Finally, the study concluded in the last section.

2 Methodology

2.1 Materials selection process

The treated and untreated (UT) short PALF, kenaf and mengkuang fibres were used in this experiment. The treated fibres were soaked in 1% and 5% sodium hydroxide (NaOH) concentration for 1 hour, respectively. Then, these fibres were rinsed with tap water to remove all traces of the NaOH. The fibres were left to dry in aging oven for 24 hours at 90°C.

2.2 Composites preparation

The hand lay-up technique is used to prepare the composites. The epoxy resin (DER 331) and hardener (Jointmine 905-3) was mixed with the ratio of two to one (2:1) and stirred with the short fibres. The resin is then poured into the 200 mm x 200 mm x 3 mm mold (see Fig. 1). The composites were left for curing at room temperature for 24 hours. The volume fraction used in this study for UT fibres were 5% and 20% volume fraction. Treated fibres are using 10% volume fraction for both 1% and 5% NaOH concentration.

Fig. 1. Mould with dimension 200 mm x 200 mm x 3 mm.

2.3 NaOH treatment

The PALF, kenaf, and mengkuang fibres were soaked in 1% and 5% concentration of aqueous NaOH, separately. These fibres were left in aqueous NaOH for 1 hour at room temperature. After that, the fibres were washed by using tap water, rinsed until there are no traces of any residual NaOH. Then, these fibres were left to dry in aging oven for 24 hours at 90°C.

2.4 Thermogravimetric analysis (TGA)

The TGA Q500 V6.7 Build 203 machine is employed to conduct the TGA. The samples weighing 5.0 mg were subjected to a maximum temperature 600°C. The heating rate of the samples was set to 10°C/min. The final residue of the samples was recorded on the set of degradation temperature. The weight loss of the samples has been documented in response to increasing temperature.

2.5 Differential scanning calorimetry analysis (DSC)

Differential Scanning Calorimetry analysis was operated by using DSC Q1000 V9.9 Build 303 machine. Seven samples with the weight of 2.30 mg were used and analyzed dynamically at a heating rate of 10°C/min. The maximum temperature was set up to 600°C under a nitrogen atmosphere.

3 Results and discussion

3.1 Thermogravimetric analysis (TGA)

The Fig. 2 (a) up to (c) showed the results of TGA curves of PALF, kenaf and mengkuang fibres, each. In Fig. 2(a), it revealed that the TGA curve for treated and untreated (UT) PALF. The specimens show two stages of weight loss process. The first phase of transition for 1% NaOH PALF shows temperature 18°C to 175°C with weight loss 10.7% (see Table 1). De Rosa et al. [17] have studied the thermal behavior of okra fibre and revealed that the initial weight loss happened due to the vaporization from the fibre. The second stage of weight loss occurred at 265°C to 595°C with the maximum peak (T_m) is 341°C. This phase indicated the weight loss due to decomposition of cellulose [17]. T_m for 5% PALF UT was 115°C in first transition and 344°C in the second transition. Meanwhile, the 20% PALF UT produce 9.2 % of residual weight after heating from 10°C to 600°C.

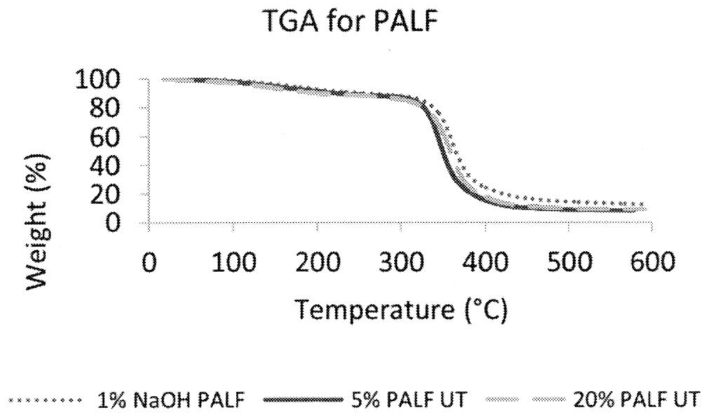

(a)

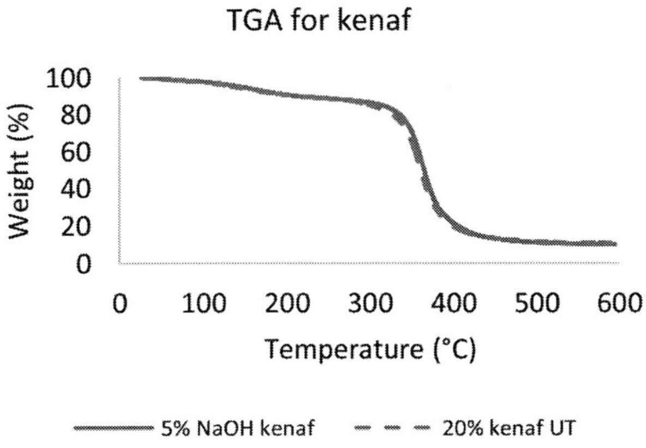

(b)

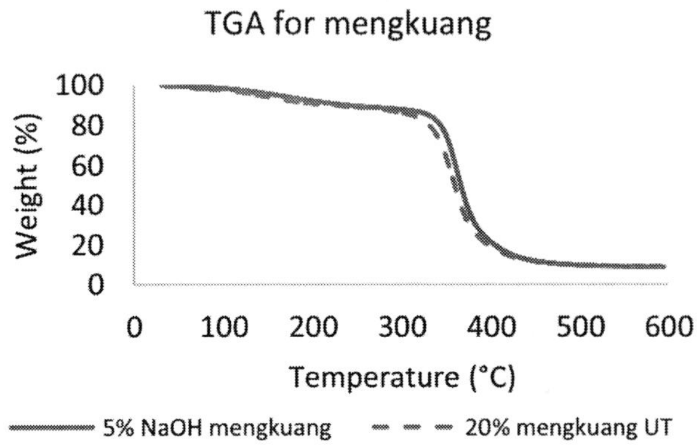

(c)

Fig. 2. Thermogravimetric analysis curves of a) PALF b) kenaf fibre c) mengkuang fibre

The Fig. 2(b) and 2(c) showed the TGA curves of kenaf and mengkuang fibres, respectively. The decomposition of these fibres befalls in two main stages. From these figures, it clearly shows that the fibre with alkali treated has a higher value as compared to untreated fibre. The 5% NaOH kenaf have a temperature range from 26°C - 165°C and 252°C - 595°C while the T_m for these steps were 124°C and 342°C, respectively. The 20% kenaf UT shows lower value with T_m 115°C and 340°C for respectively first stage transition and second stage transition. One possible explanation is that the chemical treatment on the surface of the fibre removed a certain population of hemicelluloses and lignin constituents enhanced the thermal stability of kenaf fibre [18].

The degradation temperature for mengkuang fibre of alkali treated fibre and untreated fibre was 345°C and 339°C, respectively. However, this value is greater than reported by Sheltami et al. [15] who have compared the degradation temperature between raw, alkali-treated and bleached mengkuang fibre. The reason is that the removal of hemicellulose component, lignin, and wax in the NaOH treatment of the mengkuang fibre. This situation showed a strong agreement with Sheltami et al. [15] that stated lower degradation temperature of mengkuang fibre due to remained hemicellulose after alkali treatment.

Table 1. Results of TGA on treated and untreated PALF, kenaf fibre and mengkuang fibre

Sample	No. of transition	Transition temperature (°C)			Weight loss at transition (%)	Residual weight (%) at 600°C
		T_i	T_m	T_f		
1% NaOH PALF	1	18	113	175	10.7	13.3
	2	265	341	595	76.1	
5% PALF UT	1	34	115	174	11.7	8.4
	2	275	344	595	79.9	
20% PALF UT	1	22	117	149	11.3	9.2
	2	230	338	595	79.5	
5% NaOH kenaf	1	26	124	165	11.5	10.2
	2	252	342	595	78.2	
20% kenaf UT	1	30	115	155	10.6	10.9
	2	240	340	595	78.5	
5% NaOH mengkuang	1	30	115	175	11.3	8.8
	2	275	345	595	79.9	
20% mengkuang UT	1	34	117	150	10.7	9.0
	2	230	339	595	80.2	

Table 1 shows the results of TGA for PALF, kenaf, and mengkuang fibres. The weight of the residual of 1% NaOH PALF is the highest; followed by kenaf and mengkuang fibre. The possible reason is that 1% NaOH concentration optimized the hemicellulose content in the fibre. It showed a strong agreement with the established works that the hemicellulose is responsible for the thermal degradation of fibre [13].

Additionally, the results of 5% NaOH concentration kenaf and mengkuang fibres showed that among these fibres, 5% NaOH kenaf fibre showed a greater value as compared 5% NaOH mengkuang fibre. The reason is that hemicellulose content in kenaf fibres are lower than mengkuang fibre [15, 19].

3.2 Differential scanning calorimetry (DSC)

Fig. 3 summarized the value of DSC analysis. The glass transition temperature (T_g) of treated PALF, kenaf fibre and mengkuang fibre showed are lower than the untreated (UT) fibre. The T_g of 20% mengkuang fibre UT outperformed the other samples. On the other hand, the T_g of 1% NaOH PALF is the lowest value. The reason is that the NaOH treatment reduced the formation of hydrogen. It presents a strong correlation with the established studies on kenaf-filled chitosan biocomposites [20]. The result from the study shows a greater maximum peak on a sample with higher hydrogen content.

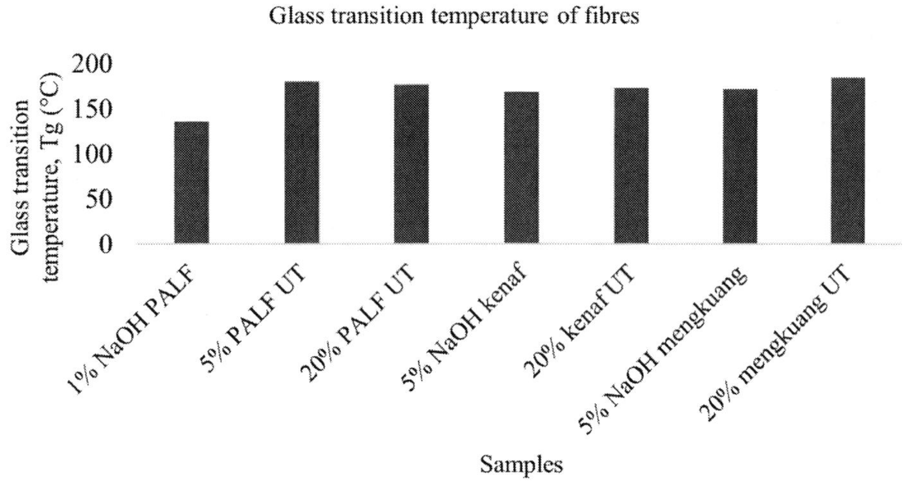

Fig. 3. Glass transition temperature, T_g of PALF, kenaf and mengkuang fibre.

4 Conclusion

The finding of this study showed that NaOH treated fibre specifically the kenaf and mengkuang produced higher maximum peak temperature than the UT fibres. However, for glass transition temperature, the treated PALF, kenaf and mengkuang fibre, the outcome showed lower value as compared to the untreated fibre. Further study needs to explore on physical properties testing such as the water absorption, melting point and density. This paper provide the DSC analysis results that can only provide glass transition temperature. The investigation of thermal properties for PALF, kenaf and mengkuang fibres could bring an improvement of using natural fibre composites for the usage of producing the sports equipment.

The author would like to thank the Universiti Malaysia Pahang for the equipment provided and Ministry of Science and Technology (MOSTI) for financial support under the E-Science grant (RDU 140502).

References

1. L. Brief, *Opportunities in natural fiber composites* (Lucintel LLC, Irving, 2011)
2. H. Alamri, I.M. Low, Polym. Test., **31**, 620 (2012)
3. Z.N. Azwa, B.F. Yousif, Polym. Degrad. Stabil., **98**, 2752 (2013)
4. A. Naughton, M. Fan, J. Bregulla, Compos. Part. B-Eng., **60**, 546 (2014)

DOI: 10.1051/matecconf/20167801042

5. T. Hojo, Z. Xu, Y. Yang, H. Hamada, Energy Procedia, **56**, 72 (2014)
6. A.R.S. Neto, M.A. Araujo, R.M. Barboza, A.S. Fonseca, G.H. Tonoli, F.V. Souza, Ind. Crop. Prod., **64**, 68 (2015)
7. Z.N. Azwa, B.F. Yousif, *Thermal degradation study of kenaf fibre/epoxy composites using thermo gravimetric analysis* (MPC, Sydney, 2013)
8. K. Anbukarasi, S. Kalaiselvam, Mater. Design, **66**,321 (2015)
9. K.P. Shadangi, K. Mohanty, Renew. Energ., **63**, 337 (2014)
10. J.P. Siregar, M.S. Salit, M.Z.A. Rahman, K.Z.H.M. Dahlan, Pertanika J. Sci. Technol., **19**, 161 (2011)
11. F. Vilaplana, A. Ribes-Greus, S. Karlsson, Anal. Chim. Acta, **604**, 18 (2007)
12. N. Saba, M. Paridah, M. Jawaid, Constr. Build. Mater., **76**, 87 (2015)
13. H.M. Akil, M.F. Omar, A.A.M. Mazuki, S. Safiee, Z.A.M. Ishak, A. Abu Bakar, Mater. Design, **32**, 4107 (2011)
14. O. Faruk, A. K. Bledzki, H.P. Fink, M. Sain, Prog. Polym. Sci., **37**, 1552 (2012)
15. R.M. Sheltami, I. Abdullah, I. Ahmad, A. Dufresne, H. Kargarzadeh, Carbohyd. Polym., **88**, 772 (2012)
16. B.F. Yousif, A. Shalwan, C.W. Chin, K.C. Ming, Mater. Design, **40**, 378 (2012)
17. I.M. De Rosa, J.M. Kenny, D. Puglia, C. Santulli, F. Sarasini, Compos. Sci. Technol., **70**, 116 (2010)
18. M.M. Kabir, H. Wang, K.T. Lau, F. Cardona, Compos. Part B-Eng., **43**, 2883 (2012)
19. M.A. Fuqua, S. Huo, C.A. Ulven, Polym. Rev., **52**, 259 (2012)
20. N.M. Julkapli, H.M. Akil, Polym.-Plast. Technol., **49**, 147 (2010)

MATEC Web of Conferences **78**, 01043 (2016) DOI: 10.1051/matecconf/20167801043
IConGDM 2016

Structural Analysis of Composite Laminates using Analytical and Numerical Techniques

Divya Sanghi[1,*], *Divyansh* Singh [1], and *Vijay* Gautam[1]

[1]Delhi Technological University, Delhi-110034, India

Abstract. A laminated composite material consists of different layers of matrix and fibres. Its properties can vary a lot with each layer's or ply's orientation, material property and the number of layers itself. The present paper focuses on a novel approach of incorporating an analytical method to arrive at a preliminary ply layup order of a composite laminate, which acts as a feeder data for the further detailed analysis done on FEA tools. The equations used in our MATLAB are based on analytical study code and supply results that are remarkably close to the final optimized layup found through extensive FEA analysis with a high probabilistic degree. This reduces significant computing time and saves considerable FEA processing to obtain efficient results quickly. The result output by our method also provides the user with the conditions that predicts the successive failure sequence of the composite plies, a result option which is not even available in popular FEM tools. The predicted results are further verified by testing the laminates in the laboratory and the results are found in good agreement.

1 Introduction

Composites have been utilized to solve challenging technological problems since a very long time, but the true potential of composite structures and technology began to be exploited with the breakthroughs in automobile and aircraft industry where every ounce counts and for each pound of weight cut, the world goes faster and greener by a mile.

The term "composite" refers to any material constituting two or more substances with significantly different properties - physical and chemical, which remain integrated yet distinct in the final load bearing structure. Normally they are used in the form of layers of woven fibres or flat tapes wetted in a resin system. These materials have advantages [1] in comparison to metal that they can be easily moulded into complex shapes and curves providing smooth aerodynamic finishes. In terms of strength; careful design, appropriate material selection and cautious fabrication of composites can yield much stronger, stiffer and lighter structures than similarly loaded metallic parts.

However the design of a fibre-reinforced composite structure is considerably more difficult than that of a metal structure, principally due to the difference in its properties in different directions. Every composite structure consists of different layers called plies which can have different material properties and thickness. Therefore a composite structure can have several stacking sequences which can differ with the number of layers, material

* Corresponding author: divyasanghi.93@gmail.com

properties and orientation of each layer. The design consists of various stages; the pre-design stage is the most crucial stage as it involves material selection and preliminary composite laminate analysis to determine possible stacking sequence of a laminate taking in account different variables. This data can be then further modelled into complex geometries for final detailed analysis.

Since most of the existing design tools are less developed for composites than for conventional materials, experimental testing is still widely used to validate design and analysis models. Some FEM [2] solutions do offer composite analysis capability, but the analysis is not simple as with isotropic materials and usually involve high computational time and cost. It is also required that the models for analysis are perfectly tuned and validated against the experimental data.

2 Methodology

The main objective is to develop a MATLAB program on the basis of classical laminate theory which can be utilized to determine a preliminary ply layup order for the initial design stage. The results will act as a feeder data for the further detailed analysis. The MATLAB program will provide a user friendly interface to evaluate different stacking sequences in a shorter time and predict the results with reasonable accuracy without experimental cost of fabricating and testing the samples. Further, we have compared the MATALB results with experimental and FEM values to determine percentage of error and analyse the accuracy of the code as well.

2.1 Analytical Analysis

A laminate is constructed by stacking a number of such laminas in the direction of the lamina thickness. The design and analysis of such laminated structures require the knowledge of stresses and strains in the laminate. Macro-mechanics is the approach used to predict the strength and stiffness of such composite structures, based on the properties of each lamina.

The MATLAB code [3] for macro-mechanical analysis of a composite laminate is constructed using a plane stress assumption for a lamina. The code uses equations based on the classical lamination theory to compute stress and strain variations across the laminate thickness. It uses engineering elastic constants of the material used in each lamina and its orientation to form individual stiffness matrices. Each of these matrices is then utilized to form the extensional, bending and coupling stiffness matrices of the entire laminate. For a given laminate which is subjected to in-plane loads [4]' namely axial and shear forces, and bending and twisting moments, the code can predict the values of local and global stresses, and local and global strains, at every co-ordinate within the laminate [5]· The deflection values for this code are then compared with the experimental and FEA results from ABAQUS.

2.2 Numerical Analysis

The analysis of the Composite Laminate was carried out in ABAQUS 6.14.1 using the following procedure:-

- The model was made using AUTOCAD in .dxf format and a planar shell of the sub-size dimension was created

- A pin of diameter 10 mm was then modelled using the 3D extrude option.
- The parts were appropriately partitioned for Mesh refinement and application of boundary conditions.
- The material properties were then added in the properties module, the material was given the properties of a Lamina and the fail stresses for each were given to separately calculate the failure of the different layers in the lamina.
- Further for a uniform mesh global and local seeds were created and then the assembly was made
- Now the boundary conditions are applied to various points, edges and partitions of the assembly
- The maximum loads as per the experimental values were applied.

2.3 Experimental Analysis

Along with FEA [6] analysis, testing of the laminates has been conducted so that the accuracy of the MATLAB program can be further verified.

2.3.1 Fabrication of Test Samples

One of the complexities involved in dealing with composite materials is the initial capital required to set up the advanced manufacturing techniques. Hence vacuum bagging using hand layup technique was used to fabricate [7] the samples for testing. Firstly E-glass fabric of 400 GSM and BD fibres was obtained locally along with the epoxy resin that would form the matrix of the composite. The epoxy selected was Araldite laminating epoxy due to its exceptional shear strength properties as well as being of medium viscosity allowing easy fibre penetration in vacuum moulding.

- The fabric was cut to shape and was laid out on an aluminium flat panel, whose top surface on which the moulding was to be done had been waxed to prevent the composite from sticking to the mould itself.
- Epoxy was then applied liberally to the layers and finally, after the last layer the peel ply (fibreglass coated teflon) was placed before the bleeder cloth. The edges of the aluminium panel was sealed using vacuum sealant tape and after connected a plastic bag with a vacuum pump the whole panel was sealed.
- The curing time of the epoxy is around 2 to 3 hours at 25 degree Celsius. After approximately 6 hours the plastic bag was opened and the laminate was removed off the panel, its GSM measured and labelled.
- The laminates were then cut according to **ASTM standard** for tensile tests using water abrasive jet cutting, utilising a cutting speed of 400mm/min.

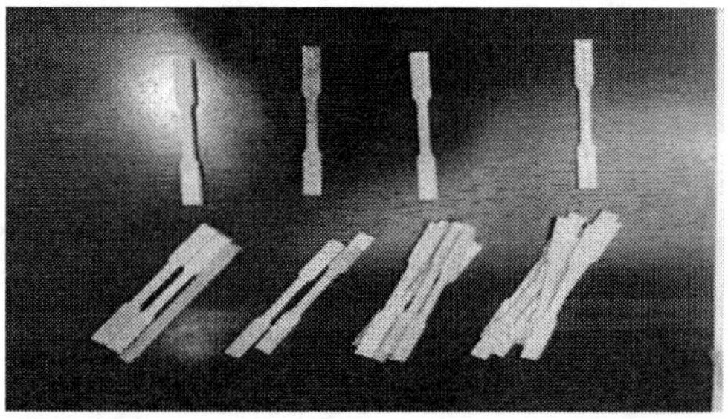

Fig. 1. Prepared samples for testing.

Table 1. Details of the tested samples.

S.No.	Sample ID	Ply Layup		Thickness(mm)
		No. of Layers	Orientation	
1	2-90-0	2	0/90	0.85
2	2-90-1	2	0/90	0.85
3	2-45-0	2	45/-45	0.85
4	2-45-1	2	45/-45	0.85
5	4-90-5	4	0/90	1.36
6	4-90-6	4	0/90	1.36
7	4-90-7	4	0/90	1.36
8	4-45-5	4	45/-45	1.36
9	4-45-6	4	45/-45	1.36
10	4-45-7	4	45/-45	1.36
11	6-90-2	6	0/90	1.81
12	6-90-3	6	0/90	1.81
13	6-90-4	6	0/90	1.81
14	6-45-2	6	45/-45	1.81
15	6-45-3	6	45/-45	1.81
16	6-45-4	6	45/-45	1.81

2.3.2 Testing of the samples

The samples fabricated were of three different thicknesses and fibre orientation of 90 degrees and 45 degrees. This would allow the MATLAB code to be verified for different fibre layups and also to help us in studying the effect of the number of layers and orientation on the laminate's strength. These samples were then tested under tensile loading using the machine which provided us with load versus displacement curves for each of the test samples. The maximum load capacity of the machine was 50KN which limited the thickness of the samples that could be tested and hence the maximum thickness was limited to 2.0 mm. The loading speed employed was 5 N/sec.

 DOI: 10.1051/matecconf/20167801043

Fig. 2. Tensile test on the UTM machine.

3 Results

All 16 samples were tested on the UTM under tensile tests. The next step was to apply force with successive increments on the samples and the corresponding deflection data was then recorded. The force versus Deflection graph is plotted for each sample and a comparison was drawn between the values of the experimental analysis, analytical analysis and FEA analysis.

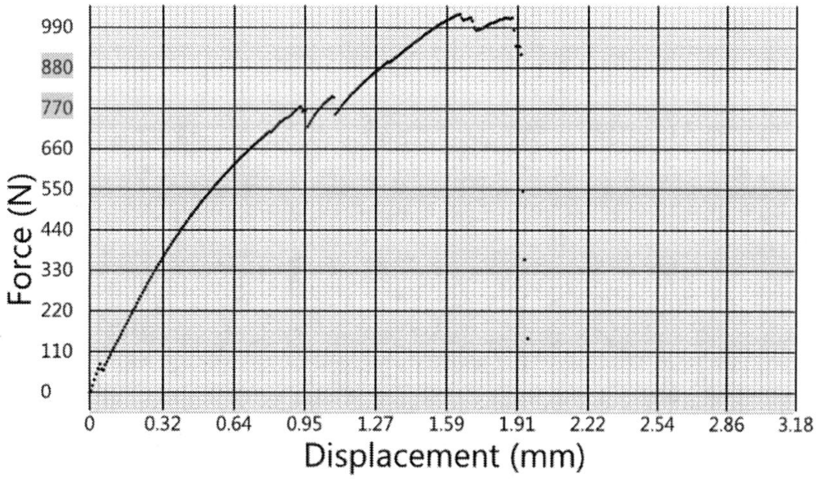

Fig. 3. Force versus Deflection Graph of a 2 layer sample.

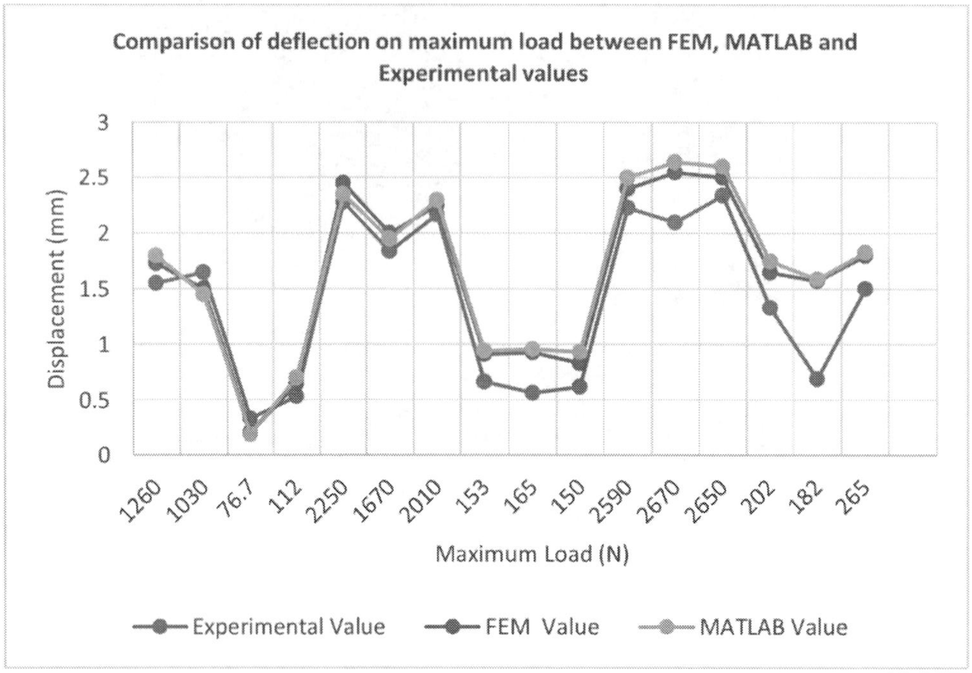

Fig. 4. Comparison of the FEA, MATLAB and Experimental Deflection Values.

The results showed that the FEA and MATLAB code values were consistent which proves the credibility of the MATLAB code. Also, it was found that the experimental results of the samples of same thickness were consistent. MATLAB and Experimental values showed a variance which was because of the defects caused during fabrication. It was also observed that the experimental and FEA results showed similar results for the region of failure at the maximum load.

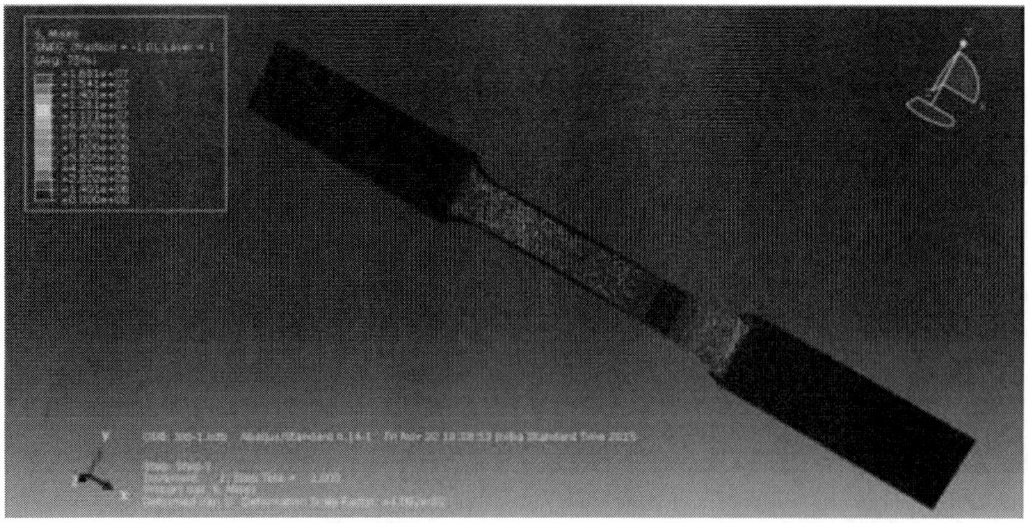

Fig. 5. Simulation results of the FEA analysis.

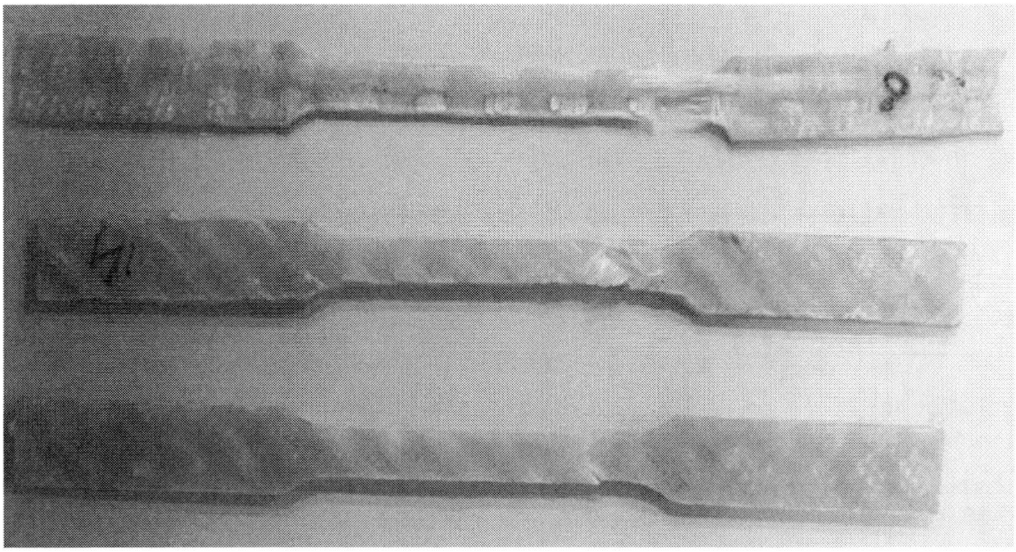

Fig. 6. Tested Samples under UTM machine.

4 Conclusion & future work

The study shows that the analytical calculation using the MATLAB code provides results with reasonable accuracy as compared to FEA and Experimental results. Further, it can be modified in a way that it can account for better accuracy of the experimental results. Work in underway so that the code can be implemented for 3-D figures and not just for 2-D laminates, such that more complex analysis of circular composite rods can also be done. Further Experimental tests need to be conducted on more samples varying in thickness and materials to analyse the scope of the MATLAB code.

References

1. http://www.autospeed.com
2. P. Roberts, *ABAQUS Lecture IV*, 2005
3. Y. Liu, *Introduction to Finite Element Methods*, *CAE Research Laboratory*, University of Cicinnati, 128, 2003
4. A.T. Nettles, NASA, 1994
5. G.H. Staab, *Laminar Composites,* Butterworth-Heinemann, Technology & Engineering, (1999)
6. H. Kim, *ABAQUS/CAE Tutorial*, 3 (2004)
7. http://alexpb.com

Simulation on Toxic Gases in Vehicle Exhaust Equipped with Modified Catalytic Converter : A Review

A.M. Leman[1,*], *Fakhrurrazi* Rahman[1] , *Afiqah* Jajuli[1] , *Winardi* Sani[1], *Supa'at* Zakaria[1] and *Dafit* Feriyanto[1]

[1] Faculty of Engineering Technology, Universiti Tun Hussein Onn Malaysia (UTHM), Parit Raja, Batu Pahat, 86400 Johor, Malaysia.

Abstract. Air pollution and global warming is a major issue nowadays. One of the main contributors to be the emission of harmful gases produced by vehicle exhausts lines. The harmful gases like NOx, CO, unburned HC and particulate matter increases the global warming, so catalytic converter plays a vital role in reducing harmful gases. Catalytic converters are used on most vehicles on the road today. This research deals with the gas emission flow in the catalytic converter involving the heat transfer, velocity flow, back pressure and others chemical reaction in the modified catalytic converter by using FeCrAl as a substrate that is treated using the ultrasonic bath and electroplating techniques. The objective of this study is to obtain a quantitative description of the gas emission in the catalytic converter system of automobile exhaust gas using ANSYS Software. The description of the gas emission in the catalytic converter system of automobile exhaust gas using ANSYS Software was simulated in this research in order to provide better efficiency and ease the reusability of the catalytic converter by comparing experimental data with software analysing data. The result will be expected to demonstrate a good approximation of gas emission in the modified catalytic converter simulation data compared to experimental data in order to verify the effectiveness of modified catalytic converter. Therefore studies on simulation of flow through the modified catalytic converter are very important to increase the accuracy of the obtained emission result.

1 Introduction

The development of new technology that aims in lowering environmental impact has become one of the most interesting and challenging issues in automotive engineering. Researchers must contribute new ideas to further decrease exhaust emission for automobiles. A serious issue that is always been debated among the environmentalists over the decades and recent years is air pollution. One of the main contributors is said to be the emission of harmful gases produced by vehicle exhaust lines [1]. Vehicles natural gas engine are facing a problem with the engine combustion as the in-cylinder produced high

* Corresponding author : mutalib@uthm.edu.my

unburned hydrocarbon (HC) and also carbon monoxide (CO) gas emission through the vehicle's exhaust [2]. The emission of nitrogen oxide (NO_x) is depending on cylinder fuel combustion; either the emission is increasing or decreasing [3]. However, the pollutant gaseous can be reduced either by the thermal or catalytic system.

Mansha M et al. stated that there are two methods that were usually being practiced to reduce the harmful adapted from the emission gases of automobiles exhaust which is by improving the automobile's engine technology and the composition of fuels. The new improvement of alternative fuels composition not only creating better combustions for vehicles exhaust but also might lower the emission [4]. The other method is after treatment process of exhaust gases which mostly vehicle used a catalytic converter. The catalytic converter is commonly installed as a device in the exhaust system of an automobile to convert toxic exhaust gas into less harmful gas. Catalyst metals are introduced in the catalytic converter to increase the rate of destruction of toxic gases such as carbon monoxide, nitrogen oxide and unburned fuel resulted from incomplete combustion of fuel [5]. It is a component that has become very effective in reducing the pollutants emission from the vehicles exhaust in the last 15 years. The chemical kinetics for the various heterogeneous reactions in a catalytic converter depend on several factors and have to be determined accurately for good qualitative and quantitative predictions by the model. The reaction rates are different for each wash coat formulation and cannot be determined a priori because (1) the kinetics are very sensitive to the wash coat formulation, (2) catalytic processes suffer from aging and deactivation, (3) the reaction path depends on the possible presence of species in real exhaust gas that are not present in controlled experimental studies, and (4) the kinetics is sensitive to feed composition and temperature [6]. The three pollutants of concern emitted from the spark ignition engines are unburned HC, CO and NOx (currently regulated). Unburned HC and CO are generated due to inefficient combustion while the thermal NOx is formed when nitrogen reacts with excess oxygen at higher temperature (>180^0K) in the combustion process and is the predominant form produced by Zeldovich Mechanism in combustion engines [7].

The effectiveness of catalytic converters is depending on the temperature gas temperature which is usually less effective during cold start [8] [9]. Thus, in order to shorten the time for the catalytic converter reaches light-off temperature, detailed analysis using computational software must be done.

2 Literature Review

2.1 Catalytic Converter

A catalytic converter is a device used to reduce the pollutant gaseous from the internal combustion of vehicles engine. It is placed between the tailpipe of vehicle exhaust and engine manifold. The effect of the combustion in the engine will produce pollutant gaseous which is flowing out of the engine to be pass through the tailpipe and undergo a chemical reaction with the catalyst within the porous wash coat in the converter to produce a harmless gaseous before the harmless gaseous is exposed to the atmosphere. The catalytic converter is a stainless steel container mounted somewhere along the exhaust pipe of the engine and inside the container is a porous ceramic structure through which the exhaust gas flows [10]. In most converters, the ceramic is a single honeycomb structure with many flow passages. The passages comprise of many shapes, including square, triangular, hexagonal and sinusoidal [1].

Early converters used loose granular ceramic with the gas passing between the packed spheres. Since it is difficult to keep the spheres in place, many converter developers opted for ceramic monolith which offers various advantages. Among these advantages are smaller volumes, lower mass and greater ease of packaging [11]. Catalytic Converters are important post-combustion after treatment devices mounted in the exhaust system of engines to reduce engine out exhaust emissions and are classified as Two-Way or Three-Way. Two-Way Catalytic Converter works on two gases, CO and unburned HC while the NOx is controlled through exhaust gas recirculation (EGR) and by retarding the ignition timing [4].

2.2 Catalyst Material

Catalysts used in catalytic converters are from various types of materials. Mostly developed catalyst is Platinum Group Metal as the main role for this group is to oxidise carbon monoxide (CO) and unburned hydrocarbon (HC) exist from the engine combustion. [12] state that the effectiveness of platinum (Pt) is under an oxygen-excessive condition which it is often being the metal of choice for the diesel application while for the vehicles that powered by petrol, platinum (Pt) and palladium (Pd) are suitable materials for catalyst development. Platinum (Pt) has several advantages to be highlighted which are; the high melting point, the interaction with poisons are limited to the metal surface such as sulphur compounds and platinum also can be efficiently recycled. However, there are recent studies that show the negative effects of the platinum (Pt) usage. Shahar and Majib (2007) [13] state that platinum (Pt) content of road dust which is partially soluble as the dust may dissolve in water, ground and also the food chain. On the other hand, Platinum Group Metal also has some side effects which can lead to asthma, hair loss effect, dermatitis and other critical health problem in human [14]. The installation of the converter using Platinum Group Metal as a catalyst is quite expensive in term of its fabrication process, which also produce exhaust combustion with a high concentration of plumbum or lead.

The stated issue lead the researchers to construct new material for catalyst development in order to replace the Platinum Group Metal (PGM). According to Bagus et al., (2015) [15] in their studies on the new development of catalyst using Manganese Coated Copper (CuMn), it is able to reduce the emission of carbon monoxide (CO). The decreasing of carbon monoxide pollutant gaseous is related to the changing of vehicles rpm. The analysed data taken from the experiment that has been run with the new development of catalytic converter shows that by increasing the cell numbers in the catalyst would reduce the carbon monoxide (CO) emission significantly. However, the usage of Manganese Coated Copper (CuMn) as a catalyst for the converter only highly effect in reducing the carbon monoxide (CO) gas emission. The other gaseous such as nitrogen oxide (NO_x) and unburned hydrocarbon (HC) is not being covered by this type of catalyst.

The usage of PGM is keep raising and demand also increasing which makes the concentration of these types of metal increased in the environment. According to Shahar and Majib (2007) [13], the concentration of Platinum (Pt) and Rhodium (Rh) has been taken upon road dust sample and the ratio range from the sample is in the range of 2.1 :1 to 7:1, where the ratio these elements in the catalytic converter usage is 7:1. Kalam et al., (2009) [2] has also come with new ideas in the development of catalytic converter that used the material of titanium dioxide (TiO_2) and cobalt oxide (CoO) as the catalyst material. The method is by using wire mesh substrate treatment immersed into a preparation of HCL solution. Titanium oxide (TiO) has its own properties as Bode et al., [16] state that the titanium is widely used in the automotive exhaust system as an oxidizing agent with low-cost fabrication. This new development of catalytic converter is tested with the original catalytic converter (OEM catco) that gives a positive feedback on the gaseous emission. The result concludes that the conversion efficiency of wire mesh catalyst and original catalytic converter (OEM catco) is 82% and 42%.

The usage of wire mesh in developing the substrate for the catalyst coating has its side effect. Even though the wire mesh will contribute to excellent mass and heat transfer with lower pressure drop, but the drawback of the wire mesh leads to the clogging effects because of high dust concentration, especially in solid fuel combustion. Another research was investigated by Kamariah et al., (2007) [17] using the woven stainless steel wire mesh to support the catalyst in reducing nitrogen oxide (NO_x) gas emission. The mixture of Silica quartz, SiO_2, and alumina, Al_2O_3 is used for the catalyst support. The area of woven stainless steel wire mesh (WSSWM) must be fully coated to avoid porosity on the catalyst surface as shown in Figure 1.

Fig. 1. Porosity on uncoated wire-mesh substrate.

In this study, a new type of catalytic converter based on nickel oxide (NiO)/ gamma-alumina (γ-Al_2O_3) materials has been selected to develop a catalyst to oxidise and reduce the pollutant gas emission. The advantage of the nickel type base catalyst is this material has high activity and firmness [22]. Nickel oxide catalyst (NiO) is a low-cost fabrication types, resistance to high temperature and domestically available in the market [18].

2.3 Emission Gas Analysis and Simulation

Simulation using software is quietly related with the mathematical models and codes parameter. According to Froment et al. [19] the use of the mathematical models in simulating models of catalytic reactors can give advantages where the formulation of such model is well documented. Kuipers et al., (1998) [20] support the above statement by stating that the development of the Computational Fluid Dynamics (CFD) codes is a useful introduction to the community. The evolution of the CFD Software might be used in various types of engineering field. Eriksson et al., (2004) [21] used the FLUENT Code to investigate the problems of particles that are overheating in a gas polymerization reactor. The investigation is done by the translation from the problems into a model of the moving gaseous particles.

3 Methodology

3.1 Catalyst Preparation

Nickel electroplating process is for depositing nickel onto FeCrAl. Electroplating is conducted on FeCrAl after new ultrasonic treatment. FeCrAl with the various sizes of 2 cm x 1 cm, acted as a cathode with the various sizes acted as a cathode while a nickel plate substrate acted as an anode with the size of 4 cm x 1 cm [22] (Sebayang et. al, 2012). Kalam et al., (2009) [2] prepared the catalyst by adding 10 gm of TiO_2 with sodium silicate in order to get 10% TiO_2 slurry before continuing with homogenization process that ends with drying process to get the catalyst slurry. The Same step is used by Syed et al., (2015) [23] by adding 10 gm Al_2O_3 with sodium silicate in order to get Al_2O_3 nanoparticles slurry. After the homogenization process takes place, the dip coating process is done before drying the catalyst in oven about 5 hours at 120^0C.

3.2 Mathematical Model for Simulation

A simulation for the gas emission flow distribution in the modified catalytic converter using ANSYS Software to be compared with the manual experiment. The simulation steps involved several previous steps such as developing of mathematical equation (thermodynamic) using for energy and mass balance, heat and mass transfer coefficient.

Thermodynamic reactions in one-dimensional system can be presented in a form of equations of mass balance of certain exhaust components and energy balance in gaseous phase and solid phase (on surface of catalyst walls). There is taken into account four exhaust components. Additionally, heat conduction can occur in the solid phase. These led to internal energy changes in the solid phase. The energy balance in the gas and solid phases can be expressed by Eq. 1 and 2, respectively [24]:

$$-\rho_g C_{\rho g} v_m \frac{dT_g}{dx} + \frac{4}{D_H} h\left(T_s - T_g\right) = 0 \tag{1}$$

$$\frac{\partial}{\partial x}\left(k_s \delta_w \frac{\partial T_s}{\partial x}\right) - h\left(T_s - T_g\right) - \sum_i \Delta H_{R,i} R_{i,s} = \delta_W \rho_S C_{PS} \frac{\delta T_S}{\delta T} \tag{2}$$

ρ_g and ρ_S = Gas density and solid phase density

$C_{\rho g}$ and C_{PS} = Constant pressure heat capacity in gas and solid phase, respectively in J kg^{-1} K^{-1}

v_m = Mean mass average velocity in $msec^{-1}$

h = Heat transfer coefficient in W m^{-2} K^{-1}

T_s and T_g = Temperature in gas and solid phases, respectively in K

δ_W = Reactor wall thickness in m and $\Delta H_{R,i}$ is enthalpy of reaction I in J mol^{-1}

3.3 Simulation Result

Refers to the result from the previous study, Mesut et al., (2013) [25] describe that the velocity for the exhaust emission gas using CFD analysis varies between 5.78 m/s and 0.92 m/s. The increasing of velocity as expected is because of the reduction in exhaust cross section. According to Mohan et al., (2013) [26] the simulation tested on the velocity of 22.6 m/s in the converter, it shows that the pressure change rapidly in the middle section of the converter while the pressure drop can be high which due to the inertial and viscous resistance of the porous media. Simon et al., (2010) [27] had simulated the pressure in catalyst and manifold of the vehicle exhaust in four selected design. The result concludes that catalyst yields a pressure drop about 10kPa where the four types of manifolds generate total pressure drop around 14.6, 12.2, 10.8 and 14.3 kPa.

4. Conclusion

The modification of the catalytic converter using FeCrAl as a substrate coated with γ-Al2O3 by using the ultrasonic and electroplating technique may improve the conversion of the pollutant. The data gained from the emission testing will be compared to the simulation of ANSYS FLUENT software in order to get the correspondent relation between the data. On the other hand, the simulation also will determine the effectiveness of the coated materials to the exhaust emission control and also influence to the lifetime of the materials.

This work was financially supported by Ministry of Education through FRGS Vote 1612 and ORICC, Universiti Tun Hussien Onn Malaysia (UTHM) for the management of the research grant and Centre for Graduate Studies – UTHM.

References

1. A.K.M. Mohiuddin, M. Nurhafez, International Journal of Mechanical and Materials Engineering, **2**, 1 (2007)
2. M.A. Kalam, H.H. Masjuki, M. Redzuan, M.A. Fuad, M. Mohibah, K.H. Halim, M.I. Mahlia, A. Ishak, A. Yusoff, Sadhana, Acad. P. Eng. S., **34**, 467 (2009)
3. S. Chuepeng, American Journal of Applied Sciences **9(8)**, 1225 (2012)
4. M. Mansha, A.H. Qureshi, I.A. Chaudry, E.M. Shahid, J. Qual. Technol., **9(1)**, 57 (2013)
5. K. Hakan, Ind. Lubr. Tribol., **53(6)**, 237 (2001)
6. S.R. Koteswara, R. Imam, K. Ramanathan, S. Pushpavanam, Ind. Eng. Chem. Res., **48(8)**, 3779 (2009)
7. M. Mansha, E.M. Shahid, A.H. Qureshi, Pak, J. Engg. & Appl. Sci., **11**, 114 (2012)
8. N. Gallopoulos. SAE Technical Paper, **920721** (1992)
9. D. Boam, T. Clark, K. Hobbs.SAE Technical Paper, **950930** (1995)
10. V. Ganessan, *Internal Combustion Engine* (McGraw Hill, New Delhi, 2004)
11. R.M. Heck, R.J. Farrauto, *Catalytic Air Pollution Control* (John Wiley & Sons, New York, 2009)
12. S.E. Golunski, Platinum Metals Review., **51(3)**, 162 (2007)
13. H. Shahar, Amran Ab. Majib, Malaysian Journal of Analytical Sciences, **11(1)**, 145 (2007)
14. K. Ravindra, L. Bencs, R.V. Grieken, Sci. Total Environ., **318(1-3)**, 1 (2004)
15. R.M. Bagus Irawan, P. Purwanto, H. Hadiyanto, Procedia Environmental Sciences, **23**, 86 (2015)
16. H. Bode, W. Maus, H. Swars, *17th International Engine Symposium, Vienna,* (1996)
17. Kamariah Noor Ismail, Ku Halim Ku Hamid, Sharifah Aishah S. A. Kadir, Mohibah Musa, Robert Mikhail Savory, The Malaysian Journal of Analytical Sciences, **11(1)**, 246 (2007)
18. A. Firdianto, *Ultrasonic Treatment With Nickel Electroplating Combined With Oxidation For Developing Gamma-Al_2O_3 Washcoat On Fe-Cr-Al Substrate* (Diss UTHM, Batu Pahat, 2012).
19. G.B. Froment, K.B. Bischoff, *Chemical Reactor Analysis and Design* (Wiley, New York, 1990)
20. J.A.M. Kuipers, W.P.M. Van Swaaij, Advances in Chemical Engineering, **24**, 227 (1998)

21. J.G. Eriksson, T.F. Mckenna, Ind. Eng. Chem. Res., **43**, 7251 (2004)
22. Darwin Sebayang, Yanuandri Putrasari, Sulaiman Hasan, Mohd Ashraf Othman, Pudji Untoro, *Preparation of NiO Catalyst on FeCrAl Substrate Using Various Techniques at Higher Oxidation Process, Electroplating* (INTECH, Crotia, 2012)
23. C. Syed Aalam, C.G. Saravanan, C. Mohamed Samath, International Journal For Research In Emerging Science And Technology, **2(7)**, 17 (2015)
24. R.E. Hayes, S.T. Kolaczkowski, *Introduction to Catalytic Combustion* (CRC Press, Amsterdam, 1998).
25. Mesut Durat, Zekeriya Parlak, Murat Kapsiz, Adnan Parlak, Ferit Fiçici, Journal of Thermal Science and Technology, **33(2)**, 89 (2013)
26. K. Mohan Laxmi, V. Ranjith Kumar, Y.V. Hanumantha Rao, International Journal of Engineering Research and Applications, **3(3)**, 518 (2013)
27. M.S. Simon Martinez, D. Ruben, A. Leal-Garza Fausto, Sanchez. Cruz, Esteban Baez Villarreal, Journal of KONES., **17(4)**, 303 (2010)

Catalytic Converter Developed By Washcoat Of γ-Alumina On Nickel Oxide (Nio) Catalyst In FeCrAl Substrate For Exhaust Emission Control: A Review

*A.M.*Leman[1,*], *J.*Afiqah[1], *Fakhrurrazi* Rahman[1], *Dafit* Feriyanto [1], *Supa*'at Zakaria[1], and *R.*Rahmad[1]

[1]Faculty of Engineering Technology, Universiti Tun Hussein Onn Malaysia (UTHM) Parit Raja, Batu Pahat, 86400 Johor, Malaysia.

Abstract. Automobile exhaust emission control is one of the trending issues in automobile research field. The existing catalytic converter using the noble metals of platinum (Pt), palladium (Pd) and rhodium (Rd) recently were in limited supply and higher in cost. There is a need for the automotive industry to produce ultra-low emitting vehicles at a reasonable cost. The objective of this study is to investigate the effectiveness of methods of fabrication of modified catalytic converter by approaching FeCrAl as a substrate which treated using ultrasonic bath technique to improve the exhaust emission control. The modified catalytic converter preparation will involve the ultrasonic bath process of FeCrAl foil which has fabricated as metallic monolith coated by γ-Al_2O_3 powder. Nickel as catalyst material will be prepared using electroplating process. The oxidation test will be conducted using a tube and automatic furnace in temperature of 1100^0C for 100 hours. Mitsubishi 4G93 1800cc Petrol E.F.I with a multi -gas analyzer equipped with a hydraulic dynamometer will be used for emission measurements of HC, CO, and NOx in varying speed and load for both conditions with and without catalytic converter. The result will expect the γ-Al_2O_3 as the washcoat material that fully embedded to FeCrAl substrate with the combination of ultrasonic and electroplating technique will effectively convert the CO, NOx and HC to CO_2, NO_2 and H_2O which means that catalytic converter is effective to improve exhaust emission control of diesel engine. The FeCrAl substrate as a metallic catalytic converter which coated by γ-Al_2O_3 using ultrasonic and nickel-electroplating technique may improve the exhaust emission control.

1 Introduction

Recent haze crisis was among the most terrific impact resulting from a critical level of urban air pollution in Malaysia. Exhaust emission from automobile has become the largest source that contributes to air pollution. Each automobile could release nearly 200 different components in its exhaust gases into the atmosphere [1]. In recent years, many researchers

* Corresponding author: mutalib@uthm.edu.my

indicate that exposure levels have increased gradually worldwide, especially in rapidly large populations of industrializing countries [2]. The vehicle population can be estimated to grow close to 1300 million by the year 2030 [3].

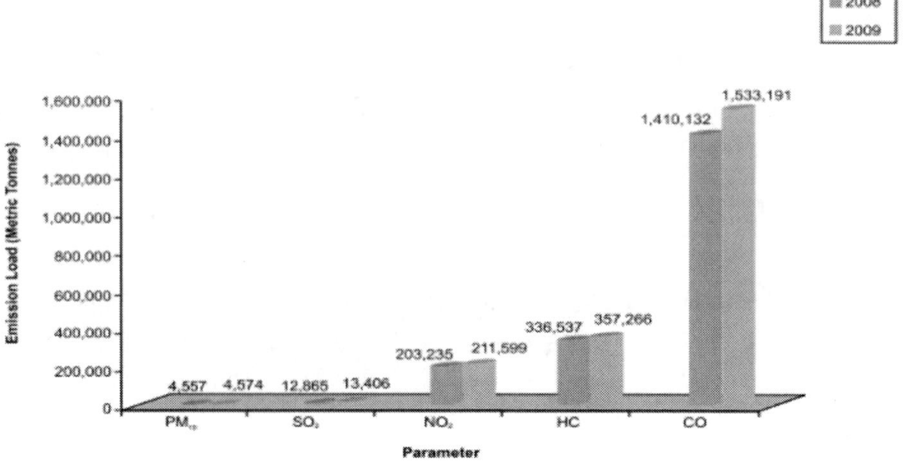

Fig. 1. Air Pollutant Emission Load from Motor Vehicles in 2008-2009 [4].

At the normal engine operating conditions, the typical exhaust gas composition are: carbon monoxide (CO, 0.5 vol.%), unburned hydrocarbons (HC, 350 ppm), nitrogen oxides (NOx, 900 ppm) hydrogen (H_2, 0.17 vol.%), water (H_2O, 10 vol.%), carbon dioxide (CO_2, 10 vol.%), oxygen (O_2, 0.5 vol.%) [5].

The catalytic converter is a tool built-in into the exhaust system of an automobile that functioning to cut down the amount of CO, NOx and unreacted HC in automotive emissions [6]. This catalytic converter is mounted along the exhaust pipe of the engine and has a porous ceramic structure through which the exhaust gas flows into the container [7]. Out of various existing technologies for the control of vehicle exhaust emission, catalytic converter has been found to be the best option. These gases are eliminated by the basic reactions occur inside a catalytic converter through oxidation for CO and HC and reduction for NOx.

Most of the catalytic converter used today consists of a monolithic structure which is ceramic honeycomb or metallic matrix. Monolith is known as multichannel which the exhaust gas flow with a high volume flow rate [8]. The schematic conversion of Three Way Catalyst (TWC) in Figure 2 shows the simultaneous reaction of oxidation and reduction in TWC in which the most common type of catalytic converter used [9, 10].

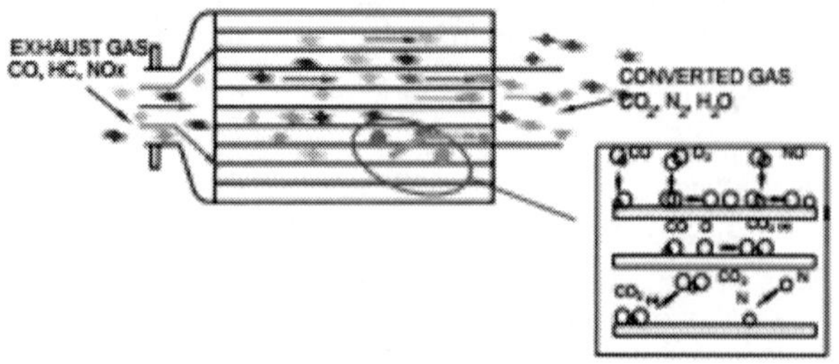

Fig. 2. Schematic of a three way catalyst converter [11].

The existing of excellent oxidation catalyst materials was usually based on the precious metal (Pt, Pd, and Rd). However, those materials are expensive and limited supply [12, 13]. In response to this matter, many researchers investigate an alternative catalyst such as the cheaper ranges of oxides (e.g. CuO, V_2O_5, NiO,MoO_3, and Cr_2O_3) to replace the use of precious metal [14].

2 Non-Noble Catalyst Development

The development of catalytic converter has evolved over the last three decades since the introduction of the first catalyst on diesel cars. There were many research works were made to investigate the best material for the replacement of the noble metals. These noble metals also could contribute to converter meltdown, carbon deposit, catalyst fracture and also poisoning in some cases [15].

The latest studies of catalyst performance were done by Ranganathan et al. [16]. They have studied the performance of two types converters which are the metal oxide such as titanium dioxide (TiO_2), cerium oxide (CeO_2) and copper nitrate ($Cu(NO_3)_2$), Zirconium dioxide (ZiO_2) with wire mesh substrate. Both the catalyst material used in this is inexpensive in comparison with conventional catalyst such as noble metals like Platinum (Pt), Palladium (Pd) and Rhodium (Rh). The study shows a positive result that is the new two catalyst catalytic converters are effective than the present Original Engine Manufacture catalytic converter. The emission from exhaust is rectified up to 20 % than the OEM fitted values.

Studies by Irawan et al. [17] have investigated the new optimum design of Manganese-Coated Copper catalytic converter. Catalytic Converter design and modification of catalytic materials can be an alternative to overcome the high air pollution problem from the transportation sector, especially particular CO exhaust emissions from gasoline motors. The result shows the increases of catalyst amount cells decreased the concentration of carbon monoxide exhaust emission. The limitation of this study is that they have only focused to reduced exhaust emission of carbon monoxide in order to find the optimum design of the catalytic converter.

The use of different catalyst of copper oxide (CuO), cerium oxide (CrO_2) and zirconium dioxide (ZrO_2) was presented by Walke et al. [18]. The purpose of this study is to determine the emission characteristics of a compression ignition engine. The result showed that all the three catalyst reduce the CO emission. The catalyst (zirconium dioxide + cerium oxide), reduce NOx emission. However, the brake thermal efficiency decreases with the use of all catalyst in the pollutant conversion.

A study on the development and performance analysis of nickel-based catalytic converter was conducted by Narendrasinh et al. [19]. This study introduces a non-noble (nickel) based catalytic converter to reduce diesel engine exhaust emission. The nickel was used as the oxidizing agents and stainless steel as a substrate. The result shows that nickel works in a limited proportion as a catalyst for the conversion of pollutants. Nickel has reduced HC by 40 % while CO by 35 %. Thus, this study introduce the use of new catalyst of nickel oxide (NiO) to full washcoat material on the catalyst and substrate using electrochemical and modified surface treatment via ultrasonic bath technique. It purposed to improve the exhaust emission control system. NiO created through plating Ni to the substrate FeCrAl and NiO existed through oxidization in high temperature. This approach is completely creatively new in order to replace the existing of excellent catalyst materials of precious metal of Pt, Pd, and Rd.

3 Methods and Materials

3.1 Preparation of Modified Catalytic Converter

There are several methods for coating the fabrication of modified catalytic converter such as ultrasonic bath and electroplating technique. Ultrasonic bath is conducted to FeCrAl foil which have fabricated as metallic monolith coated by γ-Al_2O_3 powder [20]. The samples are cleaned in water and then proceed in methanol for 5 minutes. During the ultrasonic process, the frequency of 35 kHz and ultrasonic times of 4.5 hours are imposed. That holding time is selected according to Leman et al. [21] that the ultrasonic time for 4.5 hours has most effective to achieve high thermal stability of the metallic material.

The samples are immersed into the beaker with γ-Al_2O_3 powder and ethanol with the concentration of 20 g/l. Nickel electroplating process is conducted according to Putrasar et al. [22] for depositing nickel onto FeCrAl. Electroplating is conducted on FeCrAl after new ultrasonic treatment. The electroplating process is conducted through some components such as electrolyte, anti-pitting agent, anode and cathode. The schematic diagram of the electroplating process is shown in Figure 3. Sulphamate type is used as the electrolyte medium. The electrolyte prepared with distilled water, at a constant temperature of 40-60^{0}C, and pH value of solution adjusted to 2.5-4.5 using HCl and NaOH reagent. The electrolyte agitated using a magnetic stirrer.

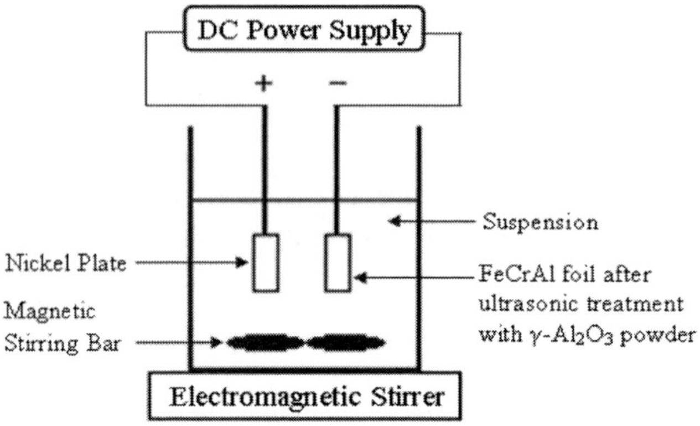

Fig. 3. Schematic diagram of electroplating process.

The oxidation test of coated FeCrAl is conducted according to Badini and Laurella [23], to identify the characteristics of alumina scales grown on a FeCrAl alloy. The cyclic oxidation test is conducted using tube furnace and automatic furnace in temperature of 1100^{0}C. The test was carried out for 100 hours which divided into 5 cycles [22]. The oxidation cycle were illustrated in Figure 4. The holding time for each cycle was 20 hours for each cycle with the heating and cooling rate of 5^{0}C/minutes.

DOI: 10.1051/matecconf/20167801045

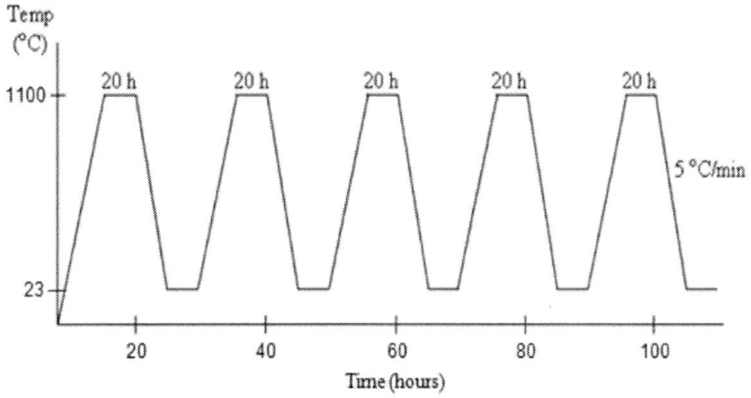

Fig. 4. Cyclic oxidation of coated Fe-Cr-Al at temperature of 1100^{0}C for 100 hours [23].

3.2 Exhaust Emission Testing

Testing for the effectiveness of the prepared catalyst can be observed by undergo the emission testing. The exhaust gas emission was measured by using multi-gas analyzer equipped with a hydraulic dynamometer for emission measurements of HC, CO, and NOx shown in Figure 7 [24]. According to the experimental studies by Merkisz [25] the function of gas analyzer is to allowed simultaneous measurement of the mass intensity of exhaust of harmful pollutant concentration. The emission test will be conducted at each operating point in varying speed and load for both conditions with and without a catalytic converter. An experimental study by Alves et al. [26] observed that dynamometer gives an accurate reading of the engine's power, speed, torque, exhaust temperature, etc during the emission testing. A probe was used to measure the gases at the inlet and outlet region, along with the thermocouple, to attain temperature values for ambient and the exhaust gas itself.

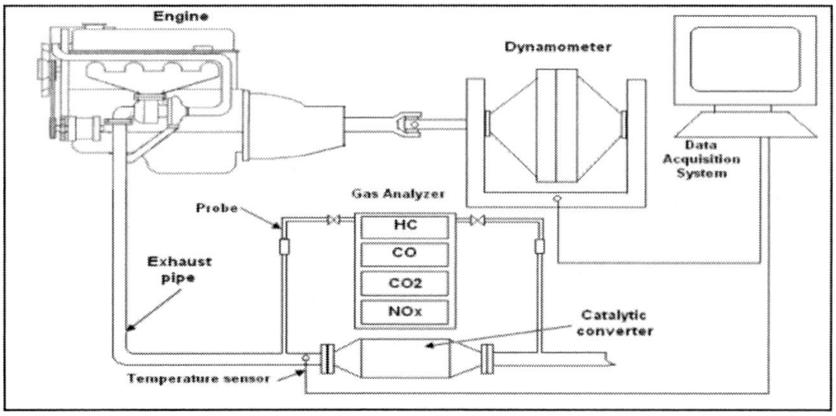

Fig. 7. Emissions measureme nts set-up [25].

4 Experimental Results

The aim of the experiment is to reduce the exhaust emission from the effectiveness of fabrication method of modified catalytic converter by approaching FeCrAl as a substrate which treated using ultrasonic bath technique. According to the emission testing that was conducted by Mizanuzzaman et al. [27], the carbon monoxide (CO2) level emitted from the engine without catalytic converter is above 7 ppm. The level is decreasing which is lower than 7 ppm when catalytic converter is applied to the vehicles. Kretzschmar [28] found that the emission of the carbon monoxide (CO) during the cold start are higher than the normal operation of vehicles which is 5 to 40 times higher.

5 Conclusion

The FeCrAl substrate as a metallic catalytic converter which coated by γ-Al_2O_3 using ultrasonic and nickel-electroplating technique may improve the exhaust emission control which tested using Mitsubishi 4G93 1800cc Petrol E.F.I. It will become new contribution to the automobile research field.

This work was financially supported by Ministry of Education through FRGS Vot 1612 and ORICC, Universiti Tun Hussein Onn Malaysia (UTHM) for the management of the research grant and Centre for Graduate Studies – UTHM.

References

1. I.V. Mishakov, A.A. Vedyagin, A.M. Volodin, M.S. Myakisheva, Chemistry for Sustainable Development, **19**, 91 (2011)
2. WHO, World Health Organization Fact sheet N313. *Air Quality and Health.* http://www.who.int/mediacentre/factsheets/fs313/ en/index.html (2013).
3. M.N. Rao and H.V.N Rao. *Air pollution* (Tata McGraw-Hill publishing company limited, New Delhi, 2003).
4. Department of Environment, *Malaysia Environmental Quality Report* (Ministry of Natural Resources and Environment, Putrajaya, 2009).
5. S. Pate Bharat, D. Patel Kuldeep, International Journal of Applied Engineering Research, **7**, 0973 (2012)
6. D. Sebayang, S.H. Amirnordin, P. Untoro, H. Abd Rahman, *Malaysian Technical University Colleges Annual Conference on Engineering and Technology (MUCET)*, (Malaysia, 2006)
7. V. Ganesan, *Internal Combustion Engines. Second Edition* (McGraw Hill, 2004)
8. M. Mansha, E.M. Shahid, A.H. Qureshi, Pakistan Journal of Engineering and Applied Sciences, **11**, 114 (2012)
9. D. Sebayang, P. Untoro, S.H. Aminordin, H. Abd Rahman, *World Engineering Congress* (Institute of Engineers Malaysia (IEM), Malaysia, 2007)
10. R.M. Heck, R.J. Farrauto, S.T. Gulati, *Catalytic Air Pollution Control Commercial Technology. 3rd edition* (John Wiley & Sons, Inc., USA, 2009)
11. J.C. Prince, C. Treviño, M. Díaz, *Proceedings of the World Congress on Engineering* (International Association of Engineers, London, 2008)
12. G.C. Koltsakis, A.M. Stamatelos, Prog. Energy Combust. Sci., **23**, 1 (1997)
13. M. Benson, C.R. Bennett, J.E. Harry, M.K. Patel, M. Cross, Resour Conserv Recy. **31**, 1 (2000)
14. S. Kolaczkowski, *Structured Catalysts and Reactors 2nd edition* (CRC Press, Taylor & Francis Group, 2006)
15. A. Chirag and P.R. Pravin, International Journal of Advanced Engineering Research and Studies, **1**, 118 (2015)
16. M. Ranganathan, S.A. Renald Remo, U. Kishore, S. Yuvaraj, S. Arun, *International Conference on Advance Research and Innovation in Engineering, Science, Technology and Management* (ARS College of Engineering, Maraimalai Nagar, Chennai, 2015)

17. R.M. Bagus Irawan, P. Purwanto, H. Hadiyanto, Procedia Environ Sci. **23**, 86 (2014)
18. P.V. Walke, N.V. Deshpande and A.K. Mahalle, *Proceedings of the World Congress on Engineering* (International Association of Engineers, 2008)
19. R. Narendrasinh Makwana, M. Chirag Amin, K. Shyam, International Journal of Advanced Engineering Technology, **4**, 10 (2013)
20. F. Ade, D. Sebayang, P. Yanuandri, S.B. Hasan, A. Bin Otman, P. Untoro, Adv. Mat. Res., **181**, 501 (2012)
21. A.M. Leman, D. Feriyanto, M.N.M. Salleh, I. Baba, AMM, **815**, 193 (2015).
22. Y. Putrasari, P. Untoro, H. Naili, S. Hasan, D. Sebayang, MEV, **1**, 53 (2012).
23. C. Badini, F. Laurella, Surf. Coat. Technol., **135**, 291 (2001)
24. A.K.M. Mohiuddin, M. Nurhafez, International Journal of Mechanical and Materials Engineering (IJMME), **2**, 1 (2007)
25. J. Merkisz, J. KONES., **18**, 253 (2001).
26. C.A. Alves, A.I. Calvo, D.J. Lopes, T. Nunes, A. Charron, M. Goriaux, P. Tassel, P. Perret, International Scholarly and Scientific Research & Innovation, **7** (2013)
27. M.D. Mizanuzzama, Amitava Ghosh Dastidar, Mohammad Rajib Uddin Rony, Reasat Azam, M.D. Almostasim Mahmud, Int. J. Eng. Res. Appl., **3**, 313 (2013)
28. J.G. Kretzschmar, Air Quality, 9 (1997)

Indoor Air Contaminant Adsorption By Palm Shell Activated Carbon Filter – A Proposed Study

A.M Leman[1,*], *Supa'at* Zakaria[1], *M.N.M* Salleh[1], *Dafit* Feriyanto[1], *N.M* Sunar[1], and *N.*Misdan[1]

[1] Faculty of Engineering Technology, Universiti Tun Hussein Onn Malaysia (UTHM), Parit Raja, Batu Pahat, 86400 Johor, Malaysia.

Abstract. Indoor air contaminant is a public issue. High Volatile Organic Compound (VOC), Carbon monoxide (CO), Carbon dioxide (CO_2), and particulate matter is becoming main issue that needs to solve. Therefore, this study focus on improving indoor air quality by using activated carbon (AC) for Ventilation and Air-Conditioning (VAC). It investigated because AC is widely explored but developing AC as a filter for VAC is not developed yet. The AC prepared by physical and chemical activation process and combination both of process and it was activated by H_3PO_4 and NaOH. Characterization and analysis process are consists of water content, ash content, bulk density, adsorption capacity, iodine number and indoor air filtering analysis. Treated activated carbon potential in achieving higher surface area of the structure to the range of 950 to 1150 m^2/g for gas phase application. The higher surface area will adsorb more air pollution. Maintained properties of activated carbon such as hardness, density, pore, extractable ash, particle size (12 by 40 mesh) and pH are becoming the main concern in achieving high quality of activated carbon.

1 Introduction

Demanding of fresh, healthy and comfortable indoor environment is essential since most people spend their time in indoor as compared to outdoor [1]. The increasing concern towards indoor air quality (IAQ) have gained the suitable technique on mitigating the indoor air contaminant [2] which can effect on health and working performance [3]. Volatile Organic Compound (VOC), Carbon monoxide (CO), Carbon dioxide (CO_2), and particulate matter defined as the major contribution of indoor air contaminant. There are several VOCs in indoor air environment such as BTX (benzene, toluene xylene), trichloroethylene and dichloromethane [4]. During long term exposure may pose adverse health effect related to human respiratory systems (such as asthma, throat irritation, lung cancer), headache, poor memory, eyes, nose and also skin irritation [5, 6]. In order to decrease the pollutant, some researchers have been conducting their methods for controlling

* Corresponding author: mutalib@uthm.edu.my

the VOC such as adsorption, condensation, Photocatalytic oxidation (PCO), negative air ions (NAIs) and non-thermal plasma (NTP) [4]. In recent years, adsorption in bulk separation or purification process has an innovative treatment process in environment application. Adsorption method is effective at low concentration level which is Part per million (ppm). Larger adsorption capacity is achieved by larger surface area of the filter material and their performance in both equilibrium and kinetics.

Generally the concentration of indoor air contaminant is higher than outdoor because it's released by human activities, building materials, furniture, carpets, paints, cleaning product, etc. [7]. Therefore, the maintenance of the indoor air quality is quite needed in order to provide sustainable and healthy environment [8].

There were some approaches methods that has been used for controlling indoor air pollutant such as ventilation and air cleaning method [9]. Mechanical Ventilation and Air-Conditioning (MVAC) system have been improved by reducing indoor particulate level. However, this technique is not promising to diminish of contaminant gases and require more energy consumption [10]. Air purification through activated carbon adsorption technique being most common air cleaning method especially for VOC and other polluted gases [11]. There are several adsorption media which commercially available or existing in the market such as Activated Carbon Fiber (ACF), Silica gel, zeolite, alumina and Granular Activated Carbon (GAC). According to some literature that the constant temperature and relative humidity in a real building for various condition is may vary significantly between months, seasons and years [12, 13]. However, the filter in the market has not been evaluated yet which led to a real VOC contaminant is real building is not recorded [13]. One of the technology which effective and reliable for removing impurities is by using activated carbon filter. Activated carbon is one of the cheapest and popular material as water purification, cleaned/desorption which used hundred or thousand times [14]. There are many advantages of activated carbon such as high carbon content, high density and low ash content, high absorption capacity, for dissolved organic, chlorine and can be used as specific application [14-17]. The other superiority of activated carbon is a reasonable large surface area, have numerous pore network (Figure 1) as a transportation media of the molecule to the interior and it's more stable performance in dynamic condition [18]. Influencing by high pollutant in our environment and many superiority of activated carbon, Therefore, the activated carbon is urgently needed in filtering the pollutant into fresh air. The principle of filtering process of activated carbon is to convert the pollutant to the CO_2, H2O as shown in Figure 2.

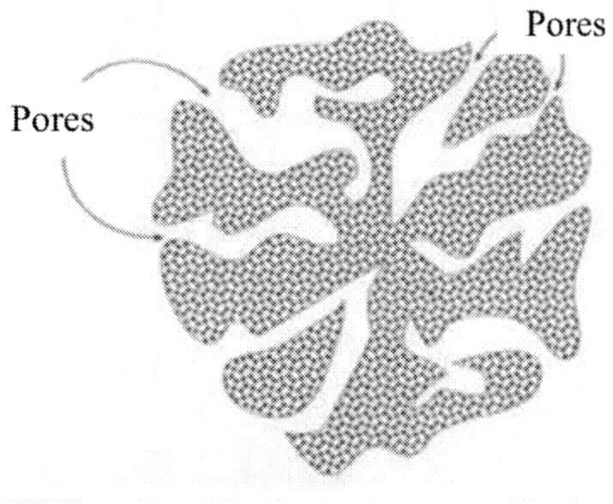

Fig. 1. Numerous pores of the activated carbon [18].

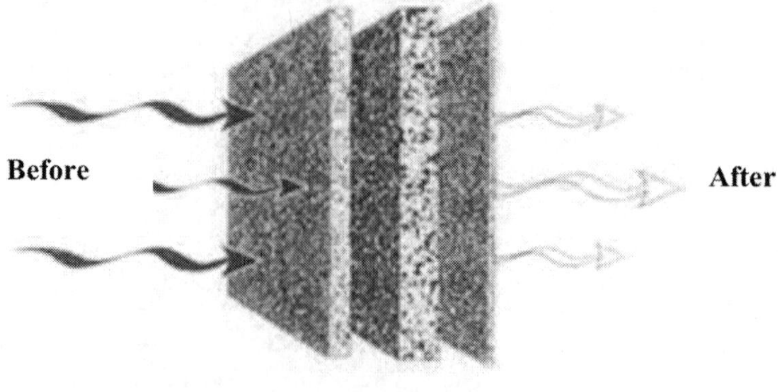

Fig. 2. Air before and after filtered by activated carbon [8, 13].

AC as an indoor air filter in order to improve indoor air quality and to reduce the cancer risk and non-cancer risk health problem. There are many researchers' studies on increasing the pore development of palm shell activated carbon [8, 13, 14, 19-22]. The pore development on activated carbon can be increased by undergoing physical and chemical treatment. In Malaysia, the palm shell is abundant due to there are 4,853,766 hectares oil palm forest in 2010 [23]. Therefore, palm shell AC (Figure 3) is very potential to explore in Malaysia.

Fig. 3. Palm shell activated carbon [23].

According to Abdullah and Wahid [23], their AC has several base properties such as bulk density (0.4-0.47), moisture content (3 %), ball-pan hardness (90-95 %), ash content (5-7 %), iodine number (950-1100 mg/g) and pH value (10).

2 Surface Modification of Palm Shell Activated Carbon

Surface modification is conducted using physical and chemical treatment. That treatment is conducted to achieve good pore structure and good adsorption properties.

2.1 Physical Activation Process

Physical activation of AC is consists of 2 steps which are hydrogen and oxygen content elimination by using pyrolysis carbonaceous material at inert atmosphere and at high temperature. Second steps are chars activation at high temperature and in the presence of

CO_2 as oxidizing gases [24]. According to Abechi et al. [18], that highest physicochemical properties are signed by the sample which treated at a temperature of 1000^{0}C, holding time of 45 min, Yiel of 19.25 % and bulk density of 0.62 gcm^{-3} and its micostructure analysis as shown in Figure 4.

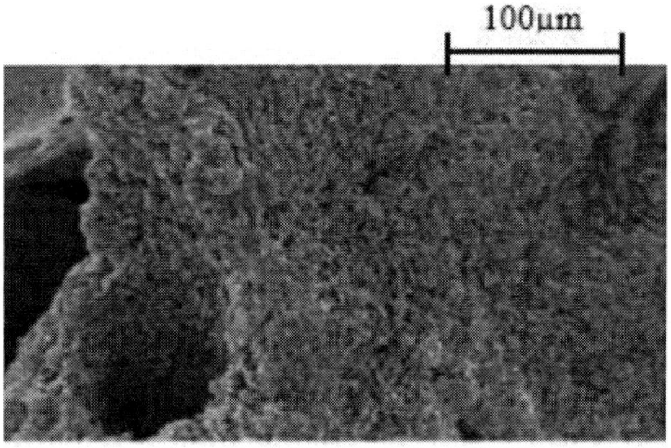

Fig. 4. Microstructure of treated AC in temperature and holding time of 1000^{0}C and 45 min [18].

Figure 4 shows a highest adsorption capacity with surface area of 217 gm^{-2}, micropore volume (Vμ) is 11 x10^{-2} cm^{-3}g^{-1} and mesopore volume of 1 x10^{-2} cm^{-3}g^{-1}. During physical activation process, There are 2 major reactions as in Eq. 1 and Eq. 2 [18].

$$C + CO_2 \rightarrow 2CO \quad (1)$$

$$C + H_2O \rightarrow CO + H_2 \quad (2)$$

Physical treatment or thermal activation is conducted by carbonization of palm shell at the temperature range of 500 to 900 ^{0}C in order to remove volatile metter. Many researchers used different temperature in thermal activation such as 500^{0}C [25, 26], 600^{0}C [26, 27], 700^{0}C [26, 28], 800^{0}C [18, 26, 29], 900^{0}C [18, 24], 1000^{0}C [18] and 1200^{0}C [24]. The influence of the temperature in enlargement of pores and to enhance the adsorption solvent by increasing activation temperature [30]. The normal physical activation time for palm shell activated carbon is 1-8 hours [31]. Moreover, physical activation process also varied by relative pressure (P/Po) in the range of 0 to 1 with the highest adsorption capacity is signed by A1 sample of 80 cm3/g STP in 1 relative pressure like shown in Figure 5 [18].

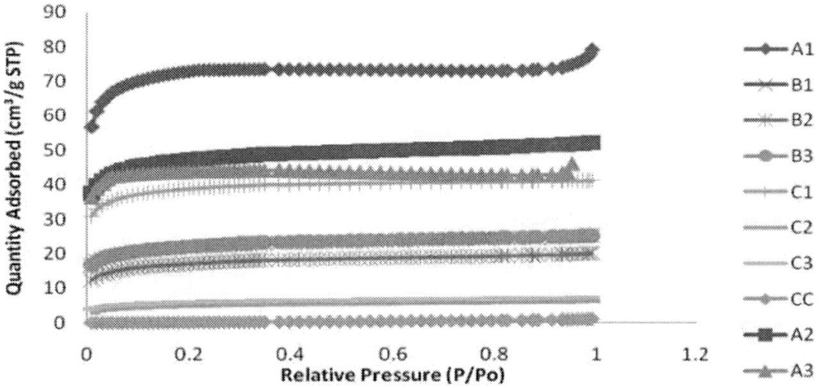

Fig. 5. Adsorption capacity of AC in different relative pressure [18].

2.2 Chemical Activation Process

The chemical activation process of AC covers the precursor treatment with chemical agent and followed by heat treatment which influence to the carbonization process and porosity generation [26, 32]. There are some dehysrating agent such as phosporic acid (H_3PO_4), zing chloride (ZnCl2) and natrium hydroxide (NaOH) [26]. The chemical analysis of palm shell is shown in Table 1.

Table 1. Chemical analysis of oil palm shell as activated carbon material [29]

Approximate analysis (wt%)		Element analysis (wt%)		Content (wt%)	
Fixed carbon	18.7	C	50.01	Cellulose	29.7
Ash	7.96	H	6.85	Halo cellulose	47.7
Moisture	1.1	N	1.9	Lignin	53.4
Volatile	72.46	O	41.15		

There are many researchers which conduct the chemical activation process in producing activated carbon using various chemical agents as listed in Table 2.

Table 2. Chemical agent for activating charcoal

Chemical agent	Composition
H_3PO_4	85 % [24]
$ZnCl_2$	65 % [26]; 98% [24]
NaOH	1000 mg/L [33]; 10 to 300 mg/L [34]
KOH	50 % [18]
HNO_3	1000 mg/L [33]
K_2CO_3	1000 mg/L [33]
H_2SO4	10 to 300 mg/L [34]
$NaHCO_3$	10 to 300 mg/L [34]
NaCl	10 to 300 mg/L [34]

3 Research Approch

AC is produced by some methods and it have explored by many researchers. The activation process is divided into 2 types which are physical (various temperature between 500 to 900^0C [25] and time of 1-8 hours [31] and chemical activation as listed in Table 2. It has conducted using various temperature, and chemical agent. According to Niya et al., (2011) [24] H_3PO_4 is used as chemical agent which heat treatment under temperature of 900 and 1200^0C. Abechi et al. [18] also activate the palm shell using a chemical agent of KOH and treated under temperature of 800, 900 and 1000^0C. Palm shell was burned in various temperature using commercially burner that unidentified reference of burner design [23, 33, 34]. The burner may not frequently release the heat flow and air flow which cause overheated of the burner and it led to a high volume of its content and decreasing adsorption properties of activated carbon. After burned, palm shell is cruised and sieved in various sizes such as 1–2 mm [24], 1.18 mm [18], 1.0–2.0 mm [35], 250–300 μm diameter [19] and 425 μm [36]. There is some limitation of those researches in combining the activation process, and in burning process. That process is strictly controlled in order to achieve high quality activated carbon based on the adsorption capacity/ surface area, hardness, density, pore, extractable ash, particle size and pH.

Research Approach is more to innovative technology and treatment process. It involves the burning process which is a fabrication process based on air and heat flow calculation in order to achieve the prevalent air and heat distribution. It conducted to reduce the possibility buner in producing high value of ash content. In this study the activated carbon size is divided into 2 types which are powdered AC and granular AC. Powdered AC especially used in lab scale and it most recommended for batch treatment of liquid (water treatment plant). It caused by slower diffusion through particles and high pressure loss across the bed. Granular activated carbon is used in pilot plant and industrial sector and in this case, it applied to air filter because better mechanical by sacrificing some efficiency. Testing process include iodine number which typically in the range of 500–1200 mg/g, water content is performed under temperature of 100^0C, ash content is performed using atmospheric furnace with various temperature of 650^0C, 700^0C and 850^0C and bulk density analyzed through drying under temperature of 100^0C for 1 hour and weighed. Adsorption analysis is conducted using dipping methods where the activated carbon is dehydrated in a convection oven at 150^0C. Indoor air quality improvement is conducted by AC filter (Figure 6) which performed by a pilot-scale acrylic chamber with a volume of 1 m3 and temperature as well as the humidity in the chamber were maintained at 20–26^0C and 40–70 %, respectively.

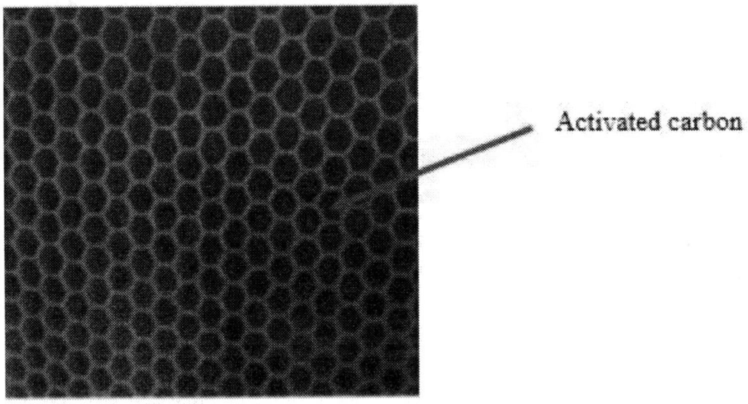

Fig. 6. Concept of AC filter [37].

4 Conclusion

AC filter has been developed by previous researcher but it still very challenging to explored. This research, proposed AC air filter completed by effective technique for improving indoor air quality. This paper proposes an innovative method inactivate an activated carbon by physical, chemical and combination of physical and chemical with new treatment parameter. Approach technology for burning process is one of the innovative technique in achieving high quality of activated carbon.

The authors would like to thank the Ministry of Higher Education Malaysia for FRGS fund vot 1216, ORICC, Universiti Tun Hussein Onn Malaysia (UTHM) and Centre for Graduate Studies – UTHM.

References

1. M. Frontezak, P. Wargocki, Build. Environ., **46**, 922 (2011)
2. C.K.H. Yu, M. Li, V. Chan, A.C.K. Lai, Build. Environ., **76**, 73 (2014)
3. B.F. Yua, Z.B. Hua, M. Liua, H.L. Yang, Q.X. Kong and Y.H. Liu, Int. J. Refrig., **32**, 3 (2009)
4. D. Das, V. Gaur, N. Verma, Carbon, **42**, 2949 (2004)
5. W.K. Jo, H.H. Chun, Aerosol. Air Qual. Res., **14**, 347 (2014)
6. Z. Eugenija, E.S. Neil, M. Jadranka, P.C. Jasna, D.J. Jagoda, M.P. Branka, Biol., **111**, 37 (2009)
7. B. Guieysse, C. Hort, V. Platel, R. Munoz, M. Ondarts, S. Revah, Biotechnol. Adv., **26**, 398 (2008)
8. J.Y. Yoo, C.J. Park, K.Y. Kim, Y. Son, C. Kang, J.M. Wolfson, I. Jung, S. Lee, P. Koutrakis, J. Hazard. Mater., **289**, 184 (2015)
9. E. Kabir, K.H. Kim, Asian J. Atmos. Environ., **6**, 137 (2012)
10. F. Haghighat, C.S. Lee, B. Pant, G. Bolourani, N. Lakdawala, A. Bastani, Atmos Environ., **42**, 8176 (2008)
11. F.I. Khan, A.K. Ghoshal, J. Loss Prev. Process Ind., **13**, 527 (2000)
12. C.H. Reed, S.J. Nabinger, S.J. Emmerich, Build. Environ., **43**, 368 (2008)
13. Gallego, E. Roca, F.J. Perales and J.F. Guardino, Build. Environ., **67**,14 (2014)
14. I. Adedayo, A.O. Inegbenebor, H.I. Boyo, ARPN Journal of Science and Technology, **2**, 2225 (2012)
15. Y.B. Onundi, A.A. Mamun, M.F. Al Khatib and Y.M. Ahmed, Int. J. Environ. Sci. Tech., **7**, 751 (2010)
16. W.B. Wan Nik, M.M. Rahman, A.M. Yusof, F.N. Ani, C.M. Che Adnan, *Proceedings of the 1st International Conference on Natural Resources Engineering & Technology, Putrajaya, Malaysia,* 646 (2006)
17. N.H. Hasan, M.R. Said, A.M. Leman, International Journal of Environmental Engineering Science and Technology Research, **1**, 28 (2013)
18. S.E. Abechi, C.E. Gimba, A. Uzairu and Y.A. Dallatu, Research Journal of Chemical Sciences, **3**, 54 (2013)
19. T.A. Metts, S.A. Batterman, Chemosphere, **62**, 34 (2006)
20. S. Wu, B. Dong, T. Qiao, J. Zhang, Zhejiang Univ Sci., A, **9**, 1576 (2008)
21. A.A. Ismaiel, M.K. Aroua, R. Yusoff, Chem. Eng. J., **225**, 306 (2013)
22. M. A. Sidheswaran, H. Destaillats, D.P. Sullivan, S. Cohn, and W.J. Fisk, Build. Environ., **47**, 357 (2012)
23. R. Abdullah, M.B. Wahid. *World palm oil supply, demand, price and prospect: Focus on Malaysian and Indonesian palm oil industry,* (Malaysian Palm Oil Board (MPOB), 2010).
24. A.A. Niya,W.M. Ashri, W. Daud and S. Farouq, Eng. Res. Des., **8**, 657 (2011)
25. J. Guo, W.S. Xu, Y.L. Chen, A.C. Lua, J. Colloid Interface Sci., **281**, 285 (2005)
26. A.B.M. Noor, M.A.B.M. Nawi, J. Phys. Sci., **19**, 93 (2008)
27. D. Adinata, W.M.A.W. Daud, M.A. Kheireddine, Bioresour. Technol., **98**, 145 (2007)
28. I.A.W. Tan, A.L. Ahmad, B.H. Hameed, Bioresour. Technol., **100**, 1494 (2009)
29. W.M.A.W. Daud, W.S.W. Ali, Bioresour. Technol., **93**, 63 (2004)
30. I.A.W. Tan, A.I. Ahmad, B.H. Hameed, Colloids Surf., A **318**, 88 (2008)
31. A. Arami-Niya, W.M.A.W. Daud, S.M. Farouq, J. Anal. Appl. Pyrolysis., **89**, 197 (2010)
32. A. Afrane, Microporous Mesoporous Mater., **112**, 284 (2008)

33. W.K. Koo, N.A. Gani, M.S. Shamsuddin, N.S. Subki, M.A. Sulaiman, J. Trop. Resour. Sustain., Sci. **3**, 54 (2015)
34. T.S. Anirudhan and S.S. Sreekumari, J. Environ. Sci., **23**, 1989 (2011)
35. S. Sethupathi, M.J.K. Bashir, Z.A. Akbar, A.R. Mohamed, Waste Manage. Res., **33**, 303 (2015)
36. E.C. Okoroigwe, C.M. Saffron, P.D. Kamdem, J. Chem. Eng. Mater. Sci., 5, 1 (2014)
37. Suki, *Honeycomb activated carbon air filter*, (Dongguan ICM Environment Tech Co., Ltd., 2015)
38. DOSH. *Industry Code of Practice on Indoor Air Quality-Department of Occupational Safety and Health Ministry of Human Recourses Malaysia*, (JKKP DP(S), 2010)

MATEC Web of Conferences **78**, 01047 (2016) DOI: 10.1051/matecconf/20167801047
IConGDM 2016

Flow Characteristics of Multi-circular Jet Plate in Premix Chamber of Air-Assist Atomizer for Burner System

Shahrin Hisham Amirnordin[1,*], *Amir* Khalid[1], *Isma Adzrai* Ismail[1], *Ronny* Yii Shi Chin[1], and *Mas* Fawzi[1]

[1]Combustion Research Group, Centre for Energy and Industrial Environment Studies (CEIES), Faculty of Mechanical and Manufacturing Engineering, Universiti Tun Hussein Onn Malaysia, Parit Raja, 86400 Batu Pahat, Johor, Malaysia

Abstract. The flow characteristics of multi-circular jet (MCJ) plate in the premix chamber of an atomizer were investigated using Computational Fluid Dynamics. Multiphase volume of fluid behavior inside the chamber was determined via steady simulations. The Eulerian–Eulerian two-fluid approach was used for execution mixing of diesel fuel and air. Spray simulation using the discrete phase with injection was generated from the nozzle hole into the ambient atmosphere. The behavior of three MCJ plates in the premix chamber was studied numerically. Results illustrated that plate open area, A_e, influenced the turbulence inside the chamber. MCJ 3, which had the lowest open area, generated the highest flow velocity and turbulence kinetic energy compared with MCJ 1 and 2. The MCJ plates could increase the turbulence in the premix chamber and contribute to the combustion efficiency.

1 Introduction

In recent years, premix combustion technology has received particular attention in response to strict emission regulations introduced by the authorities. The reduction in emission, especially nitrogen oxides (NO_x), can be observed when a conventional burner is replaced by a premix burner. The thermal efficiency of a premix burner system can also be optimized because of its capability to operate under very low excess air. In a premix burner, enhanced mixing between fuel and oxidizer can be achieved by the presence of a turbulence generator. Many techniques have been explored in generating turbulence, including the application of compact active grids [1], fractal grids [2–5], and multi-circular jet plates [6].

Among the turbulence generators, fractal grids have attracted much research because of their high fluid mixing efficiency. Theoretically, a fractal grid differs from a standard grid because it creates turbulence by exciting many length scales of various sizes simultaneously, rather than a single one. Krogstad [4] revealed the strong dependency of turbulence on the grid geometry. Soulopolous *et al.* [7] used the fractal grid to generate turbulence in the premixed flame. The results illustrated the advantages of a fractal grid,

* Corresponding author: shahrin@uthm.edu.my

which created more intense turbulence and changed the combustion regime. Verbeek *et al.* [3] studied the fractal grid with the aim of increasing the turbulent and combustion rate in a low swirl burner. An increase in turbulence intensity was observed when the flow of the multi-grid was compared with the standard grid.

Other types of turbulence generators are multi-circular jet (MCJ) and non-circular jet plates. One of the earlier studies was conducted by Coppola and Gomez [6], who investigated the effects of geometric parameters on two families of high-blockage plates (i.e., circular and non-circular). Hot wire anemometry was used on the turbulent free jet with measured parameters of axial velocity, turbulence intensity, and turbulence length scale of corresponding Reynolds numbers. Several characteristics of the plates were identified for each group. In MCJ plate, the geometry was defined as the hole diameter, d; the radial position of the holes, R_s; the number of holes, N; and the plate open area, $\%A_e$ (i.e., the area of circular openings). The MCJ was observed to perform well in achieving a high turbulence level at a given flow rate and preserving radial uniformity. From the findings, the turbulence generation through high-blockage perforated plates was a viable way to generate intense turbulent field with certain precautions on instabilities because of the openings, shape, and position. However, the effects of the MCJ plates on the flow inside the premix chamber of the atomizer need to be investigated further to determine the influence on spray formation.

This study analyzed MCJ plates inside the premix chamber of an air-assisted atomizer using Computational Fluid Dynamics (CFD), which enables the detailed investigation of the fluid flow being conducted. Three different types of MCJ plates with different patterns and geometries were studied. The results were analyzed in terms of fluid flow in the premix chamber and the behavior of atomization nozzle, such as the velocity and turbulence kinetic energy.

The rest of the paper is organized as follows. The next section describes the numerical calculation, the boundary conditions, the physical model, and the meshing of the calculated domain. In the results and discussion, air flow structure, exit velocity profile, and turbulence kinetic energy are discussed. The last section explains the conclusion from the analysis of the results.

2 Methodology

Calculation methods. Calculations were performed using ANSYS Fluent commercial CFD software. The equation involved was the fully 3D, steady, compressible, Navier–Stokes equation. The solver used the implicit pressure-based algorithm. Two phases of fluid (diesel in liquid and air) were used in this work, which required the multiphase volume of fluid model. The turbulence model employed was the transition four-equation shear stress transport model. Discretization involved the PISO scheme with spatial discretization of the first-order upwind scheme. The calculation involved species transport using air as a primary phase and diesel liquid as a secondary phase.

Boundary conditions. The inlet air and inlet fuel used mass flow inlet conditions, which define the mass flow rate, flow direction, and angle. In this study, the air mass flow rate was fixed at 0.00393 kg/s, whereas the inlet fuel mass flow rate was set at 0.00026934 kg/s [8]. The output flow was set at 1 bar, which is the standard atmospheric condition. The material of the solid wall was aluminum with the boundary of the wall for solid-fluid coupling, whereas the wall was specified as stationary wall motion (no slip condition). The discrete phase model was used for atomization characteristics with breakup as a physical model, and injection type was a solid cone with droplet particle. The effects between the droplet and flow were obtained from this model.

Physical model and mesh. The position of the MCJ plate in the injector is illustrated in Fig. 1(a). The premix chamber was the location for fuel and air mixing before spraying through the nozzle holes out into the environment. It was 6.70 mm in radius and 35.66 mm in length, which resulted in a volume of 4098.15 mm^3. The top of the chamber contained eight injector nozzles with 1 mm in diameter and 45° angle. The geometric model of the domain is illustrated in Fig. 1(b).

Fig. 1(c) displays the mesh of the MCJ plate in the mixing chamber. The tetrahedral unstructured grids were employed because of the complexity of the structure. The grid independence study was conducted for three sets of number of elements (i.e., 187,669; 212,181; and 287,350). The maximum error for the velocity inside the exit nozzle of the local position was 1.05%, and the relative errors in the remaining calculations were about 1%. Therefore, the selected numerical scheme applied in this work was 212,181 elements. The actual physical appearance of the plates is described in Table 1, which shows the terms used in the plates; the percentage of a plate's open area with respect to the plate's total area, $\%A_e = 100 \times (A_e/A_p)$, where A_p is the plate's total area.

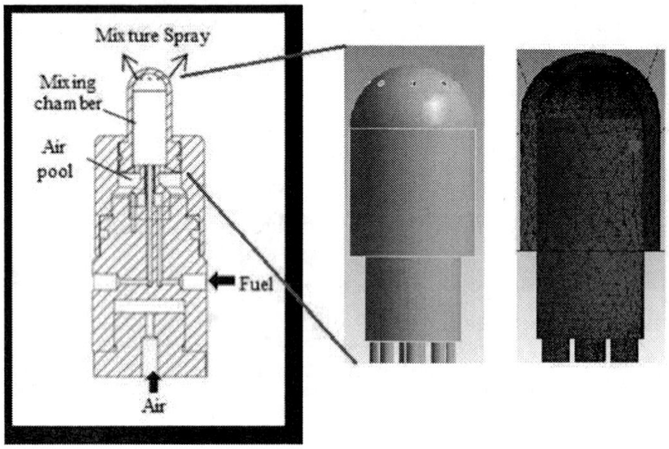

(a) (b) (c)

Fig. 1. The premix injector showing (a) the MCJ plate location (b) the geometrical model and (c) the meshing

Table 1. The geometry of (a) MCJ 1 (b) MCJ 2 (c) MCJ 3

	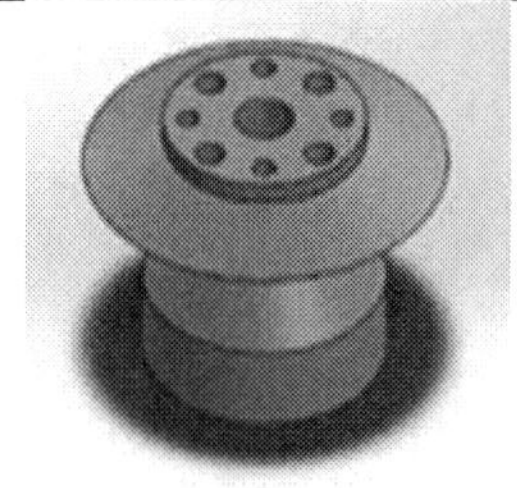	
(a) $\%A_e$: 17.82 Diameter 2 x 8 holes Standard regular holes	(b) $\%A_e$: 13.92 Diameter 2 x 4 holes Diameter 1.5 x 4 holes	(c) $\%A_e$: 10.03 Diameter 1.5 x 8 holes

3 Results and discussion

Results of mixing chamber flow simulation. The flow of the mixture from the uniform grid (MCJ 1) is shown in Fig. 2. The streamline velocity from the fuel and air inlet (at the bottom) exited through the nozzle (at the top) of the injector. The figure indicates the development of velocity from the bottom part of the chamber (near inlet) before achieving its high velocity through the atomizing nozzles [8].

Fig. 3 exhibits the nozzle exit velocity profile of all the plates and the velocity across the chamber at the exit plane, with 30 mm downstream of the plate. The profiles showed that the flow was almost symmetric with three peaks of the velocity, and the central velocity was lower than its left and right peaks. This difference was due to the large opening of inlet fuel at the center of the chamber and lower mass flow rate compared with the air inlet at its peripheral [9]. At this location far from the inlet, the velocity was close to the bulk velocity in the chamber. MCJ 3 displayed the highest velocity compared with MCJ 1 and MCJ 2. This result was related to the open area, $\%A_e$, which demonstrated that MCJ 3 had the lowest value of 10.03 compared with MCJ 1 (17.82) and MCJ 2 (13.92).

Fig. 4 represents the turbulence kinetic energy of mixing fluids along the premix chamber of the injector. The turbulence kinetic energy of all the plates increased toward its peak at 6 mm from the inlet; this value decreased and became constant before it sprayed out through the nozzle. Based on the graph, MCJ 3 flow inside the injector displayed the highest turbulence kinetic energy because of its low open area compared with MCJ 1 and MCJ 2.

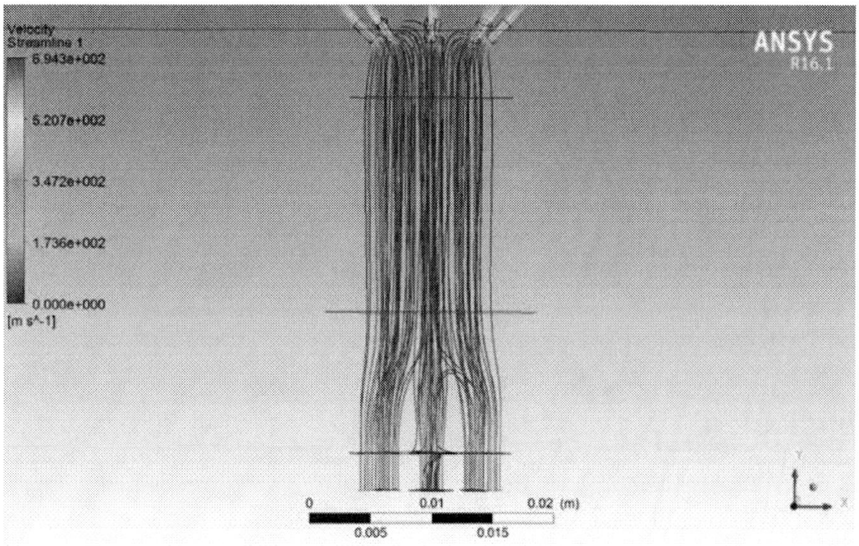

Fig. 2. The velocity streamlines of mixture from the inlet and exit through the nozzle of injector for MCJ1

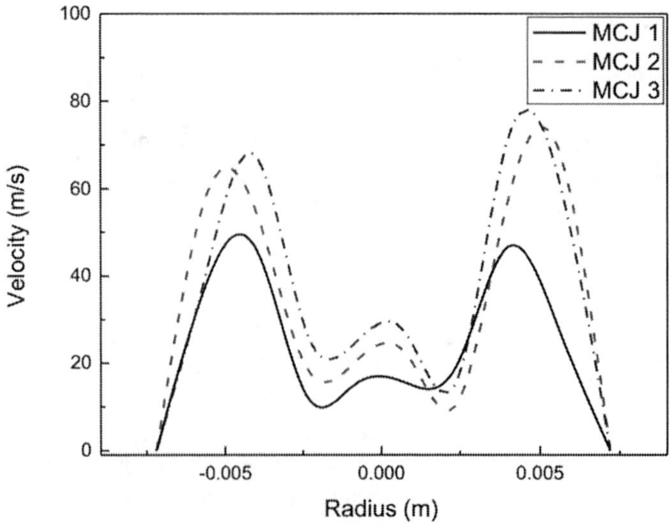

Fig. 3. The nozzle exit velocity profiles of all the plates

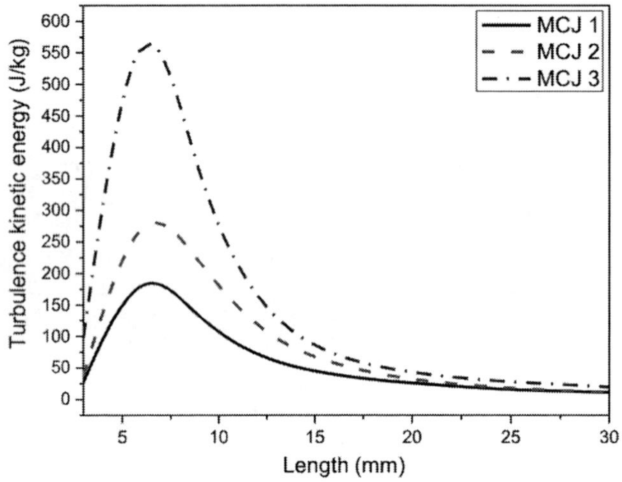

Fig. 4. The turbulence kinetic energy in the premix chamber

4 Conclusion

In this paper, the turbulence generated from MCJ plates was investigated numerically using ANSYS Fluent. Its influence on the flow behavior in the premix chamber of the atomizer was also studied. The flow characteristics were mainly influenced by the geometry of the MCJ plate. The open area, $\%A_e$, significantly affected the turbulence, which could improve the mixing of fuel and air inside the premix chamber. Overall, MCJ 3 with 10.03$\%A_e$ exhibited the highest performance among the three plates.

The authors would like to thank the Centre for Graduate Studies, Universiti Tun Hussein Onn Malaysia and Ministry of Higher Education, Malaysia, for supporting this research.

References

1. A.A. Verbeek, *Efficiently generated turbulence for an increased flame speed* (University of Twente, Netherland, 2014)
2. R.E. Seoud, J. C. Vassilicos, Phys. Fluids., **19**, 105108 (2007)
3. A.A. Verbeek, T.W.F.M. Bouten, G.G.M. Stoffels, B.J. Geurts, T.H. Meer, Combust. Flame, **162** (2015)
4. P. Krogstad, Int. J. Heat Fluid. Fl., **35** (2012)
5. H. Suzuki, K. Nagata, Y. Sakai, R. Ukai, Phys. Scripta, **T142**, 014069 (2010)
6. G. Coppola, A. Gomez, EXP. Therm. Fluid Sci., **3**, 1037 (2009)
7. N. Soulopoulos, J. Kerl, F. Beyrau, Y. Hardalupas, A.M.K.P. Taylor, Fluid. Dyn., (2010)
8. R.Y.S. Chin, A. Khalid, A.M.A. Ghani, I.M. Mahad, S.H Amirnordin, (IRIS, Putrajaya, 2015)
9. N. Soulopolous, J. Kerl, T. Sponfeldner, F. Beyrau, Y. Hardalupas, A.M.K.P. Taylor, J. C. Vassilicos, Fluid. Dyn. Res., **45**, 061404 (2013)

Ballistic Impact Response of Woven Hybrid Coir/Kevlar Laminated Composites

A.R Azrin Hani[1,2,*], *N.S* Azman[1], *R.* Ahmad[3], *M.* Mariatti[4], *M.N* Roslan[1,2], and *N.* Marsi[1,2]

[1]Faculty of Engineering Technology, University Tun Hussein Onn Malaysia, 86400 Parit Raja, Batu Pahat, Johor, Malaysia.
[2]Advanced Technology Centre, Faculty of Engineering Technology, University Tun Hussein Onn Malaysia, 86400 Parit Raja, Batu Pahat, Johor, Malaysia.
[3]School of Mechanical Engineering, Engineering Campus, Universiti Sains Malaysia, 14300 Nibong Tebal, Pulau Pinang, Malaysia.
[4]School of Materials and Mineral Resources Engineering, Engineering Campus, Universiti Sains Malaysia, 14300 Nibong Tebal, Pulau Pinang, Malaysia.

Abstract. The effects of different laminated hybrid composites stacking configuration subjected to ballistic impact were investigated. The hybrid composites consist of woven coir (C) and woven Kevlar (K) layers laminated together. The samples of woven coir were prepared using handloom device. The composites were produced by stacking the laminated woven coir and Kevlar alternately with the presence of the binder. The samples were tested under ballistic impact with different stacking configuration. The results obtained had successfully achieved the National Institute of Justice (NIJ) standard level IIA with energy absorption of 435.6 kJ and 412.2 kJ under the projectile speed of between 330 m/s and 321 m/s respectively. Samples that having Kevlar layer at the front face and woven coir layer as back face achieved partial penetration during projectile impact. This orientation is proven to have good impact energy absorption and able to stop projectile at the second panel of the composites.

1 Introduction

Body armour is a protective garment used to protect the body from any projectile penetration. According to Cavallaro [1], the earliest invention that used body armour concept is using animal skin as well as Natural Fibre (NF) made from thatch, cotton, and silk that is woven to make a protective garment. The concept is then evolved using a more advanced material such as iron, copper, steel that made into the plate to protect the body. Hence, the fibres being used need to be as strong as it could be. Today, Synthetic Fibre (SF) has manipulated the production of body armour with its good mechanical properties. The most commonly used SF in body armor is the high performances fibre such as Kevlar and Twaron from aramid family, Ultra High Molecular Weight Polyethylene (UHMPE) and nylon.

* Corresponding author: azrin@uthm.edu.my

The advanced research was carried out to improve the quality of body armor performances and trying to replace the SF used with NF that has a great potential, equal to the SF. Based on the report by Wambua [2], NFs have the economical value since it is from natural resources, renewable, biodegradable, recyclable and have low health risk when inhaled compared to GF and SF. However, the NF contain cellulose, hemicelluloses, pectins and lignin and are rich in hydroxyl groups which making the characteristics of NF hydrophobicity. With this character, it reduces the interfacial adhesion of the NF. Cellulose is often found as a relatively high modulus, fibril component, of many naturally occurring composites, such as wood, where it is found in association with lignin [3] According to the research of Yousif et al. [4] highlighted that this distinct value is due to its weakness in the interfacial adhesion of NFs with synthetic matrices. So in order to improve this issue, surface modification needs to be done.

Meon et. al [5] have conducted research on the potential Kenaf fibre with the treatment of NaOH. This research was inspired by the previous research done by Mwaikambo and Ansell [6] where they treated NFs such as hemp, sisal, jute and kapok fibres with various concentration of caustic soda. Their aim was to remove the surface impurities in order to enhance the surface adhesion characteristic. With the treatment of alkali, the NF tends to have a better thermo physical properties. It is proven that NaOH improves the interlocking and chemical bonding between the fibre and matrices which resulting in a better mechanical property.

The effects of different laminated stacking configuration subjected to ballistic impact were investigated. The composites consist of woven Kevlar (K) and woven treated coconut coir (C). A follow-up investigations were also done to assess blunt trauma effect.

2 Experimental Method

2.1 Material

Plain weave Kevlar 29 was used in this research. It was taken from China Beihai Fiberglass Co. Ltd., China. The yarn size is 1100 dtex whereas the fabric weight is 200 g/m^2. The coconut coir yarn used in this experiment was provided by Btex Engineering Ltd., India. The yarns were weaved into the plain woven structure formed into composite panel and tested it under ballistic test. The density of coir as stated by Akil et.al. [3] is within the range of 1.15 – 1.46 g/cm^3.

The matrix used was epoxy DER 332 of density 1.16 g/mL. Jeffamine D-230 hardener of density 0.948g/mL was used as curing agent. Both the resin and hardener were supplied by Penchem Technologies Sdn. Bhd. Malaysia.

2.2 Sample Preparation

The process started with the preparation of dry fabric/mat of plain woven coir. Woven coir as a composite reinforcement was produced with a self-designed handloom [7]. The woven fabrics were then soaked as part of the treatment process in sodium hydroxide (NaOH) solutions of concentrations 6%. The manufacture of structural composites involved the compression moulding method. Fibre–matrix ratios were fixed at 3:7 by weight percentage. The mould was closed and placed in a pressing device overnight for curing. The composites were then analysed in terms of thickness, and weight. The composites were labelled and stacked into one panel which consists of three same composites. The stacking configuration was varied to investigate its effect on ballistic impact.

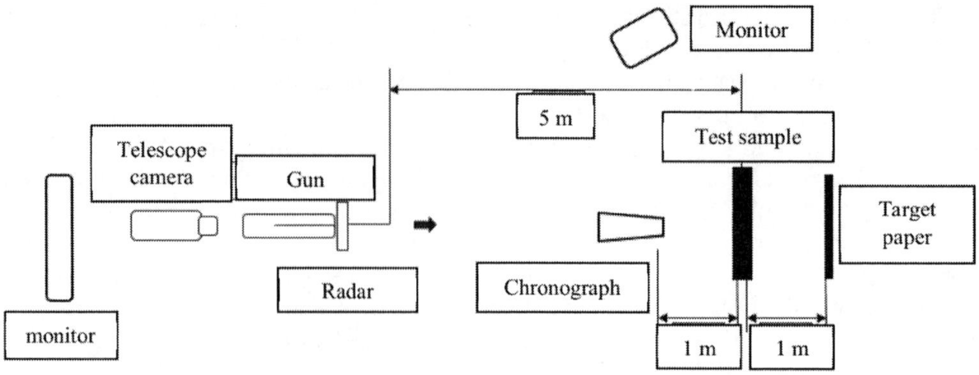

Fig. 1. Schematic diagram of testing apparatus in STRIDE, Batu Arang.

2.3 Testing Method

Samples were tested at Science and Technology Research Institute of Defence (STRIDE), Batu Arang, Selangor for high-velocity impact test. The assessments involve the determination of ballistic limit, damage mode, and blunt trauma deformation. The tests were performed guided by NIJ Standard 0101.04 [8] level IIA with 9 mm calibre FMJ RN 8.0g. Figure 1 shows the schematic diagram of the testing apparatus. The samples to be tested were cut into 310 × 205 mm dimensions, and thickness within the range of 7 mm to 9 mm for one composite panel (each containing 4 layers of woven laminated coir/ Kevlar). It was then stacked into 3 sets of panel (12 layers of woven laminated coir/Kevlar). The ballistic impact was conducted using high-velocity impact gun muzzle.

3 Results and Discussion

3.1 Ballistic Impact Response

Table 1 shows the results of ballistic impact response of composites composed of different stacking sequence. The first shot was using only 2 panels. The top surface of the panel is Kevlar while the back surface is woven coir. The results were a complete penetration. The velocity used to be 368 m/s which categorized under NIJ-Standard level II. However the panel is too thin and the velocity of the projectile was too high that the panel could not stop the projectiles. The structure of woven coir which placed as the back face of the panel was dispersed and break into small pieces. This was due to the structure of the panel where the composites were not properly dried due to the environmental factor and mould used. However, the semi-dried panel was observed to be good in energy absorption. This was proven by the 3, 4 and 5 shots in this test.

For the second shot, another panel was added. Hence given a total of 3 panels in a sample with a total weight and thickness of 1.89 kg and 25.07 mm respectively. The projectile speed was recorded as 378 m/s which satisfied threat level II. The stacking orientation for the panel was KKCC-KKCC-CCKK where Kevlar was placed on the front and back face of the panel. However, complete penetration occurred which makes the researcher decided to reduce the speed of the projectile. On the third shot, the striking velocity or the impact velocity was reduced to 330 m/s. The velocity of the projectiles was reduced by decreasing the propellant charged weight of total 350mg. The more charged weight used, the higher the velocity of the projectile. The stacking orientation of the panel for this shot was KKCC-KKCC-KKCC with Kevlar on the front face, and woven coir on the back face of the panel. The energy absorption for this shot was 438 kJ. This panel achieved the desired result with partial penetration condition. Stacking sequence or orientation was believed to be the main factor that can arrest the projectile. The placement of Kevlar on the front face of every panel slowed down the velocity of the projectiles while the presence of the woven coir at the back absorbed the energy released. It was observed that the fracture mode on Kevlar layer was small and greater delamination occurred. Delamination is believed due to the dissipation of energy at Kevlar layer during projectile impact. Aktas et. al [9] clearly stated that in hybrid composites, woven structure placed on the outer layer gave a better result. This works well for coconut coir that was weaved into the plain woven structure and proven to withstand high impact energy.

Table 1. Ballistic test results of woven coir/Kevlar composites.

Shots number	**Stacking orientation**	**Striking Velocity 1 (m/s)**	**Striking Velocity 2 (m/s)**	**Striking Energy (kJ)**	**Penetration (CP/PP)**	
					CP	**PP**
1	2 panels *KC + KC	369.8	368.0	544.0	√	
2	3 panels *KC + KC + CK	375.8	378.0	574.0	√	
3	3 panels *KC + KC + KC	328.5	330.0	438.0		√
4	3 panels *KC + KC + KC	321.5	321.0	414.0		√
5	3 panels *KC + KC + CK	348.7	348.0	487.0		√
6	3 panels *KC + KC + CK	343.8	343.0	473.0	√	

*Each panel consist of 2 layers of woven Kevlar (K) and 2 layers of woven coir (C).
*CP=complete penetration, PP=partial penetration

From the fourth shots, the same orientation used was gave a similar trend with the previous shot (shot 3). The striking velocity recorded was 321 m/s and partial penetration was achieved with 414 kJ of energy being absorbed by the composites. By maintaining the type of stacking configuration and the range of projectile striking velocity in shot number 5, it also produced the same ballistic response with the previous shot number 3 and 4. Partial penetration occurred. With this trend, it is proven that this orientation exhibited better ballistic energy absorption. However, an attempt to increase the projectile velocity will need modification on the panels thickness and orientation. With the higher projectile striking velocity (as referring to shot number 6 from Table 1), the panel shows complete penetration. It can be concluded that the hybrid of woven coir and Kevlar laminated composites has high potential as applications for high protective materials at lower cost.

3.2 Blunt Trauma

Blunt trauma test was done using composite panel where the stacking orientation was the same as the successful panel with partial penetration in ballistic test in the previous section. The backing material was Roma Plastilina with the temperature of 40°C. The results show that the projectile was successfully stopped and embedded inside the panel. Blunt trauma depth of penetration was measured using Vernier Calliper. The depth of penetration is called as back face signature (BFS).

A shot was launched and the results of BFS for blunt trauma assessment are tabulated in Table 2. From Table 2 it is obvious that area of the deformation was 21.16 cm^2. The bigger area represents the energy being dissipated in a wider area to stop the projectiles. The depth of penetration or so called BFS also shows a very good result with the length of 6.79 mm, whereas the limit given by standard in designing armor panel must be less than 24 mm.

Table 2. Blunt trauma assessment result.

Orientation	Length horizontal x, (cm)	Length vertical y, (cm)	Area of Deformation (cm^2)	Penetration	
				CP/ PP	BFS (mm)
3 panels: *KW + KW + KW	4.6	4.6	21.16	PP	6.79

3.3 Damage Assessment

The partial penetration (PP) samples are referring to shot number 3, 4 and 5 as detailed in Table 1. Damage is obviously seen on the back face of the panel. For shot number 3, the damage mode on the first panel was small. This was expected due to the impact of the projectile which was still in high velocity. When the projectile reaches the second panel, the damage mode was huge. The bigger the damage mode the higher the energy absorption. At this stage, the projectile stop and the woven coir were broken into pieces due to high impact. The projectiles were found crumpled right at the very end of the second panel on the structure of woven coir. Even though it stops on the second panel, there was still huge damage done on the third panel. The layer of Kevlar on the third panel does not penetrate through, however, the woven coir structure ruptured on the outside.

The similar damage mode can be seen for shot number 4. The similarity might be due to the same orientation used where the Kevlar as the front face and woven coir as the back face of the panel. From the results and observation, this consistency proved that this structure was strong enough to stop the projectiles. The damage on the first panel (as predicted) was small, both for Kevlar and woven coir. Whereas for the second panel, larger damage mode can be seen on the woven coir but not on the Kevlar structure. Based on this observation, woven coir is proven to have a better energy absorption. It has similar damage mode with shot number 3. On the third panel, Kevlar layer does not break, but the energy was absorbed by the woven coir on the back face of the panel.

A different condition was observed for shot number 5. The energy absorbed by the panel was 487 kJ, which was considered high. The plate achieved the condition of partial penetration but Kevlar which was placed on the back face of the panel peeled off from the panel as clearly seen in Table 3. Kevlar was expected not to be properly stuck to the woven coir during the fabrication which causing it to be separated when high impact applied. The damage mode on the third panel of the woven structure was the largest among the 3 shots of PP condition. The higher velocity of bullet resulting in this condition.

DOI: 10.1051/mateconf/20167801048

Table 3. Visual damage assessment of partial penetrated (PP) samples.

Shot number	First panel		Second panel		Third panel	
3	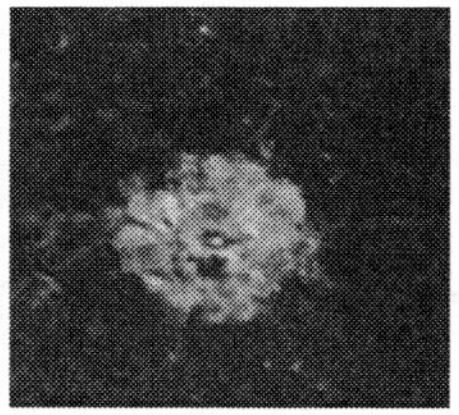	Bigger matrix crack and fiber break	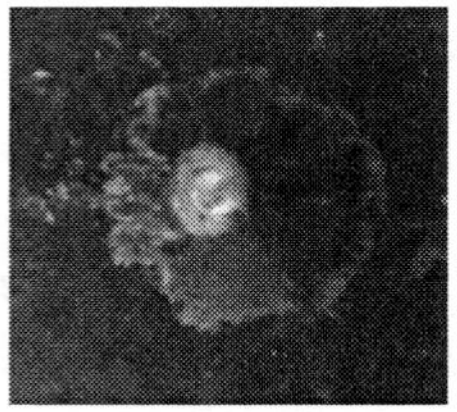	Bullet stops on the matrix crack with medium size of fabric deformation		No penetration of bullet, small crack on the matrix and fibre does not completely break
4	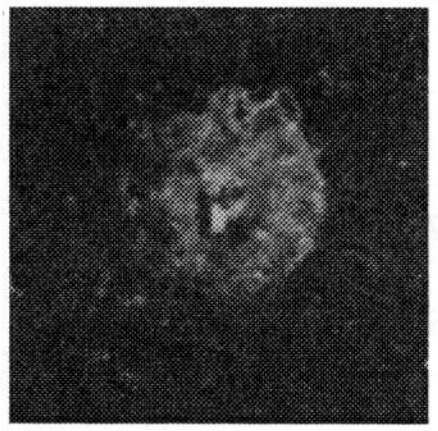		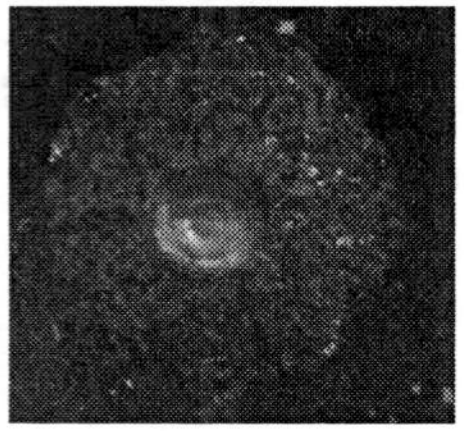	Bullet stops on the matrix that crack with bigger fiber deformation		No penetration of bullet, the matrix crack was slightly bigger and a little fiber break occurred
5	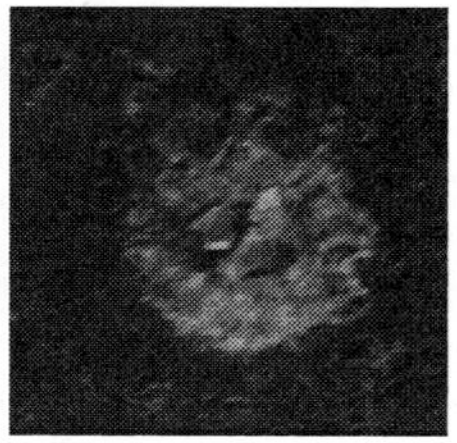		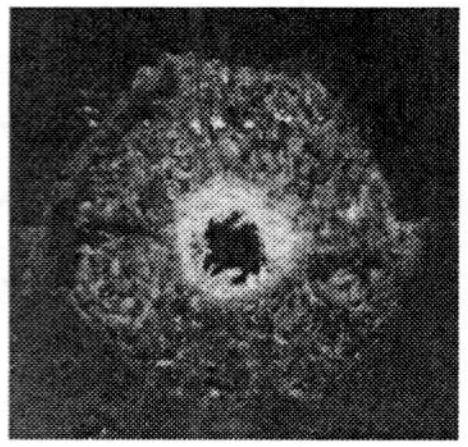	Larger matrix crack and large fiber deformation due to higher impact load	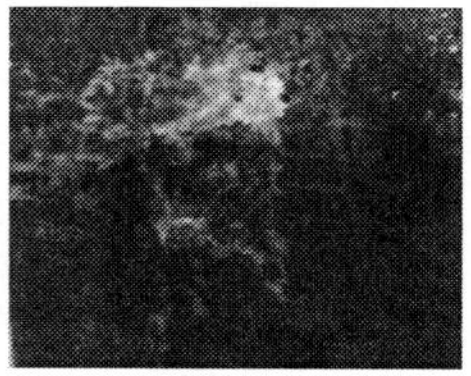	Kevlar was peeled off from the composites, bigger matrix crack and fibre break.

4 Conclusion

It can be concluded that the generated information from this research has achieved the objective which is to investigate the effects of different laminated coir/Kevlar hybrid composites stacking configuration subjected to ballistic impact. The best stacking sequence would be KKCC-KKCC-KKCC with the ballistic limit of 348 m/s. The same orientation successfully undertook blunt trauma test with BFS depth of 6.79 mm. This study has high impact in reducing the usage of the expensive and non-biodegradable product, Kevlar to more than half of the normal usage in the production of armour panel.

The authors gratefully acknowledge Advanced Technology Centre (ATC), Faculty of Engineering Technology, Universiti Tun Hussein Onn Malaysia (UTHM) for supporting this research. The authors also would like to express sincere appreciation to the staffs of Science and Technology Research Institute of Defence (STRIDE) for the generous assistance.

References

1. P.V. Cavallaro, *Soft body armor : An overview of materials, manufacturing, testing and ballistic impact dynamics* (Naval Undersea Warfare Center Division, NUWC-NPT Technical Report, 2011)
2. P. Wambua, J. Ivens, I. Verpoest, Compos. Sci. Technol., **63(9),** 1259 (2003)
3. H.M. Akil, M.F. Omar, A.A.M. Mazuki, S. Safiee, Z.A.M. Ishak, A. Abu Bakar, Mater. Design, **32**, 4107 (2011)
4. B.F. Yousif, A. Shalwan, C.W. Chin, K.C. Ming, Mater. Design, **40**, 378 (2012)
5. M.S. Meon, M.F. Othman, H. Husain, M.F. Remeli, M.S.M. Syawal, Procedia Engineering, **41**, 1587 (2012)
6. L.Y. Mwaikambo, M.P.Ansell, Macromol. Mater. Eng., **272(1)**, 108 (1999)
7. A.R. Azrin Hani, M.F. Shaari, N.S. Mohd Radzuan, M.S. Hashim, R. Ahmad, M. Mariatti, The International Journal of Automotive Engineering, **8**, 1197 (2013)
8. NIJ Standard-0101.04, *Ballistic resistance of personal body armor* (US Dept. Just. Washington, DC, 2001)
9. Aktas, M. Aktas, F. Turan, Compos. Struct,, **103**, 119 (2013)

Effect of Calcination Temperature on Morphological and Topography of Nickel-Alumina Thin Film

Khairul Ilman Sarwani[1,*], *Rais Hanizam* Madon[1], *Siti Aida* Ibrahim[1], and *Mas* Fawzi[1]

[1]Energy Technologies Research Group (EnRG),Center for Energy and Industrial Environment Studies (CEIES), Faculty of Mechanical and Manufacturing Engineering, Universiti Tun Hussein Onn Malaysia, 86400 Parit Raja, Batu Pahat, Johor, Malaysia

Abstract. Dip coating process promises good potential of nickel-alumina catalyst deposition on metal substrate for various applications especially in gas conversion reaction. This study was conducted to investigate the effect of different calcination temperature on nickel-alumina catalysts thin film formation. Four different calcination temperature were used, which are 300°C, 400°C, 500°C and 600°C. The calculation process was conducted for a duration of 90 minutes. The deposited thin films were characterized using Atomic Force Microscopy (AFM) and X-ray diffraction (XRD) equipment. The AFM result showed that the surface roughness of the nickel-alumina increase proportionally from 56 to 275 nm when the calcination temperature increased from 300 to 600°C. From an observation at high calcination temperature, the atom of grains assisted diffusion at the crystallite point causing grain with lower surface energy become larger. As the calcination temperature increase, the surface profile becomes rough and uneven representing high surface roughness. Thus, the effect of calcination temperature greatly influences the surface roughness of the nickel-alumina thin film.

1 Introduction

Catalysis is a process to accelerate chemical reaction using a catalyst that is not consumed by the overall reaction. Chemical reaction could be initiate with low amount of energy in presence of catalysts. Currently steam methane reforming has been widely used to produced hydrogen because of its economic feasibility [1]. Among many catalysts available for steam methane reforming, nickel-based catalysts meet these criteria for a whole range of chemical reaction [2]. Moreover, the ability of nickel to absorb large amount of hydrogen help increased the reaction efficiency in the microreactor.

Nickel-based catalyst prepared using sol-gel with alumina support have high hydrogen selectivity. The properties of alumina support are influence by the aluminas treatment temperature [3]. Furthermore, the activity of nickel catalyst supported with alumina made on sol-gel method is influence by the calcination temperature during preparation [4]. Nickel-alumina thin film surface roughness is one of the main potential enhancement factor

* Corresponding author: ilmansarwani@gmail.com

for the high methane conversion and hydrogen yield [4,6]. It was studied that thin film thickness and roughness is proportional to the substrate's roughness. The formation of the final prepared catalyst depends on the calcination parameter such as time and temperature [6]. An active catalyst is usually finely dispersed metal over support porous alumina. Employed catalyst coated on the plates substrate reduces mass and heat transfer restriction compared with pellet catalyst and can improve its effectiveness for further intensification. The presence of catalyst will lead to faster reaction at certain operating temperature condition, by reducing the amount of energy required to initiate chemical reaction [4]. Furthermore, nickel-alumina prepared by sol gel and dip coating method has been reported to produce more stable support structure and higher nickel dispersion [7]. These dipping process is crucial in determining the properties of coated thin film nickel-alumina catalyst on top of substrate's surface area.

This study aims to observe the effect of calcination temperature on the formation of nickel alumina thin film by dip coating method. Using fixed nickel-alumina sol-gel concentration, various calcination temperatures were tested. By comparing the results obtained from previous literature, it was expected to yield the critical information regarding calcination temperature effects for more optimized nickel-alumina catalyst preparation.

2 Materials and Methods

The substrate used are stainless steels grade 304 with a specific dimension of 15 mm width x 60 mm length x 0.5 mm thickness. Sol-gel solutions were prepared using Nickel(II) nitrate hexahydrate – $Ni(NO_3)_2{\cdot}6H_2O$ and Nitric Acid 65 % – HNO_3 obtained from R&M Chemicals, Aluminum isopropoxide – $Al[OCH(CH_3)_2]_3$ from Acros Organics, and Ethanol 99.8 % obtained from ChemAR System.

Using Skymen JP-020S ultrasonic cleaner the substrates were cleaned to remove dust and contaminants on the substrate surface. Cleaned stainless steel was polished using abrasive paper. Later, the substrates were sanded with abrasive paper grit 400 to increase it roughness for better adhesion. Each substrate was polished in one way direction for 30 cycles.

The solution was prepared by dissolving 22 g Aluminum isopropoxide with 5 ml distill water, 5 ml Nitric acid and 90 ml ethanol. Then the solution was stirred at 60°C until homogenous for one hour. Next, 10 g of Nickel (II) nitrate hexahydrate was added slowly the mixture was stirred continuously at 60°C until homogenous which took another one hour. The nickel-alumina solution was used directly for dip coating process without undergoing aging process.

The dip coating schematic diagram was shown as Fig. 1. PTL-MM01 Dip Coater was used for dip coating process with dipping and withdrawal speed range of 1-200 mm/min. For dip coating sector, the dipping and withdrawal speed used was 160 mm/min. The 304 stainless steel plates were dipped and then withdrawn from the bath at a prescribed withdrawal velocity. For this experiment, the heat treatment used was isothermal annealing by using Carbolite RHF 14/3 box furnace at operating temperature of 300°C, 400°C, 500°C, 600°C for 90 minutes.

In this research, physical tests had been carried out to investigate the characteristic of the deposited thin film catalyst. The substrates were characterized using Atomic Force Microscopy, AFM (XE-100 Park Systems) which provided a 3D profile of the surface in a Nano scale and X-ray diffraction (XRD) to identify unknown crystalline materials on the coated film. The equipment used was Panalytical X'Pert^3 Powder and analized using Highscore software.

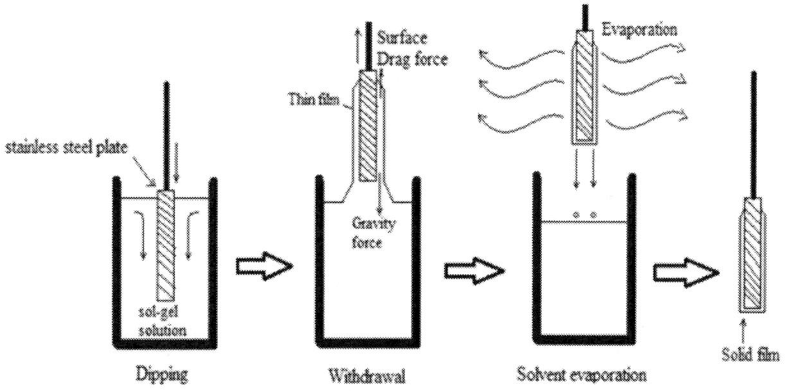

Fig. 1. Schematic diagram of dip coating process.

3 Results and Discussions

Fig. 2 shows an AFM image of nickel-alumina thin films calcination from 300 to 600°C. Based on the result obtained shows the surface of nickel-alumina forming the shape of the hill and valley. According to Kumar et al. [8] as the surface roughness increase, charged oxygen–related species were absorbed on the crystalline film for alumina formation. Due to the high calcination temperatures, there were high potential for the migration of grain's boundaries and straightly causing coalescence of more nickel-alumina grains. The nucleation rate of the particles increased rapidly as the calcination temperature increase causing larger grain. Parra et al. also mentioned when the temperature increase the supersaturation of the reaction products also increase to accelerated the crystal core forming of the reaction within short time [9]. Furthermore, at high temperature stage, the atom of grains is assisting diffusion at the crystallite point to occupy the correct site in the crystal and causing grain with lower surface energy become larger. With temperature continuing to rise, rapid formation of crystal nucleus were expected which control the morphology and structure of the final nickel-alumina film. Crystallite size, particle size and particle morphology are strongly dependent on heating temperature through dip coating, the grain size increase as calcination temperature increase.

Using AFM, the surface roughness was characterize by obtaining the roughness parameters R_a, R_q and maximum elevation. The surface roughness's of the catalyst prepared play an important role in Steam Methane Reforming (SMR) performance. Measuring the surface roughness of the prepared catalyst is very important to determine hydrogen yield [6]. Fig. 3 represent an analysis of the roughness of each film varying from calcination 300°C - 600°C. At 300°C showing only peak without valley represent smooth surface profile indicating low roughness value. Apparently, from 400°C to 600°C showing more peak and valley represent roughness of the film layer. As the calcination temperature increase the surface profile become more rough and uneven representing high surface roughness.

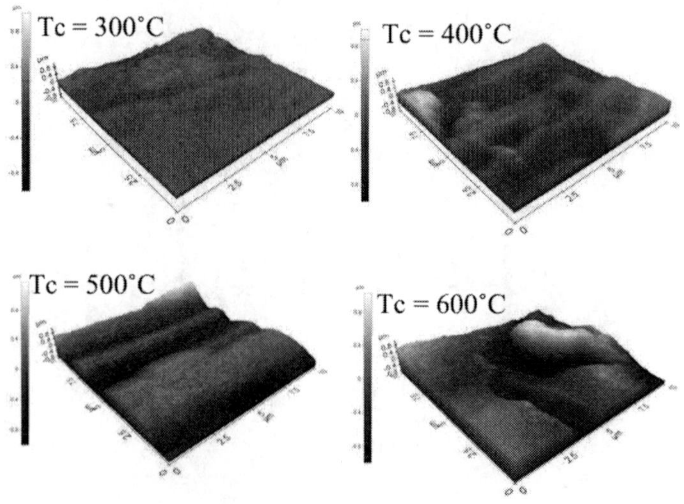

Fig. 2. AFM image of nickel-alumina for calcination temperature.

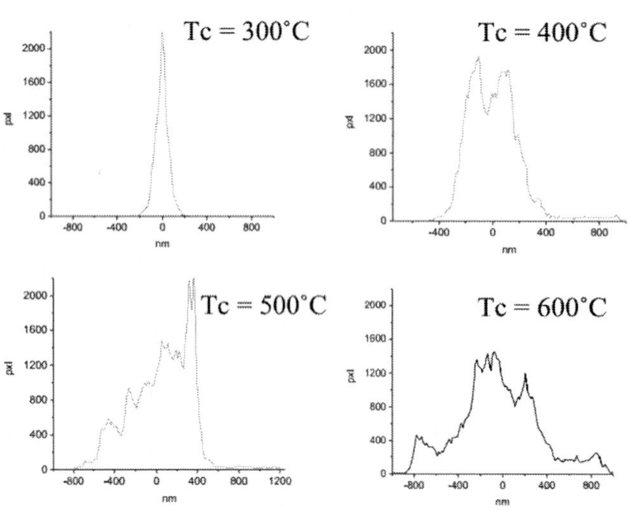

Fig. 3. AFM surface profile analysis of nickel-alumina thin films.

Table 1 shows the average roughness (R_a) had the same trend as the ten-point mean height (R_z) for all prepared nickel alumina thin film. Surface roughness of the nickel-alumina increase from 56 to 275 nm as the calcination temperature increased from 300 to 600 °C. From Table-1 shows high value R_a, R_z also has high value because of strong dependence of R_z on the peak height or valley depth. From Gaussian distribution of asperity height show that the ratio R_q to R_a should be 1.25 [8]. As shown in Table-1, the maximum value of R_q/R_a is 1.32, which reasonably closed to the value of 1.25 indicating the statistical relationships for surface roughness are applicable. In Table 1, negative value of skewness (R_{sk}) indicate that the valleys are dominant over scanned area and positive values show that the peaks are dominant on the surface. In this case, the negative value indicates crack on the film layer. The defect caused by the adatoms trapped at these sites forming nuclei for subsequent growth process called heterogeneous nucleation. The nuclei will then grow due to further addition of adatoms and the island up to that size can dissolve again as stated by Bäumer et al. Since the values were low, the cracks were minor [8].

Table 1 also showing the value of R_{ku} had more than 3 indicating high peaks and low valley due to strong relation between R_z and R_a .

Table 1. Roughness parameters of nickel-alumina thin films annealed at different temperatures.

Calcination Temperature (◦C)	R_a nm	R_q nm	R_z nm	R_q/R_a	R_{sk}	R_{ku}
300	56	74	581	1.314645	-0.006	3.746
400	134	177	1322	1.320427	-1.288	6.564
500	244	302	2007	1.237705	0.123	3.275
600	275	355	1898	1.290909	-0.213	3.132

Based on Table 1 the average roughness (Ra) and ten-point height (Rz) were plotted into graph as shown in Fig. 4. The value Rz also coFuld be used as the thicknesses of the thin film since it calculate the difference between highest peak and lowest valley. Rz values in Fig. 4 showed as the calcination temperature increase the thickness of the thin films also increase. The highest thickness obtained is at temperature 500˚C, the thickness drop at 600˚C. It was the evident that the particles become flatter when the adhesion energy increases. Apparently, the value of surface roughness did not affected by thickness of the thin film as shown in Fig. 4 , the surface roughness of the thin film still increasing even at 600˚C.

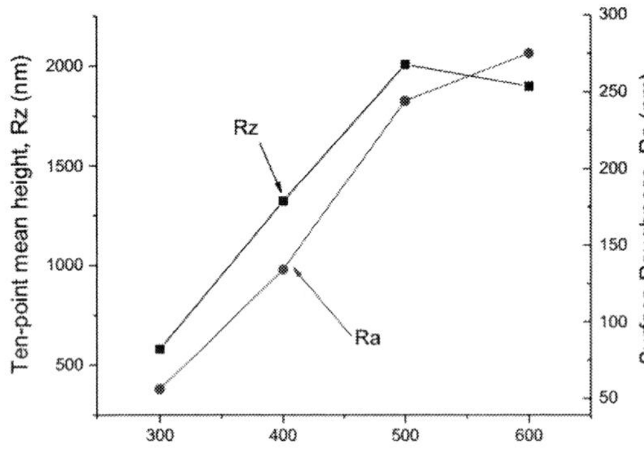

Fig. 4. Thin film surface roughness (Ra) and ten-point mean height (Rz) measured using AFM.

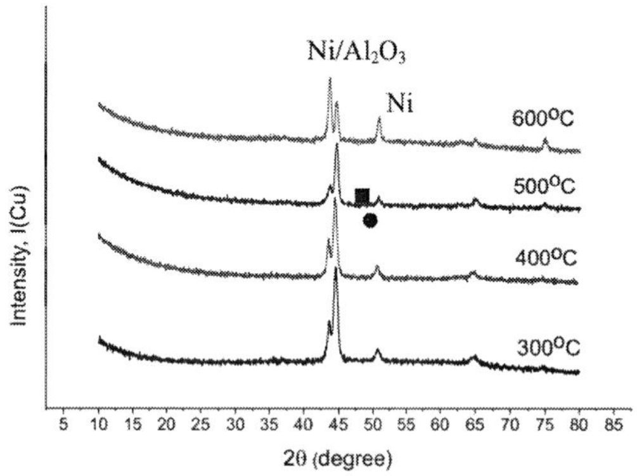

Fig. 5. XRD patterns of catalyst calcined at various temperature

All calcined catalysts were examined by XRD in order to investigate the phase development. Fig. 5 present similar pattern for catalysts calcined at 300, 400˚C, 500˚C and 600˚C. It appeared that there is not much changein particle size with calcination temperature. It can be seen that Nickel phase and nickel – alumina phase were presents on the substrates. As the temperature increase the amount of nickel decreasing to form nickel – alumina. This may indicates that temperature play an important role for the formation of nickel – alumina.

4 Conclusions

This study had shown that the surface roughness of the nickel-alumina increase proportionally from 56 to 275 nm as the calcination temperature increased from 300 to 600 °C. In addition, the calcination temperature would stimulate the migration of grain's boundaries and straightly causing coalescence of more grains. As observe from the surface profile, it has become more rough and uneven which representing high surface roughness accordingly to the increment of calcination temperature. The XRD result also shown that as the calcination temperature increase the formation of nickel alumina would greatly become higher. Therefore, a high temperature during the catalyst preparation is expected to increase the yield of the steam methane reforming process.

The author would like to thank the Ministry of Education for partly supporting this research under the Fundamental Research Grant Scheme (FRGS) vot. 1421.

References

1. Y. Bang, S. Park, S.J. Han, J. Yoo, J.H. Song, J.H. Choi, K.H. Kang, and I.K. Song, Appl. Catal. B-Environ., **180**, 179 (2016)
2. A. Ishihara, T. Imai, T. Hashimoto, H. Nasu, Fuel Process. Technol., **136**, 34 (2015)
3. Z. Yaakob, A. Bshish, A. Ebshish, S.M. Tasirin, F.H. Alhasan, Materials, **6(6)**, 2229 (2013)
4. B. Bej, N.C. Pradhan, S. Neogi, Catal. Today, **207**, 28 (2013)
5. S. Allahyari, M. Haghighi, A. Ebadi, Chem. Eng. Process. Process Intensif, **86**, 53 (2014)
6. A.S.A. Al-Fatesh, A.H. Fakeeha, Journal of Saudi Chemical Society, **16(1)**, 55 (2012)
7. M. Zangouei, A.Z. Moghaddam, M. Arasteh, Chem. Eng. Res., **14(2)**, 97 (2010)
8. B. Rajesh Kumar, T. Subba Rao, Dig. J. Nanomater. Bios. **7(4)**, 1881 (2012)
9. M.R. Parra, F.Z. Haque, J. Mater. Res. **3(4)**, 363 (2014)

MATEC Web of Conferences **78**, 01050 (2016) DOI: 10.1051/matecconf/20167801050
IConGDM 2016

A Comparative Study of an Lpg- Spark Ignition Engine using Liquid Sequential Injection Technique

Mohd Mustaqim Tukiman[1,*], *Norrizal* Mustaffa[1], *Mas* Fawzi[1] , *Shahrul Azmir* Osman[1], and *Rais Hanizam* Madon[1]

[1]Automotive Research Group (ARG) , Center for Energy and Industrial Environment Studies (CEIES), Faculty of Mechanical and Manufacturing Engineering, Universiti Tun Hussein Onn Malaysia, 86400 Parit Raja, Batu Pahat, Johor, Malaysia

Abstract. The development of liquid phase Liquefied Petroleum Gas (LPG) as an alternative fuel for the gasoline fuel is becoming popular in spark ignition engines. In this research, an experimental work was conducted on a1.6L 4-cylinder multi-port injection engine. The engine was retrofittedwith a Liquid Sequential Injection (LSI) LPG system. Its performance in terms of brake power,brake torque, brake specific fuel consumption (BSFC) and exhaust gas emissions (CO, CO_2, HC, O_2, and NOx) were evaluated from 1500rpm to 4000rpm with half-open and wide-open throttle positions. In comparison with gasoline, the brake torque and brake powerfor LPG were higher, but the BSFC and exhaust emission of CO and $CO_2$for LPG are lower.

1 Introduction

Liquefied petroleum gas (LPG) is a clean alternative fuel for vehicle compared to the conventional fuels such as gasoline and diesel. In general, conventional fuels has high carbon ratio, low octane number and may gives higher emission levels compared to LPG [1, 2]. In year 2000, more than 6 million vehicles worldwide used LPG [3]. It showed that the technologies in LPG vehicles are getting good acceptance from automotive industries and the consumers. LPG has two phases; gas and liquid. From previous studies, LPG development is divided into five generations, which the 1st-4th generation used gas phase for the fuel delivery system while the latest system uses liquid phase.

According to the Material Safety Data Sheet (MSDS) of LPG in Malaysia, it contains 60% butane and 40% propane. Generally, the composition of LPG varies followed by season, country and characteristics of the supply crude oil, the refining process and the cost refined product. Therefore there is no constant standard value for the LPG composition[4].Table 1 shows the composition of LPG as an alternative fuel of few countries.LPG is suitable as an alternative fuel for internal combustion engine (ICE). This is due to the higher Research Octane Number (RON) of LPG as compared with gasoline as

* Corresponding author: mohdmustaqimakim@gmail.com

shown in Table 2.The liquid LPG produced lower carbon emission due to the lower carbon content in LPG as compared to the gasoline fuel[5, 6].The liquid LPG injection promised to give high resistance knocking phenomena and then the system can use a higher engine compression ratio thus the efficiencies of engine was increased [3].

The liquid phase is more effective than gas phase because a large amount of heat was absorbed into the intake air manifold due energy required to change the phase. This liquid phase is vaporized quickly, but at the same time it works as the cooler on of the mixture inside the intake manifold. As consequences, the engine received a greater mass of fuel-air mixture because greater density of cooler air and resulting volumetric efficiency and torque output were increased [4].The purpose of this study is to evaluate the performance and emission of gasoline and liquid LPG as fuel of a 4-cylinder 1.6L displacement engine using multi-point port injection (MPI).

Table 1. Composition of LPG by country [7].

Country	Propane-butane (%) by volume
Belgium	50-50
France	35-65
Italy	25-75
Spain	30-70
United Kingdom	100-0
Germany	90-10
Malaysia	40-60

Table 2. Selected specifications of LPG and gasoline [8].

Property	LPG	Gasoline
RON	106-111	92-95
Specific Gravity	0.51	0.74
Stoichiometric A/F Ratio	15.7:1	14.7:1
Relative CO_2 per kJ	0.885	1.0

2 Experimental setup

All experiments were conducted on a *Proton Gen 2 (S4PH)*1.6L vehicle. Table 3 shows the specification of the test vehicle engine. The fuel delivery type of this engine is a multi-point port injection (MPI) and a prototype liquid LPG injection were used. The LPG injectors were mounted on the intake manifold that close to the inlet valve. The fuel pressure was regulatedin the range of 9 to 10bar to ensure that the LPG is in the liquid phase during injection process.The system was calibrated before running the experiments to ensure the quantity of injection LPG are adequate. The injection timing for LPG was emulated from the stock electronic control unit (ECU) and this system was controlled by the additional LPG ECU system.

The emissions of carbon dioxide (CO_2), carbon monoxide (CO), hydrocarbon (HC), oxygen (O_2) and oxides of nitrogen (NO_X) emission were measured using*Autocheck* gas analyzer. At the same time the data ofthe air fuel ratiowere also recorded. Next, the test vehicle was coupled to a *Dynapack* chassis dynamometer in order to measure the engine performance. The *Bosch* scan tool V1.2 management system was connected to the test vehicle during testing to enablereal time data to be recorded and monitored for each test condition.The fuel consumption was measured using *Ono Sokki: FZ-2100* mass flow meter for both fuels used. A Schematic diagram of the experiment is shown in Fig. 1.The experiments were carried out at two different throttle valve positions; with half open throttle (HOT) valve position and wide open throttle (WOT) valve position. The engine speed for both throttle valve positions were tested from 1500 rpm until 4000rpm with 500rpm increment. The LPG used composes of 40% propane and 60% butane which were produced by Petroliam Nasional Berhad (PETRONAS).

3 Result and Discussion

Fig. 2 shows the performance of both tested fuels at two different throttle valve positions, HOT and WOT valve position. The brake torque, brake power and brake specific fuel consumption were plotted against the engine speed.The brake power of LPG and gasoline were found increase for both HOT valve position and WOT valve position. LPG produced higher brake power at WOT valve position as compared to the gasoline, meanwhile at HOT valve position brake power produced by both of fuels showed nearly similar figure except at 4000rpm where the brake power value of gasoline dropped. This is due to the fact that the LPG in liquid phase has a higher thermal efficiency than gasoline. Futhermore the temperature mixture was decreased and therefore the density and volumetric efficiency was inversely proportional condition to give positive effect [9].

The brake torque produced by LPG was compared with the gasoline fuel for both throttle valve position. The value of brake torque LPG was increase compared to the gasoline fuel.This is because the LPG cooled surrounding intake air inside intake manifold and results on the greater density [10].BSFC variation of the tested fuels is shown in Fig. 2. For both throttle valve position, BSFC of gasoline was decreased with the increased of engine speed. Meanwhile, for the LPG, the BSFC was almost consistent throughout the engine speed. The BSFC of LPG was lower than gasoline except at 3500rpm and 4000rpm engine speed for tested throttle valve position. This is because the liquid LPG has higher pressure in fuel lines, hence the higher pressure fuel of LPG was released in the cylinder and resulted on lower fuel consumption of LPG itself as compared to gasoline fuel [8, 12].

Table 3. Specification of the test vehicle.

Type model	Proton Gen2 (S4PH)
Total displacement	1.6L
Number of cylinders	4
Orientation	East - West
Valve train	DOHC 16 V
Compression ratio	10.0: 1
Bore x stroke	76mmx 88mm
Power	82KW@6000 RPM
Torque	109lb-ft@ 4000 RPM
Fuel/ system	Gasoline/ multipoint injection

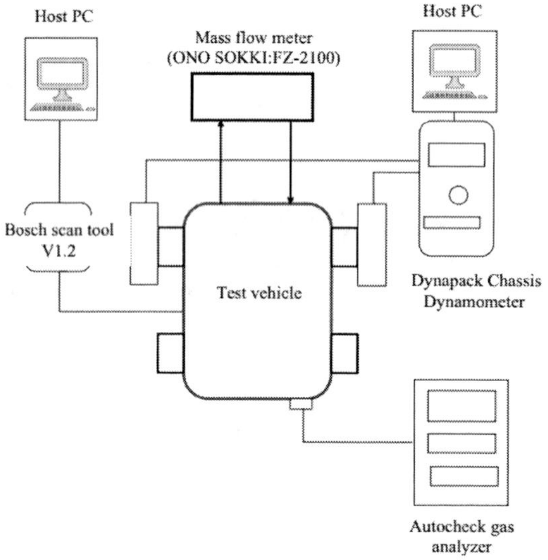

Fig.1. Schematic diagram of theexperimental setup.

Fig. 3 shows the effects of emissions on SI engine fuelled by LPG and gasoline at HOT valve position to WOT valve position. The figure illustrated the emission of CO, CO_2, HC, O_2, and the NO_x.From Fig 3, the CO emission of LPG at the both throttle valve position showed a similar pattern. At HOT valve position, the CO emission produced near similar value with WOT valve position, in the range of 1500rpm up to 3500rpm. However, at higher engine speed, 4000rpm, CO emission of gasoline yielded higher value compared to LPG. Since the carbon-hydrocarbon ratio of LPG is low and the LPG burn effectively with more homogeneous mixture[12].The trend showed CO_2 emissions of gasoline produced higher value as compared to the LPG for the all engine speed tested. This was attributed to the fact that carbon-hydrogen ratio gasoline is largest value than the LPG [12].In addition, gasoline has complete combustion compared with the LPG for these experiments.The HC emission of LPG was found higher than gasoline for both HOT valve position and WOT valve position. This is due to gasoline burned completely comparedwith LPG. The HC emission produced at WOT valve position almost constant for both LPG and gasoline. The LPG was produced high HC compared with gasoline, this may be due to high temperature combustion and non- homogeneity of mixture [12].

The NOx emission of LPG at HOT valve position shows a different pattern as compared to the gasoline fuel and the maximum NOx produced was at 2500rpm as per shown in Fig. 3. At WOT valve position, the NOx production of LPG was higher than gasoline fuel but at 4000rpm the value is nearly zero.It showed that the LPG in liquid phase has higher peak temperature combustion that contributed the increased in NOx emission [1, 3, 13].Throughout the experiments at WOT valve position, the AFR was maintained at lean condition until 3500rpm. Meanwhile at 4000rpm the AFR changed to rich condition for both of fuel. This is due to the ECU setting of vehicle used and SI engine technology itself. At WOT valve position, the AFR was remained at rich condition and the AFR value became richer as the engine speed increased.

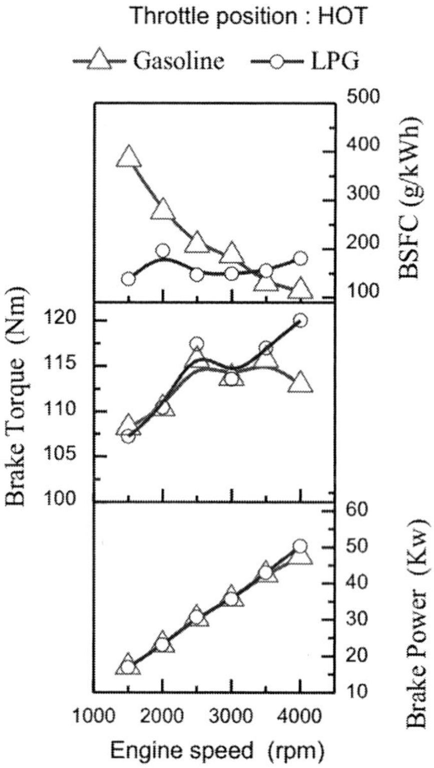

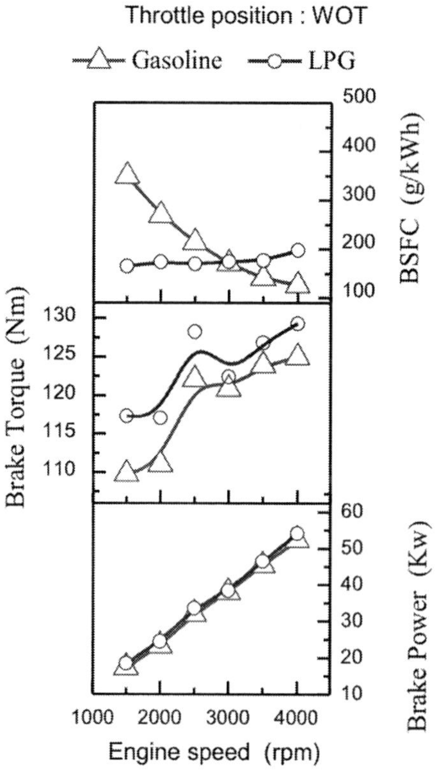

Fig. 2. Performance of LPG and gasoline at different throttle position.

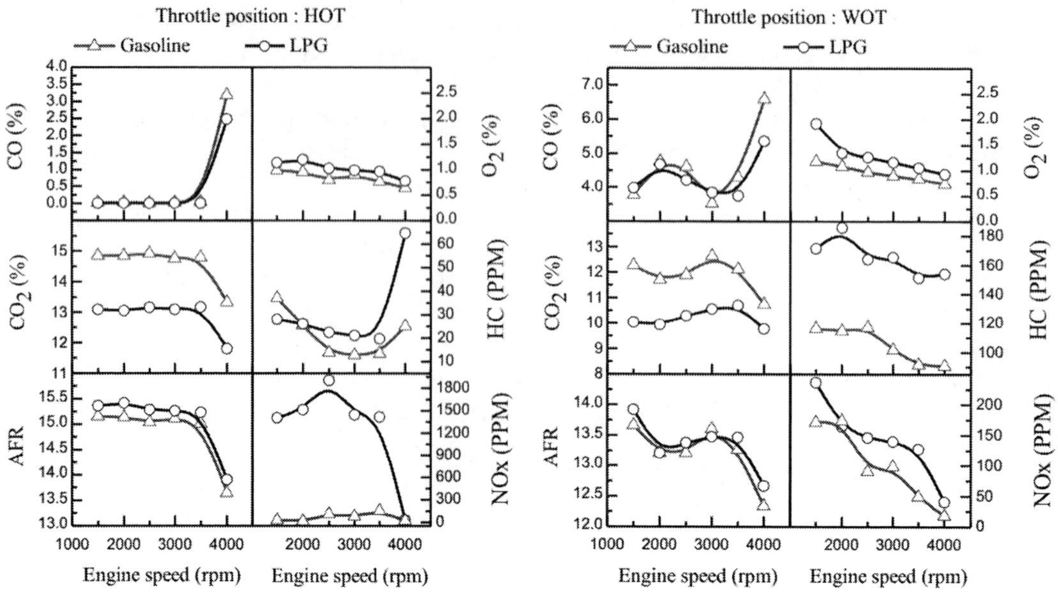

Fig. 3. Emissions of LPG and gasoline at different throttle valve position.

4 Conclusions

In a nutshell, performance of the liquid LPG injection was better than gasoline fuel.The BSFC for liquid LPG injection was significantly lower than gasoline fuel. The emission of CO was also lower as compared to the gasoline fuel. This result shows that LPG is capable to reduce the pollution of the environment. NOx emission produced by LPG is relatively higher than gasoline for both throttle valve position of which may due to the high temperature combustion in the engine cylinders.To reduce the NOx emission of liquid LPG injection, adjustment of the injection duration may be required in order to gives optimum amount of LPG injected in cylinder during combustion.

The author would like to thank the Ministry of Education for partly supporting this research under the Contract Grant, VOT: U427, Short Term Grant (STG) VOT U383 and Centre for Graduate Study, Universiti Tun Hussein Onn Malaysia.

References

1. H.C. Watson, P.X. Phuong, *SAE Technical Paper*, (2007)
2. K. Yeom, J. Jang, C. Bae, Fuel, **86**, 494 (2007)
3. K. Kang, D. Lee, S. Oh, C. Kim, *SAE Technical Paper*, (2001)
4. P. Baker, H. Watson, *SAE Technical Paper*, (2005)
5. M. Campbell, L.P. Wyszynski, R. Stone, *SAE Technical Paper*, (2004)
6. A. Boretti, *SAE Technical Paper*, (2011)
7. M. Ihsan Karamangil, Energy policy, **35**, 640 (2007)
8. M. Masi, Energy, **41**, 252 (2012)
9. A. Boretti, H.C. Watson, *SAE Technical Paper*, (2009)
10. M.A. Khan, H. Watson, P. Baker, G. Liew, D. Johnston, *SAE Technical Pape*, (2006)

11. H.S. Tira, J.M. Herreros, A. Tsolakis, M.L. Wyszynski, Energy, **47**, 620 (2012)
12. M. Gumus, Fuel Process. Technol., **92,** 1862 (2011)
13. J.A. Caton, M. Mcdermott, R. Chona, *SAE Technical Paper*, (1997)

Feasibility Study on IPMC Actuated Contractile Water Jet Thruster Body Contraction

Muhammad Farid Shaari[1,*], *Z.* Samad[2], *M. H.* Othman[1], and *O.M.F.* Marwah[1]

[1]Faculty of Mechanical and Manufacturing Engineering, Universiti Tun Hussein Onn Malaysia, Parit Raja, 86400 Batu Pahat, Johor, Malaysia.
[2]School of Mechanical Engineering, Universiti Sains Malaysia, 14300 Nibong Tebal, Pulau Pinang, Malaysia.

Abstract. This research was conducted to investigate the feasibility of Ionic Polymer Metal Composite (IPMC) as the actuator for Contractile Water Jet Thruster (CWJT). The thrust that generated from the accelerated volume of ejected fluid depends on the IPMC actuator contraction performance. During this investigation, the variation of contraction frequency and nozzle aperture influence on the body deformation was carried out. The results of the experiments exhibit that the body contraction decreased tremendously from 0.05 Hz to 0.15 Hz. The body contraction has no apparent decrement for more than 0.15 Hz of contraction frequency. Besides, larger nozzle aperture has relatively more body contraction than smaller nozzle aperture. These results show that the fixed actuation force and the contraction frequency have a significant influence on the CWJT body contraction.

1 Introduction

Contractile Water Jet Thruster (CWJT) is a thruster that inspired from the squid contractile mechanism [1]. The ejected water jet produces thrust in one direction and accelerates the thruster body in the opposite direction. It has low noise propulsion compared to the electric motor driven blade thruster if being actuated by smart actuator [2]. The function enables the CWJT to be adapted to quite or high sensitivity underwater task. IPMC utilizes low voltage and able to work in the water. However, it has a low actuation force which is between 5 gf to 10 gf, depending on the type of electrode, actuator length and thickness [3, 4]. Thus, a feasibility study had been conducted on the contraction behaviour of the CWJT. The purpose of this study is to observe the CWJT volume contraction performance over time. In this research, two parameters were included in the observation which are the contraction frequency and the nozzle aperture. The nozzle aperture of the CWJT was varied in certain sizes and the IPMC actuator will contract the CWJT at different frequencies. The experiment was repeated for another nozzle apertures. Generally, the result of this study determined the relation between the volume contraction and the actuation frequency as well as the nozzle aperture. The relation between these parameters had been presented in graph

* Corresponding author: mdfarid@uthm.edu.my

form in order to define the most influential parameters on volume contraction. The next section discusses the experimental setup and the result will be thoroughly discussed in the result and discussion section. Finally, the finding of this research is concluded in the last section.

2 Materials and Method

The CWJT was developed by using forms polyethylene (PE) film, which was then attached to a 3 mm thick acrylic plate and a nozzle. Generally, it has half tapered shape where the nozzle was placed at the sharp edge. The wider opening of the film was the inlet of the fluid. The PE film had been shaped by blow heating the PE film on a 3D printed mould. The heater temperature was recorded at 120°C. The formed film was attached to the nozzle and both of them were then pasted on an acrylic plate. The IPMC actuators were fixed to the gold plated electrodes and its tip was pasted on the PE film (Figure 1). The gold plated electrodes were connected to a power supply (GW Instek PSM-3004) and the drive voltage was fixed at 4 V to ensure an optimum underwater IPMC actuation [5]. The contraction frequency, f_c was determined by setting the IPMC actuation cycle time, T_a in Arduino software version 1.0.5. The actuation cycle time is a time taken to trigger the IPMC actuator to make a complete upward and downward motion as presented in Figure 2. The ejected fluid from the actuation was calculated from the volume difference of this displacement. The reversible IPMC actuation was obtained by using H-bridge L239D and Arduino UNO microcontroller. There were 10 sets of contraction frequencies between 0.05 Hz and 0.50 Hz had been put understudies. This frequency range had been selected based on the optimum IPMC actuator character [6, 7]. Table 1 shows the sets of the contraction frequency that utilized in the experiments. The relation between contraction frequency and the actuation cycle time can be presented as Equation 1.

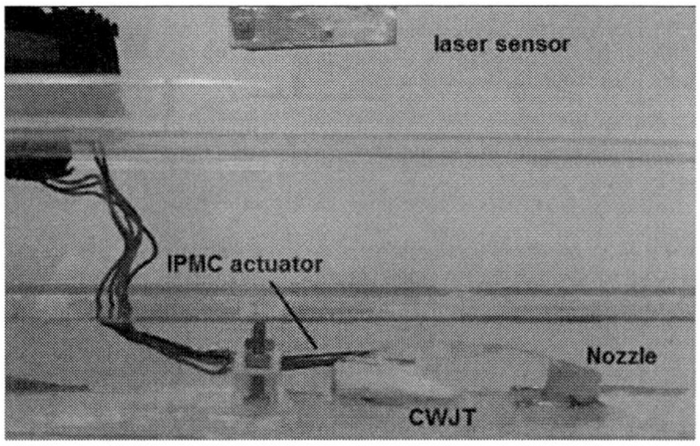

Fig. 1. Experimental setup.

 DOI: 10.1051/matecconf/20167801051

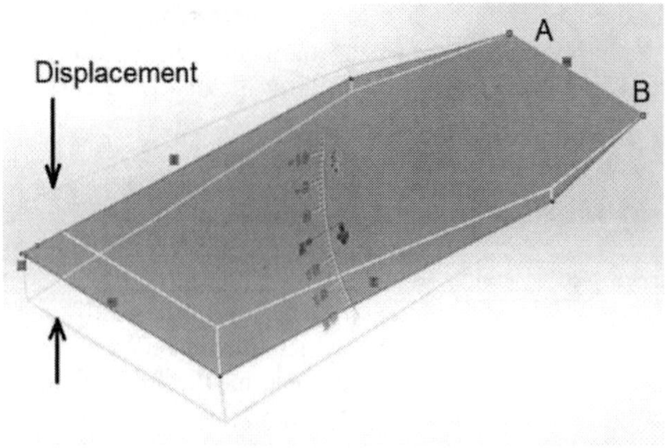

Fig. 2. Actuation of the IPMC.

Table 1: Frequency setting.

1 cycle time (s)	Frequency (Hz)
2	0.500
4	0.250
6	0.167
8	0.125
10	0.100
12	0.083
14	0.071
16	0.063
18	0.056
20	0.050

$$f_c = \frac{1}{T_a} \quad (1)$$

Observation on the body displacement was measured using laser displacement sensor (Panasonic HG-C1100). The laser was pointed to the fluid inlet area to measure the maximum displacement of the CWJT as shown in Figure 2. The obtained data were recorded in Signal Express software through National Instruments DAQ card NI USB 6009. Once the displacement data were recorded for all contraction frequencies, the experiment was repeated for another nozzle aperture size between 4 mm and 12 mm. The volume difference, because of the actuation was calculated by applying Pappus-Guldinus Theorem as described in Equation 2,

$$V_c = 2\pi \bar{y} A_c \quad (2)$$

where V_c is the contraction volume, $\tilde{y}$ is the distance between the centroid and the rotation axis which is in this case is the line along point A and point B in Figure 2. A_c is the rotating surface which is affected by the actuation. 2π refers to the rotation angle which means complete 360° angle. Therefore, the displacement data gained in the experiment utilized to determine the angle of rotation. This calculation had been achieved by using tangent formula whereas the axis of rotation. The acquired degrees were then converted into radian degree before it could be applied in the Pappus Guldinus Theorem.

3 Results and Discussion

The relation between CWJT displacement and nozzle aperture is depicted in Figure 3. The results show that the CWJT body displacement increases as the nozzle aperture became wider. However, the displacement increment is quite small which is between 11 % and 70 %, depending on the contraction frequency. Higher contraction frequency has low displacement while lower contraction frequency has higher displacement. This is due to the character of the IPMC actuator itself [8]. Lower contraction frequency allows more water swell into the one side of the IPMC to actuate it. Higher contraction frequency has limited time to absorb water for actuation. The results also show that the displacement increment trend increases exponentially as the contraction frequency decreases. This displacement data had been utilized to measure the CWJT volume contraction using Equation 2. Based on the calculated results, the relation between contraction volume and the contraction frequency had been identified as exhibited in Figure 4.

Generally, the graph shows that the relation between CWJT contraction volume and the contraction frequency has the same trend which is an exponential decay trend. The contraction volume decreases tremendously started from 0.05 Hz to 0.15 Hz. The reduction of contraction volume within this range is approximately 6 to 7 times. The highest record of the contraction volume was 717 mm^3 at 0.05 Hz. At the contraction frequency of more than 0.15 Hz, the volume contraction started to decrease in constant linear trend. At this range, the contraction volume was recorded between 100 mm^3 and 300 mm^3. In this graph, the nozzle aperture had shown no significant influence on the contraction volume except for contraction frequency within 0.2 Hz to 0.3 Hz. Based on this observation, it could be concluded that the contraction frequency has more impact on volume contraction if relatively compared to the nozzle aperture size despite of the nozzle aperture increases by almost 67 % from 4 mm diameter to 12 mm diameter. This condition occurred as the consequence of the low actuation force by the IPMC actuators. In spite of the significant influence by the contraction frequency, others vital factors for volume deformation were the actuation acceleration and force. Thrust, which can be derived from the momentum equation is highly affected by the rate of ejected volumetric fluid.

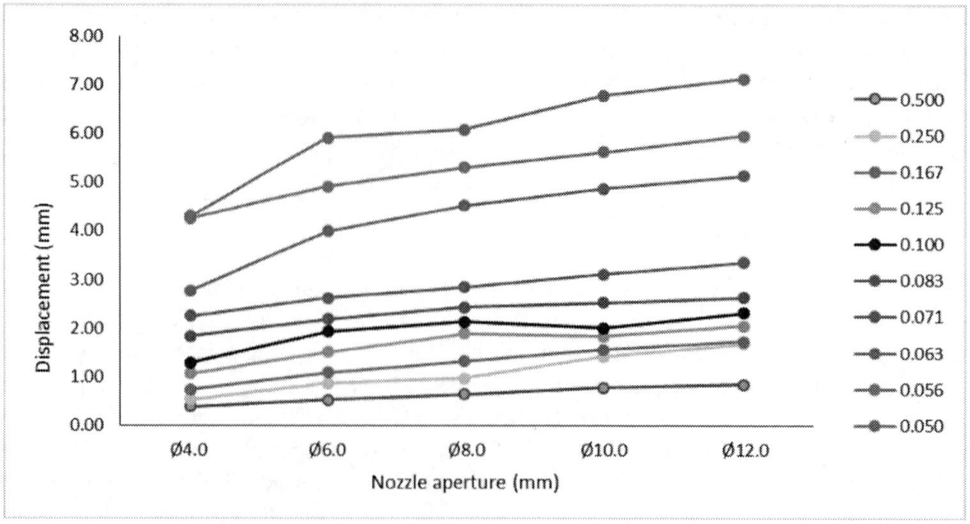

Fig. 3. CWJT body displacement measured by laser sensor at different nozzle aperture and contraction frequency.

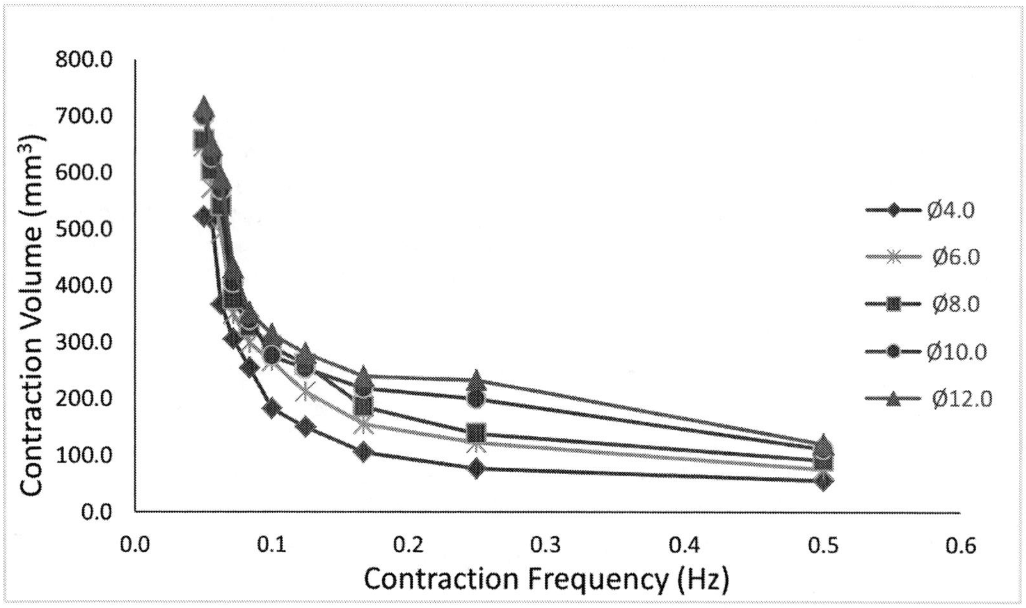

Fig.4. Contraction volume at different contraction frequency.

4 Conclusion

In this research, the measurement of contraction volume is essential to determine the feasibility of the CWJT contraction. The relationship between the contraction frequency and the contraction volume has shown that there are tremendous decreases of contraction volume between 0.05 Hz and 0.15 Hz. On the other analysis, nozzle aperture has no significant influence on the contraction volume as its value has not changed much despite of the nozzle aperture difference. Meanwhile the contraction frequency depends on the

IPMC actuation force. These two parameters were measured by single actuation force. The result might change if higher IPMC actuation force will be applied.

The author would like to compliment Universiti Tun Hussein Onn Malaysia for sponsoring this project via IGSP grant vot no. U257 and providing facilities to complete this research.

References

1. M.F. Shaari, Z. Samad, Applied Mechanics and Materials, **490**, 1099 (2014)
2. M.F. Shaari, C.J. Jun, Z. Samad, *IEEE International Conference on Robotics Biomimetics, & Intelligent Computational System* (2013)
3. A. Barber, J. Najem, L. Donald, J. Blotman, EAPAD, **7976**, (2011)
4. L. Shi, S. Guo, S. Mao, M. Li, K. Asaka, Int. J. Adv. Robot. Syst., **10**, 1 (2013)
5. M.F. Shaari, S.K. Saw, Z. Samad, Applied Mechanics and Materials, **575**, 716 (2014)
6. C.K. Chung, P.K. Fung, Y.Z. Hong, M.S. Ju, C.C.K. Lin, T.C. Wu, Sensor Actuators B. Chem., **117**, 367 (2006)
7. S.G. Lee, H.C. Park, Surya D. Pandita, Y. Yoo, Int. J. Control Autom. Syst., **4**, 748 (2006)
8. E. Malone, H. Lipson, Rapid Prototyp. J., **12**, 244 (2006)

Computer Aided Facial Prosthetics Manufacturing System

H.K. Peng[1,*], *M.N.* Salleh[1] , and *A.F.* Aiman[2]

[1]School of Technology Management and Logistics, Universiti Utara Malaysia, 06010 Sintok, Kedah, Malaysia
[2]School of Manufacturing Engineering, Universiti Malaysia Perlis, 02600 Arau, Perlis, Malaysia

Abstract. Facial deformities can impose burden to the patient. There are many solutions for facial deformities such as plastic surgery and facial prosthetics. However, current fabrication method of facial prosthetics is high-cost and time consuming. This study aimed to identify a new method to construct a customized facial prosthetic. A 3D scanner, computer software and 3D printer were used in this study. Results showed that the new developed method can be used to produce a customized facial prosthetics. The advantages of the developed method over the conventional process are low cost, reduce waste of material and pollution in order to meet the green concept.

1 Introduction

Nowadays, face is important appearance for everyone. However, it is common that some people have facial deformities or disfigurement. These facial deformities or disfigurements are caused by serious accidents like burns, diseases like cancer, drug side effect or congenital facial deformities. The deformities make those people cannot live as a normal person because of their face. Their abnormal appearances make them unable to mix with the normal people or even in society. The deformities can be reduced through plastic surgery by replacing tissues from own skin. Another method is using facial prosthetics. Facial prosthetics is an artificial device that is used to alter the outward appearance of a person's face or head.

2 Problem statement

Current method to develop the facial prosthetics is complicated because each facial prosthetic is customized for the patient. First, a mold was made according to the affected area. The shape of the prosthetic is carefully sculpted to ensure a high quality of natural contour and fit the deformity. After the sculpture complete, a durable mold was created to

* Corresponding author: kok_peng0710@hotmail.com

produce the prosthetic part. Every detail was copied to the mould to ensure it is precise and fit to the face. Due this complicated fabrication process, specialists are needed to execute the fabrication [1, 2] . In addition, a handmade silicone ear or nose can costs up to $4000, and requires 5 to 7 hospital visits for customization. The process takes 5–10 weeks in total [3].

Recently, 3D printing technology has emerged in additive manufacturing process for facial prosthetics fabrication. 3D printing technology has proven that it is capable of manufacturing precise and affordable facial prosthetics [4, 5]. However, even though the cost of facial prosthetics can be reduced using the new technology, the cost of buying the technology is still out of reach for many developing countries. Most of the research mentioned above employed excessively high cost of 3D-scanning technology and 3D printing machine.

3 Research Objectives

The main aim of this research is to investigate a new affordable method to fabricate a customized facial prosthetics. Thus, the objectives of this research are;

I. To acquire facial area 3D data
II. To select the affected area and design the prosthetics.
III. Using additive manufacturing to create the prosthetics

4 Methodology

Method to develop the facial prosthetic can be categorized into three stages which are data acquisition, facial prosthetics 3D model development and 3d printing of facial prosthetics.

4.1 Data acquisition

This section discussed the method to acquire 3D data from a face.

4.1.1 Subject Preparation

At this stage, a human face pattern with some disruptive anatomy was used as a subject. A model constructed from Plaster of Paris (P.O.P) material was used to replicate the affected area as shown in Fig. 1.

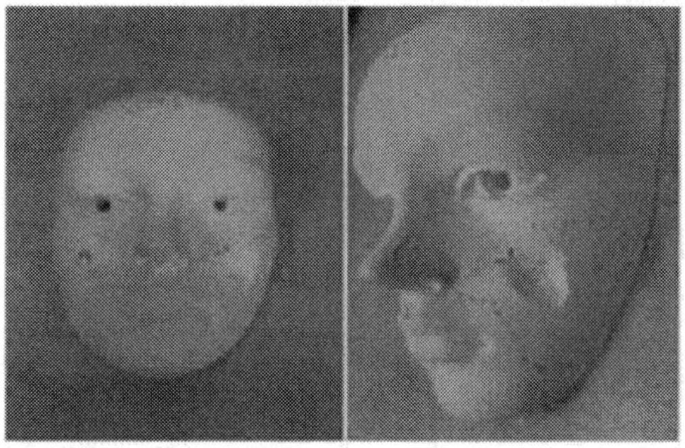

Fig. 1. Human face pattern.

4.1.2 Scanning Process

Scanning process is the important stage to acquire the digital data from the subject. A Cubify Sense 3D scanner equipped with the 3D system Sense software was used to scan the subject. The result of the scanning process is shown in Fig. 2.

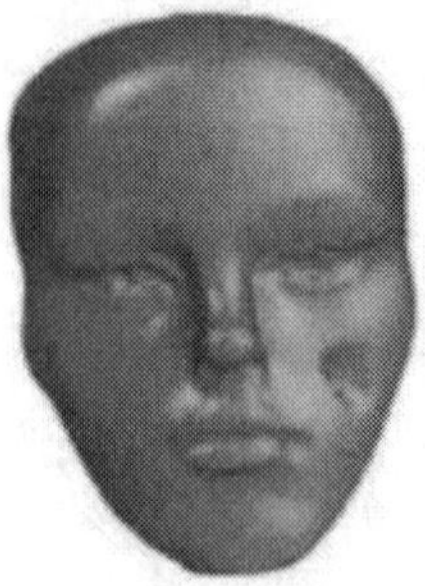

Fig. 2. Scanning result of face model.

4.2 Facial prosthetics model development

In this section, methods to develop the facial prosthetic model are discussed.

4.2.1 Software preparation

In the research, software Geomagic Sculpt developed by 3D System was used to construct the facial prosthetics due to the fast and accurate virtual sculpting software platform. This software was specifically designed for creating and modifying the 3D model that used for 3D printing.

4.2.2 Development of facial prosthetics model

In this section, the steps of facial prosthetics development were shown in details. The first step in this process is to import the previous 3D scanning result. The scanning model is imported to the Geomagic Sculpt for modifying using option. After that, curve line was created around the deformity and split. Then, subD surface is used to cover the deformity as shown in the red colour in Fig. 3. After the adjustment, the subD surface needs to be changed to become meshed and applied on the damaged face. The results of the development of facial prosthetics are perfectly smooth as shown in Fig. 4.

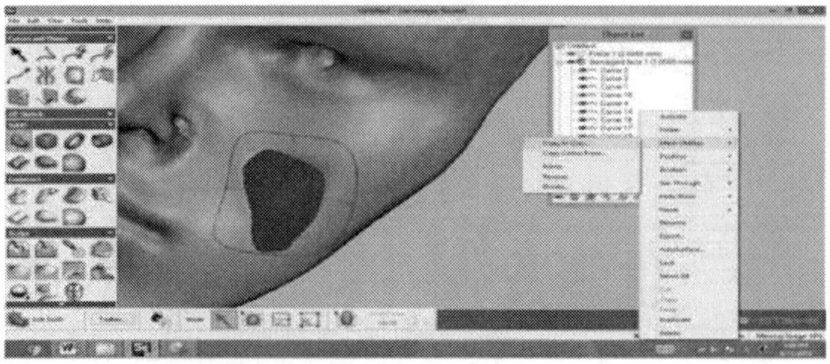

Fig. 3. The facial prosthetics model development.

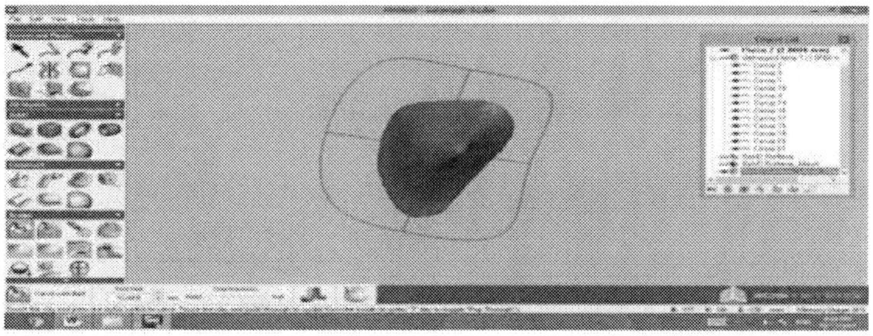

Fig. 4. 3D facial prosthetic model.

4.2.3 Result

Result shows the new method can be used to develop facial prosthetics. The part was fully covered the damaged face and fit the deformity shown in Fig. 5.

4.3 3D Printing Of Facial Prosthetics

This section discussed the process of 3D printing for the facial prosthetics model.

4.3.1 3D printer preparation

3D printer is the most important tool in this project because it was used to fabricate the developed facial prosthetics into a tangible part. In this project, the 3D printer that was used to print out the developed facial prosthetics is Flashforge Creator Pro Dual Head 3D Printer. Material used for the facial prosthetic is Polylactid Acid (PLA).

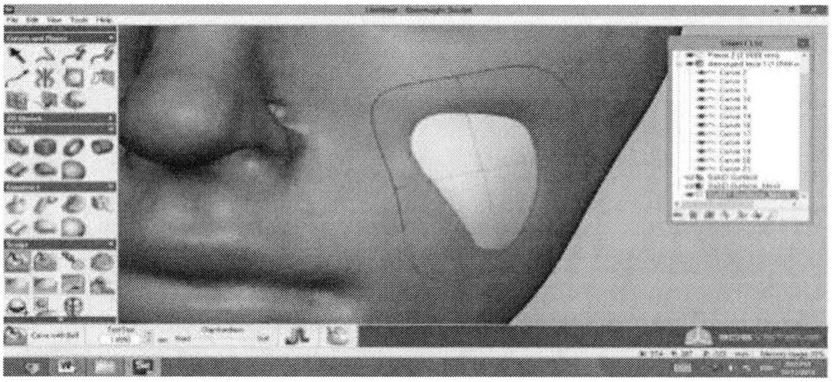

Fig. 5. 3D facial prosthetic model fits perfectly into the damaged area.

4.3.2 3D Printing Process of Facial Prosthetics

The result from the previous 3D facial prosthetics model was exported to STereoLithography (STL) format into the MakerWare software. MakerWare can be used to set the 3D printer according to the user preference, e.g. to control the temperature of the extruder, monitor the part, adding rafting and support to make sure the part is printed in perfect condition. After the parameter setting, the print file was exported and saved in X3G format that can only be read by the 3D printer.

The printing process took about 20 minutes for the prosthetics part. Post processing was done to eliminate the raft and support materials. The finished facial prosthetics part covers up the deformity with a good fitting as shown in Fig. 6.

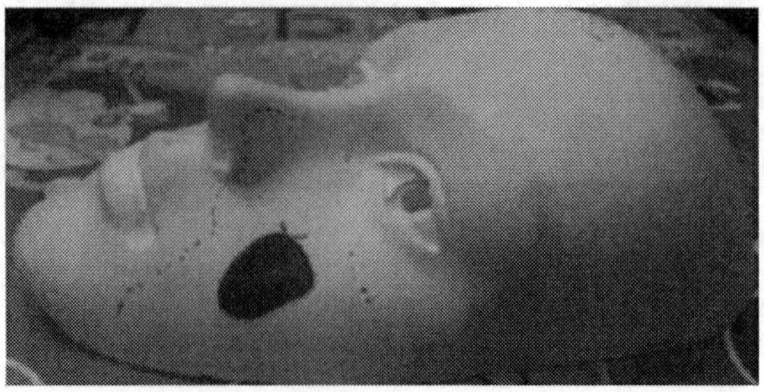

Fig. 6. Facial prosthetic fill into the damaged area.

5 Discussion

This project successfully identified a new method to develop the facial prosthetics using low cost 3D scanner and low cost 3D printer. The objective of this project to develop the facial prosthetics using 3D printing and 3D scanning apparatus was achieved successfully. Through this project, it brings benefit to manufacturers and doctor in term of the development of facial prosthetics for medical purposes.

The first advantage of the method is low production cost. With the low cost 3D scanner combined with low cost 3D printer, the facial prosthetics cost was reduced compared to the conventional method. The second advantage was time saving. For a professional user, they may be just consumed one day for the development of the facial prosthetics with the method in this project. For the conventional method, it takes time to finish the facial prosthetics because the process was complicated. Mostly the facial prosthetics fabrication is about one month. The third advantage of this project was the method was fully customization. There were different deformities or disfigurement in different people. So, the facial prosthetics have to fit perfectly according to patients. The fourth advantage of this method was a green process. With the use of 3D printing technology, there was a zero fuel emission. It reduces the air pollution. The raw material that used in this project was PLA, which is a thermoplastic and made from corn starch, sugarcane and some polymers. Therefore, the facial prosthetics were environmentally friendly and it can be recycled.

The disadvantage of the method is the level of expertise to design the facial prosthetic model. Skilled technician still in processing the 3D model was important in the process. It needed some knowledge on the function of the software so that the functions can be used in the optimum way.

6 Conclusion

This project has successfully identified a new low cost method to fabricate a customized facial prosthetics. The method has been explained and elaborated in the paper. For future research, researchers are suggested to apply the developed method on the real human face.

References

1. J. Wolfaardt, A. Sugar, G. Wilkes, Int. J. Oral Maxillofac. Surg., **32**, 121 (2003)
2. G.T. Grant, Aita Holmes, P.C. Liacouras, J. Garnes, W. Wilson O., J. Prosthet. Dent., **114(1)**, 13 (2015)
3. H. Yong, G. Xue, J. Fu, Sci. Rep., **4** (2014)
4. R. Bibb, D. Eggbeer, P. Evans, Rapid Prototyping J., **16(2)**, 130 (2010)
5. G.T. Grant, P. Liacouras, G.F. Santiago, J.R. Garcia, M. Al-Rakan, R. Murphy, C.R. Gordo, Ann. Plast. Surg., **72(6)**, 720 (2014)

Improvement on Transportation Safety on Bus by Installing A Speed Limiter in Conventional Engine to Reduce Speed Rate

Hadi Pranoto[1,2,*] , *A.M.* Leman[2] , *Darwin* Sebayang[1] , and *D.* Feriyanto[2]

[1]Faculty of Engineering, Department of Mechanical Engineering, Mercubuana University, Jakarta, Indonesia
[2]Faculty of Engineering Technology, Universiti Tun Hussein Onn Malaysia (UTHM), 86400 Parit Raja, Batu Pahat, Johor, Malaysia

Abstract. Driving is one of the conditions when the driver should be given special attention to make the motor vehicle in driving way and the driver should be in a good condition. However, the problem is in a certain condition the driver lost their control speed due to their target and minimize the driving time and it led the high potential to accident. Therefore, speed limiter which applied in the bus is urgently needed to reduce the accident and improve their awareness of road safety. Developed speed limiter is completed by fuel cut-off system to prevent the engine and maintain the speed. Limitation of speed be adapted by government regulation. From the results show that the highest and average speed of 136 km/h and 123.5 km/h is observed by bus speed prior to use speed limiter. After speed limiter applied in the bus, the approved maximum speed is 90 km/h. Those data approve that the speed limiter can reduce 83% from the top speed before speed limiter applied.

1 Introduction

National Safety Program. Safety driving is a based principle of transportation. In Indonesia, safety driving almost never happen which cause of road accidents. Based on the report of Indonesia Police department in 2010 that approximately of 31.234 peoples death by road accident. It means that there is 3-4 people death/hours. Decreasing PDB (gross domestic income) of 2.9-3% is one of the issues that influenced by road accident [1]. The driving safety has global issues from past decades and it estimates that 1.3 million people killed due to road accident [2-3]. It become biggest disaster when it not solved. Those data are approved by WHO that since 2030, Highest number of victims are not caused by disease while it caused by road accident [4-6]. There are some factors that led the accident (Fig. 1) and it will become main resources to produce road safety guideline in order to reduce the victims [7, 8].

* Corresponding author: hadi.pranoto@mercubuana.ac.id

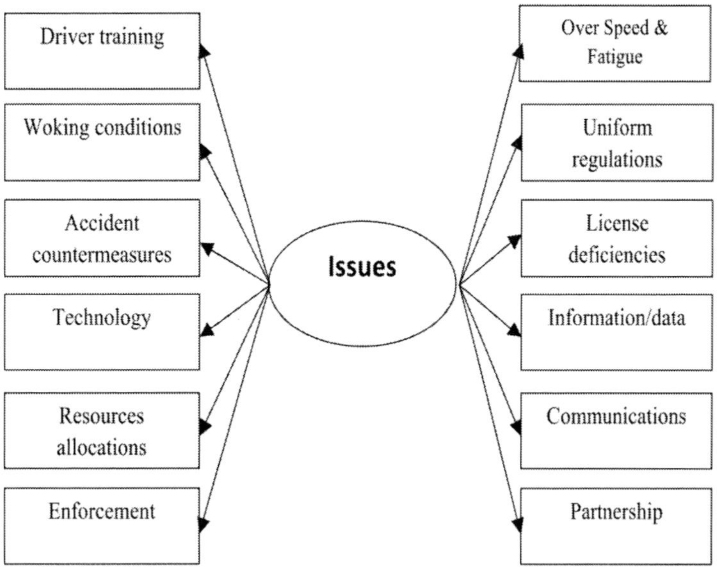

Fig. 1. Issues of bus accident on the highway [8].

Fig. 1 shows that over speed and fatigue are the most factors that cause the road accident. If approved by the number of bus accidents in West Java Region (DKI Jaya, West Java, Banten province), by the 2010 to 2015 has recorded that approximately of 17% and 20% of bus accident is caused by over speed and fatigue, respectively.

The regulation of motor vehicle. Based on the regulation that issued by Ministry of Transportation, Republic of Indonesia, No.111 mention about regulation for speed of motor vehicle [9 - 11]:

a. The minimum speed is 60 km/hours and the highest speed is 100 km/hour for a highway.
b. The highest speed in urban area is 80 km / hours.
c. The highest speed in central of urban area is 50 km/hours.
d. The highest speed in settlement place is 30 km/hours.

Through this regulation, the government aimed that it can restrict the speed of motor vehicle. Department of Police mention that at 2011-2013, bus has become most of transport accidents and it recorded that close to 9.278 cases by that time [12].

2 Methodology

Installation of speed limiter on the bus. This paper discusses and reviews how to control the speed of the bus before installing speed limiter. There are 300 buses with operating in different route and it divided by 2 classes of routes which are urban route, province route (Java, Bali and Sumatra island). The speed limiter is installed in the engine and it controlled and adapted by government regulation. The basic principle of speed limiter is signed when the bus in over speed. It will automatically cutoff the fuel which caused the bus speed in operated under the top speed regulation. Schematic diagram of speed limiter is shown in Fig. 2.

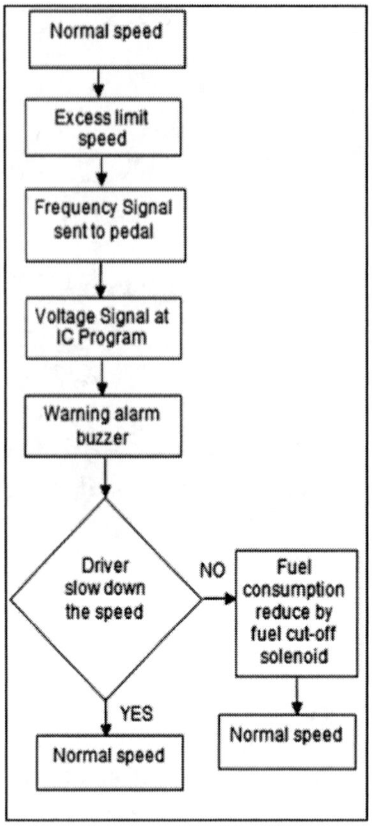

Fig. 2. Schematic diagram of speed limiter.

3 Engine speed limiter installation in bus

The modification of the speed limiter installation has been done by improving on electrical wire which placed and connected with the machine. It controlled and developed based on government regulation depending on the urban or highway speed. The speed limiter installation procedure is shown in Fig. 3 and Fig. 4 where it consists of buzzer, ECU, speed sensor, engine and speed limiter control box.

The working of the speed limiter mounted on the machines are inter-city buses, when the vehicle speed exceeds a specified speed limit speed sensor on the transmission output will send frequency signal to accelerator pedal, then the frequency signal is transformed into a voltage signal at IC program. The voltage signal is transformed into the comparators components, forwarded to relay and followed by sound a warning alarm buzzer. When the driver still performs additions speed, fuel shut off solenoid will reduce the fuel consumption into the engine and it will reduce the speed/rpm to the normal speed. Meanwhile, when the driver is slow down the speed, the buzzer alarm will be stop and back to the normal speed.

 DOI: 10.1051/matecconf/20167801053

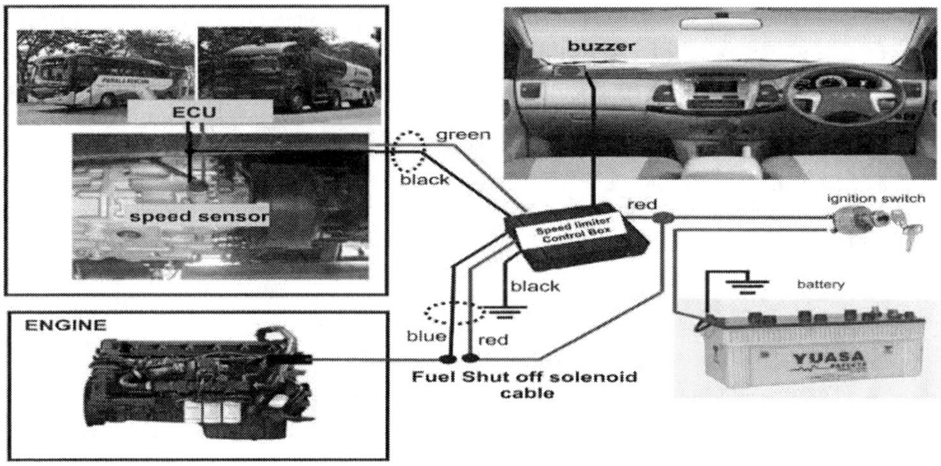

Fig. 3. Diagram Installation engine bus.

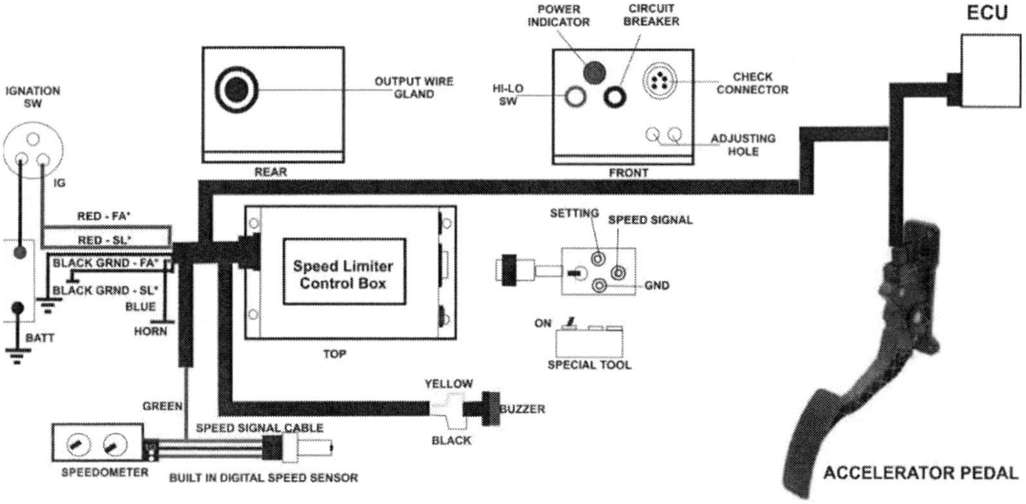

Note:
RED-FA= RED cable line-for fatigue alarm
RED-SL= RED cable line-for speed limiter
BLACK-GRAND FA= BLACK cable line-for fatigue alarm
BLACK-GRAND SL= BLACK cable line-for speed limiter

Fig. 4. Speed limiter on the engine bus scheme.

4 Development speed limiter

The speed of the bus before installing speed limiter. The data on the speed of the bus is monitored by GPS and they have monitored several buses such buses with ID No. is HT 002 ,HT 006, HT 634, HT 674,and HT 691. The speed limiter is not installed on the bus. The highest speed of 136 km/h for HT 691, and the speed flat is 123,5 km/h for HT 674 has been observed and it will very serious thing when that speed is operated in urban areas. The lowest speed is shown by HT 002 with the speed of 120 km/h. The different speed between that bus is influenced by driver behavior or attitude and their time target in order to receive more income. The detail speed data for buses in different route (Jakarta, central java, north java, and Bali) is listed in Table 1. That data is summarized from the daily speed which detail shown in Fig. 5Speed of the bus before installing speed limiter. The data on the speed of the bus is monitored by GPS and they have monitored several buses such buses with ID

No. is HT 002 ,HT 006, HT 634, HT 674,and HT 691. The speed limiter is not installed on the bus. The highest speed of 136 km/h for HT 691, and the speed flat is 123,5 km/h for HT 674 has been observed and it will very serious thing when that speed is operated in urban areas. The lowest speed is shown by HT 002 with the speed of 120 km/h. The different speed between that buses is influenced by driver behavior or attitude and their time target in order to receive more income. The detail speed data for buses in different route (Jakarta, central java, north java, and Bali) is listed in Table 1. That data is summarized from the daily speed which detail shown in Fig. 5.

Table 1. The average speed of the bus before being installed speed limiter.

No	ID Vehicle	Tipe Vehicle	Average Speed Bus (Km / hour)	Top speed (Km / hour)
1	HT 002	Bus	110,2	120
2	HT 006	Bus	100,5	122
3	HT 634	Bus	87,2	121
4	HT 674	Bus	99,2	124
5	HT 691	Bus	123,5	136

The speed of the bus after install speed limiter. The comparison between the speed of bus before and after installing speed limiter has successfully done. A speed limit that allowed by government regulation and the company when the driving in highway is 100km/h and in urban street not more than 80km/h. Base of GPS data, when the bus installs by speed limiter in the bus id of HT 002, HT 006, HT 634, HT 674 and HT 691. It shows that highest speed is 90km/h and average speed is 80km/h. The data monitored by 30 days as shown in Fig. 6 and summarized in the Table 2. Top speed allowed when the speed limiter applied is 90 km/h and the lowest speed of 85.4 km/h is shown by HT 634. After installation speed limiter can reduce and make consistent speed of the bus close to 80km/h. It can be improved to blow off to the driver how to drive, when the speed of the bus lock off in 80km/h the alarm will be on in 80db to considering to the driver to take care driving the bus in the highway or urban street. When the drivers try to do acceleration or over speed in driving almost 90km/h, the driver must to strip off the accelerator to off the speed limiter alarm. A speed limiter alarm will be on because the bus is over speed as sign that the driver must reduce the speed.

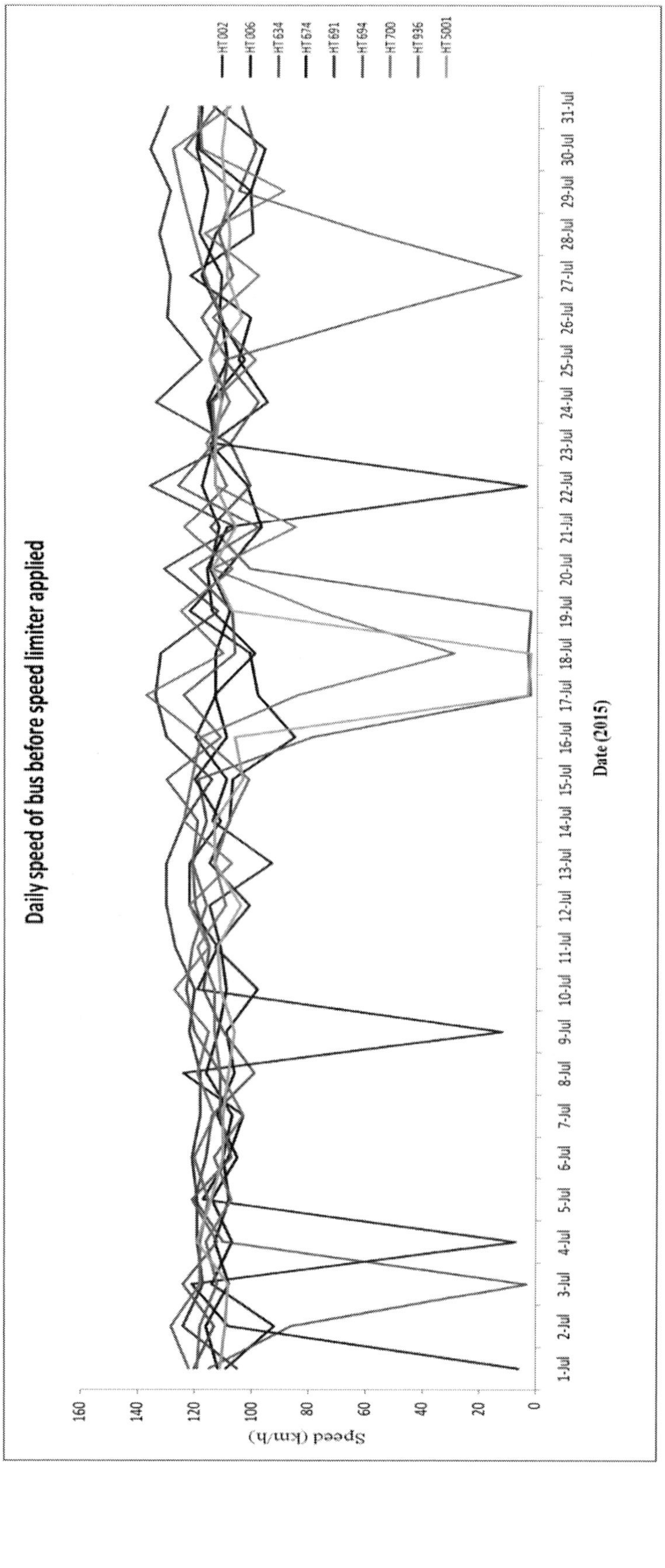

Fig. 5. Bus speed before being installed speed limiter.

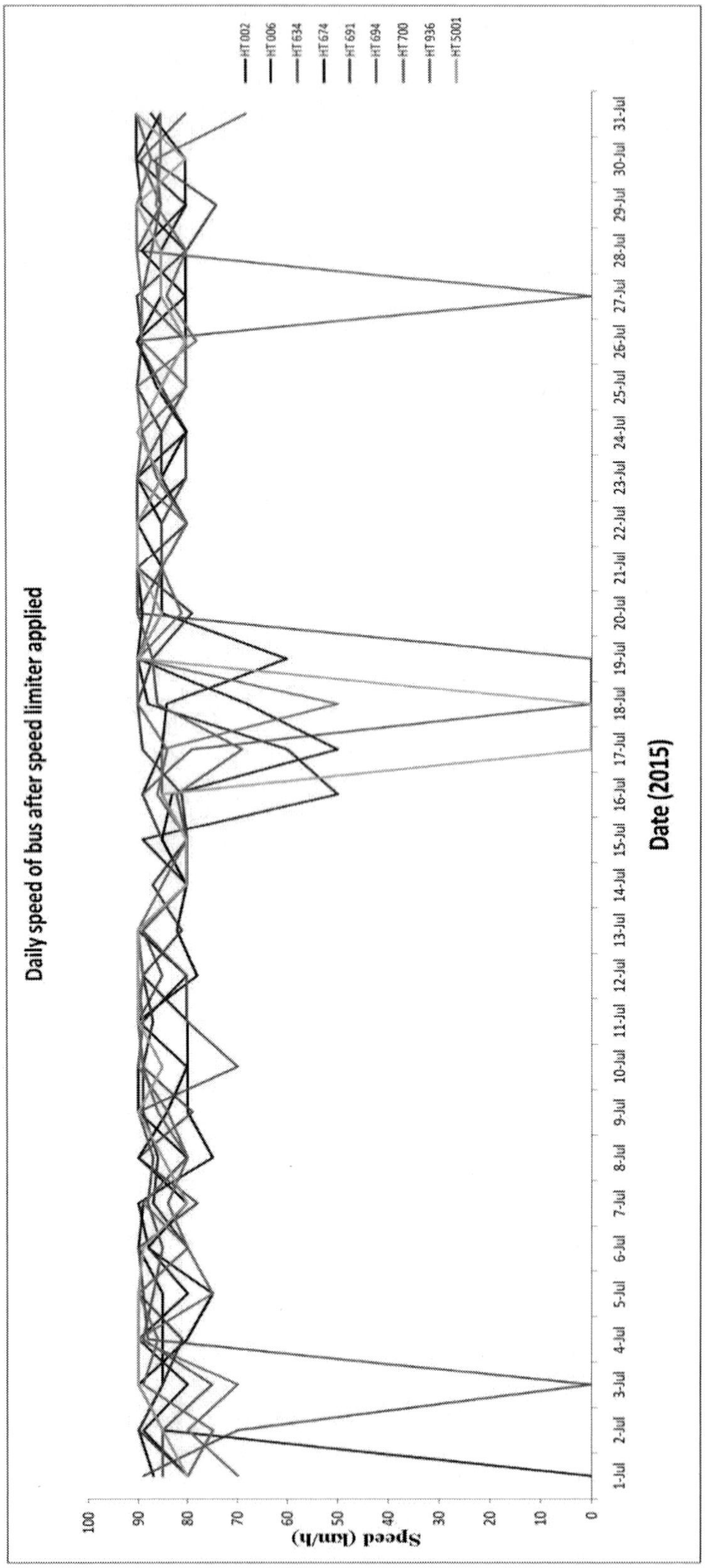

Fig. 6. Bus speed after being installed speed limiter.

Table 2. The average speed of the bus after being installed speed limiter.

No	ID Vehicle	Type Vehicle	Average Speed Bus& truck (Km/h)	Top speed (Km /h)
1	HT 002	Bus	86,3	90
2	HT 006	Bus	87,4	90
3	HT 634	Bus	85,4	90
4	HT 674	Bus	86,6	90
5	HT 691	Bus	88,1	90

The improvement and analysis after install speed limiter. Improved analysis of speed limiter usage is detail listed in Table 3 and Figure 7. Efficiency improvement is indicated that it effectively to reduce the speed bus even the driver speedup or uncontrolled the speeding bus. The drivers have to strip off the accelerator if they want the speed back to normal and they can push the accelerator back again. The speed report data day by day after install speed limiter become more controlling, it's based on a GPS report in bus id HT.002, HT 006, HT 634, HT 674 and HT 691 the highest speed of the bus is 124 km/h when the bus on rushing with average speed is 99 km/h, Base on that table's data, speed limiter can control until 98% speed of the bus in highway or urban street.

Table 3. Improved use of speed limits

No	ID Vehicle	Type Vehicle	Average speed before installed speed limiter (km/h)	Average speed before istalled speed limiter (km/h)	Average speed after installed speed limiter (km/h)	The average speed after installed speed limiter (km/h)	Improvement (%)
1	HT 002	Bus	110	110	86	24	78%
2	HT 006	Bus	101	101	87	13	87%
3	HT 634	Bus	87	87	85	2	98%
4	HT 674	Bus	99	99	87	13	87%
5	HT 691	Bus	124	124	88	35	71%

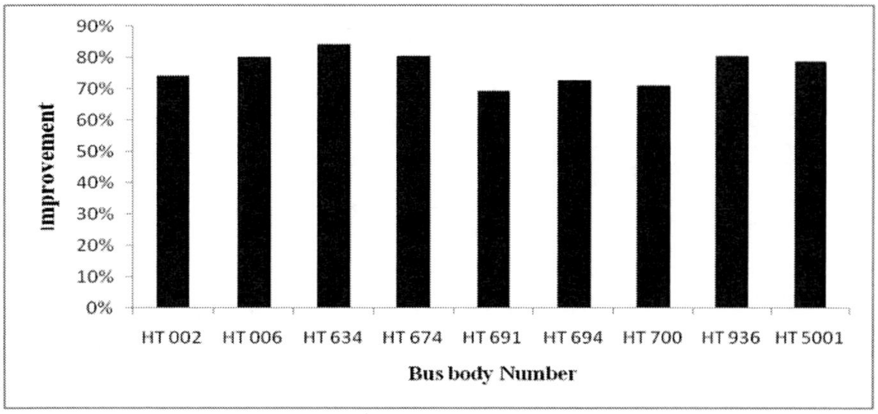

Fig. 7. Improvement use of speed limiter.

5 Conclusion

The embracing of the paper is speed limiter can be an effective appliance way for a transportation company, especially for protecting their bus machine, their human resource and their environmental. When the bus on rushing in the highway or urban way, the driver potentially speed up the bus. It can be seen on GPS tracking data before installing speed limiter, the highest speed of 136 km/h on the highway and 99 km/h in urban street. After speed limiter installs on the bus machine, the speed reduces significantly. Based on the data summary of GPS tracking, while the bus rushing on the highway become 88 to 90 km/h and on the urban street become 70 km/h. The improvement percentage of speed limiter appliance until 83 % and it is prospecting for reduce speed of the bus. It approved that this speed limiter is most recommended to apply in public bus transportation. It will potential to reduce the accident rate and reduce the victims.

The authors would like to thank the Mercubuana University for morale support and thanks to Universiti Tun Hussein Onn Malaysia (UTHM).

References

1. *Rencana Umum Nasional Keselamatan (RUNK),* (Republic of Indonesia: Technical Report, 2011)
2. E.M. Ossiander, P. Cummings, Accid. .Anal. Prev., **34**, 13 (2002)
3. M.M. Gallagher, C.M. Sewell, S. Flint, J.L. Herndon, H. Graff, J. Fenner, H.F. Hull, Journal of American Medical Association, **262**, 2243 (1989)
4. D.B. Brown, S. Maghsoodloo, M.E. McCardle, J. Safety Res., **21**, 125 (1990)
5. A.J. McKnight, T.M. Klein. Transport Results Records, **1281**, 71 (1990)
6. C. Lave, P. Elias, Accid. Anal. Prev., **26**, 49 (1994)
7. United States. Federal Motor Carrier Safety Administration. *Future Truck and Bus Safety Research Opportunities*, **38**, (2006)
8. B. S. Dhillon, *Safety and Human Error in Engineering Systems* (CRC Press, 2012)
9. Constitution Number 22, clause 203, Republic of Indonesia (2009)
10. Regulation Manistry of environment (KLH), **74** (1999)
11. Regulation Manistry of transportation, Republic of Indonesia, *Procedures for Determining Vehicle Speed Limit*, **111**, (2015)
12. Policy as expressed in an internal memo No. 154/F00010/2013-S0 on 8 April 2013 than stated on HHSE Guidance, PT. Pertamina Persero (2013)

MATEC Web of Conferences **78**, 01054 (2016) DOI: 10.1051/matecconf/20167801054
IConGDM 2016

Comparison of Impact Duration Between Experiment and Theory From Charpy Impact Test

N.B. Muhammad Said[1,*], *M.B.* Ali[1], *K.A.* Zakaria[1], and *M.A.M.* Daud[1]

[1] Faculty of Mechanical Engineering, Universiti Teknikal Malaysia Melaka,Hang Tuah Jaya, 76100 Durian Tunggal, Melaka, Malaysia.

Abstract. This study presents the comparison of impact duration between experiment and theory from impact signal through a Charpy test. Recently, the number of accidents on the highway has been increased and it depends on the impact duration of material that have the ability to provide adequate protection to passengers from harmful and improve occupant survivability during crash event. Charpy impact test was implemented on different material and thickness but at the same striker velocity. Impact signal is obtained through the strain gauge that has been installed to striker hammer and connected to frequency data acquisition system. Collected signal is then analysed to identify the time period during impact before fractured. Result from both experiment and theory shows an increment to the impact duration as thickness is increased. Charpy test shows that aluminium 6061-T6 has a higher impact duration compared to carbon steel 1050.

1 Introduction

Impact on a body is occurring in a very short period of time with very large of force is imposed. According to Oztas [1], vehicle collision on highways can be classified as head-on collision, front-to-back collision and eccentric collision and impact of that collision is strongly depends on the vehicle velocity, mass and material of the vehicle is made of. In vehicle collision, crashworthiness is an aspect that always becomes major attention in the automotive field because its ability to protect passengers from damage and improve occupant survivability during crash event. Previous study by Bhanage et al. [2] states that crashworthiness which includes the ability of the car to withstand minor accidents with little damage becomes a trend nowadays in the automotive industry. In engineering structures, strength and toughness often combined to indicate the amount of energy absorbed during the deformation and fracture.

In order to meet safety regulations, designers have to improve the structural design that can be applied in vehicle crashworthiness. However, the design for crashworthiness is a tough task in a design process often involves non-obvious decisions since the structure has to be strong enough in some zones to prevent passengers, yet sufficiently compliant in other

* Corresponding author: nurlaelamuhammadsaid@yahoo.com

zones to absorb the crash energy [12]. Vehicle that is made from material with high impact duration is more safe since there are extended crush spaces for car protection which can reduce the damage either to car structure or occupant [7].

Some ways may apply for improving the safety condition in vehicles. Samer et. al [3] recommends that improvements on the security level of vehicle structure need to be achieved in order to enhanced the capabilities of energy absorption of vehicles. Good and high quality of energy absorber should be designed to dissipate the energy of impact irreversibly via plastic deformation of metallic structure. Most of the energy absorbers were developed using metallic thin-walled structure since they tend to deform plastically due to elastic-plastic behaviour [4].

Charpy impact test is widely applied for many years in the engineering field since it is easy to conduct, rapid test method, reliability and low cost. This test often implemented to study the failure behaviour besides of investigates the toughness of a material. Some aspects may affect the failure behaviour of Charpy impact test such as temperature, notch position, material properties and impact velocity.

Jang et al. [5] remarks that experimental and numerical studies were performed to study the effects of notch position on the failure behaviour and energy absorption when the Charpy V-notch impact test is made at temperature of 1°C. Furthermore, Shende et al. [6] claims ductile-to-brittle transition curves may be obtained when the results of a number of tests presented in different temperatures is plotted. When the temperature is reduced through the range of transition, the fracture surface changes from one having a 'fibrous' or 'silky' appearance with much distortion at the sides, to one of completely crystalline appearance with negligible distortion.

Nowadays, vehicle collision has been increase per year was reported. Research on those phenomena is studied continuously in order to improve the performance of vehicles on the road. Great performances provide sufficient security or reduce the severity to occupants if a sudden collision happens. Thus, this study is conducted as a guideline for the next research to investigating the effect of parameters (materials, thickness, velocity) where it is one of the best method to be conducted in order to enhance the impact duration, so can reduce the risk during collision. The main purpose of this study is to compare the impact duration obtained from Charpy impact strain signal with theory. Since there are many factors that influence the impact duration, hence changes in the material and specimen thickness are two parameters that contribute in affecting the value of time required during impact before fracture occurs.

2 Theory

Based on Ali et al.[7], the total impact duration, t_0 can be determined from the formula shown as follow.

$$t_o = 2.94\left(\frac{5}{4Mn^1v^{1/2}}\right)^{2/5} \quad (1)$$

where,

total mass, $M = \frac{1}{m_1} + \frac{1}{m_2}$

m_1 = mass of impactor and m_2 = mass of specimen

v = velocity during impact

and $$n^1 = \frac{4\sqrt{R_1}}{3\pi(k_1+k_2)}$$

where, $k_1 = \frac{1-v_1^2}{\pi E_1}$ and $k_2 = \frac{1-v_2^2}{\pi E_2}$

v_1 *and* v_2 = Poisson ratio of striker and specimen respectively
E_1 *and* E_2 = Young Modulus for striker and specimen respectively
R_1 = radius of a spherical impactor

Based on Rechnitzer et al. (Accident Research Centre) [8], impact is considered to be essentially plastic when two vehicles that approaching each other in opposite directions do not bounce apart and continue together with a combined mass in a straight line after impact occurs. Impact duration between cars and heavy vehicles can be performed by using equation 2 below.

$$t = \frac{2s}{V_c} \qquad (2)$$

where,
t = impact duration
s = car crush (m) or overall deformation
Vc = closing velocity, $V_1 + V_2$
V_1 = velocity of truck/heavy vehicle (m/s)
V_2 = initial velocity of car (m/s)

3 Methodology

The figure 1 below shows the process involved in determining the impact duration. This study is started with material selection and preparation. Then, Charpy test was conducted to collect impact strain signal. Lastly, data on strain signal is analysed and compared to the theory.

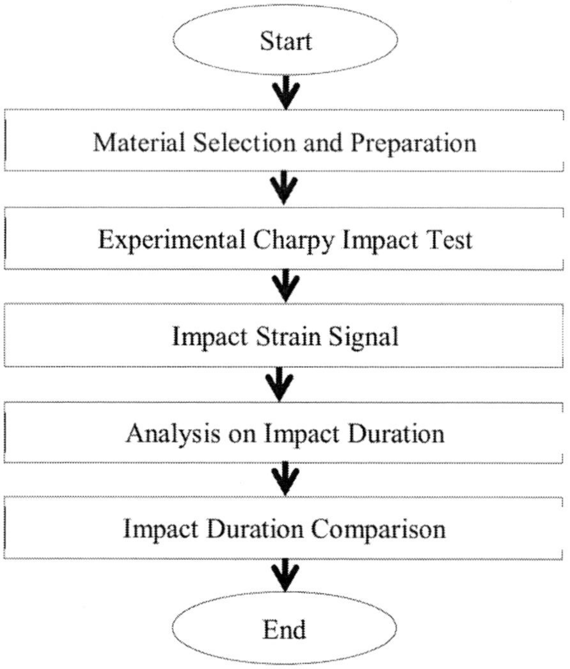

Fig. 1. Process in Impact Duration Determination.

3.1 Material selection

Since the uses of aluminium and steel are widely used in engineering application, aluminium 6061-T6 and carbon steel 1050 were chosen for conducting the Charpy impact test. Both of these specimens were selected by considering their potential properties such as high strength, durable, good toughness, cost effective and corrosion resistance. The compositions of these materials are shown in Table 1 and Table 2. Impactor of machine is made from steel and assumed to be a rigid body since it is not deformable. Young's modulus, density and Poisson's ratio for that impactor are 7.86x103 kg/m3, 210 GPa and 0.32 respectively. Table 3 represents the physical and mechanical properties for aluminium 6061-T6 and carbon steel 1050.

Table 1. Chemical composition of 6061 aluminium alloy [9]

Element	Fe	Cu	Si	Zn	Mn	Mg	Cr	Ti	Al
Mass [%]	0.16	0.19	0.71	0.04	0.02	0.94	0.08	0.03	Bal

Table 2. Chemical composition of carbon steel 1050 [10]

Element	C	Mn	Si	Fe
Mass [%]	0.55	0.65	0.25	98.55

Table 3. Properties of Aluminium 6061-T6 and Carbon steel 1050 [11]

Properties	Aluminium 6061-T6	Carbon steel 1050
Density [kg/m^3]	2700	7860
Young's Modulus [GPa]	70	210
Poisson's Ratio	0.35	0.32
Yield strength [MPa]	292	660
Ultimate strength[MPa]	328	675

3.2 Specimen design and preparation

The design of specimen is prepared by referring to ASTM E23. Figure 2 below shows the standard geometry for V-notch Charpy impact specimen. All dimensions are measured in unit of millimetre (mm). Since thickness is one of the manipulated variables, the other thickness of the specimen is 5 mm by reducing its width from 10 mm.

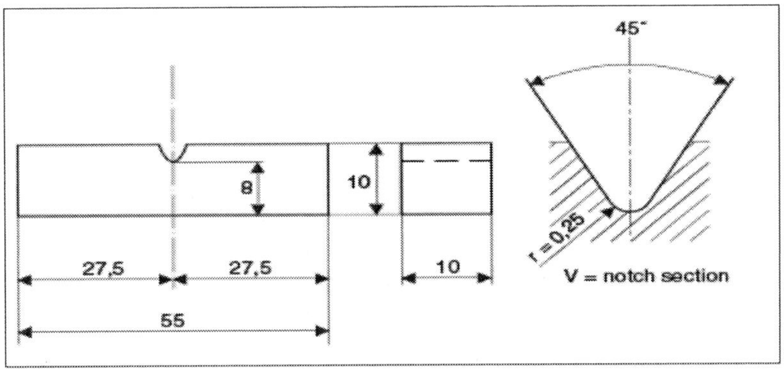

Fig. 2. Standard dimension of Charpy specimen.

3.3 Equipment for testing

Figure 3 shows equipment that was used in order to collect the information on the strain signal. Strain gauge was installed to the right and left striker surface that has radius of 2 mm. In order to prevent the short circuit occurrence, direct contact between the chip surface of strain gauge and striker is avoided. In this study, frequency of 50 kHz is selected, means that 50 data will be collected per second. After connection of all equipment is done, calibration on the Charpy machine is implemented by swing the pendulum without specimen on the test piece support. If the dial gauge on the Charpy machine shows zero reading, there is no problem for the machine to be used for data collection. The Charpy impact test is carried out under ambient temperature at a constant impactor velocity of 5.18 m/s.

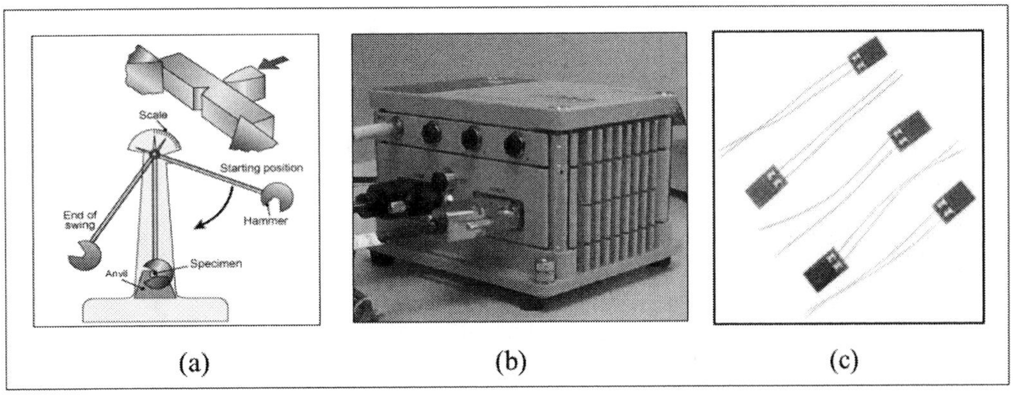

Fig. 3. Equipment for Charpy impact test in collecting strain signal: (a) Machine of Charpy impact (b) Data acquisition system (c) Strain gauge

3.4 Impact duration analysis

Collecting data from the experimental Charpy impact test is analysed by plotting a graph of strain against time. Then, average impact duration is calculated and compared to the formula that has been stated in Equations 1 and 2.

 DOI: 10.1051/matecconf/20167801054

4 Result and discussion

The signal obtained from a Charpy test is analysed by plotting a graph of strain versus time. The graphs are shown as below.

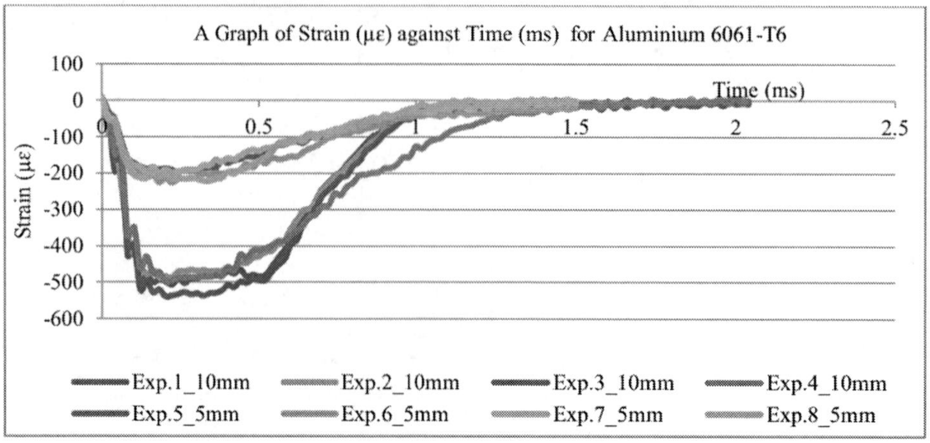

Fig. 4. Distribution of strain (με) against time (ms) for Aluminium 6061-T6 at different thickness

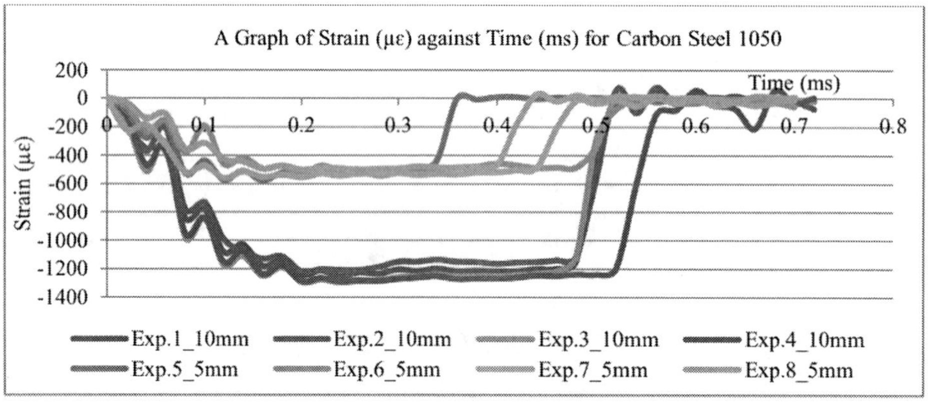

Fig. 5. Distribution of strain (με) against time (ms) for Carbon Steel 1050 at different thickness

Based on the graphs above, for each material strain due to impact is directly proportional to the time taken during impact occurs. The range of impact duration for aluminium 6061-T6 occurs between 1.1ms to 1.5ms and 0.38ms to 0.60ms for carbon steel 1050. For both materials, the highest impact duration was found in the thicker specimen. Charpy test results show an increment in impact duration when the thickness of the specimen was increased which is about 24% for aluminium and only 12% of carbon steel. Aluminium 6061-T6 has a higher average impact duration compared to Carbon steel 1050. It is because aluminium is more ductile than steel thus it has more elastic and plastic region in a stress - strain curve before fracture.

Impact duration from theory is calculated by using equation (1) and equation (2). By using equation 1, impact duration for aluminium has been increased about 32% and 28% of carbon steel when the thickness is doubled from 5 mm. Identical pattern was found in impact duration when equation 2 is used. The impact duration has increased about 115% for aluminium while 118% of carbon steel if the thickness of specimen is increased. The comparison of impact duration between experiment and theory from previous study is summarized in Table 4 and Table 5 shows the percentage of increment in impact duration when the thickness of the specimen is raised. Results of theory show that carbon steel has a higher impact duration when compared to aluminium. It shows there is significant difference regarding the comparison between material changes in the scope of impact duration. Hence, more studies have to carry out to compare the time of impact between experiment and theory.

Meanwhile, changes in specimen thickness show a correlation if result from experiment and theory was compared. All cases indicate impact duration has been increased with thickness. This because thicker material requires more energy to fail and tends to absorb more energy once impact force is exerted and it will consequently affect the impact duration.

Table 4. Impact duration of experiment and theory

Material	**Impact Duration [ms]**					
	Experiment		**Theory [Equation 1]**		**Theory [Equation 2]**	
	5mm	**10mm**	**5mm**	**10mm**	**5mm**	**10mm**
Aluminium 6061-T6	1.1500	1.4300	0.0496	0.0654	0.1004	0.2163
Carbon Steel 1050	0.4900	0.5500	0.0597	0.0764	0.2542	0.5541

Table 5. Percentage of Increment on Impact Duration

Cases	**Increment of impact duration percentages from 5 mm to 10 mm [%]**	
	Aluminium 6061-T6	**Carbon Steel 1050**
Experiment (Charpy Test)	24 %	12 %
Equation 1	32.5 %	28 %
Equation 2	115 %	118 %

5 Conclusion

In both cases experiment and theory, the impact duration was increased as the specimen thickness is increased. This shows that the correlation between experiment and theory in term of impact duration is existed. It is because thicker structure requires more energy due to impact before fracture. In terms of material changes experimental result shows that aluminium 6061-T6 able to extend the time during impact occurs compared to carbon steel 1050. However, that result against with a theory where the impact duration of carbon steel is higher than aluminium. Hence, further studies on comparison of impact duration between experiment and theory need to reviewed continuously as a basis to derive a new formula or theory that link to the experiment in order as a guideline to reduce the risk during collision.

The authors would like to thank to the reviewer community who greatly help to produce this manuscript. Special thanks to financial support under grant of FRGS/2/2013/TK01/FKM/03/3/F00173.

References

1. A.G. Oztas, Turkish J. Eng. Env. Sci., **23**, 455 (1999)
2. A. Bhanage, N. Satonkar, P. Deshmukh, R. Sundge, International Journal of Engineering and Technology, **6**, 2129 (2014)
3. F. Samer, A. Abdullah, J.O. Sameer, International Journal of Advanced Engineering and Nano Technology, **2**, 1 (2015)
4. L.K.O.K. Hao, A. Radzi, A. Ghani, P. Edi, K. Yusuf, WSEAS Transactions on Applied and Theoretical Mechanics, **4**, 95 (2009)
5. Y.C. Jang, J.K. Hong, J.H. Park, D.W. Kim, Y.Lee, J. Mater. Process. Tech., **201**, 419 (2008)
6. C.J. Shende, A.R. Sahu, A.V. Deshmukh, Int. J. Mech. Eng. & Rob. Res., **4**, 350 (2015)
7. M.B. Ali, S. Abdullah, M.Z. Nuawi, A.K. Ariffin, Indian J. Eng. Mater. S., **20**, 504 (2013)
8. G. Rechnitzer, C. Powell, K. Seyer, *Performance Criteria, Design and Crash Tests of Effective Rear Underride Barriers For Heavy Vehicles*, 218 (SAE International, 2001)
9. S.H. Lee, Y. Saito, T. Sakai, H. Utsunomiya, Mat. Sci. Eng. A-Struct., **325,** 228 (2002)
10. S.K. Jaiswal, T. Sharma, M. Rajesh, V. Kumar, *Study of the Hardness and the Microstructure of AISI 1050 Medium Carbon Steel after Heat Treatment Processes* (ELK Asia Pacific Journals, Delhi,-Special Issue)
11. M.B. Ali, S. Abdullah, M.Z. Nuawi, A.K. Ariffin, Appl. Mech. Mater., **165**, 182 (2012)
12. K. Hamza, K. Saitou, *Design for Crashworthiness of Vehicle Structures via Equivalent Mechanism Approximations and Crush Mode Matching* (ASME, Anaheim, 2004)

MATEC Web of Conferences **78**, 01055 (2016) DOI: 10.1051/matecconf/20167801055
IConGDM 2016

A Preliminary Study of Knowledge, Attitude and Practices of Pesticide Use Among Oil Palm Workers in Johor

S.N. Marina Mior[1,2,*], A.M Leman[1], M.R. Baharudin[3], R. Masripan[1], M. Faazir T.[2],M. Ifwat A.[2]

[1]Faculty of Engineering Technology, Universiti Tun Hussein Onn Malaysia (UTHM),Parit Raja, Batu Pahat, 86400 Johor, Malaysia.
[2]National Institute of Occupational Safety and Health (NIOSH), Lot 1, Jalan 15/1, Section 15, 43650 Bandar Baru Bangi, Selangor, Malaysia
[3]Faculty of Medicine and Health Science, Universiti Putra Malaysia,43400, UPM Serdang, Selangor, Malaysia.

Abstract. Agricultural sector is a fundamental activity that has been practiced since the civilization of Babylonian, which also has been proven by the glory of Nile River Valley and Mekong Valley. History has shown that the migration of investors and foreign labors in Malaysia in 1960 was to be harvested fortune from the land due to its fertility and suitability. As a developing country, agriculture is one of the important sectors in Malaysia that contributes towards the national economy development. The mentioned sector has indirectly provided benefits both local and global perspectives either in terms of business opportunity as well as employment opportunity. In general, this study aims to assess the level of knowledge, practices and attitudes in the use of pesticides in oil palm plantation workers in Johor. The findings of this study will be used as reference in the farming industry, particularly for the awareness and understanding of good farm management practices and safety for agricultural workers by providing information regarding precautionary measures. This study will engage oil palm workers in Johor which are tasked to carry out pesticides handling activities. A total of 138 respondents were selected for the study. The methodology approaches that will be used in this study are questionnaire and interview. The results of this study will be used as reference in the farming industry, particularly the awareness and understanding of good farm management practices and safety for agricultural Workers By Informing Them Of A Precautionary Measure.

1 Introduction

As a developing country, agriculture is one of the important sectors in Malaysia that contributes towards the national economy development. The mentioned sector has

* Corresponding author: nazhatulmarina@gmail.com

indirectly provided benefits both local and global perspectives either in terms of business opportunity as well as employment opportunity. Based on an economic census of the plantation in 2011 by the Department of Statistics Malaysia, there is a total of 390,708 workers in the agricultural sector, which experiencing a growth of 8.5% since 2005, and the employment of oil palm plantations alone consists of the total of 300,465 workers to cover 4,960,853 million hectares [1].

According to [2], the plantation area of oil palm and rubber were 86% of the whole main plantation area all over Malaysia in the year 2000. Due to the rubber price volatility, the plantation of oil palm has dominated the agricultural sector in order to replace most of the rubber plantations, supported by a wide range of purposes that can be exploited from oil palm to generate innovative products. Moreover, Johor has been identified as the state with the highest public workers in the oil palm plantation sector in Peninsular Malaysia. This scenario definitely demonstrates the major contribution of oil palm plantation industry towards the economic growth.

Agriculture is one of the earliest professions practiced and involves a unique system of works. This sector involves extensive use of operating techniques ranging from conventional processes to the application of modern and complex technologies. [3] stated that the working conditions in a country, whether developed or developing, will vary depending on the methods of work. Therefore, the development in these agricultural activities that carried out by the workers has created many risks to human health, and one of them is caused by chemical hazard existed from the uses of various types of insecticides or pesticides. Based on the statistics disclosed by the Department of Occupational Safety and Health (DOSH) in the year of 1991, the agriculture industry is recorded with the second highest reported occurrence of accidents among all sectors of the occupation industry. In 1990, World Health Organization (WHO) which based in Geneva explained that most of the pesticide is a chemical used in agriculture sector to control insects and weeds to assist in increasing the production of crops. The consumed organic products nowadays which potentially contain the excess of several different pesticides might be the cause of the illness. Latest investigation has discovered that a moderate to low awareness among farm workers towards the utilization of pesticide residue in soil, in the air, on plants and in groundwater [4].

In Peninsular Malaysia, there are approximately 4.3 million farmers are located all over 1.5 million hectares of rubber plantations and 0.6 million hectares of oil palm plantations. This amount does not include other agricultural areas that were also using pesticides. In 1991, a report released by the Department of Plant and Machinery Malaysia, which act as the body that implements the Occupational Safety and Health, has shown the accident occurrence rate from improper use of pesticides are four times higher than the accidents that occur in other industries. Therefore this study is planned to be conducted to determine the knowledge, attitudes and practices on the use of pesticides among oil palm plantation workers in the state of Johor.

2 Literature review

In Southeast Asia, the countries have improved agriculture from traditional rice cultivation into a more complex market oriented crop cultivation to meet the demands. Therefore, the agriculture activities as well as land area keep increasing and providing job opportunities, economic growth and varieties crop production [5]. According to Sime Darby, Oil World announced that palm oil was the highest type of oil consumed in 2011 [6]. In Malaysia, specifically for oil palm plantations, chlorpyrifors, a type of pesticide was widely used in order to reduce the population of bagworms, ceterpillars and rhinoceros beetles that may attack the plantations [7].

A record in 2011 stated that there are 426 palm oil mills in Malaysia with the capacity of processing 99.85 million tonnes of Fresh Fruit Bunches (FFB) annually. Indonesia and Malaysia altogether account for 85% of global oil palm production covering almost 15 million hectares in 2009. Palm oil is derived from the plant's fruit, which produces two types of oils: crude palm oil (CPO), which comes from the mesocarp of the fruit, and palm kernel oil, which comes from the seed in the fruit. Most CPO is used for food products, while most palm kernel oil is used in non-edible products such as detergents, cosmetics, plastics, as well as a broad range of other industrial and agricultural chemicals [8].

Instead of pesticide usage, Integrated Pest Management (IPM) relies on barn owls or snake to reduce rat populations are an excellent idea and need to be implemented widely [9]. 70% of Indian population rely on agricultural activity [10] as the emphasis was given highly in agriculture after the green revolution happened. The concerning factor is that Indian farmers apply pesticides according to the pre-determined calender rather than applying based on the requirement basis. A study conducted in Ubon Ratchani, a district in Thailand, found out that Parkinson Disease patients was recorded through their background working history as farmers, who were exposed to paraquat, organo-chlorines as well as other pesticides [11].

Farmers use highly hazardous-monochrotofos, carbofuran and methidation, while 11.11% of farmers still use PIC(Prior To Inform) pesticides which is prohibited [12]. 61% of the Indonesian farmers also endured minor poisonings while 14.4% suffered major poisonings symptoms [13]. A survey conducted in Malaysia shows that 14.5 % of farmers in Cameron Highlands had symptoms of pesticide poisoning [6]. The most common improper usages of pesticides are the usage of extra volumes or concentrations therefore lead to the probability of pesticide exposure. They also preferred to create their own pesticides by mixing several pesticides together hesitated referring to the proper usage and label rate guideline [14].

In Malaysia, Heavy metal industries and pesticides usage caused devastating impacts but was regarded insignificant on the environment [6]. A study by [15] reported that the majority of the farmers, 86% of them was using pesticides; therefore only 14% of farmers were not using the pesticides. 31.38% of pesticides users knows the name of the pesticide they used, that is Foliden, DAP bio bullet for instances. Therefore, 68.62% of pesticide users did not know even the name of their pesticide they used. A Similar study was conducted in rural areas of Ahmednagar of India revealed, 50% of the respondents stored the pesticide in the field at the farmhouse, and 20% at home. In Western Uttar Pradesh and Ahmednagar showed majority 34% of respondent had used mask/hand gloves and 81% were using mask followed by 67% who used gloves [16].

In Malaysia, in order to reduce the usage and application of rodenticides and pesticides, Integrated Pest Management (IPM) with the biological control would be very useful. Using barn owls for instance, yet the cost and efficiency is not that easy to be determined though biological solution will be very environmental-friendly method. Above all, Malaysia government has been very committed in providing good safety workplace condition [17].

3 Methodology

Simple random technique will be used in selecting the respondents among the farm workers. The data will be collected in December 2014 using structured questionnaires. The questionnaires will be provided in dual languages, Bahasa Malaysia and English Language to overcome communication issue. The questionnaire contained four sections. The first section consists of demographic information, the second section of the questionnaire consists of knowledge related questions, the following section of the questionnaire consists of practice related questions and the final section was designed to assess participant's attitude of farm workers. The questionnaire adapted from Skim Amalan Ladang Baik (SLAM). The sample size of the respondents (n) is about 217 agricultural farm workers (oil palm plantation in Johor) based on the Ministry of Plantation Industry and Commodity Malaysia database. The data will be analyzed using SPSS version 20.

$$n = \frac{z\,(P - (1 - P))}{d} \tag{1}$$

where,

z = 95% confidence interval
P = anticipated population proportion (90%) (M.M Yassin et. All)
d = absolute precision (± 5%)
= 217 farm workers

4 Result And Discussion

The totals of 217 oil palm workers in Johor were given a set of questionnaire. We narrow down our research from 217 respondents to 132 respondents as we intended to investigate a better target group which hands on directly handle the pesticide. The amounts of 132 respondents are obtained from Part A: Demoghraphy Analysis question number 14. By the data obtained from the respondents, the total Knowledge, Attitude, and Practice (KAP) score for each respondent are counted by comparing to the answer scheme provided that had been synchronized to the Skim Amalan Ladang Malaysia (SALM). SALM is a national programmed made by the Ministry of Agriculture Malaysia, MoA, to recognize and certify farms that are engaged in commercial fruits and vegetables production, which adopt agricultural practices that are environmentally-friendly, sensitive to workers' welfare, and yield quality products that are safe for consumption. However, this programme was revised in 2005. Furthermore, SALM-certified farms are entitled to use the "Malaysia Best" logo, which provides an opportunity to brand products in the marketplace [6].

Referring to a journal written by [6], he classified the score into High, Fair Practice and Poor into the range of scores as described in Table 1. The total score for each Knowledge, Attitude, and Practice (KAP) for each 132 respondents are totaled as well as their total score. If the score is more than 80% of the total question, it is classified as high score, while a score between 80% and 50% is classified as fair practice, while a score below 50% is classified as poor.

The high or good level is highlighted with green colour, while the intermediate or fair practice is highlighted with yellow colour, and the one that is highlighted in red colour, indicating the poor level of score in each component.

Table 1. Average Mark for Each KAP and Total Mark of Respondents.

Total	Score Description	Knowledge	Attitude	Practice	Kap_Overall
High	TOTAL > 80 %	32	66	101	54
Fair Practice	50% < TOTAL < 80%	96	64	30	78
Poor	TOTAL < 50%	4	2	1	0
		132	132	132	132

In Table 1, in the Knowledge Column, the number of respondents who scored more than 80% is 32 respondents out of 132 respondents, while 96 respondents scored between 50% to 80% and there are 4 people who scored below 50%. For Attitude Column, the number of respondents who scored more than 80% is 66 respondents out of 132 respondents, while 64 respondents scored between 50% to 80% and there are 2 people who scored below 50%. For Practice Column, the number of respondents who scored more than 80% is101 respondents out of 30 respondents, while 64 respondents scored between 50% to 80% and there are 1 person who scored below 50%.

In Figure 1, we may observe that only 24% of respondents are in good score, i.e. more than 80%, while 73% of them are in the intermediate class and 3% of them are in poor class. Therefore, most of them are in intermediate class and resulting the average mark for knowledge is in intermediate class with a score range between 50%-80 %. In Figure 2, specifically 50% of the respondents scored more than 80%. For intermediate class, of range of score 50-80%, there are 48% of respondents score in this range while 2% of respondents score at the poor level.

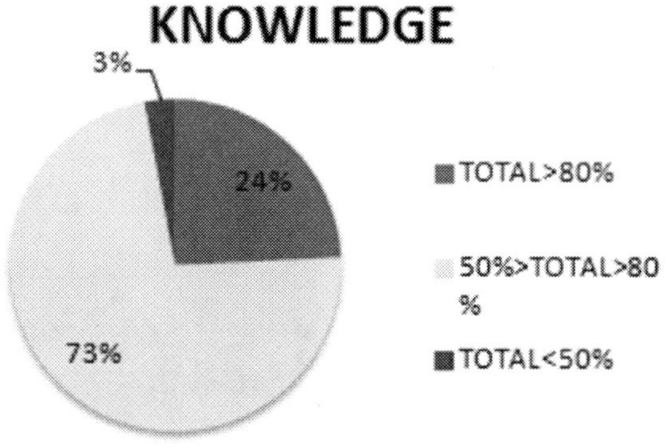

Fig. 1. The Percentage of Knowledge.

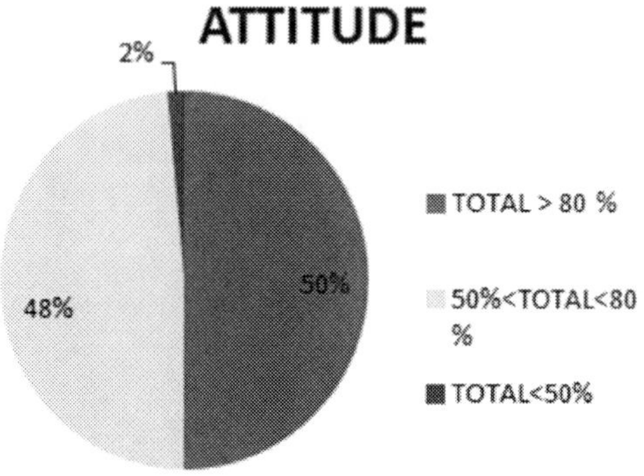

Fig. 2. The Percentage of Attitude.

In Figure 3, the figure shows that there are 76% of respondents scored more than 80% in practice part, and for intermediate level, there are 23% of respondents score between 50% to 80%, while just 1% out of 132 respondents scored less than 50%. As the majority of them scored at a good level (highlighted green) in the pie chart, therefore the average of practice is more than 80%.

After the scored has been calculated for each of 132 respondents, the total score of each respondent is calculated shown in Figure 4. It was found that only 41% of respondents fall in the good range with a score of more than 80% of total question in the questionnaire, while 59% is in the intermediate level and gladly, no respondent falls in the range of poor level, with scores less than 50%. This is the inference of the average total score falls in the range of 50-80%, as most of the respondents are in the intermediate score.

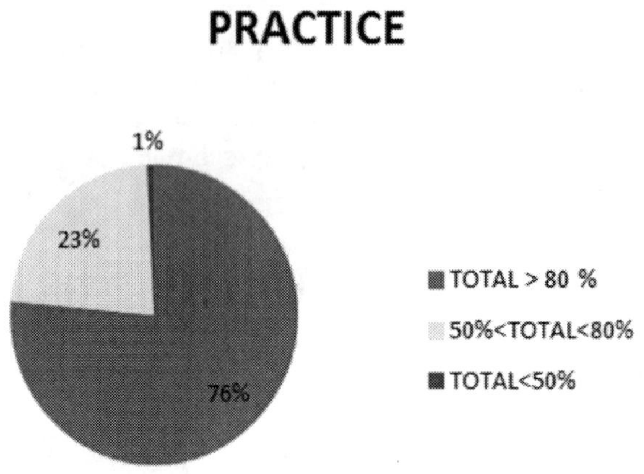

Fig. 3. The Percentage of Practice.

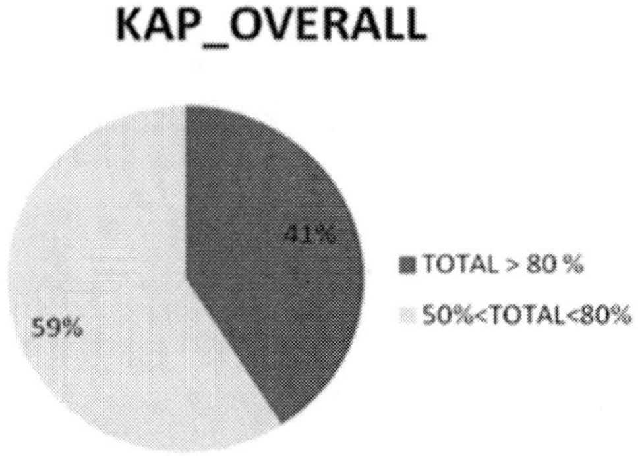

Fig. 4. The Percentage of Overall KAP.

5 Conclusion

By observing Table 1 and Table 2, as well as Figure 1, Figure 2 and Figure 3, we may observe that almost 25% score good in knowledge, 50% in attitude and almost 75% score good in practice component. This captured the attention that palm oil workers has low in knowledge, intermediately good in attitude, but surprisingly excellent in practice of handling pesticide sprayers in palm oil plantations in Johor. By the data obtained from demography analysis, as most of the respondents' age ranges from 18-35 years old, therefore they have more years to involve in agricultural activities and are prone to determine the agricultural safety and health at workplace. Therefore, more impact knowledge-focused programs need to be arranged and implemented to inculcate the better awareness of pesticides among palm oil workers, as without knowledge, people may become less careful and therefore are exposed to pesticide poisoning and other long term exposure illness in the future.

The authors would like to thank NIOSH for grant funding. Faculty of Engineering Technology, Universiti Tun Hussein Onn Malaysia (UTHM) and Centre for Graduate Studies – UTHM.

References

1. Garis Panduan Keselamatan dan Kesihatan Pekerjaan Dalam Pertanian, Department of Occupational and Health Malaysia (DOSH), 2011
2. J.K. Mohamad Khan, M. Munauwar, A.A. Fadzli Shah, A. Hassan, Universiti Utara Malaysia, 2002.
3. V. Forastieri, *The International Labour Organization (ILO) Programme on Occupational Safety and Health in Agriculture*, ILO, Geneva, (1999)
4. J.M. Safi, Egyptian journal occupational Medicine, **19**, 267 (1995)
5. J. Chalermphol, G.B. Bastakoti, R.C. Bastakoti, Procedia Environmental Sciences, **20**, 418 (2014)
6. S. Bahri, *Occupational safety and health in Commodity Agriculture : Case Studies From Malaysian Agricultural Perspective*, (2014)
7. H. Muhamad, B.H. Zainudin, N.K. Abu Bakar, Food Chemistry, **134(4)**, 2489 (2012)
8. J. Janick, History of Agricultural and Horticultural Technology in Asia. Purdue University, 1 (2013)
9. I. Comte, F. Colin, J.K. Whalen, O. Grunberger, J.P. Caliman, Advances in Agronomy, **116**, (2012)
10. M.G.C. Yuantari, C.A.M. Van Gestel, N.M. Van Straalen, B. Widianarko, H.R. Sunoko, M.N. Shobib, Environmental Monitoring and Assessment, **187(3)**, 142 (2015)
11. S. Norkaew, S. Lertmaharit, W. Wilaiwan, W. Siriwong, M.G. Robson, Roczniki Panstwowego Zakladu Higieny, **66(1)**, 21 (2015)
12. ISBN 92 4 156139 4, World Health Organization (1990)
13. F. Hidayat, T. Khamidi, S. Wiyono, Jurnal Bumi Lestari, **10(1)**, 1 (2010)
14. N. Taneepanichskul, S. Norkaew, W. Siriwong, Roczniki Panstwowego Zakladu Higieny, **65(4)**, 291 (2014)
15. D. Neupane, E. Jors, L. Brandt, Environmental Health, **13(1)**, 1 (2014)
16. R. Ranjan, K. Neupane, A.S. Wantamutte, B. Banjade, N. Kushwaha, R. Neupane, (2014). International Journal of Interdisciplinary and Multidisciplinary Studies, **1(5)**, 202 (2014)
17. H. Zulkifli, M. Halimah, K.W. Chan, Y.M. Choo, W. Mohd Basri, Journal of Oil Palm Research, **22**, 887 (2010)

DOI: 10.1051/matecconf/20167801056

Increasing ABB FlexPicker Robot's Degree of Freedom (DOF) using Flexible End Effector

M. H. Nordin[1,*], *K.* Selvaraju[1], and *M.* Fathullah [2,3]

[1]Department of Robotics and Automation, Faculty of Manufacturing Engineering, Universiti Teknikal Malaysia Melaka (UTeM), Hang Tuah Jaya, 76100 Durian Tunggal, Melaka, Malaysia
[2]Center of Excellence Geopolymer and Green Technology (CEGeoGTech), Universiti Malaysia Perlis (UniMAP), 01000 Kangar, Perlis, Malaysia.
[3]School of Manufacturing Engineering, Universiti Malaysia Perlis (UniMAP), Pauh Putra Campus, 02600 Arau, Perlis, Malaysia

Abstract. The commissioning of a high speed parallel robot such as the ABB Flexpicker Robot into the industrial environment provides greater flexibility in picking, packaging or palletizing tasks. However, current configuration of this robot only allows up to four-degree-of-freedom (DOF) which restricted its function to perform simple or complex automated assembly task. Therefore, the possibility of increasing one additional degree-of-freedom for the parallel robot was investigated. By attaching a flexible vacuum hose to the robot's end effector, it can be controlled by using a servomotor that can manipulate its angle of suction up to 90 degrees. With this new configuration, one DOF has been introduced for this robot. Several tests had been conducted to evaluate the flexibility and performance of the end effector which includes varying loads and angles. From the analysis, it is found that the end-effector can lift weights up to 1kg due to sufficient vacuum pressure; however, it could perform bending up to 90 degrees for maximum load of 600 grams only.

1 Introduction

Parallel robots are a relatively young type robot that has been developed over 25 years ago and has been implemented in productions for about 15 years. Parallel robots are robots with closed-loop mechanism and consist of three or more rotary or prismatic axes that functional parallel to each other. Parallel robots are designed for high-speed applications, manufacturing, packaging, material handling and assembly purpose. A majority of the high-speed robot is based on delta design which has been developed in 1988 by the Ecole Polytechnique Fédérale de Lausanne (EPFL). Delta architecture robots are highly dynamics due to its lightweight and parallel design. The first commercial Delta robots were produced

[*] Corresponding author: hisham@utem.edu.my

by BOSCH and ASEA Brown Boveri, (ABB) companies. However, ABB FlexPicker is the most sold and fastest commercial parallel robot in the world today [1][2].

Delta robot - FlexPicker IRB 340 (as shown in Figure 1) by ABB that was installed in Robotic Lab, Faculty of Manufacturing Engineering, UTeM, is one of the top most leading robot for industrial use for picking up lightweight products. This robot consists of 4 degree of freedom (DOF); Scara motions which are well adapted to pick and place tasks. All the arms were linked in parallel with three DOF of translation to carry an object from one point to another and a theta axis that enables rotation about a given axis in world coordinates.

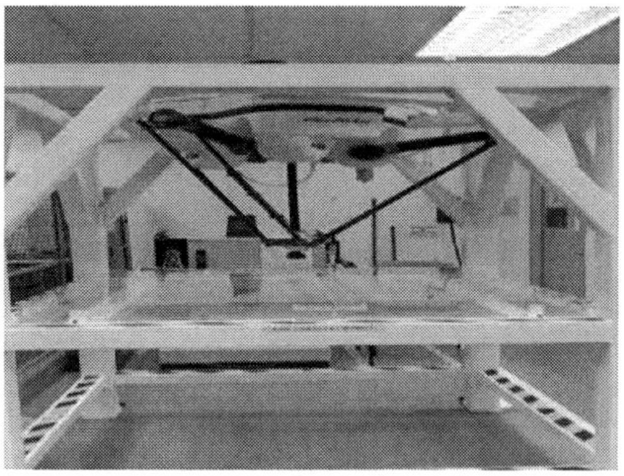

Fig. 1. ABB Delta robot – FlexPicker IRB 340 (Robotic Lab, FKP).

The aim of this project is to go one step further and to show that is indeed possible to increase the degree of freedom of the ABB FlexPicker robot by at least one degree of freedom for automated product assembly task. An end effector has been designed to increase the degree of freedom of the robot. The current robot in investigation uses Vacuum gripper (shown in Figure 2) with a fixed end effector. It consists of four DOF whereby three translations in x, y, z and one rotation about a given axis. The vacuum gripper which was attached to the fixed end-effector provides minimal flexibility of movement. For example, it only picks and place an object in a vertical direction and does not provide gripping at any angles.

Fig. 2. End-effector and vacuum gripper.

End-effector is one of the most important factors for increasing flexibility and degree of freedom of a robot. There are many types of end-effector being used in current industries for ABB FlexPicker such as mono pick, multi-suction cup, magnetic grippers and parallel grippers [3]. However, all of this end-effector can only be moved in rotational motion as additional to the x, y, and z-motions. Therefore, an initiative to design a flexible end-effector was taken to solve this problem so that the robot can perform gripping at vertical direction with a specified angle. This flexible end-effector is believed to increase the robot's degree of freedom by at least one degree, thus could increase the flexibility and improving the productivity.

2 Design overview

Figure 3 shows the selected design of the end-effector that has been fabricated and investigated in this project. Table 1 indicates the name of the parts in this design. This design is basically powered by a servo motor. In general servo motor is a rotary type motor that can be coded to output a signal that allows the shaft of the motor to revolve to a specified angular position. Therefore, when the servo motor rotates, it pulls the joint and eventually forces the flexible tube towards the direction of the servo motor. The top ring holder is used to hold the middle part of the flexible tube in stationery, thus the upper part is able to stretch and bend to the desired angle. In this case, the bending angle of the tube can be controlled by manipulating the angular rotation of the servo motor.

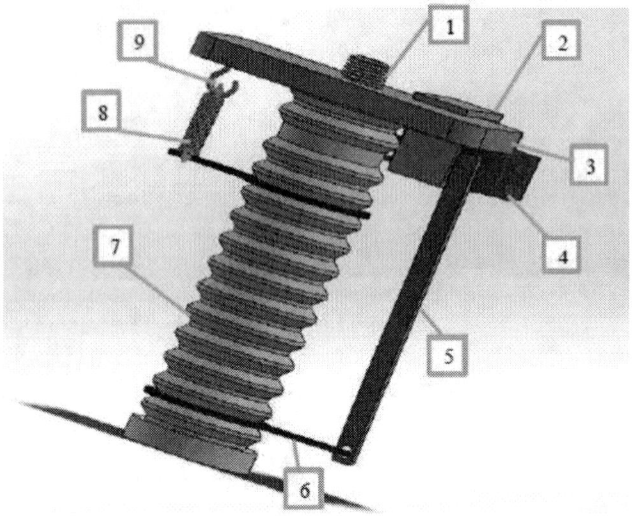

Fig. 3. Final design of end-effector.

Table 1. List of parts.

No	Name of Parts
1	Hollow screw
2	Base
3	Servo motor
4	Joint
5	Ring connector
6	Flexible hollow tube
7	Rigid holder

DOI: 10.1051/matecconf/20167801056

Among the advantages of using this design as an effector for the ABB Flexpicker are because of its durability, incur minimal noise, simple design yet flexible and easier to bent by using the servo motor. However, the servo motor is not as fast as a pneumatic cylinder, nevertheless, due to its lightweight property, the servo motor is more desirable to be attached to the ABB robot as to decrease additional payload.

3 Experiment details

The prototype of the end-effector was attached to the ABB FlexPicker by screwing it firmly to the robot, replacing the original suction cup as shown in Figure 4. A pre-programmed microcontroller (Arduino) was used in this experiment to control the angle of the servo motor, initially from 0° to 90°. By controlling the angle of the servo motor, the angle of the hose can be found by measuring the angle between both ends of the hose as illustrated in Figure 5. Initially, the attached end-effector was tested without any bending to determine the maximum load it can withstand. The suction force provided by the robot is constant at 0.4 MPa. Next, without any load, the servo motor was operated for different specified angle to investigate the flexibility of the end effector and the resulting angle of bending for the vacuum hose. The resulting hose bending angles were then recorded.

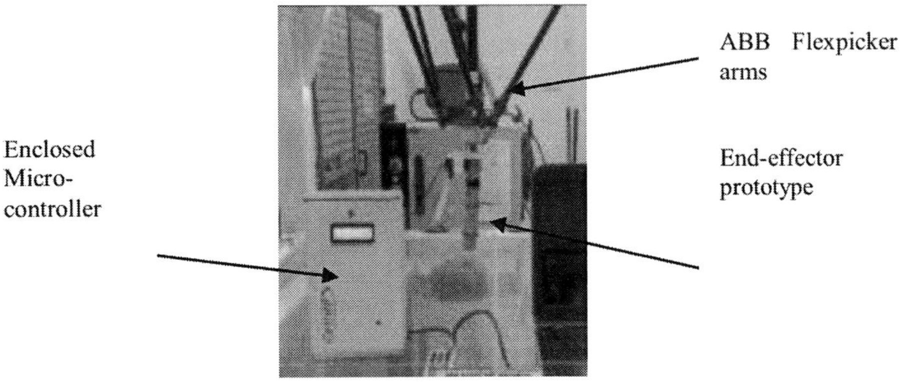

Fig. 4. End-effector attached to the ABB Flexpicker robot arms.

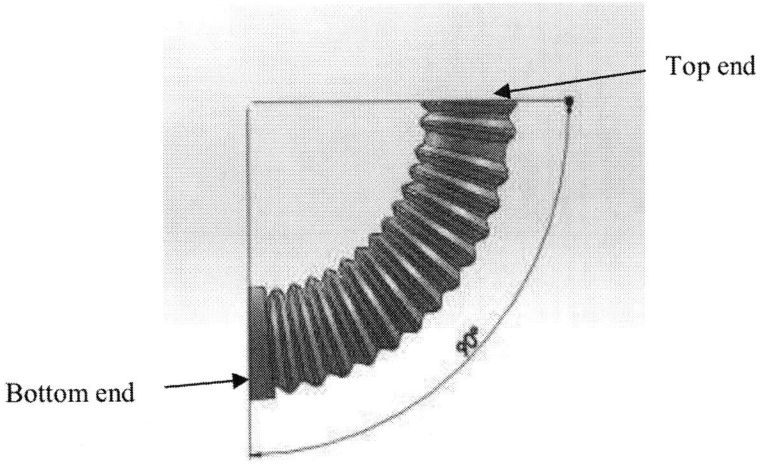

Fig. 5. Measuring hose bending angle.

The experiment was then repeated again but with varying loads attached to the end of the suction hose. Based on the maximum allowable load that was determined from the previous experiment, the load was varied gradually with 100 g increment. The result was then recorded and analyzed for the bending motion.

4 Results and discussion

The first test to determine the maximum load the end-effector can withstand resulting in failure where the suction force from the robotic system had crushed the vacuum hose inwards, regardless of the load weight. To strengthen the hose, metal rings were inserted at every crevice of the hose and the test was resumed. It was found out that the maximum load it can hold are approximately 1 kg.

According to the obtained result, the lifting force induced by the ABB FlexPicker is sufficient to hold the maximum weight of 1 kg. The results show the load can still be gripped by the supplied vacuum pressure. To proof this, a simple calculation was done by calculating the holding force required by a maximum load of 1 kg and the maximum lifting force offered by the system at 0.4 MPa vacuum pressure.

The actual holding force required to carry 1 kg of the load by the flexible end-effector can be calculated by using the following equation [4]:

$$F_H = m \times (g + a) \times s \qquad (1)$$

Where:

F_H = Holding force (N)
m = Mass (kg)
g = Acceleration due to gravity (9.81 m/s²)
a = Acceleration of the system (m/s²)
s = Safety factor (1.5 for linear and 2 for rotary movement)

Maximum acceleration values of ABB FlexPicker robot is 100 m/s^2 [5]. Thus, for lifting 1 kg of load,

$$F_H = 1\text{kg} \times (9.81\text{m/s}^2 + 100\text{ m/s}^2) \times 2$$
$$= 219.62\text{ N}$$

Maximum suction force offered by the system;

$$F_S = P \times A \qquad (2)$$

Where:

F_S = Suction force (N)
P = Vacuum pressure (Pa)
A = Size of suction cup (m^2)

The area of flexible hollow tube;

$$A = \pi r^2$$

$$= 3.142 \times 0.03^2$$
$$= 0.00283 \text{ m}^2$$

Thus, the suction force;

$$F_S = 0.4 \text{ MPa} \times 0.00283 \text{ m}^2$$
$$= 1130.9 \text{ N}$$

Comparing both the actual holding force (F_H) and suction force (F_S);

$$1130.9 \text{ N} > 219.62 \text{ N}$$
$$F_S > F_H$$

From the calculation, it is proven that the result obtained from the weight lifting test on the gripping force exerted by the ABB FlexPicker is sufficient to hold the maximum weight of 1 kg. However, in the case where the holding force is perpendicular to the gravitational force, the equation are given as below [4]:

$$F_H = (m/\mu) \times (g + a) \times s \quad (3)$$

Where:

F_H = Holding force (N)
m = Mass (kg)
μ = Frictional coefficient
g = Acceleration due to gravity (9.81 m/s^2)
a = Acceleration of the system (m/s^2)
s = Safety factor (1.5 for linear and 2 for rotary movement)

Empirical frictional value varies with different surface are shown below:

- Oily: 0.1
- Wet: 0.2 – 0.3
- Wood, metal, glass, stone: 0.5
- Rough: 0.6

Assuming frictional coefficient, μ to be 0.5, thus;

$$F_H = (1\text{kg}/0.5) \times (9.81\text{m/s}^2 + 100 \text{ m/s}^2) \times 2$$
$$= 439.24 \text{ N}$$

Therefore, it can be deduced that extra holding force is required for the case of perpendicular lifting for the current system. However, since the suction force provided by the system is higher, the 1 kg object can still be handled effectively by the end-effector.

For the bending test, the angle of bending achieved by the flexible end effector for a maximum of 90 degrees motor rotational angle is tabulated in Table 2. Without applying any load, as the angle of servo motor rotation increases, the actual bending of the vacuum hose decreases near to 8% towards desired angle of 90 degrees. The reason for this difference is due to the extra force required to overcome the stiffness of the vacuum hose which increases upon increasing angle of bending.

The angle of bending was then recorded for all loads and from the result, it is found that the maximum load that the servo motor can support is 600 gram and beyond that, the servo motor starts to jerk at a certain angle. For example, for 100 gram load, the average angle of bending achieved for servo motor at 90 degrees angle of rotation is just 79.3 degrees. The angle of bending decreases when the load increases due to the weight effect of the gravitational force. To achieve 90 degrees of hose bending for 100 gram to 600 gram load, the motor angle of rotation should be increased more than 90 degrees.

Table 2. Resulting vacuum hose angles.

Motor rotation angle, Ø	Pipe bending angle, β										
	0 g	100g	200 g	300 g	400g	500 g	600 g	700 g	800g	900 g	1000 g
10	10.7	11.3	9.3	7.7	7.7	7.0	6.7	6.7	6.0	4.7	4.0
20	20.3	20.7	17.7	17.3	15.7	13.7	10.7	10.7	9.7	8.7	8.3
30	30.3	31.0	28.3	24.3	25.0	22.3	20.7	20.3	20.7	19.7	18.0
40	39.3	35.7	36.0	33.7	34.3	31.7	31.7	31.3	30.0	29.0	0.0
50	49.0	48.3	42.3	42.0	41.0	40.0	39.7	39.7	40.3	39.0	Jerk
60	57.0	55.7	52.0	51.3	53.0	51.0	50.3	50.0	48.3	Jerk	Jerk
70	66.7	64.7	58.7	57.3	56.7	55.3	54.3	53.7	54.0	Jerk	Jerk
80	75.0	72.3	65.3	63.0	60.3	60.7	60.7	60.0	Jerk	Jerk	Jerk
90	83.0	79.3	71.3	69.3	77.7	69.7	69.3	Jerk	Jerk	Jerk	Jerk

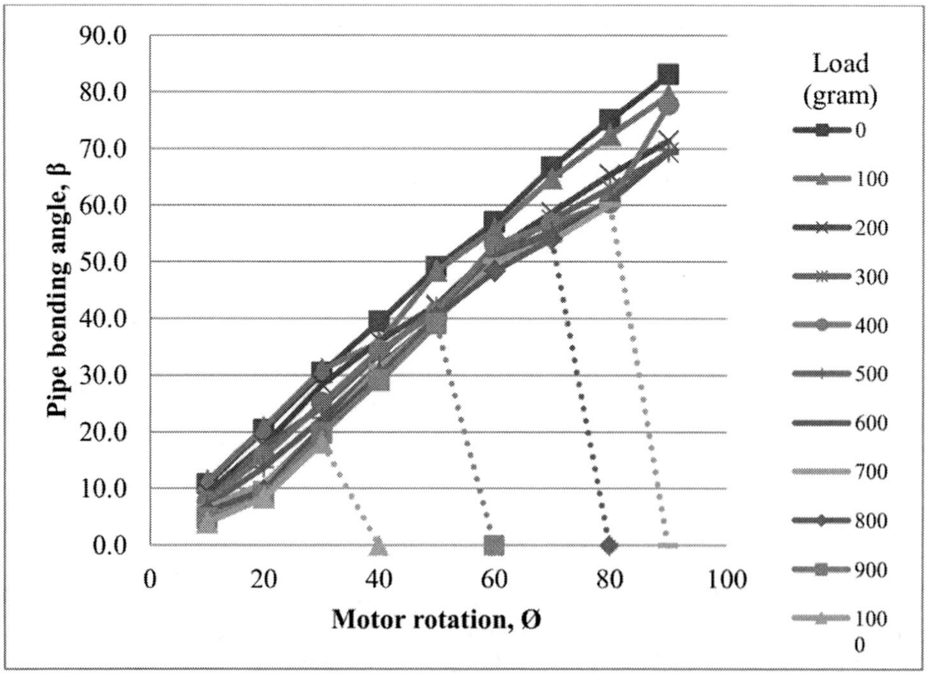

Fig. 6. Graph of servo motor rotational angle, Ø versus vacuum hose bending angle, β.

The main reason that the servo motor starts to jerk at a certain angle after implementing 600 gram of load was due to insufficient torque to overcome the gravitational force as well as the bending resistance stiffness of the hose. The specified servo motor holding torque is 1.7 N.m (17 kg.cm), which signify that a combination of gravitational force and bending resistance (stiffness) is at equilibrium with the motor holding torque. From the graph in Figure 6, the resulting bending angle for different motor rotation and different load can be estimated.

5 Conclusion

An additional one-degree-of-freedom has been introduced to the 4-DOF ABB FlexPicker robot by attaching a flexible vacuum hose as an alternative end effector. The end effector has the capacity to lift up to 1 kg of load in vertical motion, however, it can only lift up to 600 g with bending angle of 69.3 degrees. To increase the bending angle of the vacuum hose, the rotational angle of the servo motor can be increased by minor modification to the design of the end effector simply by removing the physical barrier that restricts the movement of the joint. Finally, further works are needed in order to increase the lifting capacity of the servo motor without sacrificing the weight ratio of the end effector with respect to the total maximum payload of the robot.

The experimental works were conducted at Robotic Laboratory, Faculty of Manufacturing Engineering, Universiti Teknikal Malaysia Melaka (UTeM) under grant no: PJP/2015/FKP(3C)/S01434.

References

1. M.Martin, K. Bernd, G. Harald, *Conference ISR ROBOTIK,Munich,* 225 (2014)
2. J. Brinker, B. Corves, *The 14th IFToMM World Congress* (2015)
3. K. Appu, *Robotics.* I.K. International Publishing House Pvt. Ltd., 13, (2007)
4. http://www.avs-yhtiot.fi/sites/default/files/pdf/7.01.05_alipaine_ohje.pdf
5. *IRB 340 Industrial Robot Datasheet*, ABB Automation Technologies (2005)
6. FESTO, *Basic Principle of Vacuum Technology* (2015)
7. BOSCH, *Packaging Line Automation using Delta Robots* (2015)
8. I. Bonev, *Delta Parallel Robot – the Story of Success* (2001)
9. H. Wu. (2010), *Developing an Integrated System for Automating Picking and Sorting using an ABB FlexPicker Robot,* Master's Thesis. Massey University (2010)

MATEC Web of Conferences 78, 01057 (2016) DOI: 10.1051/matecconf/20167801057
ICon*GDM 2016*

The Development of Plant Maintenance Scheduling Via Inventory System for Sustainable Plant Operation

Roslizan Masripan[1], *A. M.* Leman[1,*], *I.* Baba[1], *Dafit* Ferianto[1], *Supaat* Zakaria[1], *Fakrurrazi* Rahman[1]

[1]Faculty of Engineering Technology, Universiti Tun Hussein Onn Malaysia (UTHM), Parit Raja, 86400 Batu Pahat, Johor, Malaysia

Abstract. Industrial sector becomes the main concern for developing country. By the time, it was increased rapidly. However, there are many problems observed such as maintenance scheduling, stock inventory and supply chain. Therefore, this research develops new inventory system to develop sustainable plant operation with a high capability to plant operation especially to stock inventory of machine component. In also required green application with minimised used on paper. This system is developed using Radio Frequency Identification (RFID) for inventory control which integrated with web-based system. This system consists of several modules such as station module, item module and item request module and report of critical stock in the store. This system can be controlled from a hand-phone with internet connection or automatic alert such as Short Massage Send (SMS) and email. The developed system is very effective in monitoring the stock material through the barcode, supply chain and worker performance as well as to reduce the lead time for maintenance activities of the company through sustainable plant operation.

1 Introduction

It is widely acknowledged in literature that the design of Inventory Management Systems is of significant importance for the overall performance of manufacturing companies. Although different definitions exist, Inventory Management Systems are generally described as integrated systems designed to integrate, standardise and automate decision processes related to the management and control of inventories [1]. Inventory Management Systems in other words, provide the information to decision-makers within organisations at a strategic, tactical and operational level in order to support inventory decisions [2-3].

* Corresponding author: mutalib@uthm.edu.my

DOI: 10.1051/matecconf/20167801057

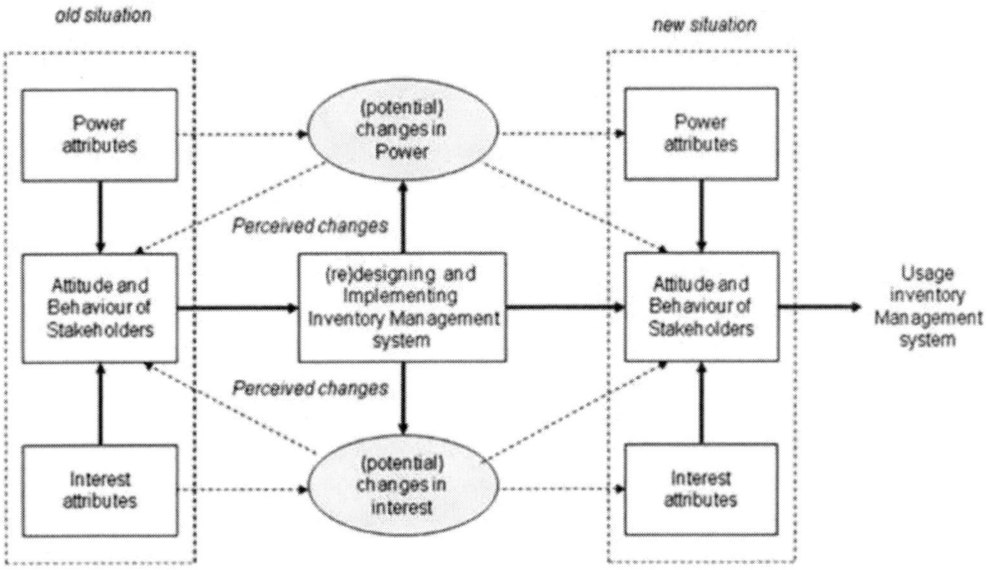

Fig. 1. Conceptual model of the implementation and usage of inventory systems network [1].

With the concern and support to The Internet of Things (IOT) industry constantly strengthening, the IOT has made full use of next generation IT technologies among the industries and has been widely applied in smart transportation, logistics management, environmental protection, public safety, smart household and other fields. In the aspect of logistics management, it can greatly reduce labour force, improve the efficiency of the operation speed and storage, and also decrease logistics costs. Meanwhile, it combined with a supply chain management system effectively and suppliers, manufacturers, customers are strongly united by Information technology, which can reduce inventory risk effectively [4].

The industrial production index has been increased for last few years. However, it is in-line with the problems observed. Nowadays, there are many companies which not completed by inventory system that can affect to the bad maintenance schedule, high cost and low profit of company [5]. The development of inventory system integrated with web-based is quite needed to solve the problem by improving communication in related staff. The inventory system basically developed using RFID system where the main components of RFIDare shownn in Fig. 2.

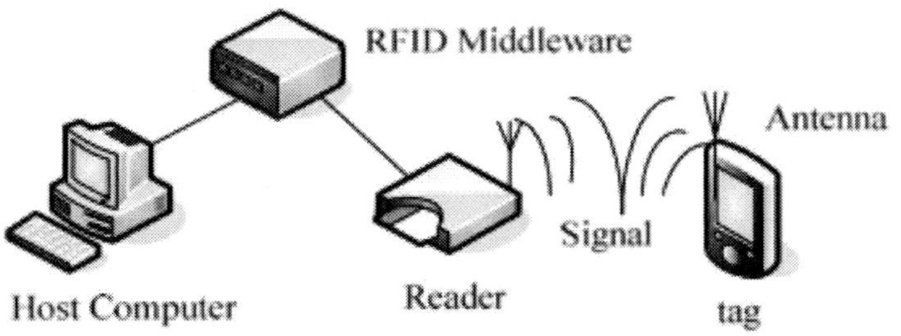

Fig. 2. The main components of an RFID system [4].

The several influences of the inventory system may caused by an unsuitable inventory system such as quantity of the components or stock is not updated automatically, delay for receiving components and unaware of the store keeper in component supply [6-7]. Therefore, this study aims to develop a new inventory system which integrated with report management for improving maintenance scheduling. This research is believed will sustain the plant production and improve the income of the company.

1.1 Maintenance types.

There are three types of maintenance in use in the company such as [8]:

i. Preventive maintenance, where equipment is maintained before breakdown occurs. This type of maintenance has many different variations and is subject of various researches to determine best and most efficient way to maintain equipment. Recent studies have shown that Preventive maintenance is effective in preventing age related failures of the equipment. For random failure patterns which amount to 80% of the failure patterns, condition monitoring proves to be effective.
ii. Operational maintenance, where equipment is maintained by using.
iii. Corrective maintenance, where equipment is maintained after break down. This maintenance is often more expensive because the worn equipment can damage other parts and cause multiple damage.

1.2 Integrated optimal reliable design, production, and maintenance planning.

In multipurpose process plants, which are characterised by sharing different resources (equipment, manpower etc.) for production, unplanned equipment shutdown could affect the timely production of products and hence process profitability. Many approaches have been proposed to ensure high equipment availability by combining design, production, and scheduling frameworks with a maintenance optimisation framework. In these approaches, the initial reliability characteristics of equipment, which also determine equipment availability, are considered fixed by problem definition. In this work, a combined design, production and maintenance planning formulation for multipurpose process plants is extended to incorporate the reliability allocation problem at the design stage and the equipment initial reliabilities are made and can be improved by increasing the reliability of equipment and/or adding redundancy like shown in Fig. 3 [9].

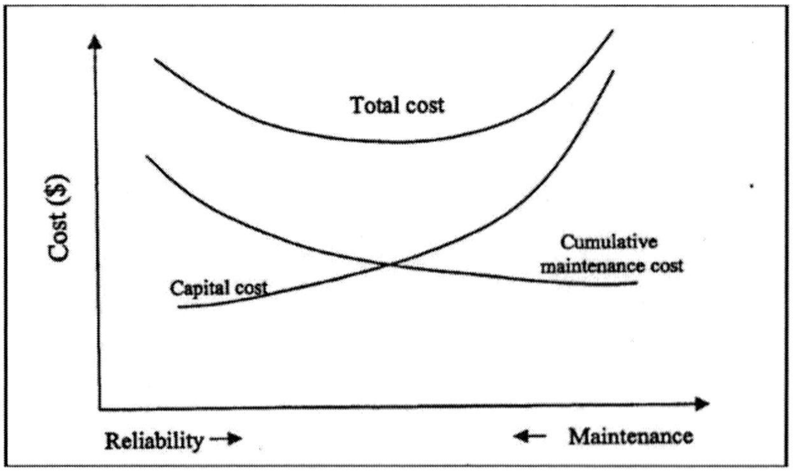

Fig. 3. Reliability versus maintenance costs to achieve a high availability [9].

1.3 Inventory theory

The goal of inventory usually comprises elements such as re-order level and re-order quantities which minimise total inventory cost (Fig.4), subject to a given service level. Total inventory cost normally ordering, setup cost, unit purchase costs, holding costs and shortage costs.
The Economic Order Quantity (EOQ), where inventory costs are minimised for independent demand. According to Sanjoy et al., (2015) [10] study on the classic economic lot size model, which recommends an optimal production batch by a trade-off of the inventory holding costs against production change-over costs.

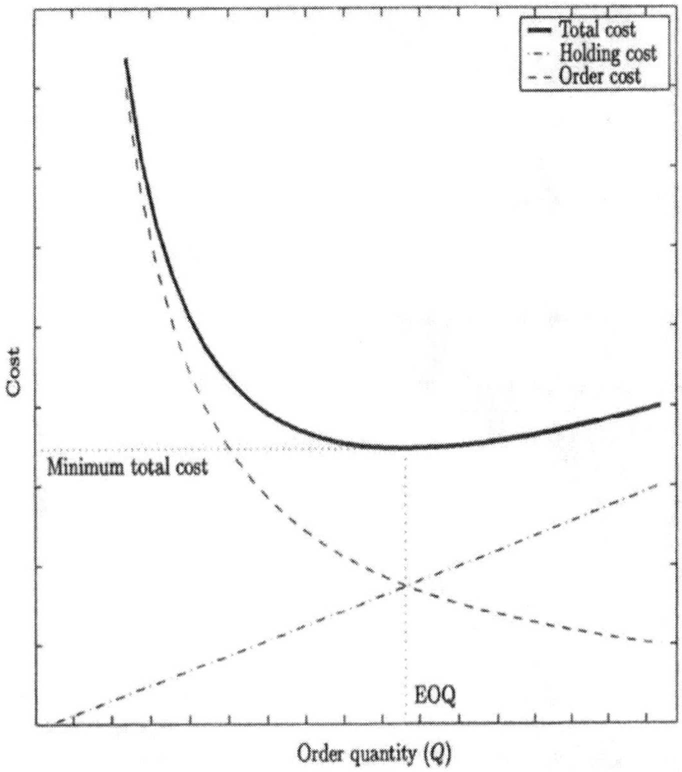

Fig. 4. A graphical depiction of the concept of an economic order quantity with alocation cost [10, 11].

1.4 Stock control and inventory

Stock control is used to show how much stock you have at a point in time, and how you keep track of it. It applies to every item you use to produce from raw materials to finished goods and whether it is a product or service. It covers stock at each stage of the production process, from purchase and delivery to using and re-ordering the stock [11].

1.5 Using RFID for inventory control, stock security and quality management.

Radio Frequency Identification (RFID) allows a business to identify individual products and components, and to track them throughout the supply chain from production to point-of-sale. An RFID tag is a tiny microchip, plus a small aerial, which can contain a range of digital information about the particular item. Tags are encapsulated in plastic, paper or similar material, and fixed to the product or its packaging, to a pallet or container, or even to a van or delivery truck. The tag is integrated by an RFID reader, which transmits and receives radio signals to and from the tag. Readers can range in size from a hand-held device to a 'portal' through which several tagged devices can be passed at once, eg on a pallet. Readers can be placed at different positions within a factory or warehouse to show when goods are moved, providing continuous inventory control. The benefits of this system are more efficient stock control and improved security make it particularly attractive to retailers, wholesalers or distributors [5].

2 Methodology

This research is conducted at Sungai Ambat Palm Oil Millstore department and involves of selected major key machinery which will give a significant effect to the maintenance and operation of the mill. The system inventory is developed to assist Sungai Ambat Management to improve maintenance scheduling to avoid major breakdowns which lead to profit lost to the company via reducing of inventory cost, time to locate and count materials.

2.1 Pre-trial / Dry –run

The coding system has given an impact the whole inventory system. The dry – run test, including coding the part, testing the scanner and system and software like shown in Fig. 5.

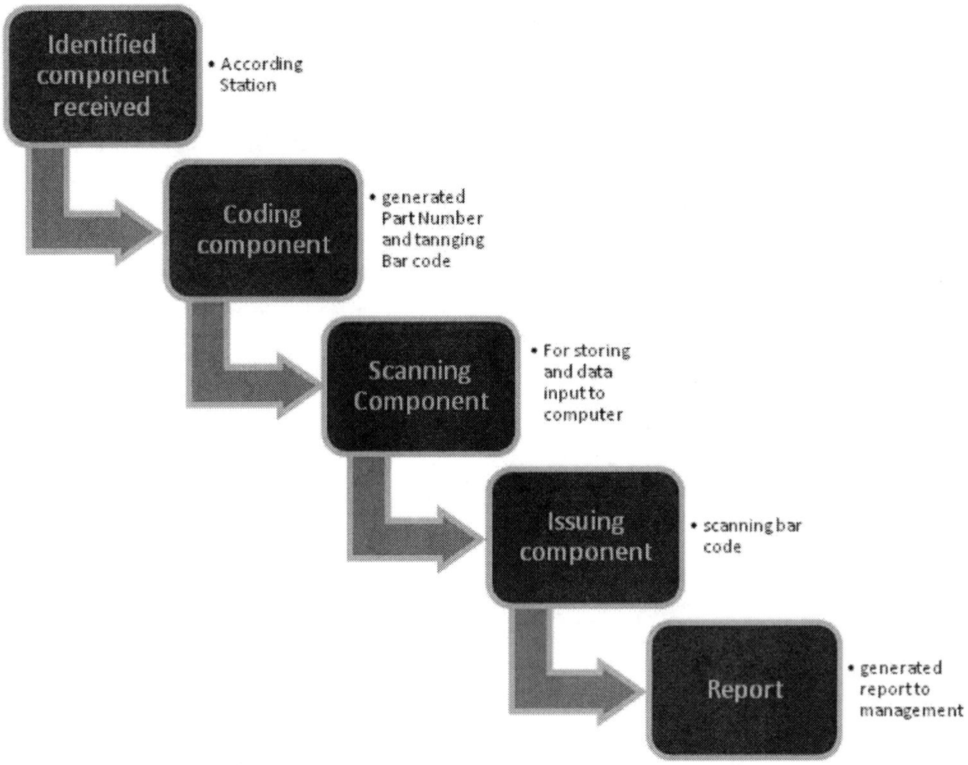

Fig. 5. Flow of pre-trial/dry-run process.

2.2 Testing

Testing will carry out on barcode itself. The simulation will carry out for few selected items to determine the equipment which reads the barcode function according with system like shown in Fig. 6.

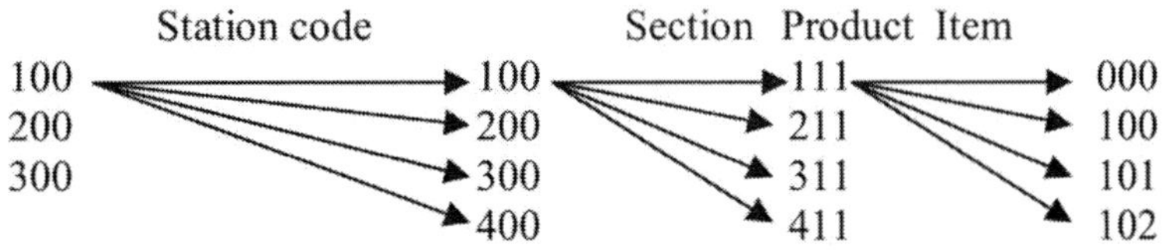

Fig. 6. Tagging system – bar coding system.

2.3 Synchronization

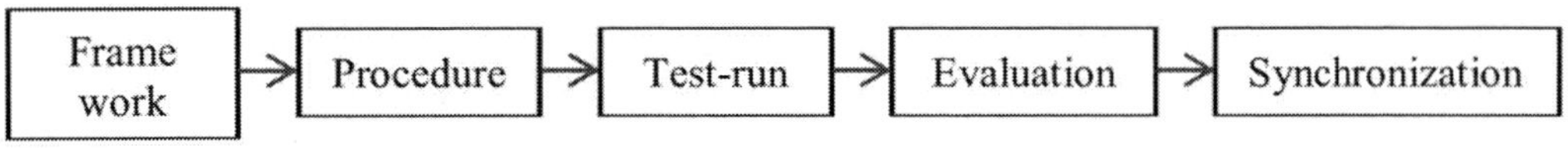

Fig. 7. Synchronization process of the inventory system.

2.4 Oil Palm Plant Inventory System (OPIS)

This Oil Palm Plant Inventory System, called OPIS. This software provides business solutions for managing simple purchase order and stock inventory in the plant. Inventory management is the management of that list of material in stock. Inventory is a list of goods in stock i.e. total amount of materials contained in a unit or store. Buffer stock should be kept in order to meet the uncertainties in the market, i.e. in the demand of goods as well as to meet the time lag from supplier to user. Inventory can be of different types like raw material inventory, work in progress inventory, finished goods inventory and goods for resale inventory. Different components of an inventory management system are barcode scanner, barcode printer, barcode label, inventory software, etc. Inventory management software works along with barcode for inventory tracking and control.

2.5 Web-based inventory system

Internet-enabled electronic links between supply chain partners have allowed better coordination and collaboration among the various supply chain segments. Meanwhile, at the front of the supply chain, increasingly sophisticated point-of-sale systems can capture product demand patterns. More accurate sales-forecasting tools take some of the guesswork out of production and reduce the need for large inventory safety stocks. Tracking and tracing tools are also available to follow orders across borders and through the hands of different supply partners. The mechanism of tracking system is shown in Fig. 8.

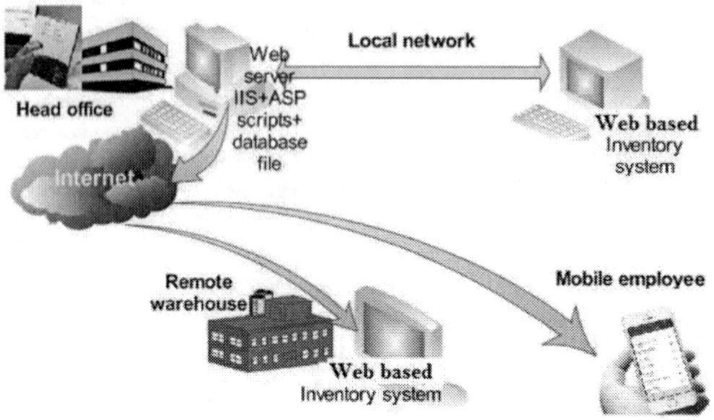

Fig. 8. Mechanism of the inventory system.

2.6 New inventory system

This system is easy to operate from anywhere and anytime control system because it controlled using the internet, alert system, SMS and Email.

2.7 Database connection

Name of the database that is designed as shown in Fig. 9, for SPMP is inventory.

```
Dim sql As New SqlConnection("Data Source=EIGHT\FARESSA;Initial Catalog=inventori;Integrated
```

Fig. 9. Code to Connect to Database.

2.8 Log in

OPIS enables two types of user login which are Kerani and Admin as shown in the Fig. 10.

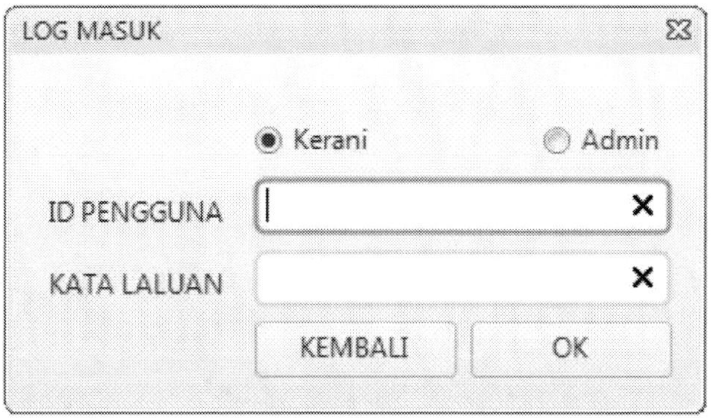

Fig. 10. Log In Interface.

2.9 Item module

Item module as shown in Fig. 11 is used by secretary to register new item and generating barcode for each item.

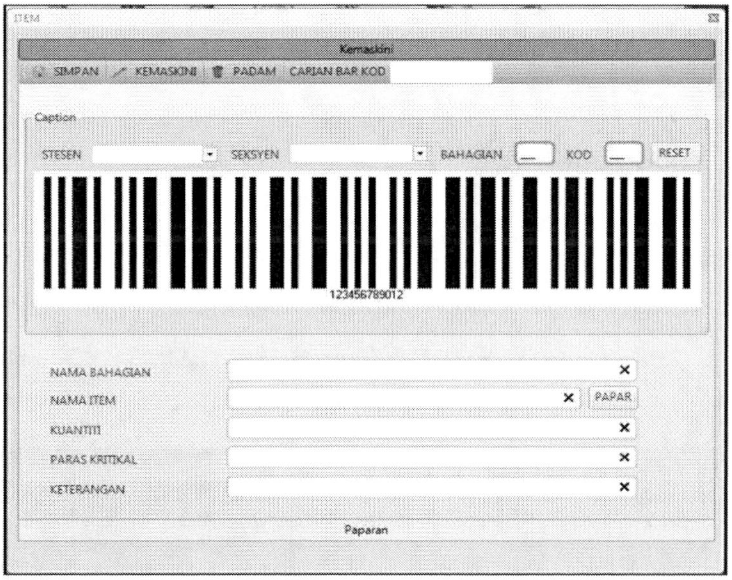

Fig. 11. Item Form Interface.

2.10 Item request module

The item request module will be accessed by secretary to manage stock in and stock out of an item. The request form must be signed by a supervisor before proceeding with this module. Staff id is required to identify each transaction occurred. This form will calculate the stock left after being requested as shown in Fig. 12.

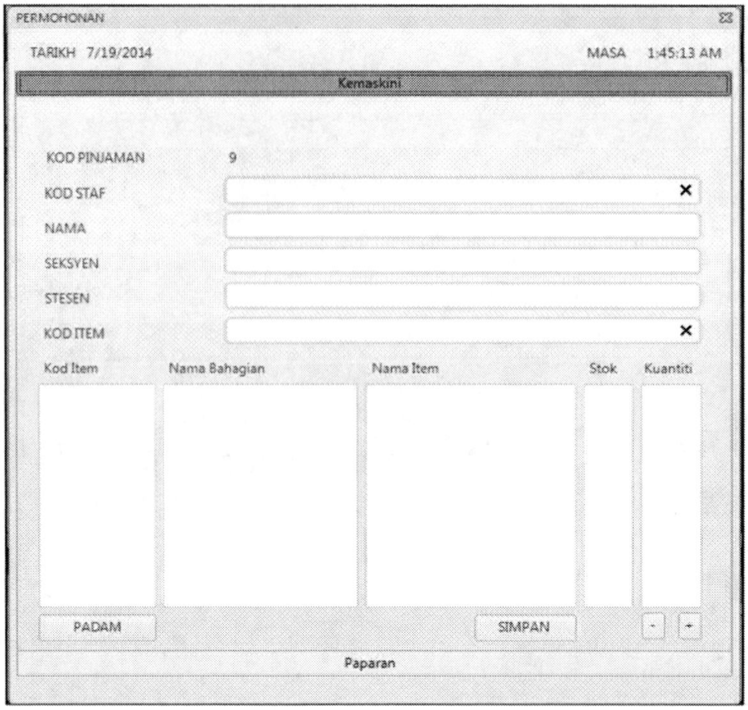

Fig. 12. Item Request Interface.

2.11 Critical stock report

OPIS also can generate critical stock reports if the current stock below the critical level as shown in Table 1.

Table 1. Critical Stock Report.

LAPORAN HARIAN STOK YANG MENCAPAI PARAS KRITIKAL

30-Jul-2014

Kod Bar	Keterangan	Kod Bahagian	Nama Bahagian	Kod Item	Nama Item	Kod Stesen	Kod Seksyen	Paras Kritikal	Stok Semasa
		0		0		0	0	0	0
101100111333	WARNA HIJAU, DIBUAT DARI KAYU JATI	111	MEJA	333	ALAS MEJA	101	100	6	0
101100123441	GG	123	JAM TANGAN ROLEX	441	AW	101	100	7	3
200102123445	EVEREADY	123	JAM TANGAN	100	BATERI	200	102	50	3
101100111	H	111	MEJA	119		101	100	5	5
101100111332	EVEREADY	123	JAM TANGAN ROLEX	123	BATERI	101	100	50	49

2.12 Notification

There are three types of notification for the critical stock value which are alert, SMS and Email. OPIS will display alert to the user each time user logs onto the system as shown in Fig. 13. OPIS also will send SMS to administrator to inform user about the critical value stock. Partial code to send SMS and email.

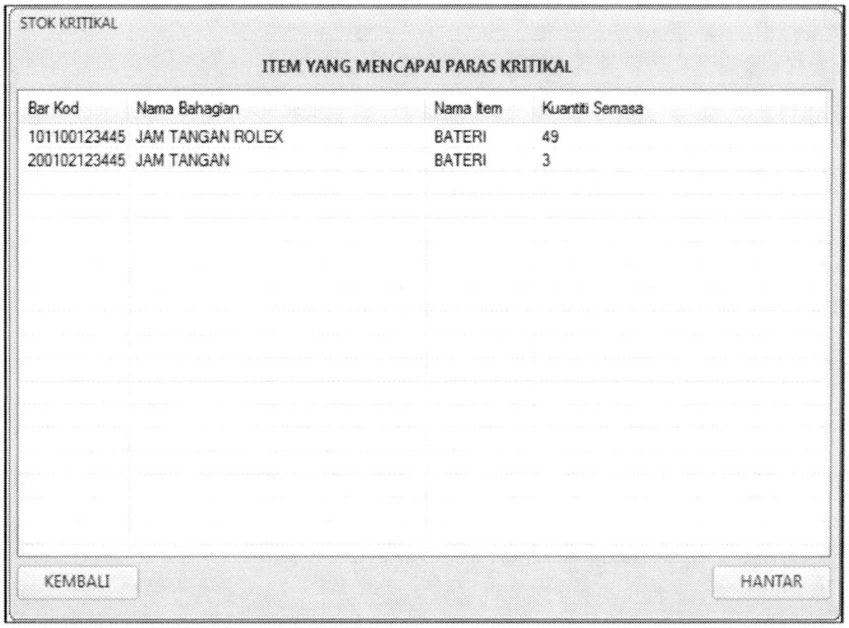

Fig. 13. Alert of Critical Stock.

3 Conclusion

New developing inventory system has been successfully developed by RFID system. It gives significant effect to the supply chain, stock monitoring, maintenance scheduling, and worker performance. In conclusion, inventory control system will be improved store system at Kilang Sawit RISDA Sg. Ambat (KSSA). Besides, by implementing this system, the work load of staff will be reduced and their job become easy and more productive.

The authors would like to thank to Universiti Tun Hussein Onn Malaysia (UTHM) and Centre for Graduate Studies – UTHM as well as Kilang Sawit RISDA Sg. Ambat (KSSA).

References

1. J. Vries, Int. J. Prod. Econ., **143**, 233 (2013)
2. N.N. Nagarur, Hu, Tai san, Baid, K. Nirmal, Ind. Manage. Data Syst., **94** (9), 22 (1994)
3. Chen, H.M. Chen, Chang, C. Sian, International Journal of Business Information Systems, **4(1)**, 105 (2009)
4. X. Jing, P. Tang, *Research and Design of The Intelligent Inventory Management System Based on RFID* (Sixth International Symposium on Computational Intelligence and Design, 2013)

5. K.H. Donselaar, R.A.C.M.B. Eindhoven, Int. J. Prod. Econ., **143**, 440 (2013)
6. D.P. Warsing Jr., W. Wangwatcharakul, R.E. King, Comput. Oper. Res., **40**, 2786 (2013)
7. S. Axsater, S. Viswanathan, Eur. J. Oper. Res., **221**, 340 (2012)
8. X. Liu, W. Wang, R. Peng, Reliab. Eng. Syst. Safe, **137**, 76 (2015)
9. D. Fu gui, L. Hui mei, L Bing de, Systems Engineering Procedia, **4**, 298 (2012)
10. S.K. Paul, R. Sarker, D. Essam, Comput. Ind. Eng., **84**, 101 (2015)
11. Z. Wang, F.T.S. Chan, Systems, **45**(2), 326 (2015)

Energy Efficiency: The Implementation of Minimum Energy Performance Standard (MEPS) Application on Home Appliances for Residential

K.A Rahman[1, *], *A.M* Leman[2], *L.H* Mansor[2], *M. N. M* Salleh[3], *M.Z.M* Yusof[1], and *M.* Mahathir S.S [2]

[1]Faculty of Mechanical & Manufacturing Engineering, Universiti Tun Hussein Onn Malaysia, 86400 Batu Pahat, Johor, Malaysia
[2]Faculty of Engineering Technology, Universiti Tun Hussein Onn Malaysia, 86400 Batu Pahat, Johor, Malaysia
[3]Faculty of Computer Science & Information Technology, Universiti Tun Hussein Onn Malaysia,86400 Batu Pahat, Johor, Malaysia

Abstract. Generally, Minimum Energy Performance Standard (MEPS) has been widespread across the country especially developed country. However, most consumers do not even know about the MEPS. Without sufficient knowledge, much energy have been wasted before this. The aim of this study is to review the implementation of MEPS of Asia country and to compare electricity consumption of home appliances with star rating and without star rating. In order to fulfil the objectives of the study, the equipment must be chosen correctly and must be learned properly. The home appliances that will be used also need to be chosen so that the comparison between the appliances will be matched correctly. To understand the results, the analysis was done using graphs and table. The purpose of using graph and table is to understand the comparison between appliances more clearly. The results show that home appliances with MEPS is more efficient on energy saving rather than without MEPS. This is the evidence as a method to educate a consumer on energy saving.

1 Introduction

The biggest contributor to high levels of CO_2 emissions is the energy industry, which accounts for 41% of emissions, followed by the transport industry (23%) and the manufacturing industry (20%) in 2009 [1]. Accurate forecasting of energy consumption in a commercial building, including residential building is an important strategy in achieving the goal of reducing energy demand, as well as to improve Energy Efficiency (EE). To implement this strategy, many methods and indicators have been proposed to monitor and measure energy performance in buildings. The EE improvement is related to a cost-effective measure in order to achieve sustainable energy development [2]. The most frequently tools that are used in any energy efficient and conservation program for

* Corresponding author : khairunnisa7482@gmail.com

appliances are Minimum Energy Performance standards (MEPS) and energy labelling [2,3,4,5]. MEPS specifies the minimum level of energy performance for electrical appliances such as lighting and cooling and those electrical appliances must meet the requirement before they can be offered for sale or used for commercial purposes [7]. MEPS are an effective way to increase the EE of products [8]. MEPS constitute the most cost-effective policy option to transform markets toward more energy efficient products. When effectively applied, in the context of an integrated policy approach, MEPS encourage manufacturers to improve the efficiency of their products [9].

Most of electricity bills soaring for a house nowadays due to the wastage of electricity. This is because consumers did not know how to use it efficiently. Generally, most houses use old electrical appliances and most of these appliances do not have energy star rating - MEPS. This is because electrical appliances with star rating a bit costly compared with electrical appliances without star rating. Although electrical appliances with MEPS a bit expensive, it is better for a long term where it is similar with high electric bills that paid every month. The purpose of this research is to test EE on MEPS application that can help us in determining the efficiency of energy on electrical appliances especially on home application. Furthermore, the comparison also will be made between the home appliance with a star rating and without a star rating to help consumers make a choice wisely. In addition, consumers can calculate by its own to differentiate which one is better whether buy the home appliances with MEPS with high price and can save the electricity for a long term or vice versa. Besides, the EE knowledge can be used for practice in order for reducing the cost of electricity especially for residential bulding.

2 Literature Review

2.1 Minimum Energy Performance Standards (MEPS)

Accurate forecasting of energy consumption in a commercial building is an important strategy to achieve the goal of reducing energy demand, as well as to improve EE. To implement this strategy, many methods and indicators have been proposed to monitor and measure energy performance in buildings. EE standards and labelling has been introduced to bring more benefit for the most major classes especially for consumers [9]. Although EE standards and labelling are now commonly used around the world, the MEPS approach still not implemented widely. However, in recent years, it become among the popular strategy to be implemented in order to improve EE. Malaysia become one of the developing country to adopt this approach. MEPS are incorporated in the amendments of the Electricity Regulations 1994 and has been gazetted by the Minister of Energy, Green Technology and Water Malaysia on 3rd May 2013. Before MEPS, a Voluntary Energy Efficient Appliances and Equipment standards and labelling program for Energy Efficient Appliances and Equipment was conducted from 2009 to 2012 [10]. Currently, MEPS only implemented to five domestic electrical products which is air conditioner, refrigerator, television, fan and lighting. Mahlia et. Al [11] proposed to introduce labels to convince consumers to buy EE appliances. At the same time also influence the manufacturers to produce more EE products into the market.

MEPS used to determine the minimum level of energy performance that electrical appliances must meet before they can be offered for sale or used for commercial purposes [12]. The function of MEPS is to protect the consumers from products that have low energy performance [13]. The efficiency level is shown by using the star rating label. This label must be affixed onto the electrical appliances before it can be sold in the market [10].

Based on Electricity Regulation 1994 (Amendments 2013) Regulation 101A (3) stated, "Any equipment that meets all the requirements of efficient use of electricity under sub regulation (1) shall be affixed with an efficiency rating label in such form and manner as may be determined by the Commission." Figure 1 shows the EE label specification including energy rating, appliance type, appliance energy rating, information on brand and model, energy consumption (in kWh/year) and testing standards used.

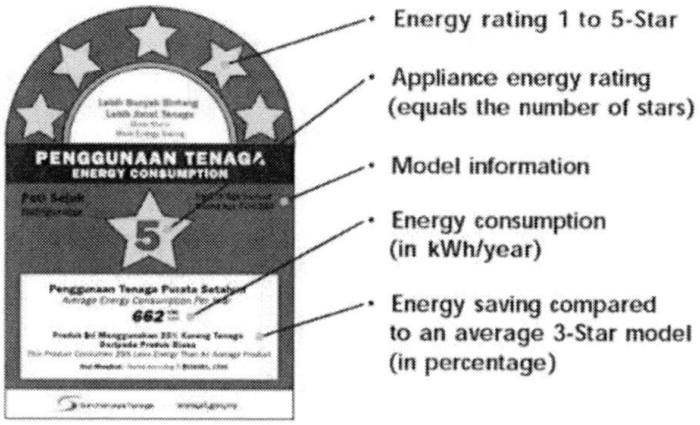

Fig. 1. The Energy Efficiency Label Specification by the Energy Commission.

2.2 Minimum Energy Performance Standards in Malaysia

Malaysia already has been implemented MEPS for selected electrical appliances (five appliances). Different types of electrical application will use a different method. The label verification of the implementation of the EE in Malaysia was conducted voluntarily by the Energy Commission that is operated through Ministry of Energy, Green Technologies and Water (KeTTHA) to amend the Electricity Supply Regulations 1994.

2.2.1 Implementation of MEPS in Asian Countries

Most of Asian countries have implemented MEPS for household appliances, equipment and lighting including Singapore, Indonesia, China and South Korea [8]. China is the most numerous target products with EE standards with around 26 target products, followed by South Korea and Japan, where both countries have the same target products that is 21 target products.

The development of MEPS for several numbers of products was handled by The Energy Commission in the Ministry of Economic Affairs (MOEA). In most cases, the energy testing were more details on Chinese Taipei for the Chinese National Standard (CNS) where the MEPS requirements are published by the MOEA. Chinese Taipei had introduced the usage of the MEPS application for the water chillers in January 2003 and was continued for the second phase in January 2005 [14]. Mahlia et. al [15] had conducted a study about the energy savings potential in Malaysia. The study indicates that implements the EE standards for room air conditioners and household refrigerators-freezers will increase the energy saving potential. There are also several research were conducted on electric motor on EE standards [16,17]. In China, EE standards for central air-conditioners and refrigerators-freezers were reported by [9,10] on the potential energy savings and environmental benefits. A comprehensive guidebook on Global Standards and Labels for appliances were developed by [13]. It contains the details of the standards, labels and test procedure for appliances in residential, commercial and industrial sectors.

2.2.1 Tariff for Tenaga Nasional Berhad (TNB) Malaysia

Tariff divided into a few sections such as a tariff for industrial, commercial and others. Every tariff has a different value and category. Table 1 shows electricity tariff in 2014 for domestic consumer. The tariff is used in calculating the monthly electricity to estimate the electricity consumption.

Table 1. Electricity Tariff in Malaysia.

Tariff Category	**Unit**	**Current Rate (MYR) (effective 1st Jan 2014)**
Tariff A – Domestic Tariff		
For the first 200 kWh (1 - 200 kWh) per month	cent/kWh	21.80
For the next 100 kWh (201 - 300 kWh) per month	cent/kWh	33.40
For the next 300 kWh (301 - 600 kWh) per month	cent/kWh	51.60
For the next 300 kWh (601 - 900 kWh) per month	cent/kWh	54.60
For the next kWh (901 kWh onwards) per month	cent/kWh	57.10
**The minimum monthly charge is MYR3.00*		

3 Methodology

A few selections of electrical items that are refrigerators, air conditioner system, fans, electric lamp and television are selected to review and analyze the advantages of using MEPS. The criteria taken into account in this selection are the electrical goods which have MEPS and basic electrical items commonly used at home. The equation 1 below shows the calculation star index for refrigerator. Equation 2 shows how to obtain Energy Efficiency Factor (EEF) value and equation 3 and 4 shows EEF average for one door and two door refrigerator.

$$\text{STAR Index} = \left(\frac{\text{EEF}_{\text{tested}}}{\text{EEF}_{\text{average}}} - 1\right) \times 100\% \tag{1}$$

Where,

$$Energy\ Efficiency\ Factor\ (EEF) = \frac{Adjusted\ Volume(litre)}{Energy\ consumed\ per\ day(kWh)} \tag{2}$$

$$EEF_{average(1\ door)} = 1.37V_{adj} - 63.3 \tag{3}$$

$$EEF_{average(2\ door)} = 0.409V_{adj} + 119.5 \tag{4}$$

Similar to the refrigerator calculation, equation 5 below shows how to determine star index for television. The different calculation showed in equation 6 that is to find the EEF value and equation 7 to calculate Annual Energy Consumption.

$$STARindex = \left(\frac{EEF_{tested}}{EEF_{average}} - 1\right) \times 100\% \tag{5}$$

$$Energy\ Efficiency\ Factor(EEF) = \frac{Screen\ Area(cm^2)}{Annual\ Energy\ Consumption(kWh)} \quad (6)$$

$$Annual\ Energy\ Consumption(kWh) = 0.365[(P_{on} \times 5) + P_{ps}(19 - T_{as}) + (P_{as} \times T_{as})] \quad (7)$$

Where,
$P_{on} = Power\ at\ on\ mode\ (W)$

$P_{as} = Power\ at\ active\ stanby\ mode\ (W)$

$P_{ps} = Power\ at\ passive\ stanby\ mode$

$T_{as} = Time\ on\ active\ standby\ mode(hour)$

$EEF\ Average\ determined\ as\ y = 0.002x + 11$

The air conditioner has two cases that are non-inverter air conditioner and inverter air-conditioner. Equation 8 shows the calculation to find EER value for non-inverter conditioner while equation 9 shows the calculation weighted EER for inverter air conditioner.

For non-inverter air conditioner,

$$ER = \frac{Cooling\ Capacity(Btu/h)}{Input\ Power(W)} \quad (8)$$

For inverter air conditioner,

$$Weighted\ EE = \left(0.4 \times \frac{100\%\ Load\ Cooling\ Capacity}{100\%\ Load\ Power\ Input}\right) + \left(0.6 \times \frac{50\%\ Load\ Cooling\ Capacity}{50\%\ Load\ Power\ Input}\right) \quad (9)$$

There is a single calculation for the fan, which is Coefficient of Performance (COP). This equation used to measure the ratio of air deliver and the input power. Equation 10 shows the formula of COP.

$$Coefficient\ of\ Performance(COP) = \frac{Measured\ Air\ Deliver(\frac{m^3}{min})}{Input\ Power\ (W)} \quad (10)$$

4 Result And Discussion

4.1 Refrigerator Analysis

The analysis has been done on several types of refrigerator: 200-299 litres, 300-399 litres and 400-499 litres in three conditions of refrigerator with inverter and MEPS, refrigerator with inverter only and refrigerator without inverter and MEPS. Figure 2(a) shows the energy consumption for refrigerator with capacity 200-299 litres, the highest energy consumption per year is refrigerator without inverter and MEPS which is 445 kWh/yr followed by refrigerator with MEPS only 318 kWh/yr while the lowest energy consumption is refrigerator with an inverter and MEPS which is 263 kWh/yr.

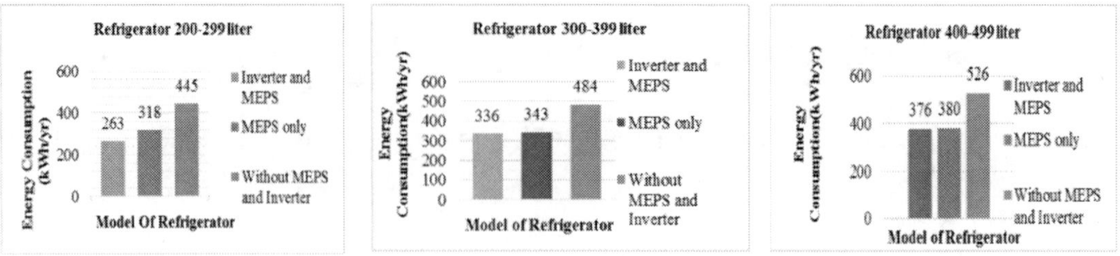

Fig. 2. Refrigerator (a) 200-299 litres, (b) 300-399 litres, (c) Refrigerator 400-499 litres.

Figure 2 (b) shows the energy consumption for refrigerator with capacity 300-399 litre, the highest energy consumption per year is refrigerator without inverter and MEPS which is 484 kWh/yr followed by refrigerator with MEPS only which is 343 kWh/yr. The lowest energy consumption is refrigerator with inverter and MEPS which is 336 kWh/yr. Figure 2(c) shows the energy consumption for refrigerator with capacity 400-499 litre, the highest energy consumption per year is refrigerator without inverter and MEPS which is 526 kWh/yr followed by refrigerator with MEPS only which is 380 kWh/yr and the lowest energy consumption is refrigerator with an inverter and MEPS which is 376 kWh/yr. Results show that the use of refrigerator with inverter and MEPS is more effective to reduce energy consumption in residential.

4.2 Air Conditioner Analysis

Analysis of the air conditioner is divided into two types that is air conditioner with 1 Horsepower (HP) and air conditioner with 1.5 Horsepower (HP). Each air conditioner is divided into three types, which is air conditioner with inverter and MEPS, MEPS only and without both the inverter and MEPS. For air conditioner analysis, it divided into two types which is air conditioner with 1 Horsepower (HP) and 1.5 Horsepower (HP). Figure 4(a) shows the energy consumption for air conditioner with 1 Horsepower (HP), the highest energy consumption per year is air conditioner without inverter and MEPS which is 3024 kWh/yr followed by air conditioner with MEPS only which is 2292 kWh/yr. The lowest energy consumption per year is air conditioner with both inverter and MEPS which is 1562 kWh/yr. Figure 4(b) is for air conditioner with 1.5 Horsepower (HP), the higher energy consumption per year is air conditioner without inverter and MEPS which is 4630 kWh/yr followed by air conditioner with MEPS only which is 3419 kWh/yr. The lowest energy consumption per year is air conditioner with both inverter and MEPS which is 1935 kWh/yr.

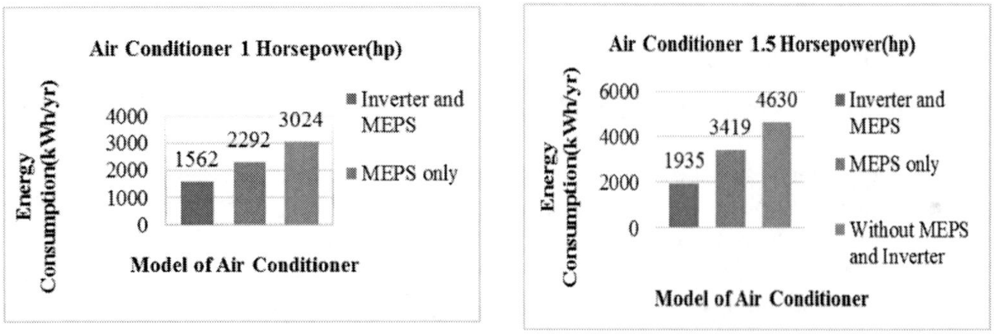

Fig. 4. Energy Consumption for Air Conditioner, (a) 1 Horsepower (HP), (b) 1.5 Horsepower (HP).

4.3 Fan Analysis

Analysis of the fan is divided into two types fan that is table fan and stand fan. Each fan is divided into two types, which is fan with MEPS and fan without MEPS. For analysis of fan also divided into two types which is stand fan and table fan. Figure 5(a) shows the energy consumption for stand fan, the highest energy consumption per year is stand fan without MEPS which is 158 kWh/yr and the lowest energy consumption per year is stand fan with MEPS which is 132 kWh/yr. Figure 5(b) shows the energy consumption for table fan. The highest energy consumption per year is stand fan without MEPS which is 175 kWh/yr and the lowest energy consumption per year is stand fan with MEPS which is 146 kWh/yr.

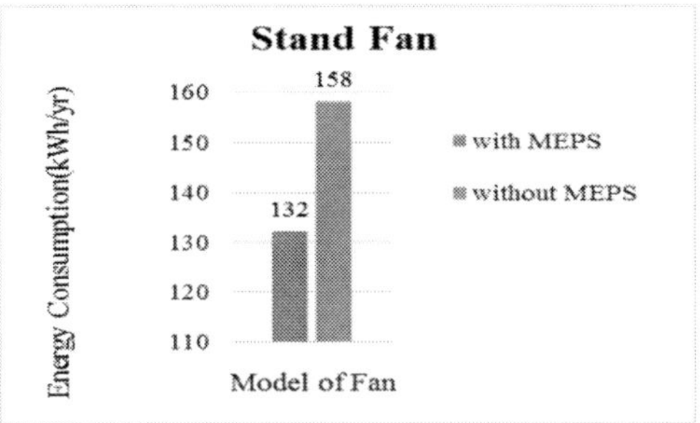

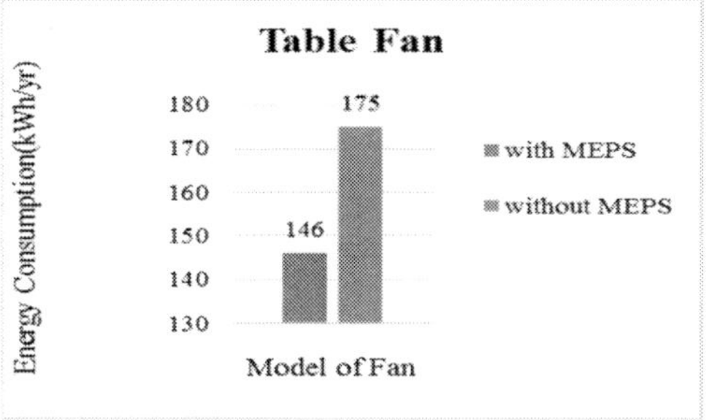

Fig. 5. Energy Consumption, (a) Stand Fan, (b) Table Fan.

4.4 Lamp Analysis

Analysis for the lamp is divided into three types such as fluorescent lamp, energy saving lamp and Light Emitting Diode (LED) lamp. Fluorescent lamp and energy saving lamp used 18 Watts of power while LED lamp only used 9 Watts of energy. The brightness for all three types of lamps are almost the same. In addition, LED lamp are brighter even though only use energy 9 Watts. For lamp analysis, it is divided into three types which is LED lamp, energy saving lamp and fluorescent lamp. Based on Figure 6, the highest energy consumption per year is fluorescent lamp which is 569.4 kWh/yr followed by the moderate is 551.2 kWh/yr. The lowest energy consumption per year is LED is 511 kWh/yr .

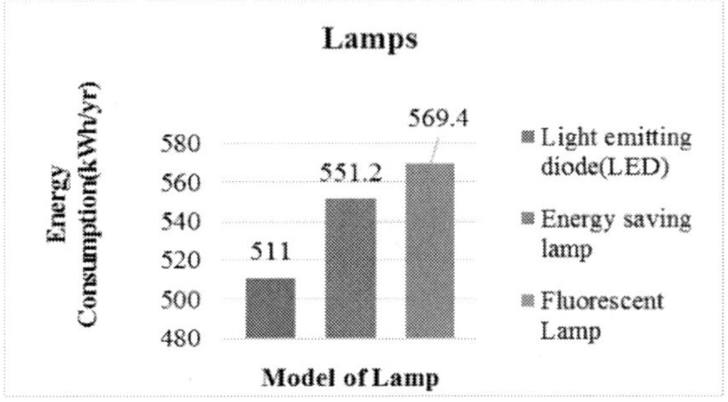

Fig. 6. Energy Consumption for Lamps.

4.5 Television Analysis

Analysis of television is divided into three types which is Liquid Crystal Display (LCD), plasma and LED. Each types of television is divided into two types which is with and without MEPS. For analysis of television, it is divided into three types which is LCD television, Plasma television and LED television. Figure 7(a) shows the results for LCD television which indicates the highest energy consumption per year LCD television is without MEPS which is 84.73 kWh/yr while the lowest is LCD television with MEPS which is 64 kWh/yr.

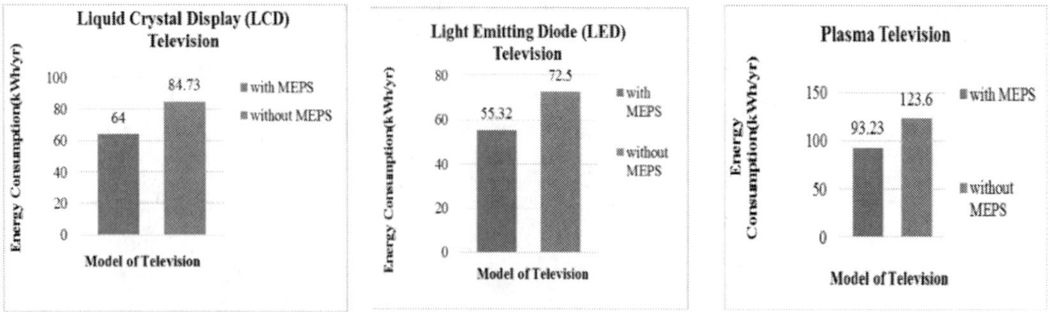

Fig. 7. (a) Liquid Crystal (LCD) TV, (b) Light Emitting Diode (LED) TV, (c) Plasma TV.

Furthermore, Figure 7(b) shows the measurement for LED television. The highest energy consumption per year is LED television without MEPS which is 72.5 kWh/yr followed by the lowest is LED television with MEPS which is 55.32 kWh/yr. Figure 7(c) shows the energy consumption for plasma television. The highest energy consumption per year is plasma television without MEPS which is 123.6 kWh/yr while the lowest is LED television with MEPS which is 93.23 kWh/yr. Table 2 shows the summary of all the measurements on five types of electrical appliances which using the EE approach (MEPS) and do not use it. Results show that the implementation of EE approach can reduce the energy consumption for five types of electrical appliances.

Table 2. Summary of Measurement on Five Electrical Appliances.

No	Electrical Appliances	Types			No	Electrical Appliances	Types		
		kWh/yr					kWh/yr		
		Without Inverter and MEPS	MEPS	Inverter and MEPS			Without MEPS	MEPS	-
1	*Refrigerator*				3	*Fan*			
	200-299Ltrs	445	318	**263**		Stand Fan	158	**132**	-
	300-399Ltrs	484	343	**336**		Table Fan	175	**146**	-
	400-499Ltrs	526	380	**376**	4	*Television*			
2	*Air-conditioner*					LCD	84.73	**64**	-
	1.0HP	3024	2292	**1562**		LED	72.5	**55.32**	-
						Plasma	123.6	**93.23**	-
	1.5HP	4630	3416	**1935**	5	*Lamp*	Fluorescent	Energy Saving	LED
						kWh/yr	569.4	551.2	**511**

5 Conclusion

EE is doing more with less energy benefits to consumers, country, and to the global. This study has successfully shown the home electrical appliances with MEPS can save the energy consumption. MEPS application has helped us in determining the efficiency of energy on electrical appliances especially on home application. In addition, the benefits of energy efficiency are numerous. Used energy more efficiently can reduce expenses, improves the economy, protect the environment, improves national security and it will enhance the quality of life.

The authors would like to express their gratitude to the University Tun Hussein Onn Malaysia (UTHM) and the Ministry of Education Malaysia for Hadiah Latihan Persekutuan (HLP) Scholarship for the study (author 1).

References

1. http://www.iea.org
2. H. Khatib, Energy policy, **48**, 737 (2012)
3. H.P. Siderius, Energy Policy, **70**, 1 (2014)
4. H.P. Siderius, Energy Policy, **59**, 762 (2013)
5. L. Harrington, G. Wilkenfeld, Energy and Buildings, **26**, 81 (1997)
6. K.A. Rahman, M.Z.M. Yusof, M.N.M. Salleh, A.M. Leman, *Engineering Technology International Conference*, 2015
7. K.A. Rahman, M.Z. Yusof, M.N. Salleh, A.M. Leman, *Technology and Innovation National Conference Proceedings*, 2015
8. K.A Rahman, M.Z. Md Yusof, M.N. Mohd Salleh, A.M. Leman, Chem. Eng. Trans., **45**, 1663 (2015)
9. X. Shi, Appl. Energy, **133**, 135 (2014)
10. http://www.st.gov.my/
11. T.M.I. Mahlia, H.H. Masjuki, I.A. Choudhury, Energy Convers. Manag., **43**, 743 (2002)
12. CSA-C743-02 Performance Standard for Rating Packaged Water Chillers, Canadian Standards Association, 2002
13. S. Wiel, J.E. McMahon, CLASP, 2 (2005)
14. R. Saidur, T.M.I. Mahlia, M. Hasanuzzaman, Energy Sci. Res., **27**, 175 (2011)
15. T.M. Mahlia, H. Masjuki, R. Saidur, M. Amalina, Energy Policy, **32**, 1819 (2004)
16. A.T. De Almeida, P. Fonseca, P. Bertoldi, Energy, **28**, 673 (2003)
17. A.G.P. Garcia, A.S. Szklo, R. Schaeffer, M.A. McNeil, Energy Policy, **35**, 3424 (2007)
18. W. Lu, Energy Policy, **35**, 1709 (2007)
19. W. Lu, Energy Policy, **34**, 1583 (2006)

Influence of Synchronized Dead Point Elimination Crank on Cyclist Muscle Fatigue

Khadijah Akmal Abdul Aziz[1,*], *Ahmad Faizal* S[1]., *Norasmadi* Abdul Rahim[1], *Mohd. Hanafi* Mat Som[1], and *Hamzah* Sakeran[1]

[1]School of Mechatronic Engineering, University Malaysia Perlis, Perlis, Malaysia

Abstract. The aim of this study was to investigate the influence of newly proposed bicycle's crank to crank angle setting on the Vastus Lateralis (VL) and Bicep Femoris (BF) muscle activity during cycling. Procedures of Conconi Test were used throughout the experiment for the data collection purpose. The muscles activities were recorded using surface electromyography and software LabChart7. The raw data were further processed in time (Root-Mean-Square, RMS) and frequency (Mean Power Frequency, MPF) domain. It was found that 0° crank to crank setting (similar to conventional crank to crank angle setting) caused the prime mover VL (Normalized RMS = 0.119) to fatigue more than BF (Normalized RMS = 0.102). This setting is expected to decrease the cycling performance. In addition, -5° is the best crank to crank angle setting that causes least fatigue to both VL and BF. In short, to increase the cycling performance by avoiding the fatigue to the main muscles, -5° is the suggested as setting angle for the proposed crank design.

1 Introduction

Cycling is one of the most famous sport worldwide, which is suitable for all ages and providing a lot of benefits such as good for health and also for the environment. One of the main components that make the bicycles move is the crank which can be manipulated to improve the cycling performance especially during competition. Several strategies to optimize the cycling power are the alteration of the crank arm length, chairing geometry and position of the chainring rotation axis [1]. Generally, the conventional cranks are separated at 180° and were reported to produce minimal power when one of the legs reaches Bottom Dead Centre (BDC) while the other leg is at Top Dead Centre (TDC) [2]. Those two points so called dead points because throughout the circular motion of cycling, TDC and BDC produce minimum tangential force [3]. Previous research found that by avoiding both leg to reach TDC and BDC simultaneously, the net cycling power will be increased and thus enhance the cyclist performance [4]. Thus, the proposed crank is able to prevent the legs reaching TDC and BDC simultaneously. In this study, the performance of the cyclists was measured based on the level of cyclist's muscle fatigue. This can be achieved by analyzing the muscles activity in time and frequency domain which were

* Corresponding author: ahmadfaizal@unimap.edu.my

recorded using surface electromyography. Therefore, in this research, the influence of the proposed crank on the cyclist muscles fatigue is investigated.

2 Methodology

2.1 Subjects

This study involved 10 normal and healthy subjects, which consisted of 5 males and 5 females (age 23.10 ± 0.32 years; weight 57.1 ± 12.84 kg; height 1.63 ± 0.12 m; Body Mass Index (BMI) 21.26 ± 2.52 kg/m^2). The subjects were healthy without any muscular, neurological and tendineous injuries. Each subject read and signed the written informed consent form. The details about the study and the procedures were presented to them clearly before the experiments were conducted.

2.2 Experimental Equipment

The brand of the bicycle used in this study is Polygon Helios 100. An interactive smart trainer with an electrobrake system, Tacx Vortex Smart T2180 was attached to the rear tire of the bicycle to provide suitable training resistance to the cyclist. The proposed crank is shown in Figure 1.

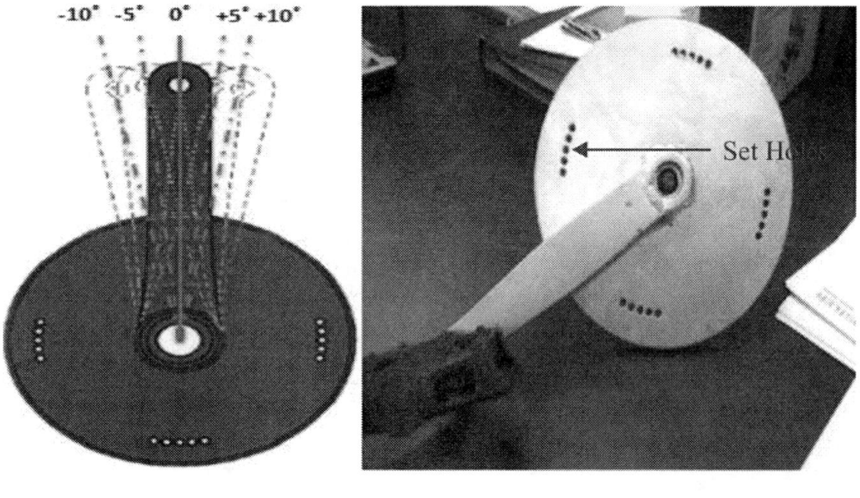

Fig. 1. Proposed Crank System.

The Polygon Helios 100 bicycle installed with the proposed crank on the left side of the bicycle. The proposed crank has two pieces of circle shape disc where one disc (no crank attaches to the disc) is fixed to the bottom bracket of the bicycle and the other (crank was welded to the disc) is freely rotated and could be fixed by screwing through few 'set holes' to the first disc for a desired crank angular position. The crank able to be set for ±10° maximum with the addition of 5° back and forth from inertial 0° of TDC point. The 0° resembles the conventional crank which the left and right cranks are separated at 180°. An EMG (ML865, ADInstrument) was used to monitor muscles activity. EMG signals were amplified and recorded at a sampling rate of 2000 Hz.

2.3 Experimental Protocol

The electrode placement on the quadriceps (VL) and hamstring (BF) of the subjects are according to the Surface ElectroMyoGraphy for the Non-Invasive Assessment of Muscles (SENIAM) recommendations. The subjects followed the Conconi Test throughout the cycling activity. The subjects started cycling after a warmed-up of 5 minutes with a workload of 70 W. Subjects kept increasing the speed until no longer able to continue the cycling experiment (fatigue). Breathing difficulties or 'burning muscles' are the examples of physical indications of near-maximal power [10]. During cycling protocol, the subject need to increase both cadence and force. This will resemble the actual cycling conditions when cyclists reach their maximal speed at a cadence exceeding 160 revolutions per minute [11].

2.4 Data Analysis

MATLAB R2013a was opted to filter the raw data using Fourth order Butterworth band pass filter (20Hz -450 Hz). The peak value represents the dynamic maximum voluntary contraction (MVC). For normalization of the data, the filtered data were divided by the peak value. Then, Root-Mean-Square (RMS) method is used for the muscle fatigue analysis. The RMS is in the time domain. Next, the data is further analysed in the frequency domain. Then, Mean Power Frequency (MPF) which is the centroid frequency of the power spectrum being used in this study [12].

3 Result and discussion

Firstly, the activity of BF and VL were investigated individually to understand the effect of varying crank to crank angles on the activity of those muscles. Besides, this analysis is crucial to examine the relationship of those muscles and to decide the best angle to avoid the severe muscle fatigue.

In Fig. 2, 0° crank to crank angle (similar to conventional crank) shows the least amount of RMS value for both VL and BF. However, only at 0° and -10°, the RMS value of VL is higher than BF. For the other angles (+10°, +5°, -5°), the RMS value of BF is higher than VL. Generally, from these two observations, 0° and -10° cause the prime mover muscle, VL fatigue more. This occurrence should be highly avoided because prime mover is responsible for the specific movement and antagonist muscle helps the prime mover to accomplish the movement.

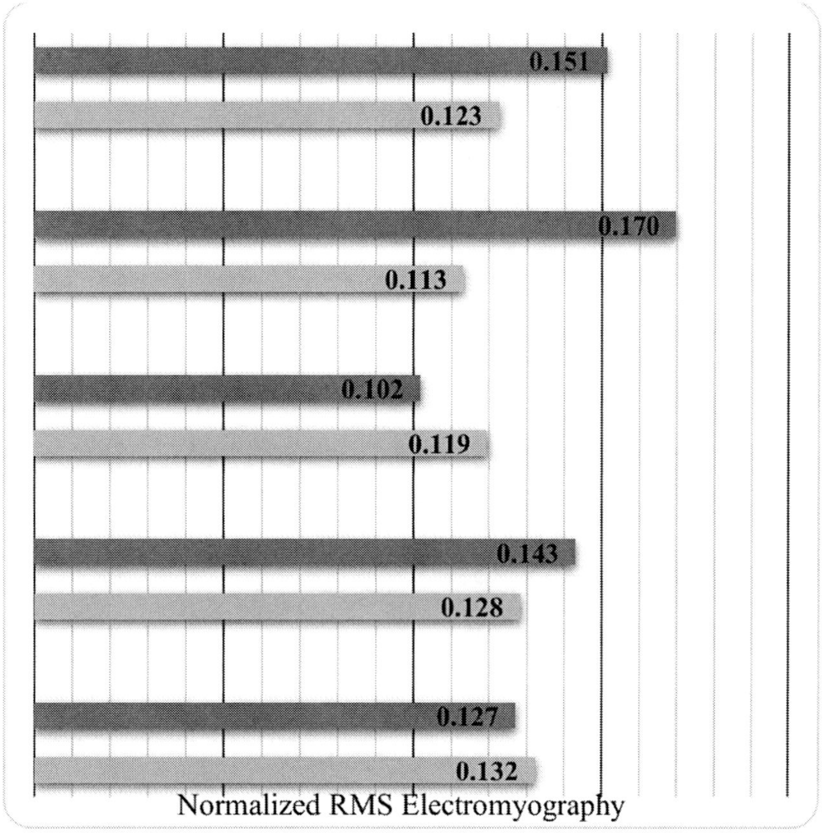

Fig. 2. RMS of VL and BF for 5 different crank angle.

In addition, the antagonist muscle, BF consumes much more energy at +10°, +5° and -5° during the pulling phase. In Fig. 2, it is shown that at least only one muscle was affected as the crank to crank angles being varied. This gives a good impact to the cyclist because the proposed crank design did not severely affect both antagonistic-pair muscles simultaneously, which could lead to the failure of cycling activity in both propulsive and pulling phase. In short, BF shows higher RMS value at +10°, +5° and -5°. However, at 0° and -10°, VL is having higher RMS value.

Next, the analysis of EMG signal in time domain on the effect of muscle fatigue to both males and females was conducted. Overall, females show the high RMS value for both muscles and for the proposed crank to crank angles as shown in both Fig. 3 and Fig. 4. Females have the higher rate of fatigue levels compared to the males. Apparently, the lifestyle of the females differs from the males. The males quite fit and used to exercise regularly compared to females.

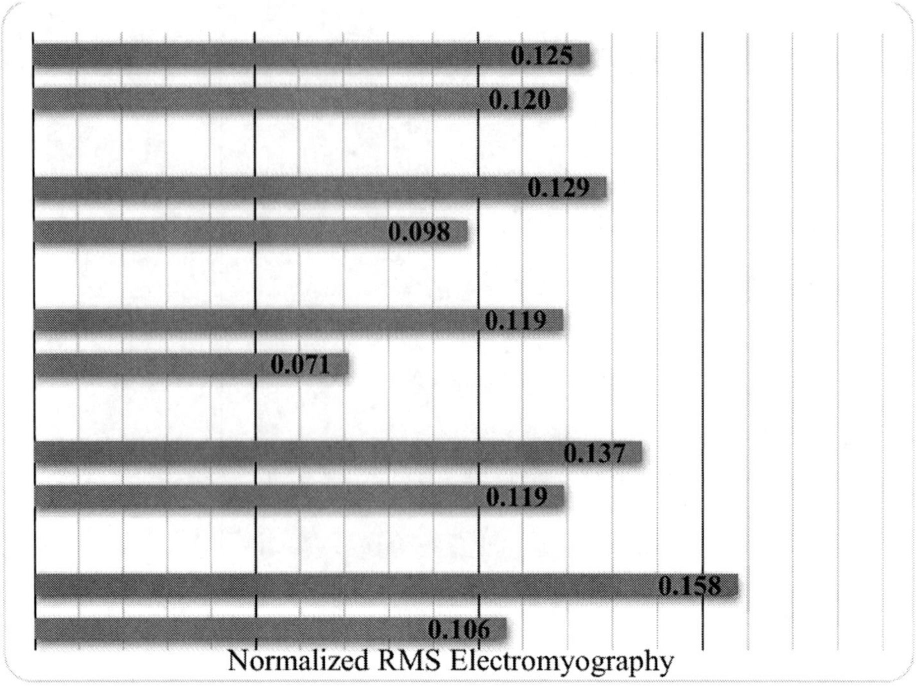

Fig. 3. RMS value of VL activity for male and female.

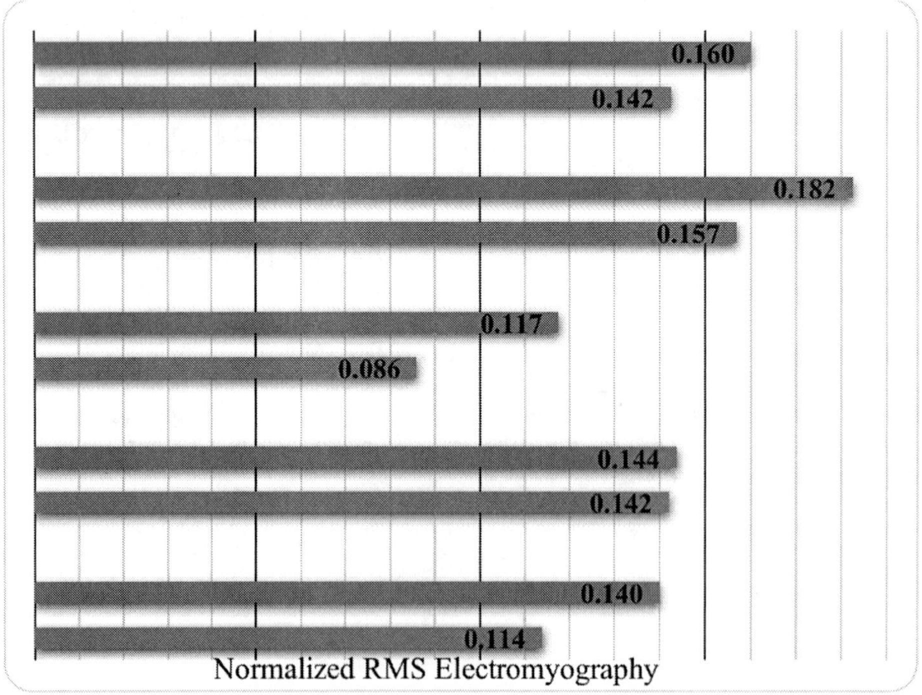

Fig. 4. RMS value of BF activity for male and female.

4 Conclusion

From the analysis, the best crank to crank angle settings that causes least fatigue to both VL and BF of the subject and does not severely affect the prime mover muscle is -5°. Conversely, 0° and -10° are the worst angles because they cause fatigue to the prime mover or so called as agonist muscle (VL). Generally, females tend to get fatigued easier in every each experiment. Male able to cycle in a longer period of time compared to the female.

References

1. R.C.H. So, J.K.F. Ng, G.Y.F. Ng, Phys. Ther. Sport, **6(2)**, 89 (2005)
2. V. Parnabas, T. Wahidah, N.M. Abdullah, *Proceedings of the International Colloquium on Sports Science, Exercise, Engineering and Technology*, Singapore, 2014
3. A. Gefen, M. Megido-ravid, Y. Itzchak, M. Arcan, Gait & Posture, **15**, 56 (2002)
4. A. Lucia, J. Balmer, R.C.R. Davison, M. Perez, A. Santalla, P. M. Smith, Int. Journal of Sports Medicine **25(7)**, 479 (2004)
5. F.M. Giovanni Grazzi, Ilario Casoni, Gianni Mazzoni, Int. Journal of Sports Medicine **17(7)**, 509 (1996)
6. A. Ignjatovi, P. Hofmann, D. Radovanovi, Physical Education and Sport, **6**, 1 (2008)
7. S.R. Alty, A. Georgakis, *Signal Processing Conference*, 1387 (2011)
8. I. Stirn, T. Jarm, Vojko Strojnik, Kinesiologia Slovenica, **14(1)**, 28 (2008)
9. C. Fritschi, L. Quinn, J. Psychosom. Res., **69(1)**, 33 (2010)
10. A. Saito, K. Watanabe, H. Akima, Hum. Mov. Sci. **40**, 14 (2015)
11. D.M. Rouffet, C.A. Hautier, J. Electromyogr. Kinesiol., **18**, 866 (2008)
12. L. Fedorowich, K. Emery, B. Gervasi, J.N. Cote, J. Electromyogr. Kinesiol., **23(5)**, 1183 (2013)
13. W.J. Albert, A.T. Wrigley, R.B. McLean, G.G. Sleivert, Dynamic Medicine, **5(2)**, 1 (2006)
14. A. Phinyomark, S. Thongpanja, H. Hu, INTECH, Croatia, 2012

Influence of Initial Crank Positions and Jaw Clenching Activity on Cycling Performance

Puteri Umi Farhana Mohd Zaaim[1,*], *S.* Ahmad Faizal[1], *Mohd Hanafi* Mat Som[1], *Hamzah* Sakeran[1], *Wan* Mohd. Zuki[2]

[1]School of Mechatronic Engineering, University Malaysia Perlis, Pauh, Malaysia
[2]Institute of Engineering Mathematics, University Malaysia Perlis, Pauh, Malaysia

Abstract. The purpose of this study was to identify the effects of three initial crank positions and jaw clenching activities on seated cycling performance. 10 normal subjects (5 females, 5 males) completed six separate occasions of Wingate Test with three different initial crank positions (0°, 47°, 71°) and two modes of jaw activities (jaw clenching and jaw loose). The findings suggest that, crank angle of 71° has the highest power output compare to other angles. The combination of 71° crank angles and jaw loose the jaw shows the best cycling performance.

1 Introduction

Cycling is a competitive physical activity which could be categorized in racing and non racing categories. The racing categories includes road cycling, track cycling, time trialling, cyclo cross, mountain bike racing, bicycle motocross (BMX) and cycle speedway while the non racing categories comprise of artistic cycling, cycle polo, freestyle BMX and mountain bike trials.

Several factors were reported to influence the racing cycling performance including but not limited to environmental, mechanical, strategy and human factors [1]. As for human factors, body position, configuration and orientation were assessed by various researchers as to improve the cyclist performance. The influence of body configurations on cycling performance were reported to receive large attention by researchers [2][3].

The cyclist leg segment position influenced the moment generated at pedal axle and also force generated by the muscles. Padulo et. al have studied the effect of initial crank positions on cyclist speed and acceleration in track cycling performance. It was found that the acceleration with standing position and initial crank position of 71° was about 35 % greater than that with seated position with an initial crank angle of 47° [4].

Other than initial crank position, other human factors e.g. jaw clenching activities were reported to improve cycling performance as well. Clenching the jaw, grasping firmly the hands and attempting to inhale out through a closed mouth at the same time, was reported to expand knee power by 14.8% that may influence the cycling performance [5]. However, to the best of author knowledge, there is no study combines both effect of cranks initial

* Corresponding author: puteri.farhana91@yahoo.com

position and jaw clenching activities on seated cycling performance. Therefore, in this project, we investigate the influence of three initial crank positions, level of comfort and jaw clenching and jaw loose activities on seated cycling performance.

2 Methodology

2.1 Subject

In this project, 10 (5 males and 5 females) university students (age is between 23.5±1.5 years old and height is 170.5±11.5cm tall and weighing 66.5±11.5kg.) volunteered as participants. Subjects were healthy without any muscular, neurological and tendineous injuries. All subjects was voluntarily agreed to participate in the study and signed the consent form prior to participation.

2.2 Experimental equipment

A kinetic fluid cycle trainer, Tacx Vortex Smart model T2810 was used to acquire the speed, cadence and power output during cycling. It also functions to hold the bike upright and to provide the desired resistance to the back wheel. A goniometer (Lafayette Gollehon Extendable Goniometer, model 01135) was used as a part of test for measuring the angle of knee when subject sit on the adjusted bicycle seat. This instrument was used to adjust bicycle seat height based on Holmes method, where the knee angle must be in the range between 25° to 35° during sit on saddle seat. The goniometer was also used to set the initial crank position of the cyclist. Cycling data from TACX Vortex were transmitted into a tablet through ANT+ wireless networking protocol.

2.3 Experimental procedure

The experiment began by measuring subject's height and body mass for calculating the Body Mass Index (BMI). The bicycle seat height was adjusted according to the Holmes method. Subject seated on the bicycle before start the experiment. Then, subject start cycling according to Wingate test method with combination of with or without jaw clenching activities in a separate session. The initial crank positions setting start with 0°, followed by 47° and 71° in each cycle of experiment. The mean power outputs were recorded and analyze offline using Microsoft Excel 2010.

2.4 Data analysis

All statistical analyses were performed using SPSS version 17.0, (SPSS Inc., Chicago, IL). The data of the protocols were tested for normality and homogeneity of variance and turned out to be normally distributed. Therefore, the analyses of differences in power outputs and comfort ratings between initial crank positions and jaw clenching activities were detected using 2 Way ANOVA. Results for all tests were considered significant at $p < 0.05$.

3 Result and discussion

3.1 Power output at difference crank angles

Figure 1 shows the average power output for the three initial crank positions. The bar graph showed that the 71° crank position had the highest overall ranking among the three initial crank positions. Table 1 shows the result of ANOVA test between relative power output and the different initial crank positions. For all subjects, ANOVA test revealed a non significant difference in cycling relative power output due to the different settings of initial crank positions (p = 0.910). Padulo. J. et. al. found that the acceleration at initial crank position of 71° was about 35% greater than that with seated position with an initial crank angle of 47° [4]. In evaluating the effect of the initial crank position on cyclist performance, perhaps, it is better to opt for acceleration or speed rather than the power output. In this study, although the cyclist show higher power output during the initial cycling stage at different initial crank position; the analysis using average power output resulted in minimal difference of power output at different initial crank position. Perhaps, it is better to measure the instantaneous rather than mean power output at specific experiment time.

Table 1. ANOVA Test Between Mean Power Output and Different Initial Crank Angular Positions

	Sum of Squares	df	Mean Square	F	Sig.
Between Groups	0.180	2	0.090	0.094	0.910
Within Groups	54.407	57	0.955		
Total	54.588	59			

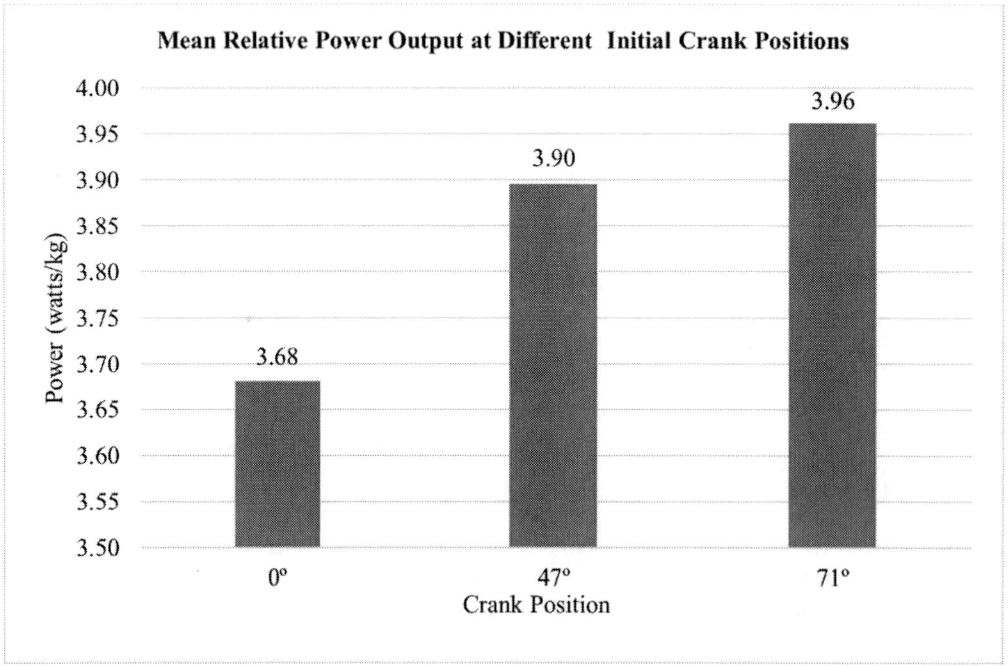

Fig. 1. The average power output between the initial crank angular positions.

3.2 Power output for jaw clenching and jaw loose activities

Table 2 presented the result of ANOVA tests for the jaw clenching and jaw loose activities. There is statistically significant difference in the jaw clenching and jaw loose jaw activities in terms of power output. As shown in Figure 2, the jaw jaw loose mode generate more power output compared to jaw clenching. However, according to William et. al., clenching the jaw, grasping firmly with the hands and attempting to inhale out through a closed mouth, at the same time, expands knee power by 14.8% when performing leg extension using Biodex System Dynamometer [5]. In this project, confirmation of clenching and jaw loose activities were performed using hand (palpation on jaw clenching) before the subjects start pushing the pedal downstroke, and thus, none could confirm that they definitely clenched their jaw when pushing the pedal towards bottom dead centre. There is a possibility that subject clenched their jaw in the unclenched jaw mode too. In future, method of confirming the clenching activities (using electromyography signal of jaw and etc) should be introduced to obtain a better results.

Table 2. ANOVA Test for Jaw Clenching and Jaw Loose Activities.

	Sum of Squares	df	Mean Square	F	Sig.
Between Groups	5.279	1	5.279	6.209	0.016
Within Groups	49.309	58	0.850		
Total	54.588	59			

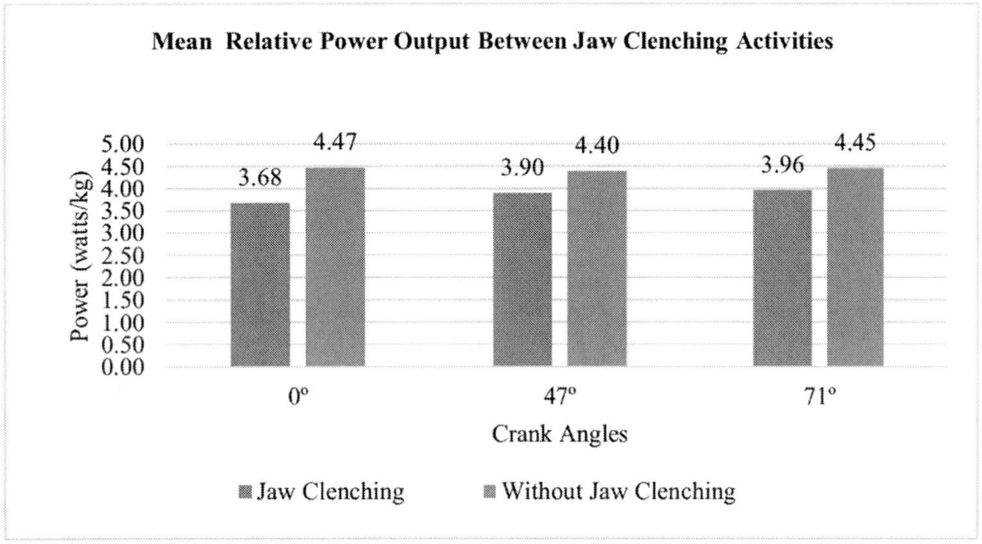

Fig. 2. Average power output between jaw clenching and jaw loose.

4 Conclusion

The main objective of this study is to assess the influence of crank's initial position on cycling performance. In this project, three types of initial crank positions, and the activities of the jaw clenching and jaw loose during cycling in seating position were utilized to understand their effects on cycling performance. There is no significant difference ($p>0.05$) in result for power output with different initial crank positions. The reason probably is due to the difference in power output value between the initial crank positions is not so large. There is a significant difference between jaw clenching and jaw loose activities during seated cycling in term of power output. The subjects generate power output higher when cycling with jaw loose jaw. In conclusion, from a performance standpoint, crank angle of 71° has the highest power output compare to other initial crank positions. Not only that, the combination of 71° crank angles and without jaw clenching could improve cycling performance.

5 Recommendation

More experiments to be carried out to have a deeper knowledge on the influence of initial crank angular position on cycling power generation. It is also advisable to utilize motion capture system to understand lower limb joint power and moment produce at different initial crank positions. Additionally, seat height, body posture and seat position (front and back) of the cyclist also should be considered to improve the cyclist performance. In addition, the difference between the genders should also be taken into account due to the difference in strength of both genders.

References

1. Danny Too, Sport Med., **10 (5)**, 286 (1990)
2. W.W. Peveler, J. Strength Cond. Res., **22 (4)**, 1355 (2008)
3. Bini RR, Hume PA, Croft JL, Procedia Engineering, **13**, 51 (2011)
4. J. Padulo, G. Laffaye, W. Bertucci, Sport Science Health, **10**, 189 (2014)
5. P. William Ebben, H. Leigh David, F. Geiser Christoper, Med. Sci. Sport Exer., **40 (10)**, 1805 (2008)
6. N.P. Craig, K.I. Norton, Sport Med., **31 (7)**, 457 (2001)
7. J.A. Bousie, P. Blanch, T.G. Mcpoil, B. Vicenzino, J. Sci. Sport, **16(1)**, 60 (2013)

MATEC Web of Conferences **78**, 01061 (2016) DOI: 10.1051/matecconf/20167801061
IConGDM 2016

Kaolin Geopolymer as Precursor to Ceramic Formation

Nur Ain Jaya[1,*], *Mohd Mustafa* Al Bakri Abdullah[2], *Che Mohd Ruzaidi* Ghazali[1,2], *Mohammed* Hussain[3], *Kamarudin* Hussin[1,2], and *Romisuhani* Ahmad[1]

[1]Center of Excellence Geopolymer and Green Technology, School of Materials Engineering, Universiti Malaysia Perlis (UniMAP), 01000, P.O Box 77, D/A Pejabat Pos Besar, Kangar, Perlis, Malaysia.
[2]Faculty of Engineering Technology, Universiti Malaysia Perlis (UniMAP), 01000, P.O Box 77, D/A Pejabat Pos Besar, Kangar, Perlis, Malaysia.
[3]King Abdul Aziz City Science & Technology (KACST), P.O. Box 6086, Riyadh 11442, Kingdom of Saudi Arabia.

Abstract. This paper introduced the potential application of kaolin geopolymer as ceramic precursor. This is one of the alternatives to produce high strength ceramic at a slightly lower temperature. Upon sintering the conversion of geopolymer to ceramic occur. The kaolin used were characterized using XRF and has plate-like structure upon investigating through microstructural analysis. Geopolymer mixture is produced using 12 M NaOH molarity with the Na_2SiO_3/NaOH ratio of 0.24. The sintering temperature used were ranging from 900 °C to 1200 °C. The flexural strength showed the highest value of 88.47 MPa when sintered at 1200 °C. The combination of geopolymerization and sintering has attributed to the strength increment as temperature increased. The density is observed to increase with increasing sintering temperature due to the appearance of the close pores in the structure. Sintering of the geopolymer resulted in the formation of liquid phase, which enables the joining of particles to produce dense microstructure.

1. Introduction

The research into geopolymers was led by the needs of new heat resistant materials after various catastrophic fires in France in 1970. Geopolymers are hard, inorganic and stable up to 1400 °C and non-inflammable [1]. Geopolymerization is carried out by putting the aluminosilicate materials in contact with the alkaline activator solution which produce the polymeric chain of Si-O-Al-O bonds [2]. Geopolymers are amorphous to semi-crystalline three-dimensional silico-aluminate materials synthesized at low temperature in a short time. However, with the increasing temperature crystalline phase begins to appear [3]. The geopolymeric materials are considered as polymer but also geopolymer. Hence, it transforms, polycondense and adopt a shape rapidly at low temperature while also hard,

* Corresponding author: nurainjaya@email.org

weather resistant and can withstand high temperatures [4]. As a result of excellent properties, geopolymers have been investigated for potential application in refractory and insulating areas industries.

In geopolymerization, an activator such as alkaline solution aids in the dissolution process of raw material by dissolving the soluble amorphous content of aluminosilicate materials in the first step of the polymeric reaction to form geopolymer precursors [5]. In this study, sodium silicate and sodium hydroxide were used as alkaline activator. The molarity of sodium hydroxide were fixed at 12 M because it provides sufficient alkalinity to activate the kaolin and released Al^{3+} and Si^{4+} ions to participate in a geopolymerization process [6]. Apart from geopolymerization component, sintering played an important role to produce geopolymer ceramic. Sintering can be defined as the thermally activated adhesion processes which produce the growth of contacts between particles and their coalescence [7]. Sintering close some of the open pores, and cause the water absorption rate to decrease while increase the strength produced [8]. Geopolymer technology can be used to produce high performance ceramics at a slightly lower temperature compared to normal ceramics such as kaolin. Over the years, kaolin has been a very important ingredient in ceramics industry. However, the use of kaolin in producing ceramics through the conventional method requires high temperature (up to 1400 °C) processing to obtain high performance.

Geopolymers are known to have inflammability at high temperature during sintering. Thus, their behaviour at elevated temperature exposure is becoming an interest. Earlier studies on the thermal behaviour of geopolymers includes high-temperature reactions, phase changes on melting and thermal expansion properties of sodium polysialates and potassium sialate geopolymers [9, 10]. Later, their behaviour at elevated temperature in the means of mechanical properties is becoming an interest. The mechanical properties of geopolymers are said to improve greatly when exposed to elevated temperatures [11]. Several researchers studied on the thermal behaviour of geopolymers prepared from different sources such as metakaolin [12-15] and fly ash [16, 17]. Despite being used in refractory and high temperature applications, geopolymers is gaining interest as precursors to ceramic formation. Cesium based geopolymers often used to produce pollucite ceramic [18, 19] and potassium based geopolymers as precursor to leucite ceramic [20, 21]. However, limited research was founded on using kaolin as the source materials and sodium based geopolymers. This study has used kaolin as the source material to study the effect of sintering temperatures on the mechanical strength and microstructure evolution of the geopolymer ceramics formed.

2. Experimental method

2.1 Materials

Kaolin used in this study was supplied by Associated Kaolin Industries Malaysia as Si-Al sources materials. The chemical composition of kaolin was determined by X-ray fluorescence. The particle size distribution of aluminosilicate materials was obtained using a Malvern particle size analyser as the mean particle size d (50) is 9.465 μm. Sodium hydroxide (NaOH) powder used to be of caustic soda micropearls, 99 % purity with brand name of Formosoda-P made in Taiwan. Sodium silicate (Na_2SiO_3) solution was supplied by South Pacific Chemicals Industries Sdn. Bhd. Malaysia with a chemical composition of 30.1 % SiO_2, 9.4 % Na_2O and 60.5 % H_2O.

2.2 Sample Preparation

Distilled water was used to prepare 12 M NaOH in a volumetric flask. The alkaline activator solution was formed by mixing Na_2SiO_3 to NaOH solution at a ratio of 0.24 until a clear solution was obtained for 24 hours. In order to produce geopolymer paste, kaolin powder was mixed with the alkaline solution with 1.0 solids-to-liquid ratio. The mixture then were mixed well for a few minutes and poured into HDPE mould. The samples were put into the oven at 80 °C for 24 hours for the curing process. After that, the samples were taken out of the oven and were crushed using mechanical crusher to make it into powder. To obtain fine powder, the samples were sieved manually to pass 150 micron sieve. In order to produce geopolymer ceramic, the sieved powders were then compacted using uniaxial isostatic pressing at 4.5 tons for 2 minutes. The final green body was sintered in a high temperature furnace to 900 °C - 1200 °C with 3 h soaking time and a cooling and heating rate of 5 °C/ min. All samples were sintered on an alumina plate as it can withstand high temperature firing.

2.3 Tests and Analysis Method

The density of sintered samples was measured using the pycnometry (AccuPyc II 1340 He pycnometer, Micromeritics, GA, USA). Flexural strength measurements of kaolin geopolymer ceramics were conducted on specimens (7 mm x 5 mm x 52 mm) using a three-point-bending fixture on an instron-500 tester, with a span length of 30 mm at a crosshead speed of 0.3 mm/min. A minimum of five samples for each group was subjected to flexural strength testing in accordance with ASTM C1161-02c [22]. Scanning electron microscopy (SEM) was used to study the microstructure of the raw kaolin and the effect of sintering temperatures on kaolin geopolymer ceramic.

3. Results and discussion

3.1 Raw materials characterization

Table 1 tabulates the chemical composition of raw kaolin obtained by x-ray fluorescence (XRF) analysis. The main elements of kaolin were SiO_2 and Al_2O_3 make up the 86.8 wt. % of the total composition. The silica content contributes to the formation of Si-O-Si bonds, which responsible for the mechanical strength of the geopolymer. Traces of other compounds also can be found in kaolin such as potassium oxide (K_2O), iron oxide (Fe_2O_3), titanium oxide (TiO_2), zirconium oxide (ZrO_2) and manganese oxide (MnO_2). The alumina and silica were held together by van der Waals and hydrogen bonding indicated by the layered structure as shown in Figure 1.

Table 1. Chemical composition of kaolin as determined by XRF

Chemical	Wt (%)
SiO_2	54.4
Al_2O_3	32.4
Fe_2O_3	4.32
TiO_2	1.33
Na_2O	0.30
ZrO_2	0.07
K_2O	5.58
MnO_2	0.10
LOI	1.74

 DOI: 10.1051/matecconf/20167801061

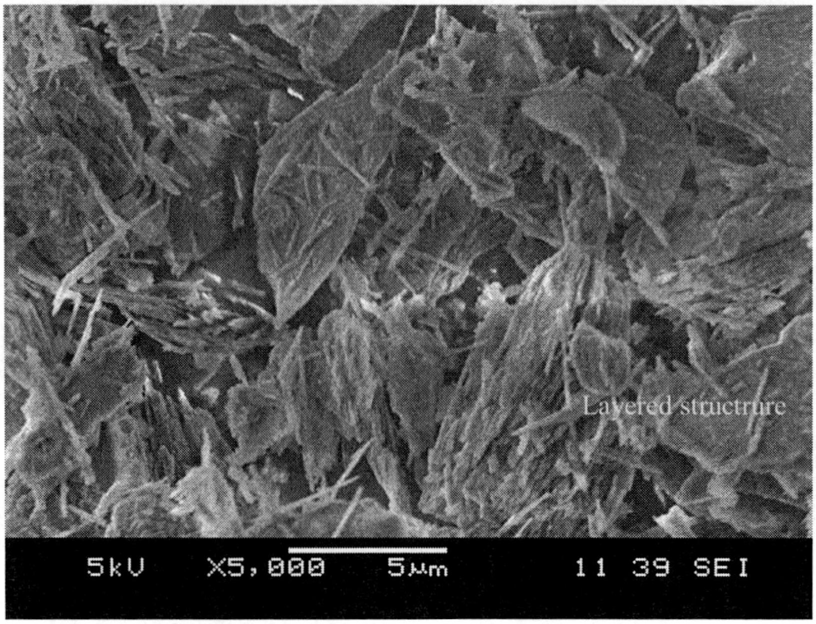

Fig. 4. Microstructure of raw kaolin

3.2 Physical observation and density

Figure 2 shows the physical observation of the kaolin geopolymer ceramic sintered at five different temperatures. All of the samples showed discolouration after sintering. The milky white colour of the samples sintered at 900 to 1200 °C resembles the colour of the pure kaolin used as the aluminosilicate source in this work. At lower temperature, the surface of the samples was rough and at higher temperature the surface become glazed. This can be observed when the samples were sintered at 1200 °C where it showed fully glazed surface which resembles glass. Increased the temperature to 1250 °C causes melting of the geopolymer which gives no definite shape. The colour also changed from milky white to brown due to the burn of geopolymer body. This indicates that temperature above 1200 °C is not suitable for sintering due to the exceeding melting point of KGC.

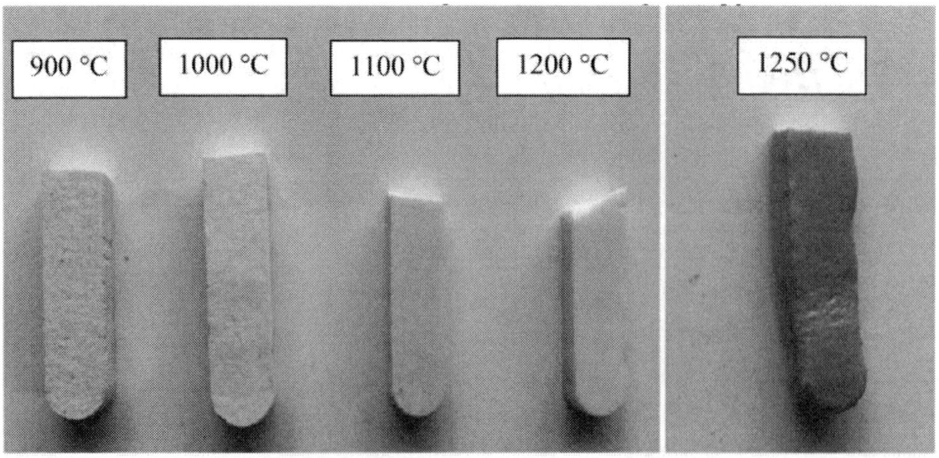

Fig. 5. Physical observation of KGC at different temperature

Figure 2 shows the density for KGC at different sintering temperature. The density measured decreased with increasing sintering temperature. Sample sintered at 900 °C has the highest density of 2.69 g/cm^3 and samples sintered at 1200 °C has the lowest density of 2.13g/cm^3. The formation of large pores created at 1200 °C contributes to the lowest density. Sintering at low temperature cause the formation of small pores in geopolymer sample. However, increasing the sintering temperature linearly decreased the density due to the closed pores that present in the structure. The density of the sample with closed pores were measured with the density of the pores whilst for sample with open pores, the density measured were included the material density. This is true as at 900 °C the microstructure analysis showed the existence of open pores and showed a high density compared to samples sintered at 1200 °C.

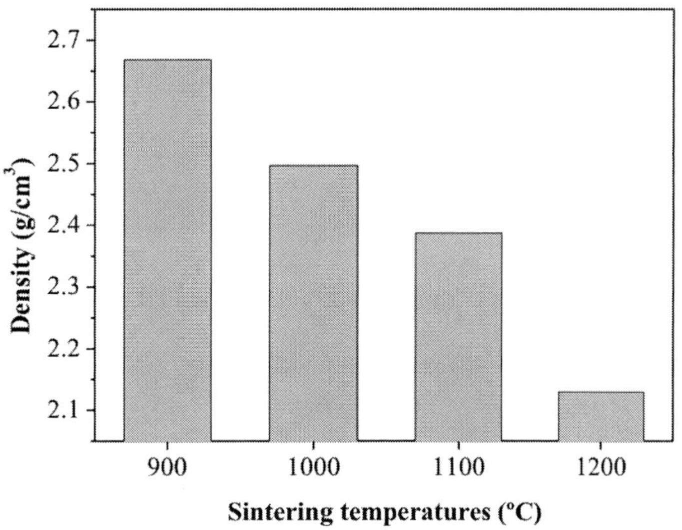

Fig. 6. Density of KGC with different sintering temperatures

3.3 Flexural strength

Figure 3 shows the flexural strength of kaolin geopolymer at four different sintering temperature. From the observation, increasing the sintering temperature increase the flexural strength. Heating the kaolin geopolymers to 900 °C resulted in the lowest flexural strength of 37.11 MPa. The low temperature of 900 °C is chosen based on Lee and Yeh [22] who reported the minimum firing temperature for clay-based ceramic is 900 °C. This is due to the transformations of fine quartz particles above this temperature. The low temperature of 900 °C were not sufficient to bring the particle to diffuse well which cause low strength. Hence, produced low point of contact between the particles which later result in the formation of the open and interconnected pores as shown in Figure 5. The high strength achieved at 1200 °C is due to the particle that well diffused together to form larger sintered area to form a smooth geopolyer matrix. Kong et al. [23] have reported that the strength increment in geopolymer might be contributed by the combination of polymerization and sintering. This is due to the portions of unreacted particles in geopolymer body remained from the dissolution process. Exposure to temperature caused the sintering of these particles which leads to increasing strength by forming stronger bonds and countering the thermal damage at elevated temperatures [15]. The chemical composition of kaolin geopolymer ceramic determined by XRF in table 2 shows the increased percentage of Na which were assumed to play important role in the geopolymer bonding.

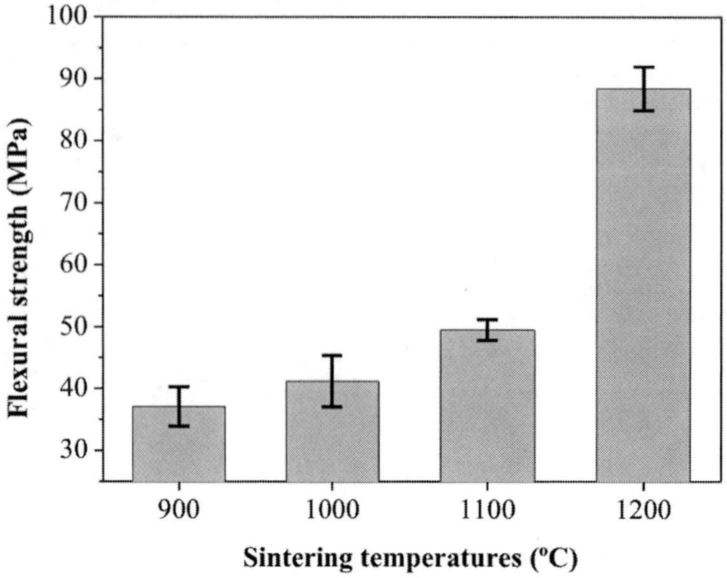

Fig. 7. Flexural strength of KGC at different sintering temperatures

Table 2. Chemical composition of kaolin geopolymer ceramic as determined by XRF

Chemical	Wt (%)
SiO_2	54.0
Al_2O_3	27.7
Fe_2O_3	5.04
TiO_2	1.23
Na_2O	4.00
ZrO_2	0.10
K_2O	5.41
MnO_2	0.11
LOI	1.74

3.4 Microstructural analysis

The microstructure of geopolymer ceramic after exposure to temperatures of 900 °C, 1000 °C, 1100 °C and 1200 °C are shown Figure 5. The micrographs show the microstructure development of kaolin geopolymer sample as the temperature increases. The particles observed as plate which forming a layer-like structure at temperatures of 900 °C and 1000 °C which resembles the microstructure of pure kaolin but with smooth geopolymer matrix. The structure becomes more compacted in matrix form (smoother geopolymeric gel) at 1100 °C and 1200 °C. The appearance of pores become well-shaped and homogeneously distributed as temperature rises indicate a better binding between kaolin and alkaline activator. Sintering the geopolymer resulted in the formation of liquid phase, which enables the joining of particles and the transformation of plate-like structure to dense microstructure [24]. This could explain the great advance in flexural strength. This study has shown that higher temperature aids in increasing the consolidation and facilitating a fairly uniform microstructure.

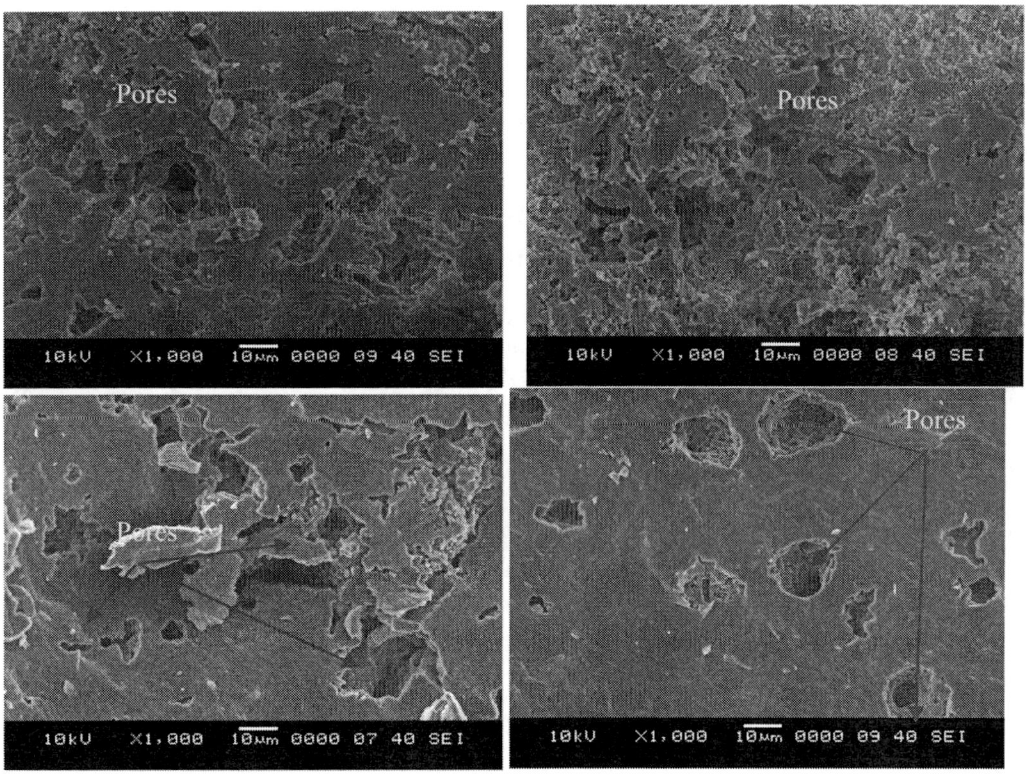

Fig. 8. Microstructure analysis of kaolin geopolymer ceramic with different sintering temperatures

4 Conclusions

This paper summarizes the work done on the synthesis of the geopolymer ceramics using kaolin as precursor material. The effect of sintering temperatures was studied on the final properties of kaolin geopolymer ceramics. Sintering temperature has a significant effect on the mechanical strength of the ceramics. From the results, it showed that sintering at 1200 °C gave highest flexural strength of 88.47 MPa. Increasing the sintering temperature increase the flexural strength of the ceramics. Sintering of the geopolymer enables the joining of particles and the transformation of plate-like kaolin structure to dense microstructure. Therefore, this indicates that high temperature were required to consolidate the geopolymer samples to achieve remarkable strength.

The authors would like to express gratitude KACST for funding this study through collaboration between KACST- UniMAP.

References

1. J. Davidovits, J. Therm. Anal, **37**, 1633 (1991)
2. D. Dimas, I. Giannopoulou, D. Panias, J. Mater. Sci., **44**, 3719 (2009)
3. J.E. Oh, P.J.M. Monteiro, S.S. Jun, S. Choi, S.M. Clark, Cement Concrete Res., **40**, 189 (2010)
4. J. Davidovits, J. Mater. Educ., **16**, 91 (1994)
5. C.D. Atiş, E.B. Görür, O. Karahan, C. Bilim, S. İlkentapar, E. Luga, Constr. Build. Mater., **96**, 673 (2015)
6. C.Y. Heah, H. Kamarudin, A.M.M.A. Bakri, M. Bnhussain, M. Luqman, I.K. Nizar, Int. J. Min. Met. Mater., **20**, 313 (2013)
7. E. Olevsky, A. Molinari, Int. J. Plasticity, **16**, 1 (2000)
8. X. Wang, Y. Jin, Z. Wang, R.B. Mahar, Y. Nie, J. Hazard. Mater., **160**, 489 (2008)
9. V.F.F. Barbosa, K.J.D. MacKenzie, Mater. Ress. Bull., **38**, 319 (2003)
10. V.F.F. Barbosa, K.J.D. Mackenzie, Mater. Lett., **57**, 1477 (2003)
11. A.V. Riessen, W. Rickard, J. Sanjayan, *Geopolymers: Structure, processing, properties and industrial applications* (Woodhead Publishing Limited, UK, 2009)
12. P. Duxson, G.C. Lukey, J.S.J.v. Deventer, J. Non-Cryst. Solid, **352**, 5541 (2006)
13. P. Duxson, G. C. Lukey, J.S.J.v. Deventer, J. Mater. Sci., **42**, 3044 (2007)
14. P. Duxson, G.C. Lukey, J.S.J.v. Deventer, J. Non-Cryst. Solid, **353**, 2186 (2007)
15. D.L.Y. Kong, J.G. Sanjayan, K. Sagoe-Crentsil, J. Mater. Sci., **43**, 824 (2008)
16. T. Bakharev, Cement Concrete Res., **36**, 1134 (2006)
17. J.G.S. Daniel L.Y. Kong, J.G. Sanjayan, Kwesi Sagoe-Crentsil, Cement Concrete Res., **37**, 1583 (2007)
18. J. L. Bell, P. E. Driemeyer, W. M. Kriven, J. Am. Ceram. Soc., **92**, 607 (2009)
19. P. He, D. Jia, M. Wang,Y. Zhou, Ceram. Int., **36**, 2395 (2010)
20. J.L. Bell, P.E. Driemeyer, W.M. Kriven, J. Am. Ceram. Soc., **92**, 607 (2009)
21. N. Xie, J. L. Bell, W. M. Kriven, J. Am. Ceram. Soc., **93**, 2644 (2010)
22. V.G. Lee, T.H. Yeh, Mater. Sci. Eng., **485**, 5 (2008)
23. D.L.Y. Kong, J.G. Sanjayan, K. Sagoe-Crentsil, Cement Concrete Res., **37**, 1583 (2007)
24. B.K. Ngun, H. Mohamad, S.K. Sulaiman, M.Y.M. Sulaiman, T. Isobe, K. Okada, J. Ceram. Process. Res., **13**, 547 (2012)

MATEC Web of Conferences **78**, 01062 (2016) DOI: 10.1051/matecconf/20167801062
IConGDM 2016

Palm Ash as an Alternative Source for Silica Production

Faizul Che Pa[1,*] , *Abdullah Chik*[1] , and *Md. Fazlul* Bari[1]

[1]Centre of Excellence Geopolymer & Green Technology,
School of Materials Engineering, Universiti Malaysia Perlis,
P.O Box 77, D/A Pejabat Pos Besar, 01007 Kangar, Perlis, Malaysia.

Abstract. Agricultural wastes such as palm ash and rice husk have a possibility to be used as a useful renewable source for the production of silica (SiO_2). Extensive researches have been carried out to extract silica from agricultural wastes such as rice husk, due to silica as a useful raw material for industrial application. In this study, the environmentally benign and economically effective process to produce SiO_2 materials from palm ash has been established by using citric acid leaching, not the conventional strong acids. Results showed that silica can be extracted from palm ash using the citric acid leaching method under the optimum extracting conditions with 700 ^{0}C of solution temperature, 60 minutes of reaction time and concentration of citric acid of more than 2 %. The purity of silica extracted is more than 90 %.

1 Introduction

Silica is one of the basic materials & the valuable inorganic multipurpose chemical compounds. Silica is occurring naturally as quartz, sand or flint. It can exist in gel, crystalline & amorphous forms. Nowadays, most silica was produced from quartz or sand by the extraction process. Sodium silicate, the precursor for silica production is currently manufactured by smelting quartz sand with sodium carbonate at 1300 °C [1]. However, manufacturing of high purity silica is energy intensive. Silica also can be found in agricultural wastes such as palm ash & rice husk.

Extensive researches have been carried out to extract silica from agricultural wastes. Leaching is a proper route to extract the silica. Leaching generally refers to the removal of a substance from a solid via a liquid extraction media. The desired component diffuses into the solvent from its natural solid form. The temperature, contact time and solvent selection are three important parameters in leaching process. The temperature can be adjusted to optimized solubility and mass transfer [2-5].

Organic acid is an alternative chemical to replace the strong acid in the leaching process. This is because the organic acid has low level of hazardous compared to the strong acid. Researchers all over the world are focusing on ways to establish an environmentally

* Corresponding author: faizul@unimap.edu.my

 DOI: 10.1051/matecconf/20167801062

benign process to produce silica from agricultural wastes by using organic acid rather than using strong acid.

2 Materials And Method

2.1 Palm Ash

Palm oil is an important export commodity in tropical countries and especially in Malaysia due to its wide spectrum of an acknowledged usability such as precursors of food products and biofuel. Extraction of the oil from fresh oil palm fruitlets requires separation of the fruitlets from empty fruit bunches prior to further processing. These empty fruit bunches, which consists of fibres and shells, are often used as boiler fuel by palm oil mills to produce steam for electricity generation and palm oil extraction. Combustion produces approximately 5 % of palm ash [6]. Because of limited uses for palm ash, it is currently disposed of as landfill; this could lead to environmental problems in the future. Palm ash has about 46.0 % silicon element. It is possible to extract silica from palm ash.

2.2 Leaching Process

Palm ash from local palm oil mill was used as raw materials to produce silica materials. Citric acid powders were dissolved in the distilled water at the ambient temperature to prepare citric acid solutions. 10 to 30 g palm ash (having a mean particle size of 75 μm) was put into 500 ml citric acid solution in the beaker. Due to the chelate reaction between–COOH groups and metallic impurities strongly depend on The concentration and temperature of the citric acid solution, and stirring time in the solution were selected as the operating parameters. The concentration was controlled from 1 to 6 % by changing the mixing ratio of citric acid powders and distilled water. The beaker was placed on the hot plate magnetic stirrer, and the solution temperature was changed from 30 to 70 ^{0}C. The reaction time recorded was 30 to 90 minutes. After the acid leaching process, the water rinsing treatment of the palm ash was carried out with the distilled water at room temperature to remove the citric acid content of the ash. The materials were dried at 60 ^{0}C for 60 min in the oven, and then combusted at 800 ^{0}C for 30 min in the tube furnace. The combustion temperature of 800 ^{0}C was used in this study to prevent the crystallization of amorphous silica contained in the ash [5,7].

3 Result And Discussion

In hydrometallurgical extraction, many leaching agents are used such as HCl, HNO_3 and H_2SO_4. However, most of these leachants cause environmental pollutions. Organic acids such as citric, acetic or oxalic acids can be attractive leaching agents as the leaching is carried out at moderate acidic conditions [8].

3.1 Palm Ash

Photographs of the as-received palm ash and pulverized palm ash samples obtained from the local oil palm company are shown in Figure 1. The colour of pulverised palm ash used to be dark grey.

Fig. 1. Digital photo of (left) as-received palm ash (right) pulverized palm ash

Typically, palm ash is characterized by a spongy and porous structure in nature, its main components are in the angular and irregular form with a sizable fraction showing cellular textures. The scanning electron microscope (SEM) images (Figure 2) revealed that palm ash used has medium-sized particles with crushed shape structure. The main components of palm ash are the oxides of Si, Al and various metallic oxides. Energy dispersive X-ray spectroscopy (EDX) analysis (Figure 3) confirms the presence of Si and Al.

Fig. 2. SEM images of ground palm.

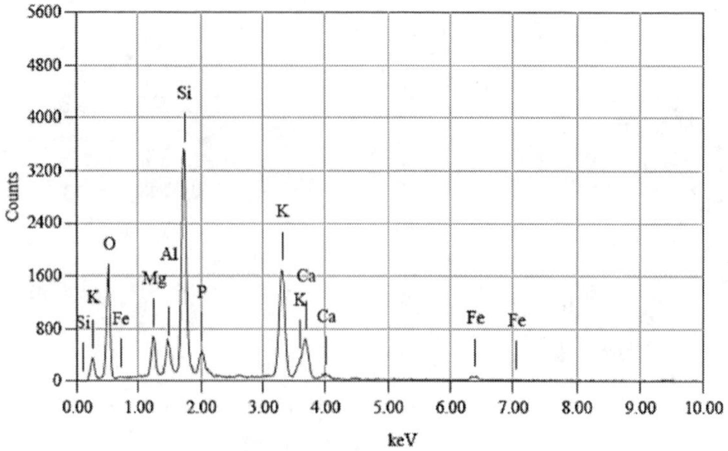

ZAF Method Standardless Quantitative Analysis
Fitting Coefficient : 0.5088

Element	(keV)	mass%	Error%	At%	Compound	mass%	Cation	K
O K*	0.525	43.22	0.95	60.72				30.6971
Mg K*	1.253	4.24	0.37	3.92				3.4948
Al K*	1.486	3.42	0.36	2.85				3.2200
Si K*	1.739	18.87	0.30	15.10				20.7402
P K*	2.013	3.08	0.40	2.24				3.6937
K K*	3.312	17.85	0.48	10.26				25.5637
Ca K*	3.690	7.35	0.60	4.12				10.1255
Fe K*	6.398	1.96	1.23	0.79				2.4650
Total		100.00		100.00				

Fig. 3. EDX result of ground palm ash.

The X-ray powder diffraction (XRD) pattern (Figure 4) indicated that a high intensity peak was observed at 26.7° (2θ) corresponding to the quartz present in palm ash used in this research, while the low intensity peaks can be assigned to other oxides present in the palm ash. An amorphous hump was observed between 30° and 35°, this may be due to the presence of amorphous glassy material [9]. A high intensity peak was observed at 22.5° corresponding to the silicates present in both forms of palm ash while the low intensity peaks can be assigned to other metal oxides present in the palm ash [10]. Phase analysis by x-ray diffraction shows that quartz (SiO_2;), calcite ($CaCO_3$) and kalsilite ($KAlSiO_4$) in palm ash mainly exists as crystalline substance.

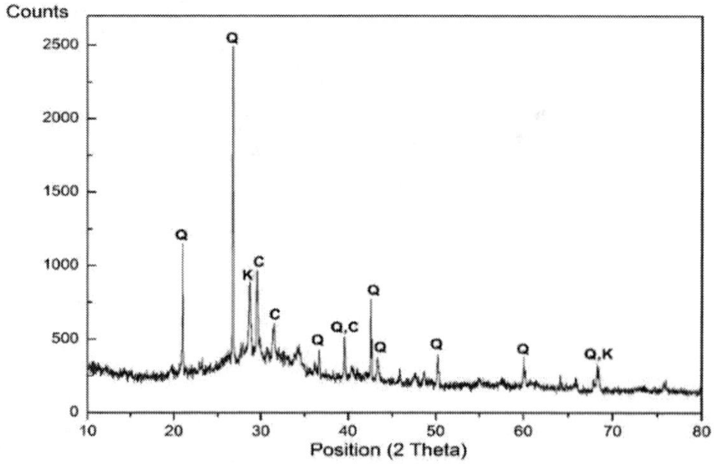

Fig. 4. XRD pattern of the palm ash. Q = SiO_2, C= $CaCO_3$, K = $KalSiO_4$.

The present of different elements concentration in palm ash is affected by the area of plantation, geographical distribution, quantity of fertilizers used, and agronomy used for the growth of palm ash [11].

3.2 Effect of Acid Concentration.

In general, the SiO_2 content increases with increasing acid concentration. The purity of silica (SiO_2) dramatically increased from just 45.5 % in palm ash up to 92 % when employing the citric acid solution with a concentration of 2 % and higher (Table 1).

Table 1. Chemical compositions of palm ash via the citric acid solution leaching treatment obtained at various concentrations.

Acid Concentration (%)	SiO_2	MgO	Al_2O_3	P_2O_5	K_2O	CaO	Fe_2O_3
Palm Ash	45.50	3.20	5.40	5.38	23.30	12.80	3.26
1	66.70	6.00	-	2.88	12.00	6.62	4.10
2	89.00	1.00	-	-	4.22	1.30	2.55
3	92.00	-	-	-	3.58	1.22	2.13
4	92.00	-	-	-	3.56	0.85	3.07
5	92.00	-	-	-	3.54	0.80	2.97
6	92.00	-	-	-	2.84	1.00	2.53

The content of metal oxide impurities also reduced when the concentration of the solution was increased. It means the use of the citric acid leaching treatment obviously caused the reduction of the total content of metal oxide impurities and resulted in a high-purification of silica materials contained in the palm ash. Concerning the content of metal elements remained in the ashes, the use of citric acid solution is remarkably effective to reduce them.

In particular, Al_2O_3 has been completely removed from the palm ash by using the citric acid solution leaching with 1 % or more concentration, while P_2O_5 has been totally taken away from the palm ash using 2 % of citric acid solution. Furthermore, it is obvious that the citric acid solution leaching treatment is also effective in removing K and Ca elements from palm ash. The content of K_2O and CaO extremely reduced and the reduction rate is 48.5 - 87.8 % and 48.3 - 93.8 % respectively by comparing the remained content of palm ash before and after citric acid solution leaching treatment. The content of MgO in palm ash, even so, was entirely removed with concentration of 3 % and higher.

Thus, it shows that citric acid leaching treatment is significantly useful and effective to remove impurities and increased the purity of silica in palm ash. Organic acids such as citric acid have carboxyl groups and therefore can act as chelating agents. These carboxyl groups, tend to donate protons (H+), resulting in the negatively charged carboxyl group that is capable of forming stable complexes with several metal ions. The impurities were removed from the ash via chelate reaction between carboxyl groups (-CHOOH) and the metal elements [2-5]. Therefore, it is possible that these negatively charged carboxyl groups are able to form stable complexes with Ca+ and K+ cations present, resulting in their removal from the palm ash.

3.3 Effect of Reaction Temperature

The use of high temperatures can lead to an increase in the leaching performance as shown in Table 2.

Table 2. Chemical composition of treated palm ash via citric acid solution treatment obtained at various reaction temperatures.

Solution Temp. (^0C)	Acid Conc. (%)	SiO_2	MgO	Al_2O_3	P_2O_5	K_2O	CaO	Fe_2O_3
Palm Ash	-	45.50	3.20	5.40	5.38	23.30	12.80	3.26
30	3	87.00	1.80	-	-	4.96	1.35	2.78
50	3	90.00	1.20	-	-	4.17	1.07	2.43
70	3	92.00	-	-	-	3.58	1.22	2.13
90	3	92.13	-	-	-	3.50	1.19	2.07
30	2	83.00	2.20	-	-	6.18	1.73	3.36
50	2	87.00	2.00	-	-	4.90	1.44	3.18
70	2	89.00	1.00	-	-	4.22	1.30	2.55
90	2	88.00	1.10	-	-	4.87	1.38	2.82

With increased in the temperature, most of the impurities are decreased and the purification of silica increased up to 92 % takes place after leaching at 70 ^0C. In the case of Al_2O_3 and P_2O_5, even the temperature of 30 ^0C and 2 % citric acid concentration leaching treatment is enough to remove Al and P impurities in palm ash. While MgO content of the ashes was fully taken out of palm ashes at a temperature of 70 ^0C and 3 % citric acid concentration. On the other hand, other impurity elements such as K, Ca and Fe discharged into the acid solution during the acid leaching process gradually increase with increasing the reaction temperature.

An increase in temperature of leaching treatment process speeds up the chemical reaction. The rising temperature will be affected on kinetic energy to molecules. The molecule will move faster, collide more frequently and causing the reaction rate of temperature in the general to increase. It indicates that using higher temperature in the process will improve the leaching result. The palm ash used in this research work has been pulverised to increase the leaching rate. Theoretically, the smaller size of the material, the easier to be heated and mass transfer, the higher leaching rate [12].

The use of high temperatures can lead to an increase in the leaching performance, but the use of high temperatures may be limited with organic acids due to their low boiling temperatures and decompositions [13]. Higher temperature tends to vaporize organic acids and water and contaminates the vapour phase stream with mainly carbon dioxide [13]. Higher temperatures tend to decrease the solubility of the reaction by-products in water and it makes the separation difficult. Furthermore, in the industrial processes at relatively higher temperatures, organic acids may cause a little corrosive effect.

3.3 Effect of Reaction Time

The purity of silica increased with the increasing of reaction time as shown in Table 3. SiO_2 content in palm ash increased dramatically from 45.5 % to 87.0 % after 30 minutes leaching process. When the reaction time increased from 30 minutes to 60 minutes, the purity of silica slightly increased by about 5 %. However, no significant increase in silica purity was observed at a longer reaction time than 60 minutes. Beyond this range the effect is negligible, so the reaction time of 60 minutes was confirmed to be the optimal reaction time. Other investigators observed the value ranges from 45 to 60 min for effective dissolution [13].

Table 3. Chemical composition (%) of palm ash after citric acid solution treatment obtained at various reaction times.

Reaction Time (minute)	SiO_2	MgO	Al_2O_3	P_2O_5	K_2O	CaO	Fe_2O_3
Palm Ash	45.50	3.20	5.40	5.38	23.30	12.80	3.26
30	87.00	1.50	-	-	5.68	1.49	2.65
60	92.00	-	-	-	3.58	1.22	2.13
90	92.00	-	-	-	3.25	1.16	2.22
120	92.30	-	-	-	3.01	0.99	2.37
180	92.40	-	-	-	2.90	0.82	2.43

Studies done by Ukiwe and Luke [14] appears to suggest that organic acids such as citric and oxalic acids seemed to be more promising as chemical extracting agents for removal of heavy metal from sludge at contact time of 60 minutes, since these acids are biodegradable and can attain a higher metal extraction efficiency at this contact time compared to other contact times.

3.5 Effect of Solid/liquid Ratio.

Table 4 shows that in the leaching process, lower solid/liquid ratios (0.02:1 and 0.04:1) lead to achieving the higher purity of silica. The silica content in treated palm ash was raised by up to 92 %. When the solid/liquid was increased from 0.04:1 to 0.06:1 until 0.10:1, with the other conditions unchanged, the purity of silica content in treating palm ash decreased to 83 %.

Table 4. Chemical composition (%) of palm ash after citric acid solution treatment obtained at various solid/liquid ratio.

Solid/liquid Ratio	SiO_2	MgO	Al_2O_3	P_2O_5	K_2O	CaO	Fe_2O_3
Palm Ash	45.50	3.20	5.40	5.38	23.30	12.80	3.26
0.02:1	91.00	-	-	-	3.76	0.96	2.86
0.04:1	92.00	-	-	-	3.58	1.22	2.13
0.06:1	87.00	-	-	-	5.28	3.59	2.65
0.08:1	85.00	-	-	-	5.61	4.51	3.71
0.10:1	83.00	-	-	-	6.05	4.98	3.64

Table 4 shows that the contents of K_2O and CaO in treated palm ash were decreased up to 84.6 % and 92.5 %, respectively, when the solid/liquid ratios were 0.04:1 and below. When the solid/liquid ratio increased to 0.10:1, the contents of K_2O and CaO were decreased only about 74.0 % and 61.0 %, respectively. It happens because of not enough of carboxyl groups (-CHOOH) that act as chelating agents to promote stable complexes with Ca+ and K+ cations present in palm ash. The impurities were removed from the ash via chelate reaction between carboxyl groups and the metal elements [2-5].

The limitations associated with the poor dissolution of metal elements by citric acid at the solid/liquid ratio higher than 0.04:1, indicate a lower purity of silica. A solid/liquid ratio 0.04:1 by using 3 % citric acid was found to be the most promising and suitable condition to produce high purity of silica from palm ash.

3.6 Comparison between Organic Acid and Strong Acid

Hydrochloric acid solution is also used as the comparison in the leaching treatment. 1, 2, 3 and 4 % of hydrochloric acid and citric acid solution was used in this study. The reaction temperature was controlled at 70 °C and 60 minutes reaction time. Figure 4 shows that the content of silica in palm ash treated with hydrochloric acid was much higher than silica content in palm ash treated with citric acid.

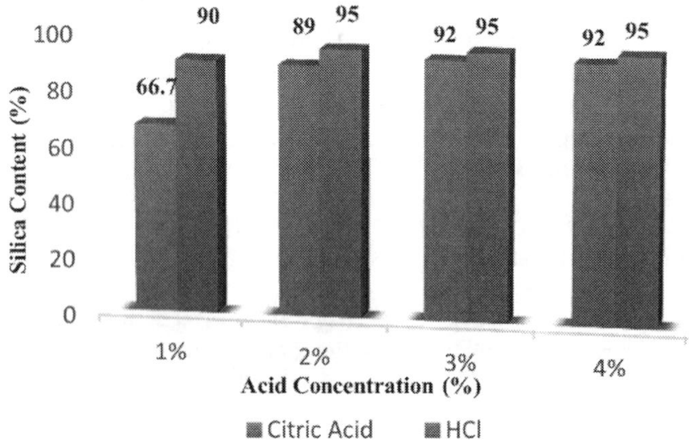

Fig. 4. Silica content (%) versus acid concentration

Two factors, mainly account; type of acid used in the leaching process & carbon content in palm ash. The strong acid almost fully ionized when mixed with water to produce hydrogen ions from water, whereas organic acid only partially dissociate into ions and leaving an increasing percentage of untreated molecules in the solution. In the leaching process, H+ ion should be the unique responsible for the dissolution reaction [8]. Anions such as NO^{3-}, SO_4^{2-} and Cl- should be non-effective. However, it is worth noting that inorganic acids completely dissolved in water, which means in case HCl and HNO_3 that the same mol number of H+ were present in the solution leading normally to the same dissolution efficiency of metal oxide. Moreover, H_2SO_4 at 0.5 produced 1 M of H^+ and in spite of that, lower dissolution efficiency of H_2SO_4 was observed compared to HCl. These results indicate that leaching is not a simple process and that anions may play an important role in the rate of metal oxide dissolution [8].

It is reported in Table 1 that metal impurities such as K and Ca are included in palm ash. These metal impurities cause the remained carbons in ashes, which are originated in palm ash. As mentioned before, palm ash is a by-product from burning process in palm oil plantation, which palm shell, fibre and empty fruit bunches of palm burn at a temperature about 800-1000°C. The eutectic reaction between the metals and SiO_2 element occurs in the burning process of palm ash, and the carbon remains in the melts SiO_2 [2-5]. Once carbon is entrapped in these melt, it cannot be easily removed because of the intimate contact between silica and carbon in palm ash (some of the silica might also be bonded to the organic compound in palm ash) [15]. XRD patterns of treated palm ash (Figure 5) show that carbon still exists after treated the palm ash with 2-3 % citric acid at a solution temperature of 70 ^{0}C for 60 minutes.

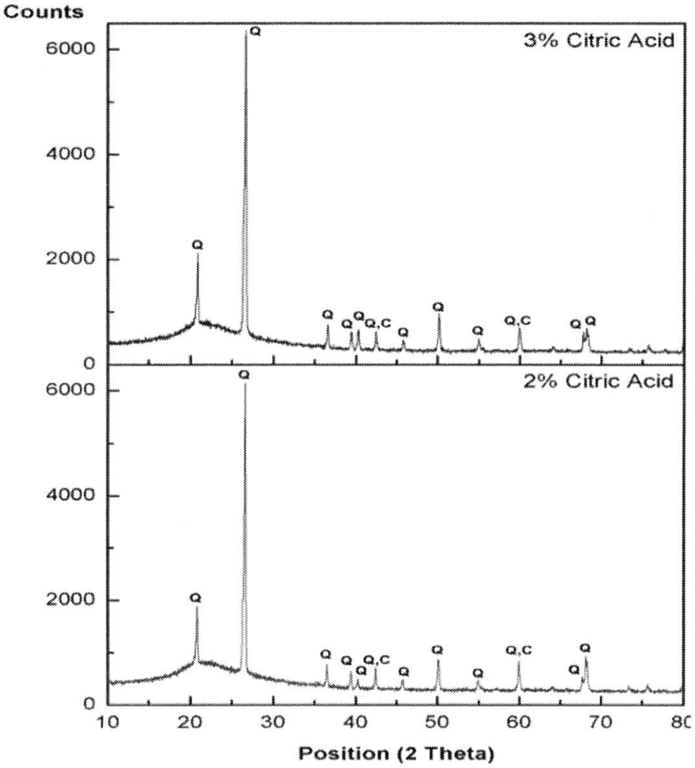

Fig. 5. XRD Pattern Of The Treated Palm Ash. Q = SIO_2, C = CARBON.

4 Conclusion

It was possible to obtain silica from palm ash using the organic acid leaching method. It consists of a citric acid leaching and the combustion process. Citric acid leaching treatment is significantly useful and effective to remove the impurities and increased the purity of silica in palm ash.

The authors would like to thank the Center of Excellence Geopolymer & Green Technology, School of Materials Engineering, Universiti Malaysia Perlis and Malaysian Government for financing this project through the Research Acculturation Collaborative Effort (RACE) grant scheme (9017-00011).

References

1. U. Kalapathy, A. Proctor, J. Shultz, Bioresour Techno., **73**, 257 (2000)
2. J. Umeda, K. Kondoh, J. Mater. Sci., **43**, 7084 (2008)

3. J. Umeda, K. Kondoh, Trans. JWRI, **37**, 13 (2008).
4. J. Umeda, H. Imai, K. Kondoh, Trans. JWRI, **38**, 13 (2009)
5. J. Umeda, K. Kondoh, Ind. Crops. Prod., **32**, 539 (2010)
6. C.P. Faizul, C. Abdullah, B. Fazlul, Adv. Mat. Res., **626**, 997 (2013)
7. C.P. Faizul, C. Abdullah, B. M. Fazlul, Key Eng. Mater., **673**, 13 (2016)
8. N. Habbache, N. Alane, S. Djerad, Tifouti, Chem. Eng. J., **152**, 503 (2009)
9. P. J. Williams, J. J. Biernacki, C. J. Rawn, L. Walker, J. Bai, ACI Mater J. **102**, 330 (2005)
10. N.F. Zainudin, K.T. Lee, A.H. Kamarudin, S. Bhatia, A.R. Mohamed, Sep. Purif Technol., **45**, 50 (2005)
11. K.Y. Foo, B.H. Hameed, J. Hazard. Mater., **172**, 523 (2009)
12. X. Fan, W. Xing, H. Dong, J. Zhao, Y. Wu, B. Li, W. Tong, X. Wu, Int. J. Nonferrous. Metall., **2**, 63 (2013)
13. M. Gharabaghi, M. Irannajad, M. Noaparats, Hydrometallurgy, **103**, 96 (2010)
14. N.U. Luke, I.I. Chinonye, Terr. Aquat. Environ. Toxicol., **5**, 73 (2011)
15. S. Yafei, Z. Peito, S. Qinfu, Microporous Mesoporous Mater., **188**, 46 (2014)

Effect Of Crumb Rubber On Compressive Strength Of Fly Ash Based Geopolymer Concrete

Ahmad Azrem Azmi[1,*], *Mohd Mustafa Al Bakri* Abdullah[2,3], *Che Mohd Ruzaidi* Ghazali[2,1], *Andrei* Victor Sandu[4,3], and *Kamarudin* Hussin[2,1]

[1]Center of Excellence Geopolymer and Green Technology, School of Materials Engineering, Universiti Malaysia Perlis (UniMAP), 01007, P.O Box 77, D/A Pejabat Pos Besar, Kangar, Perlis, Malaysia.
[2]Faculty of Engineering Technology, Universiti Malaysia Perlis (UniMAP), 01007, P.O Box 77, D/A Pejabat Pos Besar, Kangar, Perlis, Malaysia.
[3]Faculty of Technology, UniversitasUbudiyah Indonesia (UUI), JlnAlue Naga DesaTibang, Banda Aceh, Indonesia
[4]Gheorghe Asachi Technical University of lasi, Faculty of Materials Science and Engineering, Blvd. D. Mangeron 71, 700050, lasi, Romania

Abstract. This study explores the effect of different percentage of crumb rubber on compressive strength of fly ash based geopolymer concrete. This research attempted to produce rubberized geopolymer concrete as an environmentally friendly, lightweight and durable product. Crumb rubber with ranged size from 73 μm to 375 μm was used in order to replace fine aggregates in geopolymer concrete. The replacements of crumb rubber were 0%, 5%, 10%, 15% and 20% in the fly ash based geopolymer concrete. The ratio of fly ash to alkaline activator was 2.5 and the ratio of Na2SiO3 to NaOH was fixed at 2.0. After the curing process, the strengths of the samples were tested on days 7 and 28. The results show that there is a reduction in all compressive strength for crumb rubber mixture, but still higher than normal rubberized concrete. Rubberized geopolymer concrete is a suitable solution in some non structural applications.

1 Introduction

Geopolymer is known for its durability and environmental sustainability. Sustainable practice of industrial ecology demands the use of byproduct of one industry as source materials for other industrial application. This will help in complying with environmental parameters for both the industry. Environmental concerns compelled us to reduce the material consumption by enhancing the durability and resource efficiency using advanced technology [1-2]. In general, geopolymers as a class of inorganic polymer are formed by reaction between an alkaline solution (e.g., sodium hydroxide and sodium silicate) and an aluminosilicate source (e.g., metakaolin, fly ash, and slag) [3]. Geopolymer or alkali-

* Corresponding author: mustafa_albakri@unimap.edu.my

activated binder is produced by activating aluminosilicate source materials with highly alkaline activator solution at ambient or high temperature. A highly alkaline solution such as sodium hydroxide (NaOH) and sodium silicate (Na2SiO3) was usually used as alkaline solution [4-5].

Fly ash normally produced from burning anthracite or bituminous coal that meets the applicable requirements for this class. This class of fly ash has pozzolanic properties [6]. Fly ash is suitable for use as a geopolymer source material because it consists mostly of glassy, hollow and spherical particles. Fly ash-based geopolymer cement and concrete have been studied extensively, and they are well known for their properties, which are better than those of normal concrete due to their lower creep, lower shrinkage, better fire and acid resistance, and resistance to sulfate attack [7].

Geopolymer concrete is concrete which does not utilize any Portland cement in its production. Geopolymer concrete is being studied extensively and shows promise as a substitute to Portland cement concrete. Research is shifting from the chemistry domain to engineering applications and commercial production of geopolymer concrete [8]. Producing a typical car tyre uses about 40% natural rubber and 60% synthetic rubber that needs to be separated from the tyre's steel casing before it can be used as a rubber crumb. The products obtained from scrap tyres are classified as whole scrap tyres, slit tyres, shredded/chipped, ground and rubber crumb. Rubber crumb particle sizes range between 4.75mm to less than 0.075 mm [9] and are irregularly shaped, torn particles due to the micro-mill process, they are subjected to during the manufacturing process [10-12]. The recycling will reduce the waste and contribute to a greener environment.

Over the past few years, a number of researchers have studied on incorporated the waste tires into cement based materials [13-18]. However, this paper presents the study of early age, strength of geopolymer concrete with recycled rubber. This study used crumb rubber as an aggregate to produce rubberized geopolymer concrete. By using rubber waste to produce lightweight geopolymer concrete is expected to more durable, less expensive (low material cost and easy to manufacture) and absorb higher energy under impact [19].

2 Raw materials

In this experiment, crumb rubber from automotive and truck scrap tires, in order to replace fine aggregates. It has particles ranging from 73 µm to 375 µm. Crumb rubber has a lower unit weight compared to mineral aggregates. The addition of crumb rubber to the mixture reduced the unit weight of the geopolymer concrete mixture. The replacements of crumb rubber in the mixture were 0%, 5%, 10%, 15% and 20%.

Table 1. The composition of fly ash.

Elements	% mass
Al_2O_3	23.40
SiO_2	50.00
SO_3	0.08
K_2O	1.41
CaO	5.06
TiO_2	1.60
MnO	0.22
Fe_2O_3	17.29

Low calcium class F fly ash (ASTM Class F) used in this research was obtained from the Sultan Abdul Aziz Power Station in Kapar, Selangor, Malaysia. It was used as a source material in the geopolymer concrete mixture. Table 1 shows the composition of fly ash, as determined by X-Ray Fluorescence (XRF).

The brand named Formosoda-P of sodium hydroxide (NaOH) powder with pure 99% was obtained from the Formosa Plastic Corporation, Taiwan used to produce the NaOH solution. Sodium silicate (Na2SiO3) with chemical composition of 30.1% SiO2, 9.4% Na2O and 60.5% H2O (modulus, SiO2/Na2O = 3.2) was obtained from South Pacific Chemicals Industries Sdn. Bhd. (SPCI), Malaysia. It has a specific gravity at 20 °C = 1.4 g/cm3 and viscosity at 20 °C = 0.4 Pa s.

3 Mixing methode

In order to produce 12 M NaOH solution [20], 480g of NaOH powder were dissolved in 1 L distilled water and allowed to cool down at room temperature. The alkaline activator was prepared by combining of NaOH and Na2SiO3 one day prior to use. The ratio of Na2SiO3 to NaOH was fixed at 2.0. The ratio of fly ash to alkaline activator was 2.5 and was kept fixed for all mixtures. Crumb rubber and fly ash were mixed in a dry state for 30 seconds, then alkaline activator were mixed into the mixtures using a hand mixer for another 10 minutes.

4 Casting and curing

The mixture of crumb rubber and fly ash were mixed well with alkaline activator for 10 minutes using a hand mixer, forming a homogeneous mortar. The fresh geopolymer paste filled crumb rubber were poured into 50 mm cube moulds and compacted as described in ASTM C109. The samples were sealed with a thin plastic layer to prevent moisture loss and left cured at room temperature (open air) before demoulding.

5 Testing

5.1 Compressive strength test

Compressive strength test of geopolymer mortar was performed using GoTech Universal testing Machine according to ASTM C 109/C. Three samples of each percentage of crumb rubber were tested at 7 and 28 days.

6 Result and discussion

Table 2 and Figure 1 shown the compressive strength of the rubberized geopolymer concrete decreased with increasing crumb rubber content. The addition of crumb rubber in rubberized geopolymer concrete weakened the compressive strength of rubberized geopolymer concrete. As expected, the higher the crumb rubber content in the mix, the higher the reduction in compressive strength.

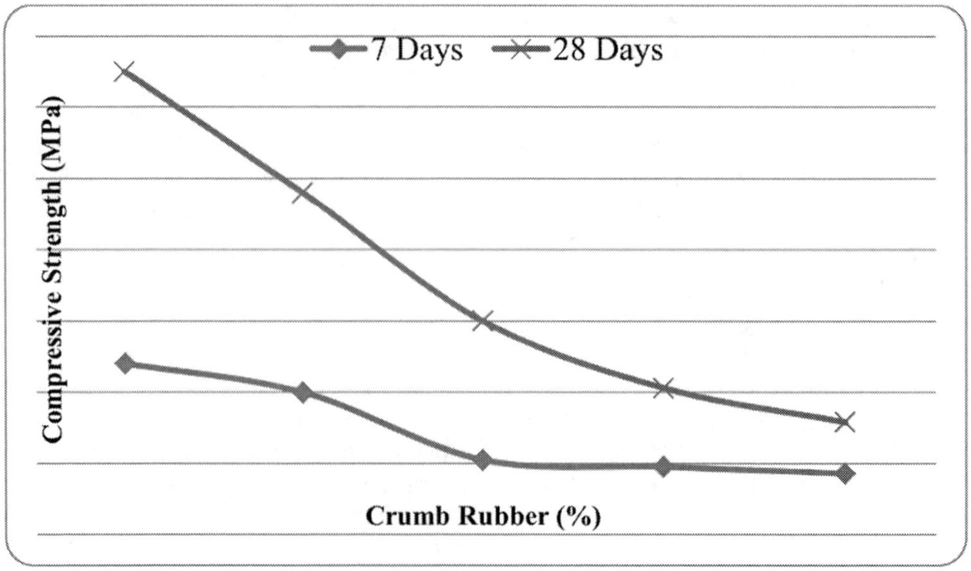

Fig. 1. Compressive strength of rubberized geopolymer concrete at 7 and 28 days.

Table 2. Compressive strength of the rubberized geopolymer concrete at 7 and 28 days.

Mix	Rubber %	Compression strength(N/mm2)	
		7 Days	28 Days
1	0	24.0	65.0
2	5	20.0	48.0
3	10	10.5	30.0
4	15	9.5	20.6
5	20	8.5	15.8

The performance of rubberized geopolymer concrete is quite encouraging up to 15%. The compressive strength of the rubberized geopolymer concrete decreases about 60% when 15% crumb rubber was added. The results of compressive strength of rubberized geopolymer concrete mixtures are affected by the size, proportions, and surface texture of rubber particles, and type of cement used in such mixtures [19]. This is due to the increased porosity or weakness points in the geoploymer concrete mix which is lack of bonding between crumb rubber and fly ash with alkaline activator and lead to high internal stress that are perpendicular to the direction of applied load [21]. With the increase in void content of the concrete, there will be a corresponding decrease in strength. In addition, the failure of the samples also due to the crumb rubber being more elastically deformable than the matrix. The cracks begin at the softest areas when the samples were loaded. The site of the inclusion of rubber is where these sites appear [19].

7 Conclusion

The test results of this study show that crumb rubber can be added in geopolymer concrete without any difficulties. With the addition of the crumb rubber, the reduction in strength cannot be avoided. The compressive strength of the rubberized geopolymer concrete decreases about 60% when 15% crumb rubber was added. For large amount of crumb rubber in the mixture, the compressive strength gain rate is lower than ordinary concrete. Although compressive strength reduction occurred due to the amount of crumb rubber in the geopolymer concrete, these concrete still can be utilized in many engineering applications.

A special thanks to the Center of Excellence Geopolymer and Green Technology (CEGeoGTech) and the School of Materials Engineering, Universiti Malaysia Perlis (UniMAP) for providing research facilities and technical assistance for their help and cooperation in research work.

References

1. A. Bhowmick, S. Ghosh, Int. J. Eng. Res. Appl., **3(1)**, 168 (2012)
2. J.T. Gourley, Journal of the Australian Ceramics Society, **50(1)**, 102 (2014)
3. J. He, *Synthesis and Characterization of Geopolymers for Infrastructural Applications*, PhD Thesis, Nottingham University, 2005
4. Z. Yahya, A.M.M. Al Bakri, H. Kamarudin, I. Khairul Nizar, A.R. Razak, Materials Science Forum, 803(2014)
5. M.M.A. Abdullah, K. Hussin, M. Bnhussain, K.N. Ismail, W.M.W. Ibrahim, Int. J. Pure Appl. Sci. Technol., **6(1)**, 35, 2011
6. M.M. Al Bakri, H. Mohammed, H. Kamarudin, I.K. Niza, Y. Zarina, Journal of engineering and technology research, **3(1)**, 1(2011)
7. Abdullah, M. M. A. B., Hussin, K., Bnhussain, M., Ismail, K. N., Yahya, Z., Abdul Razak, R., Int. J. Mol. Sci., **13(6)**, 7186 (2012).
8. S. Jaydeep, B.J. Chakravarthy, International Journal of Engineering Trends and Technology, **4(10)**, 4614 (2013)
9. S.D. Parveen, A. Sharm, International Journal of Emerging Technology and Advanced Engineering, **3(3)**, 192 (2013)
10. A. Richardson, K. Coventry, U. Dave, J. Pienaar, Journal of Green Building, **6(1)**, 83 (2011)
11. H. Maraghechi, I. Fotovat Ahmadi, S. Motahari, J. Mech. Mater. Struct., **6(9)**, 1283 (2012)
12. K.M. Kotresh, M.G. Belachew, International Journal of Scientific Engineering and Technology, **3**, 433 (2014)
13. I.B. Topcu, Cement and Concrete Research, **25(2)**, 304 (1995)
14. H.A. Toutanji, Cem. Concr. Compos., **18(2)**, 135 (1996)
15. A. Benazzouk, O. Douzane, T. Langlet, K. Mezreb, J.M. Roucoult, M. Queneudec, Cem. Concr. Compos., **29(10)**, 732 (2007)
16. S. Sgobba, G.C. Marano, M. Borsa, M. Molfetta, *2nd International Conference on Sustainable Construction Materials and Technologies*, 2010
17. A.A.Yasin, Contemporary Engineering Sciences, **5(10)**, 473 (2012)
18. Y. Park, A. Abolmaali, P. Aswath, S.Lee, *The 2015 World Congress on Advanced in Structural Engineering and Mechanics,* 2015
19. M.M.A.B. Abdullah, S.N.F.S. Adam, M.A. Bakar, K.W. Leong, *Proceedings of the 1st International Conference On Sustainable Materials*, 2007
20. A.M. Al Bakri, H. Kamarudin, M. Bnhussain, I.K. Nizar, A.R. Rafiza, Y. Zarina, Rev. Adv. Mater. Sci., **30**, 90 (2012)
21. P. Nithiya, G. Portchejian, International Journal of Structural and Civil Engineering Research, **3(3)**, 63 (2014)

Influence of Activated Carbon Particles on Intermetallic Compound Growth Mechanism in Sn-Cu-Ni Composite Solder

M.I.I. *Ramli*[1,*], M.A.A. *Mohd Salleh*[1], M.N. *Derman* [1], R.M. *Said* [1] and N. *Saud*[1].

[1]Center of Excellence Geopolymer & Green Technology (CeGeoGTech), School of Materials Engineering, Universiti Malaysia Perlis (UniMAP), Taman Muhibbah, 02600 Jejawi, Arau, Perlis, Malaysia.

Abstract. The influence of Activated Carbon (AC) particles on mechanical properties of Sn-Cu-Ni-xAC solder joint was investigated. Five different Activated Carbon (AC) percentage addition (0 wt. %, 0.25 wt. %, 0.5 wt. %, 0.75 wt. %, and 1.0 wt. %) were prepared via powder metallurgy (PM) technique. Interfacial IMC thickness measurement and shear strength results showed that with thinner IMC layer (by increasing amount of wt.% of AC), the higher the shear strength of the joint. It is believed that the AC particles suppresses the interfacial IMC growth and thus improves the shear strength.

1 Introduction

The eutectic Sn-0.7Cu solder alloy has been used as one of the candidates to replace the traditional Sn-Pb solder. However, Sn-0.7Cu has been reported to have poor mechanical properties and wettability compared to lead free solder alloys containing silver (Ag) [1]. Because of that, a small amount of nickel (Ni) were added into Sn-0.7Cu solder and has improved the fluidity of Sn-0.7Cu alloy [2]. As a result, Sn-Cu-Ni solder alloy is also considered as a low cost solder alloy which has an equal attractive as Sn-Pb solder for wave, iron and reflow soldering process [3].

The fast changing technology and increasing miniaturization of electronic devices place challenges for obtaining successful component joint. Therefore, solder with improved mechanical properties have to be developed to suit with rapid advance in the electronics industry. By these reasons, many researchers investigate and develop a solder material to meet the requirement of the modern solder. Some attractive and potentially available methods of enhancing solder joint performances are by adding reinforcements to solder alloys, to form composite solders [4]. In general, composite solder should improve the mechanical properties of a solder joint.

Major studies have proven that composite technology improves the reliability of solder joints. Ceramic reinforcement particles has been developed by Nai et al. [5] by using titanium diboride (TiB_2) and multi walled carbon nanotubes (MWCNTs). They reported

[*] Corresponding author : mohdizrulizwan@gmail.com

that mechanical properties of both composites had majorly improved. Other ceramic reinforcing material such as Si_3N_4 has also been studied by Ramli et al. [6] and Mohd Salleh et al. [7]. They also reported that the mechanical properties of the composite solder have improved significantly. During soldering, the intermetallic compound layers are formed due to an interfacial reaction between solder alloy and metal substrate. It is well known that the intermetallic compound is necessary in every solder joint because it determined the strength and wettability of a solder. However, thicker intermetallic is avoided due to brittle nature and poor interfacial bonding of the formation thicker intermetallic layer. A thin, continuous, and uniform intermetallic compound (IMC) layer between a solder and the substrate material is an essential requirement for good bonding [8].

This research discusses the effect of AC additions on the Sn-Cu-Ni solder alloy in enhancing its solder joint strength. The intermetallic compound and shear strength of the composite solder joint were investigated.

2 Experiment Details

Sn-Cu-Ni powder alloys were mixed with different weight percentage (0, 0.25, 0.5, 0.75 and 1.0 wt. %) of the Activated Carbon in an airtight container using a roller blender rotated at a speed of 200 rpm for 1 hour. Sn-Cu-Ni powders were supplied from Nihon Superior (M) Sdn Bhd with an average particle size of 20-38 μm. The mixed powders were then weighed approximately 1g each in forming a green compact pellet disc. Each of the solder mixtures was pressed using Specac 15-ton Manual Hydraulic Press and uniaxially compacted in a 12-mm diameter mold. The compacted pallets were sintered with microwave sintering to reach the temperature 185 °C. SiC was used as the microwave susceptor material [9]. The sintered samples were then placed on Cu substrate with placing a small amount of no-clean flux and subsequently heated in a reflow oven to form solder joints.

For futher analysis, the reflowed samples were mounted, polished and etched with a solution of 93% methanol + 5% HNO_3 + 2% HCl to reveal a clear microstructure for imaging. Interfacial IMC layer was observed using Optical Microscope (ECLIPSE L300N) and Scanning Electron Microscope (SEM). The IMC thickness measurement was carried out by dividing the IMC area (A) by its length (L). The average thickness of the IMC layer was taken determined at 10 positions of the interface using the micrographs. To determined the solder joint strength, a single-lap shear solder joint sample were prepared by bonding the composite solder to the Cu-substrate (PCB FR-4). The 12mm diameter of solder samples with a thickness of approximately 0.5mm were sandwiched with the Cu-substrate. The reflow soldering process was used using the F_4N reflow oven. The specifications of the shear test were based on the ASTM D1002. Shear tests were then conducted by using Instron Machine.

3 Result and Discussion

3.1 Intermetallic Compound

Fig.1 show the thickness for IMC layer of Sn-Cu-Ni-xAC composite solder. By increasing the reinforcement addition, the IMC layer thickness showed a decreasing trend. The thinnest IMC layer was achieved with the addition of 1.0 wt. % reinforcement in Sn-Cu-Ni-xAC. AC addition has not just reduce the IMC thickness, but has also altered the IMC layer shape. The IMC layer has change from needle scallop shaped to continuous shaped as shown in Fig 2. The changes in IMC layer shape (Fig. 3) are considered as desired changes since the needle scallop-shaped induce brittle fracturing at the interface and provides crack thus decreasing the solder joint reliability [10].

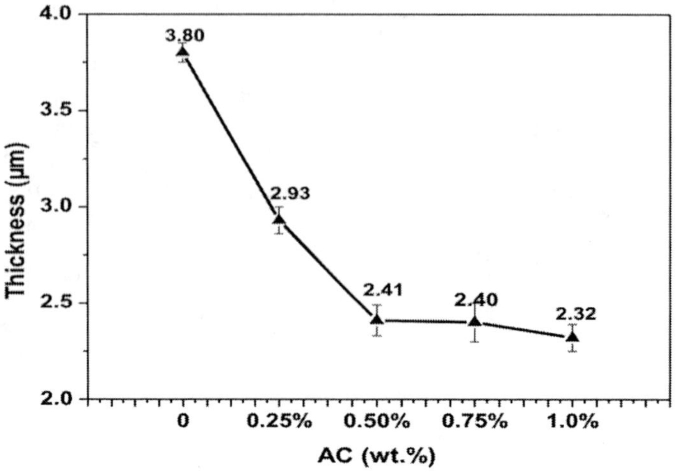

Fig.1. IMC layer thickness of Sn-Cu-Ni-xAC composite solder joint.

The observation was in conjuction with the observation done by S.L.Tay et. al [10]. They found that a scallop-shaped intermetallic layer was formed after reflow but changed to a planar type when Co particle addition in SAC solder. The addition of Ni nanoparticles also influence the thickness and shape of IMC layer as reported by S.L.Tay et al.[11]. They also found the morphology of Cu_6Sn_5 IMC changes from a scalloped structure to a planar type after reflow while the growth of $(Cu,Ni)_6Sn_5$ enhanced but reduced in Cu_3Sn [11]. They suggested that the Co and Ni containing solder entrapped in the IMC layer thus changes the interfacial IMC.

According to the EDX result using line and area scanning mode in Fig. 4 and Table 1, it shows that AC particles exist between solder and substrate interfaces (line 001). Scanning mode (Area 002) also detects the existence of reinforcement particle in that area. It supports that reinforcement particles act to retard the growth of IMC when they distributed homogeneously at the diffusion path for IMC growth.

The formation of IMC layer is mainly due to the diffusion of atoms from the substrate into the solder matrix. Diffusion rates control the IMC growth that solder/IMC interface energies control the coarsening of the IMC grain and nucleation kinetics [12]. IMCs form rapidly at the interface and act as the bonding materials between the solder and the substrate. Due to the present of reinforcement in the solder matrix, it could alter the diffusion rate at the interfacial layer. Reinforcement particles are believed to act as a barrier that block the diffusion of Cu and Sn from reacting to each others.

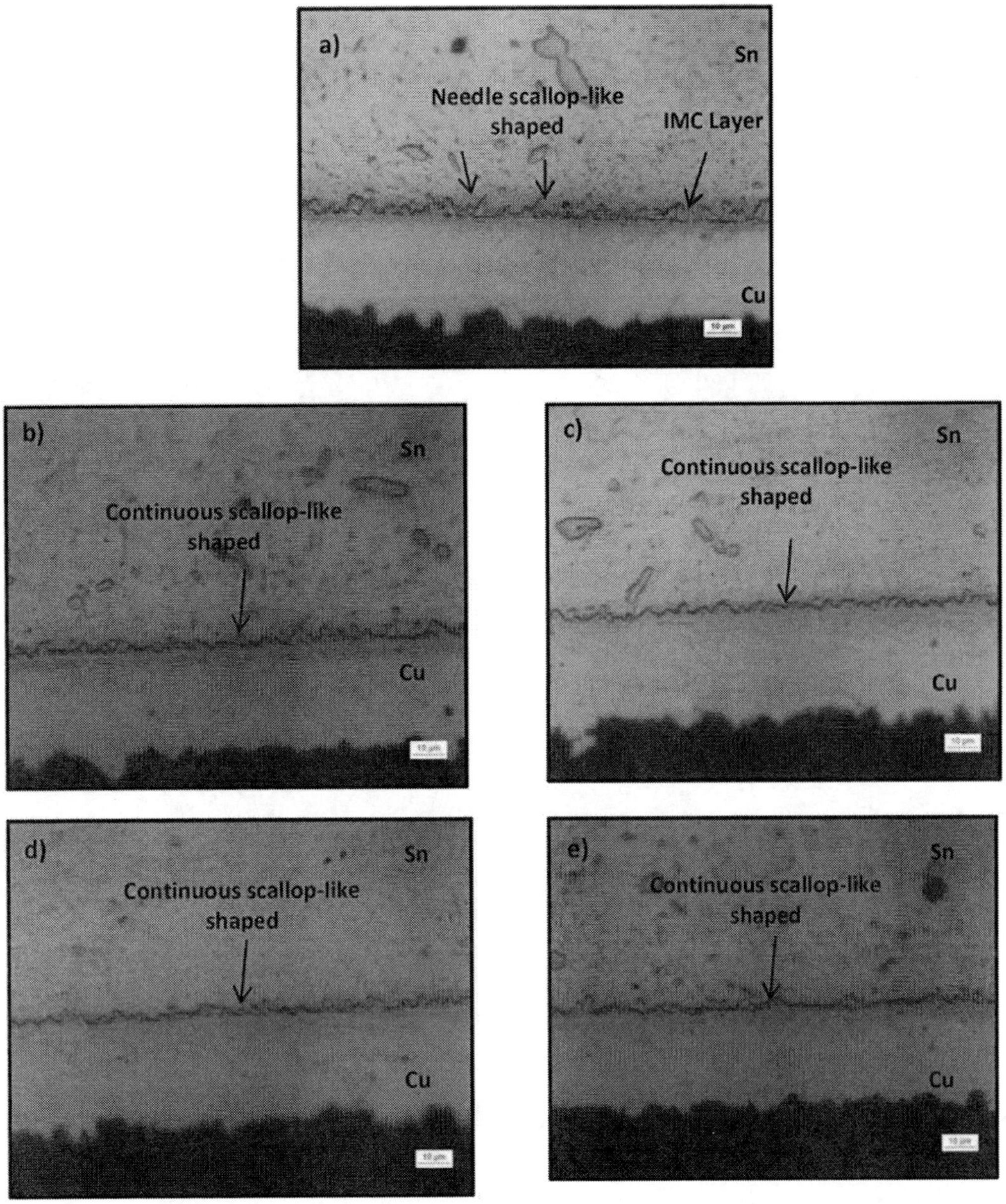

Fig.2. Micrograph of cross-sectional view of interfacial IMC (a) Sn-Cu-Ni, (b) Sn-Cu-Ni-0.25AC, (c) Sn-Cu-Ni-0.5AC, (d) Sn-Cu-Ni-0.75AC and (e) Sn-Cu-Ni-1AC.

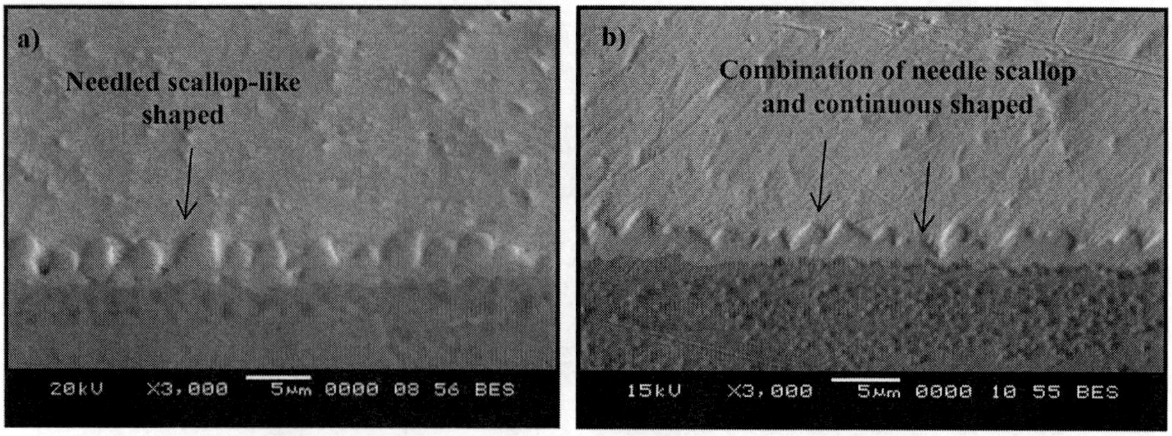

Fig.3. SEM Micrograph of cross-sectional view of interfacial IMC a) Sn-Cu-Ni b) Sn-Cu-Ni-1.0AC.

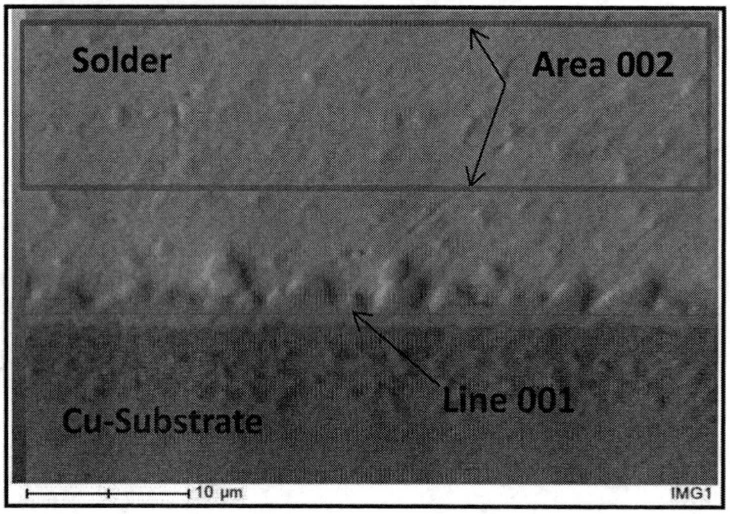

Fig. 4. Cross-sectional view of IMC layer Sn-Cu-Ni-1.0AC.

Table 1. EDX analysis of IMC layer Sn-Cu-Ni-1.0AC.

	Element	Weight Percent (%)	Atomic (%)
Line 001	C	2.24	13.52
	Sn	47.71	29.22
	Cu	50.05	57.26
Area 002	C	1.20	10.68
	Sn	98.23	88.37
	Cu	0.57	0.95

3.2 Shear strength

Strength of the solder joint basically depends on the IMC formation at the interface [13]. Although the intermetallic compound is necessary for solder joint, excessive intermetallic thicknessis need to be avoided because of their brittle properties. As shown in Fig. 5, the shear strength value increase linearly with increases weight percentages of AC particles. As the weight percentages of AC approach 1.0 wt. %, the maximum value of shear strength has reached 13.966 MPa. The factor that attributed to the increases of shear strength was cause by the decreases of IMC thickness and dislocation interaction to slide or slip. When stress being applied to the composite solder joint, the existence of AC will resist the grain boundary sliding.

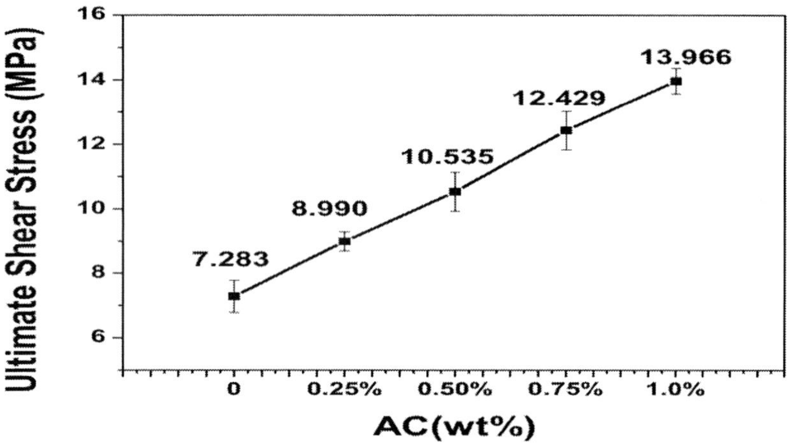

Fig. 5. Shear strength of Sn-Cu-Ni-xAC composite solder joint.

4 Conclusion

In this research, the influence of AC particles addition on intermetallic compound and shear strength of Sn-Cu-Ni lead free solder are studied and following conclusion can be made:

a) The addition of AC as reinforcement reduces the intermetallic compound (IMC) thickness and suggested that AC act as a barrier to retard the growth of the formation IMC layer.
b) The addition of AC particle improved the shear strength of the Sn-Cu-Ni/xAC solder joint.

The authors gratefully acknowledge the financial support from Nihon Superior Co. Ltd., Japan under Grant No. 9008-00003 and the School of Material Engineering, University Malaysia Perlis (UniMAP) for supporting this research effort through materials and facilities. The support of Associate Professor Kazuhiro Nogita from the University of Queensland for his valuable advice throughout this study is greatly appreciated.

References

1. G. Zeng, S. Xue, L. Zhang, L. Gao, J. Mater. Sci-Mater. El., **22**, 565 (2011)
2. T. Ventura, C.M. Gourlay, K. Nogita, T. Nishimura, M. Rappaz, A.K. Dahle, J. Electron Mater., **37**, 32 (2008)

3. Z. Liang, T. Lei, G. Yonghuan, S. Lei, M. Yong, J. Rare Earth, **32**, 1184 (2014)
4. Y. Tang, G.Y. Li, Y.C. Pan, Mater. Design, **55**, 574 (2014)
5. S.M.L. Nai, J. Wei, M. Gupta, Thin Solid Films, **504(1)**, 401 (2006)
6. M.I.I. Ramli, N. Saud, M.A.A.M. Salleh, M.N. Derman, R.M. Said, N. Nasir, Appl. Mech. Mater., **754**, 518 (2015)
7. M.A.A.M. Salleh, A.M.M. AlBakri, M.H. Zan. Hazizi, F. Somidin, N.F.M. Alui, Z.A. Ahmad, Mat. Sci. Eng. A-Struct., **556**, 633(2012)
8. J. Shen, Y.C. Chan, Microelectron Reliab., **49**, 223 (2009)
9. S.M.L. Nai, J.V.M. Kuma, M.E. Alam, X.L. Zhong, P. Babaghorbani, M. Gupta, J. Mater. Eng. Perform, **19(3),** 335 (2010)
10. S.L. Tay, A.S.M.A. Haseeb, M.R. Johan, Solder Surf. Mt. Tech., **23(1),** 10 (2011)
11. S.L. Tay, A.S.M.A. Haseeb, M.R. Johan, P.R. Munroe, M.Z. Quadir, Intermetallics, **33**, 8 (2013)
12. L.M. Lee, A.A. Mohamad, Adv. Mater. Sci. Eng., **2013(4)**, 1(2013)
13. M.O. Alam, H. Lua, C. Bailey, Y.C. Chan, Comp. Mater. Sci., **45**, 576 (2009)

Review on Potential of Geopolymer for Concrete Repair and Rehabilitation

A.Z. Warid Wazien[1,2,*], *Mohd Mustafa Al Bakri* Abdullah[1,2], *Rafiza* Abd. Razak[1,2], *Mohd M.A.Z.* Remy Rozainy[3], *Muhammad Faheem* Mohd Tahir[1], *M.A.* Faris[1], and *Hazamaah* Nur Hamzah[1]

[1]Center of Excellence Geopolymer and Green Technology (CEGeoGTech), School of Materials Engineering, Universiti Malaysia Perlis (UniMAP), P.O Box 77, D/A Pejabat Pos Besar, 01000 Kangar, Perlis, Malaysia
[2]Faculty of Engineering Technology, Universiti Malaysia Perlis (UniMAP), P.O Box 77, D/A Pejabat Pos Besar, 01000 Kangar, Perlis, Malaysia
[3]School of Civil Engineering, Universiti Sains Malaysia (USM), Engineering Campus, 14300 Nibong Tebal, P. Pinang, Malaysia

Abstract. Cracking, spalling, surface deterioration, seepage and other concrete damage of the existing infrastructure have raised concern among residents, local authorities and developers. Dealing with concrete infrastructure rehabilitation is an important issue due to most of infrastructures today are concrete based. To simply patch up the spalling concrete as temporary solution and sooner or later, the cracks will return to haunt. This paper reviewed the current research and studies on concrete repair materials, highlighting the properties of geopolymer. It covers geopolymer repair materials which addressed in the field of concrete infrastructure rehabilitation. Geopolymer had good repair characteristics and displays the potential as an excellent repair material.

1 Introduction

Nowadays, concrete infrastructure rehabilitation is very costly due to various factors. For example, in Malaysia, almost every concrete bridges require maintenance in order to repair and strengthening the structure with an insufficient annual budget [1]. The existing infrastructures were given a little attention to durability issues long years ago and this is one of the factors affecting the damaged concrete structures. However, the continuous development among researchers is a good opportunity to take in order to solve the current issue towards implementing sustainable and cost effective repair material.

Geopolymers or known as alkaline activated material is a newly developed binder which is made from alumina silica source materials activated with an alkaline solution [2-6]. Investigations and studies related to the use of geopolymers in the field of concrete infrastructure rehabilitation are required [7].

* Corresponding author: warid.wazien92@yahoo.com

Instead of complying an excellent mechanical properties as repair material, geopolymer also shows the ability and promoted an excellent bond strength to the old concrete, environmental friendly and even more cost effective. In this preliminary study, the papers related to the used of geopolymer as repair material were reviewed.

Bond strength between repair material and substrate is the crucial property of repair material. Some researchers normally will investigate and make a comparison between geopolymer mortar and commercial products [8-11]. Slant shear test is one of the approached method highlighted by the previous researchers in order to evaluate the performances of the repair material [12]. However, bond strength depends not only on the characteristics of the repair material but also on the surface roughness of concrete substrate. Roughness or called as surface treatments will increase the bonding between repair material and the substrate.

Some researchers published their studies on bonding of repair materials to a concrete substrate where the preparation of the substrate surface with different techniques is mentioned [13,14]. However, the differences also been observed on the concrete mix of the repair materials, the age of the specimens, the curing scheme, and the repair technique itself. The purpose of this paper is to give an overview of the most recent studies using geopolymer as a repair material at different condition and approached.

2 Geopolymer repair material

In recent years, the rapid growth in research and development related to geopolymer has indicated the potential to implement at various applications and so as in the field of concrete infrastructure rehabilitation. A number of researchers explored the geopolymer as a repair material. The bond strength of the geopolymer is reported higher bonding strength than comparable portland cement based repair [4], commercial repair products [5,6] and repair epoxies [7]. Moreover, geopolymeric binders had very high bond strength even at early ages. Fly ash based geopolymer shows the potential to be developed as an excellent repair materials so far. In addition, the interface zones between between substrate and geopolymer were homogenous at contact zone. Instead of engineering properties is comparable to the existing repair material, the cost of the geopolymer is also reported to be the cheapest [5].

However, the value of geopolymer as repair material could be enhanced and applied in a various applications. Geopolymer is used as repair and strengthening material in the forms of slurry for crack injection, mortar for section restoration, jacketing, filling and etc, and as concrete for section restoration and jacketing. Different application required different properties of repair material [11]. However, good bonding between repair materials and concrete substrate is very important in the concrete repairs.

3 Patch repair

This method is widely used to restore the original conditions of the concrete structures. The compatibility between repair material and concrete substrate required some specific requirement (Table 1). The method to patch the spalling concrete starts with preparing the concrete surface and end up with curing process (Fig 1).

Table 1. Compatibility between substrate and repair material [15]

Properties	Relationship between repair mortar (Rm) and concrete substrate (Cs)
Compression, tension and flexure strength	Rm ≥ Cs
Modulus in compression, tension and flexure	Rm ≈ Cs
Coefficient of thermal expansion	Rm ≈ Cs
Adhesion in tension and in shear	Rm ≥ Cs
Curing and long term shrinkage	Rm ≥ Cs
Strain capacity	Rm ≥ Cs
Creep	Dependent on whether creep causes undesirable or desirable effects
Fatigue performance	Rm ≥ Cs

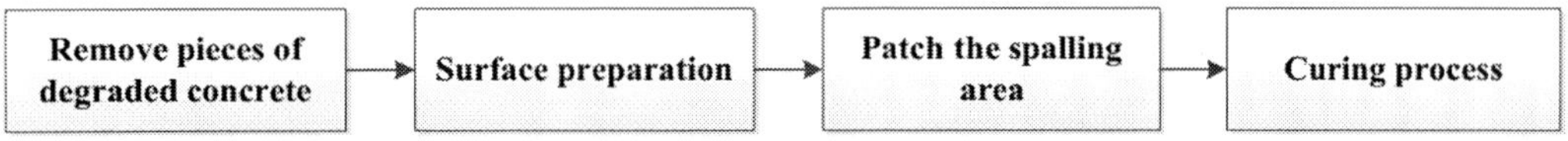

Fig. 1. General patch repair procedure

Roughning the surface of the concrete substrate is the factor affecting the performance of repair material instead of cleaning the concrete substrate surface. The different types of surface preparations has been investigated to improve the bonding between substrate and repair material. The bonding strength between repair material and concrete substrate was evaluated using slant shear test method as per ASTM C882 (Fig 2).

The bond strength (S) for the slant shear strength was calculated by dividing the maximum load (P) by the bond area (A). The ACI Concrete Repair Guide specifies the acceptable bond strength for repair work should be in the ranges between 6.9 – 12 MPa [16]. Instead of the shear strength value, the bond failure modes can be categorised into several types (Fig. 3).

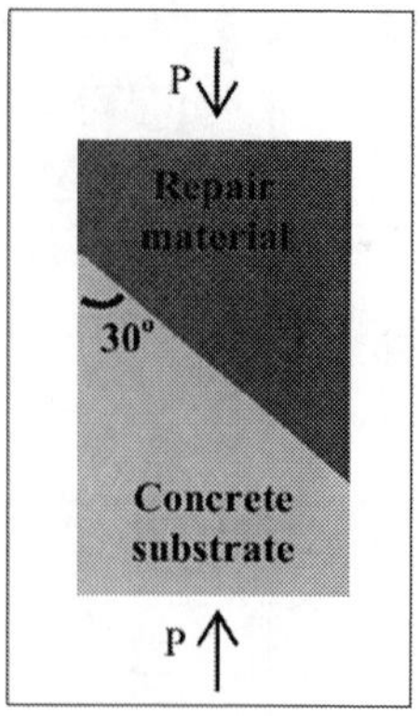

Fig. 2. Schematic diagram of the slant shear test

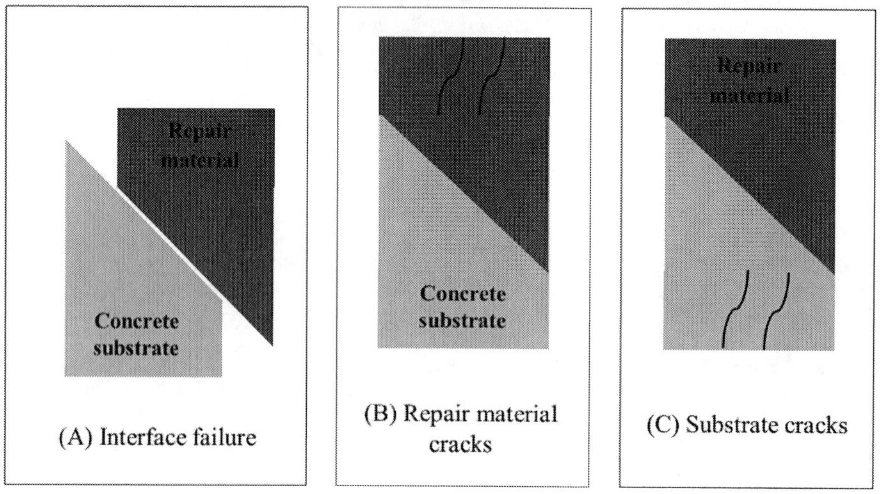

Fig. 3. Failure type

4 Surface preparation

In the field of rehabilitation and strengthening the bond strength between the substrate and repair material generally presents a weak bond [12]. A good bond is one of the main requirements for successful repair. However, the surface preparation of the concrete substrate will courage and enhance the bonding [17].

Some researchers reported that sand blasting is the best surface preparation methods. However, sand-blasting was attributed to the polishing effect. Other surface preparation such as grinding, wire-brushing, drill holes, and etc is done by some researchers. Bassam et al. [18] indicate that low bond strengths were obtained with substrate surfaces treated by this method except for sand blasting. Chemical products are also used to increase the substrate surface roughness. Pacheco et al. [9] mentioned an acid etching method is used to prepare the bond surface, but carefully washed is required to ensure the removal of $CaCl_2$ which results from the reaction between HCl and $Ca(OH)_2$.

Table 2. Different surface preparation of the concrete substrate

Repair materials	Samples	Description
Ultra High Performance Fiber Concrete [18]	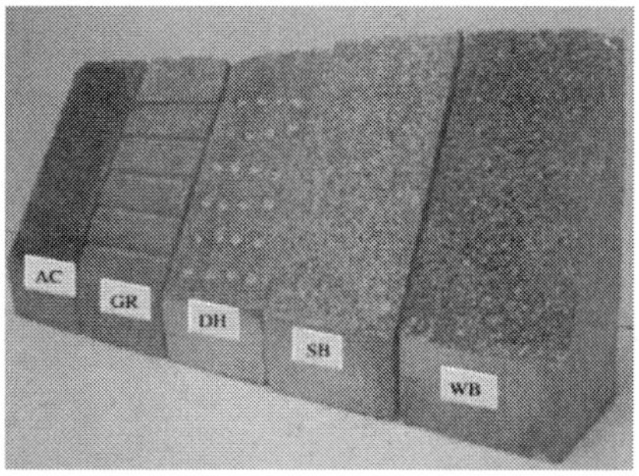	Slant shear test specimens with five different surface textures namely sand blasting, wire brushing, drill holes and grooves. Sand blasting shows the best.
Geopolymer [9]	 	Substrate concrete with etched surface and cast against metallic formwork. Bond strength is not affected by low roughness surface treatment of concrete substrate.
Portland cement [19]		Smooth as-sawn, wire-brushed, hand-chiselled, and acid etching were used in order to prepare the surface of old concrete samples. Acid etching is the best.

5 Conclusion

The aims of this paper are to overview the current research of repair materials. Thus, with the sustainable and comparable properties, geopolymer display a high potential for improvement and developed as an excellent repair material. Continuous investigation related to the used of geopolymer is required in order to champion in the concrete infrastructure rehabilitation.

The author would like to express his gratitude to the Center of Excellence Geopolymer & Green Technology (CEGeoGTech) for funding the research project.

References

1. Seminar on Practical Approaches towards Assessment, Repair and Strengthening of Crack Damaged Concrete Structures, *Practical Guideline for Investigation, Repair, and Strengthening of Cracked Concrete Structures* (Japan Concrete Institute, 2015)
2. M.M.A. Abdullah, H. Kamarudin, M. Bnhussain, I. Khairul Nizar, A.R. Rafiza, Y. Zarina, Adv. Mat. Res., **328**, 1475 (2011)
3. Y. Zarina, A.M. Mustafa Al Bakri, H. Kamarudin, I.K. Nizar, A.R. Rafiza, Rev. Adv. Mater. Sci., **34**, 37 (2013)
4. W.K. Part, M. Ramli, C.B. Cheah, Constr. Build. Mater., **77**, 370 (2015)
5. S.K. Nath, S. Maitra, S. Mukherjee, S. Kumar, Constr. Build. Mater., **111**, 758 (2016)
6. I. Khairul Nizar, A.M.M. Al Bakri, A.R. Rafiza, H. Kamarudin, A. Abdullah, Y. Zarina, Key Eng. Mater., **594**, 985 (2013)
7. F. Pacheco Torgal, Z. Abdollahnejad, S. Miraldo, S. Baklouti, Y. Ding, Constr. Build. Mater., **36,** 1053 (2012)
8. S. Hu, H. Wang, G. Zhang, Q. Ding, Cem. Concr. Compos., **30**, 239 (2008)
9. F. Pacheco Torgal, J.P. Castro Gomes, S. Jalali, Constr. Build. Mater., **22**, 154 (2008)
10. S. Songpiriyakij, T. Pulngern, P. Pungpremtrakul, C. Jaturapitakkul, Mater. Design, **32**, 3021 (2011)
11. T. Phoo-ngernkham,V. Sata, S. Hanjitsuwan, C. Ridtirud, S. Hatanaka, P. Chindaprasirt, Constr. Build. Mater., **98** 482 (2015)
12. A. Momayez, M.R. Ehsani, A.A. Ramezanianpour, H. Rajaie, Cement Concrete Res., **35**, 748 (2005)
13. G. Xiong, J. Liu, G. Li, H. Xie, Cement Concrete Res., **32**, 1877 (2002)
14. E.N.B.S. Júlio, F.A.B. Branco, V.D. Silva, Constr. Build. Mater., **18**, 675 (2004)
15. D.R. Morgan, Constr. Build. Mater., **10**, 57 (1996)
16. G. Chynoweth, R.R. Stankie, W.L. Allen, R.R. Anderson RR, W.N. Babcock, P. Barlow, Concrete Repair Manual, ACI Committee, **546**, 287 (1996)
17. S. Austin, P. Robins, Y. Pan, Cement Concrete Res., **29,** 1067 (1999)
18. B.A. Tayeh, B.H. Abu Bakar, M.A. Megat Johari, Y.L. Voo, Procedia Eng., **54**, 554 (2013)
19. K. Behfarnia, H. Jon nesari, A. Mosharaf, Asian Journal of Civil Engineering, **6**, 267 (2005)

Effect of Glass Reinforced Epoxy (GRE) pipe filled with Geopolymer Materials for Piping Application: Compression Properties

Mohammad Firdaus Abu Hashim[1,*], *Mohd Mustafa Al Bakri* Abdullah[1,2], *Che Mohd Ruzaidi* Ghazali[1,2], *Kamarudin* Hussin[1,2], *Mohammed* Binhussain[3] and *Mohd Firdaus* Omar[1]

[1]Center of Excellence Geopolymer & Green Technology (CeGeoGTech), School of Material Engineering, Universiti Malaysia Perlis (UniMAP), Perlis Malaysia.
[2]Faculty of Engineering Technology, Universiti Malaysia Perlis (UniMAP), Perlis, Malaysia.
[3]King Abdul Aziz City Science & Technology (KACST), Riyadh, Kingdom of Saudi Arabia.

Abstract. The aim of this paper is to achieve the highest compressive strength of glass reinforced epoxy pipe with the geopolymer filler content of weight percentage that were used in glass reinforced epoxy pipe. The samples were prepared by using the filament winding method. The effect of weight percentage of geopolymer materials in epoxy hardener was studied under mechanical testing, which is using the compression test. A series of glass reinforced epoxy pipe and glass reinforced epoxy pipe filled with 10 - 40 weight percentage geopolymer filler which is white clay were prepared. The compression strength of the glass reinforced epoxy pipe filled geopolymer materials is determined using Instron Universal Testing under compression mode. It was found that compressive strength for samples with white clay geopolymer filler are much higher compare to glass reinforced epoxy pipe without geopolymer filler. Moreover, the compressive strength of glass reinforced epoxy pipe filled with white clay geopolymer filler was increased from 10 wt% to 30 wt% of geopolymer content. However, the compressive strength of glass reinforced epoxy pipe with white clay geopolymer filler suddenly decreased when added to 40 wt%. The results indicated that the blending of geopolymer materials in epoxy system can be obtained in this study.

1 Introduction

Filament winding is a type of composite manufacturing process, where a controlled amount of resin and oriented fibers are wound around a rotating mandrel and cured to produce the required composite part. It was initially used to produce pressure vessels, water and chemical tanks. The process of filament winding has evolved to be the preferred, and most cost effective method, for producing pressure retaining structures from fiber reinforced polymeric (FRP) composites such as piping and tanks for the transportation/storage of

* Corresponding author: firdaushashim88@yahoo.com

fluids [1]. According to material properties combinations have been and yet being extended by the advance of the composite, which is a multiphase material that shows a proportion of the properties of the forming phase so that a better mixture of properties is achieved [2]. Filament winding also can be described as the process of winding fiber material and resin around a shape known as a mandrel to create composite product. The process of filament winding is typically used to produce circular composite products with a hollow core. The filament winding process can consume many different fibers and resins to achieve desired characteristics of the finished component.

The outstanding achievements in inorganic chemistry made through geopolymerization include mineral polymers which termed as polysiatate or geopolymers, making a great possibility to produce composite materials not only with excellent mechanical properties such as lightweight and high strength but also with ideal fire resistant, non-toxic fumes and smokes, and resisting all organic solvents [3-5]. For several years, researchers have been enthralled by the studies on geopolymers since they offer benefits regarding low energy consumption and the absence of CO_2 emissions in the preparation process [6]. Geopolymers are an example of the broader class of alkali-activated binders, which also includes alkali-activated metallurgical slag and other related materials [7]. Composites based on clay have been more widely studied since clay materials are naturally available and contribute to excellent mechanical properties of either thermoplastic or thermosets matrices [8].

In this research, GRE pipe which is diglycidyl ether of bisphenol A (DGEBA) epoxy were mixed with white clay geopolymer filler which became one of the greatest current interests in thermoset-based clay composites since it offers low cost, ease of processing, fine adhesion to many substrates, and good chemical resistance for a wide range of applications [9]. White clay and kaolin are linked with kaolinite. Kaolinite clay is one of most versatile industrial minerals which mostly used as ceramic raw material, coating and filler pigment for paper [6]. Therefore, these materials which fly ash, kaolin, and white clay has wide potential to be used as a source material to react with liquid alkaline activator within sodium silicate solution (Na_2SiO_3) and sodium hydroxide (NaOH) mixture [10].

2 Material and Experimental Details

Materials. The epoxy resin used was Diglycidyl Ether of Bisphenol-A (DGEBA) supplied by Euro Pharma Sdn Bhd and hardener Isophorondiamine (IPDA) was obtained from Dr Rahmatullah Holdings. The raw materials geopolymer filler was produced by Saudi local based materials. Saudi local based material which is White Clay was used in preparation of geopolymer. Geopolymer paste is formed by alkaline activator to induce the silicon and aluminum atoms in the material [11]. An alkaline activator liquid used in this research is a combination of sodium hydroxide (NaOH) with 4 Molar and sodium silicate (Na_2SiO_3).

Table 1. Geopolymer Formulation Design.

Ratio = 1	**Total weight**			
Material	**Material (g)**	**Alkaline Activator (g)**		
		Na_2SiO_3	**NaOH**	
				Molarity
White Clay	400	200	200	4 M

Alkaline activator was prepared by mixing NaOH and sodium silicate before mixed with raw materials to increase the reactivity of solution with concentration 4 M of NaOH to find the optimum concentration. The ratio of raw materials to alkaline activator and sodium silicate to NaOH is 1. The geopolymer pastes then were cured in oven for 24 hours to 48 hours at 60°C – 80°C and were taken out to be crushed by using ring mill. Crushed geopolymer filler were sieved using sieve size 150 µm.

Experimental procedure. The Epoxy Geopolymer was prepared according to the epoxy geopolymer of formulation in the Table 2 below by mechanical mixer using blade stirrer. The obtained mixing epoxy geopolymer were cured with cycloaliphatic amine curing agent IPDA. Epoxy and geopolymer materials were mixed for about 2 hours to make it completely homogeneous and followed by curing agent/hardener for about 5 minutes. After mixing epoxy hardener with geopolymer, the resin is then poured into the resin tank.

Continuous glass fibers (E Type) were impregnated ("wet-out") with geopolymeric resin by means of homemade "impregnation machine" as filament winding technique. The velocity of the fiber during the impregnation process was chosen based on the best penetration on geopolymer resin into the fibers [12]. The winding speed is controlled to generate the desired winding angle patterns. The velocity of the fiber feeding in the resin tank will depend on the mandrel rotational speed. This will be selected based on the best impregnation of geopolymer resin into the fibers.

After the appropriate number of layers has been applied according to fully wounded, filament winding samples were allowed to be cured at the mandrel in the room temperature for 1 day. After the curing process, the sample was ready to test. In order to determine the mechanical properties, several tests are performed on composite structure such as compressive test and elasticity modulus.

Table 2. Epoxy Geopolymer Formulation Design.

Geopolymer Materials (4 M)	Epoxy + Hardener (%)	Geopolymer (%)
White Clay	100	0
	90	10
	80	20
	70	30

3 Result and discussion

3.1 Compression

Compression test was performed according to the ASTM D3410 using Instron Universal Testing Machine. The compression test consists of deforming a cylindrical hollow specimen to produce a thinner cylinder hollow of large diameter. The compression test is a convenient method for determining stress strain response. It may be suitable to use the compression test because the specimen is easy to make and does not need a large amount of materials. In this section, the compressive test results for the entire sample present in three major categorized according to the compression properties. There were compressive strength, compressive strain and modulus. Every compression properties show the performance of the sample.

Figure 1 below is shown the typical curves from the compression strength tests for epoxy hardener and epoxy hardener filled with 10 wt% - 40 wt% of white clay geopolymer filler. It is clearly indicated that epoxy with 30 wt% of white clay geopolymer filler shows the highest compression strength compare to others composition and geopolymer filler. Compression strength of the epoxy hardener filled with geopolymer filler of white clay are found to be exciting when increased slightly with the addition of up to 20 wt% and 30 wt%

of geopolymer materials and it was supported by the rule of mixture theorem given that the addition of micro particles offer greater stiffness than epoxy matrix [13]. However, further increase of geopolymer filler loading at 40 wt% of white clay resulted in a slight decrease of about 2 % in strength; this may be due to smaller agglomerated and disordered nature of clay.

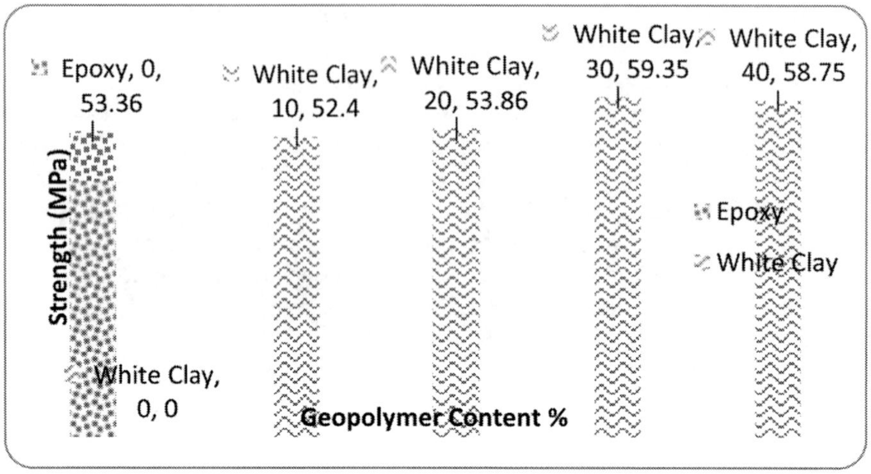

Fig. 1. Compressive Strength of Epoxy Geopolymer 0 wt% - 40 wt%.

The compression properties of GRE pipe filled with geopolymer were observed from the optimum loading of geopolymer filler at 30 wt% of the geopolymer filler loading for 4 M white clay geopolymer filler, which contributed to enhancements in the interfacial bonding between the filler and the matrix, thus increasing the surface area of matrix and filler interaction. This leads to a good stress transfer from the matrix to the micro-filler, thus resulting in improved compressive strength [14]. Compressive modulus of stiffness materials in the static bending condition is tabulated in table 3. It can be seen that the addition of epoxy-based geopolymer filler decreased from 0 wt% of geopolymer filler to 10 wt% but tend to increase the modulus elasticity respectively.

Table 3. Compression Properties of Epoxy Hardener and Epoxy Geopolymer.

	Compressive Strength (MPa)				Compressive Strain (%)				Modulus Elasticity (MPa)			
	10	20	30	40	10	20	30	40	10	20	30	40
Epoxy	53.36				0.04				1681.28			
White Clay	52.40	53.86	59.35	58.75	0.09	0.07	0.06	0.06	1008.82	1290.72	1240.95	1236.88

4 Conclusion

In this study, it was found that geopolymer composite can be as a filler in the piping system application through filament winding technique that not only environmentally product but also can be ease the production cost of the product. Glass Reinforced Epoxy (GRE) pipe filled with white clay based-geopolymer was developed on epoxy resin with different weight percentage of the geopolymer material. The experimental result shows the performance of the product through the compression test. Based on the results of the compression test, the product from white clay with 30 wt% shows the highest strength compare to the others. While on the 10 wt% of geopolymer filler show the lowest strength even compare with GRE without geopolymer filler. Due to the good properties of the waste materials based geopolymer has bigger potential to be a matrix filler of composite with glass fiber in filament winding technique.

We would like to extend our appreciation to the Center of Excellence Geopolymer & Green Technology (CEGeoGTech), King Abdulaziz City for Science and Technology (KACST) and all the people who helped to ensure that the successful completion of this study.

References

1. P. Mertiny, F. Ellyin., Composites, Part A, **33** (2002)
2. P. Karpuz, Middle East Technical University (2005)
3. J. Davidovits, Franca: Institut Geopolymere (2008)
4. J. Davidovits, , *Proceedings of 2005 Geopolymer Conference.* (2005)
5. L. Sheppar, *Proceedings of the 105th.* (2009)
6. M. Wang, D.P. H Jia, , and Y. Zhou., Mater. Lett., **64** (2010)
7. Z. S Yunsheng,. C.Wei, Qianli, and C. Lin., J. Hazard. Mater., **143** (2007)
8. M. Alexandre, P. Dubois., Mater. Sci. Eng., **28** (2000)
9. Rattanasak, U.P. Chindaprasirt., Miner. Eng., **22** (2009)
10. J. Temuujin, A.V. Riessen, K. MacKenzie, Constr. Build. Mater., **24** (2010)
11. D. Tran, D. Kroisová, P. Louda, O. Bortnovsky, P. Bezucha., Manuf. Eng. Mater. Process., **37** (2009)
12. D. Tran, P. Louda, O. Bortnovsky, P. Bezucha, *The 2nd RMUTP International Conference Green Technology and Productivity* (2010)
13. B. Nuhiji, D. Attard, G. Thorogood, T. Hanley, K. Magniez, J. Bungur, B. Fox., Materials, **6** (2013)
14. H. Alamri, I.M. Low, Mater. Des., **42** (2012)

A Review on the Manufacturing of Lightweight Aggregates Using Industrial By-Product

Aiman Mahmad Nor[1,*], *Zarina* Yahya[2], *Mohd Mustafa Al Bakri* Abdullah[2], *Rafiza* Abdul Razak[2], *Januarti* Jaya Ekaputri[3], *M. A.* Faris[1], and *Hazamaah* Nur Hamzah[1]

[1]Center of Excellence Geopolymer and Green Technology, School of Materials Engineering, Universiti Malaysia Perlis (UniMAP), 01007, P.O Box 77, D/A PejabatPosBesar, Kangar, Perlis, Malaysia.
[2]Faculty of Engineering Technology, Universiti Malaysia Perlis (UniMAP), 01007, P.O Box 77, D/A PejabatPosBesar, Kangar, Perlis, Malaysia.
[3]Department of Civil Engineering, ITS, Kampus ITS Surabaya 60111, Indonesia

Abstract: The waste materials from industrial are one of the concern issues in sustainable development. Recycle and reduce the waste material is one of the strategies to produce the better sustainable system. One of the methods is by using the waste materials to produce artificial aggregate. There are wide applications of artificial aggregate, for instance construction, agriculture, geotechnical fill, waste water, drainage and many more. Physical properties such as bulk density, specific gravity, unit weight, porosity and water absorption are important criteria. The performance of the artificial aggregate varies, depending on the application.

1 Introduction

Aggregate are materials that have a vast application in several industries. Many industries are dependent on natural resources to gain their aggregate. Thus, the manufacture of lightweight artificial aggregate is developed in order to avoid exhausting of natural resources. Nowadays one of the major issues that needs to be a concern is to minimize and recycling waste materials from industry. The waste materials from industry can be used in producing lightweight aggregate for the construction industry. The lightweight aggregate also has an application on geotechnical fill, insulation products, soil engineering, hydro-culture, drainage, roof gardens and filter [1] Lightweight aggregate can be produced either from natural rock or manufacture. The typical properties that important for lightweight aggregate are the weight, strength, thermal and acoustical, high fire resistance toughness and depending on the application other properties such as water absorption shrinkage, bonding with cement, elastic properties and abrasion resistance are also considered [2]. There are generally four groups of lightweight aggregate type[2]:

* Corresponding author: aimanorja@gmail.com

1. Natural lightweight aggregate: natural rock that had been crushed, for instance, pumice, scoria, tuff, breccias, and volcanic cinders.
2. Manufactured structural aggregates: by processing shale, clay, or slate into rotary kilns or sintering machine between 1000°C– 1200°C.
3. Manufactured insulating ultra-lightweight aggregates: by-product lightweight aggregates: a waste material like heavy metal sludge, mining residues, palm shell, paper sludge, pet bottles, sewage sludge, steel slag, bottom ash, fly ash, marine clay.

This paper focuses on the method of producing lightweight artificial aggregate using any by-product and their performance and characterization. The manufactured lightweights from natural source materials that most used and most valued in industry are from shale and expanded clay [2]. However, for any manufactured aggregate or artificial aggregate, the key physical properties are bulk density, specific gravity, unit weight, porosity, and water absorption.

Table 1. Standard to Determine the Properties of Aggregate.

Properties Of Artificial Aggregate Lightweight Concrete	Specification and standard
Bulk Density And Void In Aggregate	ASTM C 29
Specific Gravity and Absorption	ASTM C127
Sieve Analysis	ASTM C136
Lightweight Aggregates for Concrete, Mortar And Grout	EN - 13055-1
Lightweight Aggregates for Structural Concrete	ASTM C330
Heat Flow Meter Insulation Test	ASTM C518
Aggregate Impact Value (AIV)	BS 812-112:1990
Aggregate Abrasion Value (AAV)	BS 812-113:1990

2 Method of Production Artificial Aggregate

Basically, production of artificial aggregate is using the same principle which is mixing of raw materials, agglomeration, hardening or binding of the particles and then further processing like curing and sintering. Desire size of artificial aggregate is obtained according to the application by agglomerate any raw waste materials during the production process of artificial aggregate [3]. A process such as sintering, autoclaving and cold bonding are the general 3 methods to produce artificial aggregate after the mixing of raw materials, agglomeration, hardening or binding of the particles. The common method to obtain a lightweight artificial aggregate is through sintering process. There also another researcher such as [4] which has reported on manufacture and performance of cold bonded lightweight aggregate using alkaline activators for high-performance concrete.

2.1 Sintering.

Previous research had used the sintering method for producing lightweight aggregate in a furnace up until 1000^0C - 1200^0C to fuse particles at the points of mutual contact [5-9]. The raw materials usually are pelletizing by mixing with 20 – 30 % water as a binding agent to get the desire and consistency size and dried at 110^0C. There are different types of palletizer machine that can be used during the agglomeration process by granulation, such as disc or pan type, drum type, cone type and mixer type [10]. The plate revolution speeds during pelletize process using rotating disk are also affecting the physical and mechanical properties of the aggregates. These changes are due to increases coalescence process of the particle up to a limit value [11]. The sintering method required high energy, but has high engineering properties depending on agglomerated material properties and process efficiency [12]. There are also fly ash aggregate that using the 2 steps of thermal treatment where the first at 750^{0}C for 10 -15 minutes and second the expansion step at 1150^{0}C-1175^{0}C for 10-15 minutes [6]. The two-step of high heat treatment give more porous and property similar to commercial lightweight aggregate compared to one step of heat treatment.

2.2 Autoclaving.

The autoclaving process is where the chemical such as the cement and lime or gypsum are mixed with the source material in agglomeration stage [13]. After that, the sample area exposed to the autoclaving or cure in pressurized saturated steam at a temperature of 140^{0}C for several hours.

2.3 Cold Bonding.

Meanwhile, the cold bonding method is normal water curing at ambient temperature to bonding the mixing material [5]. The bonding achieved from this method are more rigid and higher drying shrinkage. The disadvantages of this method can be overcome by using compaction during agglomeration techniques. The compaction agglomeration technique is by using uni-directional piston type compaction, roll pressing, extrusion and pellet mills. The fly ash class c aggregate in the study [14] shows that increasing the duration of normal water reduce the water absorption and have close properties with autoclave due to dense microstructure formation.

2.4 Geopolymer.

There also researcher to implement geopolymerisation process bonding during agglomeration by granulation of manufactured aggregates[4,15-18]. Geopolimerisation is an inorganic polycondensation reaction which is yielding three-dimensional zeolitic framework to produce geopolymer cement [16]. The hardening mechanism of geopolymeration process is when the aluminium and silicate oxides in any raw material react with alkali polysilicates. The physical and mechanical properties of the geopolymer artificial aggregate are also affected by the molarities and quantity of the alkaline activator use. The geopolymer fly ash shows an increase in strength after exposing to 1000^{0}C [20]. The porosity of the artificial aggregate is decreasing after the heat treatment is increasing [15]. The source materials also had a huge role in geopolymer properties according to the reactivity and chemical composition. Lower compressive strength can be seen in cold bond geopolymer artificial aggregate using only fly ash show compared with the geopolymer lightweight aggregate blends with ground granulated blast slag and rice husk ash [4].

3 Physical Properties of Artificial Aggregate

3.1 Specific gravity.

Generally, the different source materials to produce artificial aggregate had a different density and specific gravity. The cold bond metakaolin aggregate size of 20 – 10 mm had a specific gravity of 1.83 where as the cement act as the binder [21]. The specific gravity of cold bonded aggregate of increases when the volume of cement proportion uses to produce the aggregate increase [22]. Other than that, the specific gravity of 100% ground granulated blast furnace slag cold bonded process aggregate is higher from the fly ash which is 1.97. The Blaine fineness of fly ash class F also effects on the specific gravity of cold bond artificial aggregate [22]. The saturated surface dry specify gravity of fly ash geopolymer artificial aggregate is 1.89 [18]. Other than that, the lower unit weight of cold bonded geopolymer lightweight aggregate can be obtained by using rice husk ash as source material [4].

The specific gravity of sintered fly ash aggregate is increased from 1.51 to 1.92 when it is exposed higher temperature to 1200^{0}C from 1100^{0}C [23]. The density of the sintered fly ash aggregate also is increase at a higher temperature from 1075^{0}C to 1200^{0}C [6]. The sintered fly ash aggregate specific gravity or density start to decreasing until the temperature are 1075^{0}C -1100^{0}C and then increase when exposed to higher temperature because of the close to open porosity of the fly ash aggregate. The specific gravity of sintered sludge ash is decreased as well when the temperature increasing up to 10900C and start to increase when it exposed to 1100^{0}C [7]. However, the time of the sintered process also affects the specific gravity of the sludge ash aggregates. The specific gravity of sludge ash artificial aggregates is lower when the pellet sintered for 30 minutes than 10 minutes exposed to the same temperature [7].

The composition of the raw material of the sintered artificial aggregates also affects the density. The increasing of the glass content in the aggregate will increase the density [9]. The high calcium content in glass powder act as fluxes which lower the melting point [24]. Thus, the density is increased when it is exposed certain temperature because of the fluxing agent in the aggregate melt and decrease the void of the aggregate. However the density is decreased as the temperature is very high such as the aggregate is melting and bloating. The sintered artificial aggregate from class F fly ash obtained from different power plants also had different density [25].

3.2 Water Absorption.

The water absorption capacity of the aggregate is contributed by the open and closed porosity of the artificial aggregates. The addition of lime and cement affect the physical properties of the fly ash cold bonded aggregates. The water absorption of the fly ash aggregates is decreased by 4-9 %with addition of lime and cement [12]. The water absorption turns out to be lower if the cement content is increased because of the high rate of the pozzolanic reaction produce calcium silicate hydrate making the structure more dense [22]. The addition of ground granulated blast slag also decreases the water absorption of the cold bond geopolymer artificial aggregate [4]. The increase duration of water curing of the cold bonded, steam and autoclaved artificial aggregates shows decreases in water absorption and porosity of the aggregate [14]. In general, the autoclaving aggregates have lower water absorption than steam curing because of the lower porosity.

The agglomeration of class F fly ash aggregates through compression method that been sintered shown that the water absorption and porosity are decreased when the soaking time and temperature are increased from 0.5-2 hours 1000^0C-1150^0C [25]. The porosity and water absorption of fly ash particles reported by [10] are lowest when the specimen are sintered at 975^0C-1050^0C. The maximum value of water absorption is obtained when the artificial aggregate is sintered at 1150^0C and then decreased at a higher temperature. The decreasing of the water absorption is due to decrease of open porosity of the surface of the void by less viscous liquid phase. The sintered artificial aggregates reported by [24] show that the water absorption is high when the specific gravity and have low strength due to porosity. The additions of binder also decrease the water absorption as the aggregate becoming denser, discontinuous porosity and small pore size [24].

3.3 Morphology.

The aggregate morphology can be analysed by a microscopic study using scanning electron microscope (SEM) analysis.

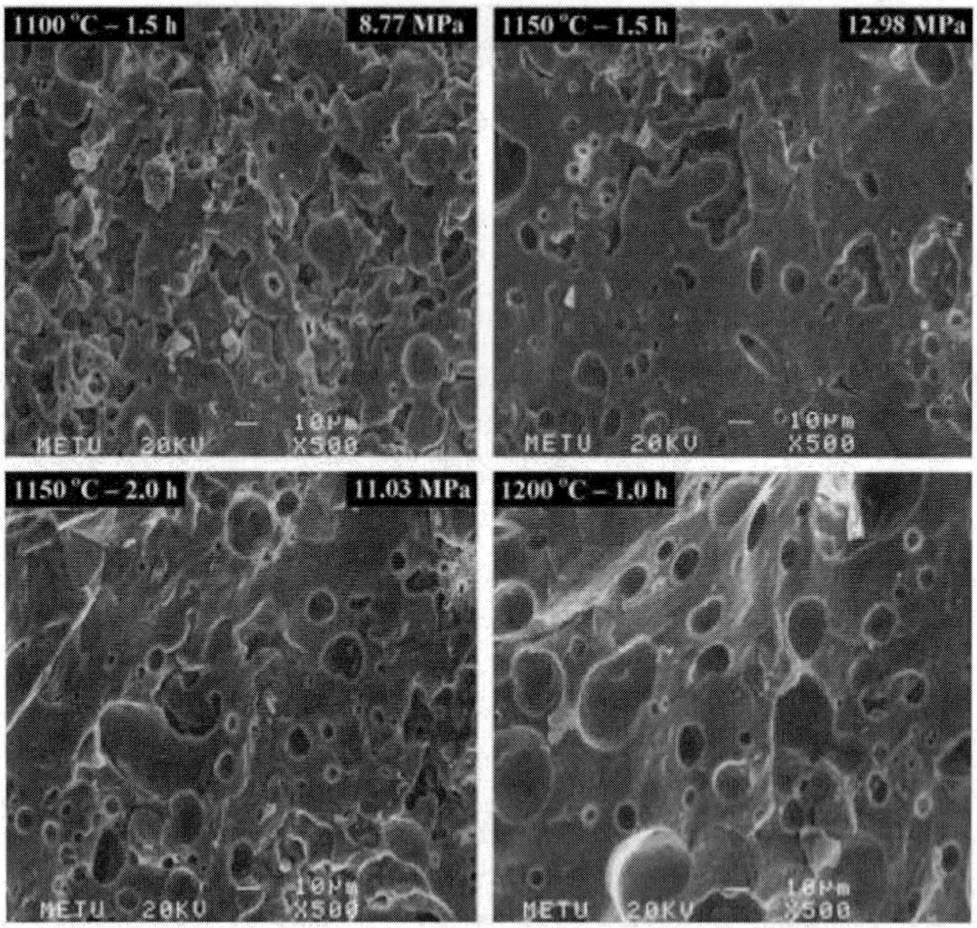

Fig. 1. The SEM images of fly ash sample for different sintering temperature and soaking time [25].

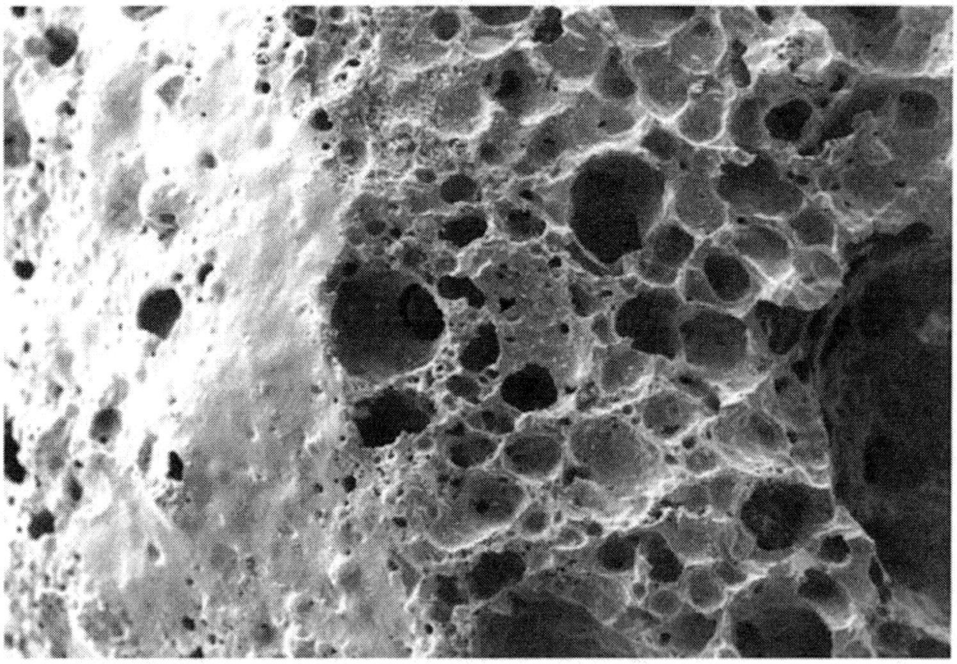

Fig. 2. The morphology of the fly ash lightweight aggregate [6].

The porosity and the formation of the binder can be seen by using SEM analysis. The formation of the pore and binder can be analysed corresponding to the various applications of the manufactured aggregate.

4 Application

The aggregate has been an important material in various industries as it can be used in may apply depending on the their properties. The by-product such as fly ash been mostly researched to be use for structural concrete. The artificial lightweight aggregate as well can be grouped as a granular thermal insulation product as it has good thermal insulation.

4.1 Lightweight concrete.

The construction industries is very important as all other industry will need infrastructures in order for development to occur. The important materials in the infrastructures are concrete. The lightweight concrete had been proving more advantages than normal weight in terms of structural load-bearing, as acoustic, thermal insulation and less cost for transportation and building as the dead load from superstructure are decreased[26-28]. There are several ways to produce a light weight concrete. A common method to produce lightweight building materials is by usage of lightweight aggregates [27]. The density of a concrete are reduced because of the air void exist in the lightweight aggregates that replaced the fine aggregate and normal weight course aggregate. Foamed concrete also categorised as lightweight concrete, which is produced by the use of foaming agent to entrap large volume of air into a cement paste or mortar [29].

The concrete performance is affected by several aggregate properties like particle shape and surface texture, relative density, bulk density, the strength of lightweight aggregates, total porosity, grading, moisture content and absorption, modulus of elasticity of lightweight aggregate particles [30]. The aggregate properties affect the performance of the concrete in workability, proportions of mixing, density, and strength.

4.2 Insulation.

The thermal conductivity coefficient is the basic term for classifying the thermal insulation materials. The unit of the thermal conductivity coefficient is a Kelvin meter per watt W/m K which can be defined as the amount the heat to pass through a certain unit area and thickness in unit time when the temperature difference between two surfaces is 1 Kelvin. The thermal conductivity of a lightweight aggregate concrete is lower than normal concrete by nearly 50% because of the decrease of the unit weight [31]. The thermal conductivity of normal concrete is in the range of the 0.62 to 3.3 W/m K, meanwhile the lightweight concrete are 0.4 – 1.89 W/m K [32].

5 Conclusion

The industrial waste can be utilised as aggregate as the aggregate has a very large application. The understanding the properties of by the product artificial lightweight aggregate are needed so that the performance can be better than natural aggregate. The use of artificial can provide more sustainable to this world and maybe better performance to traditional system which use the natural aggregate.

References

1. C.R. Cheeseman, A. Makinde, S. Bethanis, Resour. Conserv. Recy., **43,** 147 (2005).
2. A.L. Bush, D.P. Bryan, D.R. Hack, Struct. Eng., **88,** 14 (2010).
3. P. Priyadharshini, G.M. Ganesh, S. Sathi, Int. J. Earth Sci. Eng., **05,** 540–546 (2012).
4. C.L.A. Bui, Hwang, C. Chen, K. Lin, M. Hsieh, Constr. Build. Mater., **35,** 1056–1062 (2012).
5. J.M. Bijen, Int. J. Cem. Compos. Light. Concr., **8,** 191–199 (1986).
6. M. Aineto, A. Acosta, J.M. Rincon, *World of Coal Ash Conference (WOCA)* (2005).
7. J.I. Bhatty, K.J. Reid, Waste water. Res., **7,** 363–376 (1989).
8. C.R. Cheeseman, G.S. Virdi, Resour. Conserv. Recycl., **45,** 18–30 (2005).
9. I. Kourti, C.R. Cheeseman, Resour. Conserv. Recycl., **54,** 769–775 (2010).
10. A.Sivakumar, P.Gomathi, J. Civ. Eng., **3,** 42 (2012).
11. F.Colangelo, R. Cioffi, Materials (Basel),. **6,** 3139 (2013).
12. G. Baykal, A. G. Döven, Resour. Conserv. Recycl., **30,** 59 (2000).
13. I. B. Topçu, T. Uygunoglu, Build. Environ., **42,** 4108 (2007).
14. R. Manikandan, K. Ramamurthy, Cem. Concr. Compos., **30,** 848 (2008).
15. A. Abdullah, M. M. A. B. Abdullah, K. Hussin, C.M. Ruzaid, Y. Zarina, M.A.A.M. Salleh, N. Zakaria, M.F.M. Tahie, Key Eng. Mater., **594-595,** 151–155 (2013).
16. R. A Razak, M.Mustafa, A. L Bakri, K. Hussin, K. N. Ismail, Rev. Chim., **7** , 828 (2014).
17. R.Razak, A.M Mustafa Al Bakri, K. Hussin, K.N. Ismail, D. Hardjito, Y. Zarina, Int. J. Mol. Sci., **16,** 11629 (2015).
18. P. Gomathi, A. Sivakumar, Arch. Civ. Eng., **60,** (2014).
19. J. Davidovits, First Int. Conf. Alkaline Cem. Concr., 131 (1994).
20. M.M.A.B. Abdullah, Fly Ash Porous Material using Geopolymerization Process for High Temperature Exposure. Int. J. Mol. Sci., **13,** 4388 (2012).

21. D.V.B. Desai, A. Sathyam, D.Sireesha, IOSR J. Mech. Civ. Eng., **11,** 30 (2014).
22. M. Gesoglu, E. Güneyisi, H. O. Oz, Mater. Struct., **45,** 1535 (2012).
23. N.U. Kockal, T. Ozturan, J. Hazard. Mater., **179,** 954 (2010).
24. N.U. Kockal, T. Ozturan, Cem. Concr. Compos., **33,** 61 (2011).
25. I. Acar, M.U. Atalay, Fuel, **106,** 195–203 (2013).
26. R. Arellano Aguilar, O. Burciaga Diaz, J. I. Escalante García, Constr. Build. Mater., **24,** 1166 (2010).
27. O. Abdulkareem , M. Abdullah, K. Hussin, K. Ismail, M. Binhussain, Materials, **6,** 4450 (2013).
28. A.R. Rafiza, Adv. Mater. Res., **626,** 892–895 (2012).
29. M.M.Al Bakri Abdullah, *et al.* Int. J. Mol. Sci., **13,** 7186–7198 (2012).
30. J.P. Ries, *et al. Guide for Structural Lightweight-Aggregate Concrete Reported by ACI Committee* 213. (2010).
31. S. Akcaozoglu, K. Akcaozoglu, C. D. Atis, Composites Part B, **45,** 721–726 (2013).
32. T.S. Yun, Y.J. Jeong, T. S. Han, K.S. Youm, Energ Buildings **61,** 125–132 (2013).

Properties of Hooked Steel Fibers Reinforced Alkali Activated Material Concrete

M. A. Faris[1,*], *Mohd Mustafa Al Bakri* Abdullah[2,1], *Khairul Nizar* Ismail[3], *Ratnasamy* Muniandy[4], *Aiman* Mahmad Nor[2,1], *Ramadhansyah* Putra Jaya[5] and *Waried Wazien* A. Z.[2,1]

[1]Center of Excellent Geopolymer and Green Technology, School of Materials Engineering, University Malaysia Perlis, Perlis, Malaysia.
[2]Faculty of Engineering Technology, Universiti Malaysia Perlis (UniMAP), Perlis, Malaysia.
[3]School of Environmental Engineering, University Malaysia Perlis, Perlis, Malaysia.
[4]Department of Civil Engineering, University Putra Malaysia, Selangor, Malaysia.
[5]Faculty of Civil Engineering, Department of Geotechnics and Transportation, Universiti Teknologi Malaysia, Johor Bahru, Malaysia

Abstract. In this study, alkali activated material was produced by using Class F fly ash from Manjung power station, Lumut, Perak, Malaysia. Fly ash then was activated by alkaline activator which is consisting of sodium silicate (Na_2SiO_3) and sodium hydroxide (NaOH). Hooked end steel fibers were added into the alkali activated material system with percentage vary from 0 % - 5 %. Chemical compositions of fly ash were first analyzed by using x-ray fluorescence (XRF). All hardened alkali activated material samples were tested for density, workability, and compression after 28 days. Results show a slight increase of density with the addition of steel fibers. However, the workability was reduced with the addition of steel fibers content. Meanwhile, the addition of steel fibers shows the improvement of compressive strength which is about 19 % obtained at 3 % of steel fibers addition.

1 Introduction

A new kind of materials with cement characteristics known as alkali activated material has been developed in recent years. The production of alkali activation material usually undergoes geopolymerization process where the glassy constituent was altered by a chemical process in well compacted and cement featured composites [1]. Alkali-activated material achieved excellent performance in many areas such fire and acid resistant, early mechanical strength, repairing materials, coating materials, and building materials. The development of alkali activated material as cement in building materials have been started since the 1940s [2]. Industrial trials have confirmed that these kind of materials is good and suitable to be used as cement in building materials [3]. Similar to OPC concrete, alkali-activated material concrete also highly brittle which exhibit poor tensile strength. Many researchers have been done in order to improve the brittleness of alkali activated material

* Corresponding author: meorfaris88@gmail.com

[4-6]. To improve the ductility, fibers inclusions in brittle alkali activated material matrix is an effective method to be applied.

The function of steel fibers is making a bridge between the cracks in the matrix and transfers the applied load to the matrix so that fiber reinforced concrete has better post-crack behavior than normal concrete [7]. The inclusions of steel fibers proved to increase the impact resistance and higher the flexural and tensile strengths [8]. P. S. Song et al. [9] and M. Valle et al. [10] conclude a few factors will enhance the properties of concrete including fiber volume, fiber surface bonding characteristics, and matrix strength. The inclusion of steel fibers was reported to improve the fatigue and dynamic resistance of OPC concrete through the high bond between steel fibers surface and cement paste that has been hydrated during the hardening process [11, 12]. However, there has an issue that needs to consider with the additions of fibers which is the workability of OPC concrete will become lower. This low workability issue influence to higher the bleed water at the top of concrete [13]. However, detailed data regarding alkali activated material reinforced with steel fibers still less and need to be discovered more in order to strengthen the theory. Therefore, this paper goal to investigate the performance of alkali activated material with the additions of steel fibers which is hooked end steel fibers.

2 Materials and Methodology

Alkali-activated material concrete reinforced was produced by mixing between alumino-silicate raw material which is fly ash, alkaline activator, fine aggregate, coarse aggregate and fibers. Raw material (fly ash) was taken from Manjung power station, Lumut, Perak, Malaysia. This fly ash was identified as Class F according to ASTM C618 after chemical composition (Table 1) was analyzing by using XRF. According to chemical composition from Table 1, ratio between silica and alumina (SiO_2/Al_2O_3) of fly ash is 2.64. Meanwhile, total composition of SiO_2 + Al_2O_3 + Fe_2O_3 is 72.98 % with the total CaO is equal to 18.10.

Alkaline activator is consisting of a mixture between (Na_2SiO_3) and (NaOH). The Na_2SiO_3 obtained from South Pacific Chemicals Industry Sdn. Bhd. (SPCI), Malaysia with a ratio of SiO/Na_2O is equal to 3.2. Detail specification of Na_2SiO_3 is summarized in Table 2. Meanwhile, the NaOH pellet is obtained from Formosa Plastic Corporation, Taiwan with a concentration of 12 M (after diluted with distilled water). Details specification of NaOH is summarized as in Table 3.

Table 1. Chemical composition of fly ash.

Component	**SiO_2**	**Al_2O_3**	**Fe_2O_3**	**TiO_2**	**CaO**	**MgO**	**SO_3**	**K_2O**	**MnO**
Percentage (%)	38.80	14.70	19.48	1.02	18.10	3.30	1.50	1.79	0.16

Table 2. Specification of Na_2SiO_3.

Items	**Water (H_2O)**	**Silica**	**Sodium oxide**	**Viscosity at 20 °C**	**SiO_2/Na_2O**	**Molecular weight**	**Specific gravity at 20 °C**
Specification	60.5 %	30.1 %	9.4	0.4 Pa	3.2	122.06 g/mol	122.06 g/mol

Table 3. Specifications of NaOH.

Items	Physical form	Color	Molecular weight	Water solubility	NaOH content (purity)
Specifications	Pellets	White	40 g/mol	100 %	>99 %

Steel fibers used in this study were 3D Dramix Hooked end 80/60 BG produced by Bekaert which conforms to ASTM A820. The steel fiber dimensions and material properties are given details in Table 4. The total amount of steel fibers additions were 0 %, 1 %, 3 %, and 5 % from the total weight of concrete.

Table 4. Properties of steel fibers.

Items	Fiber type	Diameter, d (mm)	Length, 1 (mm)	Aspect ratio, 1/d	Tensile strength (N/mm^2)	Young modulus (N/mm^2)
Specifications	Hook	0.75	60	80	1.225	210

Type of fine aggregate used in this experiment was river sand with a maximum size of 4.75 mm. Meanwhile, coarse aggregate used was limestone with a maximum size of 20 mm. A portion of fine aggregate used in this experiment was 730 kg/m^3, while coarse aggregate used were 1093 kg/m^3.

Sample cube of dimension 100 x 100 x 100 mm were cast for the production of alkali activated material concrete. Each sample was demoulded after 24 hours and cure at room temperature. After 28 days, all hardened alkali activated material samples were tested for density, workability, and compression. Workability was measured by slump test, while compression was conducted by using universal testing machine (UTS).

3 Results and Discussions

Figure 1 shows a result of density for different steel fibers addition starting from 0 %, 1 %, 3 %, and 5 %. The density of all samples seems equivalent to that of conventional concrete which is around 2400 kg/m^3 [14]. Figure 1 shows alkali activated material concrete without steel fibers indicates the lowest density which is 2410 kg/m^3, followed by 1 % of steel fibers addition (2447 kg/m^3), 3 % steel fibers addition (2461 kg/m^3), and 5 % of steel fibers additions shows the highest density (2466 kg/m^3). The additions of steel fibers seem slightly increase the density of alkali activated material concrete.

DOI: 10.1051/matecconf/20167801068

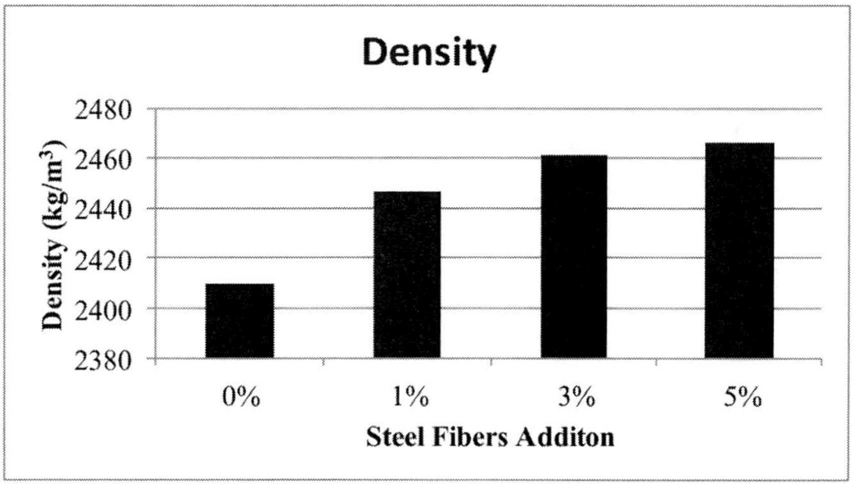

Fig. 1. The density of alkali activated material with the weight percentage of steel fibers addition.

Workability of alkali activated material concrete was measured in terms of a slump by using conventional slump cone apparatus. Slump test was conducted by following ASTM C143. All the mixtures were generally appeared cohesive and shiny due to the effect of reaction in the mixing of alkali activated material. Figure 2 shows the slump test result of alkali activated material concrete without steel fibers was 100 mm. The slump of alkali activated material were decreased with the increasing of steel fibers addition which are 88 mm (1 % steel fibers), 75 mm (3 % steel fibers), and 55 mm (5 % steel fibers). This is shown that the existence of steel fibers will reduce the workability of alkali activated material concrete. The workability of concrete will significantly decrease due to increasing of friction between the steel fibers and the matrix where the fundamentally related theory have been clearly explained by Ranjbar et al. in their research of steel and polypropylene fiber reinforced geopolymer concrete [15].

The effect of steel fibers additions which are 0%, 1 %, 3 % and 5 % to the compressive strength of alkali activated material concrete was plotted in a graph as in Figure 3. From the graph, 3 % of steel fibers addition shows the highest compressive strength which is 69.40 MPa, followed by 1 % of steel fibers (62.41 MPa), 5 % of steel fibers (59.61 MPa), and the lowest value was 0 % of steel fibers (58.32 MPa). This result indicates the contribution of steel fibers addition in order to enhance the compressive strength of alkali activated material concrete. From the graph, the increasing of steel fibers additions will slightly increase the compressive strength of alkali activated material until reached 3 % which is at the optimum value.

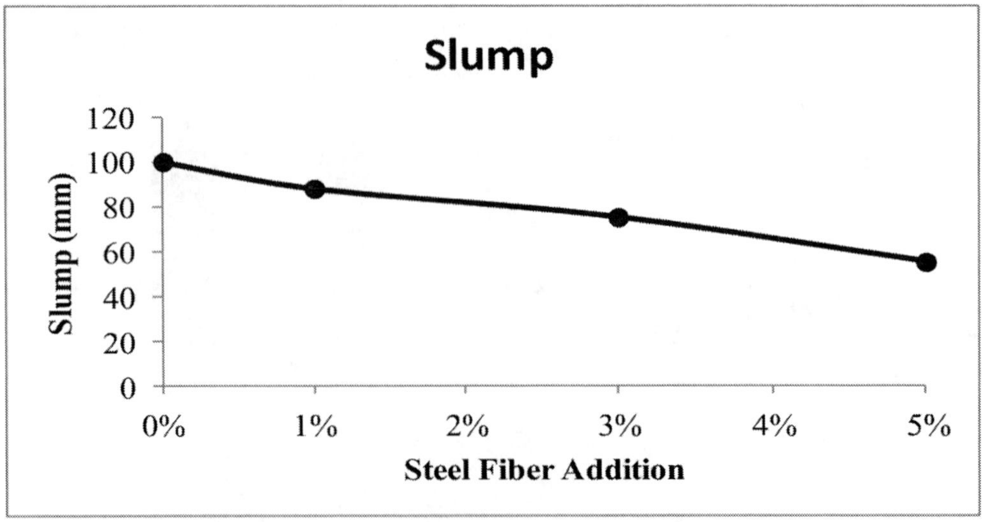

Fig. 2. Slump test of alkali activated material with the addition of steel fibers.

However, steel fibers addition higher than 3 % indicates a reduction in terms of compressive strength. The increasing of compressive strength is due to the functioned of steel fibers addition to improved post-cracking behavior. This is supported by the previous researchers where Cuenca et al. conclude steel fibers act to bridge the crack spot if the total amount of steel fibers is sufficient [16]. The addition of steel fibers in alkali activated material concrete give minor effect to the enhancement of alkali activated material concrete which is about 19 % increment at 3 % steel fibers additions (optimum).

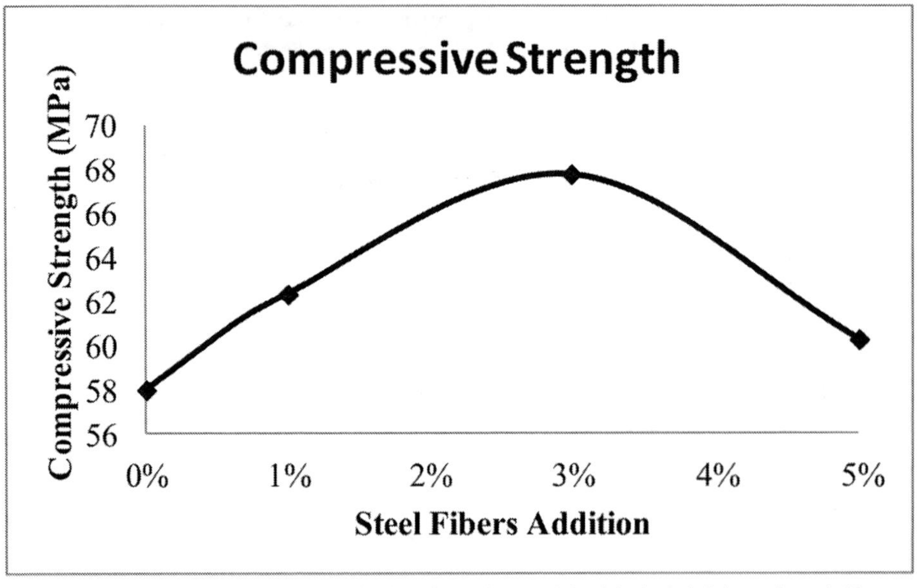

Fig. 3. Compressive strength of alkali activated material with the addition of steel fibers.

4 Conclusion

The inclusion of hooked steel fibers doesn't give significant effect to the density of alkali activated material concrete. There is slightly increasing of density with the increasing of hooked steel fibers addition. Meanwhile, the inclusion of hooked steel fibers give influence to the workability of alkali activated material concrete where the addition of hooked steel fibers reduced the workability of alkali activated material concrete. Additions of hooked steel fibers improved the compressive strength until reached at optimum value which is at 3 % and over than that, the compressive strength slightly getting lower.

Authors gratefully acknowledge the financial support and facilities in order to complete this research provided by Centre of Excellent Geopolymer & Green Technology (CEGeoGTech) and University Malaysia Perlis (UniMAP).

References

1. G. Gorhan, G. Kurklu, Compos. Part B-Eng., **58**, 371 (2014)
2. M. Roy Della, Cement Concrete Res., **29**, 249 (1999)
3. A. Fernandez Jimanez, A. Palomo, C. Lopez Hombradoz, ACI Mater. J., **2**, 106 (2006)
4. S. Karunanithi, S. Anandan, Advances in Civil Engineering, **2014**, 1 (2014)
5. A. Hutagi, R.B. Khadiranaikar, R. Shemblle, Journal of Civil Engineering Technology and Research, **2**, 15 (2014)
6. J. Venkateswararao, R.K. Srinivasa, K. Rambabu, ARPN Journal of Engineering and Applied Sceience, **8**, 800 (2013)
7. Y.Y. Lu, N. Li, S. Li, Constr. Build. Mater., **95**, 74 (2015)
8. H.T. Wang, L.C. Wang, Constr. Build. Mater., **38**, 1146 (2013)
9. P.S. Song, S. Hwang, Constr. Build. Mater., **19**, 669 (2004)
10. M. Valle, O. Buyukozturk, ACI Mater. J., **90**, 122 (1993)
11. S. Wei, G. Jianming, Y. Yun, ACI Mater. J., **93**, 206 (1996)
12. W. Yin, T. Hsu, ACI Mater. J., **92**, 71 (1995)
13. I. Topcu, V. Elgun, Cement Concrete Res., **34**, 275 (2004)
14. R.C. Dorf, *The Engineering Handbook, Second Edition* (CRC Press New York, 2004)
15. N. Ranjbar, S. Talebian, M. Mehrali, Compos. Sci. Technol., **122**, 73 (2016)
16. E. Cuenca, J. Echegaray Oviedo, P. Serna, Compos. Part B-Eng., **75**, 135 (2015)

Morphology and Properties of Geopolymer Coatings on Glass Fibre-Reinforced Epoxy (GRE) pipe

Noor Fifinatasha Shahedan[1,*], *Mohd Mustafa Al Bakri* Abdullah[1], *Che Mohd Ruzaidi* Ghazali[2], *Mohammed* Binhussain[3], *Mohammed* Al Husaini[3], *Kamarudin* Hussin[2] and *Shamala* Ramasamy[1]

[1]Center of Excellence Geopolymer and Green Technology (CEGeoGTech), School of Materials Engineering, Universiti Malaysia Perlis (UniMAP), 01007, P.O Box 77, D/A Pejabat Pos Besar, Kangar, Perlis, Malaysia.
[2]Faculty of Engineering Technology, Universiti Malaysia Perlis (UniMAP), P.O. Box 77, D/A Pejabat Pos Besar, Kangar, Perlis 01000, Malaysia.
[3]King Abdul Aziz City Science & Technology (KACST), P.O. Box Riyadh 11442, Kingdom of Saudi Arabia.

Abstract. Geopolymer coatings were coated on glass fibre-reinforced epoxy (GRE) pipe by using kaolin, white clay and silica sand as source materials and sodium hydroxide (NaOH) and sodium silicate (Na2SiO3) as alkaline solution. The microstructure and mechanical property of geopolymer coating on GRE pipe were methodically investigated through morphology analysis, and flexural strength test. The result indicates the microstructure and interfacial layer between geopolymer coating and GRE pipe significantly influence the mechanical property of geopolymer coating. However, different source materials gave different microstructure and property in geopolymer coating.

1 Introduction

Geopolymer has received much attention as green alternative to ordinary Portland cement (OPC), with most of the work focused on broad applications in transportation, emergency repairs, metallurgy, coating, membrane materials and nuclear waste disposal [1-5]. The workability of geopolymer and manufactured at normal temperature has made geopolymer increasingly being used in fireproofing covers, mending interfaces, coating pipelines and floors, and as adhesives [6, 7]. Geopolymer has displayed an opportunity all together to improve both environmental and engineering performance compared to traditional technology [8, 9]. Due to the excellent resistance to sulphate and seawater attack [10, 11], geopolymer as a novel coating material has been studied for protective marine concretes [12, 13]. Geopolymer coating application has been studied to advance the performance in aircraft [14], construction [15]. Besides, geopolymer have properties of anti-aging and anti-

* Corresponding author: sfifinatasha@yahoo.com

ultraviolet capability that make geopolymer suitable for fabricating inorganic exterior wall building coating [16, 17].

In geopolymer coating, geopolymer source material used as aluminosilicate source to produce geopolymer coating material by using sodium hydroxide (NaOH) and sodium silicate (Na_2SiO_3) as alkaline activator [18-21]. Geopolymer source materials such as fly ash, kaolin, silica fume, ground granulate blast slag (GGBS), boiler ash and volcano ash reactivated via alkaline solution has potential to be use as geopolymer coating. Previous research in geopolymer studied has proved that fly ash was possesses good quality mechanical properties and durability in aggressive environments [19] while silica fume has been research as one of the source materials by adding the silica fume content and test based on the geopolymer porosity, water absorption and strength [22]. Besides that, kaolin as source material in geopolymer has improved mechanical strength [23]. According to the past research, GGBS activated with alkaline solution can be used to make slag concrete with satisfactory workability and strength properties [24].

Hence, these geopolymer coating can be one of the solution to reduce the risk or lack of workability such as on glass fibre-reinforced epoxy (GRE). As the use of these material increases, so does the need to define reliable design limits in terms of performance. Most important arguably, is to predict the long term behaviour of these pipes (20–30 years) under complex loading based on the data recorded from short term failure tests. GRE pipes are normally designed to withstand high pressure. However, as with all pipelines, GRE pipes face significant risks of external interference of pipeline failures and has to spend millions of money in maintenance. Therefore, by applying geopolymer coating it can improvise the mechanical and physical properties of the GRE pipe based industrial needed. The potential of source materials geopolymer such as kaolin, white clay and silica sand has been study for the synthesis of geopolymer in geopolymer coating application.

2 Experimental Setup

Kaolin, white clay and silica sand from Saudi Arabia have been used as raw materials in geopolymer coating paste. Raw materials sample is sieved at 100 μm.

For alkaline activator solution, sodium hydroxide (NaOH) and sodium silicate (Na2SiO3) was prepared according to the Table 1. NaOH in pellet from with (99 %) purity and Na2SiO3 consists of Na2O = 14.7 %, SiO2 = 29.4 % and H2O = 55.9 %.

2.1 Sample preparation

Geopolymer coating paste was prepared by mixing aluminosilicate source materials with alkaline activator solution. Kaolin, white clay and silica sand from Saudi Arabia act as aluminosilicate source materials has been sieved at 100μm in this study. Chemical composition of kaolin, white clay and silica sand as measured by XRF is as follows (Table 1). Dissolution of sodium hydroxide (NaOH) one litre of distilled water in a volumetric flask to obtain a 8 M concentration of NaOH. The ratio of solid/liquid in geopolymer coating paste was fixed at 1.0 [25] for all mixtures with various Na_2SiO_3/NaOH ratio is studied. Sodium hydroxide (NaOH) and sodium silicate (Na_2SiO_3) as alkaline activator was mixed together just before it was mixed with the aluminosilicate source materials which were kaolin and white clay. After that, the alkaline solution was added to kaolin powder and mixed for 2 – 5 minutes to get homogenous mixture. The same step was conducting to the white clay and silica sand powder. Geopolymer coating paste then was applied to the GRE pipe by brushing method. Thickness of coating varied between 0.3 – 0.5 mm. Then sample GRE pipe that has been coated is placed at 75°C in the oven for 48hours. Surface treatment is applied on the GRE pipe sample before coated with geopolymer. GRE pipe sample has been cut with dimension 2.5 cm x 8 cm was cleaned with abrasive paper (grade 80) then washed with detergent, distilled water and methyl ethyl katone (MEK).

Table 1. The chemical composition raw materials of kaolin, white clay and silica sand from

Chemical Composition	Kaolin (%)	White Clay (%)	Silica Sand (%)
SiO_2	50.8	50.8	97.39
Al_2O_3	39.3	39.7	1.88
CaO	0.43	0.37	0.041
Fe_2O_3	2.21	2.13	0.090
MgO	0.45	0.40	0.41
ZrO_2	0.69	0.57	0.0459
TiO_2	5.73	5.6	0.090
K_2O	0.06	0.05	0.027
V_2O_5	0.11	0.11	0.001

3 Mechanical Property of Coating

3.1 Flexural Strength

Flexural strength is important to GRE pipe to overcome the significant risks of external interference of pipeline failures in the ground level based on industrial needed and how it improved with the applied of geopolymer coating. The result stated the highest flexural strength of kaolin, white clay and silica sand geopolymer coating sample was at 0.40 – 0.60 ratio of Na_2SiO_3/NaOH as shown in Figure 1. The coated GRE pipe sample with geopolymer coating has significantly improved the result of flexural strength up to 60 % compare to control sample (without coating) which was 18.57 MPa to the same conditions. The Na_2SiO_3/NaOH ratio was importance for the geopolymerization process and acted as alkaline activator, binder and plasticizer or dispersant [26]. Previous study Na_2SiO_3/NaOH ratios affected the workability of the kaolin geopolymer [25]. The workability of the kaolin geopolymer mixes decreased with increasing in the ratios.

 DOI: 10.1051/matecconf/20167801069

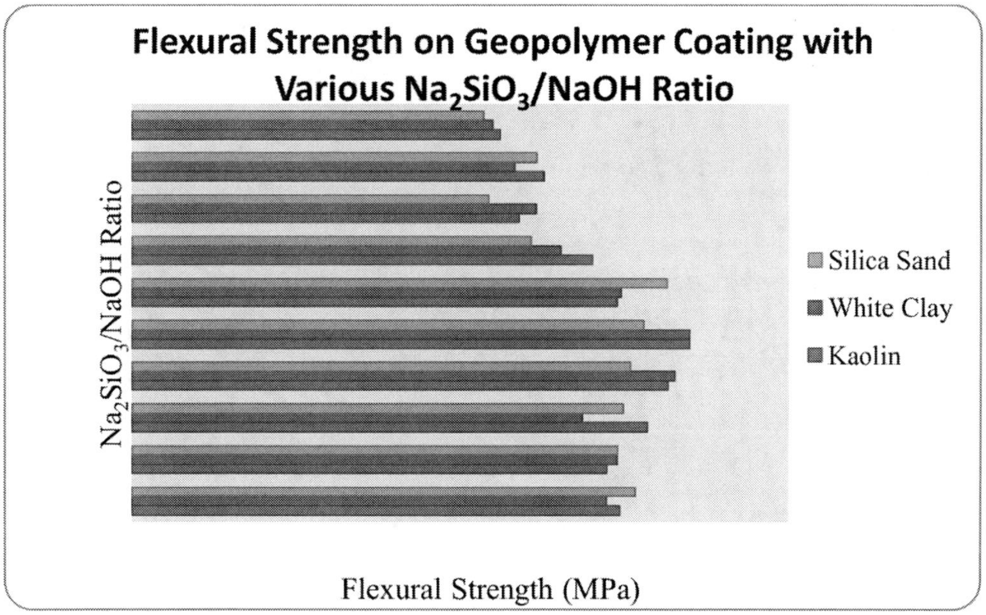

Fig. 1. Effect of $Na_2SiO_3/NaOH$ ratio to flexural strength in geopolymer coating formulation

4 Morphology Analysis

4.1 Scanning electron microscopy

The determination morphology characterization of kaolin, white clay and silica sand has been analyzed on raw materials and geopolymer coating paste. The sample is selected for the best flexural strength kaolin, white clay and silica sand geopolymer coating.

Scanning electron microscopy (SEM) image of raw materials kaolin and white clay demonstrated kaolinite particles of reveal irregular platelets that are arranged in face-to-face patterns as shown in Figure 2 (a) and Figure 3 (a). Pseudo hexagonal kaolinite and some rolled and rough edged kaolinite particles are also observed. The small platelets of submicron size of kaolin and white clay show less than 2 μm. According to the past research, the particle size in the coating application is mainly under 2 μm which lead to the bigger surface for geopolymerization process [27]. However, the rounded corners and spherical quartz particle of different sizes appears in raw materials silica sand can be observed in Figure 4 (a). Kaolin, white clay and silica sand are suitable to be use as main sources materials in geopolymer coating application because of the particles with broken edges appear contributed to the increase of the total surface charge and the reaction during geopolymerization [28].

The microstructure of the best design all geopolymer coating shows the formation of large compacted area compare to the un-reacted particle. The SEM image analysis of kaolin and white clay based geopolymer coating (Figure 2 (b) and Figure 3 (b)) obviously shows the raw particles has been stimulated by the alkaline activator solution with alters at the edges of the plate-like particles and became more compact [28, 29]. The microstructure of silica sand based geopolymer coating showed more homogeneous structure with more intervening materials was examined (Figure 4 (b)). The form of compacted structure proved that the geopolymerization reaction occurred at almost all part of raw particle in silica sand.

Fig. 2(a). SE image of the raw material kaolin: pseudo hexagonal kaolinite in face to face arrangement (→); rolled and rough edged kaolinite (- - -➔); and booklets of kaolinite (-·-·➔)

Fig. 2(b). SE image of the kaolin based geopolymer: compacted area (→); un-reacted particles (- - -➔)

Fig. 3(a). SE image of the raw material white clay: pseudo hexagonal kaolinite in face to face arrangement (→); rolled and rough edged kaolinite (- - -➔); and booklets of kaolinite (-·-·➔)

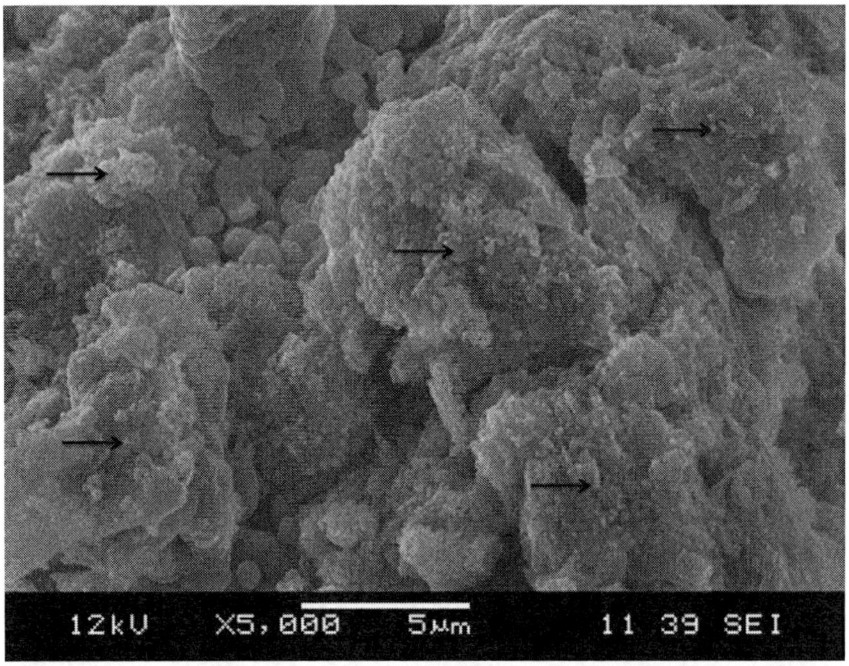

Fig. 3(b). SE image of the white clay based geopolymer: compacted area (→)

DOI: 10.1051/matecconf/20167801069

Fig. 4(a). SE image of the raw material silica fume: rounded corners (→); spherical quartz particle (- - -➧); and stack (-·-·➧)

Fig. 4(b). SE image of the silica fume based geopolymer: compacted area (→) and pore area (- - -➧)

DOI: 10.1051/matecconf/20167801069

4.2 Interfacial study on layer between geopolymer coating and GRE pipe

The optical microscope is used in interfacial study to observe the GRE pipe sample with and without geopolymer coating. Figure 5 (a) is shows the GRE pipe sample without geopolymer coating after flexural strength testing. The interface between kaolin, white clay and silica sand geopolymer coating with the best strength in flexural and GRE pipe sample is observed under optical microscope to distinguish the different. The optical microscope image in Figure 5 (b) – (c) shows the damage and crack on GRE pipe sample is decreased compare to the GRE pipe sample without geopolymer coating. Davidovits [30] propose by varied the Si/Al ratio, it is possible to regulate the properties of geopolymer performance. Therefore, the best mix design for geopolymer coating form rigid final product, improve mechanical properties and formed better interfacial in geopolymer coating.

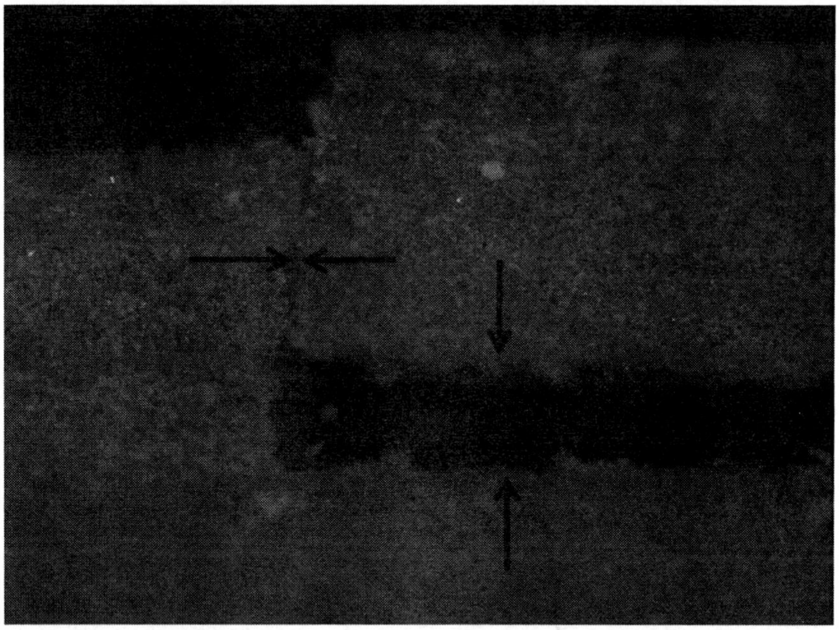

Fig. 5(a). The optical microscope image for GRE pipe sample after flexural testing

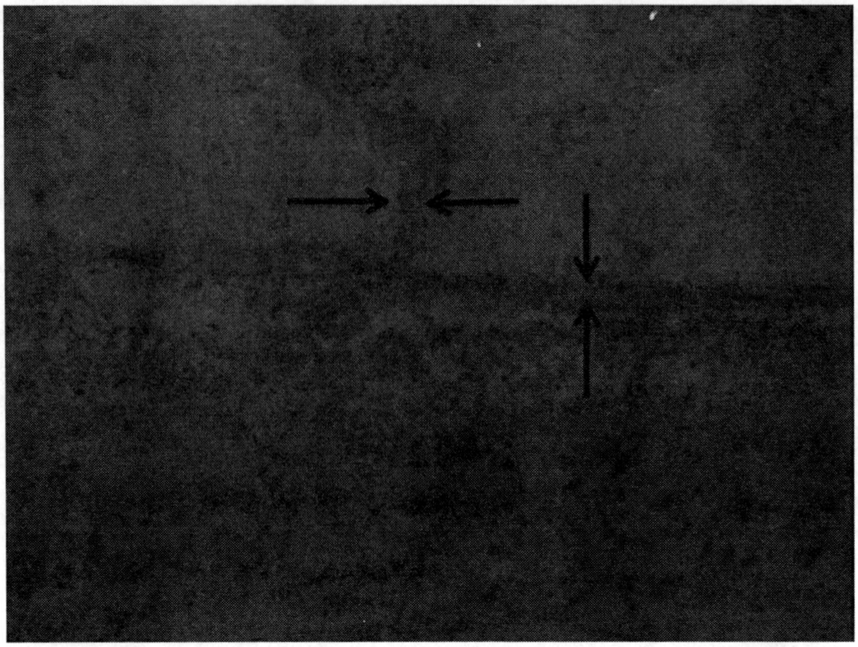

Fig. 5(b). The optical microscope image between kaolin geopolymer coating and GRE pipe sample after flexural testing

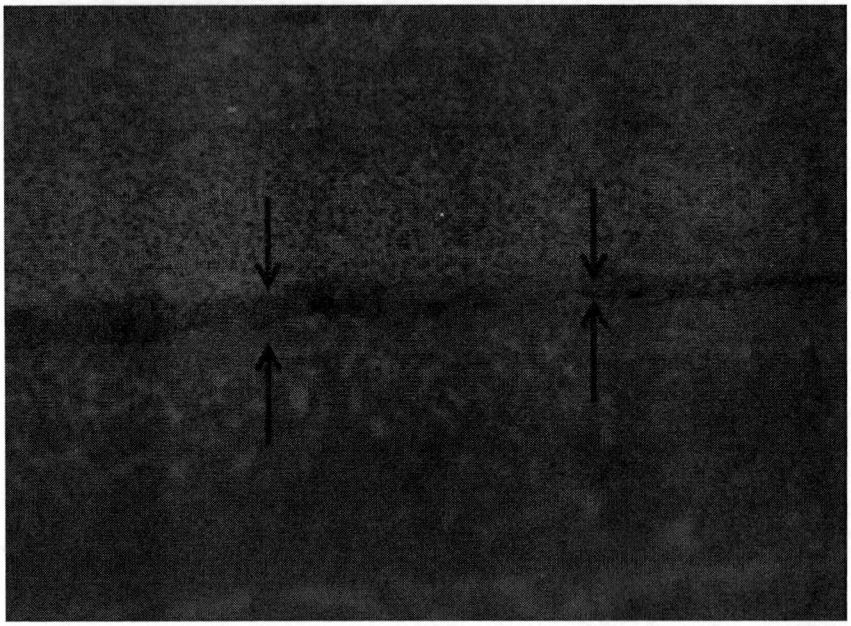

Fig. 5(c). The optical microscope image between white clay geopolymer coating and GRE pipe sample after flexural testing

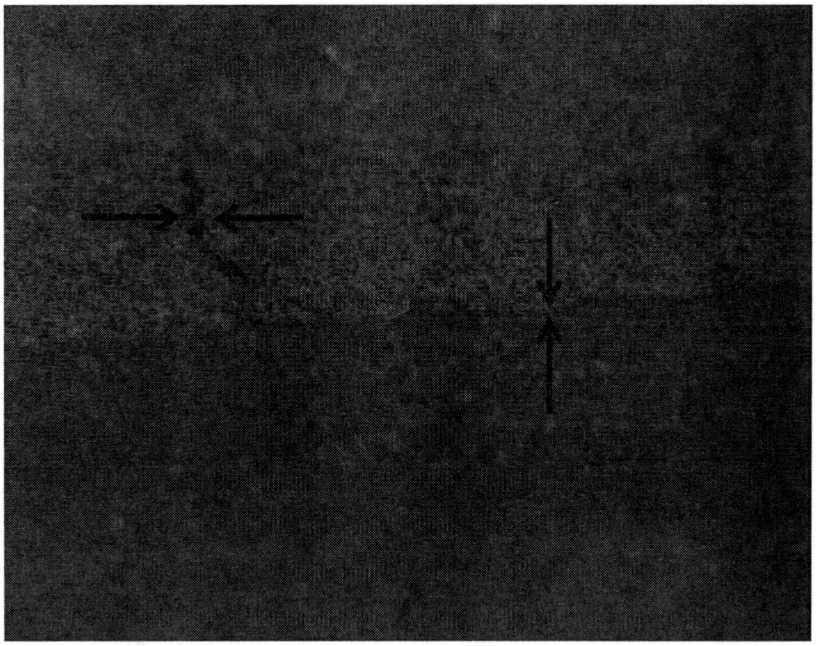

Fig. 5(d). The optical microscope image between silica sand geopolymer coating and GRE pipe sample after flexural testing

5 Summary

As conclusion, the Na_2SiO_3/NaOH ratio play importance role in the mix design for geopolymer coating form rigid final product, improve mechanical properties and formed better interfacial in geopolymer coating. However, different source materials gave different microstructure and property in geopolymer coating.

This work is supported by Centre of Excellence Geopolymer System Research @UniMAP. Also, the authors of the present work wish to dedicate their great thanks to KACST for funding this study through the collaboration between UniMAP-KACST.

References

1. J.G.S. Van Jaarsveld, J.S.J. Van Deventer, A. Schwartzman, Miner. Eng., **12**, 75 (1999)
2. H. Xu, J.S.J. Van Deventer, Int. J. Miner. Process., **59**, 247 (2000)
3. Y. He, X. Cui, X. Liu, Y. Wang, J. Zhang, K. Liu, J. Membr. Sci., **447**, 66 (2013)
4. J.H. Zhang, Mater. Lett., **116**, 167 (2014)
5. Z. Zhang, K. Wang, B. Mo, X. Li, X. Cui, Energy Build., **87**, 220 (2015)
6. J. Davidovits, J. Therm. Anal., **37**, 1633 (1991)
7. J. Davidovits, *Geopolymer Chemistry and Applications 2nd edition* (Geopolymer Institute, 2008)
8. A. Palomo, M. W. Grutzeck, M.T. Blanco, Cem. Concr. Res., **29**, 1323 (1999)

9. P. Duxson, A. Fernandez Jimenez, J. Provis, G. C. Lukey, A. Palomo, J. S. J. Van Deventer, J. Mater. Sci., **42**, 2917 (2007)
10. T. Bakharev, Cem. Concr. Res., **35**, 1224 (2005)
11. A. Fernandez-Jimenez, A. Palomo, Woodhead publishing Limited and CRC Press LLC, 167 (2009)
12. Z. Zhang, X. Yao, H. Zhu, Appl. Clay Sci., **49**, 1 (2010)
13. Z. Zhang, X. Yao, H. Zhu, Appl. Clay Sci., **49**, 7 (2010)
14. G. Bierwagen, D. Tallman, Prog. Org. Coat., **41**, 201 (2001)
15. P. Balaguru, Stephen Kurtz, and Jon Rudolph, *Geopolymer for repair and rehabilitation of reinforced concrete beams* (St Quentin, France, Geopolymer Institute, France, 1997)
16. H.F. Liu, X.M. Cui, W.P. Zhang, W.G. Hua, S.H. Qiu, Rare. Metal. Mat. Eng., 341 (2008)
17. X.F. Li, X.M. Cui, S.D. Liu, B.H. Mo, L. Cui, Key Eng. Mater., **537**, 261 (2013)
18. J. Temuujin, A. Minjigmaa, W. Rickard, M. Lee, I. Williams, A. Van Riessen, Appl. Clay Sci., **46**, 265 (2009)
19. J. Temuujin, A. Minjigmaa, W. Rickard, M. Lee, I. Williams, A. Van Riessen, J. Hazard. Mater., **180**, 748 (2010)
20. A. M. Mustafa Al Bakri, H. Kamarudin, M. Bnhussain, I. Khairul Nizar, W. I. W. Mastura, Journal of Asian Scientific Research, **1**, 247 (2011)
21. A. M. Mohd Mustafa Al Bakr, J. Liyana, H. Kamarudin, H. Mohammad, C. M. G. Ruzaidi C, A. Izzat, Adv. Mat. Res., **686**, 227 (2013)
22. D. Dutta, S. Thokchom, P. Ghosh, S. Ghosh, J. Eng. Appl. Sci., **5**, 74 (2010)
23. H. Xu, J. S. J. Van Deventer, G. C. Lukey, Industrial & Engineering Chemistry Research, **40**, 3749 (2001)
24. E. Douglas E., A. Bilodeau, J. Brandsterr, Cem. Concr. Res., **21**, 101 (1991)
25. C.Y. Heah, H. Kamarudin, A.M. Mustafa Al Bakri, M. Binhussain, M. Luqman, I. Khairul Nizar, Y.M. Liew, Constr. Build. Mater., **35**, 912 (2012)
26. K. Komnitsas, D. Zaharaki, Miner. Eng. **20**, 1261 (2007)
27. L. Huggenberger, A. Manfred, H. H. Köster, *Pigment Coating and Surface Sizing of Paper* (Fapet Oy, Jyvaskyla, 2000)
28. Y.M. Liew, H. Kamarudin, A.M.M. Al Bakri, M. Bnhussain, M. Luqman, I. Khairul Nizar, C.M. Ruzaidi, C.Y. Heah, Phys. Procedia., 22, 312 (2011)
29. R.A. Fletcher, K.J.D. MacKenzie, C.L. Nicholson, S. Shimada, J. Eur. Ceram. Soc., **25**, 1471 (2005)
30. J. Davidovits, *First international conference on alkaline cements and concretes,* **1**, 131 (1994)

An Investigation of TiO_2 Addition on Microstructure Evolution of Sn-Cu-Ni Solder Paste Composite

Norainiza Saud[1,*], *Rita* Mohd Said[1], *Mohd Arif Anuar* Mohd Salleh[1], *Mohd Nazree* Derman[1], *Mohd Izrul Izwan* Ramli[1], and *Norhayanti* Mohd Nasir[1]

[1]Center of Excellence Geopolymer & Green Technology (CeGeoGTech), School of Materials Engineering, Universiti Malaysia Perlis (UniMAP), Taman Muhibah, 02600 Jejawi, Arau, Perlis, Malaysia

Abstract. In this research, varying fraction of titanium oxide (TiO_2) reinforcement particles was successfully incorporated into Sn-Cu-Ni solder paste in an effort to study the influence of TiO_2 addition on microstructure evolution of Sn-Cu-Ni solder paste composite. Sn-Cu-Ni solder paste composite was produced by mixing TiO_2 particle with Sn-Cu-Ni solder paste. The microstructure analysis was carried out by Scanning Electron Microscopy-Energy dispersive X-ray (SEM-EDX). The addition TiO_2 particle helps to refine the bulk solder microstructure and suppress the intermetallic compound (IMC) formation at the interface as will be discussed further.

1 Introduction

In microelectronic manufacturing industry lead solder paste widely used as electronic assembly due to their excellent reliability such as low melting point, better wettability, good mechanical, fatigue and creep properties [1, 2]. Unfortunately, lead is a known as toxic metal in electronic circuit. Since most of the electronic circuit waste could not be recycled and thus they will be dumped in the landfill. This phenomenon will cause lead pollution to the environment, especially soil pollution. Due to the band of lead solder paste usage in electronic assemblies, many research now focusing on lead free solder paste as interconnect material [3].

However, there are another challenge related to lead-free solder paste such as solder melting temperature, processing temperature, wettability, mechanical and thermo-mechanical fatigue (TMF) behaviors [4]. An attractive and potentially viable method of enhancing lead-free solder paste reliability is adopting composite approach to fabricate solder paste composite [5]. Composite solder is solder alloy which is intentionally incorporated with reinforcement particles [6]. Tang et al. [7] have studied the effect of adding TiO_2 nanoparticles on microstructure, microhardness and tensile properties of Sn-3.0Ag-0.5Cu-x TiO_2 composite solder. The microstructure of solder composite becomes

* Corresponding author: norainiza@unimap.edu.my

more uniform and the size of Ag_3Sn and average grain size were decreasing. Apart from that, microhardness and tensile properties of composite solders have been improved.

In the current study, the influence of TiO_2 incorporated on Sn-Cu-Ni solder paste properties have not been studied in detail yet. The experimental data for this solder paste composite is still lacking. Therefore, an established data is needed. This research study was made to investigate the effect of various weight percentages of TiO_2 particles on the microstructure evolution and mechanical properties of Sn-Cu-Ni solder paste composite.

2 Experimental procedures

Sn-Cu-Ni solder paste (Nihon Superior Co. Ltd, Japan) with 20 to 38 μm particle size range and titanium oxide (TiO_2) reinforcement particulates (44 μm particle size from Sigma Aldrich) were mechanically mixed together in a various percentage (0, 0.25, 0.50, and 0.75 wt. %) to form composite Sn-Cu-Ni solder paste. The solder paste and reinforcement powder were pre-weight on a balance weight and mechanical mixing Sn-Cu-Ni solder paste with TiO_2 particles for 30 minutes to ensure the uniform distribution of the reinforcements. To produce a solder joint, about 0.1g of solder paste composite was then deposited on copper substrate and reflow soldering. After the reflow, the solder joint was then cleaned to remove flux residues.

The reflowed sample was first cross-sectioned perpendicularly to solder-copper interface of the solder joint and mounted in mounting cup using epoxy resin before metallographic sample preparation (grinding, polishing and etching using dilute solution of 5 vol. % nitric acid (HNO_3), 2% hydrochloric acid (HCl) in 93 vol. % methanol (CH_3OH)). Wetting angles were measured under optical microscope (ECLIPSE L300N) and using J-Image software.

The morphology of interfacial IMCs layer was observed by scanning electron microscope (SEM) JEOL JSM 6460LA coupled with energy dispersive X-ray spectroscopy (EDS). The IMC thickness was calculated by dividing the area of the IMC by its length and then takes averaged over five SEM images taken at a magnification of 3000x. J-image software was utilized to conduct the measurements of IMCs thickness form for as reflowed samples.

3 Results and discussion

Fig. 1 shows SEM image of the microstructure of monolithic and solder paste composite. The microstructure for both monolithic and solder paste composite composed of two regions which are the dark region known as eutectic region and light region as β-Sn. The eutectic region consists of Cu_6Sn_5 IMC phase. Microstructural observation shows that with the increasing amount of reinforcement particles up to 0.75 wt. %, the size of Cu_6Sn_5 IMCs were refined.

The refinement of Cu_6Sn_5 IMC phases is in conjunction with the theory of heterogeneous nucleation [8] During soldering, reinforcement particles may be finely dispersed in the molten solder and precipitate on the top of the substrate. These areas then become the heterogeneous nucleation sites for Cu_6Sn_5 IMC. With an increasing amount of reinforcement particles, it will result in decreasing the surface energy of Cu_6Sn_5 grains. These situations will then result in decreasing the growth velocity of Cu_6Sn_5 grains. According to heterogeneously nucleation theory, the Cu_6Sn_5 will nucleate on reinforcement surface in order to reduce the thermodynamic barrier [9]. The increasing amount of reinforcement particles in the solder has caused more reinforcement particles precipitate on the top of the substrate, thus produce more nucleation site. It is believed that the more production of the nucleation site, the more Cu_6Sn_5 refinement process has occurred.

Previous studies reported that reinforcement particles may attribute in alter the microstructure become finer and thus improved the properties[8-10]. The effect of TiO_2 particles on Sn-0.3Ag-0.5 solder resulting in smaller grain size with an increase in TiO_2 proportion. The refinement effect of Si_3N_4 was also found by Salleh et al. with the addition of Si_3N_4 into monolithic solder, the β-Sn grains refine greatly and the size of Cu_6Sn_5 IMCs decrease [9].

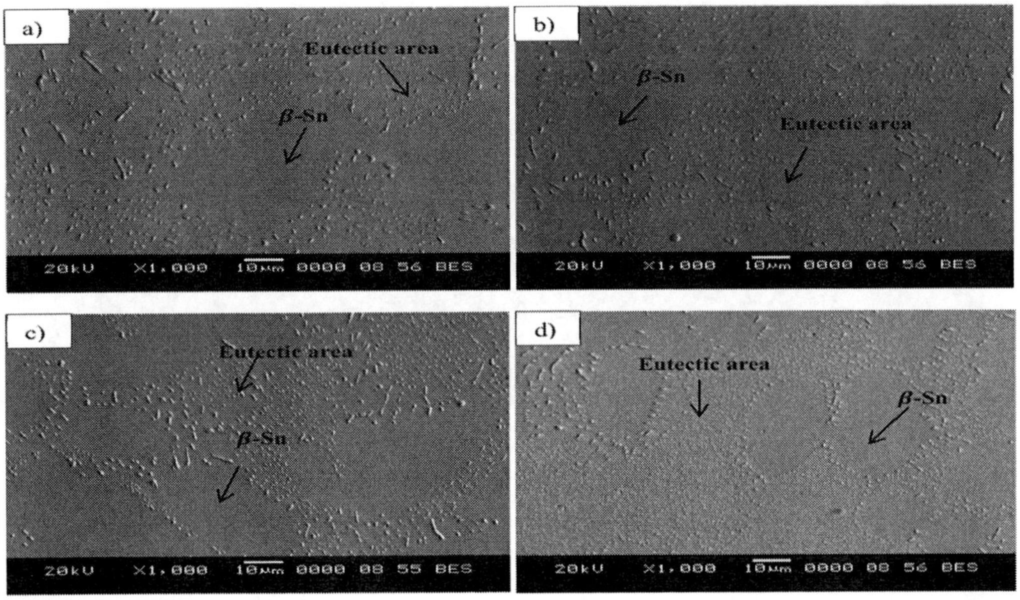

Fig. 1. Microstructure of Sn-Cu-Ni solder paste reinforced with TiO_2 at different percentage (a) 0 %, (b) 0.25 %, (c) 0.50 % and (d) 0.75 %.

4 Intermetallic compound formation

The intermetallic compounds (IMCs) layer for monolithic and solder paste composite is shown in Fig. 2. It can be seen that the scallop-like IMCs layer was observed along Cu substrate interface. The Cu_6Sn_5 was detected by SEM-EDX analysis, while Cu_3Sn phase was not detected. The thin Cu_3Sn phase was hardly to be detected for first reflow cycle and accelerated test such as thermal aging can possibly reveal this phase. It can be observed that the morphology of the interfacial Cu_6Sn_5 layer does not change too much with TiO_2 particle addition into Sn-Cu-Ni solder paste. The scallop type of IMC layer was not changed even when TiO_2 particle added. However, the addition of TiO_2 particles suppressed the growth of interfacial Cu_6Sn_5 layer. Fig. 3 shows the average IMC layer thickness for both monolithic and Sn-Cu-Ni solder paste composite with various weight percentages.

The reduction on interfacial IMC layer thickness is due to the presence of TiO_2 particles and the mechanism of suppression on this IMC thickness was discussed by Haseeb et al. [11]. They suggested that the decreasing in the thickness of interfacial IMC is due to discrete particle effect that preferentially being absorbed at the grain boundaries of IMC scallops. However, in this research, TiO_2 particles are believed has been distributed homogeneously at the diffusion path for the IMC growth thus suppressed the growth of the IMC layer.

DOI: 10.1051/matecconf/20167801070

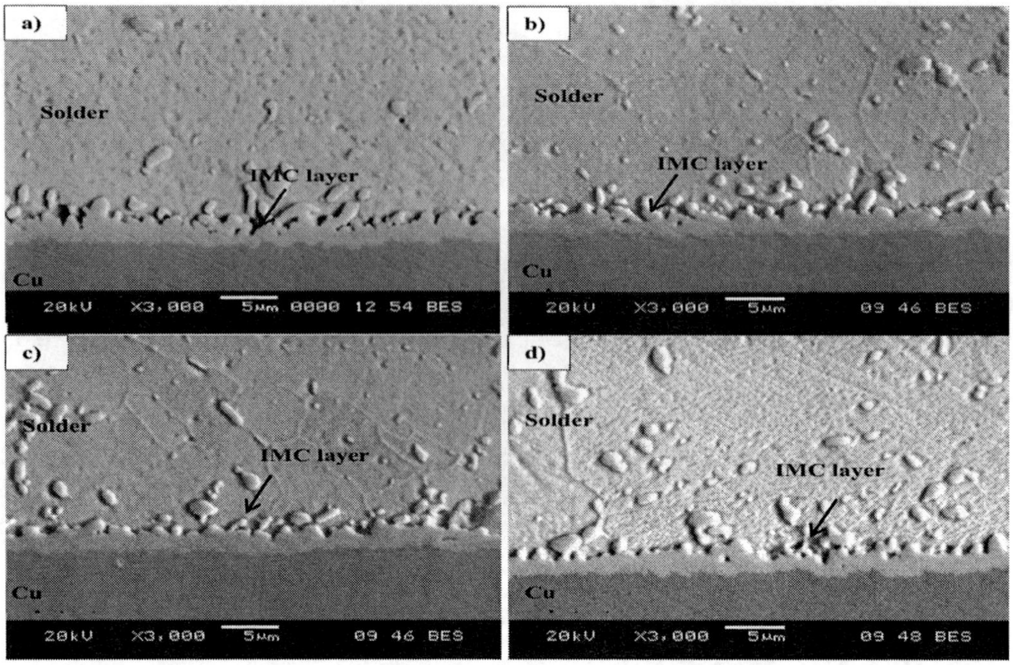

Fig. 2. Scanning electron microscope (SEM) images of the Intermetallic compound (IMC) formation for (a) 0, (b) 0.25, (c) 0.50 and (d) 0.75 of TiO_2 solder composite.

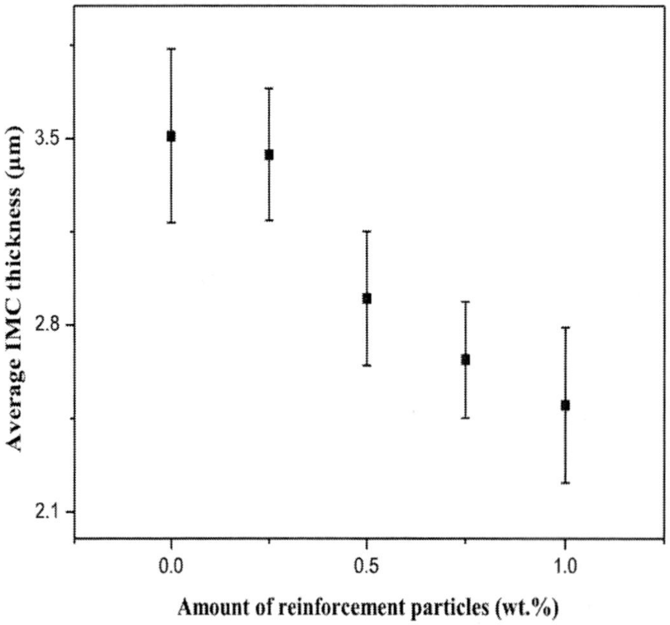

Fig. 3. Average intermetallic compound layer thickness of Sn-Cu-Ni solder paste reinforced with TiO_2 particles.

5 Conclusion

The effect of TiO_2 particles addition on microstructure evolution of Sn-Cu-Ni-TiO_2 solder joints was investigated. The addition of TiO_2 particle helps to refine the Cu_6Sn_5 IMC microstructure for bulk solder composite Sn-Cu-Ni solder paste. Besides, the addition of TiO_2 particles into Sn-Cu-Ni solder paste has caused decreases in thickness and diameter of the IMC at the interface. Apart from that, the presence of TiO_2 particles is believed to retard the IMC growth of solder paste composite resulting in a thinner IMC.

References

1. Satyanarayan, K.N. Prabu, Adv. Colloid Interface Sci., **166** (2011)
2. L. Li, R. Yang, J. Kai Lin, Int. Sym. Adv. Pkg. Mat., 112 (2001)
3. S.L. Tay, A.S.M.A. Haseeb, Mohd Rafie Johan, P.R. Munroe, M.Z. Quadir, Intermetallic, **33**, 8 (2013)
4. F.Guo, *Lead-free Electronic Solders* (2006)
5. S.M.L. Nai, J. Wei, M. Gupta, Thin Solid Films, **504**, 401 (2006)
6. J. Shen, Y.C. Chan, Microelectronics Reliability, **49**, 223 (2009)
7. Y. Tang, G.Y. Li, Y.C. Pan, Mater. Design, **55**, 574 (2014)
8. M.A.A.M. Salleh, A.M. Mustafa AlBakri, M.H. Zan@Hazizi, F. Somidin, N.F. Mohd Alui, Z.A. Ahmad, Mat. Sci. Eng. A-Struct., **556**, 633 (2012)
9. A.A. El-Daly, W.M. Desoky, T.A. Elmosalami, M.G. El-Shaarawy, A.M. Abdraboh, Mater. Design, **65**, 1196 (2015).
10. Y. Tang, G.Y. Li, Y.C. Pan, J. Alloy. Compd., **554**, 195 (2013)
11. A.S.M.A. Haseeb, M.M. Arafat, M.R. Johan, Mater. Charact., **64**, 27 (2012)

Correlation of the Na_2SiO_3 to NaOH Ratios and Solid to Liquid Ratios to the Kedah's Soil Strength

Hazamaah Nur Hamzah[1,*], *MohdMustafa Al Bakri* Abdullah[1,2], *Heah* Cheng Yong[1], *Mohd Remy Rozainy* Mohd Arif Zainol[3], *Aiman* Mahmad Nor[1,2], and *Warid* Wazien A.Z.[1,2]

[1]Center of Excellence Geopolymer and Green Technology (CEGeoGTech), School of Material Engineering, University Malaysia Perlis (UniMAP), P.O Box 77, d/a Pejabat Pos Besar, 01000 Kangar, Perlis, Malaysia.
[2]Faculty of Engineering Technology, Universiti Malaysia Perlis (UniMAP), P.O. Box 77, D/A Pejabat Pos Besar, 01000 Kangar, Perlis, Malaysia.
[3]Schools of Civil Engineering, USM, Engineering Campus, 14300 Nibong Tebal, Seberang Perai Selatan, Pulau Pinang, Malaysia.

Abstract. Geopolymer was used for the soil stabilization of Kedah's soil at different ratios of solid to liquid and Na_2SiO_3 to NaOH in order to achieve the desired compressive strength. The geopolymerization process which produces an aluminosilicate gel was occurred due to the mixing of Kedah's soil and fly ash with Na_2SiO_3 and NaOH. Soil stabilization by geopolymer was synthesized by the activation of fly ash and Kedah's soil with Na_2SiO_3 and NaOH at different ratios of solid to liquid (1.5, 2.0, 2.5 and 3.0) and Na_2SiO_3 to NaOH (0.5, 1.0, 1.5, 2.0, 2.5 and 3.0) at a specific constant concentration of NaOH solution of 6M. The compressive strength up to 5.12 MPa was obtained at 3.0 of solid to liquid ratio and 2.5 of Na_2SiO_3 to NaOH ratio in 7 days curing at room temperature.

1 Introduction

One of the first steps in the design of civil engineering is the investigation of soil conditions on the site of the proposed structure. Engineering properties important not only as a basis for the project but also as a construction material for many structures, including roads embankments, earth dams, and other types of grading projects. Soil is a collection of solid parts and is not bound with each other (including possibly organic material) between the cavities of the material filled with air and water [1]. In the field of civil engineering, soil defined as all material in the earth's crust which is not consolidated and consider that the rock is an aggregate of minerals which are connected by a wide range of great strength, while the soil is the natural particle that can be destroyed with a low strength. In other words, soil is loose material outside layers of rock, which is comprised of a group of mineral grains of various sizes and shapes as well as the content of organic matter, water

* Corresponding author: hazamaahnur@yahoo.com

and air. In the civil construction frequently encountered problems in soft soil type, among others are the low bearing capacity and settlement is high if subjected to load. Because of the soft soil has low permeability, the time required for consolidation is longer. Soil consolidation is the process where the bulk soil volume is reduced under loading due to flow of pore water. Then slowly the pressure is transferred to the effective stress soil, so soil has increased in terms of soil strength and bearing capacity. Thus, if the construction is to be built on soft soil, the construction process should be done after soil has been consolidated completely. This is to avoid large settlements on the building and obtain the soil bearing capacity that is good enough. Therefore, we need a method of soil improvement to overcome this problem.

Soil stabilization can be done either only stabilization of the ground surface or material in the soil [2]. Nowadays chemical stabilization with cementitious materials has been broadly practiced and implemented to improve the physical characteristics of the problematic soil [3]. Developing of geopolymer technology by Davidovits [4, 5] offers an attractive solution for soil stabilization regarding the emission of carbon dioxide to the atmosphere that produced by Ordinand Portland Cement (OPC) in conventional method [6, 7]. Soil stabilization by using geopolymer have been popularly applied in construction field such as fly ash [8] rice husk ash [9], lime [10] and metakaolin [11]. Dali Bondar et. al. [12] stated that highest compressive strength can be achieved when the molarity of the solution is between 5 and 7.5 M. The ratios of fly ash to alkaline activator and Na2SiO3 to NaOH can be influenced the compressive strength of geopolymer. A study conducted by D. Hardjito [13] showed that the use of a Na2SiO3 to NaOH ratio of 0.4 resulted in lower compressive strength, whereas a ratio of 2.5 gave the highest compressive strength. In addition to obtain the best compressive strength, the researchers used a ratio of sodium silicate to NaOH in the range of 0.4 to 2.5 and a ratio of fly ash to alkaline activator in the range of 2.5 to 3.3 [14].

Thus, the combination is relevant for precast products used for practical purposes. The purpose of this study was to identify the mix design proportions that can produce optimum compressive strength of geopolymer soil stabilization.

2 Experimental Method

2.1 Raw Material

Materials used in this study include soil, fly ash, sodium hydroxide and sodium silicate.

2.1.1 Soil

Laboratory tests were performed on the reddish colour clayey soil with high amounts of oxides as residual laterite soil recovered in Kedah, Malaysia [15]. This soil is found commonly in tropical regions [16] and it can be considered as representative of typical soils from the north of Malaysia. After being submitted to characterisation analysis, this laterite soil classified as MH according BS classification [17] and the properties described in Table 1 [18]. It was oven dried and sieved down by using 300 µm sieve before being thoroughly mixed. Additional characterisation included X-Ray Fluorescence (XRF) to determine chemical composition of this soil (Table 2).

Table 1. Physical properties of Kedah's soil.

Properties	**Percentage (%)**
Sand	11
Silt	52
Clay	37
Specific gravity	2.63
Optimum Moisture Content	35
Maximum Dry Density	1.3
Liquid Limit	77
Plastic Limit	42
Plasticity Index	35

Table 2. Chemical composition of Kedah's soil.

Chemical composition (oxides)	**Percentage (%)**
SiO_2	66.50
Al_2O_3	21.50
Fe_2O_3	8.24

2.1.2 Fly ash

The fly ash that used in this study had low calcium content (class F) that obtained from the Manjung Power Station in Lumut, Perak. Its characterisation (Table 3) was carried by X-Ray Fluorescence (XRF) [18].

Table 3. Chemical composition of fly ash.

Chemical composition	SiO_2	Al_2O_3	Fe_2O_3	TiO_2	CaO	MgO	Na_2O	K_2O	P_2O_5	SO_3
Percentage (%)	52.11	23.59	7.39	0.88	2.61	0.78	0.42	0.80	1.31	0.49

2.1.3 NaOH

NaOH can be used to make geopolymer material as the liquid medium mixture. It was obtained from the Formosa Plastic Corporation, Taiwan. The specification of NaOH is shown as Table 4.

Table 4. Specification of Sodium Hydroxide.

Items	**Specifications**
Physical Form	Pellets
Colour	White
NaOH Content	More than 99%
Water Solubility	100%
Molecular Weight	40 g/mol

 DOI: 10.1051/matecconf/20167801071

2.1.4 Na_2SiO_3

Na_2SiO_3 is also used as ingredients for making material where obtained from the South Pacific Chemicals Industries Sdn. Bhd. (SPCI), Malaysia. The specification of Sodium Silicate is tabulated in Table 5.

Table 5. The specification of Sodium Silicate liquid.

Items	Specifications
Water (H_2O)	60.5%
Silica	30.1%
Sodium Oxide	9.4%
SiO_2/Na_2O Ratio	3.2
Molecular Weight	122.06 g/mol
Viscosity at 20°C	0.40 Pa·s
Specific gravity at 20°C	1.40 g/cm^3

2.2 Sample Preparation

Geopolymer material mixture is manufactured by trial and error which is expected to obtain the optimum compressive strength value of the geopolymer material. It is by regulating the solid and liquid materials. The geopolymerization process for soil stabilization was synthesized by the activation of fly ash and soil with alkaline solution at solid to liquid ratios of 1.5, 2.0, 2.5, and 3.0 and Na2Sio3/NaOH solution of 0.5, 1.0, 1.5, 2.0, 2.5 and 3.0. Then the mix design proportion of the entire manufacturing of geopolymer material is determined by the following ratios in Table 6.

Then geopolymer soil stabilization was produced by mixing the materials. First of all, sodium hydroxide must be dissolved early in distilled water based composition that has been planned for the concentration was kept constant at 6M. Then sodium silicate put into a sodium hydroxide solution which had been dissolved in distilled water and the solution is allowed to stand for 24 hours. After that, the solution is mixed with fly ash and soil and it has been stirred evenly so it gets homogeneous mixture and good workability.

2.3 Moulding, Curing and Unconfined Compressive Strength (UCS) Testing

Material that has been cast into a cylindrical mould in three layers was left to stand for 24 hours before being extruded from the mould. After extruded from the mould, the samples are wrapped and left for curing at room temperature in 7 days before the compressive strength test needs to be run. A low testing speed (0.25mm/min) was used to mitigate any influence of this parameter on the test result.

Table 6. Mix design details.

Solid/Liquid	Na_2SiO_3/NaOH	Solid		Liquid	
		Fly Ash (g)	Soil (g)	NaOH + H_2O (g)	Na_2SiO_3 (g)
1.5	0.5	157.5	892.5	466.67	233.33
1.5	1.0	157.5	892.5	350.00	350.00
1.5	1.5	157.5	892.5	280.00	420.00
1.5	2.0	157.5	892.5	233.33	466.67
1.5	2.5	157.5	892.5	200.00	500.00
1.5	3.0	157.5	892.5	175.00	525.00
2.0	0.5	157.5	892.5	350.00	175.00
2.0	1.0	157.5	892.5	262.50	262.50
2.0	1.5	157.5	892.5	210.00	315.00
2.0	2.0	157.5	892.5	175.00	350.00
2.0	2.5	157.5	892.5	150.00	375.00
2.0	3.0	157.5	892.5	131.25	393.75
2.5	0.5	157.5	892.5	280.00	140.0
2.5	1.0	157.5	892.5	210.00	210.00
2.5	1.5	157.5	892.5	168.00	252.00
2.5	2.0	157.5	892.5	140.00	280.00
2.5	2.5	157.5	892.5	120.00	300.00
2.5	3.0	157.5	892.5	105.00	315.00
3.0	0.5	157.5	892.5	233.33	116.67
3.0	1.0	157.5	892.5	175.00	175.00
3.0	1.5	157.5	892.5	140.00	210.00
3.0	2.0	157.5	892.5	116.67	233.33
3.0	2.5	157.5	892.5	100.00	250.00
3.0	3.0	157.5	892.5	87.50	262.50

3 Results and Discussion

The compressive strength for geopolymer soil stabilization at different ratios of solid to liquid and Na_2SiO_3 to NaOH is shown as Figure 1. The optimum ratio of Na2SiO3 to NaOH (2.5) contributed to a maximum compressive strength for all ratio of solid to liquid which 1.5, 2.0, 2.5 and 3.0 are obtained 0.34 MPa, 1.23 MPa, 2.83 MPa, and 5.12 MPa, respectively. Based on the figure below, the pattern of the graphs was steep rise for each group of bar graph. The value of compressive strength tended to decrease after ratio of Na_2SiO_3 to NaOH of 2.5 for all ratios of solid to liquid.

Among the ratios of solid to liquid, the maximum compressive strength (5.12MPa) was obtained by the ratio of solid to liquid at 3.0. This shows that the increasing the ratio of solid to liquid resulted in an increase the compressive strength of geopolymer soil stabilization. Geopolymer gels increase could influence the increasing of compressive strength with the curing age and fill the pores in geopolymer samples, thus leading to the formation of a denser structure. The solid to liquid ratios, by mass should be approximately 3 to produce fly ash-based geopolymers to allow for geopolymerization process and alkaline activator solution formed a thick gel instantaneously upon mixing with source material [19]. In this case the activator solution is also very influential in the polymerization process and it's also one of the factors that determine the compressive strength. However,

DOI: 10.1051/matecconf/20167801071

when the ratio was increased to 3.5, the compressive strength of soil stabilization cannot be done due to low workability and non-homogeneous geopolymer mixture.

The optimum compressive strength was obtained at ratio of Na_2SiO_3 to NaOH at 2.5 for every solid to liquid ratios as shown in result. As a reason, the compressive strength of geopolymer soil stabilization seems more affected on Na_2SiO_3 to NaOH ratio compared to solid to liquid ratio. This study showed that the ratio of solid to liquid and the ratio of Na_2SiO_3 to NaOH were influenced the compressive strength of geopolymer soil stabilization.

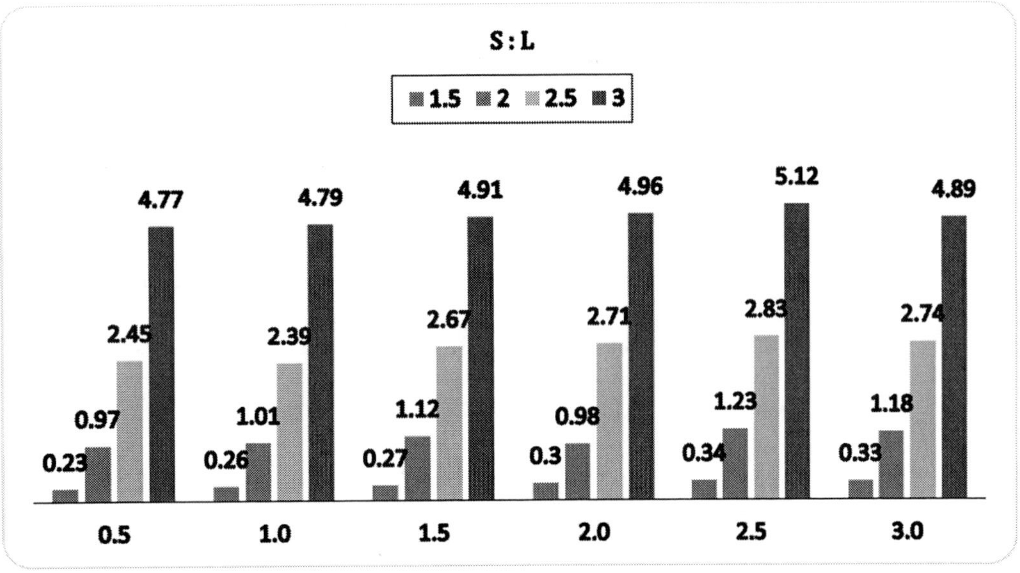

Fig. 1. The compressive strength for geopolymer soil stabilization at a different ratios of solid to liquid and Na_2SiO_3 to NaOH.

4. Conclusion

From the analysis of compressive strength results of fly ash based geopolymer soil stabilizer, the compressive strength was optimized at 2.5 of Na_2SiO_3 to NaOH ratio and 3.0 of solid to liquid ratio with 5.12 MPa. The compressive strength of Kedah's soil was increased as the increases of solid to liquid ratios. This might due to the fine size of fly ash particle which it can play a role like an OPC and the dominant of SiO_2 and Al_2O_3 in terms of chemical composition. So that the fly ash is suitable for use as a raw material to produce geopolymer material and feasible to be applied as a stabilizer in the soil. The amount of waste can be reduced and resulting in green and environmentally friendly when using fly ash in the production of soil stabilizer.

We would like to appreciate Centre of Excellence Geopolymer and Green Technology (CeGeoGTech), School of Material Engineering, Universiti Malaysia Perlis.

References

1. H.A. Verhoef, *Methods in Applied Microbiology and Biochemistry*, (Ed. A. Kaufmann, New York, 1994)
2. N. Cristelo, S. Glendinning, T. Miranda, D. Oliveira, R. Silva, Constr. Build. Mater., **36**, 727 (2012)
3. O.S. Olaniyan, R.A. Olaoye, O.M. Okeyinka, D.B. Olaniyan, IJCEE-IJENS **11,** 6 (2011)
4. J. Davidovits, *Chemistry of Geopolymeric Systems Terminology* (The Geopolymer Institute, France, 1999)
5. J. Davidovits, *Properties of Geopolymer Cements*, (Kiev, Ukraine, 1994)
6. B.V. Rangan, D. Hardjito, S. E. Wallah, D. M. J. Sumajouw, *World Congress Geopolymer* (Geopolymer Institute, France, 2005)
7. S.P. Shah, K. Wang, *Critical Issues of Sustainable Development and Emerging Technology for "Green" Concrete* (International Workshop on Sustainable Development and Concrete Technology, China, 2004)
8. N. Cristelo, S. Glendinning, L. Fernandes, A.T. Pinto, Acta Geotech., **8**, 395 (2013)
9. R. Dayakar Babu, K. Ramu, S. Durga Prasad, K Ashok Kumar, *Geotechnical Advances and Novel Geomechanical Applications* (IGC, India, 2013)
10. J.D. McKinley, H. R. Thomas, K.P. Williams, J.M. Reid, Eng. Geol., **60,** 181 (2001)
11. M. Zhang, H. Guo, T. El-Korchi, G. Zhang and M. Tao, Constr. and Build. Mater., **47**, 1468 (2013)
12. D. Bondar, C.J. Lynsdale, N.B. Milestone, N.Hassani, A. A. Ramezanianpour, Cem. Concr. Compos., **33**, 251 (2011)
13. D.Hardjito, S. E. Wallah, D. M. J. Sumajouw, B. V. Rangan, ACI Mater. J., **101**, 467 (2004)
14. D. Hardjito, B.V. Rangan, *Research Report GC 1* (Faculty of Engineering, Curtin University of Technology, Australia, 2005)
15. Commission on Engineering Geological Mapping, IAEG. (1981)
16. A. Marto, N. Latifi, H. Sohaei, EJGE., **18** 521 (2013)
17. British Standards Institution, *Classification Tests* (BS1377, London, 1990)
18. H. N. Hamzah, M. M. Al Bakri Abdullah, C. Y. Heah, M. R. R. Arif Zainol, K. Hussin, Mater. Sci. Forum, **841**, 51 (2016)
19. H. Xu, J.S.J. Van Deventer, Int. J. Miner. Process., **59**, 247 (2000)

MATEC Web of Conferences **78**, 01072 (2016) DOI: 10.1051/matecconf/20167801072
IConGDM 2016

Highly Response and Sensitivity Chitosan-Polyvinyl alcohol Based Hexanal Sensors

Nur Zuraihan Abd Wahab[1,*], *Irwana* Nainggolan[1], *Tulus Ikhsan* Nasution[2], *Mohd Nazree* Derman[1], *Devi* Shantini [1]

[1]School of Materials Engineering, Universiti Malaysia Perlis 02600 Jejawi, Arau, Perlis, Malaysia.
[2]Physics Department, Faculty of Mathematic and Natural Science, University of Sumatera Utara, Medan, Sumatera Utara 20155, Indonesia.

Abstract. This work is to study the sensing properties of chitosan-polyvinyl alcohol film sensors upon hexanal gas exposure using vapour sensing technique. The sensor fabrication process was done using electrochemical deposition method by deposit the sensing materials on the gold patterned electrode with chitosan/PVA. The response value of chitosan-PVA film sensors towards hexanal was taken as an output voltage using electrical testing method. In this study, 1.75% of chitosan with 5% of PVA were mixed with the ratio of (95:5) was tested upon exposure to hexanal gas. The concentration of the hexanal was varied as 10 ppm, 20 ppm, 30 ppm. It was found that the chitosan-PVA film sensors showed fast response, stable, good stability, good recovery, repeatable and good sensitivity towards hexanal exposure. The morphology of the pure chitosan and chitosan-PVA was analyzed by scanning electron microscope (SEM) and the interaction between chitosan and PVA was examined by Fourier Transform Infrared Spectroscopy (FTIR). The FTIR results indicate the changes in characteristics of the spectral peaks due to the formation of the intermolecular bonds between chitosan and PVA. The SEM morphology of the composites showed flat smooth surface that be a sign of uniform distribution of chitosan and PVA mixture throughout the films.

1 Introduction

Food deterioration that occurred effect the quality and decrease the shelf life of the food itself. This is caused by the lipid oxidation that occurred during food storage. This process results in altereration of flavour, texture and nutrition value in the food [1]. In addition, the autooxidation in wood pellets will released hexanal as a major emitted gas during storage [2]. Besides that, a study reported that they develop hexanal as an odour reference standard for sensory analysis of drinking water. The off flavour in drinking water will reduce intake drinking water among peoples and less of confidence in the water utilities. The off-flavour may be caused by contaminants in water sources, corrosion of metal pipes and new plumbing materials. Hexanal was chosen as good candidates for odor reference standard because it was stable during the sampling process and reproducible results [3]. Hexanal also

* Corresponding author: nurzuraihan90@gmail.com

used as food additive and it was detected in lung cancer blood [4,5]. So, it is important to have a device to detect hexanal gas in order to be applied in any fields that required to monitor the concentration level of hexanal. Current equipment to detect the concentration of hexanal was gas chromatography (GC) that has drawbacks such as expensive, large space used, required specialist to operate the equipment and high in maintenance utilities. Thus, a device with low cost, easy to operate and user friedly needed to overcome the drawbacks of GC. On the other hand, shrimps, lobster, crabs or krills were exoskeleton creatures that have chitin, one of the most plentiful renewable organic resources in nature. Derivation of chitin will produce polyaminosaccharide compound that also known as chitosan [6]. Chitosan was second most abundant polymer in nature after cellulose. Chitosan is a polysaccharide composed of B-(1,4)-linked D-glucosamine residues (deacetylated unit) with a variable number of randomly located N-acetyl-glucosamine groups (acetylated unit) [7]. It consist of two types of reactive functional groups amino (NH2) at C-2 and hydroxyl (OH) at C-3 and C-6 position on the backbone along with interdispersed acetamido groups [8]. Chitosan has excellent characteristics including good chemical inertness, high mechanical strength, biodegradability, biocompatibility, high-quality film forming properties, and low cost [9]. Polyvinyl alcohol (PVA) is a non-toxic, water soluble polymer having a strong film forming ability with very high dielectric strength, good charge storage capability, high mechanical tensile strength and dopant dependent electrical and optical properties [10]. PVA is an inexpensive hydrophilic semi-crystalline polymer. This polymer has excellent film forming properties and can produce effective electrolytes after doping with potassium hydroxide (KOH) solution [11]. During to the good properties of PVA, it was believed that it has ability to influence the sensing performance of chitosan film sensor. In this study, we investigated the capability of Chitosan-PVA compared to pure Chitosan as sensing materials upon different of concentration hexanal. The electrical testing was done in order to check the performance of the sensor which including response, stability, sensitivity, repeatability and recovery.

2 Experimental

Chitosan solution was prepared by dissolving 1.75 g of chitosan powder (low molecular weight, Sigma Aldrich)) in 100 ml of 2% aqueous Acetic Acid (99.9% purity, HmBG). The mixture was stirred continuously at room temperature for 24 hours. Polyvinyl alcohol (Average MW = 200000, Sigma Aldrich) was dissolved in water and stirred for 1 hour at temperature 80°C. The PVA solution was added dropwise in the chitosan solution that prepared from previous procedures. The mixture was stirred continuously at room temperature for 1 hour. Chitosan-PVA solution will form into film by electrochemical deposition process by a piece of patterned electrode was dipped into chitosan/PVA solution and was subjected to required deposition voltage and deposition time. The deposited chitosan-PVA film were hard baked on vacuum oven with temperature of 100°C to remove excess water that existed on the film. Then, the sample was cooled at room temperature. The electrical testing was tested in order to check the performance of the sensor which including stability, sensitivity, output voltage, repeatability and recovery. The chitosan-PVA film sensor was placed in a chamber and connected with power supply and the output was recorded by Digital multimeter (Model : Sanwa CD771). The sensor was exposed to hexanal and dry air alternately. The output voltage was recorded for every 30 seconds for 5 minutes to hexanal vapor and 5 minutes to dry air to complete one cycle.

DOI: 10.1051/matecconf/20167801072

3 Results and Discussion

Fig. 1 shows graph response of CS-PVA film sensor towards hexanal exposure. The response of the sensors during hexanal exposure was taken as an output voltage. As recorded, the response value of CS-PVA increased when exposed to the hexanal. The response of CS-PVA film sensor increased fastly within 5 minutes for 10 ppm, 20, ppm and 30 ppm respectively. CS-PVA film sensors gave good response during 5 cycles of measurement when exposed to hexanal vapour. In addition, the Chitosan-PVA composite film sensors yielded repeatable readings at when hexanal vapour was exposed within 5 cycles. The recovery of CS-PVA film sensor was satisfactory since the output response recovered gradually when hexanal was replaced by dry air. Fig. 2 shows the root mean square (R2) of CS-PVA was fit to regression line which is 0.9979. This indicates that the response of CS/PVA towards each hexanal concentration was around the average value. Fig 3 and Fig 4 shows the operating temperature (°C) and humidity (%) values that were taken during the measurement of the CS-PVA film sensor at room temperature. The range of the operating temperature was between 26°C-33°C during the measurement was collected. This indicates that CS/PVA film sensors able to operate at room temperature without any fluctuation readings.

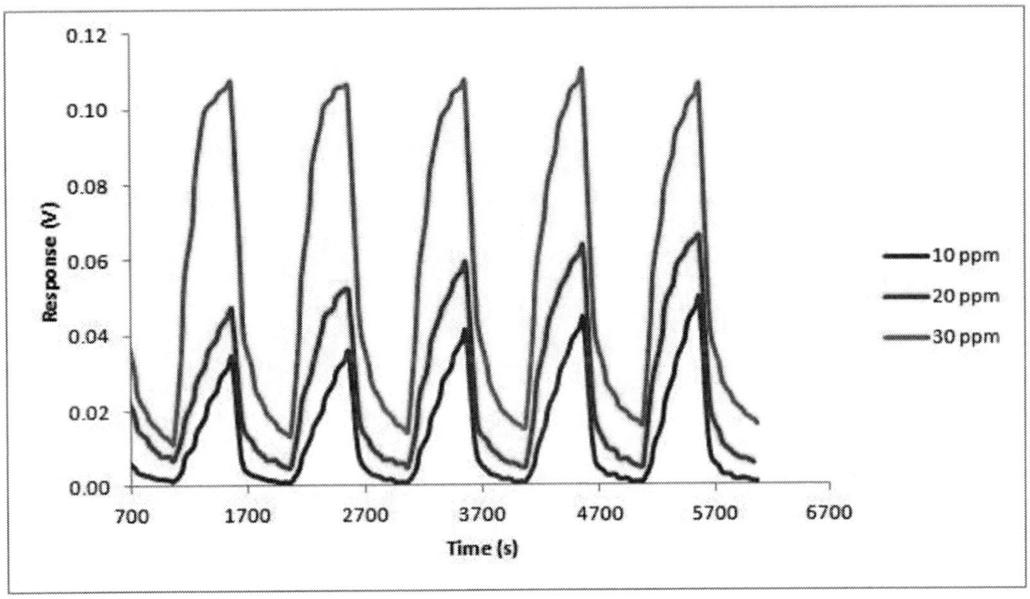

Fig. 1. Response (V) of Chitosan/PVA Sensor towards Hexanal.

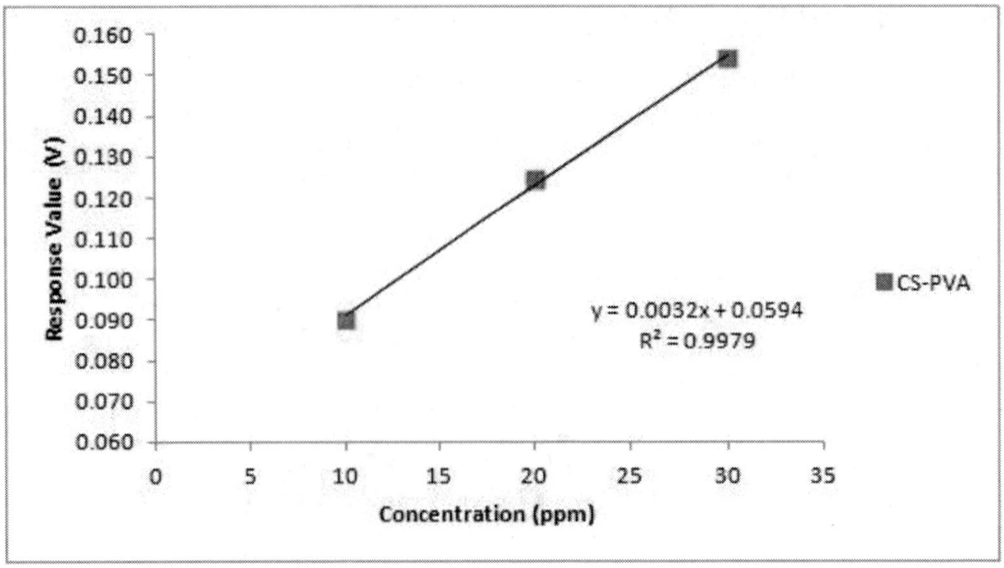

Fig. 2. Maximum Response of CS/PVA against concentration of Hexanal (ppm).

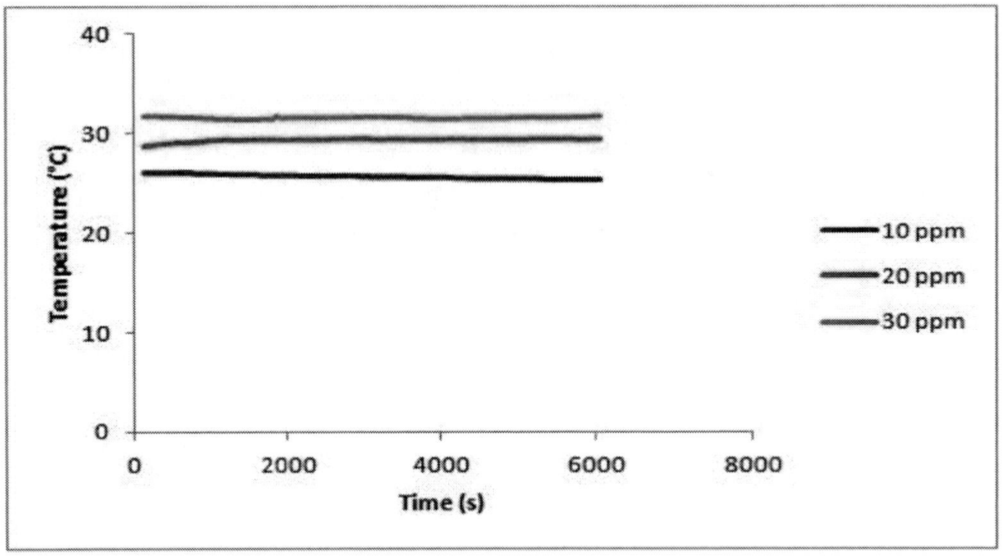

Fig. 3. Operating Temperature (°C) of CS/PVA against concentration of Hexanal.

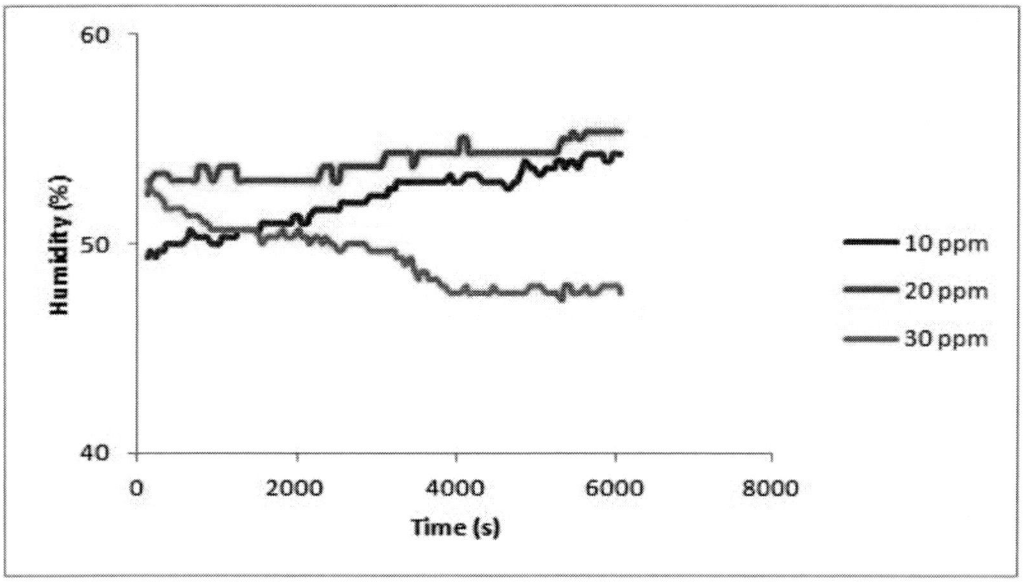

Fig. 4. Humidity (%)of CS/PVA against concentration of Hexanal.

4 Sensing Mechanism

The sensing mechanisms of the polymer sensor films occurred when the composite polymer is exposed to water vapour. The water vapor will be chemically adsorbed (chemisorption) at the activated sites of the polymer composites. The water molecules will be dissociated at the vapor stage as shown in Eq. 1.

$$H_2O \longleftrightarrow H^+ + OH^- \quad (1)$$

The hydroxyl group (as shown in Eq. 2) of each water molecule will be adsorbed at the chitosan cations and providing mobile protons. This interaction will form a layer surrounding the chitosan cations. The subsequent water vapor will form a second layer on the first layer. This process will form a number of stack layers. This will lead the vapor molecules have more mobile protons resulting the increasing in the conductivity values as well as increase the output voltage during the electrical testing [12]. During recovery process, oxygen molecules will be chemisorbed onto the surface of the chitosan particles when the sensor is exposed to normal air. Under certain input voltage, the amount of free electrons moving randomly in the conduction band is reduced as the chemisorbed oxygen traps them and transfers them from one particle to another. This results in an increase in resistance and a decrease in output voltage of the chitosan film until the oxygen species are saturated [13].

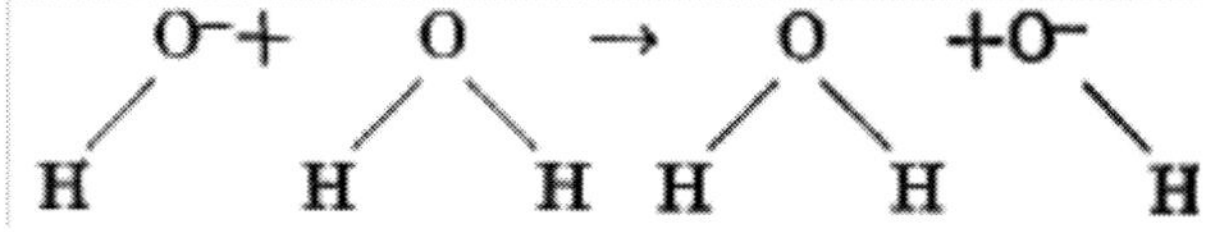

(2)

When hexanal gas vapour is exposed on the surface of the CS-PVA film sensor, the gas molecules of hexanal will make a contact with chemisorbed oxygen on the film sensor surface as shown in Eq. 3. This will make the electrons release freely within the CS-PVA surface. The equation involved as follows:

$$C_6H_{12}O + O_2^- \text{ (adsorbed)} \rightarrow C_6H1_2O_2 + H2O + e^- \quad (3)$$

The contact reaction between the analyte gas and sensor surface will release electrons and water molecules.These free electrons that occurred in conduction band will influence the increment of the response of the sensor [2].

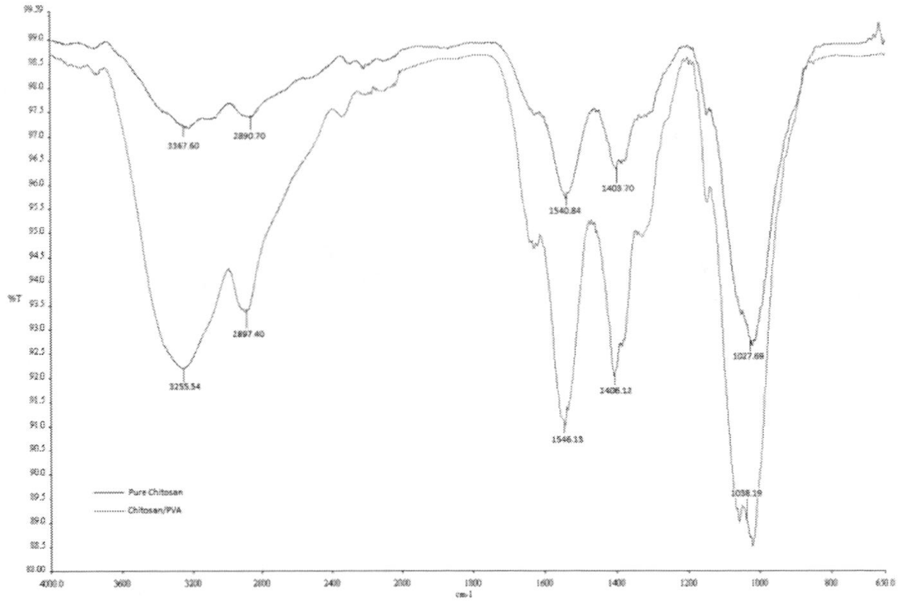

Fig. 5. FTIR Spectra of Chitosan Film and Chitosan-PVA Film.

Pure Chitosan films showed peak at 3367.60 cm-1 that indicated for the –OH and NH2 stretching. Measurement spectra of 2890.70 cm-1 for –CH stetching, 1027 cm-1 was noted for C-O-C stretching, 1540.84 cm-1 for –NH bending of NH2 (amide II) and 1403.70 cm-1 for –CH wagging coupled with –OH groups of chitosan. The addition of PVA in the chitosan caused a decrease in the intensity of the band at 1540.84 cm-1 of chitosan. CS-PVA composite exhibits an absorption band around 3255.54 cm-1 and 2897.40 cm-1 due to the –OH stretching and –CH2 asymmetric stretching from PVA [14].

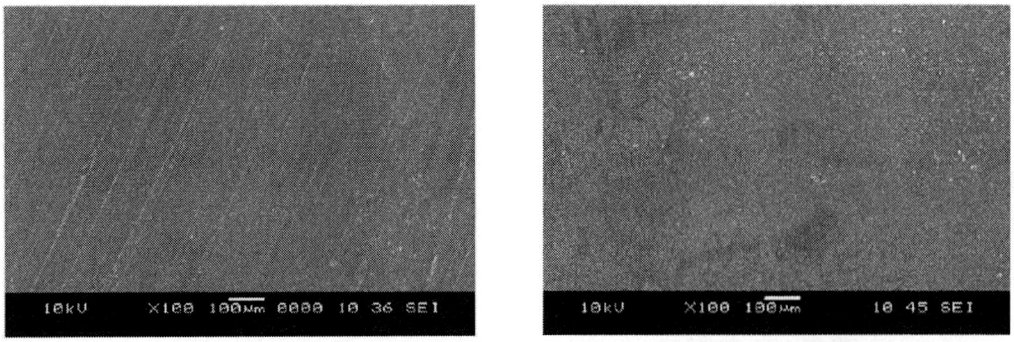

Pure Chitosan (a)	Chitosan-PVA (b)

Fig. 6. SEM image of pure chitosan film (a) and chitosan-PVA film (b).

Based on the surface morphology obtained from SEM, chitosan film shows a smooth surface with some straps existed as depicted in Fig. 6. Meanwhile, the surface of the composite film made of chitosan and PVA is more homogenous and flat surfaces that indicates the uniform distribution of chitosan and PVA in the films. This uniformity characteristics may caused by the interaction of functional groups of –NH2 in chitosan and –OH in PVA [15].

5 Summary

Chitosan-PVA film was found as a good sensing material to sense various concentration of hexanal. chitosan-PVA film sensors showed good sensing properties which include good response, stability, sensitivity, repeatability and recovery properties toward hexanal vapor exposure. The low cost production and ease of fabrication process make chitosan-PVA sensors become advantageous to be applied in various fields.

References

1. P.G. Maher, Y.H. Roos, K.N. Kilcawley, M.A.E. Auty, M.A. Fenelon, Food Sci. Technol., **63**, 1069, (2015)
2. D. Shantini, I. Nainggolan, T.I. Nasution, M.N. Derman, R. Mustaffa, N.Z.A. Wahab, Sensors, 1 (2016)
3. P. Omur Ozbek, A.M. Dietrich, Water Res., **42**, 2598 (2008)
4. T. Fadida, A. Selilat-Weiss, E. Poverenov, Food Hydrocolloid, **48,** 213(2015)
5. N. Li, C. Deng, X. Yin, N. Yao, X. Shen, X. Zhang, Anal. Biochem., **342**, 318, (2015)
6. P. Wydro, B. Krajewska, K. Ha, Biomacromolecules, **8**, 2611 (2007)
7. P.P. Zuo, H.F. Feng, Z.Z. Xu, L.F. Zhang, Y.L. Zhang, W. Xia, W.Q. Zhang, Chem. Cent. J., **7**, 1, (2013)
8. S. Kumar, J. Koh, Int. J. Biol. Macromol., **70**, 559, (2014)
9. I. Nainggolan, D. Shantini, T.I. Nasution, and M.N. Derman, Adv. Mater. Sci. Eng., **70**, 1, (2015)
10. N. Kulshrestha, B. Chatterjee and P.N. Gupta, Mater. Sci. Eng., B, **184**, 49, (2014)
11. G. Liao, C. Yang, C. Hu, Y. Pai and S. Jessie, J. Membr. Sci., **485**, 17 (2015)
12. H. Farahani, R. Wagiran and M. N. Hamidon, Sensors, **14**, 7881, (2014)
13. T.I. Nasution, I. Nainggolan, S.D. Hutagalung, K.R. Ahmad and Z.A. Ahmad, Sensor Actuat B-Chem, **177**, 522, (2013)
14. A.M. Pandele, M. Ionita, L. Crica, S. Dinescu, M. Costache and H. Iovu, Carbohydr. Polym., **102**, 813 (2014)
15. E.A. El-hefian, M.M. Nasef and A.H. Yahaya, Journal of Chemistry **7**, 4 (2010)

Influence of Difference Solders Volume on Intermetallic Growth of Sn-4.0Ag-0.5Cu/ENEPIG

O. *Saliza Azlina* [1,*] and A. *Ourdjini* [2]

[1] Faculty of Mechanical and Manufacturing Engineering, Universiti Tun Hussein Onn, 86400 Batu Pahat, Johor, Malaysia,

[2] Department of Mechanical Engineering, Faculty of Engineering, University of Ottawa, Canada

Abstract. In recent years, portable electronic packaging products such as smart phones, tablets, notebooks and other gadgets have been developed with reduced size of component packaging, light weight, high speed and with enhanced performance. Thus, flip chip technology with smaller solder sphere sizes that would produce fine solder joint interconnections have become essential in order to fulfill these miniaturization requirements. This study investigates the interfacial reactions and intermetallics formation during reflow soldering and isothermal aging between Sn-4.0Ag-0.5Cu (SAC405) and electroless nickel/immersion palladium/immersion gold (EN(P)EPIG). Solder diameters of 300 μm and 700 μm were used to compare the effect of solder volume on the solder joint microstructure. The solid state isothermal aging was performed at 125°C starting from 250 hours until 2000 hours. The results revealed that only (Cu,Ni)6Sn5 IMC was found at the interface during reflow soldering while both (Cu,Ni)6Sn5 and (Ni,Cu)3Sn4 IMC have been observed after aging process. Smaller solder sizes produced thinner IMC than larger solder joints investigated after reflow soldering, whereas the larger solders produced thinner IMC than the smaller solders after isothermal aging. Aging duration of solder joints has been found to be increase the IMC's thickness and changed the IMC morphologies to spherical-shaped, compacted and larger grain size.

1 Introduction

During the last two decades surface mount technology (SMT) has become the dominant manufacturing technology for electronic products, mainly because of its cost-effective high-volume production capability. Thus, it has made possible the manufacturing of extremely high-density portable electronics with good quality and reliability [1]. Generally, smaller electronic devices are faster because of the signals do not have to travel as far within the device. Packing more functionality into a smaller or same-sized device reduces the cost of electronics. With this continuous miniaturization and increasing functionality of electronic components, the microelectronic industry keep moving towards fine pitch devices. Area

* Corresponding author : salizaz@uthm.edu.my

array components have evolved to meet the demands of the industry and the chip scale package (CSP) is therefore becoming one of the main products in the integrated circuit (IC) industry [2].

Besides, this chip scale package (CSP) mostly use flip chip technology where solder bumps are used as interconnects between the chip and substrate. The substrate provides the connection from the chip to the exterior of the package. In flip chip technology, solder bumps are grown on pads deposited throughout the chip area. The number of solder bumps and the number of solder pads depend on the diameter of each solder bump and pad pitch respectively. The chip is placed face down onto a substrate with matched pad locations [3].

This study is focused to the interfacial reaction between different sizes of Sn-4.0Ag-0.5Cu (SAC405) lead-free solders and EN(P)EPIG (electroless nickel/electroless palladium/immersion gold) surface finish. Besides that, the effect of different aging duration on interfacial reaction also has been investigated.

2 Methodology

The copper substrate (FR-4) were cut into 45 x 50 x 1 mm and then a pretreatment process was conducted to remove oxides and activate the copper substrate surface before the electroless plating process is begun. The Ni-P layer was plated at 85□C then the deposition of a layer of electroless palladium will be plated and followed by gold deposition using immersion plating. Then, the samples were laminated with solder mask to restrict the molten solder from flat spreading during reflow. Next, the solder mask and patterned film was cured by ultraviolet (UV) light in order to produce small openings. After curing samples, a thin layer of no-clean flux was applied onto the substrate to remove the oxide layer and also to improve the wetting of molten solder during reflow. The substrates were manually arranged the solder balls with a diameter of 300μm and 700μm in several rows. The samples then was reflowed in a furnace with temperature setting at 230°C and followed by aging treatment at 125°C for 250 hours, 1000 hours and 2000 hours. Characterisation of samples involved both at top surface and cross section of solder joints. Optical microscopy, scanning electron microscopy (SEM) equipped with energy dispersive x-ray analysis (EDX) and image analyzer were used for characterizing the intermetallics.

3 Results and Discussion

Fig.1 shows SEM top surface micrographs of the interfaces between the Sn-4.0Ag-0.5Cu (SAC405) and EN(P)EPIG pads just after reflow soldering. It can be clearly seen that the dominant intermetallic compound (IMC) formed was identified as $(Cu,Ni)_6Sn_5$ and exhibited rod-like and/or needle-shape morphology for the smaller and larger solder size. The present results are consistent with previous research [4]. The cross section view revealed that the IMC layer is formed at the SAC405/ EN(P)EPIG interface has a typical scallop-like morphology as represented in Fig 2. The thickness of this IMC layer for the Ø300μm solder was found to be 0.71μm which is relatively thinner than that recorded in the Ø700μm solder at 1.71μm.

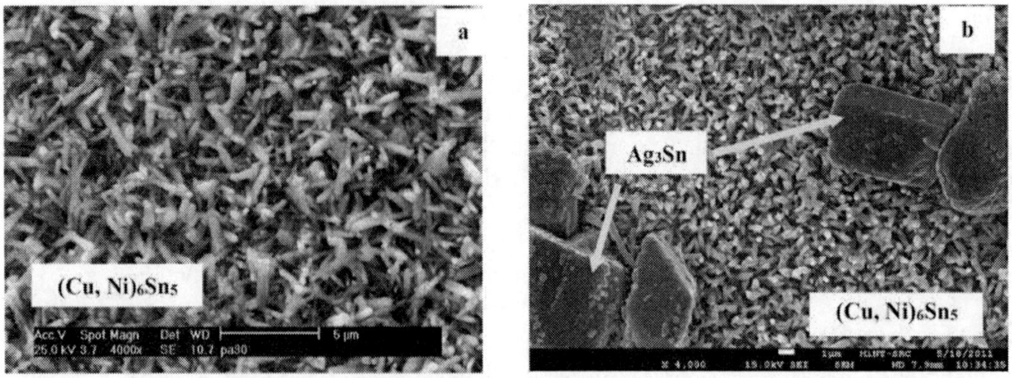

Fig.1. SEM top surface morphology of EN(P)EPIG reflowed with Sn-4Ag-0.5Cu solder at near center region with solder size: (a) Ø300µm and (b) Ø700µm

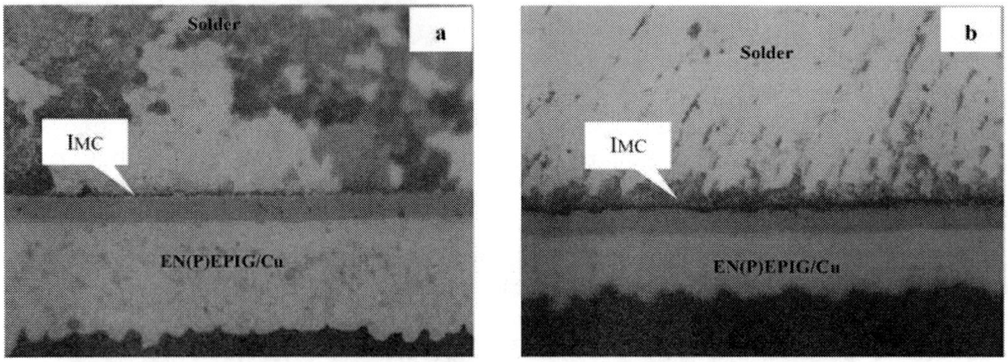

Fig.2. Cross sectional view of Sn-4Ag-0.5Cu/EN(P)EPIG for reflowed sample with magnification 500x using solder size (a) Ø300µm and (b) Ø700µm

Fig.3 shows the SEM micrographs of the interface as function of aging time on the growth of IMC. It is clear that a new IMC layer was created below the (Cu,Ni)6Sn5 layer and EN(P)EPIG/Cu substrate. EDX analysis was confirmed this IMC as (Ni,Cu)3Sn4 layer. During the solid-state aging treatment, the IMC will keep growing by inter-diffusion process between the Sn in solder and EN(P)EPIG substrate. Since the (Cu,Ni)6Sn5 growth is suppressed by a lack of Cu atoms will caused the formation of (Ni, Cu)3Sn4 is formed at the interface.

The scallope-type of (Cu,Ni)6Sn5 IMC which formed during reflow and the (Ni,Cu)3Sn4 during aging, both grew thicker and became continous layers after aging. Moreover, both IMCs also are proportional to an aging time and caused the coarsened and flattened IMC [5]. The reason of this planarization process can happened has been detailed discussed in previous paper [6].

DOI: 10.1051/matecconf/20167801073

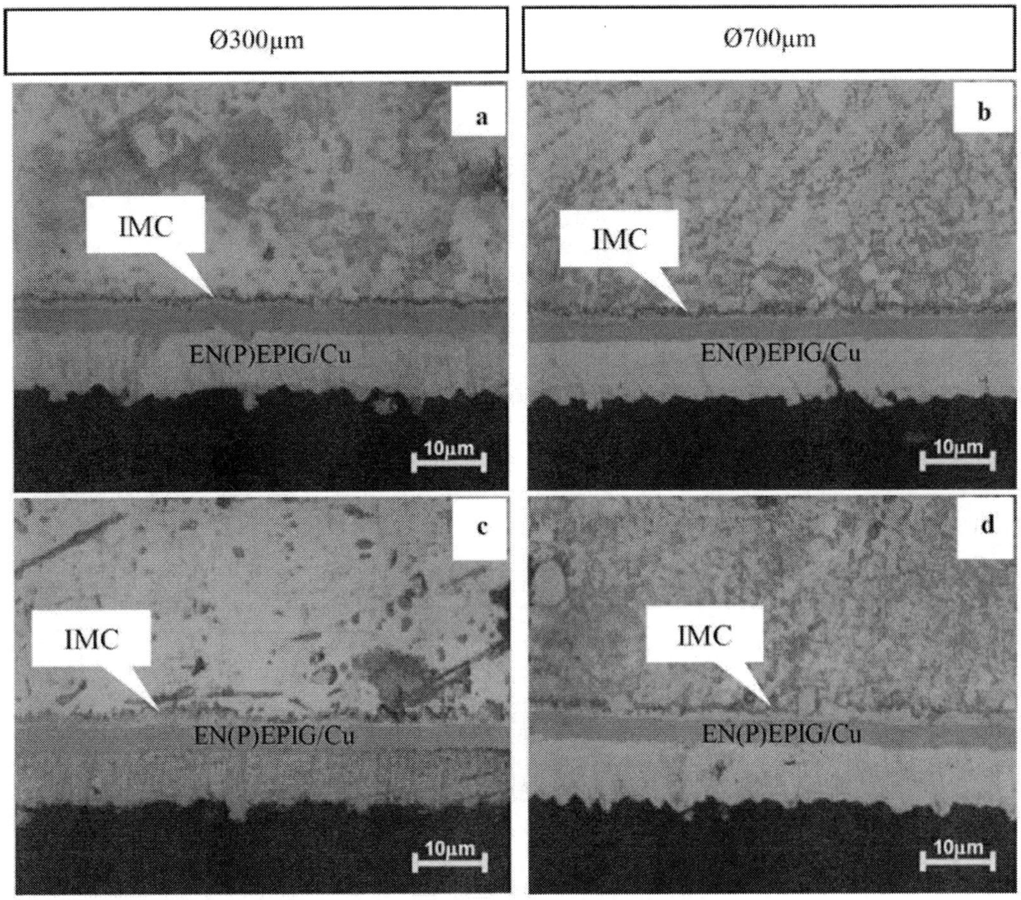

Fig.3. Cross sectional view of SAC405/EN(P)EPIG at 125°C for Ø300µm and Ø700µm for aged samples using optical microscope (a, b) 250 hours and (c, d) 2000 hours

The IMC thickness measured in both solder joints sizes was found to increase with increasing aging time up to 2000 hours as shown in Fig. 4. However, when comparing the solder size, the larger solder of Ø700µm exhibited thinner IMC for all aging times compared with the smaller solder size of Ø300µm (Fig. 4). This dissolution phenomena of metals involved are related to the ratio of solder volume to contact pad area (V/A). Besies, this explains why in larger solders the IMC layer grew slower than in smaller solders during aging treatment. Thinner IMC layer means a shorter diffusion path in high V/A samples (larger solders) [7, 8]. Hence, higher amounts of Sn and Cu atoms could diffused through the IMC barrier to speed up the growth of IMC. Moreover, with increasing the V/A ratio the diffusion distance for Cu to saturate the molten solder increases, and as a result the intermetallics will grow at a slower rate.

DOI: 10.1051/matecconf/20167801073

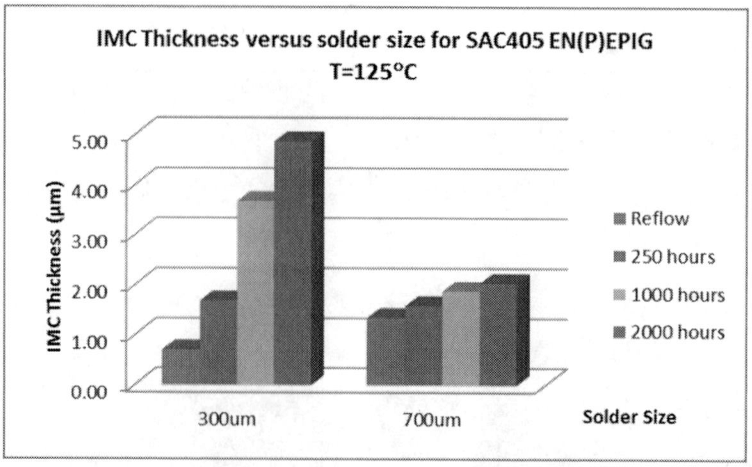

Fig.4. Intermetallic compound (IMC) thickness versus solder size between SAC405 solder and EN(P)EPIG exposed at 125°C

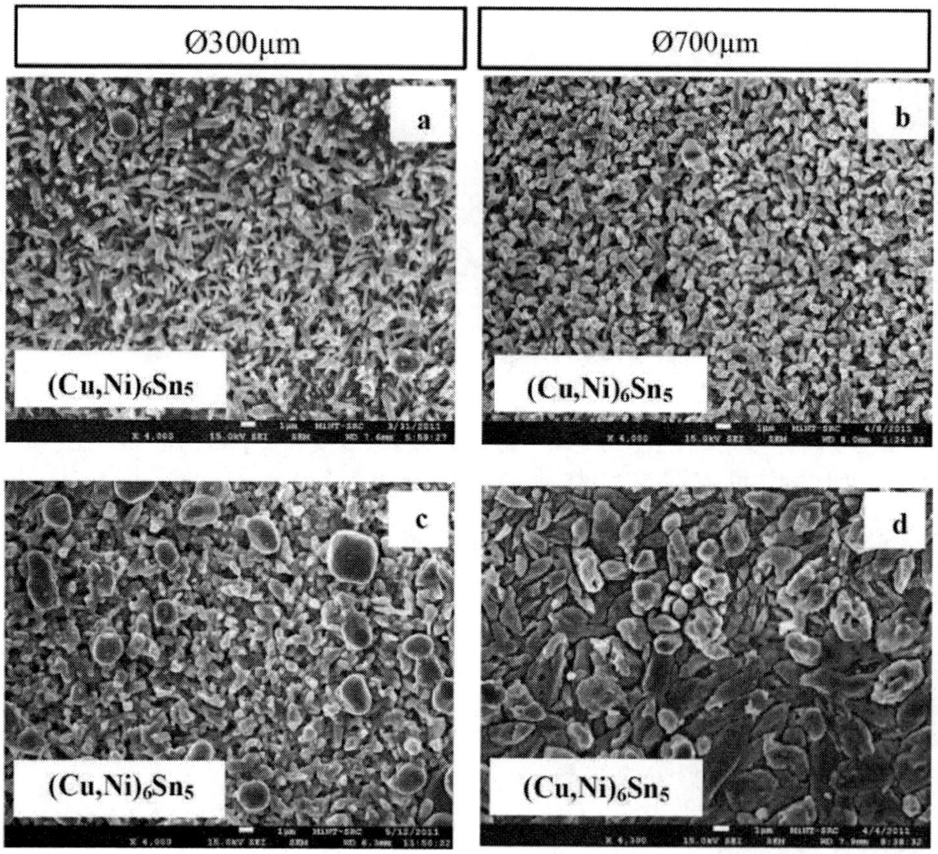

Fig.5. SEM top surface of EN(P)EPIG aged at 125°C with Sn-4Ag-0.5Cu solder (a,b) aging 250 hours and (c,d) aging 2000 hours

Fig.5 illustrates the IMC morphology of top surface examination during isothermal aging process for both solder sizes. The coarsening of (Cu,Ni)6Sn5 IMC grains was pronounced after aging for 2000 hours compared to the 250 hours sample. The rod-like (Cu,Ni)6Sn5 grains have disappeared and tend to have stout-rod-like morphology. In addition, the grain size of (Cu,Ni)6Sn5 IMC has slightly increased when aging duration and temperature was gradually increased. This also can be clearly observed on solder size of Ø700μm especially when aged for 2000 hours. The aging duration results showed an increase of IMC's thickness and changed the IMC morphologies to spherical-shaped, compacted and larger grain size.

4 Summary

The interfacial reaction between SAC405 lead-free solders and EN(P)EPIG surface finish has been examined. The results revealed that only (Cu,Ni)6Sn5 IMC was found at the interface during reflow soldering while both (Cu,Ni)6Sn5 and (Ni,Cu)3Sn4 IMC have been observed after aging process. Smaller solder sizes produced thinner IMC than larger solder joints investigated after reflow soldering, whereas the larger solders produced thinner IMC than the smaller solders after isothermal aging. Aging duration of solder joints results in an increase of IMC's thickness and changed the IMC morphologies to spherical-shaped, compacted and larger grain size.

This work was financially supported Ministry of Education, Malaysia through Fundamental Research Grant Scheme (Vot 1493) in Universiti Tun Hussein Onn Malaysia.

References

1. T. Laurila, V. Vuorinen, M. Paulasto-Kröckel, Mat. Sci. Eng. R., **68**, 1 (2010)
2. M.K. Huang, C. Lee, P.L. Wu, S.R. Tzan, Solder. Surf. Mt. Tech., **17**, 3 (2005)
3. K.K. Muchherla, J.G. Kumar, J.M.W. Roveda, Microelectron. Eng., **114**, 57 (2014)
4. J.W. Yoon, B.I. Noh, S.B.J. Jung, Electron. Mater., **40**, 1950 (2011)
5. D. Li, C. Liu, P.P. Conway, Mater. Sci. Eng., A, **391**, 95 (2005)
6. M.A. Azmah Hanim, A. Ourdjini, O. Saliza Azlina, Siti Rabiatul Aisha, International Journal of Automotive and Mechanical Engineering (IJAME) **8**, 1348 (2013)
7. W. Choi, *Reliability of Pb-Free Solders in Electronic Packaging Technology,* (Doctor
8. Philosophy, University of California, 2002)
9. M. Schaefer, W. Laub, R.A. Fournelle, J. Liang, *Design Reliability Solder Interconnects, ed. R.K. Mahidhara, D.R. Frear, S.M.L. Sastry, K.L. Murthy, P.K. Liaw and W.L. Winterbottom* (Warrendale, PA: TMS, 1997)

Thermal properties of ethyl vinyl acetate (EVA)/ montmorillonite (MMT) nanocomposites for biomedical applications

A.M. Alakrach[1,*], *Azlin F.* Osman[2], N.Z. Noriman[1], *B.O.* Betar[3], and *Omar* S. Dahham[1]

[1]Faculty of Engineering Technology (FETech), Universiti Malaysia Perlis (UniMAP), 01000, P.O.Box, D/A Pejabat Pos Besar, Kangar, Perlis, Malaysia
[2]Center of Excellence Geopolymer and Green Technology (CEGeoGTech), School of Materials Engineering, University Malaysia Perlis, 02600 Perlis, Malaysia.
[3]Nanotechnology & Catalysis Research Centre (NANOCAT), University of Malaya, Kuala Lumpur 50603, Malaysia

Abstract. The viability of nanocomposites comprising Ethyl Vinyl Acetate (EVA) filled montmorillonite (MMT) nanoclay as candidate materials of biomedical devices was investigated. EVA/MMT nanocomposites were prepared by incorporating the ratios 0, 1, 3 and 5% of organoclay MMT to EVA copolymer. In vitro biostability of the neat EVA and EVA nanocomposites was compared and assessed by exposing the materials to oxidizing and hydrolytic agents for 4 weeks at 37°C. The thermal properties of the neat EVA and EVA nanocomposites nanoclay filled were studied by using thermogravimetric analysis (TGA). TGA results indicate that the EVA nanocomposite sample containing 1 wt% MMT exhibits higher T_{onset} and significant reduction in the rate of mass loss as compared to the neat EVA and other nanocomposites.

1 Introduction

In today's world, medical devices contribute significant role in enhancing the quality and efficacy of healthcare. While several strategies are being sought for upgrading the quality of health care, however, medical device designers are facing an issue with a limited number of off-the-shelf materials that can be designed for biomedical devices, especially those implantable materials for long term application [1].

EVA is a synthetic random copolymer of hydrophobic ethylene and hydrophobic vinyl acetate monomers. It is one of the most used polymers for a wide area of applications. Due to its unique and wide range properties, researchers have discovered the promised prospects of this polymer as well as its composites in healthcare and medical applications [1]. Generally, EVA has been accepted for biomedical applications due to its random structures, which offer high ozone resistance, weather resistance, and exceptional mechanical

* Corresponding author: abdoakrash@hotmail.com

properties [2]. Nowadays, EVA comes into its emerging role as a controlled release excipient for biological systems [1].

The incorporation of nanoclay or layered silicate into polymers to obtain nanocomposites were one of the important studies since 1950. However, wide interest in the nanocomposite research area has started over forty years later. That was when the research team at Toyota Central Research & Development Co., added MMT (5%) as nanofiller into the Nylon-6 matrix, which caused an increase in the strength of the material [3]. Depending on nanoclays type, the low aspect ratio type ones such as hectorite may have an average platelet length that is approximately 30 nm. However, the larger type such as fluoromica may have an average platelet length up to 2000 nm.

Nanocomposites incorporating nanoclays as filler have been produced mostly for industrial, automotive and packaging purposes [4]. However, less attention has been paid for biomedical applications. Styan (2006) demonstrated the permeability of the thermoplastic polyurethane matrix to water and oxygen molecules was successfully reduced by the addition of the organically modified clay [5]. In a more recent publication, Styan et.al (2012) highlighted that the addition of MMT modified with amino undecanoic acid resulted in enhanced biostability of this particular matrix. They hypothesized that the presence of the impermeable organoclay layers led to a 'barrier effect', which can restrict the access of degradative species to the polymer structure, thus reducing the nanocomposite's oxidative degradation rate [6]. Andriani et.al studied the in vitro biostability of polyurethane containing organically modified clay (organoclay) with different types of surface modification. They found that the nanocomposite incorporating the most hydrophobic organoclay resulted in greater biostability by hindering greater extent of matrix surface re-structuring upon oxidative exposure [7]. These findings show that the barrier properties of the polymers may be impacted by the organoclay, thus can be controlled to enhance the biostability. These properties are highly needed for materials insisted for use in biomedical applications.

An investigation on in vitro biostability and biocompatibility of ethyl vinyl acetate (EVA) nanocomposites incorporating organically modified montmorillonite (organo-MMT) as a new material for biomedical applications [8]. They studied the effect of oxidizing and hydrolytic agents on the morphology and mechanical properties of EVA/MMT nanocomposites. In this study, the thermal behaviour of EVA/ MMT nanocomposites prepared by melt compounding process and exposed to H_2O_2 as an oxidizing agent was investigated as a new candidate of biomedical applications.

2 Experimental

2.1 Materials

EVA was supplied by UBE-Maruzen Polyethylene Co. Ltd., Tokyo, Japan and commercially known as UBE EVA V215. The weight percentage of vinyl acetate is 15%, with the rest is ethylene. Organically modified montmorillonite (organo-MMT), which contains 35–45 wt% dimethyl dialkyl (C14–C18) amine as an organic surfactant, was manufactured by Sigma-Aldrich (USA) and supplied by Zarm Scientific and Supplies Sdn. Bhd. Hydrogen peroxide H_2O_2, 30–32 % solution (Qrec®), was supplied by Qrec (Asia) Sdn. Bhd. Phosphate-buffered saline (PBS) tablets were manufactured by Sigma-Aldrich and used as the hydrolytic agent after being dissolved in distilled water.

2.2 Preparation of the samples.

The samples were prepared by using a Brabender Plasticorder machine manufactured by Lab Tech Co. (LZ80). EVA was mixed with different ratios of organo-MMT nanofiller (0, 1, 3, and 5%) at160 °C prior to drying the starting materials for 24 hours at 50 °C. The resulting nanocomposite samples were then compressed into 1 mm thick sheets using compression moulding machine model GT-7014-H30C from GOTECH Co. The samples after that were cut accordingly for testing and analysis.

2.3 Thermal Gravimetric Analysis (TGA).

TGA analysis was carried out by Perkin Elmer Pyris Diamond TG/DTA Thermogravimetric Differential Thermal Analyzer. 10 ± 1 mg of each sample was subjected to analysis of temperature ranging from 30 - 650 °C, heating rate of 20 °C min^{-1} under nitrogen atmosphere which was pumped into the system at a flow rate of 30 ml min^{-1}.

3 Results and discussions

TGA was employed to study and compared the thermal stability of neat EVA and EVA nanocomposites. The thermal stability in this study is based on the temperature when the neat EVA and EVA nanocomposites start to degrade (T_{onset}) and the rate of mass loss. Figure 1 and Figure 2 display the TGA and DTG curves of the neat EVA and EVA nanocomposites, before and after H_2O_2 treatment, while Table 1 summarizes the thermal degradation onset temperature (T_{onset}) of all the materials.

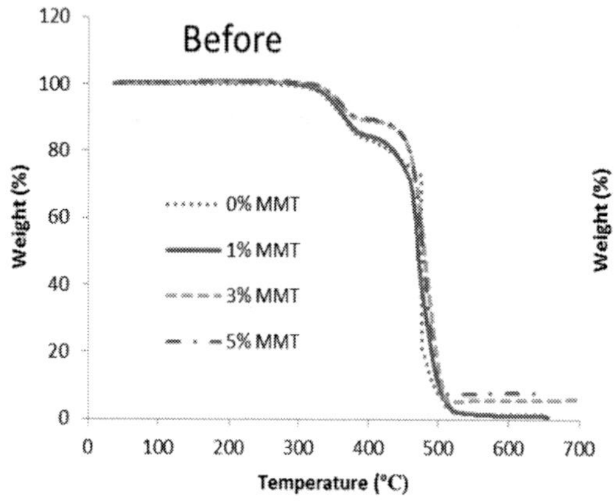

Fig. 1. TGA curves of EVA and EVA nanocomposites before and after H_2O_2 exposure at 37°C.

The intensity of the derivative thermogravimetric (DTG) peaks depends on the amount of degradation product released, which is attributed to the degree of interaction between the EVA polar and nonpolar segments, and also the EVA-nanoclay interactions. Based on Figure 1, the decomposition of neat EVA copolymer and EVA nanocomposites took place in two steps. The first mass loss step between 200 and 300°C was due to deacetylation process, where the release of gaseous acetic acid and formation of carbon-carbon double bonds along the polymer backbone occurred. The second mass loss step (between 400 and 500°C), was due to the oxidation and volatilization of hydrocarbons resulting from the decomposition of the EVA copolymer backbone [9].

For TGA of neat EVA and EVA nanocomposites before the exposure, T_{onset} of a first degradation step of EVA nanocomposites occurred at lower temperatures as compared to neat EVA, and this was due to the degradation of the organic surfactant [9]. The degradation of dimethyl dialkyl (C14-C18) amine, that used to surface modified the MMT accelerated the degradation of acetic acid. The T_{onset} of second degradation step of EVA nanocomposites incorporating 3 and 5 wt% MMT occurred at a higher temperature than that of the neat EVA and EVA containing 1 wt% MMT.

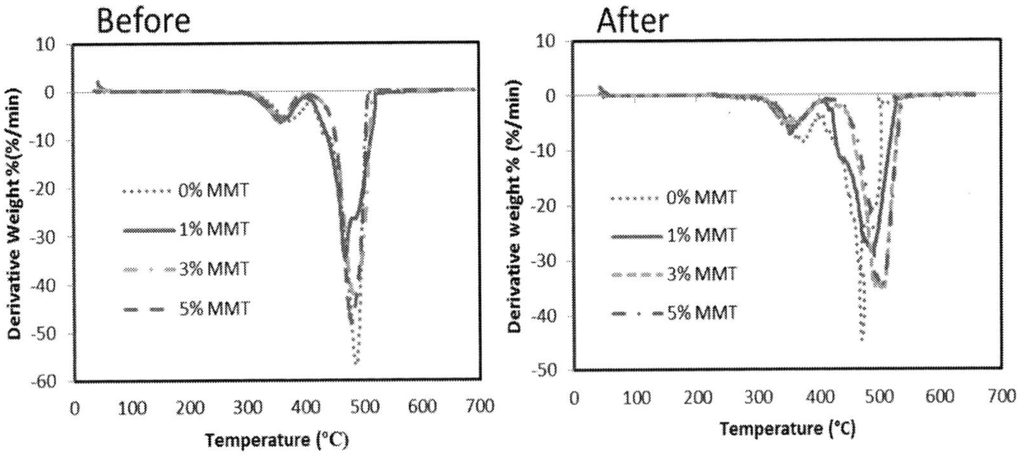

Fig. 2. DTG curves of neat EVA and EVA nanocomposites before and after H_2O_2 exposure.

Table 1. Thermal degradation onset temperature of EVA and EVA nanocomposites before and after the in vitro treatment in oxidative agent (H_2O_2) at 37°C.

Material	**T_{onset} (°C)**			
	First step mass loss		**Second step mass loss**	
	Before	**After**	**Before**	**After**
0%	265	240	420	405
1%	240	235	415	410
3%	255	235	425	425
5%	264	230	430	445

TGA of neat EVA and EVA nanocomposites after in vitro exposure to H_2O_2 at 37 °C show that T_{onset} of first degradation step for the neat EVA and EVA nanocomposites was seen to occur at lower temperature after the in vitro exposure due to degradation process. The in vitro exposed nanocomposites exhibit the same thermal behaviour as the non-exposed nanocomposites, as they show lower Tonset than the neat EVA. The Tonset of second degradation step for the neat EVA and EVA containing 1 wt% MMT was seen to occur at lower temperature after the in vitro exposure. This is expected to happen as a result of polymer degradation upon oxidative exposure. However, the incorporation of 1 wt% MMT into the EVA matrix resulted in an enhanced thermal stability, as the nanocomposite exhibits higher Tonset and significant reduction in the rate of mass loss. This suggests that the incorporation of 1 wt% MMT contributes to biostability enhancement of the EVA matrix, which is in good agreement with SEM analysis and tensile test results. The well dispersed and exfoliated organo-MMT may restrict the entrance of the oxidant molecules into the polymer chains and prevent them from undergo a more severe degradation process. It is interesting to note that the nanocomposites with 3 and 5 wt% MMT show an anomalous thermal behaviour. No reduction in thermal stability upon the in vitro exposure occurred, and in the case of nanocomposite with 5 wt% MMT, the increase in Tonset was observable. As opposed to the previous XRD, TEM, SEM and mechanical performance data, these TGA results suggest that improved thermal stability was obtained with the incorporation of 5 wt% MMT. Thus, further studies are needed to elucidate this phenomenon.

References

1. B.D. Ratner, A.S. Hoffman, F.J. Schoen, J.E. Lemons, *Biomaterials science: An introduction to materials in medicine* (Academic press, 2004)
2. H. Deka, Polym. Degrad. Stabil., **95(9)**, 1509 (2010)
3. Y. Kojima, A. Usuki, M. Kawasumi, A. Okada, Y. Fukushima, T. Kurauchi, O. Kamigaito, Journal of Materials Research, **8(05)**, 1185 (1993)
4. B. Finnigan, D. Martin, P. Halley, R. Truss, K. Campbell, Polymer, **45(7)**, 2249 (2004)
5. K. Styan, *Polyurethane organosilicate nanocomposites for novel use as biomaterials* (The University Of New South Wales, 2006)
6. K.E. Styan, D.J. Martin, A. Simmons, L.A. Poole-Warren, Acta Biomater., **8(6)**, 2243 (2012)
7. Y. Andriani, I.C. Morrow, E. Taran, G.A. Edwards, T.L. Schiller, A.F. Osman, D.J. Martin, Acta Biomater., **9(9)**, 8308 (2013)
8. A.F. Osman, RSC Advances, **5(40)**, (2015)
9. R. Wilson, T.S. Plivelic, A.S. Aprem, C. Ranganathaiagh, S.A. Kumar, S. Thomas, J. Appl. Polym. Sci., **123(6)**, 3806 (2012)

The influence of Rice Husk Fiber on The Properties of Epoxidized Natural Rubber/ Rice Husk Compounds

Omar S. Dahham[1], *N.Z.* Noriman.[1,*], *S.T.* Sam.[2], *N.* Marwa. Al-Samarrai [1], *Z.* Shayfull [3,4], *A. M.* Alakrach[1], and *S. A.* Abduati Elnaid[1]

[1]Center of Excellence Geopolymer and Green Technology (CEGeoGTech), Faculty of Engineering Technology (FETech), Universiti Malaysia Perlis (UniMAP), Level 1 Block S2, UniCITI Alam Campus, Sungai Chucuh, Padang Besar, 02100, Perlis, Malaysia
[2]School of Bioprocess Engineering, Universiti Malaysia Perlis (UniMAP), Kompleks Pengajian Jejawi 3, 02600 Arau, Perlis, Malaysia
[3]School of Manufacturing Engineering, Universiti Malaysia Perlis (UniMAP), Kampus Pauh, 01000 kangar, Perlis, Malaysia
[4]Green Design and Manufacturing Research Group, Center of Excellence Geopolymer and Green Technology (CEGeo Tech), School of Materials Engineering, Universiti Malaysia Perlis (UniMAP), Perlis, Malaysia

Abstract. In this work, curing characteristics, tensile and physical properties of epoxidized natural rubber/ rice husk (ENR-50/RH) compounds were investigated. Different RH loading (10, 20, 30, 40 and 50 Phr) and size (fine size at 100-300 μm and coarse size at 5-10 mm) were prepared and used. Results indicated that the scorch time (t_2) and cure time (t_{90}) became shorter with increasing RH content. In contrast, minimum torque (M_L) and maximum torque (M_H) increased with increasing RH content in the rubber compounds. Hardness and crosslink density showed improvement with increasing RH content. Tensile strength (Ts) and elongation at break (Eb) decreased slightly as RH content increased. However, the fine size of RH recorded better overall properties compared to the RH coarse size at same loading the rubber compound.

1 Introduction

Epoxidized natural rubber (ENR) is a modified natural rubber (NR), which is derived from cis-1, 4-isoprene. The monomer isoprene (C) and epoxidized isoprene (E) are randomly distributed along the polymer chain [1]. ENR also be known as good compatibilizer for incompatible rubber blend as well as processing aids. With incorporation of ENR 50, the rubber blends inherent improved process ability, stiffness, resilience and also shown excellent oil resistance, reduced air permeability, good damping and wet grip performance [2]. The using of natural rubber by itself in the production of desired products can affect economical and environmental aspect. To minimize these issues, it will be useful if the

* Corresponding author: niknoriman@unimap.edu.my

rubber is compound with filler, as it could be improved the properties of rubber compound. The conventional filler, such as carbon black and silica have been used widely with rubber, particularly natural rubber for improving of rubber properties. However, the environmental issues and the relatively high cost of this filler urged the scientists to find alternative material.

Currently, natural fillers, such as natural fiber were used to substitute the conventional filler. Natural fibers are produced from agriculture technology as waste products such as grass, leaves, stems, seeds, wood and other form. Most of these fibers comprise cellulose, waxes, lignin and some water-soluble compounds in their structures [3]. In Malaysia, rice husk (RH) is one of largest waste of natural fiber that produced after rice harvest. RH could be classified as usable material for renewable energy as it can be a road to the creation of a green economy.

However, most of RH been wasted by burning, which caused gases emissions such as carbon monoxide and carbon dioxide and made environmental issues. Recently, RH is attracting much attention and has become a significant part in the industry to over come the environmental pollution problem. One of the first trying that used RH fiber as a reinforcing filler in different types of polymer was reported by Haxo and Mehta. They found that RH could be effectively used as a partial replacement for finer and more reinforcing blacks [4]. Other researcher, such as Ishak et al. studied the potential of rice husk ash after burning (black and white ash) as fillers in epoxidized natural rubber (ENR). They found that the potential of rice husk ash, particularly the white ash, could be exploited further by controlling particle size and particle size distribution, improving filler dispersion and also its surface functionality [5].

Here in, we report the influence of different loading and size of Rice Husk Fiber on the cure characteristics, tensile and physical Properties of Rice Husk/Epoxidized Natural Rubber Compounds.

2 Experimental

2.1 Materials and Samples Preparation

Epoxidized natural rubber with 50% epoxidation (ENR-50) were supplied by Rubber Research Institute of Malaysia (RRIM) as a raw material. Zinc oxide, stearic acid, sulfur and N-cyclohexyl-2-benzothiazole sulfonamide (CBS) were supplied by Anchor Chemical Co. (M) Ltd. Rice husk fibers (RH) were supplied by Bernas Malaysia Sdn. Bhd., Simpang Empat, Perlis. All ingredients that used in this research are listed in Table 1. RH fibers were cut and dried under sun exposure for 3 days before use. Two different sizes of RH (150-300 μm and 10-20 mm) were prepared using Crusher model RT34 (Chyun Industrial Co. Ltd.) and lab sieves respectively. Two-roll mill X (S) K - 160 X 320 was utilized to mix rubber with other materials based on ASTM D 3184-89. Rheometer (MDR 2000) was used for cure characteristics according to ASTM D 2240-93. Electrically hot press machine was utilized for rubber compounds compression using cure time at 160 ^{0}C and under 30 tonne pressure.

2.2 Measurement of Properties

Dumbbell-shaped mold was utilized to form species for tensile test, which was done according to ASTM D412 using universal testing machine (Instron 5582). Shore-A Durometer (Asker, Kobunshi Keiki Co. Ltd) was utilized according to ASTM D-2240-81 to evaluate the hardness values of species. Toluene solvent was used for swelling test according to ASTM D3616 and swelling test data were used to calculate the crosslink density of rubber compounds using Flory-Rehner equations [6].

Table 1. The formulation of ENR-50/RH compounds.

Ingredients	phr					
	R0	**R10**	**R20**	**R30**	**R40**	**R50**
ENR-50	100	100	100	100	100	100
ZnO	5	5	5	5	5	5
Stearic Acid	2	2	2	2	2	2
CBS	1	1	1	1	1	1
Sulfur	2	2	2	2	2	2
RH fs cs	0	10	20	30	40	50

3 Results and Discussions

3.1 Cure Characteristics

The variation in scorch time (t_2), cure time (t_{90}), minimum torque (M_L) and maximum torque (M_H) values of ENR-50/RH compounds are shown in Table 2 respectively. It is clearly seen that t_2 became shorter as RH content increased, which mean that the compound of rubber needed less time to reach to vulcanization onset time. This could be attributed to the mobility restriction of the rubber matrix with the mechanical restrain. Nonetheless, RH (cs) shows longer t_2 than RH (fs) at same RH loading indicating that RH (cs) tendency needed longer time than RH (fs) to coalesce. t_{90} shows same t_2 tendency. Thus, the increasing of RH content accelerated the vulcanization process of the rubber compound. However, RH (cs) exhibits longer t_{90} than RH (fs) ar same RH loading. This might due to the poor dispersion of RH (cs) hat caused poor interaction between ENR-50 and RH (cs).

In contrast, the values of M_L and M_H increased as RH increases. This was due to the increasing of viscosity as RH content increases, which in turn led to decrease the chain mobility inside rubber compounds. Thus the processability of the compounds became more difficult. RH (fs) exhibits lower M_L and M_H than RH (cs) at same RH loading. This was due to better RH (fs) dispersion, which made better compounding process [7].

Table 2. The cure characteristics of ENR-50/RH compounds.

RH		t_2 (min)	t_{90} (min)	M_L (dNm)	M_H (dNm)
R0	Control	2.0	9.5	4.2	20.5
R10	fs	0.80	6.25	5.56	26.0
	cs	1.10	6.75	6.0	28.1
R20	fs	0.65	6.10	7.99	29.2
	cs	0.82	6.35	11.29	30.7
R30	fs	0.37	5.24	11.04	31.2
	cs	0.50	5.79	12.41	33.4
R40	fs	0.27	5.15	12.82	35.1
	cs	0.34	5.65	13.5	37.2
R50	fs	0.20	5.06	13.44	38.8
	cs	0.27	5.45	13.99	40.8

3.2 Tensile and Physical Properties

Table 3 shows the tensile strength (Ts), elongation at break (Eb) and modulus at 100% elongation (M100) of ENR-50/RH compounds. It was observed that the increasing of RH content in ENR-50/RH compounds led to gradual decreases on Ts. Azmi et al. attributed the decreasing of Ts for natural fiber /rubber compounds to the increasing of the stress transmission difficulty from rubber matrix to the filler [8]. However, RH (fs) was higher Ts than RH (cs) at same RH loading. This was due to better RH (fs) distribution in the rubber compounding compared to RH (cs). The percentages Eb decreased as RH loading increased. This decreasing made the compounds stiffer and harder because the mobility of the chain inside rubber compound became more difficult. However, RH (fs) shows higher Eb than RH (cs) due to the good corporation of RH (fs) with ENR-50 which in turn led to increase the viscosity of rubber vulcanized and became harder to be fractured. The M100 values increased continuously as RH content increased in the rubber compounds.

According to Sharifah et al. research, M100 value is directly proportional to the crosslink density [9], which mean M100 increases was due to the crosslinking density improvement as shown in Figure 1. This indicated that the incorporation between RH as filler and ENR-50 as matrix enhanced the stiffness of the compound. Additionally, RH (fs) recorded higher M100 values than RH (cs) at same RH loading due to better RH (fs) distribution that gives larger surface area to contact with ENR-50.

Hardness of ENR-50/RH compounds shows same trend of crosslink density, which increased as RH loading increased (Figure 2). Moreover, RH (fs) was higher hardness than RH (cs) at same RH loading, which indicated that the incorporation of RH (fs) with ENR-50 matrix has improved the stiffness and brittleness of the rubber compounds more than RH (cs) [10].

Table 3. The tensile properties of ENR-50/RH compounds.

RH		Ts (MPa)	Eb (%)	M100 (MPa)
R0	Control	13.0	966	0.50
R10	fs	12.2	960	0.85
	cs	8.9	841	0.75
R20	fs	10.6	901	1.55
	cs	7.2	802	1.45
R30	fs	8.3	814	1.85
	cs	5.7	712	1.75
R40	fs	5.2	715	2.50
	cs	2.6	614	2.35
R50	fs	2.4	619	2.70
	cs	1.8	511	2.60

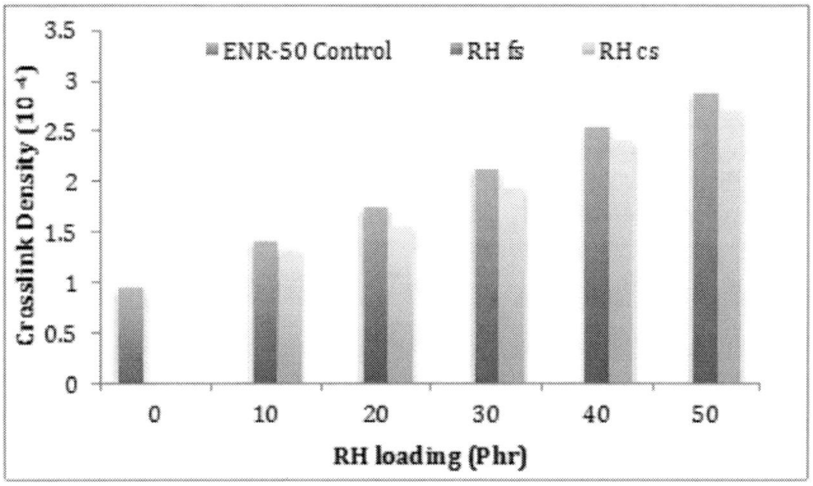

Fig. 1. The Variations of Crosslink density of ENR-50/RS compounds.

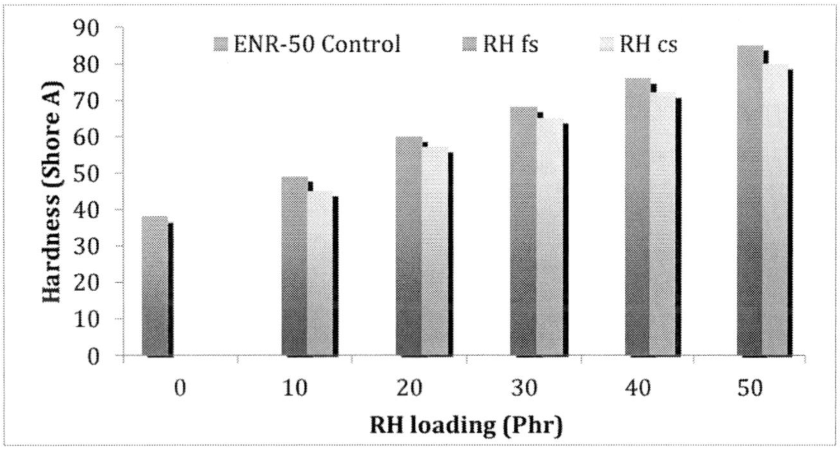

Fig. 2. The Variations of Crosslink density of ENR-50/RS compounds.

4 Conclusion

Generally, the increasing of RH into ENR-50 gave positive impacts on the cure characteristic, which reduced the scorch and cure time of the ENR-50/RH compounds. Additionally, the maximum torque, modulus and physical properties such as crosslink density and hardness were improved. Tensile strength and elongation at break slightly decreased as RH increased. However, RH (fs) shows better overall properties than RH (cs) at same loading.

The authors would like to acknowledge the Fundamental Research Grant Scheme (FRGS) Phase 1/2015 (9003-00523) under Ministry of Higher Education, Malaysia.

References

1. R. Hamzah, M.A. Bakar, M. Khairuddean, I.A. Mohammed, R. Adnan, Molecules, **17**, 10974 (2012)
2. N.R. Munirah, N.Z. Noriman, S. T. Sam, C.M. Ruzaidi, M.F. Omar, M.Z. Salihin, Applied Mechanics and Materials, **679**, 207 (2014)
3. D.N. Saheb, J.P. Jog, Adv. Polym. Tech., **18**, 351 (1999)
4. H.E. Haxo Jr, P.K. Mehta, Rubber Chem. Technol., **48**, 271 (1975)
5. Z.M. Ishak, A.A. Bakar, Eur. Polym. J., **31**, 259 (1995)
6. P.J. Flory, J. Rehner, Journal Chemistry Physics, **11**, 512 (1943)
7. I.L. Hakim, N.Z. Noriman, S.T., Sam, S.A.B. Shahnaz, S. Ragunathan, Advanced Materials Research, **795**, 550 (2013)
8. A.A. Azmi, N.N. Zulkepli, M.R. Nordin, M.M.A.B. Abdullah, I. Sandu, M.A.A.M, Salleh, I. Hanafi, Materiale Plastice, **50,** 175 (2013)
9. H.A. Sharifah, Composite Science Technology, **65**, 525 (2005)
10. O.S. Dahham, N.Z. Noriman, S.T. Sam, M.F. Omar, A. Alakrach, Appl. Mech. Mater., **754**, 693 (2015)

MATEC Web of Conferences **78**, 01076 (2016) DOI: 10.1051/matecconf/20167801076
IConGDM 2016

The Effects of Trans- Polyoctylene Rubber (TOR) as a Compatibilizer on The Properties of Epoxidized Natural Rubber/Recycled Silicone Catheter (ENR-25/rSC) Vulcanizate

Omar S. Dahham[1], *N.Z.* Noriman.[1,*], *S.T.* Sam.[2], H. Rosniza[1], *N.* Al-Samarrai Marwa [1], *Z.* Shayfull [3,4], and *A. M.* Alakrach[1]

[1]Center of Excellence Geopolymer and Green Technology (CEGeoGTech), Faculty of Engineering Technology (FETech), Universiti Malaysia Perlis (UniMAP), Level 1 Block S2, UniCITI Alam Campus, Sungai Chucuh, Padang Besar, 02100, Perlis, Malaysia.
[2]School of Bioprocess Engineering, Universiti Malaysia Perlis (UniMAP), Kompleks Pengajian Jejawi 3, 02600 Arau, Perlis, Malaysia.
[3]School of Manufacturing Engineering, Universiti Malaysia Perlis (UniMAP), Kampus Pauh, 01000 kangar, Perlis, Malaysia.
[4]Green Design and Manufacturing Research Group, Center of Excellence Geopolymer and Green Technology (CEGeo Tech), School of Materials Engineering, Universiti Malaysia Perlis (UniMAP), Perlis, Malaysia

Abstract. In this study, the influence of Trans- Polyoctylene Rubber (TOR) as a compatibilizer on the cure characteristics, tensile and physical properties of ENR-25/rSC vulcanizate were determined. Five different loading of TOR (2, 4, 6, 8 and 10 phr) were prepared and added into the vulcanizate. Results indicated that the scorch time (t_2) and cure time (t_{90}) bacame shorter as TOR increased, while minimum torque (M_L) and maximum torque (M_H) increased. The incorporation of TOR with the vulcanizates enhanced the tensile strength (Ts), modulus (M100) crosslinking density and hardness values. However, the elongation at break percentage of compatibilized vulcanizates became lower than uncompatibilized vulcanizates.

1 Introduction

In general, the blending of two or more kinds of polymers could be a beneficial method for the preparation and improvement of materials with properties superior to those of individual constituents [1]. The main purpose of polymer blending is to modify the processing characteristics and also to improve the physical and mechanical properties. In addition, polymer blending could minimize the cost of the final product. Nonetheless, most of polymer blends have poor adhesion between polymer matrix and disperse, which in turn

* Corresponding author: niknoriman@unimap.edu.my

could lead to poor interfacial adhesion between the phases. Therefore, the compatibilization could be the suitable technique to minimize the problems in polymer blending [2].

Compatibilization is described as a process that uses to overcome the high interfacial tension between the elastomeric and thermoplastic phases. The compatibilization process could achieve either by adding a component, called a compatibilizer, or by improving the interaction of the polymeric components chemically or mechanically. This process could present a smooth transfer of stress between different phases and allow the product to resist the failure under multiple stresses [3]. Compatibilizers are macromolecular species that used for interfacial improvement between different polymer blends.

Trans- polyoctylene rubber (TOR) was used broadly as a compatibilizer for heterogeneous polymer blends, particularly incompatible rubber blends. TOR has a low molecular weight made from cyclo octane by metathesis polymerization. It has crystallizability, which depend on the cis-trans ratio of the double bond. Therefore, it can present a suitable processability in the rubber processing with a temperature range (100-150 ^{0}C) and a good collapse resistance below the melting temperature due to recrystallization [4]. Pena Bizi et al. have studied the effects of TOR on the properties of NR/SBR blends. Results showed the addition of TOR in blends has improved the hardness, decrease in tear resistance and no effect on the resilience [5]. Our previous work investigated the influence of TOR on the Properties of Acrylonitrile Butadiene Rubber/Recycled Natural Latex Gloves (NBR/NRL-G) Vulcanizates. The results indicated that the presence of TOR had improved curing characteristics, tensile, physical and morphological properties of the rubber vulcanizate [6].

Here in, we studied the effect of similar compatibilizer (TOR), on the cure characteristic, tensile and physical properties of epoxidized natural rubber/recycled silicone catheter (ENR-25/rSC) vulcanizate.

2 Experimental

2.1 Materials, Preparation and Testing

Table 1 shows the materials that used in this work. Epoxidized natural rubber with 25 epoxidation (ENR-25) was provided from Malaysian Testing Laboratory Sdn. Bhd. While Sulphur, N-cyclohexyl-2-benzothiazole sulphonamide (CBS), Stearic acid, zinc oxide (ZnO) and trans polyoctylene rubber (TOR) were supplied from Anchor Chemical Co. (M) Ltd. Recycled Silicone Catheter (rSC) was supplied from Teleflex Sdn. Bhd. as rejected Silicone Catheters from the factory.

Two-roll mill machine X (S) K - 160 X 320 was used for rSC mastication and then Grinding process was conducted using Grinder RT 34 to form rSC with (1-2mm) particle size. The mixing of rSC with ENR-25 and chemicals was accompanied using same two-roll mill according to ASTM D 3184-89. Monsanto Moving Die Rheometer (MDR 2000) was utilized for Cure characteristics based on ASTM D 2240-93 at 160 ^{0}C. Electrically hot press machine was utilized for rubber vulcanizate molding based on respective cure time at 160 ^{0}C and under 30 tonne pressure. Dumbbell-shaped mold was utilized to form tensile test species and then universal testing machine (Instron 5582) was used according to ASTM D412. Shore-A Durometer (Asker, Kobunshi Keiki Co. Ltd) was used for hardness test. Toluene solvent was used for swelling test according to ASTM D3616 and Flory-Rehner equation was used for crosslink density calculations using swelling test data [7].

Table 1. The formulation of ENR-25/rSC/TOR vulcanizates.

Ingredients	**phr**					
	R0	R10	R20	R30	R40	R50
ENR-25	100	100	100	100	100	100
ZnO	5	5	5	5	5	5
Stearic Acid	2	2	2	2	2	2
CBS	1	1	1	1	1	1
Sulfur	2	2	2	2	2	2
rSC	10	10	10	10	10	10
TOR	0	2	4	6	8	10

3 Result and discussion

3.1 Cure Characteristics

The cure characteristics (scorch time t_2, cure time t_{90}, minimum torque M_L and maximum torque M_H) of ENR-25/rSC/TOR vulcanizates are listed in Table 2 respectively. It is observed clearly that t_2 and t_{90} became shorter with increasing TOR content in the vulcanizates. This was because of the characteristic nature of TOR as a compatibilizer. Furthermore, TOR also could act as an unsaturated rubber, which led to increasing of bonding between TOR and elastomer in short time. M_L shown decreasing as TOR content increased. This could be due to the plasticizer effect of TOR, which in turn led to reduce the viscosity of the vulcanizates. Consequently, the processability of the vulcanizates had improved.

In contrast, M_H shown an opposite trend, which increased as TOR content increased in the vulcanizates. This was contributed to the unsaturated structure of TOR, which reacted with the elastomer and improved the crosslink density of the vulcanizates [8].

Table 1. The cure characteristics of ENR-25/rSC/TOR vulcanizates.

TOR	**t_2 (min)**	**t_{90} (min)**	**M_L (dNm)**	**M_H (dNm)**
R0	1.12	7.1	6.9	23.9
R2	0.50	5.8	5.1	27.1
R4	0.45	5.1	4.8	28.2
R6	0.40	4.2	4.4	29.0
R8	0.30	3.5	3.8	32.4
R10	0.20	2.8	3.1	35.8

3.2 Tensile Properties

The effects of tensile strength (Ts), elongation at break (Eb) and modulus (M100) on the ENR-25/rSC/TOR vulcanizates are shown in Figure 1, 2 and 3 respectively. The values of Ts increased linearly as TOR content increased. The increasing of Ts was attributed to the higher rSC dispersion in NR-25 matrix as TOR increased, which enhanced the interfacial adhesion of the rubber vulcaniziate. M100 values shown same Ts tendency.

The higher incorporation between rSC and ENR-25 matrix as TOR increased led to increases on M100 values. Eb exhibited decline with the presence of TOR. The increasing of TOR content hindered the flow and mobility of rubber vulcanizates. Therefor, the vulcanizates flexibility decreased [9].

 DOI: 10.1051/matecconf/20167801076

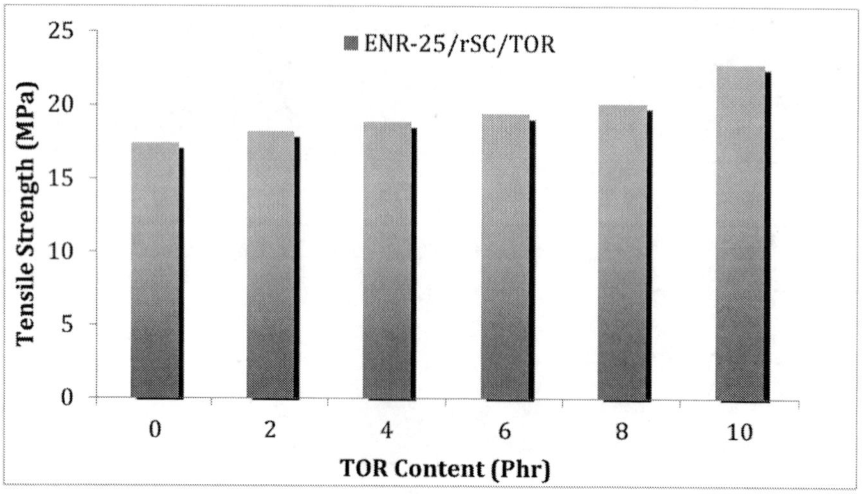

Fig. 1. The variation of Tensile Strength of ENR-25/rSC/TOR vulcanizates.

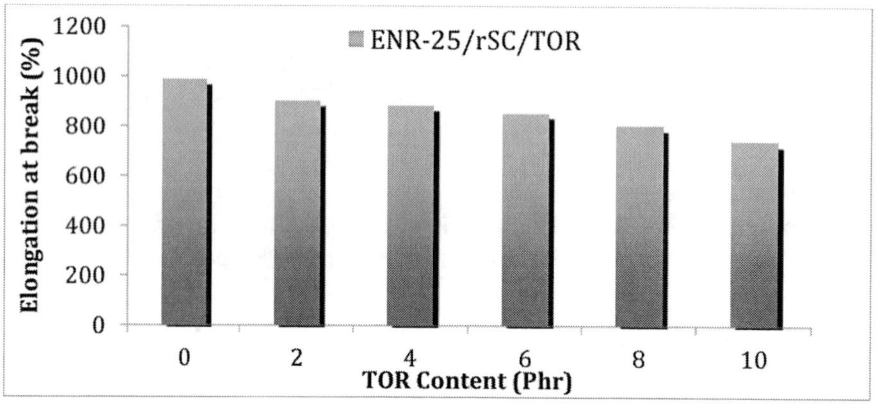

Fig. 2. The variation of Elongation at break of ENR-25/rSC/TOR vulcanizates.

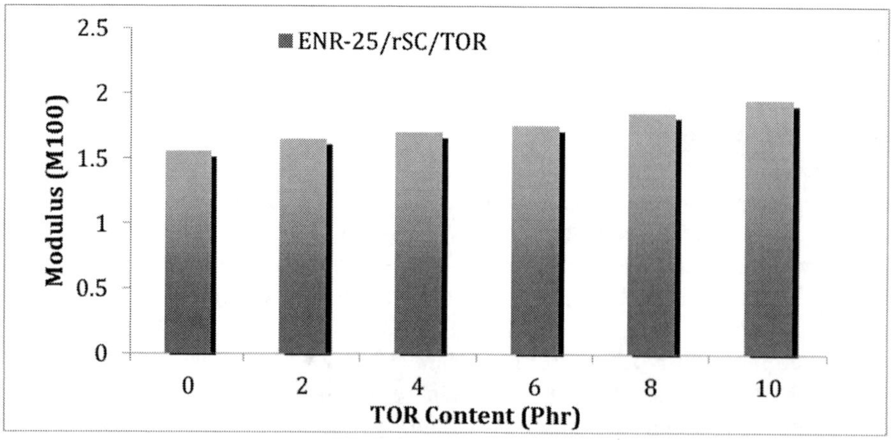

Fig. 3. The variation of Modulus of ENR-25/rSC/TOR vulcanizates.

3.3 Physical Properties

The effects of TOR compatibilizer on the physical properties (crosslink density and hardness) of ENR-25/rSC vulcanizates are displayed in figure 4 and 5 respectively. It's clearly observed that both crosslinking density and hardness values are increased as TOR increased in the vulcanizates. The increasing of crosslinking density led to decreases in Eb, which could be due to the existence of crosslinked precursors, various additives and unreacted curative in the vulcanizates with TOR.

Nah et al. stated that the increasing of hardness values as TOR content increased was might due to the fact that the relative degree of crosslinking was higher than for the other rubber components [10]. Therefore, both crossliniking density and hardness of the vulcanizates are improved.

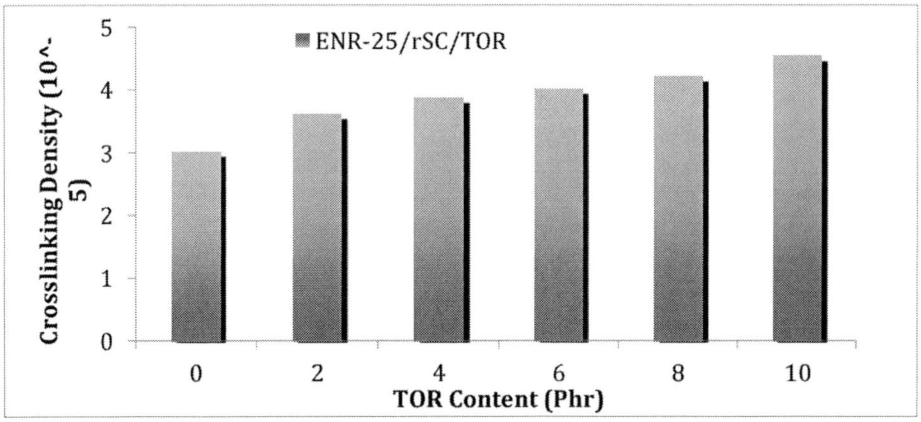

Fig. 4. The variation of crosslinking density of ENR-25/rSC/TOR vulcanizates.

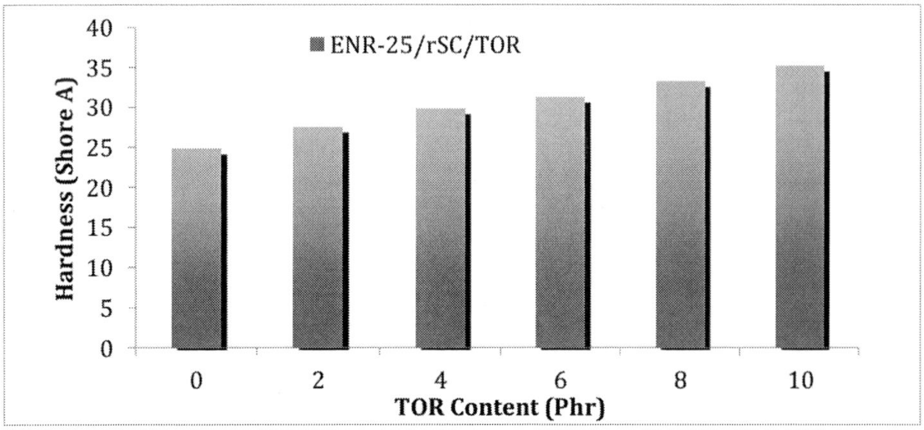

Fig. 5. The variation of hardness of ENR-25/rSC/TOR vulcanizates.

4 Conclusion

Generally, the existence of TOR as a compatibilizer made improvement on overall properties of ENR-25/rSC/TOR vulcanizates. The increasing of TOR in the vulcanizates has shortened the scorch and cure time and increased the toque. Moreover, The tensile and physical properties of the vulcanizates have increased as TOR content increased.

The authors would like to acknowledge the Fundamental Research Grant Scheme (FRGS) Phase 1/2015 (9003-00523) under Ministry of Higher Education, Malaysia.

References

11. N.Z. Noriman, H. Ismail, A.A. Rashid, Polym. Test., **29**, 200 (2010)
12. Paul, D. R, Polymer Blends, **1**, 353 (2012)
13. A. Tedesco, R.V. Barbosa, S.M.B. Nachtigall, R.S. Mauler, Polym. Test., **21**, 11 (2002)
14. M. Awang, H. Ismail, M.A. Hazizan, Polym. Test., **26**, 779 (2007)
15. C.M.P. Bizi, N.R. Demarquette, J. Appl. Polym. Sci., **109**, 445 (2008)
16. O.S. Dahham, N.Z. Noriman, H. Kamarudin, S.T. Sam, M.F. Omar, A.M. Mustafa Al Bakri, H. Rosniza, J. Appl. Polym. Sci., **815**, 59 (2015)
17. P.J. Flory, J. Rehner, Journal Chemistry Physics, **11**, 512 (1943)
18. N.Z. Noriman, H. Ismail, A.A. Rashid, J. Appl. Polym. Sci., **126(S2)**, (2012)
19. N. Hayeemasae, I. Surya, H. Ismail, Int. J. Polym. Anal. Charact., 1 (2016)
20. C. Nah, S.C. Han, B.W. Jo, W.D. Kim, Y.W. Chang, J. Appl. Polym. Sci., **86**, 125 (2002)

MATEC Web of Conferences **78**, 01077 (2016) DOI: 10.1051/matecconf/20167801077
IConGDM 2016

Dimensional Accuracy and Surface Roughness Analysis for AlSi10Mg Produced by Selective Laser Melting (SLM)

K. Kamarudin[1,*], M.S. Wahab[1], Z. Shayfull[2], Aqeel Ahmed[1], and A.A. Raus[1]

[1]University Tun Hussien Onn Malaysia, Parit Raja, 86400 Batu Pahat, Johor.
[2]Green Design and Manufacture Research Group, Center of Excellence Geopolymer and Green Technology (CEGeoGTech), Universiti Malaysia Perlis, 01000 Kangar, Perlis, Malaysia.

Abstract. Selective Laser Melting (SLM) is an Additive Manufacturing (AM) technique that built 3D part in a layer-by-layer method by melting the top surface layer of a powder bed with a high intensity laser according to sliced 3D CAD data. AlSi10Mg alloy is a traditional cast alloy that is broadly used for die-casting process and used in automotive industry due its good mechanical properties. This paper seeks to investigate the requirement SLM in rapid tooling application. The feasibility study is done by examining the surface roughness and dimensional accuracy as compared to the benchmark part produced through the SLM process with constant parameters. The benchmark produced by SLM shows the potential of SLM in a manufacturing application particularly in moulds.

1 Introduction

Selective Laser Melting (SLM) is a layer-wise material addition technique that allows generating complex 3D parts by selectively consolidating successive layers of powder material on top of each other, using thermal energy supplied by a focused and computer controlled laser beam [1-4]. The competitive advantages of SLM are geometrical freedom, mass customisation and material flexibility [5].

Aluminium-Silicon alloys are characterised by sound castability, great weld ability and excellent corrosion resistance. Due to their attractive combination of mechanical properties, high heat conductivity and low weight, the Al-Si alloys found a large number of applications in automotive, aerospace and domestic industries [2]. SLM allows parts to be built additively to form near net shape components rather than by removing waste material. Traditional manufacturing techniques have a relatively high set-up cost (e.g. for creating a mould) while SLM has a high cost per part mostly because it is time-intensive and it is advisable if only very few parts are to be produced. Much of the part from SLM technologies is on lightweight fabrication that applied for aerospace. This technology is able to manufacture multifaceted shapes where traditional manufacturing constraints, such

* Corresponding author: khairu@uthm.edu.my

 DOI: 10.1051/matecconf/20167801077

as tooling and physical access to surfaces for machining and restrict the design of components [6].

Additive Manufacturing (AM) also results in reduction of emission and very low wastage because it able to recycle the powder waste through the processes itself. Lots of energy and resources are consumed to produce tools like dies and moulds [7]. Besides, AM techniques provide almost unchallenged freedom for design without the need for part-specific tooling [8].

To employ SLM into a manufacturing technique, the laser melted parts have to comply the strict material requirements regarding mechanical and chemical properties and the process must guarantee the high accuracy and appropriate surface roughness [9]. In this study, benchmark product was produced by SLM with constant parameters. The process requirements for rapid tooling application was characterised and the feasibility study also done by examining the surface roughness and dimensional accuracy. The benchmark produced by SLM shows the potential of SLM in a manufacturing application particularly in moulds.

2 Methodology

AlSi10Mg is a typical casting alloy with virtuous casting properties and is typically used for cast parts with thin walls and complex geometry and it provides good strength, hardness and dynamic properties and is therefore also used for parts subject to high loads [10].

SLM is an AM process used a laser beam which selectively melts and fuses accumulating layers of powder to build solid metal parts [11]. SLM equipment technology provides a laser system, a set of optical laser beam focusing, a powder feeding system (loader and roller or coater) and a control centre as shown in Figure 1. The SLM is a cyclical process which consists of three steps which repeated until the end of the construction process. First, a recoater applies an even coating of metal powder in layer thicknesses of 20 μm to 100 μm. Then, the exposure will solidify the powder with a laser beam. The absorption of the laser radiation causes the metal powder to heat up above the melting temperature of the metal. This causes the blending of the exposed areas of the current layer and the solidified areas of the layer beneath it through metallurgic melting.

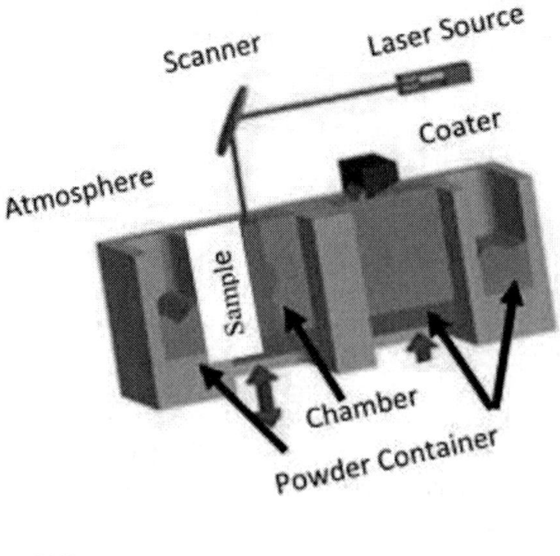

Fig. 1. SLM equipments [11]

The AlSi10Mg in a powder form was sieved into 63 microns, while the loose density of 7.68 g/cm^3 and chemical composition as shown in Table 1 supplied from the LPW Technology Limited UK. For this study, the default SLM parameters shown in Table 2 were used to produce a benchmark sample.

Table 1. Chemical composition of the investigated AlSi10Mg alloy (wt. %).

Si	Fe	Mn	Mg	Ni	Zn	Pb	Sn	Ti	Al
9.92	0.14	0.00	0.29	0.00	0.001.	0.00	0.00	0.01	Balance

Since, the good surface roughness could be an important requirement in rapid tooling, a roughness analysis has been performed. The surface roughness depends on many factors: material, powder particle size, layer thickness, laser and scan parameters, scan strategy and surface post-treatment [12]. The roughness of top and two side surfaces (Figure 4) were measured along different directions as shown in Figure 2 (a). Figure 2 (b) shows a benchmark model consisting of different features such as stairs, cylinders, circumferential surface, rectangular cuts and etc. was developed to test the influence of sloping angle, dimensional analysis, measurement accuracy and the difference between top and bottom surfaces (Figure 4). The sloping angle of this benchmark top plane are ranging from 0° to 90° and bottom planes are ranging from 30° to 90°. On the other hand, the sloping angle for the horizontal holes changes continuously. All surface roughness tests were investigated by using roughness measuring meter and mean R_a and R_z values were calculated with a cut-off length of 2.5 mm, according to the DIN 4768 standard [12].

Table 2. SLM parameters for producing benchmark for dimensional accuracy and surface roughness.

Sample Number	SLM Parameter	Setting Values
1	Laser Power	350 watt
2	Scanning Speed	1650 mm/s
3	Hatch Distance	0.13 mm
4	Layer Thickness	30 Micron
5	Laser Beam Spot	80 Micron
6	Laser Beam Offset	0.025 mm

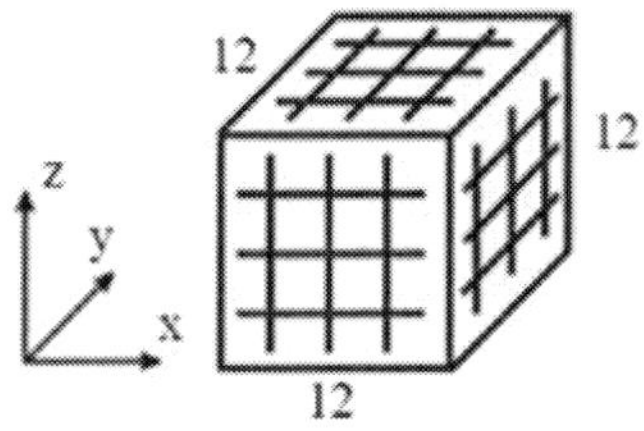

(a)

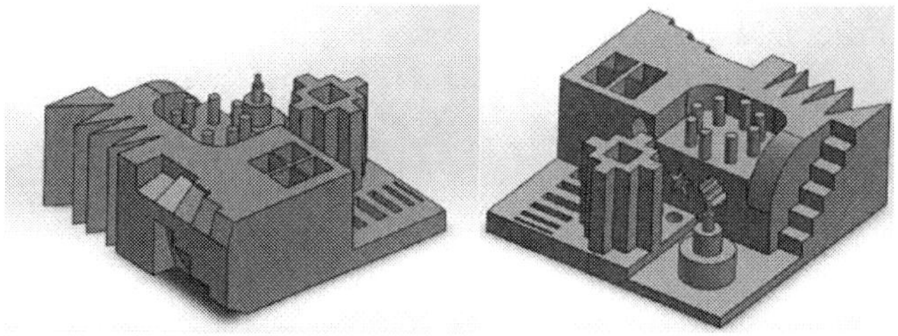

(b)

Fig. 2. (a) Indication of surface roughness measurements on blocks; (b) Benchmark model with different features and sloping angles for top and bottom planes

The benchmark model as shown in Figure (2b) was characterised to investigate the process accuracy (in x, y, and z directions), accuracy of measurement for cylinders, stairs, rectangular cuts and other angle features. The presence of 2 mm thin wall is to indicate the warpage due to thermal stresses. Otherwise the features as Table 3 was developed to evaluate the feasible precision and resolution of the process.

 DOI: 10.1051/matecconf/20167801077

Table 3. Developed features on the benchmark models

Features	Range (mm)
Small holes	0.5 – 3 (diameter)
Small slots	0.5 – 3 (thickness)
Small Cylinders	1 – 5 (diameter)
Thin walls	0.5 – 3 (thickness)

All geometrical features of benchmarks produced were measured three times by tactile probes on a Numerical Control 3D coordinate measurement machine (CMM) as shown in Figure 3.

Fig. 3. Numerical control 3-D coordinate measuring machine (CMM)

3 Results and Discussion

The breakthrough of SLM as a Rapid Manufacturing technique will depend on the reliability, performance and economical aspects like production time and cost [13]. These factors cannot be characterised in general, but some were investigated in this paper for the mould application. The results of the roughness analysis on the benchmark model as shown in Figure 2(b) are presented in Table 4.

The building orientation of side surface and top surface are difference where the top surface is built in x-y direction while side surface was built in x-z direction (Figure 4). This difference should give a difference value for surface roughness, however both top and side surfaces not present significant differences regarding measurement direction.

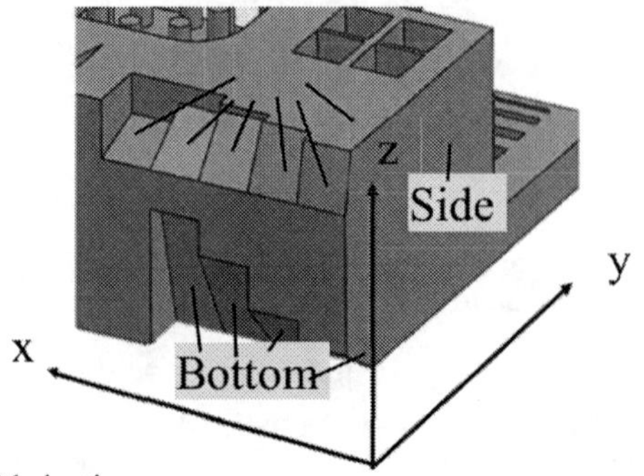

Fig. 4. Orientation of benchmark fabrication.

In spite of a lower powder particle size, the AlSi10Mg samples showed higher roughness in R_z value (Table 4), it is because of physical properties of the material where the melt pools are more stable for Silicon.

In order to have appropriate reading of average surface roughness the measurements were taken at angles, curves and other features at the top and bottom (Figure 4) of the benchmark while on the straight face of the side of it (Figure 4), as shown in Table 4.

Table 4. Average surface roughness for top, down and side of the benchmark

No	Average	SLM AlSi10Mg						
		Top			Bottom			Side
		45°	60°	75°	45°	60°	75°	
1	Ra (μm)	3.35	2.28	3.43	9.20	8.32	7.19	2.29
2	Rz (μm)	15.05	12.16	15.63	42.16	43.61	34.96	11.45

The surface roughness depends on many factors such as material, powder particle size, layer thickness, laser and scan parameters, scan strategy and surface post-treatment [14]. The surface roughness of a sloping plane depend on the sloping angle because of the stair effect due to the layer wise production. In addition, roughness of top surfaces differ strongly from the roughness of bottom surfaces. So the values of roughness in this work are varied by varying the angles of the benchmark because of the layer thickness and the particle size of the AlSi10Mg.

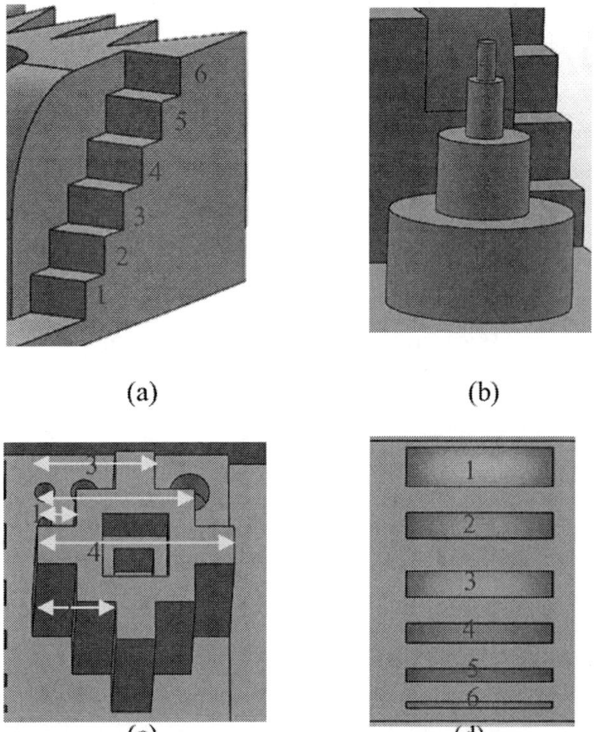

(a) (b) (c) (d)

Fig. 5 (a) to (d). Benchmark of different parts for dimensional analysis

The dimensional accuracy analysis of SLM benchmark model was shows in Table 5 to Table 8. Figure 5 (a) to 5 (d) indicate the different shape feature of benchmark such 90° edge (stair) (Figure 4a), cylinders (Figure 5b), reactangular (Figure 5c) and variate rectangular slot (Figure 5d).

Mean value and maximum deviations between measured and designed dimensions are stated absolutely (mm) and relative to the nominal dimension (percentage). All small features of the benchmark shown on Figure 5 (a) to (d), were built successfully. The highest deviation percentage is obviously shown in table 5(b) with 11.66%. However, others value are show the value between 0.10% to 3.30%. While the negative value for deviation are indicate the actual sample is expand over the designed dimension and contrary with that, the positive value of deviation show the part was shrinkage. The dimensional accuracy for this study is possible to accept for mould fabrication due and this trend is valuable for mould designer to predict the mould design.

Table 5. Dimension measurements of stairs as shown in Figure 5(a)

Stairs No.	1	2	3	4	5	6
Designed Dimensions(mm)	3.000	6.000	9.000	12.000	15.000	18.000
Reading 1	3.044	5.986	9.018	12.083	15.104	18.127
Reading 2	3.011	5.996	9.042	12.085	15.126	18.184
Reading 3	3.038	5.996	9.026	12.063	15.091	18.194
Mean Value	3.031	5.992	9.028	12.077	15.107	18.168
Deviation (%)	-1.470	0.240	-0.190	-0.690	-0.690	-0.710

Table 6. Dimension measurements of cylinders as shown in Figure 5(b)

Cylinder No.	1	2	3	4
Designed Dimensions in mm	10	5	1.5	0.5
Reading 1	9.854	4.898	1.460	0.432
Reading 2	9.862	4.892	1.460	0.451
Reading 3	9.871	4.897	1.470	0.442
Mean Value	9.862	4.896	1.460	0.442
Deviation (%)	1.380	2.080	2.530	11.660

Table 7. Dimension measurements of the rectangular slot with different relating position as shown in Figure 5(c)

Designed Dimensions (mm)	**1**	**2**	**3**	**4**	**5**
Reading 1	2.989	5.945	8.887	11.865	14.876
Reading 2	2.997	5.976	8.879	11.880	14.866
Reading 3	2.990	5.974	8.876	11.878	14.883
Mean value	2.992	5.965	8.881	11.874	14.875
Deviation (%)	0.270	0.920	1.320	1.050	0.830

Table 8. Dimension measurements of variable rectangular slots as shown in Figure 5(d)

Rectangular Slot Number	1	2	3	4	5	6	7
Designed Dimensions in mm	3	2	2	1.5	1	0.5	8
Reading 1	2.998	2.028	1.994	1.499	1.056	0.426	8.005
Reading 2	2.963	2.037	1.996	1.497	1.032	0.394	7.994
Reading 3	3.002	2.021	1.998	1.493	1.010	0.409	7.974
Mean Value	2.988	2.029	1.996	1.496	1.033	0.410	7.991
Deviation (%)	0.400	-1.450	0.200	0.250	-3.300	18.000	0.113

4 Conclusion

AlSi10Mg has been characterised with rapid manufacturing technique by SLM technology. After analysis of the roughness and dimensional accuracy, it can be concluded that the values of roughness on the bottom are higher than the top and side of the benchmark because of the starting of the SLM process the thermal conductivity influence on the layers of the features of the benchmark.

For the mould manufacturing from AlSi10Mg by SLM process, there are many positive and negative variations in the measured values and designed values of the benchmark, nevertheless the range of the deviation is acceptable.

Lastly SLM can produce even a complex shape of geometry, which proved that SLM enables an efficient production of mould manufacturing and complex parts with strong economic, good surface, dimensional accuracy and potential.

The authors would like to acknowledge the financial support from Universiti Tun Hussein Onn Malaysia for this work. (UTHM) and Kolej Kemahiran Tinggi Mara Kuantan for the technical support.

References

1. J.P. Kruth, CIRP Annals-Manufacturing Technology, **40**, 603 (1991)
2. K. Kempen, L. Thijs, J. Van Humbeeck, J.P. Kruth, Phys. Procedia, **39**, 439 (2012)
3. I.P. Ilyas, *Production of plastic injection moulding tools using selective laser sintering and high speed machining*, (University of Leeds, 2007)
4. J.P. Kruth, CIRP Annals-Manufacturing Technology, **40**, (1991)
5. J.P. Kruth, P. Mercelis, J. Van Vaerenbergh, L. Froyen, M. Rombouts, Rapid Prototyping J., **11(1)**, 26 (2005)
6. A.J. Pinkerton, Opt. Laser Technol., **78(8)**, 25 (2016)
7. I. Tolosa, F. Garciandía, F. Zubiri, F. Zapirain, A. Esnaola, Int. J. Adv. Manuf. Tech., **51**, 639 (2010)
8. C. Tuck, R. Hague, International Journal of Mass Customisation, **1(2-3)**, 360 (2006)
9. H. Pohl, A. Simchi, M. Issa, H.C. Dias, *Solid freeform fabrication symposium, University of Texas at Austin*, 366 (2001)
10. E. Louvis, P. Fox, C.J. Sutcliffe, J. Mater. Process. Tech., **211(2)**, 275 (2011)
11. S.L. Campanelli, A. Angelastro, A.D. Ludovico, N. Contuzzi, *Capabilities and performances of the selective laser melting process*, (INTECH Open Access Publisher, Italy, 2010)
12. B. Vandenbroucke, J.P. Kruth, Rapid Prototyping J., **13(4)**, 196 (2007)

13. N. Hopkinson, R. Hague, P. Dickens, *Rapid Manufacturing: An Industrial Revolution for the Digital Age* (John Wiley & Sons, 2005)
14. D. Buchbinder, H. Schleifenbaum, S. Heidrich, W. Meiners, J. Bultmann, Phys. Procedia, (12), 271 (2011)

The Influence of Selective Laser Melting Parameters on Density and Mechanical Properties of AlSi10Mg

A. A *Raus*[1,*], M. S. *Wahab*[1], Z. *Shayfull*[2], K. *Kamarudin*[1], and M. *Ibrahim*[1]

[1]Universiti Tun Hussein Onn Malaysia, Parit Raja, Batu Pahat, Johor, Malaysia
[2]Green Design and Manufacture Research Group, Center of Excellence Geopolymer and Green Technology (CEGeoGTech), Universiti Malaysia Perlis, 01000 Kangar, Perlis, Malaysia.

Abstract. Selective Laser Melting (SLM) is one of the most effective powder bed technique in the additive Manufacturing (AM) which able to fabricate functional metal parts directly from 3D Computer Aided Design (CAD) file data. In this paper, the influence of SLM parameters, such as laser power, scanning speed and hatching distance on the density of AlSi10Mg samples are investigated using one factor at a time (OFAT). Furthermore, the optimum results are used to fabricate samples for hardness, tensile strength, and impact toughness test. It is revealed that AlSi10Mg parts fabricated by SLM achieving the best density of 99.13% at the value of 350 watts laser power, 1650 mm/s scanning speed and hatching distance 0.13mm, whereby resulted comparable and even better mechanical properties to those of conventionally HDPC A360F and HDPC A360T6 alloys although without any comprehensive post processing methods.

1 Introduction

Selective Laser Melting (SLM) is considerably one of the sufficient technique in the additive Manufacturing (AM) procedure in which a component is fabricated with a layer by layer technique through melting the top surface of a powder bed with a high intensity laser according to sliced 3D Computer Aided Design (CAD) data. Thin layers of material are spread intricately and stacked on the top of the building platform. A laser beam then melts and fuse the layer of powder metal, referring to the generated slice. After fabricating a layer was completed, the build piston will lower down the build platform and the following layer of powder is spread. The fabrication process repetitions continue from bottom to top until the part is completed [1-4]. The SLM process is controlled by the set of parameters listed in Table 1 that has a major influence on the quality of the final part [5]. The foremost build parameters involved in the process are laser power, hatch spacing, scanning speed, and layer thickness.

* Corresponding author: gd150034@siswa.uthm.edu.my

Table 1. Controlling parameters in SLM process [5]

Process Parameters			
Laser-related	**Powder-related**	**Scan-related**	**Temperature-related**
• Laser power(watt) • Spot Size(μm) • Pulse Duration(μs) • Pulse frequency(Hz)	• Particle size(μm) • Particle shape & distribution • Powder bed density (g/cm³) • Layer Thicknes(μm) • Material properties	• Scan Speed (mm/s) • Scan Spacing (μm) • Scan Pattern	• Powder bed temperature(°C) • Powder feed temperature(°C) • Temperature uniformity(°C)

Besides the influence of processing parameters, powder characteristics are of significant concern. The shape, size, particle distribution and powder chemical composition have a strong effect on the melting and fusing process. Powders with low oxygen, content high flow rate and high packing density are preferred [6-9]

Aluminium-Silicon alloys are very light metal and owned a high strength, good weld ability and exceptional corrosion resistance. Due to their attractive combination of mechanical properties and high thermal conductivity, Al-Si alloys have been used in many industrial applications such as aerospace, automotive, and moulding industries [3,10-13]. By adding magnesium in aluminium-silicon alloys will increase the mechanical properties and improved the ductility and modulus of elasticity properties. Hence AlSi10Mg contains 0.3 to 0.5 wt% Mg, it is harden able through T6 cycle of solution annealing, quenching and age hardening heat treatment process to improve the mechanical properties [14-15].

The main goal of this paper is to investigate the density and mechanical properties such as hardness, tensile strength and impact toughness of AlSi10Mg samples fabricated by SLM and compare those properties to the high pressure die cast (HDPC) A360 alloy. The results will be used to predict and to facilitate the manufacturing process for mould fabrication.

2 Experimental works

The SLM system SLM 125 HL (SLM Solutions) as shown in Figure 1(a), was used to manufacture all samples in this study. The machine system is equipped with a 400 watt fibre laser , 80 μm laser beam diameter and has a building chamber of 125 mm × 125 mm × 125 mm as shown in Figure 1(b). Argon is used as an inert gas in the building chamber. The stripe scanning strategy with 45° orientation and 90° rotated incrementally on the next layer was used. The platform building was heated at 150 °C to avoid the build samples from warping due to non-uniform thermal expansion at elevated temperature. Constant thickness layer at 30μm has been set by lowering the build platform. An aluminium platform is placed on the building platform (x-y table) and levelled.

(a) (b)

Fig. 1. (a) SLM 125HL 20121213-EN, (b) Build chamber SLM system SLM 125 HL

The AlSi10Mg in powder form having sieving powder of 63 micron, a loose density of 7.68 g/cm^3 and chemical composition as shown in Table 2 supplied from the LPW Technology Limited UK.

Table 2. Chemical composition of the investigated AlSi10Mg alloy (wt. %).

Si	Fe	Mn	Mg	Ni	Zn	Pb	Sn	Ti	Al
9.92	0.137	0.004	0.291	0.004	0.001	0.004	0.003	0.006	Balance

The samples 10 mm x 10 mm x 10 mm in cubic shape were fabricated in 3 x 11 matrix by varying one factor (laser power/scanning speed/hatching distance) and the remaining two parameters constant at the values shown in Table 3 to find the relative density by using Archimedes method.

Table 3. Parameters selection using OFAT for Selective Laser Sintering Machine

Row No.	Laser Power	Scanning speed	Hatching distance	No. of Sample
1	320 – 380 W	1650 mm/s	0.13 mm	11
2	350 W	1550 -1750 mm/s	0.13 mm	11
3	350 W	1650 mm/s	0.10 – 0.15 mm	11

In order to fabricate the mechanical properties test samples in optimum condition, the experiment was conducted using the optimised parameters of the density test. A laser power of 350 watt scan speed of 1650mm/s and a hatching distance of 0.13 mm was used to fabricate the samples as shown in Figure 2.

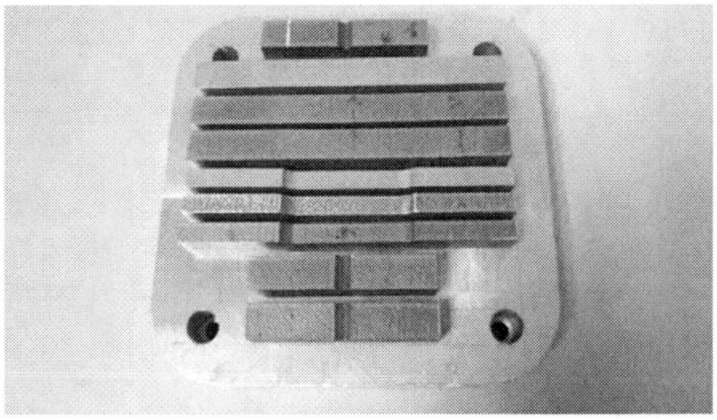

Fig. 2. The arrangement of samples on the SLM platform

To determine the mechanical properties, hardness, tensile, and Charpy impact tests were performed. The value of the hardness tests was measured using Shimadzu HMV-2T E under the load of 2.942 N (HV0.3). The measurement was taken from the top and side surfaces at 9 different locations for general characterisation of the microstructure as shown in Figure 3 (a) and (b).

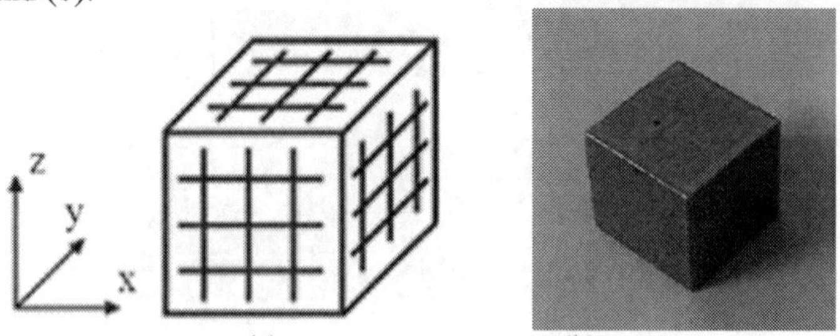

(a) (b)

Fig. 3. (a) Indication of different locations for hardness test on blocks; (b) (10x10x10)mm cube produced by SLM for hardness test.

ASTM E8 standard tensile test samples with the 25 mm gauge length were built on the platform in the x-z axis. Precision machining and polishing by abrasive papers were carry out on the gage length to provide the notch effect. A Gotech testing machine with 0.5 mm/s test speed was used to measure the tensile strength. Samples for transverse rupture strength test were prepared according to the ASTM E23 Type A [16].

3 Results and discussion

3.1 Morphology of AlSi10Mg powder particles

Figure 4 shows the Field Emission Scanning Electron Microscope (FESEM, JSM-7600F) micrograph of AlSi10Mg powder as in recieved form. The AlSi10Mg particles are not spherical and size ranges from 5 to 80 μm. The appearance of the particles is irregular morphology, with many small irregular satellite particles agglomerated to bigger particles creating some clusters of about 60 to 90 μm, and this could be detrimental for the final density of the SLM parts, considering that the layer thickness employed in this research is of 30 μm. Regardless of the irregular powder morphology, the powder had an acceptable flowability and Hauser's ratio for SLM.

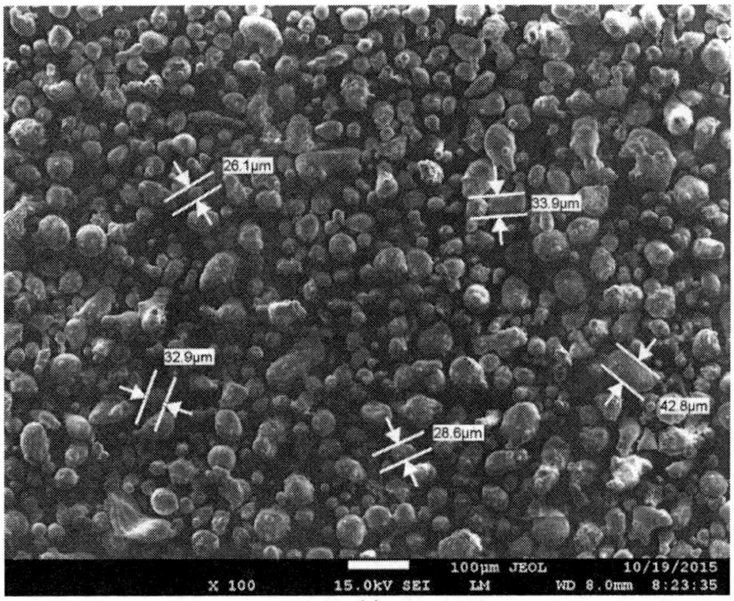

(a)

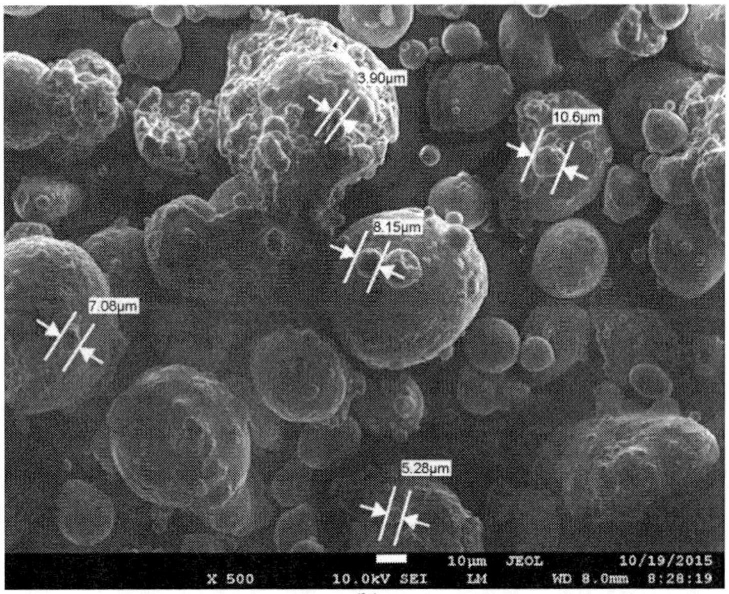

(b)

Fig. 4. Shows the Field Emission Scanning Electron Microscope (FESEM, JSM-7600F) micrograph of AlSi10Mg powder (a) x100, (b) x500

3.2 Density Study

In this study the Archimedes standard method was used to determine the density of the SLM processed AlSi10Mg samples and compared with the distilled water in the pycnometer. Table 4 and Figure 4 shows the values for measuring density using an OFAT technique on laser power, scanning speed and Hatching distance.

DOI: 10.1051/matecconf/20167801078

Table 4. Density measurement by Archimedes method

Parameter	Density #1	Density #2	Density #3	Density #4	Density #5
320-380W	98.70%	99.03%	98.91%	98.88%	99.07%
1550-1750 mm/s	98.88%	99.05%	98.96%	99.04%	99.11%
0.10-0.15 mm	98.92%	98.84%	98.78%	99.11%	99.13%

Density #-6	Density #7	Density #8	Density #9	Density #10	Density #11
98.88%	98.83%	98.85%	98.67%	98.56%	98.54%
99.00%	98.86%	99.08%	98.93%	98.94%	98.87%
99.10%	99.04%	98.90%	98.79%	98.78%	98.65%

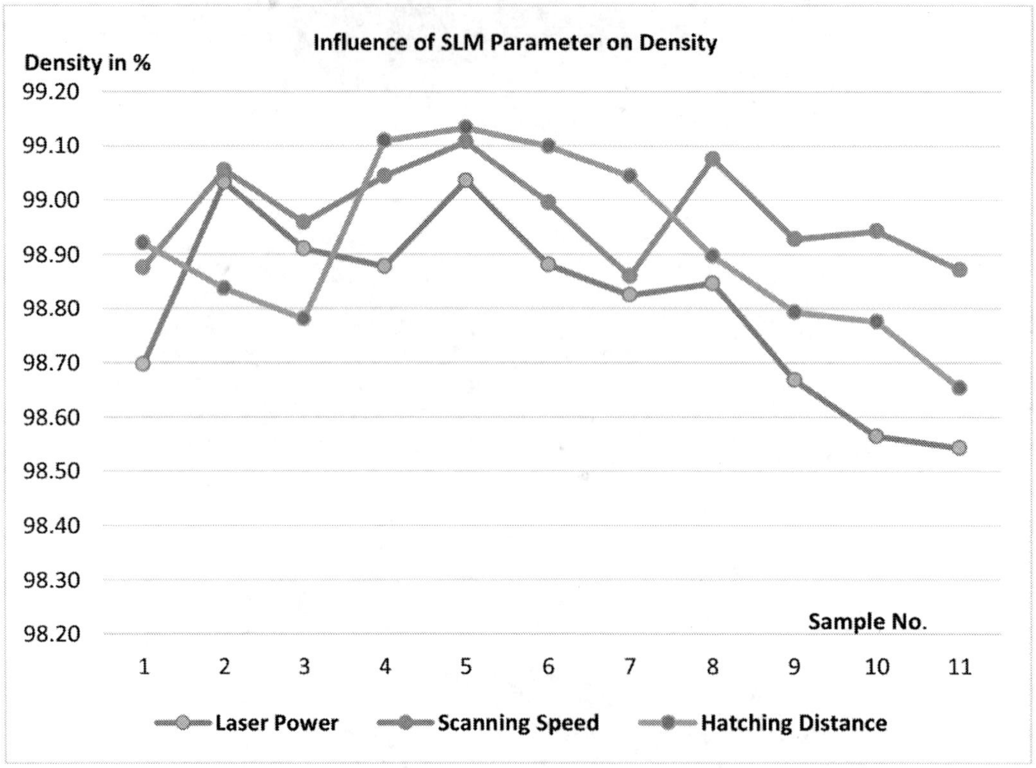

Fig. 5. Density with SLM Parameters (Laser power, Scanning speed and Hatching distance.)

At the variation of laser power from 320 to 380 watts with an interval of 6 watts for each sample and where the scanning speed and hatch distance remain constant at 1650 mm/s and 0.13 mm respectively, the highest density value of 99.07% occurred at sample No. 5 with the laser power of 350 watts. Meanwhile, at the constant laser power and a hatching distance at 350 watts and 0.13 mm, respectively, with the variation of scan speed of 1550 to 1750 mm/s and equal intervals of 20 mm/s, the density increase slightly to 99.11%. While the values of density measurement with the range of hatching distance from 0.1 mm to 0.15 mm by the intervals of 0.05 mm and remained the laser power and scanning speed constant at 350 watt and 1650 mm /s respectively, gave preeminent result of 99.34% at hatching distance of 0.13 mm.From the trend of the results from experiment, it shows that the optimum value of density occurred at 350 watts laser power, 1650 mm/s scanning speed and 0.13 mm hatching distance, whereby proven by the machine manufacture as the optimum parameters for AlSi10Mg. From the graph presented in Figure 4, it also can be noticed that the minimum value of density occurred at the highest value of each parameter. Whereby high input of the laser power may result, excessive energy density that leading to oxidation and pore formation. Meanwhile the increasing porosity with the increases of the hatching distance is due to the reduces of the intra-layer overlap and the part is held together mostly through interlayer bonding. Differently for the scanning speed effect, where metallurgical pores occurred at a slow scanning speed which the gases trapped inside the molten pool consolidation. Meanwhile keyhole pores formation arisen when the scanning speed is too fast due to rapid solidification of the alloy without completely filling the gaps with molten alloy [5].

3.3 Mechanical properties

Hardness is well recognised as a preliminary indication for mechanical properties. By using the Micro-Vickers hardness testing machine, the measurements on 3 test cube samples with dimensions of 10 mm ×10 mm × 10 mm were conducted at 10 different locations of the top surface (x-y direction) and side wall surface (y-z direction). The mean value of the hardness on both surfaces direction is listed in Table 5.

Ultimate Strength, Yield Strength and Elongation (%) results were achieved from the tensile tests according to ASTM E23 standard. Three samples of AlSi10Mg fabricated by SLM were inspected in the x-z direction at optimised parameters. The mean values are listed in Table 5 while the stress-strain curve is presented in Figure 5. As the reference and comparision to the current research work, the value of mechanical properties such as Ultimate Strength, Yield Strength and Elongation (%), Hardness and Impact energy of High pressure die cast (HPDC) alloy A360F, HDPC alloy A360T6 [17] are also listed in the table 5.

Both HPDC alloys are considered to give the optimum mechanical properties for industrial application. This is the reason to examine the mechanical properties of AlSi10Mg sample fabricated by SLM in the x-z direction for further investigation and practicability in the manufacturing and other industrial applications.

Table 5. Mechanical properties of SLM AlSi10Mg samples and HPDC alloy A360F and A360T6

Material	Direction	Yield Strength (MPa)	Ultimate Tensile Strength (MPa)	Elongation at Break (%)	Hardness (HV)	Impact energy (J)
AlSi10Mg	X-Z	242 ± 5	412 ± 2	6.34 ± 0.3	139 149	2.5 ± 0.5
A360F[17] A360T6[17]	-	160-185 285-330	300-350 330-365	3-5 3-5	95-105 130-133	2.5 ± 3.0

Both values of Vickers hardness measurements on the top surface (x-y direction) and side wall surface (y-z direction) are greater than HPDC alloy A360F and A360T6. The Vickers hardness values were improved 32% for the top surface (x-y direction) and 42% for the side wall surface (y-z direction) compared to HPDC alloy A360F. On the other hand, the Vickers hardness value were also improved by a small margin of 4% for the top surface (x-y direction) and 12% for the side wall surface (y-z direction) as compared to HDPC alloy A360T6. The yield strength value of the AlSi10Mg samples produced by SLM is 31% higher compared to HPDC alloy A360F, but 27% lower from the HDPC alloy A360T6.Whereas the value of the ultimate tensile strength of SLM AlSi10Mg samples is higher when compared to the both HPDC alloys. The elongation at break is nearly 2% higher as compared to the HPDC alloys in both conditions. Although the strength and hardness values were higher than the HPDC alloys, still the result from the charpy impact energy of this experiment is comparable to both HPDC alloys.This experiment result trend has also been proven by K.Kempen et al [18] in his experiment using the same material processed through SLM although the samples were fabricated in different scanning stratiges and x-y and z building directions.

Overall, from the experimental results, it can be observed that the AlSi10Mg samples produced by SLM in x-z direction possessed higher mechanical properties as compared to the HPDC alloy A360F and HDPC alloy A360T6. It can be concluded that AlSi10Mg product produced by SLM can be practically employed in the manufacturing and other industrial applications.

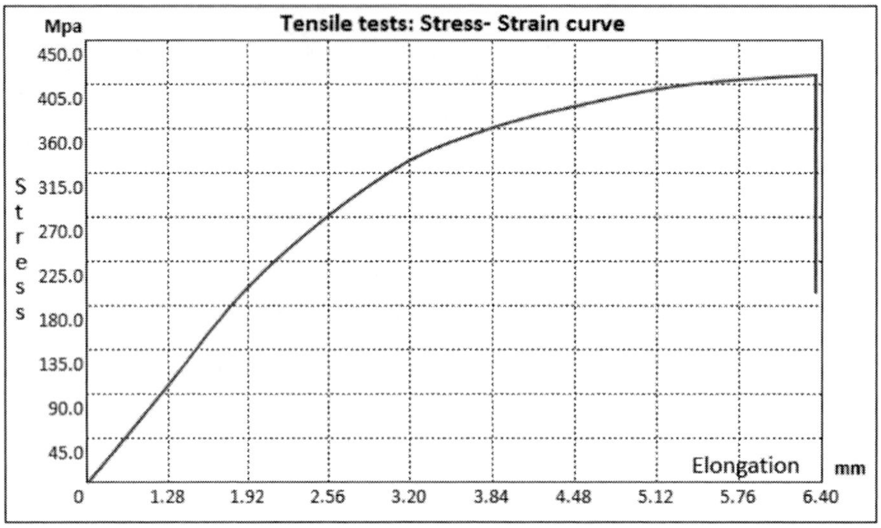

Fig. 5. Stress-Strain curves for SLM AlSi10Mg Samples built along X-Z direction

4 Conclusion

After characterisation and analysis of the powder size, morphology and chemical composition through to the assessment of mechanical properties of AlSi10Mg SLM as fabricated samples, it can be concluded that:

- The SLM process is controlled by the set of parameters involved in the process such as laser power, scanning speed, hatching spacing, and layer thickness. Furthermore, with the optimised application of scanning strategy will increase the part density from 99.34 % to 99.8%.
- The OFAT method has been used to evaluate the influence of process parameters on the porosity of SLM processed AlSi10Mg shows the trends of porosity in the SLM fabricated samples.
- As a comparison of the properties of a the HPDC alloy A360F to the HDPC alloy A360T6, AlSi10Mg SLM samples demonstrated slightly higher values of hardness, yield strength, ultimate tensile strength and elongation at break, while for the Charpy impact energy test, there is comparable, although with a slightly lower value and can

be concluded that it can be practically employed for varieties industrial applications even in very challenging like aerospace, medical and manufacturing sectors.

- Upon observation of the part density and mechanical properties of the SLM processed AlSi10Mg samples, it can be determined that the samples possessed a comparable and even better mechanical properties to those of conventionally HDPC A360F and HDPC A360T6 alloys although without any comprehensive post processing methods.

This research work is financially supported by Research Innovation Commercialisation and Consultancy Management (ORICC), University Tun Hussein Onn Malaysia (UTHM).

References

1. C.Y. Yap, C.K. Chua, Z.L. Dong, Z.H. Liu, D.Q. Zhang, L.E. Loh, S.L. Sing, Appl. Phys. Rev., **2(4)**, (2015)
2. X. Yan, P. Gu, CAD Comput. Aided Des., **28(4)**, 307 (1996)
3. N.P. Karapatis, J.P. Van Griethuysen, R. Glardon, Rapid Prototyping J., **4(2)**, 77 (1998)
4. K.V. Wong, A. Hernandez, ISRN Mech. Eng., **16**, (2012)
5. N.T. Aboulkhair, N.M. Everitt, I. Ashcroft, C. Tuck, Additive Manufacturing, **1**, 77 (2014)
6. D. Manfredi, F. Calignano, M. Krishnan, R. Canali, E.P. Ambrosio, E. Atzeni, Materials, **6(3)**, 856 (2013)
7. E. Louvis, P. Fox, C.J. Sutcliffe, J. Mater. Process. Tech., **211(2)**, 275 (2011)
8. Y. Wang, J. Bergström, C. Burman, Mater. Sci. Eng. A, **513**, 64 (2009)
9. E. Brandl, U. Heckenberger, V. Holzinger, D. Buchbinder, Mater. Des., **34**, 159 (2012)
10. B. Vandenbroucke, J.P. Kruth, Rapid Prototyp. J., **13(4)**, 196 (2007)
11. D. Manfredi, F. Calignano, E. P. Ambrosio, M. Krishnan, R. Canali, S. Biamino, M. Pavese, E. Atzeni, L. Luliano, P. Fino, C. Badini, La metallurgia italiana, **105(10)**, 15 (2013)
12. M. Rivette, P. Mognol, and J.Y. Hascoet, Rapid Prototyp. J., **19(2)**, 77 (2013)
13. A. Simchi, Mater. Sci. Eng. A, **428(1)**, 148 (2006)
14. S. Menargues, E. Martín, M.T. Baile, J.A. Picas, Mater. Sci. Eng. A, **621**, 236 (2015)
15. J. Rajabi, G. Branch, J. Rajabi, E. Alibeiki, G. Branch, M. Nekoei, S. Branch, M. Rajabi, G. Branch, M. R. Meschian, M. Branch, Academic Research International, **2(3)**, 114 (2012)
16. ASTM E23 - 16b, *Standard Test Methods for Notched Bar Impact Testing of Metallic Materials*, (ASTM Standards 2002)
17. L. Roger, *Technical data sheets for heat treated aluminium high pressure die castings*, (CSIRO Light Metals Flagship Technical data 2008)
18. K. Kempen, L. Thijs, J. Van Humbeeck, J.P. Kruth, Phys. Procedia, **39**, 439 (2012)

Investigation on Microstructure of Heat Treated High Manganese Austenitic Cast Iron

A.K. Muzafar[1,*], *M.M.* Rashidi[1] , *I.* Mahadzir[1] , and *Z.* Shayfull[2]

[1]Faculty of Mechanical Engineering, Universiti Malaysia Pahang, 26600 Pekan, Pahang, Malaysia.
[2]School of manufacturing Engineering, Universiti Malaysia Perlis, Kampus Tetap Pauh Putra, 02600 Arau, Perlis Malaysia.

Abstract. The effect of manganese addition and annealing heat treatment on microstructure of austenitic cast irons with high manganese content (Mn-Ni-resist) were investigated. The complex relationship between the development of the solidification microstructures and buildup of microsegregation in Mn-Ni-resist was obtained by using microstructure analysis and EDS analysis. The annealing heat treatment was applied at 700°C up to 1000°C to investigate the effect of the annealing temperature on the microstructure. This experiment describes the characterization of microsegregation in Mn-Ni-reist was made by means of point counting microanalysis along the microstructure. With this method, the differences of silicon, manganese and nickel distribution in alloys solidified in the microstructure were clearly evidenced. The results show microstructure consists of flake graphite embedded in austenitic matrix and carbides. There is segregation of elements in the Late To Freeze (LTF) region after solidification from melting. Manganese positively with high concentration detected in the LTF region. As for heat treatment, higher annealing temperature on the Mn-Ni-resist was reduced carbide formation. The higher annealing temperature shows carbide transformed into a smaller size and disperses through the austenitic matrix structure. The size of carbide decreased with increasing annealing temperature as observed in the microstructure.

1 Introduction

Austenitic cast irons also known as Ni-resist are an important casting materials and their use is justified by the specialist of mechanical properties which can be achieved, associated with their outstanding abilities on wistanding the effects of corrosion [1], heat and wear. These materials have been used for more than 50 years, dating back to early of 1930 [2]. As the nickel content varying from 13 to 37 wt% as one of the main element, it is the most widely utilized materials in corrosive environments, due to its excellent resistance to corrosion [3-6].

Austenitic cast irons microstructure containing high composition of austenitic matrix. This matrix appears contributed by the influence of nickel contained in the composition that acts as austenite matrix promoter. Nickel forms a continuous series of solid solutions

* Corresponding author: mrashidi@ump.edu.my

with iron and has an almost unlimited solubility in austenite. It will be in austenitic matrix when high temperature and also at room temperature which make it successfully avoid matrix volume changes. This phenomenon occurs due to the volume changes which occur because of matrix phase changes between ferrite and austenite. Ni-resist able to stay consistence in austenitic matrix at all temperatures and have avoided this type of transformation [7]. This can be achieved by suitable addition of alloying elements, the time–temperature–transformation (T–T–T) curve is generally pushed to the right, make solidification curve at room temperature possible to miss the 'nose' of the T–T–T curve [1, 8, 9]. Nickel broadens the temperature range in which austenite exists and makes the austenitic matrix stable at room and lower temperatures [10]. The austenite matrix in these alloys is a metallic phase with an FCC lattice, consisting of a solid solution based on iron which all the alloying elements enter.

Although, numerous data have been published on the advantages of the Ni-resist it has a major drawback on the production cost. The high price is contributed by the nickel element cost [11]. Some efforts have been made to reduce nickel content in Ni-resist iron. Recently, evidence discovered that production of austenitic ductile cast iron having wholly austenitic matrix only by using 13.5 % Ni for ductile form graphite [12]. This condition suppose only applied for ASTM A436 rather than ASTM A439. On the other hand, there are several investigation purposes the use of other element as replacement for nickel such as manganese and copper. They have been the object of recent studies [13-16]. Furthermore, it is also believed that manganese is better austenite promoter than copper [15].

Manganese seem to be most promising alternative to be used as it also austenitic promoter despite manganese also suspected to interrupt with the nodularisation during solidification [17], acts as sulphur controller [18] and also well known as carbide promoter [19-21]. Although some of the nickel in these alloys can be replaced by manganese, the structure will still remain austenitic matrix. The concentrations of this element in the austenite usually differ from the average concentration in the cast iron. This is due to the limited solubility of the manganese in the liquid or solid solution which is partially soluble in it and furthermore encourages the formation of carbide. The existence of carbide is very undesirable element in producing iron which limits the possibility of their being added to austenitic cast iron but demanded when related for wear ability purpose. Manganese segregates severely in inter graphite area and be in last to freeze (LTF) region [3, 7, 21, 22]. This region is enriched with positive segregation elements such as Mn and Cr [23].

The production of austenitic cast iron with high manganese content is limited by several restrictions. As for all cast irons, the Mn-Ni-resist microstructure is affected by the manganese content and cooling rate during solidification. Larger manganese content promotes the precipitation of eutectic carbide throughout austenite matrix and expected to reduce martensite start line (Ms) temperature as reported in previous report [24]. The presence of carbide in the matrix microstructure is predicted strongly detrimental to Mn-Ni-resist mechanical properties and processing.

There are two different approaches can be used to obtain carbide free iron. The first approach would be to avoid carbide precipitation during solidification. This can be achieved by maximizing the efficiency of inoculation. Good inoculation technique managed to reduce carbide precipitation during solidification and improves its mechanical properties. The second approach would be to eliminate the carbide precipitated during solidification by means of a dissolution heat treatment.

Now attention turned to improvement associates with the application of heat treatment. It is believed that annealing able to minimize the formation of excessive carbide in microstructure cause by manganese. This information based on previous study [25]. Annealing process is a heat treatment which consists combination of operation for the purpose of obtaining certain desirable conditions or properties. Through this refining process will produce a mechanical structure or properties that are superior in term strength and toughness.

This study attempts to combine manganese and reduced nickel as an austenitic promoter in Ni-resist and at the same time use annealing heat treatment to reduce side effect of the manganese on the microstructure. This was aimed to investigate the effect of manganese on flake graphite of Mn-Ni-resist and annealing temperature on the microstructure of Mn-Ni-resist. The annealing temperature varies from 700°C up to 1000°C. The result also will show optimum annealing temperature required to alter the microstructure and mechanical properties. Analysis was concentrated on evaluating the effect of temperature on the graphite, carbide of the modified austenitic cast iron.

2 Experimental procedure

2.1 Material preparation

The studied alloy was prepared in a 100 kg-capacity frequency induction furnace. Initial charge materials were pig iron, pure nickel, steel scrap, Ferro manganese and austenitic iron returns. Then the materials heated until molten state. Alloying elements such as 99.0 wt% pure nickel and fe-45.0 wt% Mn were added to the molten alloy so as to increase and control Ni and Mn percentage content respectively. The mould was made by using Y block shaped pattern with minimum thickness of 30mm was prepared by using a pattern as in Figaccording to ASTM standard for austenitic cast iron [26, 27] in the green sand mould. Inoculation process with 0.5 wt% Fe-Si was introduced to melt by using late inoculation in the mould technique [21, 28, 29].

Table 3. Chemical composition (wt. %) of the base metal.

Elements	C	Si	Ni	Mn	S	Fe
Composition wt.%	4.5	1.81	10.00	4.98	0.082	balance

This procedure ensures that the inoculation level high and reduce magnesium fading phenomena during casting. The chemical composition of casting product was shown in Table 6.

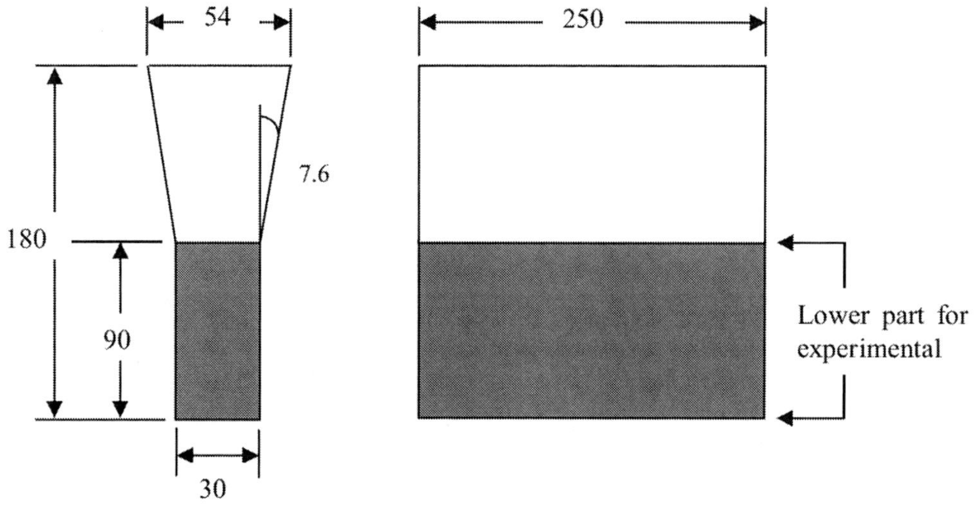

Fig. 1. Dimensions of the Y-block castings used in this experiment. Test specimens cut and machined from the lower part of the Y-block (dimensions in mm).

2.2 Heat treatment process

The heat treatment schedule outlined in Table was carried out. All dog bone samples then were placed in a box-type resistance furnace heated according to its annealing temperature for annealing and 927°C for quenching. The holding time is 3 hours was counted when it reached the desired temperature. Austenitizing time less than 2 hour was not selected because the complete transformation of the as-cast structure into austenite required at least 2 hours [30]. Annealing process Annealing was held at 700°C, 800°C, 900°C and 1000°C for 3.0 h. Then the samples cooled down to room temperature by furnace cooling respectively. For the purpose of easy analysis, as shown in Table , every heat treatment process was named according to its annealing temperature and heat treatment condition.

Due to slight oxidation of the surface of cast iron, there is every possibility of scale formation on this surface during transfer of the samples to the oil tank or cooling to room temperature and making the hardness and tensile value vary. Moreover, the specimen will not also be gripped properly in the machine. The specimens were polished to remove the scales from the surface to avoid these difficulties.

2.4 Microstructure analysis

The metallographic analysis of the specimens was carried out in the round shaped specimen obtained from the broken halves of the tensile specimen. Metallographic samples were sectioned, ground and polished with 1 to 6μm grade diamond paste, rinsed in distilled water and degreased with ethanol. The etch solution of 3 % Nital was used for investigation of microstructure of samples. The microstructure was examined using optical microscope (OM) and scanning electron microscope (SEM) equipped with energy dispersive X-ray spectrometry (EDS). The carbide area fraction (CAF) was obtained by image analysis using images analyzer ImageJ [31]. The retained austenite content of the specimens was measured through X-ray diffraction by a diffractometer.

Table 4. Chemical composition of raw material.

Material	Elements (wt. %)										
	C	Si	Mn	P	S	Mg	Ni	Ca	Cr	R.E	Fe
Pig iron	4.5	1.73	0.309	0.180	0.150	0.310	0.236	-	-	-	balance
Steel	0.19	0.15	0.54	0.90	0.02	-	-	-	-	-	balance
Nickel	-	-	-	-	-	-	99.0	-	-	-	balance
FeMn	-	-	86.00	-	-	-	-	-	-	-	-
Inoculant	-	70.00	-	-	-	-	-	2.0	-	-	balance

3 Result and discussion

3.1 Microstructure analysis

Fig presents microstructure images for each heat treated sample (as-cast, 700°C, 800°C, 900°C, 1000°C) under examination at 10x magnifications with temperature respectively. Metallographic observation shows that all samples are in austenitic microstructure with single bright matrix and the graphite precipitated in the form of flake graphite type VII or more specific type B (rosette pattern) of flake graphite as according to ASTM 247 [32]. High nickel content has produced austenitic matrix as reported in literature [12].

Fig (a) are in as-cast condition generally, shows dark flake graphite distributed and embedded throughout in the austenite matrix. Austenitic irons must contain sufficient amount of nickel and manganese in order to promote austenitic characteristic. Literature indicates that 13.5 wt% Ni together with presence of 5.5 wt% Cu for standard types of austenitic gray cast iron [26]. Through this study, the minimum Ni (10wt.%) used with addition of manganese (6 wt.%) to assist in promoting austenite matrix which is lower than the previous study and ASTM A436. Even though the combination of higher manganese wt% and less nickel wt% stabilized the austenite matrix, it is appears that those combination encourage carbide formation as reported previously [8, 14, 15].

In Mn-Ni-cast iron carbide formation was detected in area inter flake graphite region where located at grain boundary. Although, it is difficult to distinguish the boundary or frame of the iron in the microstructure, the location of carbide believed presence in grain boundary as stated by Jiyang [33] this was evidently supported by

Fig(a). As for carbide formation, this occurrences indicates segregated carbide suppress the existences of free graphite as concludes by latest investigation [21]. The formations of the carbide in this region push the graphite aside. It is making the length between graphite increases. This area is enriched with positive segregation elements such as Mn, as shown in

Fig and

Fig. Generally, they was segregated at alloyed iron grain boundary and identifies as the last to freeze (LTF) region. During the solidification of iron melt, manganese was pushed away from the solid phase (graphite) which is first nucleate and solidifies. Manganese become the last element solidifies after the nucleation of graphite. It has been grouped and segregated away from the graphite. Besides, carbides in cast iron are also obviously situated between austenite dendrites, see

Fig .

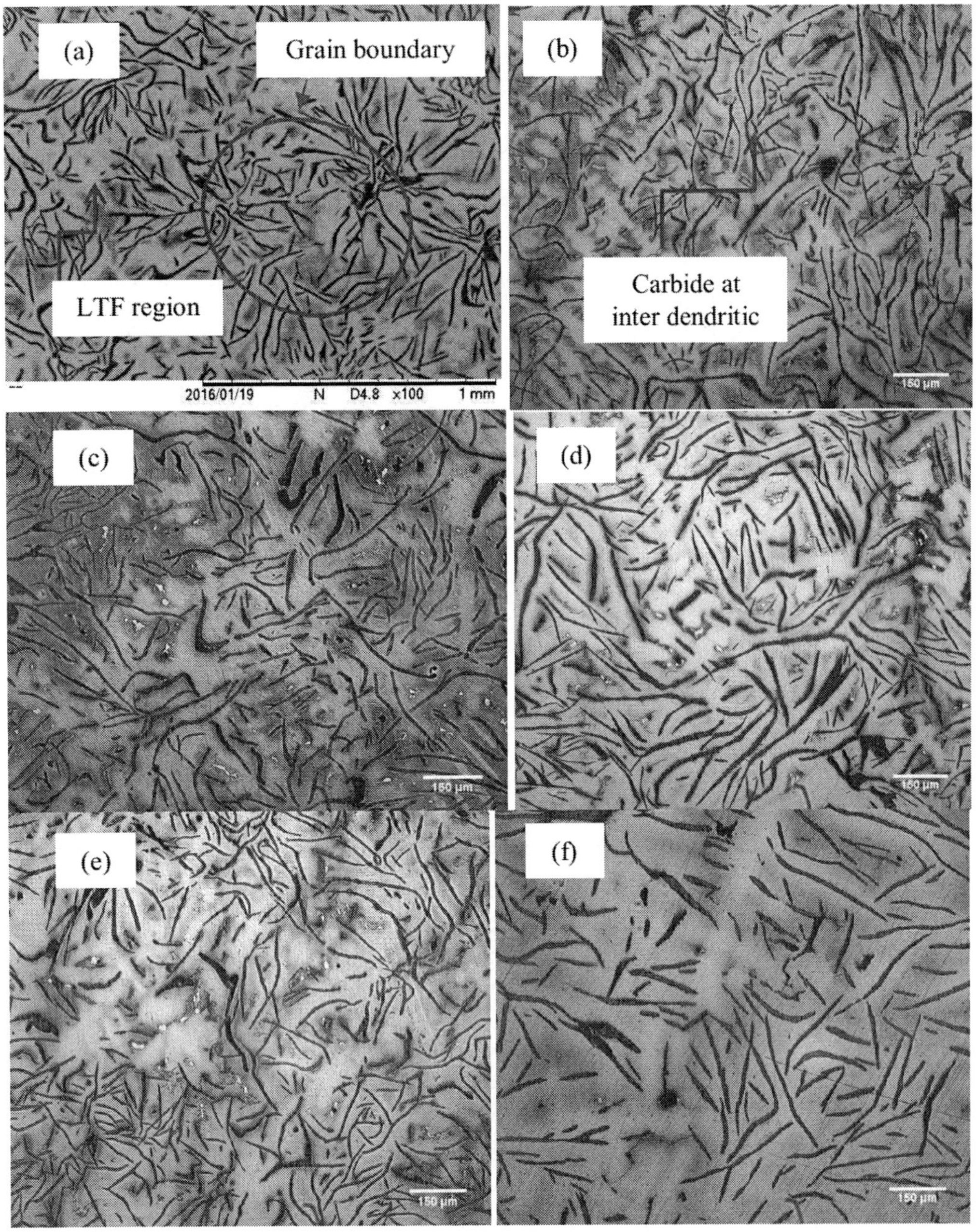

Fig. 2. Microstructure of the modified M-N-resist under 10X magnification: (a) as-cast 100X magnification (b) As- cast condition, (c) Annealing 700°C, (d) Annealing 800°C (e) Annealing 900°C (f) Annealing 1000°C.

Fig (a) to (e) were exemplifier the sequence of carbide formation throughout the microstructure. Carbide was observed dissolves progressively. The rate of carbide dissolution was slow at low temperature but very fast at high temperature. For all cases the dissolution was successfully decrease carbide formation depending on the holding temperature compared to as-cast condition. It was found that the microstructure of the original cast iron with a structure of carbide is higher than that of cast iron with the same chemical composition with a structure of carbide obtained by heat treatment. Increasing annealing temperature shows decreasing pattern of the carbide appearance in the microstructure. As the quantitative observation as recorded in

Table , area fraction of carbide decreasing. As-cast iron with 0.54 % carbide decreases to 0.51 % for sample 700°C, 0.49 % for sample 800°C and 0.32 % for sample 900°C. Surprisingly, a very fast rate of dissolution of carbides was observed in the sample 1000°C as compared to the carbide dissolution of sample 900°C. Very high drop reading showed from 0.32 % to 0.06 % may result from the contribution of near to transition region line A1 which is in the partial melting region. This lower A1 effect by the presences of manganese as austenite stabilizer also lowered the temperatures A1 [24] and believed to drop near to 1000°C. This discovery supported the previous investigation by Ahmadabadi and Feghahati [34] which reported that a heat treatment involving an austenitising at temperature partial melting was sufficient and higher rate to reduce free carbides present in samples.

Table 3. Quantitative microstructure analysis.

Condition / Sample name	Carbide			
	Area average (μm^2)	**Percentage (%)**	**Total carbide area (μm^2)**	**Number of carbide**
As-Cast	55.34	0.54	6918.38	125
700°C	45.40	0.51	6583.50	145
800°C	38.73	0.49	6312.24	163
900°C	34.92	0.32	4155.21	119
1000°C	24.56	0.06	835.13	34

As shown in

Fig the shape and distribution of the carbide particles are irregular. The carbide disperses through the matrix as the temperature increases forms. Increasing the holding temperature of the annealing decreases the carbide size prior to solidification and as a sequence, the tendency of the carbide to formed dispersed carbide is high. Carbide dispersed but still grouping at the LTF region and believed to slowly dissolve into austenite as the temperature of annealing increase. This is evidently shown in Table , a result of the number and small size carbide grouping in the inter graphite region. The growth of small dispersed carbide becomes the dominant pattern for entire matrix.

3.2 SEM analysis

The concentration of the elements is unevenly distributed in the microstructure. This phenomenon was known as micro segregation.

Fig and Table 5 shows the location of the probe and element of silicon, manganese and nickel in the region between graphite. It shows increasing reading pattern for manganese at a certain point. Overall, the highest reading is at point C and point D as evident of micro segregation of manganese at the LTF region. When this element is too much enriched in the LTF, thus expectedly leading to formation of carbide ($Mn_{23}C_6$) as be evidenced by Rashidi and Hasbullah [7] and also predicted earlier by Fallon [35] and Morrison [36]. Furthermore the detriment effect will exceed positive effect. Point B and E exhibit decreasing pattern of reading as the probe located near to graphite. Nickel reading shows two increasing pattern and peak reading at point B and E. This pattern indicates that nickel is positive at region between LTF and graphite. In LTF region nickel shows a negative segregation reading contrast to manganese. As expectedly, same graph pattern happened to silicon for what happen to nickel which is positive segregate at region between LTF and graphite.

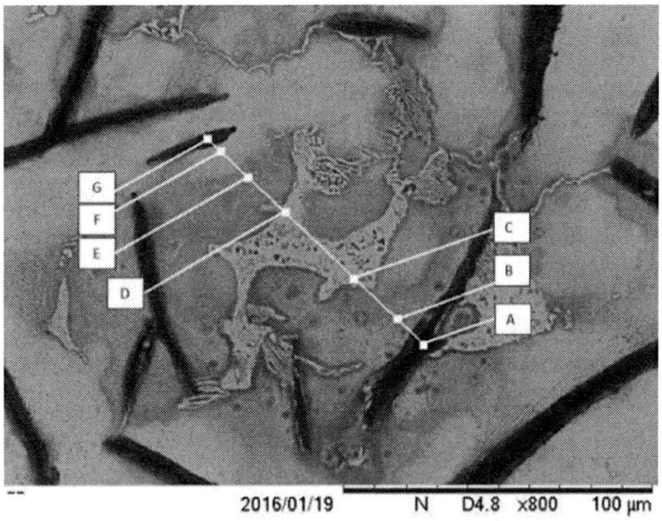

Fig. 3. Microstructure of high manganese austenitic cast iron with carbide presence

Table 5. Element constituent in the microstructure.

Elements	Electron probe location						
	A	B	C	D	E	F	G
Nickel (Ni)	5.094	6.466	3.269	6.594	7.287	3.470	0.231
Manganese (Mn)	2.728	6.616	18.007	12.585	6.138	6.016	0.428
Silicon (Si)	0.762	1.834	0.034	0.752	1.224	0.863	0.772

4 Conclusion

Base on the above results and discussion, the following conclusions are made:

1. Formation of carbide due to element segregation of manganese in inter graphite region throughout microstructure.
2. Increasing the annealing temperature reduce more segregated carbide in the LTF region in microstructure resulting in improvement in tensile strength and lowering the hardness value.
3. Higher hardness is seen for the sample having higher carbide volume in the matrix.
4. Less precipitation of carbide, $Mn_{23}C_6$ phase observed at grain boundaries of Mn-Ni-resist by increasing annealing temperature. There is possibility that annealing temperature minimize the segregation existence in alloyed iron. Segregation is the prime factor that influenced precipitation of carbide, $Mn_{23}C_6$ phase.
5. Fast rate of dissolution of carbides was observed in the sample 1000°C as compared to the carbide dissolution of sample 900°C due to the partial melting state which increase carbide dissolution rate.

The authors would like to express his sincere thanks to Universiti Malaysia Pahang (UMP) and ministry of education (Malaysia) for providing laboratory facilities and financial assistance under project no. RDU 140135.

References

1. M.M. Rashidi, I.M. Hasbullah, Mater. Design, **51,** 861 (2013)
2. R.J. McKay, Ind. Eng. Chem., **28,** 1391 (1936)
3. F. Alabbasian, S.M.A. Boutorabi, S. Kheirandish, Mat. Sci. Eng. A, **651**, 467 (2016)
4. Y.A. Alzafin, A.H.I. Mourad, M.A. Zour, O.A. Abuzeid, Eng. Fail. Anal., **16**, 733 (2009)
5. D. Zeng, C. Xie, Q. Hu, K.C. Yung, Scripta. Mater., **44**, 651 (2001)
6. Y.L. Yang, Z.Y. Cao, Z.S. Lian, H.X. Yu, Adv. Mat. Res., **194-196**, 95 (2011)
7. M.M. Rashidi, and I.M. Hasbullah, Mat. Sci. Eng. A, **574**, 226 (2013)
8. R.D. Forrest, Foundry Trade J. Int., **5,** 138 (1983)
9. A.I. Yatsenko, B.F. Martsiniv, Met. Sci. Heat. Treat., **10**, 890 (1968)
10. L.I. Aptekar', Y.E. Abramenko, Met. Sci. Heat. Treat., **19**, 280 (1977)
11. T. Oshima, ISIJ Int., **47(3)**, 359 (2007)
12. N. Fatahalla, A. AbuElEzz, M. Semeida, Mat. Sci. Eng. A, **504**, 81 (2009)
13. A. Janus, K. Granat, Arch. Civ. Mech. Eng., **14**, 602 (2014)
14. M.M. Rashidi, I.M. Hasbullah, Z. Shayfull, Key. Eng. Mat., **594-595**, 98 (2013)
15. G.J Cox, Foundryman., **81**, 453 (1988)
16. M.P. Shebatinov, Y.E. Abramenko, Met. Sci. Heat. Treat., **28**, 425 (1986)
17. A. Roula, G.A. Kosnikov, Mat. Lett., **62**, 3796 (2008)
18. R. Gundlach, M. Meyer, L. Winardi, Int. J. Metalcasting, **9**, 69 (2015)
19. K.N. Murthy, P. Sampathkumaran, S. Seetharamu, Wear, **267**, 1393 (2009)
20. X. Yaer, K. Shimizu, H. Matsumoto, T. Kitsudo, T. Momono, Wear, **264**, 947 (2008)
21. M.M. Rashidi, I.M. Hasbullah, Mat. Sci. Eng. A, **597,** 395 (2014)
22. M.N. Ahmadabadi, E. Niyama, M. Tanino, T. Ohide, Metall. Mat. Trans. A, **25,** 911 (1994)
23. A. Owhadi, J. Hedjazi, P. Davami, M. Fazli, J.M. Shabestari, Mat. Sci. Tech., **13,** 813 (1997)
24. R.K. Dasgupta, D.K. Mondal, A.K. Chakarabarti, A.C. Ganguli, Int. J. Cast. Met. Res., **25,** 239 (2012)
25. D.I. Stanchev, V.I. Rublev, Y.G.Sednev, A.M. Filatov, Met. Sci. Heat Treat., **18,** 514 (1976)
26. ASTM A436, ASTM Int.,(2001)
27. ASTM A439, ASTM Int.,(2001)
28. T. Skaland, ELKEM ASA Res., 1 (1999)
29. M. Chisamera, I. Ripson, S. Stan, E. Stefan, G. Costache, China Found., **6,** 145 (2009)
30. S.K. Putatunda, Mater. Design, **24,** 435 (2003)
31. C.A. Schneider, W.S. Rasband, K.W. Eliceiri, Nat. Methods, **9,** 671 (2012)
32. ASTM A247, ASTM Int.,(2001)
33. Z. Jiyang, China Found., (2009)
34. M.N. Ahmadabadi, S. Feghahati, Int. J. Cast. Met. Res., **20,** 53 (2013)
35. M. Fallon, BCIRA Tech., **9315,** 8 (1998)

MATEC Web of Conferences **78**, 01080 (2016) DOI: 10.1051/matecconf/20167801080
IConGDM 2016

A Review on Level of Specific Absorption Rate Due to High Power Transmission Lines: Analysis toward Human Position Posture

Z. Ghazali[1,*], *M. H.* Mat[2] , *H. A.* Rahim[3] , and *A.* Syafiq[1]

[1]School of Electrical System, UniMAP, Perlis, Malaysia
[2]Center for Diploma Studies, UniMAP, Perlis, Malaysia
[3]School of Computer and Communication, UniMAP, Perlis, Malaysia

Abstract. The main contribution of this project is the development of a homogeneous model of a man to presents the specific absorption rate (SAR) due to high power transmission line. As a low frequency application under high power transmission line of 50 Hz in electrical engineering, to studies the influence of human's posture on specific absorption rate. This project designs two types of human body which one design uses most cylinder block and another design use brick block where both blocks have a different value of mesh cells. For each design has four types of posture are standing, sitting, arms up and arms out by using Computer Simulation Technology (CST) Studio Software. This analysis does toward for four types of the human position postures because each posture has different value of specific absorption rate (SAR) based on the size of the mesh cells of the design. Based on two designs of the human body, the lowest of the mesh cells value will reduce time to simulate SAR. For each posture has different value of SAR for each part of the human body because the whole human body has different types of tissues. Therefore, the CST studio software uses extremely to simulate the SAR value toward human position posture due to high power transmission line.

1 Introduction

This project discusses about level of specific absorption rate (SAR) due to high power transmission line in Malaysia where uses 50 Hz of frequency to simulate the SAR value toward human body. The International Commission on Non-Ionizing Radiation Protection (ICNIRP) and the Council for European Union recommended permissible limit to exposure of electric shown in Table 1.

* Corresponding author: zulaikha_ghazali@yahoo.com

Table 1. ICNIRP limit and the Council for European Union exposure limit [1].

Power line operating frequency	50 Hz
	Electric field (V/m)
Public exposure limit	5000
Occupational exposure limit	10000

This analysis was done toward on four types of the human position postures which is standing, sitting, arms up and arms out posture. Each posture has different SAR value for each part of the human body because the whole human body have different types of tissues [2]. The ICNIRP have defined SAR with unit watt per kilogram (W/Kg) and reference level of field strength from exposure to EMF to protect workers and public avoid health effects [3-5]. SAR is the quantity of how much power of an electromagnetic wave is absorbed per unit mass by an absorber such as a human body. In other hand, SAR also means a measure of the rate at which energy is absorbed by the human body when exposed to a radio frequency (RF) electromagnetic field. SAR is derived from the square of the (direct) electric field strength (E) in tissue [6]:

$$SAR = \frac{P_{ab}}{\rho} = \frac{\sigma E^2}{2\rho} = \frac{J^2}{2\rho\sigma} \tag{1}$$

where P is the power loss density, E is electric field strength, J is the current density, σ is conductivity (S/m) and ρ is the density (Kg/m^3) Of the tissue interest [7-9]. For frequencies up to a few GHz, as applied in wireless communications, SAR is normally averaged over either 1 or 10 g. American National Standard Institute (ANSI) and Federal Communication Commission (FCC) has determine the SAR limit to 1.6 W/Kg per 1 g of tissues. ICNIRP and IEEE has impose the maximum of 2 W/Kg absorbed per 10 g of tissue [10, 11].

2 Numerical study

2.1 Power transmission lines.

Electromagnetic fields (EMF) are generated in the vicinity of power transmission lines [12]. Now, the places use the 50 Hz frequency tend to use 220-240 V and for 60 Hz tend to use 100-127 V [2, 13, 14]. At extremely low frequencies (ELF) the radiating field of a source is negligible. In practical exposure situations, radiation is absolutely negligible in the ELF range. Radiation only becomes dominants at distances that are large compared to the wavelength. In the free space the wavelength is related to the frequency by the formula,

$$\lambda = \frac{c}{f} \tag{2}$$

where λ is wavelength (km), c is the speed of lights ($3x10^8$ m/s) and f is frequency (50 Hz). At 50 Hz, the wavelength is very long, 6000 km. At 60 Hz, the wavelength is very long, 5000 km. As a conclusion, the smaller value of the frequency, the harder of the simulation cause the wavelength become larger with higher mesh cell [15].

2.2 Simulate specific absorption rate (SAR) using Computer Aided Design (CAD) software.

Power absorption under the whole body resonant conditions is also strongly dependent on the posture of the human body. For this simulation has developed human body models for any posture based on a CAD software product. This software includes a statistical data set of anthropomorphic dimensions and can change the posture of various human bodies. After the posture of the human bodies is modified, a newly developed software module outputs the voxel data with any voxel resolution. The models described so far in this section are homogeneous because only the outline of the human bodies can be obtained with the CAD software. This is a significant shortcoming because an internal inhomogeneous structure can affect electromagnetic power absorption in the human body [16].

The FDTD method was used to calculate Whole Body Averaged Specific Absorption Rate (WBASAR). The change of the postures from arms down to arms up increased the WBASAR at most of frequencies of interest. SAR increase due to raising arms was more dominant at low frequencies and in the older models [17]. A change in the posture of the human body can significantly affect the way in which it absorbs radio frequency EMFs [8].

2.3 Hugo model simulate with poser software and manipulate in MATLAB.

In order to perform electromagnetic simulations, the dielectric properties allocated to various tissues and organs in the frequency ranges should be concerned. The properties of Hugo model's tissues and organs at 50 Hz frequency which will be used for skin refer to Table 2. While Figure 1 show various type of posture of human body.

Table 2. The dielectric properties of Hugo model tissue at 50 Hz[18].

Relative permittivity	Conductivity (S/m)
1136	0.00020

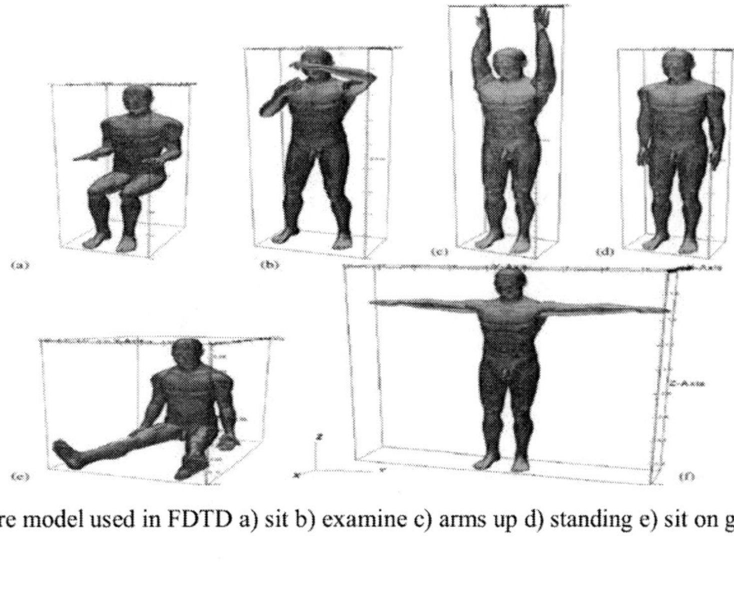

Fig. 1. Posture model used in FDTD a) sit b) examine c) arms up d) standing e) sit on ground f) arms out [18].

Before the simulation is performed formally, study the influence of the voxels size and mesh cells to the final results is necessary to perform the convergence calculation. Table 3 shows how whole body current changes with the voxels size and the number of mesh cells when voltage is fixed to be 220 V. Using three mm or four mm voxel size human body model requires much more mesh cells, and then needs much more memory and hard to do the simulation [19].

Table 3. The number of mesh cell depend on voxel size[19].

Voxel size (mm)	Number of mesh cell	Body current (mA)
8	11,476,290	49.3
6	11,674,530	56.3
5	12,072,489	58.5
4	14,479,368	65.5
3	35,123,088	66.6
2	116,385,060	65.8

2.4 Influence of specific absorption rate (SAR) of the human body using computer simulation technology (CST) microwave studio software for frequency 70 mhz under communication system.

Figure 2 shows the localized SAR (averaged on 10 g mass) distribution in the human body with sitting and standing posture at 70 MHz. For the human body with sitting posture in Figure 2, most of the SAR value is absorbed near different kinds of joints such as ankle, knee, hip joint, wrist, elbow and neck. The area near the nose and the fingers also has a little higher localized SAR. For the human body with standing posture in Figure 3, higher SAR values mainly appear in the legs, especially near the ankle and knee joint [19].

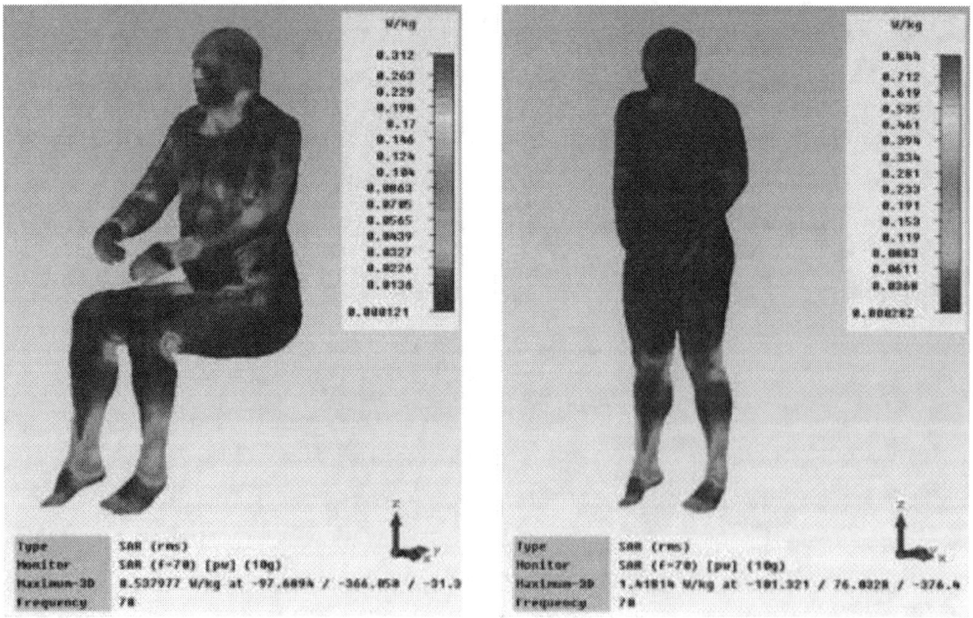

Fig. 2. The SAR distribution in the human body at frequency 70 MHz a) sitting posture b) standing posture[19].

3 Conclusion

In this simulation, a homogeneous model of man is developed to display the mesh cells value. Based on two designs from the brick block and cylinder block, the cylinder block has higher value of the mesh cells than the brick block. This is because the cylinder block needs detail the voxels size to produce cylinder curve. The smaller of the voxels size value, the higher of the mesh cell value. This project was needed to simulate the specific absorption rate (SAR) to four types of posture. This analysis was done on four types of human position postures which is standing, sitting, arms up and arms out posture. Each posture has different value of SAR based on the size of the mesh cells of the design.

References

1. P. Akinlolu, A. Kazeem, American Journal of Engineering Research (AJER) **5**, 154 (2015)
2. W. headquartes, L*egrand power guide*, 1, (2009)
3. ICNIRP, *Guidelines for limiting exposure to time varying electric, magnetic, and electromagnetic fields (up to 300GHz)*, (ICNIRP Guidelines, 1998)
4. *IEEE-SA Standards Board*, 1, (1998)
5. A. Cortel, N. Varsier, V. Dronne, O. Colas, M. F. Wong, E. Nicolas, F. Jacquin, J. Wiart, *Proceeding of The International Union of Radio Science* (2005)
6. T. Wessapan, S. Srisawatdhisukul, P. Rattanadecho, Int. Commun. Heat Mass, **38**, 255 (2011)
7. G.C.R. Melia, *Electromagnetic Absorption by the Human Body from 1 to 15 GHz*, (The University of York, 2013)
8. G.V. Cossio, Giesen, Araya, Tohme, Holmberg, Bressmann, *Radiofreq. Electromagn.* **XXXIII**, 81 (2012)
9. A.K.M.F. Hoque, Am. J. Phys. Appl., **1**, 104 (2013)
10. Hamzah Asyrani Sulaiman, Mohd Azlishah Othman, Mohamad Zoinol Abidin Abd. Aziz, Mohd Fareq Abd Malek, *Theory and Applications of Applied Electromagnetics: APPEIC 2014*, (Springer, 2015)
11. M.H. Mat, M.F. Abd Malek, W.G. Whittow, R. Bibb, J. Electromagnet Wave., 29, 514 (2015)
12. ICNIRP, *Health Phys.*, 74, 494 (1998)
13. *The World Wide Voltage/Frequency Guide*, (http://www.hightechcords.com)
14. Safety Guildelines and Electrical Requirement, *Measurement*, 429905, 1.
15. P. M. Kanarev, *J. Theor.* **5**, 1 (2004)
16. S. Watanabe, N. Hirose, T. Nagaoka, N. Hatakenaka, Y. Tanaka, M. Takahashi, Y. Suzuki, M. Taki, J. Wang, O. Fujiwara, and Y. Yamanaka, Int. Symp. Electromagn. *Compat.*, **2**, 797 (2004)
17. A. Lee, H. Choi, General Assembly and Scientific Symposium, 2011 XXXth URSI, 5 (IEEE, 2011)
18. T. Uusitupa, I. Laakso, S. Ilvonen, K. N. Sar, T. Uusitupa, I. Laakso, S. Ilvonen, and K. Nikoskinen, Phys. Med. Biol., 55, 1157 (2010)
19. J. Gao, *Generation of Postured Voxel-based Human Models Used for Electromagnetic Applications*, (Technische Universität, Darmstadt, 2012)
20. T.M. Uusitupa, S.A. Ilvonen, I.M. Laakso, K.I. Nikoskinen, Phys. Med. Biol., 53, 445 (2008)

MATEC Web of Conferences 78, 01081 (2016) DOI: 10.1051/matecconf/20167801081
IConGDM 2016

Effect on Mechanical Properties of Heat Treated High Manganese Austenitic Cast Iron

A.K. Muzafar[1,*], M.M. Rashidi[1] , I. Mahadzir[1] , and Z. Shayfull[2]

[1]Faculty of Mechanical Engineering, Universiti Malaysia Pahang, 26600 Pekan, Pahang, Malaysia.
[2]School of Manufacturing Engineering, Universiti Malaysia Perlis, Kampus Tetap Pauh Putra, 02600 Arau, Perlis Malaysia.

Abstract. This work presents an attempt to study the effect of manganese addition and heat treatment on higher carbon austenitic cast iron to form high manganese austenitic cast iron with reduced nickel content (Mn-Ni-resist) on mechanical properties. The combination on microstructure (microsegregation), mechanical properties and the relationship of heat treatment on the alloy were analyzed. For this purpose Mn-Ni-resist (4.50C, 2.64Si, 6.0 Mn, 10 Ni) was melted and cast in the form of Y-block test pieces. Four different heat treatment procedures were applied to the as-cast to investigate the effect of alloy modifications on Mn-Ni-resist. Optical and scanning electron microscopies were used for microstructure investigation. To determine the mechanical properties tensile test and hardness test were carried out. The result indicates both composition and heat treatment affect the performance of Mn-Ni-resist intensively. Microprobe analysis shows some silicon segregation near the graphite and practically little segregation of manganese. The increase in manganese contents developed some fractions of segregated carbide structures in LTF region located at austenite eutectic cell frame, which caused the tensile properties to drop in a small range. Application of annealing heat treatment gradually changed the carbide formation, so is the material's strength.

1 Introduction

Recently, researchers have shown an increased interest in austenitic cast iron study. Due to constant demand for high performances materials austenitic cast iron has been chosen for various engineering applications as substituted material to steel, as it offer an outstanding combination of properties in withstanding the effects of corrosion, heat and wear [1-5].

The high composition of austenitic matrix in microstructure is contributed by the influence of nickel contained in the composition that acts as austenite matrix promoter [6]. According to ASTM standard, with nickel composition 13.5 wt% up to 36 wt%, it already manages to produces full austenitic structure in the microstructure [7, 8]. However Fatahalla et. al (2009) claims that with only 13.5 wt% was enough to produce austenitic

* Corresponding author: mrashidi@ump.edu.my

ductile iron [1] while to produce austenitic ductile cast iron need at least 18 wt% Ni. This special alloy will be in austenitic matrix when at high temperature and also at room temperature which makes it successful avoid matrix volume changes. Despite its advantages, the used of nickel as main austenite promoter suffers from major drawbacks associated to its high price production.

Generally, the use of cast iron with high nickel content is technically as well as economically inexpedient. It can be successfully replaced nickel in austenitic cast iron with higher manganese or copper content. Since copper and manganese also can acts as austenitic promoter. Usually, austenitic cast iron with manganese addition is complexly alloyed, together with the austenitic base the structure strongly tend to produce carbide ($Mn_{23}C_6$) through the microstructure [9] while austenitic cast iron with copper addition improves corrosion resistance in mildly acid solution and also interfere with the magnesium treatment used to produce spheroidal graphite [10]. Furthermore, it is also believed that manganese is better austenite promoter than copper [11]. At present, Mn was used solely for alloying purposes to improve DNR impact toughness property instead of austenitic matrix stabilizer and does not contribute to the reduction use of Ni. Only few of publications on optimize the use of manganese in Ni-resist. Up to date, there is no generally accepted classification of austenitic manganese cast iron in international specification standard like ASTM or ISO.

There are several attempts to utilize the use of manganese as austenitic promoter in producing Ni-resist and also using heat treatment to obtain the austenitic microstructure. The use of nickel in Ni-resist can be reduced to the range of 8-10 wt% with assistance of 3-4 wt% of manganese with good properties of ductile. This finding was supported by recent investigation by Rashidi and hasbullah [12] which is studied about the effect of manganese (10wt%, 8-11wt%Mn) on ductile iron with optimization of inoculation to reduce the carbide formation while Janus and Kurzawa [13] made some effort in investigate effect of nickel equivalent on austenite transition ratio in Ni-Mn-Cu cast iron using heat treatment with manganese and copper content with low austenite stability. However, investigation does not stop here, there are improvement field still open for new discovery on reducing production prices for austenitic ductile iron by manipulating the alloying element and heat treatment and microstructure. It is of interest to work out and study types of cast iron in which the austenite is stabilized by manganese.

From the previous literature, using high manganese content in Ni-resist was invited carbide formation as has been found and recorded in since the early of 1900 by Mckay [14]. Carbide formation very undesirable and was normally avoided in foundry. There are two different approaches can be used to obtain carbide free iron. The first approach would be to avoid carbide precipitation during solidification. This can be achieved by maximizing the efficiency of inoculation. Good inoculation technique managed to reduce carbide precipitation during solidification and improves its mechanical properties. The second approach would be to eliminate the carbide precipitated during solidification by means of a dissolution heat treatment.

It is technically difficult to obtain graphite of such favorable shape in Mn-Ni-resist as in gray or high-strength cast iron, especially with high carbon content. If austenitic cast iron is to find widespread application in industry, its properties and structure must be further investigated because these kinds of cast iron are very promising materials. They are cheap because they are alloyed with the abundantly available manganese, and they have valuable technological and operational properties. The present investigation was therefore an attempt undertaken to examine the influence of manganese and annealing heat treatment to discover the effect on mechanical properties. This research will concentrate on producing austenitic cast iron with high manganese content iron and application of annealing procedure on it. Tensile tests were performed to determine if the deformation-induced transformation was strain-induced; fracture toughness tests were performed to see if the transformation was stress-induced.

2 Experimental procedure

2.1 Material preparation

The studied alloy was prepared in a 100 kg-capacity frequency induction furnace. Initial charge materials were pig iron, pure nickel, steel scrap, Ferro manganese and austenitic iron returns. Then the materials heated until molten state. Alloying elements such as 99.0 wt% pure nickel and fe-45.0 wt% Mn were added to the molten alloy so as to increase and control Ni and Mn percentage content respectively. The mould was made by using Y block shaped pattern with minimum thickness of 30mm was prepared by using a pattern as in **Fig** according to ASTM standard for austenitic cast iron [7, 8] in the green sand mould. Inoculation process with 0.5 wt% Fe-Si was introduced to melt by using late inoculation in the mould technique [4, 15, 16].

This procedure ensures that the inoculation level high and reduce magnesium fading phenomena during casting. The chemical composition of casting product was shown in Table 6.

Table 6. Chemical composition (wt.%) of the base metal

Elements	C	Si	Ni	Mn	S	Fe
Composition wt.%	4.5	1.81	10.00	4.98	0.082	balance

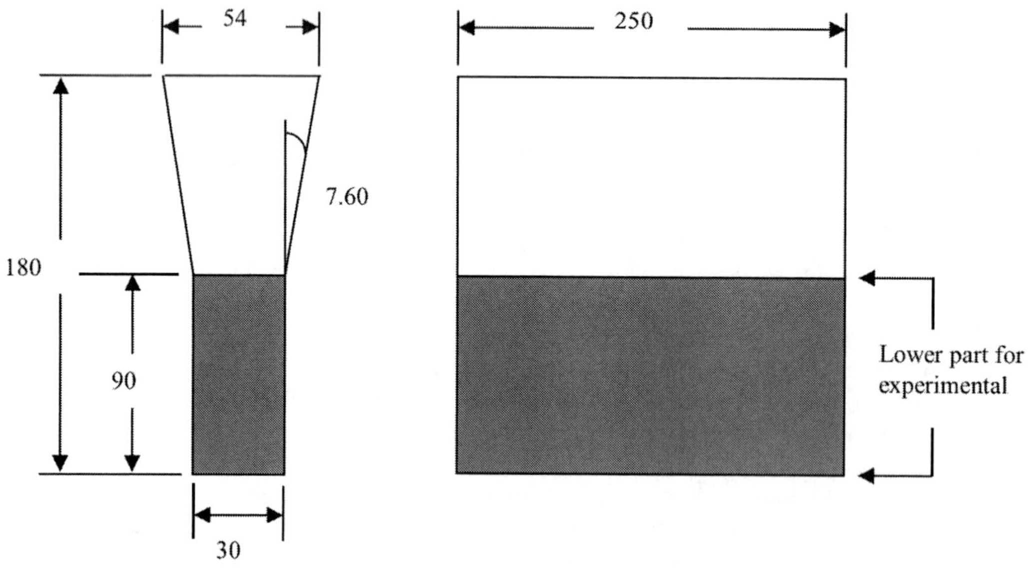

Fig. 9. Dimensions of the Y-block castings used in this experiment. Test specimens cut and machined from the lower part of the Y-block (dimensions in mm).

2.2 Heat treatment process

The heat treatment schedule outlined in Table 7 was carried out. All dog bone samples then were placed in a box-type resistance furnace heated according to its annealing temperature for annealing and 927°C for quenching. The holding time is 3 hours was counted when it reached the desired temperature. Austenitizing time less than 2 hour was not selected because the complete transformation of the as-cast structure into austenite required at least 2 hours [17]. Annealing process Annealing was held at 700°C, 800°C, 900°C and 1000°C for 3.0 h. Then the samples cooled down to room temperature by furnace cooling respectively. For the purpose of easy analysis, as shown in Table 7, every heat treatment process was named according to its annealing temperature and heat treatment condition.

Due to slight oxidation of the surface of cast iron, there is every possibility of scale formation on this surface during transfer of the samples to the oil tank or cooling to room temperature and making the hardness and tensile value vary. Moreover, the specimen will not also be gripped properly in the machine. The specimens were polished to remove the scales from the surface to avoid these difficulties.

2.3 Mechanical test

Tensile test were carried out using universal testing machine 50 kN was used according to American Society of the International Association for Testing and Materials standard ASTM E8/E8M [18] at room temperature. CNC lathe machine was used to shape the cast alloy specimen to produce dog bone shape as shown in Fig. **10**. The samples were further examined using scanning electron magnetic (SEM) to assess the microstructure after fracture. Samples from the specimens of 700°C, 800°C, 900°C and 1000°C soaking temperature were taken for the hardness test. The Brinell hardness test machine was used to determine the macro hardness. Five reading was taken from each sample and averaged.

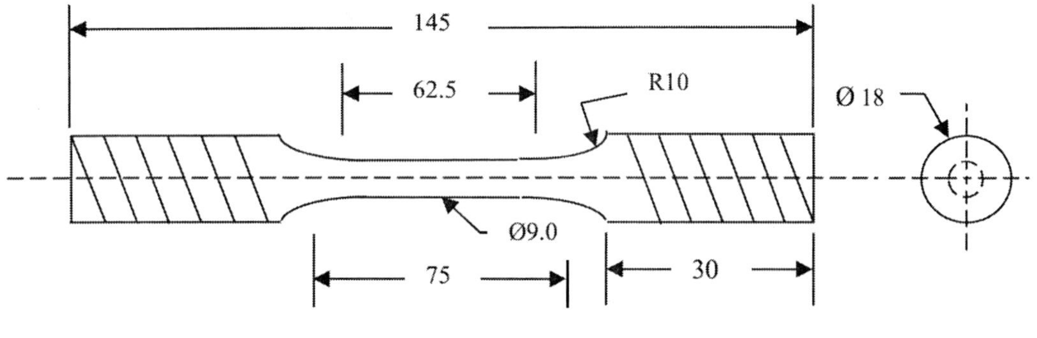

Fig. 10. Dimension of tensile test specimen.

2.4 Microstructure analysis

The metallographic analysis of the specimens was carried out in the round shaped specimen obtained from the broken halves of the tensile specimen. Metallographic samples were sectioned, ground and polished with 1 to 6μm grade diamond paste, rinsed in distilled water and degreased with ethanol. The etch solution of 3% Nital was used for investigation of microstructure of samples. The microstructure was examined using optical microscope (OM) and scanning electron microscope (SEM) next analyzed using images analyzer ImageJ [19].

Table 2. Chemical composition of raw material

Material	Elements (wt. %)										
	C	Si	Mn	P	S	Mg	Ni	Ca	Cr	R.E	Fe
Pig iron	4.5	1.73	0.309	0.180	0.150	0.310	0.236	-	-	-	balance
Steel	0.19	0.15	0.54	0.90	0.02	-	-	-	-	-	balance
Nickel	-	-	-	-	-	-	99.0	-	-	-	balance
FeMn	-	-	86.00	-	-	-	-	-	-	-	-
Inoculant	-	70.00	-	-	-	-	-	2.0	-	-	balance

3 Result and discussion

3.1 Tensile properties

Table 7 demonstrates the influence of heat treatment and annealing temperature in the range from 700°C to 1000°C in tensile properties of Mn-Ni austenitic cast iron for all heats. It is apparent from the table that when annealing temperature was introduced, ultimate tensile strength and elongation was increasing as well. This is probably due to a decreases of carbide phase amount at LTF region between graphite which restructuring itself by influenced of elevated temperature during annealing. Another possible explanation of the resulted phenomenon would be more scattered carbide in free form smaller in size situated in LTF region in contrast with as-cast condition. This claim was reiterated by Felon (1993) that heat treatment results in precipitation of small carbides and can reduce the proof strength values [20]. It is supposed to solidifies as big size carbide as shown in Fig. **12**. Most probably smaller carbide situated at LTF region propagated less crack during load applied. Thus the tensile result is much better. Hence, this will encourage slightly higher tensile characteristic that responsible for the overall material strength.

Table 7. Mechanical properties of high manganese austenitic cast iron

Materials	Room temperature test		
	Ultimate tensile strength (MPa)	Elongation (%)	Macro hardness (HRB)
As-cast	60.10	1.04	102.53
Annealing 700°C	66.67	1.14	100.93
Annealing 800°C	66.63	1.34	93.63
Annealing 900°C	71.76	1.74	90.48
Annealing 1000°C	75.03	1.92	80.55

Fig. **11** (a)(b) illustrates clearly the graph trend of the result recorded in Table 7. It is evident that material which was annealed was relatively gaining higher tensile strength (75.03 MPa) compared to as-cast condition (60.10 MPa). It is cannot be argued that higher value of Mn-Ni-resist tensile strength is well related to its annealing temperature by controlling its carbide formation. This result supported the previous report in microstructure analysis [21] as lesser carbide formation with higher annealing temperature.

so less carbide formation with increase annealing temperature successful increase the mechanical properties compare to Mn-Ni-resist in as-cast condition.

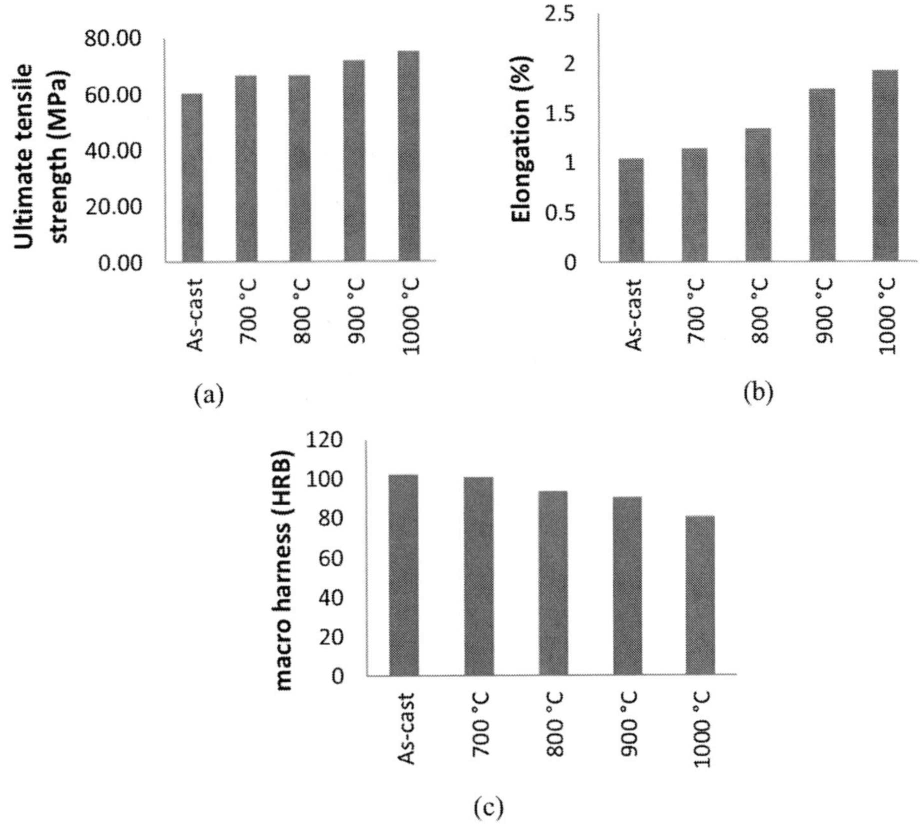

Fig. 11. Comparison of tensile properties and macro-hardness of the alloyed iron: (a) ultimate tensile strength, (b) elongation, (c) hardness

Apart from that, the influence of elements solid solution (manganese in particular) in alloyed iron which affected its mechanical properties could treat by annealing process. As a result both strength and elongation was increased. Higher annealing temperature influencing the tendency of carbide in Last to Freeze (LTF) region to form scattered. LTF crystallisation condition is different compared to mother iron liquid, where manganese as positive segregation elements concentrated more in bigger quantity. Elemental segregation influencing solidification structure which later effects solid phase transformation. Thus it is possibly exerts important effects on the mechanical properties.

3.4 Hardness test

Fig. **11** illustrates the influence of varying annealing temperature on hardness values in all of samples. This curve reveals that hardness is slightly decreased with increasing in annealing temperature. The highest value of hardness is for the 700°C sample corresponding to the large fraction of carbide in its structure compared to other samples. While the lowest hardness reading (sample 1000°C) shows its carbide content is much lower in the matrix, resulting in lower macro-hardness. It can be suggested that the macro-hardness of the alloyed iron is primarily affected upon addition of the heating temperature and phase constituent in the microstructure. According to ASTM-A436 standard the hardness values for Ni-resist (Type 1) have been reported in the range of 131 up to 149 Brinell. In this study, the hardness values were in the range of 80 up to 101 Brinell for annealed samples. It seems the relationship between the tensile properties and macro-hardness of alloyed iron is shown that good annealing heat treatment practice evidently results in decreasing hardness values and increase tensile properties. It is advocates that the formation of carbide due to element segregation governs for the most part the average macro-hardness obtained.

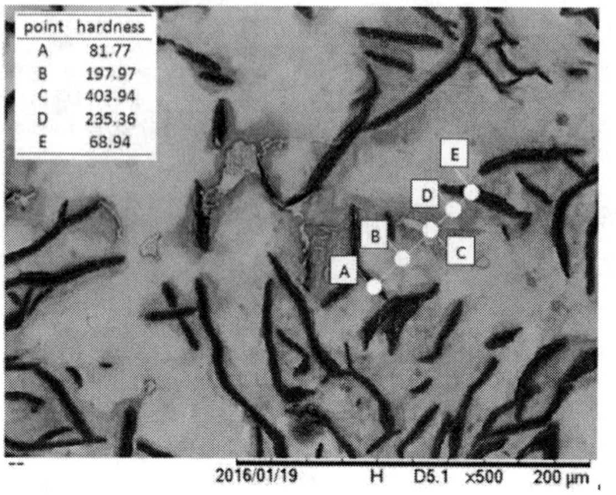

Fig. 12. Micro hardness values in the microstructure.

Fig. **12** shows the differences in micro hardness in an alloyed iron microstructure. The existence of $Mn_{23}C_6$ in the sampling microstructure explained the greater differences in the micro-hardness value discovered. $Mn_{23}C_6$, found at the last to freeze (LTF) region, has the highest micro-hardness value, averaging at 400HB and above as illustrate in **Error! Reference source not found.**. The behavior of the micro-hardness value is due to segregation of the element factor where positive segregation factors such as manganese accumulated more. Manganese strongly tends to form bonding with carbon to precipitate as carbide. This proves by the highest quantitative value of micro-hardness value in this region. It developed into a complete altered phase compared to the austenitic structure around the graphite area. The low micro-hardness at area between carbide area and graphite is contributed by accumulation of negative segregation factors such as nickel and silicon which developed the austenitic structure when alloyed with Ferum. The micro hardness value for this region was reduced by approximately 50% compared to the carbide region.

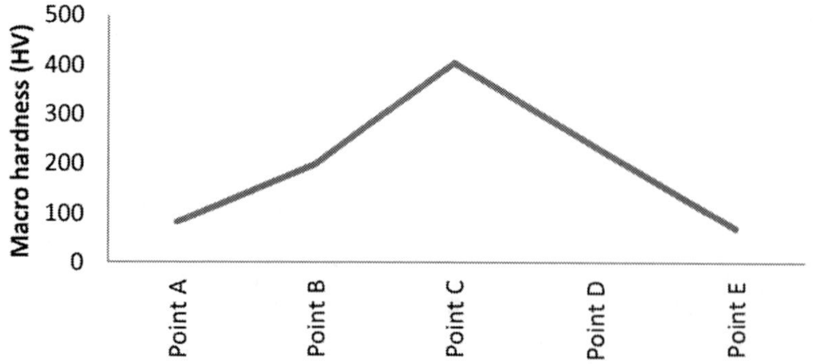

Fig. 5. Micro-hardness.

4 Conclusion

Base on the above results and discussion, the following conclusions are done:

- Formation of carbide due to element segregation of manganese in inter graphite region throughout microstructure.
- Increasing the annealing temperature increase tensile strength and lowering hardness value.
- 700°C has the lowest tensile test properties but highest macro hardness while 1000°C evidently has the highest tensile strength with lowest macro hardness.
- Higher hardness is seen for the sample having higher carbide volume in the matrix.
- There is possibility that annealing temperature minimize the carbide segregation existence in alloyed iron. Segregation is the prime factor that influenced precipitation of carbide and mechanical properties.

The authors would like to express his sincere thanks to Universiti Malaysia Pahang (UMP) and ministry of education (Malaysia) for providing laboratory facilities and financial assistance under project no. RDU 140135.

References

1. N. Fatahalla, A. AbuElEzz, M. Semeida, Mater. Sci. Eng., A **504**, 81 (2009)
2. Y.L. Yang, Z.Y. Cao, Z.S. Lian, H.X. Yu, Adv. Mater. Res., **194-196**, 95 (2011)
3. A. Janus, K. Granat, Arch. Civ. Mech. Eng., **14**, 602 (2014)
4. M.M. Rashidi, I.M. Hasbullah, Mater. Sci. Eng., A, **597**, 395 (2014)
5. Alabbasian, F., S.M.A. Boutorabi, S. Kheirandish, Mater. Sci. Eng., A **651**, 467 (2016)
6. R.D. Forrest, Foundry Trade J. Int., **5**, 138 (1983)
7. *ASTM A436*, (ASTM Int., 2001)
8. *ASTM A439*, (ASTM Int., 2001)
9. M.M. Rashidi, I.M. Hasbullah, Mater. Des., **51**, 861 (2013)
10. R. Covert, *Nickelinstitute.org*, (Nicle Institute, 1998)
11. G.J Cox, Foundryman., **81**, 453 (1988)
12. M.M. Rashidi, and I.M. Hasbullah, Mater. Sci. Eng., A **574**, 226 (2013)
13. A. Janus, A.K., **13**, 53 (2013)
14. R.J. McKay, Ind. Eng. Chem., **28**, 1391 (1936)
15. T. Skaland, *ELKEM ASA Research*. (1999)
16. M. Chisamera, I. Ripson, S. Stan, E. Stefan, G. Costache, China Foundry **6**, 145 (2009)
17. S.K. Putatunda, Mater. Des., **24**, 435-443, (2003)
18. *ASTM E8/E8M, A.*, (ASTM Int., 2001)
19. C.A. Schneider, W.S. Rasband, K.W. Eliceiri, Nat. Methods, **9**, 671 (2012)
20. M. Fallon, BCIRA Tech., **9315**, 8 (1998)
21. M.M. Rashidi, I.M. Hasbullah, Z. Shayfull, Key. Eng. Mater., **594-595**, 98 (2013)

Behavior of CuPb12Sn6 Alloys subjected to Heat Treatments

Dragoş Cristian Achiţei[1,*] , *Mirabela* Georgiana Minciună[1], *Mohd Mustafa* Al bakri Abdullah[2,3] , *Andrei* Victor Sandu[1] , *Michal* Szota[4] and *Petrică* Vizureanu[1]

[1]Department Technologies and Equipments for Materials Processing, Gheorghe Asachi Technical University of Iasi, D Mangeron Street, no 41A, Iasi
[2]Faculty of Engineering Technology, Universiti Malaysia Perlis, Kangar, Perlis, Malaysia
[3]Center of Excellence Geopolymer & Green Technology (CEGeoGTech), School of Materials Engineering, Universiti Malaysia Perlis, Kangar, Perlis, Malaysia
[4]Czestochowa University of Technology, Institute of Materials Science and Engineering, 19 Armii Krajowej Av., Czestochowa, Poland

Abstract. The paper present the structural modifications induced by 3 heat treatments: homogenization, recrystallization and quenching, respecting the specific parameters of copper based alloys. The obtained modifications by heat treatments were highlighted by micro-hardness measurements, with Vickers method, and by electronic structural microscopy, at various magnitudes.

1 Introduction

Copper is an excellent electrical conductor. Most of its uses are based on this property or the fact that it is also a good thermal conductor. However, many of its applications also rely on one or more of its other properties. For example, it wouldn't make very good water and gas pipes if it were highly reactive.

Copper is low in the reactivity series. This means that it doesn't tend to corrode. Again, this is important for its use for pipes, electrical cables, saucepans and radiators.
However, it also means that it is well suited to decorative use. Jewellery, statues and parts of buildings can be made from copper, brass or bronze and remain attractive for thousands of years.

Copper and its alloys, such as brass, are used for jewellery and ornaments. They have an attractive golden colour which varies with the copper content. They have a good resistance to tarnishing making them last a long a time.

Copper can be recycled without any loss of quality. 40% of the world's demand is met by recycled copper.

The alloying elements contribute to increase of their resistance properties, concomitant with the decrease of plasticity properties. The principal deficiencies of pure copper, like low castability, low corrosion resistance, mechanical properties at small values, can be removed partially or totally by alloying.

* Corresponding author: mirabela.minciuna@yahoo.ro

Due his physical-chemical and mechanical properties, the copper and his alloys have much utilization in electrical engineering, chemical industry, aeronautics, oil industry, food industry (under forms of laminated, drawn and cast products).

Copper have 2 categories of alloys: **brasses** (Cu-Zn alloys and other elements) and **bronzes** (Cu-Sn alloys and other elements).

The equilibrium diagram of Cu-Sn alloys system is like diagrams with partial solubility in solid state and with peritectic and eutectoid transformations. The maximum solubility, in solid states, of Sn in Cu us by 15.8% at temperature by 520°C, and is lowest than in the case of zinc and varies in function of temperature:

- At start, increase from 13.5% (at 768°C) up to 15.8 % (at 520°C).
- Decrease to 13.9% at ambient temperature.

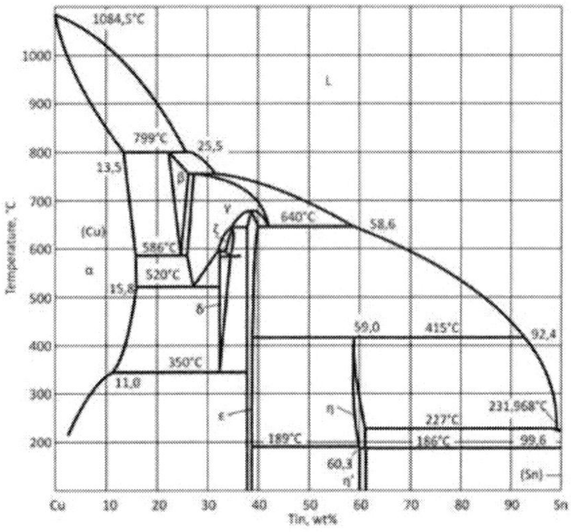

Fig. 1. The Cu-Sn equilibrium diagram.

In the diagram can observe the following intermetallic phases:

- α phase – solid solution by substitution, isomorphic with copper, which crystallize in cubic system with centered faces;
- β phase – solid solution, stabilized at high temperatures, up to 586°C, based on electronic compound Cu_5Sn, with $\frac{n_e}{n_a} = \frac{3}{2}$ ratio;
- γ phase – stabilized only the high temperature, up to 586 °C, resulted by redistribution of the atoms in crystalline network of γ phase;
- δ phase – represent the electronic compound $Cu_{31}Sn_8$, cu cubic crystalline network, with electronic concentration by 21/13.

2 Experiments

2.1 Determination of chemical composition

The determination of alloying elements percents was made on Foundry Masters optical spectrometer, using copper base for analysis. The experimental tests were realized on samples which were prepared by polishing on abrasive papers.

The analyzed alloy is part of Cu-Pb-Sn alloys class (alloyed bronze) and is fall in the standard limits of CuPb12Sn6 alloy.This type of alloy present a structure formed by copper grains, in which are dispersed lead particles.

Whereas the mechanical resistance of binary bronzes is low, it is used lead bronzes which are alloyed with tin, nickel and zinc. The tin increase the mechanical resistance, the zinc hardened the base mass, and nickel smooth the repartition of lead particles.

The Cu-Pb-Sn antifriction bronzes are sufficient by mechanical resistant; to be use like bearings, without protected enclosures by steels, when support pressures by 6000 N/cm^2. These bronzes have high corrosion resistance, and good usage resistance. This alloy is use at plain bearings for calenders, vehicles, mill, bearings, bolts for pistons, bushings for wheel drive, brake discs, usage resistant parts, corrosion resistant parts in water and armatures resistant in acid medium.

2.2 The heat treatments applied to Cu-Pb-Sn alloys

The usually heat treatments for Sn bronzes are recrystallion (hardened parts) and homogenization (cast parts).

After homogenization the values of mechanical and plasticity characteristics recorded important increases, and from structures of dendrite segregations is obtained monophasic and biphasic homogeneous structures.

In the case of pressure armature, due to constructive configuration of parts and elaboration conditions of alloys, in some areas can appear cracks, these being noseals areas at 8…16 atmosphere pressure.

The bronze quenching is similar to martensitic quenching of steels, obtaining structure with acicular aspect, similar with martensitic structure.

The recommended values of parameters by heat treatments for CuPb12Sn6 alloy is presented in Table 1. Based on these values, we adopted the adopted parameters used in experimental tests.

Table 1. The parameters of heat treatments recommended for CuPb12Sn6 alloy

The heat treatment	The heating temperature, [°C]	The maintaining time ,[h]	The cooling medium
Homogenization	700 ... 750	1 ... 2	Free air
Recrystallization	500 ... 675	1 ... 2	Free air
Quenching	> 600	0.5 ... 1	water

The heat treatments effectuated in the experimental study, were made in an electric furnace, with resistors, type UTTIS, with fixed hearth.

The adopted parameters (Table 2), for the 3 heat treatments, were controlled with a specific software, type Protherm 100.

Table 2. The parameters of heat treatments adopted for CuPb12Sn6 alloy

The heat treatment	The heating temperature, °C	Heating speed, °C/min	The maintaining time, h	The cooling medium
Homogenization	750	10	1	With the furnace
Recrystallization	650	10	2	Free air
Quenching	630	10	1	water

2.3 The microstructure analysis by electronic microscopy

The analysis were made on a electronic microscope, SEM model VEGA II LSH, manufactured by Tescan Company of Czech, coupled with a EDX detector, type QUANTAX QX2, manufactured by ROENTEC Company of Germany.

The SEM images were obtained after the samples were previously prepared by polishing on abrasive papers and moistened felt with solution by Al_2O_3 powder.

Subsequent, the samples were attacked with a chemical reactive, with the following composition: 5g ferric chloride + 30ml HCl + 100ml water, at a maintaining time between 5 – 10 seconds.

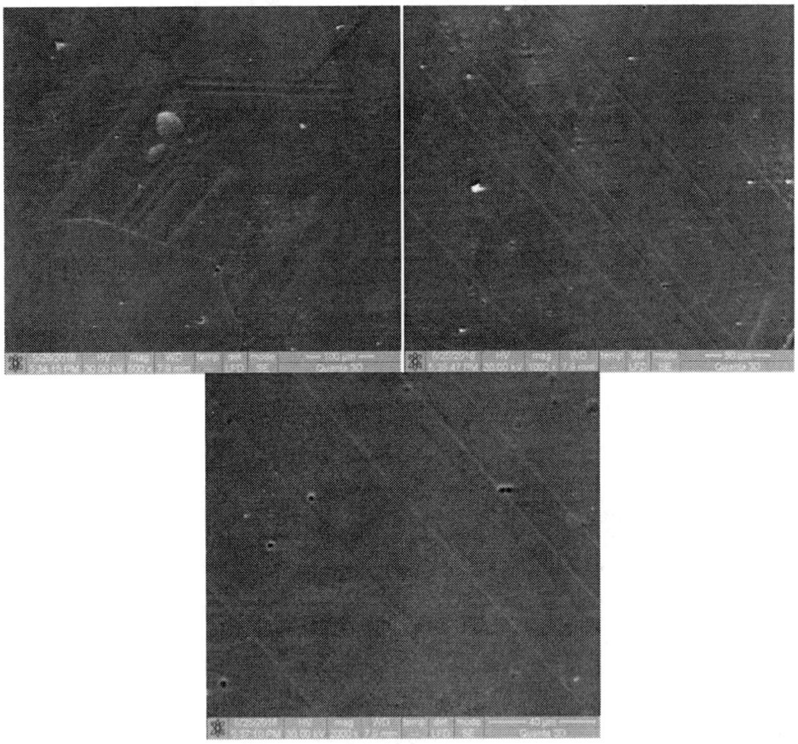

Fig. 2. SEM Microstructure – CuPb12Sn6 alloy – after homogenization heat treatment

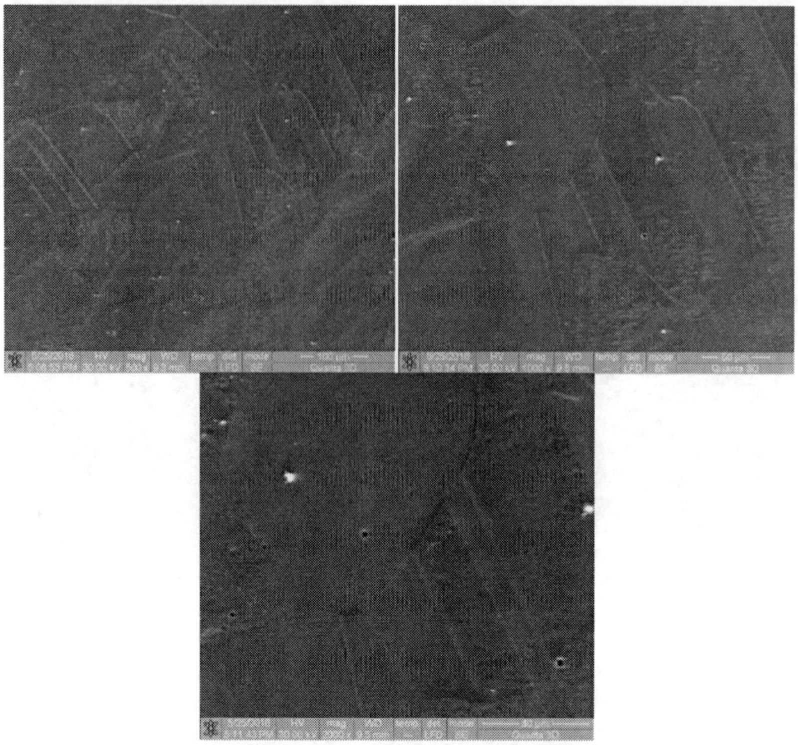

Fig. 3. SEM Microstructure – CuPb12Sn6 alloy – after recrystallization heat treatment

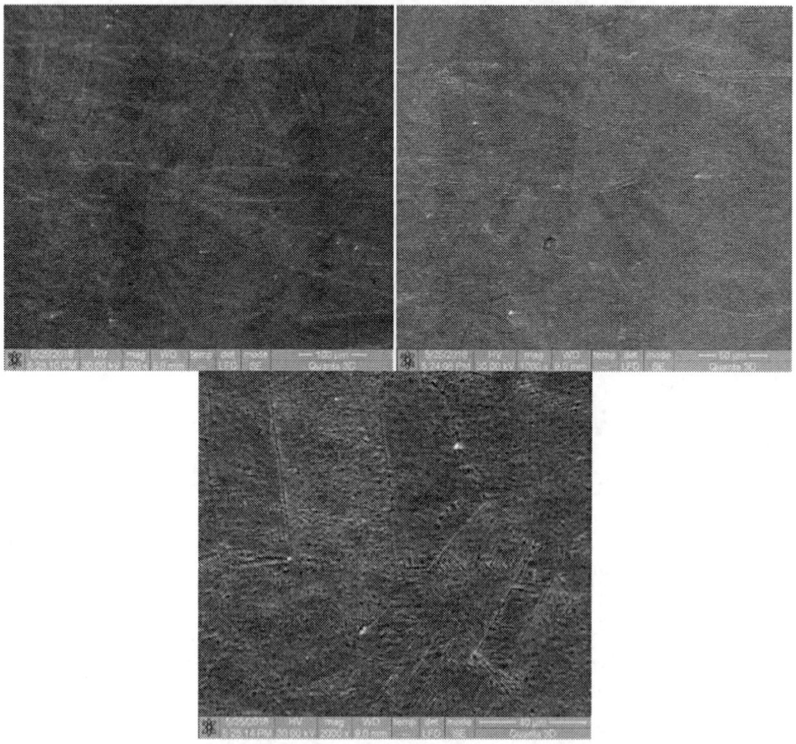

Fig. 4. SEM Microstructure – CuPb12Sn6 alloy – after quenching heat treatment

The microstructure obtained after applying the heat treatments by homogenization (Figure 2) and recrystallization (Figure 3), present a polyhedral homogeneous structure, intertwined by macle.

After quenching, from the microstructure images can observe a solid solution δ (light color) and eutectoid (α + δ).

2.4 The determination of micro-hardness characteristic

Micro-hardness measurements were performed on PMT3 micro-hardness device, with standard measure of hardness HV 50, using a force measuring 50 g and a measurement time of 15 seconds.

Were conducted a total of 3 measurements for alloy (Table 3), with the following measuring conditions: temperature 28°C (reference temperature: 23 ± 5°C) and humidity of 62 %.

Table 3. Micro-hardness of the CuPb12Sn6 bronzes

Nr.	N [div]	P [g]	HV	average value
After homogenization				
1	79	50	160,23	148,05
2	86	50	135,20	
3	82	50	148,72	
After recrystallization				
1	77	50	168,66	158,01
2	79	50	160,23	
3	83	50	145,15	
After quenching				
1	76	50	173,13	166,96
2	74	50	182,61	
3	83	50	145,15	

The obtained values for micro-hardness measurement have specific values of copper alloys, fact which permits an ease processing by cutting, to obtain finite parts with various configurations.

The obtained values for samples in quenching state have a medium micro-hardness by 167 HV50, fact which permit the use of these for manufacturing of supplies parts used in automotive industry, like bushings with collar.

3 Conclusions

By homogenization can modify the properties and structure obtained after the cast.

By recrystallization can remove the effects of cutting processing. The structural modifications induced will permit the supplementary processing up to final form of the part, or the final thermal processing.

The obtained structures after quenching, also average values of micro-hardness, permit utilization of studied alloys for manufacture the automotive parts. After a quenching, the structure is thus modified that is obtained a base mass formed preponderant from copper base compounds, but can appear lead grains. The hardness of ($\alpha + \delta$) eutectoid make that alloyed bronzes with Sn and Pb, in the cast state, to be proper for manufacturing of bearings. The mechanical properties demonstrated by micro-hardness analysis on treated samples, offer to bushings like advantages an easy processing by: casting, forging, cutting, rectification. The principal economical condition is that the final part to present a low cost, the bushings being considers the suppliers' elements in automotive industry.

References

1. V.J. Keast, J. Ewald, K.S.B. De Silva, M.B. Cortie, B. Monnier, D. Cuskelly, E.H. Kisi, J. Alloys Compd., **647**, 129 (2015)
2. A. Fattah-Alhosseini, A.H. Taheri, J Manuf Process, **20**, 98 (2015)
3. O.A. Chikova, O.P. Moskovskikh, V.S. Tsepelev, J. Phys. Chem., A, **90**, 783 (2016)
4. C. Nobel, U. Hofmann, F. Klocke, D. Veselovac, H. Puls, Wear, **344**, 58 (2015)
5. H.S. Wang, H.G. Chen, J.W. Gu, C.E. Hsu, C.Y. Wu, J. Alloys Compd., **633**, 59 (2015)
6. G. Enciu, I. Paunescu, Mat. Plast., **52**, 564 (2015)
7. M.G. Minciuna, P. Vizureanu, Metal. Int., **18**, 123 (2013)
8. M.G. Minciuna, P Vizureanu, D.C. Achitei, N. Ghiban, A.V. Sandu, N.C. Forna, Rev. Chim., **65**, 335 (2014)
9. T. Konkova, S. Mironov, A. Korznikov, G. Korznikova, M.M. Myshlyaev, S.L. Semiatin, Mater. Des., **86**, 913 (2015)
10. C. Liang, M. Ma, D. Zhang, Mater. Res., **18**, 158 (2015)
11. D.C. Achiţei, P. Vizureanu, A.A. Minea, M.M. Al Bakri Abdullah, M.G. Minciună, A.V. Sandu, Appl. Mech. Mater., **657**, 412 (2014)
12. V.P. Paun, N.I.C.A.N.O.R. Cimpoesu, R.H. Cimpoesu, G.V. Munceleanu, N. Forna, M. Agop, Mat. Plast., **47**, 158 (2010)
13. I. Carcea, M. Gherghe, *Editura Performantica*, (Iaşi, 2009)
14. G. Barbu, *Tehnologia turnării*, (Editura Universitas XXI, Iaşi, 2008)
15. P. Vizureanu, *Sisteme termice inteligente*, (Editura P.I.M., Iaşi, 2013)
16. F. Diaconescu, I. Carcea, *Obţinerea pieselor prin turnare în forme speciale*, (Editura Performantica, Iaşi, 2012)

Analysis of Shrinkage on Thick Plate Part using Genetic Algorithm

S.N. Najihah[1], *Z.* Shayfull[1,2,*], *S.M.* Nasir[1,2,3], *Mohd Sazli* Saad[1,2], *M.M.* Rashidi[4], *M.* Fathullah[1,2] and *N.Z.* Noriman[5]

[1]School of Manufacturing Engineering, Universiti Malaysia Perlis, Kampus Tetap Pauh Putra, 02600 Arau, Perlis, Malaysia.
[2]Green Design and Manufacture Research Group, Center of Excellence Geopolymer and Green Technology (CEGeoGTech), Universiti Malaysia Perlis, 01000 Kangar, Perlis, Malaysia.
[3]Centre For Diploma Studies, Universiti Malaysia Perlis, Malaysia.
[4]Faculty of Mechanical Engineering, Universiti Malaysia Pahang, 26600 Pekan, Pahang, Malaysia.
[5]Center of Excellence Geopolymer and Green Technology (CEGeoGTech), Faculty of Engineering Technology (FETech), Universiti Malaysia Perlis (UniMAP), Level 1 Block S2, UniCITI Alam

Abstract. Injection moulding is the most widely used processes in manufacturing plastic products. Since the quality of injection improves plastic parts are mostly influenced by process conditions, the method to determine the optimum process conditions becomes the key to improving the part quality. This paper presents a systematic methodology to analyse the shrinkage of the thick plate part during the injection moulding process. Genetic Algorithm (GA) method was proposed to optimise the process parameters that would result in optimal solutions of optimisation goals. Using the GA, the shrinkage of the thick plate part was improved by 39.1% in parallel direction and 17.21% in the normal direction of melt flow.

1 Introduction

Injection moulding categorised as one of the most manufacturing process that used for producing plastic products. Differ from other, forming process, injection moulding products as well as the materials are significantly affected by the moulds quality [1]. Today, injection moulding product serves range from children's toys to automobile accessories. The process of injection moulding produces a variety of thermoplastic product with tight tolerance, high production volume and complex dimension [2]. There are many different ways of moulding such as blow moulding, injection moulding, rotational moulding and compression moulding. Each method has their benefits in the manufacturing of specific item [3].

There is a common coincidence of defects arise in injection moulding from the imperfection in mould design and inappropriate of material compound. Product quality will be lower caused by inappropriate clamping force, mould seal clearance and melting temperature along with nonuniform setting. In addition, inappropriate mass to mass ratio of

* Corresponding author: shayfull@unimap.edu.my

low grade polymer is a non-conformance in the production of quality product. The most common injection moulding defects are warpage and shrinkage [4].

The webpage is one of the significant factors that affecting product quality, especially to some product that developed with thin, short, light and small design concept. Warpage become important to improve the quality of that particular design concept. Warpage is caused by uneven shrinkage of plastic parts [5]. Meanwhile, shrinkage in the mould happens is caused by thermal changes of the plastics while it is in the mould. The part can be rejected due to too much or too little shrinkage. In fact, uneven shrinkage result warpage to the moulded part. The nonuniform shrinkage gives bending or twisting out the shape, not just its dimensions, but also the angles and contour [6]. In order to avoid the shrinkage phenomenon in moulded parts, many previous researchers use parameter optimisation method. Optimisation method help to analysed the injection moulded problem and initiated various solutions [7-8]. One of the optimisation methods that widely used is a Genetic Algorithm (GA) [9].

Ozcelik and Erzurumlu [10] investigated the efficient warpage reduction on thin shell plastic parts using Finite Element (FE) analysis, statistical Design of Experiment (DOE) method, RSM and GA. A thin shell plastic part was moulded by ABS material. The FE analyses were simulated for combination of process parameters organised based on statistical full factorial experimental design. The X, Y and Z dimension are considered as process conditions dimensional parameters influencing warpage. The mould temperature, melt temperature, injection time and injection pressure were taken as constant that effecting minimum warpage. The RSM was used to create a predictive model for warpage in terms of the critical process parameters. Then, the RSM model was coupled GA in order to find the optimum process parameter values. The maximum warpage on the model was 0.0977 mm and after optimisation the warpage was reduced to 0.0582 mm by 40.4% when a maximum webpage was considered.

Ozcelik and Erzurumlu [11] studied for reducing the warpage of thin shell plastic parts using the ANOVA, ANN and GA. At first, a model of PC button for the washing machine was meshed with MoldFlow software for plastic flow analysis. Then, Taguchi method was applied for identifying the significant processing parameters which are the melt temperature, mould temperature, packing time, packing pressure, and cooling time before optimizing the minimum warpage. The ANOVA results 33.7, 21.6, 20.5, 16.1, 5.1, 1.5 and 1.3%, respectively, to packing pressure, mould temperature, melt temperature, packing time, cooling time, runner type and gate location influence warpage. ANN created a predictive model for warpage and unites with GA for optimum process parameter value. For GA, the maximum warpage on the model before optimisation is 1.703 mm reduced to 0.819 mm when considering maximum warpage. The warpage reduced by 51% after optimisation.

Changyu et al. [12] studied the possibility of the modelling as well as estimating the injection moulded parts quality and improving the process conditions in order to enhance the part quality by using the combining ANN and GA method. A top cover of an industrial refrigerator was moulded by ABS material. At first, the Computer Aided Engineering (CAE) analysis was applied to find the range of each variable. Then, a model of Back Propagation (BP) neural network was used to design the complex nonlinear relationship between quality indexes and process conditions of the injection moulded parts. Based on ANN model, a GA was used in the process conditions optimisation with the fitness function. The combination of the ANN / GA method then was used in the process optimisation for the part model in order to enhance the quality index of part's volumetric shrinkage variation. As the result, ANN gives 0.487 of volumetric shrinkage, lower than all the test sample and train sample.

There are some researchers used multi objective optimisation methods in order to reduce the defect in producing quality parts. In the injection moulding process, the multi objective optimization is used occasionally. Using a set of same input parameter, the multi objective method may help optimising more than one output like warpage and shrinkage [13].

Hakimian and Sulong [14] studied on reducing warpage and shrinkage using three different materials which were amorphous PC/ABS blend, amorphous Polyphenylene Ether/Polystyrene (PPE/PS), and crystalline Polyoxymethylene (POM) filled with glass fibres. A micro gear with 5.5 mm overall length, 18 gear teeth and 0.6mm thickness was simulated using Autodesk MoldFlow with PC/ABS, PPE/PS and POM material. The parameter which is the cooling moulding temperature, melting temperatures, temperature, packing pressure, packing time, injection pressures, and fibreglass percentages were assigned as an important factor. Three interactions factor from those important factors were investigated. The injection parameters' effects on warpage and shrinkage at different fibreglass percentages and cooling temperatures were analysed according to the Taguchi method. As the result, the PC/ABS and PPE/PS composites had 0.0051 mm of minimum warpage and shrinkage was 2.2886%, respectively.

In this study, the Genetic Algorithm (GA) is used as the optimisation method for minimising shrinkage with the help of Research Surface Methodology (RSM) for building up a mathematical model.

2 Methodology

2.1 Simulation through Autodesk Moldflow Insight 2012

Thick plate part, 4 mm thickness with single gate is chosen as a case study to minimise the shrinkage shown in Figure 1. The sprue, runner, and gate as well as the cooling channel are designed before the mould boundary is created. The thick plate part was moulded using ABS material produced by Polylac PA-777B manufactured by Chi Mei Corporation. The simulation of fill analysis, fill + pack analysis and cool analysis are simulated using Autodesk Moldflow Insight (AMI) 2012 to analyse the flow of material into mould cavities. The recommended results of AMI software are shown in Table 1. This result is used as before optimisation with shrinkage value in parallel and normal direction of 1.085mm and 0.07mm, respectively.

Fig. 1. Meshed model of thick plate part

Table 1. Results recommend by Moldflow simulation

Process Parameter		**Value**
Melt temperature (°C)		245
Mould temperature (°C)		65
Fill/ injection	Time (s)	2.75
	Velocity profile (mm)	21
Packing	Time (s)	16
	Pressure (Mpa)	50
Velocity/Pressure (V/P) switch over (mm)		21
Cooling time (s)		15.7
Coolant inlet temperature (°C)		50
Shrinkage (mm)	Parallel direction	1.085
	Normal direction	0.07

Based on the previous researches, there are four factors that being chosen as the variable factors that influence the shrinkage in injection moulding which are the mould temperature (A), melt temperature (B), packing pressure (C) and packing time (D). Through this simulation, using the result obtained, the ranges of the factors were investigated before being used through cool (FEM) + fill + pack + warp analysis for identifying the shrinkage in x and y direction. The ranges of these factors are summarised in Table 2.

Table 2. Range of variable factors

Variable Factors	Ranges
Mould temperature (°C)	40-90
Melt temperature (°C)	220-270
Packing pressure (MPa)	30-70
Packing time (s)	7.3-17.3

Through cool (FEM) + fill + pack + warp analysis, four nodes of each two cavities are selected as the part where the shrinkage measurement is taken. The nodes, including the parallel (y) side and normal (x) side of the thick plate part like Figure 2 below. The parallel and normal nodes for cavity A are 4852, 3091 and 4339, 5245, respectively. Meanwhile, for cavity B, the parallel and normal nodes are 26031, 3385 and 25518, 26424, respectively. In cool (FEM) + fill + pack + warp analysis, the value of mould temperature, melt temperature, packing pressure and packing time used are generated from the data of 30 levels of face-centered Central Composite Design (CCD) of Response Surface Methodology (RSM).

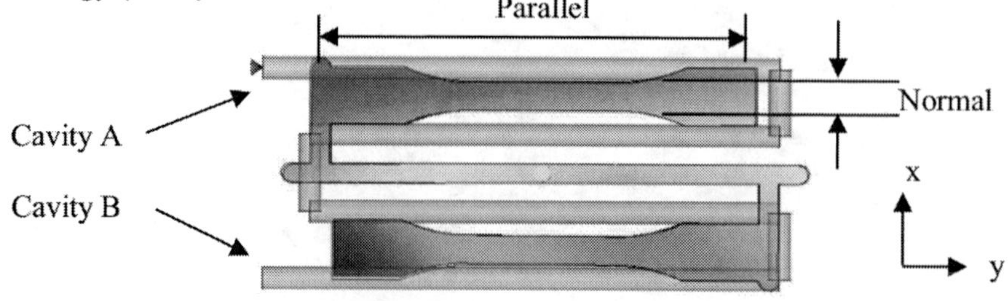

Fig. 2. Measurement method of shrinkage in parallel and normal direction

2.2 Simulation through Response Surface Methodology (RSM)

RSM is one of the methods that widely used for prediction and optimisation [15-17]. In this study, before optimisation using the Genetic Algorithm (GA) method, the RSM used was to find out the mathematical model in parallel and normal direction. The CCD model with quadratic process order, backward and alpha, α equal to 0.05 augment the factorial design was used in RSM. There are 30 runs experiments generated by specified condition of face-centered CCD design can be seen in Table 3.

Table 3. Result of face-centered CCD design

Run	A (°C)	B (°C)	C (MPa)	D (s)	Parallel (mm)	Normal (mm)
1	90	270	30	17.3	1.365	0.090
2	40	270	70	7.3	0.830	0.050
3	65	245	50	12.3	1.120	0.070
4	65	245	50	12.3	1.120	0.070
5	90	270	70	17.3	1.020	0.060
6	40	220	30	17.3	0.735	0.045
7	90	270	30	7.3	1.370	0.090
8	90	220	70	7.3	1.080	0.070
9	40	270	70	17.3	0.850	0.050
10	65	245	50	12.3	1.120	0.070
11	90	270	70	7.3	0.970	0.060
12	40	220	70	17.3	0.850	0.050
13	90	220	70	17.3	0.970	0.060
14	40	220	30	7.3	1.165	0.075
15	90	220	30	7.3	1.255	0.080
16	90	220	30	17.3	0.370	0.685
17	40	220	70	7.3	0.845	0.055
18	40	270	30	17.3	1.250	0.085
19	40	270	30	7.3	1.270	0.090
20	65	245	50	12.3	1.015	0.070
21	65	245	30	12.3	1.160	0.080
22	65	270	50	12.3	1.010	0.070
23	65	245	50	17.3	1.115	0.070
24	65	245	50	12.3	1.015	0.070
25	65	245	50	7.3	0.975	0.070
26	65	220	50	12.3	0.935	0.065
27	65	245	70	12.3	0.810	0.060
28	90	245	50	12.3	0.965	0.070
29	40	245	50	12.3	1.025	0.070
30	65	245	50	12.3	1.015	0.070

All of the variable factors were established and represented as Equation 1 for shrinkage in parallel direction, $S_{parallel}$ and Equation 2 for shrinkage in normal direction, S_{normal} of melt flow.

$$S_{parallel} = 0.42568 + 6.11594e^{-0.003}B + 0.040482C - 0.23397D - 2.25625e^{-0.004}BC + 7.32500e^{-0.004}BD + 8.15625e^{-0.004}CD \quad (1)$$

$$S_{normal}^{-1.53} = 50.94176 - 0.48740A + 0.84762C \quad (2)$$

where A is mould temperature (°C), B is melt temperature (°C), C is packing pressure (MPa) and D is packing time (s).

2.3 Simulation through Genetic Algorithm (GA)

Genetic algorithms known as search algorithms built to imitate the exact principles of biological evolution under natural genetic system. GAs are also recognised as stochastic sampling techniques, and regularly used to solve difficult problems under the terms of objective functions that hold 'bad' properties, such as multi-modal, discontinuous, non-differentiable and many more. These specific algorithms maintain and manipulate a population of solutions and implement their search for finer solutions based on 'survival of the fittest' strategy. GA solves linear and non-linear problems by considering all regions of the state space and utilizing encouraging areas through mutation, crossover, and selection process put in application to individuals in the population [18].

Using the range factor in Table 2, the value of each variable factor needs to encode into binary strings. This binary string is called as chromosomes. Each set of chromosomes constitutes a population and evolved during several generations or iteration. The chromosomes undergo the selection, based on the fitness, crossover and then the mutation for generating new generations (offspring).

Each solution given by fitness value and the higher fitness value gives chance for an individual to be selected in GA selection. Once the solution being selected, it will undergo the crossover. Crossover is the method that randomly selects one cut-point and to generate the offspring, the right parts of two parents are exchanged. In mutation, location is selected randomly and the bit number of a chromosome is flipped. Then, using a certain fitness criteria, the chromosomes are evaluated. The best ones are kept and the others are discharged. This process is repeated until a chromosome has the best fitness and it will be taken as the optimum solution.

The mathematical model generated from RSM technique is used through the GA method. The objective of this study is to minimise the shrinkage on the thick plate, put in both parallel and normal direction of melt flow. In this study, population size of 50, number of generations of 50, crossover rate of 0.40, mutation rate of 0.01, and bit number for each variable of 18 and 19 were employed.

3 Results and Discussion

After executing ANOVA, it was found that the packing pressure is the most significant factors that contribute shrinkage in both parallel and normal direction, followed by packing time, mould temperature and melt temperature. This result supported by Bushko and Stokes [19] which reported that the packing pressure resulting in minimising the shrinkage in both in-plane and through thickness direction. Ozcelik and Sonat [20] investigated that the most significant parameter for PC/ABS material of cell thin shell phone cover is packing pressure. Subsequently, Huang and Tai [21] claimed that the packing pressure is the most influential factor that affects warpage after executing the ANOVA.

Figure 3 shows the variable factors result obtained through the evolution of generations for shrinkage in parallel direction of melt flow. The whole experimental design variable factors' are optimised by the basic coded GA by defining the population size to 50. The optimised variable factors results are as following: the mould temperature is 79.45 °C, the melt temperature is 225.43 °C, the packing pressure is 31.76MPa, and the packing time is 17s which summarised in Table 4. The minimum shrinkage under the optimised variable factor conditions is 0.66127 mm based on Figure 4. The optimised shrinkage shows a reduction from the initial shrinkage before optimisation which at 1.085 mm based on Table 1. The shrinkage is improved by 39.1% after optimisation.

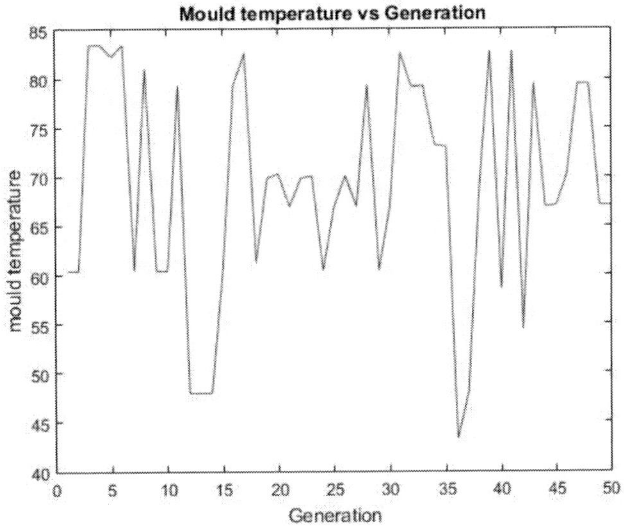

(a)

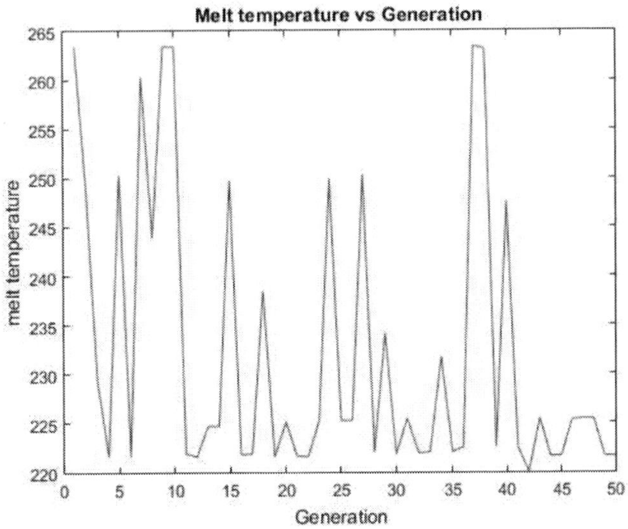

(b)

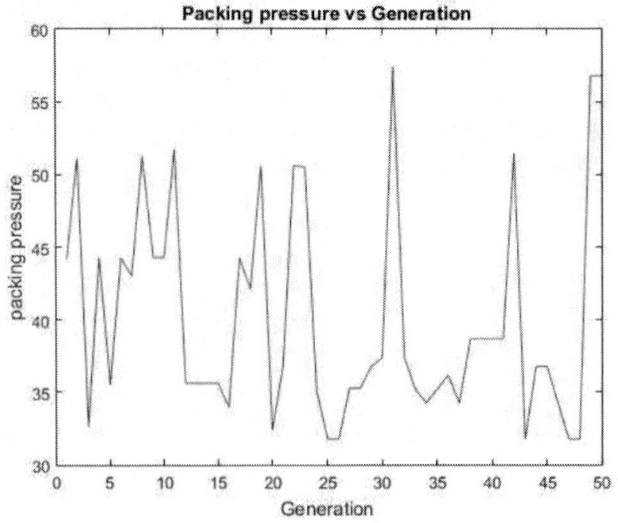

(c)

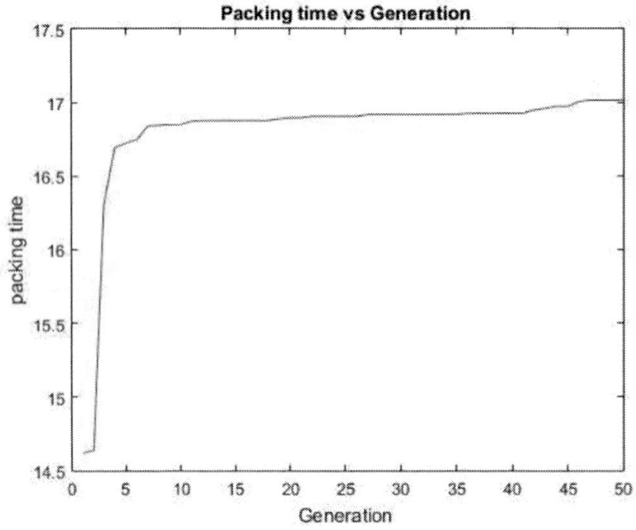

(d)

Fig. 3. (a) Mould temperature versus generation, (b) Melt temperature versus generation,(c) Packing pressure versus generation (d) Packing time versus generation

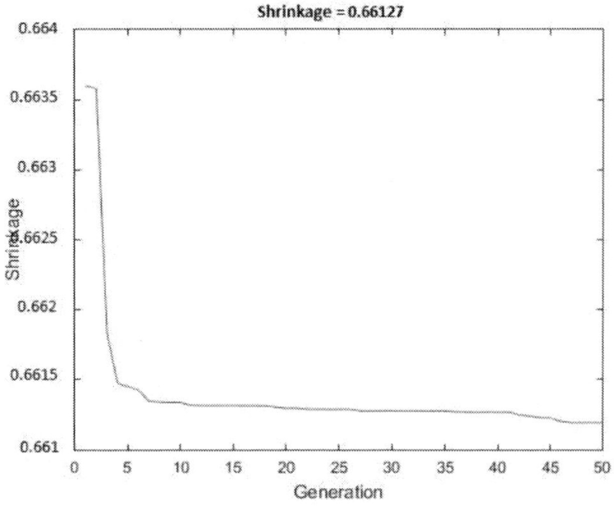

Fig. 4. Evolution of generations for the shrinkage defect

Table 4. Optimal values generated through GA method

Variable Factor	After Optimisation
Mould Temperature (°C)	79.45
Melt Temperature (°C)	225.43
Packing Pressure (MPa)	31.76
Packing Time (s)	17
Shrinkage (mm)	0.66127

Meanwhile, for the shrinkage in normal direction, Figure 4 shows the variable factors result obtained through the evolution of generations. The whole experimental design variable factors' are optimised by real coded GA by defining the population size to 50. The optimised variable factors results are as following: the mould temperature is 40.72°C, the melt temperature is 231.8°C, the packing pressure is 57.6MPa, and the packing time is 17.3s which summarised in Table 5. The minimum shrinkage under the optimised variable factor conditions is 0.057949 mm based on Figure 6. The optimised shrinkage shows a reduction from the initial shrinkage before optimisation which at 0.07 mm based on Table 1. It can be concluded that the shrinkage is improved by 17.21% after optimisation.

.

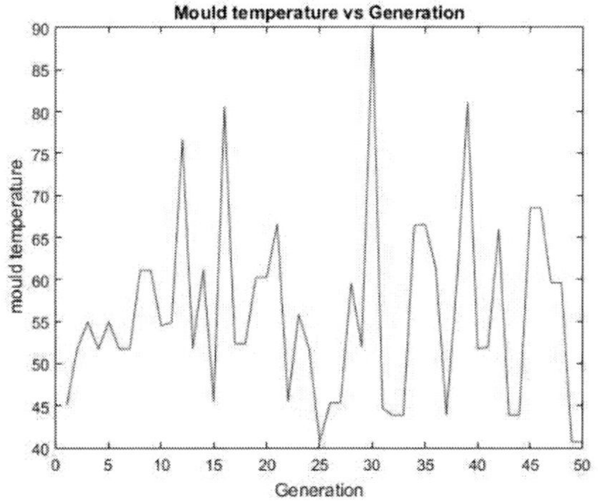

(a)

DOI: 10.1051/matecconf/20167801083

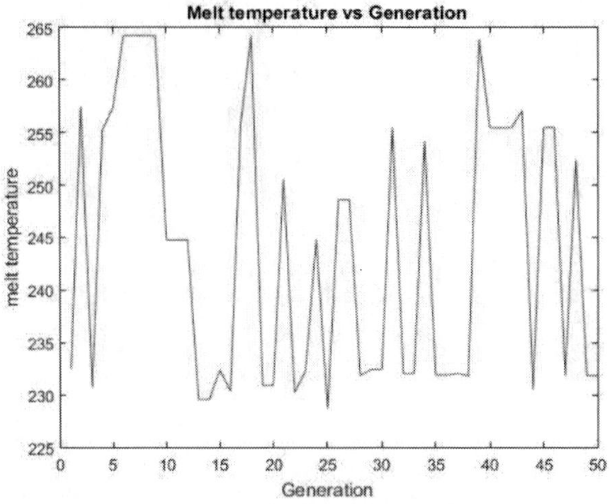

(b)

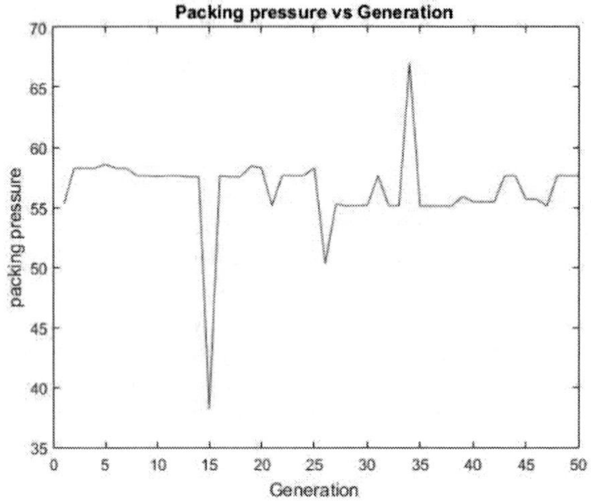

(c)

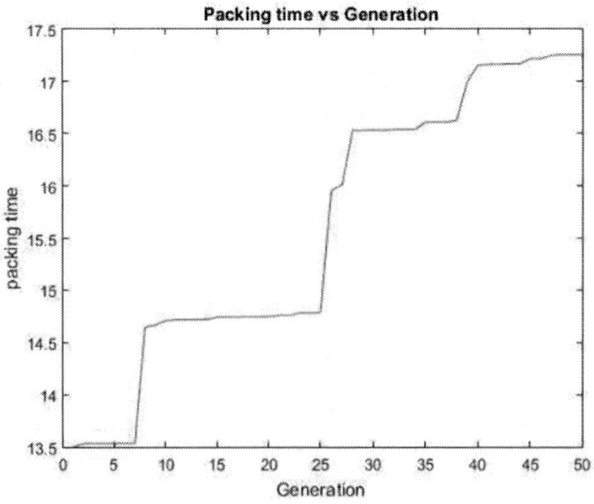

(d)

Fig. 5. (a) Mould temperature versus generation, (b) Melt temperature versus generation, (c) Packing pressure versus generation (d) Packing time versus generation

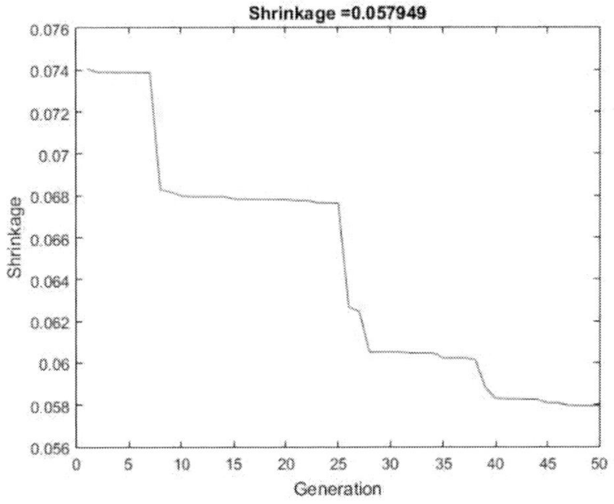

Fig. 6. Evolution of generations for the shrinkage defect

Table 5. Optimal values generated through GA method

Variable Factor	After Optimisation
Mould Temperature (^{o}C)	40.72
Melt Temperature (^{o}C)	231.8
Packing Pressure (MPa)	57.6
Packing Time (s)	17.3
Shrinkage (mm)	0.057949

Based on Table 6, it can be shown that the variable factors for parallel and normal direction varies after optimisation. The value of shrinkage before optimisation for parallel and normal direction is taken from Table 1 which recommended by AMI software. Both shrinkage values in parallel and normal direction on the thick plate part is minimised through the GA method by 39.1% and 17.21%, respectively.

Table 6. Comparison before and after optimisation

	Before Optimisation		After Optimisation	
Parameters	**Parallel**	**Normal**	**Parallel**	**Normal**
Mould Temperature (^{o}C)	65	65	79.45	40.72
Melt Temperature (^{o}C)	245	245	225.43	231.8
Packing Pressure (MPa)	50	50	31.76	57.6
Packing Time (s)	16	16	17	17.3
Shrinkage (mm)	1.085	0.070	0.66127	0.05795

4 Conclusion

In this study, best injection moulding process conditions are determined to enable minimum shrinkage. In finding optimum values, power of Finite Element (FE) software Moldflow, and GA is exploited. GA is an appropriate method to obtain the global optimal solution of the complex non-linear problem. Mould temperature, melt temperature, packing pressure, and packing time are considered as process parameters. FE analyses are conducted in combination of variable factors designed using statistical CCD face-centered experiment design. The experimental design is integrated with an effective genetic algorithm to find the optimum process parameter values. GA optimisation reduces the minimum shrinkage for parallel and normal direction to 0.66127 mm and 0.057949 mm, respectively. The findings in this study also indicate that packing pressure is the most significant factors. The optimisation methodology in this study can be employed to improve the shrinkage in the thick plate part.

References

1. S.H. Tang, Y.M. Kong, S.M. Sapuan, R. Samin, S. Sulaiman, J. Mater. Process. Technol., **171**, 259 (2006)
2. S.S. Korke, *Analysis and measurement of warpage for linear flow of amorphous and crystalline materials*, (University of Massachusetts, Lowell, 1997)
3. P.P. Shinde, S.S. Patil, S.S. Kulkarni, International Journal of Advanced Engineering Research and Studies, (2014)
4. T. Samson, *Simulation and Design of a plastic injection mold-A Joint mold for credit card and USB holder*, (Arcada, Helsinki, 2011)
5. Y. Gao, X. Wang, Int. J. Adv. Manuf. Tech., **37**, 953 (2008)
6. Z. Kamal, *Trouble Shooting In Plastic Injection Molding Machines*, (NJIT, New Jersey, 1993).
7. B. Ozcelik, Int. Commun. Heat. Mass., **38**, 1067 (2011)
8. C.Y. Wu, C.C. Ku, H.Y. Pai, Int. J. Adv. Manuf. Tech., **52**, 131 (2010)
9. F. Yin, H. Mao, L. Hua, Mater. Design, **32**, 3457 (2011)
10. B. Ozcelik, T. Erzurumlu, Int. Commun. Heat. Mass., **32**, 1085 (2005)
11. B. Ozcelik, T. Erzurumlu, J. Mater. Process. Technol., **171**, 437 (2006)
12. C. Shen, L. Wang, Q. Li, J. Mater. Process. Technol., **183**, 412 (2007)
13. S.M. Nasir, K.A. Ismail, Z. Shayfull, N.A. Shuaib, International Journal of Mechanical Engineering, **7**, 977 (2013)
14. E. Hakimian, A. B. Sulong, Mater. Design, **42**, 62 (2012)
15. N.N. Aimi, H. Anuar, M.R. Manshor, W.B.W. Nazri, S.M. Sapuan, Ind. Crop. Prod., **54**, 291 (2014)
16. G. Chi, S. Hu, Y. Yang, T. Chen, Chem. Eng. Res. Des., **90**, 1235 (2012)
17. A.M. Zain, H. Haron, S. Sharif, Expert. Syst. Appl., **37**, 4650 (2010)
18. R.H. Christopher, A.J. Jeffery, G.K. Michael, *A Genetic Algorithm for function optimization: a MATLAB implementation* (NSCU, Raleigh, 1995)
19. W.C. Bushko, V.K. Stokes, Polym. Eng. Sci., **35**, 365 (1995)
20. B. Ozcelik, I. Sonat, Mater. Design, **30**, 367 (2009)
21. M.C. Huang, C.C. Tai, J. Mater. Process. Technol., **110**, 1 (2001)

Shrinkage Analysis on Thick Plate Part using Response Surface Methodology (RSM)

M. Isafiq[1] , *Z.* Shayfull[1,2,*] , *S.M.* Nasir[1,2,3] , *M.M.* Rashidi[4] , *M.* Fathullah[1,2] and *N.Z.* Noriman[5]

[1]School of Manufacturing Engineering, Universiti Malaysia Perlis, Kampus Tetap Pauh Putra, 02600 Arau, Perlis, Malaysia
[2]Green Design and Manufacture Research Group, Center of Excellence Geopolymer and Green Technology (CEGeoGTech), Universiti Malaysia Perlis, 01000 Kangar, Perlis, Malaysia.
[3]Centre For Diploma Studies, Universiti Malaysia Perlis, Malaysia.
[4]Faculty of Mechanical Engineering, Universiti Malaysia Pahang, 26600 Pekan, Pahang, Malaysia.
[5]Center of Excellence Geopolymer and Green Technology (CEGeoGTech), Faculty of Engineering Technology (FETech), Universiti Malaysia Perlis (UniMAP), Level 1 Block S2, UniCITI Alam.

Abstract. The work reported herein is about an analysis on the quality (shrinkage) on a thick plate part using Response Surface Methodology (RSM). Previous researches showed that the most influential factor affecting the shrinkage on moulded parts are mould and melt temperature. Autodesk Moldflow Insight software was used for the analysis, while specifications of Nessei NEX 1000 injection moulding machine and P20 mould material were incorporated in this study on top of Acrylonitrile Butadiene Styrene (ABS) as a moulded thermoplastic material. Mould temperature, melt temperature, packing pressure and packing time were selected as variable parameters. The results show that the shrinkage have improved 42.48% and 14.41% in parallel and normal directions respectively after the optimisation process.

1 Introduction

Injection moulding is a very tough and complex process. This process is related to many fields of engineering such as fluid dynamics, static and dynamic, heat transfer, friction, material polymer and theory of control [1].

The term injection moulding is an oversimplified description of a quite complicated process that is controllable within specified limits. Melted or plasticized plastic material is injected by force into a mould cavity. The process is one of the most economical methods for mass production from simple to complex products. The machine parameters in injection moulding are among important factors that need to be considered to ensure the moulded part is produced in good quality.

Shrinkage is a very common defect in the injection moulding process. The value of shrinkage needs to be minimised to the lowest possible value to obtain better quality of moulded parts produced. Shrinkage occurs due to the relaxation of the polymer chain

* Corresponding author: shayfull@unimap.edu.my

molecules during the molten plastic cools down in the mould. A large value of mould temperature will make a large volume of shrinkage to the part due to the increase of the molecular energy. To overcome this issue, therefore some machine parameters need to be considered.

Previous researches highlighted on some machine parameters such as mould temperature (°C), melt temperature (°C), cooling time (s) and packing pressure (MPa) to overcome the shrinkage defect thus producing a better quality of the moulded parts.

Chen et al. [4] proposed a mathematical model to reduce the value of shrinkage variation of the thin surface moulding part using RSM. A thin shell part was used as a case study. The specimen was moulded using material from Acrylonitrile butadiene styrene (ABS). Four parameters such as melt temperature, injection pressure, packing pressure and packing time were selected as a variable parameter. From the results, shrinkage had reduced 0.981% after optimization.

Mostafa et al. [5] attempted to minimise the volume of shrinkage and warpage on a moulded fuel filter using polyamide. Mould temperature, melt temperature and injection pressure were considered as variable parameters. The response surface model was interfaced with an effective Genetic Algorithm (GA) to determine an optimum value of process parameters. The results of ANOVA and confirmation experiments showed that the quadratic models of the shrinkage and warpage were fairly well fitted with the experimental values. The values of shrinkage presented the reduction of 26.8% after optimisation.

Similarly, in different paper, Kurtaran and Erzurumlu [6] reported about their investigation of the efficiency of integrating finite element analysis, statistical design of experiment method, response surface methodology, and GA to minimise warpage on lamp base from the ceiling of bus moulded using acrylonitrile butadiene styrene (ABS). Analysis of variance (ANOVA) was used to find the significance parameters on the warpage. Mould temperature (°C), packing time (s), packing pressure (MPa) and cooling time (s) are selected as variable parameters. As a result, the values of warpage reduced 37.8% after optimisation.

In this paper, the effects of the process conditions on the shrinkage of injection moulded parts produced using ABS material were determined by the RSM method based on the input of variable parameters of the injection moulding process such as coolant inlet temperature (°C), melt temperature (°C), packing pressure (MPa) and cooling time (s). In order to minimise such defect in plastic injection moulding, Design of Experiment (DOE) with full factorial design and RSM were applied. The significant factors that contribute to the shrinkage defect were identified from DOE analysis. Furthermore, an optimal set of process conditions can be predicted from a polynomial regression equation that had been modelled from list of experiments [7]. Thick plate part was selected as a case study and Autodesk Moldflow Insight (AMI) 2012 was used to simulate and identify the shrinkage on the moulded parts.

2 Methodology

2.1 Simulation

The thick plate part as shown in Figure 1 moulded using Acrylonitrile Butadiene Styrene (ABS) was used as a case study. Thickness of the part is 4 mm and diameter of cooling channel is 8 mm. The 3D meshed model with the thickness of 2.5 mm was applied on the moulded parts using Autodesk Moldflow Insight (AMI) 2012. 3D modelling of the part and feed system was exported into AMI as well as construct the appropriate diameter, position, and cooling channels layout. The simulation starts with the fill analysis and the ram

DOI: 10.1051/matecconf/20167801084

position was determined accordingly. The inputs in this stage was melt temperature (°C) and mould temperature (°C), while the output was set as fill time (s) and shear rate (s^{-1}) before performing the Fill + Pack analysis. From the analysis, an appropriate fill time, packing time and packing pressure were determined. The input in this stage was mould temperature (°C), melt temperature (°C) and ram position (mm). The next stage was cool (FEM) analysis to obtain an appropriate cooling time (s).

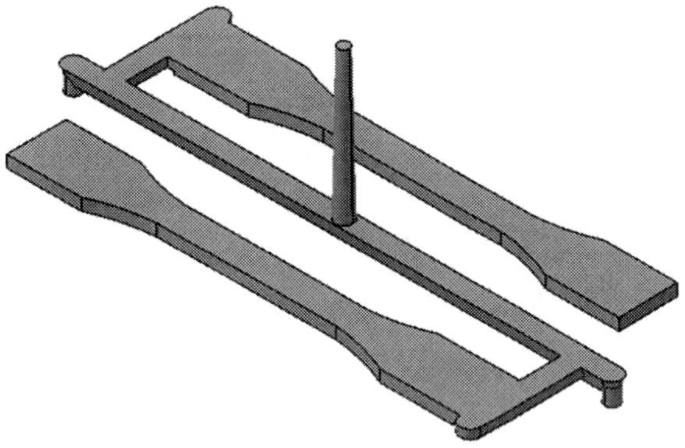

Fig. 13. Thick plate part with a gating system

2.2 Mould and thermoplastic material.

Mould Steel P20 was selected as a mould insert in this study. The mould insert was meshed using 3D mesh to simulate in Cool (FEM) analysis using AMI software. ABS material Polylac PA-777B manufactured by Chi Mei Corporation was used as a thermoplastic material to mould the part. The properties of thermoplastic and mould insert material are shown in Table 1 and 2 respectively.

Table 8. Physical and mechanical properties of ABS material.

Properties	Value
Specific heat, C_p (J/kg°C)	1464.7
Elastic modulus, E (MPa)	2.24×10^3
Poisson's ratio, υ	0.392
Thermal conductivity, *K* (w/m°C)	0.178
Melt density, (g/cm^3)	0.96

Table 9. Properties of P20 mould steel

Properties	Value
Mould density, (g/cm^3)	7.8
Mould specific heat (J/kgs)	460
Thermal conductivity, *K* (w/m°C)	29
Elastic modulus, *E* (MPa)	2.0×10^5
Poisson's ratio, υ	0.33
Mould coefficient of thermal expansion (1/C)	1.2×10^{-5}

2.3 Injection moulding machine

In the simulation, the specification of injection moulding machine Nessei NEX1000 has been set in the AMI software. The specification of the injection moulding machine used is shown in Table 3.

Table 3. Injection moulding machine specification.

Description	**Value**
Maximum machine injection stroke, (mm)	75
Maximum machine injection rate (cm^3/s)	185
Machine screw diameter (mm)	28
Machine pressure limit (MPa)	243
Maximum machine clamp force (tonne)	80

2.4 Design of Experiment (DOE)

Two level full factorial design with four factors was selected as an experimental design to estimate the main effects and interactions on the shrinkage of the moulded part using Design Expert 7.0 software. The factorial design was conducted to observe the curvature in RSM where the second order model was proposed if the curvature is significant. Mould temperature (°C), melt temperature (°C), packing pressure (MPa) and packing time (s) are four independent factors that are considered in this study. Levels for each factor are illustrated in Table 4.

Table 4. Factors and levels for DOE.

Factors	**Level**	
	Minimum	**Maximum**
Mould temperature, A (°C)	40	90
Melt temperature, B (°C)	220	270
Packing pressure, C (MPa)	30	70
Cooling time, D (s)	8	12

2.5 Response Surface Methodology (RSM)

RSM method is widely used as a tool for modelling the prediction and optimisation in various areas [7-9]. It is an empirical model to define the relationship between the various processing of variable parameters and responses according to desire criteria and searches for the significance of these process parameters in the coupled responses [10].

As much as 30 runs of experiments involved in this study were generated using the specified conditions, according to the rotatable Central Composite Design (CCD) design to minimise the shrinkage on the thick plate part. In the determination of the shrinkage on the moulded parts, the lowers of shrinkage are the better the indication of the response characteristics. Therefore, the desired responses are seen as the smaller-the-better characteristic and influence each other relatively [11].

2.6 Shrinkage measurement.

Shrinkage on the thick plate part can be calculated according to the Eq.1 in parallel to the melt flow direction, S_p and Eq. 2 for normal to the melt flow direction, S_n. The average of shrinkage was measured for both of cavities.

$$S_p = 100\frac{l_c - l_1}{l_c} \tag{1}$$

$$S_n = 100\frac{b_c - b_1}{b_c} \tag{2}$$

where, l_c is distance before deformation (mm) and l_1 is distance after deformation (mm). b_c is distance before deformation (mm) and b_1 is distance after deformation (mm). The average value of shrinkage was calculated from shrinkage in normal and parallel directions to the melt flow for both of cavities.

3 Results and Discussion

The simulation conducted using DOE indicated that melt temperature is the main factor contributing to the shrinkage defect. This result is in line with Altan [12] who found that the packing pressure is the most significant factor for shrinkage and followed by melt temperature. Besides, Azaman et al. [13] also reported that packing pressure has a significant effect on shrinkage where the higher packing pressure resulted a lower shrinkage and vice versa. This trend also agreed by Mehat et al. [14] who indicated that shrinkage significantly reduced by increasing the packing pressure. On the other hand, the cooling time was agreed by Azaman et al. [13] to have less effect on the shrinkage on the moulded parts.

The regression model from Table 5 shows the relationship between the shrinkage and all input parameters which are mould temperature (A), melt temperature (B), packing pressure (C) and packing time (D) was established by Design Expert software and represented as Eq. 3 for shrinkage for parallel and Eq. 4 for shrinkage for normal,

$$\begin{aligned} S_p = 0.42568 &+ (6.11594\, x\, 10^{-3} x\, B) + (0.040482C) - (0.23397D) \\ &- (2.2562510\, x\, 10^{-4} x\, BC) + (7.325\, x\, 10^{-4} x\, BD) \\ &+ (8.15625\, x\, 10^{-4} x\, CD) \end{aligned} \tag{3}$$

$$\begin{aligned} S_n = 14.58 - (2.01 * A) - (0.079 * B) + \\ (2.89 * c) + (0.15 * D) \end{aligned} \tag{4}$$

where A is mould temperature (°C), B melt temperature (°C), C packing pressure (MPa) and D is packing time (s).

The probability of P-value obtained from ANOVA is less than 0.05 indicating that the model is very significant for both types of shrinkage as shown in Tables 6 and 7. Among the four independent variables, the effect of packing time is the least significant on the shrinkage. Eq. 3 and Eq. 4 was applied to calculate the values of the shrinkage.

Table 5. Parameter setting in DOE

Run	Setting parameters for injection moulding simulation				Average simulation value of shrinkage (%)	
	Mould temperature (°C)	Melt temperature (°C)	Packing Pressure (MPa)	Packing time (s)	Parallel	Normal
1	90	270	30	17.3	1.365	0.090
2	40	270	70	7.3	0.830	0.050
3	65	245	50	12.3	1.120	0.070
4	65	245	50	12.3	1.120	0.070
5	90	270	70	17.3	1.020	0.060
6	40	220	30	17.3	0.7350	0.045
7	90	270	30	7.3	1.370	0.090
8	90	220	70	7.3	1.080	0.070
9	40	270	70	17.3	0.850	0.050
10	65	245	50	12.3	1.120	0.070
11	90	270	70	7.3	0.970	0.060
12	40	220	70	17.3	0.850	0.050
13	90	220	70	17.3	0.970	0.060
14	40	220	30	7.3	1.1650	0.0750
15	90	220	30	7.3	1.255	0.080
16	90	220	30	17.3	0.370	0.685
17	40	220	70	7.3	0.845	0.055
18	40	270	30	17.3	1.250	0.085
19	40	270	30	7.3	1.270	0.090
20	65	245	50	12.3	1.015	0.070
21	65	245	30	12.3	1.160	0.080
22	65	270	50	12.3	1.010	0.070
23	65	245	50	17.3	1.115	0.070
24	65	245	50	12.3	1.015	0.070
25	65	245	50	7.3	0.975	0.070
26	65	220	50	12.3	0.935	0.0650
27	65	245	70	12.3	0.810	0.060
28	90	245	50	12.3	0.965	0.070
29	40	245	50	12.3	1.025	0.070
30	65	245	50	12.3	1.015	0.070

Table 6. ANOVA of response surface model for parallel.

Source	Sum of Squares	df	Mean Square	F Value	p-value P model > F	
Model	0.86	6	0.14	9.77	<0.0001	significant
B-Melt temp	0.17	1	0.17	11.36	0.0028	
C-Packing pressure	0.16	1	0.16	11.16	0.0030	
D- packing time	0.08	1	0.08	5.79	0.0250	
BC	0.20	1	0.20	13.91	0.0012	
BD	0.13	1	0.13	9.16	0.0062	
CD	0.11	1	0.11	7.27	0.0132	
Residual	0.32	22	0.01			
Lack of fit	0.31	18	0.01	8.43	0.0259	Not significant
Pure error	8.26 x 10^{-3}	4	2.06 x 10^{-3}			
Cor total	1.19	29				

Table 7. ANOVA of the response surface model for normal.

Source	Sum of Squares	df	Mean Square	F Value	p-value P model > F	
Model	223.43	4	55.86	7.40	0.0005	significant
A-Mould temp	72.58	1	72.58	9.62	0.0049	
B-melt temp	0.096	1	0.09	0.013	0.1090	
C-packing pressure	150.34	1	150.34	19.93	0.0002	
D-packing time	0.41	1	0.41	0.054	0.4186	
Residual	181.05	24	7.54			
Lack of Fit	181.05	20	9.05			Not significant
Pure Error	0.00	4	0.00			
Cor Total	404.92	29				

4 Optimisation

To minimise the shrinkage on moulded parts, the optimal parameters should be determined. Design Expert software has been used in this study to utilise the results from the RSM method to obtain the minimum value of shrinkage. The best results for shrinkage after optimisation is presented in Table 8.

Table 8. The results of shrinkage after optimisation.

Factors	Value
Mould temperature, A (°C)	90
Melt temperature, B (°C)	270
Packing pressure, C (MPa)	70
Packing time, D (s)	7.31
Shrinkage, (%)	Parallel :0.66 Normal :0.08

5 Conclusions

The results obtained from this study are helpful in improving the quality and productivity of moulded parts. RSM method is one of the optimisation method which can be used to obtain the best setting parameters to minimise the shrinkage on moulded part. The mould temperature, melt temperature, packing pressure and packing time have been considered as variable parameters in this paper. As a result, packing pressure is concluded to be the main factor that highly affects the shrinkage of the thick plate part. The shrinkage on the thick plate parts was reduced 42.48% and 14.41% in parallel and normal directions to the material flow respectively, after optimisation using RSM.

References

1. C.K. Kwong, G.F. Smithb, W.S. Lauc, J. Mater. Process. Technol., 63, 463 (1997)
2. H. Zhou, *Computer modeling for injection molding simulation, optimization and control* (Wiley, 2013)
3. D.V. Rosato, D.V. Rosato, M.V. Rosato, *Plastic product material and process selection handbook*, 85 (Elsevier, 2004)
4. C.C. Chen, P.L. Su, Y.C. Lin, Int. J. Adv. Manuf. Tech., 45, 1087 (2002)
5. T.C. Chang, Polym. Eng. Sci., 41, 5 (2001)
6. D.V. Rosato, D.V. Rosato, M.G. Rosato, *Injection molding handbook* (Springer, 2000)
7. N.N. Aimi, H. Anuar, M.R. Manshor, W.B.W. Nazri, S.M. Sapuan, Ind. Crops Prod., 54, 291 (2014)
8. G. Chi, S. Hu, Y. Yang, T. Chen, Chem. Eng. Res. Des., 90, 1235 (2012)
9. A.M. Zain, H. Haron, S. Sharif, Expert. Syst. App., 37, 4650 (2010)
10. D.C. Montgomery, *Design and analysis of experiments* (Wiley, 2001)
11. W.C. Chen, D. Kurniawan, G. L. Fu, Adv. Mater. Res., 4, 1220 (2012)
12. M. Altan, Mater. Design, 31, 599 (2010)
13. M.D. Azaman, S.M. Sapuan, S. Sulaiman, E.S. Zainudin, A. Khalina, Mater. Design, 52, 1018 (2013)
14. N.M. Mehat, S. Kamaruddin, A.R. Othman, Adv. Mat. Res., 5, 2135 (2012)

Analysis of Strength on Thick Plate Part using Genetic Algorithm Optimisation Method

S.M. Azlan[1] , *Z.* Shayfull[1,2,*] , *S.M.* Nasir[1,2,3] , *Mohd Sazli* Saad[1,2] , *M.M.* Rashidi[4], and *M.* Fathullah[1,2]

[1]School of Manufacturing Engineering, Universiti Malaysia Perlis, Kampus Tetap Pauh Putra, 02600 Arau, Perlis, Malaysia.
[2]Green Design and Manufacture Research Group, Center of Excellence Geopolymer and Green Technology (CEGeoGTech), Universiti Malaysia Perlis, 01000 Kangar, Perlis, Malaysia.
[3]Centre For Diploma Studies, Universiti Malaysia Perlis, Malaysia.
[4]Faculty of Mechanical Engineering, Universiti Malaysia Pahang, 26600 Pekan, Pahang, Malaysia.

Abstract. This study focuses on the optimisation of the injection moulding parameters to maximise the strength ofmoulded parts using a simulation software. The moulded parts were injected with Acrylonitrile-Butadiene-Styrene (ABS) whereas mould temperature, melt temperature, packing pressure and packing time were selected as variable process parameters. The polynomial model obtained using Design of Experiment (DOE) was integrated with the Response Surface Methodology (RSM) and Centre Composite Design (CCD). The RSM was supported with Genetic Algorithm (GA) to anticipate the optimum value of processing parameters with the highest strength. It was found that strength of the parts can be improved 2.2% using the methodology reported herein.

1 Introduction

Injection moulding is among the most important polymer processing methods to produce plastic parts [1]. There are several defects on moulded part that normally appear during the injection moulding process is taking place. Among the defects are warpage, shrinkage, and weld line which influence the cosmetic quality on the surface as well as on the strength of the part. Processing parameters such as melt temperature, mould temperature, packing pressure and packing time are the main factors affecting the quality of the plastic parts produced by the injection moulding process [2-7].

Based on literature reviews (which will be discussed below), many researches have studied to improve the warpage and shrinkage by utilising Genetic Algorithm optimisation method. Park and Nguyen [8] investigated the car fender-based moulded with materials fiber-reinforced plastic, Noryl GTX810 in injection moulding process. Response Surface Methodology and Non Dominated Sorting Genetic Algorithm II (NGSA II), the multi-object optimisation method was used to find optimal values of clamping force and warpage defect. The value of clamping force was reduced 10.7% (from 4291 tonne to 3831 tonne)

* Corresponding author: shayfull@unimap.edu.my

and warpage value was reduced also 10.7% (from 3.577 mm to 3.194 mm). The results showed that the proposed method can be applied to optimise the processing parameters in solving the trade-off between consumption of energy (clamping force) and the quality of the product (warpage) produced.

Moreover, Kurtaran et al. [9] investigated on minimisation of warpage by optimising the processing parameters in injection moulding processes. A bus ceiling lamp base moulded by Acrylonitrile-Butadiene-Styrene (ABS) was used as a case study. Finite Element (FE) model of the lamp base was performed using a commercial software MoldFlow based on hybrid finite element. The injection time of 4.74 s, injection pressure of 15 MPa, packing pressure of 60 MPa, packing time 8 s and cooling time are 15 s were set during analysis. The result showed that, the warpage of the bus ceiling lamp base was reduced to 46.5% after optimisation using Genetic Algorithm (GA).

Furthermore, Yoshiko et al. [10] investigated on improvement of mechanical properties moulded by green-composites based on Polylactic Acid TE-8000 (PLA) with jute fibers. A tensile test was conducted using a universal testing machine (Autograph AG-1, Shimadzu Co.) based on the Japanese Industrial Standard (JIS) K71. It was found that, the strength of PLA reduced from 49.8 MPa to 28 MPa, but impact strength had increased more than 60% with the addition of the 50 % of jute fiber.

Another investigation was made by Chen and Kurniawan [11] who optimised the processing parameters of multiple quality characteristics in the plastic part through two stage optimisation system. In the first stage optimisation, S/N ratio and GA were used to reduce the variance of quality characteristic. In the second stage optimisation, the S/N ratio with hybrid PSO-GA were used to find an optimal setting of processing parameter for quality characteristic and stability of the process. The specimens were moulded using Victor Taichung VS-80 injection moulding machine. The Mitutoyo digital slide caliper and a NV300T laser measurement machine were used to measure the length and warpage on the moulded parts. The processing parameters such as melt temperature, injection velocity, packing pressure, packing time and cooling time were selected as variable parameters. As a result, the proposed optimisation system successfully found that the optimal parameters setting to obtain the dimensions as per designed and reducing warpage 48.5% (from 0.198 mm to 0.096 mm).

GA solves the optimisation problem by using simulating the biological evolution process, Darwin's theory of survival of the fittest [14]. The solution starts with a set of potential solution that refers as population or chromosomes. Chromosomes are in the form of bit strings and are generated randomly. The chromosomes evolve during a several generations and the generation correspond to the optimisation iteration. New generation has generated by using the mutation and crossover technique. The crossover involves splitting two chromosomes and then combining one-half of each chromosome with the other pair and mutation involves flipping a single bit of a chromosome. The chromosome has evaluated using a certain fitness criteria. This process in GA was repeated until the best fitness of chromosome had been chosen. In this study, the population size of 50, crossover rate of 1.0, mutation rate of 0.1, bit number for 4 variables of 19 except for packing time is 17 and the numbers of generations of 100 are employed. Therefore the practises of trial and error method can be reduced in order to find an optimal setting of processing parameters. At the same time the material and labour cost and time were used for mould testing to obtain the mechanical strength required by the customers can be reduced.

In this study, an experimental work was conducted to determine and to assist the injection moulding industries in order to gain an optimal setting of processing parameters with the highest strength of the moulded parts. In the beginning, Response Surface Methodology (RSM) was used to obtain the metahematical model. RSM is an empirical modelling approach used to determine the relationship between various processing parameters and responses with the various desired criteria and searches for the significance of these processing parameters in the coupled responses [12]. It is a sequential experimentation strategy to build and optimise an empirical model. Therefore, RSM is a collection of mathematical and statistical procedures which can be used for modelling and analysis the problems where the desired response is affected by several variables. The mathematical model of the desired response to the several independent input variables was gained using the experimental design and applying the regression analysis [13]. The mathematical model obtained from RSM was then used in GA to solve an optimisation problem in order to find the highest strength of the thick plate parts.

2 Methodology

This study focused on the strength effect based on a specified ranges of parameters set. The tests were conducted using Genetic Algorithm method towards the tensile strength of plastic part which conducted experimentally to obtain an optimal setting of processing parameters and maximum tensile strength. There are a few factors that influence the mechanical strength of plastic product produced from injection moulding process such as shape of product, plastic material, types of material for mould inserts, injection moulding machine, processing parameters, cooling channels design, types of coolant and room temperature [15].

This study can be divided into three phases which are selection of part design used, simulation the flow of the selected plastic material into mould cavities using Autodesk Moldflow Insight 2012 (AMI) and then experimental works using injection moulding machine and conducting tensile testing using Universal Testing Machine (UTM). Based on the simulation results, the set of parameters as shown in Table 1 had been obtained from Fill, Fill + pack and Cool (FEM) analyses were used to generated two-level full factorial design and will be analysed using Design of Experiment (DOE) to obtain the parameters setting before optimisation.

Table 1. Summarise results from AMI 2012 analysis

Parameter Setting	Value
Melt temperature (°C)	245
Mould temperature (°C)	65
Fill time (s)	2.75
Packing time (s)	12.3
Packing Pressure (MPa)	50
Cooling time (s)	15.7
Coolant inlet temperature (°C)	50

The design of thick plate part with the dimensions of 152 mm x 20 mm x 4 mm and the fabricated mould are shown in Figure 1 and 2, respectively. The thick plate parts were moulded using ABS material using 80 Tonnage injection moulding machine, NEX1000 as shown in Figure 3. Table 2 shows the technical specifications for the injection moulding machine used in this study.

DOI: 10.1051/matecconf/20167801085

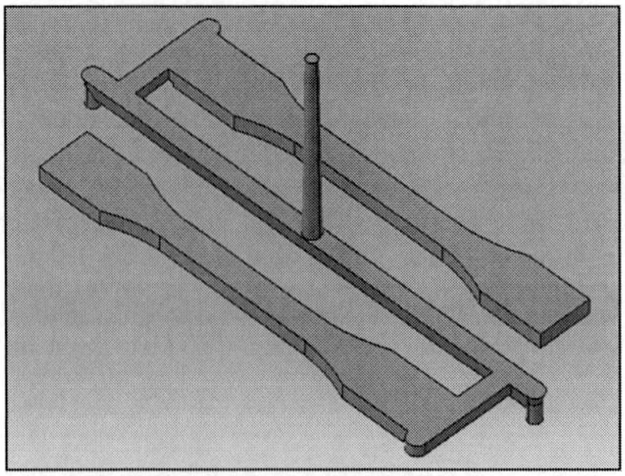

Fig. 1. Thick plate part with gating system.

Fig. 2. An injection mould for thick plate part specimen.

Table 2. Technical specifications for injection moulding machine, Nissei NEX1000.

Injection Unit	Screw diameter, mm	28
	Injection capacity, cm^3	69
	Plasticising capacity, kg/hr	28
	Injection pressure, MPa	243
	Packing pressure, MPa	205
Ejection Unit	Clamping force, tf	80
	Clamping stroke, mm	300
	Mould thickness, mm	200 – 385
	Ejector force, tf	2
	Ejector troke, mm	75
Other	Heating capacity,kW	8.36
	Machine dimensions (L x W x H), m	4.19 x 1.12 x 68
	Machine weight, tonne	3.7

Fig. 3. Injection Moulding Machine, Nissei NEX1000.

2.1 Thermoplastic material

ABS material, Poyolac manufactured by Chi Mei Corporation has been used as a thermoplastic material in this study. The properties of plastic and mould material are shown in Table 3.

Table 3. Material properties of the plastic resin.

Plastic material	**ABS**
Trade name/Grade	Polylac PA-777B
Supplier	Chi Mei Corporation
Mould shrinkage (ASTM D638)	0.4 – 0.6%
No flow melt temperature (°C)	132
Minimum melt temperature (°C)	220
Maximum melt temperature (°C)	270
Minimum mould temperature (°C)	40
Maximum mould temperature (°C)	90

2.2 Design of experiment (DOE)

In this study, two-level full factorial designs with four factors was selected as an experimental design using Design Expert 7.0 software. The factorial design was conducted to observe the significant of curvature while the second order model will propose should the curvature is significant. All variable parameters such as mould temperature, melt temperature, packing pressure and packing time are independent factors considered in this study. Table 4 shows the levels for each variable factor.

Table 4. Factors and levels for DOE

Factors	**Level**	
	Minimum	**Maximum**
Mould temperature, A (°C)	40	90
Melt temperature, B (°C)	220	270
Packing pressure, C (MPa)	30	70
Cooling time, D (s)	7.3	17.3

2.3 Responses surface methodology

The parameters setting in this study was generated using the specified conditions according to the rotatable Centred Composite Design (CCD) in obtaining the maximum strength from the moulded structure. The parameters of the experiment in this study are shown in Table 5. The curvature is significant, thus the factorial design is augmented to CCD with alpha, α value is 0.05 to an adequate quadratic model of the RSM. The empirical model in this study was obtained using the quadratic model or also known as the second-order polynomial regression model. The necessary information to construct the response model are generally accumulated by the value obtained from simulation works such as mould temperature (°C), melt temperature (°C), packing pressure (MPa) and packing time (s). The factorial portion of CCD is a full factorial design with all combinations of the genes at two levels (high, +1 and lower, -1) and collected of the eight axial points and six central points (coded level 0) which is the midpoint between the high and low points.

Table 5. Parameters setting in DOE

Run	Setting parameters for injection moulding simulation			
	Mould temperature (°C)	Melt temperature (°C)	Packing Pressure (MPa)	Packing time (s)
1	90	270	30	17.3
2	40	270	70	7.3
3	65	245	50	12.3
4	65	245	50	12.3
5	90	270	70	17.3
6	40	220	30	17.3
7	90	270	30	7.3
8	90	220	70	7.3
9	40	270	70	17.3
10	65	245	50	12.3
11	90	270	70	7.3
12	40	220	70	17.3
13	90	220	70	17.3
14	40	220	30	7.3
15	90	220	30	7.3
16	90	220	30	17.3
17	40	220	70	7.3
18	40	270	30	17.3
19	40	270	30	7.3
20	65	245	50	12.3
21	65	245	30	12.3
22	65	270	50	12.3
23	65	245	50	17.3
24	65	245	50	12.3
25	65	245	50	7.3
26	65	220	50	12.3
27	65	245	70	12.3
28	90	245	50	12.3
29	40	245	50	12.3
30	65	245	50	12.3

3 Result and Discussion

Based on the analysis was conducted using Response Surface Methodology (RSM), the results of Analysis of Variance (ANOVA) are shows in Table 6. It can be seen that, the melt temperature is the most significant factors that affected to the strength of the moulded parts. This result in line with Ozcelik et al. [16] which found that the melt temperature and packing pressure have the greatest effect on the weld line and the tensile strength of the moulded parts. This statement also agreed by Mathiavanan & Parthasaraty [17] which reported that melt temperature and packing pressure is the most influence on the strength of the moulded parts. In addition, Gou et al. [18] also reported that melt temperature has a significant effect on strength of the moulded parts.

Table 6. ANOVA results from DOE

Source	Sum of Squares	df	Mean Square	F Value	p-value P model > F	
Block	6.68	1	6.68			
Model	10.82	7	1.55	11.77	< 0.0001	Significant
B-Melt temperature	6.15	1	6.15	46.85	< 0.0001	
C-Packing Pressure	0.66	1	0.66	5.04	0.0357	
D- Packing Time	3.423×10^{-6}	1	3.423×10^{-6}	2.606×10^{-5}	0.9960	
BD	0.76	1	0.76	5.80	0.0253	
CD	1.88	1	1.88	14.28	0.0011	
B^2	0.61	1	0.61	4.65	0.0427	
D^2	1.36	1	1.36	10.32	0.0042	
Residual	2.76	21	0.13			
Lack of fit	2.76	17	0.16	197.60	< 0.0001	Not significant
Pure error	3.281×10^{-3}	4	8.201×10^{-4}			
Cor total	20.27	29				

Based on the parameter setting generated by DOE in Table 5, it can be seen that the parameter setting of run number 24 is identical with the parameters setting used in simulation works as shown in Table 1. The value of the ultimate tensile strength of the run number 24 is 38.4626 MPa as shown in Table 7. Therefore, the value of tensile strength of run number 24 (before optimization) was used to compare the tensile strength obtained after the optimization took place. In this study, a mathematical model was obtained for the use in the MATLAB software by using Design Expert software and RSM in order to maximise the tensile strength on the moulded parts produced with an optimal setting of processing parameters. The mathematical model obtained is shown in Equation 1 below:

$$\begin{aligned} Strength = 75.49189 - (0.34642x\, B) + (0.051703C) + (1.24240D) \\ - (1.74507\, x\, 10^{-3} x\, BC) - (3.42434\, x\, 10^{-3} x\, BD) \\ + (7.03051\, x\, 10^{-4} x\, B^2) - (0.026168\, x\, D^2 \end{aligned} \quad (1)$$

Where A is mould temperature (°C), B is melt temperature (°C), C is packing pressure (MPa) and D is packing time (s). The ranges of A, B, C and D were selected based on the recommended values obtained from simulation works using Autodesk Moldflow Indisght (AMI) 2012 as shown in Table 4.

Table 7. Summary results of Ultimate Tensile Strength for thick plate part.

Run	Ultimate Tensile Strength, σ_{uts} (MPa)
1	39.1373
2	40.0161
3	40.0576
4	40.0576
5	38.5548
6	40.0100
7	38.5973
8	40.5696
9	38.4682
10	40.0576
11	40.4294
12	40.4855
13	40.2298
14	39.8732
15	39.7466
16	40.2719
17	40.2022
18	39.1512
19	38.3533
20	40.0576
21	38.9426
22	38.9745
23	38.7382
24	38.4626
25	38.0670
26	40.0179
27	38.5780
28	38.8477
29	38.5396
30	38.3816

3.1 Optimisation through Genetic Algorithm

In this study, Genetic Algorithm (GA) method was used to optimise the processing parameters to mould a thick plate parts with the highest tensile strength. The equation (from RSM) was written in MATLAB programming language as shown in Equation 2 based on the Analysis of variance (ANOVA) model to solve the optimisation problem. The maximum strength is defined in the standard optimisation problem format that can be solved by a numerical optimisation algorithm. Standard optimisation problem definition requires an objective function to be minimised or maximised and constraint functions to be satisfied in terms of optimisation parameters. For a thick plate part model, an optimisation problem can be defined as follows:

To find: A, B, C, D
To maximise: Strength (A, B, C, D) (2)

Within parameter ranges:

$$40°C \leq A \leq 90°C$$
$$220°C \leq B \leq 270\ °C$$
$$30\ MPa \leq C \leq 70\ MPa$$
$$7.3s \leq D \leq 17.3s$$

Table 8. Comparision of processing parameters setting before and after optimisation

		Before Optimisation	After optimisation
Factors	Mould temperature, °C	65	51.8521
	Melt temperature, °C	270	248.0365
	Packing pressure, MPa	30	31.5412
	Packing time, s	12.3	7.3212
	Ultimate Tensile Strength, MPa	38.4626	39.3063

The results in Table 8 shows that the values of the UTS increased from 38.4626 to 39.3063. In other words, the maximum strength on the thick plate parts improved approximately 2.2% after the optimisation using GA method and the iterations of the optimisation history was demonstrated in Figure 4.

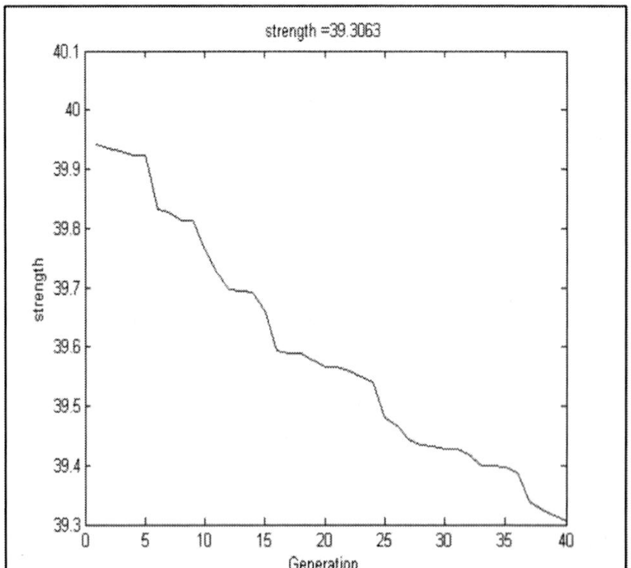

Fig. 4. The values of strength was obtained in MATHLAB software.

4 Conclusion

An efficient optimisation method using GA and RSM was useed to estimate an optimal solution of injection moulding process in order to maximise the strength on the thick plate part. The mathematical model of the strength has been carried out to correlate the dominant of the injection moulding parameters of the mould the thick plate part using ABS material. To obtain the maximum strength, the appropriate process condition parameters were determined based on the significant factor results as shown in Table 4. Mould temperature, melt temperature, packing pressure and packing time were considered as a variable processing parameters. The anaylses showed that the strength of the thick plate part can improve 2.2% after the optimisation. An optimal solution has been predicted based on a polynomial model created using RSM with CCD. This indicates that an optimisation method used in this study can be employed to improve the strength of the moulded parts required the strength for fuctionality.

References

1. E. Sachs, E. Wylonis, S. Allen, M. Cima, H. Guo, Polym. Eng. Sci., **40**, 5 (2000)
2. A.I. Isayav, *Injection and Compression Molding* (Marcel Dekker, New York, 1987)
3. G.U. Yuanxian, L.I. Haimei, S. Changyo, Adv. Polym. Technol., **20**, 1 (2001)
4. T.J. Wang, C.K. Yoon, SPE ANTEC Proceedings, **1**, 687 (2000)
5. S.J. Lui, Polym. Eng. Sci., **3**, 5 (1995)
6. B.H. Lee, B.H. Kim, Polym. Plast. Technol. Eng., **34**, 5 (1995)
7. H. Ming Chih, T. Ching-Chih, Process. Technol., **1**, 10 (2001)
8. H.S. Park, T.T. Nguyen, J. Comput. Des. Eng., **1**, 4 (2014)
9. H. Kurtaran, B. Ozcelik, T. Erzurumlu, J. Mater. Process. Tech., **169**, 314 (2005)
10. Y. Arao, T. Fujiura, S. Itani, T. Tanaka, Compos. Part B Eng., **68**, 1 (2015)
11. W. Chen, D. Kurniawan, Int. J. Precis. Eng. Manuf., **15**, 1583 (2014)
12. R.H. Myers, D.C. Montgomery, *Response Surface Methodology: Process and Product Optimization Using Designed Experiments* (Wiley, New York, 1995)
13. D.C. Montgomery, *Design and Analysis of Experiments* (Wiley, New York, 2001)
14. D.E. Goldberg, *Genetic Algorithms in Search, Optimization and Machine Learning* (Addison-Wesley, Boston, 1989)
15. Z. Shayfull, M. Fathullah, S.M. Nasir, N.A. Shuib, M.S. Abdul Manan, International Review of Mechanical Engineering, **6**, 1 (2012)
16. B. Ozcelik, E. Kuram, M. M. Topal, Int. Commun. Heat Mass Transf., **39**,2 (2012)
17. D. Mathivanan, N. S. Parthasarathy, Int. J. Adv. Manuf. Technol., **43**, 654 (2009)
18. W. Guo, L. Hua, H. Mao, **3**, 4 (2014)

MATEC Web of Conferences **78**, 01086 (2016) DOI: 10.1051/matecconf/20167801086
IConGDM 2016

Factorial Analysis on Flooded Soil Recovery Using Soil Containing Arbuscular Mycorrhizal Fungi

N. Zainol[1,*], *N.* Thangaperumal, *N.H.* Zahari and *N.H.* Aziz

[1]Faculty of Chemical & Natural Resources Engineering, Universiti Malaysia Pahang, 26300 Gambang, Pahang, MALAYSIA.

Abstract. This paper presents the factorial analysis on flooded soil (FS) recovery using soil containing Arbuscular Mycorrhizal fungi (SA) with the help of a selected host plant, *Allium cepa* L. (onion plant). In order to study the most contributing factor for the soil recovery, five factors were selected namely; pH of soil, water content, SA/FS ratio, light supply and depth of soil. This was to study the ability of SA to recover the nutrients (nitrogen, phosphorus and potassium) in the FS. Experimental design table was constructed using the concept of two levels factorial by employing Design-Expert® software (Ver 7.1.6). Soil sample was collected after 12 days and analyzed for nitrogen, phosphorus and potassium concentrations using HACH Spectrophotometer. Later, the data was analyzed in the named software to determine the main effect of respective factors and interaction between factors. The results showed that water content, light supply and depth of soil were the important factors for nutrients recovery. The best condition for FS recovery using SA was at: pH of the soil (4), water content (28 ml), SA/FS ratio (1:5), depth of soil (5 cm) and with light supply.

1 Introduction

Rainy seasons in the eastern part of Peninsular Malaysia which occurs in the period between October and March always bring about the devastating effect not only to the people but also to the ecosystem. This effects on may aspects of industries, especially on agricultural soil since agriculture, occupies a large proportion of the landscape. Flooded soils create significant challenges for agricultural lands. The floods have many direct impacts, with the most prominent one being the flooded soil syndrome where the soils losses their beneficial fungi which mobilize soil-based plant nutrients. Flooding and long periods of waterlogging have resulted in the depletion of nutrients. Prolonged flooding reduces the concentration of nitrogen (N), phosphorus, (P) and potassium, (K). N appears

[*] Corresponding author: azwinaz@gmail.com

to have been denitrified and lost from the system. Since P uptake depends on microbes in most plant, prolonged waterlogging has reduced microbial activity which in turn affects the absorption of P into the plant roots [1]. Soil K is less available in the soils that remain wet since wet soils are more prone to compaction, which restricts plant root growth and uptake of soil K.

Arbuscular Mycorrhizal fungi (AMF) is a soil borne fungi found in almost any habitat worldwide. Symbiotic associations between AMF and plant roots are widespread in the natural environment and can provide a range of benefits to the host plant. These include improved nutrition, enhanced resistance to soil-borne pests and disease, improved resistance to drought, tolerance of heavy metals and better soil structure [2]. AMF enlarges the soil volume from which nutrients can be taken up, via an extensive mycelium network, enabling host plants to access more resources. As a consequence, AMF enhances uptake of nutrients, particularly phosphorus [3], and may allow for a reduction of the amount of fertilizers applied [4]. Hence, this study attempts to manipulate the ability of AMF in facilitating nutrient uptake to recover the infertile flooded soil with the help of onion plant as a host.

2 Materials and Methods

2.1 Experimental set-up

The onion plant (*Allium cepa* L.) was selected as a host plant in this study. The soil containing AMF (SA) was collected from the domestic vegetable garden and kept in the ambient temperature. Then, simulated flooded soil (FS) was prepared by submerging the agriculture soil in the water for two weeks. The mixed soil was prepared by mixing the SA and FS according to the experimental design table (Table 1). Onion bulb was planted in the mixed soil for twelve days and planted in ambient temperature. There were 16 pots that represented 16 runs in experimental design table.

Table 1. Experimental design table for two levels factorial analysis.

Run	Factor A: pH of soil	Factor B: Water content	Factor C: SA/FS ratio	Factor D: Light supply	Factor E: Depth of soil [cm]
1	8	36	1:5	Absent	3
2	4	36	1:5	Absent	5
3	4	28	1:5	Present	5
4	8	36	1:1	Absent	5
5	4	36	1:1	Absent	3
6	8	28	1:5	Present	3
7	4	36	1:1	Present	3
8	8	28	1:1	Absent	5
9	8	36	1:5	Present	5
10	4	28	1:1	Absent	5
11	4	28	1:1	Present	3
12	8	36	1:1	Present	3
13	8	28	1:1	Present	5
14	4	36	1:5	Present	3
15	8	28	1:5	Absent	5
16	4	28	1:5	Absent	3

2.2 Sample Analysis

Soil sample near the onion plant roots was analysed for the N, P, K concentrations by using HACH Spectrophotometer. In order to test potassium content, Potassium 1, 2, 3 reagent powder pillows were used as a reagent by using tetraphenylborate method (Method 8049). Potassium in the sample reacts with sodium tetraphenylborate to form an insoluble white solid. The amount of turbidity produced was proportional to the potassium concentration. Nitrogen was analysed using persulfate digestion method (Method 10071). An alkaline persulfate digestion converts all form of nitrogen to nitrate. Sodium metabisulfite was added after the digestion to eliminate halogen oxide interferences. Nitrate then reacted with chromotropic acid under strongly acidic conditions to form a yellow complex. The measurement wavelength was 420nm for the spectrophotometer. For the phosphorus concentration, the potassium persulfate powder pillow was used with sodium hydroxide as a standard solution (Method 10127). Phosphate that present in organic and condensed inorganic forms must be converted to reactive orthophosphate before analysis. Organic phosphates were converted to orthophosphate by heating with acid and persulfate. Orthophospate reacted with molybdate in an acid medium to produce a phosphomolybdate complex. In the presence of vanadium, yellow vanadomolybdophosphoric acid forms. The intensity of the yellow colour was proportional to the phosphate concentration.

2.3 Statistical Analysis

Analysis of variance (ANOVA) was performed by using the Design-Expert® software (Ver 7.1.6). Data for the analysis were concentrations of Nitrogen (N), Phosphorous (P) and Potassium (K) in the soil from HACH Spectrophotometer.

3 Results and Discussion

3.1 Nitrogen

Data from the analysis showed that N concentration ranged from 1.1667 to 11.9333mg/L. N is a limiting nutrient in many terrestrial ecosystems [5]. AMF can transfer inorganic N ($NO3-$ or $NH4+$) to their host plant [6] as it can pull up the nutrients from deep down the soil for the host plant. These ions can readily move to the root via diffusion with the presence of water. Plants that form the AMF symbiosis frequently occur in ecosystems with high nitrification rates [7]. However, N limitation can occur even in these ecosystems [8] and nitrate acquisition via AMF might offer an advantage to the host plant in highly competitive environments, as in the analogous case of root proliferation in patches of organic material [9]. From Pareto Chart in Figure 1 it shows that the interaction between factor C (SA/FS ratio) and factor D (light supply) was the highest, meaning interaction between CD gave the most contribution to N recovery in FS with 35.22%. This was followed by only factor D with 23.07%. It clearly indicates that light supply plays a very important role for the host plant to pull up N.

Table 2 shows the ANOVA for N recovery in FS. The p-value of main effects (CD) is smaller than 0.0001 implies the significance of the interaction factors. The 2-way interactions give R2 = 0.9955 show their significance at 99.55% confidence level.

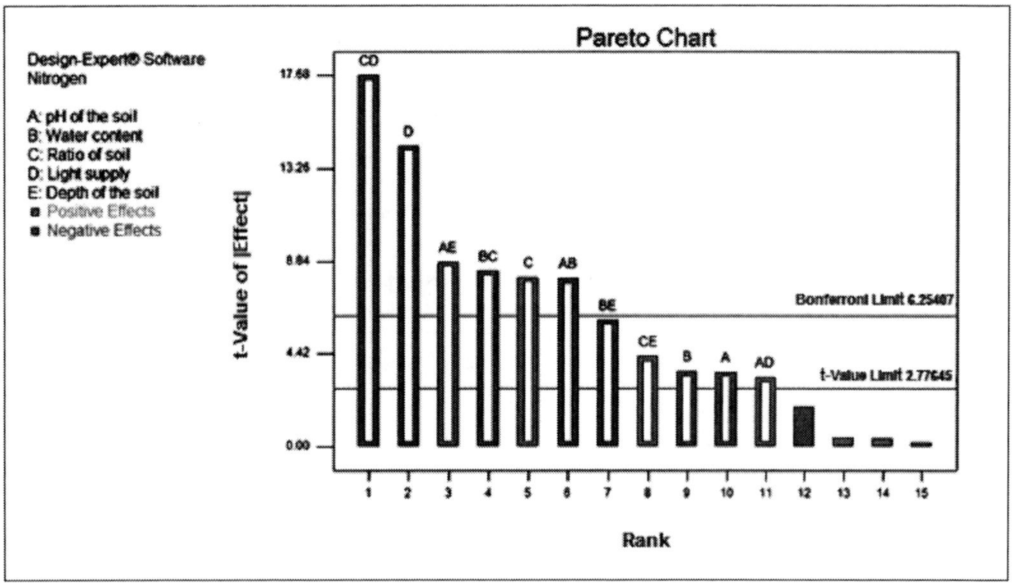

Fig. 1. Pareto chart for nitrogen.

Table 2. ANOVA table for nitrogen.

Source of variation	Sum of squares	Degree of freedom	Mean square	F value	p value Prob>F
Model	138.17	11	12.56	80.29	0.0004
A, pH of soil	1.94	1	1.94	12.38	0.0245
B, Water content	1.98	1	1.98	12.68	0.0236
C, Soil ratio	10.19	1	10.19	65.11	0.0013
D, Light supply	32.02	1	32.02	204.64	0.0001
AB	9.98	1	9.98	63.76	0.0013
AD	1.67	1	1.67	10.66	0.0309
AE	12.08	1	12.08	77.18	0.0009
BC	10.94	1	10.94	69.96	0.0011
BE	5.64	1	5.64	36.05	0.0039
CD	48.88	1	48.88	312.45	<0.0001
CE	2.86	1	2.86	18.29	0.0129
Residual	0.63	4	0.16		
Correlation Total	138.8	15			
R^2	0.9955				
Adj. R^2	0.9784				

3.2 Phosphorus

The result shows that P concentration was between the ranges of 17.23 to 26.03 mg/l. According to Shukla et al [10], both AMF and their host plants need P, and work hard to accumulate P. One of the main functions of AMF is to extract P from the soil. But when soluble P is abundant, the host plants tend to increase their restrictions to AMF, resulting in lower levels of root colonisation. Therefore, under conditions of high available P, the restrictions are increased. This has been proved by Shukla et al [10] who studied different P levels on AMF colonisation on different plants. For phosphorus recovery, the interaction of factor B (water content) and E (depth of soil) was the highest at 18.63% contribution

(Fig. 2). Both of these factors were associated with the soil condition. Factor D (light supply) and interaction between factor A (pH of soil) and B were at the same contribution around 16%. As they were all positive effects, the P recovery increased with the contribution of respective factors.

Table 3 summarizes the ANOVA (F-test) and p-value for P recovery in FS. It was observed from the ANOVA analysis that the confidence level was greater than 95% while the p-value of the model was 0.0009. The model with the p-value below 0.05 was statistically significant, which implied that the model was suitable for this experiment. The p-value of main effects (BE) is 0.0008 implies the significance of the interaction factors.

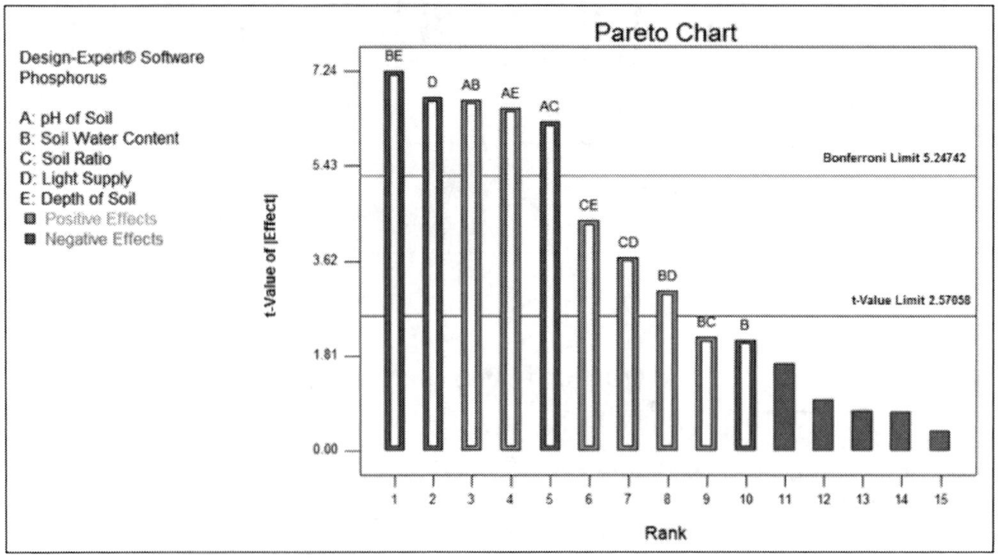

Fig. 2. Pareto chart for phosphorus.

Table 3. ANOVA table for phosphorus.

Source of variation	**Sum of squares**	**Degree of freedom**	**Mean square**	**F value**	**p value Prob>F**
Model	114.40	10	11.40	27.61	0.0009
B, Water content	1.85	1	1.85	4.45	0.0886
D, Light supply	18.85	1	18.85	45.49	0.0011
AB	18.56	1	18.56	44.79	0.0011
AC	16.34	1	16.34	39.42	0.0015
AE	17.11	1	17.11	42.74	0.0013
BC	1.94	1	1.94	4.67	0.0830
BD	3.84	1	3.84	9.26	0.0287
BE	21.70	1	21.70	52.37	0.0008
CD	5.64	1	5.64	13.61	0.0142
CE	7.98	1	7.98	19.26	0.0071
Residual	2.07	5	0.41		
Correlation Total R^2 Adj.R^2	116.47 0.9822 0.9466	15			

3.3 Potassium

The result shows that K concentration was between the ranges of 3.5 to 8.7 mg/L. As per study, potassium is one of the most important macronutrients for all organisms, especially plants. Maintaining the elevated K concentration in plants cells is vital for the smooth running of such physiological processes [11, 12]. Potassium ions are abundant in soil; however, the availability is very low due to their strong mineral adsorption. According to this research, initially the simulated flooded soil did not contain any nutrients (N, P and K). AMF was the intermediate to absorb K from the soil, indirectly helped to gain the K in the soil. The depth of the soil and water content were the most important factors in K recovery (Fig. 3). K concentration decreased as both of these factors increased.

Table 4 shows the ANOVA analysis for K recovery in FS. It was observed from the ANOVA analysis that the confidence level was 99.92% while the p-value of the model was less than 0.0001. The model with the p-value below 0.05 was statistically significant, which implied that the model was suitable for this experiment. The p-value of main effects (CD) is less than 0.0001 implies the significance of the interaction factors.

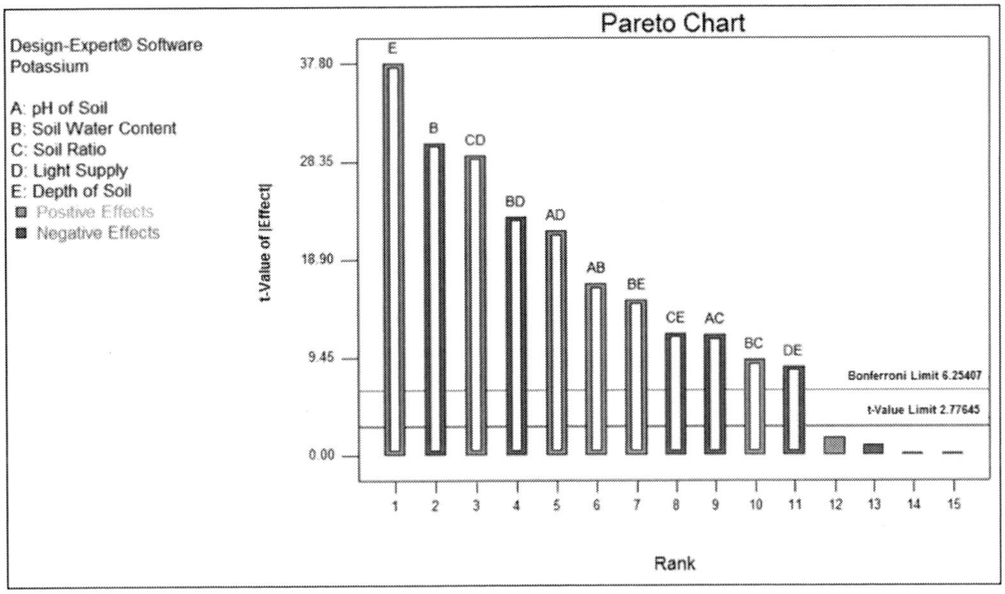

Fig. 3. Pareto chart for Potassium.

Table 4. Analysis of variance table for K.

Source of variation	Sum of squares	Degree of freedom	Mean square	F value	p value Prob>F
Model	48.84	11	4.44	463.30	<0.0001
B, Water content	8.70	1	8.70	908.10	<0.0001
E, Depth of soil	13.69	1	13.69	1428.55	<0.0001
AB	2.61	1	2.61	272.73	<0.0001
AC	1.28	1	1.28	134.03	0.0003
AD	4.48	1	4.48	467.51	<0.0001
BC	0.81	1	0.81	84.52	0.0008
BD	5.06	1	5.06	528.27	<0.0001
BE	2.15	1	2.15	224.47	0.0001
CD	8.03	1	8.03	837.70	<0.0001
CE	1.32	1	1.32	138.00	0.0003
DE	0.69	1	0.69	72.46	0.0010
Residual	0.038	4	9.583E-0.03		
Correlation Total R^2 Adj.R^2	48.88 0.9992 0.9971	15			

3.4 Best condition for flooded soil recovery

The best condition selected by design expert is summarized in Table 5. The N, P and K are at their maximum concentration in a soil condition suggested by Design-Expert® software (Ver 7.1.6).

Table 6. Best soil condition for AMF to propagate as suggested by Design-Expert® software (Ver 7.1.6).

pH of soil	Water Content	SA:FS	Light Supply	Depth of Soil
4	28	1:5	Present	5 cm

4 Conclusions

This work demonstrates the feasibility of using experimental design tools to screen significant influential factors affecting the propagation of AMF in plant roots, which in turn increase the fertility of the soil itself. The best condition of the soil for AMF in taking up nutrient at their maximum rate was identified and the result can be further used to conduct an optimization work.

References

1. T.T. Kozlowski, BioScience, **34**, 162 (1984)
2. P. Gosling, A. Hodge, G. Goodlass, G.D. Bending, Agr. Ecosyst. Environ., **113**, 7 (2006)
3. D.S. Hayman, B. Mosse, New Phytol., **70**, 19 (1971)
4. R.G. Linderman, E.A. Davis, HortTechnology, **14**, 196 (2004)
5. P.M. Vitousek, R.W. Howarth, Biogeochemistry, **13**, 87 (1991)
6. M. Govindaraju, P.E. Pfeffer, H. Jin, J. Abubaker, D.D. Douds, J.W. Allen, H. Buking, P.J. Lammers, Y. Shachar-Hill, Nature, **435**, 819 (2005)
7. D.J. Read, Experientia, **47**, 376 (1991)
8. D. Tilman, Ecol. Monogr., **57**, 189 (1987)

9. A. Hodge, D. Robinson, B.S. Griffiths, A.H. Fitter, Plant Cell Environ., **22**, 811 (1999)
10. A. Shukla, S. Lahudkar, G. Durairaj, S.R. Bhaumik, Biochemistry, **51(2)**, 706 (2012)
11. B. Benito, R. Haro, A. Amtmann, T.A. Cuin, I. Dreyer, J. Plant. Physiol., **171**, 723 (2014)
12. E. Adams, R. Shin, J. Integr. Plant Biol., **56 (3)**, 231 (2014)

Kinetic Study on Flooded Soil Recovery Using Soil Containing Arbuscular Mycorrhizal Fungi

N. Zainol[1,*], N. Thangaperumal, N.H. Zahari and N.H. Aziz

[1]Faculty of Chemical & Natural Resources Engineering, Universiti Malaysia Pahang, 26300 Gambang, Pahang, MALAYSIA.

Abstract. The purpose of this research was to determine the kinetic parameters for flooded soil recovery via soil containing Arbuscular Mycorrhizal fungi (AMF). The general procedures of this experiment started by preparation of simulated flooded soil (FS) and soil containing AMF (SA). Mixed soil was prepared by mixing FS and SA with ratio 1:1. Onion plant was chosen as a host plant and planted in the mixed soil for 14 days. The plantation was conducted in ambient temperature. The nutrients (nitrogen, phosphorus and potassium) concentrations in the soil were tested using HACH Spectrophotomer. The Michaelis-Menten equation was used to study the nutrients recovery in soil. The Lineweaver-Bulk plot was used to solve the Michaelis-Menten equation. From the experiment conducted, the maximum nutrient uptake (Vmax) and bonding affinity (Km) obtained for nitrogen (N) were 6.28mg/l.d and 82.17 mg/l, for phosphorus (P) were 9.80 mg/l.d and 60.96 mg/l.d and for potassium (K) were 0.07mg/l.d and 4.55mg/l. By comparing the result with other researcher, it showed that the Vmax and Km of nitrogen (N) and phosphorus (P) obtained were higher than other research. This was because the onion required a high level of N and P in the soil compared to other host plant.

1 Introduction

Flood has many direct impacts especially the deposition of sand and debris on productive lands, erosion of agricultural soils, anaerobic condition and this phenomenon called fallow soil syndrome. There was concern that crop production problems would develop when subsequent crops were planted to previously flooded areas. One of symptoms of flooded soil syndrome is phosphorus deficiencies developing in crops even when adequate soil phosphorus concentrations are measured. When the soil was flooded, the nutrients in the soil leached and removed out from the soil profile. Though, most of the farmers used chemical method such as fertilizer to germinate the soil nutrients back. However, this method brings more harm than good. The excess fertilizer ends up moving downwards to

* Corresponding author: azwinaz@gmail.com

the soil and lead to contamination of drinking water supplies and eutrophication of water bodies [1]. Therefore, the biological method was used to recover back all the nutrients inside the soil via Arbuscular Mycorrhizal fungi (AMF). According to [2], AMF has the ability to maintain the structure of soil and catching the nutrients from leaching out from the soil by binding it to the host plant's root. AMF is the most common type of fungus root that occurs in most soils [3]. However, AMF needs a host plant for them to survive well in the soil. They cannot propagate well in soil alone. According to Smith and Read [3], the host plants provide all the organic carbon (C) requirements of the fungi, so that AMF symbioses are mutualistic, based on an exchange of plant C for soil phosphorus (P) and other nutrients such as nitrogen (N), potassium (K), calcium (Ca) and others. It can be concluded that the symbiotic relationship between the AMF and host plant occurred when the fungi receives food carbohydrates from the plant and the plant will receive nutrients especially phosphorus from the AMF. Hence in this study the best method to recover flooded soil was via AMF by helped from the host plant. To fully understand the kinetic reactions of nutrients uptake by AMF, the Michaelis-Menten equation was used. In microbial competition theory, the Michaelis-Menten equation is used to examine the relationship between the rate of nutrients uptake and AMF. The kinetic parameters Vmax and Km determined the maximum of nutrients uptake by AMF and the bonding affinity between the AMF and the nutrients. This kinetic reaction also can be used to design equation for rate law and also used to scale up the operation. Therefore, the main objectives of this study were to use SA to recover the flooded soil and determine the kinetic parameters of nutrients uptake by AMF.

2 Materials and Methods

2.1 Experimental set-up

The soil containing AMF (SA) was collected from the domestic vegetable garden and kept in the ambient temperature at Universiti Malaysia Pahang laboratory. Then, simulated flooded soil (FS) was prepared by submerging the agriculture soil in the water for two weeks. A 200g of mixed soil was prepared by mixing the SA and the FS according to the ratio of 1:1. Onion plant *(Allium cepa* L.) was selected as a host plant in this study. Onions were bought from the convenient store. The onion was planted from the bulb using mixed soil in 14 pots. These 14 pots represented 14 days of reaction time for FS recovery. The plantation was conducted in the ambient temperature and atmospheric pressure. The sample of soil was collected daily (24 hours), then, the sample was kept in -20°C freezer to stop the activity of fungi in the soil and to avoid the denaturation of the soil contents [4].

2.2 Kinetic study

In order to study the kinetic of nutrients uptake by AMF, the Michaelis-Menten equation was used [5].

$$\mathrm{v} = \frac{\mathrm{Vmax}\, S}{\mathrm{Km} + S} \tag{1}$$

Where v is the rate of nutrient uptake, Vmax is a maximum rate achieved by AMF and [S] is the concentration of nitrogen (N), phosphorus (P) or potassium (K), Km is the bonding affinity between the AMF and which the reaction rates half of the Vmax and [S] is the initial nutrient concentration. In order to solve the above equation, the Michaelis-Menten equation was arranged linearly as shown below.

$$\frac{1}{V} = \frac{Km}{Vmax}\left[\frac{1}{S}\right] + \frac{1}{Vmax} \tag{2}$$

According to Lineweaver-Bulk plot, the 1/V against 1/S was plotted. From the graph, the value of Vmax and Km were obtained. The 1/Vmax value obtained from y-intercept while $\frac{Km}{Vmax}$ from the gradient of graph. From the calculation, the Km and Vmax determined the rates of N, P, and K uptake by AMF.

2.3 Sample Analysis

Soil sample near the onion plant roots was analyzed for the N, P, K concentrations by using HACH Spectrophotometer. In order to test potassium content, Potassium 1, 2, 3 reagent powder pillows were used as a reagent by using tetraphenylborate method (Method 8049). Potassium in the sample reacts with sodium tetraphenylborate to form an insoluble white solid. The amount of turbidity produced is proportional to the potassium concentration. Nitrogen was analysed using persulfate digestion method (Method 10071). An alkaline persulfate digestion converts all form of nitrogen to nitrate. Sodium metabisulfite is added after the digestion to eliminate halogen oxide interferences. Nitrate then reacts with chromotropic acid under strongly acidic conditions to form yellow complex. The measurement wavelength is 420nm for spectrophotometer. For the phosphorus concentration, the potassium persulfate powder pillow was used with sodium hydroxide as a standard solution (Method 10127). Phosphate that present in organic and condensed inorganic forms must be converted to reactive orthophosphate before analysis. Organic phosphates were converted to orthophosphate by heating with acid and persulfate. Orthophospate reacts with molybdate in an acid medium to produce a phosphomolybdate complex. In the presence of vanadium, yellow vanadomolybdophosphoric acid forms. The intensity of the yellow colour is proportional to the phosphate concentration.

3 Results and Discussion

3.1 Nutrients Profile

The concentrations of nutrients (N, P and K) in mixed soil (FS+SA) started to increase throughout the growing stages of onion plant. Based on Figure 1, on second day the nitrogen concentration started to increase gradually until the day 11. This stage was known as the growth stage. According to research by Feldman [2], the AMF has the ability to maintain the structure of soil and catching the nitrogen from leaching out from the soil. Therefore, the nitrogen content in the soil keeps increasing. As the height of host plant increased, the nitrogen content in the soil also increased. This was probably due to the release of nutrients by AMF during the growth of host plant [6]. The AMF started to bind with the host plant's root and triggered the accumulation of nitrogen in the soil. However, it started to decrease starting from day 12 till day 14. This happened because the AMF consumed the nitrogen and converted it into protein and supply to the host plant as food nutrients. AMF in the soil has the ability to increase nitrogen in the soil through decomposition of organic matter [2]. They also can decrease the amount of nitrogen in the soil by consuming it and convert it into protein. The presence of AMF was known to increase proportional to the availability of nutrients in soil [7-9].

While for the potassium (K), the concentration profile also has quite the same trends as N profile. According to Jeff [10], K is immobile in the soil, meaning they don't move readily with water. Therefore, the K concentration was increasing because the AMF in the soil more tightly bind with nutrient. However, at day 13, the K content started to decrease. This is because the AMF started to supply the potassium in the soil to the host (onion plant).

The phosphorus (P) concentration in the soil (Figure 1) also showed the same increasing pattern but in a fluctuated trend. This is because arbuscules in the AMF symbiosis undergo a cycle of growth, degradation, senescence and recurrent growth [10]. Phosphorus is extremely immobile in the soil and is likely to stay wherever it is placed unless moved by erosion or crop export [10] High soil P levels have been found to be detrimental for the symbiotic association of AMF with plants. These fungi need the root exudates to their vegetative growth. AMF symbioses can increase the spatial availability of P, extending the nutrient absorptive surface by formation of mycorrhizal hyphae. In the symbioses, nutrients are transferred by AMF via their extensive mycorrhizal mycelium to plants while in return the fungi receive carbon from the plant. AMF not only influence plant growth through increased uptake of nutrients (e.g. P, N, and K), but may also has non-nutritional effects in terms of stabilization of soil aggregates and alleviation of plant stresses caused by biotic and abiotic factors [3]. Overall, it can be concluded that, the AMF can increase the nutrients content in the flooded soil.

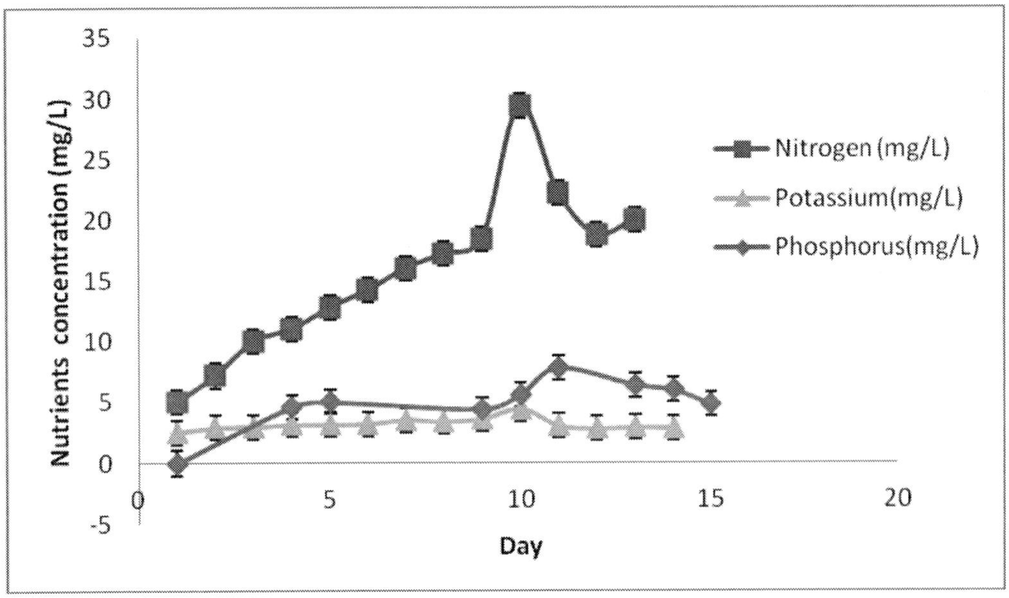

Fig. 1. Nutrients profile (N, P and K) in FS mixed with SA.

3.2 Kinetic Parameters

Kinetic study indicates the presence of high affinity transport systems for the uptake of N, P and K in the SA. N, P and K concentrations were fitted to Michaelis-Menten equation. The experimental data were plotted using Lineweaver –Bulk plot with the aid of Excel software showing r^2 more than 0.8 as shown in Figure 2, 3 and 4. Based on Figure 2, the value of Vmax obtained for N was 6.28 mg/l.d and the Km was 82.17mg/l. For K, the Vmax and Km value obtained from the Figure 3 were 0.07 mg/l.d and 4.55 mg/l respectively while for P, based on the Figure 4 the Vmax obtained was 9.80 mg/l.d and Km was 60.96 mg/l. The

 DOI: 10.1051/matecconf/20167801087

kinetic data for the nutrients uptake for this study was summarized in Table 1. The Km represented the binding affinity between the AMF and the nutrients while Vmax determined the maximum nutrients uptake by AMF. From the results, it shows that K has the lowest value of Km (highest affinity) followed by P and N. P and K were both the micronutrients that immobilize in the soil, therefore the binding between the K and AMF was stronger compared to N. From the observation of the macronutrients (N, P and K) uptake, it shows that the Vmax of the P uptake by AMF was the highest compared to N and K. This is because the higher P acquisition by mycorrhizal plants was generally attributed to production of large amounts of alkaline and acid phosphates by AMF during the course of infection that help the host plant in P absorption, existence of surface P on AMF that enable them to obtain soil P more readily than non-mycorrhizal roots [12]. This proved the ability of AMF to encourage absorption or uptake of P in the soil as earlier reported by Smith and Read [3] which stated that AMF is very effective in helping the plants to absorb P from the soil and invariably prevents P runoff that leads to eutrophication (undesired biological growth and productivity).

The increment of P uptake by AMF was well documented and considered as one of the reasons for the enhancement of the host plant growth under various soil conditions. The benefit of the mycorrhizal effect even in the soils with high levels of P provided evidence that AMF contributed efficiently to P absorption and its transportation in the plant. The increase in the P capacity of the plants can be attributed to the large increase in soil exploration area (more than one hundred times) due to mycorrhizal external mycelium production, its greater efficiency of nutrient absorption and higher affinity (lower *Km*) for P uptake from the soil solution. More recent work demonstrated that C also acts as a trigger for fungal N and K uptake and transport [11]. N in the soil had high mobility to move to plant roots through mass flow, therefore it had been suggested that the AMF played little role in N uptake. Matsumura et al., [13] George et al. [14] and Hajiboland et al. [15] also suggested that N uptake by AMF itself was usually low compared to the total demand by plants. N, P and K were important elements in the soil, however only small amount of K and N being uptake by AMF compared to P. This is because the uptake of the nutrients dependent on the type of host plant needs. From this case, it can be assumed that, P was more important to the onions, therefore the P uptake by the AMF was the highest.

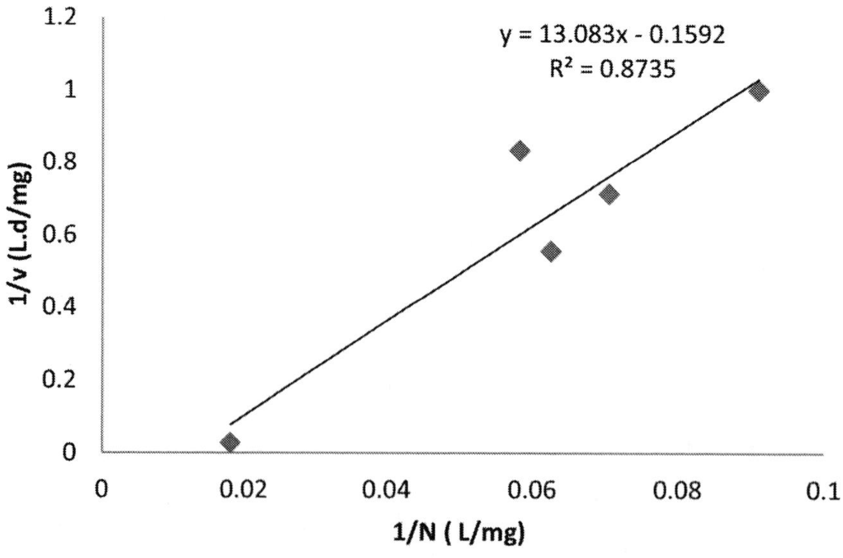

Fig. 2. Nitrogen (N) uptakes by AMF.

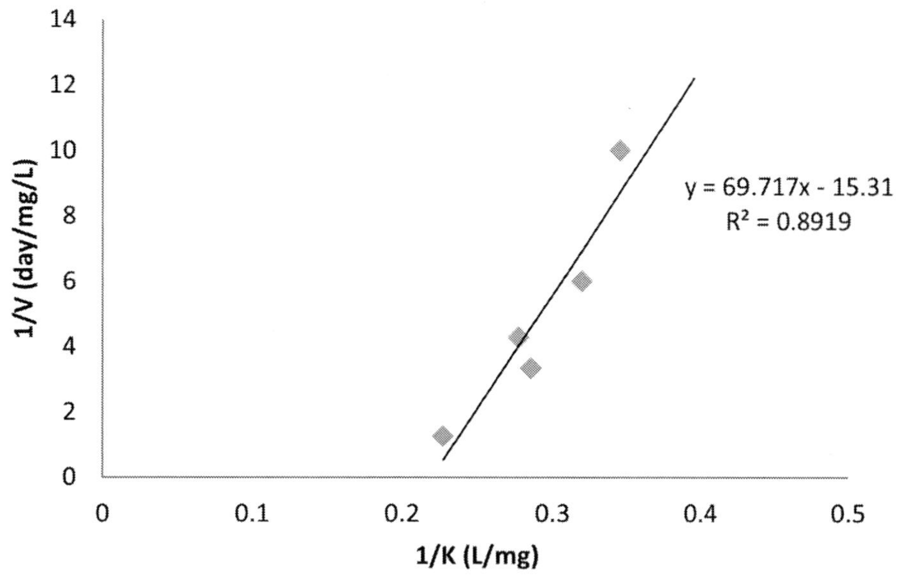

Fig. 3. Potassium (K) uptakes by AMF.

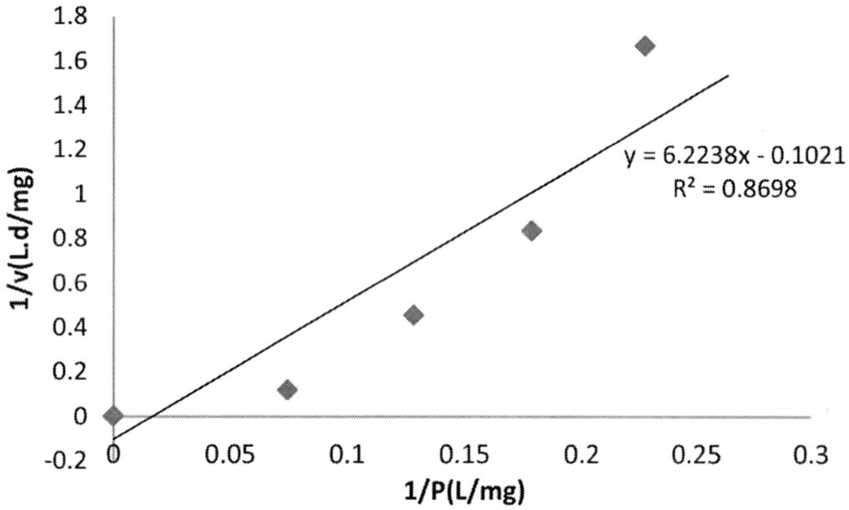

Fig. 4. Phosphorus (P) uptakes by AMF.

Table 1. Kinetic data for nutrient uptake.

Elements	Vmax (mg/l.d)		Km (mg/l)	
	This study	Other researcher [16,17]	This study	Other researcher [16,17]
Nitrogen	6.28	2.28	82.17	61.27
Potassium	0.07	0.83	4.55	7.91
Phosphorus	9.80	3.6	60.96	37.92

The kinetic parameters of nutrients uptake by AMF from other researcher also been summarized in Table 1. From the result, it shows that the Km and Vmax values obtained for the N and P from this study were higher compared to the others. This was because the host plant (onion plant) required a high level of nutrients in the soil compared to other host plant. However, the Vmax and Km obtained for K was lower compared to others because the K needed by the host plant was just in small amount.

4 Conclusions

Kinetic study showed P has the highest maximum nutrient uptake by AMF compared to N and K. The values of Vmax and Km obtained for P was 9.80 mg/l.d and 60.96 mg/l respectively which were slightly higher compared to other researcher. This was because the host plant required high P nutrients than usual. Result from this study proved that application of soil containing AMF (SA) increased nutrient (N, P and K) concentrations in the flooded soil. From a practical point of view, SA can be a valuable option for farmers to improve the sustainability of agro ecosystem by reducing the use of chemical fertilizer that gives environmental problem.

References

1. S.F. Bender, M.G. Heijden, J. Appl. Ecol., **52**, 228 (2014)
2. E.V. Feldman, *Student Pulse, 7,* 9 (2015)
3. S.E. Smith, D. J. Read, *Mycorrhizal Symbiosis*, (Third Edition. Academic Press, London, 2008)
4. M. Ayesu, G. Gyabaah, J. Ecol., **4**, 299 (2014)
5. F. Ardestani, Int. J. Energy Environ., **2**, 117 (2011)
6. A. Jessen, Plant Soil, **70**, 155 (1983)
7. S.F. Wright, A. Upadhyaya, Soil Sci., **161**, 575, 1998
8. M.C. Rillig, P.D. Steinberg, Soil Biol. Biochem., **34**, 1371 (2002)
9. G.W.T. Wilson, C.W Rice, M.C. Rillig, A. Springer, D.C. Hartnett, J. Ecol., *12,* 452 (2009)
10. B. Jeff, *Mind your P's and K's in Soils and crops*, (The Samuel Roberts Noble Foundation, Inc. 2011)
11. C.R. Fellbaum, E.W. Gachomo, Y. Beesetty, S. Choudhari, G.D. Strahan, P.E. Pfeffer, K. Toby, B. Heike, Plant Signal Behav., **7 (11)**, 1509 (2012)
12. G.R. Venkateshwar, C. Manoharacharyl, B.R. Rajeswara, Journal of Science, **131**, 47 (2011)
13. A. Matsumura, S. Taniguchi, K. Yamawaki, R. Hattori, A. Tarui, K. Yano, H. Daimon, Am. J. Plant Sci., **4,** 2290 (2013)
14. E. George, H. Marschner, I. Jakobson, Crit. Rev. Biotechnol., **15**, 257 (1995)
15. R. Hajiboland, N. Aliasgharzad, R. Barzeghar, Plant Soil Environ., **55**, 93 (2009)
16. T.J. Albaugh, H.L. Allen, T.R. Fox, Tree Physiology, **28**, 1083 (2008)
17. P.M. Bradley, J. Ecol., **71**, 282 (1990)

Design of Controls, Monitoring and a Energy Storage System for a Energy Harvesting Gymnasium Equipment

V. Thiruchelvam[1,*], *C.M.Y.* Hammad[1], *S.A.* Medni[1]

[1]School of Engineering, Asia Pacific University, Institute of Engineers Malaysia (IEM), Kuala Lumpur, Malaysia.

Abstract. The energy harvesting development was to monitor, control and optimize the output power from the gym equipment. The designed utilized Arduino and LabVIEW based Battery Energy Storage System (BESS) that incorporated switchover capability according to the load power. The manual gears-shifting mechanism was improved with the aid of a DC geared motor and L298N controller to be automatically controlled by the user's input via a keypad.

1 Introduction

The gym weight-lifting machine is a potential application to use human's exercising energy to generate power. The key area for power generation stability is based on an efficient Battery Energy Storage System (BESS), monitoring system, gears-shifter controlling system, and the brand-new concept of fully DC gymnasium. Currently, there are several ESS technologies such as: superconducting magnetic energy storage (SMES) [1], batteries (BESS) [2], electrochemical double layer capacitors (EDLC) [3], compressed air energy storage (CAES) [4] and flywheels [5]. The most promising solution in the renewable energy storage area is that NiCd and Li-ion batteries are preferred due to their high energy density, power capability and accepted economic value [6]. The proposed research carries a comparison between the power losses in different topologies of multilevel inverters. It analyzed dominant multilevel inverters; power losses due to the semiconductors were first calculated and then simulated to confirm the findings. There are four types of multilevel inverters that were used to conduct the research. Having a minimum of 8 switching modes within the circuit increases the total power losses [7].

* Corresponding author: dr.vinesh@apu.edu.my

2 Methodology

2.1 Battery Energy Storage System (BESS)

Fig 1 shows the block diagram of the BESS with switchover capability; two batteries, two SPDT relays, voltage sensor and Arduino Uno. The source block refers to the output power from the generator connected to the weight-lifting machine, and the load block is meant to be any DC device. Battery 1 is under charging mode and Battery 2 is discharging to the load. The voltage sensor on the load side will continuously detect the voltage across the discharging battery and send the information to the controller. When the voltage sensor detects 10V, the switchover between the batteries will occur. This is controlled by the Arduino which decides the modes of both relays when detecting the switching condition from the voltage sensor. After switching, Battery 1 will be connected to the load side and under discharging mode while Battery 2 is charging from the source. A Graphical User Interface (GUI) is also designed using LabVIEW and connected to the Arduino output to monitor the batteries operation. The main reason of integrating a switching feature in the BESS is to avoid heating-up the batteries from overcharging that can affect the lifespan of the batteries.

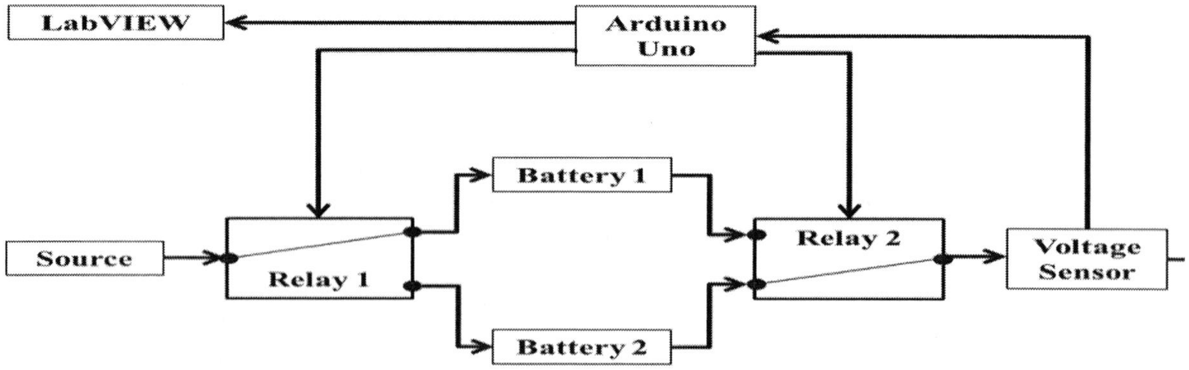

Fig. 1. BESS Block Diagram.

2.2 Gears-Shifting Mechanism Controlling System

The second objective of the research was to control the gears shifting mechanism which is coupled to the generator. The aim was to simulate the user's desired kilograms to be lifted by selecting the equivalent gear. The gears shifter is controlled by constructing an Arduino-based system which consists of a potentiometer the L298N controller, a DC geared motor and keypad. The Arduino reads the potentiometer value and sends a command to L298N Motor Controller to rotate the DC motor clockwise or counter clockwise. The shaft of 12V DC motor is attached to the 10kΩ potentiometer, which means any rotation in the motor shaft will cause a change in the potentiometer, in other words, it is used as a locator of the DC motor shaft's position. According to Table 1, when the Arduino receives a potentiometer value within the expected range, the motor's rotation will stop.

Table 1. Key and Corresponding Potentiometer Range.

Keypad Key	Kilograms	Potentiometer Range
1	21.2	646 – 663
2	22.9	566 – 585
3	25.5	508 – 525
4	27.3	463 – 482
5	29.9	400 – 425
6	32.8	300 – 345

2.3 Derivation of Theoretical Running Time

In the testing section, several experiments were conducted to determine the actual running time of the AC and DC devices when using Lead Acid or Li-Ion batteries. Before the testing, the theoretical expected running time of each device was calculated. The devices are 2Ah Lead Acid and Li-Ion batteries, 20W CFL and LED, 30W AC and DC exhaust fans, and an inverter. Eq 1 was to find the running time of each device in minutes when supplied by the 2Ah 12V batteries, until the battery voltage reduces to 10V which is the BESS switching voltage, thus the discharged voltage (V_{dis}) was 2V. Eq 2 was to determine the Battery Energy (E_B) in Wh which is to be substituted in Eqp 1; battery capacity (C_B) value is 2Ah.

Device Running Time (T) = [Battery Energy (E_B) / Load Power (L)] x 60 (1)

Battery Energy (E_B) = V_{dis} x C_B (2)

E_B = 2V x 2Ah = 4Wh

Before determining the running time of the AC devices, the current consumption of the DC-to-AC inverter from the 12V 2Ah batteries should be calculated by using Eq 4. The inverter consumption will vary according to the supplied load current, therefore the current for the CFL and AC exhaust fan is found using Eq 3 then substituted in Eq 4 and Eq 5. Eqp 5 is used to find the total AC load power consumption when both inverter and AC load are supplied by 2Ah battery; calculated values for CFL only.

Load Current (I_L) = Load Power (P_L) / Load Voltage (V_L) (3)

I_L, CFL = 20 / 12 = 1.667 A

Inverter Current (I_{inv}) = Load Current / Inverter Efficiency (4)

I_{inv} (CFL) = 1.667 / 0.90 = 1.852 A

Total AC Load Power Consumption = Battery Voltage (V_B) x (I_{inv} + I_L) (5)

Total AC Load Power (L_{TOT}), CFL = 12 x (1.852 + 1.667) = 42.228W

The running time of each AC/DC devices can be found by substituting the total AC/DC loads power and the Battery Energy (E_B) in Eq 1; calculated values for LED only. Table 2 is a summary for the derived running times for all devices.

 DOI: 10.1051/matecconf/20167801088

CFL (T_{CFL}) = [4 Wh / 42.228 W] x 60 = 5.68 Minutes
LED (T_{LED}) = [4 Wh / 20 W] x 60 = 12 Minutes

Table 2. Theoretically Derived Running Time for Devices.

Device	Running Time (Minutes)
CFL	5.68
LED	12
AC Exhaust Fan	4
DC Exhaust Fan	8

3 Design Implementation

3.1 Battery Energy Storage System (BESS)

Fig 2 shows the BESS circuit with detailed connections. It is important to note that the SPDT relays are connected to the same digital pin in Arduino (Pin 12). The relay needs a signal from Arduino to switch, whether it is HIGH or LOW, thus completing the full circuit. Relay 1 accepts one input and can connect to two outputs, but relay 2 can receive two inputs and connect to only one output. This results in relay 2 being connected in the opposite direction of SPDT default connection. When the Arduino sends digital signal to Pin 12, both relays will switch and complete different circuits charging and discharging circuits as desired.

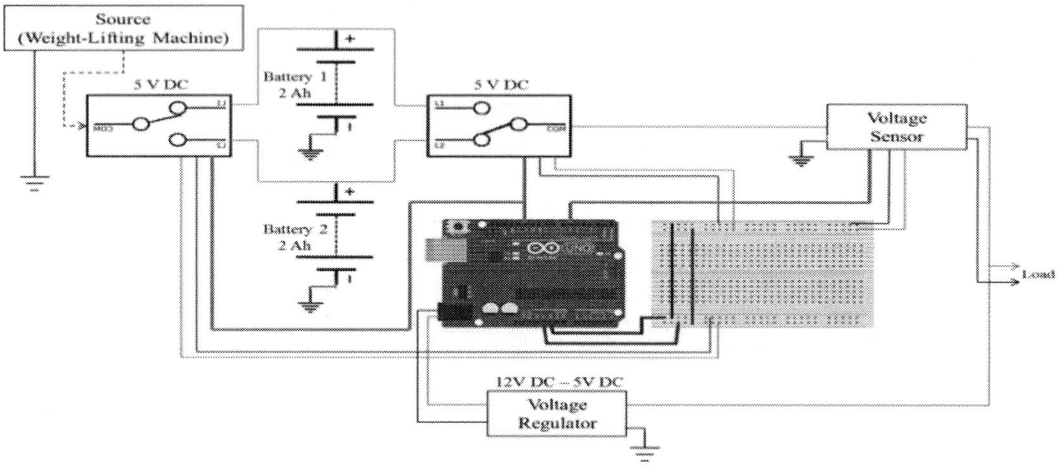

Fig. 2. BESS Circuit.

The Arduino code was developed to read voltages from 0 to 5 volts, while the voltage sensor is reading from 0 to 12 volts. When the Arduino receives the voltage from the voltage sensor, it converts the voltage from 0 to 5 volts. The Arduino has a 10-bit analog-to-digital converter that maps the analog input (0 to 5V) to 1024 (2^{10}) integer values from 0 to 1023. Voltage Sensor variable will store the integer value (0 – 1023), and ActualV variable is the voltage that is actually sensed by the voltage sensor. In ActualV formula, the ratio (5/1023) is multiplied by Voltage Sensor integer which gives a voltage range from 0 to 5 volts. To convert the voltage to 12V range it is multiplied by (2.4) which is the conversion ratio (12/5). When testing the code, the ActualV variable doesn't exactly show the correct voltage as compared to the multimeter reading (connected to the tested battery terminals), thus a calibration was needed. This was modified by multiplying 2.0797 to Voltage Sensor value after several trials to find the suitable value to correct the code. LabVIEW was used to develop a GUI to monitor the conditions of the BESS two batteries by using VISA Resource block as shown in Fig 3.

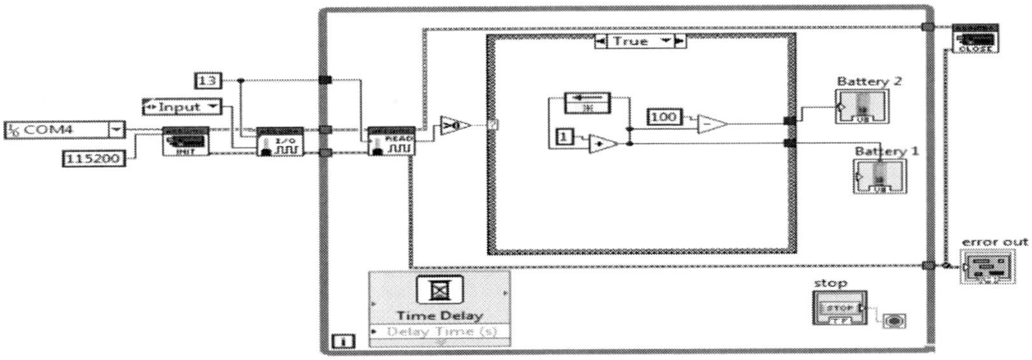

Fig. 3. BESS Block Diagram in LabVIEW.

3.2 Gears-Shifting Mechanism Controlling System

As shown in Fig 4, the constructed circuit for the gears-shifter controlling is mainly dependent on the L298N motor controller module and Arduino. The program consists of initialization of 3x4 keypad and L298N, then switch-case statement that includes six cases according to Table 1, and three if-statements in each case. To read the potentiometer value, the variable PMvalue holds the analog value from Arduino analog pin A0. Three if-statements are there to test the potentiometer value, if within or equal to 646 – 663 the code will go to MSG1 label to show the confirmation message. If the PMvalue is less than 646, the counter clockwise function will be called, the potentiometer value will be displayed and delaying of 1500ms between each change. Then, the code will go to label readPM which reads the new potentiometer value after the DC motor rotation caused by the counter clockwise function. When the PMvalue is greater than the acceptable range, the only difference will be calling the clockwise function.

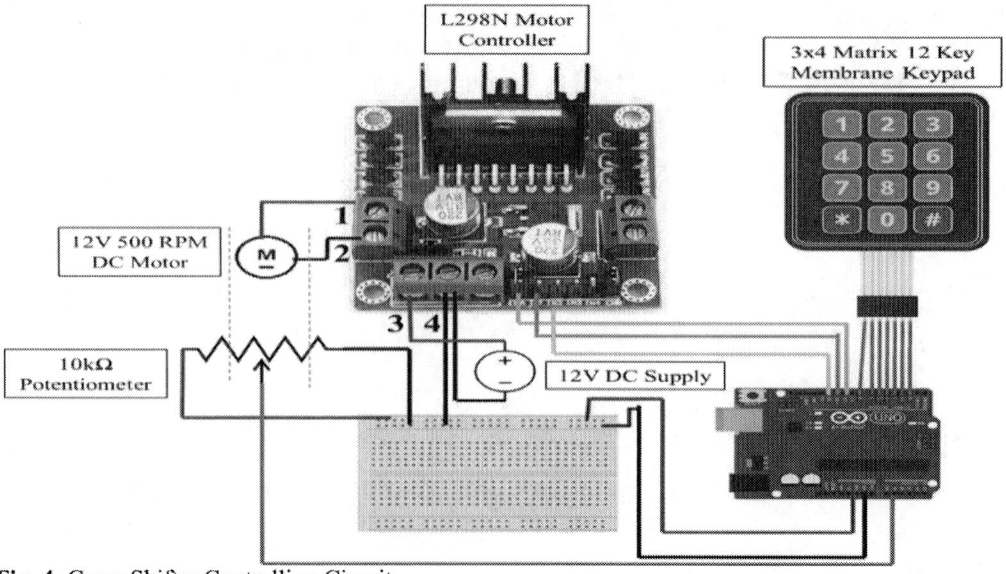

Fig. 4. Gears Shifter Controlling Circuit.

4 Conclusions

The Battery Energy Storage System (BESS) productive design was attained by developing a battery switchover controlling and monitoring scheme with the aid of Arduino and LabVIEW programming. For the Gears-Shifting Mechanism Controlling System, the prototype was efficiently designed using a Arduino-based scheme with a DC geared motor and the L298N Motor Controller Module.

References

1. M.M. Aly, M. Abdel-Akher, S.M. Said, T. Senjyu, Int J Elec Power, **83**, 485 (2016)
2. L.Yang, N. Tai, C. Fan, Y, Meng, Renewable Energy, **95**, 202 (2016)
3. B.E. Wilson, S. He, K. Buffington, S. Rudisill, W.H. Smyrl, A. Stein, J. Power Sources, **298**, 193 (2015)
4. G. Venkataramani, P. Parankusam, V. Ramalingam, J. Wang, Renew. Sustainable Energy Rev., **62**, 895 (2016)
5. K. Chatzivasileiadi, *7th International Renewable Energy Storage Conference, Germany*, 34 (2012)
6. G.S. Perantzakis, C.A. Christodoulou, K.E. Anagnostou, S.N. Manias, IET Power Electron., **7 (11)**, 2743 (2014).
7. S.B. McLaughlin, D.G. de la Torre Ugarte, C.T. Garten, Jr., L.R. Lynd, M.A. Sanderson, V.R. Tolbert, D. D. Wolf, Environ. Sci. Technol., **36**, 2122 (2002)

The Impacts of Platinum Diffusion to the Reverse Recovery Lifetime of a High Power Diode Devices

C.M. Cheh[1], *M.K.M.* Arshad [2,*], *R.A.* Rahim[1], *C.* Ibau[1], *C.H.* Voon[1], *R.M.* Ayub[1,2], *U.* Hashim[1,2]

[1]Institute of Nano Electronic Engineering, Universiti Malaysia Perlis (UniMAP), Perlis, Malaysia
[2]School of Microelectronic Engineering, Universiti Malaysia Perlis (UniMAP), Perlis, Malaysia

Abstract. The reverse recovery lifetime of a diode is one the key parameter in power electronics market. To make a diode with fast switching speed, diodes are often doped with impurities such as gold and platinum to improve its lifetime. In this works, we present the reverse recovery lifetime improvement of a power rectifier diode through platinum diffusion in the intrinsic region in between P-N junction using Design of Experiment (DOE) approach. A commercial available power rectifier is used in this study. We factored in the temperature and thermal diffusion time during the platinum diffusion process in our DOE. From results, DOE 2 (with shorter thermal duration and high temperature for diffusion) is selected based on meeting requirement for forward voltage and reverse recovery specifications i.e. forward voltage at 1.8V and reverse recovery time at 27ns.

1 Introduction

The evolution of semiconductor has made power diode with high switching speeds are widely used in variety of applications such as switching the power suppliers, circuit protection, etc. There are quite a number of studies for the past decades to explain the behaviour of P-N junction diodes and also power P-I-N diode [1-14]. In diode characteristics, one of the key parameter to determine the performance of the diode is, its switching speed capability. Switching speed is defines as the time for a diode to recovers from an off-state to on-state.

Fig. 1 shows the example of reverse recovery lifetime (TRR) of a diode. In another way, TRR is also refer as the time it takes for the diode to switch from forward bias to reverse bias. The reverse recovery time is also the combination of the storage time and recovery time. During forward bias, there is a build-up of minority carriers at the edge of the depletion region. The minority will diffuse to the opposite material during reverse bias and

* Corresponding author: mohd.khairuddin@unimap.edu.my

triggering the storage time. Once the minority carriers migrate they will diffuse and recombine which it is known as the recovery time.

In the past, there are various studies to discuss the reverse recovery of the diode [15-19]. Impurities such a gold and platinum can be diffused into the device body or the device body can be irradiated [20]. Irradiation is preferable due to inexpensive and is a cleaner technology. Irradiation is done at high level of energy around 1.5 MeV to 12MeV to alter the device structure [21-23]..Those methods are introduced to create recombination centre that stimulate the carrier recombination and depleting the semiconductor device of mobile minority more rapidly, making the device capable to switch from off state to on state quickly. One of common trade off to have fast switching diode is the increase of leakage current and also the increase in the on state of voltage. So, to find a balance in between the main electrical parameter is mandatory when designing and fabricating the power diode. Several papers discusses about the combination of few methods to control the lifetime of a diode as narrated elsewhere [24,25].

Both platinum and gold materials are used as a reverse recovery lifetime modifiers but platinum diffusion is more commonly used due to the following factors. 1) Gold as a lifetime modifier will normally exhibit higher leakage currents compared than platinum doped diode, especially at elevated temperatures. 2) Platinum cost is lower than that of gold price. In this paper, experimental studies are conducted to find the optimum process conditions for the platinum doped method through the DOE approach on the impact of thermal duration and temperature of diffusion process. The evaluation methods are discussed in methodology section. The results of the evaluation with the discussion are further elaborate in results and discussion section.

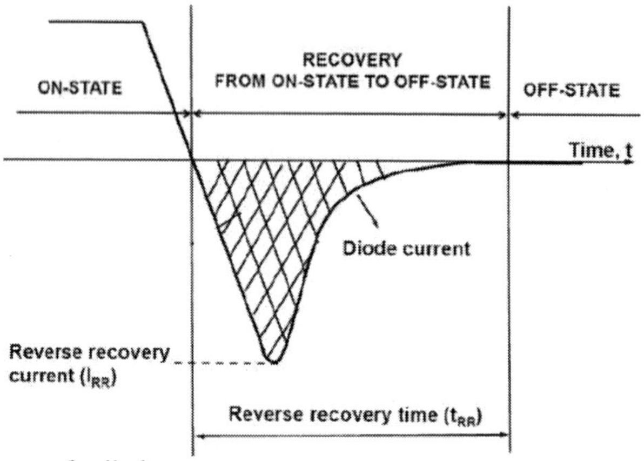

Fig. 1. The TRR diagram of a diode.

2 Methodology

In this research, the main interest of study is to evaluate the influences of platinum diffusion to the reverse recovery of a power rectifier. The evaluations are basically summarized in the Table 1, where eight Design Of Experiment (8 DOE) methods are applied in exploring the relationship between input (*X*) and response (*Y*). The input response of platinum diffusion which includes temperature of the platinum diffusion and the duration of the heat process. The output response is reverse recovery time of a power rectifier.

Table 1. Platinum diffusion DOE Matrix.

DOE	Thermal duration,[min.]	Temperature, [°C]
1	0	-1
2	-1	1
3	0	1
4	0	-1
5	0	1
6	1	-1
7	1	1
8	-1	-1

The evaluation begins with fabrication of power diodes in the fab for platinum diffusion process. For thermal durations, it is represented by 0 (the standard duration), -1 (shorter duration) and 1 (longer duration). Meanwhile for temperature matrix, 1 represent high temperature and -1 represent low temperature. The fabricated devices on the wafers are then tested electrically to determine the TRR value. The reverse recovery time of the power rectifier is then analysed in the analytical software to study the relationship between input responses(X)and output responses (Y). Take note that the platinum diffusion process need to be carried out at the middle of fabrication since the temperature diffusion process is exceeding 800 °C. Such high temperature during the diffusion process might melt the other metals of the power rectifier.

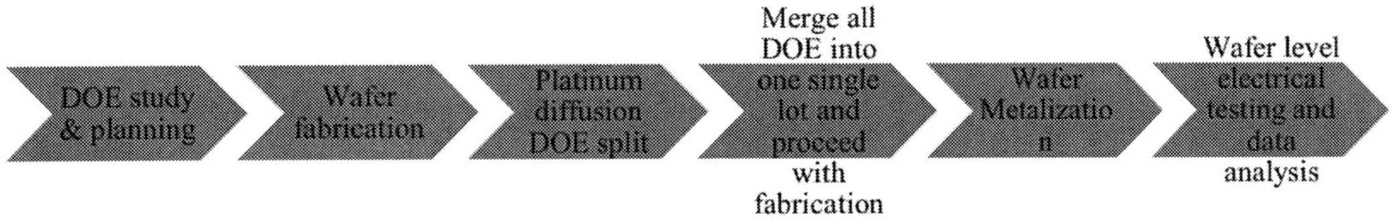

Fig. 2. The overall process flow of the evaluation.

3 Results and Discussion

The main objective of our work is to find a trade-off relationship between forward voltage and reverse recovery. Thus, both reverse recovery and forward voltage parameters have to take into consideration when designing a low forward voltage and fast reverse recovery time of a power rectifier. The forward voltage and reverse recovery of power rectifier is inversely proportional. When designing a fast reverse recovery time of a power rectifier, one should also consider the impact of lifetime to a forward voltage of the power rectifier. After completion of the eight DOE evaluations and the fabrication process, the wafers are then proceeds with electrical testing to obtain the electrical result.

Fig.3 demonstrates the relationship between reverse recovery time and forward voltage of a power rectifier. The unit for forward voltage is in voltage (V) and the reverse recovery time is in nanosecond (ns). The graph shows a negative slope, representing that the higher the forward voltage, the faster the recovery time of a power rectifier will be. Thus, it is essential to find a balance between forward voltage and reverse recovery lifetime when designing a good power rectifier.

 DOI: 10.1051/matecconf/20167801089

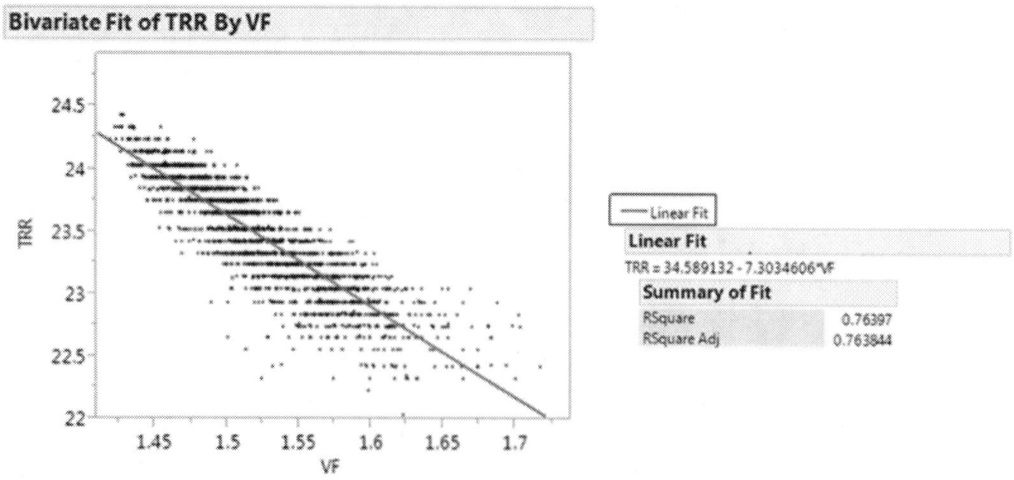

Fig.3. Relation plot between reverse recovery time (TRR) to forward voltage (VF) of a power rectifier.

Fig.4a shows the forward voltage (VF) plot for each of the DOE experiments. In each of the DOE, there are three wafers in total, which is the ideal wafer quantity to check for the variation between wafer to wafer that undergo the same process and condition. For this particular device, the forward voltage upper spec is 1.8V. In general, DOE number 1, 4, 6 and 8 give a superior performance of low forward voltage. Low forward voltage means lower power dissipation when the power rectifier is turn on. While in DOE number 3, 5 and 7 have higher VF distribution compared than that other DOE, with some of the points have exceed the upper spec limit of the VF. Higher VF is not desirable in the power rectifier application as it will dissipate more power during rectifier turn on. DOE 2 have a forward voltage distribution that is below the upper spec limit and have a minimum variation between wafers to wafer. In summary, if forward voltage is the only parameter that concerns, DOE number 1, 2, 4, 6 and 8 are fit for the change.

Fig.4b shows the reverse recovery time of all the 8 DOE experiments. Similar wafer splits and quantities were used as previously explained. For this device, the upper specification for reverse recovery time is 27nanosecond. One can see that, DOE number 1, 4, 6 and 8 considered fails the specification where some of the readings exceed 27ns. DOE number 4, 6, 7 and 8 have most significant variation between wafer to wafer. This is not desirable as the variation is unpredictable. The wide variation may cause additional yield loss reflecting the process is unstable. On the other hand, DOE number 2, 3, 5 and 7 meet the specification where none of the readings exceed the upper spec limit. With respect to that, DOE 2 and DOE 5 provide small variation wafer to wader and fulfill TRR reading specification.

 DOI: 10.1051/matecconf/20167801089

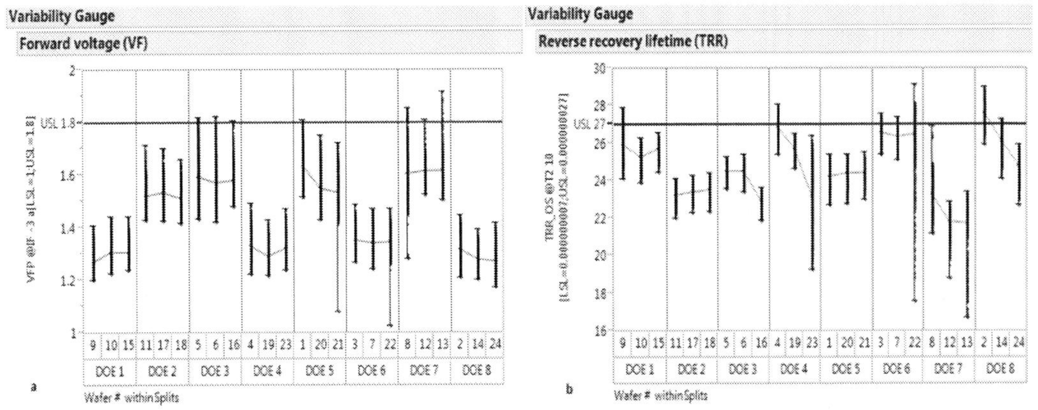

Fig. 4. a) the forward voltage (VF), b) the reverse recovery lifetime (TRR) of all 8 DOE evaluations.

The summary for each of the DOE have been tabulated in the Table 2. In conclusion, only DOE number 2 matches the requirement of forward voltage and reverse recovery time of the power rectifier. The variation between wafer to wafer in a same run is minimized compared to other splits. Although the forward voltage of the DOE number 2 is slightly higher than DOE number 1, 4, 6 and 8, but the reverse recovery time of the DOE number 2 is the best among the rest.

DOE number 1 although processed a good VF, but the reverse recovery time of a diode is not meeting the spec. The same observation goes to DOE number 4, DOE 6 and DOE 8. Despite of DOE 3 and DOE 5 gives a fast reverse recovery lifetime, but they both fail to meet the forward voltage spec at 1.8V. DOE number 7 is the worst, where it fails both the reverse recovery lifetime and forward voltage. In conclusion, DOE number 2 is provides the most optimum condition to produce a fast recovery lifetime and also meeting the forward voltage spec.

Fast recovery lifetime will give a high forward voltage as drawback. In this particular power rectifier, it has both tight tolerance to forward voltage and reverse recovery lifetime. The main purpose of this DOE is to explore a robust condition for platinum diffusion in order to process a good power rectifier that meets both VF and TRR requirements. From the distribution of both Fig. 4a and Fig. 4b, we can conclude the distribution pattern justified the hypothesis in Fig. 3, where the higher the forward voltage, the faster the recovery time of a power rectifier.

Fig. 5 shows the uniformity of the selected DOE 2 for its forward voltage and reverse recovery time. From Fig. 6, we can see there is a circular shape trend within a wafer. For forward voltage (VF), wafer edge have higher VF trend and the VF drop as it go to the centres of the wafer. The same trend observe for reverse recovery time (TRR) too where we can see a gradual increase in TRR from wafer centre to wafer edge. The uniformity for both forward voltage and reverse recovery time is acceptable where the variation within a wafer is less than 3ns for reverse recovery time and 0.2 V for forward voltage.

Table 2. DOE matrix for platinum diffusion study and summary of the results.

DOE	Thermal duration, [min.]	Temperature, [°C]	Forward voltage, VF	Reverse recovery lifetime, TRR	Conclusion
1	0	-1	√	x	Fail
2	-1	1	√	√	Selected
3	0	1	x	x	Fail
4	0	-1	√	x	Fail
5	0	1	x	√	Fail
6	1	-1	x	x	Fail
7	1	1	x	x	Fail
8	-1	-1	√	x	Fail

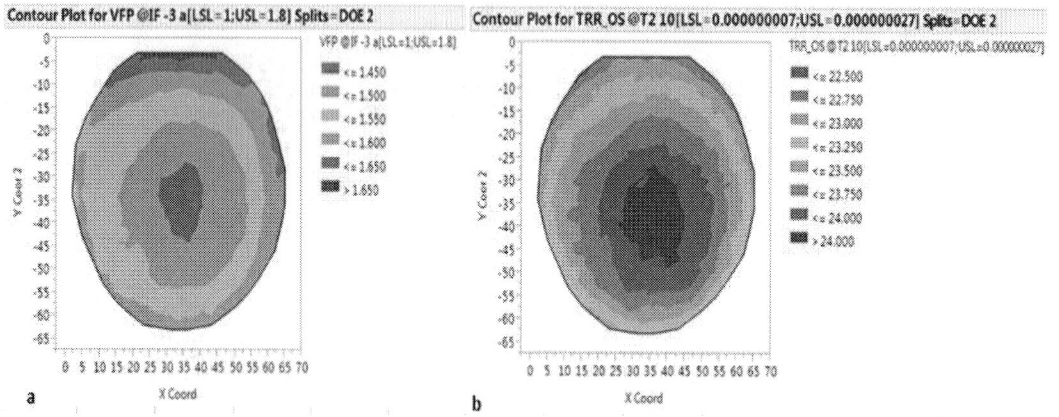

Fig.5. The contour plot of DOE number 2 to check for uniformity within a wafer for a) forward voltage (VF) and b) reverse recovery time (TRR).

4 Conclusions

Through the understanding of reverse recovery lifetime of a diode, we have further explore the details working principle of platinum diffusion towards to the influence of reverse recovery lifetime of a diode. Platinum diffusion is the optimum method to control the TRR of diode compared with gold diffusion due to it is more economical and lower leakage current drop. In this paper, we have explained the details of platinum diffusion factors that influence the output response of reverse recovery lifetime of a diode. DOE number 2 provides the optimum matrix to process good forward voltage and fast reverse recovery time for a power rectifier where it meets the given specification for VF below 1.8V and TRR below 27 ns.

References

1. S. Sze, *Physics of Semiconductor Devices*, (Wiley, New York 1969)
2. R. Kingston, Proc. IRE, **28**, 829 (1954)
3. J.L. Moll, *Physics of Semiconductor*, (McGraw-Hill, New York 1964)
4. R.N. Hall, Power Rectifiers and Transistors, **80**, 1512 (1952)
5. S.K. Ghandhi, *Semiconductor of Power device*, (Wiley, New York 1977)
6. S. Member, *IEEE Trans. on Electron Devices*, **43**, 142 (1996)

7. B. Mazhari, M. Sinha, J. Dixit, *IEEE Conference on Electron Devices and Solid-State Circuits* (2005)
8. M. Tsukuda, Y. Sakiyama, H. Ninomiya, M. Yamaguchi, *21st International Symposium on Power Semiconductor Devices & IC's* (2009)
9. B.J. Baliga, *Modern Power Devices*, (John Wiley & Sons Inc., 1987)
10. W.M. Jubadi, S.N.M. Noor, IEEE Symp. Ind. Electron. Appl., 428 (2010)
11. E. Abiri, M.R. Salehi, S. Kohan, M. Mirzazadeh, IEEE *Second Pacific-Asia Conf. Circuits, Commun. Syst.*, 60 (2010)
12. W. Jablonski, *12th International Conference on Microwaves and Radar, MIKON'98*, 69 (1998)
13. S. Sawant, B. Baliga, *Power Semiconductor Devices and ICs, 1999. ISPSD'99. Proceedings, IEEE The 11th International Symposium*, 153 (1999)
14. N.I. Shuhaimi, M. Mohamad, W.M. Jubadi, R.Tugiman, N. Zinal, R.M. Zin, *IEEE International Conference on Semiconductor Electronics*, 12 (2010)
15. E. Napoli, A.G.M Strollo, P. Spirito., *IEEE Transactions on Power Electronics 14*, 25 (1997)
16. A.G .Rojas-Hernández, A. Vera-Marquina, A. Garcia-Juárez, *IEEE Electronics, Robotics and Automotive Mechanics Conference (CERMA)*, 389 (2012)
17. V.A Temple, F.W. Holroyd, *IEEE Transactions on electron Devices*, **30(7)**, 782 (1983)
18. J. Vobecky, S. Member, V. Zahlava, P. Hazdra, *IEEE Electron Device Letters*, **35(3)**, 375 (2014)
19. B.J. Baliga, E. Sun , IEEE Trans. Electron Devices, **24**, 685 (1977)
20. E. Iliescu, V. Banu, A. Niculescu, *12th International Conference on High-Power Particle Beams*, **2**, 2 (1998)
21. R.O. Carlson, Y.S. Sun, H.B. Assalit, IEEE Trans. Electron Devices, **24**, 8 (1977)
22. W. Itthikusumarn, W. Jakphet, W. Titiroongruang, *International Symposium on Next-Generation Electronics (ISNE)*, 7 (2015)
23. D. Hu, J.K.O. Sin, B. Kang, X. Cheng, Y. Wu, S. Xie, *IEEE Southern Power Electronics Conference SPEC.* (2007)
24. J. Vobecky, P. Hazdra, *European Conference on Power Electronics and Applications*, (2005)

Review of Dolomite as Precursor of Geopolymer Materials

E.A. Azimi[1,*], *M.M.A.B.* Abdullah[1,2], *L.Y.* Ming[1], *H.C.* Yong[1], K. *Hussin*[1,2], *I.H.* Aziz[1]

[1]Center of Excellence Geopolymer and Green Technology, School of Materials Engineering, Universiti Malaysia Perlis (UniMAP), P.O. Box 77, D/A Pejabat Pos Besar, 01000, Kangar, Perlis, Malaysia
[2]Faculty of Engineering Technology, Universiti Malaysia Perlis (UniMAP), P.O. Box 77, D/A PejabatPosBesar, Kangar, Perlis 01000, Malaysia

Abstract. Geopolymer is an environmentally friendly cementitious binder that does not require the existence of ordinary Portland cement (OPC). Geopolymer has many excellent advantages, including high early strength, low shrinkage, good thermal resistance and good chemical resistance. Previous commonly used materials include fly ash, clay and slag. The used of dolomite as precursor material in geopolymer field is still new and at the early stage of study. Only a few researchers have done studies on dolomite in geopolymer. Dolomite ($CaMg(CO_3)_2$) is abundant and generally inexpensive natural minerals. The possible use of these bulk calcium carbonate materials in improving the mechanical properties of geopolymers will therefore be of great interest. This paper summarizes some research outcomes on dolomite in geopolymer along with the potential of dolomite as geopolymer composites.

1 Introduction

The demand of the usage of concrete continues to increase along with the increasing demand of infrastructures. Traditionally, ordinary Portland cement (OPC) has been used as the binder in concrete. However, cement production involves the emission of considerable amount of greenhouse gases. Therefore, the development of alternative binders that can utilize industrial by-products is considered vital in order to help reducing the carbon footprint produced during cement production [1].

Geopolymers are class of binder materials that is formed by the activation of aluminosilicate materials with alkaline or alkaline-silicate solutions and it was first named and introduced by Davidovits[2]. Geopolymers which are also known as an inorganic polymers or alkali-activated binder have rapidly gained worldwide interests [3]. Geopolymers are amorphous to semi-crystalline three-dimensional silica-aluminate materials. They are prepared by mixing and activating the aluminosilicate sources such as fly ash, metal slag, kaolin or metakaolin and dolomite with alkaline liquid (sodium

* Corresponding author: emyaizat@gmail.com

hydroxide and/or sodium silicate) and curing at a moderate temperature [4]. Geopolymers are classified as one of the green materials that contain many exceptional properties such as high compressive strength, low shrinkage, fire and acid resistant [5]. These geopolymer materials are able to offer comparable performance to the traditional cementitious binders in a wide range of applications, but with the additional advantage of lower greenhouse emissions [6]. Geopolymer technology is also increasing attention because it provides viable economical alternative to organic polymers and inorganic cements in diverse applications, such as refractory adhesives and materials[7]. These concern are also due to their excellent mechanical strength, adhesive behavior, long-term durability and exceptionally high thermal and chemical stability [8].

Silica (Si) and aluminum (Al) contents are the important compoundsin the raw material to be used and turned into geopolymer. The higher the silica and aluminum contents, the effectiveness of the geopolymers is higher. Besides, Davidovits[9] noted that the increasing in Si:Al ratio can lead to the improvement of the fire and heat resistant characteristics. One of the studies done by Temuujin et al. [10] stated that fly ash geopolymers with Si/Al molar ratio of 3.5 displayed strong adhesion to steel substrates and has promising fire resistant characteristics.

The use of dolomite in geopolymer is still new and still at the early stage of study. The current work therefore investigates the effect of the addition of larger quantities of dolomite on geopolymerization, geopolymer microstructure, long-term strength development and shrinkage properties of geopolymer mortars [11].In this paper, the properties and the suitability of dolomite as geopolymers' precursor materials are reviewed.

2 Methodology

2.1 Alkaline Solution

Alkaline solution is one of the important parts in producing geopolymer. The main criteria for developing stable geopolymer are that the source materials must be highly amorphous and possess sufficient reactive glassy content, low water demand and be able to release aluminum easily. The alkaline activators such as sodium hydroxide (NaOH), potassium hydroxide (KOH), sodium silicate (Na_2SiO_3) and potassium silicate (K_2SiO_3) are used to activate aluminosilicate materials[12, 13]. Upon mixing alkaline activator with the aluminosilicate-reactive material (raw material),alkaline activating solution dissolve the Si and Al ions from aluminosilicate material to form geopolymer precursors and finally alumino-silicate material[14].

The most commonly used alkaline activators are NaOH and Na_2SiO_3solutions. Compared to NaOH solution, KOH solution produced a greater level of alkalinity. But in reality, it has been found that NaOH possesses greater capacity to liberate silicate and aluminate monomers [15]. The potassium-based alkaline solutions are usually very costly and are not suitable for economic viability for the production of geopolymer materials. The sodium-based alkaline solutions are preferred and in such cases the formation of geopolymer matrix is also depended on its curing temperature. The curing of the product above normal atmospheric temperature is not a flexible option to adopt in commercial practice. Therefore, the selection of chemicals solution in alkaline group for geopolymerization reaction at atmospheric temperature has been the focus of research [16]. Usually this alkaline activator is prepared by mixing water, NaOH and Na_2SiO_3 but other alkali metal systems or mixtures of different alkalis also can also be used. The solution strictly needs to be concentrated to avoid the crystallization of zeolite as the end product rather than an amorphousgeopolymer[17, 18].

Joshi &Kadu[19] prepared the alkaline activator using commercially available Na_2SiO_3 liquid and NaOH pellets which contain Na_2O = 14.61%, SiO_2 = 25.18% and water = 59.99%. The alkaline solution was prepared with NaOH concentration varying from 8M to 16M and the mass ratio of Na_2SiO_3 to NaOH varied from 1.75 to 3.0. They found that the geopolymer with 14M and solid/liquid ratio 2.0 perform the optimum mechanical properties compared to others A study by Gharzouni et al. [20] stated the type of alkaline solution is important. The research found that the geopolymers based on a mixture of $K_2SiO_3$and KOH display higher mechanical properties compared to those based on $Na_2SiO_3$and NaOH or K_2SiO_3/NaOH mixtures.

2.2 Manufacturing of geopolymer

Geopolymer was usually prepared by mixing alumino-silicate or raw material with the alkaline activator solution. The raw materials previously were ground in a ring mill until fine particles were obtained [21]. The mixtures of alumino-silicate and alkali activator solution then formed slurry and was cured for 24 hours. Normally, the dissolution of sodium hydroxide was modified in one liter of distilled water in a volumetric flask to obtain a optimum molar concentration of NaOH[22]. Mustafa et al. [23] in his research mixed the raw material with an alkaline activator in the mixer. Further, the geopolymer was placed in the 50x50x50 mm^3 mold and were cured at different curing temperatures.

The ratio of the raw material with the alkaline activator plays an important role in forming the geopolymer[24]. The Si/Al composition for geopolymer prepared from raw material were 2.5 and 3.5 respectively [25]. At the same time, Cheng et al. [26], used SiO_2/Al_2O_3 ratio in the range of 3.16-3.46. Mustafa et al. [27], in his research used 2.5 as ratio of Na_2SiO_3/NaOH. Normally the ratio of solid to liquid (S/L) started from 0.6 and above.

2.3 Dolomite

Dolomite is one of the most common carbonate minerals in the geologic record. It is an anhydrous carbonate mineral composed of calcium, magnesium, and carbonate, ideally $CaMg(CO_3)_2$. The word dolomite is also used to describe the sedimentary carbonate rock, which is composed predominantly of the mineral dolomite (also known as dolostone). Dolomite is one of mineral resource that can be used as raw material ingeopolymer composites[28].The mineral dolomite will crystallize in the trigonal-rhombohedral system. It forms white, tan, gray, or pink crystals. Dolomite is a double carbonate, which having an alternating structural arrangement of calcium and magnesium ions. It does not rapidly dissolve in dilute hydrochloric acid as calcite does [29]. Table 1 display the chemical composition of dolomite. It contains small amount of alumina oxide (Al_2O_3) andsilicon dioxide(SiO_2) composition that make it possible to be raw material for geopolymer It also contain high calcium oxide (CaO) and magnesium oxide (MgO) composition.

Table 1. Chemical composition of dolomite[29].

CaO	SiO_2	Al_2O_3	Fe_2O_3	MgO	K_2O	Na_2O
33.4	2.5	0.7	0.3	17.1	0.1	0.1

*loss of ignition: 45.8

3 Results and Discussion

3.1 Properties of Geopolymer Based on Dolomite

3.1.1 Compressive strength

The properties of dolomite as substitution in metakaolingeopolymer were studied by Rodriguez-Blanco et al.[29]in terms of compressive strength. Fig.1 shows the compressive strength development of geopolymers with 20% replacement of metakaolin by dolomite.The activator modulus (Ms) of alkaline activator was also varied. The compressive strength of all samples increased gradually with ageing. The highest compressive strength (45-55 MPa) is achieved with alkaline activator of Ms = 1.5. The strength development of the dolomite-containing systems was also significantly slower, particularly in samples with alkaline solution of Ms = 1.2 samples. From the compressive strengths of dolomite-containing binders are found to be consistently lower than that of the calcite-containing binders at all alkaline conditions studied.

Zarina et al.[30] investigates the influence of dolomite on the mechanical properties of boiler ash based geopolymer pastes. It was found that the addition of dolomite decreased the 28-day strength of boiler ash geopolymer. The geopolymer sample without addition of dolomite showed the maximum compressive strength (19.4 MPa) at 28 days testing. Meanwhile the addition of 4% of dolomite into geopolymer paste achieved only maximum compressive strength of 7.3 MPa.

The dolomite particles used were larger than the calciteparticles, however their morphologies were not dramatically different. These differences are sufficient to cause the significant differences between the two minerals in geopolymerization as observed. The consistently lower compressive strength achieved by dolomite-containing systems is therefore most likely attributed to either a lower dissolved calcium level (calcite=53.5 while dolomite=33.4) or differences in surface properties of calcite and dolomite (the dolomite used here has a particle size roughly twice that of the calcite)[31, 32], which may affect the binding of the minerals to geopolymer gel.

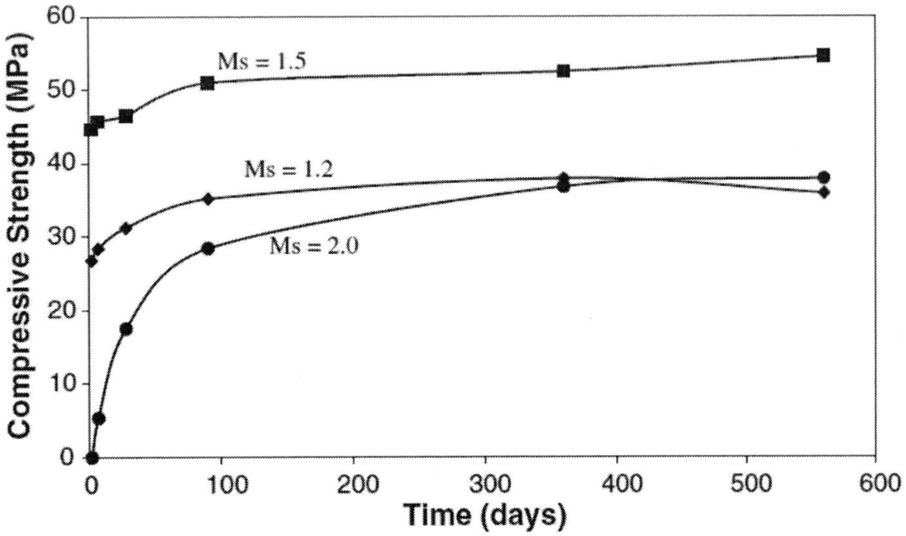

Fig.1. Compressive strength development of geopolymers with 20% dolomite and 80% metakaolin with varying Ms of alkaline activators [11, 29].

3.1.2 X-ray diffraction analysis

Fig.2 shows the XRD diffractogram of metakaolin-dolomite geopolymers. It was revealed that geopolymeric phase is the dominant product formed during alkali activation. This is again identified through the presence of an amorphous hump at around 20–35° 2θ, regardless of the alkalinity of the alkaline activator used [29]. There was also a significant amount of unreacted dolomite remained in the system. Thediffractograms exhibit slight reaction of dolomite, evidenced by diminution of the dolomite XRD peaks, which are more notable at Ms = 1.2 than Ms = 2.0 [33].

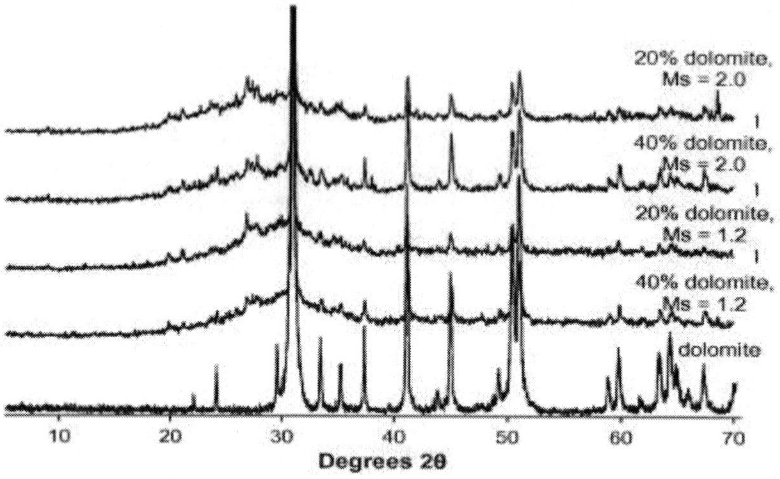

Fig.2. X-ray diffractograms of metkaolingeopolymers with 20% or 40% replacement by dolomite [29].

3.1.3 Effect of magnesium content in Dolomite

It has previously been noted that it is possible for magnesium ions to replace some fraction of the calcium ions in calcium silicate hydrate gels, or in fact to form magnesium silicate hydrate gels [34]. However, the fact that no distinct C–S–H phase has been observed throughout this study indicates that this mechanism is unlikely to be significant here for magnesium. It is therefore possible that Mg^{2+} plays a similar role to Ca^{2+} in situations where it is found within the geopolymeric gel phase [35]. Magnesium salts have been observed to have much less impact on geopolymerization kinetics than do their calcium counterparts, but do show some accelerating effect in geopolymeric setting which has been ascribed to nucleation effects[36].

Bakharev[37] done the study on the durability of geopolymer materials in magnesium sulfate solutions. SEM examination revealed formation of deep vertical cracks in the samples exposed to magnesium sulfate solutions. Ion chromatography indicated migration of alkali cations into solution. The analysis also indicated a decline of [Mg^{+2}] in testing solution in the samples, which may occur due to migration of Mg^{2+} into material. Bulk chemical variation across the specimens showed an increase of Ca, Mg and S concentrations in the surface area, a slight decrease of Al content, while Si content stayed without change. Therefore, migration of Ca from inside of the specimen to the surface area, and S and Mg from the solution into aluminosilicate matrix can be suggested[38].

The SEM image of fracture surface shown formation of lightly coloured precipitates in the surface area of the samples exposed to magnesium sulfate solution. In these case, the diffusion of Mg into material occurred at the same time as migration of alkali ions from samples into solution. These processes are responsible for fluctuations of the compressive strength and final strength improvement of the samples when immersed in the magnesium sulfate solution. Ca^{+2} and Mg^{+2} ions can be accommodated in aluminasilicate gel as network-modifying cations [39].

4 Conclusions

This paper summarizes the suitability of dolomite to be used as geopolymer materials. Based on the reviewed literature above, dolomite was used as a replacement to the main raw materials (e.g. metakaolin) in the synthesis of geopolymers. There are limited studies using dolomite as the aluminosilicate materials. Hence,the used of dolomite as precursor material in geopolymer material is still new and further study must be done to investigate the properties of geopolymer system based on dolomite. Besides, the use of dolomite which is abundantand potentially offer sustainable solution and environmental friendly for extending the service life of infrastructure and maintenance cost.Development of dolomite geopolymeris less been explored. It is hoped that future research in this field will drive a new era of greener materials in the geopolymer industry.

References

1. P.K. Sarker, S. Kelly, Z. Yao, Mater. Design, **63**, 584 (2014)
2. L. Zeng, C. Dan-yang, Y. Xu, F. Chun-wei, P. Xiao-Qin, Appl. Clay Sci., **101**, 637 (2014)
3. P. Duxson, A. Fernandez-Jimenez, J.L. Provis, G.C. Lukey, A. Palomo, J.S.J. Van Deventer, J. Mater. Sci., **42**, 2917 (2007)
4. J. Temuujin, A. Minjigmaa, W. Rickard, A. Van Riessen, J. Therm. Anal. Calorim., **107**, 287 (2012)
5. J. Liyana, H. Kamarudin, A.M. Mustafa AlBakri, M. Binhussain, C.M. Ruzaidi, A.M. Izzat, Aust. J. Basic & Appl. Sci, **2**, 100 (2013)
6. M. Sofi, J. van Deventer, P. Mendis, G. Lukey, J. Mater. Sci., **42**, 3107 (2007)
7. J.L. Bell, P.E. Driemeyer, W.M. Kriven, J. Am. Ceram. Soc., **92**, 607 (2009)
8. T. Alomayri, F.U.A. Shaikh, I.M. Low, Ceram. Int., **40**, 14019 (2014)
9. J. Davidovits, J. Mater. Educ., **16**, 91 (1994)
10. J.Temuujin, A. Minjigmaa, W. Rickard, M. Lee, I. Williams, A. Van Riessen, J. Hazard. Mater., **180**, 748 (2010)
11. C.K. Yip, J.L. Provis, G.C. Lukey, J.S.J. Van Deventer, Cement Concrete Comp., **30**, 979 (2008)
12. S.H. Sanni, R. Khadiranaikar, International Journal of Research in Engineering and Technology, 2, 100 (2013)
13. B. Singh, G. Ishwarya, M. Gupta, S.K. Bhattacharyya, Constr. Build. Mater., **85**, 78 (2015)
14. A. Fernandez-Jimenez, A. Palomo, Fuel, **82(18)**, 2259 (2003)
15. P. Duxson, A. Fernandez-Jimenez, J. Provis, G. Lukey, A. Palomo, J. Van Deventer, J. Mater. Sci., **42**, 2917 (2007)
16. D. Khale, R. Chaudhary, J. Mater. Sci., **42**, 729 (2007)
17. J. Davidovits, *99 International Conference. Saint-Quentin, France* (1999)
18. D. Hardjito, *Studies of fly ash-based geopolymer concrete*, Curtin University of Technology (2005)

19. S. Joshi, M. Kadu, International Journal of Environment and Sustainable Development, **3,** 100 (2012)
20. A. Gharzouni, E. Joussein, B. Samet, S. Baklouti, S. Rossignol, J. Non-Cryst. Solids, **410**, 127 (2015)
21. C. Tippayasam, P. Keawpapasson, P. Thavorniti, T. Panyathanmaporn, C. Leonelli, D. Chaysuwan, Constr. Build. Mater., **53**, 455 (2014)
22. A.M. Mustafa Al Bakri, L. Jamaludin, H. Kamarudin, M. Binhussain, M.C. Ruzaidi, A.M. Izzat, Adv. Mat. Res., **686**, 227 (2013)
23. A.M. Mustafa Al Bakri, , H. Kamarudin, M. Binhussain, I.K. Nizar, Y. Zarina, A.R. Rafiza, Phys. Procedia, **22**, 286 (2011)
24. E. Kamseu, B. Nait-Ali, M.C. Bignozzi, C. Leonelli, S. Rossignol, D.S. Smith, J. Eur. Ceram. Soc., **32,** 1593 (2012)
25. Z. Zhang, X. Yao, H. Zhu, Appl. Clay. Sci., **49**, 1 (2010)
26. T. Cheng, J. Chiu, Miner. Eng., **16**, 205 (2003)
27. A.M. Mustafa Al Bakri, L. Jamaludin, K. Hussin, M. Bnhussain, C.M. Ruzaidi, A.M. Izzat, Int. J. Mol. Sci., **13**, 4388 (2012)
28. C.M. Pina, Am. Mineral., **100**, 1017 (2015)
29. J.D. Rodriguez-Blanco, S. Shaw, L.G. Benning, Am. Mineral., **100**, 1172 (2015)
30. Y. Zarina, H. Kamarudin, A.M.M. Al Bakri, I. Khairul Nizar, A.R. Rafiza, Key Eng. Mat., **594**, 8 (2014)
31. K. Wright, R.T. Cygan, B. Slater, Phys. Chem. Chem. Phys., **3**, 839 (2001)
32. P. Fenter, Z. Zhang, C. Park, N.C. Sturchio, X.M. Hu, S.R. Higgins, Geochim. Cosmochim. Acta, **71,** 566 (2007)
33. K. Wang, X. Hu, P. Zhao, Z. Yin, Appl. Energy., **165**, 14 (2016)
34. F.P. Glasser, J. Hazard. Mater., **52**, 151 (1997)
35. D.R.M. Brew, F.P. Glasser, Cement Concrete Res., **35**, 85 (2005)
36. W.K.W. Lee, J.S.J. Van Deventer, Cement Concrete Res., **32,** 577 (2002)
37. T. Bakharev, Cement Concrete Res., **35**, 1233 (2005)
38. D.Bonen, M.D. Cohen, Cement Concrete Res., **22**, 707 (1992)
39. D.Perera, M. Blackford, E. Vance, J. Hanna, K. Finnie, C. Nicholson, *MRS Proceedings,* **824**, *35* (2004)

LawsoniaInermis Extract Enhances Performance of Corrosion Protection of Coated Mild Steel in Seawater

H.M. Hajar[1], *F.* Zulkifli [1], *M.J.* Suriani[1], *M.G.* Mohd Sabri[2] and *W.B.* Wan Nik [1,*]

[1]School of Ocean Engineering, Universiti Malaysia Terengganu, Malaysia.
[2]School of Fundamental Sciences, Universiti Malaysia Terengganu, Malaysia.

Abstract. The investigation of coating incorporated with various percentages of henna has been conducted by means of weight loss, potentiodynamic polarization and electrochemical impedance spectroscopy method. Mild steel has been studied at ambient temperature and 50°C in order to examine the effect of temperature. The potentiodynamic polarization measurements indicate that henna extract acts as a mixed inhibitor while the corrosion rate (CR) value decreases as the current density (I_{corr}) decreases in the presence of 10% henna extract for both temperatures. Surface and protective film analysis have been carried out using Fourier transform infrared (FTIR) and scanning electron microscopy (SEM).The inhibition activity is mainly correlated with the major active functional groups of lawsone. The main components can be found in lawsone structures are phenols O-H, carboxylic acid C=O and alkenes C=C. The morphology of the uninhibited sample showsa rough surface of mild steel as a result of corrosion process and severe damage can be seen. In the presence of henna, a smooth surface of mild steel anda uniform deposition of henna were observed.

1 Introduction

The investigation of metallic structures has become important nowadays due to the industrial applications. Thus, this application processes lead to the corrosion problems. In most industries, the uses of safe constructional materials are needed. However, the amounts spent are huge. For the example of the industrial involved are oil and gas platform, ship repair, acid pickling, industrial cleaning, acid descaling, oil-well acid in oil recovery and the petrochemical processes. Due to this problem it is originates very important material and economic losses due to partial or total replacement of equipment and structures, and plant-repairing shutdowns. Corrosion is a degradation of materials' properties due to the

[*] Corresponding author: niksani@umt.edu.my

interactions with the environments, and corrosion of most metals (and many materials for that matter) is inevitable [1].

Marine environment is probably the most aggressive environment in which metals are being to operate. The degradation of metallic surfaces due to the atmospheric corrosion is a well-known problem. Many structures exposed to this problem such as bridges, storage tanks and pipelines. Marine environments also experiencing different temperature and hence, the prediction of corrosion rate, temperature and oxygen content is important parameters that need to focus [2]. For an example, a ship is having an intense contact with seawater, but the humidity and the chloride-rich ion in seawater can be different due to the environments and surroundings. There are also the other factors that influenced corrosion rate such as salinity, pH, pressure and also the presence of bacteria [3]. Since the technology has been developed, there are many methods to overcome this problem. One of the effective methods is coating with the corrosion inhibitor which is consist of a chemical compound that able to slower corrosion rate. Corrosion inhibition can be achieved by the addition of a chemical compound that inhibits the oxidation of the metal. The chemical inhibitor added to the system may be in the form of a liquid or vapor or both. The two steps involved in the action of the corrosion inhibitor are the transport of the inhibitor to the metal surface and the interaction of the inhibitor with the metal surface [4].

Mild steel is used widely for major infrastructure in or around seawater harbors, for shipping and offshore structures. The corrosion of structural steel is an electrochemical process that requires the simultaneous presence of moisture and oxygen. Protection using coatings or cathodic protection systems or both can be effective if properly maintained [5].

Nowadays, anti-corrosion paint is applied to the ship hull andthe other ship parts in order to reduce corrosion rate [6]. Organic coatings are widely used to prevent corrosion of metal structures due to their ease of application at reasonable cost. Coating efficiency is dependent on some intrinsic properties of the organic film (barrier effect and the presence of inhibitors or sacrificial pigments) and the metal or substrate interface interaction in terms of adherence [7].

Plant extracts are rich sources of ingredients which have very high inhibition efficiency, due to the presence of heterocyclic constituents like alkaloid and flavonoid [8]. The extracts from plant leaves, roots and seed have been studied and found to be cheap, environmental friendly and good corrosion inhibitors [9]. Henna consists of four main constituents which are Lawsone, gallic acid, α-D-Glucose and tannic acid. These four constituents could combine with the metal cations and absorbed on the metal surface to protect the metal from corrosion attack [10]. Henna constituent seem to meet the basic requirements for consideration as alternative corrosion inhibitors: the ions form insoluble hydroxide, which makes it possible to use them as cathode and anodic inhibitors, they are non-toxic and relatively abundant in nature [11]. The aim of the research is to investigate the corrosion performance on mild steel coated with henna extract as organic inhibitor which exposed to the seawater. FTIR, EIS, PP and SEM methods were applied in this study.

2 Methodology

2.1 Preparation of extraction

The extraction of henna was carried out by immersion of 200g of henna powder in 1200ml of ethanol for one week at room temperature. The solutions were filtered and residual has been removed from the solutions. Rotary evaporator (ROTAVAP) was used in order to get pure henna extract.

2.2 Preparation of substrate

The specimens used in this study were mild steel grades A36. These specimens were cut into 25mm x 25mm x 4mm. The specimens were polished by using a polishing machine with the abrasive paper grade 100, 120 and 180. The abrasive paper was used to remove the carbon oxide layer on the metal surface. Lastly, these specimens were cleaned with acetone and rinsed with distilled water.

2.3 Preparation paint formulation

Commercial paint of KTH 1K Alkyd Finish was used. The thinner act as solvent and control the viscosity of paint directly. The paint was mixed by following the paint formulation and has been divided into 3 portions with different concentration of henna (0%, 10% and 20%). Table 1 shows the paint formulations according to different concentrations of henna extract.

Table 1. Paint formulation for mild steel.

Type of paint	Volume of paint [ml]	Henna extract [%]	Thinner [ml]
1 (control)	300	0	30
2	300	10	30
3	300	20	30

2.4 Immersion test

Water bath model Memmert was prepared with a specific amount of fresh water. Six beakers containing seawater then placed in the water bath. The samples were coated with different concentrations of henna extract (0%, 10% and 20%) and hung inside the beaker. It was carried out for a duration of 30 days and 4 specimens from different concentrations were taken out for each 10, 20 and 30 days in enclosed condition without any disturbance. The immersion test was conducted at different temperatures (room temperature and 50°C).

3 Results and Discussion

3.1 Fourier Transform Infrared (FTIR)

Fourier Transform Infrared (FTIR) is a device used to obtain an infrared spectrum of absorption. From Figure 1, henna consists of major component which is Lawsone. The main components can be found in Lawsone structures are phenols O-H, carboxylic acid C=O and alkenes C=C. The phenolic O-H appeared at 3393.0 cm^{-1}. The aromatic alkenes C=C stretching frequency appeared at 1638.2 cm^{-1}. The C=O stretching frequency appeared at 1655.7 cm^{-1}. Table 2 summarizes the functional groups found in henna extract.

Previous work reported that major components of henna extract are hydroxyl aromatic compounds, such as tannin and lawsone [12]. Phenol group of Lawsone would donate electron to the metal to achieve its noble state of orbit. At the same time, the metal would receive the electron to become more stable. This is indirectly retard further redox reaction and could resist metal from corrosion attack [8].

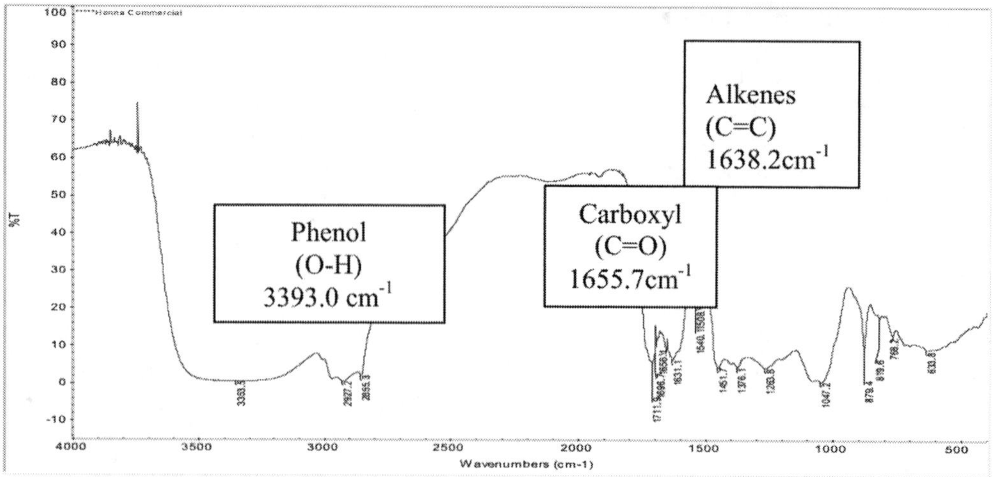

Fig. 1. Fourier Transform Infrared (FTIR) spectrum for henna.

Table 2. Main components of Lawsone identified.

Main Functional Group		Major Component
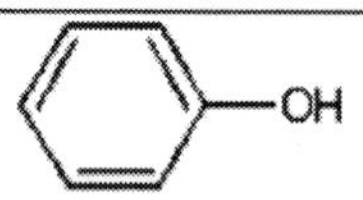	Phenol O-H	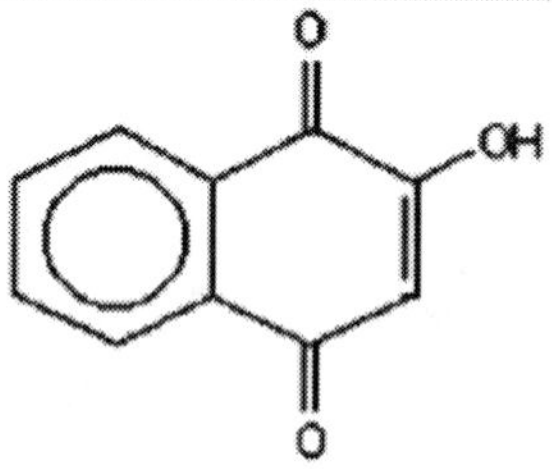
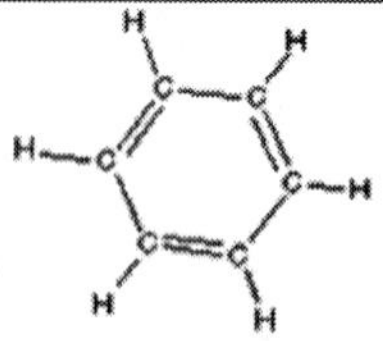	Aromatic Rings C=C	
Carboxylic Acid C=O		Structural : Lawsone Molecular Formula : $C_{10}H_6O_3$

3.2 Electrochemical Impedance Spectroscopy (EIS)

The data was presented in Nyquist plots. The x-axis indicates real impedance (Z') while the y-axis indicates imaginary impedance (Z''). Two parameters included during the test which are temperature (room temperatureand 50°C) and henna concentration (0%, 10% and 20%.).

From the Nyquist plots as shown in Figure 2, the curves were shown by capacitive semi-circle. The shape is maintained for whole concentration and both temperature which indicates no changes in corrosion mechanism during the addition of inhibitor. Double layer capacitance (C_{dl}) decreases from 8.220^{-5} F to 3.880^{-5} F for 0% of henna and 10% of henna extract respectively and increased to 5.860^{-5}F for 20% of henna extract at room temperature while C_{dl}value at 50°C shows decreasing value from $3.989E^{-4}$F to 1.352^{-4}F and increased to 1.6723^{-4}F for paint with henna of 0%, 10% and 20% respectively. The value of C_{dl} decreased due to decreasing in local dielectric constant. It shows the absorption of inhibitor takes place on the metal surface in seawater. The values of impedance parameters were tabulated in Table 3.

The potential resistance (R_p)increases as the concentration percentage increase, where at room temperature the R_p increases from 193.64kΩ to 409.96 kΩ while at 50°C, increases from 39.90 kΩ to 117.74 kΩ. However, $R_p$value decreases when the henna concentration increases to 20% with the values of 215.75kΩ at room temperature and 56.40kΩ at 50°C. Increasing R_p leads to an increase of inhibition efficiency. Based on the result, the ratio of henna extract in the paint is found limited to 10% since it was resulting in lower value of potential resistance. Thus, increasing in the percentage of henna extract caused inhomogeneity in coating sample [13].

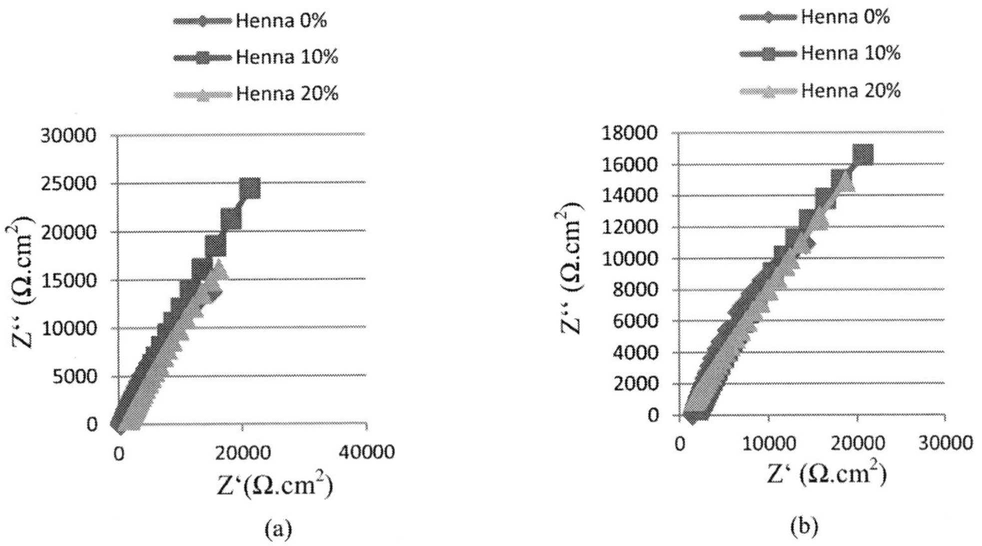

Fig. 2. Nyquist plot for 30 days (a) Nyquist plot at room temperature and (b) Nyquist plot at 50°C.

Table 3. Impedance parameter of A36 in seawater for 30 days immersion.

Type of Paint	C_{dl} (F)	R_p(kΩ)
Henna 0% Room Temperature	8.220E-5	193.64
Henna 0% 50°C	3.989E-4	39.90
Henna 10% Room Temperature	3.880E-5	409.96
Henna 10% 50°C	1.352E-4	117.74
Henna 20% Room Temperature	5.860E-5	215.75
Henna 20% 50°C	1.6723E-4	56.40

3.3 Potentiodynamic polarization

The measurements of polarization were presented in the form of Tafel plot. From Figure 3, the results show the values of corrosion potential (E_{corr}) shifted more to the positive direction as the presence of henna extract. At room temperature and at 50°C, the E_{cor}value increase as there is presence of henna which is -0.34368 V/SCE to -0.18868 V/SCE and -0.45398 V/SCE to -0.38371 V/SCE respectively. Since The E_{corr} shift toward more positive direction and indicates that henna is a mixed-type inhibitor. However, the E_{corr}value drop to -0.30495 V/SCE at room temperature and 0.38972 V as the henna percentage increase to 20%. In addition of henna percentage, the corrosion potentialE_{corr}values shift to the range (0.12-0.16) V/SCE. The higher value of corrosion potential indicates lower corrosion rate. Corrosion potential increases as the concentration of henna extract increases. The higher values of corrosion rate are due to the thin layer formed on metal surface to protect the metal from corrosion [14]. Solomon and Umoren have studied the inhibitor used was effected both anodic and cathodic sites in tafel plot with E_{corr} value towards more noble values and it suggest that inhibitor behaved as mixed-type inhibitor[15].

Table 4 shows that corrosion rate (CR) decreases when the values of current density (I_{corr}) become smaller due to the lower temperature and presence of henna extract. 10% of henna extract at room temperature has lower corrosion rate than 20% of henna extract at same temperature due to the ratio of henna extract is limited to 10%. According to Ostovari *et al.* the inhibition efficiency of mild steel increases with the increase of henna extracts concentration. As the temperature increases, the inhibition efficiency decreases [10]. Other than that, the addition of 10% henna at 50°C decreases current density from 5.71^{-5} A.cm^{2}to 1.02^{-5}A.cm^2 although it was at higher temperature. This behaviour shows that henna acts as good inhibitor for corrosion of mild steel in seawater medium.

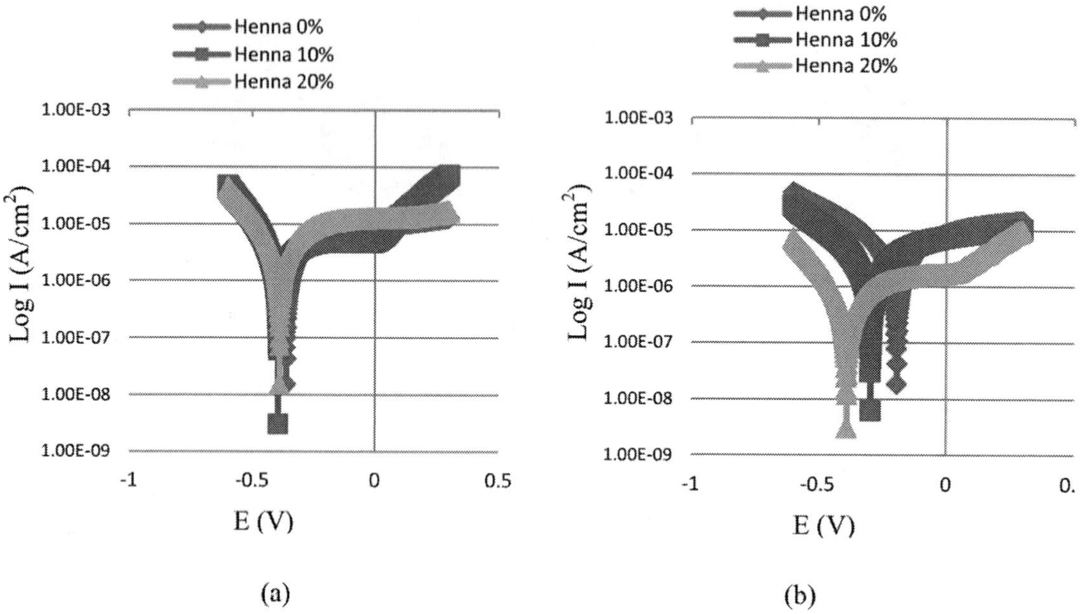

Fig. 3. Potentiodynamic Polarization of A36 in seawater for 30 days (a) at room temperature at 50°C.

Table 4. The electrochemical parameters of A36 in seawater for 30 days at different temperatures and different henna concentrations.

Type of paint	E_{corr} (V/SCE)	I_{corr} (A.cm^2)	Corrosion rate (mm/year)
Henna 0% Room Temperature	-0.34368	9.060E-6	0.105230
Henna 0% 50°C	-0.45398	5.710E-5	0.663880
Henna 10% Room Temperature	-0.18868	4.170E-7	0.0048408
Henna 10% 50°C	-0.38371	1.020E-5	0.118310
Henna 20% Room Temperature	-0.30495	5.980E-6	0.069497
Henna 20% 50°C	-0.38972	1.120E-5	0.130540

3.4 Scanning Electron Microscopy (SEM)

Scanning electron microscope (SEM) was subjected to the surface of mild steel type A36 after conducting immersion test about 30 days in the absence and presence of henna (10%) at different temperature are shown in Figures 4 and 5 respectively.

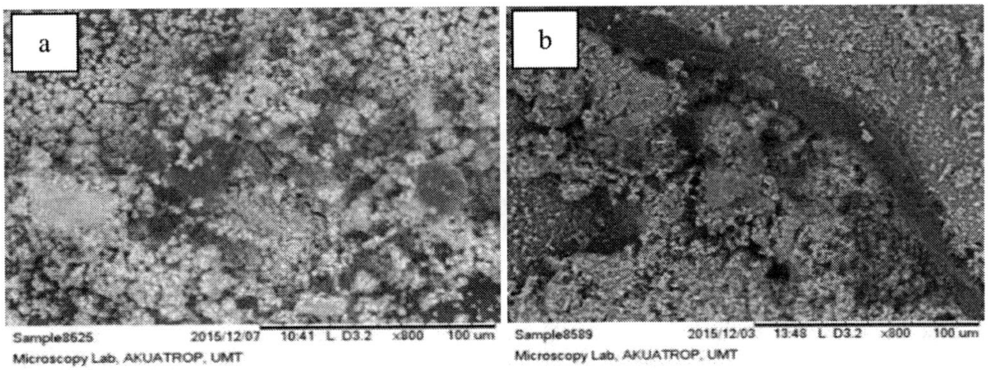

Fig. 4. SEM micrographs of the surface of A36 after 30 days immersion of paint without henna: (a) at room temperature (b) after immersion at 50°C.

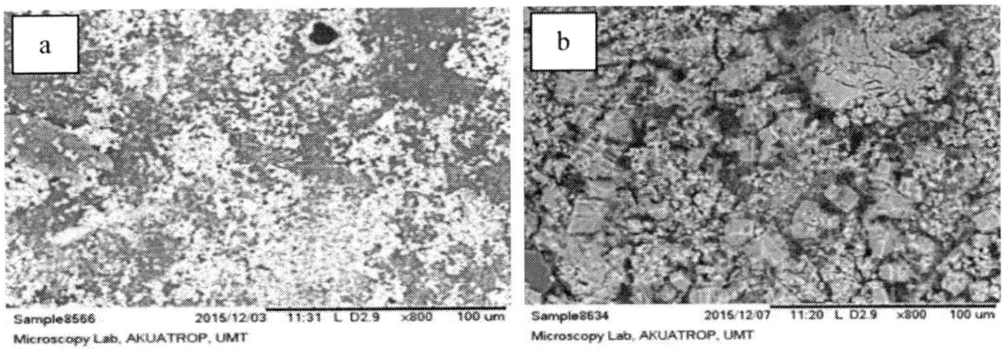

Fig. 5. SEM micrographs of the surface of A36 after 30 days immersion of paint with 10% henna extract (a) at room temperature (b) after immersion at 50°C.

Figures 4 a, and b showsthe mild steel without henna extract that have very rough surface and severely damaged in every part. Serious damage on surface of mild steel was observed when the temperature increased. The presence of henna extract makes the surface transformed into smoother and more uniform deposited. From Figure 5 a, the surface was smooth and no cracks observed. The addition of henna as an inhibitor reduced the oxidation process which led to the formation of flakes or the pitting corrosion form over the surface. However, in Figure 5 b, many cracks were observed and the surface becomes very rough. Since the ratio of henna extract is limited to 10%, Figure 5 b shows small crack on just certain part of mild steel. When the temperature increased, the surface was rough and the cracks clearly seen.

4 Conclusions

As a conclusion, inhibitors play a vital role in providing protection against corrosion. The main constituent responsible for corrosion inhibitor characteristic of henna extract is found to be Lawsone. EIS measurements show that the inhibition of corrosion on mild steel is due to the formation of protective film on metal surface and the inhibition increases as the extract concentrations increase. As presence of henna extract, the value of potential resistance (R_p) decreases with low protection towards mild steel and increases the double layer capacitance (C_{dl}). However, ratio of henna extract is found limited to 10%. Increasing in the percentage of henna extract caused inhomogeneity in coating paint sample. Besides that, the increasing in temperature caused the higher corrosion rate of mild steel. Addition of henna extract shifted the pitting potential of mild steel more positive values indicating an increased resistance to pitting corrosion. The low value of corrosion potential (E_{corr}) indicates higher corrosion rate. The values of corrosion current density (I_{corr}) increases with absence of henna extract. In the absence of henna extract, the surface of mild steel is very rough and severely damaged in every part while the surface transformed into smoother and more uniform deposited in the presence of henna extract. In a conclusion, henna extract is a good inhibitor for mild steel in seawater.

References

1. B.A Shaw, R.G. Kelly, Electrochem. Soc. Interface, **15(1)**, 24 (2006)
2. J.K. Paik, A.K. Thayamballi, Y.II. Park, J.S. Hwang, Corros. Sci., **46**, 471 (2004)
3. M.T. Gudze, R.E. Melchers, Corros. Sci., **50,** 3296 (2008)
4. V.S. Sastri, *Green Corrosion Inhibitors Theory and Practice*, (Wiley, New Jersey 2011)
5. R.E. Melchers, R. Jeffrey, Corros. Sci., **4**, 1678 (2005)
6. A.S. Yaro, A.A. Khadom, R.K. Wael, Alexandria Engineering Journal, **52**, 129 (2013)
7. M.A Hegazy, A.M. Badawi, S.S. Abd El Rehim, W.M. Kamel, Egypt. J. Pet., **22**, 549 (2013)
8. W.B. Wan Nik, F. Zulkifli, O. Sulaiman, K.B. Samo, R. Rosliza, Mater. Sci. Eng., **36**, 1 (2012)
9. V. Johnsirani, J. Sathiyabama, S. Rajendran, A.S. Prabha, *International Scholarly Research Network Corrosion ISRN*, **9**, 1 (2012)
10. A. Ostovari, S.M. Hoseinieh, M. Peikari, S.R. Shadizadeh, S.J. Hashemi, Corros. Sci., **51**, 1935 (2009)
11. M. Pourriahi, M. Nasr-Esfahani, A. Motalebi, Electronic Processing of Materials, **50**, 74 (2014)
12. Moslemizadeh, S.R. Shadizadeh, M. Moomenie, Appl. Clay Sci., **105**, 78 (2015)
13. Z.H.Z. Abidin, K.M. Nasir, S.K.M. Jamari, N. Saidon, S.V. Lee, Pigm. Resin Technol., **42**, 128 (2013)
14. N.M. Puzai, W.B. Wan Nik, F. Manuhutu, Technical Report, Final Year Project (2014)
15. M.M. Solomon, S.A. Umoren, Journal of Environmental Chemical Engineering, **3**, 1812 (2015)
16. S.N.N Jainol, W.B. Wan Nik: *Study of Lawsonia Inermis (Henna) as Anti-Corrosion Coating (Paint) For Mild Steel In Seawater*, Universiti Malaysia Terengganu (2016)

MATEC Web of Conferences **78**, 01092 (2016) DOI: 10.1051/matecconf/20167801092
IConGDM 2016

Effect of Multi Wall Carbon Nanotube Content on The Electrical and Rheological Properties of Polypropylene-based Nanocomposites

B. Afrinaldi[1,2,*], *A.L.* Juwono[2], *C.* Liza[1], *S.J.A.* Nasiri[1]

[1]Center of Polymer Technology, 460th Bld Puspiptek Area, Indonesia
[2]Physics Department, University of Indonesia, Indonesia

Abstract. Incorporating multi wall carbon nanotube (MWCNT) into the polymer can improve the electrical properties of the nanocomposites. The effects of MWCNT on the electrical and rheological properties of polypropylene/MWCNT (PP/MWCNT) nanocomposites are investigated. The nanocomposites were prepared using melt mixing technique at chamber temperature of 180°C and rotor rotation of 100 rpm for 10 minutes. Addition of MWCNT increases the electrical surface conductivity up to 10^{-7} S. The electrical percolation threshold of the nanocomposites is found to occur at 1.3 wt% MWCNT content. Viscosity of the nanocomposites increased with the addition of MWCNT. The rheological percolation threshold of the PP/MWCNT is found occur at 1.4 wt% MWCNT content. Well dispersion is observed by mean Scanning Electron Microscope (SEM) analysis.

1 Introduction

Carbon nanotubes (CNT) is a material that widely studied since their discovery by Iijima [1]. This is because the advantage of CNT which has a high aspect ratio so that increases the mechanical strength of the composites [2,3]. In addition, CNT also has other advantage such as a relatively high electrical conductivity [4].

Generally, polymer is an insulator material because it has a low conductivity (>10^{-14} S). Therefore, polymer usually used as the isolator for cable. However, by addition CNT into the polymer can increase its conductivity. Some promising applications for this material are antistatic coating, energy storage devices, and electromagnetic interference shielding [5].

Dispersion of CNTs in a polymer matrix is one of the keys to improve electrical and mechanical properties of the nanocomposites. Some techniques were using compatibilizer such as PP-g-MA [6], in-situ polymerization of polyolefin with CNT [7,8], and by modification of the CNT's surface [9].

Previous research applied the acid treatment on the CNT that can improve the percolation threshold to 1.5 wt% [10]. The other research studied the effect of glycidyl methacrylate grafted MWCNT on polypropylene nanocomposites [11].

* Corresponding author: bambang@sentrapolimer.id

In this work, the effect of MWCNT addition in nanocomposite PP/MWCNT in order to improve electrical and rheological properties with PP-g-MA compatibilizer produced by melt mixing technique is studied and investigated. The results are analyzed using insulation tester, rotational rheometer, and SEM.

2 Methodology

2.1 Materials

Polypropylene (PP HI10HO) from PT. Chandra Asri Petrochemical Tbk with MFR of 10 g/10 min was used as matrix. Maleic anhydride modified polypropylene (Epolene® E-43 Wax) from Eastman was added for improving dispersion of CNT. Antioxidant (Irganox B215 FF) from BASF was used to avoid material degradation. The multi-wall carbon nanotube (MWCNT) from Cheaptubes with outer diameter of 30–50 nm and length of 10–20 μm was utilized as the nanoscale filler.

2.2 Sample preparation and characterization

Masterbatch preparation was prepared by the addition of 10 wt% MWCNT and 5 wt% PP-g-MA in PP matrix. Masterbatch was mixed in Haake Rheomix PolyDrive at chamber temperature of 180°C and rotor rotation of 100 rpm for 10 minutes. The similar process condition was applied to fabricate PP/MWCNT nanocomposites. The masterbatch was added to obtain the composition shown in Table 1.

Table 1. Composition and abbreviations of the nanocomposites.

Composition of nanocomposites	Abbreviation
Neat PP	PP
PP + 3 wt% PP-g-MA	PP-MA3
PP + 3 wt% PP-g-MA + 0.2 wt% MWCNT	PP-MA3-CNT0.2
PP + 3 wt% PP-g-MA + 0.4 wt% MWCNT	PP-MA3-CNT0.4
PP + 3 wt% PP-g-MA + 0.6 wt% MWCNT	PP-MA3-CNT0.6
PP + 3 wt% PP-g-MA + 0.8 wt% MWCNT	PP-MA3-CNT0.8
PP + 3 wt% PP-g-MA + 1.0 wt% MWCNT	PP-MA3-CNT1.0
PP + 3 wt% PP-g-MA + 1.2 wt% MWCNT	PP-MA3-CNT1.2
PP + 3 wt% PP-g-MA + 1.4 wt% MWCNT	PP-MA3-CNT1.4
PP + 3 wt% PP-g-MA + 1.6 wt% MWCNT	PP-MA3-CNT1.6
PP + 3 wt% PP-g-MA + 1.8 wt% MWCNT	PP-MA3-CNT1.8
PP + 3 wt% PP-g-MA + 2.0 wt% MWCNT	PP-MA3-CNT2.0

Electrical conductivity of the nanocomposites was measured at room temperature with Insulation tester UNILAP ISO 5kV based on ASTM D257. Rheological properties were performed on Rotational Rheometer MARS III using oscillatory frequency sweep at strain of 0.5 and temperature of 180°C. Surface fractured of the nanocomposites was coated with Platinum and then observed using Scanning Electron Microscope (SEM) JEOL JSM-6510 at 10,000 magnification.

3 Results and discussion

3.1 Electrical properties

Figure 1 shows that the percolation threshold of nanocomposites was occurred at 1.4 wt% MWCNT. Percolation threshold was described when the conductivity value of nanocomposites increases several orders of magnitude attributes to the continuous conducting networks formation. After the percolation, conductivity of the nanocomposites was saturated due to the interconnected conductive network structure [5, 12]

The electrical percolation threshold of the PP/MWCNT nanocomposites was calculated with a power law function [2,13,14]:

$$\sigma = \sigma_0(\varphi - \varphi_c)^\beta \tag{1}$$

where σ is the electrical conductivity of the nanocomposites, σ_0 is a constant related to intrinsic conductivity, φ is the volume fraction of the filler, φ_c is a percolation threshold, and β is the critical exponent. Based on linear regression shown in Figure 2, gives intercept,(σ_0) = 1.37 x 10^{-7} S, φ_c = 1.3 wt%, and slope (β) = 0.13.

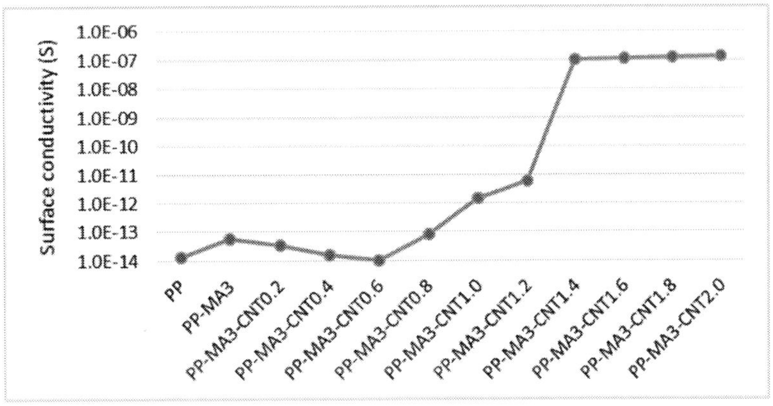

Fig. 1. Surface conductivity of PP/MWCNT nanocomposites.

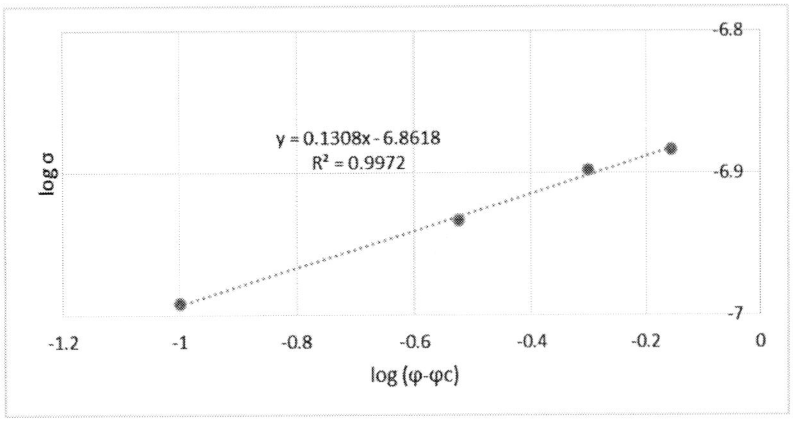

Fig. 2. Linear regression, log σ versus log (φ-φ_c).

3.2 Rheological properties

Rheology studies the flow behavior of the material under the stress applied. The rheological properties of the nanocomposites were affected by the loading, aspect ratio and dispersion of the filler, polymer molecular weight, and interfacial interaction between the filler and polymer [3].

The molten of the PP/MWCNT nanocomposites was a non-Newtonian liquid that can be shown by the increasing frequency which decreased the viscosity as shown in Figure 3. The addition of MWCNT increased the viscosity of the nanocomposites because this filler affects the molecular movement of the polymer and consequently the nanocomposites become more viscous.

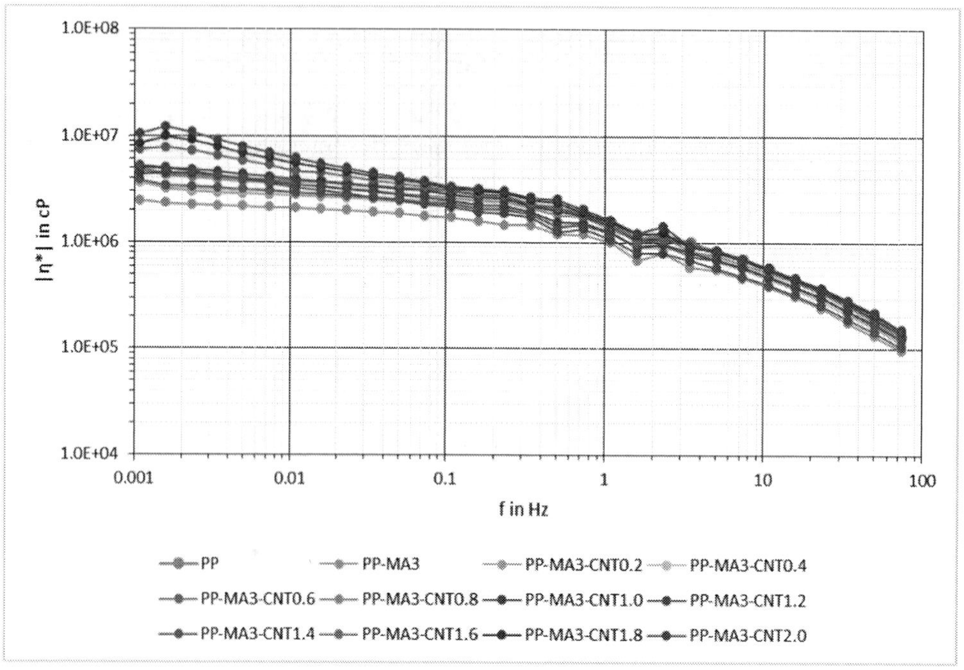

Fig. 3. Viscosity of PP/MWCNT as function of frequency at 180°C.

The storage modulus (G') is the elastic solid-like behavior and the loss modulus (G") is the viscous response of the PP/MWCNT nanocomposites. G' of these nanocomposites gradually increases with increasing frequency. This phenomenon indicated the transition of viscous liquid to solid-like behavior of the material.

Transition from rheological state to a solid-like behavior was obtained by the significant change of G' with increasing loading filler. Filler concentration in which the transition was occured is known as rheological percolation threshold [3]. Author found there was no significant increase on G' but at 1.4 wt% of MWCNT there was an indication that the percolation threshold was observed.

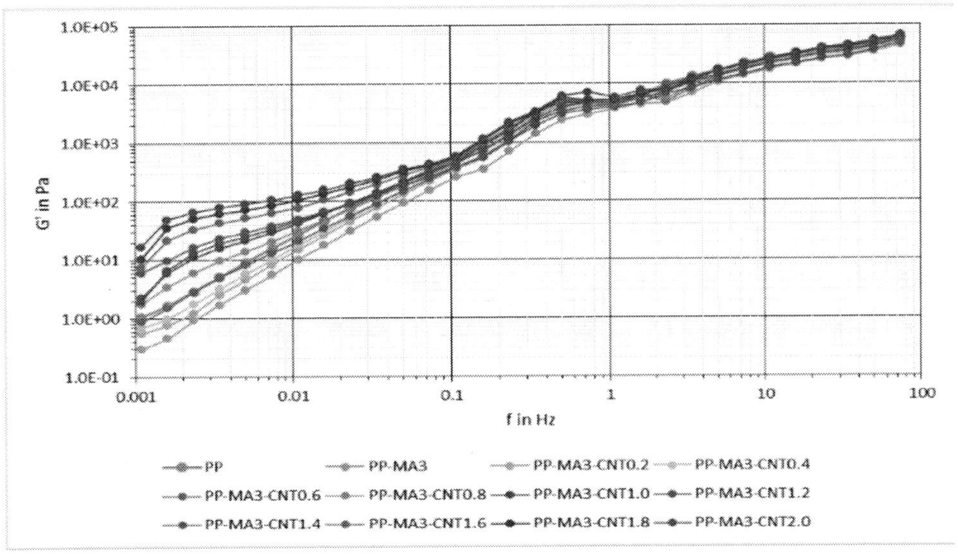

a

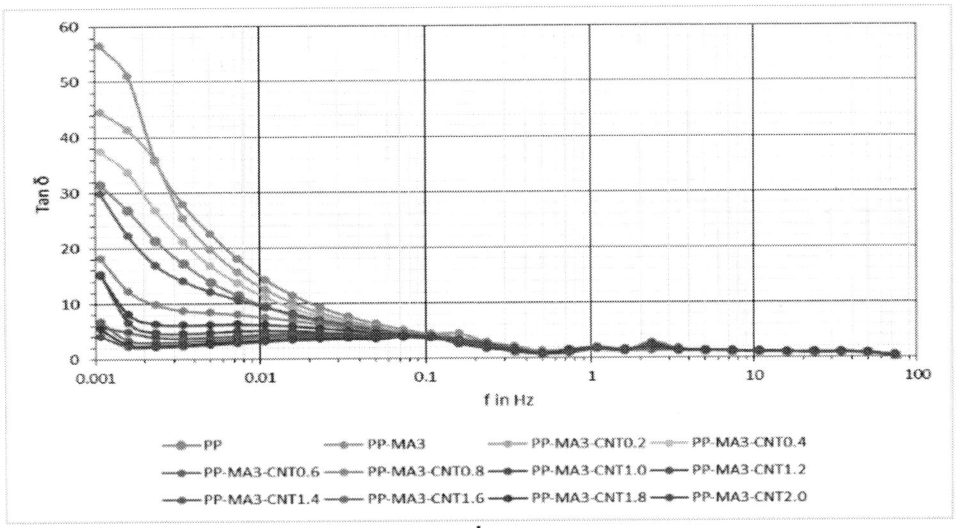

b

Fig. 4. Storage modulus (a) and phase angle (b) of PP/MWCNT as function of frequency at 180°C.

In order to have better observation of the percolation threshold, the phase angle (tan δ) was plotted against the frequency (Figure 4.b). Below the percolation threshold, tan δ increased at low frequency because of liquid-like behavior. From this observation, the rheological percolation threshold of the nanocomposites was identified between 1.2 and 1.4 wt%.

3.2 Morphology

The CNT was dispersed well in the nanocomposites as shown in Figure 5. There was no agglomeration observed in the PP/MWCNT. There were some techniques to obtain a good dispersion of CNT in polymeric matrix. One of the techniques was by applying the compatibilizer, in this work, PP-g-MA was used to disperse the CNT. In addition, high shear processing was utilized to improve the CNT dispersion. Good dispersion was achieved and expected to reduce percolation threshold of the nanocomposites.

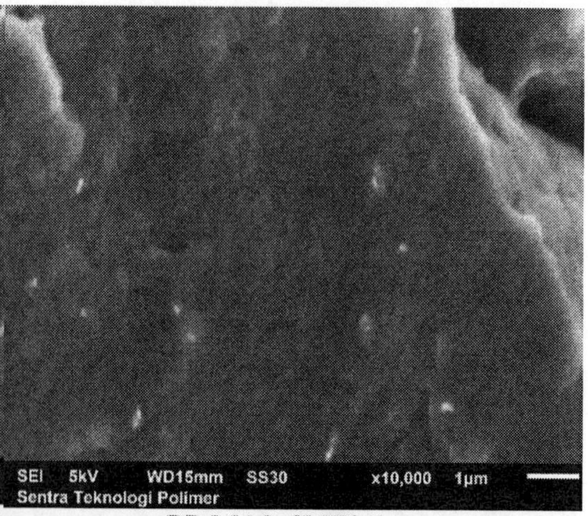

PP-MA3-CNT0.2 PP-MA3-CNT0.4

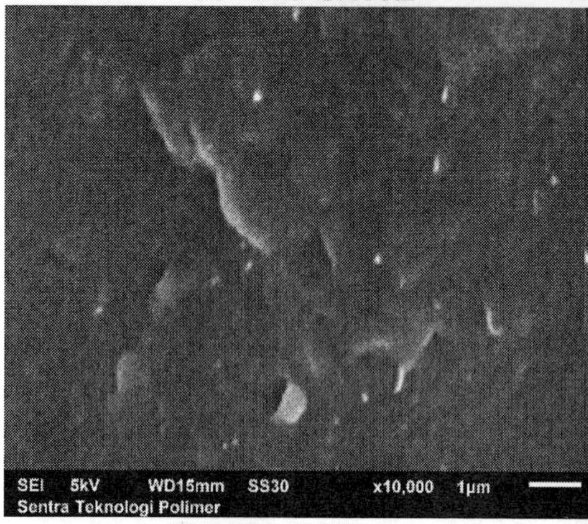

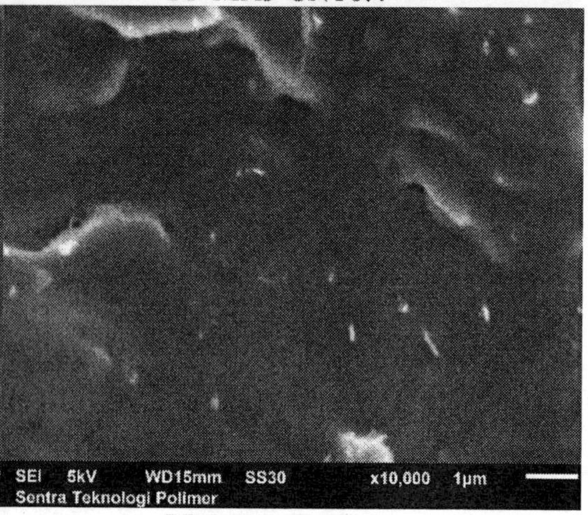

PP-MA3-CNT0.6 PP-MA3-CNT0.8

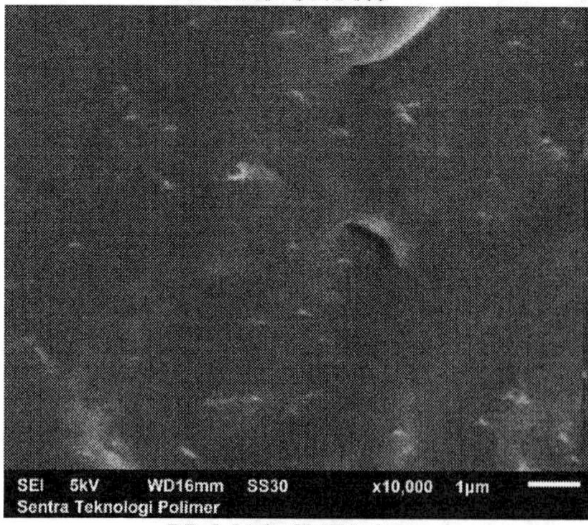

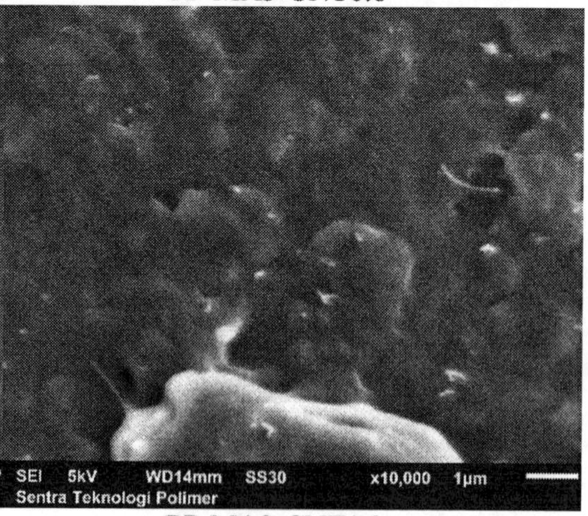

PP-MA3-CNT1.0 PP-MA3-CNT1.2

 DOI: 10.1051/matecconf/20167801092

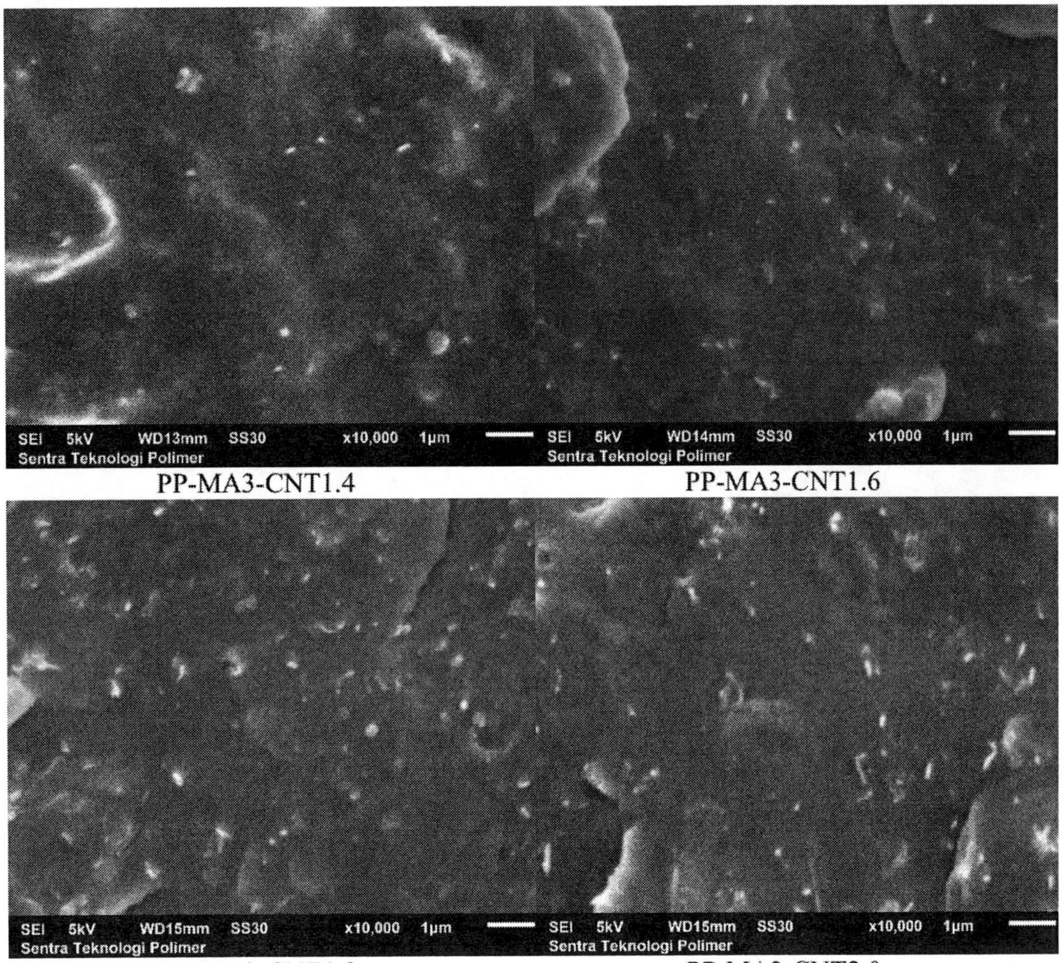

PP-MA3-CNT1.4 PP-MA3-CNT1.6

PP-MA3-CNT1.8 PP-MA3-CNT2.0

Fig. 5. Morphology of PP/MWCNT.

4 Conclusions

PP/MWCNT nanocomposites had been successfully prepared by melt mixing technique. Addition of MWCNT improved the electrical conductivity of the polypropylene. These nanocomposites had electrical percolation threshold at 1.3 wt%. From rheological observation, it was noted that the solid to liquid transition occurred at 1.4 wt% of filler loading.

References

1. S. Iijima, Nature, **354**, 56 (1991)
2. P. Verma, P. Saini, V. Choudhary, Mater. Des., **88**, 269 (2015)
3. P.C. Ma, N.A. Siddiqui, G. Marom, J.K. Kim, Compos. Part A-Appl. S., **41**, 1345 (2010)
4. I. Dubnikova, E. Kuvardina, V. Krasheninnikov, S. Lomakin, I. Tchmutin, S. Kuznetsov, J. Appl. Polym. Sci., **117**, 259 (2010)
5. G. Mittal, V. Dhand, K.Y. Rhee, S.J. Park, W.R. Lee, J. Ind. Eng. Chem., **21**, 11 (2015)

6. K. Prashantha, J. Soulestin, M.F. Lacrampe, M. Claes, G. Dupin, P. Krawczak, Express Polym. Lett., **2 (10)**, 735 (2008)
7. M. Trujillo, M.L. Arnal, A.J. Muller, St. Bredeau, D. Bonduel, Ph. Dubois, I.W. Hamley, V. Castelletto, Macromolecules, **41**, 2087 (2008)
8. R.S. Araujo, R.J.B. Oliveira, M.F.V. Marques, Macromol. React. Eng., **8**, 747 (2014)
9. R. Kotsilkova, E. Ivanov, D. Bychanok, A. Paddubskaya, M. Demidenko, J. Macutkevic, S. Maksimenko, P. Kuzhir, Compos. Sci. Technol., **106**, 85 (2015)
10. K. Zhou, S.Y. Gu, Y.H. Zhang, J. Ren, Polym. Eng. Sci., **8**, 1 (2012)
11. Y.S. Shim, S.J. Park, Carbon Lett., **11 (4)**, 311 (2010)
12. S. Maiti, N.K. Shrivastava, S. Suin, B.B. Khatua, Express Polym. Lett., **7 (6)**, 505 (2013)
13. V. Levchenko, Y. Mamunya, G. Boiteux, M. Lebovka, P. Alcouffe, G. Seytre, E. Lebedev, Eur. Polym. J., **47**, 1351 (2011)
14. W. Bauhofer, J.Z. Kovacs, Compos. Sci. Technol., **69**, 1486 (2009)

 DOI: 10.1051/matecconf/20167801093

Microstructure, Crystal structure and ionic conductivity of 3 mol % (Fe, Mn, Co, Zn) doped 8YSZ

Z.A. Ahmad[1], *B.* Johar[2,*], *S.* Tinesha[2], *S.F.* Khor[2], and *E.K.* Kok[2]

[1]School of Materials Engineering, Universiti Sains Malaysia,
[2]School of Materials Engineering, Universiti Malaysia Perlis, 02600 Kangar, Perlis, Malaysia

Abstract. Effect of 3 mol% of transition elements doped 8YSZ such as YSZZn, YSZFe, YSZCo and YSZMn on the ionic conductivity. SOFCs mostly operate at higher temperature. By substitute with dopants, it can reduce the operating temperature and costs. In this experiment, 3 mol% dopants mixed with 8YSZ and sintered at 1550 °C, hold for two hours. Crystal structure, microstructure, sintering behaviour and ionic conductivity at 300 °C were investigated. XRD demonstrates three phases (cubic, monoclinic and tetragonal) were obtained. It was confirmed that small additions of TMOs (Mn, Fe, Co and Zn) promotes densification, grain growth and ionic conductivity compared to pure 8YSZ. YSZZn obtained the highest ionic conductivity, 3.55 X 10^{-3} mS cm^{-1}at 300 °C.

1 Introduction

Pure zirconia exists in three crystallographic phases such as low temperature monoclinic phase (space group P21/c), intermediate temperature tetragonal phase (space group P42/nmc) and high temperature cubic phase (space group Fm3m) [1]. Zirconia has a fluorite type of crystal structure. During cooling process, it is about 3-5% volume expansion will form due to the transformation from tetragonal to monoclinic phase, which limit usage of ZrO_2 ceramics as structural material [2]. Monoclinic phase and tetragonal phase is not stable enough. Hence, this unstable condition was overcome by adding stabilizing agents to maintain the structure [3]. Thus, cubic ZrO_2 is stable above 2300 °C or can be stabilized by cation substitution. This is supported by Basu et al. had found that direct substitute of small amount of divalent or trivalent cations at the host lattice (Zr^{4+}) has stabilized the zirconia phase. Impurities that were commonly used are Y_2O_3, Yb_2O_3, Sc_2O_3, Gd_2O_3, Dy_2O_3, Nd_2O_3, Sm_2O_3, CaO, MgO and others [4].

Yttria doped zirconia has been widely used in variety of industrial applications [5]. Yttria stabilized zirconia (YSZ) broadly applies as a typical solid ionic conductor such as electrolyte of solid oxide fuel cell (SOFC) [3] and it was established as an oxide ion conducting ceramic[6-8]. Properties of SOFC are low thermal conduction, high ionic conductivity and chemical stability in both oxidizing and reducing atmosphere[7, 9]. Dopants plays important role to enhance the chemical and mechanical properties on

* Corresponding author: banjuraizah@unimap.edu.my

ceramic body. It can affect on the grain boundaries conductivity (GB) and grain interior (GI) during sintering process. It has been confirmed that transition metal oxides are effective sintering promoters for densification of SOFC electrolytes. As reported by Zhang and Du et al.,doping of transition metal oxides (TMOs) such as Mn and Ni was found to be an effective sintering aids for zirconia ceramic [10]. Grain interior is the disintegration of dopant in the electrolyte lattice doping can increase or decrease on grain boundary conduction. Grain growth can affect the ionic conductivity and transformation of crystal structure.Meng and Miao et al. reported that incorporation of heterogeneous and homogeneous doping. In homogeneous doping, oxygen vacancies will replace the atoms at normal lattice sites of zirconia with different of dopants [3]. A high content of dopant (Ca^{2+}, Y^{3+} and others) concentration can cause a few deformities and consequently resulted in lower ionic conductivity. Therefore, homogeneous doping has its limit to improve the ionic conductivity of YSZ. On the other hand, heterogeneous doping is insoluble in host YSZ and remains physically distinct phase. It influences the ionic conductivity of YSZ according to the space charge hypothesis [3] .

Thus, type and quantity of dopants utilisation depends on the stability limit. Kawada et al. stated that 4 mol% of Mn doped is the maximum stability limit of zirconia. Besides, Dey and Das Sharma et al. had reported that stability limit of transition metals such as iron doped with 2 mol% can increase rapidly in the sinterability of powders [4]. Additionally, Shiratori et al. developed that full stabilization and maximum solubility can occur in MgO with values of 12.1 and 14.1 mol% respectively [11]. Phase diagram can identify the solubility limit of each component which occurs at what temperature. Doping transition metals such as Zn, Fe, Mn and Co were chosen added in 8YSZ due to its ionic radius smaller than yttria (0.9 Å). Ionic radiuses of Zn, Fe, Mn and Co are 0.74 Å, 0.645 Å, 0.46 Å and 0.745 Å respectively. Therefore, a main reason of this research is to investigate the 3 mol% transition metals (Fe, Co, Mn and Zn) doped with 8YSZ by focusing on the crystal structure, microstructure and ionic conductivity at 300 °C.

2 Methodology

2.1 Sample preparation

8YSZ (size range from 0.020 to 200 μm) powder were used as a main precursors ZrO_2 (<3.8% monoclinic phase, Sulzer Metro (US) Inc) and the elemental analysis of the powders is tabulated in Table 1. Dopants manufactured from HmbG Chemicals and Sigma. Dried mass of 3 mol% of dopant (MnO, FeO, CoO and ZnO) were mixed with 8YSZ powders ground with mortar pestle for 30 minutes. Calcinations process were performed at 1000 °C for 1 hour and mixed powder with 1.5g were press into pellets of 12mm diameter with 2mm thickness at 6 tonne pressure automatically by Automatic hydraulic press, model 3890.4NE000. The pellets were sinteredat 1550 °C in air hold for two hours with heating rate at 10°C/min.

Table 1 Elemental analysis of 8YSZ.

Chemical analysis	Value (WT %)
Al_2O_3	0.02
Fe_2O_3	0.02
HfO_2	1.7
MgO	0.03
SiO_2	0.2
TiO_2	0.05
Y_2O_3	7.6
ZrO_2	90.2
CaO	0.02
Organic Solids	0.1
Uranium + Thorium	0.02

2.2 Sample characterization

Thermal analysis of doped YSZ ceramic powders were carried out using DTA/TG machine model Perkin Elmer Thermoanalyser started at room temperature until 1600°C. DTA was used to observe all the decomposition reactions (heat absorption or liberation) and determine the crystallization behavior, while TGA was used to detect the loss in mass (degradation).X-ray diffraction (XRD) patterns of the primary powders and sintered specimens were respectively measured with XRD instrument powders at 40kV and 100mA using Cu Kα radiation in air at room temperature. The diffraction intensity was measured step wise every 0.02° in the diffraction angle 2θ range between 5° and 90°. Rietveld Refinement method using XpertHighScore Plus software was used to quantify phase and refine the crystal structure. The relative densities were measured by Archimedes method. After well polished, sintered pellet ceramics sample coated with platinum (pt) using sputter coater prior characterized using the SEM. Morphology of sintered pellets were observed on different magnifications 100X, 500X, 1000X and 10,000X at 15kV by model JEOL JSM-6460LA.The ionic conductivity of sintered doped 8YSZ ceramic samples were measured by using impedance analyser, model Agilent 4294A at 300 °C. The testing was conducted at frequency range from 5000,000 Hz to 4 Hz will be measured.

3 Results and discussion

3.1 Results of DTA/TG

The experiments are carried out by differential thermal (DTA) and thermogravimetric (TGA) analysis method using a Perkin Elmer Thermoanalyser instrument. The heat flow and weight change of calcinated powder samples placed in Al_2O_3 crucibles were measured from room temperature to 1600 °C at a heating rate of 10 °C per minute. Weight gain in TG curve as shown in Fig.1 may attributes to weight gain that attributed to the oxidation [12]. YSZ, YSZMn and YSZFe gain about 0.20 mg of weight at 1500 °C. Cobalt shows an increase of weight about 0.21 mg. However, weight of dopant zinc gain about 0.205 mg at 1300 °C and loss to 0.19 mg.

DTA curves can be seen in Figure 2and as shown in the figureallundopedand doped 8YSZ samples undergoes exothermic reaction at beginning of 200 °C. YSZ, YSZCo and YSZZn undergo two endothermic peaks at each curve, while YSZFe and YSZMn undergo only one endothermic peak. It can seen clearly in Fig. 2, first endothermic peaks in DTA curve for the YSZZn happens between range 500 °C to 600 °C undergo oxidation by the air atmosphere. When temperature exceeds 550 °C and reach 600 °C, an endothermic reaction of YSZZn undergo having the maximal temperature shift at 565 °C. Simultaneously around 7% mass gain was detected. On the other hand, endothermic reaction of YSZ occurred at 600 °C. However, change in slope of YSZCo occurs at 936 °C could be due to the addition of Co^{2+} to Co^{3+} causing chemical expansion due to high temperature. In addition, YSZZn, YSZFe and YSZMn curves show a sudden change and broad endothermic reaction approximately from 1100 °C to 1250 °C. This change in the kind of thermal process is considered as occurrences of phase transformation, crystallization and structural distortion.

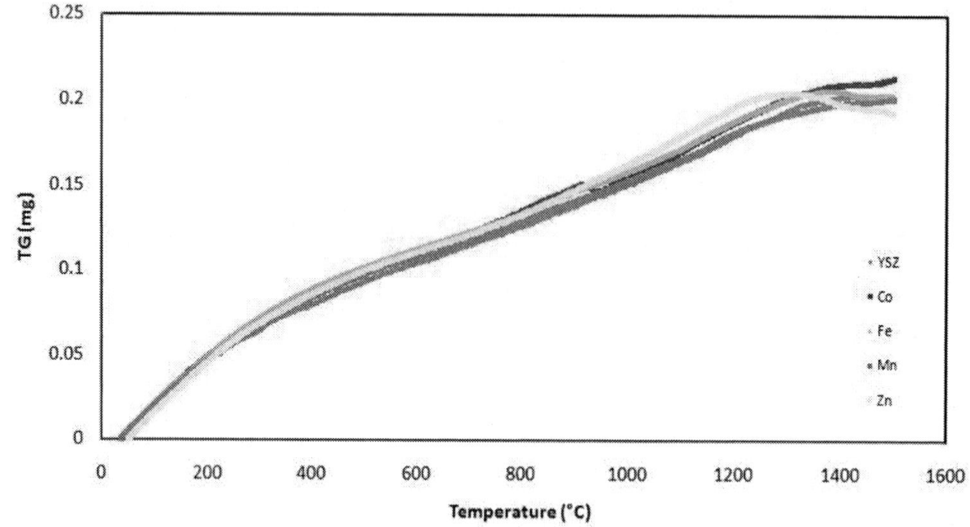

Fig. 1. TGA curve for the different transition metals doped 8YSZ. Dopants include Co, Fe, Mn and Zn.

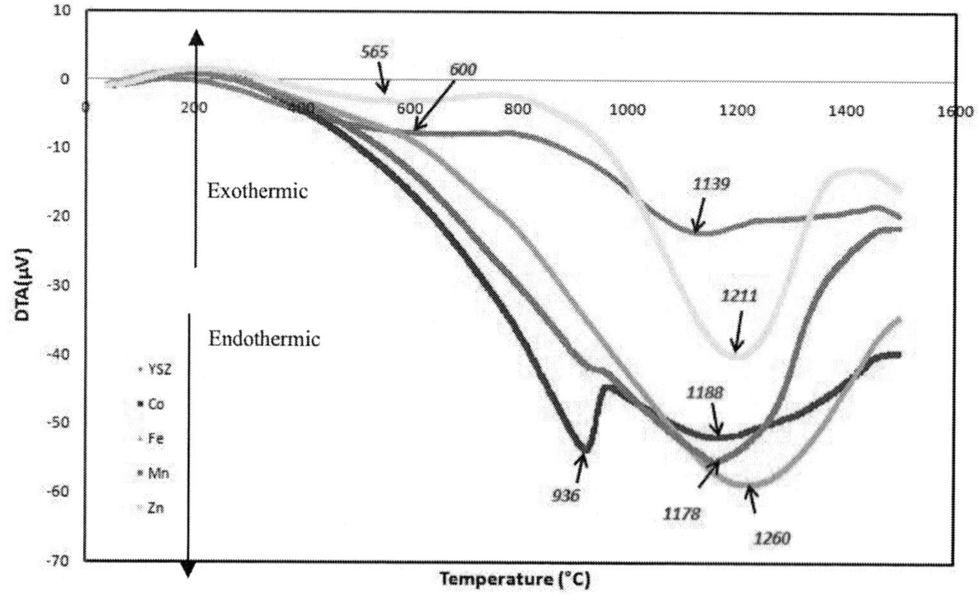

Fig. 2. DTA curve for different transition metals doped YSZ. Dopants include Co, Fe, Mn and Zn.

Two distinct features are observed from other sample: shifting in Bragg's diffraction towards higher 2θ which gives a decrease in lattice volume and appearances of shoulders in the XRD patterns corresponding to the diffraction planes indicating lower crystal symmetry [13]. Previous research mentioned that distortion of the oxygen lattice results in the tetragonal symmetry in sintered YSZ [9]. The crystalline structure of pellets was analyzed by X-ray diffraction (XRD). Figure 3 shows the x-ray diffraction of the sintered pellet of YSZMn, YSZFe, YSZCo and YSZZn along with pure YSZ. For the doped powder (up to dopant concentration of 3 mol%), all peak corresponding to the YSZ (cubic/c, tetragonal/t and monoclinic/m) are observed.Cubic phase exhibits only single peak at all positions due to broadening of peak, Z polymorph might overlap, thus crystal structure refinement using whole profile fitting method or Rietveld method to determine present either one of both polymorph.

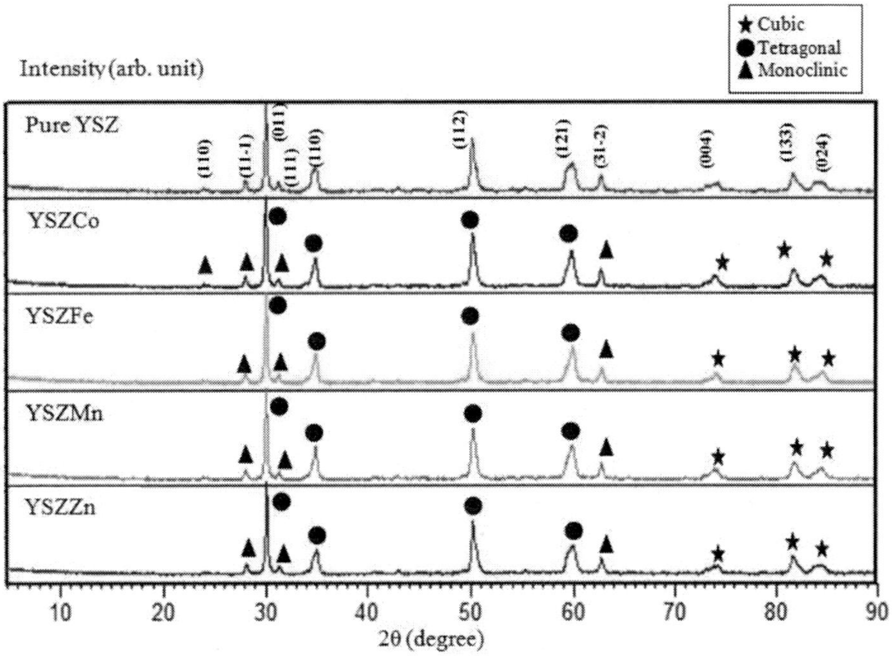

Fig. 3. Comparative XRD patterns of undoped 8YSZ and YSZCo, YSZFe, YSZMn and YSZZn.

Table 2. Refinement XRD result.

Element	Tetragonal	Cubic	Monoclinic	RWP
YSZ	65.5	26.8	7.8	13.66
YSZMn	82.4	12.4	5.2	13.25
YSZFe	77.5	17.7	4.8	13.73
YSZZn	78.9	16.1	5.0	12.80
YSZCo	80.2	14.3	5.5	13.43

Table 3 demonstrates the density of all sintered samples using Archimedes. It shows relative density of sample increasing with dopant concentration of 3 mol %. Liquid phase promotes the diffusion rate of 8YSZ matrix, enhance densification and grain growth [14].

Table 3. Result of Archimedes.

Composition	Dry mass (g)	Mass in water (g)	Bulk Density (kg/m^3)	Densification, %
Pure YSZ	1.14	0.94	5.539	75.85
3% Fe	1.49	1.22	5.143	74.13
3% Mn	1.46	1.19	5.420	83.25
3% Zn	1.47	1.19	5.306	76.32
3% Co	1.49	1.20	5.157	78.96

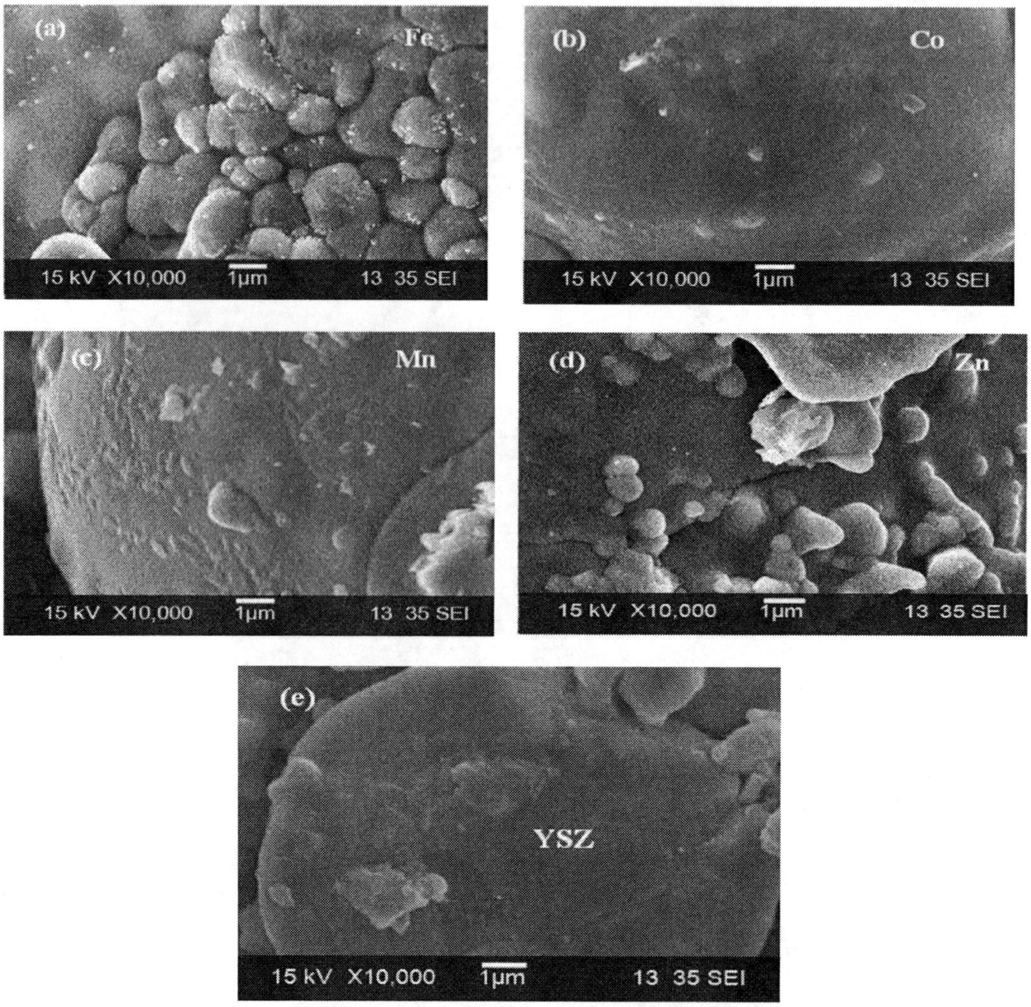

Fig. 4. The SEM micrographs showing the surface morphology of 3% of transition elements (a) YSZFe, (b) YSZCo, (c) YSZMn, (d) YSZZn and (e) undoped 8YSZ sintered at 1550 °C for 2h.

Figure 4 demonstrates the morphology of SEM images of the pellets of YSZCo, YSZMn, YSZFe and YSZZn after sintered at 1550 °C for 2h. Pure YSZ shows in Fig 4(e), YSZ in spherical shape which is similar to the microstructure previously reported by Pradhan [15, 16]. In Fig 4(a) and (d), YSZFe and YSZZn crystallite structure grows and agglomerates between the grain boundaries. In addition, YSZFe shows the least porosity and densification about 74%. Iron atoms easily trap vacancies within the grain boundary and started agglomerate along the grain boundaries. The sample of YSZCo and YSZMn were not well sintered; the micrograph revealed some open pores, showing not fully densified microstructure, it lead to relative density values being less than 100% [14]. Hence, it can conclude that sinterability of specimen prolonging the holding time, whereas ionic conductivity may improve by increasing the density.

It proposed that cause of stability of tetragonal phase in studies directly related problem statement of raw powder (8YSZ). During heat treatment of the Y_2O_3 doping either diffuse into ZrO_2 crystal lattice or segregation at grain boundary, depend how yttria atom fuses with Zr atom during synthesis process. YSZMn shows the highest porosity about 25% at 1550 °C. Undoped 8YSZ was dense with small close pore, rapid densification of MnO added specimen may attributed of liquid phase sintering. Liquid phase promotes the diffusion rate of 8YSZ matrix, enhance densification and grain growth [14]. Previous research of Vijaya Lakshmi on stabilization effect on Mn is not well established in phase diagram. Shultz and Muan reported that maximum solubility of 27 mol% of MnO in ZrO_2 at 1673 K. Mn can present in trivalent state with ionic radius of 0.065nm to have any stabilization effect [17].

Conductivity depends on the difference between size of host and dopant cation radii. High conductivity in solid oxide is associated of oxygen vacancies and dopant cations into imperfections of low mobility. Generally, concentration of dopant and size of acceptors are important in ionic conductivity of stabilized zirconia. Kosacki and Rouleau et al. had mentioned that acceptor with an ionic radius nearly similar to Zr^{4+} ion, of which Y^{3+} due to minimum activation energy. The low frequency region is undergo DC, increase in conductivity at high frequency [18]. YSZZn has highest ionic conductivity and combination of DC and AC. YSZZn has DC value of conductivity about 5 x 10^{-3} Sm^{-1} until 100K Hz. Furthermore, YSZFe stated that it has DC value of conductivity about 3 x 10^{-5} $mScm^{-1}$ until 5K Hz. YSZCo, YSZMn and YSZ curves undergo weak DC contribution among the atoms. As a result, TMOs reduces the total conductivity of pure 8YSZ of 0.5 X 10^{-6} Sm^{-1}. Consequently, 3% mol of YSZFe can reduce the GI conduction by ~27-33% [4].

There are two semicircles for the impedance plots measured at 300 °C. The left semicircles results from the grain resistance, while the right one is due to grain boundary resistance. As the measurement temperature increases, the capacitance of the grain boundary and grain become comparable, leading disappearance of one semicircle corresponding to grain. Ionic size of TMOs are small compared with yttria, as a result of TMO dissolution still present at >1550 °C which associated oxygen vacancies should be freed. From the previous research, Wang et al. proposed that addition of ZnO gives an adverse reaction to the conductivities [19]. Besides that, Shujun et al. had investigated that highest ionic value of 1 mol% of Zn doped $Ce_{0.8}Y_{0.2}O_{1.9}$ was 0.0516 Scm^{-1} at 750 °C [14].

 DOI: 10.1051/matecconf/20167801093

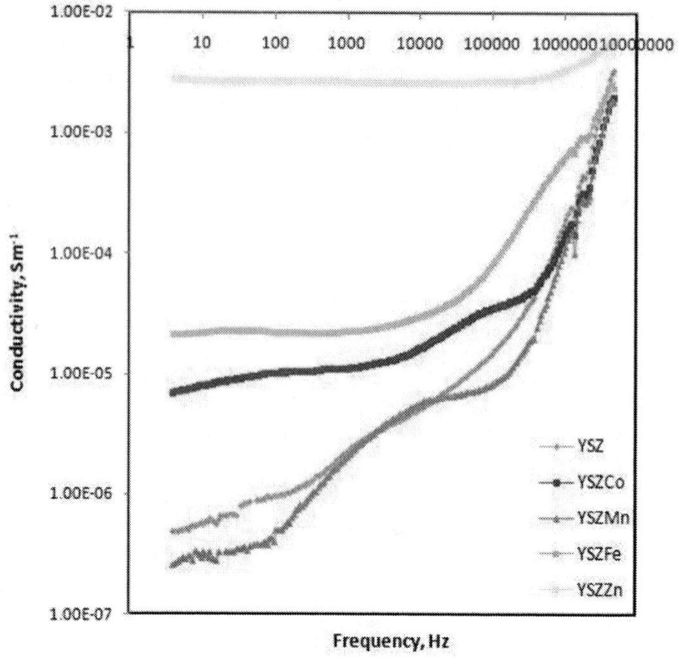

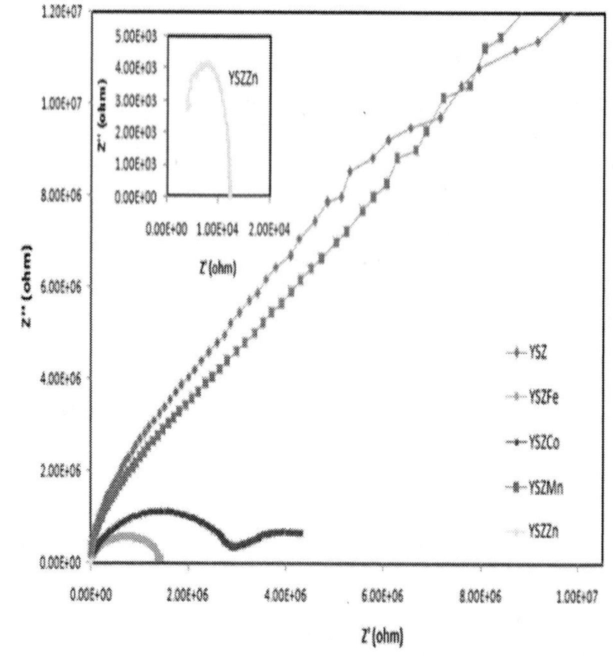

Fig. 5. Logarithmic relation between ionic conductivity (log σ) and frequency at 300 °C. Impedance diagrams recorded at 300 °C for different transition metals doped YSZ such as YSZCo, YSZFe, YSZMn and YSZZn.

In previous study, ZnO decrease the activation energy of conduction, hence it increase ionic conductivity [14]. 3 mol% ZnO of dopants can lead to increase in the sintered density value. Ionic conductivity of 3 mol% of YSZZn has highest ionic conductivity with 3.55 X 10^{-3} mS cm^{-1}. Besides that, value of conductivity of 3 mol% of cobat was 3.32 X 10^{-4} mS cm^{-1} was measured in air atmosphere. On the other hand, Fe_2O_3 cannot stabilize the cubic phase of ZrO_2 by its own. So, partial substitution of Fe^{3+} for Y^{3+} can increase the Y_2O_3

efficiency as a stabilizer of c-ZrO[20]. Ionic conductivity can be increase by the oxygen vacancies by the content of aliovalent substitution such as Fe^{3+} and Y^{3+}. Fe^{3+} ion migrate by a vacancy diffusion mechanism during sintering process. Fe^{3+} can either substitute or interstitial doped 8YSZ, this is due to the ionic radius of Fe^{3+} ions (0.55Å – 0.78Å) are far smaller than Zr^{4+} ions (0.84Å) and Y^{3+} ions (1.019Å) [7]. Furthermore, diffusion coefficient, D increases due to contribute of partially interstitial defect of Fe^{3+}. Thus, Fe^{3+} ion can weaken the bond strength of Zr-O bond. Greater activation energy is associated with a greater ionic radius. Presence of the weaken bond can led to decrease in activation energy, which partially contribute increase in the diffusion coefficient D. When temperature up to 300 °C, the value of conductivity of 3 mol% of iron is 2.67 X 10^{-3} mS cm^{-1} was measured in air atmosphere.

4 Conclusions

Themajor conclusion of the present work is to study the effect of dopants in electrolytes. YSZ solid solutions doped with small amounts of Zn, Mn, Fe and Co were investigated. It was found that the doping of 8YSZ with 3 mol% transition elements is found to be effective in lowering sintering temperature. From detailed of XRD analysis, no appreciable of new phase formation has been noticed up to 3mol%, dopants are completely dissolved in the 8YSZ. Hence, it could be that dopants mixed of metal oxides did not react with pure YSZ at 1550 °C. Pure YSZ and TMOs in this present study requires higher sintering temperature. Hence, it found that the corresponding electrochemical performances of YSZFe, YSZMn, YSZZn and YSZCo found performed better than fabricated by undoped YSZ. Based on their doping effect on grain growth and ionic conductivity, YSZZn can reduce porosity and improve sinterability to obtain dense electrolyte material.

References

1. Y. Zhou, W. Yuan, Q. Huang, W. Huang, H. Cheng, H. Liu, Ceram. Int., **41**, 10702 (2015)
2. B. Xie, C. He, P. Cai , Y. Xiong, Thin Solid Films, **518**, 1958 (2010)
3. B. Meng, Z.Y. Miao, M. Kong, X.X. Liu, J. Yu, Q.Q. Yang, Solid State Ionics, **258**, 61 (2014)
4. T. Dey, A.D. Sharma, A. Dutta, R.N. Basu, J. Alloy. Compd., **604**, 151 (2014)
5. A. Madani, A. Cheikh Amdouni, A. Touati, M. Labidi, H. Boussetta, C. Monty, Sensor Actuat. B-Chem., **109**, 107 (2015)
6. N. Matsui, M. Takigawa, Solid State Ionics, **40**, 926 (1990)
7. F. Guo, P. Xiao, J. Eur. Ceram. Soc., **32**, 4157 (2012)
8. S.K. Vijay, V. Chandramouli, S. Khan, P.C. Clinsha, S. Anthonysamy, Ceram. Int., **40**, 16689 (2014)
9. R.L. Gonzalez Romero, J.J. Melendez, D.G. Garcia, F.L. Cumbrera, A.D. Rodriguez, Solid State Ionics, **219**, 1 (2012)
10. T.S. Zhang, Z.H. Du, S. Li, L.B. Kong, X.C. Song, J. Lu, J. Ma, Solid State Ionics, **180**, 1311 (2009)
11. Y. Shiratori, F. Tietz, H.J. Penkalla, J.Q. He, Y. Shiratori, D. Stover, J. Power Sources, **148**, 32 (2005)
12. M. Jamshidijam, R.V. Mangalaraja, A. Akbari Fakhrabadi, S. Ananthakumar, S.H. Chan, Powder Technol., **253**, 304 (2014)
13. M.K. Mahapatra, N. Li, A. Verma, P. Singh, Solid State Ionics, **253**, 223 (2013)
14. S. Li, L. Ge, H. Gu, Y. Zheng, H. Chen, L. Guo, J. Alloy. Compd., **509**, 94 (2011)

15. M. Pradhan, P.C. Kapur, Pradip, Ceram. Int., **38**, 2835 (2012)
16. Z. Shu, X. Jiao, D. Chen, J. Alloy. Compd., **509**, 9200 (2011)
17. H. Misawa, K. Nakata, J. Matsuura, Y. Moriwaki, K. Kawashima, T. Shimizu, T. Shirasawa, R. Takahashi, Neurobiol. Dis., **23**, 169 (2006)
18. M. El Moudane, M. El Maniani, A. Sabbar, A. Ghanimi, M. Tabyaoui, A. Bellaouchou, A. Guenbour, Mater. Res. Bull., **72**, 241 (2015)
19. C.X. Wang, J.C. Lv, Y. Ren, T. Zhi, J.Y. Chen, Q.Q. Zhou, Z.Q. Lu, D.W. Gao, L.M. Jin, Appl. Surf. Sci., **359**, 196 (2015)
20. O. Bohnke, V. Gunes, K.V. Kravchyk, A.G. Belous, O.Z. Yanchevskii, O.I. V'Yunov, Solid State Ionics, **262**, 517 (2014)

Influence of Water Content on the Flow Consistency of Dredged Marine Soils

M. Z. Rosman[1,*], *S. F. S. M.* Johan[1], *N. A.* Rahman[1] , and *C. M.* Chan[1]

[1]Faculty of Engineering Technology, University Tun Hussein Onn Malaysia (UTHM), Beg Berkunci 101, 86400, Parit Raja Batu Pahat Johor, Malaysia

Abstract. In present time, dredged marine soils (DMS) are generally considered as geo-waste in Malaysia. It is also known to contain high value of water and low shear strength. Lightly solidified soils such as soil-cement slurry and flowable fill are known as controlled low strength materials (CLSM). On site, the CLSM was tested for its consistency by using an open-ended cylinder pipe. The vertical and lateral displacement from the test would determine the quality and workability of the CLSM. In this study, manufactured kaolin powder was mixed with different percentages of water. Cement was also added to compare the natural soil with solidified soil samples. There are two methods of flowability test used, namely the conventional lift method and innovative drop method. The lateral displacement or soil spread diameter values were recorded and averaged. Tests showed that the soil spread diameter corresponded almost linear with the increasing amount of water. The binder-added samples show no significant difference with non-binder sample. Also, the mixing water content and percentage of fines had influenced the soil spread diameter.

1 Introduction

Removal of dredged marine soils (DMS) from the sea bed is required in order to clear the passage way of ships. In Malaysia, about 300,000 m^3 volumes of DMS were gathered and removed as part of maintenance dredging [1]. Soft and problematic soils like DMS posses low shear strength and high water content. Generally, the shear strength value of DMS is less than 50 kPa [2]. With such condition, any development made on top of this unimproved soil would risk with slope failure and non-uniform settlement. In addition, transporting DMS to the designated dumping sites would cause monetary and environmental implications. With these reasons, DMS are likely to be disposed rather than to be reused.

Numerous studies show that by inducing DMS with some binders and fillers, its geotechnical properties would be improved. Other than reused as potential construction materials such as brick and cement [3,4], DMS was primarily used as reclamation fills. Binders and fillers such as cement, lime, bottom ash, fly ash and steel slag [5-8] enabled

* Corresponding author: mzawawirosman@gmail.com

DMS to be reused as reclamation fills. Large reclamation projects in Australia, Japan and Singapore had successfully utilized DMS as backfill materials [9-11].

There are several and similar terms to describe the modified fluid-like solid used as reclamation fill. *Super geomaterial* (SGM) and *Composite geomaterial* (CGM) are both mixture of dredged soil, binder, granular materials and lightweight materials [12,13]. *Self compacting material* (SCM) and *Controlled low strength materials* (CLSM) are also related where the mixture includes less binder and granular materials [14,15]. Despite the different terms used, its function as engineered and flowable fills remained the same.

This present study relates the flow consistency of soil with water content. Increasing amount of moisture would provide more fluid-like substance, thus ease the deployment of fills.

2 Methodology

2.1 Materials

Kaolin and DMS are type of soils that contain high percentage of fines and categorized as fine-grained soils. Table 1 shows the physical properties of the related samples. In this study, manufactured kaolin FM-C powder was used to resemble DMS. Different percentages of water were added into the soil sample, denoting the phase of semi-solid to liquefied form. Manufactured cement powder was also used. The addition of cement in the soil mixture was to observe the differences of flow consistency with non-solidified soil sample. Cement with percentages of 5% and 10% cement note the minimum and maximum of binder content in this study.

Table 1. Physical properties of related samples.

Samples	**Kaolin**	**Tokyo Bay A***	**Tokyo Bay B***	**Kobe***	**Okhotsk***	**Kasaoka Clay***	**Kuni-bond***
W_L [%]	79.6	114.7	112.4	108.2	85.6	55.4	133.1
WC_M [%]	60,80,100, 120,160	285	285	285	285	285	285
ρ_s [g/cm³]	2.56	2.62	2.70	2.64	2.56	2.71	2.70
Fines [%]	57	98	92	100	94	93	94

*: Samples referred to [12]

2.2 Aspect ratio

The test procedure was based on standard [16], where an open-ended cylinder was used to fill the soil mixtures. Various size and dimension of the ring pipe cylinder used in other studies were stated in Table 2. Diameter to height aspect ratio of the cylinder was used to compare the present and standard cylinder. The present ring pipe cylinder used in this study is 52 mm of diameter and 30 mm of height. As calculated, the ratio of the present cylinder size was less than the standard cylinder size ($0.00037<0.51$). Hence, the dimension of the ring pipe cylinder used in present study is permissible.

Table 2. Variation in dimension of ring cylinder used.

No.	Diameter x Height	Aspect Ratio	References
1.	50 mm x 100 mm	0.50	[12]
2.	75 mm x 150 mm	0.50	[13]
3.	76 mm x 150 mm	0.51	[16]
4.	38 mm x 76 mm	0.50	[17]

2.3 Flowability test methods.

Batches of soil mixture were mixed with $0.75W_L$, $1.00W_L$, $1.25W_L$, $1.50W_L$, $1.75W_L$ and $2.00W_L$ of its liquid limit (W_L). The variance of water content was intended to examine the behavior of soil that undergoes such levels of saturation. Another series of soil mixture incorporated with 5% and 10% of cement were also mixed with the same pre-determined moisture content. The flowability test was then prepared in two methods, namely the conventional lift method and innovative drop method.

Lift method consist of filling the soil mixture inside a ring cylinder. As the ring cylinder was slowly lifted up, the soil samples would then segregated by its own weight and form lateral displacement of soil spread. Note that before filling, the ring has been lubricated with oil or water. Due to the cohesive nature of fine grained soil, most of the soil sample adhered to the wall of the cylinder. Therefore, an innovative method that allow the soil samples to be drop at raised level was conducted. Similarly, the soil sample was filled inside the ring cylinder on a base platform. By removing the platform, the soil sample would drop at the height of 175 mm as refer to Figure. 1. It is comparable with the lift method since the soil sample for fall method segregated by gravity. Only the average value of soil spread diameter was measured and recorded.

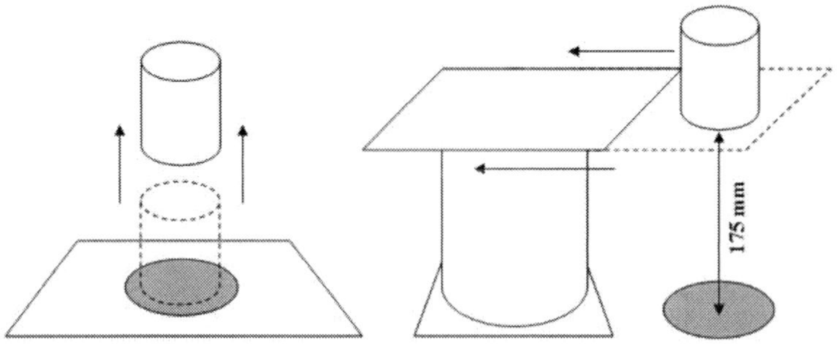

Fig. 1. Schematic diagrams of lift method [*left*] and drop method [*right*].

3 Results and Discussion

3.1 Influence of water content and binder.

The average values of soil spread diameter (S_D) were plotted against water content (WC) as shown in Figure 2. Clearly the soil spread diameter was influenced by the water content. The soil spread diameter increased with the increasing amount of moisture. The flowability of soil and binder mixture was also made in order to examine whether distinct values of spread diameter occurred. With the addition of cement, less spread diameter can be observed with the increasing dosage of 5% and 10%. However, there were no such significant differences. It is likely that the cemented soil samples reacted with water which had caused cement hydration.

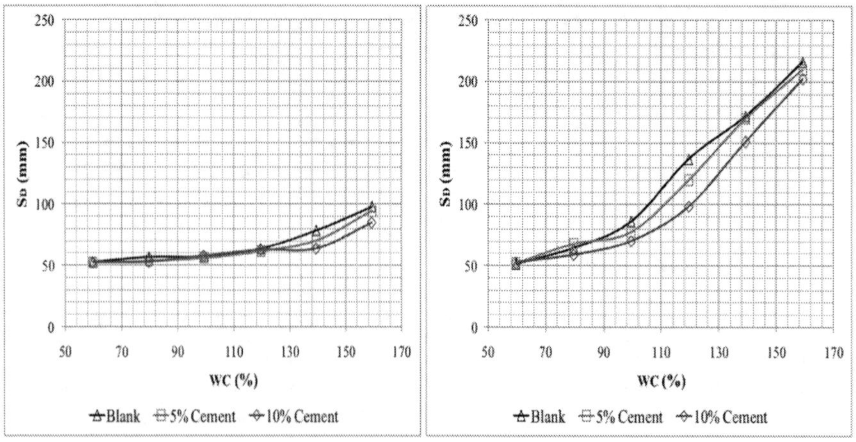

Fig. 2. Flowability test values for lift method [*left*] and drop method [*right*].

3.2 Influence of method.

Flowability by innovative fall method resulted with larger spread diameter than the conventional lift method. With more water content, the soil matrix tends to loosen and break apart. At raised level, samples by fall method were dropped by its gravitational weight. Both of the reasons contributed to the large displacement. Even so, both methods show increasing trend line in spread diameter against water content graphs.

3.3 Influence of ring cylinder.

Comparison values of data between kaolin and samples from [12] were tabulated in Table 3. The index values of mixing water content and liquid limit [WC_M/ W_L] of kaolin were predetermined. Apparently, the index values of spread diameter and initial diameter [S_D/D] were less than the others. Based from the dimension of the ring cylinders used, only the ring height for kaolin differs from the others. Despite the difference, the dimension of ring cylinders was not the main reason for the different values. Even if the ring height for kaolin sample is the same, the S_D/D value still less than the others ranged values. Illustrated in Fig.3, the normalized graph of S_D/D and WC_M/ W_L shows the distinct values of kaolin and other soil samples. According to Table 1, the mixing water content [WC_M] and percentages of fines [%] were less than the other samples. Obviously, higher water content and finer samples would produce large S_D values.

Table 3. Index values of kaolin and referenced samples.

Samples	Kaolin	Tokyo Bay A*	Tokyo Bay B*	Kobe*	Okhotsk*	Kasaoka Clay*	Kuni-bond*
WC_M/ W_L	0.75,1.00,1.25, 1.50,1.75,2.00	2.50	2.50	2.60	3.30	5.10	2.10
H [mm]	30	100	100	100	100	100	100
D [mm]	52	50	50	50	50	50	50
S_D [mm]	53,56,57,64,79, 98	475	580	490	640	660	570
S_D/D	1.02,1.08, 1.10,1.23, 1.52,1.88**	9.50	11.60	9.80	12.80	13.20	11.40

*: Samples referred to [12] **: S_D values for lift method

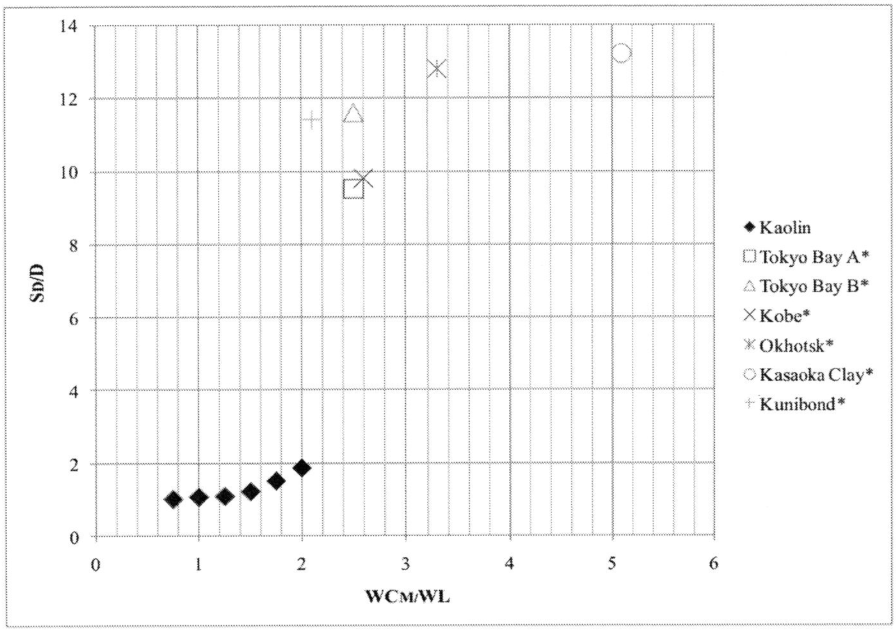

Fig. 3. Normalized results between S_D/D and WC_M/W_L.

4 Conclusion

In this study, series of soil mixture with different water percentages and binder contents were conducted. The purpose of this study is to examine the influence of moisture towards the soil samples regardless whether non-solidified or solidified. The findings are summarized as follows;

- Higher water content would resulted higher spread diameter
- By adding minimum and maximum dosage of cement namely 5% and 10%, it had affected the spread diameter accordingly. It is highly due to hydration effect between cement and water.
- Non-solidified and solidified samples do not have that significant difference of spread diameter and water content relationship.
- Despite the test methods used, both methods shows similar trend line of spread diameter and water content relationship.
- Mixing water content and percentage of fines had affected the soil spread diameter.

References

1. W.S.W. Salim, S.F. Sadikon, S.M. Salleh, N.A.M. Noor, M.F. Arshad, N. Wahid, *IEEE Symposium on Business, Engineering and Industrial Applications*, **23**, 509 (2012)
2. C.F. Chiu, W. Zhu, C.L. Zhang, Eng. Geol., **103,** 1 (2009)
3. W.S.W. Salim, N.A.M. Noor, M.F. Arshad, N. Wahid, S.M. Salleh, *Proceeding of the 2nd International Conference on Biotechnology and Environment Management* (2012)
4. T.A. Dang, S. Kamali Bernard, W.A. Prince, Constr. Build. Mater., **41**, 602 (2013)
5. A. Azhar, C.M. Chan, A.T.A. Karim, Electronic Journal of Geotechnical Engineering, **19**, 8797 (2014)

6. C.M. Chan, T. Mizutani, Y. Kikuchi, International Journal of Civil and Structural Engineering, **2,** 270 (2011)
7. H.W. Xiao, F.H. Lee, International Journal of Civil, Environmental, Structural, Construction and Architectural Engineering, **2**, 144 (2008)
8. G. Rajasekaran, S.N. Rao, Ocean Eng., **29**, 545 (2002)
9. D. Ganesalingam, J. Ameratunga, G. Schweitzer, *Proceeding of the 18th International Conference on Soil Mechanics and Geotechnical Engineering* (2013)
10. Y. Watabe, T. Noguchi, *Proceeding of Indian Geotechnical Conference* (2011)
11. A. Arulrajah, M.W. Bo, H. Nikraz, *Proceeding of the 6th International Conference on Case Histories in Geotechnical Engineering Paper* (2015)
12. S. Kataoka, T. Horita, M. Tanaka, R. Tomita, M. Nakajima, *Proceeding of the 18th International Conference on Soil Mechanics and Geotechnical Engineering* (2013)
13. Y.T Kim, C. Lee, H.I. Park, Mar. Georesour. Geotec., **29**, 1 (2011)
14. J. Kanadasan, A.F.A. Fauzi, H.A. Razak, P. Selliah, V. Subramaniam, S. Yusoff, Materials, **8**, 6518 (2015)
15. K.H. Lee, J.D. Kim, J. Civ. Eng., **17,** 674 (2013)
16. ASTM D6103-04, *Standard Test Method for Flow Consistency of Controlled Low Strength Material* (West Conshohocken 2004)
17. C.M. Chan, M.N.M. Yusof, Modern Applied Science, **9**, 13 (2015).

MATEC Web of Conferences **78**, 01095 (2016) DOI: 10.1051/matecconf/20167801095
IConGDM 2016

A study on Relay Effect via Magnetic Resonant Coupling for Wireless Power Transfer

N.A. Rashid[1,*], *M.N.M.* Yasin[1], *K.* Kamardin[2], *N.* Ramli[5], *M.* Jusoh[3], *S.A.Z.* Murad[1] , and *M.I.* Sulaiman[4]

[1]School of Microelectronic Engineering, Universiti Malaysia Perlis
[2]Computer Systems Engineering Group, Advanced Informatics School, Universiti Teknologi Malaysia
[3]School of Computer and Communication Engineering, Universiti Malaysia Perlis
[4]Universiti Kuala Lumpur, British Malaysian, Selangor, Malaysia
[5]Engineering and Built Environment, Segi University Sdn Bhd

Abstract. Wireless power transfer (WPT) transmits electrical energy from a power source to an electrical load wirelessly or without any conductors. The capability of WPT to transmit the energy is limited. Therefore, a relay was introduced to increase the distance of the WPT capabilities. The effect of the relay has been investigated to extend the energy transfer distance. The effect of relay was demonstrated by placing a relay coil between transmitter and receiver, relay biased to transmitter and placing two relay coils in the designed system. Experimental results are provided to prove the concept of the relay effect. The power transmission efficiency can be achieved up to 75% at 1 meter distance.

1 Introduction

The concept of Wireless Power Transfer (WPT) has been introduced over the century and now this technology is gaining more advances in wireless charging especially in portable consumer electronic devices, medical and industrial devices. The concept of transmitting power wirelessly has been proposed by Nikola Tesla, a pioneer electrical engineer in WPT by introducing a large scale wireless power distribution [1].

The capability of transmission power distance in WPT system is relatively short due to the effect of magnetic resonance coupling. The magnetic coupling decreases as the distance increases. Consequently, the capability of the WPT needs to be improvedto extend the distance in power transmission system. The transmission power at a longer distance can be enhanced by placing a relay coil (resonator) between the transmitter (Tx) and receiver (Rx). The resonator will receive magnetic field from the transmitter and then sending the magnetic field to the receiver. This concept is able to enhance the transmission power at a longer distance [2].

* Corresponding author: norafinie.abdrashid@gmail.com

In a previous research work [3], the researchers managed to transfer power with a little radiated loss and approximately 40% to 50% of the power source has been delivered to the load when using the resonance at both the transmitter and receiver ports.

In magnetic coupling principle, the Rx coil is not sufficient to capture the magnetic flux linkage for the power delivery process from source to the load if the coupling distance is large [4]. Therefore, by placing a relay coil between Tx and Rx coils, longer distance for the power transmission process can be achieved [5], [6]. In [7], the relay effect in the magnetic resonance coupling has been reviewed. The relay coil was designed in spiral form to reduce the volume and high Q-capacitors were placed to alter the resonant frequency. The function of the relay coil was studied in [8]. The setting of an optimal position to place the relay coil has been recommended to achieve maximum power transmission. However, the relay coil can sometimes decrease the efficiency when the coupling distance is equal between Tx and Rx coils.

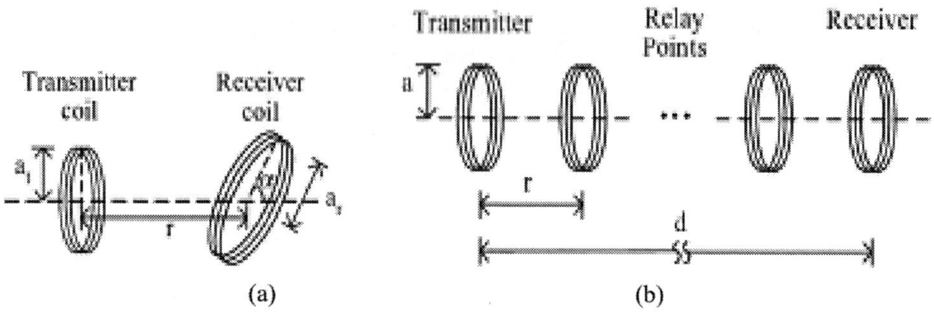

Fig. 1. (a) The typical WPT system and (b) WPT with relay coils.

In this paper, the effect of relay coil has been investigated. The relay resonator is proposed to enhance the transmission distance and improve the power efficiency to increase the performance of the WPT system. The relay effect was examined by placing one relay coil between the Tx and Rx, placing the relay coil very close to the transmitter and inserting the two relay coils between Tx and Rx. Comparison is then made between each methods that has been proposed with the typical WPT system.

2 Basic principles for mid-range WPT

2.1 The mutual coupling strength

Mutual coupling can be defined as the electromagnetic interaction between the system elements in an array. The analysis of the mutual coupling between two resonant circuits has been discussed in [1]. Mutual inductance was described as:

$$M = k\sqrt{L_1 * L_2} \quad (1)$$

where k is the coupling coefficient that is represented by the distance of the Tx and Rx coils. It can be defined as:

$$k_{xy} = \frac{M_{xy}}{\sqrt{L_{xy}}} \quad (2)$$

For the model represented by the four-coil system as depicted in Fig. 9, th e coupling coefficients were determined by k_{12}, k_{23} and k_{34}. M_{xy} is a mutual inductance between coil x and coil y.

2.2 Power transfer efficiency

The power transfer efficiency (PTE) is interrelated with the scattering S-parameter that is represented by S_{21}sincethe WPT model system contains a two-port network. The two-port network is represented as input fed at the source while the other as output fed at the load. As for the S-parameter, the PTE can be defined as $\eta = |S_{21}|^2$when the network matches at both ports [9].

PTE is an important element of the Q-factor of primary coil (Q_p) and secondary coil (Q_s). The mutual coupling of both primary and secondary coils reacts to the function alignment and distance between the coils. Thus, the power efficiency can be defined as:

$$\eta = \frac{k^2 \, Q_P \, Q_S}{1 + k^2 \, Q_P \, Q_S} \tag{3}$$

The parameter of S_{21} can be defined as the ratio of signal exiting at output port to the signal incident at input port and this can be expressed as:

$$S_{21} = 2\frac{V_L}{V_S}\sqrt{\frac{R_S}{R_L}} \tag{4}$$

S_{11} represents the reflection coefficient of port 1 and S_{21} is the transmission coefficient from port 2 to port 1 of the WPT system and expressed as:

$$S_{11}(dB) = 20\log S_{11} \tag{5}$$

$$S_{21}(dB) = 20\log S_{21} \tag{6}$$

Therefore, based on the equation in (5), the ratio of power reflection, η_{11} can be expressed as:

$$\eta_{11} = |s_{11}|^2 x100\% \tag{7}$$

and the PTE of the system is represented by η_{21}as the following:

$$\eta_{21} = |s_{21}|^2 x100\% \tag{8}$$

3 Results and discussion

Modelling and analysis of the WPT system was implemented using Computer Simulation Technology (CST) Microwave Studio simulation software. The coil for this WPT system has been constructed as shown in Fig. 2 using planar hexagon spiral shape.

DOI: 10.1051/matecconf/20167801095

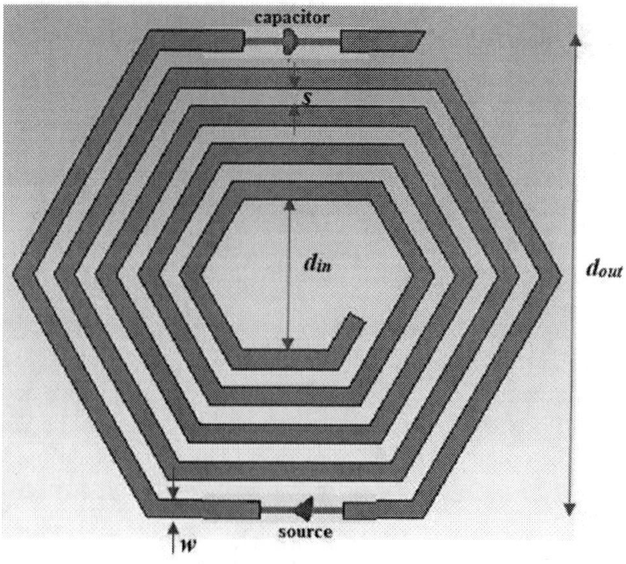

Fig. 2. Layout for LC resonator.

The coil was designed with 4.75 turns of coil, with an inner diameter (d_{in}), outer diameter (d_{out}), width (w) and turn spacing (s) of 8cm, 26cm, 1cm and 2cm respectively. The capacitance was set at 17.31pF to perform at resonant frequency of interest. Thus, making this WPT system to be well resonant at the operating frequency. The size of this model is similar for both transmitting (Tx) and receiving (Rx) inductive coils for a distance of 1 m. The design and optimization of the effect of relay coil and number of turns have been conducted for this investigation.

Fig. 3 shows the design of the Tx and Rx inductive coils at 1m distance. The system model of WPT in Fig. 3 is considered as two port network system. Therefore, the S-parameters i.e. S_{11} and S_{21} were determined to analyze this network system. S-parameter refers to the voltage travelling waves of the system which describe the relationship of the input-output between ports in the system model. S_{11} parameter can be defined as the power reflection coefficient by determining how much of the supplied power is reflected back from the system terminal, while S_{21} parameter represents the transferred power that indicates the vector of the ratio of signal exiting at output port to the incident signal at the input port.

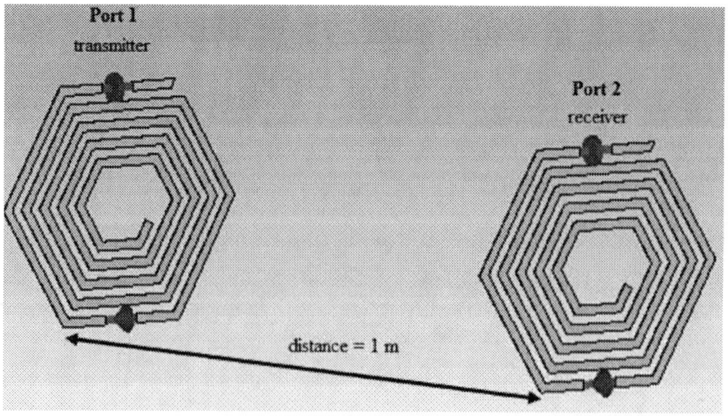

Fig. 3. Transmitter and receiver coils of atypical WPT system.

Fig. 4 shows the simulation result based on the designed system in Fig. 3. The simulation result was determined by S_{11} and S_{21}. Based on the S-parameters results in Fig. 3, the S_{11} parameter is -0.37 dB while the S_{21} parameter is -26.68 dB at the operating frequency of 38.58 MHz. The power transfer efficiency (PTE) that has been indicated with this parameter value is 4.6 % which is very low to transfer the power in the WPT system. Low efficiency will affect tthe performance of the WPT system. Therefore, such modelled system is not appropriate for the power transmission at a distance of 1m.

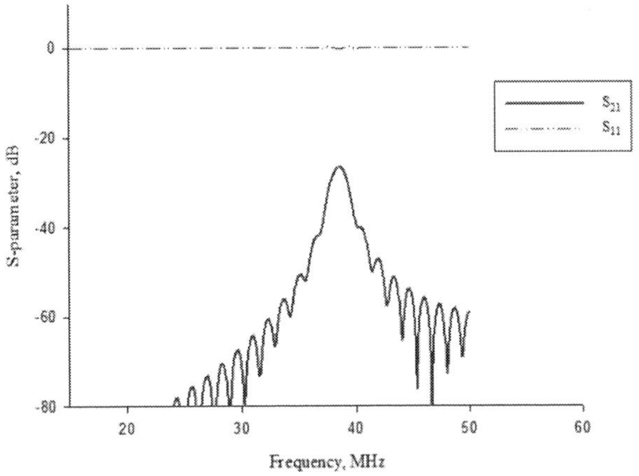

Fig. 4. Simulation results of S_{11} and S_{21} for a typical WPT system.

To overcome this issue, the relay coil was placed in between the Tx and Rx coils to be able to transfer the power for a distance of 1m. Consequently, the transmission efficiency can also be improved for a better performance. The magnetic coupling can be enhanced at a longer distance by placing the intermediate relay coil. The relay coil receives the magnetic field from the Tx and then relays the field to the Rx. Fig. 5 shows a WPT system with a resonator i.e. relay coil.

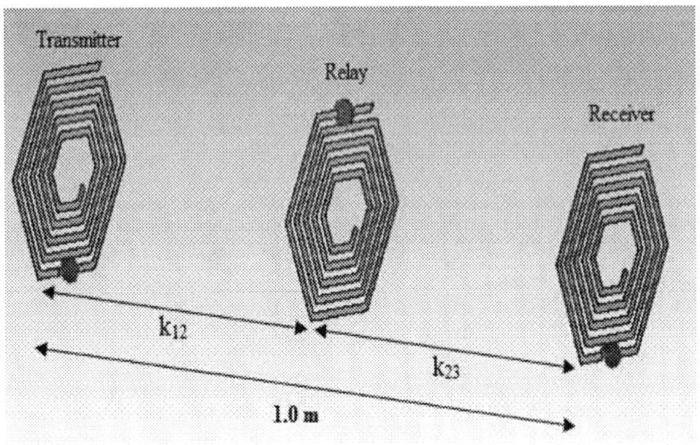

Fig. 5. WPT system with a relay coil.

The k_{12} and k_{23} represent the coupling coefficient of the system design. Coupling coefficient is referred as a dimensionless value for the coils that connects each other via magnetic field. This system was designed in a spiral form using hexagon shape. The coil in spiral shape is able to reduce the volume and a high-Q capacitor was inserted to tune the resonant frequency.

Based on the S-parameters results in Fig. 6, the S_{11} parameter is -1.90 dB while the S_{21} parameter is -13.30 dB at the operating frequency of 38.25 MHz. The PTE that has been indicated with this parameter value is 21.4 %. This result shows almost 16.8% increment of the power efficiency when inserting the relay coil compared with the typical WPT system.

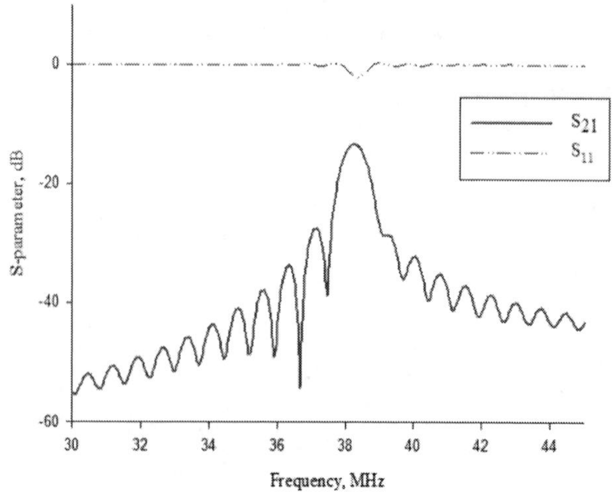

Fig. 6. Simulation results of S_{11} and S_{21} parameters with a relay coil.

To improve the power efficiency by more than 21.4 %, the relay coil is then placed very close to the transmitter as shown in Fig. 7. The experimental result is shown in Fig. 8.

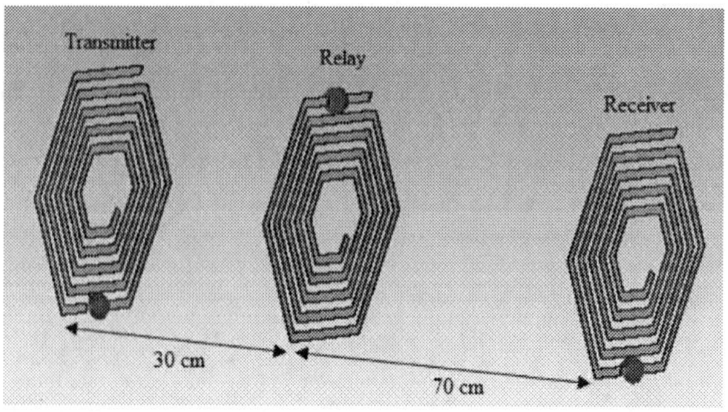

Fig. 7. Relay coil being placed very close to the transmitter.

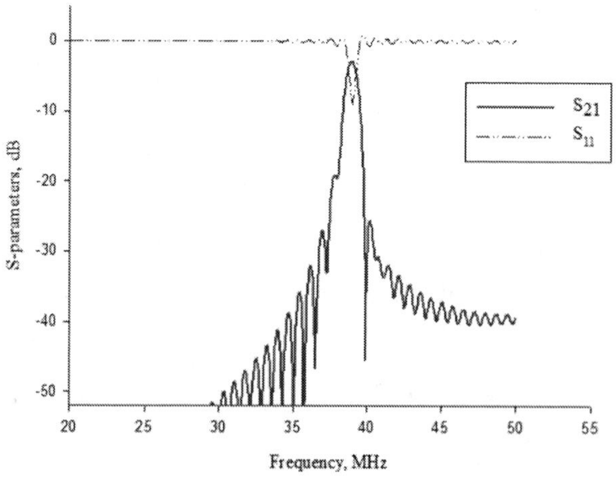

Fig. 8. Simulation results of S_{11} and S_{21} parameters when relay is biased to the transmitter.

Based on the S-parameters results in Fig. 8, the S_{11} parameter is -8.95 dB while the S_{21} parameter is -2.92 dB at the operating frequency of 38.91 MHz. The PTE that has been indicated with this parameter value is 71.4 %. This result shows almost 50 % increment of the power efficiency when placing the relay coil biased to the transmitter compared with the relay coil placed with equal distance in between the Tx and Rx of the WPT system. The relay effect in wireless power transfer can be improved with higher performance by more than 71 % improvement when inserting two relay coils between the Tx and Rx as shown in Fig. 9. The simulation result is shown in Fig. 10.

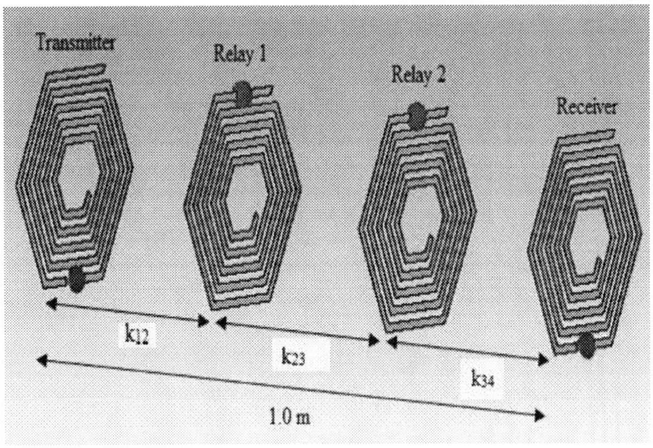

Fig. 9. The WPT system with 2 relay coils.

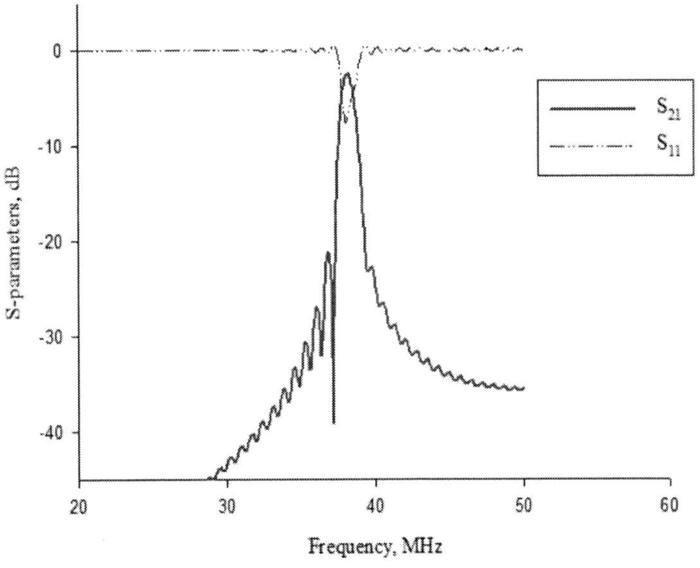

Fig. 10. Simulation results of S_{11} and S_{21} parameters when inserting 2 relay coils.

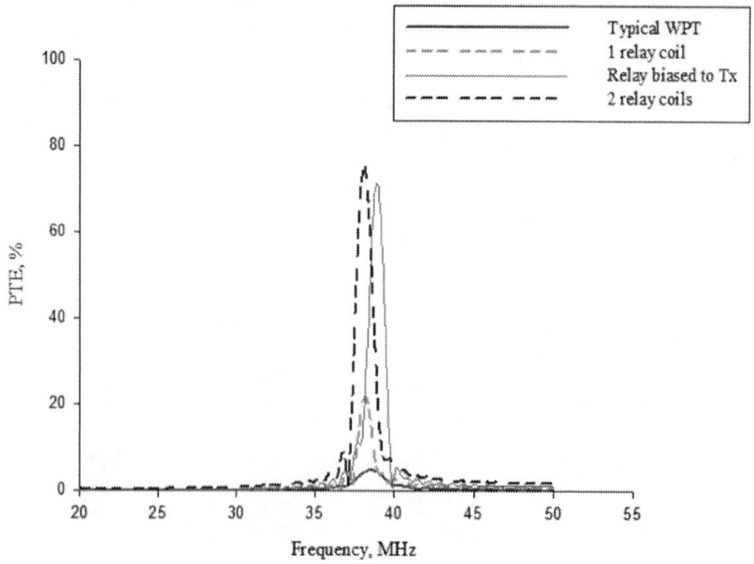

Fig. 11. Comparison of power efficiencies influenced by relay effect.

Based on the S-parameters results in Fig. 10, the S_{11} parameter is -6.98 dB while the S_{21} parameter is -2.45 dB at operating frequency of 38.15 MHz. The PTE that has been indicated with this parameter value is 75.4 %. This result shows almost 4 % increment of the power efficiency when placing the two relay coils between the Tx and Rx of the WPT system. Fig. 11 shows the comparison of the power efficiencies in WPT system that has been influenced by the relay coil effect. The power efficiency in WPT system increases when the number of relay increases.

5 Conclusions

The mid-range WPT system was designed to be applied in small device applications. The power efficiency of a typical WPT system has poor performance to transmit energy at 1 meter distance which can only reach up to 4.6 % power efficiency. Therefore, relay coil was inserted between Tx and Rx coils to extend the power transfer distance of the system. After placing the relay coil resonator, the power efficiency has shown an improvement up to 50 % increment compared to the system without relay coil. The effect of relay coil has been furtherinvestigated to increase the efficiency by positioning the relay coil close to the transmitter port. When the relay coil was placed closer to the Tx, result shows about 50 % increment of power efficiency.. In addition, by adding two relay coils between the Tx and Rx, the power efficiency was able to be increased up to 75 % at 1 meter distance of the WPT system. Thus, the relay effect is one of the solutions to enhance the distance for wireless power transmission.

References

1. S.Y.R. Hui, W. Zhong, C.K. Lee, IEEE Trans. Power Electron., **29**, 4500 (2014)
2. D. Ahn, S. Hong, IEEE Trans. Ind. Electron., **60**, 360 (2012)
3. B.L. Cannon, J.F. Hoburg, D.D. Stancil, S.C. Goldstein, IEEE Trans. Power Electron., **24**, 1819 (2009)

4. S.R. Barman, A.W. Reza, N. Kumar, M.E. Karim, A.B. Munir, Renew. Sust. Energ. Rev., **51**, 1525 (2015)
5. F. Zhang, S.A. Hackworth, W. Fu, C. Li, Z. Mao, M. Sun, IEEE Trans. Magnetics, **47**, 1478 (2011)
6. M. Kiani, U.M. Jow, M. Ghovanloo, IEEE Trans. Biomed. Circuits Syst., **5**, 579 (2011)
7. J.W. Kim, H.C. Son, K.H. Kim, Y.J. Park, IEEE Antennas Wireless Propag. Lett., **10**, 389 (2011)
8. X. Zhang, S.L. Ho, W.N. Fu, IEEE Trans. Magnetics, **48**, 4026 (2012)
9. L. Chen, S. Liu, Y.C. Zhou, T.J. Cui, IEEE Trans. Ind. Electron., **60**, 339 (2013)

A 900-2400 MHz AC-DC Rectifier Circuit for Radio Frequency Energy Harvesting

M.A. Rosli[1,*], *S.A.Z.* Murad[1], and *R.C.* Ismail[1]

[1]School of Microelectronic Engineering, Universiti Malaysia Perlis, Arau, Perlis, Malaysia

Abstract. This paper presents a 900-2400 MHz AC-DC rectifier circuit for radio frequency (RF) energy harvesting. The proposed circuit consists of a rectifier and charge pump. The multi-stage NMOS RF-DC rectifier circuit is designed to convert the AC signal to DC signal while the charge pump helps to amplify the DC amplitude. The proposed circuits are designed and simulated using CMOS 0.13-μm technology. The simulation results show that the proposed circuit able to produce 2.9 V DC voltage at 10 kΩ load with an input voltage of 150 mV.

1 Introduction

The growing of modern technologies interests people to have their own gadgets. Mobile phones, tablets, laptops and others bring a lot of advantages of their usefulness. The increment demand of wireless application devices especially mobile phones and computers shows that the important of wireless application in the whole world. The usage of these devices however required continuous supply or battery life in order to maintain the operating system. Without a doubt, the batteries have a limited life time and hazardous chemicals concern to safety matters.

Thus, the study of power energy harvesting now become an important topic in order to help the user to use the devices for a long period without worrying about the battery life. The idea is to make the devices to harvest the available surrounding power and transfer to the electrical power which can be stored in the battery. There are a lot of power energy sources lies around us such as radio frequency (RF), vibration, solar, wind and thermal [1-5]. Up to date, the present energy harvesting only generates small electrical power, which depends on technique and concept used in the system.

Among all energy harvest sources, the RF is available and easily found at any places every time. RF is an electrical oscillation, with a frequency ranging about from 3 KHz to 300 GHz and carrying an alternating current. With this concept, device like mobile phones can be used to call or access the internet connection while charging the battery at the same time.

Fig. 1 shows the block diagram of the proposed RF energy harvesting system. Basically in the system, the RF signal is received by an antenna. However, the antenna is not a part of this work, thus antenna design is neglected. The RF signal that went through antenna will be transferred to the rectifier block. In this process, the RF signal will be converted to DC

* Corresponding author: anuar_ee87@yahoo.com

voltage. The RF-DC converter should have the availability to operate at low input power levels and have a high efficiency to minimize the power loss. The DC voltage then goes to charge pump block. The charge pump is a DC-DC converter that will amplify the low input voltage to the desired output voltage. The output voltage, which should be in smooth and saturated in DC then will be stored in the capacitor. In real application, the DC voltage will be directly charge battery cell with the specific load and capacitance.

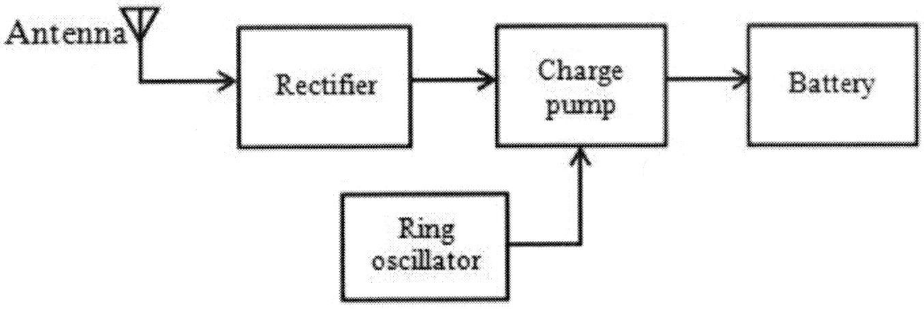

Fig. 1. Block diagram of RF energy harvesting system.

In this proposed design, the rectifier's operation is based on charging and discharging of the capacitors. Previous works mainly used the Schottky diode in order to rectify the signal [6], but the usage of diode only can be applied when the AC voltage is higher than the forward voltage. Somehow the amplitude of RF is most likely not reaching 500mV, which much lower than Schottky's forward voltage. Therefore the Schottky diode has a limitation and the study thereafter is focusing on the usage of a CMOS. Several proposed designs had been simulated before and the results proved that the usage of CMOS have much advantages compared to Schottky diode [7-13]. In this paper, selected CMOS 0.13 μm technology is proposed and implemented.

2 Methodology

2.1 Circuit implementation

2.1.1 Rectifier

A multi-stage NMOS RF-DC converter is used as the rectifier unit in the power conversion circuit. The circuit rectifies the AC input in both positive and negative cycle. As shown in Figure 2, each stage consists of NMOS transistors, a coupling capacitor and multiplying capacitor. The type of transistors and value of capacitors are selected based on the ability to convert as multi-range of RF frequency.

During the negative half cycle, drain-to-source voltage is in a negative value; making M1 is in a reverse biased condition and it is turned off. While M2 is a forward biased and it is turned on. This allows the current going through M2 and charge in C2. As the negative cycle ends, M1 becomes forward biased and turned on, allowing the current from C2 to charge through Cc. As more stages cascaded, each capacitor charged will provide a bias voltage, therefore increasing the corresponding charged current at the end until the AC signal is saturated and the ripples is minimal.

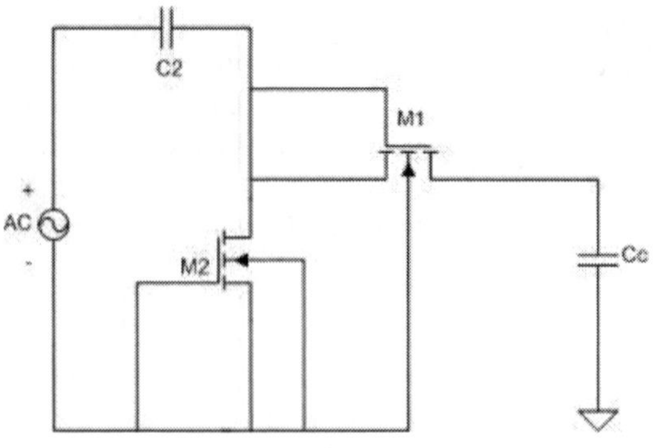

Fig. 2. NMOS RF-DC converter.

2.1.2 Charge pump

The output from rectifier is assumed to contain a small DC voltage and here come the usage of charge pump. The objective of charge pump is same as transformer, but with the external cycle pulse as a biasing. Charge pump is DC-DC converter that uses the capacitors as energy storage element to create either higher or lower voltage power source. In this design, the charge pump used the pulse generated by ring oscillator to generate the charge transfer. The pumping gain is determined by the pulse amplitude and the pumping capacitor charge. Fig. 3 shows the basic schematic of proposed charge pump used in energy harvesting system. The ring oscillator will generate two signals, θ1 and θ2 where the signals have same 50% duty cycle and frequency but with different phases. The small DC voltage from rectifier circuit will be biased by these pulses. The frequency of the pulse should be considered to minimize the external power resources. As the DC input voltage from rectifier increases, the current needed to bias the voltage will be saturated hence the efficiency of the charge pump circuit will be increased. The circuit also has to be cascaded so the output voltage can be optimized as much as possible.

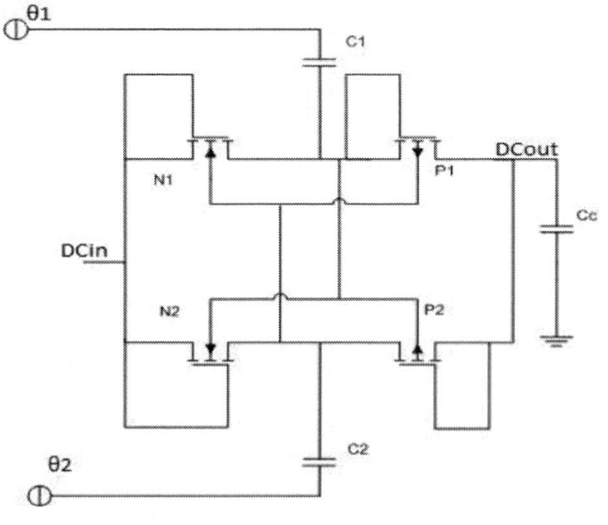

Fig. 3. Schematic of charge pump (TBC).

DOI: 10.1051/matecconf/20167801096

3 Results and discussion

Table 1 shows the data of proposed rectifier output voltage at different frequencies. The simulation is done using variation of AC input voltage, ranging from 50 mV to 500 mV with two frequencies, 900 MHz and 2400 MHz. Figure 4 shows the rectifier output voltage versus input voltage. As can be seen in Table 1 and Figure4, the rectifier circuit reacts to both input frequencies with the same output. It means the proposed rectifier design has the ability to rectify wide frequency from 900 MHz to 2400 MHz.

Table 1. Proposed rectifier output voltage with different input frequencies.

Vin (mV)	Vout (mV)	
	Freq=900 MHz	Freq=2.4 GHz
50	14.44	14.71
100	57.79	58.21
150	130.46	130.11
200	235.62	234.42
250	361.3	362.48
300	495.94	492.04
350	641.32	637.32
400	789.76	788.05
450	944.60	937.06
500	1102.4	1098.6

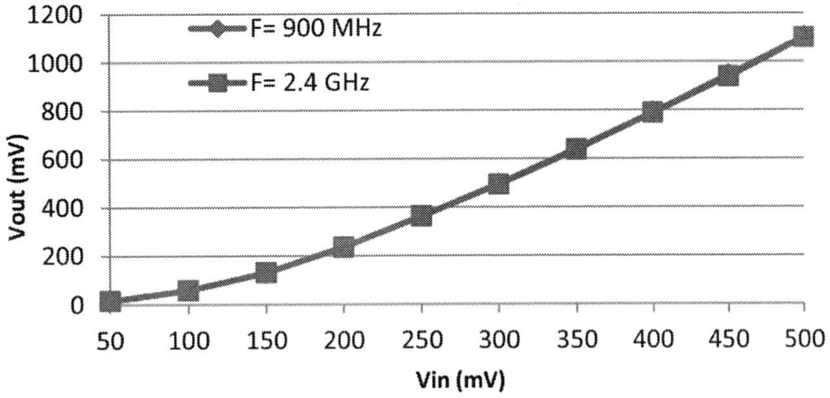

Fig. 4. Rectifier ouput voltage vs input voltage (variant input frequencies).

Table 2 shows that the proposed rectifier output voltage at different loads. The simulation is done by using different AC input voltages, ranging from 50 mV to 500 mV. The resistance loads to be tested are 10 kΩ, 50 kΩ and 100 kΩ. Table 2 and Figure5 proved that the output voltage is proportional to the value of resistance load. Fig. 5 shows the rectifier output voltage versus input voltage. It is observed that, the output voltage increased when the load resistance increases from 10 kΩ to 100 kΩ.

Table 2. Proposed rectifier output voltage with different output load.

Vin (mV)	Vout (mV)		
	Rload = 10 kΩ	**Rload = 50 kΩ**	**Rload = 100 kΩ**
50	4.50	11.63	14.71
100	17.20	45.89	58.21
150	38.07	101.98	130.11
200	67.16	181.14	234.42
250	102.40	275.98	362.48
300	140.18	380.27	492.04
350	181.18	489.60	637.32
400	225.01	605.02	788.05
450	271.97	721.40	937.06
500	317.86	844.35	1098.6

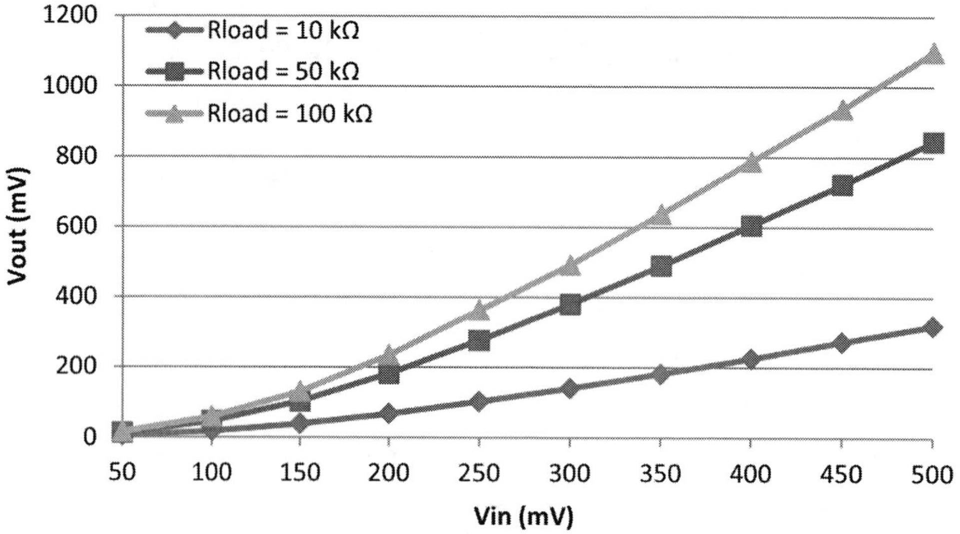

Fig. 5. Rectifier output voltage vs input voltage (variant output load).

Table 2 and Figure 5 proved that the output voltage is proportional to value of resistance load. We can see from the graph that by using Rload=100kΩ the output voltage is produced compared to Rload=10kΩ.

Table 3 shows the proposed charge pump output voltage at different loads. The data is taken using different resistance loads, 10 kΩ, 50 kΩ and 100 kΩ while the DC input voltages are 100 mV, 500 mV and 1000 mV, respectively. Charge pump output voltage versus input voltage is depicted in Fig. 6. The output voltage would be increased for higher load.

Table 3. Proposed charge pump output voltage with different load.

Vin (mV)	Vout (V)		
	Rload = 10 kΩ	**Rload = 50 kΩ**	**Rload = 100 kΩ**
100	3.39	5.91	6.44
500	3.56	6.20	6.78
1000	3.79	6.58	7.14

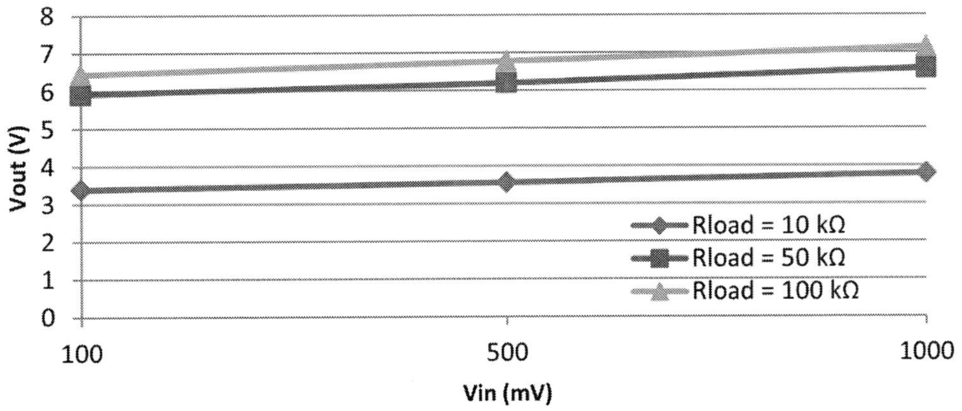

Fig. 6. Charge pump output voltage vs input voltage (variant output load).

Finally, the complete circuit is simulated with different load resistances are shown in Table 4. Figure 7 shows output voltage versus load resistance. The AC input voltage of 150 mV is used with different resistance loads at the range of 1 kΩ to 100 kΩ. As can be seen inFigure 7, the DC output voltage is proportional to the increment of the load resistance.

Table 4. Circuit output voltage vs resistance load.

Vin (mV)	Vout (V)						
	R= 1kΩ	R= 5kΩ	R= 10kΩ	R= 20kΩ	R= 50kΩ	R= 75kΩ	R= 100kΩ
150	0.544	1.991	2.914	3.715	4.385	4.558	4.648

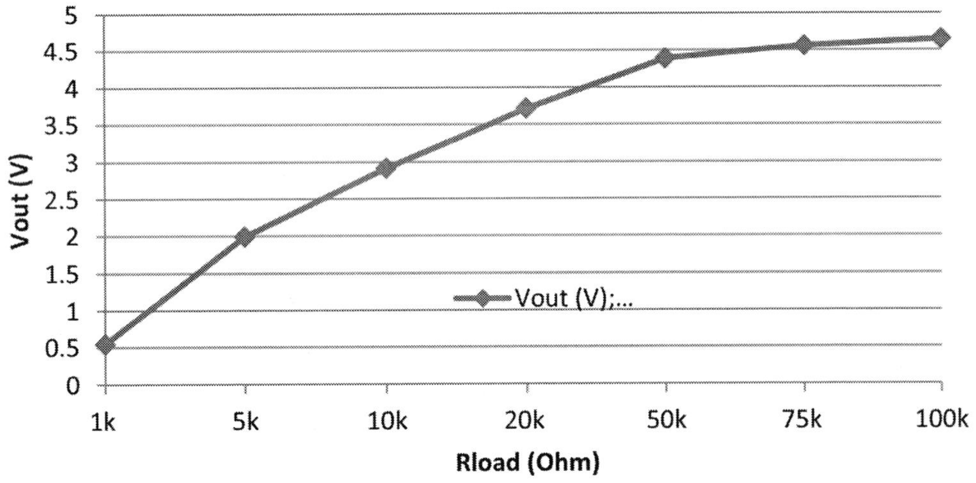

Fig. 7. Circuit output voltage vs input voltage (variant output load).

Table 5 shows the comparison between this works with previously published works. Even the resistance load can be used as higher as possible to generate higher output voltage, but it will affect the charging rate. Therefore, it can be proved that with lower load, this work can produce much higher output voltage compared to other topologies. Even [15] has the advantage in frequency range, but the range from this work also can't be compared since the objective of the study is to design a circuit for RF energy harvesting, which targeting WSN frequency that are commonly in the range from 900 MHz to 2.4 GHz.

Table 5. Comparison with other previous work.

Reference	[8]	[9]	[15]	This work
Technology (CMOS)	0.13-μm	0.18-μm	0.65-μm	0.13-μm
Topology	Cross-coupled charge pump	5-stage CMOS rectifier	3-stage Dickson charge pump	Multi-stage NMOS
Load (kΩ)	100	150	12.5	10
Input Power (dBm)	-6	-10	-10	-16.478
Frequency (MHz)	900	900	700-2400	900-2400
Output Voltage (V)	2.05	2.6	0.5	2.9

4 Conclusion

From the graphs plotted, it can be proved that the proposed AC-DC rectifier is successfully designed and simulated. The usage of multistage NMOS topology as a rectifier converts the AC signal to DC signal efficiently. The AC signals that can be converted are in wide range, from 900 MHz to 2400 MHz which is commonly used in wireless application. While the charge pump boost the DC level for higher output voltage. The proposed design AC-DC rectifier circuit is able to produce DC output voltage of 2.9 V at 10 kΩ load resistor with frequency of 900 MHz to 2400 MHz.

References

1. G. Papotto, F. Carrara, G. Palmisano, IEEE J. Solid-State Circuits, **46**, 1985 (2011)
2. E. Carlson, K. Struntz, B. Otis, IEEE J. Solid-State Circuits, **45**, 741 (2010)
3. Y. Ramadass, A. Chandrakasan, IEEE J. Solid-State Circuits, **46**, 333 (2011)
4. D. Kwon, G. Rineon-Mora, IEEE Trans. Biomed. Circuits Syst., **4**, 400 (2010)
5. M.A Abouzied, E. Sanchez Sinencio, IEEE Trans. Microw. Theory Techn., **63**, 3794 (2015)
6. D. Bouchouicha, F. Dupont, M. Latrach, L. Ventura, *Ambient RF Energy Harvesting*, Proceedings of the International Conference on Renewable Energies and Power Quality, Granada (2010)
7. Z. Hameed, K. Moez, IEEE Trans. Emerg. Sel. Topics Circuits Syst., **4**, 335 (2014)
8. K. Gudan, S. Chemishkian, J.J. Hull, *IEEE RFID Technology and Applications Conference (*2014)
9. D. Masoti, A. Costanzo, M. Filippi, A. Romani, IEEE Trans. Microw. Theory Techn., **62**, 994 (2014)
10. G.C. Lin, M.W. Lee, Y.C Hsu, *Proceeding of APMC 2012, Taiwan* (2012)
11. M. Rastmanesh, E. El Masry, *IEEE 56th International Midwest Symposium on Circuits and Systems* (2013)
12. S.S Chouhan, K. Halonen, *IEEE International Conference on Electronics, Circuits and Systems* (2014)
13. Y.J Kim, H.S. Bhamra, J. Joseph, P.P Irazoqui, IEEE Trans. Circuits Syst. II, Exp. Briefs., **62,** 1028 (2015)
14. Y. Wang, M. Sawan, *IEEE International Conference on Electronics, Circuits and Systems* (2014)
15. M.A. Rosli, S.A Zainol Murad, M.N. Md Isa, ARPN Journal of Engineering and Applied Sciences, **10**, 8326 (2015)

MATEC Web of Conferences **78**, 01097 (2016) DOI: 10.1051/matecconf/20167801097
IConGDM 2016

Adsorption Properties and Potential Applications of Bamboo Charcoal: A Review

S.S.M. Isa[1,*], *M.M.* Ramli[1], N.A.M.A. Hambali[1], *S.R.* Kasjoo[1], *M.M.* Isa[1], *N.I.M.* Nor[1], N. Khalid[1], and N. Ahmad[1]

[1]School of Microelectronic Engineering, Universiti Malaysia Perlis, Pauh Putra Campus, 02600 Arau, Perlis, Malaysia

Abstract. Bamboo charcoal was produced by pyrolysis or carbonization process with extraordinary properties such as high conductivity, large surface area and adsorption property. These properties can be improved by activation process that can be done thermally or chemically. In this paper, carbonization and activation process of bamboo, its structural and adsorption properties will be presented. Herein, the adsorption properties of bamboo charcoal that has fully utilized in solar cell as the electrode, adsorbent for water purification and electromagnetic wave absorber are reviewed.

1 Introduction

Bamboo charcoal is made up from pieces of bamboo (5 years >) which undergoes a heat treatment normally at 800-1200 °C (pyrolysis process). As early as 1486 AD, bamboo charcoal has been used in various ways due to its tremendous extraordinary properties. What makes this charcoal is so amazing is the carbonization process which creates a product with an enormous surface area to mass ratio which has high ability to attract and hold (adsorption) a wide range of materials, chemicals, minerals, radiowaves, humidity, odours and harmful substances.

Among all, absorption property of this material is well discussed and has been utilized in various applications such as household products (water filtration, humidity and odour adsorbent, air freshener, fruits and vegetables saver etc.) and health and beauty products (soap, powder, foot patch etc.). After carbonized, this material can be used as an efficient absorbent. Depends on what it adsorbs, the carbonized bamboo also can become saturated and act as the fertilizer. In another hand, the adsorbed impurities can be burned off without destroying the adsorbent property, which means it can be reused. The adsorption properties of bamboo charcoal can be improved and become a perfect absorbent when this material is further activated either thermally or chemically. No longer limited to the previous applications, recently, the bamboo charcoal has been applied as the working electrode composite in dye sensitized solar cell to increase device efficiency [1], as the absorbent for water purification [2] and microwave absorber in antenna [3].

* Corresponding author: sitisalwa@unimap.edu.my

Become the alternative to activated carbon, carbon nanotubes, graphene and other carbon based materials, this element is the most notable among readily and renewable biomass resources because bamboo is the fastest growing plants in Earth [4]. In this study, bamboo charcoal production, structural and adsorption properties specifically in dye sensitized solar cell, water purification and microwave absorber are reviewed.

2 Carbonization of camboo charcoal

Generally, Moso bamboo (*Phyllostachys pubenscens*) is used as the starting material in the production. This bamboo is cut into small pieces and washed by boiling distilled water for some hours and dried at temperature nearly 110 °C to remove moisture [5]. Then, the carbonization process is held in the oven normally at N_2 flow [2] at temperature over 800-1200 °C at longer period hours. However, bamboo charcoal also can be produced at lower temperature (500-900 °C) with different quality which has been presented in [2]. The fresh bamboo charcoal produced from this process is the raw bamboo charcoal. This material already has an extraordinary micro structure with high absorptive capacity.

Fig. 1. Bamboo (a) Moso, (b) charcoal and (c) powder.

3 Activation of bamboo charcoal

Activation of carbonized bamboo charcoal is important to improve the original structure and increase the adsorption properties. It can be mixed with CO_2 [2,6], HNO_3 [2,7], NH_3 [2,8] and many more before it is heated again at certain temperature (500-1200 °C) for some periods. The annealed bamboo charcoal need to be cooled to a temperature between 170 to 240 °C before it serves as activated bamboo charcoal. The activation step is important to increase the volume of the material, breaks some bonds of the turbostratic carbon structures that form surface functional groups and removes non-crystallized carbons from the bamboo charcoals.

4 Structure and adsorption properties

Figure 2 shows the structure of bamboo charcoal before and after activation. As shown in the figures, vascular bundles of the bamboo charcoal are arranged in various sizes. The pore characteristics of this material can be macropore (> 50 nm), mesopore (2-50 nm) or micropore (< 2nm). This pore characteristic and the resistant coefficient of this material are affected by the temperature of carbonization and activation processes. As the temperature increases, the material will change from insulator to semiconductor or even to conductor, while the pore size is getting bigger. Normal and activated bamboo charcoal have the same ability to absorb chemicals and substances from air and water, however the normal bamboo charcoal has poor holding capacity compared to the activated forms. Other than surface area to mass ratio, the excellent adsorption property is contributed by the van der Waals force [11].

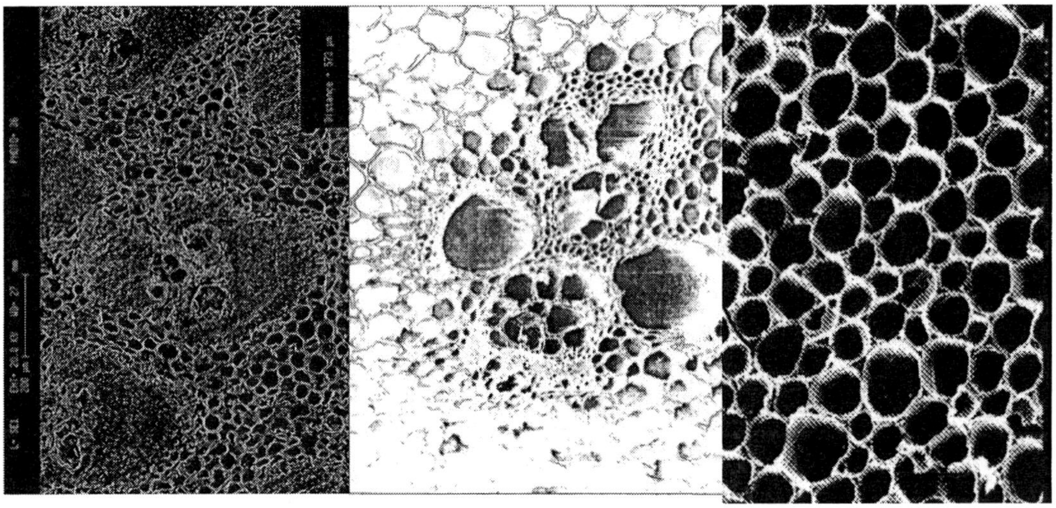

Fig. 2. Bamboo charcoal structure (a) after carbonization, (b) after activation [9] and (c) at higher magnification [10].

5 As electrodes in dye sensitized solar cell

Conventionally in Dye Sensitized Solar Cell (DSSC), carbon based material and titanium dioxide (TiO_2) were employed as counter and working electrodes. However, TiO_2 working electrodecan only absorb ultraviolet and several approaches including the use of dopants such as nitrogen [12], hydrogen [13], metal [14] and others to introduce disorder in the surface layer and narrow the band gap of TiO_2 which can improve the photocatalytic activities were done. The introduction of disorder and dopant at the surface would enhance visible and infrared absorption [13]. Several initiatives using carbonaceous materials such as graphene [15], carbon nanotubes [16] and hybrid graphene-carbon nanotubes [17] have been performed in order to increase device efficiency due to their good conductivity and high effective surface area. These materials can help the electron excitation and enhance the light harvest efficiency of the photoanode and working electrode when hybridizing this material with TiO_2.

Aside from the stated carbon based materials, mesoporous carbon such as bamboo charcoal also potentials to be used as the electrodes in DSSC. Carbonized bamboo charcoal which undergoes heat treatment between 700 to 900 °C has the resistance

DOI: 10.1051/matecconf/20167801097

coefficient between 0.12 to 11.9 Ωm which perfect as the electrode [10]. In other work, Chen et al. [13] has proved that modified TiO_2 potentials to increase solar absorption by altering the surface structure and its energy band. The black TiO_2 (refer Fig. 3(a)) can be resolved into two peaks compared to typical white TiO_2. The changes in reflectance and absorbance spectra at 806.8 nm suggest that the optical gap of the black TiO_2 was narrowed by intraband transition. Further analysis with XPS (Fig. 3(b)) also has been performed to show the changes of density of states between typical TiO_2 and modified TiO_2 and the results were illustrated in Fig. 3(c).

Recently, the best performance of bamboo charcoal in DSSC is 5.4% [1] attributed to the porous of bamboo charcoal embedded in the nano-crystalline TiO_2 which facilitates the capability of dye absorption and the generation of electrons during exposure to the light. However, the development of DSSC utilizing bamboo charcoal still in preliminary stage and more improvements need to be done.

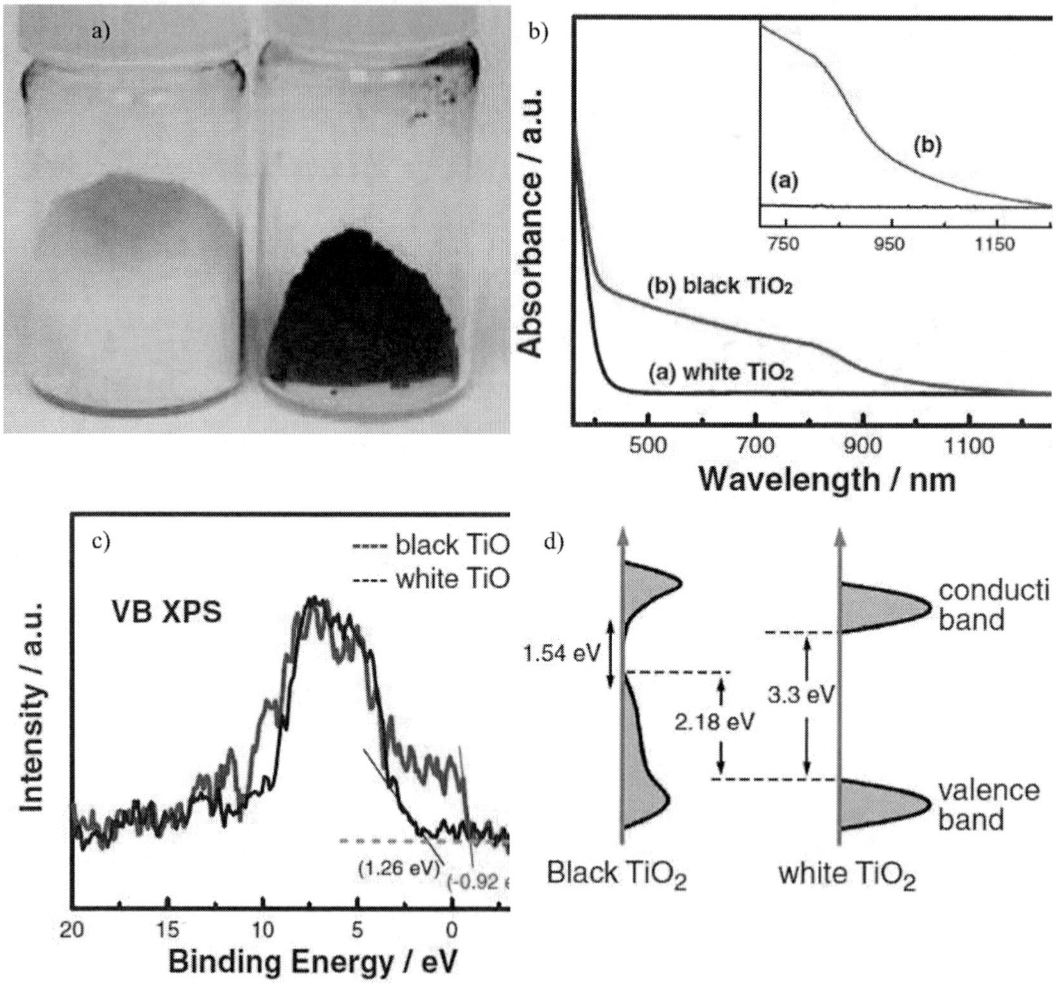

Fig. 3. Comparison between typical TiO_2 and modified (a) powder forms, (b) Spectral absorbance, (c) XPS spectra and (d) Illustration of density of states [13].

6 As an absorbent in water purification

Many industries produce volatile compounds that can pollute the environment and harm the human health. The common compounds include hydrocarbon compounds such as O_2, N_2, sulphur, halogen and others. Generally, adsorbents were employed to remove the pollutants from wastewater or waste gas. The adsorption capacity of the adsorbents comes from the affinity of molecules, and is classified into physical adsorption, chemical adsorption and catalytic action. Examples of adsorbent are activated clays, aluminium oxide based materials, silica gel, ion exchangers, magnesia based materials, activated carbon and others. Among all, the activated carbon is the most common used absorbent as it is cheaper, easier to be used and recyclable. It has superior performance in dealing with organic and toxic waste such as chrome, ozone, pesticide and others. In that matter, bamboo charcoal is a potential cheaper alternative absorbent among carbon based material such as carbon nanotubes [18, 19], graphene [19], graphene oxide [19] and others which offers a larger specific area and greater pore volume which can perform a greater degree of adsorption.

It has been shown that, the adsorption of bamboo charcoal is affected by the carbonization temperature as shown in Fig. 4(a). As the temperature increased, the pore is greater and the adsorption rate is increased. The adsorption of substances in water also depends on the way the bamboo charcoal is activated. It has been investigated that bamboo charcoal activated with HNO_3 (refer Fig. 4(b)) showed better surface area with the introduction of acidic functional groups without destruction of pore structure compared to the charcoal activated with CO_2 (refer Fig. 4(c)) which much depends on the activated temperature [2]. Author also proved that charcoal activated by NH_3 not present a promising yield although the mesopore was increased with the applied temperature. However this HNO_3 activation process posses more functional groups and resulted higher water vapour adsorption capacity at low humidity area.

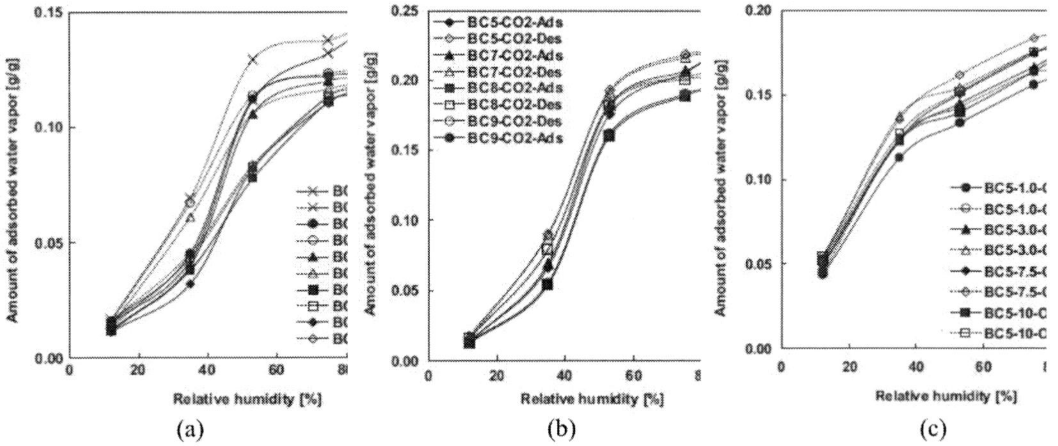

Fig. 4. Water vapour adsorption-desorption of (a) bamboo charcoal, (b) activated CO_2 and (c) activated HNO_3 [2].

7 As an electromagnetic waves absorber

Due to the extensive effort towards gigahertz (Ghz) electronic system and telecommunication devices, electromagnetic (EM) interference problems have raised vastly. EM is defined as conducted or radiated electromagnetic signals which emitted by electrical circuits which under operation, perturb proper operation of surrounding equipments which also may cause radioactive damage to living/biological species [3]. Thus, an effective shielding material in wide variety applications is needed. For this purposes, Radar Adsorbing Material (RAM) has been applied from commercial and electronic instrumentations to antenna system and military electronic devices which can be produced under different forms such as conductive paints or rubber loaded with ferrite or carbon which generally used for stealth military planes; or in foam or multilayered topologies that commonly used as liners for all enclosures in which reflection waves has to be minimized such as an anechoic chambers. But, the existing adsorption solution is still facing very low power loss.

Recently, the focus on the research on designing shielding material which works by adsorption was boosted all over the world. Most of the proposed material was based on polymeric materials in order to take the advantages of their lightness, low cost, easy shaping and others. There are three main strategies on producing the adsorbing material: 1) dispersion of metallic fillers, fibers or nanoparticles within polymer matrix which can increase the interaction with the electromagnetic radiation; 2) using conductive polymers such polyaniline (PANI) and polydimethylsiloxane (PDMS) and 3) dispersion carbon based electrically conductive polymer matrices. Based on that, carbon material like carbon black [20], carbon nanotubes [21] and graphene [22 - 24] have been used as the polymeric filler because of their extraordinary properties. However the production of these carbon materials is expensive and some of the materials are facing impedance match mechanism that may harmful to the EM adsorption properties [24]. Besides, the carbon based dielectric absorbers have narrow absorbing frequency bandwidth [25]. Therefore, a development of the EM absorbing material with low density, high thermal stability, high absorption properties, broad microwave frequency bandwidth as well as cheaper solution is highly demanded. Based on that demands, bamboo charcoal can be a magnificent alternative as this material has similar high conductivity, large surface area and extremely higher adsorption properties.

8 Conclusion

In this paper, the properties and potential applications are reviewed. The magnificent adsorption property of this material becomes the main highlight which it can be fully utilized in various potential applications such as electrode in solar cell, absorbent for water treatment and wave absorber in antenna. This adsorption property can be influenced by the carbonization temperature, the activation process and activation precursor as the pore is getting larger and the surface area is improved.

References

1. C.S. Chou, C.Y. Chen, S.H. Lin, W.H. Lu, P. Wu, Adv. Powder Technol., **26**, 711 (2015)
2. R. Wang, Y. Amano, M. Machida, J. Anal. Appl. Pyrol., **104**, 667 (2013)
3. J.M. Thomassin, C. Jerome, T. Pardoen, C. Bailly, I. Huynen, C. Detrembleur, Mat. Sci. Eng. R., **74**, 211 (2013)
4. Y.S.H. Lin, L.Y. Hsu, C.S. Chou, J.W. Jhang, P. Wu, J. Anal. Appl. Pyrol., **107**, 9 (2014)
5. M. Sakuma, Y. Amano, M. Machida, Tanso, **252**, 1 (2012)
6. T. Horikawa, Y. Kitakaze, T. Sekida, J. Hayashi, M. Katoh, Bioresource Technol., **101**, 3964 (2010)

7. P. Chingombe, B. Saha, R.J. Wakeman, Carbon, **43**, 3132 (2005)
8. R. Okayama, M.A.A. Zaini, M. Aikawa, M. Machida, H. Tatsumoto, J. Environ. Chem., **18**, 533 (2008)
9. C.H. Wang, United States Patent Application No. 11/147,371 (2005)
10. http://www.faqs.org/patents/app/20100307066
11. http://www.c60bamboo.com
12. X. Chen, S.S. Mao, Chem. Rev., **107**, 2891 (2007)
13. X. Chen, L. Liu, P.Y. Yu, S.S. Mao, Science, **331**, 746 (2011)
14. M.R. Hoffmann, S.T. Martin, W. Choi, D.W. Bahnemann, Chem. Rev., **95**, 69 (1995)
15. A.C.M.S. Esteban, E.P. Enriquez, Sol. Energy, **98**, 392 (2013)
16. Z. Yang, L. Li, H. Lin, Y. Luo, R. He, L. Qiu, J. Ren, H. Peng, Chem. Phys. Lett., (2012)
17. L.H. Chang, C.K. Hsieh, M.C. Hsiao, J.C. Chiang, P.I. Liu, K.K. Ho, C.C.M. Ma, M.Y. Yen, M.C. Tsai, C.H. Tsai, J. Power Sources, **222**, 518 (2013)
18. X. Liu, M. Wang, S. Zhang, B. Pan, J. Environ. Sci., **25**, 1263 (2013)
19. S.C. Smith, F. Ahmed, K.M. Gutierrez, D.F. Rodrigues, Chem. Eng. J., **240**, 147 (2014)
20. S. Vinayasree, M.A. Soloman, V. Sunny, P. Mohanan, P. Kurian, M.R. Anantharaman, Compos. Sci. Technol., **82**, 69 (2013)
21. Z. Liu, G. Bai, Y. Huang, F. Li, Y. Ma, T. Guo, X. He, X. Lin, H. Gao, Y. Chen, J. Phys. Chem. C., **111**, 13696 (2007)
22. V.K. Singh, A. Shukla, M.K. Patra, L. Saini, R.K. Jani, S.R. Vadera, N. Kumar, Carbon, **50**, 2202 (2012)
23. B. Wu, H.M. Tuncer, A. Katsounaros, W. Wu, M.T. Cole, K. Ying, L. Zhang, W.I. Milne, Y. Hao, Carbon, **77**, 814 (2014)
24. Y. Chen, Z. Lei, H. Wu, C. Zhu, P. Gao, Q. Ouyang, L.H. Qi, Q. Qin, Mater. Res. Bull., **48**, 3362 (2013)
25. J. Zeng, J. Xu, J. Alloy. Compd., **493**, 39 (2010)

MATEC Web of Conferences **78**, 01098 (2016) DOI: 10.1051/matecconf/20167801098
IConGDM 2016

Synthesis and Characterization of Stabilized Zirconia Toughened Alumina (ZTA) Doped Niobium Pentoxide (Nb2O5)

B. Johar[1,*],*Z. Zul.* Azhar[1], and *Z.A.* Ahmad[2]

[1]School of Materials Engineering, Universiti Malaysia Perlis, 02600 Kangar, Perlis, Malaysia
[2]School of Materials and Mineral Resources Engineering, Universiti Sains Malaysia, Engineering Campus, 14300 Nibong Tebal, Penang, Malaysia

Abstract. The main target of this research is to study the mechanical properties of the Nb_2O_5 doped with alumina-zirconia composites. Pellets form samples of ZTA (5YSZ) doped niobium pentoxide in varied amount of 0 wt%, 0.3 wt%, 0.5 wt% and 0.7 wt% were prepared by solid-state conventional method with respective 4:1 ratio or alumina and 5YSZ. The effect of different dopant addition on the morphology, phase transformation and hardness was studied. Each mixture of the compositions were calcined at 1000°C followed by pressing at 220 MPa and then they were sintered at 1650°C with heating rate 10°C min^{-1} and holding for 2 hours in air. Phase transformation, phase quantification and crystal structure analysis was revealed by using X-ray diffraction data with the help of X-Pert HighScore Plus software using Rietveld Refinement Method while microstructure was observed by scanning electron microscopy. The hardness of the composite is determined by using Vickers hardness test. It found that with 0.5 wt% niobium pentoxide addition would give a great effect on the microstructure distribution, densification and mechanical properties of the ZTA-doped sample.

1 Introduction

Zirconia toughened ceramics have proven to give abrasion, wear resistance and they can be last longer up to 10 time compared to steel tool parts [1]. Zirconia toughened alumina is the composite that involves alumina matrix and zirconia particles as reinforcement where they are enhancing each other by embedded zirconia particle into alumina matrix [2-4]. This kind of ceramics composite includes the combination of hardness provided by alumina with fracture toughness and strength from zirconia particles.

Basically, zirconia wills presence in 3 crystallographic forms, which are monoclinic, tetragonal and cubic phases [2]. Zirconia is in the monoclinic phase at ambient temperature and in tetragonal phase when heat is applied up to 1700 °C. Furthermore, at about 2370 °C, it is changes into a cubic phase [5]. The transformation of tetragonal to monoclinic phase will resulted in toughening mechanism of alumina zirconia composite materials and the system can be stabilized with the addition of some oxides materials in order to maintain the

* Corresponding author: banjuraizah@unimap.edu.my

tetragonal phase in the ambient temperature [2,6]. This is because zirconia in tetragonal phase shows excellent mechanical properties. In addition, the transformation of tetragonal phase into monoclinic phase during cooling process will cause the volume increment that may lead to failure [5,6]. Failures also may cause by the presence of porosity, inhomogeneity in phase, grain size, micro-crack and any other failure possibilities which lowering the mechanical properties of the ceramics materials [7].

Al_2O_3 matrix reinforced with ZrO_2 particles that are called as ZTA. According to Hannink et al., in their study showed that addition of the niobium pentoxide contents affect some properties such as microstructure, hardness, fracture toughness and densification of the composites [8]. Therefore, the main purpose of this research is to investigate the effect of the ZTA doped with niobium oxide which focusing on the hardness properties of alumina and zirconia ceramics materials.

Researchers have studied that the densification and mechanical properties of samples can be enhanced by the addition of Nb_2O_5 [6,12]. The enhancement of ceramics composite material can be accomplished by using different type of manipulated dopant contents to improve their mechanical properties such as hardness and fracture toughness that prompted to various applications. According to Hassan et al. the addition of 0.2 - 0.6 vol% of Nb_2O_5 showed an increment in bending strength, hardness, relative density and reduce in porosity of the ceramics materials [2]. Addition of zirconia particles in the alumina matrix will result in the both high fracture toughness and hardness compared to pure alumina and zirconia ceramics materials [15]. All the factors described above lead to this research scope. In this research, 5 mol% YSZ and alumina with 1:4 ratio varies with the addition amount of 0 wt% to 0.7 wt% of Nb_2O_5 were prepared. The target of this research is to investigate the stabilized zirconia toughened alumina (ZTA) doped with different weight percent of niobium oxide that will affect the hardness, morphology and phase transformation of the specimen.

2 Materials and methods

2.1 Materials

The main raw materials used in the present research are alumina, Al_2O_3 (MAJU SaintifikSdn. Bhd., 99% purity), 5YSZ, ZrO_2 + 5 mol% Y_2O_3 (MAJU SaintifikSdn. Bhd., purity of 96%) and niobium pentoxide, Nb_2O_5 (Sigma-Adrich Chemistry, 99.9% purity).

2.2 Methods of preparation and characterization

The initial mixtures for 5YSZ were 1:4 ratio of 20 wt% YSZ and 80 wt% Al_2O_3 respectively. These compositions was mixed with an addition of the varied amount of niobium pentoxide (0 wt%, 0.3 wt%, 0.5 wt% and 0.7 wt%). The weighted powders were mixed by using a conventional mortar and pestle for about 30 minutes on each composition and ethanol was added to facilitate the mixing process. The homogenized powder werecalcined at 1000°C for 2 hours with rate of 10°C min^{-1} followed by crushing the dried samples using mortar and pestle. Each of the compositions was hydraulically pressed at 220 MPa into a 20 mm diameter mould cavity and 3 mm thickness pellet. All the sintered pellets were sintered at 1650°C for 2 hours at heating rate of 10°C min^{-1}.

Differential thermal analysis (DTA) was used to detect the decomposition reaction especially on the heat absorption while thermal gravimetric analysis (TGA) is to measure the weight change in the tested samples as function of time or temperature. DTA-TGA was performed at 1500°C with heating rate of 10°C min^{-1} by using Pelkin Elmer

DOI: 10.1051/matecconf/20167801098

thermoanalyzer. The different phases of samples were developed by using X-ray diffraction analysis (XRD) with Cu Kα monochromatic radiation (XRD 6000, SHIMADZU). X-ray diffraction data for qualitative and Rietveld analysis were collected on the 2θ ranges from 5° to 90° with scan axis of 1°/min and a step scan of 7s. The X-ray tube was operated at 40 kV and 50 mA.

Vickers hardness values were taken by using Vickers Hardness Tester model of Mitutoyo HM-210/810-403E Japan by indentions technique at 1 kg load for 15 seconds on the sintered body. Vickers hardness values were measured by averaging 5 different readings on each of the sintered body. Bulk density of the sintered samples was calculated by using Archimedes principle and Pycnometer (AccuPcy 1330) with helped of Helium Pcynometer. All samples were coated with platinum (pt) by using sputter coater (JFC1600). Scanning electron microscopy was carried out to observe the morphology of the sintered ceramics samples by using SEM (JSM-6460LA, JEOL) and assisted with an energy dispersive X-ray (EDX).

3 Results and discussion

3.1 DTA-TGA

Thermal analysis appraises the behaviour of the sample under controlled temperature changes. TGA analysis was performed to understand the decomposition and measure the weight changes in ceramics materials over temperature and time. It is important to understand the thermal treatment and time, as it will effect or influence the ceramics composite materials such as degree of particle agglomeration and second phase size distribution.

Fig. 1 shows the behaviour for all ZTA samples, which can be seen on the TG curve patterns. The TG curve increase, which indicate that the weight gain from the beginning until the end of the tests done. A linear graph of weight gain respective to temperature trend can be seen at room temperature up to temperature of 580°C. This shown that the weight increases gradually as temperature increases. The weight gained in the TG curve may be due to the oxidation of the ZTA samples in the air atmosphere at the beginning of the analyses. Increment of ZTA samples weight was ZTA, ZTA 0.3, ZTA 0.5 and ZTA 0.7 with 0.14 mg, 0.15 mg, and 0.165 mg. A slightly weight gain rate shown at temperature above 1420 °C for all ceramics powder may be due to the oxidation reaction.

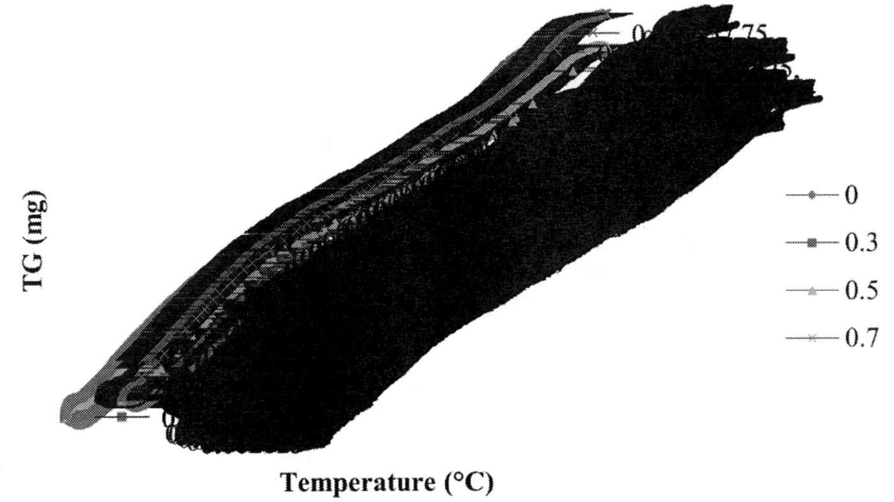

Fig. 1. TGA curve for different ZTA composition.

Fig. 2 shows DTA curve shift from the exothermic peak to the endothermic peak for all ZTA compositions at about 1100 °C endothermic peak onset. A small difference between undoped and doped ZTA samples, which is exothermic peak at temperature of 185°C while the doped compositions, is at 210°C. This shows that the ZTA powder for undoped ZTA undergo crystallization earlier compared to the doped sample. The shifted from exothermic to endothermic took place at temperature around 1100 °C and the endothermic peak for ZTA and ZTA0.7 was 1210°C while broaden endothermic peak for ZTA0.3 and ZTA0.5 powder was at temperature of 1280°C. A small decrement in weight which about 0.0952 mg to 0.0951 mg at temperature about 1030°C. This indicates that this is the onset of the endothermic for the undoped ZTA samples to melts. The same trend could be seen at temperature range of 950°C to 960°C in which the weight loss from 0.101 mg to 0.100 mg for ZTA0.3 ceramics sample. The change of the exothermic to endothermic is due to the high temperature.

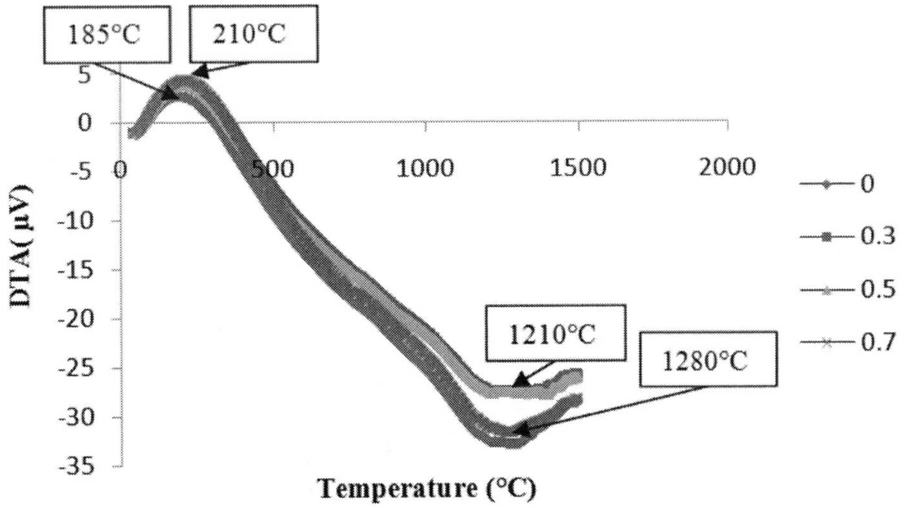

Fig. 2. DTA curve for different ZTA composition.

 DOI: 10.1051/matecconf/20167801098

3.2 XRD

X-ray diffraction (XRD) patterns of undoped and doped ZTA sintered body at 1650°C is shown in Fig. 2 and it can be observed that alumina phases present as dominant phase compared to tetragonal zirconia (t-ZrO_2) and monoclinic zirconia (m-ZrO_2). This is due to the 4:1 of alumina and YSZ used in each of the composition.

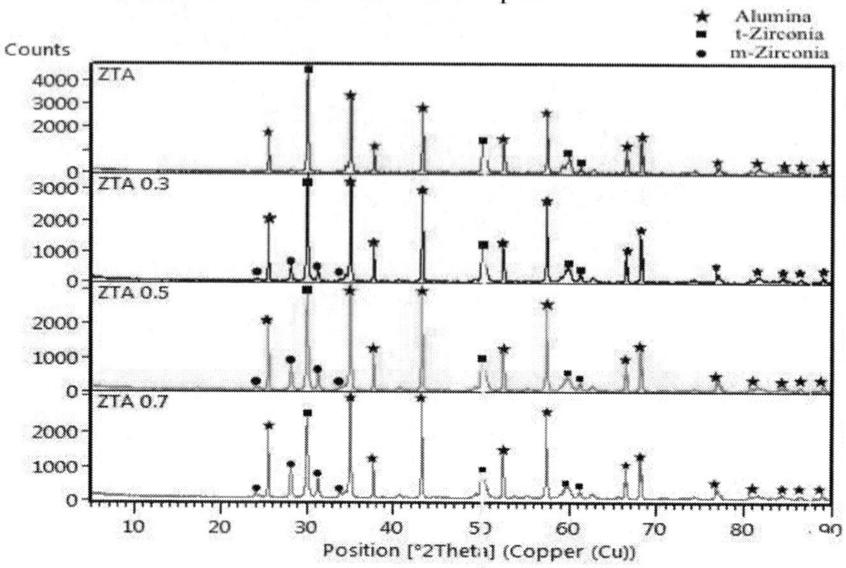

Fig. 3. XRD pattern for undoped and doped ZTA samples at 1650°C.

For the pure ZTA sample, alumina present as major phase followed by t-ZrO_2 phase while for the niobium oxide doped ZTA samples, it can be seen that at all of the doped ZTA sample, alumina phases present beside the t-ZrO_2 (strongest peak) and both of the phases are surrounded by the traces peak of m-ZrO_2 at ranges of 20 to 40 position. Hassan et al. studied on the pentavalent oxide materials doped ZTA stated that alumina phases existed as the major phases with the tetragonal phase zirconia and a minor amount of monoclinic zirconia phases in all samples [2]. In addition, Shen et al. also revealed that m-ZrO_2 formed with major t-ZrO_2 by using pentavalent oxide (Ta^+) as dopant [10].

Besides, it is also have few noticeable changes on m-ZrO_2 peak intensity are increasing from pure ZTA up to 0.7 wt% Nb_2O_5-doped ZTA. The t-ZrO_2 phases present from pure ZTA to 0.7 wt% dopant added causing the tetragonal phase of zirconia peak intensity slightly decreasing. The same agreement pattern shown by pentavalent oxide doped ZTA in the XRD result of Hassan et al.

3.3 SEM

Fig. 4 (a), (b), (c) and (d) represents the morphology of SEM images for undoped and doped ZTA samples after being sintered at 1650°C for 2h respectively. The figures show alumina grain with dissimilar grain sizes. Zirconia grains commonly appear in rounded and small in grain sizes which can be found in the intergranular position and the grain growth of the alumina grain is mainly due to the grain coalescence or fused that is correlated with the grain boundary migration [2, 11]. Fang also stated that coalescence is the other grain growth mechanism, which related to the rapid grain growth [11].

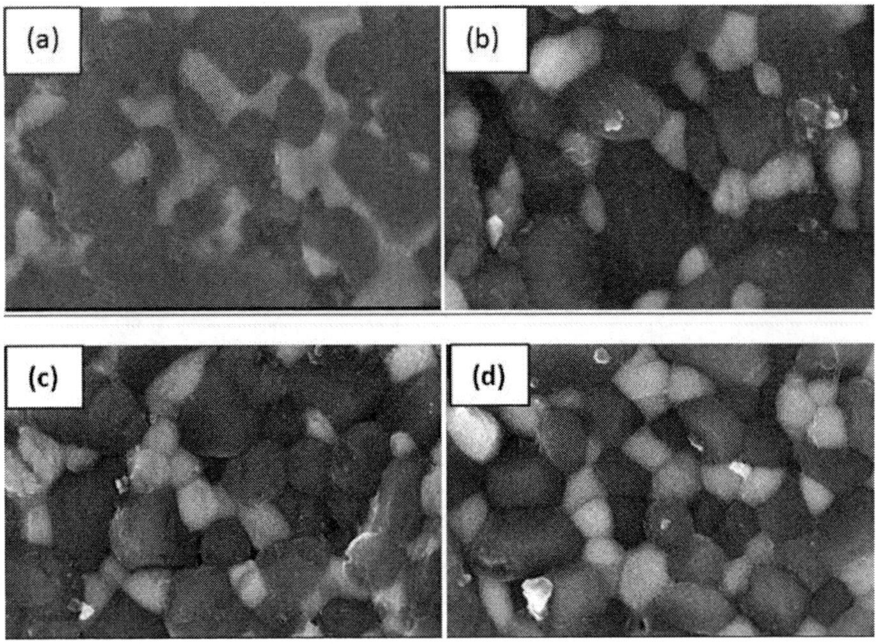

Fig. 4. SEM micrographs of undoped ZTA samples (a), ZTA0.3 (b), ZTA0.5 (c) and ZTA0.7 (d) at 10000x magnification.

In Fig. 4 (b), (c) and (d), the microstructure of doped ZTA samples distribution shows the same allocation as the undoped ZTA sample even with presence of niobium pentoxide particles. This is because the addition of Nb_2O_5 does not change the microstructure that is only up to 6 wt% niobium pentoxide addition [12,13]. Lesser porosity with good particle distribution and sizes shown by ZTA0.5 and ZTA0.7 compared to pure ZTA sample and 0.3 wt% Nb_2O_5 dopant. This behaviour is may be due to the The small dopant agglomeration of Nb_2O_5 between the grain boundaries in the doped ZTA samples which may cause the particle shape of the ZTA0.3 and ZTA0.7 tend to be rounded shapes. It also due to the increasing amount of nobiumpentoxide as dopant.

3.4 Density

The bulk density for each of the compositions is tabulated on the Fig. 5. a correlation on the bulk density has improved by addition of 0wt% up to 0.5 wt% of niobium pentoxide and decreases with the addition of 0.7 wt% Nb_2O_5. The graph has illustrated the relation between the bulk density and the Vickers hardness gained. This trend resulted from the particle size of ZTA0.7 which is higher compared to the other composition after the synthesis method. Somton Et al stated that smaller particle size resulted to the low amount of porosity and higher apparent density [14]. Besides, broaden distribution of size will cause the densification with lesser shrinkage compared to the less range size distribution [15,16].

 DOI: 10.1051/matecconf/20167801098

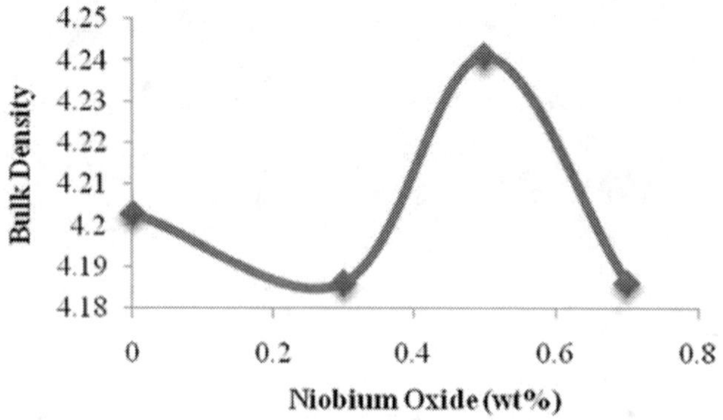

Fig. 5. Bulk Density of ZTA doped Nb_2O_5 composition.

3.5 Vickers hardness

Fig. 6 shows the resulted of the Vickers hardness, which plotted as function of dopant contents. Fig. 6 revealed that the Vickers hardness of ZTA undoped and doped niobium oxide increases with the dopant amount. But, this trend only could be seen up to 0.5 wt% of Nb_2O_5 content. The decreasing pattern presented from 0.5 wt% to 0.7 wt% addition of niobium oxide.

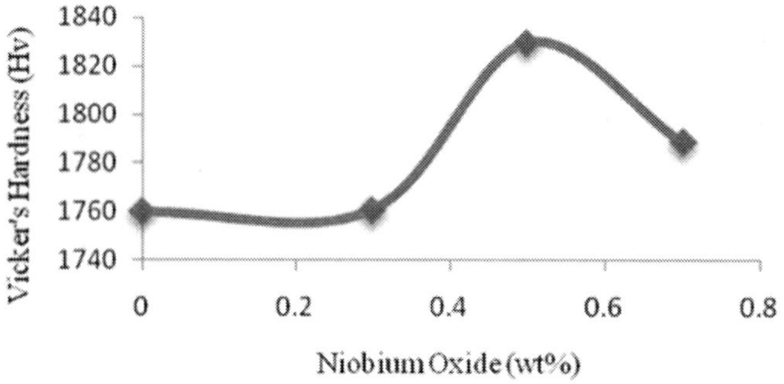

Fig. 6. Hardness obtained at different composition of ZTA.

The Vickers hardness of ZTA of undoped and doped niobium oxide is gradually increases from 1759.8 Hv (ZTA) to the highest value of 1828.62 Hv (ZTA0.5). This resulted of about 3.91% improvement compared to the undoped sample. The result also showed the slightly increased in hardness value from ZTA to ZTA0.3, which showed an increment of hardness about 0.8 Hv. While ZTA0.7, recorded a drop in hardness value from 1828.62 Hv (ZTA0.5) to 1788.6 Hv (ZTA0.7). Addition of more than 0.5 wt% niobium oxide would slightly decrease the hardness properties. This improvement is contributed by the highest densification and lowest porosity of ZTA0.5 samples. This improvement also was achieved with the presence of lowest apparent porosity and highest bulk density in ZTA0.5 sample compared to the other respective ZTA compositions. Zhang et al. mentioned that the different amounts of porosity in different particular sample could cause different hardness result and this might be contributed by sintering temperature used to eliminate porosity in the particular sample [17]. Porosity will give a great effect on the strength of the ceramics samples. Ceramics sample will be either failure in brittle manner or successfully survived from the stress applied. Rejab et al. stated that the size, shape and amount of pores in the ceramics samples would affect the strength as the cross sectional area over load lesser and caused the lower stress can supported [18]. He also added that the improvement of Vickers hardness of ZTA doped rare earth materials is because of the higher densification of the ceramics materials. It should be noted that the porosity could reduce the strength by formation of crack when it reach critical value and ceramics materials might fail without plastic deformation, so crack formed in the sample will propagate until it meet failure [2]. Therefore, it can be concluded that the addition of niobium oxide will significantly increase the hardness of the ZTA samples.

4 Conclusions

This research demonstrates the effect of Nb_2O_5 particles as dopant on the microstructure and mechanical properties of ZTA ceramics composites. It can be concluded that the addition of niobium pentoxide particles as dopant enhance the densification at 1650°C sintering temperature. The Vickers hardness increases from pure ZTA to 0.5 wt% of niobium pentoxide addition with the highest value of 1828.62 Hv and then at 0.7 wt% Nb_2O_5 addition, it caused slightly decrease on the Vickers hardness. Bulk density and Vickers hardness shows a positive relation as it shown same trend. According to scanning electron microscopy, the porosity for the ZTA0.5 is slightly lesser compared to the undoped and other doped ZTA composition.

References

1. Campbell, C. Flake, *Lighweight Materials - Understanding the basics* (ASM International, USA, 2012)
2. A.M. Hassan, S.M. Naga, M. Awaad, Int. J. Refract. Metals Hard Mater., 48, 338 (2015)
3. M. Vallet-Regi, *Bio-Ceramics with Clinical Applications* (John Wiley & Sons, West Sussex, 2014)
4. B. Basu, D.S. Katti, A. Kumar, *Advanced Biomaterials: Fundamentals, Processing and Applications* (John Wiley & Sons, USA, 2010)
5. Nevarez-Rascon, A. Aguilar-Elguezabal, E. Orrantia, M.H. Bocanegra-Bernal, Int. J. Refract. Metals Hard. Mater., 27, 962 (2009)
6. V. Tebaldo, G. Gautier, Ceram. Int., 39, 2683 (2013)
7. Nevarez-Rascon, A. Aguilar-Elguezabal, E. Orrantia, M.H. Bocanegra-Bernal, Int. J. Refract. Metals Hard. Mater., 29, 333 (2011)
8. R. Hannink, P. Kelly, B. Muddle, J. Am. Ceram. Soc., 83, 461 (2000)
9. S. M. Kurtz, S. Kocagoz, C. Arnholt, R. Huet, M. Ueno, W.L. Walter, J. Mech. Behav. Biomed. Mater., 31, 107 (2014)
10. Y. Shen, R. Leckie, C. Levi, D. Clarke, ActaMater., 58, 4424 (2010)
11. Z. Fang, *Sintering of advanced materials*, (Woodhead Pub., Cambridge, 2010)
12. S. Vidyavathy, V. Kamaraj, Modern Applied Science, 3, 102 (2009)
13. D. Almeida, C. Cairo, C. Silva, M. Nono, Journal of Aerospace Technology and Management, 2, 195 (2010)
14. K. Somton, P. Sanguanwit, K. Sujirote, R. McCuiston, Journal of Metals, Materials and Minerals, 20, 1 (2010)
15. T. Yeh, M. Sacks, J. Am. Ceram. Soc., 71, 841 (1988)
16. B. Lawn, D. Marshall, , J. Am. Ceram. Soc., 62, 347 (1979)
17. D. Zhang, T. Yoshida, T. Oekermann, K. Furuta, H. Minoura, Adv. Funct. Mater., 16, 1228 (2006)
18. N. Rejab, A. Azhar, M. Ratnam, Z. Ahmad, Int. J. Refract. Metals Hard. Mater., 36, 162 (2013)

Influence of Sintering Temperature on Crystallization Behavior of Cordierite synthesized from Non-Stoichiometric Formulation

K. K. Eing[1,*], *B.* Johar[1], *L. N.* Ho[1], and *Y. Zabar*[2]

[1]School of Materials Engineering, University Malaysia Perlis, 02600 Kangar, Perlis, Malaysia
[2]Advance Materials Department (Ceramic), ADTEC Taiping, 34600 Taiping Perak,Malaysia

Abstract. Cordierite body with formulation of non-stoichiometric composition (2.5 MgO. 1.8 Al_2O_3. 5 SiO_2) was synthesized using conventional techniques with standard raw materials. The sintering and crystallization behavior of the compositions was observed by Differential Thermal Analysis and Thermogravimetric Analysis (DTA/TG). The sequence of reaction and phase transformation was analyzed using X-ray Diffraction (XRD) technique and Rietveld structural refinement after sintering the samples at different temperature regarding the information from the DTA. The Scanning electron microscopy (SEM) was employed for morphology analysis. The DTA curve shows the crystallization temperature, T_c occur at 1259 °C.Rietveld quantitative phase analysis results reveal that α phase Cordierite constitutes up to 96.4 wt% when the samples was sintered for 2 hours at the optimal temperature of 1375 °C. The SEM micrograph revealed that the sample was heat treated at 1375 °C obtained densified body with well alignment of crystal structure.

1 Introduction

Cordierite (nominal composition 2MgO. $2Al_2O_3.5SiO_2$) know as of the useful crystalline phases in MgO-Al_2O_3-SiO_2ternary system. Recently, Cordierite has been focuses in numerous important applications due to its excellent properties, such as low thermal expansion coefficient (3.0 x10^{-6}/°C), excellent thermal shock resistance, low dielectric constant (4.0-5.0), good chemical durability, excellent refractoriness, and mechanical properties [1-3]. Moreover, cordierite has lower sintering temperature and lower density than the alumina, it also much cheaper and can be synthesized using mineral talc and kaolin [4]. Among the three cordierite polymorph, α-phase is more preference due to it has unique properties, which are very low coefficient of thermal expansion, dielectric constant and dielectric loss, density and stable at elevated temperature [5]. Although α-cordierite is a promising material for many applications, but it is difficult to crystallize and sinter high purity α-cordierite phase at low temperature because of its sintering temperature is near to

* Corresponding author: kuankok29@gmail.com

the incongruent melting point of the cordierite. It has the incongruent melting point because the solid compound of cordierite does not melt to form the liquid of its composition, but instead dissociates to form a new solid phase and the liquid.

Logvinkov et al [6] reported that the change in the crystal lattice of cordierite should be considered as changes arising from composition solid solution rather than phase transformation due to thermodynamic instability of cordierite. Additionally, Torres and Alarcon [7] mentioned that the initial crystalline phase form was cordierite within μ-cobalt and after annealing it transformed to α-cordierite. The μ-cordierite growth along the particle, surface and nucleation and form μ-cordierite dendritic, while the α-cordierite growth within the μ-cordierite dendritic and some Al and Si took place in the tetrahedral sites after long period of annealing at 1100 °C , which introduced some transformation to β-cordierite. Therefore, numerous of studies had been carried out with the purpose of accelerate phase transformation of MgO-Al_2O_3-SiO_2 with cooperation of additive [8-12]. Nevertheless, the TiO_2 was recommended as most effective crystallization catalyst in numerous papers [13-16]. The work of synthesis cordierite through solid state reaction with stiochiometric composition has been carried out by C.Liu et al [17] and founded that, 87 wt% of α-Cordierite formed at optimal sintering temperature of 1350 °C for 3 hours. However, the quantity of α-cordierite observed trending down as elevated the sintering temperature as the result of crystallization of secondary phase such as mullite.

Recently, non-stoichiometric cordierite composition ($2.8MgO.1.5Al_2O_3.5SiO2$) has been discovered to give high purity of α-cordierite phase at lower temperature using control crystallization of glass method [4]. The present of high concentration of secondary phase such as forsterite, spinel and mullite (which have different coefficient of thermal expansion and different volume) in cordierite synthesis using stoichiometric composition might affect the performance and its application especially in evaluated temperature atmosphere. The Cordierite body synthesized through glass method by using non-stoichiometric cordierite formulation $2.8MgO.1.5Al_2O_3$. $5SiO_2$ obtaining single phase α-phase Cordierite (100 wt%). However, in previous work, the cordierite with same formulation has been synthesized by solid state reaction and more than 10 wt% of secondary phase such as enstatite was observed. The formation of secondary phase was mainly contributed by excess of MgO in the system. Hence, in this present study, the Cordierite body has been synthesized through solid state reaction by alternated the mole ratio of MgO and Al_2O_3 to $2.5MgO$. $1.8Al_2O_3$. $5SiO_2$ in order to obtain high purity (>95%) stable α- cordierite.

2 Materials and method

α-Cordierite with non- stoichiometric formulation was synthesized using reagent grade oxides (MgO, Al_2O_3, and SiO_2) with 5 wt% of TiO_2 as nucleating catalyst. The magnesia was obtained for R&M Marketing, Essex, U.K, Alumina and silica were obtained from MajuSaintifikSdn.Bhd. Table 1 showed the elemental analysis of the raw materials, while Table 2 gives the details on weight precent of the non-stoichimetric cordierite formulation composition in this study. The particle size of the reagent oxide was sieved and determined by using particle size analyzer to ensure the median of the particle size d_{50} less than 30 μm. The components were weighted and mixed thoroughly by using jar mill together with 10 mm diameter alumina balls as media at rotation speed of 136 RPM for 36 hours. The batchers were compacted at 7 tones into pellet form of 20 mm in diameter by using hydraulic powder press (CARVER AUTO 3890). The green body compact were sintered at different temperature, which were 1250°C, 1275 °C, 1300 °C, 1325°C, 1350°C and 1375 °C for 2 hours at a heating rate of 10 °C/min respectively

The qualitative and quantitative crystalline phases of specimens were characterized by using X-Ray diffraction (XRD) (RigakuUltimalV) with Cu Kα radiation, in the $5° \leq 2\theta \leq$

90° range at a scanning rate of 1 ^{o}/min. Crystal structure analysis was carried out with whole diffraction powder pattern data by utilized Rietveld Refinement method using XpertHighScore Plus software. Differential Thermal Analysis and Thermogravimetric Analysis (DTA/TG) (Rigaku TG 8120) technique was employing to identify the thermal behavior of the batch. The thermal analysis was made in range 40-1500 ^oC by using corundum crucible as an inert reference material at heating rate of 10 ^oC /min. The Microstructure of the sintered samples was observed by scanning electron microscopy, SEM (Hitachi TM3000).

Table 1. Elemental analysis reagent oxides compound by XRF.

	Chemical Composition (wt%)												
	MgO	Al_2O_3	SiO_2	P_2O_5	K_2O	CaO	Fe_2O_3	NiO	ZnO	Ga_2O_3	SO_3	Cl	Na_2O
Al_2O_3	-	99.50	0.17	-	0.03	0.05	0.05	0.02	0.01	0.02	-	-	0.16
MgO	95.10	0.72	0.71	0.10	-	1.11	0.08	-	-	-	1.96	0.22	-
SiO_2	0.06	0.88	98.40	0.04	-	0.30	0.06	0.02	-	-	-	-	0.08

Table 2. Composition (wt%) of initial raw materials with formulation $2.5MgO.1.8Al_2O_3.\ 5SiO_2$.

Sample	Wt %			
	MgO	**Al_2O_3**	**SiO_2**	**TiO_2**
$2.5MgO.1.8Al_2O_3.\ 5SiO_2$	18.09	35.81	50.42	5.216

3 Results and discussion

3.1 Crystallization behavior

Differential Thermal Analysis and Thermogravimetric Analysis (DTA/TG) curves recorded on the thermal behaviour of the non-stoichimetric cordierite composition of $2.5Mg.1.8Al_2O_3.5SiO_2$ from 40 to 1500 ^oC are shown in Figure 1. As can be seen in Figure 1, the thermogravimetric (TG) curve of the sample has been segregated into 7 regions. The regions A to F was corresponding to the mass losses, while there was mass gain occur at region G which is related to the crystallization and phase transformation. At the first region, the initial weight losses about 0.73% occurs in the range between 40 and 190 ^oC which is corresponding to the elimination of water physically adsorbed when wet method mixing route was carried out. Followed by the major weight change about 3.61% losses in the temperature range of 190 to 388 ^oC, which can attributed to the dehydration of the MgO. The existence of endothermic peak in DTA plot at 588 ^oC further supports the observation. The mass losses about 1.84% and 1.17% was observed in temperature range 388 to 447 ^oC (region C) and temperature range 447 to 569^oC (region D) respectively. The mass losses occur in region C is contributed by the decomposition or organic burn out of the sample. While in region D, the small endothermic peak was observed at 564 ^oC, which is correlated to the phase transformation of silica. There is no significantly weight change in region E. However, the transition temperature, Tg endothermic peak of the sample was presence at 772 ^oC in the DTA curve at this particular peak.

The exothermic peak of was detected at 979 °C and reaching at endothermic at 1163 °C in the DTA curve and about 0.14% weight losses in region F . In this region the nucleation of glass might take place due to the nucleation temperature (T_n), is usually taken in the nucleation range, which is between T_g and softening point (T_d) [18] and it can be calculated by using formulation:

$$T_n = T_g + 2/3\,(T_d - T_g) \qquad (1)$$

Considering the endothermic peak at 1163 °C is softening point of the samples, thus the nucleation temperature, T_n of sample was located at 1132.67 °C. A shape exothermic crystallization peak was presence at 1259 °C, which is corresponding to the phase transformation of μ-Cordierite to α-Cordierite [19]. Moreover, thermogravimetric curve in region G shown mass gain causes by solid state reaction of decomposition and leading formation of α-Cordierite and leading the variation of samples density of the sample. The sequence of reaction and phase transformation occurs in region G was detailed by the X-ray Diffraction (XRD) pattern.

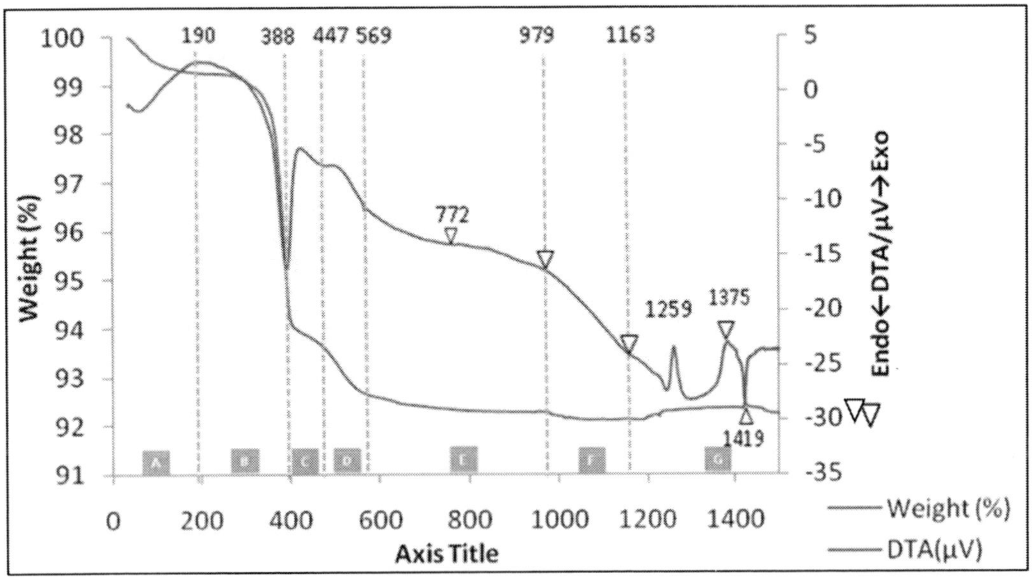

Fig. 1. DTA/TG curve for composition of $2.5MgO.\ 1.8Al_2O_3.\ 5SiO_2$.

3.2 Phase evolution

To investigate the sequence of reactions and phase transformation, the samples were heat-treated at different temperatures. According to crystallization temperature, T_c, and information of Figure 1, the samples heat treated at 1250 °C, 1275 °C, 1300 °C, 1325 °C, 1350 °C and 1375 °C.Figure 2 and Figure 3 present the phase identification of XRD patterns and Rietveld quantitative phase analysis of the samples with non-stoichiometric cordierite formulation of $2.5Mg.1.8Al_2O_3.5SiO_2$ sintered at different temperature for 2 hours as dwell time respectively. The result indicate that, the thermal treating heat have greatly impact on the purity of α-Cordierite. The α-Cordierite (ICSD # 98-004-5314) peak was observed in the entire XRD pattern, and the intensities of the corresponding peaks increases with increasing temperatures of 1250 °C-1375°C. Rietveld quantitative phase analysis results reveal that α-Cordierite constitutes up to 96.4 wt% of the samples sintered for 2 hours at the optimal temperature of 1375 °C. At 1250°C, the α-Cordierite was the dominant phase. However, the secondary phase such as Sapphirine (ICSD# 98-007-8446), Cristobalite low (ICSD# 98-001-6962), Quartz (ICSD# 98-003-5149), spinel (ICSD# 98-005-8347) and Rutile (ICSD# 98-007-6756) was observed. Furthermore, about 6.3 wt% of μ-Cordierite (ICSD# 98-010-9832) was presence at sintering temperature 1250°C due to the the uncompleted phase transformation which is related to result obtained from DTA curve in Figure 1. Further raise up the sintering temperature to 1275 °C, the completely phase transformation leads to increased amount of α-Cordierite from 52.3 wt% to 59.4 wt%. Additionally, the Magnesium Dititanate (ICSD# 98-005-3931) was observed as decomposed of Rutile. Alongside, the disintegration of minor phase such as quartz high and Cristobalite high might prop up the amount of Sapphirine. This phenomenon also can been seen in sintering temperature 1300 °C, the amount of α-Cordierite and Sapphirine increased as further decomposition of quartz and Cristobalite. Meaning that, the decomposition of quartz or Cristobalite contributed Si^{2+} and O^- in the mixed system and it might endorse the formation of α-Cordierite and Sapphirine at particular sintering temperature. As increase of the sintering temperature form 1300 °C to 1325 °C, it could be noticed the amount of Sapphirine is decrease and the Spinel peak was detected. The Sapphirine was totally vanished at 1350 °C and left behind the small amount of Spinel. Moreover, crystalline content of α-Cordierite phase has dramatically intensified to 90 wt% as disintegration of Sapphirrine. The crystalline phase content of α-Cordierite was achieved up to 96.7 wt% as decomposition of Spinel at sintering temperature of 1375 °C for 2 hours soaking time with minor secondary phase of Anosovite (Ti_3O_5) (ICSD# 98-001-1581) and Aluminium Magnesium ($Al_{0.9}Mg_{0.9}$) (ICSD# 98-006-5874). This is competitive result with stiochiometric cordierite formulation [17] [20]

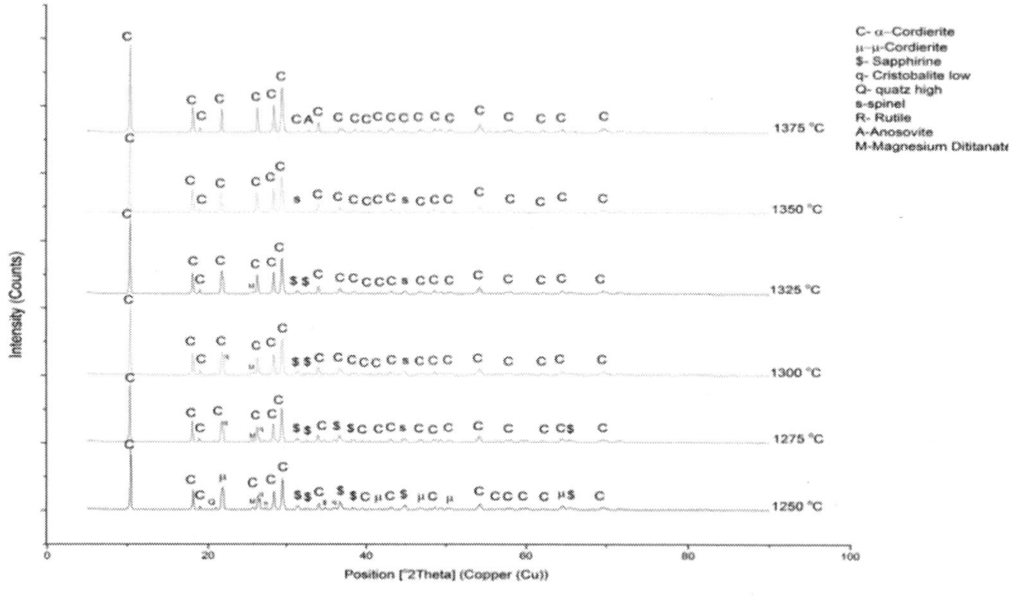

Fig. 2. XRD patterns of samples sintered at temperature ranging from 1250 to 1375 °C for soaking times of 2 hours.

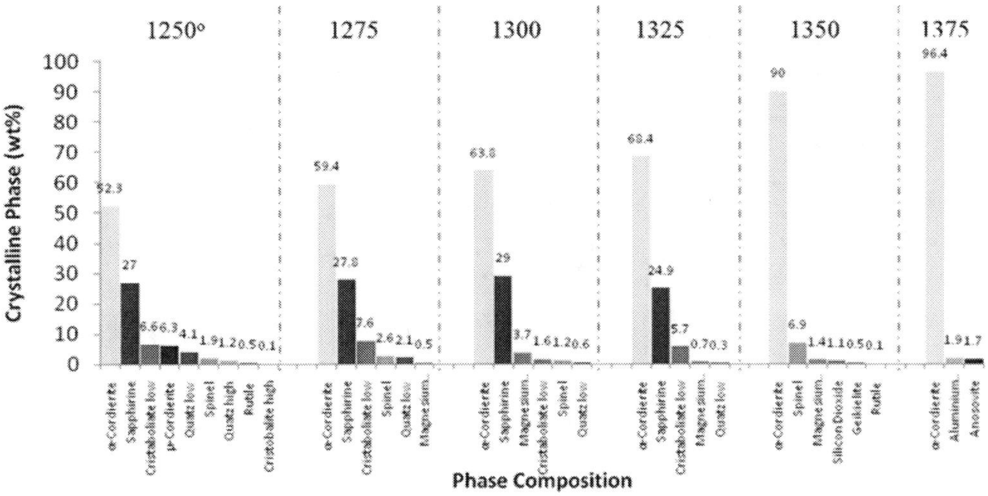

Fig. 3. Crystalline phase content of samples sintered at temperature ranging from 1250 to 1375 °C for soaking times of 2 hours.

3.3 Microstructure analysis

Figure 4 illustrated the Scanning electron microscopy (SEM) images of the samples heat treated at different temperature for 2 hours. It was found that, the sample has heat treated at 1375 °C (Figure 4f) obtained optimal morphology and fined microstructure as well as α-Cordierite hexagonal crystal structure was formed over the surface of the sample. Conversely, the formation of secondary phase induce the rough morphology as can been seen in Figure 4a. With corroborating the results obtained through XRD, in can be noticed that, the morphology of samples show better appearance more denser body as diminish of secondary phase

Figure 5, showed the comparison between the samples sintered at 1250 °C and 1375 °C. As can be seen in Figure 5a, the formation of dissimilar and complex crystal structure of secondary phase has inferior alignment with other phase and induce pore between each other. As a result, the performance in terms of mechanical strength, thermal expansion coefficient, and dielectric constant has impacted. On the other hand, the approximately single phase of α-Cordierite present well alignment of the crystal structure.

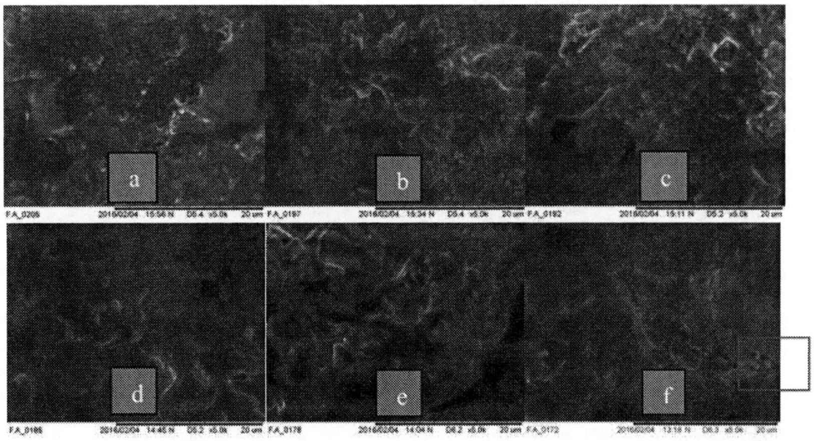

Fig. 4. SEM micrograph of samples sintering at (a) 1250 °C, (b) 1275 °C, (c) 1300 °C, (d) 1325 °C, (e) 1350 °C, (f) 1375 °C for 2 hours under magnification of 5000X.

 DOI: 10.1051/matecconf/20167801099

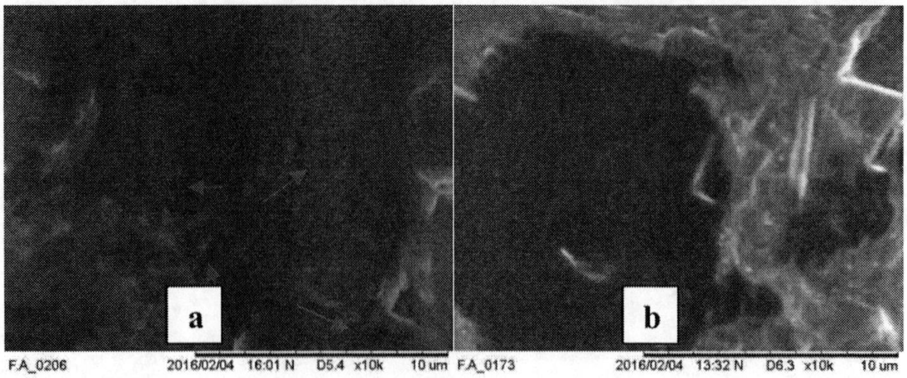

Fig. 5.The SEM micrograph of samples sintering at (a) 1250 °C, (b) 1375 °C for 2 hours under magnification of 10000X.

4 Conclusions

The high purity of α phase Cordierite has been successfully synthesized from non-stoichimetric formulation $2.5MgO.1.8Al_2O_3$. 5 SiO_2 using a conventional technique with standard raw materials. The X-ray diffraction technique and the Rietveld quantitative phase analysis results reveal that α-Cordierite constitutes up to 96.4 wt% of the samples sintered for 2 hours at the optimal temperature of 1375 °C. With corroborating of XRD patterns and DTA curve, the μ-Cordierite has been confirm completely transformed to α-Cordierite at sintering 1275 °C. SEM micrograph revealed that fully dense with well crystal structure alignment α-Cordierite was successfully crystallized.

The authors gratefully acknowledge the financial support from the fundamental Research Grant Scheme (FRGS 900300495) Universiti Malaysia Perlis and equipments support from ADTEC Taiping Malaysia.

References

1. X. Guo, N. Kazuki, K. Kazuyoshi, Y. Zhu, H. Yang, J. Eur. Ceram. Soc., **34**, 817 (2014)
2. P. Orosco, M. del C. Ruiz, J. Gonzalez, Powder Technol., **267**, 111 (2014)
3. Y. Dong, X. Liu, Q. Ma, G. Meng, J. Memb. Sci., **285**, 173 (2006)
4. J. Banjuraizah, M. Hasmaliza, Z.A Arifin, J. Alloys Compd., **482**, 429 (2009)
5. J.Banjuraizah, H.Mohamad, Z.A Ahmad, J. Alloys Compd., **509**, 7645 (2011)
6. S. M. Logvinkov, G. D. Semchenko, D. A. Kobyzeva, V. I. Babushkin, Refractories and Industrial Ceramics, **42**, 434 (2001)
7. F. J. Torres, J. Alarcon, J. Eur. Ceram. Soc., **24**, 681 (2004)
8. Z. Yuea, J. Zhoua, Z. Maa, J. Baoa, Z. Guia, L. Lia, Ferroelectrics, **262**, 31 (2001)
9. X. Hao, Z. Luo, X. Hu, J. Song, Y. Tang, A. Lu, Journal of Non-Crystalline Solids **432**, 265 (2016)
10. L. Barbieri, C. Leonelli, T. Manfredini, R. Bertoncello, Journal of Materials Science, **29**, 6273 (1994)
11. C. Leonelli, T. Manfredini, M. Paganelli, P. Pozzi, G.C. Pellacani, Journal of Materials Science, **26**, 5041 (1991)

12. H.S. Kim, R.D. Rawlings, P.S. Rogers, Journal of Materials Science, **24**, 1025 (1989)
13. M. Rezvani, B.E. Yekta, V.K. Marghussian, J. Eur. Ceram. Soc., **25**, 1525 (2005)
14. J. Yang, S.G. Zhang, B. Liu, D.A. Pan, C.L. Wu, A.A. Volinsky, Journal of Iron and Steel Research, **22**, 1113 (2015)
15. M. Fathia, A. Johnson, Dental Materials, **32**, 311 (2016)
16. K. Maeda, Y. Sera, A. Yasumori, Journal of Non-Crystalline Solids, **434**, 13 (2016)
17. C. Liu, L. Liua, K. Tan, L. Zhang, K. Tang, X. Shi, Ceramics International, **42**, 734 (2016)
18. V. I. Marghussian, M. H. D. Niaki, J. Eur. Ceram. Soc., **15**, 343 (1995)
19. LingHong, H. Zhou, C. Xu, Journal of Materials Science **13**, 381 (2002)
20. N. Obradovic, N. Dordevic, S. Filipovic, N. Nikolic, D. Kosanovic, M. Mitric, Markovic, V. Pavlovic, Powder Technol., **218**, 157 (2012)

Green Roof Technology- Mitigate Urban Heat Island (UHI) Effect

*Z.S. M.*Odli[1,*], *I.A.*Zakarya [1], *F.N.*Mohd[1], *T.N.T.*Izhar[1], *N.M.* Ibrahim[1] , and *N.* Mohamad[1]

[1]School of Environmental Engineering, Universiti Malaysia Perlis (UniMAP), Kompleks Pusat PengajianJejawi 3, 02600 Arau, Perlis, Malaysia

Abstract. Alterations on the land surfaces, which are attributed by human activities, especially in cities, cause many implications to the ecosystem. The increase of buildings in cities is reflecting the growth of human activities resulted in a significant temperature increase and warmer pattern in the urban area than the surrounding countryside. The phenomenon defined as urban heat island. This study investigates the application and efficiency of the green roof as an approach to mitigate urban heat island and reducing indoor temperature in a building. Two types of roof models, which consist of vegetative roof and non-vegetative roof, were built to investigate the efficiency of vegetated roof in reducing indoor temperature compared to the non-vegetated roof. The outdoor and indoor temperature and humidity of each roof model were monitored by using RH520 Thermo Hygrometer. The data was collected for three times in a week for 9 weeks at 9:00am to 5:00pm. It was found that the indoor average temperature data for vegetative roof could be reduced 2.4℃ from the outdoor average temperature and 0.8℃ for non-vegetative roof. The difference of temperature reduction for vegetative roof was greater than the non-vegetative roof, thus indicate that green roof was highly efficient in reducing indoor temperature and mitigate urban heat island impact.

1 Introduction

Global warming is described as when the temperature of the Earth's atmosphere is being increased. Continuous rise in global temperatures make the world warmer than it has been. Burning of fossil fuels, biomass burning, cement manufacture, cow and sheep rearing, deforestation and land-use changes are the activities that associated with the emission of greenhouse gases that posed to global warming scenario [1].

Presently, other environmental issues related to global warming and has created a concern is an urban heat island (UHI). The UHI is the phenomenon when the air temperature in the cities area is higher compared to suburban and rural areas (Figure 1) [2]. According to Davis [3], an area can be classified as undergo UHI when the ambient air temperature in downtown urban area is 2℃ – 3℃ hotter than in rural areas. Urban heat is likely to raise a problem related to energy demand, human discomfort, air quality, and heat related diseases. With the increase of urban area and large cities, the problem is getting

* Corresponding author: zaity@unimap.edu.my

more serious since the vegetative land surface is replaced with built surface. Generally, this will result in critical disruption on the role of the trees, which blocked the sun radiation during the day and hold in the heat at night and resulting to more extreme temperature swings at global level.

A study conducted by Rizwan et al. [4] shows that urban heat is mainly caused by the heat released from vehicles, power plant, air conditioner and heat from complex urban structures. Building materials such as concrete and asphalt have significantly different thermal and surface radiative properties than natural landscape, which absorb energy from the sun and convert it to sensible heat. Besides, it has been reported that UHI can be affected by certain ways like the lack of green spaces and water [5]. Another analysis made by Getter and Rowe [6], trace that the UHI phenomenon do not hit the urban only in the day, but also in the night since the absorbed heat is radiated back to the surrounding during the evening.

In response to the adverse impact of UHI towards the societies, many research works related to the studies on the concepts, methods, technologies and approach to mitigate the problem were iniated. Among the strategies that have been suggested, it is proven that increasing the green coverage with roof gardening and tree planting is one of the most effective way to mitigate the urban heat island phenomenon [7,8]. Therefore, application of green roof technology is the most attractive and relevant strategy to present the mitigation towards the UHI phenomena in the cities.

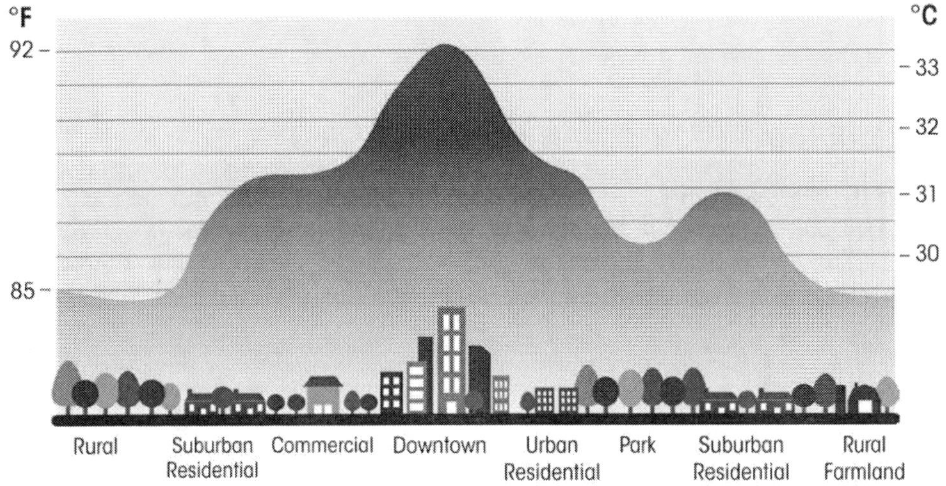

Fig. 1. Heat island profile [9].

Correspond with this issue, many related research for theses issues has been reviewed and discussed, and at the same time, the researchers have developed several recommendations on mitigation strategies to counteract the phenomenon. One of the strategies is green roof technology. Green roof technology is a technology of implementing vegetation and growing medium on the roof layer. Green roof offer high mitigation potential for high heat island problem [10].

Green roof or rooftop garden is a vegetative layer grown on a rooftop. A significance analysis from previous research stated that green roof promotes many economics, environmental, and social benefits [6]. Besides, Santamouris[10] has revealed that green roof presenting a great mitigation potential towards heat island phenomenon through vegetation usage. Firstly, plant and soils on the roof protects the surface below them by from direct sunlight radiation through shading. As a result, the temperature of the surface below the plants is reduced by the re-emitted of heat transmit into the atmosphere and the usage of sunlight by leaves for photosynthesis process [11]. Furthermore, the soils also acted to protect the underlying layers from UV radiation and strong wind [11]. However, amount of sunlight transmitted under the canopy of plants depends on plant species [12]. Secondly, green roof helps to reduce surface urban heat island through evapotranspiration. Evapotranspiration is the combination of evaporation and transpiration process is the process when liquid is converted into gas. Transpiration as explained by Monteith [13] is the condition when plants absorb water through their roots and emit it through their leaves (Figure 2).

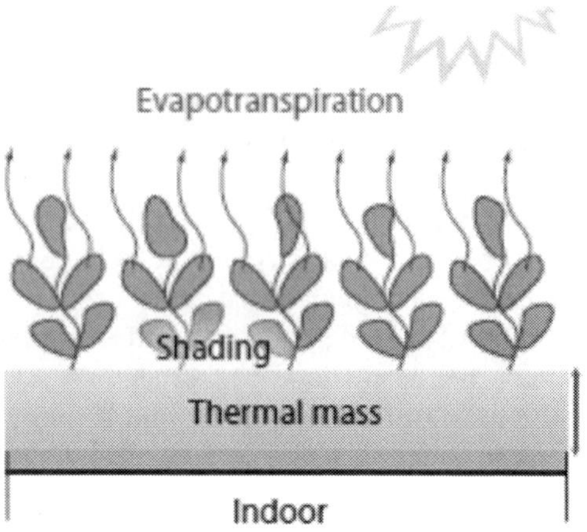

Fig. 2. Evapotranspiration and shading of a green roof [14].

2 Methodology

In this study, the observation and collection of the indoor temperature and humidity data were carried out on each roof model. The extensive roof models contain a layer of clay brick, roof, soil and vegetation. There are four extensive roof models, where three of them were planted with vegetation, which each consist a layer of Japanese grass (*Zoysiatenuifolia*), chives, joyweed (*Alternantherasessilis*) and one is non-vegetated roof. The extensive green roof has a soil thickness of 4inch (10.16cm) and total size of 144inch ($0.093m^2$). During the experimental period, a manual watering were provided with volume of 1.5L to each green roof for twice a day which at every 9:00am and 5:00pm. The roof model with and without vegetative layer are presented in Figure 3.

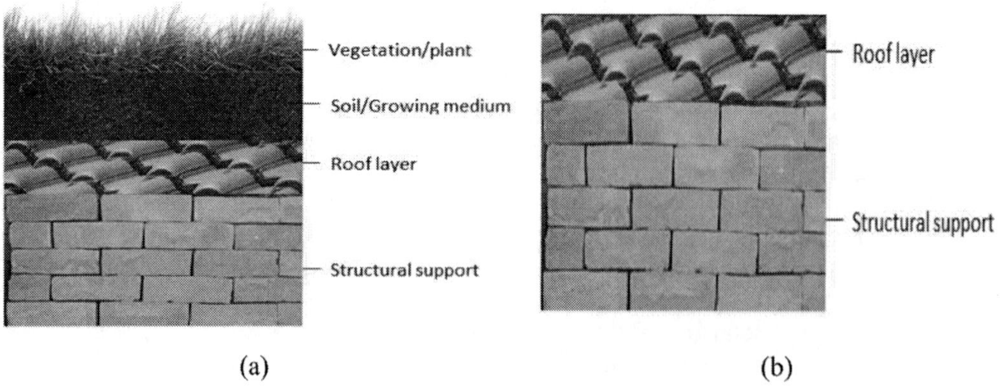

Fig. 3. The model of green roof with vegetation (a) and without vegetation (b).

In order to investigate the effectiveness of the extensive roof model in reducing heat, the outdoor and indoor temperatures were monitored. The temperature and humidity below the roof surface in the structural support column of each roof model was defined as the indoor temperature and indoor humidity, while the ambient temperature and humidity outside of the structural support column and roof surface was defined as the outdoor ambient temperature and outdoor humidity. The temperature and humidity for both indoor and outdoor were recorded every 2 hours starting at 9:00am until 5:00pm for three days in a week for nine weeks using RH520 Thermo Hygrometer.

3 Results and discussion

The experiments were monitor continuously for 9 weeks between 27th January until 25th March 2016. The monitoring results and effectiveness of green roof were discussed as the following.

3.1 Average temperatures for outdoor and indoor hardware roof models

The average temperatures at different times for both outdoor and indoor of roof model are shown in Figure 4. The average temperatures data for the ninth week's show that indoor temperature of roof models has a lower reading than the outdoor temperature especially for the vegetation roof as compared to the non-vegetated roof. Moreover, to analyse the differences in term of effectiveness of the roof models at different times of the day, the average indoor temperature for each roof model was subtracted by the average outdoor temperature for every respective monitoring time. Table 1 compares the temperature difference of each period for blank, grass, joyweed and chives roof model from the average ninth weeks monitoring period values.

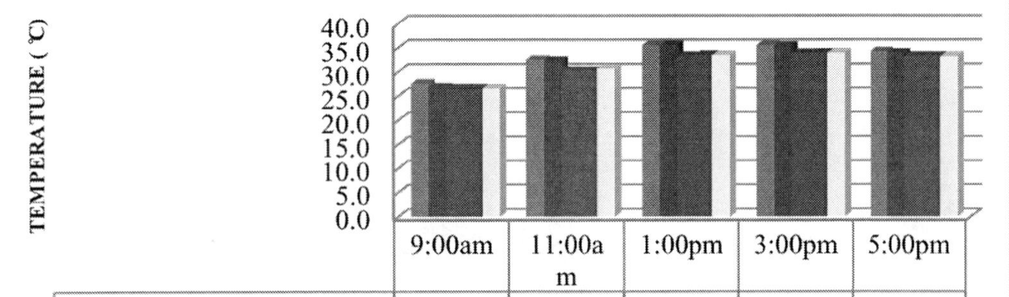

	9:00am	11:00am	1:00pm	3:00pm	5:00pm
Avg. Outdoor Ambient air	27.6	32.5	35.5	35.5	34.2
Avg. Indoor Blank Roof	26.8	32.3	35.5	35.3	34.0
Avg. Indoor Grass Roof	26.5	30.1	33.2	33.8	33.2
Avg. Indoor Joyweed Roof	26.6	30.2	33.4	33.9	33.3
Avg. Indoor Chives Roof	26.6	30.6	33.5	33.9	33.2

Fig. 4. Average temperature at different times for outdoor and indoor roof models.

Table 1 illustrates the temperature difference for each roof model. The highest difference of outdoor and indoor temperature for each period is from grass roof model, followed by joyweed, chives and blank roof models. The greatest difference of outdoor and indoor temperature was recorded at 2.4℃ from grass and joyweed roof model during 11:00am with efficiency of 7.4%, while the lowest difference is from blank roof model during 1:00pm.

 DOI: 10.1051/matecconf/20167801100

Table 1. Difference of outdoor and indoor temperature.

Time	Average outdoor temperature (℃)	Average indoor temperature (℃)				Difference of outdoor and indoor temperature (℃)			
		Blank	Grass	Joyweed	Chives	Blank	Grass	Joyweed	Chives
9:00am	27.6	26.8	26.5	26.6	26.6	0.8	1.1	1.1	1.0
11:00am	32.5	32.3	30.1	30.2	30.6	0.3	2.4	2.4	1.9
1:00pm	35.5	35.5	33.2	33.4	33.5	0.0	2.3	2.1	2.0
3:00pm	35.5	35.3	33.8	33.9	33.9	0.2	1.7	1.6	1.6
5:00pm	34.2	34.0	33.2	33.3	33.2	0.3	1.0	0.9	1.0

Based on these figure and table, the significant cooling period for the vegetative roof model was at 11:00am to 1:00pm. During this time, the vegetation and soil layer act efficiently in insulating the roof from direct radiation hit from the sun.

3.2 Average humidity for outdoor and indoor hardware roof models

Aside from temperature, humidity also influences the cooling process in the indoor roof model. Comparing the average humidity data for both outdoor and indoor humidity, simple data tabulation was made. Figure 5 shows the experimental data on average outdoor humidity and average indoor humidity of roof model. From the figure, the highest average humidity for both outdoor and indoor were recorded at 9:00am with humidity range from 73-78% while the lowest humidity range was recorded at 3:00pm with range of 41-51%. From the entire roof model, the indoor of the grass roof model reported more humid than the other three roof models and the outdoor humidity.

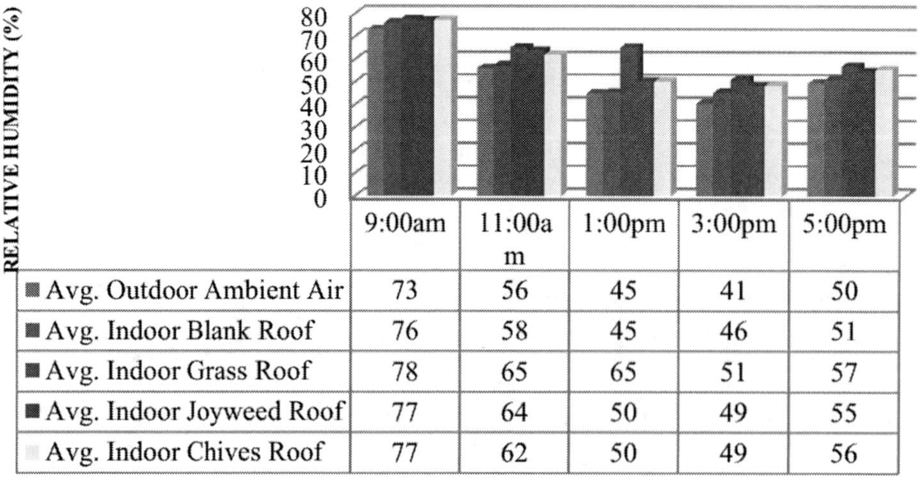

	9:00am	11:00am	1:00pm	3:00pm	5:00pm
Avg. Outdoor Ambient Air	73	56	45	41	50
Avg. Indoor Blank Roof	76	58	45	46	51
Avg. Indoor Grass Roof	78	65	65	51	57
Avg. Indoor Joyweed Roof	77	64	50	49	55
Avg. Indoor Chives Roof	77	62	50	49	56

Fig. 5. Average humidity at different times for outdoor and indoor roof models.

As can be seen from Table 2, the highest humidity difference was displayed by grass roof model at 9:00am with difference of 20% from the outdoor humidity, while for joyweed, the highest difference in humidity was achieved at 11:00am with 8% humidity difference and as for chives, the highest humidity difference wasis 8% at 3:00pm. As for blank bare roof, the indoor humidity recorded the highest difference reading at 3:00pm with 5% humidity difference with the outdoor humidity. The outdoor humidity was basically greatly influenced by the cloud which shade the sun, and wind speed of the environment.

Table 2. Difference of outdoor and indoor humidity.

Time	**Average outdoor humidity (%)**	**Average indoor humidity (%)**				**Difference of outdoor and indoor humidity (%)**			
		Blank	Grass	Joyweed	Chives	Blank	Grass	Joyweed	Chives
9:00am	73	76	78	77	77	-3	-5	-4	-4
11:00am	56	58	65	64	62	-1	-9	-8	-6
1:00pm	45	45	65	50	50	0	-20	-5	-5
3:00pm	41	46	51	49	49	-5	-10	-8	-8
5:00pm	50	51	57	55	56	-2	-7	-5	-6

Overall, these results indicate that the vegetative roof was better at reducing indoor temperature and offer significant cooling state for the indoor condition than non-vegetative roof. These finding were supported by the result from the study by Rosenweig et al. [15], where green roof provides a cooling effect to the city by 2-3K compared to cooling effect by the non-green roofs. Moreover, from this result, it was explained that grass was the best choice of plant to be used as green roof instead of joyweed and chives since grass provided high temperature and humidity difference. However, from the observation, the three types of plant used as green roof was highly sun tolerance, which can survive 3 days without water in hot day based on the Perlis climate.

4 Conclusion

Comparing the effectiveness of the green roof and non-green roof, a significant difference in indoor temperature and humidity with the outdoor temperature and humidity was found. It was confirmed that the indoor temperature of the hardware roof model for extensive green roof was lower than the indoor temperature of the hardware roof for extensive non-green roof. Extensive green roof also has higher humidity as compared to extensive non-green roof. The cooling effectiveness of the extensive hardware green roof model was significant when the ambient air temperature started to increase especially from 11:00am to 3:00pm. Besides, it is confirmed that vegetation and soil help in react as an insulator for the roof form receiving high direct radiation form the sun. To conclude, applying vegetation of roof can reduce the impact of urban heat island.

References

1. S. Solomon, D. Qin, M. Manning, Z. Chen, M. Marquis, K.B. Averyt, M. Tignor, H.L. Miller, *Climate change 2007: The physical science basis* (Cambride University Press, Cambride, United Kingdom, 2007)
2. M. Santamouris, *Solar Enery: The State o Art* (James and James Science Publishers, London, 2001)
3. M.P. Davis, N.A. Nordin, M. Ghazali, M.J. Durak, G. Reimann, *Conference on Sustainable Building South East Asia, Malaysia* (2005)

4. A.M. Rizwan, L.Y. Dennis, L.I.U. Chunho, Journal of Environmental Sciences, **20**, 120 (2008)
5. N.H. Wong, H. Feriadi, P.Y. Lim, K.W. Tham, C. Sekhar, K.W. Cheong, Building and Environment, **37**, 1267 (2002)
6. K.L. Getter, D.B. Rowe, HortScience, **41**, 1276 (2006)
7. M.A. Hart, D.J. Sailor, Theor. Appl. Climatol., **95**, 397 (2009)
8. Z. Ismail, M. Makmor, R. Hashim, R.M. Shah, N.A. Hanifah, S. Ahmad, *IEEE Colloqium on Humanities, Science & Engineering Research* (2012)
9. Ahmed Rizwan, Journal of Environmental Science, **20**,120 (2008)
10. M. Santamouris, Sol. Energy, **103**, 682 (2014)
11. Green Roofs, *Reducing Urban Heat Island: Compendium of Strategies* (U..S. Environmental Protection Agency's Office of Atmospheric Programs, U.S., 2015)
12. R. Arabi, M.F. Shahidan, M.M. Kamal, M.F.Z.B. Jaafar, M. Rakhshandehroo, Current World Environment, **10**, 918 (2015)
13. J. Monteith, Symp. Soc. Exp. Biol., **19**, 205 (1965)
14. K. Liu, B. Bass, *Performance of Green Roof Systems* (National Research Council Canada, Toronto, Canada, 2005)
15. Rosenzweig Cynthia, William Solecki, Ronald Slosberg, *A report to the New York State Energy Research and Development Authority* (2006).

The effect of ZnO addition on the phase transformation, microstructure, and ionic conductivity of 8YSZ ceramics

B. Johar[1,*], and *Y*. Zabar[1]

[1]School of Materials Engineering, Universiti Malaysia Perlis, 02600 Kangar, Perlis, Malaysia

Abstract. In this paper the effect of zinc oxide (ZnO) addition on the phase transformation and ionic conductivity of 8 mol % Yttria-stabilised Zirconia (8YSZ) is investigated. Pure 8YSZ and ZnO doped YSZ ceramics are prepared using the solid state reaction method sintered at 1550°C for 2 hours. The X-ray Diffraction (XRD) results reveal the presence of tetragonal, cubic and monoclinic phase of Zirconium dioxide (ZrO_2) for all undoped and doped YSZ sintered samples. The phase stability of tetragonal YSZ was found to be increased with the increase in ZnO addition. Minor fraction of monoclinic phase was found in pure YSZ sintered sample and the amount of monoclinic phase decreased with the increasing amount of Zn after sintering at 1550°C for 2 hours. The fraction of cubic phase was also found to decrease with the increase in Zn concentration. The highest ionic conductivity of $1.03X10^{-3}$ S cm^{-1} was obtained at room temperature for samples with 3 mol% ZnO. Pure 8YSZ sintered sample on the other hand, yielded $9.88x10^{-4}$ S cm^{-1}.

1 Introduction

Solid oxide fuel cells (SOFCs) have attracted much attention in recent years due to their various potential advantages, including a wide variety of available fuels, inexpensive technology, good durability and highly efficient energy conversion [1–3]. Yttria-stabilized zirconia (YSZ) is the traditional electrolyte of the cell system. 8 mol% of Yttrium oxide (Y_2O_3) is one of the most explored compositions where the doping provides sufficient oxygen vacancies in fluorite ZrO_2 lattice thus, to ensure high ionic conductivity (σ) up to 0.1 S/cm. In addition, it also provide a stable cubic structure at 1000 ◦C as well as to obtain sufficiently high current densities and good power output.). Such high temperatures accelerate degradation of the fuel cell system and often lead to problems, such as solid reactions between the components, thermal degradation as well as thermal expansion mismatch. Furthermore, if the operation temperature of SOFC is to be decreased to 800◦C, the ionic conductivity of YSZ reduces to 0.044 S/cm which becomes insufficient for high SOFC performance due to very high ohmic losses [1].

Numerous studied have been carried out on yttria-stabilized zirconia (YSZ) as solid electrolyte materials and a key component of SOFC systems. Therefore, efforts have been

* Corresponding author: banjuraizah@unimap.edu.my

exerted to reduce the operating temperature of such systems. In addition to solve the elevated cost of the sealing materials and interconnects, as well as the durability of the cells. Some experiments have reported the effect of dopants on the sintering, electrical, microstructure and mechanical properties of YSZ [2-4]. The ionic conductivity of YSZ are resulting from oxygen vacancies which depends on the type of dopant element and their amount [5]. Some dopant might increase and decrease the ionic conductivity level even though it could enhance the densification and reduce the sintering temperature. When ceria is doped with aliovalent oxides, such as Y_2O_3 and other rare-earth oxides the ionic conductivity increases significantly. However, an increasing amount of dopants tends to form a second phase due to solubility limits, resulting in the reduction of conductivity.

ZnO is well known not only as a semiconductor but also as a probable oxygen-ion conductor due to enrichment of oxygen vacancies at higher temperature. Although Liu and Lau already found that the small addition of ZnO was effective in reducing the sintering temperature and promoting the densification rate of the 8YSZ ceramics, but they are using co-precipitation method to synthesize 8YSZ [2]. Very recently, Gao et al. [6] synthesized $(Ce_{0.84}Y_{0.16}O_{1.92})_{1-x}(ZnO)_x$ electrolytes through a solid state reaction using ZnO as the sintering aid and they found the sintering temperature decreased from 1600 °C (20 hours) to 1375 °C (20 hours). However the addition of ZnO exhibited a slight detrimental influence on their conductivities.

In this work, the effect of ZnO addition on sinterability, microstructure, phase transition and ionic conductivity in commercial 8YSZ powders were investigated using solid state reaction. The aim is to determine whether the addition of ZnO in 8YSZ with large particle size and high impurity content could also improve densification and give positive impact to ionic conductivity.

2 Methodology

2.1 Sample preparation.

ZnO (1, 2, 3 mol%) was added to the 8YSZ powders using mortar and pestle followed by calcinations process. The calcined powders were die-pressed into cylindrical pellets (12 mm in diameter) at 6 tonne pressure by Automatic hydraulic press, model 3890.4NE000. The green pellets were sintered at temperatures 1550◦C in the air for 2 hours and subsequently cooled to room temperature.

2.2 Characterization.

Phase analysis was performed on the sintered pellet. The crystalline phase of all samples was examined using XRD (XRD, D/max-III, Rigaku) with CuKα radiation with an applied voltage of 40 kV and current of 40 mA. The XRD data were collected over the range of 2h = 20°-80° with a step size of 0.02° and a count time of 0.3°/min. The microstructure of the ceramics was investigated using SEM with an applied voltage of 30 kV (model JEOL JSM-6460LA SEM) and the energy dispersive. The oxide ionic conductivity of the materials was measured using electrochemical impedance spectroscopy based on sintered pellets.. Both sides of the pellets were coated with platinum (pt) using sputter coater prior characterized as electrodes and fired at 600 ◦C for 30 min before measurement to ensure good bonding. The AC impedance spectra were made using impedance analyser, model Agilent 4294A at 300 °C. The testing was conducted at frequency range from 4 Hz to 5 MHz

3 Results and discussion

3.1 Phase composition and crystal structure

The XRD patterns of 8YSZ and ZnO doped 8YSZ samples sintered at 1550°C for 2 h are shown in Figure 1. The XRD pattern has been indexed with the help of ICDS card and it confirms that all samples mainly comprises a combination of tetragonal (ICSD 98-002-0789) and cubic phase (ICSD 98-003-2797). Closed observation on the XRD pattern revealed the presence of small fraction of monoclinic phase (ICSD 98-010-8439) phase which can be characterized by a peak at $2\theta=28.2^{\circ}$. Yttria and ZnO based phase has not been detected which confirms the well incorporation of ZnO ions into the 8YSZ matrix.

Increases the dopant concentration to 3 mol% of ZnO resulted in increasing the formation of tetragonal phase and reduces the formation of monoclinic phase. It is very difficult to distinguish between tetragonal and cubic zirconia structure. The assignment of cubic and tetragonal structures, based solely on the x-ray diffraction analysis, can be misleading because the cubic and tetragonal structures (a0=0.5124 nm for cubic, and a0=0.5094nm and c0=0.5177 nm for tetragonal structures) are very similar as demonstrated in simulated ICSD paten shown in Figure 2.

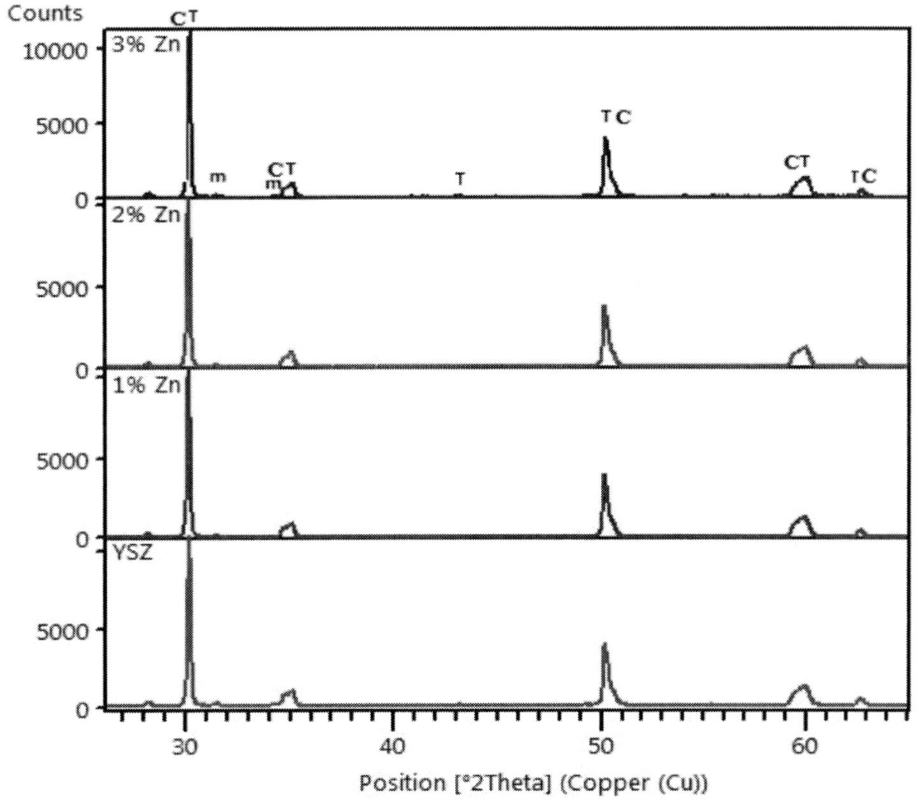

Fig. 1. XRD pattern of the 8YSZ ceramics with the addition of ZnO (1, 2, 3 mol%) sintered at 1550°C for 2 h in air (c:cubic, t:tetragonal, m:monoclinic).

The tetragonal structure can be distinguished from the cubic structure by the presence of the characteristic splitting peak at $2\theta \cong 34^{\circ}, 50^{\circ}, 59^{\circ}$, such as (002) (110), (021) (112) and (121) (013), and etc., whereas the cubic phase exhibits only single peaks at all of these positions. Due to the broadening of the peak, the two polymorph might overlap, thus crystal structure refinement using whole profile fitting method or Rietveld method to determine the present of either one or both polymorphs. Table 1 demonstrates phase quantification using Rietveld Method and presents the phase composition for all sintered undoped and doped 8YSZ samples.

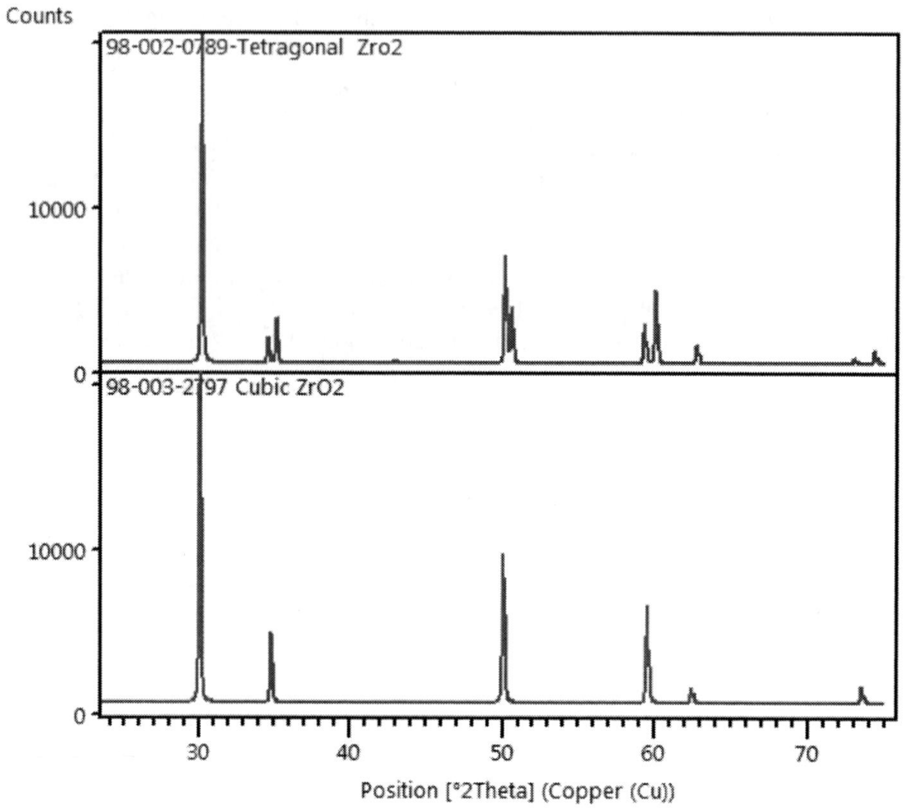

Fig. 2. Simulated pattern of ICSD data of cubic and tetragonal fluorite structure.

Table 3. Phase composition of sintered 8YSZ ceramics doped with different amount of ZnO.

% molZnO	Tetragonal	Cubic	Monoclinic	Rwp
0	65.5	26.8	7.8	13.66
1	69.2	23.2	7.6	12.87
2	70.7	22.8	6.5	12.65
3	78.9	16.1	5.0	12.80

It is proposed that the cause of the stability of the tetragonal phase in these present studies is directly related to the particle size of the raw powders (8YSZ). In particular, during the heat treatment the Y_2O_3 doping can either diffuse into the ZrO_2 crystal lattice or segregate at grain boundaries, depending on how the yttrium atoms were combined with the zirconium atoms during the synthesis process. In this context, the presence of monoclinic phase even in sintered pure 8YSZ samples could be attributed to an inhomogeneous distribution of yttrium and zirconium ions due to the conditions used in each synthesis process, so that stabilization of the tetragonal phase is not possible.

3.2 Microstructure

Figure 3 displays the SEM photographs of $(YSZ)_{1-x}(ZnO)_x$ samples sintered at 1550°C for different mol% of ZnO. The sample with 1 mol% ZnO was found to be porous despite being sintered at 1550◦C. In contrast to sample with undoped 8YSZ samples was dense with small closed pore left in the ceramic matrix (cant understand). This was resulted from the insufficient pore elimination The addition of ZnO inhibited the sintering process. ZnO has higher melting point which is 1975°C [7]. The addition of Zno does not lead to an increase in the grain size[8]. The effectiveness of sintering aids depends on several factors, such as sintering temperature, viscosity, solubility and glass wettability. The main requirement for liquid sintering is that the liquid phase should wet the grains of the ceramics. Generally, the chemical reaction between sintering aids and ceramic particles can provide the best wetting condition[9].

Although Y. Liu et al. reported that ZnO-added samples experience viscous flow sintering which promotes the diffusivity rate of 8YSZ matrix, thus enhancing the densification and grain growth however, it also greatly depends on the type of initial raw materials used. During liquid phase sintering process, both solid and liquid are subject to rearrangement because of the unbalanced capillary forces. The sintering aids with low melting temperature would act as lubricant and the rearrangement of the particles proceeds in the direction of reducing porosity. This rearrangement is driven by an imbalance capillary pressure which is a result of initial raw materials used such as (1)particle size distribution, (2) irregular particle shape, (3) local density fluctuation in the powder compact, and (4) anisotropic materials properties which limits the wetting angles between particles.

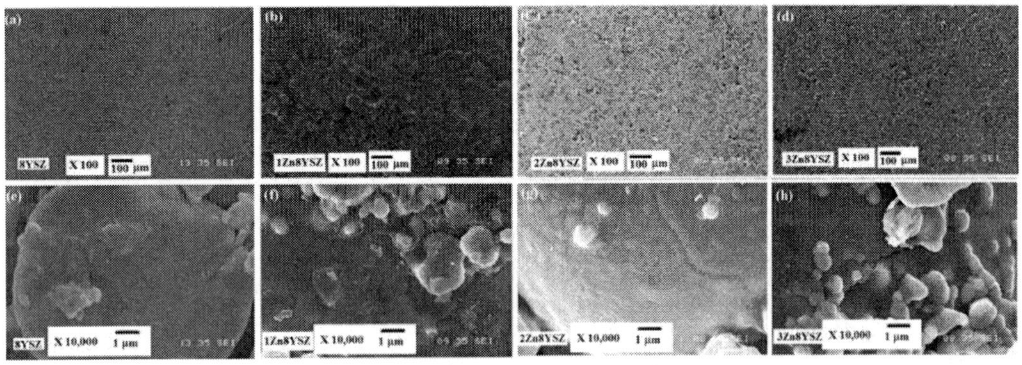

Fig. 3. SEM micrographs of ceramics samples at 1550°C with X100 magnification: undoped 8YSZ(a), with 1 mol % ZnO (b), with 2 mol % ZnO (c), with 3 mol% ZnO (d) and ceramics sintered at 1550 °C for 2 h. with X 10,000 magnification: undoped 8YSZ(e), with 1 mol % ZnO (f), with 2 mol % ZnO (g), with 3 mol% ZnO (h)

The SEM analysis shows the formation of individual particles' micro-sized spherical and irregular crystals with size of 0.5–40μm. They are partially fused to form hard agglomerates. Moreover, the structure of YSZ samples obtained at 800°C consisted of large-grained monolithic matrix, indicating good connectivity between the grains.

3.3 Ionic conductivity

Figure 4 shows the ionic conductivity of all investigated samples measured at room temperature as a function of frequency. Ionic conductivities of all undoped and Zn doped 8YSZ samples increase with increasing frequencies. Figure 5 demonstrate the ionic conductivity values of all undoped and Zn doped 8YSZ sintered samples measured at 5 MHz at room temperature. For $(8YSZ)_{1-x}(ZnO)_x$ (x = 1, 2, and 3 mol%), the ionic conductivity decreased from $9.884 \times 10^{-4}\ \Omega^{-1}\ cm^{-1}$ for undoped samples to x = 3.75 $\times 10^{-4}\ \Omega^{-1}\ cm^{-1}$ then increased to $1.03 \times 10^{-3}\ \Omega^{-1}\ cm^{-1}$ for 3 mol% ZnO addition,

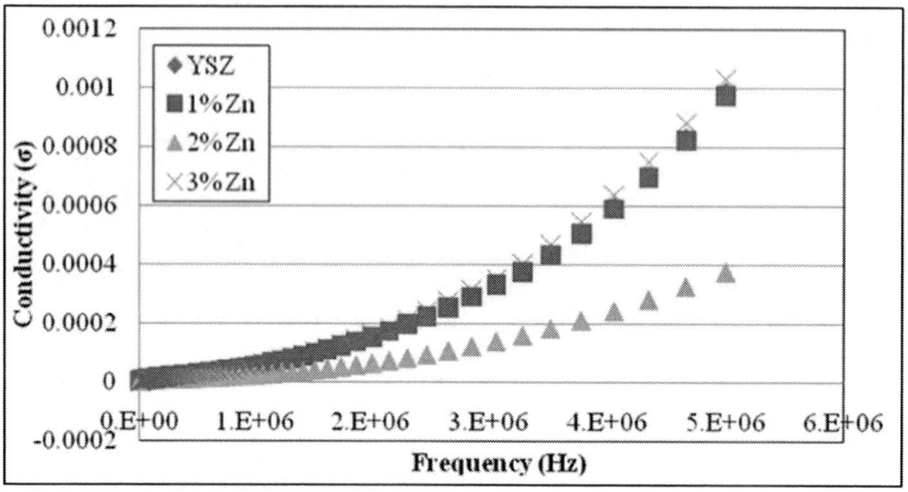

Fig. 4. Ionic conductivity of undopedand ZnO-doped 8YSZ ceramics measured at different frequency.

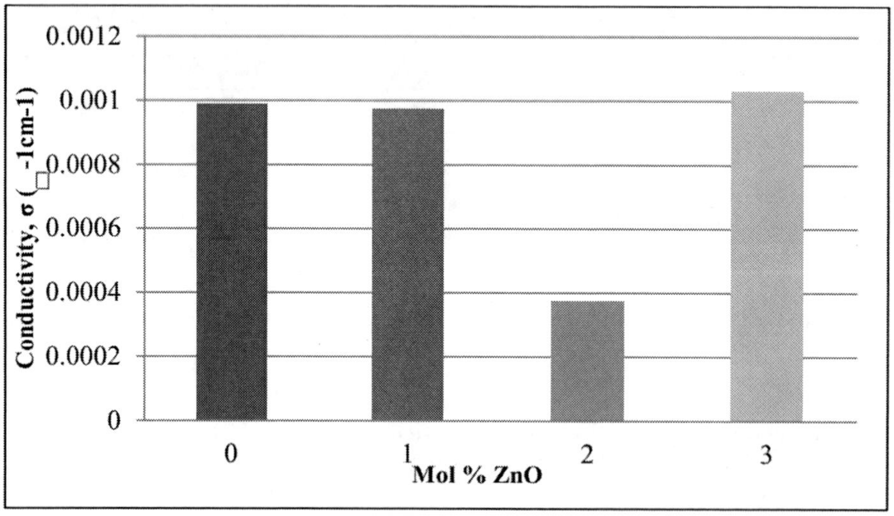

Fig. 5. Ionic conductivity of undoped and ZnO doped 8YSZ sintered samples measured at room temperature at frew 5 MHz.

The ionic conductivity of $(8YSZ)_{0.97}(ZnO)_{0.03}$ was observed to be much higher than that of pure 8YSZ samples (x = 0). The above analysis ascertained that YSZ with certain ratio of ZnO could improve the electrical property of 8YSZ electrolytes further. The study demonstrated that better electrical properties could be obtained with samples synthesized using 3 mol% ZnO .in this study. Although $(8YSZ)_{0.97}(ZnO)_{0.03}$ was still not well sintered as shown at the micrograph in Figure 3. The present of closed pores, indicate the samples are not have fully densified microstructure. Thus, it can be concluded that sinterability of the specimens can be promoted further by increasing the sintering temperature or prolonging the holding time, whereas ionic conductivity may be improved by increasing the density.

The higher ionic conductivity for the $(8YSZ)_{0.97}(ZnO)_{0.03}$ samples could probably be attributed to the higher densification (as shown in Figure 3) and the more compact arrangement of crystal grains that favored the transportation of the carriers. In this study, doping with ZnO may cause two opposite effects. On the one hand, as a sintering aid, ZnO can reduce porosity and improve sinterability to effectively obtain a dense electrolyte material, leading to a decrease in activation energy of conduction and an increase in ionic conductivity. On the other hand, ZnO also shows a detrimental influence on conductivity when the amount added is not enough to enhance desification, which leads to an increase in the activation energy of conduction and a decrease in ionic conductivity. When the ZnO content (x) was at 0.2 mol %, the latter effect may be stronger than the former, resulting in lower ionic conductivity. However, when the dopant content (x) was at 3 mol.%, the former effect may be stronger than the latter, leading to higher ionic conductivity and lower activation energy for conduction. It is known that ZnO dissolves into the ZrO_2 can create the oxygen vacancies as the addition of $Y_2O_3Y_2O_3 \rightarrow 2Y_{Zr}' + V_O^{\bullet\bullet} + 3O_O^{x}\}$ and $ZnO \rightarrow Zn_{Zr}'' + V_O^{\bullet\bullet} + O_O^{x}$ 3 mol % addition of ZnO causes an increase in bulk conductivity due to the increasing number of vacancies. If the concentration of the oxygen vacancy is too high, it might leads to the formation of high activation energy pathways for oxygen diffusion and the formation of ordered oxygen vacancy complexes at high Y concentrations thus, hinders the vacancy mobility. It has become evident that the atomic defect configuration of Y and oxygen vacancy complexes have a profound effect on the oxygen diffusion and transport in YSZ, and hence on the efficiency of solid electrolyte.

4 Conclusion

Un-doped and doped 8YSZ ceramic were successfully synthesized using the solid state reaction method. Structural characterization demonstrates the present of cubic, tetragonal and monoclinic phase for all doped undoped samples. The monoclinic phase decreased as the amount of ZnO increases. The surface textural and morphological property revealed the formation of fine grain in a matrix of large grain. The addition of ZnO on 8YSZ powders are not significantly enhanced the densification as previously reported which might due to large particle size of 8YSZ used in the present studies. For $(8YSZ)_{1-x}(ZnO)_x$ (x = 0,1,2,3 mol%), ionic conductivity was improved in comparison to pure 8YSZ, and the highest ionic conductivity was 0.001 S cm^{-1} measured at room temperature with x = 0.03. It was confirmed that an appropriate amount of ZnO could improve the ionic conductivity of 8YSZ electrolytes.

References

1. Y. Chen, N. Orlovskaya, E.A. Payzant, T. Graule, J. Kuebler, J. Eur. Ceram. Soc., **35**, 951 (2015)
2. Y. Liu, L.E. Lao, Solid State Ionics, **177**, 159 (2006)
3. K.V. Kravchyk, O. Bohnke, V. Gunes, A.G. Belous, E.V. Pashkova, J. Le Lannic, Gouttefangeas, Solid State Ionics, **226**, 53 (2012)
4. Guo F., P. Xiao, J. Eur. Ceram. Soc., **32**, 4157 (2012)
5. S. Manjunatha, M.S. Dharmaprakash, Materials Letters, 164, 476 (2016)

6. L. Gao, M. Zhou, Y. Zheng, H. Gu, H. Chen, L. Guo, J. Power Sources, **195**, 3130 (2010)
7. H.K. Kim, S.H. Lee, S.G. Lee, K.T. Lee, Y.H. Lee, Materials Research Bulletin, **58**, 218 (2014)
8. C. Peng, J. Melnik, J.L. Luo, A.R. Sanger, K.T. Chuang, Solid State Ionics, **181**, 1372 (2010)
9. Y. Pan, J. Yin, K. Zuo, D. Yao, Y. Xia, H. Liang, Y. Zeng, Ceramics International, **42**, 9222 (2016)

MATEC Web of Conferences **78**, 01102 (2016) DOI: 10.1051/matecconf/20167801102
IConGDM 2016

Fe-doped 8YSZ at different composition for solid electrolyte in solid oxide fuel cell

B. Johar[1,*], and *N.A.* Zaili[1]

[1]School of Materials Engineering, Universiti Malaysia Perlis, 02600 Kangar, Perlis, Malaysia

Abstract. Pure 8 mol% yttria stabilized zirconia (YSZ) and Fe-doped (1 mol%, 2 mol% and 3 mol%) YSZ electrolyte were prepared and sintered at 1550°C. Transition metal oxide is added into YSZ as sintering aided has a function to reduce the sintering temperature. The microstructure, crystal structure and ionic conductivity of pure YSZ and Fe-doped YSZ at different composition were investigated. The amount of cubic phase decreased as the amount of Fe increased. Fe-doped 8YSZ had higher conductivity than pure 8YSZ. The ionic conductivity of 3FeYSZ is 9.35×10^{-8} S/cm higher than 1FeYSZ which is 4.72×10^{-9} S/cm when operated at 300°C.

1 Introduction

Yttria stabilized zirconia (YSZ) is regularly utilized as electrolyte for SOFC because of its pure and high ionic conductivity at elevated temperature, chemical stability, low thermal expansion coefficient and good stability in both oxidizing and reducing gases [1,6]. YSZ has high ionic conductivity and chemical stability in oxidizing atmosphere. Zirconia (ZrO_2) is a thermal insulator with high thermodynamic and chemical stability. On 2013, Figueiredo and his team reported doping Zirconia with Yttria replaces Zr^{4+} with Y^{3+} by addition of oxygen vacancies and ionic conductivity. 8 mol% of Yttria stabilized zirconia (8YSZ) is the most preferred electrolyte material because it has excellent stability and pure ionic conductivity at typical operating temperature [21].

Transition metal oxides have proven as a good sintering promoter especially for the densification of SOFC electrolytes [1]. From the previous research by Satardekar et al. [2] and Zhang et al. [23], Fe is has been widely used because of effective dopant for YSZ electrolyte to enhance the performance of YSZ electrolyte.

Zhang et al. [23] studied that the maximum relative density which is almost ~98% of transition metal oxide of pure and doped Si-containing 8YSZ can be maintained when sintered at 1550°C. However, the lack of studied has been found for the different composition of Fe doped 8YSZ for sintering temperature at 1550°C. Therefore, in this research pure 8YSZ and different amounts of Fe (1 mol% of Fe, 2 mol% of Fe and 3 mol%) was doped in 8YSZ ceramics.

* Corresponding author: banjuraizah@unimap.edu.my

2 Experimental

Iron (Fe)-doped YSZ electrolyte powders were prepared using8 mol% yttria stabilized zirconia (8YSZ) powder and pure Fe_2O_3 powders. They were weighted according to the mole as 1 mol% of Fe (1FeYSZ), 2 mol% of Fe (2FeYSZ) and 3 mol% of Fe (3FeYSZ). The mixture of 8YSZ with Fe_2O_3 powder were ground by mortar and pestle with ethanol for 30 minutes until homogeneous mixture was obtained. The powders were then calcined at 1000°C and hold for 1 h for 10°C/min. The calcined powders were ground using mortar and pestle prior compacted in a cylindrical die of 12 mm diameter at 6 tonnes for 2 minutes. The compacted pellets were then sintered at 1550°C for 2 hours. .XRD was carried out using scan range 5-90°C and scan axis with a continuous 1°C/min, step scan of 7s and step size of 2 thetas. XRD patterns of the sintered samples were obtained using a model SHIMADZU XRD 6000 X-ray diffractometer with Cu Kα radiation of 50mA and 40kV. Morphology and microstructure of sample was investigate using SEM measurements was model JEOL JSM-6460LA. The sintered samples doped 8YSZ were coated with gold by using sputter coating before tested by the SEM to eliminate the charging effect. For the conductivity measurement, the impedance spectroscopy was used to determine the conductivity of the samples. The impedance spectroscopy measurement was carried out by impedance analyzer model Agilent 4294A. The ionic conductivity of the sintered pure and doped 8YSZ samples (coated and uncoated) were tested at high temperature which is 100°C, 200°C and 300°C. The conductivity value of coated and uncoated samples was measured at increasing temperature. The sample is fixed in the spring loaded sample holder prior dielectric properties measurement. The data was collected during heating. The measurement was conducted in air at room temperature and high temperature in a frequency range of 5000000 Hz- 4 Hz.

3 Results and discussion

3.1 Differential thermal analysis/ thermogravimetric analysis (DTA/TGA)

The ideal temperature and phase transformations of the calcined powder were studied by using DTA/TGA. The weight is monitored as a function of temperature. DTA was used to observe all the decomposition reaction and determine the crystallization behavior. Figure 1 showed exothermic peaks corresponding to various stages of weight loss were observed in DTA plot. The exothermic was occurred and the maximum rate of mass-change was 3FeYSZ which is at temperature around 1183°C while the lowest was pure YSZ which is at temperature 1238°C. The crystallization temperatures for 1FeYSZ and 2FeYSZ were 1186°C and 1223°C respectively. The crystallization temperature decreased as the composition increased due to the crystallization of certain phases.

TGA was used to detect loss in mass and quantity the change in physical or chemical properties of the materials as a function of temperature or time. Figure 2 showed weight gains occurred for all calcined powder (pure YSZ, 1FeYSZ, 2FeYSZ and 3FeYSZ) as the Tg curve increased. All samples showed almost similar trend. Fe-doped 8YSZ increase slightly uniformly up to temperature 1337°C. As the composition of Fe-doped 8YSZ increase which is 1FeYSZ, 2FeYSZ and 3FeYSZ, the weight gain decrease as 0.22mg, 0.21mg and 0.20mg respectively. Weight gains occurred due to the oxidation of the samples.

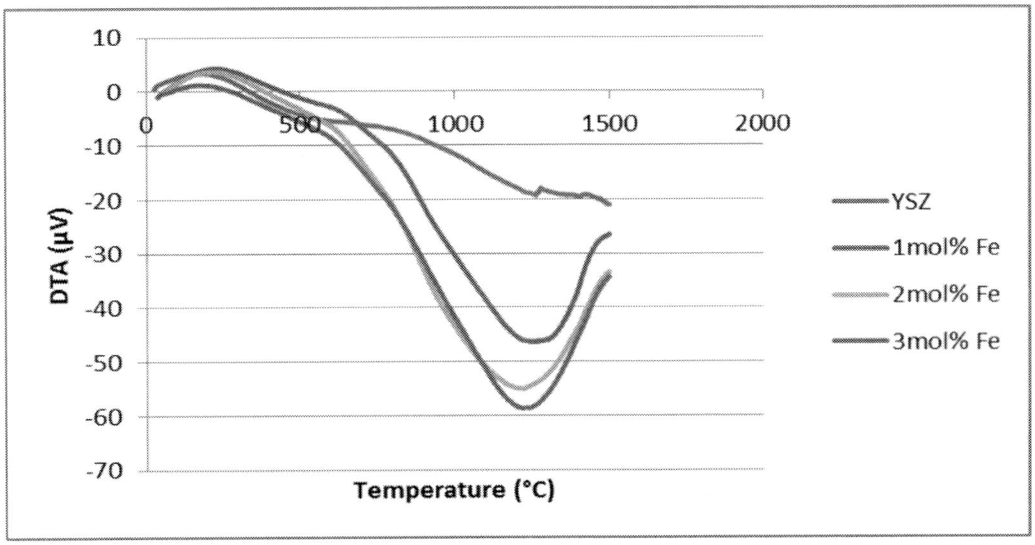

Fig. 1. The DTA graph for pure YSZ and Fe doped 8YSZ (1FeYSZ, 2FeYSZ and 3FeYSZ).

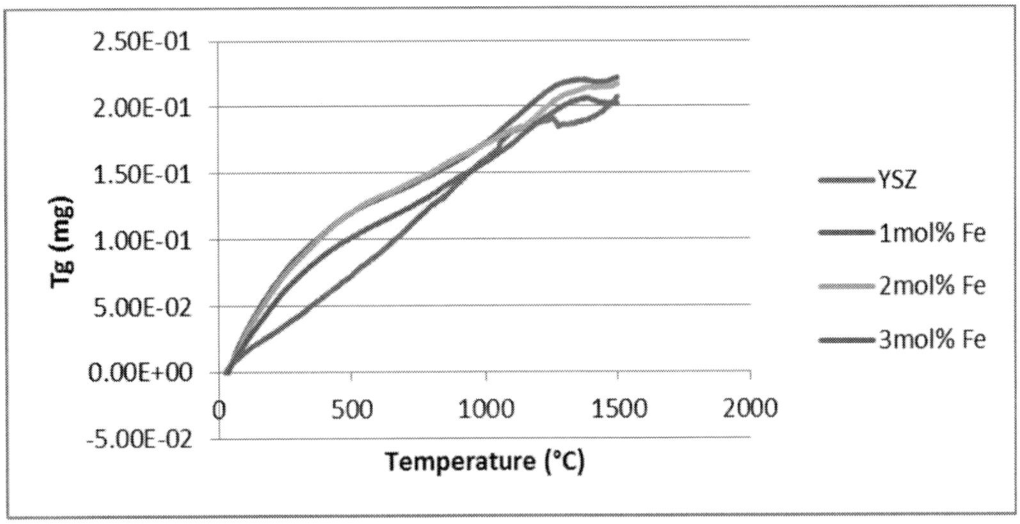

Fig. 2. The DTA graph for pure YSZ and Fe doped 8YSZ (1FeYSZ, 2FeYSZ and 3FeYSZ).

3.2 Phase composition

Figure 3 shows XRD pattern of Fe-doped 8YSZ electrolyte sintered at 1550°C at different composition. There are mixed phases corresponds to tetragonal and monoclinic phase in pure 8YSZ sample. While all tetragonal, cubic and monoclinic are present in Fe doped 8YSZ samples. No characteristic peak belong to Fe_2O_3 has been detected. The predominant phase for undoped 8YSZ samples is tetragonal phase. Rietveld quantitative phase analysis demonstrates phase composition of sample as shown in Figure 4. When the concentration of Fe increased, the proportion of tetragonal phase decreased. The addition of Fe into 8YSZ up to 1 mol% has resulted in crystallization of cubic phase, however the amount of cubic phase tend to decrease with increasing amount of Fe. The decomposition of cubic phases was expected related to crystallization of monoclinic phase. The higher the amount of Fe added into 8YSZ ceramics has distorted the crystal lattice and caused the formation of monoclinic phase. Unlike pure YSZ, high doping of Fe has retarded the formation of tetragonal or cubic phase. This is because certain concentration of dopant has scattered in the grain boundaries and resisted the formation of tetragonal phase. The dopant might also substitution or interstitial in the lattice and consequently cause distortion of lattice. The contraction or expansion of the YSZ lattice parameter can be observed by looking at changes of crystal volume as demonstrated in Table 1. It can be observed that, with the addition of Fe, the crystal volume of tetragonal phase tend to decrease. This is because Fe^{3+} ions have ionic radius which is 0.055nm smaller than Zr^{4+} ions which is 0.084nm, substitution of Zr^{4+} ions with smaller Fe^{3+} ions induce the variation in lattice parameter, thus decrease the volume of the unit cell [22]. The decreasing of crystal volume for both monoclinic and tetragonal phase proved that Fe has substitute zirconium or yttrium atom in the fluorite structure, while small expansion in cubic structure might due to the interstitial of Fe atom inside the crystal lattice. The results were in contrast with what Nakajima and his team found where they reported that the crystal structure of cubic phase was not changed by Fe doping [6].

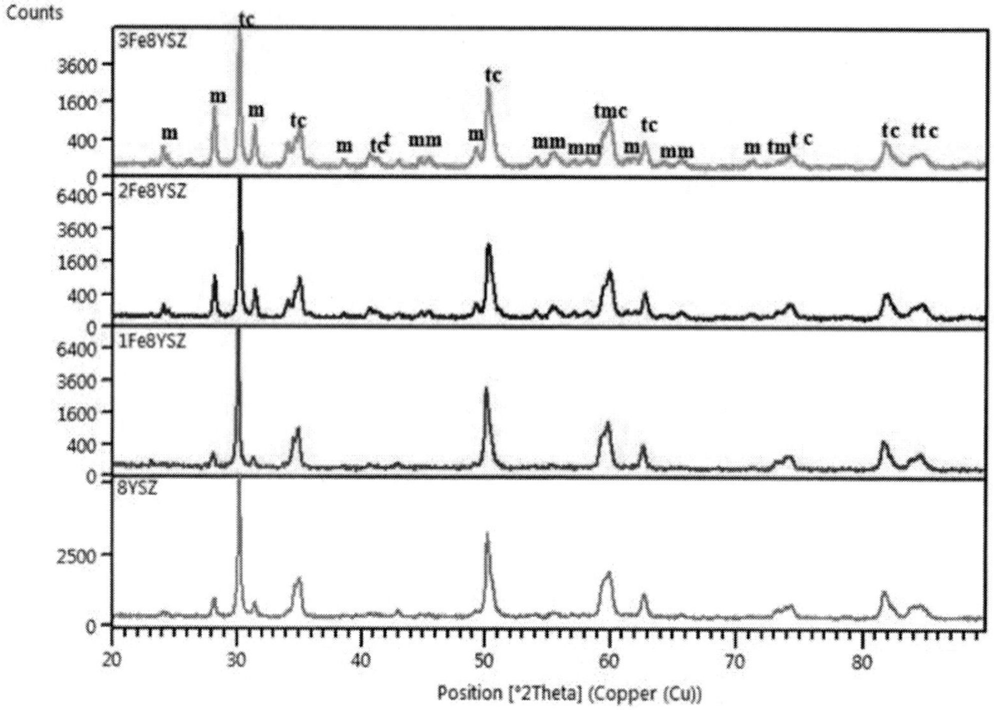

Fig. 3. XRD profile for the sintered samples of pure YSZ, Fe (1,2,3mol%) doped 8YSZ samples. (t:tetragonal #icsd no 98-002-0789, m:monoclinic #icsd no98-07-1860, c:cubic #iscd no 98-005-0481).

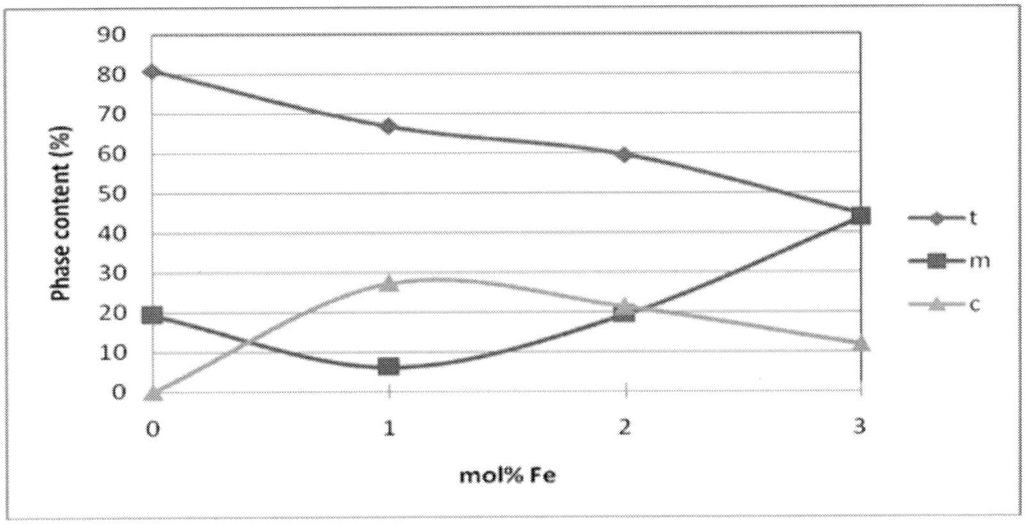

Fig. 4. Phase composition of sintered samples with doped with different amount of Fe.

Table 1. Crystal volume of phases that are present in undoped and Fe doped 8YSZ samples.

mol% Fe	Crystal volume (A°3)		
	tetragonal	monoclinic	cubic
0	67.61	142.05	0.00
1	67.32	142.00	135.10
2	67.42	141.24	135.20
3	67.40	141.20	135.28

3.3 Phase composition SEM morphology observation

The SEM morphologies of pure 8YSZ, 1FeYSZ, 2FeYSZ and 3FeYSZ sintered at 1550°C were shown in Fig. 4 (a)-(d) at magnification X10000 respectively. The pure YSZ showed the spherical shaped. The SEM micrograph indicated that dopant gave great effects on the microstructure of the 8YSZ. Addition of Fe in 8YSZ promoted the grain growth and densification. This also has resulted in non-homogeneous distribution of grain sizes. The grain size increased as the Fe concentration [2]. The figure showed that the grain size increased until 2FeYSZ and decreased when doped with 3FeYSZ concentration. It shows grain boundary with minimum grain size around 1μm and maximum grain size around 8μm.

3.4 Ionic conductivities

Impedance diagrams of samples sintered at 1550°C is shown in Figure 5 for pure YSZ, 1FeYSZ, 2FeYSZ and 3FeYSZ operated at 300°C. The trend of conductivity as a function of studied parameter showed one semicircle for the impedance plots measured at 300°C. Decrement in grain resistivity illustrated that the sintered samples had higher conductivity [24]. 3FeYSZ has smaller grain resistivity compare to 1FeYSZ. Therefore, 3FeYSZ has higher conductivity than those pure and doped sintered samples. This might be due to the Fe dopant may improve electrolyte-electrodes interfaces of SOFCs causing the power output to increase [1]. The ionic conductivity of 3FeYSZ is 9.35×10^{-8} S/cm higher than 1FeYSZ which is 4.72×10^{-9} S/cm when operated at 300°C.

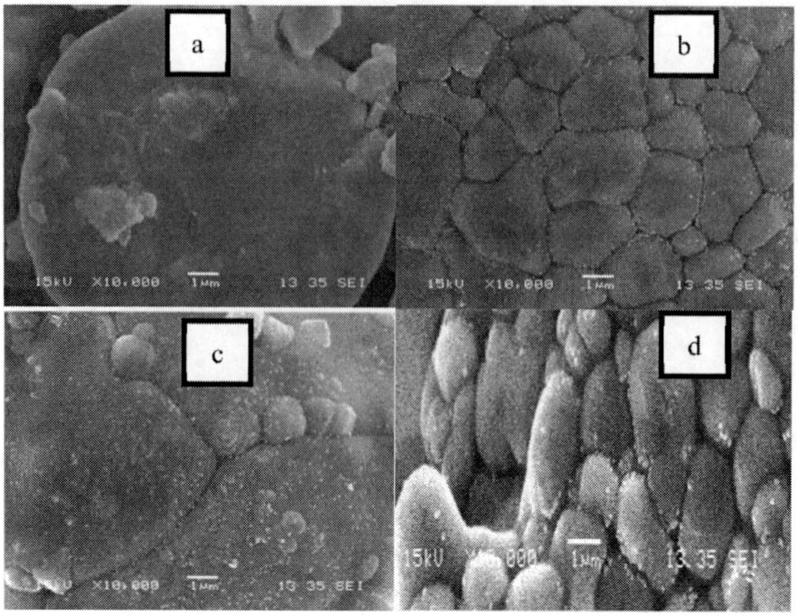

Fig. 4. SEM micrograph of pellets prepared from powder with different mol% of Fe: (a) pure YSZ, (b) 1FeYSZ, (c) 2FeYSZ and (d) 3FeYSZ.

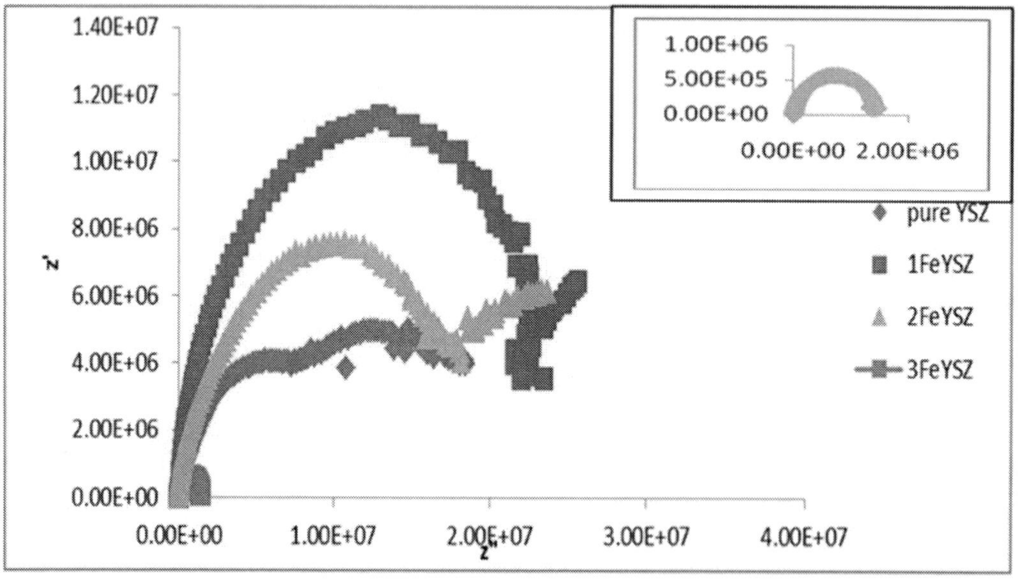

Fig. 5. Impedance spectra of sintered pure YSZ, 1FeYSZ, 2FeYSZ and 3FeYSZ operated at 300°C.

4 Conclusion

Fe-doped 8YSZ was found to be a suitable electrolyte material in solid oxide fuel cell. The crystallization temperature decreased as the composition increased. As the addition of Fe increased, the crystal volume of tetragonal phase tend to decrease because Fe^{3+} ions have ionic radius which is 0.055nm smaller than Zr^{4+} ions which is 0.084nm, substitution of

Zr^{4+}ions with smaller Fe^{3+} ions induce the variation in lattice parameter, thus decrease the volume of the unit cellThe conductivity of Fe-doped 8YSZ was found to be slightly higher than pure 8YSZ. 3FeYSZ has lower grain resistivity and thus has higher conductivity.

References

1. H. Gao, J. Liu, H. Chen, S. Li, T. He, Zhang, J. Solid State Ionics, 179, 1620 (2008)
2. P. Satardekar, D. Montinaro, V.M. Sglavo, Ceramics International, 41, 9806 (2015)
3. T. Dey, A.D. Sharma, A. Dutta, R.N. Basu, J. Alloys Compd., 604, 151 (2014)
4. In Jin Shon, In Kyoon Jeong, Jeong-Hwan Park, Byung-Ryang Kim, Lee, K.T., Ceramics International, 35, 363 (2009)
5. O. Bohnke, V. Gunes, K.V. Krachyk, A.G. Belous, O.Z.Y Anchevskii, V. Yunov, O.I.V. Yunov, Solid State Ionics, 262, 517 (2014)
6. H. Nakajima, K. Itoh, H. Kaneko, Y. Tamaura, J. Phys. Chem. Solids, 68, 1946 (2007)
7. GurpreetKaur, S. Basu, J. Power Sources, 241, 783(2013)
8. S. Bang, University of California, Irvine (2008)
9. M.L. Perry, T.F. Fuller, J. Electrochem. Soc., 149, 59 (2002)
10. S.P.S Badwal, K. Foger, In. Ceram. Int., 22, 257 (1996)
11. Chih Wei Kuoa, Y. H.S., I. Ming Hungb, Shaw Bing Wenc, M.C. Huey ErLeed., J. Alloys Compd., 8 (2008)
12. Kaiser, A.J. Feighery, D.P. Fagg, J.T. Irvine, Ionics 4, 215 (1998)
13. G.S. Lewis, A. Atkinson, B.C.H. Steele, J. Mater. Sci. Lett., 1155 (2001)
14. J. Schefold, A. Brisse, M. Zahid, J. Electrochem. Soc., 156, 897 (2009)
15. J.H. Kim, G.M. Choi., Solid State Ionics, 130, 157 (2000)
16. S. Linderoth, N. Bonanos, K.V. Jensen, J.B.B. Sorensen, J. Am. Ceram. Soc., 84, 2652 (2001)
17. H. Yakonawa, N. Sakai, T. Horita, K. Yamaji, M.E. Brito, MRS Bulletin, 30, 591 (2005)
18. J.W. Fergus, J. Power Sources, 30 (2006)
19. E.C. SubbaRao, H.S. Maiti, Solid State Ionics, 11, 317 (1984)
20. F. Gou, P. Xiao, Ceram. Soc., 32, 4157 (2012)
21. Q. Dong, Z.H. Du, T.S. Zhang, J. Lu, X.C. Song, J. Ma, Int. J. Hydrog. Energy, 34, 7903 (2009)
22. F.M.L. Figueiredo, F.M.B. Marques, WIREs Energy Environ., 2, 52 (2013)
23. T.S. Zhang, Z.H. Du, S. Li, L.B. Kong, X.C. Song, J. Lu, Solid State Ionics, 180, 1311 (2009)
24. B. Meng, Z.Y. Miao, M. Kong, X.X. Liu, Q.Q. Yang, Solid State Ionics, 258, 61 (2014)

MATEC Web of Conferences **78**, 01103 (2016) DOI: 10.1051/matecconf/20167801103
IConGDM 2016

Automatic Semiconductor Wafer Image Segmentation for Defect Detection Using Multilevel Thresholding

N.H. Saad[1,*], *A.E.* Ahmad[1], *H.M.*Saleh[1] , and *A.F.* Hasan[1]

[1]School of Microelectronic Engineering, Universiti Malaysia Perlis, KampusPauh Putra, 02600 Arau, Perlis, Malaysia

Abstract. Quality control is one of important process in semiconductor manufacturing. A lot of issues trying to be solved in semiconductor manufacturing industry regarding the rate of production with respect to time. In most semiconductor assemblies, a lot of wafers from various processes in semiconductor wafer manufacturing need to be inspected manually using human experts and this process required full concentration of the operators. This human inspection procedure, however, is time consuming and highly subjective. In order to overcome this problem, implementation of machine vision will be the best solution. This paper presents automatic defect segmentation of semiconductor wafer image based on multilevel thresholding algorithm which can be further adopted in machine vision system. In this work, the defect image which is in RGB image at first is converted to the gray scale image. Median filtering then is implemented to enhance the gray scale image. Then the modified multilevel thresholding algorithm is performed to the enhanced image. The algorithm worked in three main stages which are determination of the peak location of the histogram, segmentation the histogram between the peak and determination of first global minimum of histogram that correspond to the threshold value of the image. The proposed approach is being evaluated using defected wafer images. The experimental results shown that it can be used to segment the defect correctly and outperformed other thresholding technique such as Otsu and iterative thresholding.

1 Introduction

Semiconductor fabrication industry has greater competition where it is being influenced by factors of cost, yield, reliability and delivery time. Each process in semiconductor wafer fabrication becomes crucial including testing and quality control. In these processes, one of the main activities is defect detection. Defect detection involves imaging and analysis of the whole surface of the semiconductor wafer. The objective of this task is to provide an analysis of production batches, i.e. the numbers of good and defective wafers. Most of semiconductor manufacturer used human experts for doing the visual inspection process on the wafer surface to detect any defects [1]. By using human experts, the visual inspection

* Corresponding author: hidayahsaad@unimap.edu.my

process needs supreme concentration and the time used to inspect each wafer is limited [2]. This process can be replaced by machine vision system. In order to be adopted into machine vision system, the wafer image can be segmented into background and the defect. In defect segmentation, we are only interested at the defected portion of the image.

Over the years, extensive researches have been done for semiconductor defect detection. A semiconductor defect detection using a novel fractal encoding algorithm has been proposed by Gleason et. al [3] that emphasized non-referential defect detection (NRDD). This research used new algorithmic rule that wrote of two major steps via the use of fractal images encoding and enhanced defect boundary delineation using active contours. Fractal image encoding broke the image to different sub regions and then determined the best set of affine transformations map one area to another.

Zontak et.al [4] suggested a defect detection procedure that avoid image registration and robust to variations of the pattern. The method used is based on anisotropic kernel reconstruction of the original image by using a reference image. The sources and reference images are mapped into a feature space, where each features are estimated by the weighted sum of neighbouring characteristics reference images. They shown that the features derived from the areas of disability cannot be reconstructed from the reference image and can be identified.

P.Bourgeat studied the wafer image segmentation by the Gabor filters [5]. He developed a technique designed to train the classifier with a small set of training samples and applied to wafer inspection optical tool known as Direct-to-Digital Holography (DDH) tool. The phase information shown a strong invariance to non uniformities and can readily be used for image segmentation.

Image segmentation involves subdividing the image into areas of similar attributes. It is one of the most crucial and challenging tasks in image analysis and pattern recognition. Image segmentation methods are generally based on one of two fundamental properties of the intensity values of image pixels: similarity and discontinuity [6]. Similarity concept is focusing on partitioning the image into several different regions and there must be similar image pixels that belong to a region according to a set of predefined criteria. Discontinuity concept will partition the image on the basis of abrupt changes in the intensity values of the image. Based on these two concepts, many segmentation methods have been introduced and can be divided into six main categories which are Edge-based, Threshold based, Region-based, Neural Network based, Cluster-based and Hybrid [7].

This paper presents an efficient image segmentation method using multilevel thresholding technique based on histogram segmentation. Using this method, it is possible to determine the threshold value effectively.

1.1 Histogram thresholding

Clustering is one of the methods that can be used for segmentation. A clustering technique classifies the objects into different groups, or more specifically, partitioning of a data set into clusters (subsets) [8]. By doing that, the data in each cluster ideally shares some common trait. The histogram of an image contains some peaks, so it can be separated into a number of modes. Each mode is expected to correspond to a region, and there exists a threshold at the valley between any two adjacent modes [9]. In this project, a histogram based thresholding technique proposed by Papamarkos& Gatos known as multilevel thresholding is adopted to segment the semiconductor wafer defect from image background. In general, this technique can be divided into three main stages. At the first stage, a hill-clustering technique was applied to the image histogram in order to approximately determine the peaks location in the histogram [11]. Then, by using a linear mini-max approximation function, the histogram segments between the peaks were identified and approximated to a rational function [11]. Finally, the global minimum of each rational function was determined using one dimensional golden search minimization algorithm [10].

Consider that hill clustering gives a total number of histogram peaks equal to *P*, where $(W_n, G(W_n)), n = 1,........,P$, are the coordinates of the peaks[10]. This means that the histogram has *P-1* valleys which lie between the peaks. For the nth valley let:

- *K* be the total number of gray levels included;
- $G(w_k), k = 1,........,K$, be the values of the histogram at the $w_k$gray level
- $w_k, k = 1,........,K$, be the gray values with $W_n \leq w_k \leq W_{n+1}$

For each valley, the Linear Rational Approximation (LRA) algorithm can fit the histogram points $(w_k, G(w_k)), k = 1,........,K$ by a real function of the general form [10]:

$$R(w) = \frac{A(w)}{B(w)} = \frac{\sum_{m=0}^{N} a_m w^m}{1 + \sum_{m=1}^{M} b_m w^m} \quad (1)$$

wherea_m and b_m are the unknown coefficients, *M* and *N* are integers that define the degree of the polynomials *A(w)* and *B(w)*. Minimax criterion was used to solve this approximation problem.

The rational function *R(w)* is real and continuous. Therefore, the one dimensional golden search algorithm was applied in order to find the minimum value in the region W_n, W_{n+1}. The inputs of the golden search algorithm are only the limits of the interval which defined by W_n and W_{n+1} while the one dimensional function for each region is the real rational function, *R(w)*. To ensure that the golden search algorithm always converges to the global minimum, the following procedures were used [10]:

Step 1: Find the minimum, R_{min} of *R(w)*.
Step 2 : Define the function *Y(w)* according to the relation:

$$Y(w) = \begin{cases} R(w), if R(w) \leq R_{\min} \\ R_{\min}, otherwise \end{cases} \quad (2)$$

Step 3 : Find the minimum,Y_{min} of *Y(w)*.
Step 4 : If $Y_{min} = R_{min}$, go to step 5, otherwise put $R_{min}=Y_{min}$ and go step 2.
Step 5 : The global minimum solution is equal to R_{min}.The procedure is terminated.

From the above procedure, the minimum value of *R(w)* and its position was being determined. The minimum value of the first hill segment was taken as a threshold value of the histogram.

1.2 Defect segmentation

Fig. 1 shows the framework of automatic wafer image defect segmentation using multilevel thresholding.

DOI: 10.1051/matecconf/20167801103

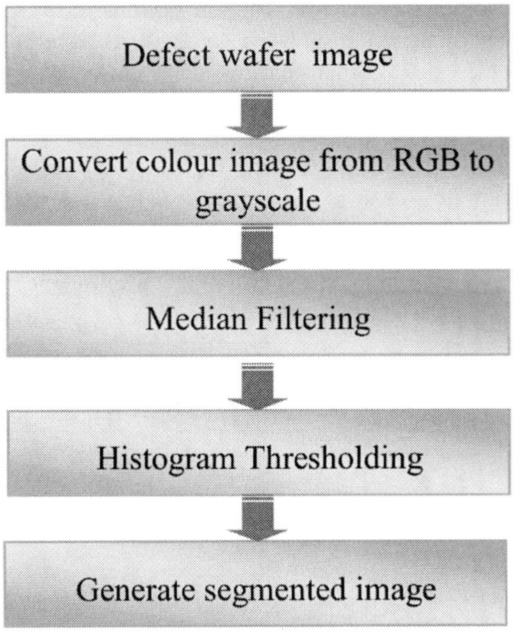

Fig. 1. Automatic Wafer Image Defect Segmentation.

Based on Fig. 1, the defect segmentation system operates by reading the input which is defected semiconductor wafer image in RGB image. Then, the RGB image was converted to gray scale image for simplifying the computational process. Each pixel in the gray scale image has intensity value from 0 (black) to 255 (white). The next step is to apply image filtering for modifying or enhancing an image to emphasize certain features or remove other features. In this project, median filtering was used to reduce the outliers without reducing the sharpness of image. The size of neighborhood used is 3x3. After completed enhancing the image, modified multilevel thresholding was used to find the threshold value for segmenting the defect from the background. The multilevel thresholding technique used histogram as a base to find the threshold value.

2 Experimental results

In order to demonstrate the performance of the proposed method, semiconductor wafer image has been used as an input. Three types of semiconductor wafer defect that commonly occur in the semiconductor industry have been chosen which are contamination, chip and crack. The images have been taken using high power microscope. The defects are shown by the black pixels in Fig. 2.

(a) (b) (c)

Fig. 2 Sample Wafer Images With (a) Contamination (b) Chip (c) Crack.

Table 1 shows the results of each steps for the automatic defect segmentation using multilevel thresholding. There are three types of defects that have been used as an input to the proposed method which are contamination, chip and crack. By referring to the imagesafter multilevel thresholding, they show that the defect had successfully segmented from the background for all three types of defect. The defects are indicated by the black pixels while the background is indicated by the white pixel. The threshold values found from multilevel thresholding using hill clustering approach are 45, 43 and 7 for each type of input image.

Table 1. Results of Each Steps on Wafer Image Defect Segmentation Using Histogram Thresholding.

Original Image	Gray scale Image	Image after Median Filtering	Image after Multilevel Thresholding	Thresholdin g Value
			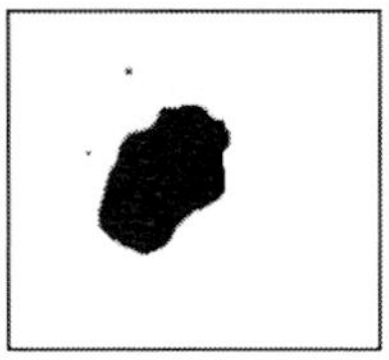	45
				43
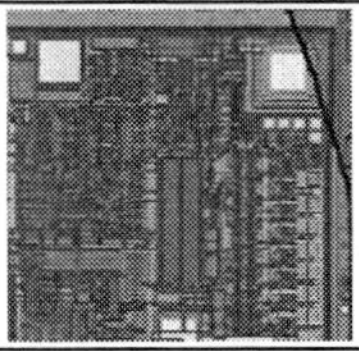	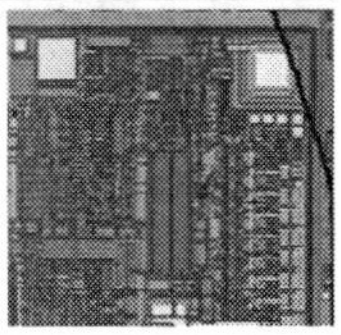	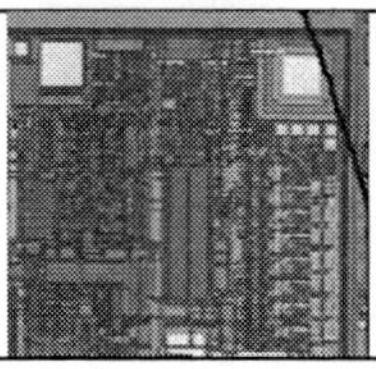	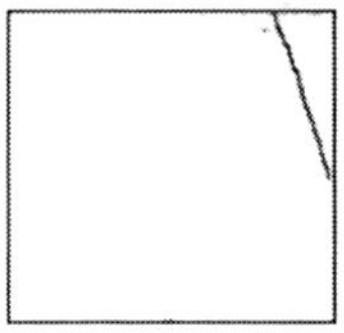	7

Table 2 shows the input image, Multilevelthresholding, Otsu thresholding and Iterative thresholding images. The threshold value computed by multilevel thresholding technique clearly segments the defect compared with other two techniques. Experimental result shows that multilevel thresholding outperforms the Otsu and Iterative thresholding in segmenting defect on semiconductor wafer image. This is due to Otsu and Iterative Thresholding techniques are more suited to the images that produce bimodal gray level histogram.

Table 2. Comparison of Thresholding Image and Value Between Multilevel, Otsu, and Iterative Thresholding Technique.

Input Image	**Thresholding Technique**	**Thresholding Image**	**Threshold Value**
	Multilevel Thresholding	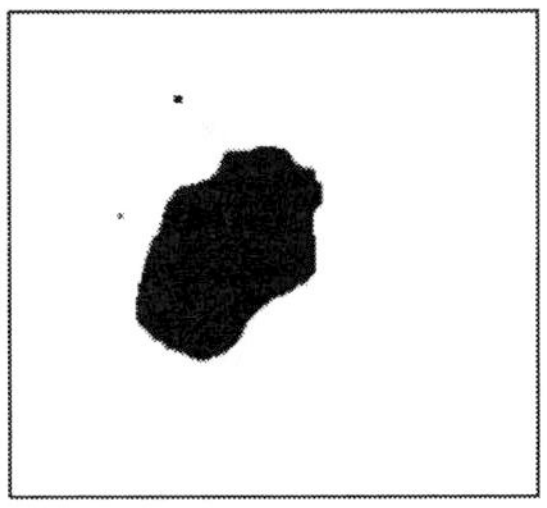	45
	Otsu Thresholding		138
	Iterative Thresholding		136

3 Conclusions

The automatic defect segmentation of the semiconductor wafer image is proposed and evaluated in this paper. The proposed approach used multilevel thresholding technique to determine the threshold value that further be used to segment the defect from image background. We have used three types of defected wafer images for the experimental observations and evaluation of the proposed method. Experimental results suggest that the proposed approach is able to segment the defected area from the background.

The authors would like to thank the staff from Microfabrication Laboratory, School of Microelectronic Engineering Universiti Malaysia Perlis for providing the sample of semiconductor wafer images.

References

1. N.G. Shankar, Z.W. Zhong, Microelectronic Engineering, 77, 337 (2005)
2. Chen, Fei-Long, Shu-Fan Liu, IEEE transactions on semiconductor manufacturing, 13, 366 (2000)
3. S. Shaun Gleason, K. Regina Ferrell, P. Thomas Karnowski, W. Kenneth Tobin, *Design, Process Integration, and Characterization for Microelectronics SPIE Proceeding*, Santa Clara, CA, US (2002)
4. Zontak Maria, Israel Cohen, Machine Vision and Applications, 21, 129 (2010)
5. P. Bourgeat, F. Meriaudeau, Jr. Tobin, K.W. Gorria, *International Society for Optics and Photonics* (2004)
6. MacQueen, James, *Proceedings of the 5th Berkeley symposium on mathematical statistics and probability* (1967)
7. Rajeshwar Dass, Priyanka, Swapna Devi, Image Segmentation Techniques, 1, (2012)
8. S.R. Dubey, P. Dixit, N. Singh, J.P. Gupta, IJIMAI, 2, 65 (2013)
9. Cheng Heng-Da, X.H. Jiang, Jingli Wang, Pattern recognition, 35, 373 (2002)
10. Papamarkos Nikos, Basilios Gatos, Graphical Models and Image Processing, 56, 357 (1994)
11. Tsai Du-Ming, Ying-Hsiung Chen, Pattern Recognition Letters, 13, 245 (1992)

MATEC Web of Conferences **78**, 01104 (2016) DOI: 10.1051/matecconf/20167801104
IConGDM 2016

Study on Modified Sand Filtration Towards Water Quality of Wet Market Waste Water

F.N.M. Saad[1,*], *M.N.* Jamil[1], *Z.S.M.* Odli[1] , and *T.N.T.* Izhar[1]

[1]School of Environmental Engineering, Universiti Malaysia Perlis (UniMAP), Kompleks Pusat Pengajian Jejawi 3, 02600 Arau, Perlis, Malaysia.

Abstract. Investigation on the potential of sand filter as a pre-treatment of waste water was done in Kangar wet market, Perlis. Besides, the best composition of filter in order to treat wastewater based on BOD, COD, SS, AN, turbidity and pH levels are further examined. In this study, there are four types of sand filter composition which the medias consist of fine sand and coarse sand while the modified sand filter are consist of sand, course sand and activated carbon prepared from rice husk and coconut shells. After 10 weeks of treatment, the results show that the concentration of BOD, COD, SS, AN, turbidity and pH were reduced up to 86%, 84%, 63%, 88%, 73%, respectively while pH nearly to neutral with 6.83. Moreover, the result also revealed that the sand filter added with rice husk almost complied with Standard B of Malaysia Environmental Quality (Sewage) Regulations 2009 as well as gives the highest number of WQI with 36.81. Overall, WQI obtained in this study are ranged from 12.77 to 36.81.

1 Introduction

Wet market is the place where people get raw materials such as fish, vegetables and chicken. Waste water generated from this premise is from different sources and activities such as cleaning of fish, vegetables and fruits as well as slaughtering poultry which used huge water usage. Then, this waste water containing high organic and inorganic matters are directly discharged into the drains without any treatment.

As said by Idris et al. [1], the wet market waste water usually contains two to three times of organic matter and solids commonly found in residential waste water, thus waste water, produce from market sources classified as "high intensity". Discharged of untreated waste water adds oxygen-demanding substances, nutrients and toxic elements into the water, thus influencing the level of parameters such as BOD, COD and TSS in the receiving water. This scenario in turn converts the streams to become unsuitable for aquatic flora and fauna. Because of this effect, discharged of grey water into the steam need to be treated to avoid or reduce that effect [1]. Therefore, an economical and easy to be used treatment should be applied in the premise in order to treat the waste water generated from wet market before it being released to receiving water.

* Corresponding author: farahnaemah@unimap.edu.my

Sand filter is one of the oldest wastewater treatment technologies. It is also a biofilm technology that has been used in waste water treatment due to its low cost and maintenance and relatively high treatment efficiency. There have been several studies carried out and reported relevant to the application of sand filter in treating waste water [2,3]. Results showed that the sand filter may reduce the concentration of water quality such as BOD, COD and SS. Therefore, this study is aimed to observe the potential of sand filter act as a pre-treatment of wet market waste water as well as to compare the quality of waste water produced from conventional sand filter with modified sand filter which added with activated carbon based on BOD,COD, suspended solid, AN, turbidity and pH.

2 Materials and methods

2.1 Sampling location and activity

Kangar wet market (6°26'11"N 100°11'33"E) was chosen as sampling location. Grab water sample was collected at the final discharge of the wet market during peak hour of wet market operation before being transfered to the laboratory for further study.

2.2 Preparation of activated carbon

Coconut shell was taken from the local shop selling coconut milk while the rice husk was collected from local rice mill. Both materials were washed with distilled water to remove impurities before being dried in the oven. Then, both coconut shell and rice husk were burned in the furnace at 300°C for 2 hours before being ground and sieved according to the desired size.

2.3 Preparation of sand filter

The sand filtrations were constructed in the transparent plastic columns. The dimension of each plastic column is 260 mm of height, 80 mm for top diameter and 24 mm for bottom diameter. A hole was created at the bottom of the plastic column which acts as a channel for effluent water sample.

Fine sand, coarse sand, as well as rice husk and coconut shell act as activated carbon, with their effective size of 0.06 mm, 2.00 mm, and 0.08 mm, respectively were used in this study. All these materials were washed by using distilled water to remove impurities and dried before being filled up in the plastic column. Plastic net was used to avoid the material from wash out.

There were four types of sand filtrations used in this study. There were sand filtration consist only fine and course sand, sand filtration with additional of rice husk, sand filtration with additional of coconut shell and sand filtration with additional of rice husk and coconut shell. The schematic diagram of all sand filters used in this study is shown in Fig. 1.

2.4 Experimental study

5 L of water sample was pumped into each of the sand filter. The flow rate for all filters was 0.13 L/ min. Then, the effluent of each sand filter was collected for further analysis. In this study, six parameters were chosen based on water quality index (WQI). The parameters are BOD, COD, SS, ammoniacal nitrogen, pH and turbidity. All procedures of experiments were according to American Public Health Association (APHA) [4].

 DOI: 10.1051/matecconf/20167801104

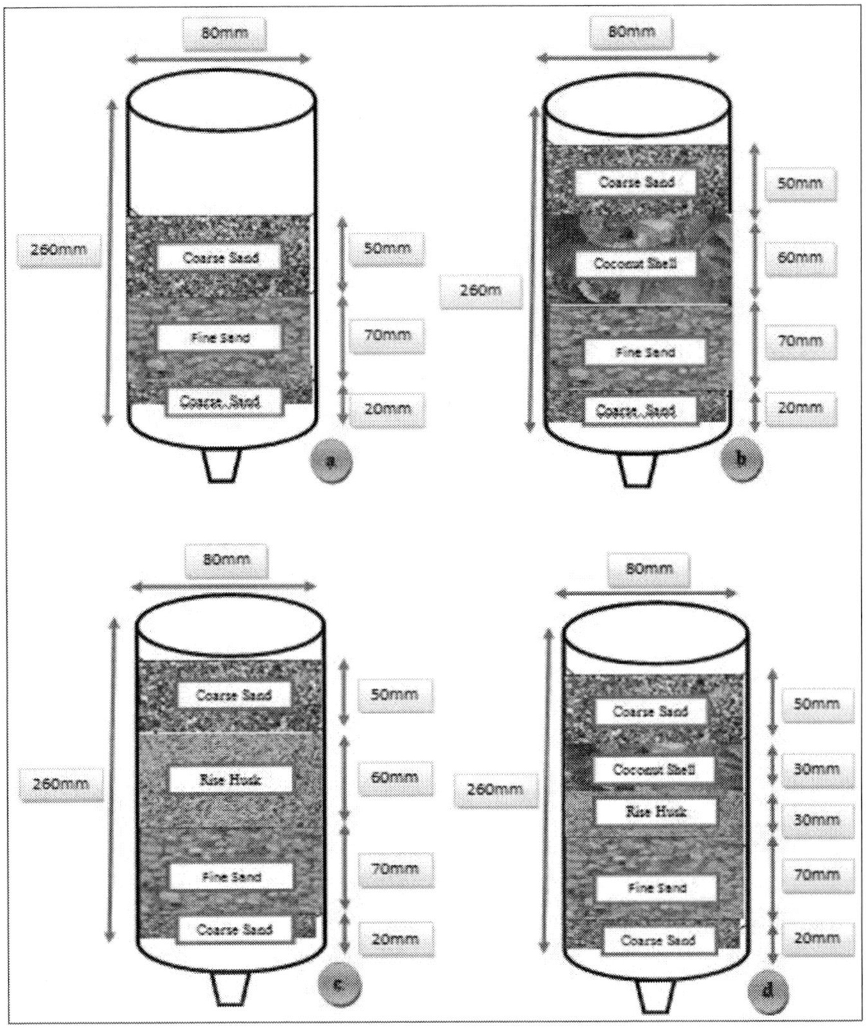

Fig. 1. Schematic diagram of (a) sand filtration, (b) sand filtration with coconut shell, (c) sand filtration with rice husk and (d) sand filtration with rice husk and coconut shell.

3 Results and discussion

The wet market waste water was collected at the final discharge of the premise before it being treated with the four types of sand filtrations. Then, the filtered waste water were analysed based on six parameters and compared with the Standard B of Malaysia Environmental Quality (Sewage) Regulations 2009 [5]. The WQI of each effluent of sand filtration is also calculated.

The concentrations of BOD, COD, SS and AN recorded by the effluent of wet market before and after being treated by four types of sand filters is shown in Fig. 2. The results of these parameters are then compared with the Malaysia's regulation.

The result shows that high BOD level is detected before the treatment is applied with 137.07 mg/L while the lowest is recorded from the sand filter added with rice husk which is 18.97 mg/L. This indicates that more than 80% of organic matter is removed. Besides that, the result also shows that only two types of sand filters are comply with the Standard B (50 mg/L) which are sand filter added with rice husk and sand filter added with both activated carbons. This suggests that the addition of activated carbon in the sand filter composition may reduce the pollutant in the water sample.

High BOD level in water sample has course a threat to the aquatic environment since it reduces the dissolved oxygen concentration to the levels that affect aquatic organisms [6]. This is due to the fact that oxygen is used by microorganisms to oxidize the organic matter to stable inorganic form [7].

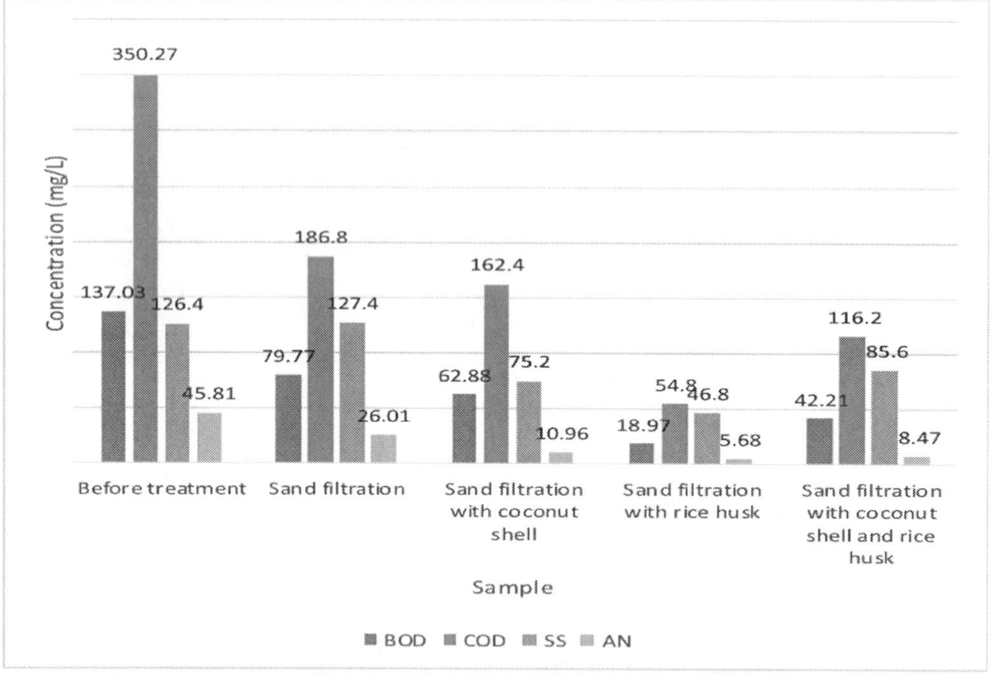

Fig. 2. Concentration of BOD, COD, SS and AN recorded by the wet market waste water before and after treatment with sand filters.

COD is the amount of oxygen required to chemically oxidize organic and inorganic matter. COD gives an estimation of the amount of organic and inorganic matter present. In Fig. 2, it shows that high level of COD recorded before the water sample is treated with 350.27 mg/L. On the other hand, all the water samples that gone through the treatments have shown a good finding where the COD concentrations recorded are lower than the Standard B (200 mg/L), ranging from 54.8 to 186.8 mg/L. This indicates that all types of sand filtrations tested in this study have a potential in removing organic and inorganic matters in the water sample with the higest percentage reduction of 84%.

Further analysis was carried out by examined the level of suspended solids in the water sample before and after the treatment with sand filters. According to Standard B of Malaysia Environmental Quality (Sewage) Regulations 2009, the limits of SS is not more than 100 mg/L. Refering to the Fig. 2, the results show that the water samples are high with SS before the treatment as well as treated with the sand filter. However, the SS concentrations recorded by sand filtrations added with activated carbon have reduced the pollutants in the effluent with the lowest concentration of 46.8 mg/L, indicating of more than 60% reduction. Excessive SS in the water sample may cause the recieving water to lose its stability in suppoting the diversity of aquatic life. It will abdsorb heat from the sun thus increase the water temperature and lower the levels of dissolved oxygen [8].

The concentrations of AN in the water sample before and after the treatment are shown in Fig. 2. The result shows that the level of AN are ranging from 5.68 mg/L up to 45.81

mg/L with the lowest and the highest AN concentrations are recorded by sand filter added with rice husk and before the water sample is treated, respectively. Although high percentage reduction is recorded with 87.6%, this concentrations are still not comply with the limit set up by the Malaysian law which is 5 mg/L. AN is very soluble in the water. High level of AN contributes to oxygen demand and nutrient loading of the water bodies, promoting toxic algal blooms and leading to a destablised aquatic ecosystem [9].

The pH of the water sample is recorded in the study before and after sand filter treatment is shown in Fig. 3. The result shows that the highest pH is 6.83 which recorded by sand filtration while the lowest is recorded by sand filtration added with both activated carbons with 6.38. Overall, it reveals that the water sample is slightly acidic. Based on the Standard B of Malaysia Environmental Quality (Sewage) Regulations 2009, all the water samples are comply with the limit (5.5-9.0). According to Morrison et al. [10], high pH values could alter the toxicity of other pollutants while low pH values may affect aquatic life and impair recreational uses of water.

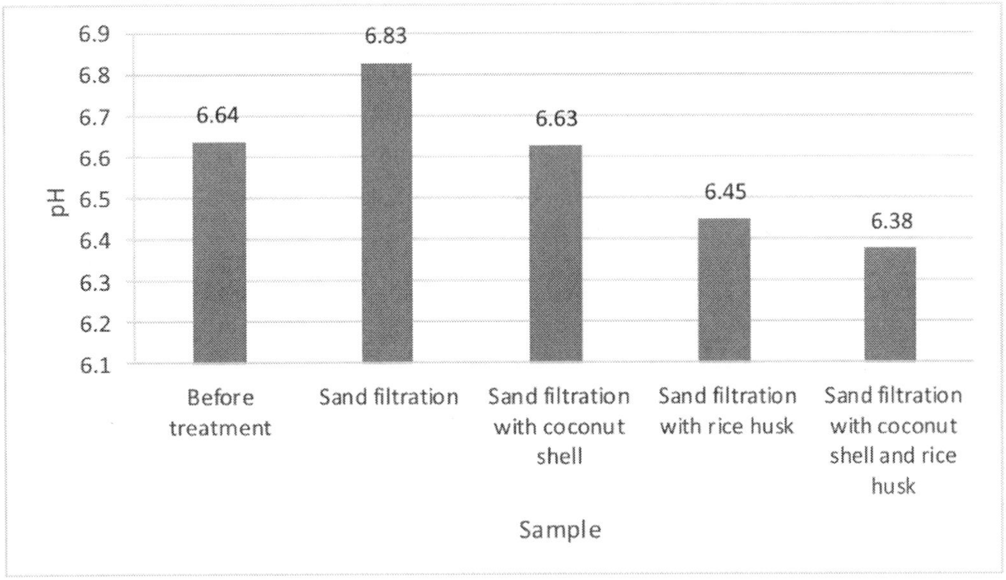

Fig. 3. pH of the water sample before and after treatment with four types of sand filters.

Fig. 4 shows the level of turbidity of the water samples recorded during the study period. It shows that high turbidity is recorded before the sample is treated with 41.68 NTU while the lowest is recorded by sand filter added with rice husk which recorded 11.12 NTU. Generally, all the effluents collected from sand filters are low which is less than 16 NTU, showing that more than 70% are reduced. As said by Ntengwe [11], turbidity is caused by the presence of suspended matter, organic particles, plankton and other microscopic organisms.

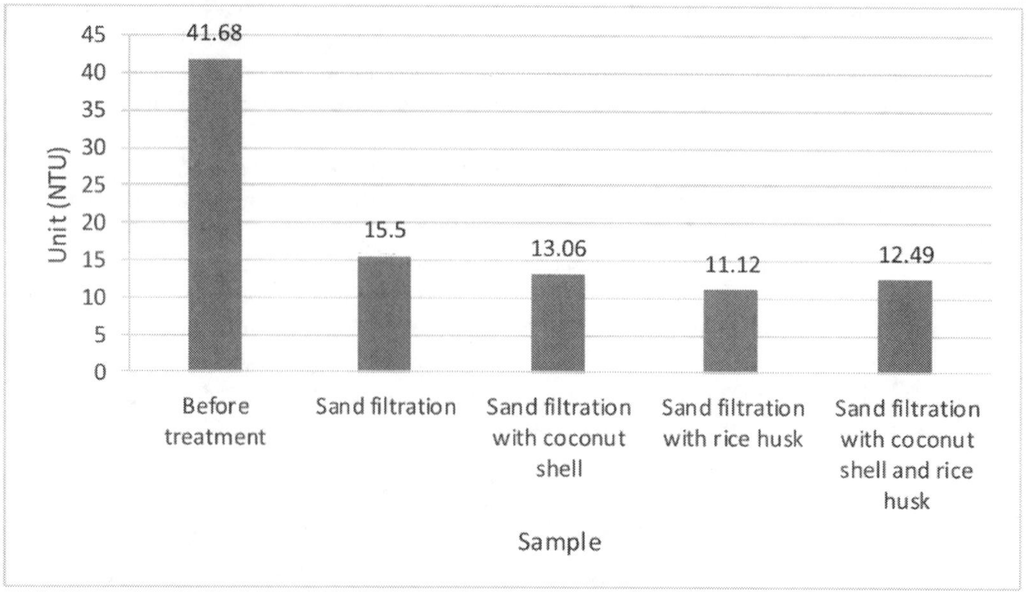

Fig. 4. The turbidity of water sample before and after sand filters treatment.

Water quality index of the water samples are also calculated in this study. Fig. 5 lists the WQI recorded by water samples before and after sand filters treatment. Overall, the result shows that only sand filtrations with rice husk falls in Class IV (31.0-51.9) which is suitable for irrigation purpose only while others fall in Class V (less than 31.0) which is not suitable for any activities. High WQI recorded by sand filter added with rice husk may be due the pores produced during activation that trapped more pollutants of the water thus filtered more efficiently. According to Kalderis et al. [12], the relative surface area (S_{BET}) for rice husk was up to 750 m^2/g while for coconut shell, the relative surface area was 546 m^2/g [13]. Hence, it gives a better result compared to the other samples.

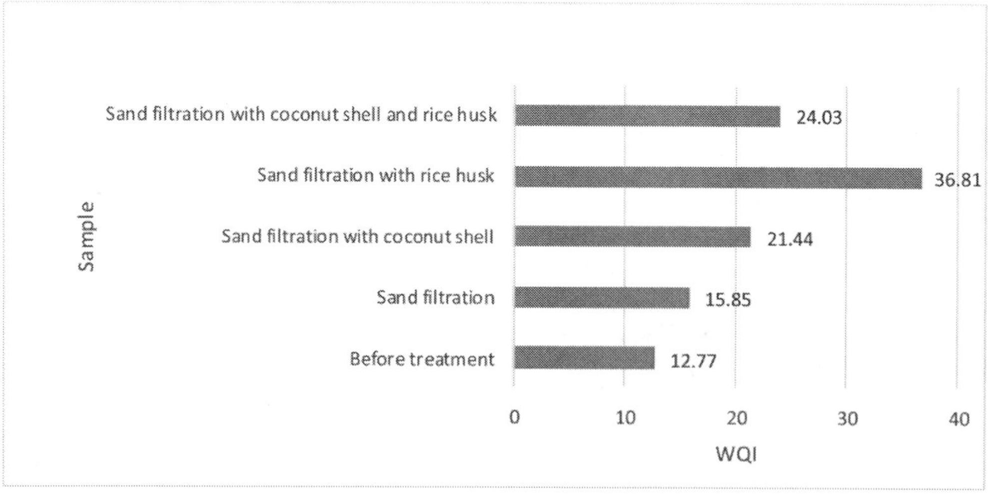

Fig. 5. Water quality index of wet market waste water before and after treatment with four types of sand filters.

4 Conclusions

This study shows that wet market waste water is high with BOD, COD, AN and SS which exceed the limit that set up by Malaysia authority. Therefore, four types of sand filters are constructed in order to treat the waste water produced. There are four types of sand filters examined in this study which one is conventional sand filter while the rest is added with activated carbon. Based on laboratory tests, the modified sand filter that added with rice husk is better than sand filtration added with coconut shell or a combination of both activated carbon. The results show that better performance in eliminating the concentration of the parameters tested is by sand filters added with rice husk where most results are comply to Standard B of Malaysia Environmental Quality (Sewage) Regulations 2009 such as the concentration of BOD, COD, SS, and pH. Besides that, this study also reveals that sand filtration added with activated carbon shows a potential in eliminating the concentration of some parameters compared to the conventional sand filters. Therefore, the addition of activated carbon in the filter can be one of the alternative for the treatment of waste water wet market.

References

1. Idris, W.N.W. Azmin, M.A.M. Soom, A.A. Mamun, *1st International Conferences on Managing Rivers in The 21st Century : Issue and Challenges* (2004)
2. R. Seswoya, A.M.M. Daud, Z.M. Ali, Z.D.M. Basri, R.N.R. Yunus, *The first attempt of chicken wastewater treatment using sand filtration* (2010)
3. J. Xi, K.M. Mancl, O.H. Touvinen, *Proceedings of the 8th International Symposium on Agricultural and Food Proceesing Waste., St. Josepth, MI* (2000)
4. American Public Health Association, *Standard Methods for the Examination of Water and Wastewater, supplement* (American Public Health Association, Washington D.C, 2000)
5. Legal Research Board (LRB), *Environmental Quality Act 1974 (ACT 127)* (Regulations, Rules and Orders, International Law Book Services, Petaling Jaya, Selangor, Malaysia, 2013)
6. S.K. Garg, Environmental Engineering, **2**, 228 (2006)
7. W. Viessman, J.M. Hammer (Harper Collins, New York, 2000)
8. S.H. Peavy, D.R. Rowe, G. Tchobanoglous, *Environmental Enginnering* (McGraw Hill, New York, 1985)
9. E.O. Igbinosa, A.I. Okoh, Int. J. Environ. Sci. Technol., **6**, 175 (2009)
10. G. Morrison, O.S. Fatoki, L. Persson, A. Ekberg, Water SA, **27(4)**, 475 (2001)
11. F.W. Ntengwe, Phys. Chem. Earth, **31**, 832 (2006)
12. D. Kalderis, S. Bethanis, P. Paraskeva, E. Diamadopoulos, Bioresource Technol., **99**, 6809 (2008)
13. E.Y. Yazici, H. Devici, I. Alp, T. Yilmaz, O. Celep, *Survival and Sustainability: Environmental Concerns in the 21st Century* (Springer-Verlag Berlin Heidelberg, 2001)

Study on Characteristics of Lightweight Aggregate Concrete Made From Foam and Ordinary Portland Cement

N.M. Ibrahim[1,*], *K.N.* Ismail[1] , *S.* Salehuddin[1] , *R.C.* Amat[1], *A.R.A.* Razak[2] , and *Z.S.* Odli[1]

[1]School of Environmental Engineering, University Malaysia Perlis (UniMAP), Kompleks Pusat Pengajian Jejawi3, 02600 Arau, Perlis, Malaysia.
[2]School Electrical System Engineering, University Malaysia Perlis (UniMAP), Pauh Putra Engineering Campus, 02600 Arau, Perlis, Malaysia.

Abstract. The production and characteristic of lightweight bubble aggregates (LBA) are presented in this paper. The LBA are produced by mixing between the foam and ordinary Portland cement according to the composition which has been set. Then, the characteristics of LBA such as density, water absorption, specific gravity, compressive strength, aggregate impact value and microscopic analysis of the LBA are analyzed. Those characteristics are identified in order to ensure that the LBA are successfully categorized into lightweight aggregate. The loose bulk density is obtained at 812.5 kg/m3 which can be categorized under lightweight aggregate group. For water absorption the value obtained is 9.7 % which is slightly higher compared to normal aggregate. Meanwhile the average specific gravity obtained for the samples of LBAis 1.75. Compressive strength for the aggregates was 17.76 MPa. The highest compressive strength for LBA foamed concrete was obtained at 25% replacement with 7.83MPa. Thus, the LBA have a significant features and characteristics that can be used as coarse aggregates in concrete.

1 Introduction

Lightweight aggregate is one type of coarse aggregate that is normally used in the production of lightweight concrete elements such as concrete block, beams, and columns as well as in slabs for precast concrete. It is proven from previous research that have been done that there are many advantages infusing lightweight aggregate in concrete which are: a) total permanent action or also known as dead load can be reduced that may results in reduced footings sizes and significant reduction in total reinforcement, lighter and smaller precast elements which needs smaller and less expensive handling and transporting element, smaller sizes of columns and slab and beam dimensions, bigger space availability and better thermal insulation, enhanced fire resistance[1].

In lightweight concrete production, the most popular method is to use lightweight aggregate (LWA). It is now common in construction industry that the lightweight concrete

* Corresponding author: norlia@unimap.edu.my

has been made using artificial lightweight aggregate for the coarse aggregate substitution and using normal weight sand for the fine aggregate. Examples of artificial lightweight aggregates such as expanded clay and shale are made because they are most suitable for the production of structural quality lightweight concrete. Commercially available, these types of lightweight aggregates are obtained through heat treatment which also known as sintering process temperature 1000–1200 °C. Due to this high temperature process, overall production process will acquire higher fuel costs. As a result, overall construction cost will increase. Therefore, it is very beneficial to find a lightweight aggregate from a natural source. However, with the increasing use of these types of lightweight aggregates, it will consequently cause a severe depletion of the natural resources. As an alternative source for lightweight aggregate, artificial lightweight aggregate is used. These artificial aggregates as alternative building material will definitely provide a significant saving in the overall cost of construction and also preserve natural environment [2].

Other than using LWA, lightweight concrete can also be produced using foamed concrete which is either made from a cement paste or mortar. Foamed concrete is generally classified as lightweight concrete, where air-voids are entrapped in mortar by suitable foaming agent. It is noted that it possesses high flow ability, low self weight, minimal consumption of aggregate, controlled low strength and more importantly excellent thermal insulation properties. With proper control of dosage of foam, a wide range of densities (1800– 400 kg/m^3) of foamed concrete can be obtained for application to structural, partition, insulation and filling grades[3].

With a wide range of applications, foamed concrete is an extremely versatile material. The properties of foamed concrete are very suitable to be used in many other applications besides for concrete production. Its application varies from road sub-bases, bridge abutments, and bridge strengthening up to road widening. It is very cost effective and especially advantageous when used for projects on soft ground. Its higher porosity effects will help to mitigate flood in urban area. To achieve a proper road structure; usually allows less excavation of the road bed, because cellular concrete will provide high strength and insulation in one layer [4].

2 Materials and methods

2.1 Preparation of samples

Ordinary Portland cement and foam are the main ingredients in producing LBA. Figure 1 shows the overall method in production of LBA. By using dilution factor of 1:2.5 where one portion of foaming agent diluted with two and half portion of water, nice and thick foam will be produced. The foam that has been produced was added into the Ordinary Portland Cement with the ratio of 1:2. The mixture will be stir thoroughly and carefully until all cement and foams are well blended. Then, the fresh paste will be formed into rounded shape lightweight aggregate. The mixture was based from the research made by[5]. Next, the LBA were placed in the oven at temperature of 105 ± 5 °C for 24 hours. LBAs were leaved at room temperature to cools it down as shown in Figure 2.

DOI: 10.1051/matecconf/20167801105

Fig. 1. The mixing process of LBA.

Fig. 2. LBA at room temperatures.

2.2 Loose bulk density test

In this study, loose bulk density was determined according to the standard of ASTM C29/C29M as the aggregate was below the maximum size, 37.5 mm. the loose bulk density is the mass of non-compacted aggregates required to fill the container of a unit volume after aggregates were batched based on volume. Loose bulk density was determined by shovelling method [6].

2.3 Water absorption test

Water Absorption was the increase in mass of aggregate due to water penetration into the pores of the particles during a prescribed period of time, but not including water adhering to the outside surface of the particles. It is conducted to calculate the change in the mass of an aggregate due to water absorbed in the pore that present in the aggregate [7]. Referring to the method in ASTM C127, 2 kg of lightweight aggregate (LWA)are washed and dried before the test sample is placed in the oven at atemperature of 105 ± 5 °C for 1-3 hours. Next, let it cools down at room temperature until it can be handled. Then the samples were immersed in water at room temperature for a period of 24 ± 4 hours. The test sample was removed from the water and wiped until it's became saturated surface dry. Then the lightweight aggregate will be weighted and recorded as B. Afterwards, LWA is placed in the oven at temperature of 105 ± 5 °C for 24 ± 4 hours and it is weighted and recorded as A. Water absorption can be calculated by knowing the difference in weight of LBA.

2.4 Specific gravity test

Specific gravity can be known as the ratio of the weight of a given volume of aggregate to the weight of an equal volume of water using an apparatus known as Densimeter. The result of specific gravity for LBA will be compared with granite.

2.5 Aggregate impact value

The aggregate impact value test is conducted to determine the strength value of coarse aggregate that is conducted by using apparatus name Aggregate Impact Value Tester. It follows the Indian Standard

2.6 Microscopic analysis for aggregate

Microscopic analysis of the lightweight aggregate are carried out by using Scanning Electron Microscope (SEM) TM3000 Table Top Microscope in order to obtain the images of the LBA before and after it was mixed in the concrete mixture. The morphology of LBA will also important to avoid segregation in concrete.

2.7 Production of foamed concrete

Foamed concrete also known as lightweight concrete is a concrete made up of mortars (cement, sand and water) and also foam. The proportion of the mortars were calculated by using foam excel calculation. Firstly, there are three parameters needed to be set. The three parameters are the water/cement (W/C) ratio, sand/cement(S/C) ratio and target density. W/C ratio, S/C ratioand target density was set to be 0.5, 1 and 1500 kg/m^3. After these threeparameters are set, the rest of other raw materials were calculated and came out with 9 kg of cement, 9 kg of sand and 4.5 kg of water and 4.23 kg of foam are needed in producing normal foamed concrete. The proportion of foam was taken from the proportion of production of foam in bubble aggregate which is 1:2.5 which one portion of foaming agent diluted with 2.5 portion of water.

Firstly, the foam is produced when the 1.19 kg of foaming agent and 2.99 kg of water let into the foam generator and 4.23 kg of foam produced is poured in the cylinder bowl. Then, the 4.23 kg of foam is mixed with 9 kg ordinary portland cement, 9 kg of sand and 4.50 kg of water in the wheelbarrow in order to cast 15 cubes of control sample. Next for the 25% substitution of lightweight bubble aggregate (LBA), it takes 4.23 kg of foam mix with 9 kg of OPC, 6.75 kg of sand, 2.25 kg of LBA and 4.50 kg of water in order to cast 15 cubes. Next for the third sample which is the 50% substitution of LBA, 4.230 kg of foam is mixed with 9 kg of OPC, 4.50 kg of sand, 4.50 kg of LBA and 4.50 kg of water. All the batches are mixed manually by hand in wheelbarrow until it is well mix as shown in Figure 3.5. The batch of each type of samples is poured into 15 cubes of 100 mm × 100 mm × 100 mm cube without using vibrator to avoid losing of bubble particle. Then after 24 hours, remold it and placed it in the curing tank.

3 Results and discussion

3.1 Density

In this study loose bulk density test is conducted. After conducting the test, it is found that the value of loose bulk density is 812.5 kg/m^3 ; the value is below the maximum value which is 880 kg/m^3 permitted in [8]. The lightweight aggregate is light due to the cellular or high internal porous microstructure, which gives this type of aggregate a low density. The lightweight aggregate is lighter due to the presence of air voids, which gives this type of aggregate a low density [9].

3.2 Water absorption

Water absorption of the LBA is 9.7%. Since the LBA have porous structure, the lightweight aggregates absorb more water than normal aggregate.After undergoing the test for 24 hours which is according to procedure in ASTM Test Method Specific Gravity and Absorption of Coarse Aggregate (C 127), lightweight aggregates will absorb from 5% to more than 25% by weight of dry aggregate while for normal aggregates, it generally absorbs less than 2% [10].

3.3 Aggregate impact value

Based on the test, the value obtained is 18.99% which falls into group of strong aggregate. Table1 shows the aggregate impact value classification.

Table 1. Aggregate Impact Value Classification [11].

value	group
<10 %	Exceptionally Strong
10 – 20 %	Strong
20-30 %	Satisfactory for road surfacing
>35 %	Weak for road surfacing

According to the result, the value obtained is 18.99% which is classified as strong aggregate because during the formation of LBA, the uniform paste is compressed before it is formed into a spherical shape. Due to compressed, some of the pore will diminish or the some of it will become smaller resulting increasing in strength of LBA which make it difficult to be crushed.

3.4 Microscopic analysis for aggregate

Based on Figure 3, pores in the LBA have a variety of shape and size. Although, the spherical shape of pores is dominating, they are also accompanied with small amount of irregular shaped pores in the aggregate. The pores are formed during the production of the aggregate as the bubbles from the foam entrapped between the cement. At the same time, sightings of cracks or fractures also can be seen which occurs resulting from the impact of a hammer as this analysis requires only a fragment of the aggregate to be enlarged using SEM.

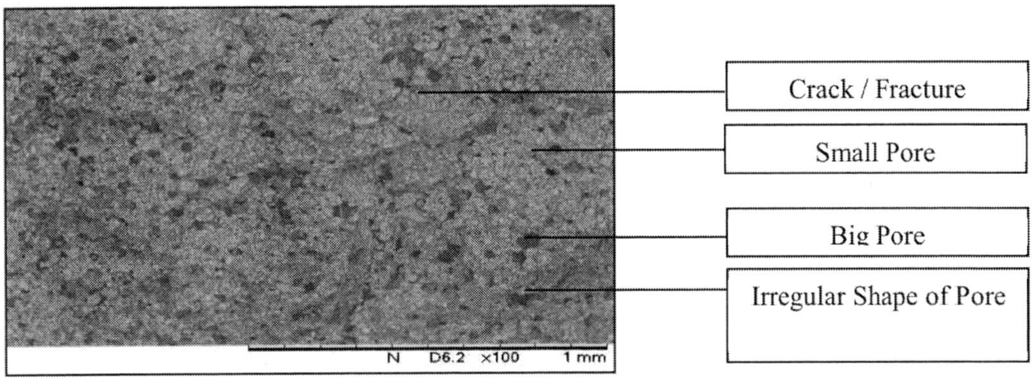

Fig. 3. Microstructure of LBA.

Figure 3 also shows that the pore in the LBA is not consistent in size. From the figure, there were many pore with different of sizes. This occurs during the production process of LBA, aluminum are not added into the mixture. Aluminum acts as a foaming agent which can create a consistent pore size [12].

3.5 Compressive atrength of LBA foamed concrete

Figure 4 shows the result of the samples where three type of concrete sample were used in this analysis which is the control sample, substitution 25% and 50% of LBA. All the samples were tested at the age of 7, 14 and 28 days by taking an average of 3 cubes for each sample. For control sample, the foamed concrete is formed without the addition of LBA. The pattern of compressive strength is increase from day 7 to until day 28 as shown in the Figure 4. On day seven, the control samples have the strength of 2.51 MPa and rose up to 2.80 MPa on day 14 which it is 11.55% increase. On day 28 it show the strength of 3.21MPa which increases 14.6% from day 14.

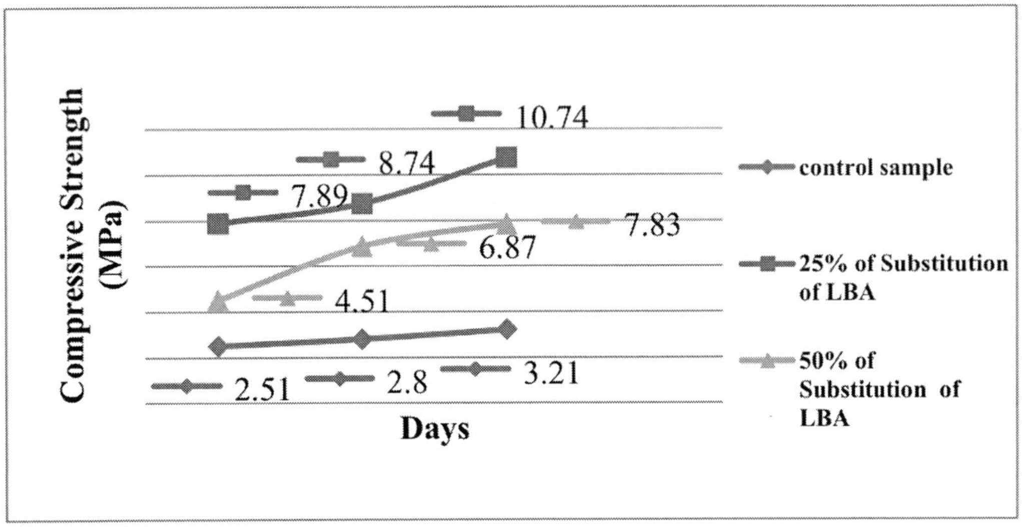

Fig. 4. Compressive Strength of LBA Foamed Concrete.

For sample with 25% substitution of LBA, the pattern of compressive strength increased with day as shown in the Figure 4.6. On day 7, the samples have the strength of 7.89 MPa which increase up to 10.77% and on 14th day reach the compressive strength of 8.74 MPa.

During the final day of 28 days, the compressive strength increased up to 10.74 MPa which rose up with 22.8% from day 14 until day 28.

The pattern for compressive strength of 50% substitution of LBA also increases with days as in Figure 4.7 which is similar to pattern of control sample and 25% substitution of LBA. The compressive strength for 7 days is 4.51 MPa and increase up to 6.87 MPa. The increase of percentage from day 7 to day 14 is 52.3% which is higher compare to control sample and sample of 25 substitution of LBA for period of 8 days. During final day which is day 28, the compressive strength of the sample rose up to 7.83 from day 14 which increased 13.9 7%.

4 Conclusions

The strength of foamed concrete with 25% substitution of of LBA is higher compare to 50% substitution of LBA and control sample with 10.738 MPa but past research has confirmed that strength of foamed concrete can achieve up to 15 MPa. In this study, the 25% substitution of LBA is the optimum percentage required in order to enhance the strength of foamed concrete. For density, the density of foamed concrete is increasing as the LBA is increasing. The results show that 50% substitution of LBA has density of 1832 kg/m^3 which is the highest, followed by 25% substitution of LBA with 1735 kg/m^3 and normal lightweight concrete with 1577 kg/m^3 All the value for the density is in the range of foamed concrete density which is below than maximum value, 1840 kg/m^3.

References

1. O. Kayali, *World of Coal Ash Conference, Lexington, Kentucky, USA* (2005)
2. P. Shafigh, M.Z. Jumaat, H. Mahmud, H, Constr. Build. Mater., 25(4), 1848 (2011)
3. K. Ramamurthy, E.K.K. Nambiar, G.I.S Ranjani, Cement Concrete Comp, 31(6), 388 (2009)
4. Nadia Salim, T.M. Dhirar, Al-Taqani, 25, 160 (2012)
5. L.Q. Wen, *Properties Of Concrete Made From Lightweight Aggregate Under Different Curing* Time, Universiti Malaysia Perlis (2014)
6. American Standard, *Standard Test Method for Bulk Density ("Unit Weight") and Voids in Aggregate* (ASTM C29/C29M, 2009)
7. American Standard, *Standard Test Method for Density, Relative Density (Specific Gravity), and Absorption of Coarse Aggregate* (ASTM 127, 2012)
8. American Standard, *Standard Specification for Lightweight Aggregates for Stuctural Concrete* (ASTM C330, 1999)
9. F. Zulkarnain, M. Ramli, *2nd International Conference on Built Environment in Developing Countries, Penang, Malaysia* (2009)
10. American Concrete Institute, *Guide for Structural Lightweight-Aggregate Concrete* (ACI 213R, 2003)
11. http://civilblog.org/about/
12. B. Middendorf, Mater. Charact., 60(7), 741 (2009)

Fabrication of AZO TCO Films by RF-sputtering and Their Physical Properties

T.S. Jang[1], and *D.C.* Oh[*]

[1]Department of Nanobiotronics, Hoseo University, Sechul-ri, Baebang-eup, Asan-si, 31499, Korea

Abstract. We report on the fabrication of Al-doped ZnO (AZO) transparent-conductive oxide (TCO) films on glass substrates by RF-sputtering, their physical properties, and the effect of thermal annealing on the AZO TCO films. The AZO films on glass substrates have a preferred orientation of the c-axis, irrespective of deposition conditions, which means that the AZO films have textured structures along the c-axis. The film thickness and surface roughness in the AZO films are proportional to plasma power and deposition time, while they are inverse-proportional to working gas ratio and working pressure. The AZO films have the optical transmittance over 80 % in the wavelength range of 400 – 1000 nm, irrespective of deposition conditions. The plasma power and the deposition time relatively give a large influence on the optical transmittance, compared to the working gas ratio and the working pressure. The AZO films deposited at room temperature have poor electrical properties, while the thermal annealing under Ar ambient significantly improves the electrical conductivity of the AZO films: an as-deposited sample has an electrical resistivity of 87 Wcm and an electron concentration of 1.3´10^{17} cm^{-3}, while the annealed sample has an electrical resistivity of 3.7´10^{-2} Wcm and an electron concentration of 1.2´10^{20} cm^{-3}.

1 Introduction

Recently, transparent conductive oxide (TCO) has attracted strong attention as the unprecedented applications such as smart phone, tablet PC, and flat panel display, etc beyond the existing devices such as solar cell and LED come on stage one after another [1-3]. AZO, Al-doped ZnO has been studied as a promising material for transparent conductive oxides (TCOs) for a long time due to the wide bandgap of 3.37 eV and the high n-type conductivity by native donor-type defects [1-3]. Thin film ZnO has the background electron concentration of high 10^{16} – high 10^{18} cm^{-3} due to native defects such as H, Zn_i, and V_O [4,5]. The inappropriate fabrication conditions of the low growth temperature that is not enough to supply the formation energy of ZnO bonds and the heterosubstrate with a large lattice mismatch of ZnO unit cells, are known to generate the various donor-type native defects unintentionally, though the background electron concentration level of the ZnO films depends on their growth techniques [4,5]. Moreover, the n-type conductivity of

[*] Corresponding author: ohdongcheol@hoseo.edu

ZnO can be effectively controlled by doping group III elements such as Al, Ga, and In. The electron concentration approaches to the level of low 10^{21} cm^{-3} and the electrical resistivity approaches to the level of low 10^{-4} Wcm in the Al-doped ZnO, Ga-doped ZnO, and In-doped ZnO films, fabricated by pulsed-laser deposition (PLD), metal-ogranic chemical-vapor deposition (MOCVD), atomic-layer deposition (ALD), DC sputter, and RF sputter [6-10]. Especially, Al-doped ZnO is very powerful and effective when used in the target sources of sputters, because the sputtering system has the merits of low cost, mass-productivity, and convenient maintenance, compared to other thin film fabrication systems such as PLD, MOCVD, and ALD. On the contrary, Ga-doped ZnO and In-doped ZnO have the difficulty that the volatile characteristics of high vapor pressure and low melting temperature in the Ga and In sources can cause the unexpected problems in the target fabrication and the sputtering process. Also, the AZO material has the merits of environment-friend and low-cost, compared to Indium-Tin-Oxide (ITO) that is widely used as a TCO material more than 100 years, because it does not have the rare-earth materials of In and Sn [1-3].

In this work, we report on the fabrication process of AZO TCO films on glass substrates by RF-sputtering, their physical properties such as surface, structural, optical, and electrical properties, and influence of thermal annealing on the electrical properties of AZO TCO films.

2 Experimental details

Four sets of AZO films as functions of plasma power, working gas ratio, working pressure, and deposition time were fabricated on glass substrates by using RF-sputter (SNTEK RSP5004) and AZO target of $ZnO:Al_2O_3$ (= 98:2 wt%). First, the plasma power (P_{plasma}) is varied in the range of 50 – 200 W under the working gas ratios ($O_2/Ar+O_2$) of 0 – 100 % and the working pressures (P_{work}) of 10 – 30 mT for a deposition time (t_{dep}) of 30 min. Second, the working gas ratio is varied in the range of 0 – 100 % under the working pressures of 10 – 30 mT at a plasma power of 100 W for a deposition time of 30 min. Third, the deposition time is varied in the range of 5 – 60 min at a plasma power of 100 W under a working pressure of 10 mT and a working gas ratio of 0 (Ar). Then substrate temperature is fixed at room temperature (RT). All fabricated samples were thermally annealed by using open-tube furnace at the temperature range of 200 – 500 °C under an Ar ambient for at a temperature of 60 min.

Table 1. Fabrication conditions of AZO films and thermal annealing conditions.

Target **Substrate**	**ZnO (98 wt%) : Al_2O_3 (2 wt%)** **Sodalime glass**
Plasma power (P_{plasma}) (W)	50, 100, 200
Working gas ratio ($O_2/Ar+O_2$) (%)	0 (Ar), 50, 100 (O_2)
Working pressure (P_{work}) (mT)	10, 20, 30
Substrate temperature (T_{sub}) (°C)	RT
Deposition time (t_{dep}) (min)	5, 15, 30, 60
Annealing temperature (T_{anneal}) (°C)	200, 400, 500
Annealing ambient	3 liter/min of Ar
Annealing time (t_{anneal}) (min)	60

The film thickness of the AZO films was investigated by wavelength variable ellipsometer. The surface morphology of the AZO films was investigated by the non-contact mode of atomic-force microscope (AFM). The structural properties of the AZO films was investigated by the q-2q scan of X-ray diffractometer (XRD). The electrical properties of the AZO films was investigated by the Hall measurement of van der Pauw type. The optical properties of the AZO films was investigated by UV-VIS spectrophotometer.

3 Results and discussion

Figure 1 shows the relationships between film thickness (or deposition rate) and plasma power, working gas ratio, working pressure, and deposition time, respectively, in the AZO films. The film thickness (or deposition rate) increases with the increase of plasma power, decreases with the increases of working gas ratio and working pressure, and is nearly constant with deposition time (for the case of the deposition rate). The change of the deposition rate in the AZO films fabricated by RF-sputter is ascribed to the change of sputter yield. The sputter yield has the direct relationship with the plasma power and the working pressure: the plasma power supplies the incident energy of sputtering particles, which is proportional to the sputter yield and the working pressure reduces the mean-free path of sputtering particles, which is proportional to the sputter yield [11,12]. Also, the sputter yield is related to the working gases and its ratio: the sputtering yield in the gas mixture including O_2 is reported to be dropped by the three mechanisms of decrease of mass transfer rate effect of target poisoning, and increase of electron affinity [11,12]. It is no wonder that the sputter yield should be independent of the deposition time.

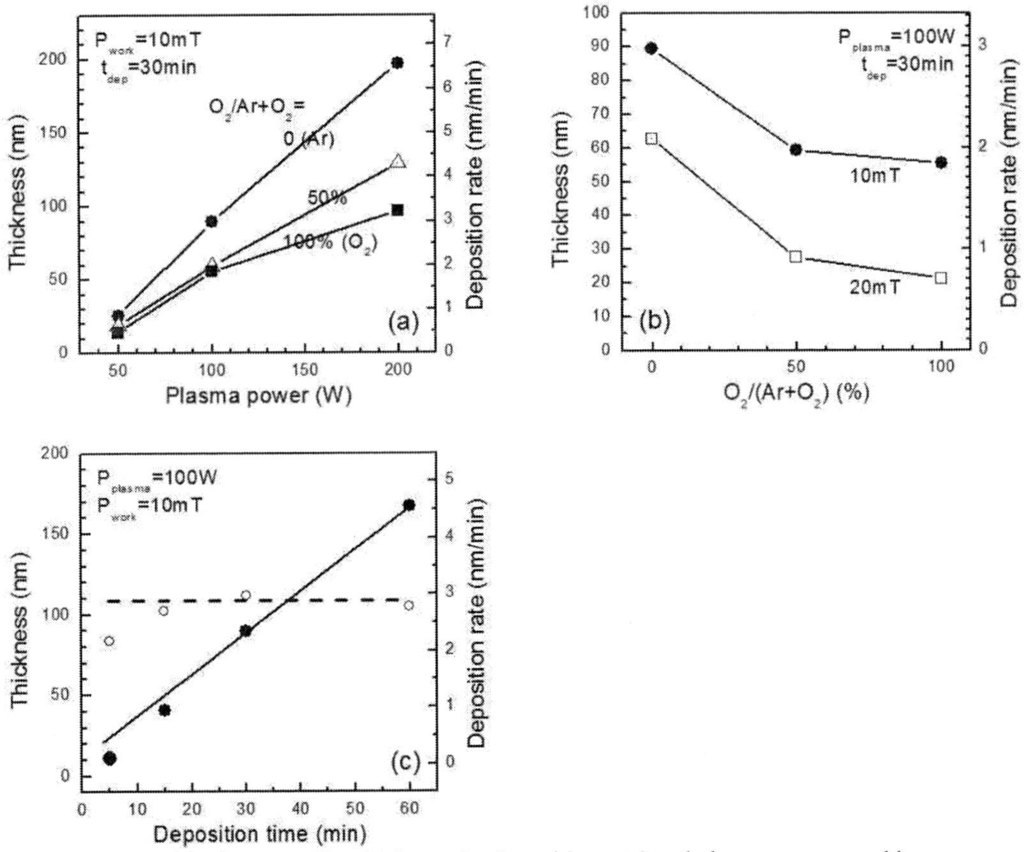

Fig. 1. Relationships between film thickness (or deposition rate) and plasma power, working gas ratio, working pressure, and deposition time, respectively, in the AZO films: (a) film thickness as a function of plasma power; (b) film thickness as a function of working gas ratio; and (c) film thickness as a function of deposition time.

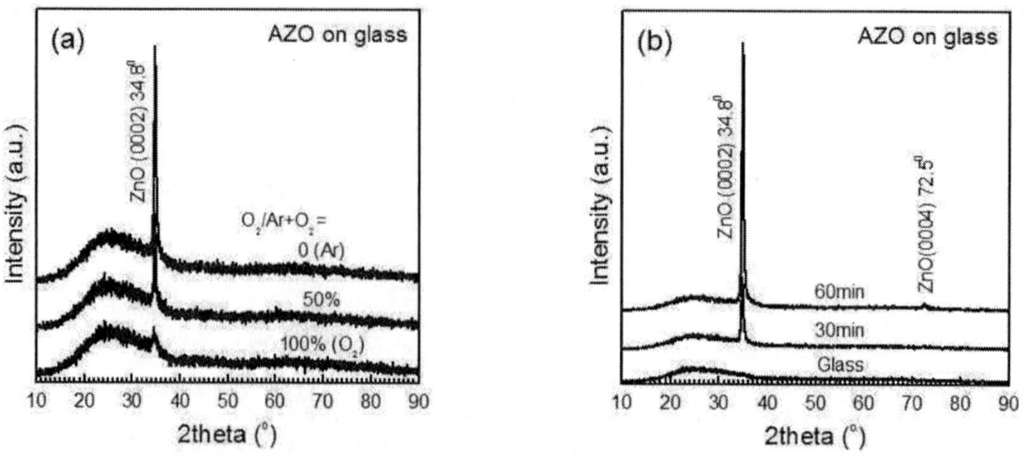

Fig. 2. Representative X-ray diffraction curves of the AZO films: (a) X-ray diffraction curve as a function of working gas ratio at a deposition time of 30 min; and (b) X-ray diffraction curve as a function of deposition time at a working gas ratio of 0 (Ar).

Figure 2 shows the representative X-ray diffraction curves in the AZO films fabricated on glass substrates by using RF-sputter. Figs. 2(a) and 2(b) show the X-ray diffraction curves as a function of working gas ratio at a deposition time of 30 min and as a function of deposition time at a working gas ratio of 0 (Ar), respectively. A broad peak in the range of 10 – 40° is assigned to the amorphous SiO_2. (The bare glass substrate has the peak of the same shape at the same position.) A peak of 34.8° is assigned to the (0002) planes of ZnO lattices [11,12]. The diffraction peak intensity decreases with the increase of working gas ratio, which means that Ar surrounding is better in the AZO film formation than O_2 surrounding. The diffraction peak intensity increases with the increase of deposition time, which is due to the increase of diffracting volume by the increased film thickness. On the other hand, in the 60-min sample the peak of ZnO (0004) planes begins to be exposed at the position of 72.5°. Also, other phases related to ZnO and AZO were not observed in the wide range of 20 – 80°, which means that the AZO films deposited on glass substrates have a preferred orientation of the c-axis [11,12]. All the other AZO films had the similar X-ray diffraction curves.

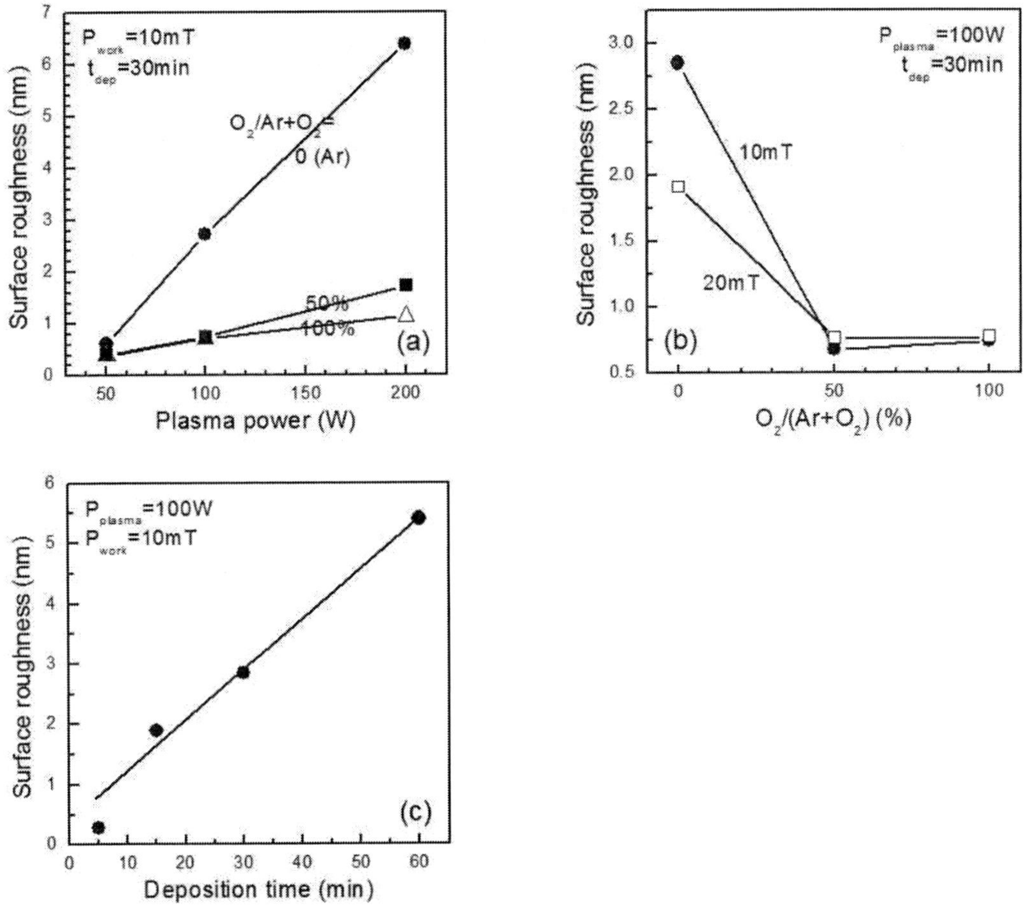

Fig. 3. Relationships between surface roughness and plasma power, working gas ratio, working pressure, and deposition time, respectively, in the AZO films: (a) surface roughness as a function of plasma power; (b) surface roughness as a function of working gas ratio; and (c) surface roughness as a function of deposition time.

Figure 3 shows the relationships between surface roughness and plasma power, working gas ratio, working pressure, and deposition time, respectively, in the AZO films. The surface roughness increases with the increase of plasma power, decreases with the increases of working gas ratio and working pressure, and increases with the increase of deposition time. These results exactly follow the relationships between film thickness and sputtering parameters in Fig. 1, which means the increase and decrease of the AZO film thickness offers the decisive factor to the increase and decrease of its surface roughness. On the other hand, Fig. 4 shows the representative surface morphologies in the AZO films fabricated on glass substrates by using RF-sputter, which is as a function of deposition time. All samples commonly have the surface structures composed of uniform nano-sized grains that mean the typical evidence of a textured growth along the c-axis [11,12], which is consistent to the results of Fig. 2. The nano-sized grains grow 3-dimensionally with the increase of the deposition time, inducing the increase of the surface roughness in Fig. 3(c). All the other AZO films had the similar surface morphologies.

DOI: 10.1051/matecconf/20167801106

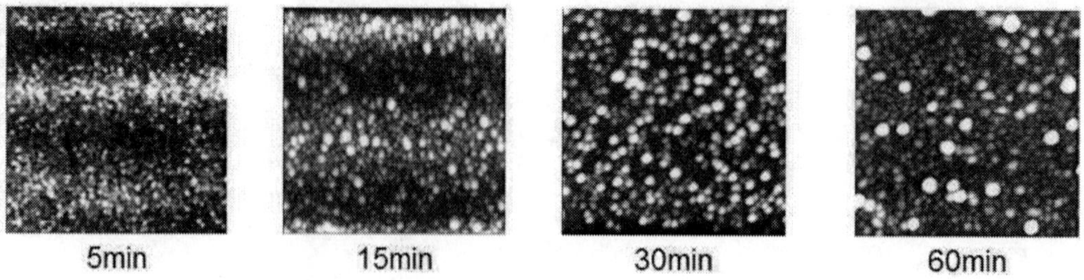

Fig. 4. Representative surface morphologies in the AZO films fabricated on glass substrates by using RF-sputter, which are as a function of deposition time.

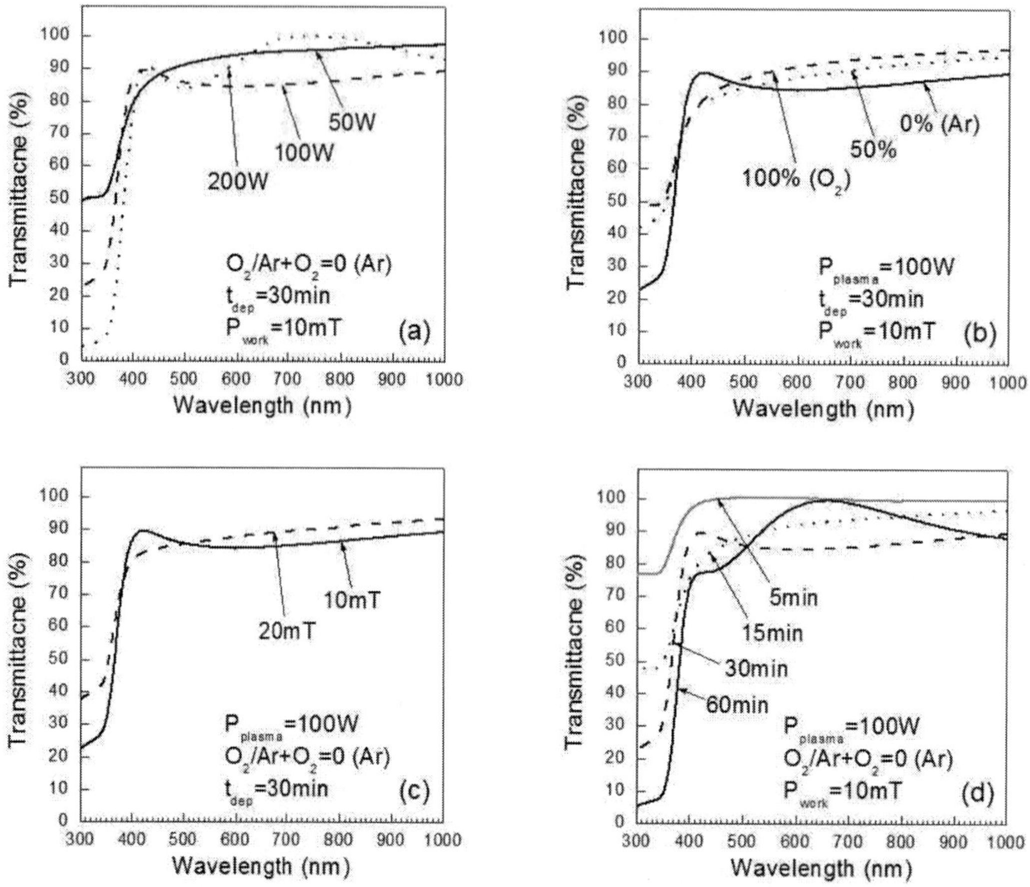

Fig. 5. Optical transmittance spectra of the AZO films as functions of (a) plasma power, (b) working gas ratio, (c) working pressure, and (d) deposition time.

Figure 5 shows the optical transmittance spectra of the AZO films as functions of plasma power, working gas ratio, working pressure, and deposition time, respectively. All samples have the high transmittance above 80 % in the wide range of near UV – near IR (400 – 1000 nm), irrespective of deposition conditions. The plasma power and the deposition time relatively give a large influence on the optical transmittance, compared to the working gas ratio and the working pressure, whose origin is related to the fact that the plasma power and the deposition time more seriously change the thickness and surface roughness of the AZO films than other sputtering parameters. The change of absorption edge in the UV region of 300 – 350 nm is due not to the shift of band edge but to the thin

 DOI: 10.1051/matecconf/20167801106

film thickness smaller than extinction length. As a result, the AZO films fabricated on glass by RF-sputtering are found to have the high optical transmittance in the wide wavelength range.

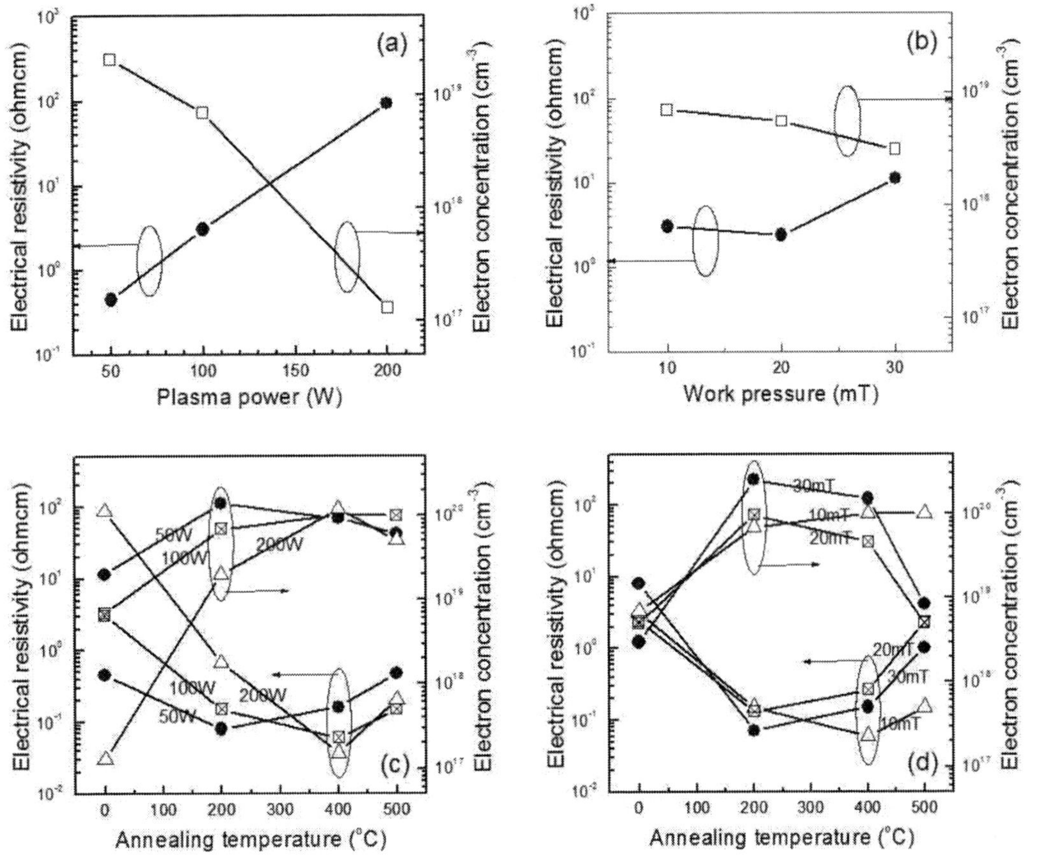

Fig. 6. Relationships between electrical resistivity (or electron concentration) and plasma power, working pressure, and annealing temperature, respectively, in the AZO films: (a) electrical resistivity and electron concentration of as-deposited AZO films as a function of plasma power, (b) electrical resistivity and electron concentration of as-deposited AZO films as a function of plasma power, (c) electrical resistivity and electron concentration as a function of annealing temperature for each plasma power, and (d) electrical resistivity and electron concentration as a function of annealing temperature for each working pressure.

Figure 6 shows the relationships between electrical resistivity (or electron concentration) and plasma power, working pressure, and annealing temperature, respectively, in the AZO films. The electrical resistivity rapidly increases with the increase of plasma power and the electron concentration is vice versa. The electrical resistivity slowly increases with the increase of working pressure and the electron concentration vice versa. Those change of electrical properties has direct relationship with the change of unintentionally generated native defects in the AZO films: the increase of the plasma power induces the increase of the native defects due to higher deposition rate, which compensate free electrons, and the increase of the working pressure induces the increase of the native defects due to shorter mean-free path, which can compensate free electrons [13,14]. Also, all AZO films have the serious dependence of the annealing temperature: an electrical resistivity of 87 Ωcm and an electron concentration of 1.3 x 10^{17} cm^{-3} is changed to an electrical resistivity of 3.7 x 10^{-2} Ωcm and an electron concentration of 1.2 x 10^{20} cm^{-3} in terms of thermal annealing. The electrical resistivity increases with the increase of annealing temperature below a critical temperature, and above the critical temperature it decreases with the increase of annealing temperature. The electron concentration is vice versa. The break point is in the range of 200 – 400 °C. The annealing temperature dependence of electrical properties in the AZO films is opposite to the phenomena found in high-quality single-crystalline ZnO films, reported previously. Those change of electrical properties by the thermal annealing has direct relationship with the formation and annihilation of the native defects: the increase of the electron concentration in the low-temperature region is due to the annihilation of the native defects, which weaken the effect of carrier compensation, and the decrease of the electron concentration in the high-temperature region is due to the formation of the native defects, which strengthen the effect of carrier compensation [4,5,15,16]. On the other hand, the relationship between electrical properties and working gas ratio in the AZO films were excluded in this discussion, because the electrical resistance was significantly increased as much as any current does not flow in the AZO films fabricated under the working gases of the mixture of Ar and O2.

4 Conclusions

In this work, we reported on the fabrication process of AZO TCO films on glass substrates by RF-sputtering, their physical properties such as surface, structural, optical, and electrical properties, and influence of thermal annealing on the electrical properties of AZO TCO films.

References

1. K. Ellmer, A. Klein, B. Rech (Springer, Berlin, 2008)
2. P. Barquinha, R. Martins, L. Pereira, E. Fortunato, *Transparent Oxide Electronics: from material to device* (John Wiley & Sons, 2012)
3. A. Facchetti, T.J. Marks, *Transparent Electronics: from synthesis to applications*, (John Wiley & Sons, 2010)
4. D.C. Oh, T. Suzuki, J.J. Kim, H. Makino, T. Hanada, M.W. Cho, T. Yao, Appl. Phys. Lett., **86(3)**, 032909 (2005)
5. D.C. Oh, S.H. Park, H. Goto, I.H. Lim, M.N. Jung, J.H. Chang, T. Yao, J.S. Song, C.H. Bae, C.S. Han, K.W. Koo, Appl. Phys. Lett., **95(15)**, 151908 (2009)
6. T. Minami, Semicond Sci. Tech., **20(4)**, 35 (2005)
7. T. Minami, K. Oohashi, S. Takata, Thin Solid Films, **193**, 721 (1990)
8. H. Agura, A. Suzuki, T. Matsushita, T. Aoki, M. Okuda, Thin Solid Films, **445(2)**, 263 (2003)
9. A. Hafdallah, F. Yanineb, M.S. Aida, N. Attaf, J. Alloy Compd., **509(26)**, 7267 (2011)
10. T.W. Nam, C.W. Lee, H.J. Kim, H.J. Kim, Appl. Surf. Sci., **295**, 260 (2014)
11. K.B. Kim, S.M. Lee, D.C. Oh, H.J. Ko, J. Korean Phys. Soc., **67(4)**, 676 (2015)
12. S.M. Lee, K.B. Kim, H.J. Ko, D.C. Oh, J. Nanoelectron Optoe, **11(2)**, 250 (2016)
13. D.K. Kim, H.B. Kim, J. Alloy Compd., **509(2)**, 421 (2011)
14. W. Yang, Z. Wu, Z. Liu, A. Pang, Y.L. Tu, Z.C. Feng, Thin Solid Films, **519(1)**, 31 (2010)
15. D.C. Oh, H.J. Ko, S.K. Han, S.K. Hong, W.G. Jeong, T. Yao, Appl. Phys. Express, **5(7)**, 075801 (2012)
16. D.C. Oh, T.Taki, I.H. Im, S.H. Park, T. Hanada, T. Yao, J.S. Song, J.H. Chang, M. Makino, C.S. Han, K.H. Koo, J. Vac. Sci. Technol. A, **26(2)**, 259 (2008)

MATEC Web of Conferences **78**, 01107 (2016) DOI: 10.1051/matecconf/20167801107
IConGDM 2016

Comparative Study About the Structural Properties of ZnO Films Grown on Si, GaAs, And Al2O3 Substrates and a GaN Template

T.S. Jang[1], *S.M.* Lee[1], *K.B.* Kim[1] , and *D.C.* Oh[1,*]

[1]Department of Nanobiotronics, Hoseo University, Sechul-ri 165, Baebang-eup, Asan 336-795, Korea

Abstract. We report on the structural properties of ZnO films grown on (001) Si, (001) GaAs, (0001) Al_2O_3 substrates and a (0001) GaN template by RF-sputtering. It is observed that all the ZnO films have textured structures with a preferred orientation of the c-axis, irrespective growth conditions, in terms of atomic-force microscope and wide-range X-ray diffractometer. However, it is found that the ZnO films in the ZnO-on-Si and ZnO-on-GaAs are consisted of various domains with different orientations, while the ZnO films in the ZnO-on-Al_2O_3 and ZnO-on-GaN are consisted of a single domain with the same orientation in terms of the φ-scan of X-ray diffractometer. Moreover, it is found that the ZnO-on-GaN has smaller edge dislocation density and screw dislocation density than the ZnO-on-Al_2O_3, which seriously depends on substrate temperature, in terms of the ω-scan of X-ray diffractometer.

1 Introduction

ZnO has attracted an attention as one of the promising materials required to realize the emitters and detectors in the ultraviolet region and the high-speed devices in the information communication system, because it has the outstanding physical properties of a wide-bandgap of 3.37 eV, an exiton-binding energy of 60 meV, a saturation velocity of 3.1×10^7 cm/sec, and a cohesive energy of 1.89 eV [1-3]. However, ZnO has an essential problem where it is still lack of the reliably available homosubstrate that thin film ZnO can be grown on, even if the substrate that a material can be grown on is an essential problem in the thin film technology. Instead, Al_2O_3, GaN, Si, and GaAs substrates, etc have been tried as the second best thing. The ZnO film of a wurzite structure is generally grown on the (0001) Al_2O_3 substrate of a corundum structure, which is the most developed substrate among the materials with the similar crystalline structure to ZnO. However, the ZnO lattice has a lattice mismatch of 30 % between the basal plane of the Al_2O_3 lattice and a lattice mismatch of 18 % between the small cell of Al atoms on the basal plane [4,5]. The GaN template, which is fabricated on the Al_2O_3 substrate by metal-organic chemical-vapor deposition (MOCVD), with the same crystalline structure as ZnO effectively can suppress the structural defects in the subsequently grown ZnO film, because the GaN lattice has a very small lattice mismatch of 1.8 % with the ZnO lattice. However, because the GaN

[*] Corresponding author: ohdongcheol@hoseo.edu

template is also grown on the Al_2O_3 substrate, the intrinsic problem derived from the Al_2O_3 substrate itself is remained [5,6]. The Si substrate of a diamond structure and the GaAs substrate of a zinc blende structure, which have the highest crystalline qualities in the surface and bulk among any other developed semiconductor substrates, have the merits of large-area wafer ability and economic feasibility. However, the two materials have the structural mismatches with ZnO, which hampers the epitaxial growth of ZnO films on the both substrates [7-10]. We note that there is not yet systematic result about the difference in the structural properties of the ZnO films grown on each substrate, referred above, though a lot of research results about the growth technique and physical properties of thin film ZnO have been reported until now. In this work, we have a comparative study about the structural properties in the surface and bulk of the ZnO films grown on Si, GaAs, and Al_2O_3 substrates and a GaN template by using RF-sputtering.

2 Experimental details

Four types of ZnO films were prepared on (001) Si, (001) GaAs, and (0001) Al_2O_3 substrates and a (0001) GaN template by RF-sputtering. Each sample was grown under each optimized condition, whose detailed conditions were reported elsewhere [5,8,10]. The ZnO-on-Si structure was grown at a substrate temperature of 500°C and an $O_2/Ar+O_2$ ratio of 80 %. The ZnO-on-GaAs structure was grown at a substrate temperature of room temperature (RT) and an $O_2/Ar+O_2$ ratio of 0 (Ar = 100 %). The ZnO-on-Al_2O_3 structure and the ZnO-on-GaN structure were grown at a substrate temperature of 700°C and an $O_2/Ar+O_2$ ratio of 80 %. The other conditions were fixed at a plasma power of 100 W, a working pressure of 10 mTorr, and a growth time of 1 hr.

The surface properties of ZnO films were measured by the non-contact type mode of atomic-force microscope (AFM). The bulk properties of ZnO films were measured by wide-range X-ray diffractometer (XRD) and high-resolution X-ray diffractometer (HRXRD). The θ-2θ scan of wide range XRD was used in order to investigate the intercompounds between ZnO films and substrates and the ZnO phases with different crystalline directions. The ω scan and φ scan of HRXRD were used in order to investigate the crystalline qualities, dislocation types, and epitaxial growth of ZnO films.

3 Results and discussion

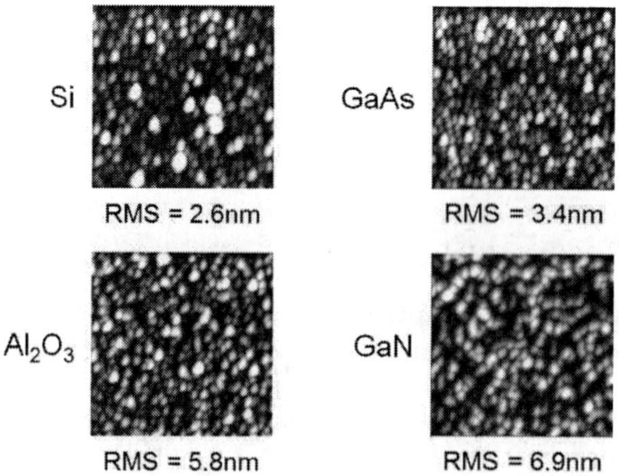

Fig. 1. Surface morphology and surface roughness (RMS) in the ZnO films grown on Si, GaAs, and Al_2O_3 substrates and a GaN template, respectively.

Figure 1 shows the surface morphology and surface roughness in the ZnO films grown on Si, GaAs, and Al_2O_3 substrates and a GaN template, respectively. All the ZnO films commonly have the similar surface morphology composed of nano-sized grains, which indicates a typical evidence of textured growth, though their details are dependent on each substrate. It is reported that the ZnO film with a preferred orientation of the c-axis has the textured surface, which has the nano-scaled surface structures of various pyramidal shapes due to the columnar growth along an axial direction [11,12]. Each ZnO film has the different grain size and surface roughness, which is ascribed to the fact that each optimized growth condition is different. It is known that i) the substrate with a different crystalline structure and a different surface state, ii) the substrate temperature that supplies the activation energy for ZnO film formation, iii) the working gas ratio that manipulates the stoichiometry of ZnO films, and iv) the working pressure and plasma power that control the growth rate of ZnO films by changing sputter yield give influences on the surface structures of the ZnO films [5,8,10]. On the other hand, hexagonal or triangular islands and step or terrace structures sometimes reported in the films grown by metal-organic chemical-vapor deposition (MOCVD) and molecular-beam epitaxy (MBE) were not observed in this research [13-16].

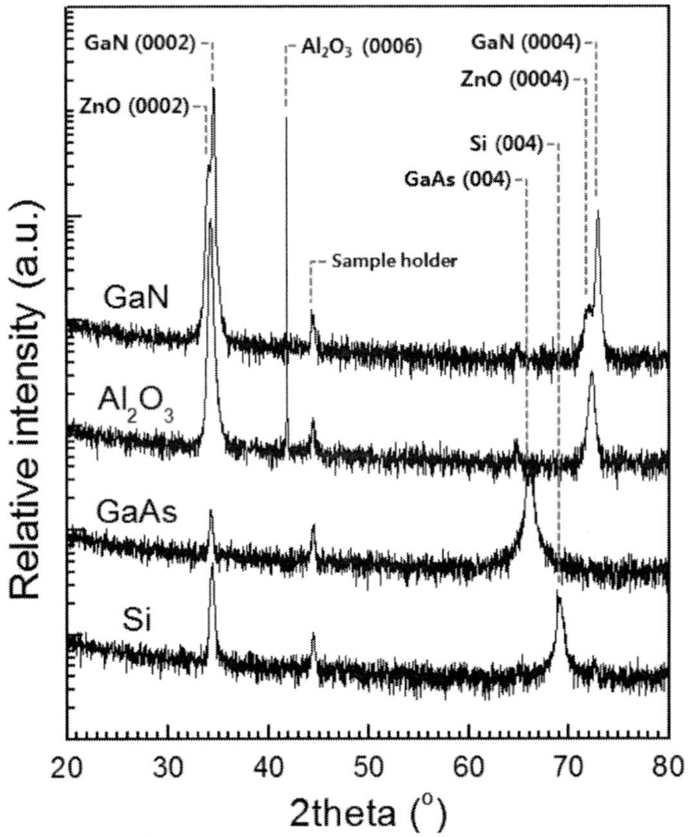

Fig. 2. X-ray θ-2θ scan diffraction curves in the wide range of 20 – 80°C in the ZnO films grown on Si, GaAs, and Al_2O_3 substrates and a GaN template, respectively.

Figure 2 shows the X-ray θ-2θ scan diffraction curves in the wide range of 20 – 80°C in the ZnO films grown on Si, GaAs, and Al_2O_3 substrates and a GaN template, respectively. The peak located at a low angle of 34.3° is assigned to the ZnO (0002) plane and the peak located at a high angle of 72.2° is assigned to the ZnO (0004) plane [5,8,10]. In the ZnO-on-Si a peak at 69.0° is due to the (004) plane of the Si substrate and in the ZnO-on-GaAs a peak at 65.9° is due to the (004) plane of the GaAs substrate. In the both samples, any other diffraction peaks do not exist beside the ZnO (0002) peaks and their crystal planes are parallel to the (004) planes of each substrate, which indicate that the ZnO films were grown with a preferred orientation of the c-axis. The peaks commonly found at 44.5° are due to a sample holder. In the ZnO-on-Al_2O_3 a peak at 41.8° is due to the (0006) plane of the Al_2O_3 substrate and in the ZnO-on-GaN the two peaks at 34.6° and 72.9° are due to the (0002) and (0004) planes of the GaN template, respectively. The peak separation between the ZnO film and the Al_2O_3 substrate is estimated to be 7.5°, which corresponds to the strain of 20.1 %. The peak separations between the ZnO film and the GaN template are estimated to be 0.48° in the (0002) plane and 0.96° in the (0004) plane, which corresponds to the strains of 1.35 and 1.16 %, respectively. In the both samples, the two ZnO films grown on the Al_2O_3 substrate and GaN template have larger diffraction intensities than those on the Si and GaAs substrates and the two ZnO films grown on the Al_2O_3 substrate and GaN template have the diffraction peaks for (0004) planes that are not exposed in those on the Si and GaAs substrates. This is ascribed to the fact that the film and substrate have the similar structure in the former while they have the different structure in the latter, which indicates that the ZnO-on-Al_2O_3 and the ZnO-on-GaN have the improved crystalline qualities, compared to the ZnO-on-Si and the ZnO-on-GaAs.

Figure 3 shows the X-ray φ-scan diffraction curves for asymmetric (10-11) planes in the ZnO films grown on Si, GaAs, and Al_2O_3 substrates and a GaN template, respectively. The φ-scan is that the sample surface rotates as a function of azimuthal angle whose axis is normal to the sample surface, while the ω scan (or θ-2θ scan) is that the sample surface rotates as a function of incident angle of X-ray beam whose axis is parallel to the sample surface [17,18]. In the cases of ZnO-on-Si and ZnO-on-GaAs there no diffraction peaks exist, while in the cases of ZnO-on-Al_2O_3 and ZnO-on-GaN there 6 diffraction peaks as a period of 60° exist. The latter is due to the six-fold symmetry of ZnO lattices, which indicates that the grown ZnO films are single domain epitaxial films with one orientations, while the former is due to the random distribution of various domains with different orientations, which indicates that the ZnO lattices are not randomly stacked along the a-axis, though they are well grown along the c-axis [17,18].

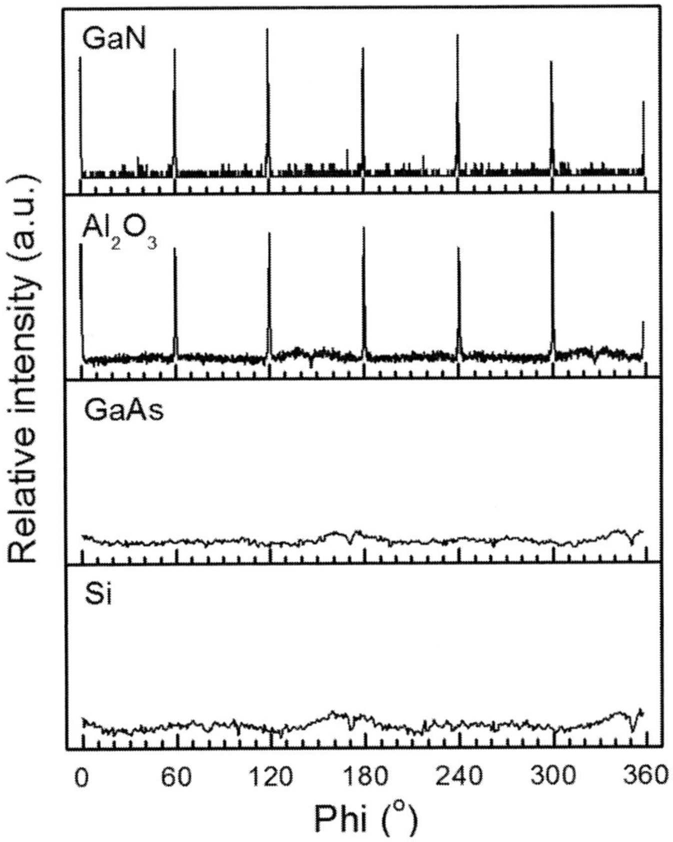

Fig. 3. X-ray φ-scan diffraction curves for asymmetric (10-11) planes in the ZnO films grown on Si, GaAs, and Al_2O_3 substrates and a GaN template, respectively.

Figure 4 shows the X-ray ω-scan diffraction curves for symmetric (0002) planes and asymmetric (10-11) planes in the ZnO films grown on Si, GaAs, and Al_2O_3 substrates and a GaN template, respectively. In the (0002) scan curves, the ZnO-on-GaN with a diffraction linewidth of 0.289° is narrower than the ZnO-on-Al_2O_3 with a diffraction linewidth of 0.480°. In the (10-11) scan curves, the ZnO-on-GaN with a diffraction linewidth of 0.643° is also narrower than the ZnO-on-Al_2O_3 with a linewidth of 0.789°. The ω scan for symmetric planes such as (0002) reflects screw dislocation densities because Burger's vector is parallel to the c-axis, while the ω scan for asymmetric planes such as (10-11) reflects edge dislocation densities because Burger's vector is normal to the c-axis [17,18]. Therefore, the ZnO-on-GaN has smaller edge dislocation density and smaller screw dislocation density than the ZnO-on-Al_2O_3, which is ascribed to the fact that the GaN lattice has a small lattice mismatch of 1.8 % with the ZnO lattice, while the Al_2O_3 lattice a large lattice mismatch of 18 % with the ZnO lattice [4-6]. This phenomenon was more clearly shown in the ZnO film growth at RT without the supply of surface reaction energy: the case of ZnO-on-Al_2O_3 had 5 times larger broadness than the case of ZnO-on-Al_2O_3 in the X-ray ω-scan diffraction curves of (0002) planes.

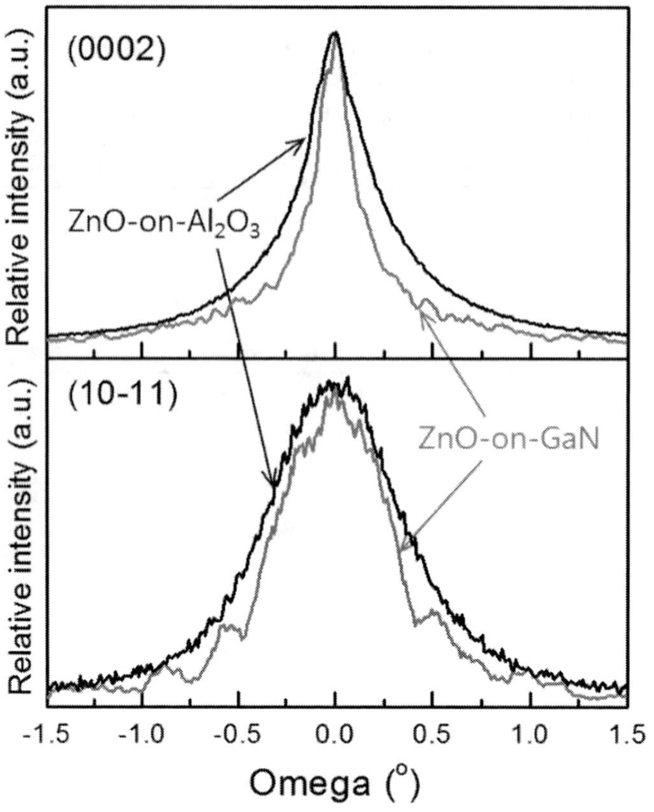

Fig. 4. X-ray ω-scan diffraction curves for symmetric (0002) planes and asymmetric (10-11) planes in the ZnO films grown on Si, GaAs, and Al_2O_3 substrates and a GaN template, respectively.

4 Conclusions

We had a comparative study about the structural properties in the surface and bulk of the ZnO films grown on (001) Si, (001) GaAs, and (0001) Al_2O_3 substrates and a (0001) GaN template by using RF-sputtering. First, all the ZnO films had textured structures with a preferred orientation of the c-axis, irrespective growth conditions. Second, the ZnO films in the ZnO-on-Si and ZnO-on-GaAs were consisted of various domains with different orientations, while the ZnO films in the ZnO-on-Al_2O_3 and ZnO-on-GaN were consisted of a single domain with the same orientation. Finally, the ZnO-on-GaN had smaller edge dislocation density and screw dislocation density than the ZnO-on-Al_2O_3, which seriously depended on substrate temperature.

References

1. C. Jagadish, S.J. Pearton, *Zinc Oxide Bulk, Thin Films and Nanostructures, Processing, Properties and Applications* (Elsevier, 2011)
2. K. Ellmer, A. Klein, B. Rech, Transparent Conductive Zinc Oxide, Basics and Applications in Thin Film Solar Cells, **104** (2007)
3. T. Yao, S.K. Hong, *Oxide and Nitride Semiconductors; Processing, properties, and applications* (Springer Science & Business Media, 2009)
4. H.F. Liu, S.J. Chua, J. Appl. Phys., **106**, 023511 (2009)

5. K.B. Kim, S.M. Lee, H.J. Ko, D.C. Oh, Submitted to Journal of Nanoscience and Nanotechnology, 2016
6. S.K. Sharma, S.E. Heo, B.H. Lee, H.H. Lee, C.M. Kim, D.Y. Kim, Curr. Appl. Phys., **14**, 1696 (2014)
7. S. Singh, R.S. Srinivasa, S.S. Major, Thin Solid Films, **515**, 8718 (2007)
8. K.B. Kim, S.M. Lee, D.C. Oh, H.J. Ko, Korean Phys. Soc., **67(4)**, 676 (2015)
9. H.F. Liu, A.S.W. Wong, G.X. Hu, H. Gong, J. Cryst. Growth., **310(19)**, 4305 (2008)
10. S.M. Lee, K.B. Kim, H.J. Ko, D.C. Oh, J. Nanoelectron Optoe., **11(2)**, 250 (2016)
11. R. Hong, H. Qi, J. Huang, H. He, Z. Fan, J. Shao, Thin Solid Films, **473(1)**, 58 (2005)
12. Z.W. Liu, C.W. Sun, J.F. Gu, Q.Y. Zhang, Appl. Phys. Lett., **88(25)**, 1911 (2006)
13. T. Ive, T. Ben-Yaacov, C.G. Van de Walle, U.K. Mishra, S.P. DenBaars, J.S. Speck, J. Cryst. Growth, **310(15)**, 3407 (2008)
14. C. Munuera, J. Zuniga-Perez, J.F. Rommeluere, V. Sallet, R. Triboulet, F. Soria, V. Munoz-Sanjose, C. Ocal, J. Cryst. Growth., **264**, 70 (2004)
15. H. Kato, M. Sano, K. Miyamoto, T. Yao, J. Cryst. Growth, **265**, 375 (2004)
16. H. Matsui, H. Saeki, T. Kawai, A. Sasaki, M. Yoshimoto, M. Tsubaki, H. Tabata, J. Vac. Sci. Technol. B, **22(5)**, 2454 (2004)
17. V. Srikant, J.S. Speck, D.R. Clarke, J. Appl. Phys., **82**, 4286 (1997)
18. T. Metzger, R. Hopler, E. Born, O. Ambacher, M. Stutzmann, R. Stommer, M. Schuster, H. Gobel, Philos. Mag., **77(4)**, 1013 (1998)

Design and Analysis of XOR Logic Gate Based on Two-Photon Absorption

N.S. Nazir[1,*], *M.H.A.* Wahid[1], *N.A.M.A.* Hambali[1], *M.M.* Shahimin[1], *N.Z.* Khairuddin[1], *N.H.* Hasanuddin[1], and *M.M.* Ramli[1]

[1]Semiconductor Photonics & Integrated Lightwave Systems (SPILS), School of Microelectronic Engineering, Universiti Malaysia Perlis, Pauh Putra Campus, Malaysia.

Abstract. Two-photon absorption with high intensity pump beam occurs in a SOA depends on fast phase change of a weak probe signal. This work analysed optical XOR logic function using two-photon absorption induced fast phase change. A rate equation for SOA and both input data signals A and B with high intensity, configured in the form of MZI has also been proved. The model shows that XOR operation at 10 Gb/s with good signal to noise ratio is obtained with high input intensities. The result on the generation of XOR indicates that operations on 10 Gb/s with a high signal to noise ratio can easily be implemented. The average input power into the SOA is 20 dbm corresponding to the peak power of 5.5 dBm at 10 Gb/s when the width of the input pulse 3.6 ps. The short narrow pulse width is utilised in the study for stronger effect of two-photon absorption.

1 Introduction

All optical logic gates are very important in signal regeneration, addressing, switching and data encoding [1].They serve a vital part in upcoming high-speed communication system [2]. In recent years, some prevalent optical logic gates systems have been studied which are Mach-Zehnder interferometer (MZI) [3, 4], dual semiconductor-optical amplifier (SOA) [1], and ultrafast nonlinear interferometer (UNI) [5]. In this work, operation of XOR logic gates based on two-photon absorption (TPA) and its induced free carrier effects (FCE) is investigated. The TPA effect is applied in XOR optical gates particularly.

When two-photon absorption of pump beam with high intensity occurs in a semiconductor optical amplifier, a fast phase change associated with weak signal probe is generated. A scheme that realizes the function of quick all optical logic gates using two photon absorption induced phase changes has been investigated [1]. The rate equation for semiconductor optical amplifier, a high intensity input data signal, has been solved which is constructed within MZI [1] configuration. Input of two-photon induced phase changes is greater than normal induced phase change due to the high intensity.

This work is dedicated to improve the performance of SOA-MZI based XOR logic gate. In order to produce output signal with fast phase change, XOR logic gate based on two-photon absorption is developed. This process occurs when two input signals of pump beams

* Corresponding author: najwasarihah@gmail.com

are injected into the configured circuit with two different or same wavelengths. This work enhanced the SOA-MZI based XOR logic gate whereby it only uses a train of pulses or CW beam as the control signal. Besides that, phase shifter is added to the circuit in order to shift the output signal for 900 because the signal travelling through the input acquires a phase change subjected to the cross phase modulation [6] .

2 Basis of two-photon absorption

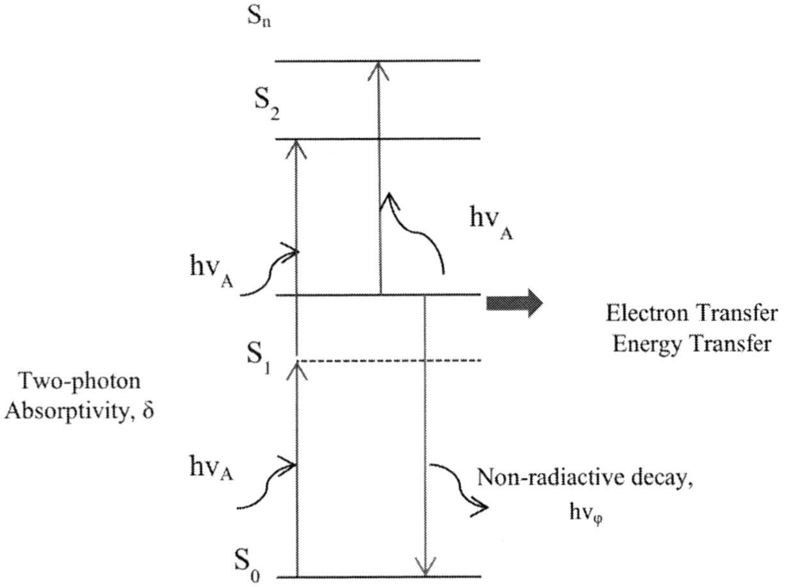

Fig. 1. Degenerated TPA Process.

Two-photon absorption (TPA) is one of the nonlinear optical phenomenon absorption processes that occur when two photons are absorbed and excited from lower state to higher state. Hence, sum of both photon energies absorbed are same with transition energy and this phenomenon happens at high optical intensities because the optical intensity proportional with absorption coefficient. The degenerated two-photon absorption process is illustrated in Fig.1 above, which two photons having equal energy hv_A are simultaneously absorbed and excited directly from the lowest state (S_0) to the second state (S_2). Photon is quickly relaxed by internal conversion to the first state (S_1) once it is in state (S_2). Then, photon returns to the ground state (S_0) by non-radiactive decay. In order to induce fast phase change in the optical XOR gate, two-photon absorption in a semiconductor optical amplifier which depends on optical intensity of pump beams is utilised. This process occurs when two input signal of pump beams at different wavelengths is supplied to the system.

3 Principle of XOR logic gate based on TPA operation

In this work, two-photon absorption with high intensity pump beam operation at 20 dBm takes place in order to induce larger phase change than regular gain so that both XOR operation and data generation at high frequency with good signal to noise ratio can be achieved. Mach-Zehnder interferometer (MZI) is used in an arm of which has a semiconductor optical amplifier (SOA) [1] and together with pump beam as the data signal is injected into each port. Also a weaker probe beam is introduced into the middle arm. Optical input data signal X and Y with two different wavelengths at λ1 (1550 nm) and λ2 (1552 nm) are pumped into the MZI meanwhile two SOAs are injected with weak probe beam of clock signals. When both inputs X and Y are equal to 0, the output is 0 because of the signal at port Z travelling through the two arms of the SOA has a difference of π when it recombines at output port [1]. An output 1 is obtained when inputs X=1 , Y=0 due to the cross phase modulation and signal travelling through the arm with signal X acquires a phase change. It also happens if X=0 and Y=1. But, the phase changes for the signal travelling in both arms are equal when both inputs A and B are equal to 1 and the output is 0 [7].

Figure 2 depicts the XOR logic gate connection schematic by using operation of SOA-MZI and truth table of XOR logic. The architecture is based on the SOA-MZI. In this case, data signals enter the device at port input X and Y with two different wavelengths λ_1 and λ_2. Control signal is split into upper and lower branch of the interferometer. After passing through two of the SOAs, the amplified signal is shifted by 90^0 and coupled out carrying the XOR operation between the two inputs.

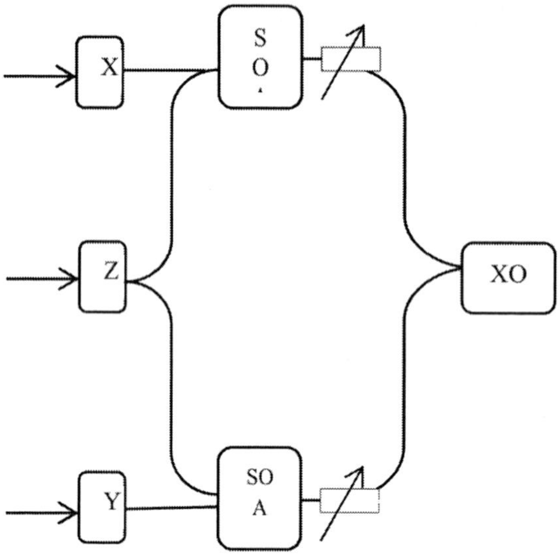

Input X	Input Y	Output XOR
0	0	0
0	1	1
1	0	1
1	1	0

(a) (b)

Fig. 2. (a) Operation of XOR logic gate using SOA-MZI, (b) Truth table of XOR.

4 Results and discussion

Figure 3 shows the input signals for input data X and Y, and the output signal from observation on optical time domain visualizer. High optical pulse signal shows data '1' whereas low optical pulse signal is '0'. This work is performed at 10 Gbit/s input data using signal sequences X (001000011) and Y (1001001000). These results illustrate both input signal X and Y with output XOR logic gate from time 2 ns until 4 ns. At time 2.4 ns both input data signal X and Y are 1, the output is 0. When both input data signals are 0 at 3 ns, the output is 0. Output signal is 0 at time 3.6 ns when input X is 0 and input Y is 1. The results are tabulated in Table 1.

Table 4. Input and output data for XOR logic gate.

Input X	Input Y	XOR output
0	1	1
0	0	0
1	0	1
0	1	1
0	0	0
0	0	0
0	1	1
1	0	1
1	0	1

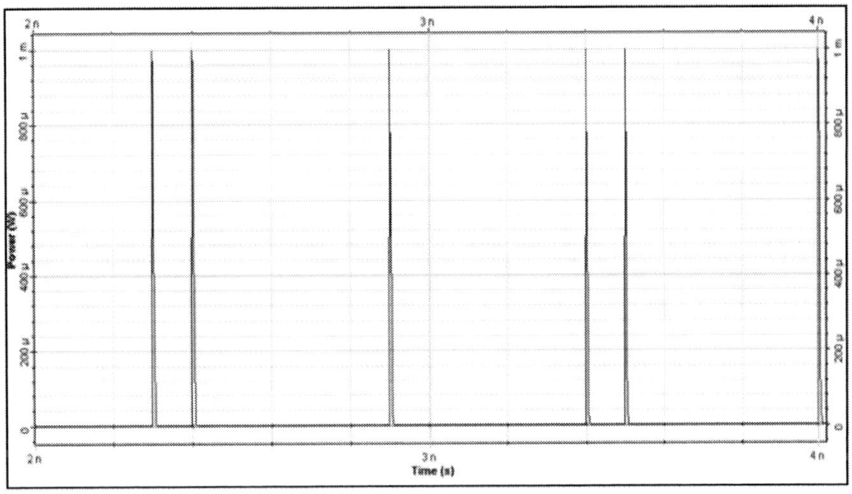

(a) Input data X is (001000011).

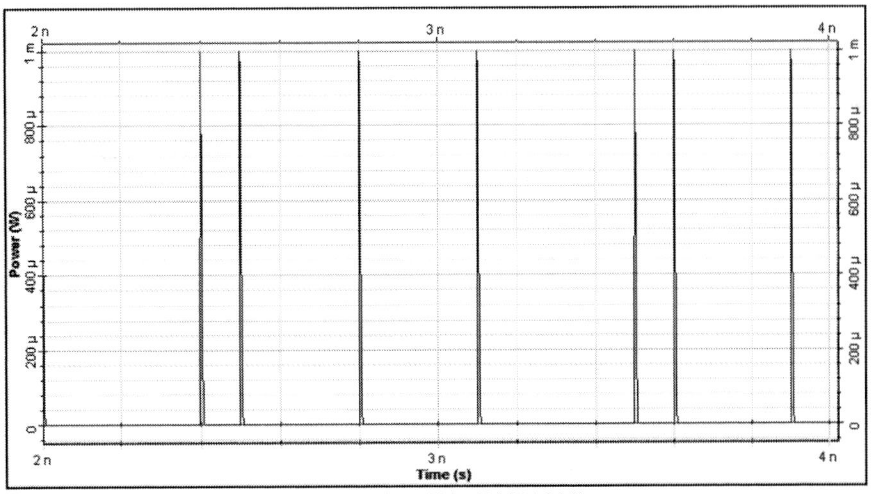

(b) Input data Y is (10010010).

 DOI: 10.1051/matecconf/20167801108

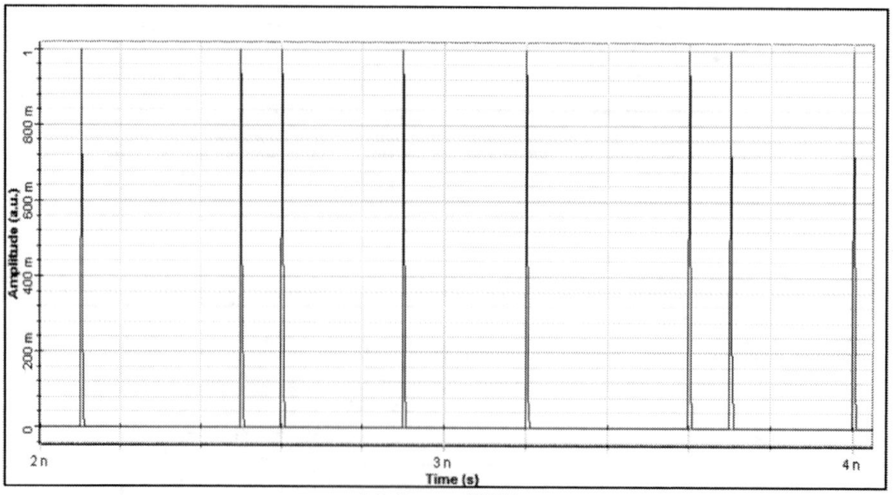

(c) Output XOR.

Fig. 3. (a) Input X, (b) Input Y and (c) Output XOR.

From the observation, the result is well simulated as desired but there is a small delay at the output because of the amplified signal at high intensity level thus induces non-linear effect by semiconductor optical amplifier (SOA). Non-linear effect mechanism occurs during the propagation of light through the SOA that induced phase changes. The non-liner effect is instantaneous. Errors in detection and delay are increased because of non-linear effects which modified the pulse carrying information in the presence of other signals taking place within the SOA's waveguided layer [8]. Another components that caused delay in output signal are the short pulse generator with narrow pulsewidth and the pulse intensity with fast rise and fall times. This generator also needs high bandwidth because its immunity to non-linear-effect [9].

4.1 Varying values of bit rates with optical power

Bit rate variation also shows good results as summarized in Table 2. Generally, it can be concluded that the output power decreases as the bit rate increases as illustrated in Fig. 4. This relation is obtained as a result of self-phase modulation whereby refractive index of a medium depends on the intensity of the light beam at high intensities. Since non-linearity induced phase change depends on optical power, and since the optical power in the pulse decreases due to attenuation, the effect of non-linearity would decrease continuously as light propagates through the system. Light with maximum power will result in high non-linear phase shift but non-linear effect would decrease as light propagates with lower power [10]. In this circuit, there are pulses propagating simultaneously at different wavelengths which are input data A at 1550 nm and input data B 1552 nm. Each wavelength can result in a change in refractive index due to the presence of non-linear effect depending on the carried power by the signal [9]. The changes in refractive index will affect the propagation of the other beam. Thus, the output is dependent on the presence or absence of the inputs.

Table 5. Table of bit rates with maximum output power (dBm).

Bit rates for input A and B (Gbit/s)	Maximum output power (dBm)
10	0.448478
20	0.278784
30	0.234797
40	0.145784

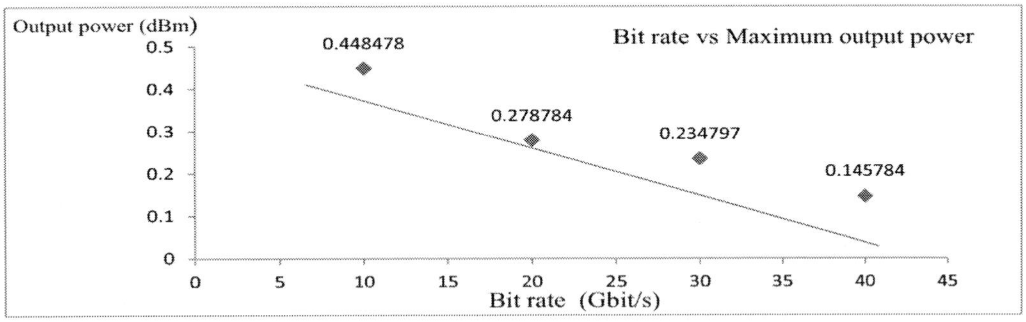

Fig. 4. Plot of bit rate versus maximum output power.

4.2 Varying bit rates with min BER

Table 3 depicts the four bit rates that are varied with the minimum recorded BER. As the bit rate increases, min BER also increases. At 10 Gbit/s, the eye is widely open due to overlapping pulses. An open eye implies that the pulses are easily resolvable [10] . But, at 20 Gbit/s, 30 Gbit/s and 40 Gbit/s, it shows there are jitters and pulses have suffered dispersion. In this case, the eye pattern deteriorates because of the time interval between adjacent pulses keeps changing randomly and remain constant [10]. Although the SOA's current is increased, it is not sufficient to counter the saturation at high speed. Eventually, the most suited operation is measured at 10 Gbit/s as shown in Fig. 5.

Table 3. Bit rates variation with the minimum observed BER.

Bit rates for input A and B (Gbit/s)	Min BER
10	2.11439×10^{-68}
20	1.30321×10^{-6}
30	7.10153×10^{-3}
40	3.45961×10^{-1}

DOI: 10.1051/matecconf/20167801108

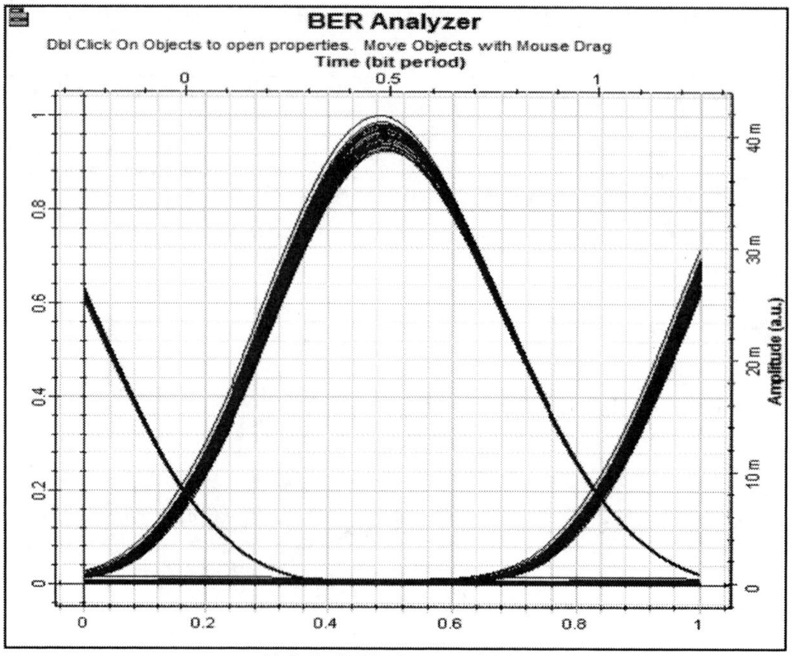

(a) Min BER for 20 Gbit/s.

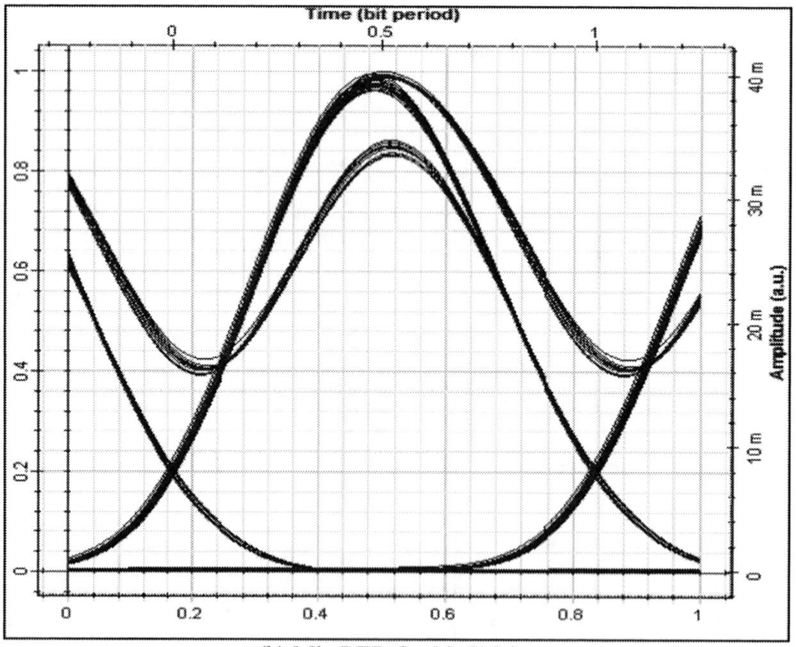

(b) Min BER for 20 Gbit/s.

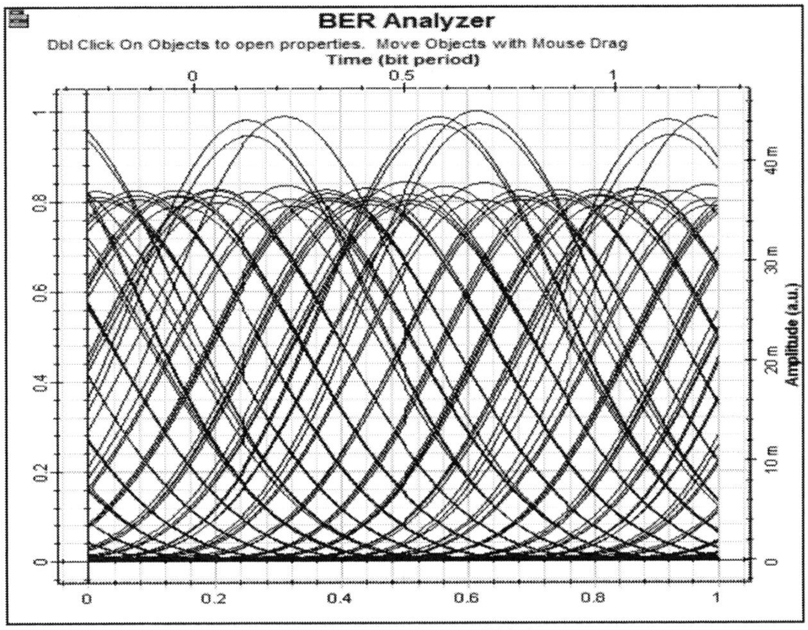

(c) Min BER for 30 Gbit/s.

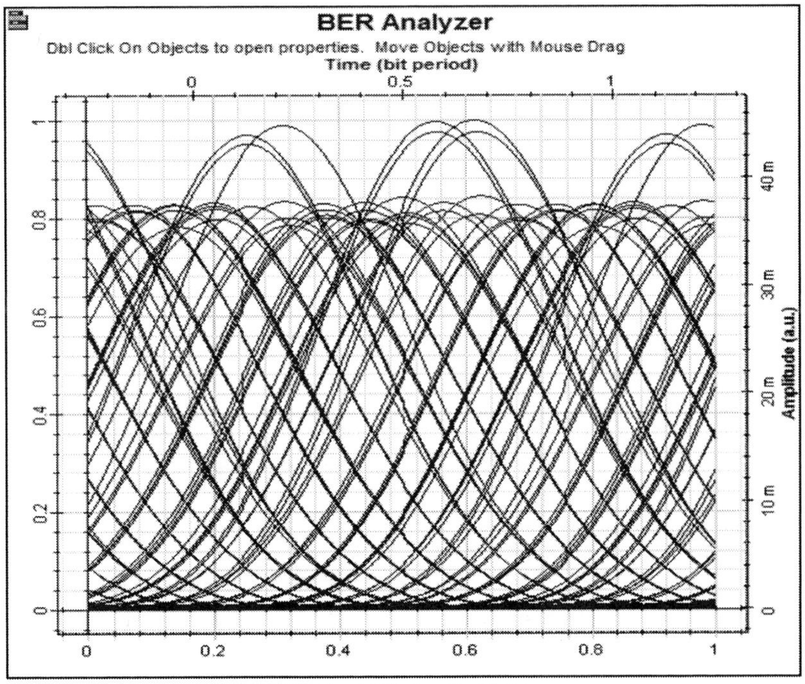

(d) Min BER for 40 Gbit/s.

Fig. 5. Eye diagrams (a), (b), (c) and (d) of minimum recorded BER.

Figures 6 & 7 show that XOR operation at 10 Gbit/s with 20 dBm input power portrays significant two-photon absorption which corresponds to a sharp peak power of 5.5 dBm when the pulse width of injected signal is ~3.6 ps .

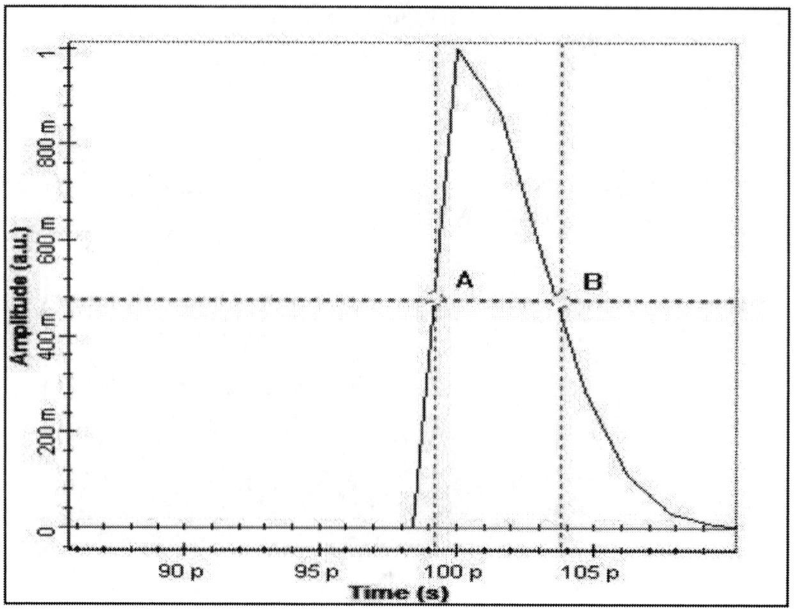

Fig. 6. The figure illustrates the pulse width of injected signal which is 3.6 ps at amplitue 500 m.

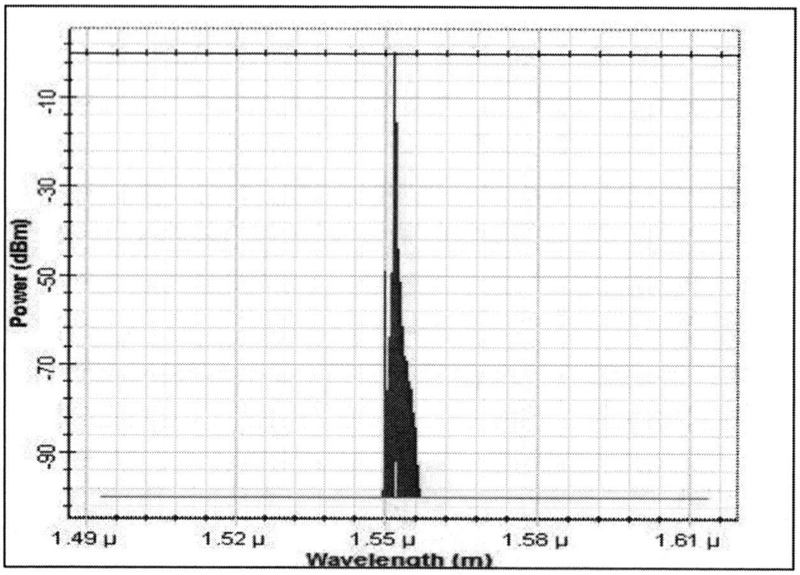

Fig. 7. The figure shows the XOR output spectrum at wavelength 1550 nm with sharp peak recorded at 20dbm power at the peak.

5 Conclusions

Design and analysis of XOR logic gate based on two-photon absorption using semiconductor optical amplifier based Mach-Zehnder interferometer is well developed and analysed. The simulated results show that the outputs are obtained as expected accordingly. Thus, validated the XOR logic gate based on two-photon absorption. Lower data rate at 10 Gbit/s describes the best performance in average which is limited by the SOA's recovery. In addition other bit rates performance also present good outputs with minimal differences. This concludes that TPA is a promising effect that can be ultralined not only in XOR but other logic gates with higher speed as well.

References

1. S. Ma, A. Kotb, Z. Chen, N.K. Dutta, *International Society for Optics and Photonics,* 7750 (2010)
2. N.K. Dutta, Q. Wang, *Semiconductor optical amplifiers* (World Scientific Pub, 2006)
3. Q. Wang, G. Zhu, H. Chen, J. Jaques, J. Leuthold, A.B. Piccirilli, N.K. Dutta, IEEE J. Quantum Elect., **40**, 703 (2004)
4. J.Y. Kim, J.M. Kang, T.Y. Kim, S.K. Han, Electron Lett., **42**, 1 (2006)
5. C. Bintjas, M. Kalyvas, G. Theophilopoulos, T. Stathopoulus, H. Avramopoulos, L. Occhi, L. Schares, G. Guekos, S. Hansmann, R. Dall Ara, IEEE Photonic Tech. L., **12**, 834 (2000)
6. K. Chan, C. Chan, L. Chen, F. Tong, IEEE Photonic Tech. L., **16**, 897 (2004)
7. Z. Li, Y. Liu, S. Zhang, H. Ju, H. De Waardt, G.D. Khoe, H.J.S. Dorren, D. Lenstra, Electron. Lett., **41**, 2 (2005)
8. N.S. Yuksek, X. Sang, E.K. Tien, F. Qian, Q. Song, O. Boyraz, *Conference on Lasers and Electro-Optics (CLEO)* (Optical Society of America, 2008)
9. Kumar, P. Syam, Marie Stanislas Ashok, and R. Subramanian, International Journal of Cloud Applications and Computing, **2**, 1 (2012)
10. K. Thyagarajan, A. Ghatak, *Fiber optic essentials* (John Wiley & Sons, **10**, 2007)

MATEC Web of Conferences **78**, 01109 (2016) DOI: 10.1051/matecconf/20167801109
IConGDM 2016

Design and Development of ZnO Based Gas Sensor for Fruit Ripening Detection

N.H. Hasanuddin[1,*], *M.H.A.* Wahid[1], *M.M.* Shahimin[1], *N.A.M.A.* Hambali[1], *N.S.* Nazir[1], *N.Z.* Khairuddin[1], *M.M.* Ramli[1] , and *S.S.M.* Isa[1].

[1]Semiconductor Photonics & Integrated Lightwave Systems (SPILS), School of Microelectronic Engineering, Universiti Malaysia Perlis, Pauh Putra Campus, Malaysia.

Abstract. This work focuses on the nondestructive fruit ripeness detection system at different maturity stages. The study involved a computational approach whereby the detector is simulated and characterized. The development of the 2 dimensional finite elements provides deeper understanding in an isotropic media of the sensor. The paper presents a sensitive layer of the zinc oxide with a thickness of 0.5 μm on LiNbO3 piezoelectric substrate for ethylene (C2H4) gas sensing application in order to distinguish maturity level of fruits. In addition the sensor acts as a better device for sensing the amount of ethylene gases that releases from mature fruits.

1 Introduction

Sensors are part of our daily life and have been applied in variety of fields and gas sensing is only one of the uses. Theoretically surface acoustic wave sensors (SAWs) categorized in microelectromechanical systems (MEMS) which depend on the modulation of acoustic surface to detect a physical phenomenon. Microelectromechanical systems (MEMS) are one of the areas that are widely used for gas sensing purpose, because of the advantages it brings in efficiency, robustness and cost effectiveness [1,2]. By introducing ethylene gaseous as an input to the sensors, receiver will receives and transduces input into a mechanical wave [1]. The sensor then transduces again the mechanical wave into electrical signal. Ethylene (C_2H_4) is fruit ripening agent. C_2H_4 gas plays an important role in the ripening process of climacteric fruits. Ethylene molecule composed from two carbon atoms bond with four hydrogen atoms. Knowing C_2H_4 gaseous concentration that been released by fruits, may distinguish the maturity level and remaining shelf life can be predicted [3].

In this research, the eigen frequency of SAW gas detector is investigated. Surface acoustic wave gas detector is based on the development of acoustic energy, where the transferring of energy is more responsive near to the surface of the detector [4]. SAW gas sensor consists of interdigitated transducers (IDT) as shown in Figure 1 which are placed on the top of substrate in order to transmit and receive acoustic waves. The IDTs layer is divided into two sets which is called the transmitter and the second set the receiver. Transmitter converts electrical signal into acoustic wave that travels along the piezoelectric

* Corresponding author: irhafika@gmail.com

DOI: 10.1051/matecconf/20167801109

substrate, then receiver converts back into electrical signal for detection and analysis purposes [5,6].

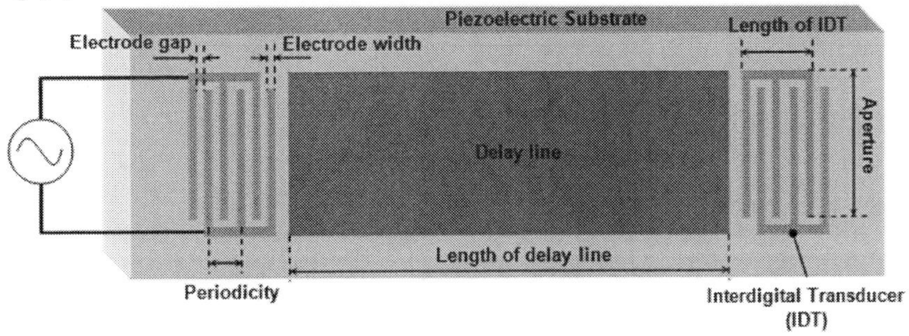

Fig. 1. Surface Acoustic Wave Sensor [1].

2 Material selection

2.1 Piezoelectric substrate material

Piezoelectric material is chosen based on the electrochemical coefficient (K^2), polarization and orientation. The extension of the wave interacts between the energy conversion that will effect in efficient measurement or (K^2) which insisted by $K^2 = 2(V_f - V_m) / Vf$ where Vf and Vm represent free surface and metalized surface phase velocities [7]. Most popular substrate materials are Lithium Niobate ($LiNbO_3$) and Lithium Tantalate ($LiTaO_3$) because of high K^2. Mode of SAW need to be considered before choosing between these two common substrate materials, because it has different acoustic properties in which each of them will provide the differences criteria with conductivity based sensitivity. In this study, XY $LiNbO_3$ is selected because its advantage that can support a single mode SAW in three directions U_x, U_y, and U_z.[6].

2.2 Interdigital transducers (IDTs) material

Aluminum has been selected as IDTs material because it is easy to deposit and cohere well with common oxide substrate compared to gold (Au) and chromium (Cr). Basically two set of IDTs commonly used in acoustic wave device, which are input and transmitter. Input work in generating energy while transmitter will convert electrical energy into mechanical energy. In the meantime output IDTs work as receiver that converts back mechanical energy into electrical energy in which at this stage occurrence of detection and analysis of ethylene gas [7].

2.3 Intermediate layer material

ZnO is chosen as intermediate layer because intermediate layer material should be dielectric layer in order to interact with piezoelectric substrate material. ZnO has low acoustic velocity about 2531 m/s compared to other dielectric material velocities. By varying the thickness of the ZnO intermediate layer it may increase effectiveness towards rate of gas absorption [5].

2.4 Sensing layer material

Sensing material is very important to make sure that the absorption occurred. ZnO nanowire is much useful in sensor application due to its good conductivity compared to others. Deformation occurs on ZnO due to the absorption of ethylene gas molecules on the nanowire which lead to alterations in conductivity [5].

3 Multiphysics modelling

Figure 2 shows a 2D schematic layer of the SAW zinc oxide based gas sensor for sensing ethylene gas that releases from fruits. The dimension of $LiNbO_3$ substrate is 30 µm in the x-axis and 4 µm in the z-axis. The dimension is based on the limitation of generated nodes number. Intermediate layer of ZnO with x-axis 30 µm and z-axis 0.2 µm is placed on the $LiNbO_3$ substrate. The ZnO sensing layer is placed on top of all substrates with 30 µm in length and 0.5 µm in height. The IDTs aluminum substrate is given by 1 µm as the width and 0.1 µm as the height, and it is defined massless electrodes in order to ignore the second order effects of the electrodes [8].

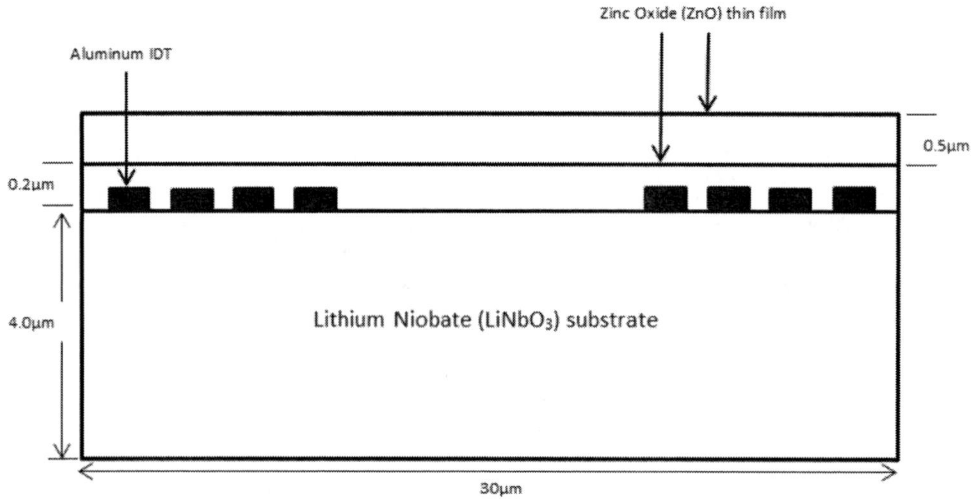

Fig. 2. SAW Sensor as the Sensing Material.

The simulations are executed using transient analysis for a simulation time of 0.1 s with 0.001 time step. Initially all the exterior condition of mechanical boundaries are set to be free exclude for the third boundary that is set to be fixed. All exterior electrical boundaries condition is fixed to be zero charge and those for transmit aluminium IDTs, electrode 1 and electrode 3 are positioned to be in positive electrical potential and zero potential is set to the second and fourth electrode. This combination of electrical boundary conditions corresponds to an open circuit configuration, which is typically suitable for sensing applications [9]. The use of periodic boundary condition implies that the frequencies of interest correspond to wavelengths that are integer fractions of the width of the geometry. The lowest SAW eigenmode has its wavelength equal to the width of the geometry, that is, 4 µm. Using this along with the Rayleigh wave velocity for the given piezoelectric substrate material, an estimation of the resonance frequency of interest can be established. The information can be used in the eigenfrequency solver, which helps it to find out the resonance frequencies close to this estimated number.

The wavelength of the substrate is described by equation 1 and the center frequency is described by equation 2 [10,11,12];

$$\boldsymbol{\lambda} = 2(W_e + W_{sp}) \tag{1}$$

$$f_0 = \frac{V0}{\lambda} \tag{2}$$

where λ is the wavelength of acoustic wave, W_e, width of each electrode and W_{sp} is the space between each electrode. The operating frequency, f_0, is determined by the propagation velocity, V_0 and the wavelength, $\boldsymbol{\lambda}$. A yz-cut $LiNbO_3$ whose Rayleigh wave velocity (vR) is around 3488 m/s. This gives an estimate of the lowest SAW frequency (f_0) to be 872 MHz.

Figure 3 shows the meshes cross section model. The maximum node size is chosen to be smaller than the wavelength to ensure the compensation of appropriate harmonics. This resulted in 3873 elements in the mesh. Meshing process is performed to determine the nodes in SAW gas sensor detecting structure which can be seen that the highest density of the nodes is directly under IDTs location.

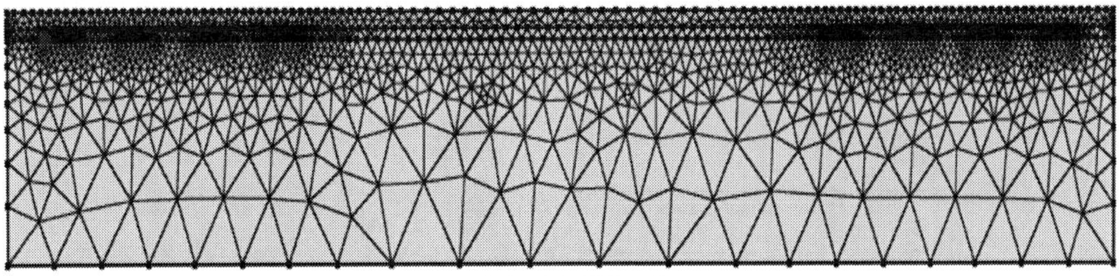

Fig. 3. Meshed image of 2D cross-section of the SAW Gas Sensor.

4 Results and discussion

Figure 4 illustrates the difference color variation due to the applied voltage on the input IDTs. It is shown that more energy at the surface rather than bulk substrate. This proves that SAW gas sensor confines more energy at the surface of piezoelectric substrate [12].

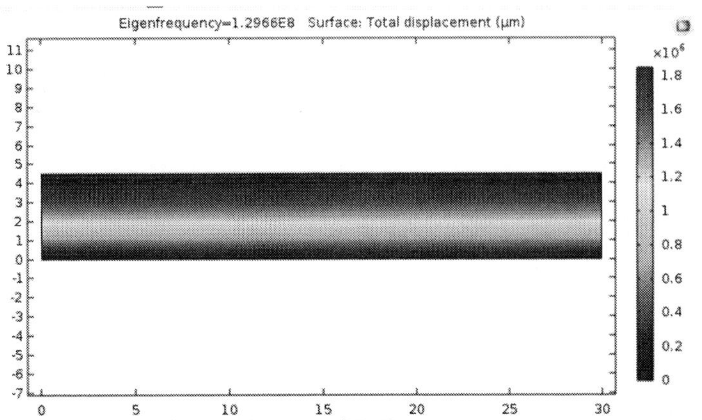

Fig.4. SAW wave propagation with the absence of C_2H_4 gas.

 DOI: 10.1051/matecconf/20167801109

Frequency dependent study is performed and produced its output as depicted in Figure 5. The acoustic wave can be observed, at the determined frequency, traveling at the top of substrate.

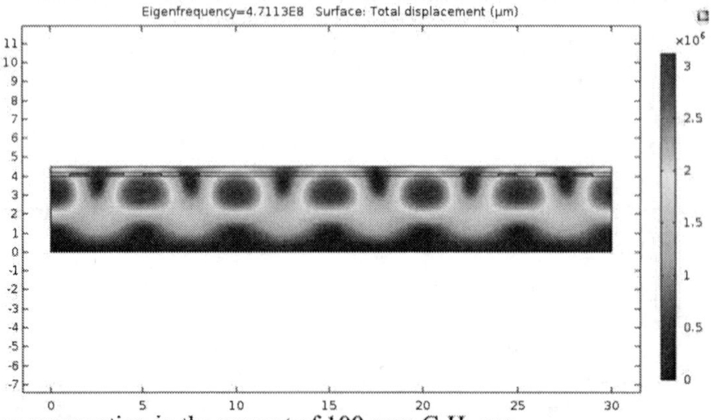

Fig. 5. SAW wave propagation in the present of 100 ppm C_2H_4 gas.

A parameter sweep of the electrode resonant frequency or eigenfrequency is done during frequency dependent study. This is done by firstly sweeping the parameter from 1.0E8 to 7.0E8 as displayed in Figure 6.

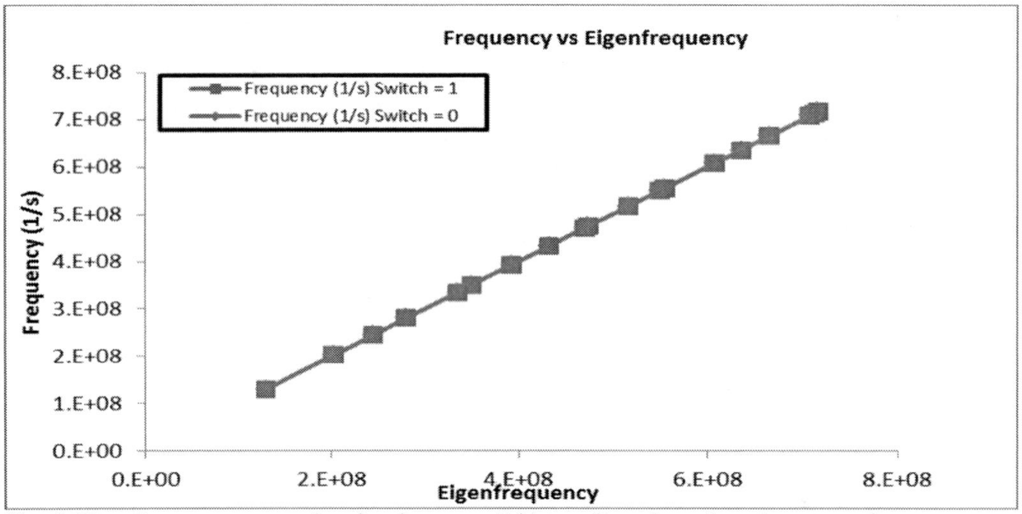

Fig. 6. Eigenfrequency sweep showing frequency(1/s) traveling along the substrate surface.

Figure 6 shows the exposure of C_2H_4 to the SAW gas detector that engaged a 2 KHz frequency shift. The frequency shift presents the existence of C_2H_4 gas releasing from fruit.

5 Conclusions

A simulation of SAW ethylene gas detector with ZnO as core material is successfully developed. An optimum frequency for this ethylene detector to be operated is 872 MHz. The x-y $LiNbO_3$ substrate is used as a piezoelectric layer as it can support a single mode SAW in three directions U_x, U_y, and U_z. The simulation is prepared in order to investigate the nature of the gas detector in the existence of ethylene gaseous. The outcome from the simulation shows that the total displacement at a determined frequency, gives a slightly changes and this concludes that with the present of C_2H_4 exposed to the surface of gas detector, frequency is shifted by about 200 Hz. Future work includes the simulation of the full sensor for ripening identification as well as the fabrication of SAW gas sensor.

References

1. D.S. Ballantine, *Acoustic Wave Sensors: Theory, Design, & Physico-Chemical Applications* (Academic Press,1997)
2. Morgan, P. David, *IEEE International Frequency Control Symposium* (1998)
3. S. Janssen, K. Schmitt, M. Blanke, M.L. Bauersfeld, J. Wöllenstein, W. Lang, Mathematical, Physical and Engineering Sciences, **372**, 20130311 (2017)
4. Y. Rao, G. Zhang, *Proceedings of the COMSOL Conference, Boston* (2007)
5. Hercules G du Plessis, *Proceedings of the COMSOL Conference in Rotterdam* (2013)
6. H. Wohltjen, R. Dessy, Analytical Chemistry, **51**, 1465 (1979)
7. Nurul Izzati Binti Mohamad Isa, *Modelling of Intermediate Layer for Love Mode SAW device thesis*, Universiti Teknologi Malaysia, Johor (2011)
8. L.J. Bie, X.N. Yan, Y. Jing, Y.Q. Duan, Z.H. Yuan, Sensors and Actuators B: Chemical, **126**, 604 (2007)
9. M.Z. Atashbar, A.Z. Sadek, W. Wlodarski, S. Sriram, M. Bhaskaran, C.J. Cheng, R.B. Kaner, K. Kalantar Zadeh, Sensors and Actuators B: Chemical, **138**, 85 (2009)
10. K. Ho, E. R. Lindgren, K. S. Rawlinson, L. K. McGrath, J. L. Wright, Sensors, **3**, 236 (2003)
11. S. Ahmadi, F. Hassani, C. Korman, M. Rahaman, M. Zaghloul, Sensors, **3**, 1129 (2004)
12. Di Francia, Girolamo, Alfano, Brigida, La Ferrara, Vera, Journal of Sensors (2009)

Development of ZnO Infrared LED and Its Emissivity

N.N.A. Saidi[1,*], *M.H.A.* Wahid[1], *P.* Poopalan[1], *N.A.M.A.* Hambali[1], *M.M.* Shahimin[1], *U.K.* Sahbudin[1], *S.N.* Ariffin[1], and *Muhammad* M. Ramli[1]

[1]Semiconductor Photonics & Integrated Lightwave Systems (SPILS), School of Microelectronic Engineering, Universiti Malaysia Perlis, Pauh Putra Main Campus, Malaysia.

Abstract. This paper provides an investigation of the design and characterization of zinc oxide (ZnO) in infrared light emitting diode (LED) application. Zinc oxide has been proved as one of the material that consists of wide range of electrical and optical properties [1,2]. A structure of ZnO infrared LED with pn junction is designed using a simulated environment. Layer of p-type doping is located at the top center of the LED while layer of n-type doping is at the bottom. This infrared LED is developed for its capability to emit electromagnetic spectrum. In order to obtain expected outputs, some parameters of the material have to be configured precisely. As a result, a current-voltage relationship and internal quantum efficiency is obtained. Besides that, emission rate and total emission rate of the design also has been successfully investigated. Finally, the properties of ZnO infrared LED are summarized and concluded.

1 Introduction

Light Emitting Diode (LED) is one of the extraordinary lighting technology that being invented to replace the traditional lighting system such as incandescent light bulb, high intensity discharge (HID) lamp and compact florescent (CFL). With the used of LED as our major lighting, greenhouse effect can be reduced and the economy growth can be sustained [3]. These benefits attract researchers to explore more regarding the characterization of LED. For your information, LED be able to produce a significantly higher light production and longer lifetime compare to a typical lighting systems. The typical light production of white LED is between 30-130 lm/W. Lifetime can be even 25 000 - 50 000 hours [3].

Unlike typical LED, infrared light emitting diode (IR LED) performing their function without emitting beautiful visible light. The IR LED emits in the infrared part of the electromagnetic spectrum and widely used in many optoelectronic applications, but perhaps the most familiar in a household setting is for used in remote controls. Although more efficient light production can be achieved using complicated design structures, simple p-n junction are still used for many IR emitters due to their low cost and ease of manufacture. As we know, p-n junctions are formed by joining n-type and p-type semiconductor material. Since the n-type region has a high electron concentration and the p-type a high holes concentration, electrons diffuse from the n-type side to the p-type side. Similarly,

* Corresponding author: atiqahsaidi@gmail.com

holes flow by diffusion from the p-type side to the n-type side. However, in the p-n junction, when the electrons and holes move to the other side of the junction, they leave behind exposed charges on dopant atom sites, which are fixed in the crystal lattice and are unable to move. As a result an electric field is formed between the positive ion cores in the n-type material and negative ion cores in the p-type material. This region is called the "depletion region" since the electric field quickly sweeps free carriers out, hence the region is depleted of free carriers.

Zinc Oxide (ZnO) is the major element in this research which exhibits the characteristic of semiconductor and piezoelectric properties. Lack in the symmetry of wurtzite combined with strong electromechanical coupling, help to produce strong piezoelectric in ZnO [4]. ZnO has a stable wurtize structure with lattice spacing a=0.325 nm and c=0.521 nm. Besides, ZnO is a polar group of 11-V1 semiconductor material with a direct wide bandgap of 3.37 eV which is suitable for short wavelength optoelectronic application [5,6]. High electron mobility, high thermal conductivity and high exciton binding energy make ZnO suitable for wide range devices [6]. ZnO is transparent to visible light and can be made highly conductive by doping.

2 Design of infrared LED

Infrared light emitting diode (IR LED) has been designed to have cylindrically symmetric geometry that made 3D visualization with a much shorter computation time in Comsol. The upper side of the p-n junction of IR LED has 50 um diameter of zinc oxide (ZnO), while the bottom side has 60 um diameter. However, due to r=0 axis is an axis of symmetry, the device geometry is represent by half of the diameter wide rectangle. The overall thickness of the LED geometry is 10 um. Since this research is investigated using p-n junction, p-type contact is designated at the center top of the IR LED while n-type contact at the bottom outside of the cylinder. The top 2.5 um of the IR LED thickness is dominated by p-doped while the bottom 7.5 um of the device is n-doped. Figs. 1 & 2 show 2D and 3D structure respectively.

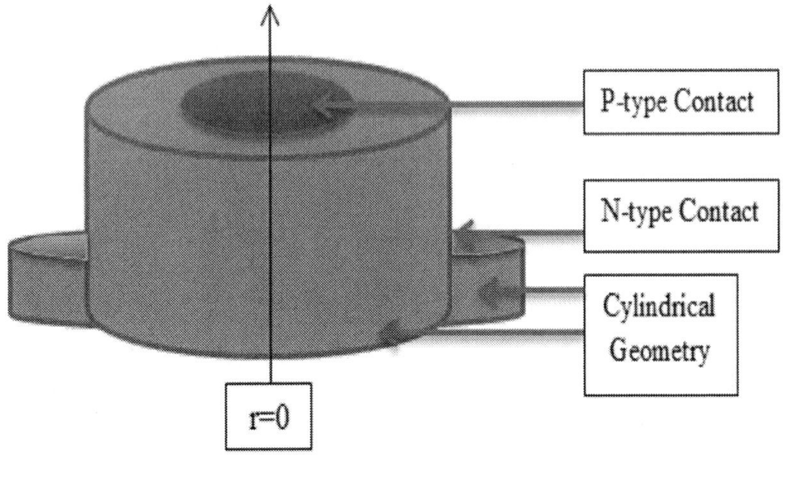

Fig. 1. Show 3D view of model.

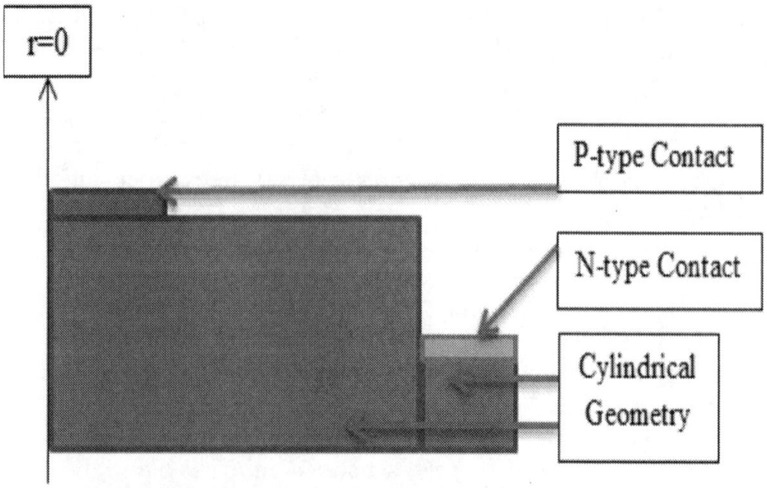

Fig. 2. Show 2D view of model.

As being explain in the introduction section earlier, the p-n junction method is used in the research due to low cost and ease of manufacturing. Hence, semiconductor (Semi) module is choosen in designing and interfacing the IR LED. The initialization study is used to refine the mesh based upon the gradient of impurity doping concentration whereas stationary study is adopted to sweep the bias applied to the device from 0 V to 1.5 V, allowing the current-voltage curve and optical emission characteristics to be calculated. The device geometry is represented by two rectangular structure represents n and p with different dimension in microns. Optical transition and Auger recombination segments are utilized for electroluminescence calculation and efficiency of the radiative emission measurement.

3 Results and discussion

Figs. 3 & 4 depict the function of mesh 1 and mesh 2. The mesh represents partition of the geometry model into small unit of simple shape. Thus, the function of mesh 1 is to increase the success rate of geometry functionality. Besides that, there are two part of the result that shows higher concentration of mesh which is at the p-type and n-type metal contact. This is because in the dark region, the mesh curvature factor is the parameter that provides the active constraint due to a circular shape that will be formed at both of the area in 3D geometry.

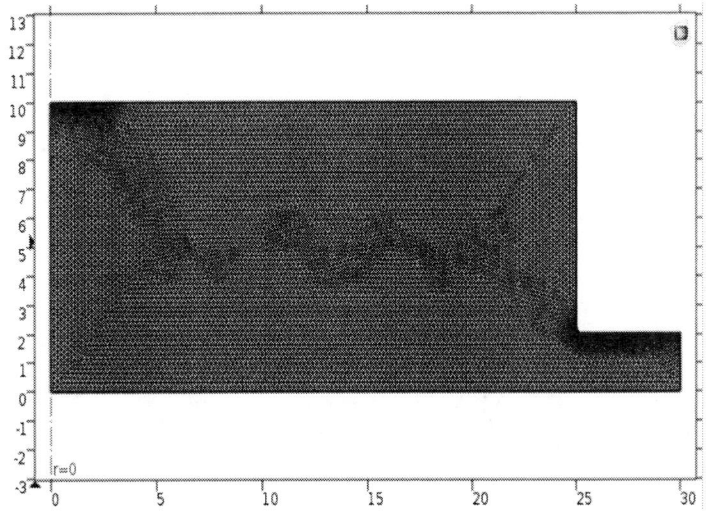

Fig. 3. Default mesh 1.

However, different things happen in mesh 2 which is the result of refined mesh output by the first study. The higher mesh densities that take place at the top of the figure show the p-n junction region. Therefore, it has being proved that p-n junction can be used in order to make Infrared LED.

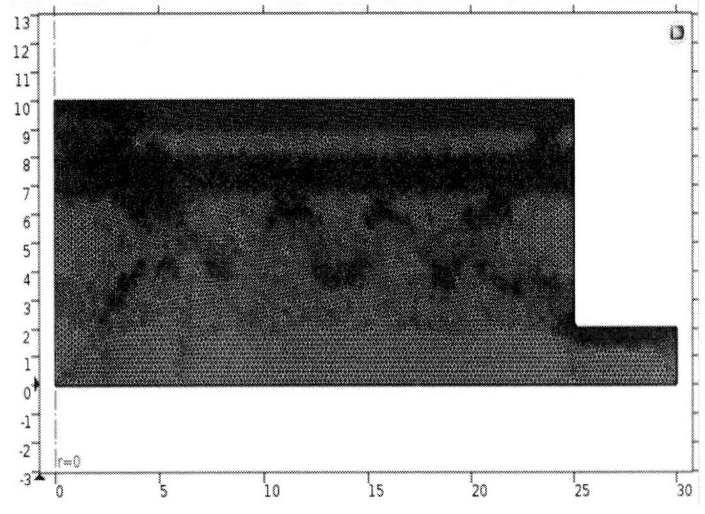

Fig. 4. Semiconductor initialization with mesh 2.

Figure 5 shows the result of doping that occur in the stimulation process. There are two kinds of doping which is p-type doping, where a regions that has extra holes and n-type doping, where a region that has extra electron. Based on the figure, the p-type surface layer is represented in blue while n-type region is represented in red. The doping gradually transitions from n-type to p-type over a range of a few micrometers (more than 4 um) to create a wide region in which light emission can occur. The variety colour plot is represented as threshold voltage and at the same time allows us to see the p-n junction. With the used of legend, the voltage range of each colour can be estimated. Based on the observation, the threshold voltage of n-type increases with doping while p-type decreases with doping.

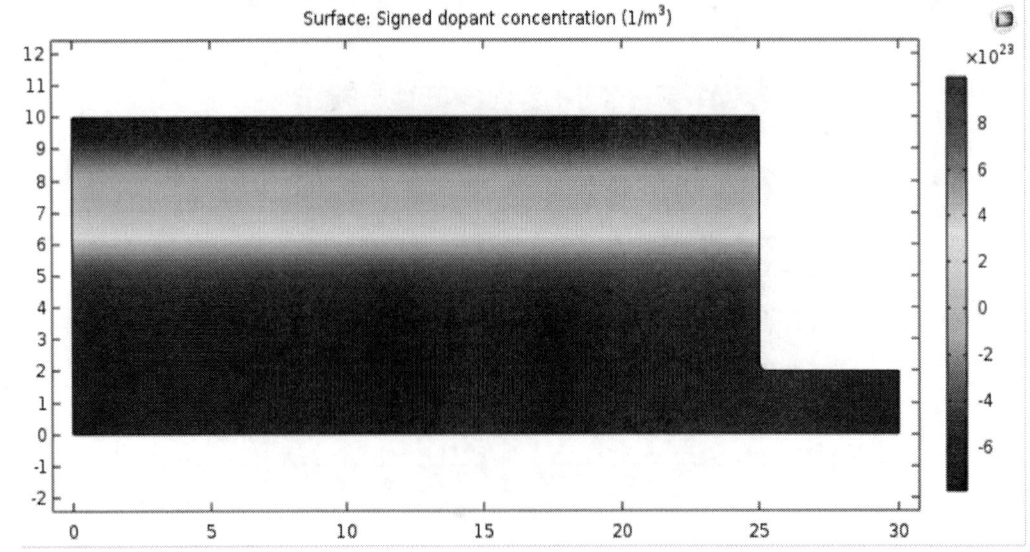

Fig. 5. Dopant Concentration.

Stationary study which is the sweeps of the biased voltage across the device from 0 V to 1.5 V is performed to prove that the device is clearly a diode. Fig. 6 illustrates the diode behavior with a turn-on voltage just around 1.1 V. After the turn-on voltage is achieved, even a small amount of voltage will give a great increase in current. It is also shown that the current is zero at any voltage below turn-on voltage whereby the diode can be viewed as a switch.

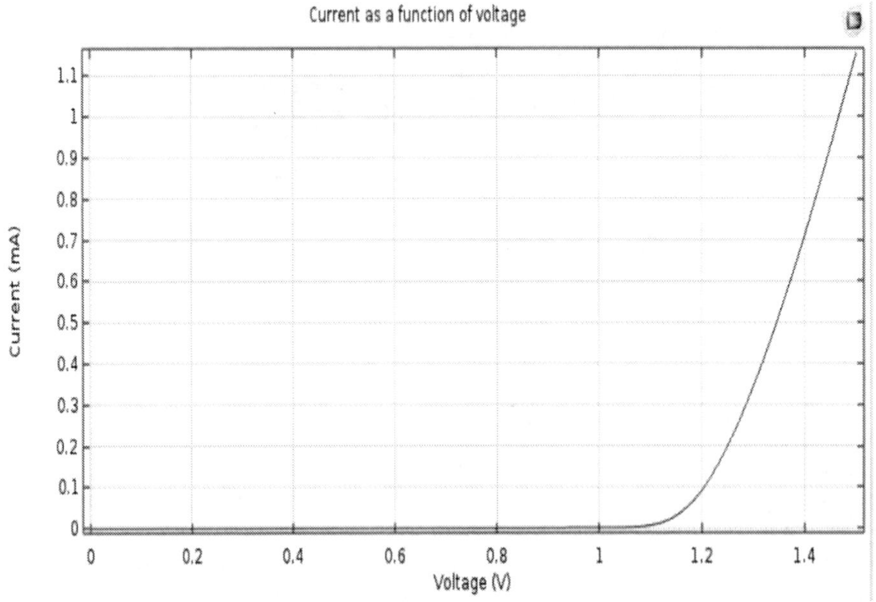

Fig. 6. Current versus applied voltage.

Figs. 7 and 8 depicts the emission rate throughout the device for a voltage of 1.5 V and 1.2 V. At the larger bias the emission has become concentrated towards the center of the devices which show the spacial distribution of the radiative emission.

DOI: 10.1051/matecconf/20167801110

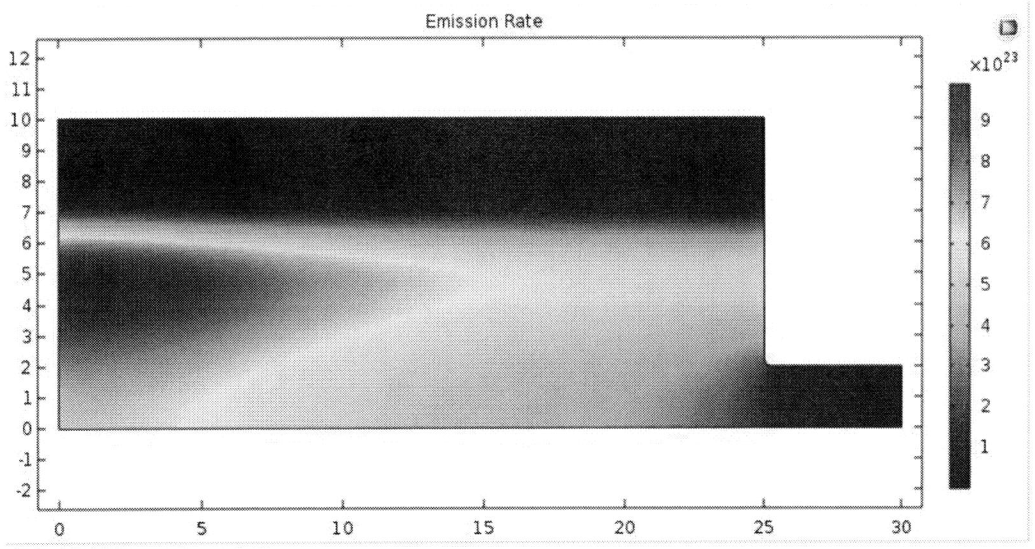

Fig. 7. Emission Rate throughout the device at 1.5 V.

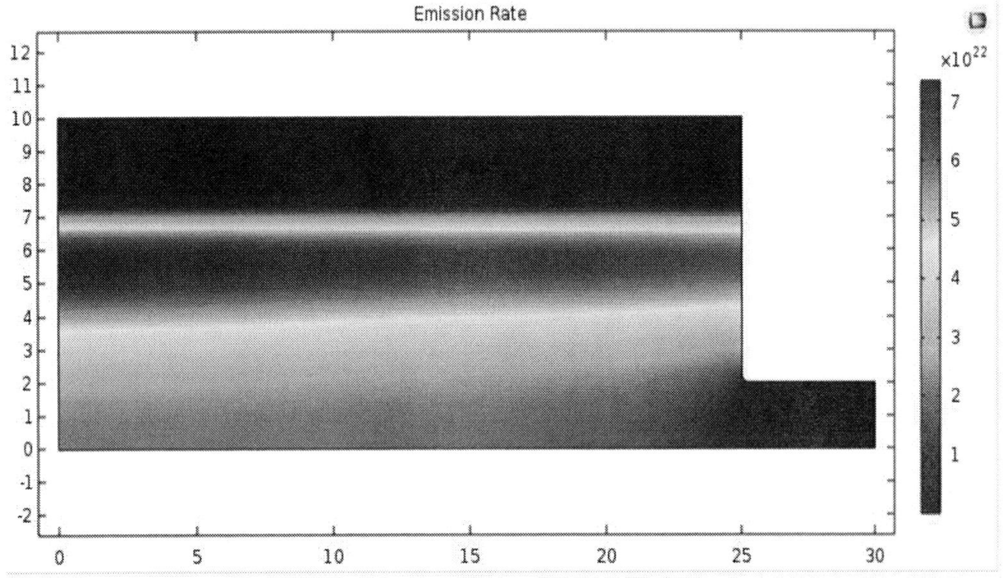

Fig. 8. Emission Rate throughout the device at 1.2 V.

Figure 9 shows the total rate of emission, integrated over the entire domain of the device, as a function of the current. The sub-linear curve is characteristic of LED droop, a phenomenon where the efficiency of LEDs decreases with increased current density. This reduction is even more apparent in Figure 10, which shows the internal quantum eficiency (IQE) as a function of the current. The IQE is the ratio of the emission rate to the carrier injection rate, and gives the proportion of injected carriers which radiatively recombine to emit light. Initially the IQE is very low, as no current flows, however as soon as current begins to flow the IQE is quite high. As the current increases the IQE drops drastically to plateau at a value of 0.981.

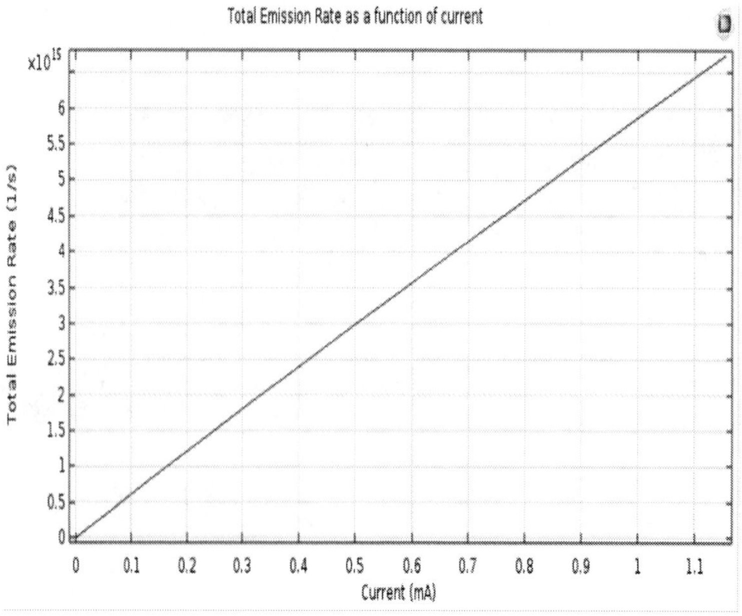

Fig. 9. Total emission rate as a function of current.

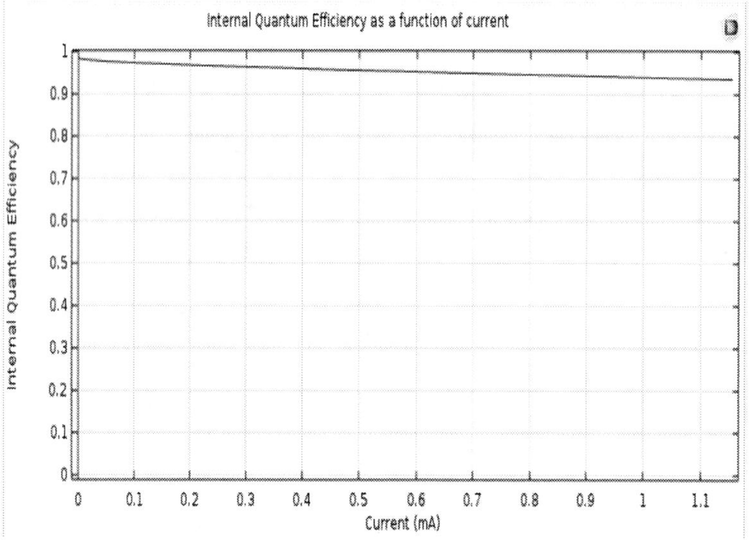

Fig. 10. Internal quantum efficiency(IQE).

In this work, Auger recombination has been included as non-radiative recombination mechanism. The rate of Auger recombination is proportional to the cube of the carrier density, whist the radiative emission rate is proportional to the square of the carrier density [7]. This means that as the carrier density is increased by the increasing current a larger proportion of the carrier recombine via the Auger mechanism, leading to the observed drastic reduction in IQE. The IQE is only a measure of the efficiency of the radiative recombination within the device. The IQE also indicates that efficient operation of LED would occur with low current below 5 mA. This would be the desirable range of operation current for a remote control device, where intensity of emission is not important compared to increased battery lifetime. This current corresponds to voltage between 1.2 V and 1.3 V and so the emission will be distributed similar to that shown in Fig. 8. However, if the LED is to be used in a situation where total brightness is more important than efficiency, it could be desirable to drive the device with a larger current. It would be advantageous to change the material, the thickness or the shape of the LED. For example, a ring shaped p-contact would produce the brightest emission on the central axis to leave the device .

4 Conclusions

As a conclusion, the infrared LED is successful developed and characterized. The result from I-V curve potrays that the p-n junction method can be used as one of the methods to build a quality and simple infrared LED. Besides that, IQE plot also indicates the efficiency of LED operation whether by using low or high current.

The authors would like to thank the School of Microelectronics Engineering, Universiti Malaysia Perlis (UniMAP) especially SPILS for their support in this work. Partial of this research is funded by FRGS 9003-00461 under Ministry of Higher Education, Malaysia.

References

1. S. Jeong, Thin Zinc Oxide and cuprous oxide films for photo voltaic application (University of Minnesota, U.S, 2010)
2. K.L. Foo, M. Kashif, U. Hashim, M.E. Ali, Adv. Mater. Research, **9**, 292 (2015)
3. A.T. Mika Maaspuro, Proceedings of the 2012 Comsol Conference, Milan, Italy (2012)
4. Z.L. Wang, Condensed Matter., **16**, 857 (2004)
5. R.Y.H.P. Zhang, Q. Chen, W.G. Feng, D. Badami, Mater. Electron, **692**, (2014)
6. A. Janotti, C.G.V.D. Walle, Report On Progress in Physics, **72**, 29 (2009)
7. O.N.M. Willander, S. Zaman, A. Zainelabdin, N. Bano, I. Hussain, Jurnal of Physics D-applied Physics, **44,** (2011)

MATEC Web of Conferences **78**, 01111 (2016) DOI: 10.1051/matecconf/20167801111
IConGDM 2016

Emission Rate Variation and Efficiency Measurement in TiO_2 Light Emitting Diode

S.N. Ariffin[1,*], *N.A.M.A.* Hambali[1], *M.H.A.* Wahid[1], *M.M.* Shahimin[1], *U.K.* Sahbudin[1] , and *N.N.* A.Saidi[1]

[1]Semiconductor Photonics & Integrated Lightwave Systems (SPILS), School of Microelectronic Engineering, Universiti Malaysia Perlis, Pauh Putra, 02600 Arau, Perlis, Malaysia.

Abstract. Titanium dioxide light emitting diode was successfully designed and characterised. An approach in developing TiO_2 was based on cylindrically symmetric configuration with shorter computation time. The main goal of the research was to look into the performance of light emitting diode by using titanium dioxide and compare its validity with the standard or common criteria. The circular chip of titanium dioxide with 60 μm diameters was made with a circular active region formed by p and n-type dopants. The drastic reduction of internal quantum efficiency droops almost to zero value even though the current started to increase by 0.002 mA. The efficient operation within the device was observed that it would occur at the lowest current 0.01 mA while the total emission rate shown linearly increased at axis (0.002 mA, 5.6 (1/s)). This light emitting diode based on titanium dioxide also enabled to operate at lower turn-on voltage as low as 0.5 V. As the increment of voltages, the distribution of emission rate became less concentrated towards the top active of p-type layer of the device. The result also compared the two meshes where it was found that the refined mesh was focused around the p-n junction.

1 Introduction

Light emitting diode (LED) is one of solid state lighting device that emits photons when the electrons and holes are recombining as electric current flow through it [1]. Previously, the fluorescent light bulbs are currently used in daily application. However, due to problem of fluorescent light bulbs in electrical power consumption, lifetime and also reliability have attracted researchers to develop light emitting diode technology as primary lighting source in future. In 1962, as the first inventor of LED, Nick Holonyak believes the time ahead in LED [1].

Various semiconductor materials that are widely used in preparing LED such as Zinc Oxide (ZnO), Gallium Nitrites (GaN), Zinc Selenide (ZnSe), and Gallium Asenide (GaAs). By using these materials, the electrical energy of LED can be converted into lighting energy as the current conducting across the p-n junction. For instance, ZnSe-based white LED was homoepatixally prepared by using ZnSe substrates via molecular beam epitaxy that was reported in [2]. As a result, the greenish-blue emission at origin 483 nm from epitaxial layer

* Corresponding author: ctnursyamimi92@gmail.com

was observed and is found that the emission wavelength of the LED and the measured chromaticity coordinate were almost independent of the injected current.

Since these materials are most common in LED preparation, the idea of using a Titanium Dioxide (TiO_2) has been proposed. This is due to low cost material, good chemical and thermal stability [3-5]. Generally, the common semiconductor that relates to the band gaps shows 3 eV to 4 eV results in ultraviolet light emitting diode (UV-LED) colour [6]. Compared to other materials as stated above, with the higher of band gap energy which is around 3.2 eV to 3.4 eV [7] and the higher of wavelength for TiO_2 that is ≤ 388 nm, hence TiO_2 has been a good candidate for the LED performance enhancement research. These advantages make TiO_2 as a potential material to be used in the UV LED applications due to bandgap energy and wavelength itself.

Two studies are reported in this work; voltage sweep studyand semiconductor initialization study. Voltage sweep study is used to investigate on the internal quantum efficiency (IQE), total emission rate, current-voltage curve and sign dopant concentration. Consequently, IQE graph is used to observe the radiative recombination in the device while total emission rate is specified to investigate the total emission in the device over time that varies with the increment of current. Also, a plot of current-voltage graph is used to determine the turn-on voltage of the device. The voltage bias is swept from 0 V to 5 V. A sign dopant concentration is used to detect the color changes around the p-n junction. Furthermore, semiconductor initialization study is used for electric potential and the drift-diffusion equations for electrons and holes in a semiconductor material. This is to refine mesh adaptively based upon the gradient of the impurity doping concentration. Two different meshes are compared in order to spot the denseness around the p-n junction. In short, these studies are focused on the internal quantum efficiency, total emission rate, current-voltage, sign dopant concentration, emission rate and mesh analysis that results in LED performances

2 Structure of LED

The model structure of LED based on TiO_2 design was illustrated in Fig. 1. The design was started by conducting a two-dimensional axisymmetric with 60 μm diameter circular chip of TiO_2 with 3.42 eV of band gap energy as the main material. A rectangular chip was designed as shown in Fig. 1 below due to the crystalline structure of semiconductor material. The active region that formed by the p and n-type doping was created in a circular pattern. By assuming the $r = 0$ axis as the axis of symmetry, the LED design was presented by a thickness of 10 μm with 30 μm wide rectangle. The size of designed LED chip was about 60×10 μm^2. N-side contact was heavily doped at the bottom of device with thickness of 7.5 μm, while 2.5 μm was doped at the top. This is to ensure the most recombination takes place in p-side. P-type contact was deposited directly on the top surface at the centre of the device with a metal disk. Also, the n contact that was etched down on the n side layer of the device was to allow electrical contact that was deposited onto n-type side of the p-n junction. An auger recombination property has been added into this LED. It functions as a non-radiative recombination and to demonstrate the behaviour of the efficiency of the radiative emission which results in IQE and total emission rate. In order to look into the current-voltage curve, voltage sweep was adopted in the study. The voltage bias across the device was swept from 0 V to 1.5 V to observe the turn-on voltage in LED performance. Apart from that, sign dopant concentration was determined within the device. The emission rate throughout the device was also measured in active region of p-n junction with different of bias voltages. Following that, the model of emission rate was obtained in three-dimensional structure. Finally, the mesh results were analysed.

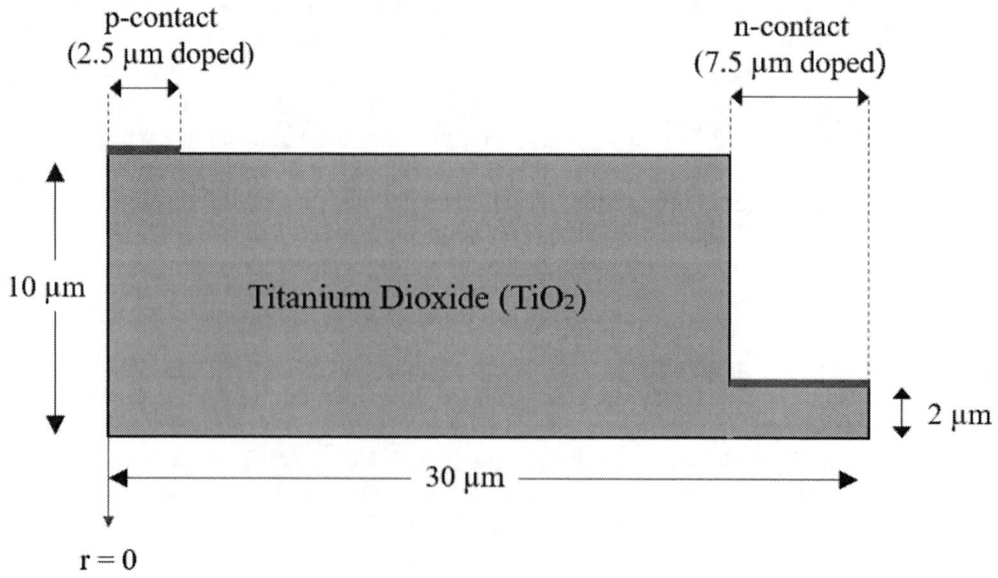

Fig. 1. The basic LED structure based on TiO_2 material

3 Results and Discussion

In order to study the performance of LED based on TiO_2, Fig. 2 was presented. The internal quantum efficiency (IQE) versus current was plotted in order to measure the efficiency of the radiative recombination of the LED device. IQE is the rate of emission variation with electrons injection. Generally, IQE gives the part of injected electrons that radioactively recombine to emit the LED light. In other words, the number of generated photons in the active region of p-n junction per electrons injection. Since the IQE only measures the efficiency of radiative recombination within the device, the actual efficiency of LED needs to consider other parameters such as size or thickness of the design.

Figure 2 depicts that IQE droops as current density increased. This IQE droops are also known as LED droops. Generally, LED droop is known as the LED efficiency decreases with the driving current [8]. LED droop can cause the amount of emitted light increases linearly with the increase of bias current. Apart from that, the drastic reduction of IQE droops almost to zero even though the current is increased by 0.002 mA. This indicates that the efficient operation of LED would occur at the lowest current 0.01 mA. This is due to the IQE results that show a decrement in LED efficiency as the current density increases in Fig. 2.

Generally, the amount of LED emission increases with bias current. Fig. 3 illustrates the evolution of total emission rate as a function of current. As a consequence, the total emission rate shows a linear increment starts from axial point of 0.002 mA, 5.6 (1/s). This indicates the total emission rate was increased with the higher current.

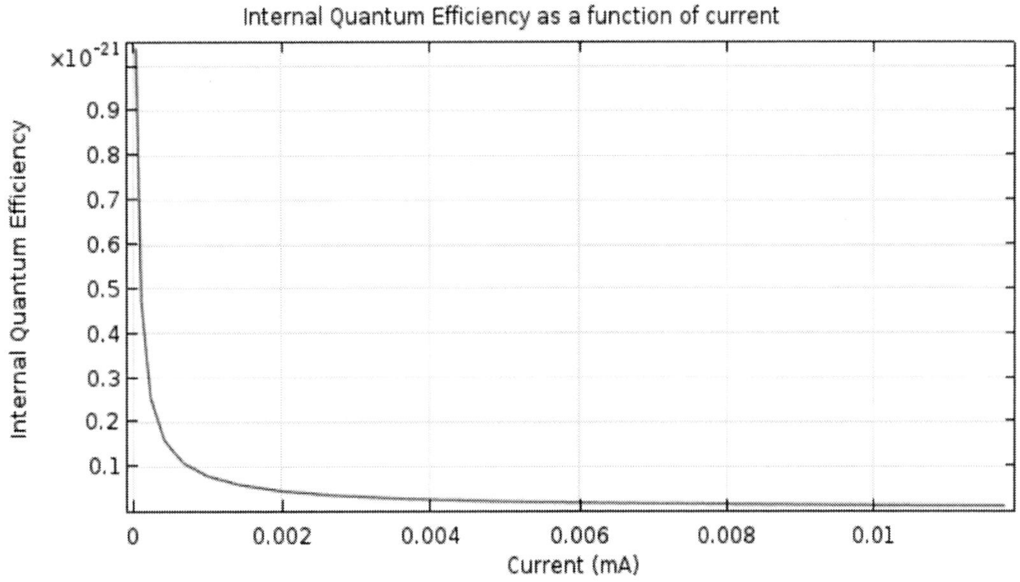

Fig. 2. The internal quantum efficiency as a function of current.

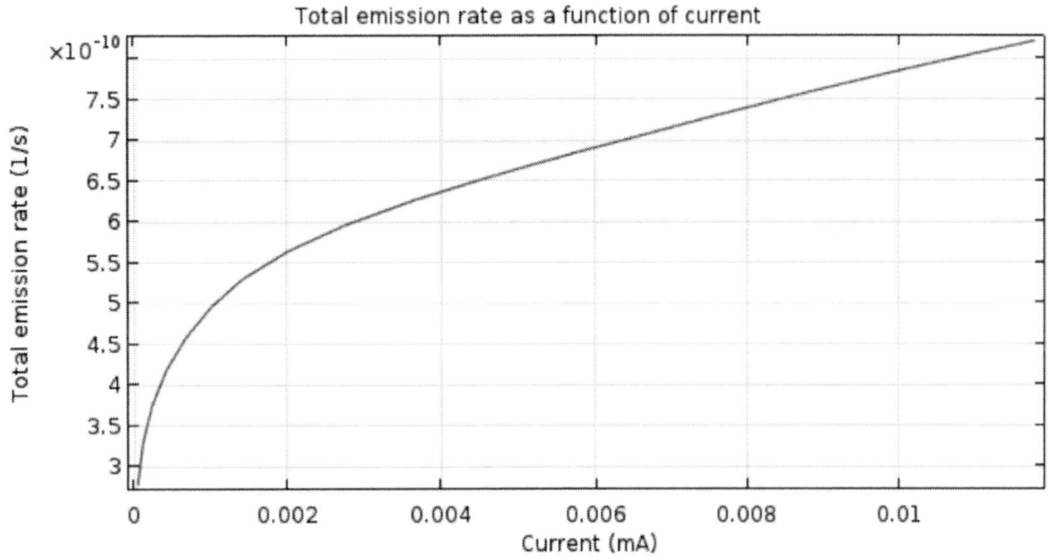

Fig. 3. Evolution of total emission rate as a function of current.

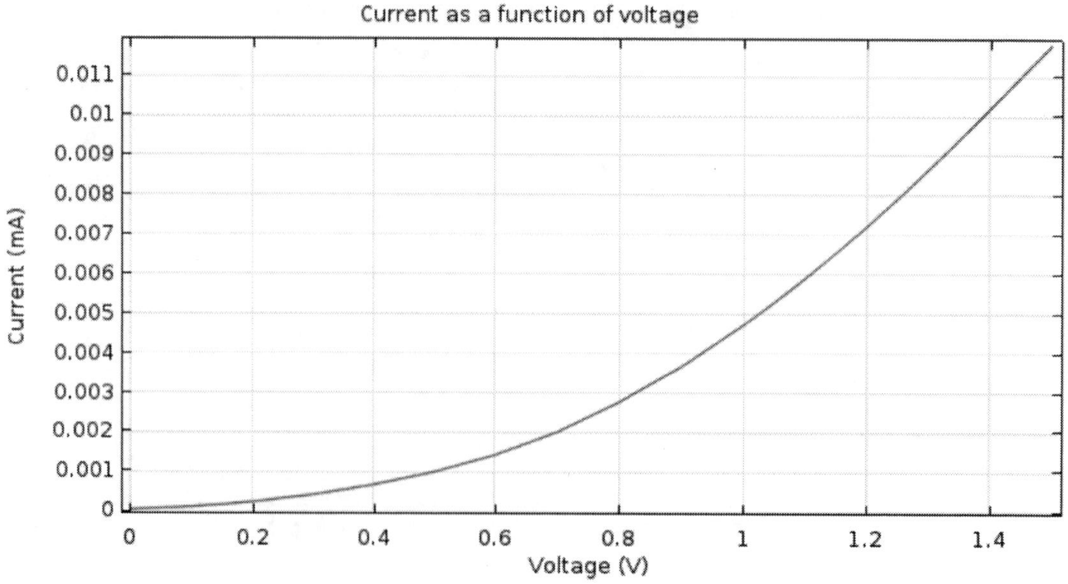

Fig. 4. Development of current as a function of voltage.

Despite of internal quantum efficiency and total emission rate that varies with bias current are important in LED designing, current versus voltage plot also plays important role in the performance of LEDs. Fig. 4 shows the development of measured current-voltage (IV) characteristics of the LED based on TiO_2. It was found that the voltage increased as the injection current increased. This shows that the LED could operate at lower of turn-on voltage as low as 0.5 V. This indicates that the LED starts to conduct current as the bias voltage is above 0.5 V. Moreover, closer to above 0.5 V of bias voltage, the current grows rapidly up to 1.5 V as a function of voltage.

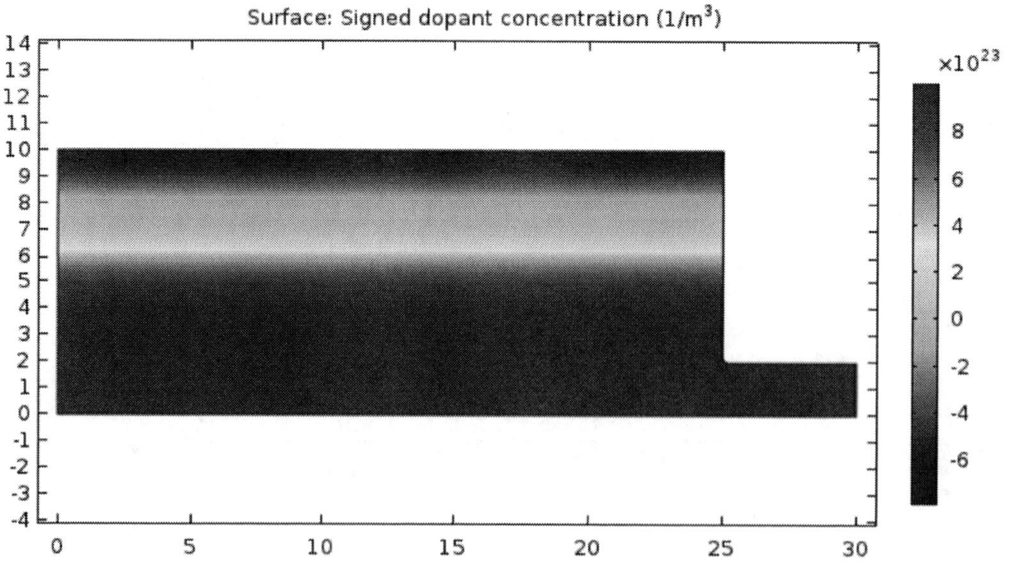

Fig. 5. Evolution of signed dopant concentration.

In the sign dopant concentration, the distribution of radiative emission within LED device was investigated with an increment of bias voltage. The sign dopant concentration relation was used to observe the changes of the spatial light emission that distributed towards the p and n-type layer of LED device. Fig. 5 shows default result obtained in a two-dimensional form of LED based on TiO_2 model where the doping concentration was transited from n-type surface layer to p-type layer in which LED light can occur. Different colours in the figure indicate the different log of dopant concentration. The blue area designates a lower of dopant concentration, while red denotes a higher of dopant concentration. This specifies higher doping concentration results at n-type layer in red while lower doping concentration at p-type layer in blue. Therefore, the changes of doping colour that transit from n-type layer to p-type layer will create a region of the light emission within the device.

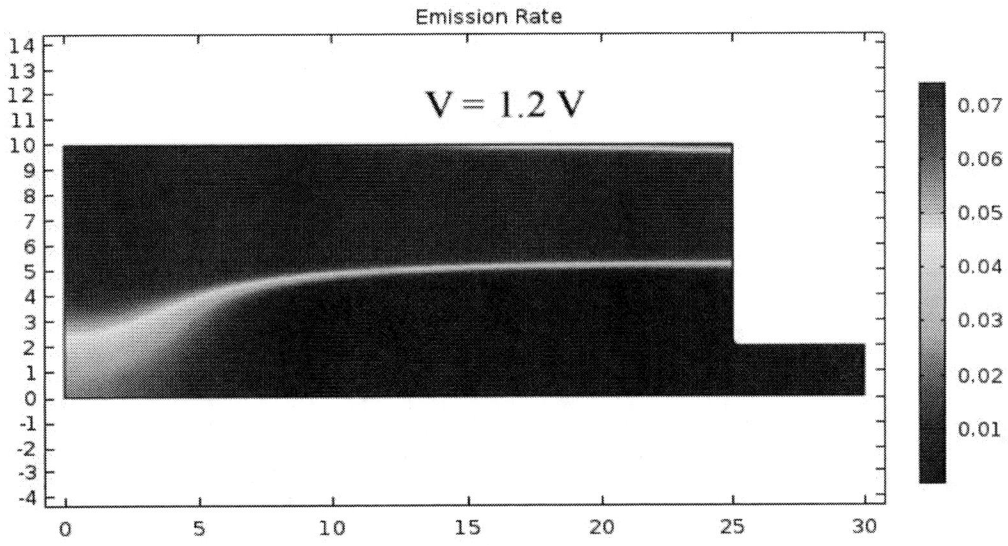

(a)

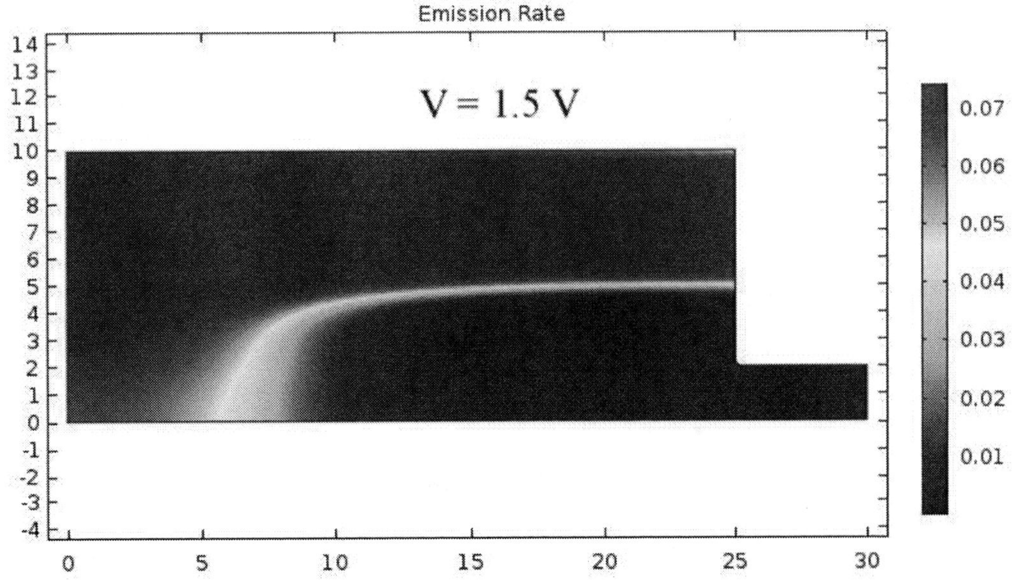

(b)

Fig. 6. Correlations emission rate at voltage bias at (a) V = 1.2 V and (b) V = 1.5 V.

As the bias voltage was applied, the emission rate showed the colour changes of doping concentration throughout the device. This correlation was depicted in Fig. 6 where the emission rate was observed at bias voltages of 1.2 V and 1.5 V. Thus, as bias voltage of 1.2 V and 1.5 V were applied, both of the colour regions of p-type layer were changed from blue to red. This denotes that the light emission within the device occurs. In addition, it can be seen that as the increment of voltage, the distribution of emission rate for both voltage biases becomes less concentrated towards the top active of p-type layer of the device.
However, even though the emission rate does not distributed towards the p-type layer of the device, the light emission of within LED device still can occur. This indicates that by increasing the current the light of the device will be emitted as well as there is increment of current and voltage. To be more apparent, Fig. 7 illustrates the spatial distribution of emission rate in three-dimensional version. The circular active region of LED based on TiO_2 model was formed.

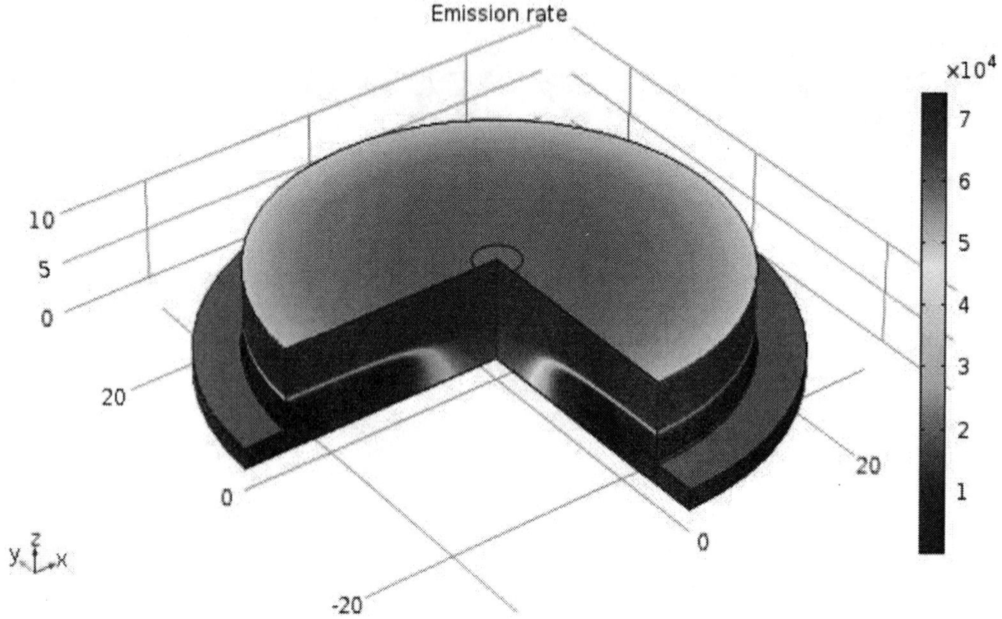

Fig. 7. Magnified view 3D model structure of emission rate.

In brief, from Fig. 2 to Fig. 7, some results have been computed to the observation that relates with the voltage study sweeps. Consequently, the LED material, high efficiency, current voltage characteristic, as well as emission rate have a certain degree of effects in producing LED device. In short, these factors play important role on LED performances. Following that, Fig. 8 represents the mesh refinement which is to observe the semiconductor initialization study. The study was focused on the observation to refine the mesh around the region of dopant concentration. Mesh is defined as a geometrical object that used as a set of finite elements in LED based on TiO_2 modelling [9]. Fine mesh was studied in order to observe the mesh in the impurity of doping concentration. This mesh provides a fine mesh at the p-n junctions which requires a higher resolution in order to see the changes of mesh. The results were compared between two meshes where the mesh was set to default mode which denoted as Fig.8 (a) and mesh that was set in fine mode indicated as Fig. 8 (b). It can be seen that the refined mesh graph was denser around the p-n junction. As a result, the gradient of mesh density increased in the p-n junction region. This indicates

DOI: 10.1051/matecconf/20167801111

the positions of the highest gradient electrostatic potential at p-n junction and the current flows were changed.

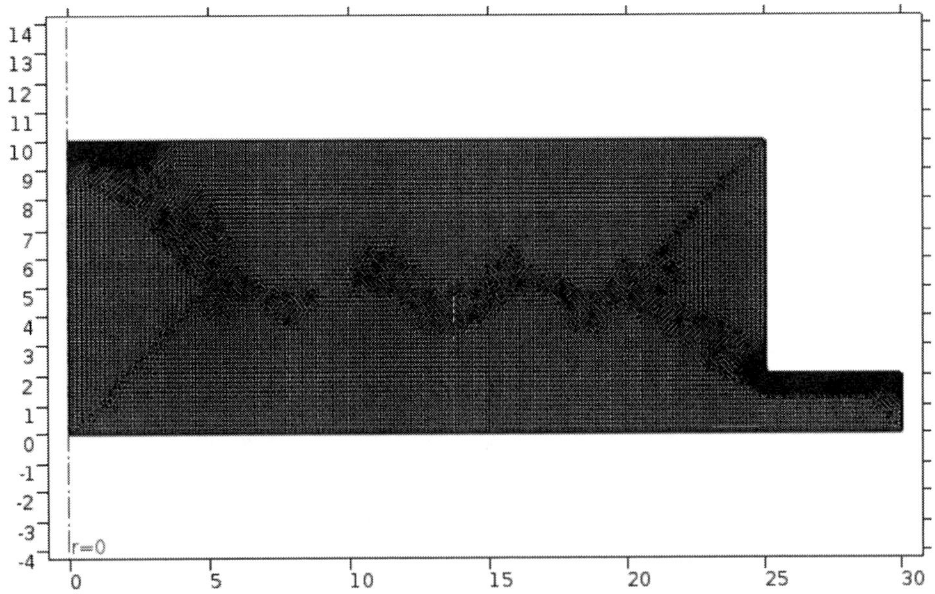

(a)

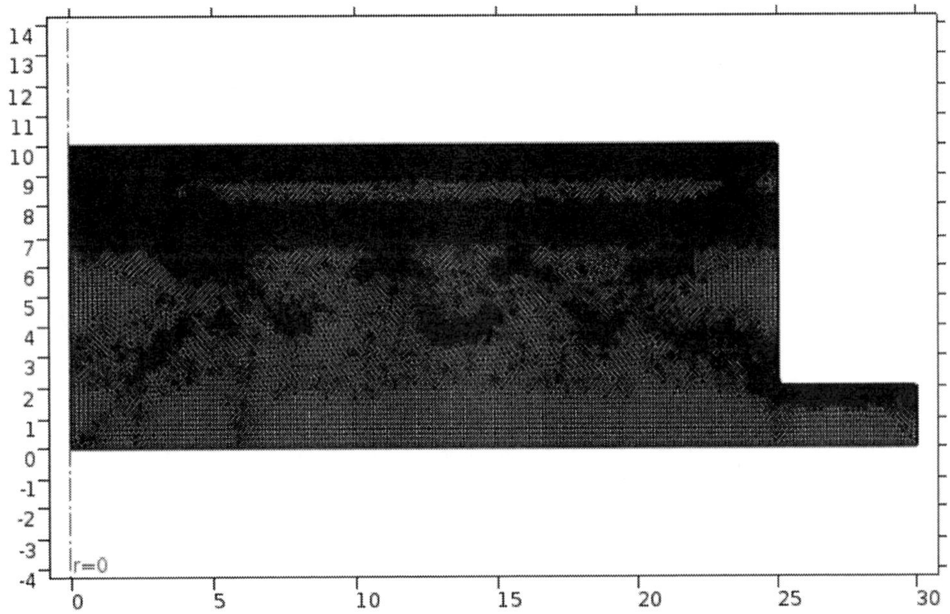

(b)

Fig. 8. Evolutions of (a) mesh 1 is the default mesh and (b) mesh 2 is the refined mesh output.

4 Conclusion

We have successfully demonstrated a LED structure based on TiO_2 material. It was found that drastic reduction of IQE droops almost to zero as the current was increased by 0.002 mA. Following that, it was observed that efficient operation of LED based on TiO_2 material

DOI: 10.1051/matecconf/20167801111

would occur at the lowest current of 0.01 mA. The total emission rate shows a linear increment axis (0.002 mA, 5.6 (1/s)). This LED based on TiO_2 design shows that the LED could operate at lower turn-on voltage as low as 0.5 V and below. In addition, it was noticed that as the voltage was increased, the distribution of emission became less intensed towards the top active of p-type layer of the device. The refined mesh was more dense around the active region of p-n junction in comparison with other mesh.

The authors would like to thank the School of Microelectronics Engineering, Universiti Malaysia Perlis (UniMAP) especially SPILS for their support in this work. Partial of this research is funded by FRGS 9003-00461 under Ministry of Higher Education, Malaysia.

References

1. S.T.R. Hetzel, G. Leising, submitted to Elektrotechnik & Informationstechnik, (2012)
2. T.K.L.S.J. Chang, Y.Z. Chiou, B.R. Huang, S.P. Chang, C.M. Chang, Y.C. Lin, C.C Wong, IET Optoelectron, **1**, 39 (2007)
3. T.S. Natarajan, M. Thomas, K. Natarajan, H.C. Bajaj, R.J. Tayade, submitted to Chem. Eng. J., (2011)
4. K. Natarajan, T.S.Natarajan, H.C. Bajaj, R.J. Tayade, submitted to Chem. Eng. J., (2011)
5. M. Izadifard, G. Achari, C.H. Langford, Catalysts, **3**, 726 (2013)
6. P. Hanselaer, A. Keppens, S. Forment, W.R.Ryckaert and G. Deconinck, Meas. Sci. Technol., **20**, 9 (2009)
7. S.V.M.K. Madhusun Reddy, A. Ramachandra Reddy, Materials Chemistry and Physics, **78**, 239 (2002)
8. C. Weisbuch, *Droop in LEDs*, Materials Department University of California Santa Barbara (2014)
9. S. Miguet, *6th International Workshop Discrete Geometry for Computer Imagery* (1996)

Investigation of All-photonic Multiple Signals Format Converter to Carrier- suppressed Signals

N.Z. Khairuddin[1,*], *M.H.A.* Wahid[1], *N.A.M.A.* Hambali[1], *M.M.* Shahimin[1], *N.S.* Nazir[1], *N.H.* Hasanuddin[1] , and *M. Muhammad.* Ramli[1]

[1]Semiconductor Photonics & Integrated Lightwave Systems (SPILS),School of Microelectronic Engineering, Universiti Malaysia Perlis, Pauh Putra Campus, Malaysia.

Abstract. Advance modulation format such as carrier suppressed return to zero (CSRZ) is used in several optical communication systems. The generation of CSRZ signal is reported in this paper by means of all-photonics format converter to its suppressed carrier signal. It examines five photonics signals of NRZ, RZ, CRZ, soliton and duobinary. All formats are successfully converted to the carrier suppressed return to zero, CSRZ. The converter utilizes the semiconductor optical amplifier in a loop mirror and the control signal which is typically at ten times higher than the input signals. The suppressed signal can reduce the peak power alteration in the transmission. It also minimizes the nonlinear effects due to unavailability of the peak carrier. The CSRZ in the WDM system with 4 and 8 channels shows soliton as the best suppressed signal within selected parameters.

1 Introduction

In recent years, several modulation formats have been suggested as alternatives to the commonly used nonreturn-to-zero (NRZ) and return-to- zero (RZ) [1,2]. The main difference between conventional RZ and CSRZ is phase shift of π between neighboring bits [1]. In its standard form CSRZ is generated by a single Mach–Zehnder modulator, driven by two sinusoidal waves at half the bit rate, and in phase opposition. An apparatus for generating a CSRZ signal is shown in some references [2,3]. The apparatus includes a mixer, a low pass filter, a driver amplifier and a single external modulator. The mixer generates a modulator input by mixing data with a half clock signal. The low pass filter band-limits the modulator input data into low frequency band data [3]. Several types of modulation formats are used in optical networks, and new type of formats are extensively studied and validated. To relief the potential of optical transmission systems and achieve higher transmission capacity, a lot of research on some special modulation formats other than traditional non return- to-zero (NRZ) and return-to-zero (RZ) has been done in recent years [4,5] including CSRZ [6].

[*] Corresponding author: zarhiedah@gmail.com

This work proposes all-photonic multiple format converter to carrier suppressed signals for RZ-to-CSRZ , NRZ-to-CSRZ, CRZ-to-CSRZ, soliton-to-CSRZ and duobinary-to-CSRZ format conversion. Carrier suppressed return to zero reduces the main carrier component, which carries the majority of the signal power and narrows the optical spectrum to a format that has a higher spectral efficiency and more tolerant to chromatic dispersion. In WDM applications, CSRZ has been shown to be more sensitive, where there are carrier components of the transmitted signals. The higher the number of channels, the more of sensitivity of CSRZ. All optical signal processing including multiplexing, demultiplexing and format conversion have recently received much interest [1,6]. Next generation communication networks are a mixture situation where different technologies come to play due to the large variety of applications and to their hierarchal configuration that implies various specifications for the networks segments. All-optical format conversion is considered to be the main technology used in optical signal processing to enhance the flexibility of optical networks [7,8].

2 Principle operation & setup

Figure 1 shows the setup of the CSRZ converter. The conversion section is the main part which contains the nonlinear loop biased in between 0 to 2π. The control beam is half the data rate with sinusoidal property. As a result the refractive index also changes sinusoidally. If the signals injected into the loop propagating in opposite directions, both will interfere constructively or destructively that may end in transmission or reflection. This is governed by the phase shift between them which later converted to amplitude modulation. Hence, a sequence of alternating bits in phase is produced at the transmission port which is CSRZ signal.

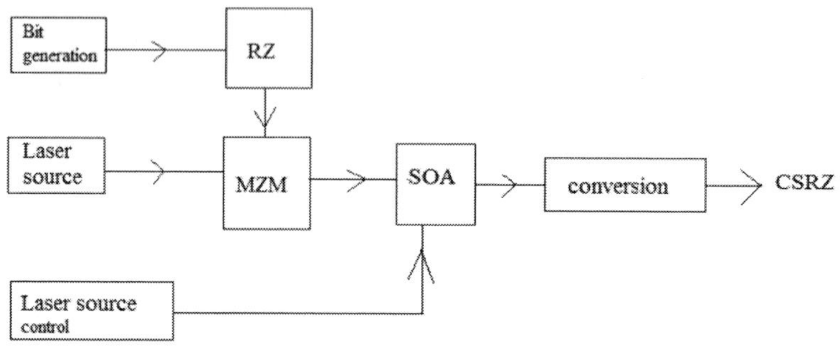

Fig. 1. Simulation setup of the converter.

The signal power is derived from NRZ, RZ, CRZ, soliton and duobinary transmitters. This signal combines in a multiplexer with control beam where the power of both signals are controlled by attenuators. Both signals are then directed to an SOA for the modulation to take place. Eventually, both signals are coupled again and the CSRZ output is observed as depicted in Figure 2. All signals are again multiplexed up to 8 channels for verification its suitability in WDM networks. Details of the eye diagrams and optical power spectra are presented in the next section.

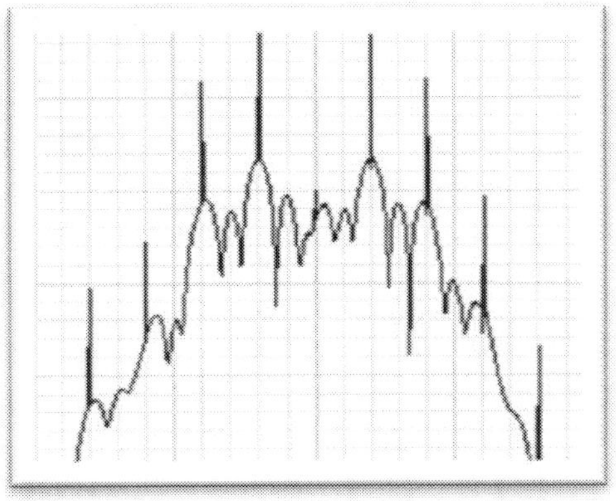

Fig. 2. Typical carrier suppressed spectrum.

3 Results and discussion

Table 1 summarizes the observation of the signals at both input and output ports. The signals are classified into original data and its converted eye and spectrum.

Table 1. Before and after conversion of each signal.

Signal	Before Conversion		After Conversion	
	BER analysis	Spectrum	BER analysis	Spectrum
NRZ	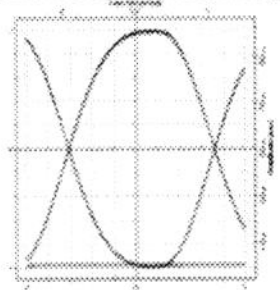	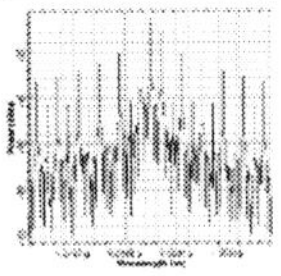	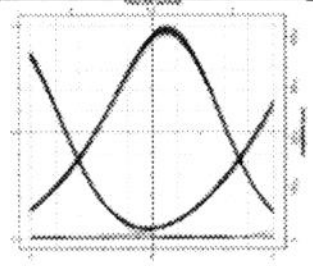	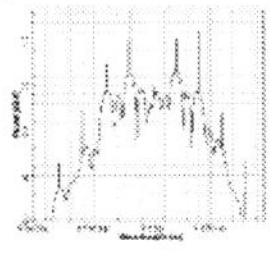
RZ	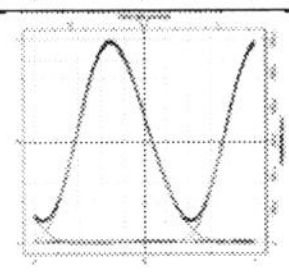	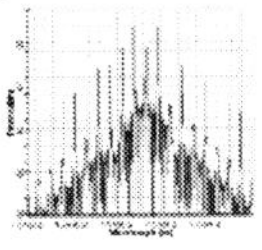	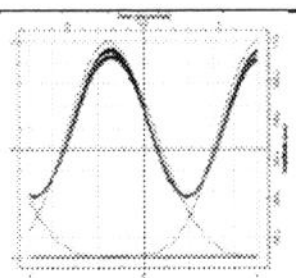	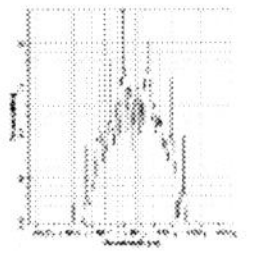
CRZ	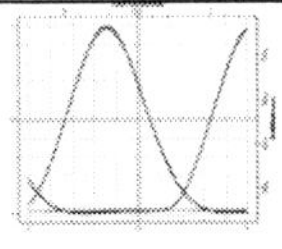	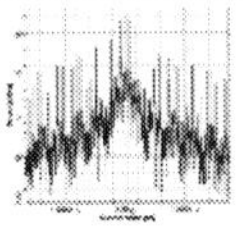	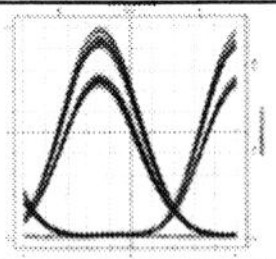	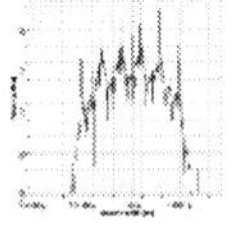
Soliton	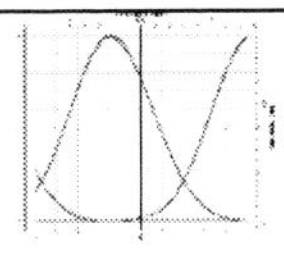	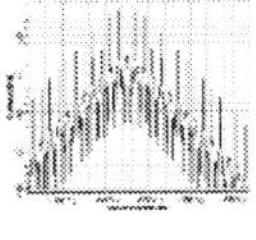	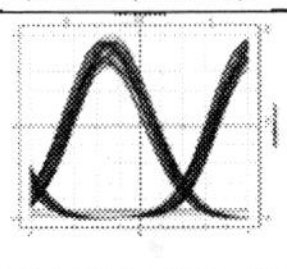	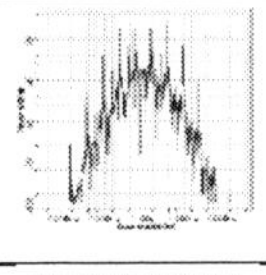
Duobinary	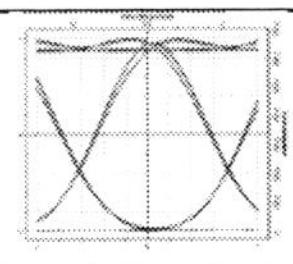	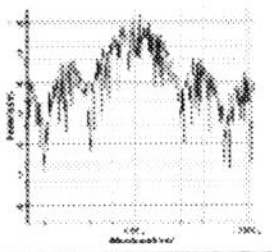	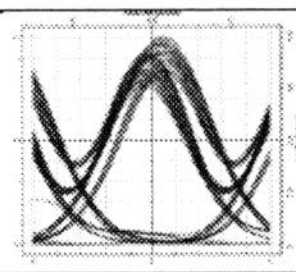	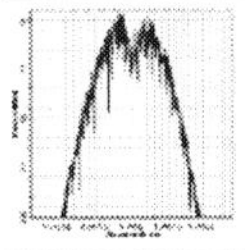

All signals are successfully converted with the same conversion parameters. The input power is maintained at 0 dBm while the control power at 10 dBm. Injected current into the SOA is configured at 250 mA for 5 dBm saturation output power. It is clearly observed that the carrier suppression denoted at the output. The level of suppression varies which is dependent on the powers and format of the signals. The RZ has shown to be the deepest carrier suppressed within the setup parameters. Changing these values might end up with different suppression ratio.

This work is extended to 4- and 8-channel WDM signals as depicted in Figure 3. Equal space between signals is used instead of random distance although it has similar possibility of conversion. It is found that the suppressed carrier signal easily generated. However, the control power needs an alteration with the number of channels increases. Table 2 & 3 depicts BER for back-to-back measurement of both WDM channels. Range of wavelengths are chosen from 1548 to 1555 nm, while the control signal is kept at 1545 nm. Some BERs are not recorded firmly due to the filtering of control power. When higher control is used, filtering selected wavelength also become more difficult. Thus, BER will give a small fluctuation in certain wavelength especially those closer to the control wavelength. Increasing the signal power will push the SOA into saturation that may also turns the BER to be worst. Properly adjusting powers and injected current could yield in better conversion.

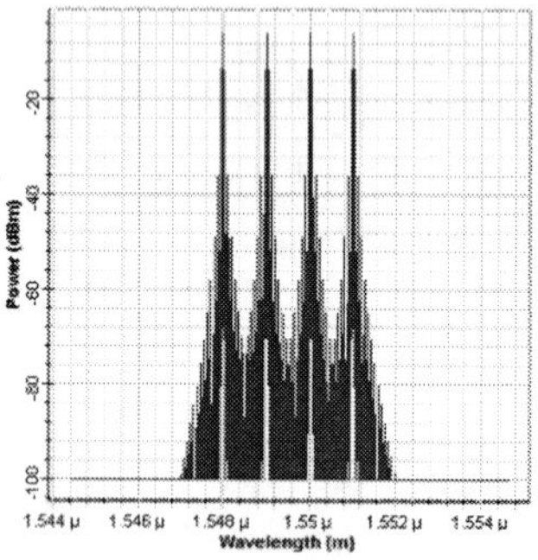

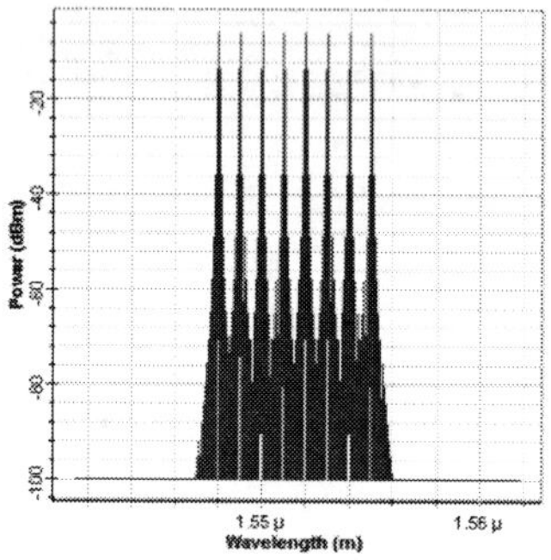

Fig. 3. Spectrum for 4-Channel and 8-Channel.

Table 2. BER for 4-channel.

Wavelength	NRZ	RZ	CRZ	Soliton	Duobinary
1548nm	2.09 e-25	1.13 e-21	2.14 e-10	1.16 e-33	5.05 e-9
1549nm	6.19 e-33	7.76 e-23	5.65 e-11	3.69 e-42	4.59 e-9
1550nm	2.57 e-37	2.16 e-21	2.24 e-10	7.98 e-40	6.52 e-9
1551nm	1.12 e-26	4.51 e-31	2.37 e-10	1.31 e-31	9.76 e-9

Table 3. BER for 8-channel.

Wavelength	NRZ	RZ	CRZ	Soliton	Duobinary
1548	2.67 e-21	3.86 e-13	5.24 e-21	3.54 e-48	2.59 e-10
1549	7.21 e-16	2.01 e-13	1.06 e-12	4.67 e-35	2.14 e-9
1550	1.45 e-16	1.87 e-13	8.40 e-14	9.47 e-37	2.21 e-8
1551	1.22 e-22	1.22 e-13	6.71 e-27	7.58 e-47	6.29 e-9
1552	1.39 e-19	2.71 e-13	2.14 e-17	2.05 e-30	5.01 e-8
1553	1.24 e-22	3.12 e-13	1.49 e-23	1.99 e-34	2.31 e-8
1554	4.08 e-23	6.35 e-11	7.01 e-22	7.31 e-31	3.21 e-9
1555	2.59 e-22	3.71 e-11	1.66 e-19	1.67 e-28	3.40 e-9

Supressed carrier signals from its origin of soliton signal beat other formats in all observation. Duobinary has the poorest eye and spectrum that resulted to be the least converted signal. In order to confirm this remark , all signals have been put in the same multiplexer with the same configuration. The soliton supersedes other signal formats. This might be the nonlinear effect is reduced due to unavailability of the peak carrier in the signal

4 Conclusion

All-optical format conversion is successfully performed using SOA in a nonlinear loop. Converted signals can be observed for NRZ, RZ, CRZ, soliton and duobinary. Carrier suppressed signal from soliton surpasses other formats with good BER at all wavelengths where asduobinary has shown the weakest likely to be converted. Multiple signals can be carrier supressed at the same time with one setup. Thus, the converter is suitable to be adopted in WDM networks.

The authors would like to thank the School of Microelectronics Engineering, Universiti Malaysia Perlis (UniMAP) especially SPILS for their support in this work. Partial of this research is funded by FRGS 9003-00461 under Ministry of Higher Education, Malaysia.

References

1. R. Ali, M.S Ali, T. Mir, B. Shabir, U.F. Lilla, Electrical and Electronics Engineering: An International Journal, **4**, 35 (2015)
2. T. Tokle, C. Peucheret, P. Jeppesen, Opt. Commun., **225**, 79 (2003)
3. D.S. Lee, M.S. Lee, S.S. Lee, J. SooKo, U.S. Patent 7 412 173 (2008)
4. Y. Yu, X. Zhang, D. Huang, Optics Express, **15**, 1 (2007)
5. Prasenjit Pal, Optical fiber communication system-theory and analysis (online) http://www.gatewayforindia.com/technology/opticalfiber.htm.
6. M.H.A. Wahid, M.M. Nahas, R.A. Ibbotson, K. J. Blow, Opt. Commun., **332**, 55 (2014)
7. A. Marculescu, S. Sygletos, J. Li, D. Karki, D. Hillerku\s s, S. Ben-Ezra, S. Tsadka, W. Freude, J. Leuthold, *Optical Fiber Communication Conference* (2009)
8. V. Sharma, A. Shrivastav, A. Jain, A. Panday, Int. J. Eng. Res. Appl., **2**, 1619 (2012)

Multi-wavelength Brillouin Fiber Laser with Fiber Bragg Grating Employs Loop Mirror

N.A. *Idris*[1,*], N.A.M.A. *Hambali*[1], A.Z. *Malek*[1], N.H.M. *Apandi*[1], M.H.A. *Wahid*[1], M.A.M. *Azidin*[1], and S.S.M. *Isa*[1]

[1]Semiconductor Photonics & Integrated Lightwave Systems (SPILS), School of Microelectronic Engineering, Universiti Malaysia Perlis, Pauh Putra, 02600 Arau, Perlis, Malaysia.

Abstract. In this paper, the development of multi-wavelength Brillouin fiber laser with fiber Bragg grating utilizing a loop mirror is successfully demonstrated. A multi-wavelength Brillouin fiber laser configuration employs 11 km length of single mode fiber as the gain medium, while the reflection of fiber Bragg grating is adopted to produce multi-wavelength signals. Based on the experimental results, it shows that the position of the fiber Bragg grating in the Brillouin fiber laser configuration plays an important role in producing the multi-wavelength signals. By using 19.0 dBm of Brillouin pump power, the highest number of Brillouin Stokes signal and tuning range obtained from this formation is about 23 and 4 nm, respectively. Furthermore, 50% output coupling ratio demonstrates the best ratio to produce Brillouin Stokes signal which gives a better performance in the Brillouin fiber laser system.

1 Introduction

Generally, the growth towards the communication is rapidly increasing with the higher demands on the data bandwidth which led to the development of the optical fiber and the exploration on the long haul transmission media [1-6]. The advancement of the optical fiber technology is continued by the invention of fiber lasers where the laser resonator is made up by selective fiber materials [7]. The advantages of fiber laser in mainly due to better reliability lifetime, high power, high stability, low amplitude noise, high output stability, immune to tough environment changes, low threshold power and narrow line width [8]. These advantages have extensive application such as optical sensing [9], optical filtering [10], narrow–bandwidth amplification [11], high-resolution spectroscopy [12], and optical fiber communications.

Brillouin fiber laser (BFL) is a type of fiber laser that utilizes stimulated Brillouin scattering (SBS) nonlinear effect. The SBS is a nonlinear process in optical fiber which involves the interaction between intense pump light and acoustic wave [13]. This process produces a unique result where Brillouin Stokes (BS) signals are generated in backward propagation at a downshifted frequency of ~ 0.08 nm (10 GHz) from the injected signal wavelength [13].

* Corresponding author: athirah.idris@yahoo.com

The implementation of the fiber Bragg grating (FBG) and loop mirror in the BFL configuration plays an important role in generating the multi-wavelength signal. This FBG is known as the optical fiber core that possess a special disorder refractive index region which it will diffracts the light ray propagation at a particular wavelength [14]. The loop mirror is a reflecting structure that acts as the mirror and help the generation of the BS signals [15]. On the other hand, this configuration has been conducted in the C-band in the range of 1530 nm to 1565 nm transmission wavelength. This band promotes the lowest attenuation and compatible system with amplifiers [16].

In this paper, the performance of multi-wavelength Brillouin fiber laser (MWBFL) with FBG utilizing loop mirror is demonstrated. The fiber loop acts as a mirror in such a way that it is able to reflect back most of the input signal. Evidently, the 50% output coupling ratio shows the best ratio of Brillouin pump power to stoke power which gives the better performance in this configuration. At 50% output coupling ratio and 19.0 dBm of BP power, 23 BS signals are produced. In addition, this configuration has been operated in the C-band region in the range of 1530 nm to 1565 nm transmission wavelength. This band promotes the lowest attenuation and compatible system with amplifiers [16].

2 Experimental setup and principle

The basic configuration of MBFL utilizing loop mirror is shown in Fig. 1. Based on this setup, the external tunable laser source (TLS) is utilized as a Brillouin pump (BP) signal. Besides that, 11 km length of single mode fiber (SMF) functions as Brillouin gain medium. A reflection of the FBG is the main factor in producing multi-wavelength in FBL configuration. The FBG has central wavelength of 1550 nm and 3 dB bandwidth of 1 nm with 90% reflectivity. The bidirectional optical coupler possesses the coupling ratio of 20%, 30%, 50%, 70% and 80%. Erbium-doped fiber amplifier (EDFA) amplifies the BP power in the range of 16 dBm to 20 dBm before going through an optical circulator. The optical circulator avoids the BP signal to be redirected to the source. An optical spectrum analyzer (OSA) monitors the output signals of MBFL.

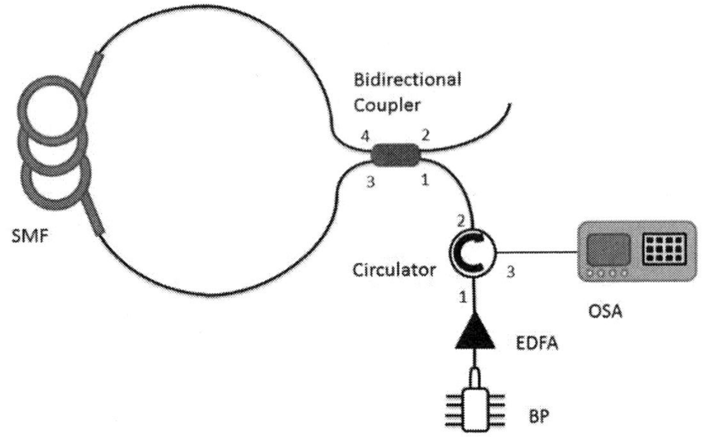

Fig.2. The basic configuration of the MBFL utilizing loop mirror.

The operation is started by the injection of the Brillouin pump (BP) signal via external TLS into the EDFA. The EDFA amplified and produced stronger BP signal in the loop [17]. Next, the signal is travelled through the optical circulator where the signals are conducted from one port to another and also has a function to prevent the signal to propagate in random directions [18]. The BP signal is then undergone the bidirectional optical coupler and propagated into the gain medium of the SMF. After achieving the

Brillouin threshold in SMF, the Brillouin Stokes (BS) signal is generated with the down shifted of ~ 0.08 nm from the initial BP signal wavelength and propagated in backward direction. This signal is the first generation of the BS signal. The second generation of the BS signal is produced when the signal propagated to the FBG reflects back into the loop mirror, thus creating the second generation of BS signal. Finally, the BS signal is directed to optical spectrum analyzer (OSA) through the bidirectional optical coupler and optical circulator. The observation and measurement of the output results is performed using the OSA with resolution bandwidth of 0.02 nm.

3 Experimental result and discussion

Producing a narrow linewidth multi-wavelength in the BFL setup of this work requires the position of FBG at different locations. Fig. 2 depicts three different locations of FBG. The FBG is placed in the loop mirror which is located after and before SMF as shown in Fig. 2(a) and Fig. 2(b), respectively. Whereas the FBG placed at the end is illustrated in Fig. 2(c).

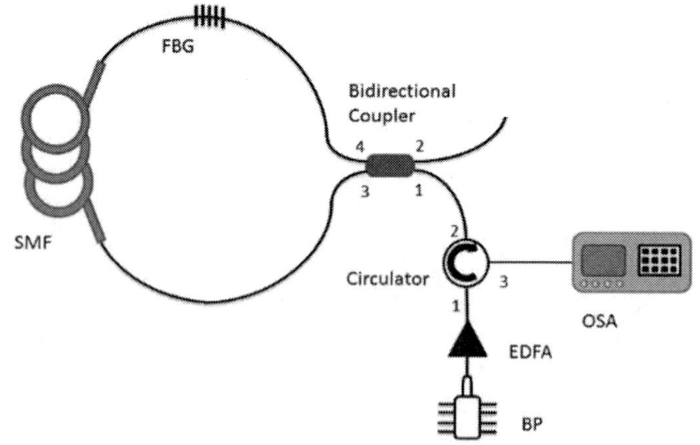

(a)

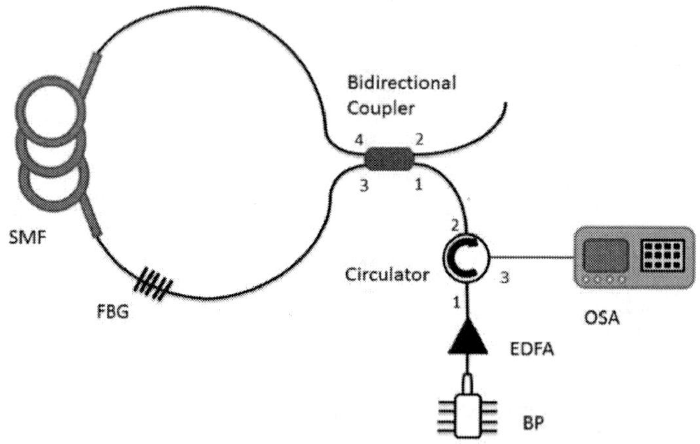

(b)

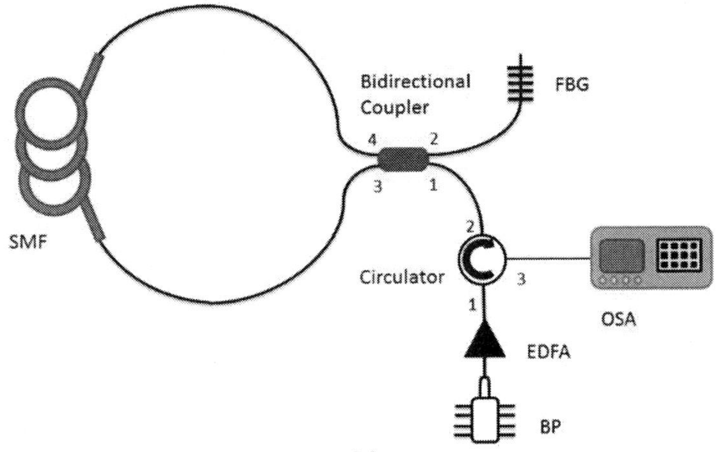

(c)

Fig. 2. The different positions of FBG in the MBFL configuration. (a) The FBG is placed in the loop mirror after the SMF. (b) The FBG is placed in loop mirror before the SMF. (c) The FBG is placed at the end of the configuration.

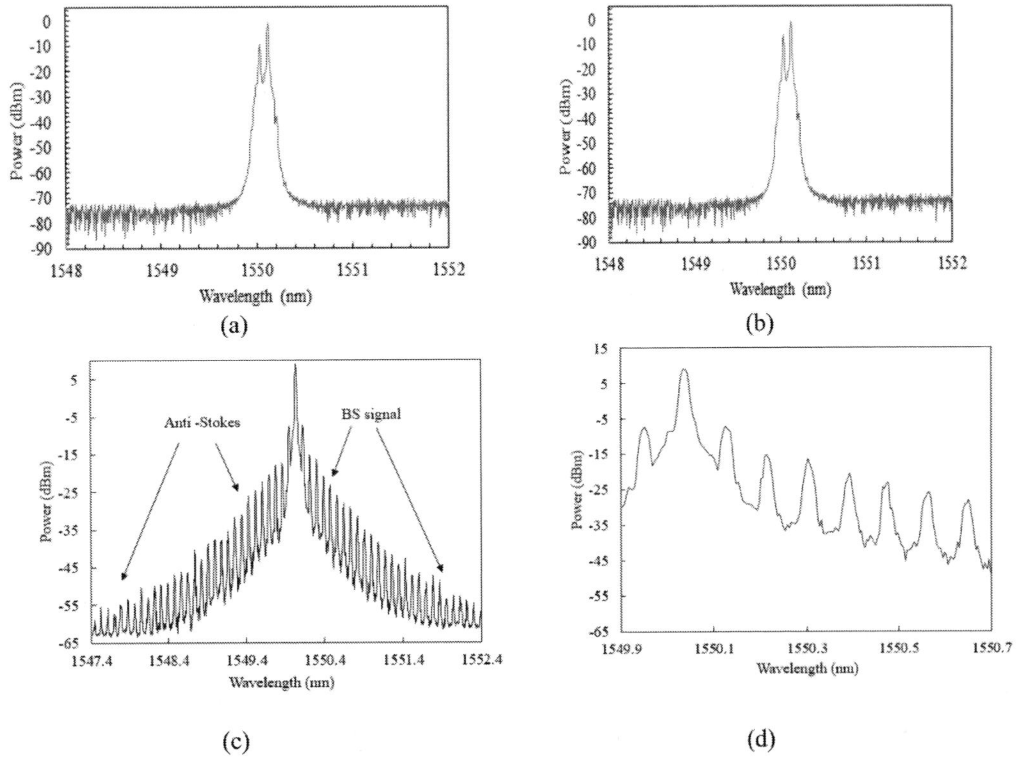

Fig. 3. The output spectra for different FBG locations in the BFL configuration. (a) The FBG is placed in the loop mirror after SMF, (b) the position of FBG in the loop mirror before SMF, (c) The position of the FBG at the end of the fiber configuration and (d) The BS signals with wavelength downshifted ~0.08 nm.

As a result, the implementations of three different locations of FBG have yielded different output spectra as shown in Fig. 3. Based on the Fig. 3(a) and Fig. 3(b), when FBG is placed in the loop mirror neither before nor after the SMF, the single wavelength is produced even though the input power has been optimized and the threshold power has been achieved. Besides that, the position of the FBG at the end of the fiber laser has successfully demonstrated the multi-wavelength generations where the signals are reflected back by the FBG into the loop mirror as depicted Fig. 3(c).

This has successfully produced the generations of the BS signals with wavelength downshifted of ~0.08 nm as shown in Fig. 3 (d). These spectra are generated at 1550 nm of BP wavelength with amplified BP power at 19.0 dBm. Characterization of the setup in Fig. 2 (c) where the FBG is placed at the end of the fiber laser is successfully obtained in this research. In order to investigate the number of BS signals, the 1550 nm of BP wavelength and amplified BP power of 16.0 dBm to 20.0 dBm are injected to the BFL configuration. From Fig. 4, 19.0 dBm BP power and 50% output ratio have been identified as the optimum parameters to produce a highest number of BS signals when compared with the other output coupling ratio. At these values, 23 BS signal are produced. This condition can be explained that when the BP power increases the number of BS signals rises up by owing to sufficient energy is provided to generate higher order BS signals generation [19]. When the BP power increases at 20.0 dBm, the number of BS signals shows slight declination to 22. It is observed at 16.0 dBm, 17.0 dBm and 18.0 dBm of BP powers, 15, 14 and 20 BS signals are generated, respectively. Meanwhile, when the values of output coupling ratio increase beyond 50%, which is at 70% and 80%, the number of BS signals is reduced. This trend is attributed from the increment of the cavity loss [20]. This factor affected the performance to generate higher number BS signals. At the lowest output coupling ratio of 20%, 1, 5, 6, 9 and 11 BS signals are generated at 16.0 dBm, 17 dBm, 18.0 dBm, 19.0 dBm and 20.0 dBm, respectively. The output spectra at different coupling (20%, 30%, 50%, 70% and 80%) when 1550 nm of BP wavelength and 19.0 dBm of BP power are injected inside the configuration is shown in Fig. 5.

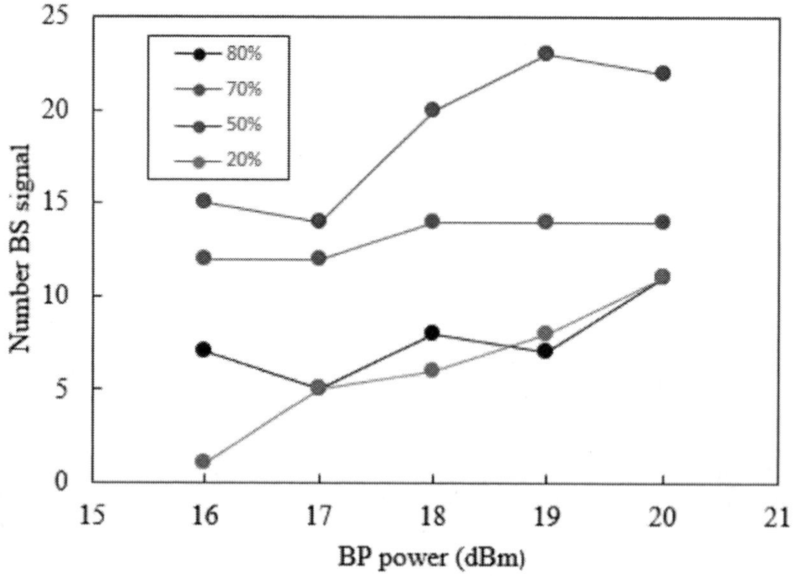

Fig. 4. Number of BS signal versus BP power at BP wavelength of 1550 nm with output coupling ratio of 80%, 70%, 50% and 20%.

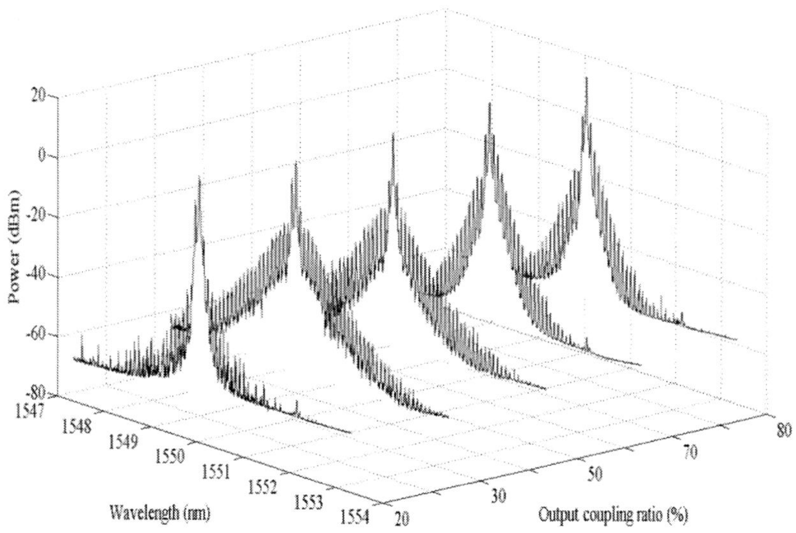

Fig. 5. The MWBFL spectra at different output coupling (20%, 30%, 50%, 70% and 80%) when 1550 nm of BP wavelength and 20.0 dBm of BP power are injected.

In order to observe the relationship between tuning range and multi-wavelength generation, produced optical spectra from 1548 nm to 1552 nm of BP wavelengths is plotted in Fig. 6. The amplified BP power is fixed at 20.0 dBm. Meanwhile, the optical spectra are extracted at 50% output coupling ratio. The MWBFL configuration has the ability to tune only 4 nm which is from 1548 nm to 1552 nm of BP wavelengths. At 1550 nm of BP wavelengths, 22 BS signals are produced. As 1550 nm is the center wavelength of the FBG properties, it is attributed to the creation of higher number of BS signals. Meanwhile, 16, 18, 19, 18 BS signals are generated at 1548 nm, 1547 nm, 1551 nm and 1552 nm, respectively.

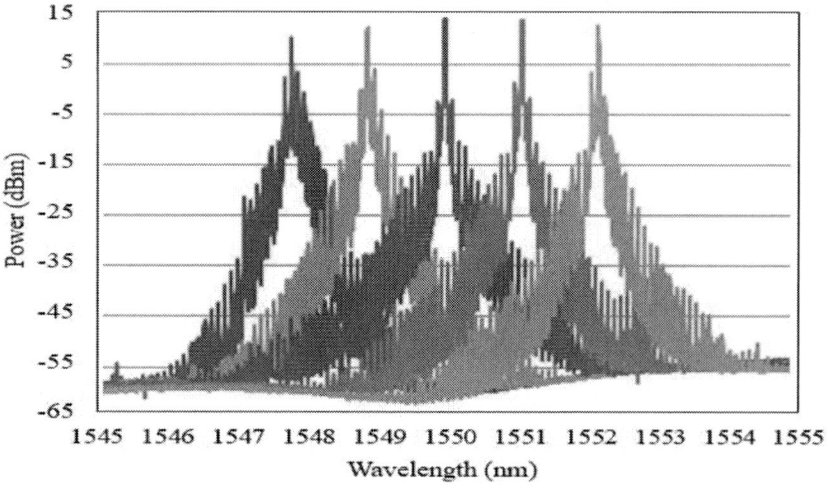

Fig. 6. Tuning ability of the MWBFL 1548 nm to 1552 nm of BP wavelength by injected 20.0 dBm of amplified BP power at output coupling ratio of 50%.

4 Summary

As presented in this paper, the development of the MWBFL with FBG utilizing loop mirror is successfully demonstrated in the C- band region of 1530 nm to1565 nm. The position of the FBG in the MWBFL configuration plays the important role in producing multi-wavelength signals. In addition, with injected BP power of 19.0 dBm, 23 BS signals and 4 nm of tuning range are obtained. Besides, the 50% output coupling ratio has the ability to generate higher number of BS signals. It can be concluded that the performance of the MWBFL utilizing loop mirror is dependent on the FBG location to produce multi-wavelength signals.

This work was fully supported by the Ministry of Higher Education, Malaysia under research grant # FRGS/9003-00532#.

References

1. N.A.M Ahmad Hambali, N. Roshidah, M. SafuanZainudin, A. ZakiahMalek, M.H.A. Wahid, M. Shakirin Abu Hurera, *International Conference on Electronic Design (ICED), Penang, Malaysia* (2014)
2. M. Shakirin Abu Hurera, *Development of L-band ring cavity Brillouin Fiber Laser,* Universiti Malaysia Perlis, Perlis, Malaysia (2015)
3. R.J. Ducar, *Particle Accelerator Conference, Accelerator Science and Technology, San Francisco, CA, US* (1991)
4. G. Snell, SMPTE Journal, **105(1)**, 4 (1996)
5. K.Kikuchi, Journal of Lightwave Technology, **34**, 157 (2016)
6. M.J. Hackerr, O. Muja, *Avionics Fiber-Optics and Photonics IEEE Conference, Annapolis, US* (2006)
7. S.W. Harun, IEEE Journal in Quantum Electronics, **20,** 1100108 (2014)
8. D. Schuocker, G. Liedl, A. Kratky, University of Technology, Vienna, Austria (2013)
9. D.Choudhury, A.K. Kar, R. Mary, IEEE Journal of Selected in Quantum Electronics, **20** (2014)
10. M.E. Fermann, I. Hartl, Nature Photonics, **7,** 868 (2013)
11. M. Giesberts, S. Nyga, O. Fitzau, B. Jungbluth, H.D. Hoffmann, O. Bang C. Larsen, Optics Express, **21**, 12303 (2013)
12. D.W. Hahn, N. Omenetto, SAGE Journal, **66**, 347 (2012)
13. H. Gong, L. Zhang, Y. Jin, J. Wang, Y. Tu, *International Conference on Electronics and Optoelectronics* (2011)
14. Z. Yusoff, M.H. Al-Mansoori, H.A. Abdul Rashid, P.K. Choudhury, M.N. Mohd Nasir, *National Conference on Telecommunication Technologies and IEEE 2008 2nd Malaysia Conference on Photonics* (2008)
15. N.H. Muhammad Apandi, *MultiwavelengthBrillouin Fiber Laser with Fiber Bragg Grating utilizing loop mirror,* Universiti Malaysia Perlis, Perlis, Malaysia (2015)
16. A. Buxens, A.T. Clausen, P. Jeppesen, H.N. Poulsen, Lasers and Electro-Optics, **39**, 277 (2000)
17. R.J. Mears, *Fifth Asia-Pacific Conference, Beijing, China* (1999)
18. Y. Makiuchi, H. Matsuura, Development of Low Loss Optical Circulator, Furukawa Electric Review, **59,** 1-4 (2002)

19. N.A.M Ahmad Hambali, M. Ajiya, M.A Mahdi, R. Sonee Shargh, Journal of Optics, **17** (2015)
20. M.A. Mahdi, M.H. Al-Mansoori, M.I. Saripan, A.F. Abas, N.A.M Ahmad Hambali, Applied Optics, **48**, 5055 (2009)

ZnSe Light Emitting Diode Quantum Efficiency and Emission Characterization

U.K. Sahbudin[1,*], *M.H.A.* Wahid[1], P. Poopalan[1], *N.A.M.A.* Hambali[1], *M.M.* Shahimin[1], *S.N.* Ariffin[1], *N.N.A.* Saidi[1], and *M.M.* Ramli[1].

[1] Semiconductor Photonics & Integrated Lightwave Systems (SPILS), School of Microelectronic Engineering, Universiti Malaysia Perlis, Pauh Putra Main Campus, Malaysia.

Abstract. ZnSe has demonstrated as a potential candidate in realizing advance LED in some appications for current and future works that utilize a cheaper preparation technique. Blue and white LEDs have been shown to spread across compound semiconductors. This II-VI compound semiconductor with a direct and wide band gap is used in the study which focused on a preparation and its characterization. The device is developed using a circular chip of ZnSe but only part of the active region is designed to allow shorter computation time. Analyses of the proposed LED are performed in an environment that allows optical transition and non-radiative recombination mechanisms. Voltage variation from 0 V to 1.5 V is maintained throughout the observation. The curent-voltage plot shows the p-n junction or diode behavior with central emissive layer. The two dimensions surface emission rate obtained indicates that voltage increment causes the emission concentration to become higher near the central p-contact. The LED efficiency is assessed in terms of internal quantum efficiency and emitting rate.

1 Introduction

Light emitting diode, LED has attracted many researchers due to its low energy consumption and long lasting operation. It has created trillion dollars business turn over in the last decade for the demand in houses, offices and vehicles. Solid state lightings, SSLs have attracted huge attention especially from wide band gap semiconductors based emitting devices. Many research works are focused in light emitting diodes, LEDs with high efficiency, high colour, low voltage and low temperature.

Researchers have been studying the physics of II-VI and III-V compound materials such as gallium nitride (GaN) and zinc oxide (ZnO) ever since the 1960s in order to develop optical devices with shorter wavelengths [1]. Zinc selenide, ZnSe is one of the II-VI compound semiconductors with wide band gap and also in the category of direct band gap. The ZnSe has wide range of visible spectrum and often used in electroluminescent display and window layers in solar cells [2]. During the 1990's, ZnSe LED studied in [3] is fabricated by applying p-type layer on the n-ZnSe. The p-n junction diode is formed by

* Corresponding author: ummikhaireah@gmail.com

DOI: 10.1051/matecconf/20167801114

introducing lithium, Li to the compensated layers by thermal diffusion. Blue light emission is seen when the structure is in forward bias at room temperature.

Methods in fabricating and producing zinc selenide LEDs have been researched and analyzed to date. Layers of ZnSe can be formed in ways such as growing them by metalorganic vapour phase epitaxy (MOVPE), physical or chemical vapour deposition and conventional molecular beam epitaxy (MBE). The fabrication of LED is done using conventional photolithographic techniques with device isolation being achieved by wet chemical etching, patterning by vacuum evaporation and lift-off.

The study presented in this paper is devoted to preparation and characterization of ZnSe LED using a simulated approach, and its comparison with fabricated works. The efficiency of LED and its' relation to LED droop is also discussed. The study is accomplished using COMSOL Multiphysics semiconductor software package.

2 LED principle operation

Light emitting diode (LED) is a p-n junction diode made up of layers of semiconductor grown on a thin wafer of crystalline material. The p-type and n-type semiconductor consist of an excess amount of free conducting holes and conduction electrons respectively. Once forward bias, V is applied, voltage drops across depletion region which allows electron and holes to diffuse across the junction. Electrons from n-side diffuse into p-side whereas holes from p-side diffuse into n-side. Electron-hole pair (EHP) recombination occurs in the depletion region which results in spontaneous emission of photon as can be seen in Fig. 1. When all free electrons and holes recombined, enough light will be emitted through the device.

A variation of light colour with specific wavelength can be produced using different kind of semiconductors. As EHP recombination occurs, each photon is emitted at different amount of energy according to the material's bandgap. Therefore, different light colour can be produced. For instance, gallium nitride (GaN) and gallium phosphide produce green LEDs, aluminium gallium indium phosphide (AlGaInP) alloys for red, orange and yellow LEDs while silicon carbide (SiC) and zinc selenide (ZnSe) produce blue LEDs. Slight changes in the composition of these materials govern the colour of the emitted light. White LED can be produced by combining blue, red and green.

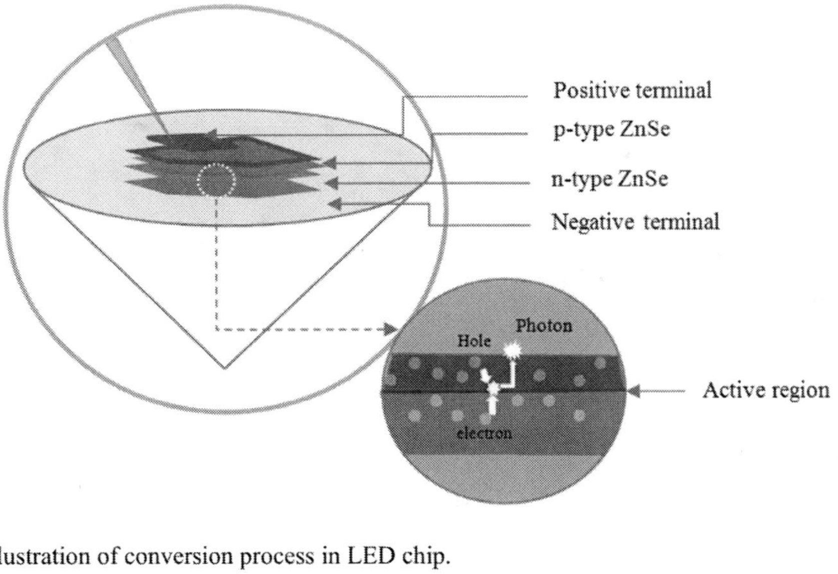

Fig. 1. Illustration of conversion process in LED chip.

3 Methodology

In this work, zinc selenide light emitting diode is evaluated and structurally designed to be cylindrically symmetrical to shorten the required computation time in the three dimension visualization. The design undergoes two different studies or experiments with different choosen parameters. First study is on semiconductor initialization that focuses on mesh and doping whereas the second study analyses the effect of applying voltage to the LED. Both studies investigate the relationship between current and voltage, the emission rate as well as the efficiency of LED in relation to bias voltage and current.

3.1 LED structure

A cross section of the active region formed by the p- and n-type doping is designed as shown in Fig. 2. The whole model is designed in two dimension axisymmetric module under semiconductor physics. The ZnSe layer is 30 μm and 10 μm in terms of width and height, accordingly. Doping profile of ZnSe is set using the analytic doping feature where the impurity type and concentration is specified. The top layer 2.5 μm of the structure is p-doped while the bottom 7.5 μm is n-doped. Part of the structure is etched to the n-type layer and n-contact is deposited on the n-type layer. Then, a 2.5 μm width p-type contact is deposited on top surface of ZnSe. Boundary for metal contact is added at both p- and n-contacts to allow electric potential to flow at the terminal.

Two additional features are included in this LED simulation. In order to allow the calculation for electroluminescence of the LED, an optical transition feature is added to the domain. Optical transition is generally, a process where an atom changes from one state of energy to another and thus, emits an electromagnetic radiation. This feature aids in realizing the radiative recombination and spontaneous emission that occurs in the semiconductor. The choosen transition model is the direct band gap which means two-band model is used to represent a direct band gap semiconductor. The spontaneous emission of photons as a result of recombination is computed and analysed.

In terms of EHP recombination, there are radiative recombination that results in photon emission and nonradiative recombination that releases energy in other forms instead of light. Auger effect occurs as EHP recombines and the released energy is passed to another electron in the conduction band which would then dissipates as heat which causes it to be listed as part of nonradiative recombinations [4]. Auger recombination is adopted in order to include nonradiative recombination mechanism and to infer the performance of LED.

Fig. 2. Cross section of the active region of ZnSe LED.

4 Results and discussion

The results obtained from the first study are the 2D meshes and doping gradient. The default mesh, with the size set to fine, is shown in Fig. 3.This simulation focuses in resolving the p-n junction region, thus, no special scales are specified for the mesh. Mesh is used to represent the solution field to the physics being solved which in this case is the doping gradient. Fig. 4 shows the solution for the first study where the mesh is refined around the doping profile. The mesh is also tighter at the area with p and n electrical contact in order to resolve the high current density and electric field effect. Fig. 5 shows the distribution of dopant throughout the structure. Red area at the bottom indicates the uniform n-doped region and the p-doped surface layer is displayed in blue. The colour that changes steadily from red to blue reflects the doping transition from n to p-type region in order to create a region where light emission can occur.

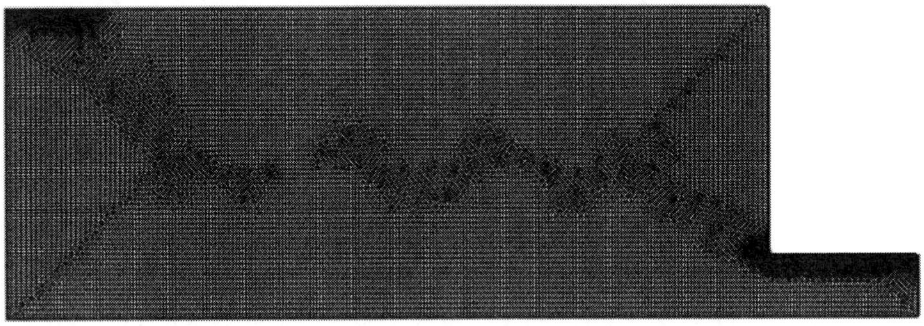

Fig. 3. Induced mesh set to Fine.

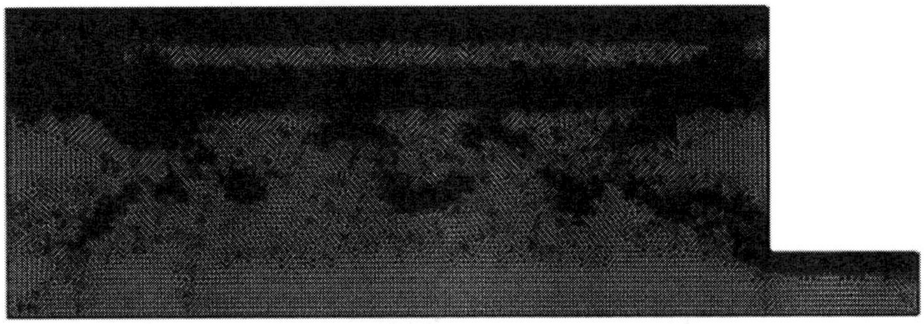

Fig. 4. Refined mesh output.

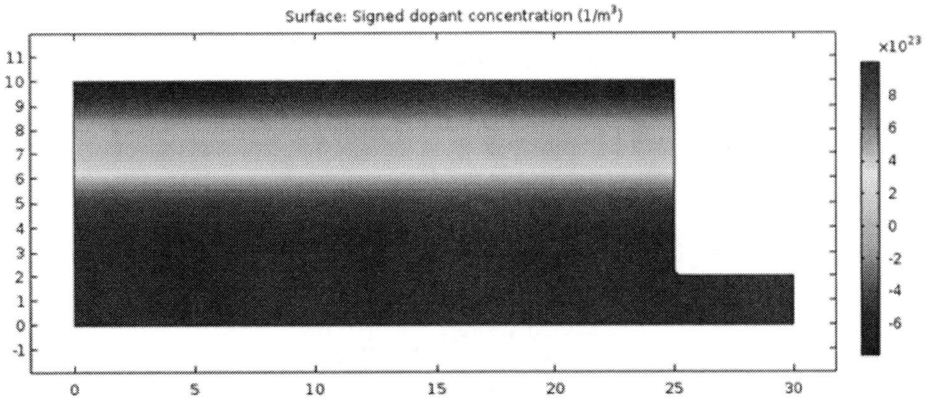

Fig. 5. Two dimension surface graph of signed dopant concentration.

Fig. 6 shows the current voltage curve of the simulated LED from the second study. The graph shows that the LED is a diode with a turn-on voltage of 1.4 V. This clearly shows that the ZnSe LED displays the expected diode behaviour as obtained in other reported researches [5,6].

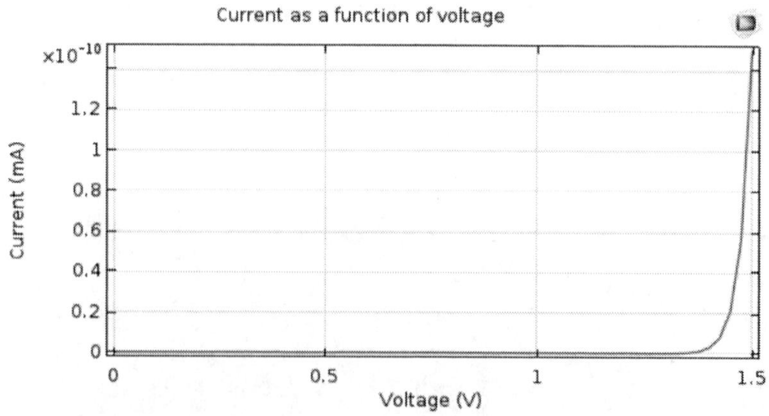

Fig. 6. Current voltage curve.

The following two figures represent the result of the two dimension emission rate across the LED structure. Observing Fig. 7 and Fig. 8 that show the emission rate at 1.0 V and 1.5 V, respectively, proves that varying the voltage affects the rate of emission. The emission can be seen to spread radially at voltage just above the turn-on voltage. However, voltage increment causes the emission to become intense at the center of the device.

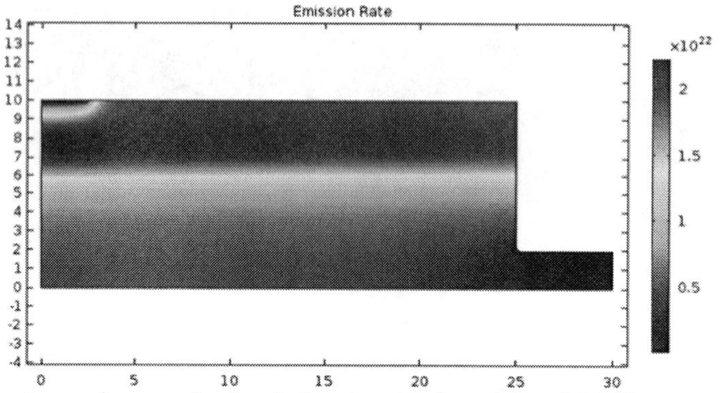

Fig. 7. Two dimension surface graph of emission rtae at voltage bias of 1.2 V.

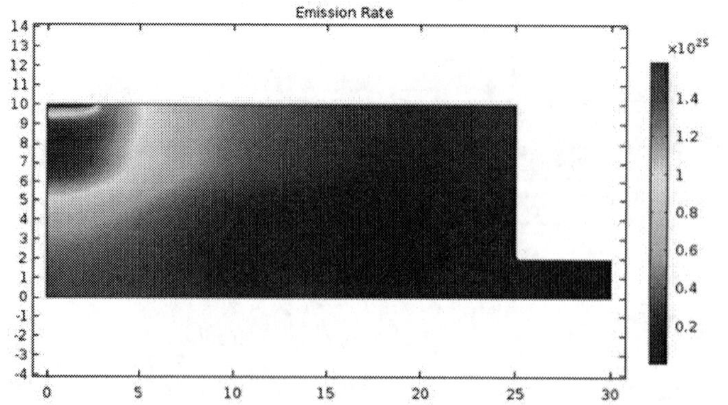

Fig. 8. Two dimension surface graph of emission rtae at voltage bias of 1.5 V.

Fig. 9 shows one-dimensional plot of the total emission rate against current. The rate of emission increases linearly at the start but it then curves sub-linearly after a few mA. The sub-linear curve is a characteristic of LED droop. As mentioned earlier, LED droop is a phenomenon that occurs in LEDs and the Auger effect is one of the reasons behind the LEDs' efficiency droop. Research done by S. Verma in [4] mentions that Auger effects gives an impact on the quantum efficiency, luminous power and electroluminescence intensity which explains the obtained results..

The decrease in efficiency of the LED can be considered by referring to Fig. 10 that shows the internal quantum efficiency (IQE) as a function of current. IQE is the ratio of emitted photon to the carriers in the active region. The IQE value starts off low but as current flows, there is an abrupt peak in IQE. At higher injection current, the IQE is reduced, thus, showing evidence of LED droop.

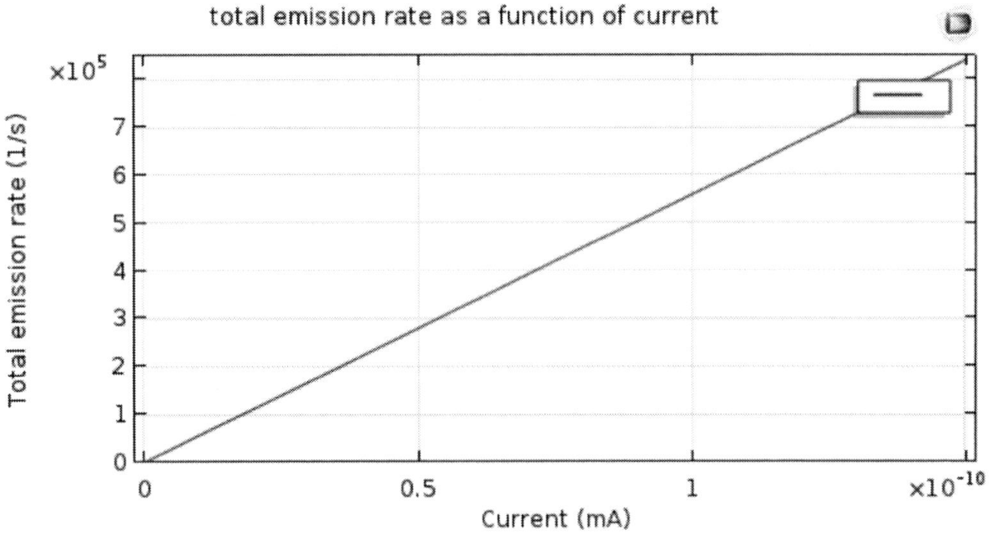

Fig. 9. Total emission rate against current.

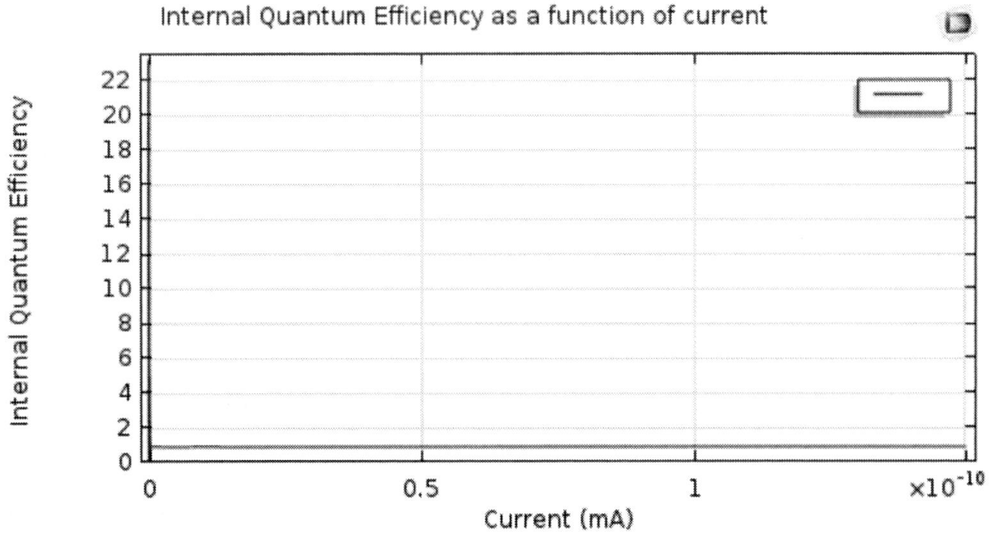

Fig. 10. Internal quantum efficiency against current.

Fig. 11 shows the cross sectional area of the LED structure. This is the 3D version of the spatial distribution of emission of Fig. 8 with 1.5 V applied bias. By observing the colour gradients, it can be concluded that the emission becomes concentrated underneath the p-contact surface.

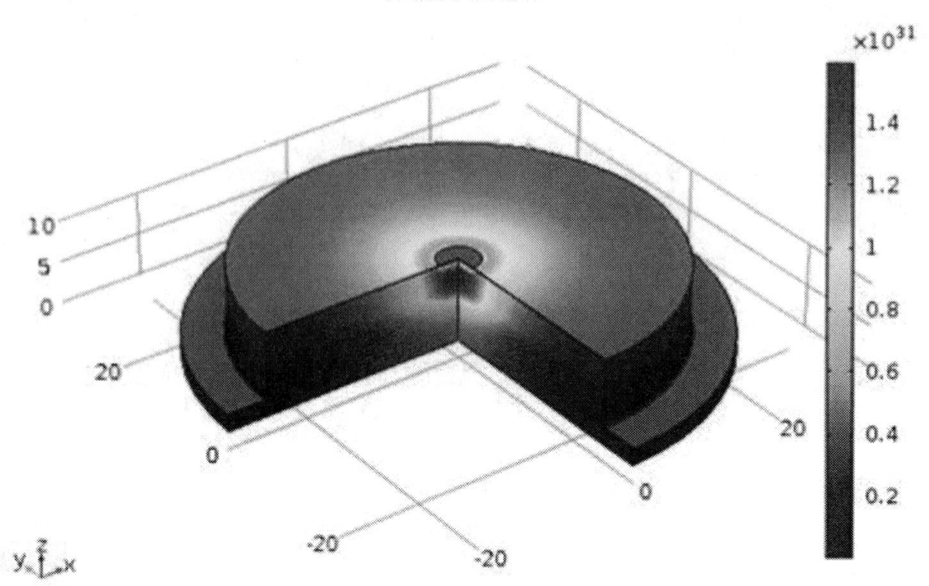

Fig. 11. 3D visualization of LED cross section.

5 Conclusions

In this work, a simple ZnSe LED is modeled and analyzed using the semiconductor module in COMSOL. The evaluated LED is made up of ZnSe layer that is p-doped and n-doped specifically. Metal contacts are deposited on those layers thus, allowing electrical current to flow through the structure. From the two studies on the structure, it is found that the IV curve displayed the expected diode behavior as the fabricated ones. Aside from that, the emission rate and the efficiency of LED is also analyzed. Eventhough it is possible to increase LED brightness by applying larger current, the efficiency of the LED will eventually decrease that end with a droop. Future study of this LED could include varying the doped material and evaluating the performance of the device in different voltages and a different nonradiative recombination mechanism.

References

1. T. Shirakawa, Mater. Sci. and Engin. B, **91**, 470 (2002)
2. S. Muthumari, G. Devi, P. Revathi, R. Vijayalakshmi, C. Sanjeeviraja, Journal of Applied Sciences, **12 (16)**, 1722 (1992)
3. A.N. Krasnov, Y.N. Purtov, Y.F. Vaksman, V.V. Serdyuk, J. Cryst. Growth, **125 (1-2)**, 373 (1992)
4. Y. Namikawa, Sei Technical Review, **72** (2011)
5. V.P. Sirkeli, O. Yilmazoglu, F. Kuppers, H.L. Hartnagel, Semiconductor Science and Technology, 30 (2015)

Clarification of Colloidal Particles in Lake and River Water Using AC Electrokinetic

M.N. Mohtar[1,2,*], *M.M.* Ramli[3], *N.A.M.* Yunus[1], *J.* Jasni[1], and *N.* Aziz[1]

[1]Faculty of Engineering, Universiti Putra Malaysia, Serdang, Selangor, Malaysia
[2]Institute of Advance Technology (ITMA), Universiti Putra Malaysia, Serdang, Selangor, Malaysia
[3]School of Microelectronic Engineering, Universiti Malaysia Perlis, Pauh Putra Campus, 02600 Arau, Perlis, Malaysia

Abstract. Scenery of clear water of a lake or river is always a fascinating view. The clarity of a water is subjected to the water free from colloidal particles. Lake or river usually have foreign colloidal particles such as sand, mud, foreign particle, etc. which make the water cloudy. Usually the cloudy water become clear because of natural sedimentation process. However it is not easy to clarify cloudy water of a lake or river and make it clear especially if the sediment of colloidal particle is influence or disturb by water current. The approach by AC Electrokinetic phenomenon able to manipulate colloidal particles in a suspension. It can separate, trap or sort colloidal particle which made the phenomenon as possible reliable option for clarifying lake or river water from colloidal particles hence make it clear water. This work will simulate the process of clarification of colloidal suspension using AC Electrokinetic phenomenon in a lab. Electrodes were fabricated on Indium Tin Oxide (ITO) coated glass slide using laser etching technique. The electrode which poses unique geometry will be able to demonstrate electric field gradient as soon as it is introduced with electrical signal. Base on the surface potential of the colloids and the surface potential of the electrode, the colloids will be manipulated. This phenomenon is known as AC Electrokinetics. This can be regarded as guided sedimentation process. The trapped colloidal particle can be now easily extracted or remove from the water thus transform the water from cloudy to clear hence complete water clarification process.

1 Introduction

Injection Crystal clear water of a river or lake is always be an attraction, be it to the local community or to the tourist. As world is threat by pollution there are not many river or lake that maintain its crystal clear water. Many efforts have been done to regain the clarity of the water.There are many techniques that have been introduced to clarify or purify water such as filtering technique, biotechnology technique (using micro bacteria), etc.[1] This paper introduced electrical approach as additional option for clarifying water by manipulating colloidal particles using electrical field. The manipulated colloids is then can be set for

* Corresponding author: nazim@upm.edu.my

removal to transform cloudy water to clear water. The phenomenon for manipulating colloids in suspension using electric field is known as AC Electrokinetics. There are two conditions where AC Electrokinetics take part, either electric field directly manipulate the colloidal particles which is known as Dielectrophoresis (DEP) or the electric field induces motion to the suspending medium where it subsequently will manipulate the colloids which is known as AC Electroosmosis (ACEO) [3].

Electrode with specific geometry needs to be fabricated in order to utilize electric field for manipulating colloids [2]. The electrodes were micro fabricated on Indium Tin Oxide (ITO) coated glass slide using laser etching technique. In previous works the micro fabrication were done using lithography techniques [3][4][5] The laser etching techniques is easy to use and applicable practically to the whole classes of materials, such as silicon, germanium, glass, oxides, nitrides, salts, alloys, binary and ternary compounds and alloys and metals. The laser etching is done using Resonetics RapidX250-L Series as shown in Fig. 1. Resonetics RapidX250-L is a laser-based laser micromachining systems designed to fabricate micro-machined components. The system is capable to fabricate up to 1μm in size.

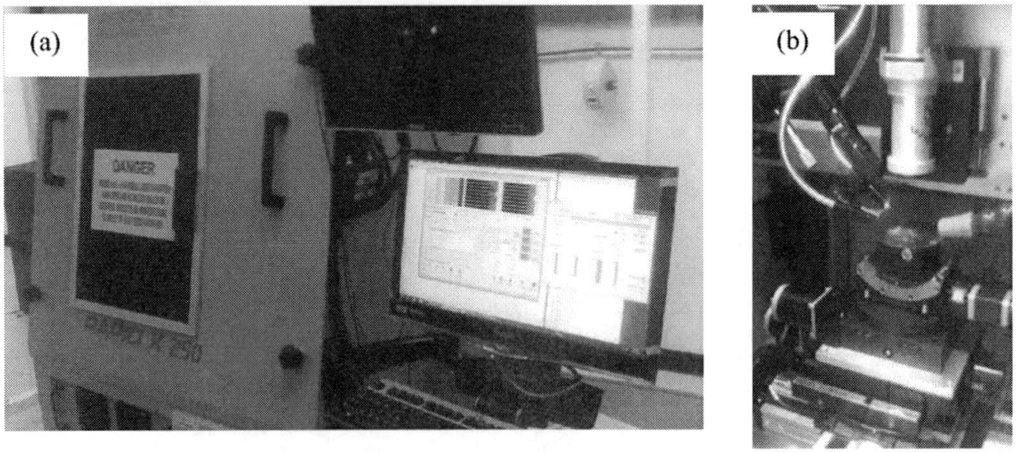

Fig. 1. (a) Resonetics RapidX250-L Series, (b) The stage where the samples were mounted.

The samples used for this work were controlled size colloids; 3.1 μm in diameter Fluorescent Polymer Microspheres purchased from Duke Scientific Corporation (Palo Alto, CA, USA), and specimens were taken from cloudy water from a lake in Serdang, Malaysia.

2 Methodology

2.1 Electrodes

The electrodes were fabricated on the ITO coated glass using laser etching technique. The geometry of the electrodes was designed using computer aided drawing (CAD) software. This work used zipper geometry as used by Mohtar et. al. [3] and Hoettges et. al. [5]. Zipper geometry is selected because the geometry will have electric potentials of opposing polarity applied allowing focusing of particles across a large area toward a central spot. The CAD drawing of zipper geometry is used by Resonetics RapidX250-L laser micromachining system in order to etch ITO on the ITO coated glass to produce the zipper electrode geometry. The ITO is used because it is transparent. This is important during the real time observation under the microscope. The fabrication process is carried out in a cleanroom because the low level of environmental pollutants such as dust, airborne microbes, aerosol particles, and chemical vapors can cause disturbance to the final results of the fabrication. Fig. 2 shows the electrode used during the experimental work. A chamber of 390 μm height is build enclosing the electrode for the optimal condition as in previous work done by Mohtar et. al. [3].

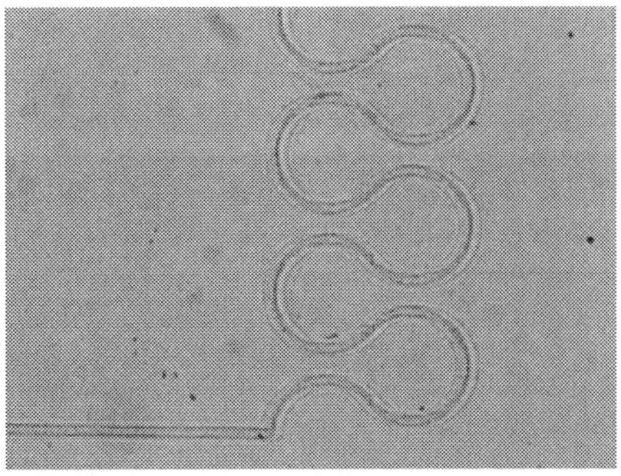

Fig. 2. Zipper Electrode 500 μm diameter, 100 μm inter electrode gap micro fabricated using laser etch technique as used in the experiment.

2.2 Sample preparation

Two set of samples were prepared for this work. Sample A is a control samples prepared using Fluorescent Polymer Microspheres, 3.1 μm in diameter purchased from Duke Scientific Corporation (Palo Alto, CA, USA). It was used because the size makes the particles easily seen through normal light microscope without the need of using fluorescent microscope. The stable properties of the latex bead acts as a constant to avoid the particles to have variables that can affect the experiment. They were supplied as dyed polystyrene microspheres in water with 6.7 x 108 beads/ml concentration. The aqueous polymer microsphere is diluted to 104 beads/ml using deionised water to get a single layer collection of particle [3]. Sample B is cloudy water taken from a lake in Serdang, Malaysia with an unknown colloids as shown in Figure 3. Both samples were shake using vortex shaker for 60 seconds before undergo ultrasonic shaker for evenly distribute the colloidal particles in the suspension.

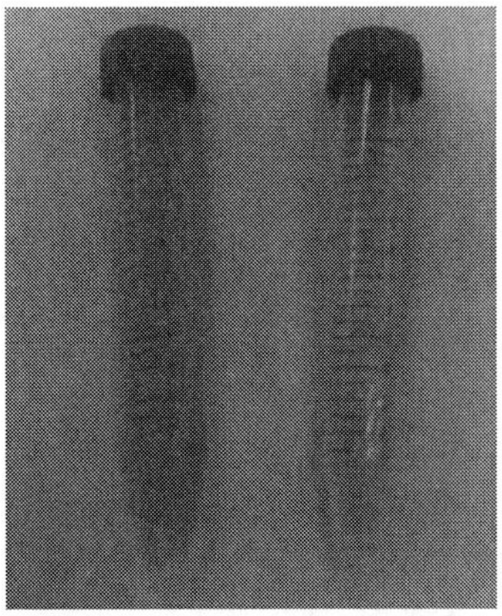

Fig. 3. Cloudy lake water taken from a lake in Serdang, Malaysia.

2.3 Particle separation

The sample (20 µl – 50 µl) was pipetted onto the electrodes array and covered with a glass cover slip thus stay in a chamber. Electric potentials were applied using a function generator giving 10 Vp-p at 1 KHz frequency for 1000 seconds. The experiment is observed using Nikon Eclipse Ni-U microscope. Particle collection was observed and image is taken using Nikon NIS-Elements BR 4.20.00 software.

3 Results and discussion

As soon as the electrical signals were introduced to the electrodes, diagonal movement of the particles toward the center of the electrodes can be observed. Image of collected particles after 1000 s for Sample A is as shown in Figure 4 and for Sample B is as shown in Fig. 5.

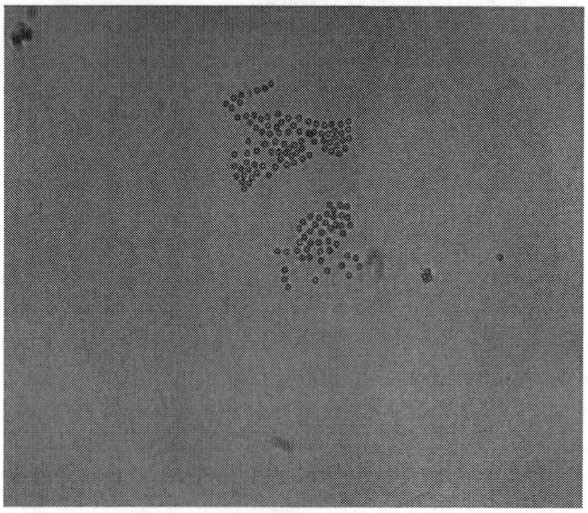

Fig. 4. Collection of particle for Sample A on a Zipper electrode 500 µm diameter, 100 µm inter electrode gap, applied to 10 Vp-p, 1 KHz, after 1000s.

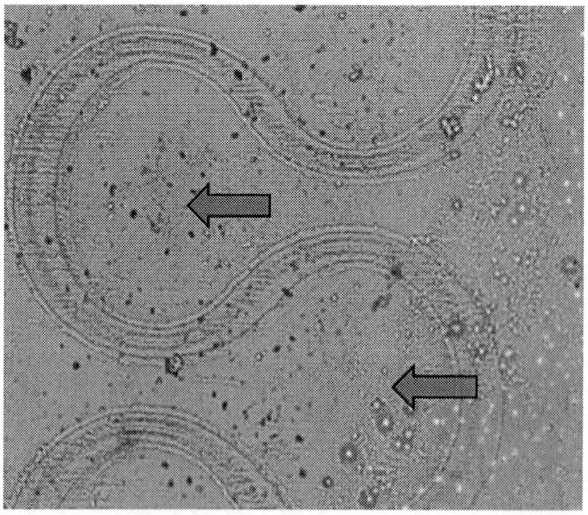

Fig. 5. Collection of particle for Sample B on a Zipper electrode 500 µm diameter, 100 µm inter electrode gap, applied to 10 Vp-p, 1 KHz, after 1000s.

The electrical signal of 10 Vp-p, 1 KHz were choose for this experiment because it was proven as the optimal electrical signal needs to be applied in previous study [3]. The vortex motions can be observed at the edge of the electrodes before the particles move diagonally towards the center of the electrodes and trapped. This suggested that the electric field induced motion to the suspension medium hence ACEO were occurred. It has also been observe that when the frequency is change to 100 Hz for both samples the particles were collected at the electrode edge. This suggested that the electric field acts directly on the particles hence DEP were occurred. This has confirmed the previous works done by Mohtar [3], Hoettges [5] and Hubner [6]. This has also means that the manipulation of colloids is selective according to the electrical potential supplied.

4 Conclusions

This work has shown that it is possible to manipulate colloidal particles in a suspension using electric field. This work has been able to trap particles of known size (3.1 μm) and unknown size (which were later found out to be 8-10 μm when measured using Nikon NIS-Elements BR 4.20.00 software) using electric field known as AC Electrokinetic. Both phenomenon either DEP or ACEO which were discussed earlier manage to trap the particles to the electrode. In future, this experiment in the lab using small scale model can be introduce for real scale study for clarification process of lake, river, etc. The selectivity of the manipulation is one of the advantages over filtering method as it can choose what to trap and remove and what to be left remain.

This work was supported by Fundamental Research Grant (FRGS 9003-00368) funded by Ministry of Education, Government of Malaysia.

References

1. N. Spahis, A. Addoun, H.Mahmoudi, N. Ghaffour, Desalination, **222**, 519 (2008)
2. H.A. Pohl, K. Pollock, J. Electrostat., **5**, 337 (1978)
3. M.N. Mohtar, K.F. Hoettges, M.P. Hughes, Electrophoresis, **35**, 345 (2014)
4. N.A. Kadri, K.F. Hoettges, M.P. Hughes, *4th Kuala Lumpur International Conference on Biomedical Engineering* (2008)
5. K.F. Hoettges, M.P. Hughes, A. Cotton, N.A. Hopkins, M.B. McDonnell, IEEE Eng. Med. Biol. Mag., **22**, 68 (2003)
6. Y. Hubner, K.F. Hoettges, M.B. McDonnell, M.J. Carter, M.P. Hughes, International journal of nanomedicine, **2(3)**, 427 (2007)

MATEC Web of Conferences **78**, 01116 (2016) DOI: 10.1051/matecconf/20167801116
IConGDM 2016

Effects of Straight and Serpentine Flow Field Designs on Temperature Distribution in Proton Exchange Membrane (PEM) Fuel Cell

Izzuddin Zaman[1], *Bukhari* Manshoor[2,*], *Amir* Khalid[2], *Laily Azwati* Mohamad Sterand[2], and *Shiau* Wei Chan[3]

[1]Structural Integrity and Monitoring Research Group, Faculty of Mechanical and Manufacturing Engineering, Universiti Tun Hussein Onn Malaysia, 86400 Batu Pahat, Johor, Malaysia
[2]Centre for Energy and Industrial Environment Studies, Faculty of Mechanical and Manufacturing Engineering, Universiti Tun Hussein Onn Malaysia, 86400 Batu Pahat, Johor, Malaysia
[3]Faculty of Technology Management and Business, Universiti Tun Hussein Onn Malaysia, 86400 Batu Pahat, Johor, Malaysia

Abstract. Proton exchange membrane fuel cells or sometimes called as polymer electrolyte membrane (PEM) fuel cells is a device for energy transformation in a changing process from one form of energy to another form of energy. It became as an alternative especially for future use in stationary and vehicular applications. PEM fuel cells provide high efficiency and power density with null emission, low operating temperature, quickly start and long life. One of the aspects that are crucial in optimizing the PEM fuel cells performance is a flow field geometry. In this paper, a simulation case of PEM fuel cells was simulated to determine effects of a straight and serpentine flow field on temperature distribution in PEM fuel cells. ANSYS Fluent software was used to simulate 3-dimensional models of single PEM fuel cells in order to determine the effects of changes in the geometry flow field on temperature distributions. Results showed that the serpentine flow field design produces a better temperature distribution along the membrane. The simulation result shows a good agreement with the experiment, thus boost a higher confidence in the results to determine the effectiveness of the flow field design in PEM fuel cells.

1 Introduction

One of the most significant aspects of developing a fuel cells design is modeling concept that calculation which makes it possibility investigate complex geometries and various circumstances by spending less time and effort from attempt investigation [1]. Basically, the operation concept of fuel cells can be distinguished as a transportation of gas-mixture and transformation of species through a process of change of electrochemical circumstances. In this process, hydrogen from anode flow channel will be transported through the Gas Diffusion Layer (GDL) toward Membrane Electrode Assembly (MEA)

* Corresponding author: bukhari@uthm.edu.my

surface. In Catalyst Layer (CL), hydrogen molecules will dissociate to proton and electrons, as the reaction of hydrogen CL can be shown as follow.

$$H_2 \rightarrow 2H^+ + 2e^- \tag{1}$$

From the above relation, water that fertilizes the MEA will hydrate the protons, hence the process of an electro-osmosis and diffusion will transport water in MEA. At the same time, the mixture of air in cathode channel will transport through the GDL toward the MEA where the reaction occurs between oxygen and proton. This reaction will produce water as in reaction shows below.

$$\frac{1}{2O_2} + 2H^+ + 2e^- \rightarrow H_2O \tag{2}$$

The oxygen received an electron from one electrode and hydrogen will supply the electron to another electrode. This flow can generate electrical power for an automobile or other electrical devices. This system saves the environment because the waste product of the PEM fuel cells only contains a water, as showed in the chemical reaction above.

In the fuel cells systems, there are several thing that needs to avoid in order to keep the performance of the fuel cells and also to evade the system from the failure. Basically, the common failure occurs in the fuel cells system includes a chemical degradation of the ionomer membrane or mechanical failure in the PEM. This failure can cause a gradual reduction ionic conductivity, increasing the total cell resistance, reduction of voltage and the power output also may lose or decrease [2, 3]. Therefore, thermal management is important to control the rate of water evaporation or condensation from the system. By control the temperature and humidity, the PEM, gas diffusion layer (GDL) and bipolar plates will all experience expansion and contraction. Resulting from the thermal expansion and contraction, hydrothermal stresses will be introduced into the cell unit during operation.

In addition, the distribution of the current flow and the reactants are not uniform in cell temperature and humidity can cause a non-uniform content of the cell and at the same time have a potential to cause a localized increase in the magnitude of the pressure. According to Hashemi *et.al* [4], temperature distribution in the fuel cells are highly dependent on the charge and also the activity during the electrochemical reaction. At high current density, the fuel cells were performed badly while the membrane thickness provide a better performance at low current density [5, 6]. Membranes have some resistance to the passage of protons from the anode to the cathode. The resistance will decrease when the membrane thickness decreased, hence the deformation due to thermal stress is low. However, the thin membrane has increased the cross gas problems, thus, will lead to a reverse flow and degradation of the temperature and membrane.

2 Simulation models

In order to increase the performance of the PEM fuel cells, one of the important aspects that needs to be considered in water management is the design of the flow channel. In this study, a numerical simulation based on 3-dimensional models will be performed. Several assumption had been considering to make sure the simulation can be performed easily. Among the assumption made are;

1) The cells were operating under the steady state condition.
2) The temperatures of cell are uniform and fix.

 DOI: 10.1051/matecconf/20167801116

Figure 1 illustrates a schematic diagram of the PEM fuel cells, which is the typical design that will use for the simulation. The main part in PEM fuel cells consist three main parts, which are an anode, a cathode and polymer electrolyte membrane. There have three distinct regions at the anode and cathode. The region consists of the gas channel, gas diffuser, and catalyst layer, as shown in the diagram.

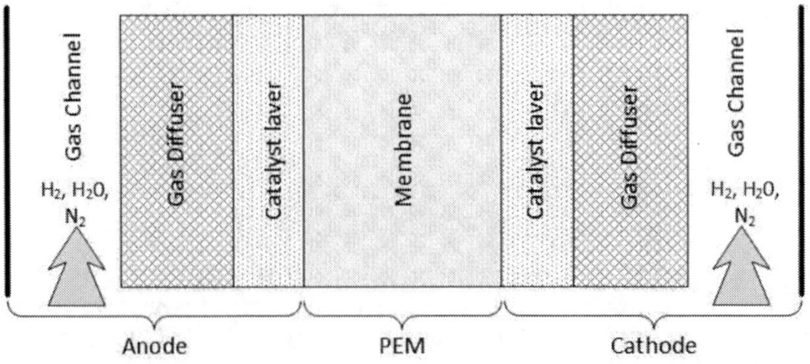

Fig. 1. Schematic diagram of a PEM fuel cell [7].

In the research of improvement of the PEM fuel cell performance, a number of experiment and numerical efforts have been devoted for it especially for water management inside the PEM fuel cells for steady state and transient condition as reported in [8-10]. However, their flow-field effect in the performance is absent from literature. To optimize the flow-field design, hopefully, the research of flow patterns inside the PEM fuel cells will give an idea to fix the lacking of absent from previous research.

3 Simulation work

As shown in Figure 1, the PEM fuel cells have a three main parts. There are an anode, PEM and cathode. For both anode and cathode, it contains a gas channel, gas diffusion and catalyst layer and both are divided by PEM at the middle of the fuel cell. Humidified air will be fed into the cathode channel, whereas hydrogen is supplied to the anode channel. For this study, the authors were concentrating on the flow channel that used to flow an oxygen and hydrogen through the PEM fuel cell. Figure 2(a) shows the view of gas flow channel in this research. There are divided into two part flows which are an anode side flow channel and a cathode side flow channel. Both channels have same parameter which is 0.3mm for the dimension channel.

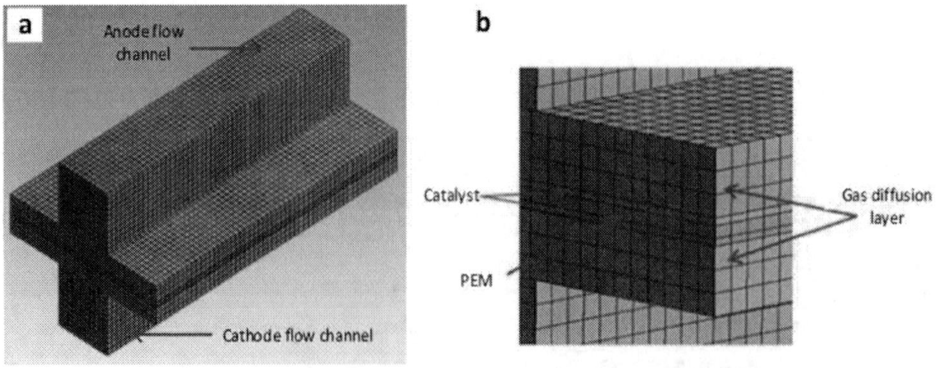

Fig. 2. (a) 3-D computational of PEMFC flow channel, (b) Computational domain of PEMFC.

Figure 2 shows the 3-dimensional meshed model of the PEM fuel cell flow channel. From the figure, it is more clearly showed the components in the system which are anode and cathode flow channel and both of the flow channel contain of gas diffusion layer and catalyst layer that separated by the PEM. The membrane (PEM) at the middle part is the main part of this system which is hydrogen separated with an electron and start to transfer to the catalyst layer at the cathode part.

The boundary conditions used are summarized in Table 1. The modeling is done with the available condition in research by Al-Baghdadi [3] and Hashemi *et. al* [4]. The simulation parameter that used in this study also referred to research that had been done by Hossain *et. al* [11], which was used an Ansys DM version 15.1 as their modeler. Ansys mashing set up as default, size function and smoothing as fixed and medium with slow transition. Fluent 3-D single precision with solver type as pressure based.

Table 1. Geometry and operational parameter

Parameter	Symbol	Value
Cell length	L_{cell}	5 mm
Cell width	W_{cell}	5 mm
Gas Channel width	W	0.3 mm
Gas Channel height	H	0.3 mm
Gas diffusion layer thickness (GDL)	δ_{GDL}	0.26 mm
Wet membrane thickness (PTFE)	δ_{mem}	0.05 mm
Catalyst layer thickness (CL)	δ_{CL}	0.0287 mm
Anode and cathode pressure	P_a and P_c	$1x10^5$ Pa
Initial temperature both inlet	K	353.15
Porosity of GDL Porosity of CL Porosity of Membrane	- - -	0.8 0.6 0.4
Operating pressure	Pa	101325
Mass flow rate: Anode Cathode	kg/s	6.0e-7 5.0e-6

Model flow is turbulent because the Reynolds number more than 4000 with fluid Hydrogen and Oxygen while at wall solid set up as Aluminum with constant temperature. However, we assume it laminar cause of the channel is very small. By using the turbulent model, the accuracy of performance will decrease. Scheme for solution method is SIMPLE method with spatial discretization Green-gauss node based; standard pressure and solution initialization is hybrid with 150 iterations.

4 Results and discussion

The simulation work had been done for two cases of flow which are serpentine flow and straight flow. The results that were come out from this research are the distribution of temperature for both cases. Figures 3(a) and 3(b) show the distribution of the local contour temperature on membrane plane surface in straight gas flow channel and serpentine gas flow channel. For the straight gas flow channel, it showed that the temperature distributes from 353.155K near at inlet anode and decrease to 353.148K near the outlet anode. However for the range of serpentine gas flow channel, the temperature distribute from 353.147K near at inlet anode while decrease to 353.132K near the outlet anode.

 DOI: 10.1051/matecconf/20167801116

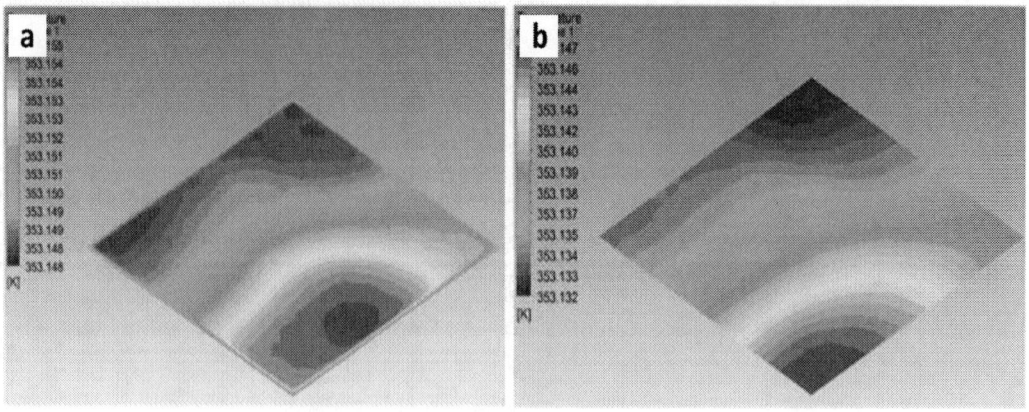

Fig. 3. Distribution of the temperature at membrane (a) straight gas channel (b) serpentine gas channel.

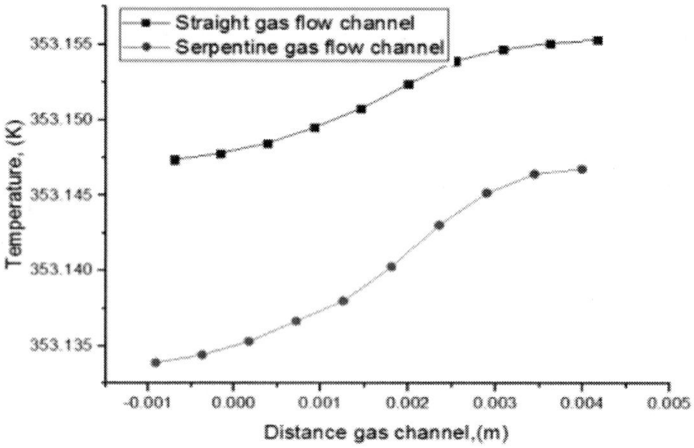

Fig. 4. The plot of temperature distribution against distance gas channel (m).

Although the different temperature change at the anode and cathode are small for both case, it was showed that the straight and serpentine gas flow channel gave a significant effect on the temperature distribution on the membrane. The small change of the temperature basically due to the thickness of membrane which is 5mm. In a real case, the membrane can be thinner and an assumption had been done for this simulation work to choose a thickness of the membrane up to 5mm so that it easy to model and simulate the problem regarding the PEM fuel cell.

From the images shown in Figure 3, the decreasing of temperature occurred from the middle of the membrane surface. This is because the hydrogen will flow through the catalyst layer from the anode side to the cathode side which is the area where the oxygen is supplied. The oxygen will combine with the hydrogen during the process and a water will be produced as a waste product. Figure 4 shows the temperature distribution for both channels of straight and serpentine flow. Since there is no chemical reaction occurred at the inlet for both channels, the temperature was higher at the location. The thickness of membrane also give effect of temperature distribution. Membrane thickness decreased, resistance decrease and temperature will drop.

5 Conclusion

As performed in this study, the simulation of two flow-field design configuration and effect between two designs on PEM fuel cell performance were presented. The temperature distribution along the membrane in PEM fuel cells for serpentine gas flow channel and straight gas flow channel were analyzed. In conclusion, the changing flow-field design or pattern of PEM fuel cell can give an effect to the temperature distribution and other characteristic that need to be consider in order to improve the performance of PEM fuel cells. The serpentine flow-field gives more uniform temperature distribution compare with straight flow-field. This is due to the shape of gas channel for serpentine is direct from inlet to outlet, compare to straight gas flow channel, the gas fed from inlet and gas distribute by channel and it maybe not constant when it distribute.

Partial support was provided by MOHE (Ministry of Higher Education, MALAYSIA) and Universiti Tun Hussein Onn Malaysia (UTHM) under the Fundamental Research Grants Scheme (FRGS), vote 1466 and Incentive Grant Scheme for Publication (IGSP), vote U247.

References

1. N. Djilali, Energy, **32**, 269 (2007)
2. V. Mehta, J.S. Cooper, J. Power Sources, **114**, 32 (2003)
3. M.A.S. Al-Baghdadi, International Journal of Energy and Environment, **5**, 1 (2014)
4. F. Hashemi, S. Rowshanzamir, M. Rezakazemi, Math. Comput. Model., **55**, 1540 (2012)
5. B.M. Koraishy, J.P. Meyers, K.L. Wood, J. Electrochem. Soc., **158**, 1459 (2011)
6. A.K. Sahu, S. Pitchumani, P. Sridhar, A.K. Shukla, Fuel Cells, **9**, 139 (2009)
7. S. Um, C.Y. Wang, K.S. Chen, J. Electrochem. Soc., **147**, 4485 (2000)
8. P.R. Pathapati, X. Xue, J. Tang, Renew. Energ., **30**, 1 (2005)
9. D.H. Jeon, S. Greenway, S. Shimpalee, J.W. Van Zee, Int. J. Hydrogen Energ., **33**, 1052 (2008)
10. S.Y. Cha, W.M. Lee, J. Electrochem. Soc., **146**, (1999)
11. M. Hossain, S.Z. Islam, A. Colley Davies, E. Adom, Renew. Energ., **50**, 763 (2013)

Parameter Estimation of Damped Compound Pendulum Differential Evolution Algorithm

Mohd Sazli Saad[1,2,*] , *Z.* Shayfull[1,2] , *S.M.* Nasir[1,2] , and *M.* Fathullah[1,2]

[1]Green Design and Manufacture Research Group, Center of Excellence Geopolymer and Green Technology (CEGeoGTech), Universiti Malaysia Perlis, 01000 Kangar, Perlis, Malaysia
[2]School of Manufacturing Engineering, Universiti Malaysia Perlis, Kampus Tetap Pauh Putra, 02600 Arau, Perlis, Malaysia

Abstract. This paper present the parameter identification of damped compound pendulum using differential evolution algorithm. The procedure used to achieve the parameter identification of the experimental system consisted of input output data collection, ARX model order selection and parameter estimation using conventional method least square (LS) and differential evolution (DE) algorithm. PRBS signal is used to be input signal to regulate the motor speed. Whereas, the output signal is taken from position sensor. Both, input and output data is used to estimate the parameter of the ARX model. The residual error between the actual and predicted output responses of the models is validated using mean squares error (MSE). Analysis showed that, MSE value for LS is 0.0026 and MSE value for DE is 3.6601x10-5. Based results obtained, it was found that DE have lower MSE than the LS method.

1 Introduction

The pendulum is a relative mass object hung from the pivot point. Naturally, when the external force acts on it, it will swing back and forth with periodic motion. This kind of phenomena usually works due to the force of *gravity*. *H*owever, in this research the pendulum motion is powered by a driven propeller. When the propeller is driven by a specific amount of voltage, it will spin gradually hence produce a thrust force that moves the pendulum at certain angle position. This kind of driven pendulum can be seen in wide applications of multi-rotor helicopters which also known as drones. One of the vital parts to be focused on is the position control of the pendulum. This part is intrinsically unsteady systems, requiring exact control so as to fly. Therefore, in order to get a better control system, the physical system needs to be modeled accurately.

Therefore, in this research, system identification is employed in order to get an accurate model of damped compound pendulum. This due to the fact that, physical modeling of the real system is not as easy to acquire. Ones, needs to have a physical system about the system such as the dynamic behavior of a brushless DC motor, electronic speed controller (ESC) and aerodynamic forces and moment generated by the propeller installed on motor. Furthermore, ESC for the motor has it own dynamic property where a precise dynamic

* Corresponding author: sazlisaad@unimap.edu.my

model is hard to derive using mathematical modeling. Parametric system identification is one of the simplest methods in system identification to be used for estimating the model from the actual input output experimental data. In practice, parametric estimation requires of three main steps, model structure selection, estimation of the model parameters and validation test. Many well-developed techniques, for instance, least square, maximum likelihood and instrumental variable.

Recently, nature inspired algorithms have become potential candidates to many parameter estimation techniques. Among them, differential evolution (DE) has been widely used in system identification and controller optimization. The prominent advantages of this technique is it can works effectively without require any search space and to be differentiable or continuous. Many researchers attempt to study the performance of DE in optimizing processes. Wenjuan Gu et. al (2016) utilized hybrid artificial bee colony algorithm combined with differential evolution to optimise the estimate parameters of the unknown fractional-order memristor-based chaotic systems. Results revealed that the unknown chaotic system can be estimated effectively with minimum residual error and higher convergence accuracy [1]. Ashik and Shahid (2016) studied the performance of DE in searching for the optimum proportional intergral (PI) controller parameters to track small variations in stand-alone solid oxide fuel cell (SOFC) system load current. The objective function was formulated based on the minimization of an eigenvalue in such a way that the error between the actual and desired eigenvalues were reduced. Comparative study showed that show that the performance of DE is superior than those of particle swarm optimization (PSO) and invasive weed optimization (IWO) [2]. Leandro et. al (2015) improved the original differential evolution (DE) in tuning the discrete VSC parameters for applied in CSTR plant. Results showd that discrete VSC parameters tuned by improve DE produced fast setpoint tracking and good disturbance rejection [3].

Over the years, parametric modeling with differential evolution (DE) algorithm has been widely used several engineering plant or system. Many researchers have employed the algorithms to estimate the unknown parameters in a parametric model structure of a particular system. The estimated model was found to be more accurate and effective than the conventional methods. AL-Khafaji and Darus (2014) applied differential evolutionary techniques (DE) and particle swarm optimization technique (PSO) to estimate the parametric model of flexible single-link manipulator system. The identified model was conducted based on the input output data from the simulation study using finite element method (FDM) based on Lagrangian approach. Mean squared error (MSE), correlation tests and one step ahead (OSA) validation test showed that DE performed better than PSO [4]. Sazli et. al (2012) proposed an alternative approach to search for the optimum parameter for the linear parametric structure controller to suppress the vibration of a flexible beam system. The objective function of the controller is the minimum value of mean square error. Simulation result depicted a sufficient vibration reduction with optimum controller setting [5]. Yousefi et. al (2008) identified the parametric model for electro hydraulic servo systems (EHSS) utilizing a novel evolutionary algorithms in particular differential evolution. The optimisation procedure was intended to seek for a global minimum in terms of mean square error (MSE) on a non-linear constraint functions with boundary limits of variables of the unknown parameters of a servo-hydraulic system. Rather than having a quick convergence profile, DE algorithm also exhibited a a good agreement between the predicted and measured value. The validity of the obtained model is guaranteed[6]. Swayamsiddha et. al (2015) investigated the performance of differential algorithm (DE) for nonlinear multiple input multiple output (MIMO) system. Convergence profile indicated that, the proposed algorithm able to prevent from trapping at local minima which cause a premature settling of the model parameters. Furthermore, with this result, it was assured that this technique can be used to solve for predicting model of real system [7]. Zakaria et.

al (2015) employed multi-objective optimization differential evolution (MOODE) to estimate the dynamic model of palm oil biodiesel engine. There were two objective function aims to be minimised, the number of terms used model structure and the mean square error between measured and estimated outputs. Validation test has shown that the proposed algorithm accurately search for the optimal parameter of the estimated model [8]. From the overall study, it is assured that DE able to find the optimal solution for various applications. Apart from that, DE is useful, reliable and offer better solution in the fields of control system and particularly in system identification.

This paper attempts to estimate the model for motorized-propeller pendulum via the experimental approach using DE as a tool for model optimisation. The input output data is collected from the experimental rig comprise of input voltage to regulate the motor speed and the output voltage taken from the position sensor. Then, the simple model structure is selected i.e. ARX model structure. The model parameters is carried out using DE algorithm to find the best estimated model that accurately represent the actual system. Finally, the obtained model is validated using mean square error and one step ahead (OSA) predicton. At the end of the study, the performance of DE is compared with conventional technique namely least square (LS).

2 Experimental setup

In the present work, the real time interacting and non-interacting fabricated system was used for collecting the input, output data. In this part will explain the parameter and function for each component/device. The setup consists of hardware and software. The hardware (Figure 1) use in this project is DC power supply, Brushless speed controller, Arduiono , Data Acquisition card National Instrument, Automatic voltage regulator, Motorized Propeller, potentiometer, and pendulum apparatus. The set-up is intended to study dynamic response of single input and single output for motorized propeller pendulum. Software use for this project to the system identification are Matlab and Simulink.

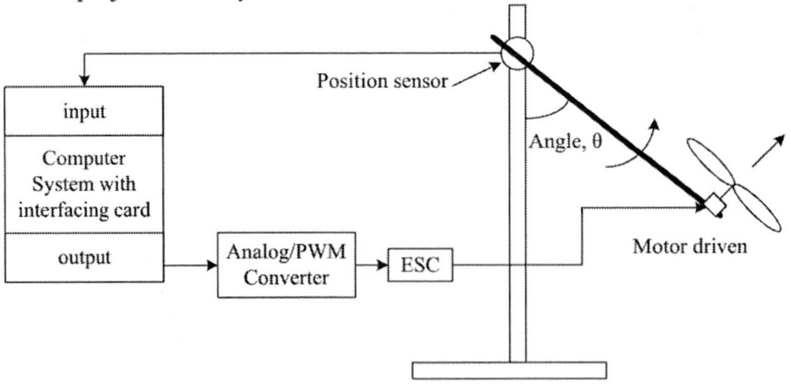

Fig. 1. Schematic diagram for motorized propeller system.

Figure 2 shows the process of system identification. Initially, data collection is adopted. The input of the system was analog output to the motor and the output was the input voltage from the position sensor. Based on the acquired input-output data, the model parameter is estimated until minimum error, e(t) is reached. The parameter of estimated model is determined using the differential evolution algorithm and compared with least square (LS) [9]. The performance of the estimated model is measured using mean square error (MSE).

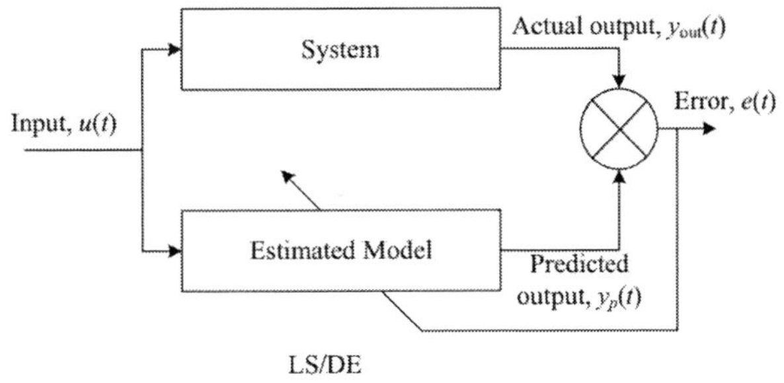

Fig. 2. The system and the model.

3 System identification

The system identification is the process to get the system model through the data input-output system (stimulus-response). The data input and output can be obtained through the experiments, simulations or directly collected from the factory. After the model of the physical system is obtained, it can be used to solve a variety of problems such as, for controlling physical systems or to predict the behaviour under different operating conditions [10].

The goal Identification System is a mathematical model based on experimental data of dynamic systems. Model should be solid and sufficient in respect of the purpose it is intended for. Identification system more systematic information from experimental observation is a key aspect of any scientific work. The information obtained led to the establishment of model systems considered. Models such as some form of pattern that explains the observed experimental results and justify predictions of future system response to be made.

In system identification there are three types of white-box model, Grey-box and black box. The most suitable identification for this project is black-box identification [11]. Black box model assumes that the system is unknown and all model parameters can be adjusted regardless of the physical background. Type of system identification that is appropriate for this project is the black-box. Most system identification algorithms are using this type of system to obtain system model. Black-box identification are rarely used in real-life control.

3.1 Model structure

In this study, the chosen method for estimating the model of driven pendulum system is single input and single output Auto-regressive with eXogenous (ARX) input model structure. ARX model is the simpliest identification technique for black-box *parametric* model. This model is developed from a linear regression analysis which is very efficient in presenting the actual dynamic systems. Moreover, the solution is unique. In other words, the solution always satisfies the global minimum of the loss function [4]. The general equation of ARX model is shown in equation (1).

$$y[k] = \frac{B(q)}{A(q)}u[k] + \frac{1}{A(q)}e[k] \tag{1}$$

The transfer function form of this ARX model is:

$$\frac{y_n}{u_n} = \frac{b_1q^{-1}+b_2q^{-2}+...b_rq^{-r}}{1-a_1q^{-1}-a_2q^{-2}-...-a_mq^{-m}} \tag{2}$$

The model parameter, a and b are estimated using Differential evolution algorithm (DE) and compared with least square (LS) method.

3.2 Least squares (LS)

The least square (LS) parameter estimator is one of the conventional estiomation algorithm. It was discovered by Karl Friedrich Gauss in 1975 for astronomical computation purposed. This method works based on minimizing the residual error between actual and estimated outputs. This method is used because it is easy to implement and suitable for higher order linear system [12]. The least squares method is also known as linear regression and equation error method.
The LS method estimates the model parameter, θ (*k*), denoted by $\hat{\theta}(k)$, such that it minimize the sum of the squares of the residual, *E*(*k*). Parameter, θ(*k*) is obtained by minimizing the cost function in equation (3).

$$J = \frac{1}{2}\sum_{K=1}^{N}[E(K)] \tag{3}$$

Consider the mathematical model written in the regression model in equation (4):

$$y(k) = x_1(k)\theta_1 + x_2(k)\theta_2 + \ldots + x_i(k)\theta_i = [x_1(k) \quad x_2(k) \quad \ldots \quad x_i(k)]\begin{bmatrix}\theta_1\\ \theta_2\\ \vdots\\ \theta_i\end{bmatrix} = \varphi^t(k)\underline{\theta} \tag{4}$$

Model parameter, *θ*(*k*) is determined such that the output calculated the right-hand side of equation (4) matches as closed as possible with the measured y(*k*) in the sense of least squares, by minimizing:

$$J = \frac{1}{2}\sum_{K=1}^{N}[y(k) - \varphi^T(k)\underline{\theta}]$$

$$J = \frac{1}{2}E(N)^T E(N) \tag{5}$$

The variables in equation 3.2 are given by:

$$\begin{aligned} E(N) &= Y(N) - \hat{Y}(N) \\ &= Y(N) - \phi(N)\underline{\theta} \\ &= [\varepsilon(1) \quad \varepsilon(2) \quad \ldots \quad \varepsilon(N)] \\ Y(N)^T &= [y(1) \quad y(2) \quad \ldots \quad y(N)] \\ \phi(N) &= \begin{bmatrix}\varphi^T(1)\\ \varphi^T(2)\\ \vdots\\ \varphi^T(N)\end{bmatrix} \end{aligned}$$

J is minimized by setting:

$$\frac{dJ}{d\theta} = -\phi^T(Y - \phi\theta) = 0 \tag{6}$$

Finally the estimated parameter is given by:

$$\hat{\theta} = [\phi(N)^T \quad \phi(N)^{-1}]\phi(N)^T Y(N) \tag{7}$$

Where,
k = discrete time index
$u(k)$ = system input at discrete-time *k*

$y(k)$ =measured system output at discrete-tike k
$\underline{\theta}(k)$ =unknown parameter vector to be estimated
$\hat{\underline{\theta}}(k)$ = estimate of $\theta(k)$
$\varphi(k)$ = vector of input and output signal, and possibly their delayed signals
$\varepsilon(k)$ = $y(k)$ - $\bar{y}(k)$ = residual
$\hat{y}(k)$ = estimated output signal.

Equation (7) is used to find the estimated parameter for batch processing of LS method. The method can be used to find the relationship between two variable of linear system.

3.3 Differential evolution (DE) algorithm

Differential Evolution (DE), proposed by Storn and Price, is metaheuristic optimisation method to seek for global optimisation of a particular feasible solutions. It has a capabilities to deal with various problem ranging from linear system, nonlinear system, handling non-differentiable, and multi-modal objective functions, with easier selection of control parameters. DE works similarly with a standard evolutionary algorithm (EA). However, there is a bit different from other EA in the mutation and recombination phases. DE uses weighted differences between solution vectors to change the population [13]. Figure 3 shows the flow of DE operations.

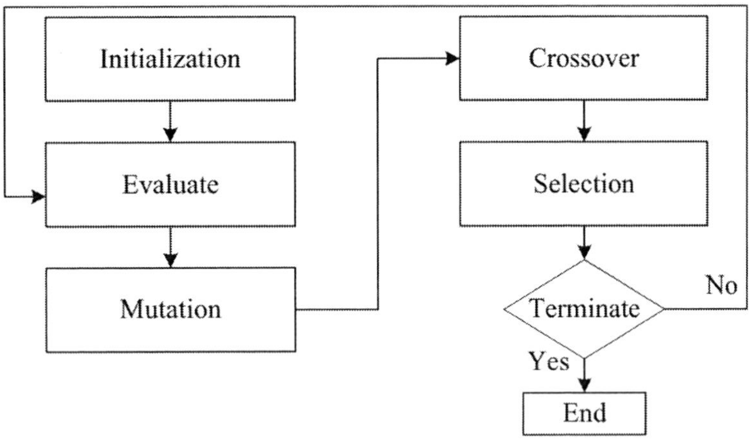

Fig. 3. Flow chart for DE.1

3.3.1 Initialization

Initialization process is performed to set appropriate parameter setting such as number of generation mutation rate, differential factor, and boundary of each solution vector. All the vector population and its fitness value randomly initialized in the given upper & lower bound.

3.3.2 Mutation

Three set of vector population is randomly selected, and the mutant vector is evaluated using equation (8). The difference between two vectors is multiply with its mutant factor, *F*.

A mutant vector is generated by:

$$v_{j,G+1} = x_{r1,G} + F.(x_{r2,G} - x_{r3,G}) \quad (8)$$

where the three distinct vectors x_{r1}, x_{r2} and x_{r3} randomly chosen from the current population other than vector $x_{j,G}$.

3.3.3 Crossover

Upon approaching the mutation operation, The trial vector $v_{i,G+1}$ is developed via crossover process between the target vector and the mutant vector, $X_{i,G}$ using equation (18). Crossover is done in order to increase the diversity of the perturbed parameters for each individual in the population [7].

$$u_{j,i,G+1} = \begin{Bmatrix} v_{j,i,G+1} & \text{if } rand \leq CR \\ x_{j,i,G} & \text{if } rand > CR \end{Bmatrix} \quad (9)$$

3.3.4 Selection

After the mutation and crossover operations, the fitness value from the trial vector and target vector is compared. The best fitness value is reserved for the next generation.

4 Results and discussion

In this project, the experiment was carried out to compare the value of mean square error (MSE) for least squares (LS) and differential evolution (DE) algorithm. Input-output data from the actuator and position sensor are used for parameter estimation in Least squares (LS) algorithm and differential evolution (DE) algorithm. Figure 4 and 5 show position sensor voltage signal and PRBS actuator signal respectively.

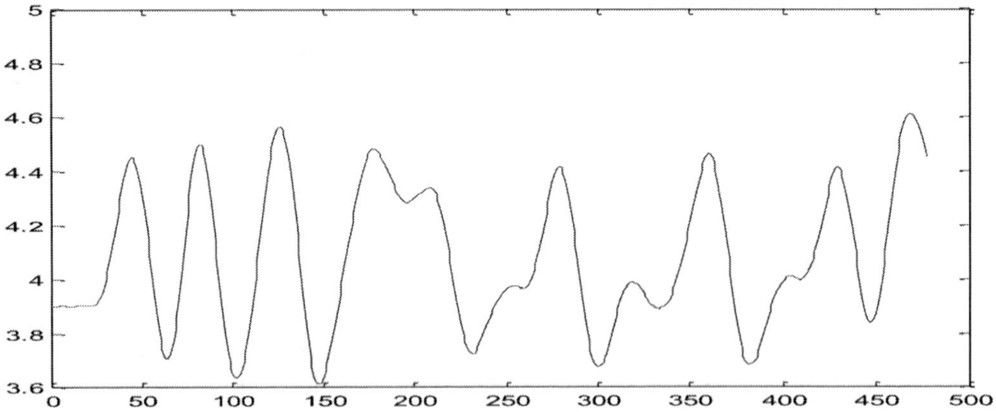

Fig. 4. Output voltage from potentiometer

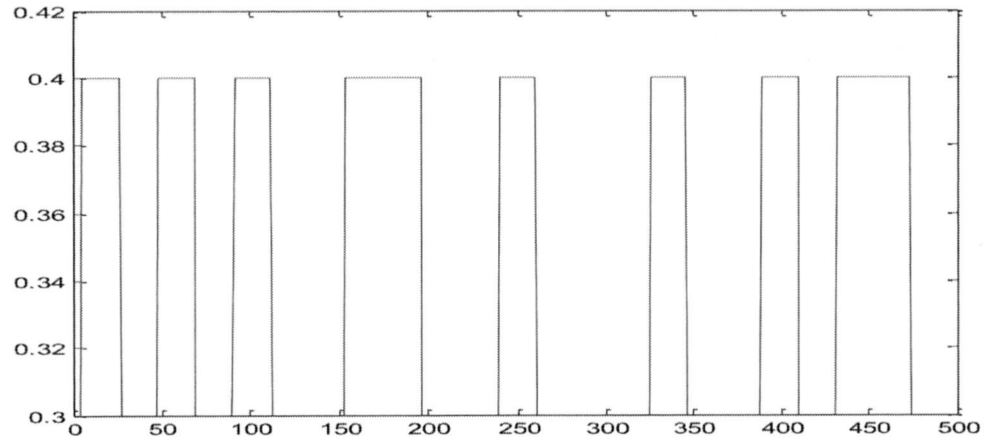

Fig. 5. PRBS input voltage to ESC

Initially for DE, the control parameter such as number of population, NP and generation, GEN is set to 30 and 100 respectively. The experiment was carried out to investigate the value of MSE by changing the crossover rate, CR between 0.1 to 1.

Table 1. The value for model parameter, crossover and MSE for NP=30.

Beta				CR	MSE(Output)	
a1	a2	b1	b2		Train	Test
-1.7556	0.7922	0.0186	-0.0121	0.1	$5.8047x10^{-4}$	$3.3974x10^{-4}$
-1.8379	0.8591	-0.0218	0.0120	0.2	$2.3507x10^{-4}$	$2.5269x10^{-4}$
-2.0206	1.0414	-0.0016	0.0088	0.3	$8.8214x10^{-5}$	$9.6176x10^{-5}$
-1.9956	**1.0116**	**-0.0021**	**0.0030**	**0.4**	**$3.2059x10^{-5}$**	**$3.8338x10^{-5}$**
-1.9713	0.9877	-0.0026	0.0042	0.5	$3.1864x10^{-5}$	$3.8973x10^{-5}$
-1.9839	1.0013	-0.0012	0.0016	0.6	$3.0536x10^{-5}$	$4.2412x10^{-5}$
-1.9815	0.9992	-0.0025	0.0031	0.7	$3.0261x10^{-5}$	$4.2477x10^{-5}$
-1.9804	0.9980	-0.0021	0.0028	0.8	$3.0223x10^{-5}$	$4.2179x10^{-5}$
-1.9806	0.9982	-0.0021	0.0028	0.9	$3.0222x10^{-5}$	$4.2178x10^{-5}$
-1.9806	0.9982	-0.0021	0.0028	1.0	$3.0222x10^{-5}$	$4.2178x10^{-5}$

Table 2. The value for model parameter, crossover and MSE for NP=100.

Beta				CR	MSE(Output)	
a1	a2	b1	b2		Train	Test
-1.9110	0.9466	0.0196	-0.0202	0.1	$2.2701x10^{-4}$	$2.3989x10^{-4}$
-2.0314	1.0450	-0.0102	0.0103	0.2	$6.1928x10^{-5}$	$4.4968x10^{-5}$
-1.9614	**0.9725**	**-0.0041**	**0.0038**	**0.3**	**$4.9184x10^{-5}$**	**$3.6601x10^{-5}$**
-1.9440	0.9616	-0.0035	0.0041	0.4	$3.7794x10^{-5}$	$4.5543x10^{-5}$
-1.9770	0.9935	-0.0016	0.0026	0.5	$3.0696x10^{-5}$	$3.9261x10^{-5}$
-1.9851	1.0023	-0.0006	0.0014	0.6	$3.0644x10^{-5}$	$4.1630x10^{-5}$
-1.9799	0.9974	-0.0028	0.0033	0.7	$3.0296x10^{-5}$	$4.2009x10^{-5}$
-1.9805	0.9981	-0.0021	0.0028	0.8	$3.0222x10^{-5}$	$4.2178x10^{-5}$
-1.9806	0.9982	-0.0021	0.0028	0.9	$3.0222x10^{-5}$	$4.2178x10^{-5}$
-1.9806	0.9982	-0.0021	0.0028	1.0	$3.0222x10^{-5}$	$4.2178x10^{-5}$

 DOI: 10.1051/matecconf/20167801117

From Table 3, when NP=100, the lowest MSE (test) value is 3.6601×10^{-5} at CR=0.3 and for NP=30 the lowest MSE value is 3.8338×10^{-5} at CR=0.4. Among these tests, the lowest value for MSE is 3.6601×10^{-5} at CR=0.3 with NP=100. This is the best parameter for the ARX model to present the dynamic driven pendulum system.

Table 3. Value of Beta, CR, NP and MSE (test)

Beta				CR	NP	MSE(test)
a1	a2	b1	b2			
-1.9956	1.0116	-0.0021	0.0030	0.4	30	3.8338×10^{-5}
-1.9614	0.9725	-0.0041	0.0038	0.3	100	3.6601×10^{-5}

4.1 Comparison between least squares (LS) with differential evolution (DE) algorithm.

The table 4 proved that the DE method produced the lowest MSE value 3.6601×10^{-5}, and parameter *a* and *b* values are a_1=-1.9614, a_2=0.9725, b_1=-0.0041 and b_2=0.0038. From Figure (99) it is clearly depicted that differential evolution (DE) algorithm performed better than least square (LS) algorithm.

Table 4. Comparison between least squares (LS) with differential evolution (DE) algorithm.

Method	**Beta**				**MSE**	
	a1	**a2**	**b1**	**b2**	**Train**	**Test**
Least Squares	2.9196	1.9717	-0.0065	0.0088	0.0057	0.0026
Differential Evolution	-1.9614	0.9725	-0.0041	0.0038	4.9184×10^{-5}	3.6601×10^{-5}

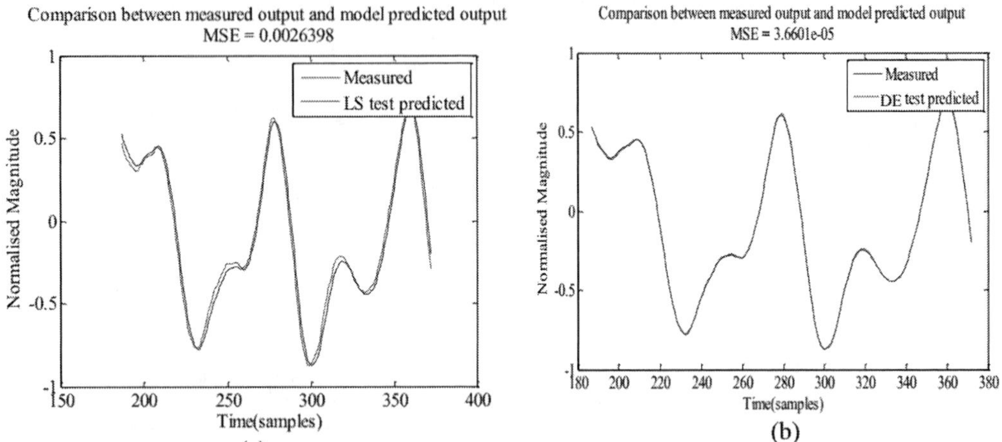

(a) (b)

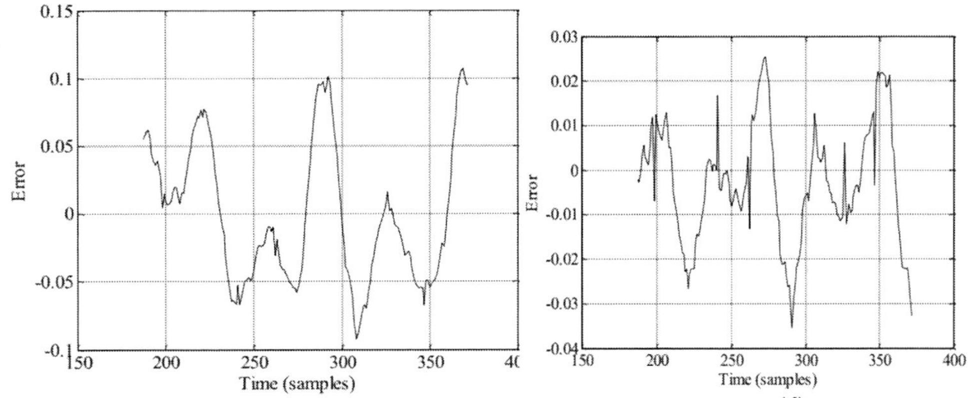

(c) (d)

Fig. 2. Comparison between LS and DE. (a) actual and predicted output using LS, (b) actual and predicted output using DE, (c) Error between actual and predicted output using LS, (d) Error between actual and predicted output using DE

The transfer function for the ARX model using differential evolution algorithm is shown in equation (10).

$$\frac{y(t)}{u(t)} = \frac{-0.0041z^{-1}+0.0038z^{-2}}{1+-1.9614z^{-1}+0.9725z^{-2}} \quad (10)$$

5 Conclusion

The estimate model for motorized-propeller pendulum was successfully developed using DE as a tool for model optimisation. The model parameters is carried out using DE algorithm to find the best estimated model that accurately represent the actual system. Validation test showed that DE is performed better than least square (LS). For future work, the identified model will be useful to study the controller performance.

The authors would like to express their gratitude to Universiti Malaysia Perlis and Universiti Teknologi Malaysia for providing facilities to conduct this research. This research was funded by the Ministry of Education Malaysia through Fundamental Research Grant Scheme No. FRGS 9003-00451.

References

1. W. Gu, Y. Yu, W. Hu, Nonlinear Dynam., **84**, 779 (2016)
2. A. Ahmed, M. S. Ullah, SpringerPlus, **5**, 1, (2016)
3. L. d. S. Coelho, H. V. H. Ayala, A. A. R. Coelho, Chem. Eng. Res. Des., **100**, 248 (2015)
4. A. AL-Khafaji, I. Darus, *Evolutionary algorithm for identification of a flexible single-link system*, WSEAS International Conference on Manufacturing Engineering, Automatic Control and Robotics (Kuala Lumpur, 2014)

5. M. S. Saad, H. Jamaluddin, I. Z. M. Darus, World Academy of Science, Engineering and Technology, International Journal of Mechanical, Aerospace, Industrial, Mechatronic and Manufacturing Engineering, **6**, 446 (2012)
6. H. Yousefi, H. Handroos, A. Soleymani, Mechatronics, vol. **18**, 513 (2008)
7. S. Swayamsiddha, S. Behera, H. P. Thethi, *Blind Identification of Nonlinear MIMO System Using Differential Evolution Techniques and Performance Analysis of Its Variants*, International Conference on Computational Intelligence and Networks (CINE, 2015)
8. R. Ahmad, M. Z. Zakaria, A. M. Nor, H. Jamaluddin, WSEAS Transactions on Systems and Control, **9**, 500 (2014)
9. J. McLellan, *Pseudo random binary sequence* (PRBS TESTING, 2005)
10. L. Ljung, *System Identification* (Prantice Hall, New Jersey, 1999)
11. M. Killian, B. Mayer, M. Kozek, Energy Build., **96**, 175, (2015)
12. J. G. Turner, B. Samanta, *Least Squares Estimation of Dynamic System Parameters using LabVIEW*, Proceedings of IEEE, (2012)
13. G. Y. Li, M. G. Liu, Int. Conf. Adv. Comput. Theory Eng. Proc., **3**, 153, (2010)

MATEC Web of Conferences **78**, 01118 (2016) DOI: 10.1051/matecconf/20167801118
IConGDM 2016

Parameter Estimation of Damped Compound Pendulum Using Bat Algorithm

Mohd Sazli Saad[1,2,*], *Luqman Nul Hakim* Mat Deri[2], *Z.* Shayfull[1,2], *S.M.* Nasir[1,2], and *M.* Fathullah[1,2]

[1]Green Design and Manufacture Research Group, Center of Excellence Geopolymer and Green Technology (CEGeoGTech), Universiti Malaysia Perlis, 01000 Kangar, Perlis, Malaysia
[2]School of Manufacturing Engineering, Universiti Malaysia Perlis, Kampus Tetap Pauh Putra, 02600 Arau, Perlis, Malaysia

Abstract. In this study, the parameter identification of the damped compound pendulum system is proposed using one of the most promising nature inspired algorithms which is Bat Algorithm (BA). The procedure used to achieve the parameter identification of the experimental system consists of input-output data collection, ARX model order selection and parameter estimation using bat algorithm (BA) method. PRBS signal is used as an input signal to regulate the motor speed. Whereas, the output signal is taken from position sensor. Both, input and output data is used to estimate the parameter of the autoregressive with exogenous input (ARX) model. The performance of the model is validated using mean squares error (MSE) between the actual and predicted output responses of the models. Finally, comparative study is conducted between BA and the conventional estimation method (i.e. Least Square). Based on the results obtained, MSE produce from Bat Algorithm (BA) is outperformed the Least Square (LS) method.

1 Introduction

A simple pendulum system naturally exhibits a periodic motion due to external forces. This system contain of particle like bob of mass fixed to a string of length hinged at the lower part [1]. In most physics courses, pendulum system can be a compulsory topic because it includes some mechanical subjects such as the period of oscillation, the simple harmonic motion the center of mass, , the acceleration of gravity, momentum, the moment of the inertia, etc.[2]. One of the popular applications of damped compound pendulum is in aerospace technology i.e. In this area, there are many commercial flight controllers available in markets including those in [3-5].

In designing a positioning control of damped compound pendulum, ones need to have a a dynamic modeling of a whole system. One key difficulty, in our view, for theoretical analysis and design of damped compound pendulum systems is to make a proper modeling

* Corresponding author: sazlisaad@unimap.edu.my

of a dynamics system [6]. This is because a typical damped compound pendulum system composed a propeller, a brushless DC motor, a micro-process based ESC (electronic speed controller), difficult to model with a set of mathematical equations. In order to solve this kind of problem, system identification technique is proposed in this research. This method is performed based on the input-output data measured from the experimental rig of damped compound pendulum, which describes the dynamic behavior of the overall system.

System identification is one of the most rudimentary necessities for many engineering and scientific applications. The aim of system identification is to search for the approximate model that represent the actual system. This technique is implemented by utilising input output data collected via phisycal system or simulation model. The identified model can be used to take care of numerous issues, for example, to design a controller in control system or to predict the future output of a given system under different working conditions [7].

Mainly there are two types of identification techniques which is parametric and non-parametric. This study focuses on system identification based on parameter identification process. Generally, its involve two steps, namely input-output data collection, selection of model structure and estimation of model parameters. Several methods have been considered for parameter estimation such as least square, maximum likelihood and prediction error method. However this method often fail in searching for the global optimum if the search space is not differentiable or linear in the parameters [8].

Nowadays, nature inspired algorithms have been widely applied in system identification [9-11]. It offers an alternative strategy that works with computational methods inspired by nature and natural ecosystem. These techniques provide more reliable and effective methodology for parametric estimation in system identification. For instance, bat algorithm gets to be one of the prominent algorithm works in optimization processes. It was invented from the echolocation behaviour of bats that also known as captivating creatures. Moreover, bats are great at finding the victims and they can recognize small size of prey and avoid obstacle in the darkness. They are the only mammals with wings and they have a capability of echolocation. They able to identify a prey and its moving speed by utilizing echolocation behaviour. This bat algorithm can see and identify nourishment sources with the assistance of echolocation. Among the bats species, microbats are an example of bat that use echolocation effectively, which is different from megabats. Microbats utilize a kind of sonar, called echolocation, maintain a strategic distance from snags, to recognize prey, and find their roosting crevices in the dark [12].

The utilisation of bat algorithm in optimisation problem has been reported in various applications. Bendjeghaba (2014) applied bat algorithm to search for optimum PID parameters in automatic voltage regulator (AVR) system. Bat algorithm was iterated to achieve optimum or near optimum controller parameters to improve voltage regulator features automatic response measures the simulation showed the effectiveness and efficiency of the proposed approach [13]. Marichelvam and Prabaharam 2013 proposed a modified bat algorithm to solve the multistage hybrid flow shop (HFS) scheduling problems. Results showed that the proposed bat algorithm was more efficient than genetic optimization algorithm [14]. Santos Coelho and Askarzadeh (2016) proposed a new enhanced differential bat algorithm (DBA) which was a swarm intelligence paradigm for minimising the total electrical power consumption of the multiple chiller systems (MCS) during 24 hours in an air conditioning system.

Their previous studies showed promising results in comparison with other optimization metaheuristics, such as particle swarm optimisation (FA), Artificial Bee Colony Algorithm (ABC), differential evolution (DE), chicken swarm optimization (CSO), and the Social Spider Algorithm (SSA) [15]. Furthermore, in the field of system identification and control engineering, Mustafa et. al (2013) applied bat algorithm in optimising the generalised predictive controller for double acting pneumatic cyclinder. The results showed that the controller tuned by bat algorithm produces a better transient response, no overshoot and less steady state error [16]. Sambariya and Manohar (2015) have studied the effectiveness of bat algorithm in model order reduction system. The numerator and denominator coefficients of a desired reduced-order system were optimized using integral square error minimization as a fitness function. Comparative study showed that bat algorithm was outperformed in terms of performance indices integral square error (ISE) [17].

In this research, bat algorithm is employed for parameter estimation of damped compound damped pendulum system.Accordingly, experimental rig is set up and prepared for data collection. Then, the information from the experiment was collected, select and satisfy the ARX model to get the model structure of the damped pendulum experiment. Afterwards, bat algorithm method is used to identify the experimental compound damped pendulum system. The estimated model is validated by mean square error and plotting the output predicted by the model and comparing it with the measured output for the validation data set.

2 Parametric identification

The parametric identification technique is a method used in system identification. This method is commonly used to estimate the parameters of a model from the physical system based on observed input-output data. Typically, system identification constituted of three main parts; data collection, selection of model structure, parameter estimation and model validation.

2.1 Model structure

The compound damped pendulum system is a single input single output (SISO). Due to simplicity of the system, an autoregressive with exogenous input (ARX) model was selected as a model structure. The ARX model can be written in the form :

$$Ay(t) = Bu(t-1) + e(t) \tag{1}$$

where $u(t)$ is the discrete input signal, $y(t)$ is the discrete output signal, $e(t)$ is the noise disturbance.
The following polynomial A and B are given by equation (2) and (3) where n_a and n_b are the positive number which define the order of polynomials,

$$A = 1 + a_1 z^{-1} \dots\dots\dots\dots + a_{na} z^{-na} \tag{2}$$

$$B = 1 + b_1 z^{-1} \dots\dots\dots\dots + b_{nb} z^{-nb} \tag{3}$$

The polynomial coefficients of equation (2) and equation (3) are the parameter to be estimated during estimation process. Therefore, it is necessary to write equation (4) using the backshift operator of z^{-1} and write the equation in the form:

$$y(t) = X(t)\boldsymbol{\theta} + e(t) \tag{4}$$

$$\boldsymbol{\theta}^T = [-a_1, \dots\dots, -a_{n_a}, -b_0, \dots\dots, b_{n_b}] \tag{5}$$

where θ is the vector of unknown parameters and $X(t)$ is the regression vector comprise of measured input and output variable.

In this experiment an ARX model by equation of transfer function:

$$y(t) = \frac{b_1 z^{-1} + b_2 z^{-2}}{1 + a_1 z^{-1} + a_2 z^{-2}} \tag{6}$$

2.2 BAT Algorithm

Bat algorithm was created by Yang in 2010. It was motivated by the echolocation conduct of microbats, with shifting pulse rates of emission and loudness. The ability of echolocation of microbats is interesting as these bats can discover their prey and segregate diverse sorts of bugs even in dark condition [20].

By looking to the natural behavior of Bats, it can fly randomly with velocity, v at position, x with a fixed frequency f_{min}, changing wavelength λ, and loudness A_0 to search for the prey. The wavelength (or frequency) of their emitted pulses and the rate of pulse emission $r \in [0,1]$ can be adjusted automatically, depending upon the closeness of their target. In this study,the loudness is assumed to vary from a large (positive) A_0 to a minimum constant value A_{min}.

Furthermore, for straightforwardness, they also use several assumptions: in general, the frequency f in a range $[f_{min}, f_{max}]$ relates to a range of wavelengths $[\lambda_{min}, \lambda_{max}]$. Moreover, they just vary in the frequency while fixed in the wavelength λ and assume $f \in [0, f_{max}]$ in their usage. This is because λ and f are related because of the way that $\lambda f = v$ is constant.

While the algorithm is executed, they utilise virtual bats to update their of their positions, x_i and velocities, v_i in a d-dimensional search space [21]. The new solutions, x_i^t and velocities, v_i^t at time step t are as follows:

$$f_i = f_{min} + (f_{max} - f_{min})\beta \tag{7}$$

$$v_i^{t+1} = v_i^t + (x_i^t - x_*) f_i \tag{8}$$

$$x_i^{t+1} = x_i^t + v_i^t \tag{9}$$

where $\beta \in [0,1]$ is a random vector coming from a uniform distribution and, x is the current global best location (solution) which is computed after comparing all the solutions among all the n bats. For the local search, once a solution is chosen among the current best solutions, a new solution for each bat is generated locally using random walk:

$$X_{new} = X_{old} + \mathcal{E} A^t \tag{10}$$

where $\mathcal{E} \in [-1,1]$ is a random number, while $A_t = \langle A_i^t \rangle$ is the average loudness of all the bats.

Finally, the loudness, A_i and the pulse rate, r_i will be updated every iteration. This is given by the following equations:

$$A_i^{t+1} = \alpha A_i^t, \gamma_i^{t+1} = \gamma_i^0 = [1 - \exp(-\gamma t)] \quad (11)$$

where α and are γ constants.

3 Experimental setup

Experiment setup of the whole system is shown in Figure 1. The system was driven by a brushless DC motor, a propeller, and electronic speed controller (ESC). The motor and its propeller were attached at the tip of the pendulum. The position of the pendulum was measured with a potentiometer sensor mounted at the suspended point of the pendulum. The data acquisition system (NI PCI-6221) was used as an interface between MATLAB program in PC with sensor and actuator. Selection of the input signal is the most prominent. It will influence the accuracy of the estimated model. The input signal should be selected in such a way that modes are effectively excited. In this study, PRBS signal was chosen as excitation input and sensor voltage as an output of the system. The position of the pendulum was displayed on the computer display panel.

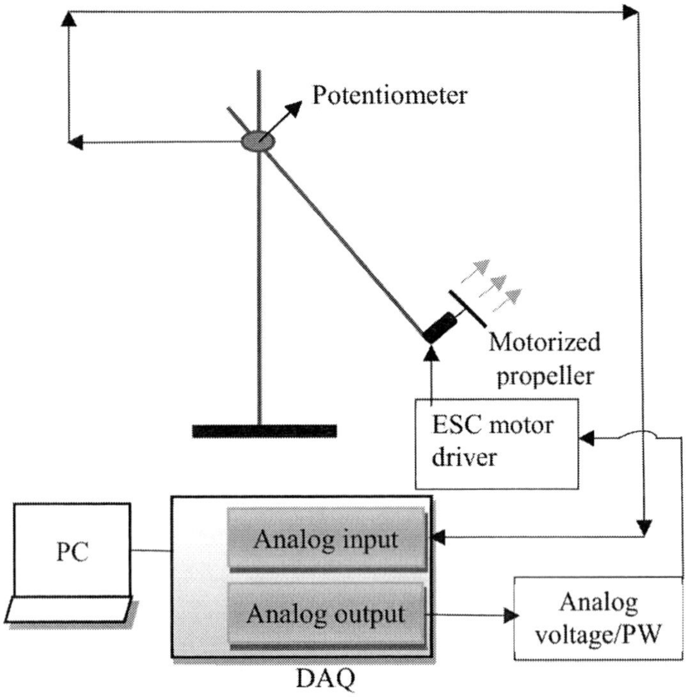

Fig. 1. Compound damped system

4 Results and discussion

Figure 2 and 3 show input-output that consists of the input voltage fed to ESC and output voltage from the potentiometer. There are 500 samples of data have been collected. It is important to make sure that data has been filtered and pre-process in order to remove any unwanted noise. Then the input-output data are divided into two parts, for estimation and validation process. During the estimation process, BA is used to search for the optimum model parameters. Then, these parameters are validated by comparing between the actual output and its predicted output. The mean square error is used to evaluate the performance of BA in searching the best parameter values for ARX model.

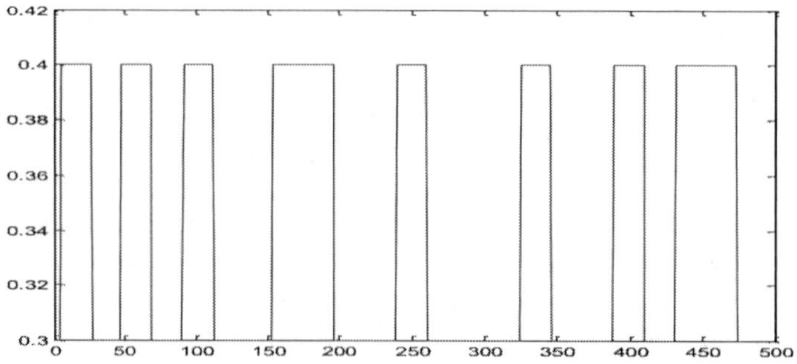

Fig. 2. PRBS input voltage to ESC

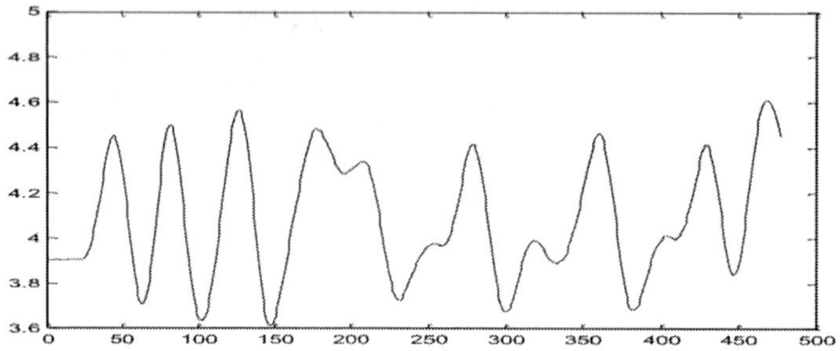

Fig. 3. Output voltage from potentiometer

4.1 Estimated Parameter

The design parameters of the bat algorithm used are presented in Table 1. In this study, minimum mean square error (MSE) is used as a fitness function, while number of generation is used as stopping criterion. The MSE is calculated from the error between actual output and predicted output.

$$mse = \frac{1}{N}\sum_{t=1}^{N}[y(t) - \hat{y}(t)]^2 \quad (12)$$

where N is the number of sampled data.

A vital approach is then conducted to get the best estimated model as the predicted output, $\hat{y}(t)$ must fit the actual output, $y(t)$ with minimum error as best as possible. Thus, BA is executed in several generations during the optimization process to get minimum MSE. It is expected that, MSE will improve gradually for every generation.

Table 1. Parameters of bat algorithm.

Population size, N	25
Number of Generation, Ngen	100
Loudness, A	0.9
Pulse Rate, R	0.1
Variation of Loudness, Alpha	0.1
Pulse Emission, Gamma	0.5

Based the Figure 4, parameter a1 and a2 start converge at generation of 90 and 93. Meanwhile, parameter b1 and b2 start converge at generation of 84. The final parameters values are taken at generation 100. At this generation, all parameters are stable. Figure 4(b) shows the minimum fitness function evaluating process using the BA method. BA converges at generation of 90 with MSE value of 0.00038285. The lower the MSE value the model is good enough to present the actual system. Figure 5 shows the error between actual and predicted output and comparison actual and predicted output. The model error is significantly reduced as BA has effectively identified the dynamic behaviour of the estimated model close to the actual output.

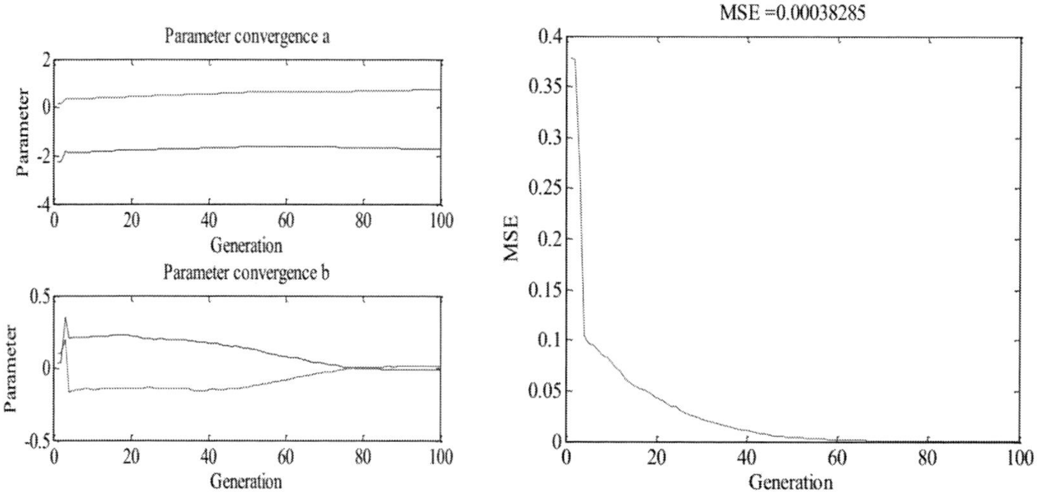

(a) (b)

Fig. 4. (a) parameter convergence a and b. (b) MSE training

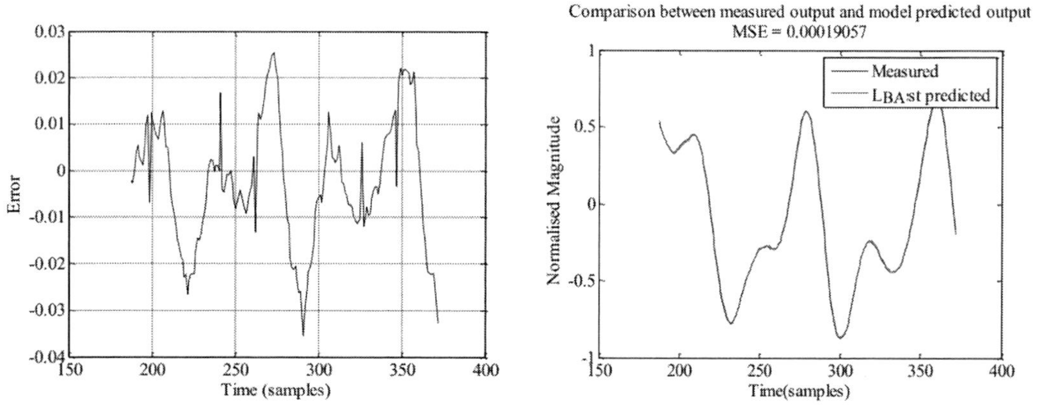

(a) (b)

Fig. 5. (a) The error for BA method and (b) Comparison between actual and predicted output.

The optimum parameters for ARX model is shown in Table 2. MSE train is the value obtained during optimisation process with BA, while MSE Test is the value obtained from model validation process.

Table 2. Estimated parameter and MSE

a1	a2	b1	b2	MSE Train	MSE Test
-1.7319	0.7482	- 0.0092	0.0099	0.0003828	0.00019057

The transfer function for the estimated model of damped compound pendulum is represented by equation (13).

$$y(t) = \frac{-0.0092z^{-1} + 0.0099z^{-2}}{1 + -1.7319z^{-1} + 0.7482z^{-2}} \tag{13}$$

4.2 Comparison the model structure LS and BA method

Comparative study is conducted to compare the performance between BA and conventional estimation technique, i.e. least square (LS). Figures 6 shows the comparison between measured output and model predicted output in terms of MSE. From the observation, the value of MSE for BA method is 0.00019057 lower than LS method. It has been shown that, the model obtained using BA has captured the dynamic characteristics of the system well. This condition proved that BA able to search for the global optimum of the given fitness function.

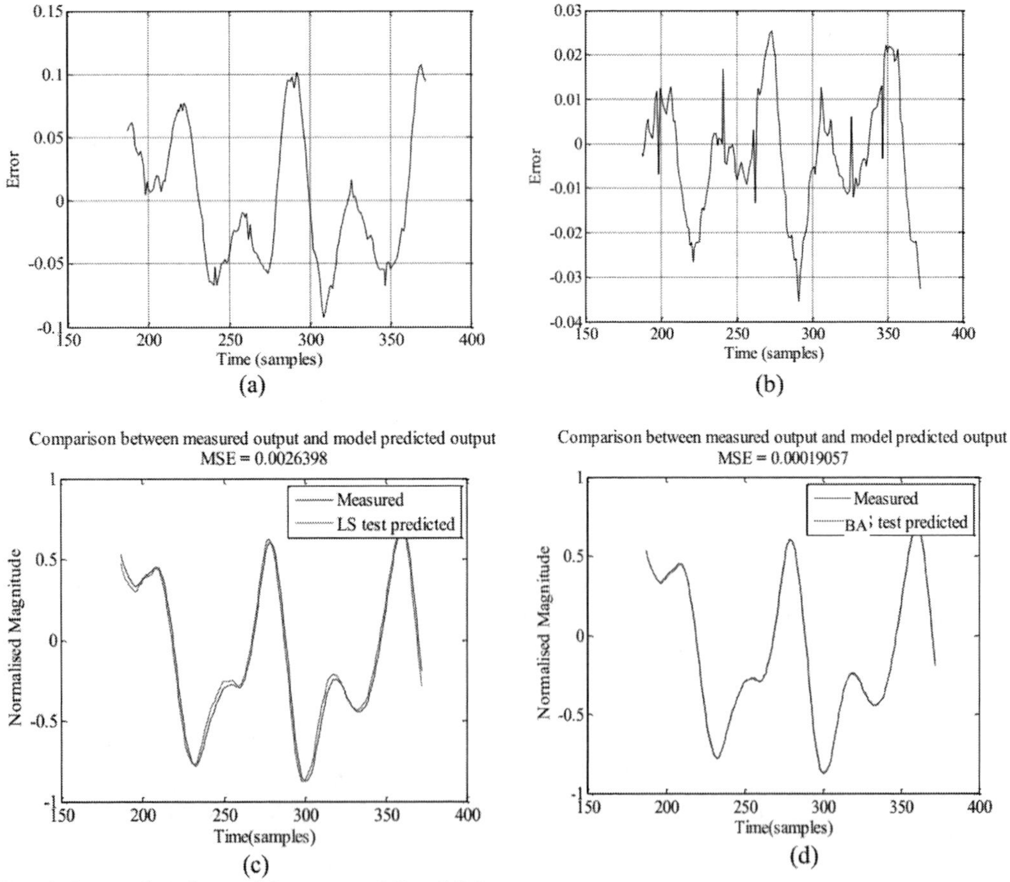

Fig. 6. Comparing the model structure LS and BA

5 Conclusion

This study shows that bat algorithm is easy to implement, and able to estimate the model of the damped pendulum system from a given input-output data. From the results, BA has been demonstrated to outperform least square method in minimizing the prediction error and resulting in a dependable level of accuracy of the approximated model. The value of MSE for Bat Algorithm method is 0.00019057 compared Least Square method is 0.0026398.

The authors would like to express their gratitude to Universiti Malaysia Perlis and Universiti Teknologi Malaysia for providing facilities to conduct this research. This research was funded by the Ministry of Education Malaysia through Fundamental Research Grant Scheme No. FRGS 9003-00451.

References

1. R. A. Serway, *Physics: For scientist and engineers with modern physics*, 468-475, 264-265 (Saunders, 1990)
2. P. F. Hinrichsen, Physics Teacher, **19**, 286-292 (1981)
3. A. Samano, R. Castro, R. Lozano, S. Salazar, J. Intell. Robot. Syst., **69**, 161 (2013)
4. Y. Naidoo, R. Stopforth, G. Bright, Int. J. Adv. Robot. Syst., **8**, 139 (2011)
5. N. Intaratep, W. N. Alexander, W. J. Devenport, S. M. Grace, A. Dropkin, *Experimental Study of Quadcopter Acoustics and Performance at Static Thrust Conditions*, 22nd AIAA/CEAS Aeroacoustics Conference (2016)
6. K. Yoshida, I. Kawanishi, H. Kawabe, *Stabilizing control for a single pendulum by moving the center of gravity: theory and experiment*, American Control Conference, **5**, 3405 (1997)
7. G. Ronquillo Lomeli, G. Rios Moreno, A. Gomez Espinosa, L. Morales Hernandez, M. Trejo Perea, Mech. Based Des. Struc., **44**, 5, (2016)
8. M. A. Hossain, M. O. Tokhi, A. J. Chipperfield, M. J. Baxter, C. M. Fonseca, N. V. Dakev, *Adaptive active vibration control using genetic algorithms*, Genetic Algorithms in Engineering Systems: Innovations and Applications (GALESIA, 1995)
9. C. Li, L. Chang, Z. Huang, Y. Liu, N. Zhang, Eng. Appl. Artif. Intel., **50**, 177 (2016)
10. B. Subudhi, D. Jena, *Evolutionary Computing Approaches to System Identification*, Handbook of Research on Computational Intelligence Applications in Bioinformatics, (IGI Global, 2016)
11. A. Rahimi, F. Bavafa, S. Aghababaei, M. H. Khooban, S. V. Naghavi, Int. J. Elec. Power., **78**, 285, (2016)
12. X. S. Yang, *A new metaheuristic bat-inspired algorithm,* Nature inspired cooperative strategies for optimization (NICSO 2010), (Springer, 2010)
13. B. Omar, I. B. Saida, *Bat Algorithm for Optimal Tuning of PID Controller in an AVR System,* International Conference on Control, Engineering & Information Technology, (2014)
14. M. Marichelvam, T. Prabaharam, ICTACT Journal on Soft Computing, **3**, 428 (2012)
15. L. dos Santos Coelho, A. Askarzadeh, Appl. Therm. Eng., 99, 834 (2016)
16. N. D. Mustafa, A. A. M. Faudzi, A. F. Z. Abidin, K. Osman, K. Suzumori, *Generalized predictive controller using Bat algorithm for double acting pneumatic cylinder*, Research and Development (SCOReD, 2013)

17. D. K. Sambariya, H. Manohar, *Model order reduction by integral squared error minimization using bat algorithm*, 2nd International Conference on Recent Advances in Engineering & Computational Sciences (RAECS, 2015)

Application of Firefly Algorithm for Parameter Estimation of Damped Compound Pendulum

Mohd Sazli Saad[1,2,*], *Mohamad Izzat* Kamal[1] , *Z. Shayfull*[1,2] , *S.M. Nasir*[1,2] , *and M. Fathullah*[1,2]

[1]Green Design and Manufacture Research Group, Center of Excellence Geopolymer and Green Technology (CEGeoGTech), Universiti Malaysia Perlis, 01000 Kangar, Perlis, Malaysia
[2]School of Manufacturing Engineering, Universiti Malaysia Perlis, Kampus Tetap Pauh Putra, 02600 Arau, Perlis, Malaysia

Abstract. This paper presents an investigation into the parameter estimation of the damped compound pendulum using Firefly algorithm method. In estimating the damped compound pendulum, the system necessarily needs a good model. Therefore, the aim of the work described in this paper is to obtain a dynamic model of the damped compound pendulum. By considering a discrete time form for the system, an autoregressive with exogenous input (ARX) model structures was selected. In order to collect input-output data from the experiment, the PRBS signal is used to be input signal to regulate the motor speed. Where, the output signal is taken from position sensor. Firefly algorithm (FA) algorithm is used to estimate the model parameters based on model 2nd orders. The model validation was done by comparing the measured output against the predicted output in terms of the closeness of both outputs via mean square error (MSE) value. The performance of FA is measured in terms of mean square error (MSE).

1 Introduction

In order to design a good control system, such as controlling the position of compound pendulum system, a suitable model always required to from the input and output of the system. System identification technique is the best choise to deals with the problem of developing a model of the real physical system based on the measured input output data. At first, when the structure of the model is identified, the parameter estimation is employed to the model's parameters. In classical approach, one needs to have a priory knowledge about the system in order to obtain the governing equations. However this method is tedious when modeling a complex dynamic system. Therefore, the black box modelling approach is an altenative technique which can be adopted for system identification. System identification is only concerned the system that has signal input and output from the real experimental data. Furthermore, it may takes less time than from physical modeling [1]. System identification methods aim to find appropriate models for the real physical systems. Over the years, many algorithms and methodologies have been developed for this purpose,

* Corresponding author: sazlisaad@unimap.edu.my

ranging from conventional method, i.e. least squares regression algorithm up to intelligent method i.e. nature-inspired algorithms.

Nature-inspired algorithms is an algorithms that developed from imitating process in nature or in other way it is inspired from the nature. These algorithms provide an efficient way to solve engineering problem especially in optimization process [2]. In the fields of system identification, many researcher applied Nature-inspired algorithms for parameter estimation in system identification. Hanim et. al [3], studied about developing self-tuning PID controller optimized for vibration suppression structure flexible manipulator using particle swarm optimization (PSO). The PSO is applied for estimating the model parameter of a flexble manipulator. Comparative study showed that PSO is outperformed the conventional method with minimum mean square error (MSE). The authors also employed PSO for PID controllers tuning for better vibration suppression. Rahimi et. al [4] estimated the parameter of chaotic behaviour in permanent magnet synchronous motor using Self-Adaptive Learning Bat-inspired algorithm (SLBA). Simulation study showed that the parameter can be estimated effectively via offline or online modes. Furthermore, it able to maintain its performance in the presence of noise. Nasir et. [5] proposed an improved version of spiral dynamic algorithm (SDA) becoming linear adaptive spiral dynamic algorithm (LASDA). This new approach employed a novel mathematical equation based on linear function. The performance of the LASDA is tested in parameter optimization for autoregressive model with exogenous input (ARX) structure for flexible manipulator. Results depicted that LASDA give better convergence speed and minimum mean square error between the actual output and predicted output.Furthermore, Ali [6] proposed proportional integral (PI) controller tuning using Firefly Algorithm (FA) for speed control of DC series motor. The results have been compared with Genetic Algorithm (GA) and Ziegler Nichols (ZN) under various operating conditions and disturbances. It is found that FA tuning PI controller produces an excellent performance due to the change in load torque, radiation, and temperature compared with GA and conventional techniques.

In this study, the Firefly algorithm has been chosen to estimate the parameter of the damped pendulum. The model validation was done by comparing the measured output against the predicted output in terms of the closeness of both outputs via mean square error (MSE) value. The performance of FA is measured in terms of mean square error (MSE). The organization of this paper is as follows. Section 2 gives the fundamental background of the system identification. Section 3 describes briefly the theory of FA. Section 4 explains the experimental setup of the compound pendulum system. Section 5 presents the parameter optimization process and the obtained results. Section 6 concludes the study.

2 System identification

System identification is modeling of a damped compound pendulum using input-output data from the experiment. Generally, the step involved in system identification is experimental data collection, model structure selection, parameter estimation and model validation. The model is developed based on auto-regressive with exogenous inputs (ARX) structure and model parameters are estimated by Firefly algorithm and least square algorithm.

2.1 Model structure

The model structure represented the mathematical model of the system to be modeled. The beam is modeled based on second order transfer function. Equations 1 below are the transfer function of damped compound pendulum bonded with actuator and sensor in continuous time and discrete time for second order system respectively:

$$\frac{y(t)}{u(t)} = \sum_{i=1}^{N} \frac{b_{1i}z^{-1} + b_{2i}z^{-2}}{1 + a_{1i}z^{-1} + a_{2i}z^{-2}} \quad (1)$$

where *y(t)* is the output signal, *u(t)* is the input signal, b_1 and b_2 are locations of zeros of the transfer function. For this study, the model is built by considering a 2nd model order for estimation method. Therefore, a model built under the assumption that the structure of the system is unknown [3]. So that the black box is used in this experiment. The input-output dataset from unknown model parameter to be used for system identification which is achieve from data experiment.

2.2 Parameter estimation

Figure 1 shows the process of system identification. Initially, data collection is adopted. The input of the system was analog output to the motor and the output was the input voltage from the position sensor. Based on the acquired input-output data, the model parameter is estimated until minimum error, e(t) is reached. The parameter of estimated model is determined using firefly algorithm [9]. The performance of the estimated model is measured using mean square error (MSE).

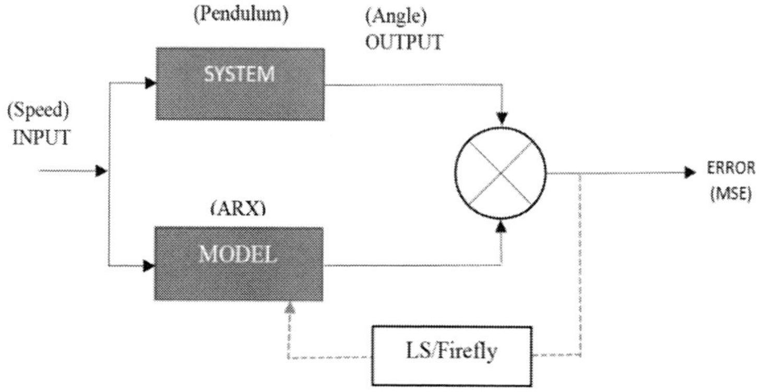

Fig. 1. Difference between the process and model output

3 Firefly algorithm

The Firefly Algorithm (FA) have been inspired by the flashing of fireflies in nature and has been proposed by Yang [7]. Each species of firefly produces its own pattern of flashes, and even complete function flashes is not known, the main goal for them is to flash to attract mates. These fireflies belongs to the family of insects that are can produce natural light to attract prey or mates. This light is found in a unique pattern and generate an amazing sight in tropical regions during the summer. The light intensity decreases as the distance increases and thus the most fireflies can communicate only up to a several hundred meters. For the implementation of the Firefly algorithm, with the objective function to be optimized the flashing light was formulated in such way that it will be associated. For simplicity, several rules are used to extend the structure of FA:

a) All fireflies are unisex and are attracted to other fireflies regardless of their sex.
b) The degree of the attractiveness of a firefly is proportional to its brightness. Their attractiveness is proportional to their light intensity. Thus for any two flashing fireflies, less bright firefly moves toward the brighter one. As brightness is proportional to distance, more brightness means less distance between two

fireflies. If any two flashing fireflies have the same brightness, then they move randomly.

c) The brightness of a firefly is determined by the objective function to be optimized

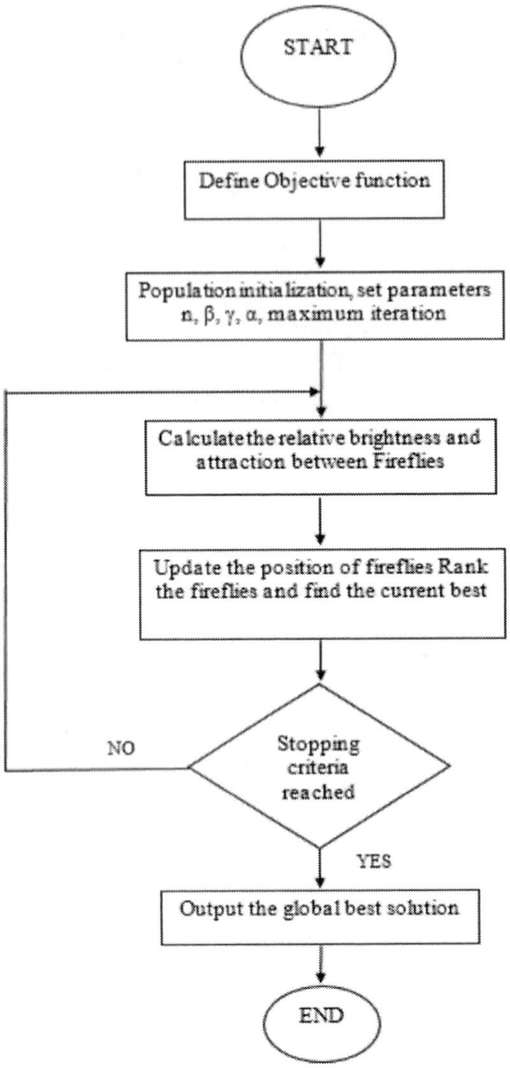

Fig. 2. Flow chart FA

3.1 The formulation

In the Firefly Algorithm has the two important information to be determine there are the light intensity variation *(I)* and the attractiveness *(β)* formulation. For simplicity, the attractiveness of fireflies is determined by the brightness with respect to the objective function [16]. For a simple case, the brightness (*I*) in a particular location is a function of its position *x* as follows:

$$I\,(x)\;\alpha\;f(x) \tag{2}$$

3.2 Light intensity and attractiveness

The β is an attractiveness factor which it ought to be reviewed or saw by different fireflies. The β is normally in the range from 0 to 1 where its determined how attracted fireflies are to others, if bringing down the level of β it will lowers the yearning for a firefly to move towards the brighter fireflies. The value of β and a random component will influent the firefly to move towards a brighter firefly. Hence, the β is depend on the distance r_{ij} between firefly i attracted with the brightness of firefly j. In addition, the attractiveness varies with the degree of adsorbption where the light intensity decreases with increasing distance from its source and depends on the propagation medium. It follows that the light intensity as indicated from equation (3) is varies depending to the inverse square distance r,

$$I(r) = \frac{I_s}{r^2} \tag{3}$$

where I_s is the light intensity at the source.
As the light intensity varies with r when the medium provided with fixed light absorption coefficient γ shown in equation (4).

$$I = I_0 e^{-\gamma r^2} \tag{4}$$

where I_0 is the original light intensity.

In order to evade the singularity at $r = 0$ occur in equation (4), the Gaussian form is developed by combining the effect of both the inverse square law and absorption as in equation (5).

$$I(r) = \frac{I_0}{1+\gamma r^2} \tag{5}$$

The adjacent fireflies later on will compute for the attractiveness is proportional to the light intensity.

$$\beta = \beta_0 e^{-\gamma r^2} \tag{6}$$

where β_0 is the attractiveness at $r = 0$. This function approximated by:

$$\beta(r) = \frac{\beta_0}{1+\gamma r^2} \tag{7}$$

The movement of a firefly towards a brighter firefly is resolved by $\beta(r)$ and its random component. The random component is key for all metaheuristic algorithms, it allows the algorithm to escape from local optimums.

3.3 Distance

The distance between any two fireflies i and j at x_i and x_j, respectively, is the Cartesian distance as in equation (8).

$$r_{ij} = |x_i - x_j| = \sqrt{\textstyle\sum_{k=1}^{d}(x_{i,k} - x_{j,k})^2} \tag{8}$$

Position Update
Position of fireflies is updated when the attractive of firefly i to one another is more than attractive firefly j. Where $x_{i,k}$ is the k^{-th} component of the spatial coordinate x_i of i^{-th} firefly. The movement of a firefly i attracted to another more attractive (brighter) firefly j in a given time step *(t)* is determined by equation (9).

$$x_i^{t=1} = x_i^t + \beta_0 e^{-\gamma r_{ij}^2}(x_i^t - x_j^t) + \propto \left(rand - \frac{1}{2}\right) \quad (9)$$

where α is the parameter controlling the amount of randomness. The randomness parameter, α is typically in the range from 0 to1, where 0 is referred to no randomness and 1 to highly random.

4 Experimental setup

The pendulum is classical control problems which have been used until today to apply in the real life most easy and stabilize the basic machine with pendulum system. By modelling and controller design, the pendulum characteristic can be easy to understand, view and design. The compound pendulum are one of a basic topic in physics because it is included some physical subjects such as the period of oscillation, center of mass, simple harmonic motion, the acceleration of gravity, moments of inertia, momentum and so on. For this project, the pendulum that used is the compound pendulum with motorized-propeller.

An experimental of a damped compound pendulum is developed in Figure 3 is for the significance test and input-output data collection. This compound pendulum consists Motorized propellers that are attached at the end of an aluminium pendulum arm. At the end of the pendulum there has motorized propeller so it can lift the pendulum if given actuator of the voltage which is able to provide thrust in a single direction. The sensor used for sensing the swing of aluminium pendulum arm when it has been applied a disturbance and signals are transmitted to computer control system via data acquisition system (DAQ). The following Table 1 below shows the properties of the aluminum damped compound pendulum:

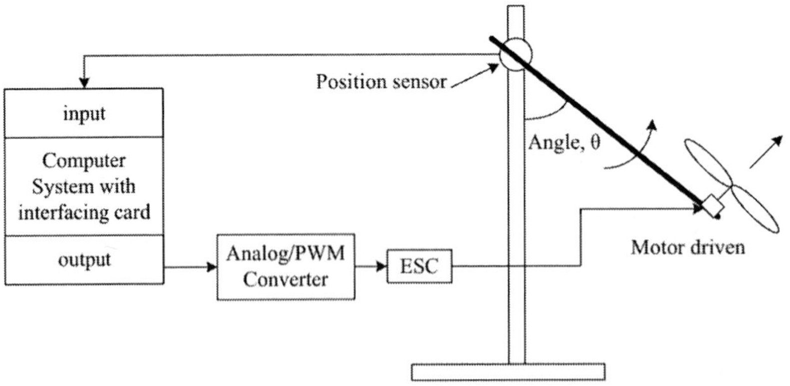

Fig. 3. Schematic diagram for motorized propeller system.

4.1 Data collection

In this experiment, the PRBS signal with the maximum length sequence of 30 with the time period of 41 sec was used during data collection of the damped compound pendulum at the (0.3v to 0.4v) and shown in Figures 4 and 5. A set of real time data were collected from the damped compound pendulum that are consist of 477 data. The set will divided to two part. The first part was used for training while second part of data used for identification activities for FA. Figure show the PRBS signal injected to actuator from computer.

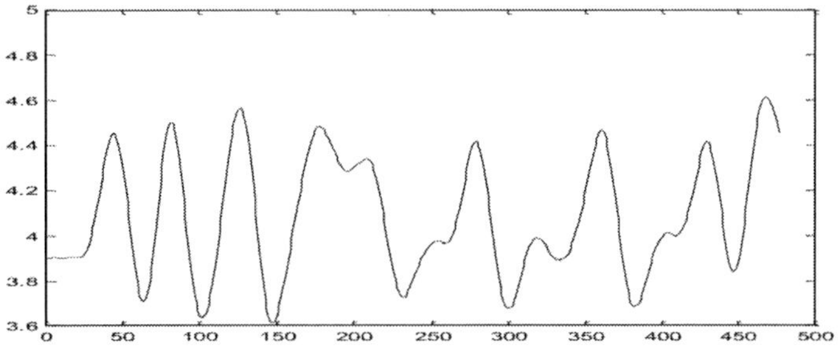

Fig. 4. Output voltage from potentiometer

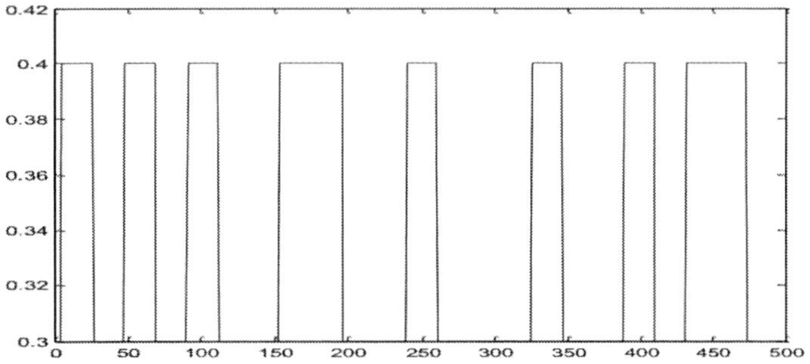

Fig. 5. PRBS input voltage to ESC

5 Results and discussions

In this study, the parameter estimation of compound pendulum is estimated using firefly algorithm in order to achieve the best value of MSE.

5.1 Analysis of results Firefly Algorithm

By using the best settings of n=18, α=0.4, β=0.3, γ=0.9 and G=100. Figures 6 and 7 show parameters a_1 and a_2 start convergence at the 55 generation while the parameter of b_1 and b_2 start convergence at the 53 generation. Both of parameters a and b converge almost at the same generation.

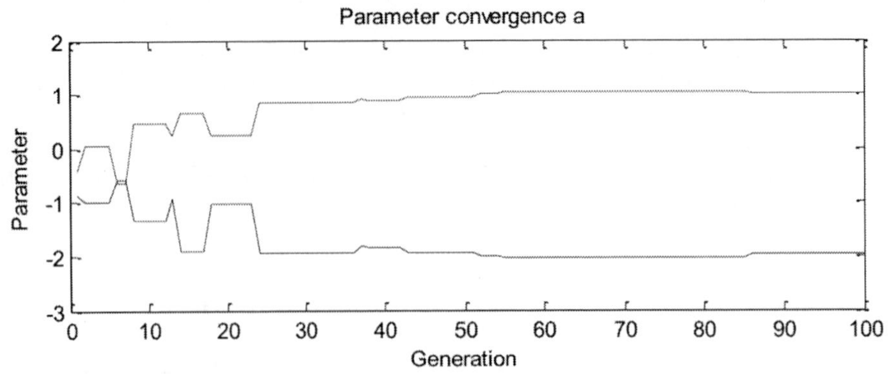

Fig. 6. Parameter convergence parameter *a*

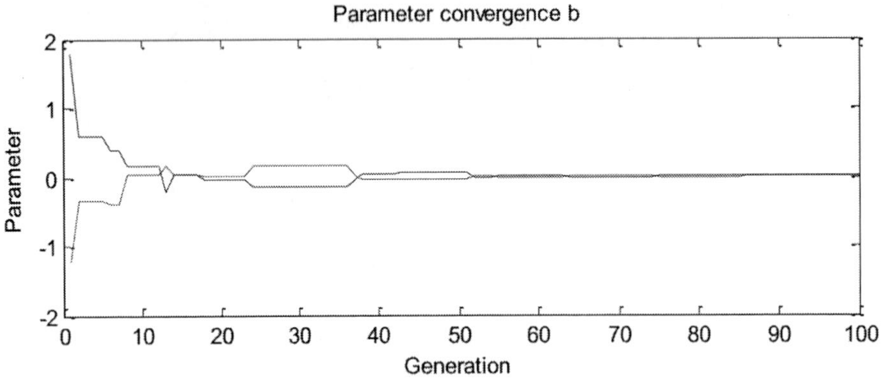

Fig. 7. Parameter convergence parameter *b*

The convergence profile during optimization process in shown in Figure 8. It is observed that the minimum fitness functions achieved at MSE value of 3.7835e-05. It can be seen that FA obtained the optimal solution after 38 generations. This shows FA produced a good convergence capability.

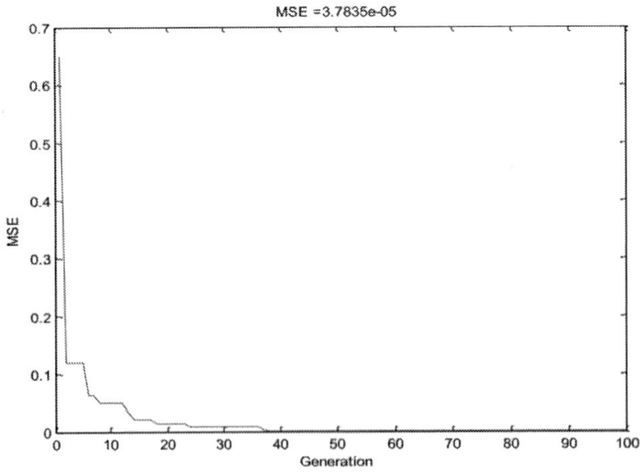

Fig. 8. Convergence profile of MSE

5.2 Model validation

After obtaining the value of a_1, a_2, b_1, b_2 from the predicted output, then these value will be validated with the measure output show as in Figure 9. It can be observed that the measured output and the model predicted output are in a good agreement during the model validate procedure. The MSE between measured output and the model predicted output is about 4.3759e-05. Thus the using firefly algorithm in the parameter estimation of pendulum compound is acceptable.

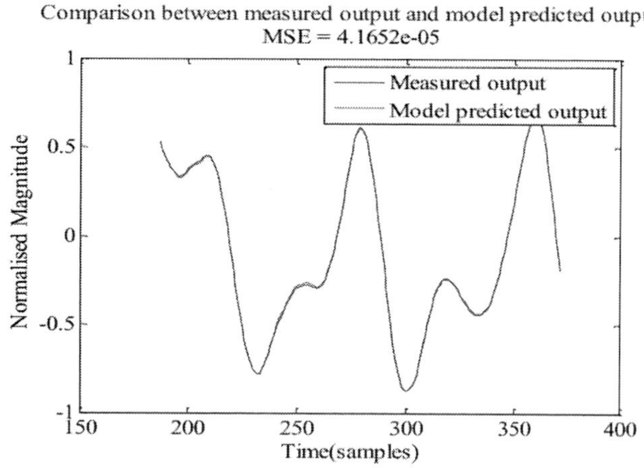

Fig. 9. The MSE of measured and model predicted output

From Figure 10, it can be observed that the error of FA is in the range of ±0.015.

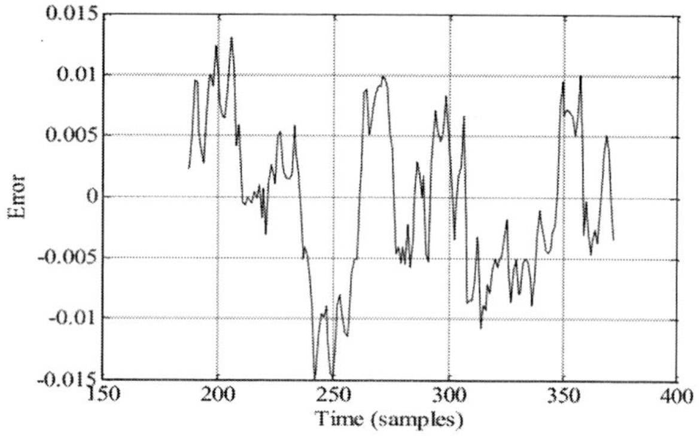

Fig. 10. The error of model predicted output

The optimal parameter can be achieved using FA algorithm is $a_1 = -1.9843$, while the value of $a_2 = 0.9994$ beside $b_1 = 0.0008$ and $b_2 = 0.0027$. The MSE obtained during training is about 3.7835e-05 while the MSE obtained during test shows the lowest value at 4.16523e-05. The overall results of FA show in Table 1.

Table 1. Summary of FA

a_1	a_2	b_1	b_2	MSE training	MSE test
-1.9843	0.9994	0.0008	0.0027	3.7835e-05	4.16523-05

Finally, the optimal parameters obtained from optimization process are substituted into the model structure a damped compound pendulum. The model is shown in the transfer function below:

$$y(t) = \frac{-1.9843z^{-1} + 0.9994z^{-2}}{1 + 0.0008z^{-1} + 0.0027z^{-2}} u(t)$$

6 Conclusion

As a conclusion, firefly algorithm firefly algorithm is performed well in terms of mean square error (MSE) with a very low value of 4.3759e-05. The error of model predicted output also show that the FA had the lowest error range that are ±0.015. It appears that the FA is the superior algorithm when it comes to problems with many local optima. Furthermore, FA produce fast convergence speed to reach the optimal solution. For future work, the identified model will be useful to study the controller performance.

The authors would like to express their gratitude to Universiti Malaysia Perlis and Universiti Teknologi Malaysia for providing facilities to conduct this research. This research was funded by the Ministry of Education Malaysia through Fundamental Research Grant Scheme No. FRGS 9003-00451.

References

1. A. Juditsky, H. Hjalmarsson, A. Benveniste, B. Delyon, L. Ljung, J. Sjoberg, Q. Zhang, Mathematical foundations, Automatica, **31**, 1725 (1995)
2. H. Zang, S. Zhang, K. Hapeshi, J. Bionic Eng., **7**, S232 (2010)
3. H.M. Yatim, I.Z. Darus, Recent Advances in Electrical Engineering Series (2013)
4. A. Rahimi, F. Bavafa, S. Aghababaei, M.H. Khooban, S.V. Naghavi, Int. J. Elec. Power., **78**, 285 (2016)
5. A. Nasir, R.R. Ismail, M. Tokhi, Appl. Math. Model., **40**, 5442 (2016)
6. E.S. Ali, Neural Comput Appl., **26**, 1321 (2015)
7. X.S. Yang, Int. J. Bio-Inspir. Com., **2**, 78 (2010)

AUTHOR INDEX

AUTHOR INDEX

AUTHOR INDEX

AUTHOR INDEX